ESSENTIALS

WORLD REGIONAL GEOGRAPHY

SECOND EDITION

ESSENTIALS OF

World Regional Geography

SECOND EDITION

CHRISTOPHER L. SALTER

JOSEPH J. HOBBS

JESSE H. WHEELER, JR.

J. TRENTON KOSTBADE

University of Missouri–Columbia

SAUNDERS COLLEGE PUBLISHING
Harcourt Brace College Publishers

FORT WORTH PHILADELPHIA SAN DIEGO NEW YORK ORLANDO AUSTIN SAN ANTONIO TORONTO MONTREAL LONDON SYDNEY TOKYO

Requests for permission to make copies of any part of the work should be mailed to: Permissions
Department, Harcourt Brace & Company, 6277 Sea Harbor Drive, Orlando, Florida 32887-6777.

Publisher: John Vondeling
Publisher: Emily Barrosse
Acquisitions Editor: Jennifer Bortel
Product Manager: Nick Agnew
Developmental Editor: Beth Rosato
Project Editor: Anne Gibby
Production Manager: Alicia Jackson
Art Directors: Joan Wendt, Lisa Caro
Text and Cover Designer: Ruth Hoover

Cover Credit: Southern Australia. Photo courtesy of David Doubilet.
Cover art and text maps: Geosystems Global Corporation
Landscape in Literature logo: Doug Plummer/Photonica
Title page: Yemen, Arab Republic. Manakha village and ancient agricultural terraces. © George Holton 1978.

Printed in the United States of America

Essentials of World Regional Geography, second edition

ISBN 0-03-018382-0

Library of Congress Catalog Card Number: 97-66890

90123456 032 10 98765

This edition of *Essentials of World Regional Geography* is dedicated to the memory of Professor Jesse H. Wheeler, Jr. This text came to life initially more than forty years ago because of his belief that American college students should be educated in the world beyond their local campuses. We both share that belief and hope that Jesse would still find this book useful in achieving such a goal.

CONTENTS OVERVIEW

APPROACH AND SCOPE

This textbook has been written to open the door of world regional geography to all who want to come into the 21st century with the geographic knowledge and cultural wisdom necessary to play the role of a thoughtful citizen. As powerful as local identity and regional connections currently are in our society, this next century will be more international than any the United States has ever experienced. This text has been written to provide access to this ever more global world, building on a base of regional Geographic Profiles and structuring learning around chapters that weave environments and peoples together in patterns of distinctive landscapes and geographic characteristics.

From the presentation of environmental fundamentals from Chapters 1 and 2 on, the text looks at the ways people have assessed their settings and resources and crafted worlds of distinct and evocative personalities over time. Seeking to illustrate the ways in which people have created the patterns of cooperation and conflict that have emerged from these blueprints of geographic development, our text presents issues that have made and will continue to make headlines in the chronicles of the next century. By constructing a story of contemporary concerns and historical themes, with the help of totally new maps and symbolic photographs, the text serves as a guide to understanding the complex and exciting world in which today's students will play out their lives.

The geography of this text is the geography of current affairs, significant historical elements, and contemporary environmental issues. It is our hope that this geography will play a creative and instructive role in preparing students to see the powerful role and responsibility they are faced with as concerned citizens of the United States—and as stewards of the geography of the future.

CHANGES IN THIS EDITION

With the untimely death of Professor Jesse Wheeler, Jr., in July 1994, this text lost the author who had been most deeply involved in its creation and original writing, and who had been instrumental in the teaching of world regional geography to more than forty thousand students for over four decades at the University of Missouri. The subsequent retirement of Professor Trent Kostbade, Jesse Wheeler's colleague and coauthor for years, provided an opportunity to reshape significant aspects of the text.

As the two new authors added to the team responsible for *Essentials of World Regional Geography,* we have incorporated a number of changes in this edition, including:

- An entirely new art program, with newly rendered maps of cartographic precision unmatched in the world regional geography realm
- Many new photographs which more closely mesh with the issues of the chapters in which they appear
- Expanded coverage of the non-Western world
- Broadened treatment of environmental and cultural outlooks
- New chapters on the Geographer's Field of Vision (Chapter 1), and Physical and Human Processes That Shape World Regions (Chapter 2), which more fully explain the spatial and environmental perspectives that geography brings to world affairs
- Updated (current as of press time) population statistics and political configurations
- Clear and engaging writing style by two professors actively teaching a world regional geography course that reaches more than 1100 students annually.

FEATURES NEW TO THIS EDITION

This second edition incorporates a number of innovations to streamline the text and expand the intellectual range of the world regional geography student. They include:

- **Geographic Profiles.** Each of the text's discussions of the eight regions of the world is introduced by a Geographic Profile chapter. This chapter lays out the basic dimensions of the region and outlines themes that provide the student with a sense of both the individuality of the region and the commonalities that create a regional cohesion.

- **Problem Landscapes.** Thirteen chapters incorporate case studies referred to as Problem Landscapes. These essays bring the student face-to-face with current points of tension in the regions being studied and represent specific areas in which features of geography are at the heart of problem situations demanding current solutions.

- **Landscape in Literature.** In order to add a new dimension to the personality of the regions and nations discussed in this text, fourteen chapters include literature selections relevant to those areas. Essays, novels, or short stories have been excerpted in such a way that students can see that the issues being discussed are not isolated in the world of the textbook but occur in the minds of the peoples of these regions again and again.

- **Definitions and Insights.** These special sections bring to center stage terms and concepts that are critical to understanding the geography of world issues.

- **Regional Perspective.** Geographic issues that work their influence across the expanse of the regions surveyed in the text are defined and made relevant in these features.

- **Environmental themes.** We have given special emphasis to issues of the environment, which are introduced and developed in Chapter 2, Physical and Human Processes That Shape World Regions, and are also referenced throughout the text.

- **End-of-chapter questions.** Thirteen chapters conclude with questions that will lead the student into further research and discovery on important issues in the Special Edition of the Rand McNally Atlas of World Geography published by Saunders College publishing.

ORGANIZATION OF THE TEXT

We have used a fairly common system of regionalization in this text, dividing the world into eight major regions. Each of these regions has certain common features that provide for internal coherence. The opening chapters of the eight regions—The Geographic Profile chapters—outline basic characteristics and describe the sorts of physical and cultural features that create a regional unity. Not every region has the same elements of continuity, but each region can be identified by similar characteristics of language, religion, geographical proximity, or some other blend of features that connect the parts. For example, not all of Latin America speaks Spanish, but the Spanish language and Spanish cultural influence serve as major features of cohesion in the Latin American world region.

Chapters 1 and 2 consider how a geographer sees the world, and how critical environmental assessment, modification, and management is to such a vision. It is in these chapters, and Chapter 2 especially, that a clear link between human-environmental modification and world regional problems is established. Such a viewpoint is essential not only for comprehending the world in which our students will live, but also for demonstrating a geographic perspective on issues that range from resources of demography to environmental perception.

Chapters 3, 4, and 5 explore Europe. In consolidating the European coverage to three chapters from the earlier five, we have focused on providing a concise historical coverage of this important region. Europe is the first region studied because of the centrality of the role of Western civilization in shaping the world in which the great majority of the readers of this text will spend their lives. Geographic dynamics of migration, diffusion, and political flux are developed in this regional unit.

In Chapters 6 and 7, the geographic nature of the Former Soviet Region is outlined. Study of this region presents the students with a first-hand analysis of a major region attempting to reshape itself, abandoning communism in favor of capitalism in fifteen nations rather than one. The tensions associated with such a major political, cultural, and economic transformation help the reader to see the pivotal role that geography plays in such a metamorphosis.

Chapters 8 and 9 take the student deep into a world that every day presents both crisis and opportunity to the United States. The Middle East, a region lived in, researched, and written about extensively by Joe Hobbs, is given crisp analysis through consideration of elements as basic as water and religion as well as complex historical features and patterns of human migration.

Chapters 10 through 14 focus on Monsoon Asia. This region is home to more than half of the world's population and is critical because of the historical significance of its many innovations and land use patterns. Also, the dramatic economic power being marshaled by the "Asian Tigers" in an effort to compete with the economic and cultural force of Europe and North America drives home the important role this region will play in the future. Kit Salter has lived, traveled, researched, and written about the world of China and East Asia, bringing an immediacy to the text's treatment of this region. Even while regional commonalities based on economic development are given focus, these chapters also craft a clear picture of unique national characteristics.

In Chapters 15 and 16, the Pacific World is highlighted as a region of extraordinary isolation that is undergoing change primarily at the hands of the advancing technologies of transport and communication. These chapters carry the student to the Southern Hemisphere and show the historic impact of sheer distance in the development of the cultural landscapes of this region.

Major changes have been introduced in Chapters 17 and 18, which study the geography of the nations in and south of the Sahara. The single chapter on South Africa in the earlier edition has been eliminated and more space has been given to the full ensemble of African nations striving for development, regional independence, and greater access to the wealth of the world.

Latin America is the focus of Chapters 19 and 20, bringing the student back to the hemisphere of our major readership. In these two chapters, attention is devoted to landscape patterns that give unique identity to different subregions within the massive Latin American region. The sometimes difficult and always dominant link between this region and North America is explored in terms of the geographic elements of migration flow and political stability.

The final region covered in the text is North America, in Chapters 21, 22, and 23. Canada continues to be given its own chapter not only because of its close links to the United States, but also because of the growing significance of the North American Free Trade Agreement (NAFTA). In giving major attention to the United States' increasingly global connections, this comprehensive study of the regional geography of North America brings to focus once again the critical geographic elements of environmental perception, resource development, urbanization, and human mobility.

The progression through the text may be taken from Chapter 1 to 23 in straight sequence, or a class may chart its own course after reading the Overview in Chapters 1 and 2. The eight Geographic Profile chapters call attention to the elements that give character and identity to each of the eight regions in the text, and the regional chapters that round out each section are fundamentally self-contained in their presentation and assumptions. A Glossary at the end of the text provides term definitions, should a class cover only a portion of the text in their particular study of world regional geography.

ANCILLARIES

This text is accompanied by a number of ancillary publications to assist instructors and enhance students' learning:

- A Pocket Study Guide aids students in self-evaluation on text material.

- A Map-Pak and Place/Name Workbook contain unlabeled maps with exercise questions.

- The Instructor's Manual contains both general and curricular suggestions for world regional geography teaching.

- A Test Bank, developed by C. L. Salter and J. J. Hobbs, contains a large number of multiple-choice questions keyed to the text.

- The ExaMaster™ Computerized Test Bank is the software version of the printed Test Bank. Instructors can create a great variety of examinations in a multiple-choice format. A command will also reformat multiple-choice questions into short-answer questions. ExaMaster™ has capabilities for recording and graphing students' grades. Available in IBM 3.5″, Macintosh, and Windows formats.

- 100 full-color overhead transparencies of maps and tables are available.

- The World Regional Geography MediaActive™ CD-ROM provides still imagery from Salter/Hobbs/Wheeler/Kostbade and other Saunders College Publishing textbooks. Available

as a presentation tool, this CD-ROM can be used in conjunction with commercial presentation packages such as Powerpoint™, Persuasion™, and Podium™, as well as the Saunders LectureActive™ presentation software. Available in both Macintosh and Windows platforms.

Saunders College Publishing may provide complimentary instructional aids and supplements or supplement packages to those adopters qualified under our adoption policy. Please contact your sales representative for more information. If as an adopter or potential user you receive supplements you do not need, please return them to your sales representative or send them to:

Attn: Returns Department
Troy Warehouse
465 South Lincoln Drive
Troy, MO 63379

ACKNOWLEDGMENTS

We wish to acknowledge the long list of reviewers whose comments helped to mold the second edition. We sincerely thank each and every one of these distinguished geographers and outstanding teachers.

Thomas Anderson	Bowling Green State University
Helen Ruth Aspaas	Utah State University
Warren Bland	Florida State University
Brian W. Blouet	The College of William and Mary
Karen DeBres	Kansas State University
Keith Debbage	University of North Carolina–Greensboro
Donald B. Freeman	York University, Ontario, Canada
Hari Garbharran	Middle Tennessee State University
Jeff Gordon	Bowling Green State University
James Harlan	University of Missouri–Columbia
Todd Heibel	University of Missouri–Columbia
Daniel J. Hammel	Illinois State University
Douglas Heffington	Middle Tennessee State University
Scot Hoiland	Butte College, California
Robert K. Holz	University of Texas–Austin
Patricia Hullet	Northern Oklahoma College
Richard H. Jackson	Brigham Young University
Scott Jeffrey	Catonsville Community College
Karen Johnson	North Hennepin Community College

Thomas Klak	Miami University, Ohio
Karen Koegler	University of Kentucky
Elizabeth Leppman	University of Georgia
Ines Miyares	Hunter College, CUNY
James L. Newman	Syracuse University
Richard Pillsbury	Georgia State University
Henry Rademacher	University of Missouri–Columbia
Rose Sauder	University of New Orleans
Joseph E. Schwartzberg	University of Minnesota
Randy Smith	Ohio State University
Joel Splansky	California State University–Long Beach
Robert C. Stinson	Macomb County Community College
B. L. Sukhwal	The University of Wisconsin–Platteville
Xiaolun Wang	Arizona State University
Barney Warf	Florida State University
Kay Weller	University of Northern Iowa
William Wyckoff	Montana State University

The authors also thank University of Minnesota professor Joseph E. Schwartzberg for an excellent, detailed review of the chapter on the Indian subcontinent in the 1995 edition of this text, which led to improvements in the present chapter.

The new authors wish to extend special thanks and appreciation to the Saunders editorial staff who worked so diligently to help them meet tight deadlines and to provide thoughtful counsel throughout the process. Sara Tenney, and then Jennifer Bortel—as acquisitions editors—did all they could to make new authors feel that we had made the right decision in stepping into the giant shoes of Jesse Wheeler and Trent Kostbade. Beth Rosato as developmental editor, had to deal with two authors new to all the tasks involved in the creation of such texts as this. She did it in good style and continuing efficiency. Thanks to Anne Gibby, project editor, for successfully guiding new authors through rough publishing terrain. Her persistent, one-paragraph, urgent, relentless missives helped us to achieve a very good book in a very short amount of time. Cathy Dexter, copy editor, worked steadfastly to meld two distinct writing and thinking styles. Where she was successful, we laud and thank her. Where it did not work, we take full responsibility for our continuing stubbornness.

Kit Salter gives a special nod to the geographers most deeply involved in making this a timely and more provocative text. Jim Harlan, a colleague at MU, has been a steady and thorough source of data, perspectives, and energy. Kit recognizes the profound devotion to world regional geography that Jesse Wheeler and Trent Kostbade, two giants in the field, have expressed through the first edition of this text. He hails the persistent wisdom and wit shown by Joe Hobbs, Anne Gibby, and Beth Rosato. Their spirit and productivity were absolutely crucial to maintaining the integrity of this book. And a special nod to Chloe of Breakfast Creek, and to all their friends in "The English Patient," who contributed so much to Kit's sanity during this bookmaking process.

Joe Hobbs thanks his parents Gregory and Mary Ann Hobbs, and Quark Expeditions, for providing him with so many opportunities to see the world. Joe is also grateful to Kit Salter for inviting him to coauthor the text, and to the College of Arts and Science at the University of Missouri for the encouragement to teach and write about world regional geography. Finally, Joe thanks his always supportive wife Cindy, his daughter Katherine, who has already spent half her life among the drafts of this book, and Joe and Cindy's newest arrival, Lily, who may grow up with the third edition of *Essentials of World Regional Geography*.

C. L. Salter
Joseph J. Hobbs
June 1997

A "Warning" to Students Assigned to Read This Text

Like all authors who have given life to a college textbook (and, in our case, taken on the monumental task of revising and reshaping this classic world regional geography text), we want you to know that the words inside are real and our perspectives genuine. Between the two of us, we have studied, traveled, or lived in almost all of the landscapes we bring you in these pages. Both of us early on made the choice to give our lives to geography because we believe that the years to come will be powerfully influenced by the issues of resources, population, migration, environmental management, and shifting spatial patterns of political and economic control.

As you launch yourself into the reading of these pages and the studying of these maps and photographs, know that these views come from real people looking at the real world—a world that will be more yours than ours simply because we started this exploration before most of you. We encourage—no, we exhort—you to see these regions and nations through the eyes of a geographer. Look for the patterns, the processes, and the spatial distributions, as well as the regional and human personalities that give identity to the worlds you are studying.

The payoff for true investigation of the issues we raise will not only likely be a better grade in the class that assigns this text, but will also be a clearer understanding of the role that each of you will play during the next five or six decades to keep our extraordinary world in productive shape for the accommodation of the generations *you* will send forward to learn and live in these same landscapes.

C. L. Salter
Joseph J. Hobbs

CHRISTOPHER L. "KIT" SALTER

Kit did his undergraduate work at Oberlin College, with a major in Geography and Geology. He spent three years teaching English at a Chinese university in Taiwan immediately following Oberlin. Graduate work for both the M.A. and the Ph.D. was done at the University of California, Berkeley. He taught at UCLA from 1968 to 1987, when he took on full-time employment with the National Geographic Society in Washington with his wife, Cathy. They were both involved in the Society's campaign to bring geography back into the American school system. Kit has been professor and Chair of the Department of Geography at the University of Missouri–Columbia since moving to the Heartland in 1988.

Themes that have made Salter glad he chose geography as a life field include landscape study and interpretation in both domestic and foreign settings; landscape and literature in order to show students that geography occurs in all writing, not just in textbooks; and geography education to help learners at all levels see the critical nature of geographic issues. He has written more than 100 articles in various aspects of geography; has traveled to a lot of the places that he writes of in this text; and has been wise enough to have a son in Spain, a daughter in the Bay Area, and a neat wife in the heart of the Heartland—Breakfast Creek, Missouri.

JOSEPH J. HOBBS

Joe Hobbs received his B.A. at the University of California–Santa Cruz in 1978, and his Ph.D. at the University of Texas–Austin in 1986. He is an associate professor at the University of Missouri–Columbia, and a geographer of the Middle East with many years of field research on biogeography and Bedouin peoples in the deserts of Egypt. Joe's interests in the region grew from a boyhood lived in Saudi Arabia and India. His research in Egypt has been supported by Fulbright fellowships, the American Council of Learned Societies, and American Research Center in Egypt, and the National Geographic Society Committee for Research and Exploration. He has served as the team leader of the Bedouin Support Program, a component of the St. Katherine National Park project in Egypt's Sinai Peninsula. He is the author of *Bedouin Life in the Egyptian Wilderness* and *Mount Sinai* (both University of Texas Press), and *The Birds of Egypt* (Oxford University Press). He teaches graduate and undergraduate courses in world regional geography, environmental geography, the geography of the Middle East, and the geography of global current events. In 1994 he received the University of Missouri's highest teaching award, the Kemper Fellowship. Since 1984, he has led "adventure travel" trips to remote areas in Latin America, Africa, the Indian Ocean, Asia, Europe, and the High Arctic. Joe lives in Missouri with his wife Cindy, daughters Katherine and Lily, and a wildlife menagerie of tortoises and Old World chameleons.

Contents

Part 4   Monsoon Asia   *257*

CHAPTER 10   A GEOGRAPHIC PROFILE OF MONSOON ASIA   258

CHAPTER 11   COMPLEX AND POPULOUS SOUTH ASIA   274

CHAPTER 12   THE TIGERS AND TRIBULATIONS OF SOUTHEAST ASIA   294

CHAPTER 13   CHINA: AMBITIOUS BLENDING OF THE PAST AND FUTURE   316

CHAPTER 14   JAPAN AND THE KOREAS: ADVERSITY AND PROSPERITY IN THE WESTERN PACIFIC   342

LIST OF MAPS

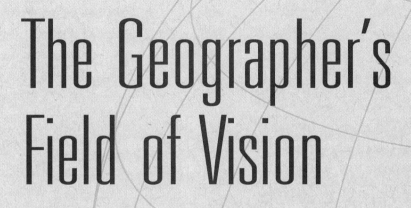

The Geographer's Field of Vision

1

THE GEOGRAPHER'S FIELD OF VISION

sh, the hero of George Stewart's *Earth Abides*, miraculously survives an unexplained plague that kills almost every human in the United States, and perhaps the world. He slowly adjusts to his solitude and undertakes a cross-country car trip in the hope of finding some other humans still alive. In the passage that follows, all he has found so far is a dog he calls Princess.

. . . he did not go any further that night. With his new-found sense of freedom he rather enjoyed merely taking additional chances. He pulled the sleeping-bag out. Unrolling it, he lay on the sand under the slight shelter of a mesquite bush. Princess lay beside him, and went soundly to sleep, tired from her run. Once he awoke in the night, and lay calm at last. He had passed through so much, and now he seemed to know a calm which would never pass. Once Princess whimpered in her sleep, and he saw her legs twitch as if she were still chasing the rabbit. Then she lay quiet; he, too, slept away.

When he awoke, finally, the dawn was lemon yellow above the desert hills. He was cold, and he found Princess close up against his sleeping bag. He crawled out just as the sun was rising. *This is the desert, the wilderness. It began a long time ago. After a while, men came. They camped at the springs and left chips of stone scattered about in the sand there, and wore faint trails through the lines of the mesquite bushes, but you could hardly tell that they had been there. Still later, they laid down railroads, and strung up wires, and made long straight roads. Still, in comparison with the whole desert, you could hardly tell that man had been there, and ten yards aside from the steel rails or the concrete pavement, it was all the same. After a while, the men went away, leaving their works behind them.*

There is plenty of time in the desert. A thousand years are as a day. The sand drifts, and in the high winds even the gravel moves, but it is all very slow. Now and then, once in a century, it may be, there comes a cloudburst, and the long-dry streambeds roar with water, rolling boulders. Given ten centuries perhaps, the fissures of the earth will open again and the black lava pour forth.

But as the desert was slow to yield before man, so it will be slow to wipe out his traces. Come back in a thousand years, and you will still see the chips of stone scattered through the sand and the long road stretching off to the gap in the knife-like hills on the horizon. As for the copper wires, they are next to immortal. This is the desert, the wilderness—slow to give, and slow to take away.

For a while the speedometer needle stood at 80, and he drove with the wild joy of freedom, fearless at the thought of a tire blowing out. Later, he slowed down a little, and began to look around with new interest, his trained geographer's mind focusing upon that drama of [humankind's] passing. In this country, he saw little difference.[1]

---

[1] George Stewart, *Earth Abides*. 1949. pp 50–51. Greenwich, Conn.: A Fawcett Crest Book.

◄ Geography and geographers are continually interested in determining what role humankind has played in changing the Earth's surface. In this scene from southern Australia, even the narrow presence of a dirt road means everything to people who live at the end of, or along, this road. The geography of connections and networks of interaction are central to human development and play a major role in the geographer's field of vision. *David Doubilet*

In this fictional drama of human life virtually vanishing from Earth, author George Stewart's hero, Ish, tries to get a sense of the scale of the unknown catastrophe that has befallen the world. Bitten by a rattlesnake while climbing in the hills east of San Francisco, Ish had drifted in and out of consciousness for a number of days while a plague was changing forever the world in the lowlands. In this excerpt, he has just begun a car trip across the United States to see if he can locate any other human survivors and to determine if the plague that appears to have eliminated humankind from California was as thorough in its impact in other parts of the country.

In the final lines of the passage, Stewart reveals that Ish had been trained as a geographer. As Ish looks out on this desert scene with its minimal signs of earlier human occupance, he begins to learn what the world offers him for survival.

Definitions and Insights

GEOGRAPHY

Geography is the study of the Earth's surface and the spatial patterns of its physical and human characteristics. Geographers determine what distinct forces of nature and society have been at work in the creation of the **landscape**—the collection of physical and human geographic features on the Earth's surface—and evaluate the role such a landscape will have on the economic and social development of the local area. Such evaluation involves **environmental perception**—our individual response to environmental features—**environmental assessment**—our determination of the value and proper management of a given environment—**geographic analysis**, and the analysis of **spatial patterns**—the distribution of given geographic phenomena—and the characteristics of **site and situation**.

Our hope in writing this text is that you, too, will begin to develop a sense of the geographer's mind and learn the richness and usefulness of a geographer's field of vision. The ability to come upon a scene—whether it be a desert landscape, an urban intersection, a broad, well-farmed valley, or even a photograph on the front page of a newspaper or an image on television—and to begin to make sense of the physical and cultural geographic elements that create this image is one of the skills we hope to achieve with this text. Ish uses such vision again and again in *Earth Abides*. We hope that you, too, will add this talent to your own learning as you come to grips with the concepts and characteristics of world regional geography at the dawn of the twenty-first century.

Geography has many perspectives, and our profession just recently created a learning model (Table 1.1) that divides geography into six realms, or the Six Essential Elements (see *Geography for Life: National Geography Standards 1994* for a full discussion of the elements and their associated Geography Standards). The geographer's field of vision draws on each of these elements in distinct ways, and the composite enables one to see the world in an ordered and logical way.

1.1   The World in Spatial Terms

We recognize that all phenomena of the world can be organized in spatial patterns. As random as the shape of a coastline might seem, there is a spatial order in the location of both physical and cultural phenomena (Fig. 1.1). As we open this text's chapters on specific regions, we will give you facts on each region's area, population, and characteristics of the natural environ-

FIGURE 1.1

This aerial shot of Iowa farmland demonstrates the blend of physical features and human landscape transformation. The differing crop patterns, the few stands of trees that remain by streams, and the trees that have been planted by settlements are all signatures of the cultural landscape. There has often been considerable human attention to the planning and shaping of the landscape, as is so evident in this scene.   *©1994 Stephen Graham/Dembinsky Photo Associates*

TABLE 1.1

The Six Essential Elements and the Eighteen Geography Standards

I. The World in Spatial Terms
Geography studies the relationships between people, places, and environments by mapping information about them into a spatial context.
The geographically informed person knows and understands
1. how to use maps and other geographic representations, tools, and technologies to acquire, process, and report information from a spatial perspective
2. how to use mental maps to organize information about people, places, and environments in a spatial context
3. how to analyze the spatial organization of people, places, and environments on Earth's surface

II. Places and Regions
The identities and lives of individuals and peoples are rooted in particular places and in those human constructs called regions.
The geographically informed person knows and understands
4. the physical and human characteristics of places
5. that people create regions to interpret Earth's complexity
6. how culture and experience influence people's perceptions of places and regions

III. Physical Systems
Physical processes shape Earth's surface and interact with plant and animal life to create, sustain, and modify ecosystems.
The geographically informed person knows and understands
7. the physical processes that shape the patterns of Earth's surface
8. the characteristics and spatial distribution of ecosystems

IV. Human Systems
People are central to geography; human activities, settlements, and structures help shape Earth's surface, and humans compete for control of Earth's surface.
The geographically informed person knows and understands
9. the characteristics, distribution, and migration of human populations
10. the characteristics, distribution, and complexity of Earth's cultural mosaics
11. the patterns and networks of economic interdependence
12. the processes, patterns, and functions of human settlement
13. how the forces of cooperation and conflict among people influence the division and control of Earth's surface

V. Environment and Society
The physical environment is influenced by the ways in which human societies value and use Earth's natural resources, while at the same time human activities are influenced by Earth's physical features and processes.
The geographically informed person knows and understands
14. how humans modify the physical environment
15. how physical systems affect human systems
16. the changes that occur in the meaning, use, distribution, and importance of resources

VI. Uses of Geography
Knowledge of geography enables people to develop an understanding of the relationships between people, places, and environments over time—that is, of Earth as it was, is, and might be.
The geographically informed person knows and understands
17. how to apply geography to interpret the past
18. how to apply geography to interpret the present and plan for the future

Source: *Geography for Life: National Geography Standards 1994.* National Geographic Research and Exploration, Washington, D.C.

ment—basic spatial elements that will help you build your own map of the regions of the world. These characteristics are fundamental to seeing the world in spatial terms.

MAPS

To think geographically is to think in terms of such spatial patterns, associations, and interconnections into which details fit. This inevitably means to think in terms of maps, for maps portray spatial patterns and associations with a clarity that is generally beyond the reach of words. Students should refer to maps frequently as they read geographic material, use outline maps for note taking, make their own sketch maps to show important spatial relationships, and try to visualize map relationships when patterns are stated in words in text material. In short, thinking geographically is to think graphically.

Definitions and Insights
SPATIAL ANALYSIS

The term **spatial** comes from the noun **space**, and it relates to the ways in which space is organized or patterned. In **spatial analysis**, geographers examine the patterns created in the distribution of phenomena (farms, auto dealerships, port cities, Baptists, etc.). In considering **spatial organization**, one studies the decision-making processes and outcomes that give such distributions their particular configuration. Important to the understanding of this term is the knowledge that almost all phenomena are characterized by a spatial organization of one sort or another. Even aspects of the world that appear random are generally bound by some system of ordered location and distribution.

A person's life is filled with numerous local spatial realities. The route you take when you walk, ride, or drive to school is built around spatial decisions: which road, which highway, what parking lot or structure? When you come into a classroom—particularly on the first day of a new term—you have a whole raft of spatial decisions to make: Sit by the aisle, sit in the back, sit by friends, sit in new territory, sit alone? In adjusting to a new campus and town, you ask yourself, Where do I eat? What street do I explore to find a bookstore, coffee house, the Salvation Army Thrift Store, a theater? When you have the leisure or the money to seek entertainment, where do you go? Life is quilted together by the sorts of spatial decisions we make, and each of those is part of our personal geography.

One reason for studying geography is to learn where things are in the world, to acquire a framework upon which countries, important cities, rivers, mountain ranges, climatic zones, agricultural and industrial areas, and other features can be arranged in relation to each other. No person can claim to be truly edu-

cated who does not carry this kind of map in his or her mind. However, it is not enough simply to know the *facts* of location; one must also develop an understanding and appreciation of the *significance* of location.

Great Britain and New Zealand: An Example of Location Dynamics. Perhaps an illustration will clarify these remarks. Let us compare the location of Great Britain with that of New Zealand (Fig. 1.2). Both are island countries. Westerly winds,

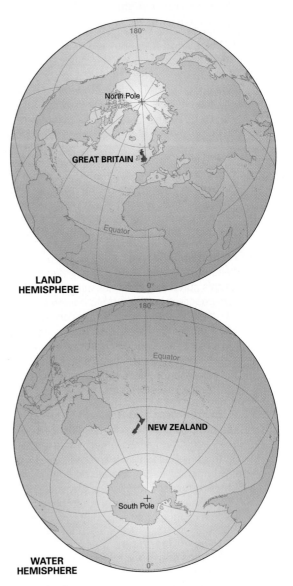

FIGURE 1.2
In the top map, note how the major land masses are grouped around the margins of the Atlantic and Arctic oceans. The British Isles and the northwestern coast of Europe lie in the center of the "land hemisphere," constitute 80 percent of the world's total land area and have approximately 91 percent of the world's population. New Zealand lies near the center of the opposite hemisphere, or "water hemisphere," which has only 20 percent of the land and 9 percent of the population.

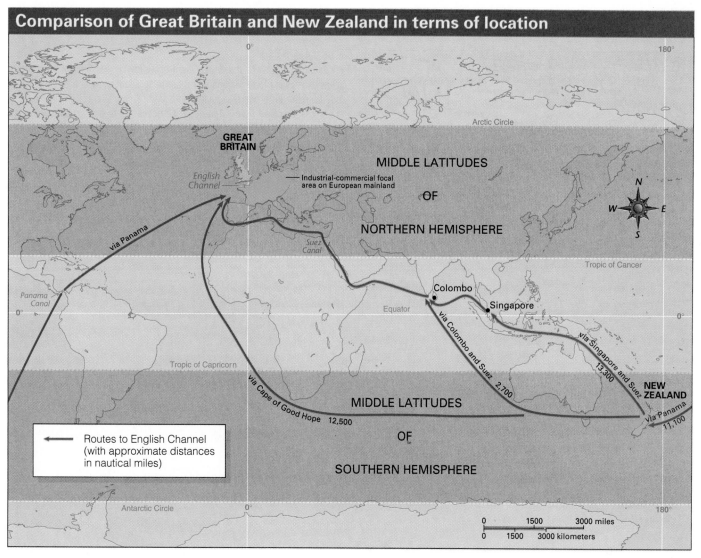

Comparison of Great Britain and New Zealand in terms of location

FIGURE 1.3
Great Britain and New Zealand compared in location.

which blow off the surrounding seas, bring abundant rain and moderate temperatures throughout the year. Their climates are remarkably similar, even though these areas are in opposite hemispheres and are about as far from each other as it is possible for two places on the Earth to be.

Great Britain is located in the Northern Hemisphere, which contains the bulk of the world's land and most of the principal centers of population and industry; New Zealand is on the other side of the equator, in the Southern Hemisphere. Great Britain is located near the center of the world's land masses (Fig. 1.3) and is separated by only a narrow channel from the densely populated industrial areas of western continental Europe; New Zealand is surrounded by vast expanses of ocean. Great Britain is located in the western seaboard area of Europe, where many major ocean routes of the world converge; New Zealand is far away from the centers of world commerce. For more than four centuries, Great Britain has shared in the development of northwestern Europe as a great organizing center for the world's eco-

nomic and political life; New Zealand, meanwhile, has existed in comparative isolation (Fig. 1.3). Great Britain, in other words, has a *central* location within the existing frame of human activity on the earth, whereas New Zealand has a *peripheral* location. Centrality of location is a highly important factor to consider in assessing the economic as well as political geography of any country, region, or other place.

Mental Maps

Our understanding of location, however, is not completely objective, for each of us has our own personal sense of geography—that is, a series of **mental maps**—which includes relative location. A mental map is the collection of personal geographic information that we use to order the images and facts we have about places, both local and distant. The mental map also serves to accommodate changes that one might imagine for a landscape (Fig. 1.4). In a course like this one, we will confront

7

FIGURE 1.4

Personal perception of the environment is widely varied. The potential future scenes for example, envisioned by three people looking at undeveloped land may be quite different. Geography is concerned with understanding where these images come from, what might be necessary to bring such images to reality, and what the environmental consequences of such development might be.   *The Key to the National Geography Standards, Illustration by Suzanne Dunaway*

FIGURE 1.5

Map of major world regions that form the basic framework of this text. Some regions overlap others. Former Soviet Region refers to the total expanse occupied by the Former Soviet Union (Union of Soviet Socialist Republics) prior to the period of dissolution that began with independence for the Baltic Republics, 1990–1991 (see Chapter 6). Names and boundaries of the 15 new independent states in the former Soviet Union are shown in Figure 6.3.

and reshape your mental map repeatedly as we introduce new images and facts about places that you have possibly never seen but about which you have some impressions. We will try to expand your geographic sense of place about a world of regions and nations, not only through this text's facts and concepts, but also through the literature we include—such as this chapter's opening excerpt from *Earth Abides*.

The Language of Maps

Because maps are so basic to all geographic understanding, it is important to learn the language of maps. **Cartographers** are the geographers and skilled draftspeople who create maps through careful design and skillful presentation of geographic information. Increasingly, **cartography**—the art and science of making maps—utilizes computer manipulation of spatial data, a field increasingly dependent on computers and known as **computer cartography**. Because the field of geography considers anything that potentially can be mapped, cartography is first and foremost a graphic portrayal of **location**. The cartographer shows where things and places on the Earth are located in relation to each other (Fig. 1.5). Thus, maps are indispensable tools for discovering, examining, and portraying spatial relationships and associations. Some maps show the **distribution** and interrelation of things that are fixed in place, such as structures and terrain features, whereas other maps portray the **flow**

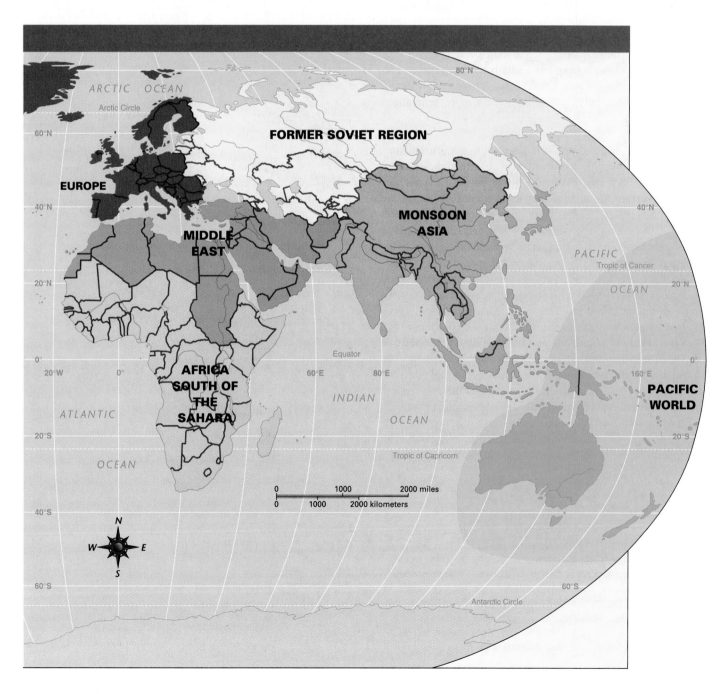

of people, goods, and ideas from place to place. A map communicates locational facts and concepts to us by means of a specialized "language" that has to be learned so that the map can be properly understood and used. Major elements of this language are **scale**, **projection**, and **symbolization**.

Scale.   A map is a reducer; it enables us to comprehend an extent of Earth-space by reducing it to the size of a single sheet. The amount of reduction appears on the scale—that is, the actual distance on the Earth represented by a given linear unit on the map. A common way of denoting scale is to use a representative fraction, such as 1/10,000 or 1/10,000,000. In other words, one linear unit on the map (for example, an inch or centimeter) equals 10,000 or 10,000,000 such real-world units on the ground. A **large-scale map** is one with a relatively large representative fraction (for example, 1/10,000 or even 1/100) that portrays a relatively small area in more detail (Fig. 4.15). A "global" or **small-scale map** has a relatively small representative fraction (for example, 1/1,000,000 or 1/10,000,000) that, in contrast, portrays a relatively large area in more generalized terms (Fig. 1.6).

Projection.   A map projection is a device to minimize distortion in one or more properties of a map (direction, distance, shape, or area). All maps inherently distort because it is not possible to represent the three-dimensional curved surface of the Earth with complete accuracy on a two-dimensional flat sheet of paper. If the area represented is very small (for example, a large-scale map), the distortion may be slight enough to be disregarded, but maps that represent larger spatial areas (small-scale maps) may introduce very serious distortions. A classic example is the well-known Mercator projection (Fig. 1.6), originally designed to aid navigation. On that projection, every straight line has a constant bearing (direction). To achieve this, both the **parallels of latitude** and the **meridians of longitude** are shown as straight lines. A globe can show both geographic shapes and locations without distortion; lines of longitude draw closer together as they go from the equator to the poles where they converge. On a Mercator map these lines remain parallel. For these reasons, a Mercator map greatly exaggerates the east-west dimension of areas near the poles. It also exaggerates the north-south dimension because the parallels are not spaced

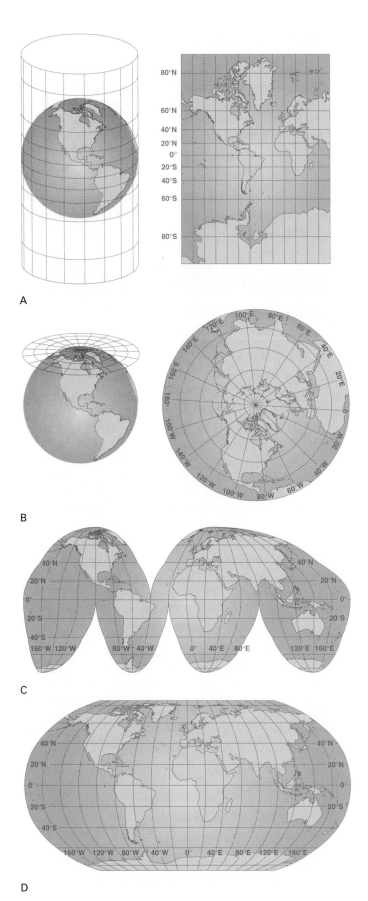

A

B

C

D

FIGURE 1.6

These assorted map projections each represent a specific goal in representing the three-dimensional surface of the Earth on a two-dimensional sheet of paper. Note the name of each different projection included in this set of map figures and try to determine what function that projection deals with. Look at the maps at the front of any good college atlas (these are from Goode's 19th edition published by Rand McNally) and you will be able to find a detailed explanation of a variety of projections.

evenly between the equator and the poles, as they are on a globe, but gradually draw farther and farther apart with increasing distance from the equator. On a Mercator map, the mathematical, or absolute, location (latitude and longitude) of every place is accurate (see Definitions and Insights, below), but areas near the poles are greatly distorted in size and shape, with the result that Greenland appears bigger than all of South America, although in reality it is only slightly larger than Mexico. There is great variety in the map projections utilized in reproducing the shape of continents on a flat surface (Fig. 1.6).

Definitions and Insights
PARALLELS OF LATITUDE AND MERIDIANS OF LONGITUDE

The term **latitude** denotes position with respect to the equator and the poles. Latitude is measured in degrees (°), minutes ('), and seconds ("). The equator, which circles the globe midway between the poles, has a latitude of 0°. All other latitudinal lines are parallel to the equator and to each other and therefore are called **parallels**. Every point on a given parallel has the same latitude. Places north of the equator are in north latitude; places south of the equator, in south latitude. The highest latitude a place can have is 90°N or 90°S latitude. Thus, the latitude of the North Pole is 90°N, and that of the South Pole is 90°S. Places near the equator are said to be in **low latitudes**; places near the poles, in **high latitudes**. The **Tropic of Cancer** and the **Tropic of Capricorn**, at 23.5°N and 23.5°S respectively, and the Arctic and Antarctic Circles, at 66.5°N and 66.5°S respectively, form convenient and generally realistic boundaries for the low and high latitudes. Places occupying an intermediate position with respect to the poles and the equator are said to be in **middle latitudes** (Fig. 1.3).

Meridians of longitude are straight lines connecting the poles. Every meridian is drawn due north and south. They converge at the poles and are farthest apart at the equator. Longitude, like latitude, is measured in degrees, minutes, and seconds. The meridian that is most often used as a base (starting point) is the one at the Royal Astronomical Observatory in Greenwich, England. It is known as the **meridian of Greenwich** or **prime meridian** and has a longitude of 0°. Places east of the prime meridian are in east longitude; places west of it are in west longitude. The meridian of 180°, exactly halfway around the world from the prime meridian, is the other dividing line between places east and west of Greenwich. At the equator, 1° of longitude is equivalent to 69.15 statute miles (111.29 km). However, because the meridians converge toward the poles, a degree of longitude at the Arctic Circle is equivalent to only 27.65 miles (44.50 km).

The combination of latitude and longitude give us **mathematical** or **absolute location**.

Symbols.   Maps enable us to extract certain items from the totality of things, to see the patterns of distribution they form, and to compare these patterns with each other. No map is a complete record of an area; instead it represents a selection of certain details, shown by **symbols**, that a cartographer records to accomplish a particular purpose. Unprocessed data must be classified in order to provide categories that the symbols will represent. The classes may be categories of physical or cultural forms (rivers, roads, settlements, and so on), aggregates such as 100,000 people or one million tons of coal, or averages such as **population density** (the number of persons per square mile within a defined area). Aggregates or averages are frequently ordered into ranked categories in a graded series (for example, population densities of 0–49, 50–99, or 100 or more people per square mile) for portrayal on the map. The categories are not self-evident but are selected by the cartographer, sometimes by elaborate statistical procedures. To represent varied phenomena, the cartographer has available a wide range of symbols: **lines**, **dots**, **circles**, **squares**, **shadings**, and others. Color, increasingly, is used in the production of maps.

Dots usually portray quantities, and the dot map is one of the most common types of maps showing distributions of people or things on the Earth (Fig. 2.16). On a dot map of population, for example, each dot represents a stated number of people and is placed as near as possible to the center of the area that these people occupy. In interpreting such a map, we are not greatly concerned with the individual dots but with the way the dots are arranged—that is, the way they cluster—and with the **pattern**—that is, the distribution of a given phenomenon—they form. Some maps show a rather even spacing of the dots, but others show the dots strongly clustered in some areas and very sparse in others, reflecting the real-world spatial distribution of the phenomenon being mapped (Chapter 2).

One complexity of cartography is that different scales and ways of showing distributions can give different impressions. For example, methods of symbolization other than dots could be used to show the distribution of world population and might convey rather different ideas from the dot map presented in Figure 2.16. Some alternative methods might include the use of a single symbol for each country, with the size of the symbol proportional to the country's population, or drawing a map on which the area of each country is proportional to the item being shown—such as caloric intake, literacy, population, or number of doctors per 100,000 population. This sort of a proportional map is called a **cartogram** (Fig. 1.7a). As you compare the distribution of any given entity by differing gradients (colors or patterns) in **choropleth maps** (Fig. 1.7b), you can see a graphic distinction from the cartogram. In looking at the two maps in Figure 1.7 you can see how the different goals of the cartographer—as determined by the style of map selected—will influence the maps chosen for a text, or for a presentation. Even if symbolization by dots is retained, one map might use one dot per 100,000 people and another show one dot per 1 million people, also producing a very different look to the two maps.

geography studies the locational associations, internal spatial organization, and functions of cities.

Geographers specializing in **cultural geography** concern themselves with places of origin ("culture hearths"), diffusion, interactions, landscape evidence, and regionalization of human cultures (see Definitions and Insights, page 16). Closely allied to cultural geography is the field of **cultural ecology**, which focuses on the relationship between culture and environment. **Social geography** deals with spatial aspects of human social relationships, generally in urban settings; **population geography** assesses population composition, distribution, migration, and demographic shifts. **Political geography** studies such topics as spatial organization of geopolitical units, international power relationships, nationalism, boundary issues, military conflicts, and regional separatism within states. **Historical geography** studies the geography of past periods and the evolution through time of such geographic phenomena as cities, industries, agricultural systems, and rural settlement patterns.

Technical specialties in geography include cartography, computer cartography, **remote sensing, quantitative methods** (mathematical model building and analysis), **air photo interpretation**, and **geographic information systems (GIS)**. The last of these—GIS, which involves the computer manipulation of spatial data—is a very successful interface between the analysis of spatial data and the computer (Fig. 1.A). There is increasing use of this technique in everything from spatial analysis to model building and analysis to sophisticated cartography. All of these technique-oriented specialties center on the cartographic (mapped) expression of spatial data while remote sensing and air photo interpretation utilize various kinds of satellite imagery or photo coverage to assess land use or geographic patterns. The expanding use of the computer gives geographers greater speed, accuracy, and facility in incorporating new and changing data into the creation of maps, graphs, and charts. This has become an area of major professional activity for geographers.

In this text we call upon these perceptions and interactions to help us understand the past as well as to chart the possible and potential courses of future events. Knowing about the spatial distribution of resources, mountain passes, ethnic populations, animal herds, and settlements, for example, means that one can achieve a clearer understanding of the past. History has been shaped by the interplay of geographic influences and human ambitions—whether at the scale of a peasant farmer trying to feed his family or of an army commander trying to expand his region of military authority. One of the prime uses of geography, then, is to be able to determine and understand the map of the locale being studied at the time of the event you are attempting to comprehend.

Understanding one place helps prepare you for seeing the meaning of yet other places. In this respect, no place is truly an island, although some locales are more insular than others; nor are places the same in the magnitude, intensity, and reach of the geographic effects they engender. A hurricane in Florida, for example, will have an impact on citrus prices in California. The return of Hong Kong to China has influenced financial markets all across the world. Such effects may be physical, economic, political, cultural, or social, and they arise from trade, investment, migration, military action (or the threat of it), and other mechanisms of spatial interaction. Especially large and complex effects are generated by such major nations as the United States, Japan, Germany, and Brazil, but many smaller and less powerful places may have an impact disproportionate to their size.

PLACE INTERACTION AND FOCAL POINTS IN UNDERSTANDING WORLD REGIONAL GEOGRAPHY

One of the most reliable uses of geography is the study of the geographic impact of places on each other. Awareness of this factor leads to enhanced appreciation of the world as an extremely complicated system in which conditions, activities, events, and changes in one part affect various other parts in diverse ways. The same mode of thought is applicable to the internal geography of countries: Within a given country, certain areas exert greater geographic impact than others. Prominent examples that appear throughout this text—to clarify regional and place identity—include:

1. major metropolitan areas;
2. major manufacturing areas, especially those that specialize in products of basic importance to the country's economy;
3. major seaports and other centers of commerce;
4. unsually productive or distinctive agricultural areas, particularly those that produce an important surplus for export;
5. areas that produce minerals on which the country's economy is very dependent;
6. areas that differ markedly in culture from the rest of the country, especially if this results in political tension;
7. areas with a pronounced strategic importance; and
8. areas of unusual poverty that consume a disproportionate share of the nation's revenues.

In this text, we treat certain areas at greater length than other areas that have a larger relative area or population but are seen as less critical in their impact on the country or the world.

All of these geographic perspectives play a role in understanding world regional geography. The role of the map, of the distributional patterns of human resources and physical properties, and of the dynamics of interaction between varied and significant players in the drama have keen importance to our making sense of the world. Many of the problems, for example, that your generation will have to deal with are problems that are evident in today's cultural landscape—that is, the landscape transformed by human action. We want you to see a utility in studying the world's regions and nations through a geographer's eyes, with a geographer's field of vision.

2

PHYSICAL AND HUMAN PROCESSES THAT SHAPE WORLD REGIONS

©1990 Tom Van Sant, Inc. / The Geo
Santa Monica, California

n the following chapters you will study the astonishing variety of environments, cultures, and events that characterize the geography of our world today. It is a complex world. You cannot learn what "makes Earth tick" without knowing a great deal about particular places. But details should not overwhelm you; in this chapter you will be introduced to the larger geographic processes and patterns that shed light on the details and make them more meaningful.

The geographer's field of vision sees physical and human processes at work around the world. We begin here by introducing climatic patterns, types of vegetation, and the flow of energy through ecosystems so that you can better understand how people around the world accommodate and often challenge nature. We introduce modern patterns of people-land associations by viewing them as the products of revolutionary changes in the past: the arrival of agriculture and industrialization. You will see where rich and poor countries are located on the face of the Earth, and understand some of the causes of their prosperity and poverty. You will see where and why populations are increasing and what the implications of that growth are. You will study human impacts on the atmosphere and consider actions to solve some of the most important global problems of our time.

2.1 Climatic Processes and Their Geographic Results

As you experience a warm, dry, cloudless summer day or a cold, wet, overcast winter day, you are encountering the **weather**—the atmospheric conditions prevailing at one time and place. The **climate** is a typical pattern recognizable in the weather of a large region over a long time. Along with surface conditions such as altitude and soil type, climatic patterns have a strong correlation with patterns of natural vegetation and, in turn, with human opportunities and activities on the landscape.

PRECIPITATION

Water is essential for life on Earth, so we begin our study of climates with precipitation. Warm air holds more moisture than cool air, and therefore precipitation is best understood as the result of processes that cool the air to release moisture (Fig. 2.1). Precipitation results when water vapor in the atmosphere cools to the point of condensation, changing from a gaseous to a liquid or solid form. The amount of cooling necessary depends on the original temperature and the amount of water vapor in air.

◄ Earth's environments as seen from space. This composite of satellite images shows forested areas in green, snowcapped mountains and masses of ice in white, largely barren deserts in yellowish-tan, and croplands, grasslands, and tundra in varied shades. Note the prominence of the tropical rain forest of the Amazon Basin in South America, the mass of high mountains in western China and its borderlands, the huge coniferous forest regions of Russia and Canada, and the east-west belt of deserts in North Africa and adjacent Asia.

World Precipitation

AVERAGE ANNUAL PRECIPITATION

Over 40 inches (100 cm)

20–40 inches (50–100 cm)

10–20 inches (25–50 cm)

Under 10 inches (25 cm)

FIGURE 2.1
World precipitation map, modified from a map by the United States Department of Agriculture.

For this cooling and precipitation to occur, generally air must rise in one of several ways. In equatorial latitudes or in the high-sun season (summer, when sun's rays strike the Earth's surface more vertically) elsewhere, air heated by intense surface radiation may rise rapidly, cool, and produce a heavy downpour of rain—an event that occurs often in summer over the Great Plains of the United States. Precipitation that originates in this way is called **convectional precipitation** (Fig. 2.2). **Orographic (mountain-associated) precipitation** results when moving air strikes a topographic barrier and is forced upward. Most of the precipitation falls on the windward side of

the barrier, and the lee (sheltered) side is likely to be excessively dry. Such dry areas—for example, Nevada, which is on the lee side of the Sierra Nevada mountain range in the western United States—are in a **rain shadow**. Rain shadows are the primary cause of arid and semiarid lands in some regions.

Cyclonic or **frontal precipitation** is generated in traveling **low-pressure cells (cyclones)** that bring air masses with different characteristics of temperature and moisture into contact (Figs. 2.2 and 2.3). A cyclone may overlie hundreds or thousands of square miles of the earth's surface. In the atmosphere, air moves from areas of high pressure to areas of low pressure.

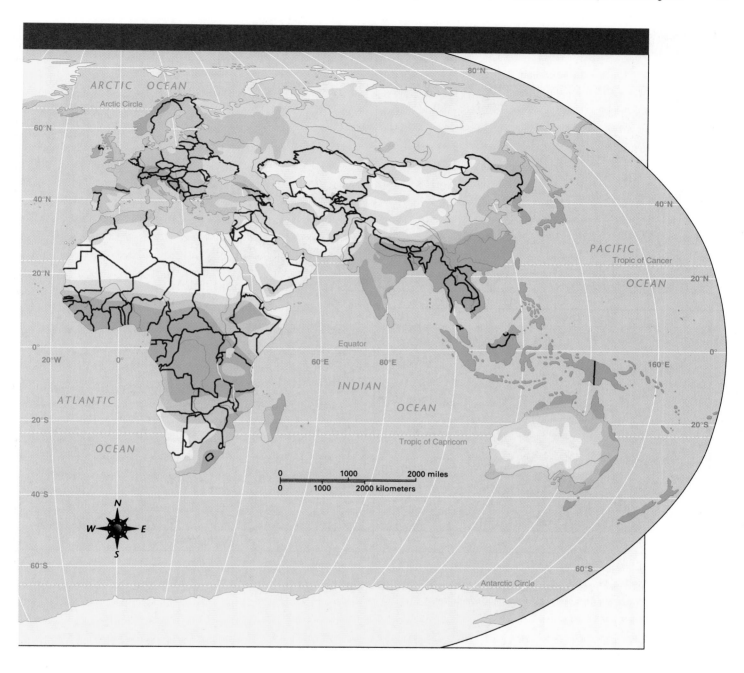

Air from masses of different temperature and moisture characteristics is drawn into a cyclone, which has low pressure. One air mass is normally cooler, drier, and more stable than the other. Such masses do not mix readily and tend to retain distinctive characteristics. They come in contact within a boundary zone 3 to 50 miles (*c*. 5 to 80 km) wide called a **front**. A front is named according to which air mass is advancing to overtake the other. In a **cold front**, the colder air wedges under the warmer air, forcing it upward and back. In a **warm front**, the warmer air rides up over the colder air, gradually pushing it back. But whether a warm or cold front, precipitation is likely

to result because the warmer air mass rises and condensation takes place.

The middle latitudes, especially in winter, experience a succession of traveling cyclones moving from west to east in an airstream called the **westerly winds** (Fig. 2.4). These cyclones rotate slowly, somewhat like whirls and eddies in a stream. Normally there are two fronts as a cyclone passes, first the warm front and then the cold front. But the cold front moves faster and eventually overtakes the warm front. The cyclone is then said to be **occluded**, and it disappears from the atmospheric pressure map.

Types of Precipitation

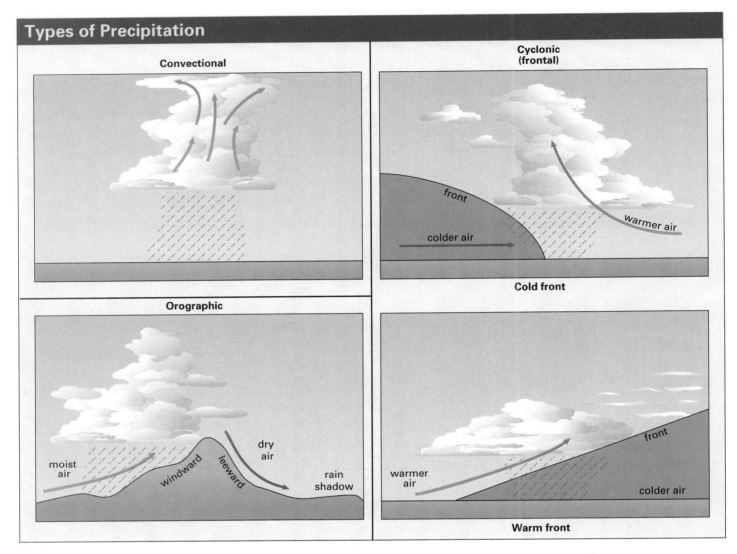

Convectional

Cyclonic (frontal)

front

colder air

warmer air

Cold front

Orographic

moist air

windward

leeward

dry air

rain shadow

warmer air

front

colder air

Warm front

FIGURE 2.2
Diagrams showing origins of convectional, orographic, and cyclonic precipitation. Note that cyclonic precipitation results from both cold fronts and warm fronts.

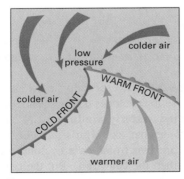

colder air

low pressure

colder air

COLD FRONT

WARM FRONT

warmer air

FIGURE 2.3
Diagram of a cyclone (Northern Hemisphere). The entire system is moving west to east in the airstream of the westerly winds.

Large areas of the Earth receive very small amounts of precipitation (see Fig. 2.1), primarily because of subsiding air masses of high atmospheric pressure. Some parts of the atmosphere generally exhibit high pressure; these are known as semipermanent **high-pressure cells**, or **anticyclones** (see Fig. 2.4). In such a high, the air is descending and thus becomes warmer as it comes under the increased pressure (weight) of the air above it. As it warms, its capacity to hold water vapor increases, its relative humidity decreases, and the result is minimal condensation and precipitation. Streams of dry, stable air moving outward from the anticyclones often bring prolonged drought to the areas below their path. Most famous are the **trade winds**—streams of air that originate in semipermanent anticyclones on the margins of the tropics and are attracted

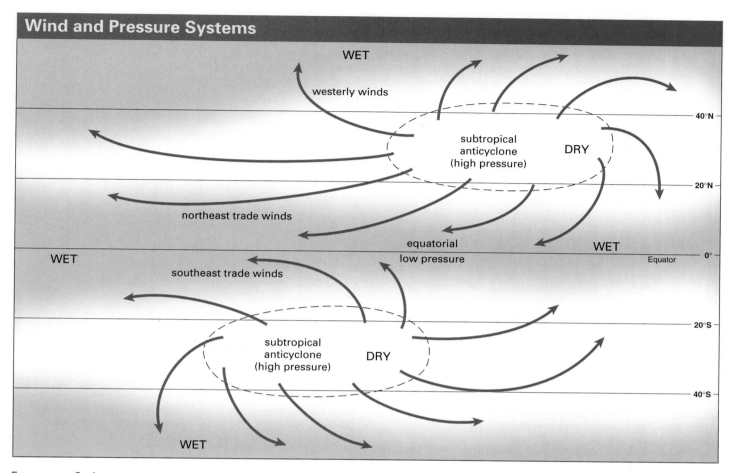

Wind and Pressure Systems

FIGURE 2.4

Idealized wind and pressure systems. Irregular shading indicates wetter areas.

equatorward by a semipermanent low-pressure cell, the **equatorial low**.

Cold ocean waters are responsible for the existence of coastal deserts in some parts of the world, such as the Atacama Desert in Chile and the Namib Desert in southwestern Africa. Here, air moving from sea to land is warmed. Instead of yielding precipitation, its capacity to hold water vapor is increased and its relative humidity is decreased; the result is little precipitation. But many areas of excessively low precipitation result from a combination of influences. The Sahara of northern Africa, for example, seems to be primarily the result of high atmospheric pressure. The rain-shadow effect of the Atlas Mountains and the presence of cold Atlantic Ocean waters along its western coast contribute to the Sahara's dryness.

TEMPERATURE

Most of the sun's visible short-wave energy that reaches the land or ocean surface is absorbed, but some of it returns to the atmosphere in the form of infrared long-wave radiation, which

generates heat and is the principal agent that warms the atmosphere. Although the sun is the initial and primary source of the heat, the air is heated mainly by re-radiation from the underlying land or water surface.

The Earth varies greatly from place to place in the total amount of energy received annually from the sun; it also varies in the resulting air temperatures. In general, absent the effects of cloud cover, the lower the latitude of a place, the more solar energy it receives annually. The major reason for this is that on average throughout the year, the sun's rays strike the Earth more nearly vertically at lower latitudes, concentrating a given amount of solar energy on a smaller extent of surface than at higher latitudes (Fig. 2.5). A second reason is that when solar rays approach the Earth at a more nearly vertical angle in lower latitudes, they must pass through a smaller thickness of absorbing and reflecting atmosphere before reaching the surface (Fig. 2.5). However, many humid areas near the equator have a relatively abundant cloud cover that reflects, scatters, or absorbs much of the incoming solar radiation and thus reduces the amount reaching the surface. The result is that the areas of highest annual solar energy actually received at the surface are in

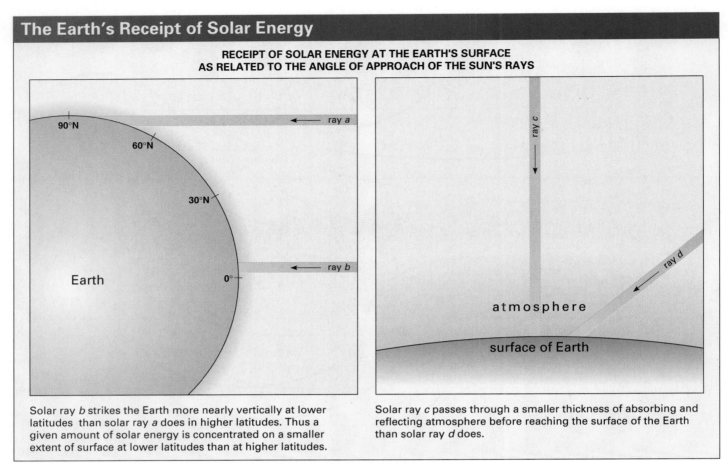

The Earth's Receipt of Solar Energy

**RECEIPT OF SOLAR ENERGY AT THE EARTH'S SURFACE
AS RELATED TO THE ANGLE OF APPROACH OF THE SUN'S RAYS**

Solar ray *b* strikes the Earth more nearly vertically at lower latitudes than solar ray *a* does in higher latitudes. Thus a given amount of solar energy is concentrated on a smaller extent of surface at lower latitudes than at higher latitudes.

Solar ray *c* passes through a smaller thickness of absorbing and reflecting atmosphere before reaching the surface of the Earth than solar ray *d* does.

FIGURE 2.5
Diagrams showing receipt of solar energy as related to Earth curvature and thickness of atmosphere.

the vicinity of 20° to 30° north and 20° to 30° south of the equator. In these subtropical zones, the average angle of the sun's rays is still close to vertical and the cloud cover is much less. The world's greatest deserts are found at these latitudes.

Great differences also exist in the world's annual and seasonal temperatures. In lowlands near the equator, temperatures remain high throughout the year, while in areas near the poles, temperatures remain low for most of the year. The intermediate (middle) latitudes have well-marked seasonal changes of temperature, with warmer temperatures generally in the summer season of high sun and cooler temperatures in the winter season of low sun (when sunlight's angle of impact is more oblique and daylight hours are shorter). Intermittent incursions of polar and tropical air masses increase the variability of temperature in these latitudes, bringing unseasonably cold or warm weather.

Also, air at lower elevations is denser and contains more water vapor and dust than air at higher elevations. Thus it ab-

sorbs more radiation and is warmer. Increasing altitude lowers temperatures on average about 3.6°F (2.0°C) for each increase of 1000 feet (305 m) in altitude. As a result, a place on the equator at an elevation of several thousand feet may have temperatures resembling those of a middle-latitude lowland.

MAJOR TYPES OF CLIMATES AND NATURAL VEGETATION

Weather reflects a great variety of local climates. Geographers group the local climates into a limited number of major **climate types** (Fig. 2.6), each of which occurs in more than one part of the world and is associated closely with other types of natural features, particularly vegetation. Geographers recognize 10 to 20 major types of terrestrial ecosystems, called **biomes**, which are categorized by dominant type of natural vegetation (Fig.

2.7). Climate plays the main role in determining the distribution of biomes, but differing soils and landforms may promote different types of vegetation where the climatic regime is essentially the same. Vegetation and climate types are sufficiently related that many climate types take their names from vegetation types—for example, the **tropical rain forest climate** and the **tundra climate**. You may easily see the geographic links between climate and vegetation by comparing the maps in Figures 2.6 and 2.7. The spatial distributions of climate and vegetation types do not overlap perfectly, but there is a high degree of correlation.

In the **ice-cap**, **tundra**, and **subarctic climates**, the dominant feature is a long, severely cold winter, making agriculture difficult or impossible. The summer is very short and cool. There is no vegetation on the ice caps. **Tundra vegetation** is composed of mosses, lichens, shrubs, dwarfed trees, and some grasses. Needleleaf evergreen coniferous trees can stand long periods of freezing weather with attendant lack of water. Thus, **coniferous forest** (often called boreal forests or *taiga*, their Russian name) occupies large areas in the subarctic climate.

In **desert** and **steppe climates**, the dominant feature is aridity or semiaridity. Deserts and steppes occur in both low and middle latitudes. Agriculture in such regions usually requires irrigation. The principal region, sometimes called the "Dry World," extends in a broad band across northern Africa and southwestern and central Asia. The deserts of the middle and low latitudes are too dry for either trees or grasslands. They have **desert shrub** vegetation, and some areas have practically no vegetation at all. The bushy desert shrubs are *xerophytic* (literally, "dry plant"), having small leaves, thick bark, large root systems, and other adaptations to absorb and retain moisture. Grasslands dominate in the more moist steppe climate, a transitional zone between very arid deserts and humid areas. The biome composed mainly of short grasses is also called the **steppe**. The steppe region of the United States and Canada originally supported both tall and short grass vegetation types, known in those countries as *prairies*.

Rainy low-latitude climates include the **tropical rain forest climate** and the **tropical savanna climate**. The critical difference between them is that the tropical savanna type has a pronounced dry season, which is short or absent in the tropical rain forest climate. Heat and moisture are almost always available in the **tropical rain forest** biome, where broadleaf evergreen trees dominate the vegetation. In tropical areas with a dry season but still enough moisture for tree growth, **tropical deciduous forest** replaces the rain forest. Here the broadleaf trees are not green throughout the year; they lose their leaves and are dormant during the dry season and then add foliage and resume their growth during the wet season. The tropical deciduous forest approaches the luxuriance of tropical rain forest in wetter areas but thins out to low, sparse **scrub and thorn forest** in drier areas. **Savanna** vegetation, which has taller grasses than the steppe, occurs in areas of greater overall rainfall and more pronounced wet and dry seasons.

The humid middle-latitude regions have mild to hot summers and winters ranging from mild to cold, with several types of climate. In the **marine west-coast climate**, occupying the western sides of continents in the higher middle latitudes (for example, in the Pacific Northwest region of the United States), warm ocean currents moderate the winter temperatures, and summers tend to be cool. Coniferous forest dominates some cool, wet areas of marine west-coast climate; examples of growth are the Douglas fir and redwood forests of the western United States.

The **mediterranean (dry-summer subtropical) climate** (named after its most prevalent area of distribution, the lands around the Mediterranean Sea) typically has an intermediate location between a marine west-coast climate and the lower latitude steppe or desert climate. In the summer high-sun period it lies under high atmospheric pressure and is rainless. In the winter low-sun period it lies in a westerly wind belt and receives cyclonic or orographic precipitation. **Mediterranean scrub forest**, known locally by such names as *maquis* and *chaparral*, characterizes mediterranean climate areas. Because of hot, dry summers, the vegetation consists primarily of xerophytic shrubs.

The **humid subtropical climate** occupies the southeastern margins of continents and is characterized by hot summers, mild to cool winters, and ample precipitation for agriculture. The **humid continental climate** lies poleward of the humid subtropical type; it has cold winters, warm to hot summers, and enough rainfall for agriculture, with the greater part of the precipitation in the summer. In middle-latitude areas with these two climate types, a **broadleaf deciduous forest** or **mixed broadleaf-coniferous forest** (as in northeastern North America) is found. As cold winter temperatures freeze the water within reach of plant roots, broadleaf trees shed their leaves and cease to grow, thus reducing water loss. They then produce new foliage and grow vigorously during the hot, wet summer. Coniferous forests can thrive in some hot and moist locations where porous sandy soil allows water to escape downward, giving conifers (which can withstand drier soil conditions) an advantage over broadleaf trees. Pine forests on the coastal plains of the southern United States are an example.

Undifferentiated highland climates exhibit a range of conditions according to altitude and exposure to wind and sun. Undifferentiated highland vegetation types differ greatly depending on altitude, degree and direction of slope, and other factors. They are "undifferentiated" in the context of world regional geography because a small mountainous area may contain numerous biomes, and it would be impossible to map them on a small scale. The world's mountain regions have a complex array of natural conditions and opportunities for human use. In climbing from sea level to the summit of a high mountain peak near the equator (in western Ecuador, for example), a person would experience many of the major climate and biome types to be found in a sea level walk from the equator to the North Pole!

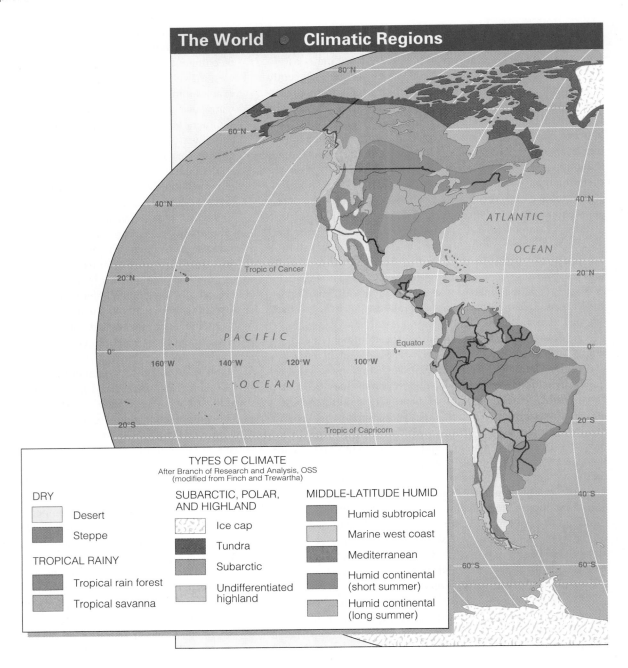

The World ● Climatic Regions

TYPES OF CLIMATE
After Branch of Research and Analysis, OSS
(modified from Finch and Trewartha)

DRY
- Desert
- Steppe

TROPICAL RAINY
- Tropical rain forest
- Tropical savanna

SUBARCTIC, POLAR, AND HIGHLAND
- Ice cap
- Tundra
- Subarctic
- Undifferentiated highland

MIDDLE-LATITUDE HUMID
- Humid subtropical
- Marine west coast
- Mediterranean
- Humid continental (short summer)
- Humid continental (long summer)

FIGURE 2.6
Map showing the world distribution of the types of climate discussed in this text.

FIGURE 2.7
World biomes (natural vegetation) map.

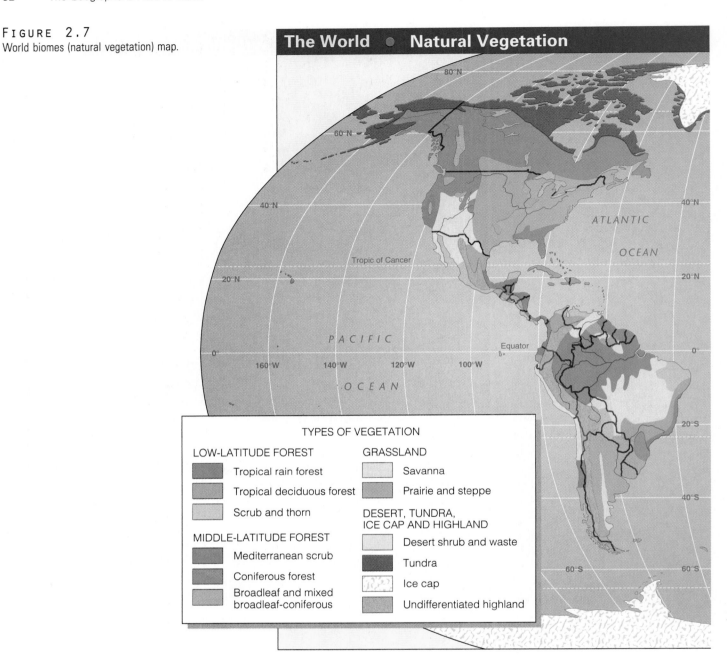

The World ● Natural Vegetation

TYPES OF VEGETATION

LOW-LATITUDE FOREST

Tropical rain forest

Tropical deciduous forest

Scrub and thorn

MIDDLE-LATITUDE FOREST

Mediterranean scrub

Coniferous forest

Broadleaf and mixed broadleaf-coniferous

GRASSLAND

Savanna

Prairie and steppe

DESERT, TUNDRA, ICE CAP AND HIGHLAND

Desert shrub and waste

Tundra

Ice cap

Undifferentiated highland

2.2 Biodiversity

Anyone who has admired the nocturnal wonders of the "barren" desert or appreciated the variety of the "monotonous" arctic can vouch for an astonishing diversity of life even in the biomes people favor least. Geographers and ecologists recognize the exceptional importance of some biomes because of their **biological diversity**—the number of plant and animal species and the variety of genetic materials these organisms contain.

The most diverse biome is the tropical rain forest. From a single tree in the Peruvian Amazon region, entomologist Terry Erwin recovered about 10,000 insect species. From another tree several yards away he counted another 10,000, many of which differed from those of the first tree. Alwyn Gentry of the Missouri Botanical Garden recorded 300 tree species in a single hectare (2.47 acre) plot of the Peruvian rain forest. Just 20 years ago, scientists calculated that there were 4 to 5 million species of plants and animals on earth. Now, however, their estimates are much higher, in the range of 40 to 80 million. This startling revision is based on research, still in its infancy, on species inhabiting the rain forest.

Such diversity is important in its own right, but it also has vital implications for nature's ongoing evolution and for people's lives on earth. Humankind now relies on a handful of

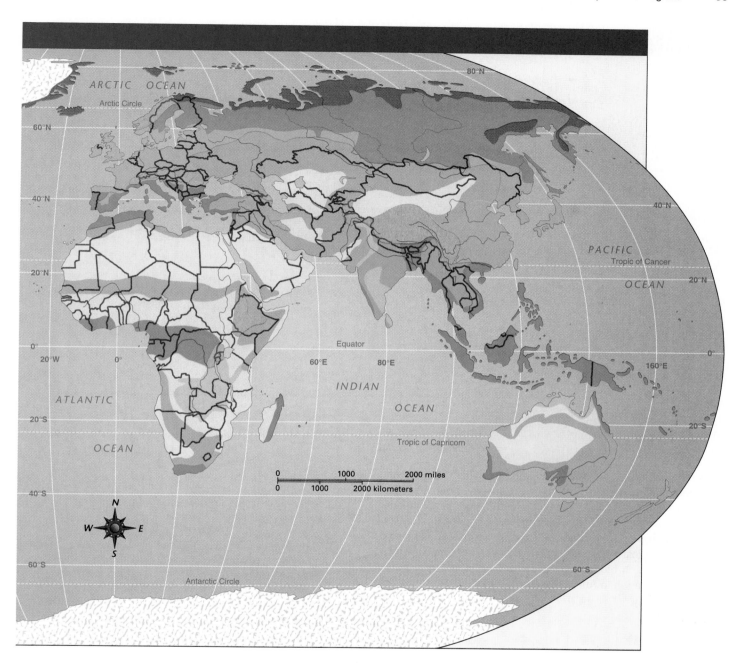

crops as staple foods. In our agricultural systems, the trend in recent decades has been to develop high-yield varieties of grains and to plant them as vast monocultures (single-crop plantings). This trend, which characterizes the **Green Revolution**, may render our agriculture more vulnerable to pests and diseases.

In evolutionary terms, we have reduced the natural diversity of crop varieties that allows nature and farmer to turn to alternatives when adversity strikes. As we remove tropical rain forests and other natural ecosystems to provide ourselves with timber, agriculture, or living space, we may be eliminating the foods, medicines, and raw materials of tomorrow—even before we

have collected them and assigned them scientific names. "We are causing the death of birth," laments biologist Norman Myers.

Regions where human activities are rapidly depleting a rich variety of plant and animal life are known as **biodiversity hot spots**, a ranked list of places scientists believe deserve immediate attention for study and conservation (Fig. 2.8). These regions are: the Choco region of western Colombia; the uplands of the western Amazon Basin in Ecuador and Peru; Brazil's Atlantic coast; the highland and lowland forest zones of Central America; the entire island of Madagascar; the eastern Himalaya Mountains; the Philippine Islands; peninsular Malaysia; northwestern Borneo; the Queensland province in northeastern

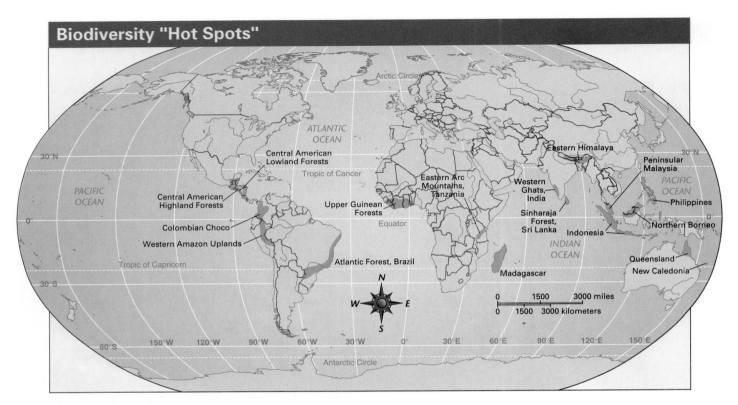

Biodiversity "Hot Spots"

FIGURE 2.8
World biodiversity hot spots.

Australia; the island of New Caledonia; the upper Guinean forest zone of West Africa; Tanzania's eastern arc mountains; the Western Ghat mountains of southern India; and the Sinharaja forest zone of Sri Lanka. By referring to the map of the Earth's biomes in Figure 2.7, you will see that most of these hot spots are within tropical rain forest areas. Many too are islands which tend to have high biodiversity because species on them have evolved in isolation to fulfill special roles in the ecosystem.

2.3   Energy and Ecosystems

In order to appreciate the complex problems affecting the global environment, the geographer's field of vision includes some of the fundamentals of **ecology**, the study of the interrelationships of organisms to one another and to the environment. Living organisms and nonliving components of the environment interact to comprise **ecosystems**, which, like climates and biomes, are classified by major types such as tundra, tropical rain forest, and steppe. Living and nonliving components in the ecosystem interact according to predictable natural patterns, especially in the flow of energy and the cycling of materials between living and nonliving forms.

Definitions and Insights
THE GAIA HYPOTHESIS

All the Earth's ecosystems together make up a vast ecosystem known as the **ecosphere** or **biosphere**. Although it is too large a system to study easily, in theory it is subject to the same laws applied to its component ecosystems, giving rise to the **Gaia Hypothesis** (named for the Mother Earth Goddess of ancient Greece). This hypothesis states that the entire global ecosystem is capable of restoring its equilibrium following any natural or human disturbance if the disturbance is not too drastic. Many environmentalists dislike the Gaia Hypothesis because of the implication that nature can correct any negative environmental impacts people's actions have on the earth, including industrial production of carbon dioxide and other "greenhouse gases."

Geographers find it useful to understand how energy flows through ecosystems because this process is central to the question of how many people the earth and its regions can support and at what levels of resource consumption. The sun is the initial source of energy for all living components of the system. In the process of **photosynthesis**, green plants use the energy of solar radiation to combine carbon dioxide (which they extract from air or water) with water (which they extract from the soil or store directly in their leaves) to give off oxygen and produce their own food supply in the form of carbohydrates such as sugars, starches, and cellulose. Plants are therefore known as **producers** and self-feeders.

In photosynthesizing, plants also provide food energy for almost all other organisms on Earth. Ultimately, the size of that photosynthetic product determines how many organisms—including humankind—the Earth can support and at what level of consumption. The world's economies depend in large part upon the current photosynthetic productivity of croplands and forests and also upon the past photosynthetic productivity of organisms whose energy is now stored as fossil fuel.

After plants produce energy in the form of food, **consumer organisms**—animals that cannot produce their own food—

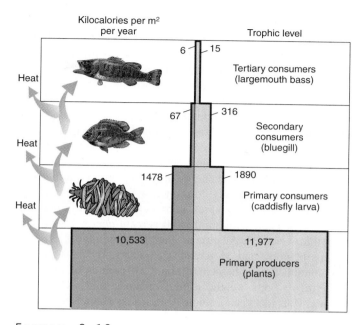

FIGURE 2.10

Another way of looking at the food chain: a pyramid of energy flow for a river ecosystem. The measurements represent usable energy available at each trophic level, and are represented as kilocalories per square meter. In each link of the food chain, about 90 percent of an organism's energy is lost as heat and feces.

pass energy through the ecosystem in a sequence of **trophic** or **feeding levels** (Fig. 2.9). The sequence, called a **food chain**, begins when **primary consumers** or **herbivores** (plant-eaters, such as rabbits and antelopes) consume green plants. They in turn are eaten by **secondary consumers** (**carnivores** or meat-eaters, such as snakes and coyotes). In some food chains, primary and secondary consumers are eaten by **tertiary consumers** (predatory or **top carnivores**, such as leopards and lions). **Decomposer** organisms such as bacteria and fungi do the important work of breaking down the organic material of dead producer and consumer organisms, thus supplying nutrients to the ecosystem.

Food chains are short, seldom consisting of more than four trophic levels, and the collective weight (known as **biomass**) and absolute number of organisms declines substantially at each successive trophic level; there are far fewer tertiary than primary consumers (there are fewer leopards than antelopes). The reason is a fundamental rule of nature known as the **second law of thermodynamics**, which states that much of the energy involved in doing work is lost to the surrounding environment. In living and dying, organisms use and lose the high-quality, concentrated energy that green plants produce. An organism loses about 90 percent of its energy as heat and feces as it passes to the next trophic level on the food chain; a graph of this energy flow resembles a pyramid and is known as the **pyramid of energy flow** (Fig. 2.10). The amount of animal biomass that can be supported at each successive level thus declines geometri-

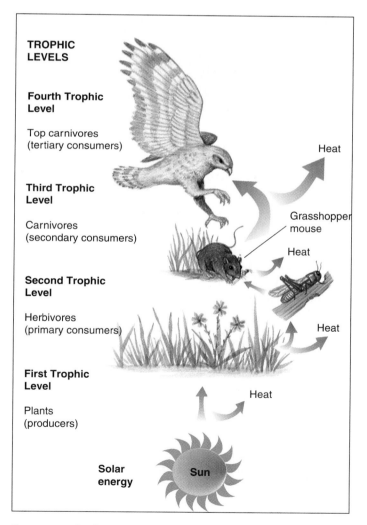

FIGURE 2.9

A food chain, made up of trophic (feeding) levels.

cally, and little energy remains to support the top carnivores. A graphic profile of biomass making up each trophic level in an ecosystem therefore also usually looks like a pyramid and is known as the **pyramid of biomass**.

The second law of thermodynamics has important consequences and implications for human use of resources. The higher we feed on the food chain (the more meat we eat), the more energy we use. The question of how many people Earth's resources can support thus depends very much on what we eat. The affluent consumer of meat demands a huge expenditure of food energy, because that energy flows from grain (producer) to livestock animal (primary consumer, with a 90 percent energy loss), and then from livestock animal to human consumer (secondary consumer, with another 90 percent energy loss). Each year the average U.S. citizen eats 100 pounds of beef, 50 pounds of pork, and 45 pounds of poultry. By the time it is slaughtered, a cow has eaten about 10 pounds of grain per pound of its body weight; a pig, 5 pounds; and a chicken, 3 pounds. Thus in a year, a single American consumes the energy captured by two thirds of a ton of grain—1330 pounds—in meat products alone. By eating grain exclusively (feeding lower on the food chain), many people could live on the energy required to sustain just one meat-eater.

Most U.S. citizens have the choice to eat vegetables and grains or meat; they are an affluent people. However, the vast majority of the world's people are too poor to feed high on the food chain. Geographers recognize distinct regional patterns in the global distribution of wealth and poverty. In order to understand these spatial patterns, and then to think about how to solve some of the major environmental and social problems they represent, it is essential to consider the history of humankind's roles, numbers, and impacts in the global ecosystem.

FIGURE 2.11
San (Bushmen) hunters setting snares in the Kalahari Desert, Botswana. Some San tribespeople are among Earth's last remaining hunters and gatherers. *(Art Wolfe/Tony Stone Images)*

2.4 Two Revolutions That Have Shaped Regions

The geographer's field of vision is necessarily historical, for it is impossible to understand the current appearance of the landscape without reference to its past. An especially useful perspective on the current spatial patterns of our relationship with Earth is to view these patterns as products of two "revolutions" in the relatively recent past: the **Agricultural** or **Neolithic (New Stone Age) Revolution** that began in the Middle East about 10,000 years ago, and the **Industrial Revolution** that began in 18th-century Europe. Each of these revolutions transformed humanity's relationship with the natural environment. Each increased substantially our capability to consume resources, modify landscapes, grow in number, and spread in distribution.

HUNTING AND GATHERING

Until about 10,000 years ago, our ancestors practiced a **hunting and gathering** livelihood. Joined in small bands consisting of

extended family members, they were nomads practicing an **extensive land use** in which they covered large areas to locate foods such as seeds, tubers, foliage, fish, and game animals. Moving from place to place in small numbers, they had a relatively limited impact on natural environments compared with our impact today. Many scholars praise these preagricultural people for the apparent harmony they maintained with the natural world in both their economies and their spiritual systems. Hunters and gatherers have even been described as the "original affluent society," because after short periods of work to collect the foods they needed, they enjoyed long stretches of leisure time. Studies of those few hunter-gatherer cultures that lingered into the 20th century, such as the San ("Bushmen") of southern Africa (Fig. 2.11), suggest that although their life expectancy was low, they suffered little from the mental illnesses and broken family structures that characterize industrial societies.

Hunters and gatherers were not always at peace with one another or with the natural world, however. With upright posture, stereoscopic vision, opposable thumbs, an especially large brain, and no rutting season, *Homo sapiens* was from its earliest times an **ecologically dominant species**—one that competes more successfully than other organisms for nutrition and other essentials of life, or exerts a greater influence than other species on the environment. Using fire to flush out game or create new pastures for the herbivores they hunted, preagricultural people, unlike other animals, shaped the face of the land on a vast scale relative to their small numbers. Many of the world's prairies, savannas, and steppes where grasses now prevail originated from hunters and gatherers setting fires repeatedly. These people also overhunted and in some cases eliminated animal species. The controversial **Pleistocene Overkill** hypothesis states that rather than being at harmony with nature, hunters and gatherers of the Pleistocene Era (1 million to 10,000 years ago) hunted many species to extinction, including the elephantlike mastodon of North America.

THE AGRICULTURAL REVOLUTION

Despite these excesses, the environmental changes that hunters and gatherers could cause was limited. Humankind's power to modify landscapes took a giant step with the domestication (controlled breeding and cultivation) of plants and animals—the **Agricultural Revolution**. Why people began to produce rather than continue to hunt and gather plant and animal foods—first in the Middle East and later in Asia, Europe, Africa, and the Americas—is uncertain. Two theories prevail. Climatic change in the form of increasing drought and reduced plant cover may have forced people and wild plants and animals into smaller areas, where people began to tame wild herbivores and sow wild seeds to produce a more dependable food

supply. A more favored theory is that their own growing populations in areas originally rich in wild foods compelled people to find new food sources, so they began sowing cereal grains and breeding animals. The latter process apparently began about 8000 B.C. in the Zagros Mountains of what is now Iran. The culture of domestication spread outward from there but also developed independently in several world regions.

In choosing to breed plants and animals, people gave up the nomadic, extensive land use of hunting and gathering for the sedentary, **intensive land use** of agriculture and animal husbandry. The new system enabled them to create large and reliable surpluses of food. Through **dry farming**, or planting and harvesting according to the seasonal rainfall cycle, population densities could be 10 to 20 times higher than they were in the hunting and gathering mode. By about 4000 B.C., people along the Tigris, Euphrates, and Nile rivers began **crop irrigation**—bringing water to the land artificially—an innovation that allowed them to grow crops year-round, independently of seasonal rainfall or river flooding (Fig. 2.12). Irrigation technology allowed even more people to make a living off the land; irrigated farming yields about five to six times more food per unit area than dry farming. In ecological terms, the expanding food surpluses of the Agricultural Revolution raised the Earth's **carrying capacity**—that is, the size of a species' population (in this case, humans) that the Earth's ecosystems can support.

This steep increase in food surpluses freed more people from the actual work of producing food, and they undertook a wide range of activities unrelated to subsistence needs. Irrigation and the dependable food supplies it provided thus set the stage for the development of **civilization**, the complex culture of urban life characterized by the appearance of writing, eco-

FIGURE 2.12
Irrigated fields on the banks of the Meander (Menderes) River in southwestern Turkey. After 4000 B.C., highly productive irrigated agriculture in Southwest Asia and North Africa sparked rapid population growth and the emergence of the world's first cities.
Richard Nowitz/FPG International

nomic specialization, social stratification, and high population concentrations. By 2500 B.C., for example, 50,000 people lived in the southern Mesopotamian city of Ur, in what is now Iraq. Other **culture hearths**—regions where civilization followed the domestication of plants and animals—emerged between 8000 and 2500 B.C. in China, Southeast Asia, the Indus River Valley, West Africa, Mesoamerica, and the Andes.

The agriculture-based urban way of life that spread from these culture hearths had larger and more lasting impacts on the natural environment than either hunting and gathering or early agriculture. Acting as agents of humankind, domesticated plants and animals proliferated at the expense of the wild varieties people came to regard as pests and competitors. Agriculture's permanent and site-specific nature magnified the human imprint on the land while the pace and distribution of that impact increased with growing numbers of people.

THE INDUSTRIAL REVOLUTION

The human capacity to transform natural landscapes took another giant leap with the **Industrial Revolution**, which began in about 1700 A.D. This new pattern of human-land relations was based on breakthroughs in technology that several factors made possible (Fig. 2.13). First, western Europe had the economic capital for experimentation, invention, and risk. Much of this money was derived from the lucrative trade in gold and slaves undertaken initially in the Spanish and Portuguese empires after 1400. Second, in Europe prior to 1500, there had been significant improvements in agricultural productivity, particularly with new tools such as the heavy plow, and with more intensive and sustainable use of farmland. Crop yields increased, and human populations grew correspondingly. A third factor was population growth itself. More people freed from work in the fields represented a greater pool of talent and labor in which experimentation and innovation could flourish. As agricultural innovations and industrial productivity improved, an increasingly large proportion of the growing European population was freed from farming, and for the first time in history there were more city-dwellers than rural folk. The process of industrialization continues to promote urbanization today.

Most geographers see population growth today as a drain on resources, but the Industrial Revolution illustrates that, given the right conditions, more people do create more resources. Inventions such as the steam engine tapped the vast energy of fossil fuels—initially coal and later, oil and natural gas. This energy, the photosynthetic product of ancient ecosystems, allowed Earth's carrying capacity for humankind to be raised again—this time into the billions.

As they began to deplete their local supplies of resources needed for industrial production, Europeans began to look for these materials abroad. As early as their **Age of Exploration**

FIGURE 2.13
The crystal soda factory in Cheshire, England. The process of industrialization which began in eighteenth century Europe has rapidly transformed the face of the Earth. *(E. Stevenson/FPG International)*

that began in the fifteenth century, Europeans probed ecosystems across the globe to feed a growing appetite for innovation, economic growth, and political power. Thus, the process of European **colonization** was linked directly to the Industrial Revolution. Mines and plantations from faraway central Africa and India supplied the copper and cotton that fueled economic growth in Belgium and England.

No longer dependent upon the foods and raw materials they could procure within their own political and ecosystem boundaries, European vanguards of the Industrial Revolution had an impact on natural environment that was far more extensive and permanent than that of any other people in history. There are many measures of the unprecedented changes that the Industrial Revolution and its wake have wrought on Earth's landscapes. Between 1700 and 1990, the total forested area on Earth declined by 20 percent. During the same period, total cropland grew 460 percent, with more expansion in the period 1950–1990 than in the 150 years from 1700–1850. Human use of energy increased more than a hundred-fold from 1700 to 1990. Today the single species of humankind uses about 40 percent of Earth's land-based photosynthetic productivity in the production of crops and exploitation of forests and other natural vegetation. Of particular interest to geographers is how the costs and benefits of such expansion are distributed in unequal patterns across the Earth.

2.5   The Geography of Development

One very notable characteristic of human life on Earth is the large disparity between wealthy and poor people, both within and between countries. At a high level of generalization, the world's countries can be divided into "haves" and "have-nots" (Table 2.1 and Fig. 2.14). Writers refer to these distinctions between countries variously as "developed" and "underdeveloped," "developed" and "developing," "more developed" and "less developed," "industrialized" and "nonindustrialized," and "north" and "south," based on the general geography of the wealthier countries occupying the middle latitudes of the Northern Hemisphere. This text uses the distinction between **more developed countries (MDCs)** and **less developed countries (LDCs)** as a framework to help explain the world's complexities. It must be emphasized, however, that as an introductory tool, this division cannot account for the tremendous variations and ongoing changes in economic and social welfare that characterize the world today. Some countries, such as the "Asian tigers," are best described as **newly industrializing countries (NICs)** since they do not fit either the MDC or LDC idealized type. We will describe these cases in the relevant discussions of the Earth's regions.

TABLE 2.1   Characteristics of More Developed (MDC) and Less Developed (LDC) Countries

Characteristic	MDC	LDC
Per capita GNP and income	High	Low
Percent in middle class	High	Low
Percent in manufacturing	High	Low
Energy use	High	Low
Percent urban	High	Low
Percent rural	Low	High
Birth rate	Low	High
Death rate	Low	Low
Population growth rate	Low	High
Percent under age 15	Low	High
Percent literate	High	Low
Leisure time available	High	Low
Life expectancy	High	Low

MEASURES OF DEVELOPMENT

It is customary to distinguish MDCs from LDCs by examining annual **gross national product (GNP)**—the total output of goods and services a country produces in a year—or by per capita gross national product, which is the GNP divided by the country's population. The gulf between the world's richest and poorest countries is startling (Table 2.2). The average per capita GNP in the MDCs is 17 times greater than in the LDCs. In 1994, with a per capita GNP of $39,850, Luxembourg was the world's richest country, while Mozambique was the poorest with just $80. These raw numbers suggest that economic productivity and income alone characterize **development**, which, according to a common definition, is a process of improvement in the material conditions of people through diffusion of knowledge and technology.

Such economic definitions reveal little about measures of well-being such as income distribution, gender equality, literacy, and life expectancy. Recognizing the shortcomings of strictly economic definitions, the United Nations Development Programme created the **Human Development Index (HDI)**, a scale that considers these attributes of quality of life. According to this index, Canada in 1995 was "the world's best place to live," although it is seventeenth in per capita GNP ranking. Following Canada, in descending order, are the United States, Japan, the Netherlands, Finland, Iceland, Norway, France, Spain, and Sweden.

On the basis of per capita GNP, 1.2 billion, or 20 percent, of the world's 5.8 billion people inhabit the MDCs. Most citizens of these countries, such as the United States, Canada, the nations of Western Europe, Japan, Australia, and New Zealand enjoy an affluent lifestyle with freedom from hunger. Employed in industries or services, most of the people live in cities rather than in rural areas. Disposable income, or money that people can spend on goods beyond their subsistence needs, is generally high. There is a large middle class. Population growth is low, as a result of low birth rates and low death rates. Life expectancy is long, and the literacy rate is high.

Life for the planet's other 80 percent, or about 4.6 billion people, is very different. In the LDCs, including most countries in Latin America, Africa, and Asia, poverty and often hunger prevail. Most people engage in subsistence agriculture, and the industrial base is small. The middle class tends to be small, with an enormous gulf between the vast majority of poor and a small wealthy elite, which owns most of the private landholdings. With high birth rates and falling death rates, population growth is high. Life expectancy is low. The literacy rate is low.

With four fifths of the world's people living in the poorer, less developed countries, it is important to understand the root causes of underdevelopment and to appreciate how wealth and poverty affect the global environment in very different but equally profound ways.

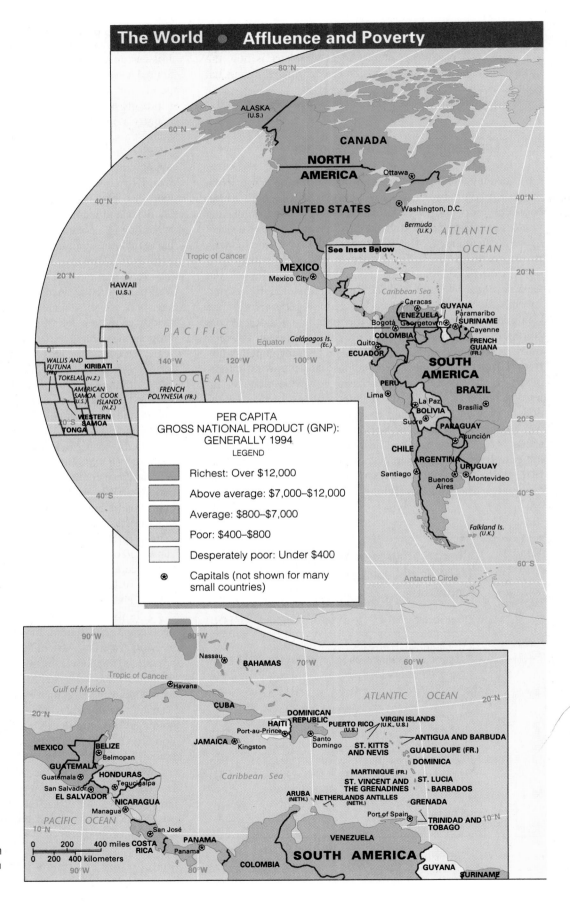

The World ● Affluence and Poverty

PER CAPITA GROSS NATIONAL PRODUCT (GNP): GENERALLY 1994

LEGEND

- Richest: Over $12,000
- Above average: $7,000–$12,000
- Average: $800–$7,000
- Poor: $400–$800
- Desperately poor: Under $400
- ⊛ Capitals (not shown for many small countries)

FIGURE 2.14

Wealth and poverty by country. Note the concentration of wealth in the middle latitudes of the northern hemisphere.

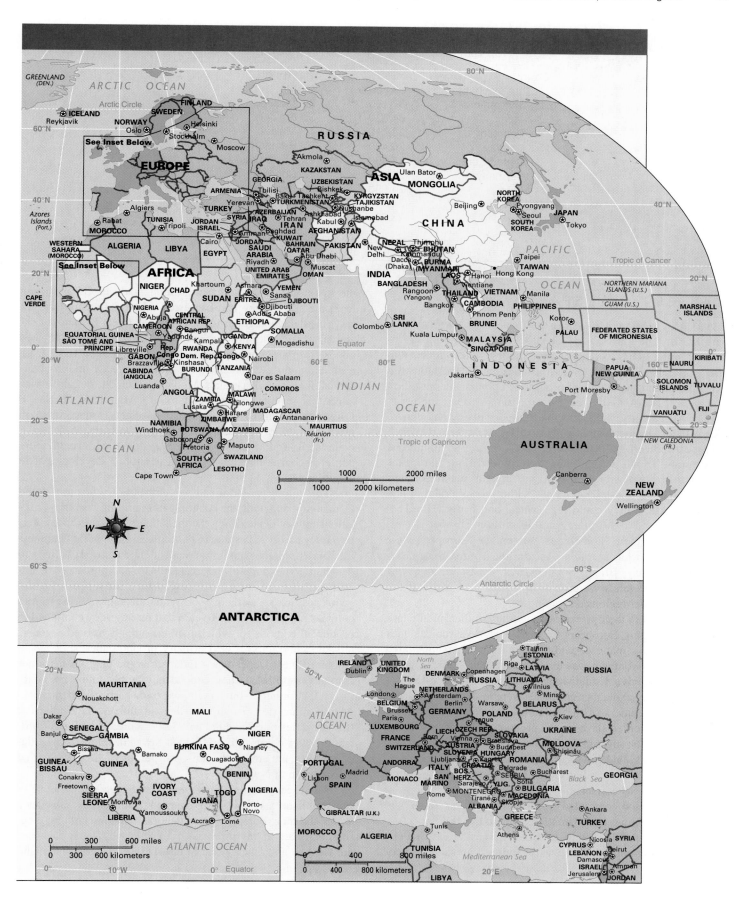

TABLE 2.2	**Top 10 and Bottom 10 Countries' Per Capita GNP (in U.S. Dollars) for 1994 (Estimated)**			
	Top Ten (Richest)		**Bottom Ten (Poorest)**	
	Country	GNP	Country	GNP
	1. Luxembourg	39,850	1. Mozambique	80
	2. Switzerland	37,180	2. Tanzania	90
	3. Japan	34,630	3. Ethiopia	130
	4. Denmark	28,110	4. Malawi	140
	5. Norway	26,480	5. Burundi	150
	6. United States	25,860	6. Sierra Leone	150
	7. Germany	25,580	7. Chad	190
	8. Austria	24,950	8. Vietnam	190
	9. Iceland	24,590	9. Nepal	200
	10. Sweden	23,630	10. Uganda	200

Source: Population Reference Bureau, World Population Data Sheet, 1996.

CAUSES OF DISPARITIES

Many theories attempt to explain the disparities between MDCs and LDCs. The most widely debated, and also the most embraced in the LDCs, is **dependency theory**, which argues that the worldwide economic pattern established by the Industrial Revolution and the attendant process of colonialism continues today. In his book *Ecological Imperialism*, historian and geographer Alfred Crosby explains the rich-poor divide and dependency theory by idealizing two very different patterns by which European powers utilized foreign lands during the Industrial Revolution. In the pattern of **settler colonization**, Europeans sought to create new or "neo-Europes" in lands much like their own: temperate mid-latitude zones with moderate rainfall and rich soils where they could raise wheat and cattle. Thus, between 1630 and 1930, more than 50 million Europeans emigrated from their homelands to create European-style settlements in what are now Canada, the United States, Argentina, Uruguay, Brazil, South Africa, Australia, and New Zealand. These lands were destined to become some of the world's wealthier regions and countries.

In contrast to their preference to settle familiar middle latitude environments, Europeans viewed the world's tropical lands as sources of raw materials and markets for their manufactured goods. The environment was too different from home to make settlement attractive. Establishing a pattern of **mercantile colonialism**, Crosby explains, Europeans were less inhabitants than conquering occupiers of the colonies, overseeing indigenous peoples and resettled slaves in the production of primary or unfinished products: sugar in the Caribbean; rubber in Latin America, West Africa, and Southeast Asia; and gold and copper in southern Africa, for example. Colonialism required huge migrations of people to extract the Earth's resources, including 30 million slaves and contracted workers from Africa, India, and China to work mines and plantations around the globe.

In the mercantile system, the colony provided raw materials to the ruling country in return for finished goods; thus people in India might purchase clothing made in Great Britain from the cotton they had harvested. The relationship was always most advantageous to the colonizer. Great Britain, for example, would not allow its colony India to purchase finished goods from any country but Britain. It prohibited India from producing any raw materials the empire already had in abundance, such as salt (India's Mohandas Gandhi defiantly violated this prohibition in his famous "March to the Sea"). Finished products (or "value-added" products) are worth much more than the raw materials they are made from, so focusing manufacturing in the ruling country concentrated wealth there while discouraging industrial and economic development in the colony.

The colony was obliged to contribute to, but prohibited from competing with, the economy of the ruling country—a relationship that dependency theorists insist continues today. Dependency theory states that to participate in the world economy, the former colonies but now-independent countries continue to depend upon exports of raw materials to, and purchases of finished goods from, their former colonizers, and this disadvantageous position keeps them poor. With independence, the former colonies needed revenue. To earn that money, they continued to produce the goods for which markets already existed—generally the same unprocessed primary products they supplied in colonial times. Dependency theorists argue that when former colonies try to break their dependency by becoming manufacturers, the former colonizers impose trade barriers and quotas to preclude that step. To support their argument, they point out that countries like Thailand and South Korea, which the Europeans never colonized, and which never developed these dependencies, are among the most prosperous LDCs or are in that select group of NICs.

Many geographers, however, view dependency theory as too simplistic and politically charged. Rather, they consider a wider and more complex set of factors, including culture, location, and natural environment, to explain why some countries are wealthy and others poor. Situated close to a great mainland with which to trade, the island of Great Britain enjoys a central location favorable for economic development.

Japan has a similar location relative to the Asian landmass. In contrast, landlocked nations such as Bolivia in South America and numerous nations in Africa, have locations unfavorable for trade and economic development, and they have not overcome this disadvantage. A superabundance of one particular resource (such as oil in the Persian/Arabian Gulf states) or a diversity of natural resources has helped some countries to become more developed than others. The former Soviet Union and the United States developed superpower status in the 20th century in large part based on the enormous natural resources of both countries. In other cases, human industriousness can help to compensate for resource limitations and promote development. Japan, for example, has a rather small territory with few natural resources (including almost no petroleum); yet, in the second half of the 20th century, it became an industrial powerhouse largely because the Japanese people united in common purpose to rebuild from wartime devastation, placing priorities on education, technical training, and seaborne trade from their advantageous island location. Conversely, cultural or political problems like corruption and ethnic factionalism can hinder development in a resource-rich nation, as in mineral-wealthy Democratic Republic of Congo.

Whether because of neo-colonialism (as many dependency theorists argue) or a more complex array of variables, many developing countries continue to rely heavily on income from the export of a handful of raw materials. This makes them vulnerable to the whims of nature and the world economy. The economy of a country heavily dependent on rubber exports, for example, may suffer if an insect pest wipes out the crop or if a foreign laboratory develops a synthetic substitute. When demand for rubber rises, that country may actually harm itself trying to increase its market share by producing more rubber, because in the process it drives down the price. If the country withholds production to shore up rubber's price, it provides consuming countries with an incentive to look for substitutes and alternative sources. The developing country is in a dependent and disadvantaged position.

ENVIRONMENTAL IMPACTS OF UNDERDEVELOPMENT

Of particular interest to geographers are the environmental impacts of relations between MDCs and LDCs, especially the impacts on the poorer countries. LDCs generally lack the financial resources needed to build roads, dams, energy grids, and the other infrastructure assets they perceive as critical to development. They turn to the World Bank, International Monetary Fund (IMF), and other institutions of the MDCs to borrow funds for these projects. Many borrowers are unable to pay even the interest on these loans. When lender institutions

threaten to sever assistance, borrowing countries often try to quickly raise cash to avoid this prospect. One method is to dedicate more quality land to the production of **cash (commercial) crops**, luxuries such as coffee, tea, sugar, coconuts, and bananas exported to the MDCs. Governments or foreign corporations often displace or "marginalize" subsistence farmers in the search for new lands on which to grow these commercial crops. In the process of **marginalization**, poor subsistence farmers are pushed onto fragile, inferior, or marginal lands which cannot support crops for long and which are degraded by cultivation. In Brazil's Amazon Basin, for example, peasant migrants arrive from Atlantic coastal regions where government and wealthy private landowners cultivate the best soils for sugarcane and other cash crops. The newcomers to Amazonia slash and burn the rain forest to grow rice and other crops that exhaust the soil's limited fertility in a few years. They move on to cultivate new lands, and in their wake come cattle ranchers whose land use further degrades the soil.

National decision makers often face a difficult choice between using the environment to produce more immediate or more long-term economic rewards. In most cases they feel compelled to take short-term profits, and by cash cropping and other strategies initiate a sequence having sometimes tragic environmental consequences. Most LDCs have resource-based economies that rely not on industrial productivity but upon stocks of soils, forests, and waters. The long-term economic health of these countries could be assured by the perpetuation of these natural assets, but to pay off international debts and meet other needs, the LDCs generally draw on their ecological capital faster than nature can replace it. In ecosystem terms, they exceed **sustainable yield** or **natural replacement rate**, the highest rate at which a renewable resource can be used without decreasing its potential for renewal.

In tropical biomes, people cut down ten trees for every one they plant. In Africa, that ratio is 29 to 1. In 1950, about 30 percent of Ethiopia's land surface was forested. Today, less than 1 percent is in forest. Such countries are ecologically bankrupt; they have exhausted their environmental capital. Many political and social crises result from this bankruptcy: the wars, refugee migrations, and famines of the Horn of Africa, for example, have underlying environmental causes. Such problems are becoming more central to the national interests of the United States and other powerful MDCs. The U.S. Central Intelligence Agency, for example, is increasingly using Geographic Information Systems (GIS) and other geographic techniques to analyze the natural phenomena and human misuses of environment that are root causes of war and threats to global security.

People in the LDCs feel the impacts of environmental degradation more directly than do people in the MDCs. The world's poor tend to drink directly from untreated water supplies and to cook their meals with fuelwood rather than fossil fuels. They are more dependent upon nature's abilities to replenish and cleanse itself, and thus they suffer more when those abilities are diminished (see Definitions and Insights, p. 44).

Definitions and Insights

THE FUELWOOD CRISIS

The removal of tree cover in excess of sustainable yield (as in highland Nepal, for example; Fig. 2.A) illustrates the many detrimental effects that a single human activity can have in the LDCs. These impacts comprise the complex phenomenon, widespread in the world's LDCs, known as the **fuelwood crisis**. As people remove trees to use as fuel, for construction, or to make room for crops, their existing crop fields lose protection against the erosive force of wind. Less water is available for crops, because in the absence of tree roots to funnel water downward into the soil, it runs off quickly. Increased salinity (salt content) generally accompanies increased runoff, so that the quality of irrigation and drinking water downstream declines. Eroded topsoil can choke irrigation chan-

nels, reduce water delivery to crops, raise floodplain levels, and increase the chance of floods destroying fields and settlements. As reservoirs fill with silt, hydroelectric generation and therefore industrial production is diminished. Upstream, where the problem began, fewer trees are available to use as fuel. People must now change their behavior. Women, most often the fuelwood collectors, must walk farther to gather fuel. Turning to animal dung and crop residues as sources of fuel deprives the soil of the fertilizers these poor people most often use when farming. Reduced food output is the result. A family may eventually give up one cooked meal a day or tolerate colder temperatures in their homes because fuel is lacking—measures that negatively impact the family's health.

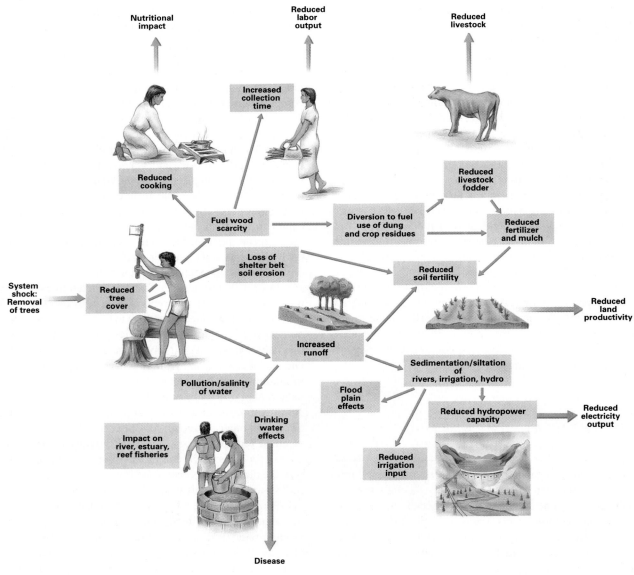

FIGURE 2.A Impacts of deforestation in the LDCs.

TWO TYPES OF OVERPOPULATION

High rates of human population growth intensify the environmental problems characteristic of the LDCs. More people cut more trees, a phenomenon that suggests there is a problem of **people overpopulation** in the poorer countries. Many persons, each using a small quantity of natural resources daily to sustain life, may add up to too many people for the environment to support.

There is another type of overpopulation, **consumption overpopulation**, that is characteristic of the MDCs. In the wealthier countries a few persons, each using a large quantity of natural resources from ecosystems across the world, may also add up to too many people for the environment to support. With just five percent of the world's population, the United States accounts for about a third of the world's annual energy consumption. These Americans use more fossil fuels, iron, and copper than do the inhabitants of Latin America, Africa, and Asia combined. One analyst calculated that in his or her lifetime, the average U.S. citizen will consume more than 250 times as many goods as the average person in Bangladesh. Such disparities suggest that, just as Bangladesh is "underdeveloped," the United States is "overdeveloped." Much of this overconsumption is unnecessary for sustenance or well-being and impacts markedly the global environment.

If the vast majority of the world's population were to consume resources at the rate that U.S. citizens do, the environmental results might be ruinous. Even if consumption levels in the LDCs do not rise substantially, the sheer increase in numbers of people in those countries suggests that degradation of the environment will accelerate in the coming decades. The basic attributes of the world's population geography provide an insight into this dilemma.

2.6   Population Geography

Earth's human population 10,000 years ago, before the Agricultural Revolution, was probably about 5.3 million. By 1 A.D. it was probably between 250 and 300 million, or about the population of the United States today. The first billion was reached about 1800. Then a staggering population explosion occurred as a result of the Industrial Revolution. The second billion came in 1930, the fourth in 1975, and in 1997 we approached six billion. As one measure of ecological dominance, humankind is now by far the most populous large mammal on earth and has succeeded, where no other animal has, in extending its range to the world's farthest corners. As Figure 2.15 reveals, some world regions have especially dense populations. Notable are China and India, where productive agriculture and a long time span of occupation are the main reasons for high density, and western Europe and the northeastern United States, where industrial productivity is the main reason.

Figure 2.16 illustrates the **population explosion** as seen in the classic J-shaped curve of exponential growth in the human population (Fig. 2.16). Most striking is the upward curve of growth after the dawn of the Industrial Revolution, with the greatest increase in the population growth rate taking place since 1950. At the current rate of 1.5 percent annual growth, our population would double in 46 years.

Dense settlement and growing numbers pose a fundamental question: Will we exceed Earth's carrying capacity for our species, and what will happen if we do? Early in the Industrial Revolution, an English clergyman named Thomas Malthus (1766–1834) postulated that human populations, which can grow geometrically or exponentially, would exceed food supplies, which usually grow only arithmetically or linearly. He predicted a catastrophic human die-off as a result of this irreconcilable equation. He could not have foreseen that the exploitation of new lands and resources, including tapping into the energy of fossil fuels, would permit food production to keep pace with or even outpace population growth for at least the next two centuries. However, the **Malthusian scenario** of the race between food supplies and mouths to feed remains a source of constant and important debate today.

POPULATION CONCEPTS AND CHARACTERISTICS

Two primary variables determine population change in a given village, city, country, or the entire Earth: birth rate and death rate. The **birth rate** is the annual number of live births per thousand people in the population. The **death rate** is the annual number of deaths among that same thousand people. The **population change rate** is the birth rate minus the death rate in that population. On a worldwide average in 1996, the birth rate was 24 per thousand and the death rate 9 per thousand. By year's end, among the 1000 people, 24 babies had been born but 9 people of varying ages had died, resulting in a net growth of 15. That figure, 15 per thousand or 1.5 percent, represents the 1996 population change rate (population growth rate) for the world. As no one leaves or enters earth, there is no need to consider a third variable, **migration**, at this scale. At the scale of a village, city, or country, migration does affect the population. Compulsory and voluntary migration are strong forces in the world's regions and countries, and this text cites frequent examples of both.

Many factors affect birth rates. Better educated and wealthier people have fewer children. Conversely, poorer, less educated people generally want and have more children. Poor parents view additional children as an economic asset rather than a burden, because they represent additional labor to work in fields or factories and will care for them in their old age. People in cities tend to have fewer children than those in rural areas. Those who marry earlier have more children. Couples

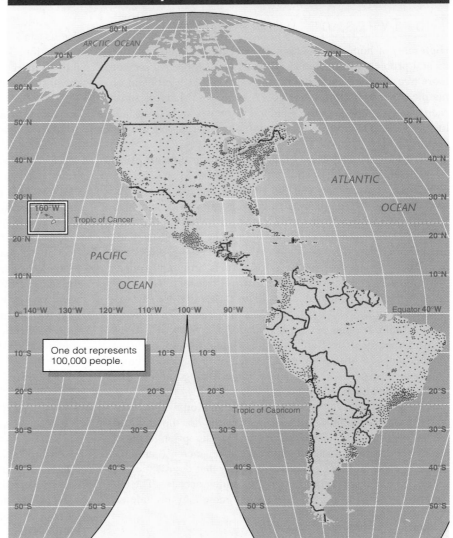

FIGURE 2.15

World population map. Each dot represents 100,000 people. The French geographer Jean Brunhes insisted that the two most significant world maps are those of population and rainfall. There are striking concentrations of people in the intensively and long-cultivated farming regions of southern and eastern Asia, and in the industrialized regions of Europe, the eastern United States, and Japan. Note the tendency of people to congregate on the margins of continents, near the sea. Populations are notably sparse in the more challenging environments of the tropical rain forest, deserts, subarctic taiga and arctic tunda.

One dot represents 100,000 people.

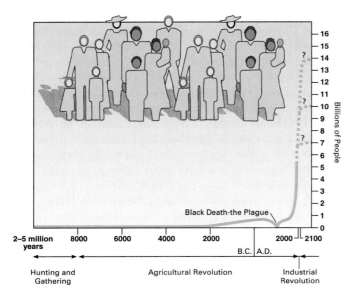

with access to and understanding of contraception may have fewer children. Cultural norms are important, because even where contraception is available, for religious or social reasons a couple may decide not to interfere with what they perceive as God's will or the social status associated with a larger family.

Death rates are affected mainly by health factors. Improvements in food production and distribution help reduce death rates. Better sanitation, hygiene, and drinking water eliminate

◀ FIGURE 2.16

The J-shaped curve tracing exponential growth of the human population.

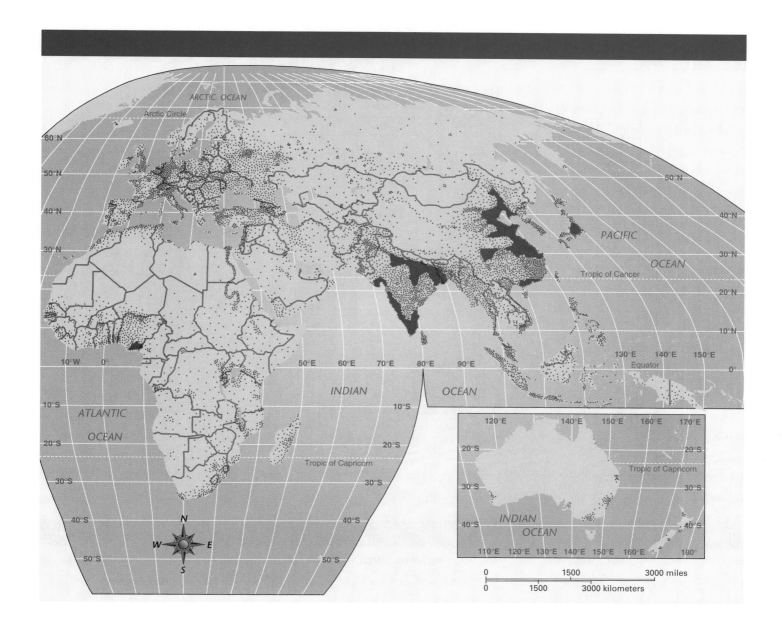

infant diarrhea, a common cause of infant mortality in LDCs. The availability of antibiotics, immunizations, insecticides, and other improvements in medical and public health technology have a marked correlation with declining death rates.

The explosive growth in world population since the beginning of the Industrial Revolution is the result not of a rise in birth rates, but of a dramatic decline in death rates, particularly in the LDCs. The death rate has fallen as improvements in agricultural and medical technologies have diffused, or spread, from the MDCs to the LDCs. Until recently, however, there were no strong incentives for people in LDCs to have fewer children. With birth rates remaining high and death rates falling quickly, the population has grown sharply.

With about 9 out of 10 babies worldwide born in the LDCs, the current and projected rates of population growth are distributed quite unevenly between the MDCs and LDCs (Fig. 2.17).

This phenomenon is apparent in the age-structure diagrams typical of these countries. An **age-structure profile** (often called a "population pyramid") classifies a country's population by gender and by five-year age increments (Fig. 2.18). One important index these profiles show is the percentage of a population under age 15. The very high proportion typical of LDCs, on average about 35 percent, is remarkable, for it means that populations will continue to grow in these poor countries as these children enter their reproductive years. The bottom-heavy, pyramid-shaped age structure diagram of LDCs contrasts markedly with the more rectangular structure of the MDCs. The wealthier countries have a much more even distribution of population through age groups, with a modest share of 20 percent under age 15. Such profiles suggest that, when migration is not considered, their population growth will be low in the near future.

Age Structure and Population of LDCs and MDCs, 1995

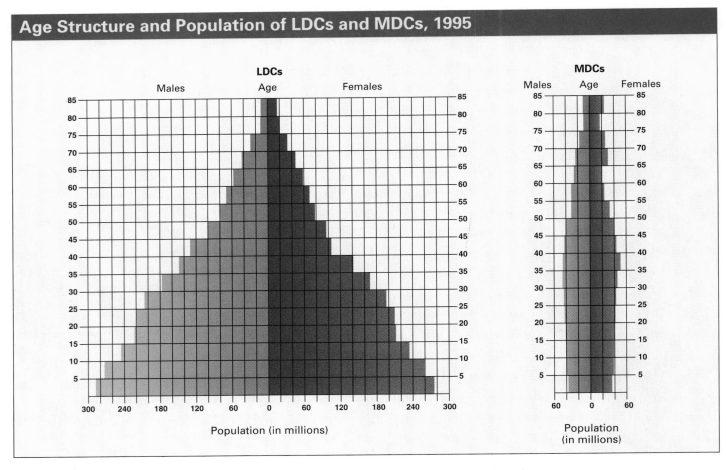

FIGURE 2.17 Age-structure profiles of the LDCs and MDCs in 1995. *(Source: Population Reference Bureau.)*

Population Age Structure Diagrams

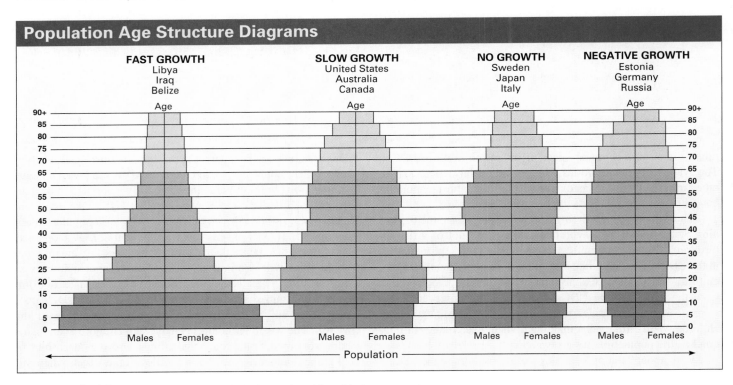

FIGURE 2.18 Age-structure profiles typical of countries with rapid, slow, zero and negative rates of population growth. *(Source: Population Reference Bureau).*

THE DEMOGRAPHIC TRANSITION: WILL THE LDCs COMPLETE IT?

The history of population change of the MDCs provides a model whose usefulness in predicting the population future for LDCs is still unknown. Known as the **demographic transition**, this model depicts the change from high birth rates and high death rates to low birth rates and low death rates that accompanied economic growth in the MDCs. Four stages comprise the model (Fig. 2.19). In the first or preindustrial stage, birth rates and death rates were high, and population growth negligible. In the second or transitional stage, birth rates remained high but death rates dropped sharply after about 1800 with the medical and other innovations of the Industrial Revolution. In the third or industrial stage, beginning around 1875, birth rates began to fall as affluence spread. Finally, after about 1975, some of the industrialized countries entered the fourth or postindustrial stage, with both low birth and low death rates, and therefore

(once again, as in stage one) low population growth. Some MDCs, including Japan, Sweden, and Italy, hover near **zero population growth** (ZPG), a rate of equal birth and death rates, and a few (in 1996, Estonia, Latvia, Lithuania, Germany, Belarus, Bulgaria, the Czech Republic, Hungary, Romania, Russia, and Ukraine) are experiencing a negative growth rate with more deaths than births.

Viewed as a group, the LDCs have proceeded from stage one to stage two of the demographic transition. With falling or already low death rates and only slightly declining birth rates, many LDCs appear to have stalled in this stage of very high population growth. There are two very distinct points of view on where these LDCs will proceed from this stage. Some observers believe that the poorer countries will follow the wealthier ones through the transition, achieving prosperity and a stabilized population. Most recognizable among these observers are the so-called **technocentrists** or **cornucopians**, who argue that human history provides insight into the future: through their technological ingenuity, people always have been

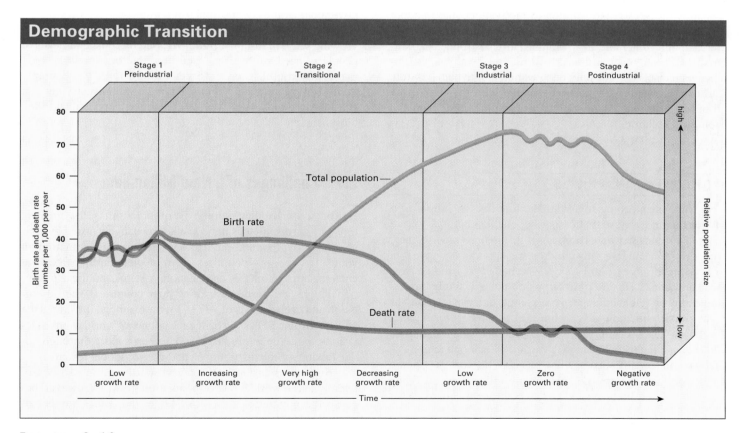

Demographic Transition

FIGURE 2.19
The demographic transition model. In the MDCs, economic development has resulted in a transition from high birth rates and high death rates to low birth rates and low death rates. Population growth rates at stages 1 and 4 are low because birth and death rates are nearly equal.

able to conquer food shortages and other problems, and therefore always will. University of Maryland economist Julian Simon, for example, argues that far from being a drain on resources, additional people create additional resources. Technocentrists thus insist that people can raise the Earth's carrying capacity indefinitely and that the die-off that Malthus predicted will always be averted. Our more numerous descendants will instead enjoy more prosperity than we do.

In contrast, the **neo-Malthusians** argue that although successful so far, we cannot indefinitely increase Earth's carrying capacity. There is an upper limit beyond which growth cannot occur, with calculations ranging from 8 to 40 billion. Neo-Malthusians insist that LDCs cannot remain indefinitely in the transitional stage. Either they must intentionally bring birth rates down and make it successfully through the demographic transition, or they must unwillingly experience nature's solution, a catastrophic increase in death rates. By either the "birth rate solution" or "death rate solution," the neo-Malthusians argue, the less developed world will confront a population crisis.

Whereas technocentrists view the equation between people and resources passively, insisting that no corrective action is needed, neo-Malthusians tend to be activists who describe dire scenarios of a death rate solution to motivate people to adopt the birth rate solution. Biologist Paul Ehrlich thinks that we are increasingly vulnerable to a Malthusian catastrophe, particularly as HIV and other viruses diffuse around the globe with unprecedented speed. "The only big question that remains," wrote Ehrlich, "is whether civilization will end with the bang of an all-out nuclear war, or the whimper of famine, pestilence and ecological collapse."[1]

LIFEBOAT ETHICS

Neo-Malthusian ecologist Garrett Hardin offers the **ethics of a lifeboat** in an essay so distressing but challenging that we are obliged to consider and respond to it. "People turn to me," writes Hardin, "and say 'my children are starving. It's up to you to keep them alive.'[2] And I say, 'The hell it is. I didn't have those children.'" He describes our world not as a single "spaceship earth" or "global village" with a single carrying capacity as many environmentalists do, but as a number of distinct "lifeboats," each occupied by the citizens of single countries and each having its own carrying capacity. Each rich nation is a lifeboat comfortably seating a few people. The world's poor are in lifeboats so overcrowded that many fall overboard. They swim to the rich lifeboats and beg to be brought aboard. What

should the passengers of the rich lifeboat do? The choices pose a dilemma.

Hardin proposes the following: There are 50 rich passengers in a boat with a capacity of 60. Around them are 100 poor swimmers who want to come aboard. The rich boaters have three choices. First, they could take in all the swimmers, capsizing the boat with "complete justice, complete catastrophe." Second, as they enjoy an unused excess capacity of 10, they could admit just 10 from the water. But which 10? And what about the margin of comfort that excess capacity allows them? Finally, the rich could prevent any of the doomed from coming aboard, ensuring their own safety, comfort, and survival.

Translating this metaphor into reality, as an occupant of the rich lifeboat United States, for example, what would you do for the drowning refugees from lifeboat Haiti?

Hardin's choice is the third: drowning. To preserve their own standard of living and ensure the planet's safety, the wealthy countries must cease to extend food and other aid to the poor, and must close their doors to immigrants from poor countries. "Every life saved this year in a poor country diminishes the quality of life for subsequent generations," Hardin concludes. "For the foreseeable future, survival demands that we govern our actions by the ethics of a lifeboat."[3]

In the U.S. and other Western countries there is now a growing sense of **donor fatigue**. Wearied by constant images of people in need, and feeling their contributions might be only marginally useful, people who can afford to give are simply tired of thinking about giving. The neo-Malthusian view that Hardin represents would see donor fatigue as beneficial to the biosphere.

2.7   Challenges of a New Millennium

Hardin assumes that people overpopulation is the greatest threat to Earth and does not consider the dangers posed by consumption overpopulation. He foresees a static situation in which the wealthy countries alone decide Earth's fate. But changes are ongoing in the economic status, political power, populations and resource uses of poor countries. The oil embargo and energy crisis of 1973–74, for example, illustrated the ability of some LDCs to change the course of world events. Decisions about the future must be based on a more thorough understanding of the distributions and uses of global resources.

Birth rates in the LDCs are now falling, and so consequently is the world's rate of population growth. However, because the population base is so large, the actual number of people added to our population is greater than at any time in the past. Several current projections (which do not consider the unpredictable effects of a devastating AIDS or other epidemic)

---

[1] Ehrlich, Paul R. 1988. "Populations of People and Other Living Things." In De Blij, Harm, ed. Earth '88: *Changing Geographic Perspectives*, pp. 302–315. Washington, D.C.: National Geographic Society.

[2] Hardin, Garrett. 1968. "The Tragedy of the Commons." *Science* **162:**1243–1248.

[3] Ibid.

place the world's population at 7 billion by 2009, 8 billion by 2020, 9 billion by 2033, 10 billion by 2046, 11 billion by 2066, and 12 billion by 2100. The World Bank projects that the population will level off at about 12.5 billion soon after that.

Considering these numbers, the geographer asks whether, how, and above all where, global resources can support them. About 11 percent of Earth's land surface is now in crops, 25 percent in pasture and rangelands, 31 percent in forests and savanna, and 33 percent is little productive (including deserts, ice caps and other harsh biomes, and paved and built-up areas). The proportion dedicated to crops grew from the Neolithic until 1981, when the total area of farmland began to decrease. The other productive areas of pasture, rangeland, forests, and savanna are also decreasing. Only the "little productive" category is growing. After a long period of growth, crop yields on existing farmlands are also diminishing. Worldwide, irrigated area per capita has been decreasing since 1978, and per capita grain production has been decreasing since 1984. In sum, just as our demands for more food and other photosynthetic products are growing, supplies are dwindling.

CONFRONTING CLIMATE CHANGE

While attempting to improve the technical means to meet the growing demand for food, we must also be concerned with the long-term effects of technology on the atmospheric processes identified at the start of this chapter. Climate change as a result of human activity could reverse efforts to improve life on the planet. This problem may become the most significant environmental challenge of the new millennium.

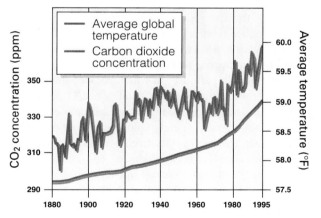

FIGURE 2.20

Industrialization and burning of tropical forests have produced a steady increase in carbon dioxide emissions. Most scientists believe these increased emissions explain the corresponding steady increase in the global mean temperature.

Until 1995, many scientists believed it was premature to state firmly that global climatic change, and in particular global warming, is now taking place as a result of human activities. The year 1995 was the Earth's warmest since records first were kept in 1856. The year also concluded three successive five year periods in which Earth's temperatures were warmer than in any previous five year period (Fig. 2.20). Until 1995, experts on climate change held out the possibility that the recent periods of warm years may have been natural variations. But in 1995, breakthroughs in computerized models of climate change prompted the United Nations-sponsored Intergovernmental Panel on Climate Change, the world's leading organization concerned with this issue, to conclude that the warming of the last century "is unlikely to be entirely due to natural causes" and that "a pattern of climatic response to human activities is identifiable in the climatological record." Increasing numbers of former skeptics now hold the view that Earth's atmosphere is growing warmer, and that people are responsible. The theory of the **greenhouse effect** (see page 52) is the most widely-adopted explanation for global warming.

In 1896, Swedish scientist Svante Arrhenius feared that Europe's growing industrial pollution would eventually double the amount of carbon dioxide in the earth's atmosphere and as a consequence raise the global mean temperature as much as nine degrees Fahrenheit (5°C.). Computer-based climate change modellers today also use the scenario of a doubling of atmospheric carbon dioxide, and they similarly conclude that the mean global temperatures might warm from three to nine degrees Fahrenheit by the year 2030. The estimated range and consequences of the increases vary widely because of different assumptions about the little-understood roles of oceans and clouds. Most models concur that because of the slowness with which the world's oceans respond to changes in temperature, the effects of this rise will be delayed.

Geographers are most concerned with where the anticipated climatic changes will occur. Computer models conclude that there will not be a uniform temperature increase across the entire globe but that the increases will vary spatially and seasonally. Differing models produce contradictory results about the timing and impacts of warming. Two teams of scientists publishing their results in the journal *Nature*, for example, first assumed a doubling of CO_2. Both then predicted rises in winter precipitation (due to increased evaporation) and in temperature, but assuming different seasonal combinations, they reached different conclusions about the effects on agriculture in the United States. One assumed warmer, drier summers and forecast that crop productivity would decline by 20 percent. The other team assumed warmer winters and forecast a 10 percent increase in crop yields.

There is general agreement that with global warming the distribution of climatic conditions typical of biomes will shift poleward. Many animal species will be able to migrate to keep pace with changing temperatures, but plants, being stationary

Definitions and Insights

THE GREENHOUSE EFFECT

Formulated in 1827, the theory of the **greenhouse effect** begins with an observation that the Earth's atmosphere acts like the transparent glass cover of a greenhouse (see Fig. 2.B) (for modern purposes, like a windows of a car). Visible sunlight passes through the glass to strike the Earth's surface. Ocean and land (the car upholstery) reflect the incoming solar energy as invisible infrared radiation (heat). Acting like the greenhouse glass or car window, the Earth's atmosphere traps some of that heat.

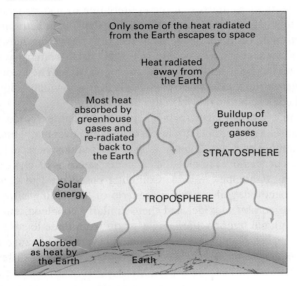

FIGURE 2.B

The greenhouse effect. Some of the solar energy reflected as heat (infrared radiation) from the Earth's surface escapes into space, while greenhouse gases trap the rest. Naturally-occurring greenhouse gases make Earth habitable, but carbon dioxide and greenhouse gases emitted by human activities accentuate the greenhouse effect, making Earth unnaturally warmer.

In the atmosphere, naturally occurring greenhouse gases such as carbon dioxide (CO_2) and water vapor make Earth habitable by trapping heat from sunlight. Concern over global warming focuses on human-made artificial sources of greenhouse gases, which trap abnormal amounts of heat. Carbon dioxide released into the atmosphere from burning coal, oil, and natural gas is the greatest source of concern, but methane (from rice paddies and the guts of ruminating animals like cattle), nitrous oxide (from the breakdown of nitrogen fertilizers), and chlorofluorocarbons, or CFCs (from cooling and blowing agents), are also human-made greenhouse gases. Continued production of these gases may have the effect of rolling up the car windows on a sunny day—increased temperatures and physical problems for the occupants result.

organisms, will not. Conservationists who have struggled to maintain islands of habitat as protected areas are particularly concerned about rapidly changing climatic conditions.

There is also general agreement that if global temperatures rise, so will sea levels as polar and glacial ice melts, and as sea-water warms and thus occupies more volume. Decision makers in coastal cities throughout the world, in island countries, and in those with important lowland areas adjacent to the sea are worried about the implications of rising sea levels. The coastal lowland countries of the Netherlands and Bangladesh have for many years been outspoken advocates of reductions in greenhouse gases. President Gayoom of the Maldives, an Indian Ocean country comprised entirely of low islands, pleaded "We are an endangered nation!"

There is not yet enough international political consensus to take the strong measures that may be necessary to prevent or greatly reduce global warming. Present efforts to reduce the production of greenhouse gases focus on the MDCs, where about 85 percent of the emissions originate. The United States alone now accounts for 25 percent of all emissions, China for 9 percent, and more than 70 LDCs together for only about 5 percent. However, this balance is likely to change as the MDCs adopt stricter emissions standards and as coal-wealthy China matures into a major industrial power, possibly with less stringent standards. In 1993, the United States joined Japan and most western European countries in a pledge to reduce CO_2 emissions to 1990 levels by the year 2000. However, by 1997 it appeared that most of these countries would not fulfill their pledges, and it was uncertain how effective even that modest reduction, if met, would be.

The LDCs which rely largely on fuelwood rather than fossil fuels release relatively little CO_2 into the atmosphere, but they do contribute to the perceived greenhouse effect. When burned, trees not only release the CO_2 which they stored in the process of photosynthesis but cease to exist as organisms which remove CO_2 from the atmosphere. However, because these countries produce relatively little CO_2, policy-makers are considering innovative ways to reward them and encourage them to develop clean industrial technologies. One system being considered is "tradable permits," in which each country would be assigned a right to emit a certain quantity of CO_2, according to its population size, setting the total at an acceptable global standard. The United States, already an overproducer by this scale, could purchase emission rights from a populous country like China, which currently "underproduces" CO_2. China would be obliged to use the income to invest in energy-efficient and non-polluting technologies. Most notable about this system is that it views global warming as a global problem.

Promising steps toward reducing another global threat, the **"ozone hole,"** have already occurred. CFCs destroy the stratospheric ozone, a layer of gas about 10–30 miles above the earth that absorbs much of the ultraviolet radiation from the sun and thus protects plants, animals, and people from the sun's harmful effects. A rise in human skin cancer in Australia recorded since

1985, and a growing incidence of blindness in sheep in Chile, are among the indications that destruction of ozone has immediate and negative consequences. Due to the Earth's rotation and climatic patterns, depletion of stratospheric ozone is concentrated at the Earth's poles, particularly over Antarctica, as "holes" in the ozone layer. Breaking up in the polar spring season, these zones elongate and drift equatorward. In the mid 1990s, scientists discovered that, as anticipated, ozone holes were beginning to creep over populated areas of northern Europe, posing a serious threat to large numbers of people, livestock and crops.

In contrast to the still-controversial geography of global warming, the geography of future ozone depletion has for many years been almost unanimously regarded as a major threat. The world's 37 major CFC-producing countries took action in 1989 with a treaty known as the **Montreal Protocol**, in which they agreed to cease production of these chemicals between the years 2000 and 2010. The MDCs which initiated the treaty agreed to cut their production sooner than the LDCs would have to. The U.S. ceased production in 1996. Many countries applauded its leadership but also reminded the U.S. that it should lead the world in a more vigorous attack on the problem of CO_2 emissions.

SUSTAINABLE DEVELOPMENT

Actions against the potential consequences of global warming and ozone depletion are part of a growing agenda aimed at reversing destructive global patterns of resource and landscape use. Within the last two decades, new concepts and tools for managing the Earth in an effective, long-term way have emerged. Known collectively as **sustainable development** or **ecodevelopment**, these ideas and techniques consider what both MDCs and LDCs can do to avert the possible Malthusian dilemma and improve life on the planet.

The World Conservation Union defined sustainable development as "improving the quality of human life while living within the carrying capacity of supporting ecosystems." Sustainable development refutes what its proponents perceive as the current pattern of unsustainable development, whereby GNP and other measures of economic growth are based in large part on the destruction of global resources. According to these measures, a country that plunders its resource base for short-term profits gained through deforestation increases its GNP and appears to be more "developed" than a country that protects its forests for long-term harvesting of sustainable yield. Deforestation appears to be beneficial to a country because it raises GNP through the production of pulp, paper, furniture, and charcoal. However, GNP does not measure the negative impacts of deforestation, such as erosion, flooding, siltation, and malnutrition.

Sustainable development is a complex assortment of theories and activities, but its proponents call for eight essential changes in the way people perceive and use their environments:

1. People must change their worldviews and value systems, recognizing the finiteness of resources and reducing their expectations to a level more in keeping with Earth's environmental capabilities. Proponents of sustainable development argue that this change in perspective is needed especially in the MDCs, where rather than trying to "keep up with the Joneses," people should try to enjoy life through more social pursuits.

2. People should recognize that development and environmental protection are compatible. Rather than viewing environmental conservation as a drain on economies, we should see it as the best guarantor of future economic well-being. This is especially important in the LDCs with their resource-based economies.

3. People all over the world should consider the needs of future generations more than we do now. Much of the wealth we generate is in effect borrowed or stolen from our descendants. Our economic system values current environmental benefits and costs far more than future benefits and costs, and thus we try to improve our standard of living today without regard to tomorrow.

4. Communities and countries should strive for self-reliance, particularly through the use of appropriate technologies. Villages, for example, could rely increasingly on solar power for electricity rather than be linked into national grids of coal-burning plants.

5. LDCs need to limit population growth as a means of avoiding the destructive impacts of "people overpopulation." Advances in the status of women, improvements in education and social services, and effective family planning technologies can help limit population growth.

6. Governments need to institute land reform, particularly in the LDCs. Poverty is often not the result of too many people on too little total land area, but of a small, wealthy minority holding a disproportionally high share of quality land. To avoid the environmental and economic consequences of marginalization, a more equitable distribution of land is needed.

7. Economic growth in the MDCs should be slowed to reduce the effects of "consumption overpopulation." If economic growth, understood as the result of consumption of natural resources, continues at its present rate in excess of sustainable yield, the Earth's environmental "capital" will continue to diminish rapidly.

8. Wealth should be redistributed among the MDCs and LDCs. Because poverty is such a fundamental cause of environmental degradation, the spread of a reasonable level of prosperity and security to the LDCs is essential. Proponents argue that this does not mean that rich countries should give

cash outright to poor. Instead, the lending institutions of MDCs could forgive some existing debts owed by LDCs, or use such innovations as **debt-for-nature swaps**, in which a certain portion of debt is forgiven in return for the borrower's pledge to invest that amount in national parks or other conservation programs. Reducing or eliminating trade barriers that MDCs impose against industries in LDCs would also help redistribute global wealth.

Some geographers and other scientists believe that sustainable development will be the "third revolution," a shift in our ways of interacting with earth so dramatic that it will be compared with the origins of agriculture and industry. The formidable changes called for in sustainable development are attracting increasing attention, perhaps because at the present there are no comprehensive alternative strategies for dealing with some of the most critical issues of our time.

Europe

1

3

A GEOGRAPHIC PROFILE
OF EUROPE

Many areas have contributed to shaping the present world but none more impressively than Europe. The influence of Europe on modern nation building, the development of science and technology, the rise of advanced capitalist economies, and expansion overseas are central to world history, even though there are positive and negative perspectives to these influences. Today, the region continues to play a vital role in the world's economic, political, cultural, and intellectual life even as it creates more internal links within the region.

3.1 Definition and Basic Magnitudes

In this book, we define Europe as the countries of Eurasia lying west of the Former Soviet Region and Turkey. Thus defined, Europe is a great peninsula of Eurasia, fringed by lesser peninsulas and islands, and bounded by the Arctic and Atlantic oceans, the Mediterranean Sea, Turkey, the Black Sea, and the Former Soviet Region (Figs. 3.1, 4.1, and 5.1). Europe achieves the label of continent largely because fundamental geographic

standards were determined and defined by Europeans when they had enormous global authority.

Understanding the geographic reality of Europe requires a grasp of some basic magnitudes that are very different from those of the United States. For example, all of Europe has an area only about two thirds as large as that of the 48 contiguous states of the United States. This means that the areas of Europe's many countries tend to be rough equivalents of those of U.S. states. One of the striking things about Europe is the large impact that societies nourished on some of these small pieces of land have had on the world.

A different set of comparative magnitudes emerges when comparing the populations of the European countries with those of the individual United States. California—with its population of nearly 33 million—would be a rather populous country if it were European, but even California is considerably exceeded in population by six European countries. While most European countries are comparable in population to relatively populous American states (Table 3.1), four major countries—Germany, the United Kingdom, France, and Italy—far outsize all others in population. Their respective populations range

◄ Europe from space. This view of Europe is a mosaic composed of a number of National Oceanic Atmospheric Administration (NOAA) weather satellite images taken on cloud-free days. The colors approximate natural tones and are a good indication of different ground cover. In western Europe, the light and mid-green colors indicate pasture and cultivated land. The bluer greens of Eastern Europe show greater aridity and a less lush vegetation. This mosaic is composed of summer shots so little snow is evident. Most evident in this graphic is the geographic fact that Europe is more a peninsula of the massive Eurasian land mass than a free-standing continent, even though convention marks Europe as one of the world's continents. However, in terms of political, economic, and social significance, it is very much a continent.

FIGURE 3.1

Europe showing political units as of early 1997. Note that in nearly all European countries the political capital is also the largest city (metropolitan area).

from about 81 million for Germany to 58 million for France, and together the four countries represent about one half of Europe's population, or nearly as many people as there are in the United States.

Another useful comparison between European countries and the United States is by density of population. Even exclud-

ing great disparities among small units such as Malta and the **microstates** of Andorra and Monaco, Europe exhibits an extreme range of population density—from over 1070 persons per square mile (over 414 per sq km) in the intensively developed Netherlands to about 7 per square mile (3 per sq km) in Iceland (Table 3.1 and Fig. 3.2). Comparisons with the United

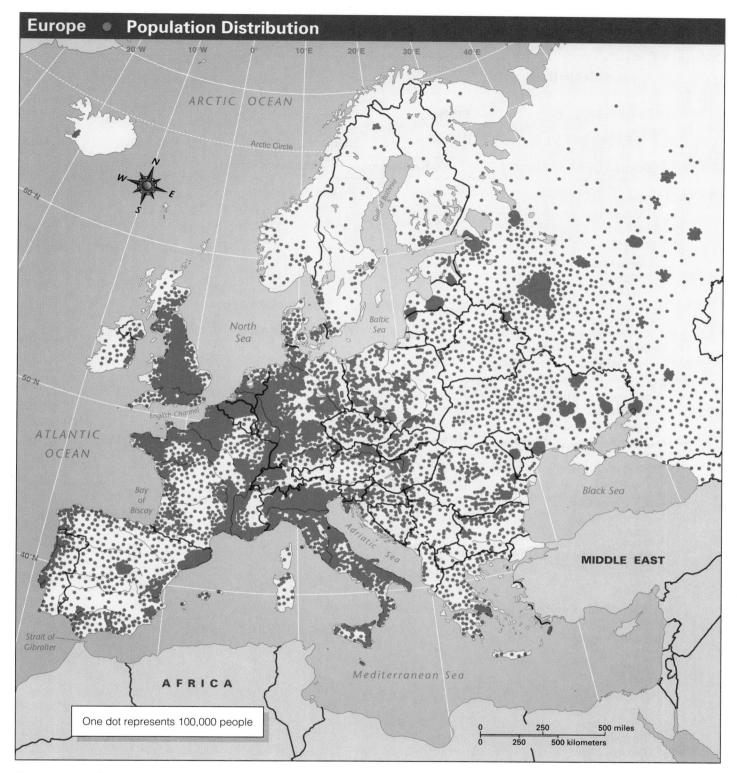

Europe • **Population Distribution**

One dot represents 100,000 people

ARCTIC OCEAN

Arctic Circle

N
W   E
S

60°N

50°N

North
Sea

40°N

ATLANTIC
OCEAN

Bay
of
Biscay

English Channel

Strait of
Gibralter

AFRICA

Baltic
Sea

Gulf of Bothnia

Adriatic
Sea

Mediterranean Sea

Black Sea

MIDDLE EAST

20°W   10°W   0°   10°E   20°E   30°E   40°E

0   250   500 miles
0   250   500 kilometers

FIGURE 3.2

The distribution of the human population is one of the most significant patterns to be
found in geography. In this map of European population distribution, the strong role of rivers,
coasts, and lowlands is evidenced. In addition, the long time concentration of urban, industrial
population on the island of Great Britain, along the coastlines of northwest Europe, and along
both margins of the Italian peninsula is apparent. The dot map clearly distinguishes high population areas from areas that are relatively less densely populated.

TABLE 3.1

Europe: Basic Data

Political Unit	Area		Estimated Population	Estimated Annual Rate of Increase
	(thousand/sq mi)	(thousand/sq km)	(millions)	(%)
British Isles				
United Kingdom	94.5	244.8	58.3	0.27
Ireland	27.1	70.3	3.6	0.33
Total	121.6	315.1	61.9	0.27
Western Central Europe				
France	211.2	547.0	58.1	0.46
Germany	137.8	356.9	81.3	0.26
Belgium	11.8	30.5	10.1	0.17
Netherlands	14.4	37.3	15.5	0.52
Luxembourg	1.0	2.6	0.4	0.57
Switzerland	15.9	41.3	7.1	0.57
Austria	32.4	83.9	8.0	0.35
Total	424.5	1099.5	180.5	0.36
Northern Europe				
Denmark	16.6	43.1	5.2	0.22
Norway	125.2	324.2	4.3	0.37
Sweden	173.7	450.0	8.8	0.46
Finland	130.1	337.0	5.1	0.30
Iceland	39.8	103.0	0.3	0.92
Total	485.4	1257.3	23.7	0.36
Southern Europe				
Italy	116.3	301.2	58.3	0.21
Spain	194.9	504.8	39.4	0.27
Portugal	35.6	92.1	10.6	0.36
Greece	50.9	131.9	10.6	0.72
Malta	0.12	0.3	0.4	0.75
Total	397.82	1030.3	119.3	0.29
East Central Europe				
Poland	120.7	312.7	38.8	0.36
Czech Republic	30.4	78.7	10.4	0.26
Slovakia	18.8	48.8	5.4	0.54
Hungary	35.9	93.0	10.3	0.02
Romania	91.7	237.5	23.2	0.09
Bulgaria	42.8	110.9	8.8	−0.25
Albania	11.1	28.8	3.4	1.16
Yugoslavia	39.5	102.3	11.1	0.79
Bosnia-Herzegovina	19.8	51.2	3.2	0.65
Croatia	21.9	56.5	4.7	0.13
Macedonia	9.8	25.3	2.2	0.90
Slovenia	7.8	20.3	2.1	0.24
Total	450.2	1166.0	123.6	0.30
Summary Total	1757.92	4553.1	509.0	0.32

Sources: *The World Almanac and Book of Facts* 1996, *Statesman's Yearbook* 1995–1996, *Britannica Book of the Year 1995*, and *The World Factbook* 1995.

Estimated Population Density		Infant Mortality	Urban Population	Arable Land	Per Capita GNP
(sq mi)	(sq km)	Rate	(%)	(% of total area)	(SUS)
616	238	7.0	92	29	17980
132	51	7.2	57	14	14060
509	196	7.0	90	26	17752
275	106	6.5	74	32	18670
591	228	6.3	85	34	16580
855	330	7.0	97	24	18040
1072	414	6.0	89	26	17940
407	157	6.6	86	24	22830
445	172	6.3	68	10	22080
246	95	6.9	54	17	17500
425	164	6.4	80	30	16947
313	121	6.8	85	61	19860
34	13	6.1	73	3	22170
52	20	5.6	83	7	18580
39	15	5.2	64	8	16140
7	3	4.0	91	1	17250
49	19	5.8	78	6	18970
500	193	7.4	68	32	17180
202	78	6.7	64	31	13120
297	115	9.1	34	32	10190
208	80	8.3	63	23	8870
2292	1155	7.7	85	38	10760
300	116	7.4	63	30	14458
321	124	12.4	62	46	4920
343	133	8.9	75	—	7350
288	111	10.0	57	—	6070
287	110	11.9	63	51	5700
253	98	18.7	55	43	2790
3	79	11.4	67	34	3830
308	119	28.1	36	21	1110
281	108	9.8	47	30	1000
162	63	11.6	34	20	—
215	83	8.4	54	32	2640
220	85	24.2	58	5	900
262	101	7.9	50	10	8110
275	106	13.2	59	38	4182
290	112	8.3	74	27	13457

States reveal about the same range, although in general the European countries are more densely populated. Table 3.1 provides a set of data that will allow a comparison within Europe and among other regions and nations as well.

This is a region worthy of some detailed study, for many of the cultural characteristics and patterns that give image, personality, and tension to the world today derive from the influences of **Western Civilization** in its many manifestations.

Definitions and Insights

WESTERN CIVILIZATION

Western civilization and related terms, such as **the West** and **Westernization**, are indispensable to a study of world geography because they connote innumerable traits that give geographic distinctiveness to places. In general, Western civilization refers to the sum of values, practices, and achievements that had roots in ancient Greece, Rome, Mesopotamia, and Palestine; that subsequently flowered in Europe; and which are still being developed and modified there and in other areas to which they have been diffused. The term incorporates a set of languages, religious practices (associated with "Western" Christian churches including the Roman Catholic church and the Protestant churches but not the Orthodox Eastern churches), systems of law, and systems of social, economic, and political organization. Among the core conceptions of modern Western civilization are strong commitments (not always effective) to education, experimental science, technological progress, economic development, democratic representative government, and explicit protection of individual rights and liberties. A strong reliance on private capitalism developed prior to and during the Industrial Revolution, although this has been modified by experiments in communism, socialism, and fascism during the 20th century. **Westernization** refers to the process whereby non-Western societies acquire Western traits, adopted with varying degrees of thoroughness. As used today, "the West" refers primarily to Europe, North America, Australia, and New Zealand, although the concept embraces other countries—such as the Latin American countries, Israel, and South Africa—to a lesser degree.

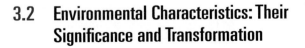

3.2 Environmental Characteristics: Their Significance and Transformation

There is a broad range of environmental settings in Europe. The application of human creativity to this varied environment is an important feature of Europe's long record of struggle and accomplishment. The roots of this development within Europe far antedate the Age of Discovery (see Chapter 1). Later, profits from exploitation of a worldwide colonial realm funded technological advances and economic growth. It should not be forgotten, however, that Europeans also exploited each other: The unbelievably primitive working conditions of early English coal mines are only one of many possible illustrations. Even considering both internal and external excesses in the marshaling of human energy to extract resources and transform the environment, we still acknowledge the ingenuity displayed over centuries by Europeans in the study and development of the region's resources.

PHYSICAL ATTRIBUTES

Irregular Outline

A noticeable characteristic of Europe is its extremely irregular outline. The main peninsula of Europe is fringed by numerous smaller ones, most notably the Scandinavian, Iberian, Italian, and Balkan peninsulas (Fig. 3.1). Offshore are numerous islands, including Great Britain, Ireland, Iceland, Sicily, Sardinia, Corsica, and Crete. Around the indented shores of Europe, arms of the sea penetrate the land in the form of significant **estuaries**, and countless harbors offer protection for shipping. This complex mingling of land and water provides many opportunities for maritime activity, and much of Europe's history has focused on maritime trade, sea fisheries, and sea power. Today, the region continues to dominate the world's sea routes.

Northerly Location

Another striking environmental characteristic of Europe is its northerly location: Much of Europe lies north of the 48 conterminous states (Fig. 3.3). Despite their moderate climate, the British Isles are at the same latitude as Canada, and Athens,

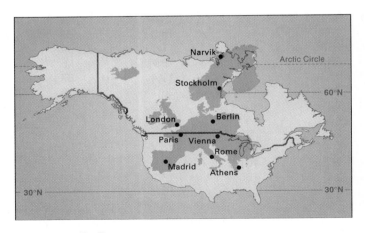

FIGURE 3.3
Europe in terms of latitude and area compared with the United States and Canada. Most islands have been omitted.

Regional Perspective

THE NORTH ATLANTIC DRIFT

One of Europe's most dramatic geographic gifts is a warm ocean current that has a strong influence on both climate and fisheries. The entire movement of this warm ocean current is known as **the North Atlantic Drift**. The portion that skirts the eastern shore of the United States before curving off toward Europe is called the **Gulf Stream**. With winds blowing predominantly from the west (from sea to land) along the Atlantic coast of Europe, the moving air in winter absorbs heat from the ocean and transports it to the land, making temperatures abnormally mild for this latitude, particularly along the coast. In the summer, the climatic roles of water and land are reversed: Instead of being warmer than the land, the ocean is now cooler; hence, the air brought to the land by westerlies in the summer has a cooling effect. The result is that only the southern fringes of Europe have hot-month temperatures averaging over 70°F (21°C).[1] This influence is carried as far north as the Scandinavian peninsula and into the Barents Sea, allowing Murmansk, Russia, to be an ice-free port even though it lies within the Arctic Circle.

[1] In general, the climate statistics used in this text are rounded data based on standard sources that report temperatures in degrees centigrade and precipitation in millimeters. Conversions and roundings may produce apparent small discrepancies in figures.

Greece, is only slightly farther south than St. Louis, Missouri. One effect of a northerly latitude that visitors to such cities as Berlin and Stockholm notice is the long duration of daylight during summer (giving a boost to the flow of tourist dollars into the economy) and the brevity of daylight in the winter.

Temperate Climate

The overall climate of Europe is more temperate than its northerly location would suggest. Winter temperatures, in particular, are mild for the latitude. For example, London has approximately the same average temperature in January as Richmond, Virginia, which is 950 miles (*c.* 1500 km) farther south. These anomalies of temperature are caused by relatively warm currents of water which originate in tropical western parts of the Atlantic Ocean, drift to the north and east, and make the waters around Europe much warmer in winter than the latitude would warrant.

The same winds that bring warmth in winter and coolness in summer also bring abundant moisture. Most of this falls as rain, although the higher mountains and more northerly areas have considerable snow. Ample, well-distributed, and relatively dependable moisture has always been one of Europe's major assets. Actually, the average precipitation in most places in the European lowlands is only 20 to 40 inches (*c.* 50 to 100 cm) per year (Fig. 3.4). In some parts of the world nearer the equator, this amount would be distinctly marginal for agriculture. But in most of Europe, the moisture is sufficient for a wide range of crops because mild temperatures and high atmospheric humidity lessen the rate of evapotranspiration.

Advantages and Limitations of a Varied Topography

Europe's topographic features are highly diversified. Each class of features—plains, plateaus, hill lands, mountains, and water bodies—is well represented (Fig. 3.5) . The whole physical assemblage, overspread with a varied plant life and enriched by the human associations and constructions of an eventful history, presents a series of distinctive and often highly scenic landscapes (Fig. 3.6).

One of the most prominent surface features of Europe is a plain that extends without a break from the Pyrenees Mountains at the French-Spanish border, across western and northern France, central and northern Belgium, the Netherlands, Denmark, northern Germany, and Poland, and far into Russia. Known as the North European Plain, it has outliers in Great Britain, the southern part of the Scandinavian peninsula, and southern Finland. For the most part, the plain is not flat but undulating or rolling. Flat stretches are generally on alluvial land. The North European Plain contains the greater part of Europe's cultivated land, and it is underlaid in some places by deposits of coal, iron ore, potash (used in the production of soaps, glass, and the manufacture of other potassium compounds), and other minerals that have been important in the region's industrial development. Many of the largest European cities—including London, Paris, and Berlin—developed on the plain. From northeast France eastward, a band of especially dense population extends along the southern edge of the plain; this concentration apparently has been true since prehistoric times. It coincides with (1) extraordinarily fertile soils formed from deposits of *loess* (see Regional Perspective, page 67), (2) an important natural transportation route skirting the highlands to

(text continues on page 66)

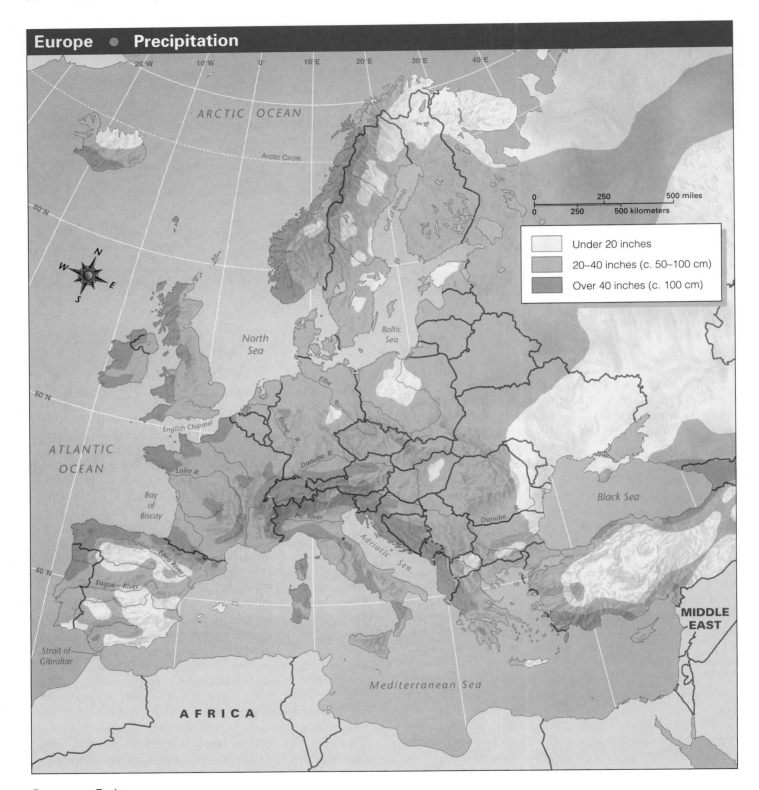

Europe ● Precipitation

Legend:
- Under 20 inches
- 20–40 inches (c. 50–100 cm)
- Over 40 inches (c. 100 cm)

FIGURE 3.4

Average annual precipitation in Europe. Most parts of Europe receive sufficient total precipitation for crop production, although in areas that have mediterranean climate, the concentration of rain occurs in the colder half of the year, thereby reducing its usefulness. Highlands along windward coasts often receive excessively heavy precipitation. International boundaries are not shown for areas outside Europe as defined in this text.

Europe • Natural Regions

Legend:
- Humid mid-latitude plains
- Humid mid-latitude hill lands (including small areas of low mountains and of plains)
- Mediterranean (dry-summer) subtropical hills, tablelands, and small plains
- Arctic tundra
- Glacially scoured subarctic plains and hills forested in conifers
- Mountains
- Approximate boundaries between marine west coast climate (west of line) and humid continental climate (east of line) between the Alps and the Baltic Sea

Selected cities are shown as reference points.

FIGURE 3.5

The "natural regions" of the map are attempts to show natural features as composite habitats. Each color-symbol represents a distinctive association of landforms and climate, with natural vegetation also stated or implied. The result is a broad-scale view of the *natural setting* for human activity in Europe.

FIGURE 3.6
The European capacity to blend water, mountains, flowers, and small scale human settlements has been a hallmark for centuries. This scene from Lake Thun and Spiez in Switzerland represents such skill. This pattern has promoted tourism for decades and decades.   *© 1995 Dembinsky Photo Associates*

the south, and (3) a large share of the mineral deposits of the region.

South of the northern plain, Europe is predominantly mountainous or hilly, although the hills and mountains enclose many plains, valleys, and plateaus. The hill lands and some low mountains are geologically old. Exposed to erosion for a long

FIGURE 3.7
The raw stone, water, snow, and thin vegetation of the Alps are a distinctive landscape feature of Europe. This scene near Grimsel Pass in the Swiss Alps in Switzerland suggests why this mountain system has been an effective geographical and political barrier for centuries   *© 1994 Dembinsky Photo Associates*

time, these uplands are often rather smooth and round in outline. But many mountains in southern Europe are geologically young, and they are often high and ruggedly spectacular, with jagged peaks and snowcapped summits. They reach their peak of height and grandeur in the Alps (Fig. 3.7).

To the north, the North European Plain is bordered by glaciated lowlands and hill lands in Finland and eastern Sweden and by rugged, ice-scoured mountains in western Sweden and most of Norway. The British Isles also have considerable areas of glacially scoured hill country and low mountains, along with lowlands where glacial deposition occurred.

3.3   Types of Climates

Climate patterns tend to be fairly uniform over wide areas of the Earth. This allows classification of the Earth's surface into climatic regions, each characterized by a particular type of climate. Certain ranges and combinations of temperature and precipitation conditions define the climate types. Associated with and strongly influenced by each type are certain vegetation and soil conditions. One common classification is shown in this text on the map of world types of climates (see Fig. 2.6). An examination of this map reveals that Europe, despite its modest dimensions, has remarkable climatic diversity, including every type of climate except desert and tropical/rainy.

MARINE WEST COAST CLIMATE

This is the type of climate in which Atlantic influences are dominant. It extends from the coast of Norway to northern Spain and inland toward central Europe. The main characteristics are mild winters, cool summers, and ample rainfall, with many drizzly, cloudy, and foggy days. Throughout the year, changes of weather follow each other in rapid succession as different air masses temporarily dominate or collide with each other along weather fronts. Most precipitation is frontal in origin or results from a combination of frontal and orographic (highland) influences (see Chapter 2). In lowlands, winter snowfall is light, and the ground is seldom covered for more than a few days at a time. Summer days are longer, brighter, and more pleasant than the short, cloudy days of winter, but even in summer there are many chilly and overcast days. The frost-free season of 175 to 250 days is long enough for most crops grown in the middle latitudes to mature, although most areas have summers that are too cool for heat-loving crops such as corn (maize) to ripen.

HUMID CONTINENTAL CLIMATES

Inland from the coast, in western and central Europe, the marine climate gradually changes. Winters become colder and summers hotter; cloudiness and annual precipitation decrease.

Influences of maritime air masses from the Atlantic diminish and are modified by continental air masses from inner Asia. At a considerable distance inland, conditions become sufficiently different that two new climate types are apparent: the humid continental short-summer climate in the north (principally in Poland, Slovakia, and the Czech Republic) and the humid continental long-summer climate in the warmer south (principally in Hungary, Romania, Serbia, and northern Bulgaria). The natural vegetation is mostly forest, and soils vary greatly in quality. Among the best soils are those formed from alluvium and loess along the Danube River.

MEDITERRANEAN CLIMATE

Southernmost Europe has a distinctive climate—the mediterranean (dry summer–wet winter) subtropical climate. This pattern of precipitation results from a seasonal shifting of atmospheric belts. In winter, the belt of westerly winds shifts southward, bringing precipitation at a time of relatively low evaporation. In summer, the belt of subsiding high atmospheric pressures over the Sahara shifts northward, bringing desert conditions. Mediterranean summers are warm to hot, and little

precipitation occurs during the summer months, when temperatures are most advantageous for crop growth. Winters are mild; frosts are few. Drought-resistant trees originally covered mediterranean lands, but little of the forest remains. It has been replaced by the wild scrub that the French call *maquis* (called *chaparral* in the United States), or by cultivated fields, orchards, or vineyards. Much of the land consists of rugged, rocky, and badly eroded slopes where thousands of years of deforestation, overgrazing, and excessive cultivation have taken their toll. The subtropical temperatures make possible a great variety of crops, but irrigation is necessary to counteract the dry summers.

SUBARCTIC AND TUNDRA CLIMATES

Some northerly sections of Europe experience the harsh conditions associated with subarctic and tundra climates. The subarctic climate, characterized by long, severe winters and short, rather cool summers, covers most of Finland, the greater part of Sweden, and some of Norway. The brief and undependable

Regional Perspective

GLACIATION

During the Great Pleistocene Ice Age (*c.* 1,000,000 to 10,000 years B.C.) continental ice sheets formed over Europe, beginning with the Scandinavian peninsula and Scotland. They moved outward to cover Finland, much of Russia, the North European Plain as far west as the eastern Netherlands, and the British Isles except for a strip of southern England. For nearly a million years, ice sheets alternately advanced and melted back as glacial and interglacial periods succeeded each other. The latest retreat of the ice possibly began around 35,000 years ago. During the periods of ice advance, much of Europe, Asia, and North America lay under "continental" glaciers. Glacial evidence suggests there were four major periods of widespread glacial coverage.

In many landscapes, the effects of glaciation are still evident. In some areas, the ice sheets scoured the preceding surface, removing the soil and gouging hollows that became lakes and swamps when the ice melted. Today, landscapes of **glacial scouring**—the erosive action of ice masses in motion—with thin soil, much bare rock, and many lakes, characterize most of Norway and Finland, much of Sweden, parts of the British Isles, and Iceland. In such areas, the ice often rearranged the preexisting drainage, leaving rivers that now follow very irregular courses and have many rapids and waterfalls, creating favorable sites for hydroelectric installations.

In other glaciated areas, **glacial deposition**—the process of offloading rock and soil in glacial retreat or lateral movement—had the predominant effect on the present landscape. As it ground along, the ice accumulated Earth material on its underside and subsequently deposited it, either as **outwash** carried by streams of meltwater from underneath the ice, or in the form of morainal material dropped in place as the ice melted. Today, the landscapes formed of outwash are likely to be flat and usually not

very fertile, whereas moraines have a more irregular surface. Moraines may, in fact, form low ridges, such as the **terminal moraines** that were formed by long-continued deposition at the front end of an ice sheet during times when melting balanced the ice movement to such an extent that the ice front remained approximately fixed.

Glacial deposits of varying thickness were left behind on most of the North European Plain. Most of the deposits were so sandy that present soils formed from them are not very fertile, although they have often been made productive by careful handling and large applications of fertilizer. However, a belt of fertile windblown soil material called **loess** deposited at the southern edge of the plain provides exceptionally fertile farmland there. This material appears to have been picked up by winds from barren surfaces after the glacial retreat and deposited where wind velocities diminished.

frost-free season, coupled with thin, highly leached, acidic soils handicaps agriculture. Human settlement is scanty, and a forest of needle-leaved conifers, such as spruce and fir, covers most of the land. In the tundra climate, which occupies northernmost Norway and much of Iceland, cold winters combine with brief, cool summers and strong winds to create conditions hostile to tree growth. An open, windswept landscape results, covered with lichens, mosses, grass, low bushes, dwarf trees, and wild-flowers. Wildlife is present, but human inhabitants are few. Agriculture in a normal sense is not feasible.

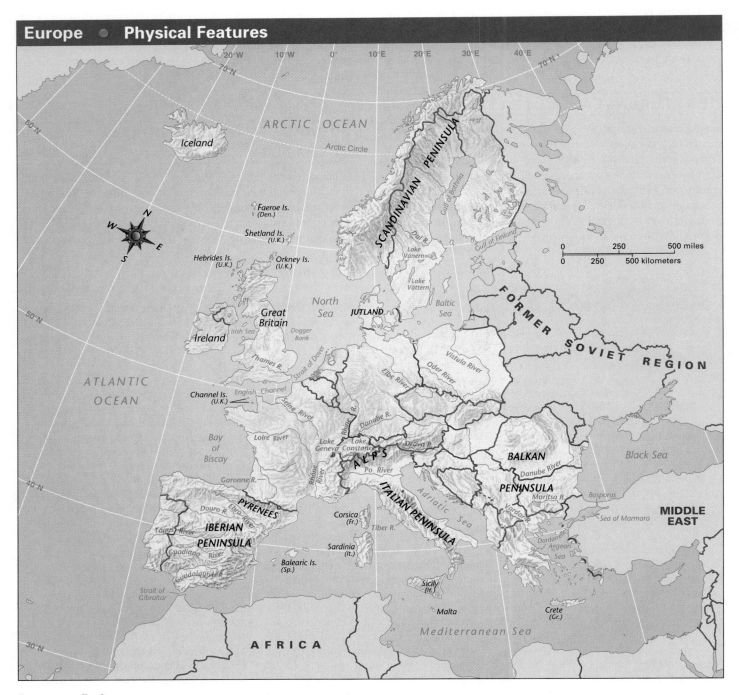

Europe ● Physical Features

FIGURE 3.8
Reference map of the main islands, peninsulas, seas, straits, and rivers of Europe. A few lakes are also shown.

HIGHLAND AND ICE CAP CLIMATES

The higher mountains of Europe, like high mountains in other parts of the world, have an undifferentiated highland climate varying with altitude and differential exposure to sun, wind, and precipitation. Given enough height, the variety can be startling. The Italian slope of the Alps, for instance, ascends from subtropical conditions at the base of the mountains to tundra and ice cap climates at the highest elevations. The ice cap climate experiences temperatures that average below freezing every month in the year, enabling ice fields and glaciers to be preserved.

3.4   The Role of Rivers

As one might expect in such a humid area, Europe has numerous river systems that are very important economically for Europe's transport, water supply, the generation of electricity, and in the creation of regional images.

Definitions and Insights

RIVER TERMS

Geographic discussion of rivers requires an understanding of certain terms: A **river system** is a river together with its tributaries. A **river basin** is the whole area drained by a river system. The **source** of a river is its place of origin; the **mouth** is where it empties into another body of water, often forming an **estuary** or a drowned river mouth. As they near the sea, many rivers become sluggish, depositing great quantities of sediment to form **deltas**, and often dividing into a number of separate channels, known as **distributaries** (Fig. 3.8). The management of a river requires a high level of cooperation throughout the river system because major change upstream—taking water out for irrigation or putting in industrial pollutants, for example—will have significant impact on downstream populations and farming patterns.

River systems provide an important role not only for transportation and fishing, but also for defining and advertising city and regional images. Such images are important for tourism, especially as the rivers course through farm land or mountainous landscapes—as is often the case in Europe.

Rivers were an important part of Europe's transport system at least as far back as Roman times and probably before; today they are still central to the transporting of large quantities of bulk cargo at low cost, mostly in motorized (self-propelled) barges. The more important rivers for transportation are largely those of the highly industrialized areas in northwestern Europe. An extensive system of canals connects and supplements the rivers.

Important seaports have developed along the lower courses of many rivers, and some have become major cities. London on the Thames, Antwerp on the Scheldt, Rotterdam in the delta of the Rhine, and Hamburg on the Elbe are outstanding examples (Fig. 3.8). Often the river mouths are wide and deep, allowing ocean ships to travel a considerable distance inland. This is true even of short rivers such as the Thames and Scheldt. Such enlarged river mouths are estuaries formed by the submergence of the lower ends of the river valleys.

The Rhine and the Danube are European rivers of particular importance, both touching or crossing the territory of many countries (see Fig. 3.8). In the case of the Rhine, these countries include Switzerland—where the river rises in the Alps—Liechtenstein, Austria, France, Germany, and the Netherlands. Highly scenic for much of its course, the Rhine River is Europe's most important inland waterway. Along or near it, a striking axis of intense urban-industrial development and extreme population density has developed. At its North Sea end, the Rhine axis connects to world commerce by one of the world's most active seaports, Rotterdam, in the Netherlands.

The Danube (German: *Donau*) touches or crosses more countries than does any other river in the world. Within Europe, the Danube is also unusual in its southeasterly direction of flow—from its source in the Black Forest of southwestern Germany to its delta on the semi-enclosed Black Sea (Fig. 3.8).

3.5   Patterns of the Past and Consequent Landscapes

Far more than any other region, Europe has shaped the human geography of the modern world. From the region's tentative beginnings in the 15th century, European seamen, traders, soldiers, colonists, and missionaries burst upon the scene, and, by the end of the 19th century, had created a world in which Europeans or their descendants were dominant.

From almost any perspective, the attempt to understand the geography of other world regions requires an assessment of Europe's influence on their development and vice versa. During the centuries that Europe was reaching out to capture and exploit colonial holdings and dominate the world, her own societies were changing and developing. Such transformations were in part a response to overseas contacts. In turn, Europe radiated its changes to the rest of the world. The entire process is modern history's most important example of spatial interaction and cultural diffusion. From it emerged the basic components of the political and economic world that we know today.

THE KALEIDOSCOPE OF LANGUAGES

Languages are one of the most highly identifiable elements in a culture group's personality. In learning a language, one also learns a great deal about tradition, history, **cultural mores**, **belief systems**, and spirit. Language also plays a major role in the tensions frequently at work in the geography of borders in Europe (see Fig. 3.A).

Europe emerged from prehistory as the homeland of many different peoples. In ancient and medieval times, certain of these peoples experienced periods of vigorous expansion, and their languages and cultures became widely diffused. First came the expansion of the Greek and Celtic peoples, and later that of the Romans, the Germanic (Teutonic) peoples, and the Slavic peoples. As each expansion occurred, traditional

languages persisted in some areas but were displaced in others. Each of the important languages eventually developed many local dialects. With the rise of centralized nation-states in early modern times, particular dialects became the bases for standard national languages.

The first millennium B.C. witnessed a great expansion of the Greek and Celtic peoples. In peninsulas and islands border-

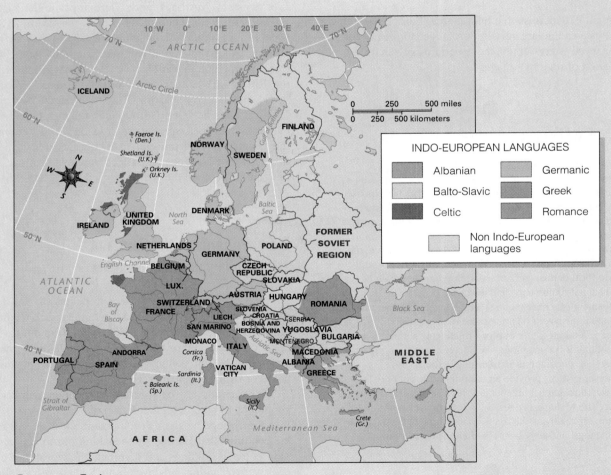

FIGURE 3.A

Branches of the Indo-European language family. Most Europeans speak languages from the Indo-European language family and the three most important branches are Germanic (north and west), Romance (south and west), and Balto-Slavic (east). The fourth major branch, Indo-Iranian—not shown in the European map here—is clustered in southern and western Asia.

ing the Aegean and Ionian seas, the early Greeks evolved a civilization that reached unsurpassed heights of philosophical inquiry and literary and artistic expression. Greek adventurers, traders, and colonists used the Mediterranean Sea as their highway to spread classical Greek civilization and its language along much of the Mediterranean shoreline, although there was already a strong Phoenician influence along the shores of the eastern Mediterranean before the Greeks arrived. Evidence of the early geographic range and subsequent influence of the Greek language and culture is apparent in the many Greek elements in modern European languages. But over time, the use of Greek in most areas disappeared as new peoples and languages expanded. Today, it remains the spoken language of Greece itself, but the dialects in use are very different from the classical Greek that was used in the writing of the major works that form part of the foundation of Western civilization (see *Definitions and Insights,* p. 62).

Europe's Celtic languages expanded at roughly the same time as Greek and, like Greek, are represented today only by remnants. The expansion of preliterate Celtic-speaking tribes took place from **hearth areas**—regions of original development—in what is now southern Germany and Austria, eventually occupying much of continental Europe and even the British Isles. Conquest and cultural influence by later arrivals, mainly Romans and Germans, eventually eliminated the Celtic languages except for a few traces that survive today in relatively remote peninsulas and islands: Breton in Brittany, Welsh in Wales, Irish Gaelic in western Ireland, and Scottish Gaelic in the northwestern Highlands of Scotland.

In present-day Europe, the overwhelming majority of the people speak **Romance**, **Germanic**, or **Slavic languages**. The Romance languages evolved from **Latin**, originally the language of ancient Rome and a small district around it. In the few centuries before and just after the birth of Christ, the Romans subdued territories extending from Great Britain to northern Africa and western Asia, and the use of Latin spread to this large empire. Latin had

by far the greatest impact in the less developed and less populous western parts of the empire. Over a long period, extending well beyond the collapse of this western part of the empire in the fifth century, regional dialects of Latin survived and evolved into the **Italian**, **French**, **Spanish**, **Catalan**, **Portuguese**, and other Romance languages of today. It's worth noting here that language frontiers do not always coincide exactly with political frontiers. For example, the French language extends to western Switzerland and also to southern Belgium, where it is known as **Walloon**.

In the middle centuries of the first millennium A.D., the power of Rome declined and a prolonged expansion by Germanic and Slavic peoples began. Germanic peoples first appear in history as a group of tribes inhabiting the coasts of Germany and much of Scandinavia. They subsequently expanded southward into Celtic lands east of the Rhine. Roman attempts at conquest were repelled, and the Latin language had little impact in Germany. In the fifth and sixth centuries A.D, Germanic incursions overran the Western Roman Empire, but in many areas the conquerors eventually were absorbed into the culture and language of their Latinized subjects.

However, the German language expanded into, and remains, the language of present-day Germany, Austria, Luxembourg, Liechtenstein, the greater part of Switzerland, the previously Latinized part of Germany west of the Rhine, and parts of easternmost France (Alsace and part of Lorraine). The Germanic languages of Europe include many languages other than German itself. In the Netherlands, Dutch developed as a language closely related to dialects of northern Germany, and Flemish—almost identical to Dutch—became the language of northern Belgium. The present languages of Denmark, Norway, Sweden, and Iceland descended from the same ancient Germanic tongue.

English is basically a Germanic language, although it has many words and expressions derived from French, Latin, Greek, and other languages. Originally, English was the language of the Germanic tribes known as Angles and Saxons who invaded England in the fifth and sixth cen-

turies A.D. The Norman conquest of England in the 11th century established French for a time as the language of the English court and the upper classes. Modern English retains the Anglo-Saxon grammatical structure but borrows great numbers of words from French and other languages. English is now the principal language in most parts of the British Isles, having been imposed by conquest or spread by cultural diffusion to areas outside of England.

Slavic languages are dominant in most of eastern Europe. The main ones today include the Russian and Ukrainian of the adjoining Former Soviet Region; Polish; Czech and Slovak; and Serbian, Croatian, and Bulgarian in the Balkan Peninsula. They apparently originated in eastern Europe and Russia, and were spread and differentiated from each other during the Middle Ages as a consequence of migrations and cultural contacts involving various peoples.

A few languages in present-day Europe are not related to any of the groups just discussed. Some are ancient languages that have persisted from prehistoric times in isolated (usually mountainous) locales. Two outstanding examples are **Albanian** (Indo-European in its origin) in the Balkan Peninsula and **Basque**, spoken in or near the western Pyrenees Mountains of Spain and France. Some languages unrelated to others in Europe do have relatives in Russia. They reached their present locales through migrations of peoples westward. The prime examples are **Finnish** and **Hungarian**, also called **Magyar**.

While it is easy to understand the importance of language as a medium of communication, it is just as essential to realize a language represents a whole universe of culture, tradition, and history. The kaleidoscope of languages in Europe has contributed to the patterns of cooperation, affiliation, tension, and distrust that characterize this region today. Language plays this same role in all other parts of the world, especially where a great number of cultures and languages are resident in a relatively small area. The current tensions that are so critical in the former Yugoslavia area derive in part from language differences within the region.

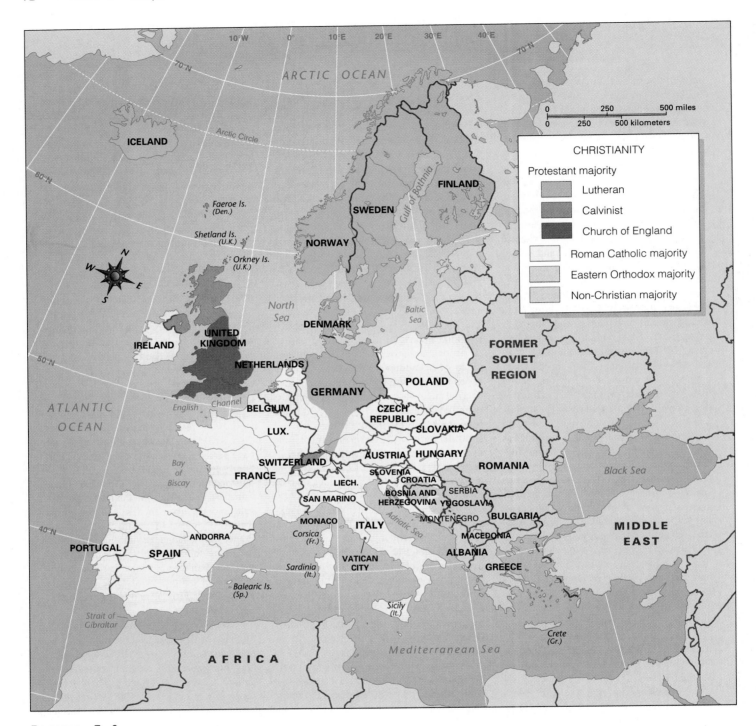

FIGURE 3.9

The religious patterns of Europe are largely expressions of distinct sects of Christianity. There are also significant areas of non-Christian populations. They are shown by the symbol of the Non-Christian majority category in the map.

RELIGIOUS BELIEF SYSTEMS

Europe's diverse patterns of religious belief systems are powerful indicators of culture, badges of nationality, and repositories of achievement. They—along with language—stand very high among the factors that differentiate Europe geographically (Fig. 3.9). During the long course of European history, both these cultural systems have played an important role in political conflict, including warfare, repression, discrimination, and terrorism.

In the Roman Empire, Christianity arose, spread, survived persecution, and became a dominant institution in the late stages of the empire. It survived the empire's fall and continued to spread, first within Europe and then to other parts of the world. In total number of adherents, the Roman Catholic Church, headed by the Pope in the Vatican, is Europe's largest religious division, as it has been since the Christian church was first established. Today, it remains the principal faith of a highly secularized Europe. The main areas that are predominantly Roman Catholic include: Italy, Spain, Portugal, France, Belgium, the Republic of Ireland, large parts of the Netherlands, Germany, Switzerland, Austria, Poland, Hungary, the Czech Republic, Slovakia, Croatia, and Slovenia.

During the Middle Ages, a center of Christianity existed in Constantinople (now Istanbul, Turkey) as a rival to Rome. From it, the Orthodox Eastern Church spread to become and remain dominant in Greece, Bulgaria, Romania, Serbia, and Macedonia, as well as in Russia and Ukraine to the east.

In the 16th century, the Protestant Reformation took root in various parts of Europe, and a subsequent series of bloody religious wars, persecutions, and counterpersecutions left Protestantism dominant in Great Britain, northern Germany, the Netherlands, Denmark, Norway, Sweden, Iceland, and Finland. Except for the Netherlands, where a higher Catholic birth rate has since reversed the balance, these areas are still mainly Protestant.

The Islamic (Muslim) faith, which is the principal religion of Albania and is an important faith in other parts of the Balkan Peninsula, plays a major influence in the contemporary struggle of European states in south-central Europe to achieve both political and religious stability. Islam was once widespread in the Balkan Peninsula, where it was established by the Ottoman Turks during their period of rule from the 14th to the 19th century. It was also the religion of the Moors, a powerful cultural presence on the Iberian Peninsula from the 8th century until the end of the 15th century (Fig. 3.10). Its presence has served as a dominant factor in the ethnic unrest in the former Yugoslavia since 1991.

Under the Roman Empire, **Jewish** minorities spread from Palestine into Europe, where they have since persisted as a significant minority population in many countries despite recurrent persecution.

SHAPING THE WORLD IN THE COLONIAL AGE

The process of exploration and discovery by which Europeans filled in the world map began with 15th-century Portuguese expeditions down the west coast of Africa. In 1488, a Portuguese expedition headed by Bartholomew Diaz rounded the Cape of Good Hope at the southern tip of Africa and opened the way for subsequent European voyages into the Indian Ocean. Then, in 1492, North America was brought into contact with Europe when a Spanish expedition commanded by a Genoese (Italian), Christopher Columbus (Cristóbal Colón), crossed the Atlantic to the Caribbean Sea. Less than half a century later, these early feats of exploration culminated in the first circumnavigation of the globe by a Spanish expedition commanded initially by a Portuguese, Ferdinand Magellan—who named the Pacific Ocean. Worldwide exploration, both coastal and inland, continued apace for centuries, with leadership being wrested from the Spanish and Portuguese by the Dutch, French, and English.

Missionaries, soldiers, and traders were seldom far behind the explorers, often traveling with the first ships. Trading posts, Christian missions, and military garrisons were established along many coasts, or, in some cases, inland. Inland posts were especially notable in Latin America, where they were often a response to the highland location of Indians who could be Christianized, robbed of precious metals, and put to work in mines or on the land. Many European outposts came to dominate the areas where they were established and eventually became bases from which complete colonial control was often extended.

FIGURE 3.10

Although the recent years of war in the former Yugoslavia center the world's attention on the tensions in east and central Europe, the Moorish influence is more widespread as this church near Oberammergau, Germany demonstrates. © 1996 Howard Garrett/Dembinsky Photo Associates

Regional Perspective

EXPLORATION AND DISCOVERY

Fundamental to all geography is the process of **exploration** and **discovery**. There has always been a fascination with what is seen when someone goes to a place that is unknown. The sorts of things learned in exploration have social, economic, and military significance. Centuries ago, explorers brought back information about routeways, plants, animals, peoples, and possible dangers in lands generally unknown to the peoples who launched the exploration. When the world being explored seemed to have a great potential for exploitation, it was claimed that a "discovery" had been made. Generally, the people being discovered did not feel "lost,"

or in need of discovery, but because exploration has often been a part of a political, economic, or military effort to gain control of some new region, the term "discovery" helped the newcomers justify their frequently harsh treatment of the peoples and places being exploited. In the **Age of Discovery** (c. late 15th through late 18th century), the scale of such exploration and discovery was greater than it had ever been before. While the indigenous peoples of the New World had themselves earlier been explorers in the Western Hemisphere, when Columbus found land in the Caribbean on October 12, 1492, he opened a new world to all of Europe. In that discovery—or arrival—there began a powerful series of

changes that could not be reversed. Native populations were diminished mightily by the introduction of diseases for which they had no immunity. As historians look more and more at the lives of common people rather than only at the social patterns and **material culture**—the artifacts of a group—of the wealthy and the politically powerful, it is increasingly clear that being "discovered" often extracts a high cultural cost. Every exploring, discovering, exploiting, and conquering people has its own range of explanations about why such geographic acts were meant to bring benefits to those being discovered as well as to those setting the discovery in motion.

Settlement by European colonists, on differing scales and at different times, took place in a great many non-European areas. In well-populated or environmentally difficult areas, European populations remained small minorities, but in other circumstances local versions of European societies took such firm root that they became numerically dominant. Today the descendants of transplanted Europeans make up most of the population in the United States, Canada, and parts of Latin America, Australia, New Zealand, and they are the financially dominant group in South Africa.

In carrying on their various overseas enterprises, Europeans not only migrated themselves but often also transferred non-Europeans from place to place. The most notable instance was the massive forced migration of slaves from Africa to the New World that began in the 16th century. Such transfers greatly influenced the ultimate racial and cultural makeup of sizable areas. Also of great importance in shaping the world's geography was the transfer of plants and animals from one place to another. A few major examples are the introduction of hogs and cattle and the reintroduction of horses to the Americas from Europe; of tobacco and corn (maize) from the Americas to Europe and other parts of the world; of rubber from South America to Asia; of the potato from the Americas to Europe, where it became a major food; and of coffee from Africa to Latin America, where it became the principal basis of a number of regional economies.

Diffusion of Foodstuffs Between New World and Old World: The Columbian Exchange

This table shows some of the most significant foodstuffs diffused from the New World to the Old World after 1492, as well as the major New World foodstuffs taken into world trade after 1492 by the ships and crews that came to the newly found lands of North and South America.

Diffused from Old World to New World	Diffused from New World to Old World
wheat	maize
grapes	potatoes
olives	cassava (manioc)
onions	tomatoes
melons	beans (lima, string, navy, kidney, etc.)
lettuce	
rice	pumpkins
soybeans	squash
coffee	sweet potatoes
bananas	peanuts
yams	cacao

A worldwide system of trade, with Europe at its core, was a major outcome of European expansion. The system principally involved the movement of raw materials and food products from the rest of the world to Europe and the return sale of European manufactured goods to non-European areas. In general, this exchange favored Europe, and wealth from the exchange mainly accumulated there, particularly in Seville, Spain during the Age of Discovery. The consequent infusions of money enabled certain European cities, particularly London, to grow and become centers of world finance. Today other centers of commerce and finance—notably in the United States, Japan, and Hong Kong—have risen to challenge Europe's dominance.

A HIGHLY DEVELOPED AND DIVERSIFIED ECONOMY

Diversity is a keynote of the European economy, just as it is of the European environment. Numerous forms of economic activity are highly developed, and much variety exists from one country and region to another. Most European countries rank high in productivity and affluence compared with other countries of the world. Western Europe was the first region of the world to evolve from an agricultural society into an industrial economy. The key series of events, commonly called the **Industrial Revolution** (see below), were followed by vast advances in productive (and destructive) knowledge, technology, and environmental transformation. Even military devastation did not stop the economic dynamism of Europe, which seized upon U.S. aid after World War II to rapidly reorganize, rebuild, and attain new heights of production and affluence.

Industrialization and Levels of Urbanization

One of the most significant facts of global geography is the high proportion of the world's manufacturing capacity and output that is European. This proportion is decreasing, but approximately one quarter to one third of the world output of most major industrial products originates in Europe. Industrialization goes far toward explaining the highly urban character of most European countries as well as their high standing in overall productivity and average income.

Europe was the place where a large-scale manufacturing industry, using machines driven by inanimate power, first arose. Notable for its surge of inventiveness, the Industrial Revolution began to be felt about the middle of the 18th century. Industrial innovations, many of which British inventors developed, made water power—and then steam power—increasingly available to turn machines in the new factories. The invention of a practical steam engine in Great Britain made coal a major resource and greatly increased the amount of power available. New processes and equipment—including the invention of a way to use coke instead of charcoal in blast furnaces to smelt iron ore—made iron relatively abundant and cheap. The invention of new industrial machinery made of iron and driven by steam

engines multiplied the output of manufactured products, principally textiles at first. Then, in the 19th century, British inventors developed processes that allowed steel, a metal superior to iron in strength and versatility, to be made on a large scale and cheaply for the first time. Such developments led the world into the modern age of massive, mechanized industrial production.

Europe has one of the highest regional averages of urban population in the world (Figure 3.11). In the United Kingdom and Belgium, more than 90 percent of the population is urban, with Belgium reaching 97 percent urban in 1996. More than 85 percent of the population of Germany, the Netherlands, Denmark, and Luxembourg resides in cities, while Sweden is 83 percent urban. These high levels of urbanization complement the industrial nature of these nations' economies, but they also have an influence on levels of population growth. These nations generally have the lowest levels of crude birth rate (the total number of live births per 1000 population per year) and often the highest per capita GNP levels.

However, as later chapters will show, the high rates of industrialization and urbanization are far from uniform in Europe. From the western and northwestern subregions toward eastern and central Europe, there are increasingly lower levels of industrial activity, higher ratios of nonurban population, and lower per capita GNP levels.

FIGURE 3.11
Urban populations far overshadow rural populations in much of Europe and the urban densities of central cities and their ring suburbs are also very high. In this scene of urban sprawl of Munich, Germany, the intermixing of three- and four-story apartments, governmental buildings, churches, and retail activity gives character to European urban space. © 1996 Howard Garrett/Dembinsky Photo Associates

Continuity and Change: Moving Toward a Postindustrial Society

Scholars have traced a slowly rising trend of European technical inventiveness since the Middle Ages, involving such things as clocks, gears, waterwheels, windmills, the invention of printing in 15th-century Germany, improvements in shipbuilding, drainage of mines by pumping to permit mining at deeper levels, improvements in artillery and other armaments, and many others. The practical improvements leading to greater production were furthered by mathematical and scientific discoveries and by the development of a scientific viewpoint toward nature and its possibilities. At the same time—still previous to the 18th century—capitalist values, practices, and institutions were evolving and becoming prominent—for example, private ownership of property; favorable attitudes toward profit making, including interest from loans; commercial banking, insurance, and credit institutions; double-entry bookkeeping; and corporate forms of business organization. Increased law and order imposed by newly centralized states—especially in England, the Netherlands, France, and Spain—protected rising commercial and industrial interests which profited further from overseas trade.

The developments spawned by the Industrial Revolution continue to the present time. Today, industrial technology has advanced from the crude steam engines and textile machines of 18th-century Britain to modern electronics, computer-controlled factory robots, and the human exploration of space. As it has advanced, it has diffused geographically. Early coal-based and steam-powered industry first spread from Britain to Belgium, and then more widely in Europe and the eastern United States. It began to expand in influence to some non-European areas like Japan in the late 19th century, and today's more modern technology is continuing its expansion with a rising tempo.

Much evidence exists that the older and more advanced industrial societies are in the process of becoming "postindustrial"; that is, they will eventually be societies in which industrial workers are as uncommon as farmers now are in most "industrial" societies. The introduction and improvement of **computer-controlled robots** (machines programmed to do specific jobs without further direction) on industrial production lines is moving this possibility along. Recent occurrence of serious **technological unemployment** (the supplanting of workers by technology) in western Europe is, in part, related to this trend. Hence, the more advanced and prosperous European countries are beset by such concerns as (1) where new jobs will come from (the "service" sectors of the economy?), (2) how to make the transition from manufacturing to service roles without too much suffering, and (3) how to allocate income between a few highly productive industrial workers who tend robots and a predominant mass of service employees whose output per worker will often be less than a robot's, and whose basic employment is increasingly at risk.

While this process is moving across northwestern Europe at the present, it is a shift in employment and manufacturing that characterizes the entire developed world. It is significant not only as an economic transformation, but also as a societal change.

Energy resources are central to any program of industrialization. When Europe began to industrialize, the energy resource situation was quite favorable. There were fewer people, the scale of production was much smaller, and raw materials were generally sufficient for the simpler requirements of the time. A relatively primitive technology was served by extensive forests, numerous small-scale sources of water power, and a large variety of minerals in small deposits found in widespread locations. When coal became the main source of power and heat, Europe—especially England—was favored by its many coal fields. These fields proved to be the main localizing factor for much of Europe's industrial and urban development. During the early period of industrialization, electricity was still in the future and coal was expensive to transport; consequently, the new factories were generally built on or near the coal fields. The working population, in turn, clustered near the factories and coal mines (Fig. 3.12) and developed a group of industrial districts that incorporated mines, factories, and urban areas.

These older districts are still very prominent on the map of Europe today. Notable examples include most major cities in Great Britain, although not London; the east-west line of cities strung across Belgium south of Brussels; the huge urban agglomeration in western Germany called the Ruhr; and the Pol-

FIGURE 3.12
The mining of coal has been an important European occupation for centuries. Seen here are miners coming off shift at a British colliery. *Jake Sutton/Gamma Liaison*

ish industrial cities near the Czech border. Economic depression and urban blight are increasingly common in these places, which had the advantage of an early start but now suffer from changed industrial circumstances, including a drastic shift in the type and availability of natural resources; increasing competition from non-European industries; competition from newer products made elsewhere (for example, new synthetic fibers competing with the cottons and woolens of old textile centers); and decreasing requirements for labor due to the automation of factories.

Industrial Resources and Landscapes Today

A few countries—notably Sweden, Finland, Norway, and Austria—currently have surpluses of wood. Hydroelectricity has been developed in the Alps, the Scandinavian Peninsula, and along the Rhone and Danube rivers (see Table 3.1). Most deposits of metallic ores are too depleted or too small to be important today. But the most far-reaching change is the altered situation of coal. Worked long and intensively, most European coal fields are now expensive to mine and yield less coal. Many were phased out as Europe shifted to cheaper imported oil for power after World War II. North Sea oil production started only in the 1970s; prior to that much of the oil used in Europe after World War II was imported from the Middle East or Russia. In a broader sense, Europe, which was once relatively self-sufficient in energy resources, must now buy and import energy and raw materials from a global base, and most European countries now enter world trade primarily to acquire energy resources.

In addition, new environmental demands have further increased the cost of coal as a fuel source because of its implication in acid precipitation as well as in the generation of smog. This shift away from relative energy self-sufficiency has had implications for economic well-being, factory employment levels, and the vitality of the economic landscape in Europe. The transition is an ongoing example of the globalization of manufacturing, trade, and political interdependency. What Europe has achieved in developing a much wider network of raw material import and export of high value-added manufactured products is a pattern that is occurring in all world regions. Because of its colonial legacy, Europe has more experience in such global trade patterns than most of the world's manufacturing centers.

A number of European countries do have mineral resources of considerable importance. The United Kingdom became a large producer and net exporter of oil and also a large producer of natural gas when its offshore fields in the North Sea were brought into production during the 1970s. Norway's oil and gas fields in the North Sea likewise yield a large production. For several decades, the Netherlands has been a major producer and exporter of natural gas. Large iron ore fields are worked in eastern France and northern Sweden. Coal from southern Poland is a mainstay of that country's economy, and Germany is still a sizable coal producer.

3.6 The Rise and Uncertain Future of European Global Centrality

What factors account for the rise of this Eurasian peninsula as the center of world trade and industrialization? Many interpretations are possible, with various scholars emphasizing such factors as the following:

1. *Capitalism.* Parts of Europe had already developed capitalist institutions by the end of the Middle Ages, and the colonial age saw a further rapid development of capitalism in the region. Profit became an acceptable motive, and the relative freedom of action afforded by the capitalistic system provided opportunities for gaining wealth by taking risks. Energetic entrepreneurs found it possible to mobilize capital into large companies and to exploit both European and overseas labor.

2. *Technology.* By the end of the Middle Ages, Europeans had reached a level of technology generally superior to that of non-Europeans with whom they came in contact during the early Age of Discovery. In particular, achievements in shipbuilding, navigation, and the manufacture and handling of weapons gave them decided advantages. Nations that had earlier made significant progress in the development of technology—such as China—were disinclined to compete at a global level during the Age of Discovery. This gave Europe a virtual free rein in energetic expansion and control of technological innovation for several centuries.

3. *Science.* Europe's scientific prowess must not be overlooked as an explanation for the region's rise to world dominance. Technology is partly applied science, and the foundations of modern science were constructed almost entirely in Europe during the centuries when European influence was becoming paramount. Until recently, the great names of science were overwhelmingly the names of Europeans. As late as World War II, the development of atomic fission in the United States was carried out by a team composed largely of prominent refugee European scientists, and Nobel Laureates continue to reflect the scientific capacities of the peoples of, and from, this region.

In the 20th century, however, the preeminence in world trade and industry of leading European countries diminished. What caused the relative decline in its fortunes? Some important reasons and circumstances include:

1. *War dislocation.* Europe suffered enormous casualties and damage in World Wars I and II, which were initiated in Europe and fought mainly there. Recovery was eventually

achieved with U.S. aid, but the region's altered position could not be reversed. The wars destroyed Europe's ability to maintain its predominance in the face of such trends as anticolonialism and thereby accelerated the capacity of the U.S. and the Former Soviet Region to rise to world power.

2. *New nationalism.* Rising **nationalism**—the quest of a nationality to have its own homeland—in the colonial world during the 20th century resulted in the virtual end of the European colonial empires within a surprisingly short time after World War II. Taking advantage of a weakened Europe and a mounting disapproval of colonialism in the world at large, one European colony after another gained independence quickly in the decades following the war. Opposition to continued European control was often spearheaded by colonial leaders who had been educated in Europe and had absorbed nationalistic ideas there; nationalism is a sentiment that has had its most pervasive expression in Europe. This same drive received an additional boost with the collapse of the Soviet Union in 1991.

3. *Ascendence of the United States and the Former Soviet Region.* Europe's predominance was seriously eroded by the rising stature of the United States and the Former Soviet Region. These enormous countries, each far larger than any European country, outstripped Europe in military power, economic resources, and world influence particularly in the rebuilding years of the 1950s–1970s.

4. *Shift in world manufacturing patterns.* Europe once enjoyed nearly a monopoly in exports of manufactured goods, but in the past three decades, manufacturing has developed in many countries outside of Europe. Japan, South Korea, Hong Kong, and Taiwan are prime examples just in East Asia alone. In the decades following World War II—although Japan began its transformation in the late 19th century—these Asian countries emerged from more traditional

Regional Perspective

EUROPE'S PROBLEMS OF ENVIRONMENTAL POLLUTION

A much-publicized problem in some European areas is **environmental pollution**. For example, pollutants pour into the Mediterranean Sea from urban-industrial districts in Europe, Africa, and Southwest Asia. Wastes discharged into this most historical of seas damage both the coastal tourist industry and fisheries. The Mediterranean, almost totally enclosed by land, cannot cleanse itself by flushing wastes into the ocean. By contrast, the North Sea is far more able to interchange water with the ocean, thanks to strong currents that move in and out through the broad opening to the Atlantic between Scotland and Norway. But even the North Sea has become alarmingly contaminated with pollutants discharged by great industrial metropolises, the undersea oil industry, and a heavy volume of shipping.

Still another famous water feature, the Rhine River, is under stress from untreated wastes. Stringent legislation requiring municipalities and industries to maintain waste-treatment plants has lessened the pollution, but the Rhine, from Basel, Switzerland, to the North Sea, is still so dangerous to human health that swimming is generally forbidden. Aquatic life in the Rhine continues to be damaged, not only by toxic wastes but also by excessive warmth when water withdrawn for industrial use is returned to the river in a heated state. Several nuclear stations, for example, use large volumes of Rhine water to cool radioactive cores and produce steam to generate electricity. The return of this water heats the river. Similarly, steel mills and coke works withdraw water to cool red-hot metal or to quench flaming coal in the coking process, and this water is still warm when it comes back to the river.

Atmospheric pollution is another widespread environmental affliction in Europe, as it is in North America and various other parts of the world where there are overgrown metropolises and industrial concentrations. One form of this pollution is known as **acid rain** (sometimes called acid precipitation). Especially publicized have been the ravages of pollutants that have killed great numbers of trees in some of Europe's finest forests. The exact mechanisms that cause the trees to die are not precisely known, but there is much suspicion that the deaths occur primarily because of industrial emissions that combine with atmospheric moisture to create acid rain. Emissions containing sulfur, when combined with precipitation, produce weak sulfuric acid, and those containing nitrogen yield weak nitric acid. Both can damage not only trees but also soils in which trees grow and aquatic life in lakes that receive the runoff. There has been much alarm about lakes that are "dying." Germany and Scandinavia are among the European areas particularly affected. Many consider Germany's great heavy-industrial district called the Ruhr to be a major generator of the pollutants that cause acid rain in European areas to the east.

Another form of atmospheric pollution caused worldwide alarm in 1986 when significant **radioactive fallout** from a nuclear disaster at the **Chernobyl** power plant in the Ukraine (western part of the Former Soviet Region) was deposited on parts of Europe as well as on a large area of the Former Soviet Region itself (Fig. 3.B). Sections of Scandinavia and Finland were particularly affected; in fact, some northerly areas used by Lapp reindeer herders became so radioactive that the wild forage could no

rural societies to become industrial forces capable of competing with Europe in markets all over the world. The United States began its industrialization earlier than Japan, but it, too, reached industrial maturity in the 20th century and also became a vigorous competitor of Europe.

5. *Energy factors.* Europe's ability to assert itself in world affairs has been weakened by the region's new dependence on outside sources of energy. The region's coal fields have become increasingly costly to exploit, and, despite the recent development of North Sea oil and gas resources (which benefit primarily Great Britain and Norway), Europe is now very dependent on Middle East oil and other imported sources of energy.

The speed of technology transfer and the complex global interdependence of capital flow makes it difficult to pinpoint current, much less future, centers of manufacturing. However, current patterns of industrial activity in Europe show that the coalescence of resources, skilled labor, cultures that promote entrepreneurship, and spatial networks of transport and communication are all still in place in this region. It is certain that Europe will continue to play a major role in manufacturing and global trade for a long time to come.

3.7   Agriculture and Fisheries

Agriculture was the original foundation of Europe's economy, and it is still a very important component. Food provided by the region's agriculture allowed Europe to become a relatively well-populated area at an early time. After about 1500, a period of steady agricultural improvement began. Introduction of important new crops such as the potato played a part, but so did such practical improvements as new crop rotations and scien-

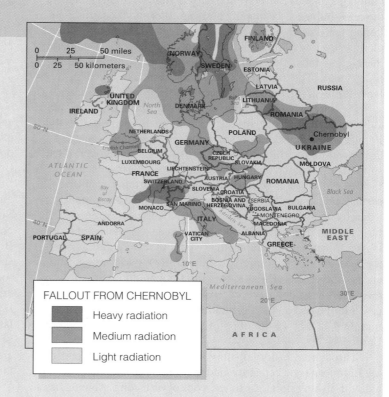

FIGURE 3.B

Extent of radiation dispersal seven days after the April 26, 1986 explosion at Chernobyl. The release of radioactive material lasted 10 days in all, and its distribution varied with changing wind patterns. Initially, southeasterly winds blew the radiation over Poland, Ukraine, Belarusa, Latvia, Lithuania, Finland, and across Norway and Sweden. The winds then changed, bringing fallout to most of Europe. The extent of the fallout was determined by whether or not rain washed the contamination from the atmosphere. Restrictions on livestock movement and slaughter were introduced in the United Kingdom and the Nordic Countries.

longer be used and the animals that had grazed on it had to be destroyed.

Many other examples of environmental damage can be drawn from Europe. For instance, old mining and industrial regions contain numerous waste heaps, sometimes centuries old, that are full of poisonous substances. Seepage of these poisons into groundwater may cause cancers and a variety of health problems. Today there is a very active environmental consciousness in Europe, as in many other parts of the world, but many problems deriving from environmental neglect are so deep-seated that solutions will be slow, very costly, and difficult to achieve. This is particularly true of former Communist bloc areas in eastern Europe.

tific advances that produced a better knowledge of the chemistry of fertilizers. The rise of industrial cities provided growing markets for European farmers, and the latter received some protection by tariffs or direct subsidies to encourage production and support incomes. Both of these governmental favors are now enjoyed by farmers in the European Union (EU). The EU has its origin in the 1967 organization of the European Economic Community, the European Coal and Steel Community, and the European Atomic Energy Committee. Today, the result of Europe's long agricultural evolution is a farming sector that stands very high in overall volume of production and in yields per unit of cultivated land. Large surpluses of many products have emerged as a result of the EU's **Common Agricultural Policy (CAP)**. The surpluses include grains, butter, cheese, olive oil, and table wines. But, as a region, Europe is by no means agriculturally self-sufficient. Imports of supplemental agricultural commodities are necessary. Increasingly, policies dealing with agricultural goods are under consideration by the **European Union** (EU) which now includes all of the earlier members of the EEC and EC. The European Union is the more common collective term for the current fifteen nations that have the most influence in shaping both agrarian and industrial marketing patterns in Europe today. Current EU nations include Germany, France, Italy, the U.K. (the four largest); Denmark, the Netherlands, Belgium, Luxembourg and Austria—sometimes called The Junior Countries; and Finland, Sweden, Ireland, Portugal, Spain, and Greece.

Pressure on European governments to eliminate farm subsidy programs meets strong resistance. Farmers—and the French are the ones most often in the limelight in this matter—are increasingly militant in their refusal to give up government supports that enable them to maintain family farms, even in the face of agricultural overpopulation. The farming tradition is seen as much more than simply working the land and producing a family income. The maintenance of these relatively small farms (often of fewer than 60 acres) is seen as a means of holding on to national traditions that would otherwise fall away in the face of increasing pressures from urban and foreign influences. There is a tension in this rural population that is felt in most European countries because of the powerful pull of the city and its jobs and opportunities, as well as the increasingly marginal nature of family farming, particularly in western Europe.

Throughout history, fishing has been an important part of the European food economy, and the coasts are still thickly dotted with fishing ports. At times, control of fishing grounds has been a major commercial and political objective of nations, even resulting in warfare. Fisheries are particularly important in shallow seas that are rich in the small organisms known as plankton. These organisms are the principal food for schools of herring, cod, and other fish of commercial value. The Dogger Bank in the North Sea and the waters off Norway and Iceland are major fishing areas, and two nearby countries—Norway

and Denmark—are Europe's leading nations in total catch. Iceland, which has few other resources and few people, ranks fourth in total catch and depends mainly on exports of fish and fish products, which represented 75 percent of all Iceland's exports by value in 1995.

3.8 Population as an Element in Europe's Continuing Global Importance

A major reason for Europe's role in the global economy lies in the region's large and technically skilled population. Europeans represent one of the great masses of world population (see Table 3.1). Europe's estimated total of 509 million people in 1996 was nearly twice that of the United States. One out of every 11 people in the world live in a space half the size of the United States. Birth rates are low in this region of relative affluence and high urbanization, and the whole region is increasing in numbers much more slowly than most of the world.

MAJOR POPULATION AXES AND THEIR SIGNIFICANCE

Except for some northern, rugged, or infertile areas, European population density is everywhere greater than the world average. However, the greatest densities appear along two axes of industrialization and urbanization near coal or hydroelectric power sources. One axis extends north-south from Great Britain to Italy (see Fig. 3.2). In addition to large parts of Great Britain, it includes extreme northeastern France, most of Belgium and the Netherlands, Germany's Rhineland, northern Switzerland, and—across the Alps—northern Italy. The second axis of dense population extends west-east from Great Britain to southern Poland and continues into Ukraine. It corresponds to the first axis as far east as the Rhineland, where it forms a relatively narrow strip eastward across Germany, western Czechoslovakia, and Poland. Although the two belts represent a minor part of Europe's total extent, they contain more large cities and a greater value of industrial output than the rest of Europe combined. Only in eastern North America and Japan are there urban-industrial bands of the same order of importance and complexity.

The European belts coincide with major routeways that were in use very early on for migration, trade, and military movement. Many cities along these routes are very old. With the rise of modern industry, coal fields along the west-east axis became important, and in the southern part of the north-south axis, the age of electricity saw the development of many industries based on the hydroelectric resources of the Alps. Both belts benefit from the relatively good soil they encompass.

3.9   The Drive for New Patterns of Regional Cooperation

Since World War II, the countries of Western Europe have moved toward greater economic, military, and political cooperation. Searching for development and security, the countries have banded together in a series of organizations designed to foster unity. In eastern Europe there were efforts to create economic and security pacts under Soviet leadership, but the collapse of Communism in that region raises the question as to whether the former Communist countries will be able to join western Europe's cooperative arrangements.

What is responsible for the new age of cooperation in a region too well known for national quarrels and strife? Important among the many motivations are: a long-standing ideal of European unity dating from the Roman Empire and the medieval Catholic Church; the experience of two World Wars (1914–1918; 1939–1945); perceived military and political threats after World War II; and a search for economic better-

ment through the concerted action of nations. After World War II, economic recovery was slow, and there were fears in western Europe of military aggression by the Former Soviet Region and/or a political takeover by powerful Communist parties in Italy and France. In the meantime, the Former Soviet Region and the Communist governments of the satellite countries east of the Iron Curtain (the term Winston Churchill devised for the western boundary of the Soviet bloc in a speech in 1956) saw U.S. military power in Europe as a major threat to be countered by economic and military cooperation under Soviet leadership. Thus the desire for economic recovery and military security in both the Western and Eastern nations set the stage for a remarkable period of supranational organization.

The alliances developed in this era of regional cooperation have played a major role in both the economic and political development of the region since the end of World War II. The collapse of the Soviet Union and the Communist satellite nations in the 1990s has changed the political fabric of the region, and opened new possibilities—and perhaps new demands—for broader regional cooperation.

Regional Perspective

NEW PATTERNS OF EUROPEAN REGIONAL COOPERATION

Following World War II, many influential people in western Europe took the view that new wars could best be avoided and more peaceful, secure, and prosperous societies achieved through a close alignment with the United States—a strategy favored by the British in particular. Coupled with this was the development of cooperation among the non-Communist European states—the alliance between France and the former West Germany being the most definitive move in that realm.

In 1948, a major step was taken when the **Organization for European Economic Cooperation (OEEC)** was formed among many European states to coordinate the use of U.S. aid proffered under the **Marshall Plan** to bolster Europe against communism. The Marshall Plan was terminated in 1952, but the OEEC continued to function. In 1960 the name changed to **Or-**

ganization for Economic Cooperation and Development (OECD). The membership now includes the United States and several other countries outside Europe.

But the most significant economic organization among Europe's western nations came to be the European Economic Community (EEC), at first called the **Common Market**, and now officially called the European Union or EU (since November 1, 1993). It was established by six countries—France, the former West Germany, Italy, Belgium, Luxembourg, and the Netherlands—under the Treaty of Rome in 1957, which in 1967, reorganized into European Community. These countries had already eliminated barriers to trade in coal and steel and now looked toward removal of all economic barriers. Later, Britain, Denmark, Ireland, Greece, Portugal, and Spain became members. By the mid-1990s other nations—Poland, Hungary, the Czech

Republic, Slovakia, Bulgaria, and Romania all indicated their desire to become part of the EU. Cyprus, Malta, and Turkey are associate members. As of 1996, the fifteen members of the European Union included the United Kingdom, Ireland, Portugal, Spain, France, Luxembourg, Germany, Italy, Greece, Austria, Belgium, Netherlands, Denmark, Sweden, and Finland.

The EEC was initially designed to secure the benefits of large-scale production by pooling the resources—natural, human, and financial—and markets of its members. Hence, tariffs were eliminated on goods moving from one member state to another, and restrictions on the movement of labor and capital between member states were greatly reduced. Monopolistic trusts and cartels that formerly restricted competition were discouraged. Meanwhile, a common

(Box continued on following page)

Regional Perspective (CONTINUED)

NEW PATTERNS OF EUROPEAN REGIONAL COOPERATION

set of external tariffs was established for the EEC's entire area to regulate imports from the outside world, and a common system of price supports for agriculture replaced the individual systems of member states. The founders of the EEC anticipated that free trade within such a populous and highly developed bloc of countries would (1) stimulate investment in mass-production enterprises, which could, wherever they were located, sell freely into all EEC countries, and (2) encourage a productive geographical specialization, with each part of the Community expanding lines of production for which it was best suited. Thus, each country might achieve greater production, larger exports, lower costs to consumers, higher wages, and a higher level of living than it could achieve on its own.

It is possible that this success would have been attained without the formal organization of an international economic community. For example, both Switzerland, which is a close neighbor of the European Union but not a member, and Norway, which voted in 1995 not to join the EU, have among the highest GNP and per capita incomes in the world. Sweden, which joined the EU in 1995, also ranks near the top of the world scale in affluence. However, Switzerland, Norway, and Sweden have all depended heavily on trade and other dealings with the Common Market, and thus their prosperity has been closely tied to that of the Market (Fig. 3.C). Six nations that were not initially enrolled in the EU—Iceland, Norway, Sweden, Switzerland, Finland, and Austria—belonged for many years to a separate trading organization called the European Free Trade Association (EFTA). It has worked closely with the Common Market and both are increasingly integrated within a single market area, particularly for nonagricultural products. The Common Market made preferential trade and economic aid treaties with many other countries—mainly ex-colonies of Common Market members—in areas outside of Europe. By late 1996, however,

three of the EFTA nations (Sweden, Finland, and Austria) had joined the EU.

The EU has an elaborate political and administrative machinery. Its policies are set by meetings of representatives of the member governments (ministers or heads of government sometimes called the European Council). The 1992 Maastricht Treaty on European Union set several review stages in motion. The late December, 1996, review of its effectiveness was Stage II of a process that is intended to be completed at the end of 1998 (Stage III). Research, monitoring, advice, and execution of policies are provided by a European Commission appointed by the members and supported by a large and very highly paid executive bureaucracy located at Brussels, Belgium. Thus, Brussels has a good claim to becoming the capital of the European Union. The Union also has a Court of Justice, located in Luxembourg City, to decide disputes over EU regulations, and a European Parliament, which meets at Strasbourg, France, on the border with Germany. The latter body has some authority over the budget of the organization, which is financed by contributions from member states and by a percentage of the revenue from import tariffs on goods entering the EU countries. A direct vote of the people in member countries elects the parliament members.

The European Union, under provisions of the Maastricht Treaty of European Union that went into force on November 1, 1993, is now trying to achieve major new steps in unification. These steps include the removal of nontariff trade barriers, such as "quality standards," that still exist between member countries and implementation of a single EU currency by 1999. Progress is slow because of questions concerning national sovereignty and enormous protest by populations all through the EU who do not want to lose the autonomy of having their own unique currency.

In the sphere of military defense, the key international body in non-Communist Europe became the **North Atlantic Treaty Organization (NATO)**. This military al-

liance was formed in 1949 and now includes the United States, Canada, a majority of the European nations west of the former Iron Curtain, and Turkey. The member countries pledged to settle disputes among themselves peacefully, to keep their individual and joint defense capacities in good order, to consider an attack upon any of them in Europe or North America an attack upon all, and to come to the aid of the country or countries being attacked. Questions arise today as to whether NATO is necessary any longer, and about what its role should be in view of the dissolution of the Soviet Union and the retreat of Soviet power from eastern Europe. NATO has responded to these issues by establishing a new Partnership for Peace program of military and political cooperation (including joint military exercises and peacekeeping operations) with Russia in response to such questions. Eighteen ex-Soviet Iron Curtain allies joined the program on June 22, 1994.

The response of the former Communist East to the formation of the Common Market and NATO was the creation in 1955 of an international economic organization called the **Council for Mutual Economic Assistance (COMECON)** and a military

FIGURE 3.C ▶

Some of the key organizations in Europe's intricate framework of political, strategic, and economic relationships. The sequence of the creation and elaboration of the various steps leading to the contemporary European Union (EU) is chronicled in this Regional Perspectives box. The critical fact to remember is that this region has a long history of conflicts that have been political, religious, and economic. The EU marks a very significant innovation in European regional cooperation.

alliance called the **Warsaw Pact**. The Former Soviet Region dominated the Communist "satellite" states in these organizations. The new eastern Europe still needs freely entered agencies of economic cooperation and collective security, but, since COMECON and the Warsaw Pact were not freely entered, they thus have been terminated. Currently, NATO and the EU are working

with a variety of potentially useful and productive collaboratives with the eastern European nations. For example, in the former Yugoslavia—once an east European nation—warring factions have thrown NATO into a new role as a peacekeeping force. The 1995 Dayton Accords tried to bring stability to a European region already known for centuries of discord.

These examples of unprecedented collaboration and economic and political cooperation among European nations reflect the changing geopolitical nature of the world today. Patterns of traditional geographic isolation, independence, and relative autonomy are being modified by these new linkages.

International Organizations of Europe, January 1997

Legend:
- European Union (EU) members
- Countries being considered for next EU enlargement
- North Atlantic Treaty Organization (NATO) members
- **(E)** European Free Trade Association (EFTA) members
- **(O)** Organization for Economic Cooperation and Development (OECD) members

*Note: Turkey was granted a customs union treaty with the E.U. in 1996.

Sources: EUROSTAT (Statistical Office of the European Communities) (1996). CIA, *The World Fact Book* (1995), *Britannica Book of the Year* 1997.

4

The Intricate Geography of Europe's Northwestern Coreland

This chapter looks at the process and implications of the Industrial Revolution, the forces of industrial transformation as shown in the literature selection—and its regional landscapes in the coreland of Europe. This monumental process of economic change led to major demographic shifts, social change, and ultimately European concerns for environmental deterioration through the forces of industrialization. To illustrate their concerns, environmentalists of this century often cite images of England and the manufacturing scenes of the Industrial Revolution as it moved across Great Britain and onto the European continent. Carry these images with you as we learn in the pages that follow of the incremental abandonment of 19th and early 20th century industrial landscapes in the coreland of northwest Europe.

4.1 The Industrial Development of West Central Europe

Three of Europe's major countries and a series of smaller nations and microstates occupy the British Isles and the west central portions of the European mainland. The United Kingdom and the Republic of Ireland share what is called the British Isles. Germany, France, their smaller neighbors of the Netherlands, Belgium, Luxembourg, Switzerland, and Austria, and the

microstates of Liechtenstein, Monaco, and Andorra may be conveniently grouped as the countries of West Central Europe. Considered as a whole, these units are highly developed, and they form Europe's economic and political core. Development is far more intensive in some areas than in others, but the overwhelming majority of people in the coreland live in areas that are among the world's most intensively developed cultural landscapes (see Fig. 4.1 and Table 3.1).

Even given the complexity of Europe, the geography of the northwestern coreland is unusually intricate. Spatial variations are great, layers of history appear at every turn, and functional linkages are extraordinarily complicated. The individual countries differ sharply in their human geographies and national perspectives, and the whole area is differentiated in a mosaic of regions and localities exhibiting great physical and cultural variety.

In the 19th and early 20th centuries, the largest coal and iron ore deposits in Europe provided bases for large-scale industrial development (Figs. 4.1 and 4.2). Consequently, a series of major industrial concentrations now stretches from the United Kingdom to Poland along the main axis of mineral deposits. Many individual industrial cities lie outside the main coal and steel axis, and some of these—such as Paris and London—are sufficiently important to be considered major industrial concentrations in their own right. In the typical industrial agglomeration, coal mining and iron and steel production have been economic foundations, with chemicals, textiles, and heavy engineering also common.

◀ The resurgence of Germany from the ruins of World War II is dramatically illustrated by the recent photo of Frankfurt-on-the-Main compared with the photo of the shattered city in 1949. *Left: H.P. Merten/The Stock Market. Right: AP/Wide World.*

FIGURE 4.1

Industrial concentrations and cities, seaports, internal waterways, and highlands in west central Europe. Older industries such as coal mining, heavy metallurgy, heavy chemicals, and textiles cluster around highland margins in the congested districts shown as "major industrial concentrations." Local coal deposits provided fuel for the Industrial Revolution in most of these districts, which have shifted increasingly to newer forms of industry as older industries have declined. But the new and more diversified industries have also proliferated in metropolitan London, Paris, and smaller industrial cities away from the old "major industrial concentrations." Relative sizes of dots and circles indicate rough groupings of industrial cities and seaports according to importance. Rivers and canals shown on the map are navigable by barges and, in some instances, by ships. The Channel Tunnel (Eurotunnel) will change forever the sense of proximity or isolation perceived by the political units on the map.

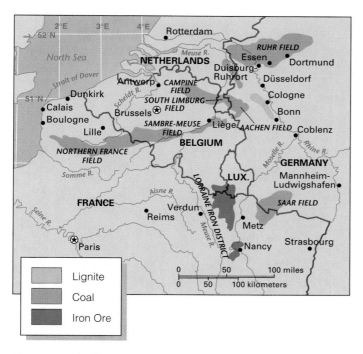

FIGURE 4.2

Major coal fields of west central Europe. Note the line of fields stretching across western Germany, the Low Countries, and northeastern France. The Northern France and South Limburg coal fields have been closed. Stars show national capital.   *After a map by the U.S. Geological Survey*

4.2 Changing Fortunes in the British Isles

Located off the northwest coast of Europe, two countries occupy the British Isles: the Republic of Ireland, with its capital at Dublin, and the larger United Kingdom of Great Britain and Northern Ireland, with its capital at London. The United Kingdom incorporates the island of Great Britain, consisting of England and Scotland, plus the northeastern corner of Ireland and most of the smaller out-islands (see Figs. 4.3 and 4.4). Altogether, it encompasses about four fifths of the area and well over nine tenths of the population of the British Isles (see Table 3.1).

POLITICAL SUBDIVISIONS AND WORLD IMPORTANCE OF THE UNITED KINGDOM

England, which lies only 21 miles (34 km) from France at the closest point (see Fig. 4.1), is the largest of the four main subdivisions of the United Kingdom (the others being Scotland, Wales, and Northern Ireland [Fig. 4.3]). These all were origi-

nally independent political units. England conquered Wales in the Middle Ages but preserved some cultural distinctiveness associated with the Welsh language.

Northern Ireland, together with the rest of Ireland, was twice conquered by England. One of the outcomes of the second conquest in the 17th century was the settlement of Scottish and English Protestants in the North, where they became numerically and politically dominant. In 1921, when the Irish Free State (later the Republic of Ireland) was established in the Catholic part of the island, the predominantly Protestant North, for economic as well as religious reasons, elected to remain with the United Kingdom. The North received a large degree of autonomy, including its own parliament in Belfast to deal with local concerns. However, direct British rule was later imposed as a result of the grim and prolonged violence that broke out between the region's Protestants and Catholics toward the end of the 1960s. Despite many attempts at a settlement, this situation continues.

Scotland was first joined to England when a Scottish king inherited the English throne in 1603, and the two became one country under the Act of Union passed in 1707. Scotland has no separate parliament, but it does have its own currency as well as special administrative agencies in Edinburgh that deal with Scottish affairs. Wales has even less autonomy than does Scotland.

The international influence once attained by the United Kingdom and the imprint it left on the world are a legacy of the 19th century. When British power was at its peak in the century, between the defeat of Napoleonic France and the outbreak of World War I (1815 to 1914), the United Kingdom was generally considered the world's greatest power. Its overseas empire eventually covered a quarter of the Earth, and the influence of English law, education, and culture spread still farther. Until the late 19th century, the U.K. was the world's foremost manufacturing and trading nation. The Royal Navy dominated the seas, and the British merchant marine carried half or more of the world's ocean trade. London became the center of a free trade and financial system that invested its profits from industry and commerce around the world.

From the late 19th century onward, the rate of economic growth in the United Kingdom tended to be slower than that of many other industrialized or industrializing countries. Thus, a relative decline in the country's economic importance set in. The free trade and financial system based in London was damaged by World War I and decline accelerated after World War II, although London remains a major world financial center. Relative economic decline has been accompanied by a drop in political stature. The large colonial countries once included in the British Empire have all gained independence, and Britain's remaining overseas possessions are scattered and small. The United Kingdom, however, is still a country of consequence in many fields. It plays a major role in the European Union and is associated with many of its former colonies in the worldwide Commonwealth of Nations.

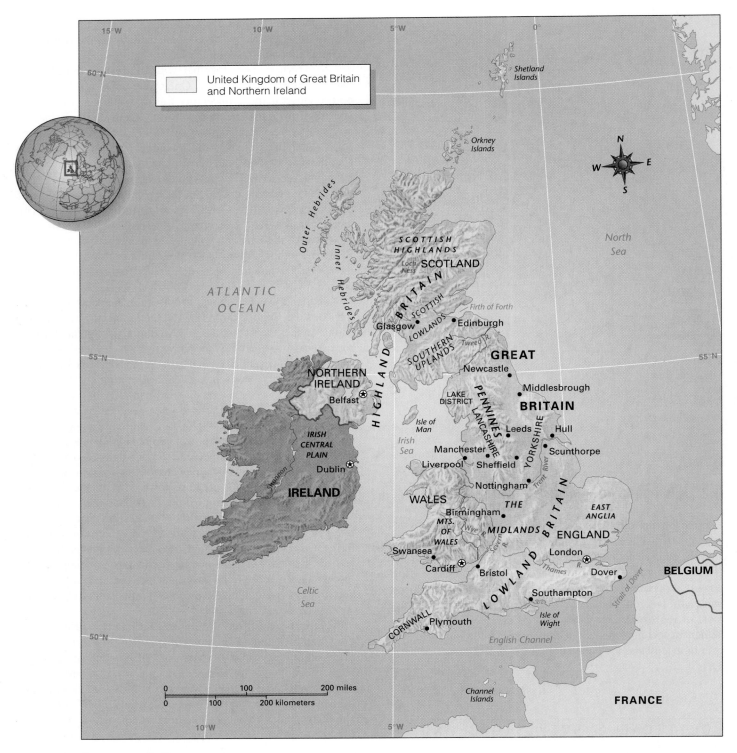

FIGURE 4.3
Highland and Lowland Britain.

THE CHANNEL TUNNEL

The 31-mile three-tube tunnel that links southern Britain with Calais, France, is also called the **Chunnel** or **Eurotunnel**. It opened for commercial travel in 1994 (a year and a half later than its scheduled opening), but discussion about the possibility of a tunnel link between the British Isles and the European continent began 140 years earlier during the reign of Britain's Queen Victoria. The estimated costs for this structure, which houses two commercial rail lines and a service shaft, was set at $7.7 billion in the initial 1987 plans. Final costs came to more than $15 billion in 1994, making it the most expensive single building project in human history. It is now possible to board a train in London and arrive in central Paris in three hours, traveling at speeds that can reach 200 miles per hour. The Chunnel opens a whole new era of speculation on the impact of reducing the insularity of Britain—a geographic characteristic that the British see as central to their traditional ability to chart a history somewhat independent of western Europe.

REGIONAL VARIETY ON THE ISLAND OF GREAT BRITAIN

The physical base of the United Kingdom is Great Britain, an island about 600 miles (somewhat under 1000 km) long by 50 to 300 miles (80 to nearly 500 km) wide. For its small size, this island provides a homeland with a notable variety of landscapes and modes of life. Some of this variety can be attributed to a broad fundamental division of the island into two contrasting areas: Highland Britain and Lowland Britain (Fig. 4.3).

FIGURE 4.4

Fields in the English Lowland normally are fenced by hedgerows, seen here in southwestern England near the mouth of the River Severn. Hedgerow trees blend in the distance to give a misleading impression of forest. This landscape was created centuries ago by the Enclosure Movement, in which large landowners appropriated unfenced land tilled by villagers. Hedges were planted so that sheep could be raised under a system of controlled breeding.   *Jesse H. Wheeler, Jr.*

Highland Britain

Highland Britain embraces Scotland, Wales, Ireland, and parts of northern and southwestern England. It is predominantly an area of treeless hills, uplands, and low mountains formed of rocks that are generally older and harder than those of Lowland Britain. Broad areas are moors tenanted by sheep farmers or uninhabited. The principal highlands include the Scottish Highlands, the Southern Uplands of Scotland, the Pennine Chain mountains and the Lake District of northern England, the mountains that occupy most of Wales, the uplands of Cornwall and Devon in southwestern England, and the mountainous and hilly rim of Ireland. The mountains are low—Ben Nevis in the Scottish Highlands is the highest at 4406 feet (1343 m)—but many slopes are steep, and the mountains often look high because they rise precipitously from a base at or near sea level. Highland Britain includes two important low-lying areas: (1) the boggy, agricultural Central Plain of Ireland, and (2) the Scottish Lowlands, a densely populated industrialized valley separating the rugged Scottish Highlands from the gentler and more fertile Southern Uplands of Scotland. The Scottish Lowlands compose only about one fifth of Scotland's area, but they incorporate more than four fifths of its population and its two main cities: Glasgow (with a population of 1.8 million in the metropolitan area, including Glasgow city [689,000] plus suburban and satellite districts nearby) on the west, and Edinburgh, Scotland's political capital (with a population of 630,000 in the metropolitan area and 435,000 in the city), on the east.[1]

[1]Unless otherwise noted, city populations in this text are rounded approximations for metropolitan areas as of the mid-1990s. They are only crudely comparable from one country to another because of variations in the accuracy of statistics and the actual definition of "metropolitan area." Such figures, based on standard sources such as *Goode's World Atlas* (Rand McNally), the *New International Atlas* (Rand McNally), the *Statesman's Year-Book* (St. Martin's Press), and the *Britannica Book of the Year*, do enable students to broadly group urban areas by orders of magnitude—an activity far more meaningful than attempts to memorize populations city by city.

Lowland Britain

Practically all of Lowland Britain lies in England and it is often called the English Lowland. It has better soils than the highlands and a gentler topography (Fig. 4.4). Most of it exhibits a mosaic of well-kept pastures, meadows, and crop fields, fenced by hedgerows and punctuated by closely spaced villages, market towns (Fig. 4.5), and industrial cities. The largest industrial and urban districts, aside from London, lie in the midlands and the north, around the margins of the Pennine Chain (Fig. 4.6). Here, clusters of manufacturing cities rise in the midst of coal fields.

GEOGRAPHIC ELEMENTS AND IMPACT OF BRITAIN'S INDUSTRIAL REVOLUTION

Development in the Industrial Revolution

Five industries were of major significance in Britain's industrial rise: *coal mining, iron and steel, cotton textiles, woolen textiles,* and *shipbuilding.* The advantage of an early start in these enterprises gave the United Kingdom industrial preeminence through most of the 19th century. Exported surpluses paid for massive food and raw material imports, and made Britain the hub of world trade and very wealthy. Current difficulties in these same industries—due to growing international competition and changing patterns of demand—are fundamental to many of the problems that Britain faces in the 20th century.

FIGURE 4.5

Early Sunday morning view of the High Street of Dorchester, located in Lowland Britain near the English Channel. Dorchester, a market town in the county of Dorset, is the "Casterbridge" of Thomas Hardy's Wessex novels. The town was founded in Roman times, and the town center exhibits varied building styles dating from different historical periods. *Jesse H. Wheeler, Jr.*

The British Coal Industry

Coal became a major industrial resource between 1700 and 1800, when it fueled the blast furnace and the steam engine. It was coked for use in blast furnaces, and coke replaced the charcoal previously used for smelting iron ore. With Britain's development of an economically practical steam engine, coal supplanted human muscle and running water as the principal source of industrial energy. Coal production rose until World War I. Then, between 1913 and the 1990s, output fell by about two thirds, the number of jobs in coal mining was cut by more than four fifths, and practically all of the coal export trade disappeared. Major factors contributing to this decline included (1) increasing competition from foreign coal in world markets, (2) depletion of Britain's own coal that was easiest and cheapest to mine (although large reserves of coal still remain), and (3) increasing substitution of other energy sources for coal—both in overseas countries and in Britain itself. In the late 1960s and the 1970s, large-scale development of newly discovered oil and gas fields underneath the North Sea revolutionized Britain's energy situation. By the 1980s, once-dominant coal's principal remaining role in Britain was reduced to supplying fuel for generating domestic electricity.

Iron and Steel

Steel was a scarce and expensive metal before the Industrial Revolution. Iron, not too cheap or plentiful itself, was more commonly used. It was made in small blast furnaces by heating iron ore over a charcoal fire, with temperatures raised by an air blast from a primitive bellows. Supplies of iron ore in Great Britain were ample for the needs of the time, but the island was largely deforested by the 18th century and charcoal was in short supply. By 1740 Britain was importing nearly two thirds of its iron from countries with better charcoal supplies, notably the American colonies, Sweden, and Russia.

This heavy import dependence then changed when 18th-century British ironmasters made revolutionary changes in iron production—namely, replacement of charcoal by coke; improvement of the blast mechanism; invention and use of a refining process called puddling to make the iron more malleable; and adoption of the rolling mill in place of the hammer for final processing. By 1855, Britain was producing about one half the world's iron, much of which was exported. Further advances in production resulted when British inventors developed the Bessemer converter and the open-hearth furnace in the 1850s. Steel, a metal superior in strength and versatility to iron, could now be made on a large scale and cheaply for the first time.

The early iron industry in Great Britain used domestic ores, but these eventually became inadequate, and Britain now relies almost entirely on high-grade imported ores. Although British inventors and entrepreneurs led the world into the age of cheap, mass-produced, and extensively used iron and steel, Britain soon lost its leadership in steel production. The United States

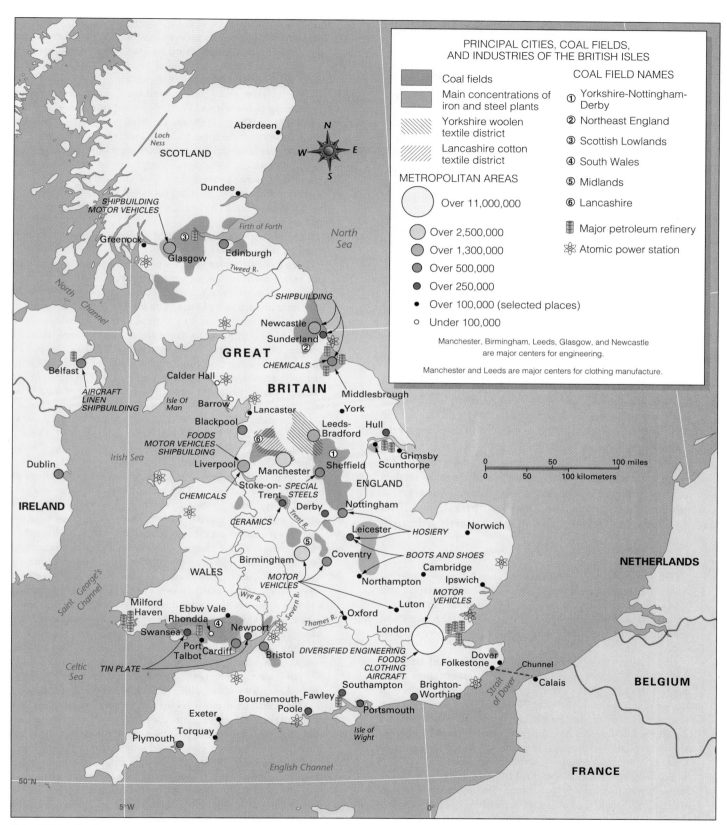

PRINCIPAL CITIES, COAL FIELDS,
AND INDUSTRIES OF THE BRITISH ISLES

Coal fields

Main concentrations of
iron and steel plants

Yorkshire woolen
textile district

Lancashire cotton
textile district

COAL FIELD NAMES

① Yorkshire-Nottingham-
   Derby

② Northeast England

③ Scottish Lowlands

④ South Wales

⑤ Midlands

⑥ Lancashire

Major petroleum refinery

Atomic power station

METROPOLITAN AREAS

Over 11,000,000

Over 2,500,000

Over 1,300,000

Over 500,000

Over 250,000

Over 100,000 (selected places)

Under 100,000

Manchester, Birmingham, Leeds, Glasgow, and Newcastle
are major centers for engineering.

Manchester and Leeds are major centers for clothing manufacture.

FIGURE 4.6
Principal cities, coal fields, and industries of the British Isles.

and Germany surpassed the British output before 1900. By 1996, the United Kingdom ranked no better than eleventh in the world.

Cotton Milling in Lancashire

During the period of Britain's industrial and commercial ascendancy, cotton textiles were its leading export by a wide margin, amounting to almost one quarter of all exports by value in 1913. The enormous cotton industry eventually became concentrated in Lancashire, in northwestern England, with Manchester (population: 2.8 million metro, 438,000 city) the principal commercial center (Fig. 4.6). Imported raw cotton was the basis for this industry, and manufactured textiles were exported through the Mersey River port of Liverpool (population: 1.5 million metro, 539,000 city), although Manchester itself became a supplementary port after completion of a ship canal to the Mersey Estuary in 1894.

A hand-labor textile industry in Lancashire dated from the Middle Ages and found itself involved in the Industrial Revolution in several lines of development. First, a series of inventors, in good part Lancashire men, mechanized and speeded up the spinning and weaving processes.

These innovations led to considerable demographic as well as economic change. In the 1845 book, *Hard Times*, author Charles Dickens painted a picture of difficult conditions for the rural men and women who came to the city in response to the new labor opportunities offered by these machines and the growing textile industry. Images like the one in the excerpt below began to unsettle the social conscience of Dickens' readers and child labor laws came under greater scrutiny. These unfair laws ultimately were changed to disallow the long and bitter days English youth had been providing to the early industrialization.

Coketown

The Fairy palaces burst into illumination, before pale morning showed the monstrous serpents of smoke trailing themselves over Coketown. A clattering of clogs upon the pavement; a rapid ringing of bells; and all the melancholy mad elephants, polished and oiled up for the day's monotony, were at their heavy exercise again.

Stephen bent over his loom, quiet, watchful, and steady. A special contrast, as every man was in the forest of looms where Stephen worked, to the crashing, smashing, tearing piece of mechanism at which he labored. Never fear, good people of an anxious turn of mind, that Art will consign Nature to oblivion. Set anywhere, side by side, the work of God and the work of man; and the former, even though it be a troop of Hands of very small account, will gain in dignity from the comparison.

So many hundred Hands in this Mill; so many hundred horse Steam Power. It is known, to the force of a single pound weight, what the engine will do; but, not all the calculators of the National Debt can tell me the capacity for good or evil, for love or hatred, for patriotism or discontent, for the decomposition of virtue into vice, or the reverse, at any single moment in the soul of one of these its quiet servants, with the composed faces and the regulated actions. There is no mystery in it; there is an unfathomable mystery in the meanest of them, for ever. Supposing we were to reserve our arithmetic for material objects and to govern these awful unknown quantities by other means!

The day grew strong, and showed itself outside, even against the flaming lights within. The lights were turned out, and the work went on. The rain fell, and the Smoke-serpents, submissive to the curse of all the tribe, trailed themselves upon the earth. In the waste-yard outside, the steam from the escape pipe, the litter of barrels and old iron, the shining heaps of coals, the ashes everywhere, were shrouded in a veil of mist and rain.

The work went on, until the noon-bell rang. More chattering upon the pavements. The looms, and wheels, and Hands all out of gear for an hour. . . .

A sunny midsummer day. There was such a thing sometimes, even in Coketown.

Seen from a distance in such weather, Coketown lay shrouded in a haze of its own, which appeared impervious to the sun's rays. You only knew the town was there, because you knew there could have been no such sulky blotch upon the prospect without a town. A blur of soot and smoke, now confusedly tending this way, now that way, now aspiring to the vault of Heaven, now murkily creeping along the earth, as the wind rose and fell, or changed its quarter; a dense formless jumble, with sheets of cross light in it, that showed nothing but masses of darkness—Coketown in the distance was suggestive of itself, though not a brick of it could be seen.

The wonder was, it was there at all. It had been ruined so often, that it was amazing how it had borne so many shocks. Surely there never was such fragile china-ware as that of which the millers of Coketown were made. Handle them ever so lightly, and they fell to pieces with such ease that you might suspect them of having been flawed before. They were ruined, when they were required to send laboring children to school; they were ruined when inspectors were appointed to look into their works; they were ruined, when such inspectors considered it doubtful whether they were quite justified in chopping people up with their machinery; they were utterly undone, when it was hinted they need not always make quite so much smoke. Besides Mr. Bounderby's gold spoon which was generally received in Coketown, another prevalent fiction was very popular there. It took the form of a threat. Whenever a Coketowner felt he was ill-used—that is to say, whenever he was not left entirely alone, and it was proposed to hold him accountable for the consequences of any of his acts—he was sure to come out with the awful menace, that he would "sooner pitch his property into the Atlantic." This had terrified the Home Secretary within an inch of his life on several occasions.

However, the Coketowners were so patriotic after all, that they never had pitched property into the Atlantic yet, but, on the contrary, had been kind enough to take mighty good care of it. So there it was, in the haze yonder; and it increased and multiplied.

The streets were hot and dusty on the summer day, and the sun was so bright that it even shone through the heavy vapor drooping over Coketown, and could not be looked at steadily. Stokers emerged from low underground doorways into factory yards, and sat on steps, and posts, palings, wiping their swarthy visages, and contemplating coals. The whole town seemed to be frying in oil. There was a stifling smell of hot oil everywhere. The steam-engines shone with it, the dresses of the Hands were soiled with it, the mills throughout their many stories oozed and trickled it. The atmosphere of those Fairy palaces was like the breath of the simoon; and the inhabitants, wasting with heat, toiled languidly in the desert. But no temperature made the melancholy elephants more mad or more sane. Their wearisome heads went up and down at the same rate, in hot weather and cold, wet weather and dry, fair weather and foul. The measured motion of their shadows on the walls, was the substitute Coketown had to show for the shadows of rustling woods; while, for the summer hum of insects, it could offer, all the year round, from the dawn of Monday to the night of Saturday, the whirr of shafts and wheels.[2]

Then, power was applied to drive enlarged and improved machines—first using falling water from Pennine streams, then steam fueled by coal from underlying Lancashire fields. In 1793, the American invention of the cotton gin, which separated seeds from raw cotton fibers economically, made cotton a relatively cheap material from which inexpensive cloth could be manufactured. Industrial growth acted like a magnet to bring farm youth to the cities in quest of factory labor. As is so powerfully chronicled by Charles Dickens in *Hard Times*, these new factory complexes that shone in the soot of the cities were difficult environments. They did, however, provide alternatives to unemployment and underemployment in the countryside, and they were seen by many youth as the only means of breaking pattern with the past and getting into the city.

Definitions and Insights

ENCLOSURE

One of the rural changes that played a role in the promotion of rural youth migration to the city was a shift in rural land use called **enclosure** (also spelled **inclosure**). This was the establishment of fences, hedges, or ditches or other barriers around the perimeters of farm fields (Fig. 4.4). The process began in the 12th century in England, but it became widespread in the mid-18th and early 19th centuries. Landowners realized that, with enclosure, they could make more money on their animal herds—especially sheep—by controlling breeding and, hence, producing better wools for the growing textile trade. Enclosure also provided abundant, cheap labor for new city industries because this act eliminated or severely restricted peasant access to lands traditionally held in **commons** for farmers who had little or no land for themselves. The commons served traditionally as unfenced lands that all farmers could run animals on, or even plant. For the farmers who had counted on access to the commons, it meant that an already marginal agricultural operation became even more untenable. It was this marginal peasant population that became so important to the "Fairy palaces" described by Dickens in *Hard Times*. These rural youth—particularly the youngest children of land-poor families—came to the factory centers and found

themselves pressed into factory work in difficult situations. Although the full set of influences on this critical demographic shift is quite complex, the act of enclosing the rural commons did lead to a major new **migration stream**—the patterns of migration between the countryside and the city—in Britain as farm youth moved to the cities. It also led to higher levels of agricultural productivity because the enclosed fields were farmed more thoroughly and intensively. Enclosure, in various forms, spread to other nations of the European coreland.

For more than a century, until World War I, Lancashire dominated world trade in cotton and cotton textiles. But 1913 saw the peak of British production; after that, the decline of the industry was nearly as spectacular as its earlier rise. The root of the trouble was increasing foreign competition. In the 20th century, country after country has surpassed the United Kingdom in cotton textile production and exports, with a resulting slide in employment and great economic distress in Lancashire. Lancashire's future as an industrial area lies mainly with other industries such as engineering (a general name for a wide range of industries that make machinery, vehicles, and tools), chemicals, and clothing.

Shipbuilding on the Tyne and Clyde

When wood was the principal material in building ships, Great Britain, as a major seafaring and shipbuilding nation, created serious difficulties for itself due to deforestation. Large timber imports were brought from the Baltic area and from North America, and the importation of completed ships from the American colonies, principally New England, was so great that an estimated third of the British merchant marine was American-built by the time of the American Revolution. During the first half of the 19th century, American wooden ships posed a serious threat to Britain's commercial dominance on the seas.

In the later 19th century, iron and steel ships propelled by steam transformed ocean transportation. Britain led the way and built four fifths of the world's seagoing tonnage by the 1890s. Two districts developed as the world's greatest centers of shipbuilding: (1) the Northeast England district, with a great concentration of shipyards along the lower Tyne River between Newcastle (population: 1.3 million) and the North Sea, and (2) the western Scottish Lowlands, with shipyards lining the River Clyde for miles downstream from Glasgow. Both districts had a seafaring tradition—Newcastle in the early coal trade and fishing, and Glasgow in trade with America—both were immediately adjacent to centers of the iron and steel industry, and both had suitable waterways for location of the yards. Their ships were built mainly for the United Kingdom's own merchant marine, as they still are, but even the minority built for foreign shipowners were an important item in Britain's exports.

[2]Dickens, Charles. *Hard Times and Other Stories.* Boston: Houghton Mifflin and Company, 1894. Pp. 67–68, 106–107.

Regional Perspective

TEXTILES AND INDUSTRIALIZATION

In the early stages of the Industrial Revolution in Europe textiles were the most important industry. The mills could accept workers at a low skill level and generally used linen or readily available imported cotton. There was always a local market for the manufactured clothing products. Factory work quickly became a goal for rural youth, and the textile industry was the most common employment target for such a migration. Bleak dormitories were constructed and work in "Coketowns" and in "fairy palaces" all across northwest Europe became characteristic of the growing industrial landscape. Today, many of the same patterns exist. Textiles still can often be supported by local or easily available fibers. The level of technology is more complex than in Dickens' time, but there are still many entry-level positions that require no prior factory experience. Domestic markets for denim, cotton goods, and increasingly fancy weaves are often protected and supported as a nation attempts to build an industrial base. In the past fifty years there has been enormous mobility of textile industries. New nations make a move toward a more industrialized future by staffing that ambition with workers who have a lower wage scale than more fully developed nations. While modern textile machine technology is important in the manufacturing of bulk goods, there continues to be a major world market for handmade textiles that come from traditions that are literally thousands of years old. The textile industry is an example of modern industry and traditional handicrafts existing side by side in countries seeking industrial development. The continued productivity of textile handicrafts has significance for the tourist trade, and is a growing source of foreign exchange.

A LAND OF CITIES

The Industrial Revolution gave a powerful impetus to the growth of cities at the same time that Britain's total population was increasing rapidly. In 1851, the United Kingdom became the world's first predominantly urban nation, and it remains so today. Most of the large cities, with the notable exception of London, are on or near the coal fields that supplied the power for early industrial growth. The most striking cluster is composed of cities associated with the coal fields that flank the Pennines to the east, south, and west—that is, in the industrial districts of Yorkshire, the Midlands, and Lancashire respectively (see Fig. 4.6). In contrast, the largest metropolitan areas that lie at a considerable distance from the coal fields—aside from London—are generally smaller and more widely spaced. Most of them are in southern England. For many years, they have tended to grow more rapidly and be more prosperous than the coal-field industrial cities, which tend to be physically unattractive and too dependent on old and failing industries. The transportability of electricity, natural gas, and oil has made it unnecessary for industries to locate near coal, and recent development tends to favor the relatively pleasant environments of southern England.

URBAN LOCATION DYNAMICS: THE LONDON EXAMPLE

Cities are almost always where they are because of the presence of some particular geographical characteristics. These might include a ford across a river, a strategic pass, the location of a resource, the intersection of significant migration paths, or even a religious or historic event. London, as one of the most dominant of the European cities, is characterized by a complex set of urban location dynamics that are worth some detailed consideration. While what follows is the story of London, it is important to realize that many aspects of this development would be important to the location and importance of Paris or Brussels or Berlin. Use this chronicle as a set of specifics that has broad implications for urban positioning across the globe.

The London metropolitan area, with more than 11 million people and 6,574,000 in the city proper, sprawls over a considerable part of southeastern England (Fig. 4.7). London is unusual because it is far from the island's major coal fields and their associated industrial centers. Its historical background is also unique in that London was already a major city before the Industrial Revolution, and its economy is not based so much on industry. Although the metropolitan area contains a great deal of manufacturing, this sector does not dominate. Instead, it shares leadership with several other major elements, some of which developed before the Industrial Revolution and have greatly expanded since then. These include commerce, government, finance and insurance, and corporate administration. Basic elements more recently developed include communications media (publishing, television, films) and tourism.

Outstanding importance as a seaport and commercial center was the original foundation for London's development. Trade generated population growth, profits, and financial operations, including insurance (the insurance business began with marine insurance covering risks to ships and cargoes). Population, trade, and wealth attracted government. Finance, trade, and government attracted corporate offices. The large, growing,

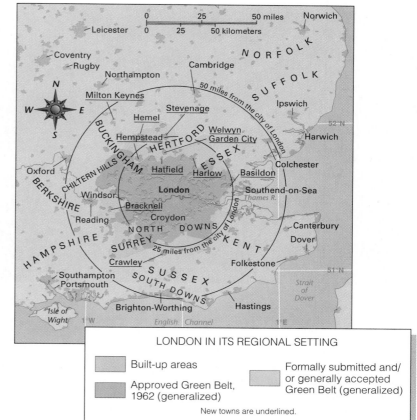

LONDON IN ITS REGIONAL SETTING

- Built-up areas
- Approved Green Belt, 1962 (generalized)
- Formally submitted and/ or generally accepted Green Belt (generalized)

New towns are underlined.

FIGURE 4.7

The maps of London illustrate the structure and setting of a major world metropolis. They may be used together with the New York maps panel in Chapter 23 to examine general concepts of urban geography, such as the Sector Model and the Concentric Zone Model (see Glossary). The City of London, built originally by Roman conquerors of Britain, contains one of the world's greatest financial districts. The Green Belt limits further urban sprawl and the New Towns are planned communities built to provide residential areas and employment for migrants from London proper in order to lessen congestion there. The West End is the governmental, hotel, and entertaining center. The Docks, built as artificial anchorages for shipping, have been closed. Various details on the maps are generalized from the *Atlas of Britain and Northern Ireland* (1963), by permission of Clarendon Press, Oxford; a government map of the Green Belt on an Ordnance Survey base, Crown copyright; and maps in J.T. Coppock and High C. Prince (ds.), *Greater London*, Faber and Faber, 1964

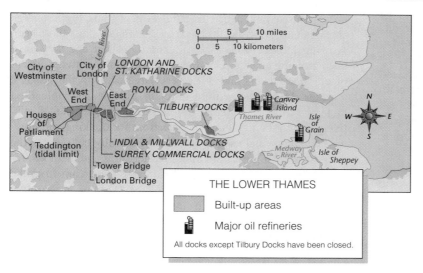

THE LOWER THAMES

- Built-up areas
- Major oil refineries

All docks except Tilbury Docks have been closed.

and increasingly prosperous population provided both a labor force and a large market for manufacturing.

London became an important seaport as early as the Roman occupation of England in the first century A.D and continues to be so today. Certain aspects of London's location and evolution as a port not only help account for the city's own development but also shed some light on port cities in general. Such factors include:

1. *Location relatively near major trading partners.* London is located in the corner of Britain nearest the continent, which was the only important area with which England traded before the Age of Discovery and which is still of major importance to British trade today.

2. *Location in a productive area.* London is located in the fertile English Lowland, which was the most productive and populous part of Great Britain in the preindustrial period,

FIGURE 4.8
London on the Thames River in southeastern England is a major node in the worldwide system of places that geographers study. Tower Bridge is at the lower left, with new London Bridge next upstream. The financial district (City of London) is at the right center, adjacent to London Bridge. *Colour Library Books*

supplying most of the wool that was Britain's main export and providing most of the British market for imports.

3. *Location facilitating transport connections with the surrounding region.* From the site of London, a major route to the interior, with branches, was provided by the Thames River and its tributaries. Before the age of the railroad and the truck, water transport had greater advantages over land transport than it has today, and these natural inland waterways were extensively used for transporting goods to and from London. During the Industrial Revolution, their utility was enhanced by canals (little used today) that connected the Thames system with other British rivers. Furthermore, London became a major junction of roads as early as Roman times, and again the city's location provided advantages. Tidal marshes originally fringed the lower Thames, forming a major barrier to transportation, but the original site of London provided firm ground that penetrated these marshes and approached the river bank on both sides. Highways could avoid the marshes by converging on London and crossing the river there.

4. *Location on a river (and estuary) permitting relatively deep penetration of ocean ships into the land.* A seaport set inland in such a manner can serve more trading territory in the surrounding countryside, within a given radius, than can a port on a comparatively straight stretch of coast or on a promontory. Dense urban development surrounds the port of London on all margins of the Thames (Fig. 4.8). Not only does the river provide good channel access for oceangoing ships, but London's network of highway connections to the rest of Britain facilitates port activities that link the island with global trade. The inland location of London's port tends to maximize the distance goods can be carried toward many destinations by relatively cheap water transportation (ocean vessels) and by minimizing the more expensive land transport.

Although London, like many other ports, originally developed well inland, later growth of the port has been downstream toward the ocean. This has been necessary to accommodate the growth in size (and hence the efficiency) of ships, because large ships cannot go as far up the river. Although ships have been growing larger for centuries, in the industrial age they have been growing at an accelerated pace, especially with the advent of oil supertankers and other very large ships in the last few decades. In addition, more efficient loading and unloading methods, such as containerization, require large areas of relatively open ground along the waterfront for tracking and moving containers, and such areas cannot generally be found in the congested older ports built up for earlier ships and methods. Instead, they must be sought in the areas downstream from the original port. Thus, a large share of the facilities in the older part of the London port have closed, with much loss of employment, whereas new port areas, using modern methods that require fewer people, are being developed far out on the Thames Estuary.

FOOD AND AGRICULTURE IN BRITAIN

One of the traditional characteristics of the United Kingdom is its large dependence on imported food. This dependence on imported food is not due to lack of efficiency in British agriculture but rather to the high density of population and the relatively high standard of living. Crop yields and the output of animal products per unit of land are among the highest in the world; the importation pattern was influenced by the success of a traditional free trade policy in the U.K. that allowed foodstuffs from Canada, Australia, the United States, and other nations into the British market with low or no tariffs. This system was in full swing in the middle of the 19th century and farmers—particularly wheat farmers—suffered from the availability of low-cost imported food. During World War II, British farmers increased their production of wheat by 35 percent because of the war's rupture of traditional dependence on oceanic trade in agricultural commodities.

Now, with the new political clout of the European Union, farmers are accustomed to EU agricultural tariffs and British subsidies for agriculture. As EU policy promotes expanded EU self-sufficiency, British farmers grow more wheat, linseed, and rapeseed—crops that had not been seen for ages in the U.K.

A major characteristic of the United Kingdom's agriculture is its high concentration on grass farming for milk and meat production. Meat and, to an even greater degree, fresh milk are traditionally more difficult and expensive to transport than grains and many other vegetable products; consequently, they afford local producers a degree of advantage over competitors located at a greater distance from the market. Land in crops is used primarily to produce barley or other supplementary feeds for dairy cattle, beef cattle, and sheep. Even so, the United Kingdom imports large quantities of meat and dairy products as well as other foods.

Total food production has increased sharply since World War II, mainly an outcome of various government policies aimed at reducing imports. A large increase in wheat and barley production has been fostered by subsidies to farmers and has been expensive. However, despite the high cost, Britain's membership in the Common Market—and now the European Union—probably will continue to foster the tendency to expand British agricultural production. Up to now, the agricultural policy of the EU emphasized subsidies to agriculture, including generally high import tariffs on farm products from outside (see Problem Landscape on page 103). As long as this policy holds, British farmers will continue to receive incentives in the form of relatively high and protected prices, and the British people, like the other peoples in the European Union, will eat more expensive food, produced to a greater degree within the home country or, at least, within the EU.

THE DECLINE OF ESTABLISHED INDUSTRIAL AREAS

The United Kingdom has experienced an almost constant succession of economic difficulties since World War I, punctuated by military and political struggles that have tended to accelerate the country's decline in power and influence. The economic difficulties are largely associated with the declining fortunes of the very industries upon which former growth was based. A num-

ber of the major industrial districts have been chronically depressed most of the time since World War I (Fig. 4.9). Overall, the economic downturn in Britain is relative rather than absolute. There has been quite a substantial increase in the standard of living over the long term, with a few sharp reversals interspersed. Except for a few promising years in the 1980s, gains have been much slower than in most industrial countries.

Britain's dependence on trade requires that British industries be competitive internationally if the country is to prosper. But for the past 80 years or so, the older industries— especially coal, cotton textiles, shipbuilding, and iron and steel—have had much difficulty in marketing their products. Consequently, their sales, output, employment, and relative wage levels have declined drastically. Whole communities in the coal-field industrial areas have experienced chronic distress because they have specialized in one or two faltering industries. When this occurs, continuing unemployment can be avoided if the community can develop or attract new industries. Britain's old industrial areas have witnessed both the development of new employment sources and migration to growing areas—mainly in southeast England around London or in such overseas countries as Australia or Canada. But most of the time, neither process has been rapid enough to avert unemployment and its depressing effect on wages. To complicate the problem, even some of Britain's industries that developed in the 20th century—such as the automobile manufacturing of the Midlands and southeast England, and the aircraft industry of southern England—have been experiencing the same competitive difficulties.

FIGURE 4.9

This derelict nineteenth-century factory building radiates the disuse and abandonment that characterize urban landscapes that have been left in the wake of new suburban development. Some of these buildings, however, are renovated by people who feel that their architecture and urban centrality make refurbishing worth the investment of capital and considerable human energy. Such an effort can promote a full scale neighborhood transformation—both in terms of appearance and social characteristics. *David Hoffman/Tony Stone Images*

Regional Perspective

THE DYNAMICS OF INDUSTRIAL AND COMMERCIAL RELOCATION

As in the case of urban location dynamics, the city of London serves well as an example of the collection of significant geographic features in the siting of an urban place. In this section, we use a British example for another common economic and geographic dynamic characterizing the past four decades: industrial and commercial relocation.

Industrial difficulties in older industrial areas and countries, of which Britain is the most outstanding example, are a product of many circumstances, but the most vital circumstance—at least for the United Kingdom—is the increasing degree of international industrial competition that developed in the 20th century, especially since World War II. Intensified competition is partly the logical outcome of the continued spread of industrialization around the world, accentuated in recent decades by the rapid improvement of communications and transportation. This brought a precipitous drop in the costs of maintaining contact with distant production facilities and markets and of shipping materials, subassemblies, and finished products for long distances. Distance from competitors no longer offers much protection, and distance from markets is no longer much of a handi-

cap. Competition has also been accentuated by the systematic and scientific development of new products having the potential to compete with and replace older ones. A 20th-century example is the replacement of older means of transport—of rail connections—by the automobile and truck, and the attendant replacement of coal by oil. Another is the development of synthetic fibers, made from wood or petroleum, to compete with cotton and wool.

In the strenuous competition for international markets, older industrial areas tend to lose to newer ones. Some major reasons include:

1. *Loss of a resource advantage*. A common example is the change for the worse in the significance of local coal. Location on a coal field was the main basis for the development of many older districts, which eventually lost the advantage because of (a) depletion of the local coal, (b) the rise of lower-cost coal fields elsewhere, or (c) replacement of the coal by oil or natural gas.
2. *Cheaper labor in newer areas*. In older districts, there is time for unionization to occur and to increase wages, whereas a new area is likely to have an unorga-

nized work force to whom even low wages seem a great improvement from urban unemployment or rural underemployment. With lower wages, prices can be lower.
3. *Technological advantages of newer areas*. Such areas can be developed with the newest and most efficient technology. In older areas, there is a large investment in older plant and equipment, which is difficult to write off, scrap, and replace with the latest technology. In some cases, strong unions may oppose the new technology because it provides fewer jobs.
4. *Technological unemployment resulting from the efforts of older industries to survive*. Even if an industry of an older area survives the competitive struggle, unemployment and hardship are likely to be visited upon the community because the industry is likely to survive only through increased mechanization, automation, and the use of robots. Production may be maintained or even expanded by this tactic, but workers will be fewer in number.
5. *Conservatism of management*. After years or decades of success, there is often a tendency for the management as

4.3   Ireland

Ireland is a land of hills and lakes, marshes and peat bogs, cool dampness and verdant grassland (Fig. 4.10). The island consists of a central plain surrounded on the north, south, and west by hills and low, rounded mountains. The Republic of Ireland, which occupies a little over four fifths of the island—and is included in the definition of the term "British Isles"—is in the midst of a transition from an agricultural to an industrial economy. Manufacturing is now a larger employer than agriculture and supplies more exports by value. But the transition is far from complete. Agriculture is still more important as an employment sector in Ireland than in most countries in western Europe, and the Republic is a considerable exporter of agricul-

tural commodities and an importer of manufactures. Another indicator of the transitional nature of this still industrializing economy is the relatively low income level of the population. In per capita income, Ireland ranks near the bottom of western Europe's countries (see Table 3.1).

Both nature and history contributed to the long delay in Irish industrialization. Some of the delay was undoubtedly due to an almost total lack of the mineral resources that facilitated early industrial development next door on the island of Great Britain. But Ireland's delayed industrialization is also a consequence of its long subordinate relationship to England. After centuries of strife, England completed an effective conquest of Ireland in the 17th century and subsequently made the island a formal part of the United Kingdom. But Ireland's status was ac-

well as the workers in an older industrial area to be somewhat resistant to necessary changes. There is a natural tendency to be complacent in the belief that procedures that worked in the past will continue to succeed in the future. Many charges of this nature have been leveled against established management teams not only in Britain, but anywhere this dislocation occurs.

As the process of dislocation takes place in one sector of the economy, there is often a parallel expanding employment sector. This is currently taking place in the *service sector*—a term embracing all workers not directly involved in producing material goods. In industrialized nations, the pattern that is ever more significant is the reduction of a manufacturing base and the expansion of the service base. Apparently this trend must continue if an increase in employment is to be enjoyed: agriculture has fallen to low levels of employment, and industrial jobs are being lost to competition and to automation. Increasingly, Britain—along with other industrial nations—is a country of white-collar workers. The expanding service sector that employs them is highly varied in its makeup, some of its major parts being government, finance, transportation, insurance, health services, management, sales, education, research,

personal services, the professions, tourism, and communications media.

There are many unanswered questions about this service-sector–oriented, "post-industrial" society toward which the industrialized Western nations are evolving. These questions focus on:

1. *The role of exports.* Will these societies be able to maintain themselves industrially with lightly manned, highly automated manufacturing industries?
2. *Capacity of the service sector.* Can the service industries provide an adequate supply of exports needed in international trade? The United Kingdom long supported itself in part by "invisible exports" such as capital, financial services, insurance services, ocean shipping and airline services, and tourism, and other Western nations are now involved in such activities. But industrial exports remain important to them. And, just as important, can service jobs be expanded rapidly enough to supply high employment in the face of declining industrial employment?
3. *Income potential in the service sector.* Can service jobs be made productive enough to provide high incomes to those holding them? There is reason to doubt the ability of many parts of the service sector to increase productivity at rates

comparable to those of large-scale and highly mechanized agricultural or manufacturing industries. If such doubts prove justified, the countries with a high dependence on services are likely to experience slow economic growth and chronic problems.

4. *Location of the service sector.* Will new service employment locate in the afflicted old industrial areas where it is especially needed, or will it grow mainly elsewhere? In Britain, the primary growth of such employment, as with employment in newer forms of manufacturing, has taken place in London and surrounding parts of southeast England rather than in the depressed coal-field industrial areas. This has accentuated the sharp contrast between the relatively prosperous London region and the rest of the country.

The answers to the foregoing questions will go far toward determining the future of Western countries, both within and outside of Europe. The United Kingdom will be a key country to watch because the process of change is relatively far along there and is proceeding rapidly.

tually colonial, and the island was governed for about two centuries under probably the harshest rule imposed on any British colonial territory. Most of the land was expropriated and divided among large estates held by English landlords. The Irish peasants became tenants on these estates, with the proceeds of their labor drained off in excessive rents and taxes. Government measures designed to favor British competitors restricted Irish trade and industry. Ireland, especially the part now constituting the Republic, became a land of deep poverty, sullen hostility, and periodic violence.

The Protestant North of Ireland—now known as Northern Ireland and still a part of the United Kingdom—was somewhat less handicapped by British policies than was the Catholic South. There was a strong element of religious discrimination

in British rule, Catholics being regarded as potentially disloyal to the Protestant crown. The Protestants were the descendants of Scottish Presbyterians and other British Protestants who were brought to the north of Ireland in the 17th century under the auspices of the British crown. They were intended to form a nucleus of loyal population in a hostile, conquered country. Protestant entrepreneurs were able to develop shipbuilding and linen textile industries in and near the North's main city, Belfast (population: 685,000 metro, 295,000 city). These industries have declined recently, but Northern Ireland has been effective in attracting new industries seeking cheap labor despite recent political violence.

In the Republic of Ireland, the recent surge of industrial development has been accomplished under a program designed to

FIGURE 4.10

The combination of abundant rainfall, rich peaty soils, and a continuing maintenance of family farms gives Ireland a lush pastoral landscape that is both agriculturally productive and attractive to tourists. This scene is from near Lough Derravaragh County, Westmeath, Ireland. *Dembinsky Photo Associates*

attract foreign-owned plants by a combination of cheap labor, tax concessions, and help in financing plant construction. Hundreds of plants have been attracted, although most of them are rather small. They are owned mainly by U.S., British, and German companies, and their major products are processed foods, textiles, office machinery, organic chemicals, and clothing. Most of the new plants have been built in and around the two main cities—Dublin (population: 1.2 million metro; 503,000 city) and Cork (population: 175,000)—and in the vicinity of Shannon International Airport in western Ireland.

With approximately 13 percent of its labor force on farms, the Republic of Ireland still depends on agriculture to a marked degree. The same physical and climatic handicaps to cultivation that affect Great Britain are present but to an even greater extent. Consequently, agriculture centers on grazing and feeding of cattle. Much of the arable land is devoted to fodder crops, which are supplemented by imported feeds. Meat, dairy products, and live feeder cattle for fattening elsewhere are major Irish exports. In addition to supplemental grain imported to feed animals, there are imports of wheat and many other human foods produced in inadequate quantities or not at all in Ireland. Although the large estates of the period of English rule are gone and the Irish farmer now tends to be a landowner, there is marked rural poverty. Farms tend to be both small and under-mechanized.

4.4   The Many-Sided Geographic Character and Personality of France

France has long been one of the world's more significant countries. It is the largest country by area in Europe (see Table 3.1), excluding Russia and Ukraine in the Former Soviet Region, and is Europe's leading nation in value of agricultural output. The country is one of the world's major industrial and trading nations and an atomic power. France has lost its former significance as a major colonial power, but it still maintains special relations with a large group of former French colonies. More than a dozen are now Departments of France and send representatives to the Assembly in Paris. Still another important foundation for France's international importance is the high prestige of French culture in many parts of the world.

France was once a very great power. In the 17th and 18th centuries, its power was a primary factor in European politics. In the early 19th century, under Napoleon I, it subjugated nearly the entire continent before it was checked by a continental coalition in alliance with Great Britain. France's power was based on the fact that the French kings unified a large and populous national territory at a relatively early date. Before Germany and Italy achieved political unification in the late 19th century, France was a larger country relative to its European neighbors than it is today.

TOPOGRAPHY AND FRONTIERS

France lies mostly within an irregular hexagon framed on five sides by seas and mountains (Fig. 4.11). In the south, the Mediterranean Sea and Pyrenees Mountains form two sides of the hexagon; in the west and southwest, the Bay of Biscay and the English Channel form two sides; in the southeast and east, a fifth side is formed by the Alps and the Jura Mountains, and, farther north, by the Vosges Mountains, with the Rhine River a short distance to the east. Part of the sixth or northeastern side of the hexagon is formed by the low Ardennes Upland, which lies mostly in Belgium and Luxembourg, but which has broad lowland passageways leading into France from Germany both north and south of the Ardennes. Aside from the large Massif Central in south central France, all the country's major highlands are peripheral. Even the Massif Central, a hilly upland with low mountains in the east, is skirted by lowland corridors that connect the extensive plains of the north and west with the Mediterranean littoral. These corridors include the Rhône-Saône Valley between the Massif to the west and the Alps and Jura to the east, and the Carcassonne Gap between the Massif and the Pyrenees. Thus, France is a country whose frontiers coincide, except in the northeast, with seas, mountains, or the Rhine and that has no serious internal barriers to movement.

The state that emerged as modern France was to a considerable degree a natural fortress, as two of its three land frontiers

FIGURE 4.11
General location map of France.

FRANCE
INDEX MAP

URBAN AREAS

⊛ Over 10,000,000 (national capital)

⊙ Over 1,000,000

● Over 600,000

○ Over 275,000

∘ Over 100,000 (selected places)

• Selected smaller places

LANDFORMS

Lowlands

Hills and uplands

Low mountains

Swiss Plateau

High mountains

City-size symbols are based on metropolitan area estimates.

lay in rugged mountain areas. The Pyrenees are a formidable barrier, lowest in the west, where the small linguistic group called the Basques—neither French nor Spanish in language—occupies an area extending into both France and Spain. In southeastern France, the Alps and Jura Mountains follow France's boundaries with Italy and Switzerland. The sparsely populated Alpine frontier between France and Italy is even higher than that of the Pyrenees: Mont Blanc, the highest summit, reaches 15,771 feet (4807 m). The number of important routes through the mountains is limited, and the Alps have been an important defensive rampart throughout France's history. The Jura Mountains, on the French-Swiss frontier, are lower but also difficult to cross, being arranged in long ridges separated by deep valleys.

FRANCE'S FLOURISHING AGRICULTURE

Agriculture is a more important part of the economy in France than it is in most highly developed countries. Since World War II, rapid economic change has been accompanied by a massive migration from farms to cities. The number of farmers has fallen to the point that only about seven percent of the country's labor force is now in agriculture (as of 1996). But this is still higher than in many developed countries, and, although their numbers have fallen, France's farmers have become more efficient. Hence, agricultural output has increased greatly, which is a common experience in industrializing countries.

Not only is agriculture important in France, but France is important in global agriculture. Despite its relatively small size on a world scale, the country is a major producer and exporter of agricultural products. Two products—wheat and wine—are of primary importance in the export trade, but France also has large exports of barley, cereal products, corn, dairy products, fruits, and vegetables. For a country with only four fifths the area of Texas to attain such a large volume of farm exports, some human and physical advantages for agriculture must obviously be present. One advantage often cited is the relatively small population compared to France's area. But one must remember that a density of 275 people per square mile is low only by west European standards; it is almost four times the density of the United States. Also, France's membership in the European Union is a major advantage. This gives French farmers free access to a huge and affluent consuming market protected by tariffs from non-EU producers. In addition, the European Union budget provides generous price supports at rates above world prices for many agricultural products, with French farmers the leading beneficiaries.

France's physical advantages for agriculture include topographic, climatic, and soil conditions. The main topographic advantage is the high proportion of plains. Large areas, especially in the north and west, are level enough for cultivation. Climatically, most of the country has the moderate tempera-

tures and year-round moisture of the marine west coast climate. The wettest lowland areas are in the northern part of the country, and it is here that crop and dairy production is especially intense. Southern France often experiences notably warmer temperatures, allowing a sizable production of corn in the southwestern plains. Productive vineyards are also concentrated particularly in the south and southwest, although important vineyard areas are found as far north as Champagne, east of Paris. Finally, northeastern France from the frontier to beyond Paris has fine soils developed from loess. As in much of Europe, these exceptionally fertile soils are used principally to produce wheat and sugar beets. This wheat belt in northeastern France is the country's chief breadbasket; the beet production from the same area provides not only sugar but also livestock feed (beet residues from processing) for meat and milk production. Wheat and beets tend to occupy the best soils because both crops are unusually sensitive to soil fertility.

FRENCH CITIES AND INDUSTRIES

The Primacy of Paris

Paris is the greatest urban and industrial center of France, completely overshadowing all other cities in both population and manufacturing development. With an estimated metropolitan population of about 10 million (and 2,152,000 in the city proper), it is by far the largest city on the mainland of Europe (excluding Moscow in Russia). Paris is located at a strategic point relative to natural lines of transportation, but it is especially the product of the growth and centralization of the French government and of the transportation system created by that government. Like London, it has no major natural resources for industry in its immediate vicinity, yet it is the greatest industrial center of its country.

The Primate City: The Example of Paris

By formal definition, a **primate city** is one which is many times larger (in population) than any other city in the country. In a more common usage, an urban center is the primate city when its demographic, economic, political, and cultural importance far outshines all other cities in a country. Such a city is considered primate by the **hierarchal rule**—meaning that its cultural and economic influence overshadows all other urban centers. In the rank order of urban center size in France (using city proper population figures and not the metropolitan conurbation in which these cities are central), Paris leads with 2,200,000; then Marseilles, 800,000; Lyon, 415,000; Toulouse, 359,000, and Nice, 342,000.

The concept of the primate city, however, is a more culturally defined term, and in that sense, there is a singularity in the authority and dominance of Paris making it an excellent example of a contemporary and popular primate city. The unusual as-

FRANCE IN THE DOCK AS FARMERS WAGE FRUIT WAR

. . . Every April at the start of the strawberry season the war against cheaper Spanish produce begins in France. Since 1993, French farmers have attacked more than 60 Spanish lorries transporting cheaper fruit and vegetables to markets within the European Union.[1]

One of the most impressive geopolitical shifts of the past half century in Europe has been the evolution of the European Union from the 1957 Treaty of Rome. The initial European Economic Community (EEC) became the European Community (EC) and now the functioning unit of great consequence in Europe is the fifteen member European Union (EU).

The EU has had an impact on all facets of the European economy. For agriculture it is an influence in the various support programs and marketing arrangements for member nations. One of the most vocal populations—in occasional support but more frequently in a role of dissent—has been the French farmers. In the decades since the 1957 Treaty of Rome, they, like farmers all throughout Europe, have undergone profound change in farm organization, levels of mechanization, size of farms, and dependence upon foreign markets. This article about the fruit war deals with some of those elements.

The EU makes a continual effort to treat the regions embraced by EU membership as one open market area. Early in European efforts to stabilize and optimize farm output, the European Community devised a Community Common Agricultural Policy (CCAP) which was intended to maximize agricultural productivity and efficiency. The benefits of disregarding national boundaries and individual support programs within the EU focused upon maximizing agricultural production in regions best suited to a particular crop or suite of crops. In broad terms, there has been a major shift in the expansion of wheat, barley, and maize (corn) production in Europe and especially in France. In vegetable production there has been a major expansion from subsistence farming and market gardening in the Paris basin to a much wider region, now linked to the Paris market by development of both highways and railsystems.

The French farmers in the article above are representative of the recently vocal farmers who find the support systems of CCAP to work sometimes to their benefit, but sometimes to their perceived detriment. In this case, French farmers found that the elimination—required by the EU— of the French tariffs put local farmers—especially near the Spanish border—at an economic disadvantage. While the importation of Spanish vegetables is not of major overall economic significance, the influence of EU policies overall has caused the French to rise in protest because of perception of cultural, as well as financial, hardship being worked upon them by the EU. Because France delivers almost a quarter of the EU's agricultural production annually, there is considerable interest in what these farmers say and do.

The cultural impact that the French fear is the further loss of the French family farm. Since 1995, more than 4 million farm jobs have been lost in France, and the country is now approximately 80 percent urban.

The average French farm size has grown to nearly 30 hectares (75 acres), with more than half of the country's farms amounting to less than 20 hectares (50 acres). Though this total is about one eighth the size of a United States farm, it does mean that it takes increasing capital to be able to stay afloat in such an operation. For the French, to see all farming becoming capital intensive, export-market-oriented, and controlled by the EU is quite unacceptable. The French continue to view themselves as an agricultural nation, holding the customs of the family farm central to their identity and their moral well-being. The economic growth resulting from EU membership has extracted its cost by increasing the migration of French youth away from family farms toward Paris and other cities. Meanwhile, the aging farm population goes further into debt purchasing new farm machinery that might enable them to increase productivity and stay competitive in the domestic, EU, and global marketplace.

The importation of Spanish vegetables is only one symptom of a larger irritation that has led French farmers to periodic protests in Paris as well as along the highways where the lorries carry produce from competing farmers.

With shifting farm technologies and national demographies, and increasing global agricultural competition, the problem landscape for the French farmer is a specter overshadowing family farms all across the globe.

[1]*The European*, 2–8 May, 1996, 1–2.

pect of the Paris example is that the primate city is more often a geographic characteristic in *developing* nations, where the dominant city is the center of everything—industry, politics, culture, and government. In France, the lesser cities have considerable industrial and cultural strength, but they all continue to exist in the shadow of Paris politically.

Paris began on an island in the Seine River which offered a defensible site and facilitated crossing of the river. Its early growth, which dates from Roman times if not before, was furthered by its location in a highly productive agricultural area and amidst navigable streams as well as land routes crossing the Seine. The rivers that join in the vicinity of Paris include the

Seine, which comes from the southeast and flows northwestward from Paris to the English Channel; the Marne, which flows from the east to join the Seine; and the Oise, which joins the Seine from the northeast (Fig. 4.12). In recent centuries, these rivers were modified and canals were provided where necessary to give Paris waterway connections with seaports on the lower Seine, with the coal field of northeastern France, with Lorraine, and with the Saône and Loire rivers (see Fig. 4.1).

In the Middle Ages, Paris became the capital of the kings who gradually extended their effective control over all of France. As the rule of the French monarchs became progressively more absolute and centralized, their capital, housing the administrative bureaucracy and the court, grew in size and came to dominate the cultural as well as the political life of France. When national road and rail systems were built in relatively recent times, their trunk lines were laid out to connect Paris with the various outlying sections of the country. The result was a radial pattern, with Paris at the hub. As the city grew in population and wealth, it became an increasingly large and rich market for goods.

The local market, plus transportation advantages and proximity to the government, provided the foundations upon which a huge industrial complex developed. Speaking broadly, this development involves two major classes of industries. On the one hand, Paris is the principal producer of the high-quality luxury items—fashions, perfumes, cosmetics, jewelry, and so

FIGURE 4.13
This aerial view of Paris suburbs shows the continuing French interest in urban greenery and parks, as well as the increasing use of high rise apartment space in the expanding city. Such growth is a continuing reminder of the forces of attraction that work to keep Paris and its environs central to French development.   *Claude Abron/Gamma Liaison*

FIGURE 4.12
Notre Dame rests on a small island in the Seine River in Paris, standing majestic in its site and a major landscape signature of this most famous European City. The Seine functions as a major transportation artery for the city and the region, but it is better known as a landscape hallmark of the French capital.   *© 1995, Dembinsky Photo Associates*

on—for which France has long been famous. The trades that produce these items are very old. Their growth before the Revolution of 1789 was based in considerable part on the market provided by the royal court; since the Revolution, it has been favored by the continued concentration of wealth in the city. On the other hand, Paris is now the country's leading center of engineering industries, secondary metal manufacturing, and diversified light industries. These industries are concentrated in a ring of industrial suburbs that sprang up in the 19th and 20th centuries (Fig. 4.13). Automobile manufacturing is the most important single industry, but a great variety of other goods is produced. In addition, the city's economic base includes an endless variety of services for a local, regional, national, and worldwide clientele.

With the more recent growth of Marseilles, Lyon, Arras, and Lille, the dominance of Paris has been steadily modified and diminished. Two ports near the mouth of the Seine handle much of the overseas trade of Paris. Rouen (population: 380,000), the medieval capital of Normandy, is located at the head of navigation for smaller ocean vessels using the Seine. However, with the increasing size of ships in recent centuries, more and more of Rouen's port functions have been taken over by Le Havre (population: 250,000), located at the entrance to the wide estuary of the Seine.

French Urban and Industrial Districts Adjoining Belgium

France's second-ranking urban and industrial area, located in the north near the Belgian border, consists of cities clustered on or near the country's leading coal field that is now closed. A bit north of the field, several cities, of which Lille is the largest, form a metropolitan agglomeration of about 1 million people. On the coal field itself are a considerable number of smaller mining and industrial centers. This region, called the Nord, is both an important contributor to, and a major problem in, the economy of France. In the 19th and early 20th centuries, it developed a rather typical concentration of the coal-based industries of that period, with close juxtaposition of coal mining, steel production, textile plants, coal-based chemicals, and heavy engineering. These have not been among France's growth industries in the second half of the 20th century, and the region suffers now with extraordinary employment problems.

Urban and Industrial Development in Southern France

France's third largest urban-industrial cluster centers on the metropolis of Lyons (French: *Lyon;* metropolitan population: about 1.3 million) in southeastern France. The valleys of the Rhône and Saône rivers join here, providing routes through the mountains to the Mediterranean, northern France, and Switzerland. Lyons, located at this junction, has been an important city since pre-Roman times. The Rhône-Saône Valley, which connects Lyons with the Mediterranean and leads northward from Lyons toward Paris, is of major importance to France, both as a routeway and as an energy producer. Since prehistoric times, the valley has been a major connection between the North European Plain and the Mediterranean. Today, it forms the leading transportation artery in France, providing links between Paris and the coast of the Mediterranean via superhighways, rails that carry some of the fastest trains in the world, and the barge waterway formed by the Rhône, Saône, and connecting rivers and canals to the north.

In addition, the Rhône-Saône Valley corridor between the Alps and Jura on the east and the Massif Central on the west also plays a prominent role in French energy production. Recently, France has stressed hydroelectric and atomic energy to compensate for the inadequacy of its fossil fuel supplies, and the Rhône Valley plays a major role in both. A series of massive dams on the Rhône and its tributaries produce about one tenth of France's electricity, and further power comes from several large nuclear plants in the Rhône Valley using uranium from the nearby Massif Central and foreign sources.

France's principal Mediterranean cities and metropolitan areas—Marseilles, Toulon, and Nice—are strung along the indented coast east of the Rhône delta. Marseilles (French: *Mar-*

seille; population: 1,230,000) ranks in a class with Lyons and Lille in metropolitan population and is France's leading seaport. It is located on a natural harbor far enough from the mouth of the Rhône to be free of the silt deposits that have clogged and closed the harbors of ports closer to the river mouth. Recently, rapid growth of port traffic and manufacturing in the Marseilles area has been particularly associated with increasingly large oil imports and related development of the petrochemical industry. Toulon (population: 437,000), southeast of Marseilles, is France's main Mediterranean naval base, and Nice (population: 516,000), near the Italian border, is the principal city of the French Riviera, probably the most famous resort district in Europe. Along this easternmost section of the French coast, the Alps come down to the sea to provide a spectacular shoreline dotted with beaches. Near Nice, a string of resort towns and cities along the coast includes the tiny principality of Monaco, which is nominally independent but closely related to France economically and administratively.

Mining and Heavy Industry in Lorraine

Lorraine, in eastern France, is a crucially important area in the French economy but one with an exceptionally severe and chronic problem of unemployment. Lorraine's importance stems from the fact that it contains the principal concentration of heavy industry in the country and has sizable resources related to this type of development. The largest iron ore production in Europe comes from ore bodies extending northward from the vicinity of Nancy (population: 329,000), past Metz (population: 193,000), and into southern Luxembourg and Belgium (see Fig. 4.2). The main output of iron and steel in France derives from plants scattered along the ore belt. Coal fields extend from Germany's Saar across the border into adjacent Lorraine.

A problem typical of that experienced by many old heavy-industrial and mining areas in western Europe and the United States is competition from new and more efficient plants and superior resources elsewhere. Lorraine iron ore is large in quantity but low in quality. It is losing markets to richer ores from overseas, and the region's coal industry has long been afflicted by the competition of imported oil and gas. Newer iron and steel plants in western Europe tend to be located on the seaboard—as at Dunkirk in northern France and on the Gulfe du Lion adjacent to Marseilles in Mediterranean France—for convenient access to ores and fuels brought in from overseas by increasingly large and efficient ships. Like other European governments with similar problems, the French government is attempting to aid distressed areas by attracting new and more diversified development there. In Lorraine, this has had limited success thus far. One problem in such areas is lack of amenities—the qualities of natural and cultural environments that people find agreeable. Belching chimneys, grimy buildings, and industrial waste piles, to say nothing of outmoded tenements and general congestion left over from a bygone age, give very negative impressions to newer industries seeking a place to locate. The tradition of highly unionized labor in the Lorraine area serves also as a deterrent.

The Dispersed Cities and Industries of Western France and the Massif Central

Western France and the Massif Central are considerably less populous, urban, and industrial than the parts of France to the northeast and east. The largest urban center, Bordeaux, has only about 700,000 people in its metropolitan area, and the next two cities, Toulouse and Nantes, have only about 650,000 and 400,000 respectively. Bordeaux is a seaport on the Garonne River, which discharges into the Gironde Estuary. The city passed its name to the wines from the surrounding region, exported through Bordeaux for centuries. Nantes is the corresponding seaport where the other main river of western France, the Loire, reaches the Atlantic. Toulouse, a very old and historic city with modern machinery, chemical, and aircraft plants, developed on the Garonne well upstream from Bordeaux, at the point where the river comes closest to the Carcassonne Gap. Lying between the Pyrenees and the Massif Central, the Gap provides a lowland route from France's Southwestern Lowland to the Mediterranean.

Recent Dynamism in French Urban and Industrial Development

France shows striking contrasts in the nature of its development before and since World War II. From the Industrial Revolution until recent decades, the country tended to fall behind its European neighbors, except for Spain and Italy, in industrialization and modernization. It changed enough to become a considerable industrial power while remaining more rural and less urban and industrial. In the decades following World War II, France's old patterns were rapidly modified with new development.

Population growth, encouraged by government subsidies for children, speeded up. Inadequate coal resources became less of a handicap as other energy sources increased in availability and importance. Protection of inefficient producers from competition was weakened or abolished by France's membership in the European Union. The French government promoted combinations of companies into larger units capable of higher efficiency and international competitiveness. France was well placed, in both site and situation, to move into newly developing industries and technologies. It had a large pool of labor available for transfer from an over-manned agriculture, a relatively modest share of its manpower and capital tied up in older industries, and an excellent educational system. Not the least of its advantages was the adoption of a future-oriented national

economic planning system following the conclusion of World War II.

FRENCH NATIONAL AND REGIONAL PLANNING

The new dynamism of France has been achieved with deliberate planning. Since 1947, the country has engaged in national and regional planning with the stated objectives of improving national output and living standards, reducing the considerable differences in prosperity of different sections of the country, and making France a pleasant place to live in a technological age. This type of government planning is envisaged as complementary to the activities of private enterprise, with the state using various means to influence the decisions of private companies while at the same time cooperating with them.

In addition to fostering overall national growth and improvement, French planning is directed toward reducing imbalances of long standing among the regions of France. In the mid-1960s, the government called particular attention to the contrasting levels of development among four major regions—the Paris region, Northeast, Southeast, and West (Fig. 4.14). The most significant contrasts were (1) between the Paris region and the rest of the country and (2) between the less devel-

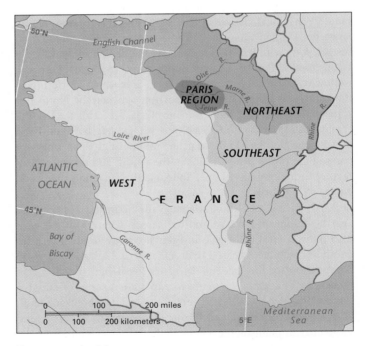

FIGURE 4.14
France, with its four regions of contrasting levels of economic development, with Paris as the highest, and declining from the Northeast to the Southeast to the West.

oped and poorer West and the rest of the country. Because modern economic growth concentrates strongly in cities, the French planners are attempting to divert development from Paris to the main cities of other regions. They use restrictions and penalties on various types of new development in the Paris region, a varied array of financial rewards for firms locating facilities in regions of greater need, and direct location or relocation of government-owned facilities in the regional centers.

An outstanding success in promoting regional centers is the rapid growth of Toulouse in southwestern France as a major European center of aircraft and aerospace industries. Within the Paris region itself, a similar policy of economic and population decentralization continues, with functions such as wholesale markets and universities moving from the old city to the suburbs and several large new planned cities being constructed in the suburbs.

Although there has been a real decentralization of growth in France during the planning period, some analysts believe that decentralization might well have occurred without any planning policy. They maintain that after manufacturing industries reach a certain stage of development, the work in them becomes so standardized and routine that little skill is required from the workforce, and such businesses then tend to relocate from industrial cities to previously nonindustrial places where lower wages can be paid to unskilled and nonunionized labor. This point of view sees French economic decentralization as a normal result of the evolution of industries, and it questions much of the planning process as an expensive irrelevancy. However valid such criticisms may be, some decentralization of jobs and prosperity in France is essential for political stability and national cohesion.

4.5   The Axial Role of Germany in Europe's 20th Century

Germany reappeared on the map of Europe as a unified country in 1990. Between 1949 and 1990, there were two Germanies: West Germany (German Federal Republic), formed from the occupation zones of Britain, France, and the U.S. after Germany's defeat in World War II in 1945; and Communist East Germany (German Democratic Republic), in territory occupied by the former Soviet Union; the urban landscape of West Berlin was a part of West Germany although the city of Berlin overall was entirely surrounded by East Germany. The withdrawal of Soviet support in 1989 led to the rapid collapse of East Germany's Communist government in 1990, the reunification of Germany, and the de facto inclusion of East Germany in the EU.

The new Germany is, by a wide margin, the most dominant country of Europe. Its population of 81 million is much greater than that of any other nation in the region. Economic consider-

ations are even more important. The former West Germany alone, with 61 million people, was Europe's leading industrial and trading state and was the principal focus of the economy of western Europe. The former East Germany was an advanced country by Soviet bloc standards but an economic disaster by West German standards. Much time and money will be required to rehabilitate Germany's East, but this area can be expected to contribute increasingly to overall German dominance in the European and global economy.

THE DIVERSIFIED GERMAN ENVIRONMENT

Germany arrived at its favorable economic position by skillfully exploiting the advantages of its centrality within Europe, together with certain key features of a diverse environment (Fig. 4.15). Broadly conceived, the terrain of Germany can be divided into a low-lying, undulating plain in the north and higher country to the south (Fig. 4.16). The lowland of northern Germany is a part of the much larger North European Plain. The central and southern countryside is much less uniform. To the extreme south are the moderately high German (Bavarian) Alps, fringed by a flat to rolling piedmont or foreland that slopes gradually down to the east-flowing upper Danube River. Between the Danube and the northern plain is a complex series of uplands, highlands, and depressions. The higher lands are predominantly composed of rounded, forested hills or low mountains. Interspersed with these are agricultural lowlands draining to the Rhine, Danube, Weser, or Elbe rivers.

The climate of Germany is maritime in the northwest and increasingly continental toward the east and south. The annual precipitation is adequate but not excessive, with most places in the lowlands receiving an average of 20 to 30 inches (*c.* 50 to 75 cm) per year. Farmers, however, must contend with soils that generally are rather infertile. Two areas are major exceptions:

1. The southern margin of the North German Plain forms a belt with soils of extraordinary fertility formed from deposits of loess. This belt continues beyond Germany both east and west.
2. The long, nearly level alluvial Upper Rhine Plain, extending from Mainz to the Swiss border and mostly enclosed by highlands on both sides, is also an area with fertile agricultural soils.

LOCATIONAL RELATIONSHIPS OF GERMAN CITIES

Within the environmental framework just described, Germany's major cities tend to be located in the most economically advantageous areas and those most strategic with respect to

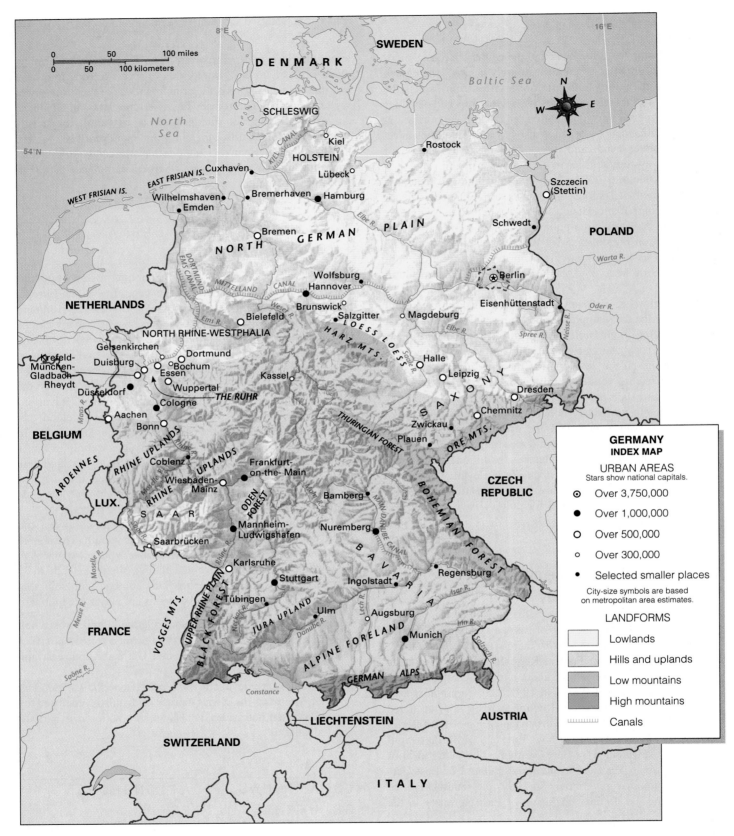

FIGURE 4.15

General location map of Germany. Heavy dashes mark the boundary between former West Germany (German Federal Republic) and the former East Germany (German Democratic Republic).

FIGURE 4.16
The meadows that flank the Wetterstein Range in Bavaria in south Germany frame the images that make this farming area so popular for hikers and tourists. The Bavarian Alps that rise in the background are part of the European mountain system that has played a major role in borders and battles for centuries in this region.   *Josef Beck/FPG International*

transportation. A string of major metropolitan areas is located in the western part of Germany, along and near the Rhine River (Fig. 4.15). This waterway has been a major route since prehistoric times. Four of the eight largest German metropolises—Düsseldorf, Cologne, Wiesbaden-Mainz, and Mannheim-Ludwigshafen-Heidelberg—are on the Rhine itself. The other four—Essen, Wuppertal, Frankfurt, and Stuttgart—are on important Rhine tributaries. This north-south belt of cities also includes several other large metropolitan centers such as Dortmund, Duisburg, and Aachen, plus numerous smaller places. The concentration of seven of the forenamed cities at or just north of the boundary between the Rhine Uplands and the North German Plain reflects additional conditions. Historically, some of the seven cities profited from their location on the agriculturally excellent loess soil just north of the Uplands. This concentration of large and medium sized urban centers differentiates the German landscape from the French world, in which Paris plays such a profoundly singular role.

Other German cities developed as early industrial centers before the Industrial Revolution, drawing on the Uplands for ores, waterpower, wood for charcoal, and surplus labor. Then, in the 19th and early 20th centuries, all seven cities grew in connection with the development of both the Ruhr coal field, located at the south edge of the Plain and east of the Rhine, and the smaller Aachen field. Except for Aachen and Cologne, they became parts of "the Ruhr," Europe's greatest concentration of coal mines and heavy industries. Still other important German

cities lie apart from the Rhine and Ruhr. These include (1) Hanover, Leipzig, and Dresden, spread along the loess belt across Germany; (2) the North Sea ports of Hamburg, located well inland on the Elbe Estuary, and Bremen, up the Weser Estuary; (3) two cities in the southern uplands—Munich, north of a pass route across the Alps, and Nuremberg; and (4) Berlin, whose large size belies its unlikely location in a sandy and infertile area of the northern plain. Berlin is mainly an artificial product of the central governments, first of Prussia and then of Germany, which developed it as their capital. Berlin's metropolitan population is 4,150,000 and Hamburg's is 2,385,000; the remaining cities previously named vary between 500,000 and 2 million.

HOW GERMANY EVOLVED: PEOPLE, MIGRATIONS, AND TERRITORIAL CHANGES

The present location of Germany and the Germans is only the latest phase in a series of large fluctuations on the map of Europe. By the first century B.C., German tribes occupied a wide expanse from the Rhine River in the west, where they confronted the Romans, to the vicinity of the Vistula in what is now Poland. In the fifth and sixth centuries A.D., the Germans broke the Rhine and Danube frontiers of the western Roman Empire and spread across western Europe. In the course of time, most Germans were absorbed into the peoples of the occupied areas, but Germanic speech was permanently extended into some areas near the German homeland. The early medieval expansion of "Germany" westward was accompanied by contraction on the east.

As the German tribes drifted west, much of the territory they abandoned was occupied by a new element, the Slavs. By about 800 A.D., Slavic peoples occupied what is now Germany to the line of the Elbe River and the Saale, as well as the Czech Republic to the south. Following the Slavic advance, a new phase of German expansionism set in. For about twelve centuries, from 800 A.D. to the 20th century, German control, influence, and population pushed eastward into territories occupied by Slavs and some other peoples in eastern Europe. Along the Baltic coast, medieval German crusaders, traders, soldiers, and settlers conquered or dominated territories as far north as the Gulf of Finland.

Farther south, centuries of German conquest and colonization eastward reached their peak in the late 1700s when the German state of Prussia, with its capital at Berlin, joined Russia and the German state of Austria in "partitioning" Poland. The latter disappeared, and German control reached almost to Warsaw. To the south, Austria built an empire between the 16th and 19th centuries that included the present Czech Republic and Slovakia, parts of southern Poland, Hungary, and parts of the northern Balkan Peninsula. Enclaves of German settlement extended much farther east than the areas actually under Prussian or Austrian political control. For example, there were sizable German settlements along the middle Volga River in Russia.

After unification in 1871, the development of a scientific, protected, and subsidized agriculture made Germany nearly self-sufficient in food despite the generally poor soils and dense population. Coal, chemical, metallurgical, and engineering industries boomed, and Germany became second only to the much larger United States in heavy industry. A world-famous educational system and scientific establishment supported these achievements. Germany's railway system, owned and operated by the state and supplemented by river improvements and canals, tied the country together and allowed a relatively wide areal spread of development and prosperity. Manipulation of railway freight rates helped bring about types and locations of development that the government desired. The German ports, especially Hamburg, became major centers of world trade, as did two foreign ports that largely served Germany: Rotterdam in the Netherlands and Antwerp in Belgium. Between 1884 and 1900, a colonial empire was acquired, mainly in Africa. Germany's colonial tenure was brief, however, as the victorious Allies took over control of the colonies at the end of World War I. In the social field, the Second Reich (1871–1919), despite its general conservatism, pioneered many of the social insurance and welfare policies that later became identified with the "welfare states" of post–World War II Europe and North America.

Defeat in World War II drastically redefined Germany's eastern boundary, split the country's remaining territory into Eastern and Western segments, and brought about a massive redistribution of Europe's German population (Fig. 4.17). All territories east of the line formed by the Oder River and its tributary, the Neisse, were transferred from Germany to Poland except for northern East Prussia, which went to Russia and remains on the map today as Kaliningrad. The occupying Allied powers soon split into two antagonistic camps, each of which sponsored a German regime of its own persuasion. In 1949, the Federal Republic of Germany (West Germany) was formed from the American, British, and French zones of occupation, and the German Democratic Republic (East Germany) was formed from the Soviet zone of occupation. The depth of antagonism between the two Germanies was well symbolized by the fortified Iron Curtain boundary between them, including the Berlin Wall between East and West Berlin.

The territorial changes following World War II were associated with the relocation of millions of Germans. Some fled westward in 1945 to escape Russian armies. Many others were forcibly expelled from eastern Europe to what was left of Germany. Within Germany itself, large westward movements of

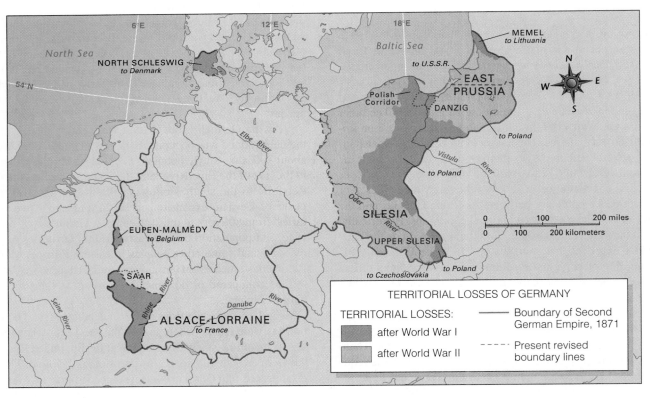

FIGURE 4.17

Territorial losses of Germany resulting from World War I and World War II. Minor frontier cessions of territory to the Netherlands, Belgium, and Luxembourg following World War II are not shown.

FIGURE 4.18

Berlin citizens from both sides of the Berlin Wall watch its demolition in June, 1990. This structure had served as one of the most powerful icons of the Cold War since its construction in 1961.   *AP/Wide World Photos*

population continued until 1961 as a result of the flight of East Germans to West Germany. Even after the boundary between East and West was closed by wires, landmines, and patrols, it was still possible to cross from East Berlin to West Berlin and thence to the main part of West Germany. By 1961, more than 13 million West Germans—almost one fourth of the country's population—were migrants from the east or the children of such migrants. A little more than one tenth of the newcomers came from East Germany, but this flow was such a drain on both the prestige and the economic strength of East Germany (and hence of the Soviet Union) that the Berlin Wall was built in 1961. This wall physically sealed off the last open boundary between East and West Germany.

Then, in less than fifty years since the end of World War II, the Soviet empire in eastern Europe collapsed. In 1989, the Soviet Union withdrew the military support that had propped up the region's Communist "satellite" governments, and popular mass demonstrations forced East Germany to dismantle the Berlin Wall (Fig. 4.18) and open the closed boundary else-

where. By the end of 1990, the Communist government of East Germany had fallen, a non-Communist interim government had negotiated unification, and East Germany had been absorbed into a reunified Germany.

GEOGRAPHIC EVOLUTION AND PROBLEMS OF GERMANY'S INDUSTRIAL ECONOMY

Germany's development into one of the world's greatest industrial areas has roots in its distant past (Fig. 4.19). Certain areas, such as the Harz Mountains, were exceptionally important centers of European mining in the Middle Ages, and many urban handicraft industries were also well established in medieval times. By about 1800, there were three German regions with unusually high concentrations of small-scale industries. One was the Rhine Uplands, in which industry spread along both sides of the river north of Wiesbaden and also away from the river in both directions, but especially toward the east. A second was Saxony, north of and including the Erzgebirge (Ore Mountains) on the German-Czech border. The third was Silesia (now in Poland), consisting of the plains along the upper Oder River—which contained the major mineral resources of Upper (southeastern) Silesia—and the Sudeten Mountains adjoining the plains. All three regions mined metallic ores, smelted the ores with charcoal, used the waterpower of mountain streams, and produced textiles as well as metals and metal goods.

In the 19th and early 20th centuries, each of the foregoing regions developed into a major center of coal-based industry. As the Industrial Revolution spread into Germany, each area discovered coal resources that were conveniently located and, in the case of the Rhine area and Upper Silesia, extremely large. The Ruhr coal field, located at the north edge of the Rhine Uplands (Fig. 4.19), is the greatest coal field in Europe, and the Upper Silesia field is also of major importance. In both areas, the abundance of coal favored the growth of large iron and steel industries, which had become increasingly dependent on imported ores. Secondary development—such as metal goods, machinery, and coal-based chemicals—evolved much more in the Ruhr than in Silesia, perhaps because of the Ruhr area's much better market position. Today, "the Ruhr" denotes an area of somewhat indefinite extent beyond the coal field proper. The "Inner Ruhr" (total population: about 5 million) denotes the more concentrated urban-industrial zone incorporating Essen, Duisburg, and Dortmund, whereas the "Outer Ruhr" includes Düsseldorf and other more widely spaced cities.

In Saxony, coal resources were much poorer. Brown coal and lignite fields were eventually developed but the heat produced is only about one quarter of that produced by bituminous coal, which yields coke for the iron and steel industry. Hence, an iron and steel complex did not arise in Saxony, although factory-type industries grew rapidly and promoted the growth of Leipzig, Dresden, and other cities. Machinery, precision instru-

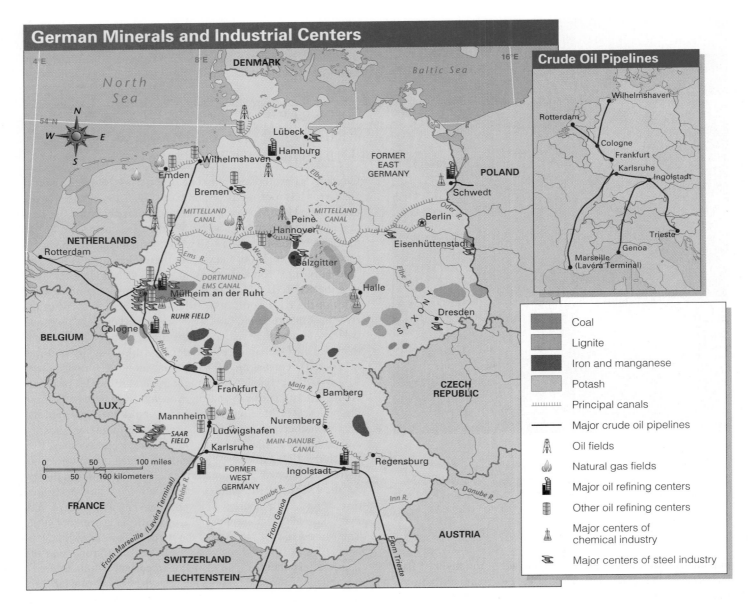

German Minerals and Industrial Centers

Crude Oil Pipelines

Legend:
- Coal
- Lignite
- Iron and manganese
- Potash
- Principal canals
- Major crude oil pipelines
- Oil fields
- Natural gas fields
- Major oil refining centers
- Other oil refining centers
- Major centers of chemical industry
- Major centers of steel industry

FIGURE 4.19

Major minerals and iron-and-steel manufacturing centers, chemical-manufacturing centers, and oil-refining centers in Germany. Dashes show the boundary between former East Germany and former West Germany.

ments, textiles, and chemicals were (and are) the main emphases. Saxony's industrial area is roughly triangular and is often called the Saxon Triangle (see Fig. 4.1).

Much of Germany's present industry, however, is not within the Ruhr or Saxony. A wide scatter of industry and of urban centers is characteristic. Many cities are large, but none is as gigantic and nationally dominant as London or Paris. The number and scatter of fairly large industrial cities apparently resulted in good part from two circumstances:

1. Germany was divided for a long time into petty states, a number of which eventually became internal states within German federal structures. These units often promoted the growth of their own capitals, and many of the large cities of today were once the capitals of independent German states.

2. German cities were able to secure power from a distance during the age of industrialization. In the sequence of German development, railways generally reached major cities before coal became practically an industrial necessity. Thus,

when coal was needed, the cities away from the fields could generally get it and continue to industrialize and grow. Waterways also helped many cities get coal; electricity then began to provide power from a distance.

Today, there is a major tendency in former West Germany, as in many other advanced industrial countries, for factory industry to shift from the cities to more suburban locations and to rural areas. The attractions appear to be lower wages, lower site costs, and environmental amenities. The shift is allowed by the wide availability of power in the form of electricity, access by superhighways, and the ability and increasing willingness of labor to commute long distances by car or rail.

The former West Germany recovered rapidly from World War II. A new currency was provided in 1948 and a government empowered to act formed in 1949. Former production levels were regained and surpassed. German cities were badly bombed in World War II (see photograph, page 84), but industrial capacity was not as badly damaged as is sometimes suggested. Various circumstances facilitated the process of economic rebuilding, not the least of which was the rather freewheeling free-enterprise approach followed by the government and massive aid received from the United States. The skills of the population were still there, and labor was cheap for a time, as workers were abundant and glad to be working again. Markets were hungry for industrial products. By the 1960s the pace of growth slowed, but it was still fast enough to provide jobs for large numbers of foreign workers who came in to take jobs vacated by upwardly mobile German workers.

Some analysts cite a further factor as an important contributor to cheap and abundant labor for a time—a rapid and large decrease in the farm population, releasing large numbers of workers to nonfarm work, which usually proved more remunerative. Germany had traditionally maintained a relatively large proportion of its people on the land by restricting imports and raising the prices of foreign food products competitive with German products. The situation then changed when the European Common Market deprived the West German farmer of protection from European competitors at the same time that industrial expansion opened new jobs. The resulting exodus of farmers from the land was so great that the West German farm labor force declined from 5.2 million to 1.7 million in the period between 1950 and 1976, freeing millions of workers for industrial and service employment. Even this decline, with the accompanying consolidation of farms, left most farms too small to provide, by West German standards, a good living for their operators.

The pace of development slowed during the 1970s and 1980s, although West Germany continued to be a very wealthy and productive country. West German coal became increasingly unable to compete with the imported petroleum which now replaces it as Germany's main energy source. New plants to generate electricity with atomic energy have sparked violent antinuclear protests, and the need for cheap energy led West Germany to quarrel with the United States in the early 1980s in order to complete a deal for Soviet natural gas from a new Russia-to-Europe pipeline. The export of German industrial products to world markets became more difficult as Japanese and other foreign competition stiffened. However, despite the various difficulties of the former West Germany's "economic miracle," it ranked with the United States and Japan as one of the world's three largest exporters of goods at the time of the 1990 reunification.

The former East Germany (1949–1990) was Communist-ruled and Soviet-dominated, but it was still German. Even Soviet exploitation and the inefficiencies of bureaucrats attempting to plan and operate a "command economy" were unable to destroy completely a historical and traditional German capacity to produce. East Germany was the most productive Communist society. But when its regime collapsed and extensive outside investigation became possible, East Germany was revealed to be a poor and inefficient country by Western standards. Many billions of dollars of investment and considerable economic pain over a number of years are estimated to be necessary if the economy of what is now the eastern part of reunified Germany is to be brought to levels of productivity and living standards comparable to those of the western part. Huge costs are involved in the cleanup of severe environmental damage incurred under Communist rule.

4.6 Benelux: Limited Resources and Well-Developed Trade Networks

Belgium, the Netherlands, and Luxembourg have been closely associated throughout their long histories. Today, they are often referred to collectively as "the Benelux Countries." An older name often applied to the three is "the Low Countries." In its strictest sense, however, this term is properly applied only to Belgium and the Netherlands because approximately the northern two thirds of the former and practically all of the latter consist of a very low plain facing the North Sea (Fig. 4.20). This plain is the narrowest section of the North European Plain (see Fig. 4.15). Large sections near the coast, especially in the Netherlands, are indeed below sea level. They are protected from flooding only by sand dunes, man-made dikes, and constant pumping.

The only land in the Low Countries with even a moderate elevation lies in the Ardennes Upland of Belgium, south of the Sambre and Meuse rivers. Much of it is an area of little relief, although in places intrenched rivers produce a more rugged terrain. To the south, the Ardennes includes the northern half of Luxembourg. Southern Luxembourg and a small adjoining tip of Belgium have a lower, rolling terrain similar to that of neighboring French Lorraine. High population densities, high pro-

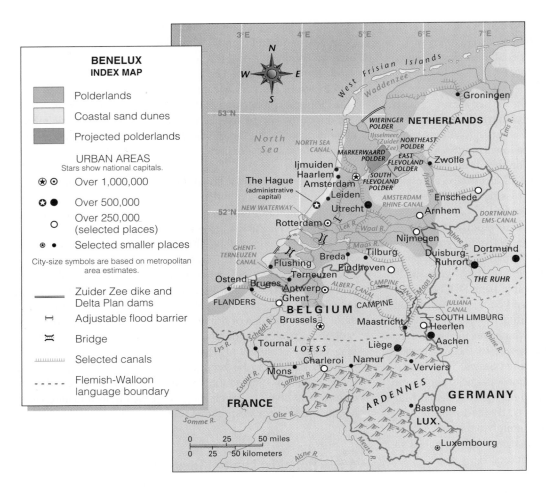

FIGURE 4.20

General location map of the Benelux countries: Belgium, the Netherlands, and Luxembourg. Note the extensive area of polder land reclaimed from the sea.

ductivity, and good incomes are outstanding characteristics of the Benelux countries. By effective utilization of such opportunities and resources as their small and crowded territories afford, the peoples of these countries are able to maintain very high standards of living (see Table 3.1). The economic life of the Benelux countries is characterized by an intensive and interrelated development of industry, agriculture, and trade. An especially distinctive feature of their economies is a remarkably high development of, and dependence on, international trade.

PORTS OF THE NETHERLANDS AND BELGIUM

Exploitation of their commercial opportunities by the Low Countries is reflected in the presence of three of the world's major port cities within a distance of about 80 miles (c. 130 km): Rotterdam and Amsterdam in the Netherlands and Antwerp in Belgium. Amsterdam (population: 1.9 million) is the largest metropolis and the constitutional capital of the Netherlands, although the government is actually located at The Hague (population: 773,000). The city was the main port of the Netherlands

during the 17th and 18th centuries and for most of the 19th century, while the Dutch colonial empire was expanding. The principal element in this empire was the Netherlands East Indies (now Indonesia), where the Royal Dutch Shell Company originated. Until the loss of Indonesia after World War II, Amsterdam built a growing trade in imported tropical specialties. Many of these imports were re-exported, often after processing, to other European countries. It also became, and remains, an important manufacturing center. Profits accumulated during these centuries supplied funds for the large overseas investments of the Netherlands and made Amsterdam an important financial center, which it continues to be.

Amsterdam also remains an important port, but it has lost much of its relative position in this respect during the past century. The original sea approach via the Zuider Zee—a landlocked inlet of the North Sea—proved too shallow for the larger ships of the 19th century. The opening in 1876 of the North Sea Canal solved this problem for a time, but even successive expansions of the canal did not give access to the sea comparable to that of Rotterdam. And the Dutch loss of Indonesia in 1949 greatly decreased the port's importance as an entrepôt, although continuing ties with its former colony have

THE VITAL ROLE OF TRADE: THE BENELUX EXAMPLE

One of the most significant driving forces in geography today is the expanding network of connectivity that links economic, political, and cultural worlds in ever more dimensions. Trade is at the very heart of this dynamic (see page 114). The Benelux countries make a useful example of the geographic factors that led to their dependence on the trading relationships they have developed. Here are four features that define their network, but that also serve as elements in parallel relationships worldwide:

1. *Trade As a Resource.* Trade is essential for such small countries to make maximum use of limited internal resources. Such resources as Luxembourg's iron ore and the Netherlands' natural gas must be either left in the ground or exported for comparatively small returns unless complementary materials needed for manufacturing can be imported.

2. *Specialization.* The need of these countries to trade matches their ability to trade. Specializing in activities that offer the greatest possibilities for effective use of limited resources, the Benelux nations are able to export large surpluses of certain manufactured goods and, in the case of the Netherlands, livestock products, vegetables, and flowers.

3. *Maximizing of Locational Factors.* The position of the Benelux countries is highly favorable for trade. They lie in the heart of the most highly developed part of Europe. Their nearest neighbors—Germany, France, and the United Kingdom—are among the world's foremost producing, consuming, and trading nations. The resulting commercial opportunities are reflected in the fact that approximately three fifths of Benelux foreign trade occurs among the three Benelux countries or between these countries and their three larger neighbors. A second significant aspect of position lies in the location of the Netherlands and Belgium at or near the mouth of the Rhine. Location where the Rhine meets the sea permits the Low Countries to handle in transit much of the foreign trade of Switzerland, eastern France, and especially Germany.

4. *Maximizing of Prior Trade Networks.* The trade of the Low Countries has profited somewhat from the fact that both the Netherlands and Belgium were able to exploit large colonial empires in the tropics. Before they became independent, the Netherlands East Indies and the Belgian Congo offered assured markets for goods and capital and allowed the home countries to act as European **entrepôts**—trading centers—for tropical agricultural products and certain minerals, particularly nonferrous metals. Wealth from this trade was important in financing industrialization, especially in the Netherlands. Today, industrial and trading specialties connected originally with this colonial trade still function in the Low Countries, although only a tiny fraction of their foreign trade is with their former colonies.

This capacity to link geographic realities with economic, political, and culture opportunities made an enormous economic difference to the standard of living of the Benelux nations. In parallel fashion, the EU has caused each member nation to exploit more fully the vital role of trade within and beyond the European region. The patterns that are emerging internationally now relate to each nation's attempt to maximize the benefits of traditional trading relationships, while attempting to find new outlets for additional trading energy. In reality, the traditional role played by the geography of location is being modified daily by the ever smoother movement of goods, information, and people across broad spatial networks of increasing interaction. The expanding global connection of traditional and current trade networks means that every trading nation will be attempting to take advantage of the geographical factors that the Benelux nations have utilized. As energy sources become more easily established away from traditional resource sites, manufacturing locales and trade centers will be established at the most trade-facilitating nodes. These spatial shifts will lead to changing urban demographics as well, not only in Europe but worldwide.

given Amsterdam a large Indonesian community that imparts a distinctive ethnic and cultural flavor to the city.

Rotterdam (population: 1.12 million metro, 560,000 city), the world's largest port in tonnage handled, is better situated for Rhine shipping than either Amsterdam or Antwerp, being located directly on one of the navigable distributaries of the river rather than to one side (see Fig. 4.20). Accordingly, Rotterdam controls and profits from the major portion of the river's transit trade, receiving goods by sea and dispatching them upstream by barge, and receiving goods downstream by barge and dispatching them to sea. Two developments are mainly responsible for the port's tremendous expansion. First was the opening in 1872 of the New Waterway, an artificial channel to the sea far superior to the shallow and treacherous natural mouths of the Rhine and the sea connections of either Amsterdam or Antwerp. Second, and more fundamental, is the increasing industrialization of areas near the Rhine (particularly the Ruhr district), for which Rotterdam has long been the main sea outlet. A large addition to Rotterdam's facilities known as Europoort lies along the New Waterway at the North Sea entrance to the Rotterdam port area and is specially designed to handle supertankers and other oversized container carriers.

Antwerp (population: 1.1 million), located about 50 miles (80 km) up the Scheldt River, is primarily a port for Belgium itself but also accounts for an important share of the Rhine transit trade. Belgium's coast is straight and its rivers shallow, so that the deep estuary of the Scheldt gives Antwerp the best harbor in the country, even though it must be reached through the Netherlands. Transit trade is facilitated by the fact that Antwerp lies slightly closer to the Ruhr than does either Rotterdam or Amsterdam.

INDUSTRIAL PATTERNS IN BENELUX

All the Benelux countries are highly industrialized. Before major discoveries of natural gas were made in the northeastern Netherlands in the 1950s, Belgium and Luxembourg were better provided with mineral resources. This helped them become somewhat more industrialized than the Netherlands. Now, however, the Netherlands, with its advantageous trade position and natural gas, is the leading country within Benelux in total value of industrial output.

Heavy Industry in the Sambre-Meuse District and Luxembourg

One of Europe's major coal fields crosses Belgium in a narrow east-west belt about a hundred miles (161 km) long (see Fig. 4.2). It follows roughly the valleys of the Sambre and Meuse rivers and extends into France on the west and Germany on the east. Liège (population: 750,000) and smaller industrial cities along this Sambre-Meuse field account for most of Belgium's metallurgical, chemical, and other heavy industrial production. A sizable iron and steel industry developed here well before the Industrial Revolution, using local iron ore and charcoal. Subsequently, local coal and the adoption of British techniques made Liège the first city in continental Europe to develop modern large-scale iron and steel manufacture. The Sambre-Meuse district also specializes in smelting imported nonferrous ores, with much of the metal being exported, though it now suffers from the customary problems of old, coal-based heavy industrial districts. Coal production has declined drastically in the face of competition from imported oil and even coal imported from better and less depleted fields in America and elsewhere. International competition has also adversely affected the iron and steel industry. Newer industrial plants have generally located in northern Belgium, where unions are weaker and wages lower. Belgium's capital and largest city, Brussels (population: 2,390,000 metro), is the leading center of a breed of new, less energy-dependent cities, where industry is able to utilize less skilled, nonunion employees.

Luxembourg's most important exports come from the iron and steel industry, carried on in several small centers near the southern border where the Lorraine iron ore deposits of France overlap into Luxembourg. Production is on a large scale, and the very small home market both necessitates and permits export of most of the metal. Luxembourg is attempting, with some success, to follow the European pattern in diversifying its manufacturing industries and exports.

INDUSTRIES IN THE NETHERLANDS

Until after World War II, the Netherlands had a smaller development of industry than might have been expected of a country in the heart of western Europe. It had no internal power resources except one small coal field (now closed) in the southernmost province called Limburg. The country's dense population provided cheap labor, and its location allowed cheap import of fuel and cheap export of products. So the Netherlands developed no heavy industrial region but had a rather scattered development of light industries such as textiles and electrical equipment. This situation changed after World War II as imported oil increasingly supplanted coal in the industries of western Europe. More oil imports enter Europe via Rotterdam-Europoort than any other port, with some arriving as crude oil and some refined further in the Rotterdam area before forwarding. This activity has generated one of the world's largest concentrations of refineries. Petrochemical plants near the refineries represent the main branch of the sizable Dutch chemical industry. Port locations and dependence on port functions also characterize many other important Dutch industries and activities, particularly the operation of a good-sized merchant fleet. Port activities reflect not only the country's present emphasis on trade but also a long maritime tradition. The Dutch played a prominent part in European discoveries and colonization overseas, and for a brief time in the 17th century were probably the world's greatest maritime power.

Certain other industries of consequence are in the major ports but also in smaller Dutch cities. Among them are the food processing, electrical machinery, textile, and apparel industries. Some food industries process imported foods for European distribution, but more of them handle products of the remarkably productive Dutch agriculture. Certain industries tend toward interior locations having relatively cheap, productive labor and good transport connections. Among the best known are the electrical and electronics industries.

INTENSIVE AGRICULTURE IN AN INDUSTRIALIZED SETTING: THE LOW COUNTRIES

Both the Netherlands and Belgium are very productive agriculturally, and are characterized particularly by extremely high yields per unit of land. In fact, yields are so high that the total output of farm products is surprisingly great for such small countries. Such productivity results mainly from the following circumstances:

1. *Fertile soils.* Parts of each country have very fertile soils, the largest such areas being the polders (diked and drained lands) in the Netherlands and the loess lands of Belgium.

2. *Farm labor.* Despite rapid decreases of farm population in recent times, there is still an intensive use of labor on the small farms that characterize these countries.

3. *Highly modified landscape.* There is an intensive application of capital in such forms as water control, agricultural machines, knowledge, and especially fertilizer. Both countries are export producers of nitrogen fertilizers made from oil, natural gas, or coal, and both countries have developed speciality crops and very productive greenhouse agriculture.

4. *Role of the EU.* Agriculture in the Low Countries has profited from Common Market tariff protection and subsidies for agricultural producers. These have stimulated spectacular increases in output, to the extent that overproduction of certain products such as milk and butter has become a more worrisome problem than food supply, even in such crowded countries as these.

5. *Trade networks.* Belgium and the Netherlands profit agriculturally from trading relationships that enable the two countries to specialize in farm commodities particularly suited to their conditions. The dense and affluent populations in and near the Low Countries provide a large market for such specialties. Requirements for other commodities such as grains are met in good part by imports.

Definitions and Insights

THE POLDER AND THE CULTURAL LANDSCAPE

One of the most definitive elements of the Low Countries landscape is the presence of land reclaimed from the North Atlantic. This pattern exists in other parts of the world, but the

FIGURE 4.A
Windmills are a common feature in the polder landscape of the Low Countries. This scene from Alkmaar, Hoorn, in the Netherlands shows the resultant fertile farmland from the poldering process. *Tony Stone Images*

FIGURE 4.B
Cheeses stacked in a marketplace at Alkmaar, Netherlands, symbolize the importance of dairy farming as one element in the rich economic diversity of the Benelux countries, Switzerland, and Austria. *Lee Foster*

leading example of this expansion of the cultural landscape is the **polder land** in the Netherlands. These are lands that have been surrounded by dikes and artificially drained (Fig. 4.A). The process of turning former swamps, lakes, and shallow seas into agricultural land has been going on for more than seven centuries. An individual polder, of which there are a great many of various sizes, is an area enclosed within dikes and kept dry by constant pumping into the drainage canals that surround it. About 50 percent of the Netherlands now consists of an intricate patchwork of polders and canals, and Belgium has a narrow strip of such lands behind its coastal sand dunes. The polders are the best agricultural lands in the Netherlands and production from them is the heart of Dutch farming. The drier polders are often used for crop farming and produce huge harvests. Most polders, however, are kept in grass, and dairy farming is the main type of agriculture (Fig. 4.B). These highly manipulated landscapes— perfect examples of a combination of tradition and modern engineering to expand a limited farmland resource—represent the conflict created by the modification of a natural landscape—the shallow seascape—into a more productive cultural landscape. In a major 1953 flood, more than 1800 people drowned, providing further evidence of nature's capacity to remind engineers of the potential danger in massive landscape transformation. Polders not only underlie much of the extremely productive agriculture of the Low Countries, but they are also central to tourist images of Holland: windmills, dykes, and gardenlike farmscapes. The entire scene, however, is always subject to threats from the sea at times of storm. Like all modified and artificial landscapes, polders require strict human maintenance to continue the unusually high levels of production that have been wrestled from nature.

Major Problems of the Benelux Nations

Population and Planning Problems

Countries as intensively developed as the Netherlands and Belgium, with high standards of living and high aspirations, must consider how development can be guided to provide an environment offering pleasant living conditions for their crowded populations. In the Netherlands, such planning problems are especially acute as this country has a population density higher than virtually any country of Europe. It also has a rapidly growing population. In this situation, the Dutch are trying to limit urban sprawl, provide more outdoor and waterfront recreational areas, and, in general, maintain a pleasant and varied environment. They are especially concerned with maintaining the open space that is semi-encircled by Amsterdam, Utrecht, Rotterdam, The Hague, and other cities. These urban areas are tending to coalesce into what is referred to as Randstad Holland—the "Ring City" of Holland.

Reclamation of New Land

Land reclamation is one response to population pressure in the Low Countries. In the past, the objective was new farm land. The past century and a half has seen much artificial improvement and expanded cultivation of infertile areas. One major reclamation work in the Netherlands was the formation of new polders from the former Zuider Zee. A massive 18-mile-long dike was completed across the entrance of the Zuider Zee in 1932, and a sizable area of sea bottom has now been reclaimed for agriculture. The remainder of the Zuider Zee has become a freshwater lake, the Ijsselmeer. However, current reclamation (centered near Amsterdam) is directed mainly toward sites for a great mix of land uses—city extensions, new cities, industries, and recreation.

Another massive and much-publicized project of the Netherlands, called the Delta Plan, is aimed primarily at flood control and only incidentally at land reclamation. Its purpose is to prevent a recurrence of the devastating flood of February 1953, when the sea broke through the dikes in the southwestern part of the country during a storm. The heart of the Delta Plan is the construction of dikes connecting the islands in the triple delta of the Rhine, Maas (Meuse), and Scheldt with each other and with the adjoining mainland. The New Waterway, giving Rotterdam access to the sea, and the Scheldt River outlet for Antwerp will remain unobstructed, but the other channels through the delta are being closed by huge dikes.

4.7   The Contrasting Development of Switzerland and Austria

The two small countries of Switzerland and Austria, located in the heart of Europe, have many environmental and cultural similarities but are remarkably different in their historical development. Switzerland represents perhaps the world's foremost example of the economic and political success of a small neutral nation, whereas Austria experienced great difficulties, although these have lessened markedly in recent years.

With respect to physical environment, Switzerland and Austria have much in common. More than half of each country is occupied by the high and rugged Alps (Fig. 4.21). North of the Alps, both countries include strips of the rolling morainal foreland of the mountains. The Swiss section of the Alpine foreland, often called the Swiss Plateau, extends between Lake Geneva on the French border and Lake Constance (German: *Bodensee*) on the German border. The Austrian section of the foreland, slightly lower in elevation, lies between the Alps and the Danube River from Salzburg on the west to Vienna on the east. The Swiss and Austrian sections are separated from each other by a third portion of the foreland in southern Germany. On their northern edges, both Switzerland and Austria include mountains that are much lower and smaller in extent than the Alps. These are the Jura Mountains, on the border between Switzerland and France, and the Bohemian Hills of Austria, on the border with Czechoslovakia.

The Disparity Between Resources and Economic Success

Historical assessment of economic success has often suggested a close link between such success and a generous resource base. In a comparison of Austria and Switzerland, however, it can be seen that natural resources and raw materials play only a partial role in their distinct level of economic achievements. Based on natural resources alone, one might expect Austria to be the more successful country economically. It has more arable land than Switzerland and more per person, primarily as a result of Austria's greater proportion of non-mountainous terrain. It also has more forested land than Switzerland, and more per person. This results mainly from the fact that the Austrian Alps have somewhat lower elevations than the Swiss Alps and hence a smaller proportion of land is above the tree line. The same contrast holds true for minerals. Austria has no mineral resources of truly major size, but it does have a varied output of minerals that are valuable collectively. Switzerland, on the other hand, is almost devoid of significant mineral resources. Both countries depend heavily on abundant hydroelectric potential, but Austria's is greater.

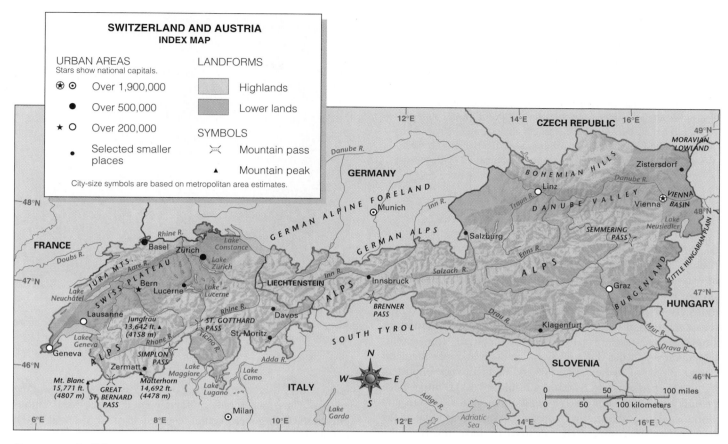

FIGURE 4.21
General location map of Switzerland and Austria.

In economic success, however, the advantage lies with Switzerland, which in 1995 had a per capita GNP of $22,080, while Austria, in contrast, had a per capita GNP of $17,500. But this must be seen in perspective, as Austria is quite well-off today, largely as a result of economic growth since World War II.

Historical developments that favored Switzerland far outweighed Austria's natural resource advantage. The primary reason seems to be the long period of peace enjoyed by Switzerland. Except for some minor internal disturbances in the 19th century, Switzerland has been at peace inside stable boundaries since 1815. The basic factors underlying this long period of peace seem to be (1) Switzerland's position as a buffer between larger powers, (2) the comparative defensibility of much of the country's terrain, (3) the relatively small value of Swiss economic production to an aggressive state, (4) the country's value as an intermediary between belligerents in wartime and a refuge for people and money, and (5) Switzerland's policy of strict and heavily armed neutrality.

During well over a century and a half of peace, the Swiss have had the opportunity to develop an economy finely adjusted to the country's potentials and opportunities. In particular, they have skillfully exploited the fields of banking, tourism, manufacturing, and agriculture. Favored by Swiss law, the country's banks enjoy a worldwide reputation for security, discretion, and service. The result has been a massive inflow of capital—including some of dubious origins—to Swiss banks. Switzerland's largest city, Zürich (population: 870,000 metro, 365,000 city), is one of the major centers of international finance, and other Swiss cities are also heavily involved.

Definitions and Insights

BUFFER STATE

In traditional political geography, the concept of the **buffer state** has been significant in diminishing the tension between adjacent, larger nations. It has also been useful in that nations without particularly rich natural resources are able to use their geographic location as a resource. In many parts of the world buffer states have unusual histories because of this role; they include Thailand in Southeast Asia, Jordan in the Middle East, and the colonial dependencies in Africa that separated white South Africa from black nationalist states in central Africa. The world's most dramatic example is Switzerland, however. Situated in the heart of western Europe, graced with few resources of either fossil fuels or metallic ores, this mountainous country has become a neutral zone for international political and financial activities for the world. Buffer states are often challenged by the strength and potential expansionist inclinations of their larger neighbors. Switzerland has been particularly successful in its ability to maintain its independence since early in the 19th century.

Switzerland has scenic resources in abundance, and the country has long been a major tourist destination. Development of winter sports makes snow an important resource, and Alpine ski resorts such as Zermatt, Davos, and St. Moritz are world famous. Less publicized is Switzerland's development of the manufacturing of delicate timepieces and high-technology instrumentation, based primarily on hydroelectricity from mountain streams and the skills of Swiss workers and management. Because most of its raw materials and fuels must be imported, industrial specialties in Switzerland minimize the importance of bulky materials and derive value from skilled design and workmanship. Reliance on hydroelectricity has facilitated the development of many small industrial centers, mostly in the Swiss Plateau. Five of the country's six largest cities are strung along the Plateau. Except for Zürich, these cities are under 500,000 in metropolitan population, although Geneva approaches that figure. The remaining city, Basel (population: 360,400), lies beyond the Jura at the point where the Rhine River turns north between Germany and France. Basel is the head of navigation for Rhine barges that connect Switzerland with ocean ports and other destinations via river and canal.

THE NATIONAL UNITY OF SWITZERLAND

The internal political organization of Switzerland expresses and makes allowance for ethnic diversity. The country was originally a loose alliance of small sovereign units known as cantons. When a stronger central authority became desirable in the 19th century, not only were the customary civil rights of a democracy guaranteed, but governmental autonomy was retained by the cantons except for limited functions specifically assigned to the central government. Although the functions allotted the central government have tended to increase with the passing of time, each of the local units (now 26 in number) has preserved a large measure of authority. Local autonomy is supplemented by the extremely democratic nature of the central government. In no country are the initiative and the referendum

FIGURE 4.22
The great achievement of the Swiss government in creating transportation routes through the rugged Alps is graphically illustrated by this view of bridges and a tunnel in precipitous terrain.   *Swiss National Tourist Office*

more widely used to submit important legislation directly to the people. Thus, guarantees of fundamental rights, local autonomy, and close governmental responsiveness to the will of the people are used to foster national unity despite the potential handicaps of ethnic diversity and local particularities.

The central government, in turn, has pushed economic development vigorously. Two outstanding accomplishments have been the construction and operation of Switzerland's railroads (Fig. 4.22) and the development of the hydroelectric-power generating system. Despite the difficult terrain, the Swiss government has built a railway network of great density, which carries enough international transit traffic to earn important revenues. The hydroelectric system, utilizing the many torrential streams of the mountains, yields about 55 percent of the country's electric output and places Switzerland very high among the world's nations in hydroelectricity produced per capita.

AUSTRIA: PROBLEMS AND READJUSTMENTS

Austria emerged in its present form as a remnant of the Austro-Hungarian Empire when that empire disintegrated in 1918 as a consequence of defeat in World War I. Austria's population is essentially German in language and culture, but a unification with Germany was forbidden by the victors. The economy was seriously disoriented by the loss of its empire, and the country limped through the interwar period until it was absorbed by Nazi Germany in 1938. In 1945, after defeat in World War II, Austria was reconstituted as a separate state but was divided into occupation zones administered, respectively, by the United States, the United Kingdom, France, and the Soviet Union. During 10 years of occupation, the Austrian Soviet Zone in the east (which included Vienna)—already badly damaged in the war—was subjected to extensive removal of industrial equipment for reparations. In 1955, the occupation was ended by agreement among the four powers and Austria became an independent, neutral state.

As the core area of an empire of 51 million people in southeastern Europe, Austria developed a diversified industrial economy prior to 1914. Its industries depended heavily on resources and markets provided by the empire. Austrian iron ore was smelted mainly with coal drawn from Bohemia and Moravia, now in the Czech Republic. Austria's textile industry specialized to a considerable extent on spinning, leaving much of the weaving to be done in Bohemia. Austrian industry was not outstandingly efficient, but it had the benefit of a large protected market in which one currency was in use. In turn, the producers of foodstuffs and raw materials had a market within the empire protected by external tariffs.

When the empire disintegrated at the end of World War I, the areas that had formed Austria's protected markets were incorporated into independent states. These states, motivated by a desire to develop industries of their own, began to erect tariff barriers, and other industrialized nations began to compete with Austria in their markets. The resulting decrease in Austria's ability to export made it more difficult for the country to secure the imports of food and raw materials that its unbalanced economy required. Such difficulties were further increased after World War II by the absorption of East Central Europe into the Communist sphere. The necessary reorientation of Austrian industries toward new markets, principally in western Europe, was not easy because Austria did not join the EU until 1995. As a consequence, Austrian exports of goods were consistently inadequate to pay for imports. However, the country has been able to compensate for this deficit by revenues from a growing tourist industry.

Austria made notable progress after World War I, and has made especially rapid progress since World War II, in building a successful economy suited to its status as a small independent state. Major elements in the economy include forest industries, tourism, some iron and steel production based on Austrian iron ore and imported ore and fuel, increased hydroelectric development, traditional manufacturing of clothing and crafts, and larger-scale engineering and textile industries. The latter industries contribute quite importantly to exports, but development is not sufficient to satisfy the internal market, and Austria is now a net importer of machinery and textiles.

Since the end of the Empire, Austrian agriculture has moved toward greater specialization in dairy and livestock farming. Such production is a logical use of the large upland areas of Austria that are best adapted to pasture. An enlarged acreage devoted to feed crops such as barley and corn accompanies the increased emphasis on animal products. But Austria is still far behind western Europe's leading countries in agricultural efficiency. Many small and poor farms still exist, and eight percent of the labor force is still in agriculture (as of 1996)—a high proportion by western European standards. Meanwhile, substantial imports of food and feed remain necessary.

A significant aspect of Austrian readjustment since World War I has been the lessened importance of the famous Austrian capital, Vienna (population: 1.9 million). Although Vienna is still by far the largest city in Austria, there has been a pronounced tendency for population and industry to shift away from the capital toward the smaller cities and mountain districts. From a position as the capital of a great empire, Vienna has regressed to the capital of a small country. Its present metropolitan population is less than the population of the city proper in 1918.

Vienna's importance, however, has persisted since Roman times and is based on more than purely Austrian circumstances. The city is located at the crossing of two of the European continent's major natural routes: the Danube Valley route through the highlands separating Germany from the Hungarian plains and southeastern Europe, and the route from Silesia and the North European Plain to the head of the Adriatic Sea. The latter follows the lowland of Moravia to the north and makes use of the passes of the eastern Alps, especially the Semmering Pass, to the south. Thus, Vienna has long been a major focus of transportation and trade and also a major strategic objective in time of war. In the geography of the late 1990s, Vienna finds itself particularly well-positioned for the potential expansion and development of the Eastern European nations and the western tiers of the states emerging from the Former Soviet Region.

QUESTIONS

1. Locate and define the geographic characteristics of the two major linear regions of highest population density that begin in The Low Countries and trend eastward.

2. What is Switzerland's annual per capita Gross Domestic Product (GDP) figure, and how does it rank worldwide?

3. What are the three most highly cultivated countries in western Europe?

4. What is the official name of the political unit in western Europe, of which London is the capital, and what are the political components that make up this unit?

5. How many of the world's slower growing countries—in terms of population—occur in Europe and what are two that have negative population growth rates?

5

EUROPE: THE NORTH, THE SOUTH, THE EAST

P resent-day Europe's northwestern coreland—the focus of Chapter 4—is ringed by three outer groups of countries: Northern Europe (sometimes called Norden), Southern Europe, and East Central Europe. Although these countries all possess distinctive traits and unique histories, today they tend to play a more minor role than do the coreland nations in influencing the other regions and nations of the world. We explore in this chapter the geographic personalities and economic and political dynamics of these nations outside the European center. Area and population data can be found in Table 3.1.

Sweden; more often it includes these countries plus Denmark; and sometimes it includes these three plus Iceland. When Finland is included in the group, the term Fennoscandia, or Fennoscandian countries, is used, thus acknowledging the cultural and historical differences between the Finns and the other proximate nations.

DOMINANT TRAITS OF NORTHERN EUROPE AS A REGION

"The North" is a good descriptive term for these lands. No other highly developed countries have their principal populated areas so near the North Pole. Located in the general latitude of Alaska, the countries of Northern Europe represent the northernmost concentration of advanced industrial societies in the world.

Westerly winds from the Atlantic, warmed in winter by the North Atlantic Drift (see page 63), moderate the climatic effects of this northern location in some areas. Temperatures average above freezing in winter over most of Denmark and along the coast of Norway. But, away from the direct influence of the west winds, winter temperatures average below freezing and are particularly severe at elevated, interior, and noncoastal northern locations. In summer, the ocean tends to be a cooling rather than a warming influence, and most of Northern Europe's Fahrenheit temperatures in July average no higher than the 50s

5.1 The Independent, Cohesive, and Prosperous Countries of Northern Europe

Denmark, Norway, Sweden, Finland, and Iceland are the countries of Northern Europe (Fig. 5.1). The peoples of these lands recognize their close relationships with each other and group their countries geographically under the regional term Norden, or "The North." A more common term to describe these countries is Scandinavia, or the Scandinavian countries. But this regional name is ambiguous: It occasionally refers to only the two countries that occupy the Scandinavian Peninsula, Norway and

◀ The Mediterranean world is ablaze with the colors of its flowers, fruit crops, and sun. This market scene in Portugal's Madeira Island shows the color and variety of local farm products, as well as the potential tourist interest in such a setting. *Hugh Sitton/Tony.Stone Images.*

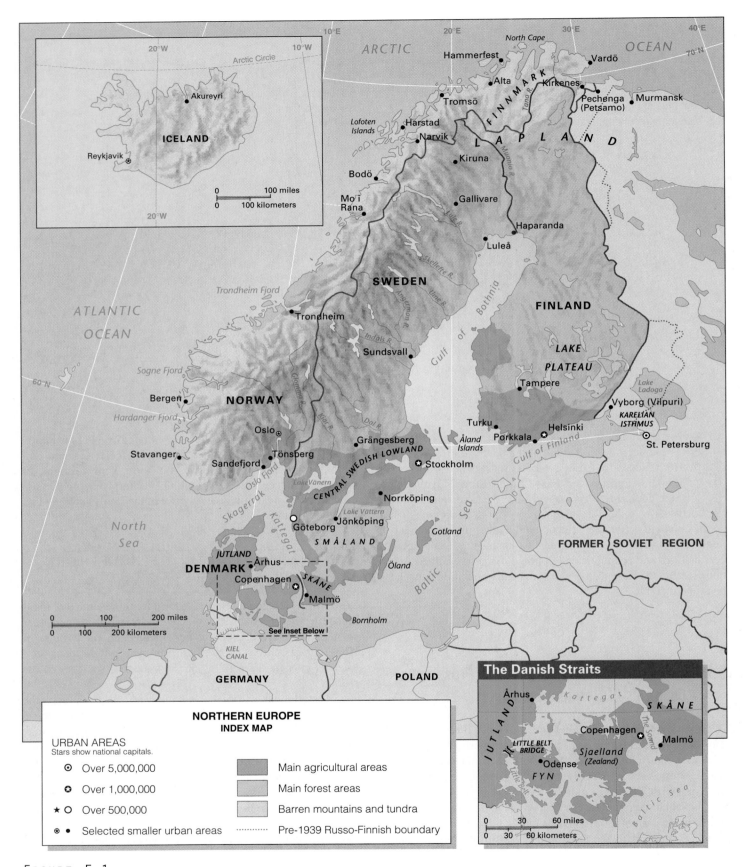

The following labels appear on the map:

Iceland inset:
Arctic Circle
Akureyri
ICELAND
Reykjavik
0 100 miles
0 100 kilometers

Main map:
ARCTIC OCEAN
North Cape
Hammerfest
Vardö
Alta
Kirkenes
Tromsö
FINNMARK
Pechenga (Petsamo)
Murmansk
Lofoten Islands
Harstad
Narvik
L A P L A N D
Kiruna
Bodö
Gallivare
Mo i Rana
Haparanda
Luleå
SWEDEN
FINLAND
ATLANTIC OCEAN
Trondheim Fjord
Trondheim
LAKE PLATEAU
Gulf of Bothnia
Sundsvall
Tampere
Lake Ladoga
Sogne Fjord
Bergen
NORWAY
Turku
Vyborg (Viipuri)
KARELIAN ISTHMUS
Hardanger Fjord
Oslo
Grängesberg
Porkkala
Helsinki
Stavanger
Sandefjord
Tönsberg
CENTRAL SWEDISH LOWLAND
Åland Islands
Gulf of Finland
St. Petersburg
Stockholm
Skagerrak
Lake Vänern
Norrköping
North Sea
Kattegat
Göteborg
Lake Vättern
Jönköping
SMÅLAND
Gotland
FORMER SOVIET REGION
JUTLAND
Århus
Öland
DENMARK
Copenhagen
SKÅNE
Baltic Sea
Malmö
Bornholm
See Inset Below
0 100 200 miles
0 100 200 kilometers
KIEL CANAL
GERMANY
POLAND

Legend:
NORTHERN EUROPE
INDEX MAP
URBAN AREAS
Stars show national capitals.
⊙ Over 5,000,000
✪ Over 1,000,000
★ O Over 500,000
⊗ • Selected smaller urban areas
Main agricultural areas
Main forest areas
Barren mountains and tundra
.... Pre-1939 Russo-Finnish boundary

The Danish Straits inset:
The Danish Straits
Århus
Kattegat
SKÅNE
JUTLAND
Copenhagen
The Sound
Malmö
LITTLE BELT BRIDGE
Odense
Sjaelland (Zealand)
FYN
Little Belt
Baltic Sea
0 30 60 miles
0 30 60 kilometers

FIGURE 5.1
Index map of Northern Europe. Small tracts of agricultural land are scattered through the areas of forest, mountain, and tundra, and smaller forest tracts occur within the agricultural areas.

or low 60s (10° to 18°C). Despite the overall moderation of the climate—compared with what might be expected from the latitude—the populations of the various countries tend to cluster in the southern sections, and all countries except Denmark have considerable areas of sparsely populated terrain where the problems of development are largely those of overcoming the rigors of a northern environment.

However, historical interconnections and cultural similarities are more important factors in the regional unity of Northern Europe than partial similarity of environmental conditions. Historically, each country has been more closely related to others of the group than to any outside power. In the past, warlike relations often prevailed, and for considerable periods some countries ruled others of the group. But since the early 19th century, relations among the Northern European peoples have been peaceful, and have come to be expressed in close international cooperation among the five countries.

Cultural similarities among the countries are many. The languages of Denmark, Norway, and Sweden descend from the same ancient tongue and are mutually intelligible. Icelandic, although a branch of the same root, is more difficult for the other peoples because it has changed less from the original Germanic language and has borrowed less from other languages. Only Finnish, which belongs to a different language family, is entirely distinct from the others. Even in Finland, however, about six percent of the population speaks Swedish as a native tongue, and Swedish is recognized as a second official language.

Among all these countries, cultural unity is also embodied in a common religion. The Evangelical Lutheran Church is the church of 88 to 93 percent of the respective populations, at least nominally. It is a state church, supported by taxes, and is probably the most all-embracing organization outside of the state. The countries of Northern Europe also exhibit basic similarities with respect to law and political institutions. They all have a long tradition of individual rights, broad political participation, limited governmental powers, and democratic control. Today, they are recognized as strongholds of democratic institutions.

Small size and resource limitations made it necessary for each of the countries of Northern Europe to build a highly specialized economy in pursuit of a high standard of living. Success in such endeavors has been so marked that these countries are probably known as much for high living standards as for any other characteristic. In all five, high standards of health, education, security for the individual, and creative achievement are characteristic.

AGRICULTURE'S KEY ROLE IN DENMARK

Denmark has the distinction of possessing the largest city in Northern Europe while, at the same time, being the most dependent on agriculture of any country in the region. The Danish capital of Copenhagen (population: 1,670,000) has one third of Denmark's population in its metropolitan area, allowing it—

like Paris—to play the role of a primate city. Denmark has a far greater density of population than the other countries of Northern Europe, a fact accounted for by the presence of Copenhagen, the greater productivity of the land, and the lack of any sparsely populated zone of frontier settlement.

Copenhagen and the Danish Straits

Copenhagen lies on the island of Sjaelland (Zealand) at the extreme eastern edge of Denmark (Fig. 5.1). Sweden lies only 12 miles (c. 20 km) away across The Sound, the main passage between the Baltic and the North seas. The city grew beside a natural harbor well placed to control traffic through The Sound. Before the 17th century, Denmark controlled adjacent southern Sweden, as well as the less favored alternative channels to The Sound—the Great Belt and the Little Belt. Tolls were levied on all shipping passing to and from the Baltic. Although the days of levying tolls are now long past, Copenhagen still does a large transit and entrepôt business in North Sea–Baltic Sea trade.

Trade led to industry, and Copenhagen is the principal industrial center of Denmark as well as its chief port and capital. Outstanding industries include the processing of both exported and imported foods, the manufacture of chemicals, brewing, and engineering.

Factors in Denmark's Agricultural Success

Denmark stands out sharply from the other countries of Northern Europe in the nature of its land and the place of agriculture in its economy. The western part of the peninsula of Jutland consists mainly of coastal dunes and sandy outwash plains deposited by glacial meltwater. Eastern Jutland and the Danish islands exhibit a rolling topography of glacial moraines. Although the sandy areas of the west are not very fertile, most have been reclaimed and are cultivated; the more fertile soils of the east support a very intensive and productive agriculture upon which the country's prosperity largely is based.

About three fifths of Denmark is cultivated—the largest proportion of any European country. It is fortunate that so much of the nation is arable, because the country is greatly lacking in other natural resources. Danish agriculture is so efficient that only five percent of the working population is employed on the land (Fig. 5.2). Agriculture, however, is basic to the country's economy. Many additional workers engage in processing and marketing the country's agricultural products and in supplying the needs of the farms. Approximately one third of Denmark's exports relate to agriculture.

Few countries or areas that depend so heavily on agriculture are as materially successful as Denmark. Danish agriculture is based on the highly specialized and carefully fostered development of animal husbandry, which was emphasized when the competition of cheap grain from overseas brought ruin to the previous Danish system of grain farming in the latter part of the 19th century. Until recently, Danish animal husbandry focused on dairy farming, but in recent years the pro-

FIGURE 5.2

Farmland in the North often comes down to the edges of the fjords that provide both transport for farm surpluses and a connection with the more densely settled world. This photo of Aurlandfjord, Norway, shows how little of the mountain flank landscape has been opened for farming. *Frank Clarkson/Gamma Liaison*

duction and sale of animals for meat has become dominant over dairying.

The following are some of the favorable factors contributing to the success of Danish agriculture and animal husbandry:

1. *Environmental setting.* Danish land and climate favor fodder crops such as barley, beets, and potatoes.
2. *Markets.* Nearby European markets are large, and the Danish membership in the European Union gives it open access to them.
3. *Prominence of agriculture.* The Danish state has long supported agriculture by encouraging family farms, financing agricultural research and education, and extending financial aid to private reclamation projects.
4. *Cooperatives.* Cooperative societies have developed to a point that practically all Danish farmers are members. The

cooperatives process and export the farmers' products, provide farm supplies at advantageous prices, and enforce rigid standards of quality control. They make the Danish farmer his own middleman to some extent and give him the benefit of large-scale marketing and buying.

5. *Education.* Probably underlying much of the success of Danish agriculture, and especially the success of the cooperative movement, is a very high level of education. Various forms of adult education supplement the traditional school system. This emphasis on the continuing education of adults is characteristic not only of Denmark but of all the countries of Northern Europe.

ECONOMIC EMPHASES IN NORWAY

Norway stretches for well over a thousand miles (*c.* 1600 km) along its west side and around the north end of the Scandinavian peninsula. The peninsula, as well as Finland, occupies part of the Fennoscandian Shield, a block of ancient and very stable metamorphic rocks. The western margins of the block in Norway are extremely rugged. Most areas have little or no soil to cover the rock surface, which was scraped bare by the continental glaciation that centered in Scandinavia. About 70 percent of Norway is classified as wasteland, 3 percent as arable or pasture land, and the remainder as forest.

The nature of the terrain hinders not only agriculture but also transportation. Glaciers deepened the valleys of streams flowing from Norway's mountains into the sea, and when the ice melted, the sea invaded the valleys. Thus were created the famous **fjords**—long, narrow, and deep extensions of the sea into the land, usually edged by steep valley walls (Fig. 5.3). It is difficult to build highways and railroads parallel to such a coast.

FIGURE 5.3

The village of Gudvagen, Norway, settled as it is on the edge of a characteristic Scandinavian fjord, has a small hinterland going upslope and along the edge of the drowned river. The real hinterland of the village is the fjord as it is used for trade, fishing, and general transportation. *© 1991, Marvin L. Dembinsky, Jr./Dembinsky Photo Associates*

Regional Perspective

NORWAY AND "NO" TO THE EUROPEAN UNION

In November, 1994, Norwegians voted, by a margin of 53 to 47 percent, *not* to join the European Union (EU). This negative vote came about because recent petroleum discoveries in Norway's sector of the North Sea have given it an economic stability and the capacity to decline the interference of the EU, as well as the benefits of membership. Sweden, during the same time period, voted *to* join the EU by the reverse percentages. In the European Economic Area (EEA) this means that only Iceland, Liechtenstein, and Norway are not members of the increasingly powerful EU.

In the case of Norway, the situation has dimensions that are unusual for Europe. Because Norway's economy is so strong and stable, there was a reluctance to allow the EU to interfere. For example, in 1995, Norway was second only to Saudi Arabia in

the export of petroleum worldwide. In economic growth, Norway experienced a rate of 4.5 percent in 1995, one of the highest in Europe. Norway's 5 percent unemployment rate is one of the lowest in Europe. And in the current effort to get the EU nations' financial houses in order in anticipation of the possibility of a common EU currency in 1999, Norway is one of the few nations that already meets the essential EU criteria—a balanced budget, modest national debt levels, currency stability, and low inflation—required for participation in the new monetary unit. Norway's status improved when its government moved into a time of budget surplus and it became a net lender in 1996.

The forces that led the battle against membership in the EU—at a time when even most of the Eastern European nations are petitioning for EU membership—were

coalitions of environmentalists, farmers, fishermen, and nationalists. Their call was to preserve the relative independence of Norway and not slip into the situation where uniquely Norwegian situations would be tampered with by bureaucrats in Brussels—the administrative "capital" of the EU. On the other hand, the people nervous about the "no" vote warned that Norway's petroleum income will surely decline in the future, and that the aging of the population will change the budget picture profoundly as ambitious promises for retirement and medical benefits become more costly. However, the farmers and fishermen wanted to retain their Norwegian state subsidies, which are even more generous than those offered by the EU. In the end, it was Norwegian suspicions of transferring political power to Brussels that resulted in the unusual "no" vote.

The fjords provide some of the finest harbors in the world, but most of them have practically no **hinterland** or productive rural landscape dependent upon the central places established on the fjords. Much of Norway's population is scattered along these waterways in many relatively isolated clusters, most of which are small and only moderately interconnected.

More than half of the population of Norway lives in the southeastern core region, which centers on the capital, Oslo (population: 725,000). Located at the head of Oslo Fjord, where several valleys converge, Oslo is the principal seaport and industrial, commercial, and cultural center of Norway as well as its largest city. In this region valleys are wider and the land is less rugged. The most extensive agricultural lands and the largest forests in the country are located here. Hydroelectricity powers sawmilling, pulp and paper production, metallurgy, electrochemical industries, and industries that produce various consumer goods for Norwegian consumption.

Apart from its scanty agricultural base, Norway has rich natural resources relative to the size of its population, which is small and increasing at a very slow rate. Key resources include spectacular scenery, water power, fish, forests, and a variety of minerals. The country's resource position was greatly enhanced by the discoveries and development, in the 1960s and 1970s, of

large oil and natural-gas deposits far offshore in the Norwegian sector of the North Sea floor. In just a few years, oil and gas exports grew from nothing to the country's largest by value, and Norway advanced from being reasonably well-off to being one of Europe's richest countries as measured by per capita GNP (see Table 3.1).

Meanwhile, older economic emphases remain important. Manufacturing is a large employer and provides exports of goods related to Norway's resources. And Norway has had a major fishing industry for centuries. An ancient seafaring tradition going back to the Vikings is expressed in one of the world's largest merchant fleets.

THE VARIED LANDSCAPES OF SWEDEN

Sweden is the largest in area and population (8.8 million) and the most diversified of the countries of Northern Europe. In the northwest, it shares the mountains of the Scandinavian peninsula with Norway; in the south, it has rolling, fertile farmlands like those of Denmark; in the central area of the great lakes, another block of good farmland occurs. To the north, between the mountains and the shores of the Baltic Sea and Gulf of Bothnia, Sweden consists mainly of ice-scoured, forested uplands. A smaller area of this type occurs south of Lake Vättern.

Economy and Resources

The high development of engineering and metallurgical industries is the feature that most distinguishes the Swedish economy from that of the other countries of Northern Europe. These industries have evolved out of a centuries-old tradition of mining and smelting. The iron ores of the Bergslagen region, just north of the Central Swedish Lowland, have been made into iron and steel since medieval times. Originally, the fuel used to smelt the ore was charcoal made from the surrounding forest. When coal and coke began to replace charcoal as fuel in Europe's iron industries, Sweden was handicapped by its scarcity of coal. But, in the age of electricity, the Swedes responded to this problem by a heavy reliance on electric furnaces using current from hydropower plants; more recently, this kind of smelting has been supplemented by a Swedish-invented oxygen process. This electric process is expensive but gives Swedes an extraordinary control of quality, and "Swedish steel" became a synonym for steel of the highest grade. The modern Swedish steel industry still centers in the Bergslagen region, but the output from that area is supplemented by production from some large conventional plants built on the Baltic and Bothnian coasts which use imported coal and coke. And although Sweden now produces large quantities of ordinary steel, the ultrafine steels and the products made from them still impart a special character to Swedish metallurgy.

Some iron ore continues to come from the Bergslagen district, but this source is overshadowed today by the output of high-grade ore, mainly for export, from Kiruna and Gällivare, mining settlements located in the northern wilderness of Sweden far north of the Arctic Circle. The ore exports move through Luleå on the Bothnian coast in summer, when the Baltic is free of ice, and through Norway's ice-free Atlantic port of Narvik the year round.

The Swedish steel industry's emphasis on skill and quality carries over to the finishing and fabricating industries that use the steel. Among Swedish specialties yielding a large volume and value of exports are automobiles (Volvo and Saab), industrial machinery (including robots), office machinery, telephone equipment, medical instruments and machinery, and electric transmission equipment. The country's reputation for skill in design and quality of product also extends to items such as ball bearings, cutlery, tools, surgical instruments, and the glassware and furniture manufactured in Småaland, the infertile and rather sparsely populated plateau south of the Central Lowland.

Except in the far north, where the forest cover is sparser and the trees are stunted, practically all of Sweden is naturally forested, mainly with coniferous softwoods. The wood-products industries based on these forests provide important exports. These industries include sawmilling, pulp milling, papermaking, the manufacture of wood chemicals and synthetic fabrics, and the production of fabricated articles such as plywood. Logging and wood industries are characteristic of most of Sweden (Fig. 5.4), but the main concentration is found in areas to the north of the Central Lowland. Here, even before the age of the truck, logs could be transported in winter by sled or in summer by floating them to mills on the numerous rivers. Most of the mills are located in industrial villages and towns that dot the coast of the Gulf of Bothnia at the mouths of rivers. Numerous hydroelectric stations supply power, and high-tension systems transmit electricity, to the central and southern parts of the country, where growing demands cannot be met by streams that are now almost completely developed for power.

Overall, Swedish manufacturing occurs principally in numerous centers in the Central Lowland. These places, ranging in population from about 125,000 to mere villages, have grown up in an area favorably situated with respect to minerals, forests, water power, labor, food supplies, and trading possibilities. The Central Lowland is the historic core of Sweden and has long maintained important agricultural development and a relatively dense population.

Unlike Denmark, Sweden has not developed a specialized export-oriented agriculture; and unlike Norway, Sweden commands enough good land to supply all but a minor share of its small population's food requirements. More than 90 percent of Sweden is nonagricultural, but there is some good land, largely in Skåane at the southern tip of the peninsula and in the Central Lowland. Skåane, with relatively fertile soils like those of adjacent Denmark, and with Sweden's mildest climate, is the prime agricultural area.

FIGURE 5.4

Logs in central Sweden. Most of Sweden's forest industries are located farther north in the zone of subarctic climate. *Jesse H. Wheeler, Jr.*

Stockholm and Göteborg

Stockholm (population: 1.5 million) is the second largest city in Northern Europe (Fig. 5.5). The location of the capital reflects the role of the Central Lowland as the center of the Swedish state and the early orientation of that state toward the Baltic and trans-Baltic lands. Stockholm is the principal administrative, financial, and cultural center of the country and shares in many of the manufacturing activities typical of the Central Lowland. But in the past century, as Sweden has traded more via the North Sea, Stockholm has been displaced by Göteborg (population: 700,000), located at the other end of the Central Lowland, as the leading port of the country, and, in fact, of all Northern Europe. In addition to its North Sea location, Göteborg has a harbor that is ice-free the year round, whereas icebreakers are needed to keep Stockholm's harbor open in winter.

The Swedish Policy of Neutrality and Preparedness.   In the Middle Ages and early modern times, Sweden was a powerful and imperialistic country. In the 17th century, the Baltic functioned almost like a Swedish lake. During the 18th century, however, the rising power of Russia and Prussia put an end to Swedish imperialism, aside from a brief campaign in 1814 through which Sweden won control of Norway from Denmark. Since 1814, Sweden has never engaged in a war, and it has become known as one of Europe's most successful neutrals. The long period of peace has undoubtedly been partially responsible for the country's success in attaining a high level of economic welfare and a reputation for social advancement. However, Sweden stays heavily armed while maintaining the nation's traditional policy of neutrality.

FINLAND AND ITS RUSSIAN CONNECTION

Conquered and Christianized by the Swedes in the 12th and 13th centuries, Finland was ceded by Sweden to Russia in 1809 and was controlled by the latter nation until 1917. Under both the Swedes and the Russians, the country's people developed their own culture and feelings of nationality. The opportunity for independence provided by the collapse of Tsarist Russia in 1917 was eagerly seized.

Most of Finland is a sparsely populated, glacially scoured, subarctic wilderness of coniferous forest, ancient igneous and metamorphic rocks, thousands upon thousands of lakes, and numerous swamps. Despite this environment, Finland was primarily an agricultural country until very recently, and the majority of the population resides in the relatively fertile and warmer lowland districts scattered through the southern half of the country. Hay, oats, and barley are the main crops of an agriculture in which livestock production, especially dairying, predominates.

To pay for a wide variety of imports, the nation depends heavily on exports of forest products. About two thirds of Fin-

FIGURE 5.5
Stockholm, Sweden's capital, has the low skyline typical of European cities. The oldest part of the city and the main harbor are in the center of the photograph.   *Peter Jordan/Gamma Liaison*

land is forested, mainly in pine or spruce, and about two fifths of the country's exports in 1996 consisted of wood products—paper, pulp, and timber. Forest production is especially concentrated in the south central part of the country, often referred to as the Lake Plateau. A poor and rocky soil discourages agriculture here, but a multitude of lakes, connected by streams, aid in transporting the logs.

Helsinki (population: 1 million), located on the coast of the Gulf of Finland, is Finland's capital, largest city, main seaport, and principal commercial and cultural center. It is also the country's most important and diversified industrial center. Hydroelectricity originally powered Finland's industries, but this source has now been surpassed by both thermal and nuclear power. The thermal stations use imported fossil fuels.

Problems of a Buffer State

For centuries, Finland has been a buffer between Russia, Scandinavia, and the former eastward reach of German power. In 1939, Finland refused to accede to Soviet demands for the cession of certain strategic frontier areas and was overwhelmed by the Former Soviet Region in the "Winter War" of 1939–1940. Then, in an attempt to regain what had been lost, Finland fought with Germany against the Soviet Union from 1941 to 1944 during World War II and again was defeated. Following the war, the economy of Finland was heavily burdened by (1) the necessity for rebuilding the northern third of the country, devastated by retreating German soldiers after Finland surrendered to the Soviets, (2) the necessity for resettling one tenth of the total population of the country after they fled as refugees from border areas ceded to the Soviets, (3) the loss of the ceded areas themselves, some portions of which were of considerable importance in the prewar economy, and (4) the necessity for making large reparations to the Soviet Union.

In spite of these difficulties, however, the Finns were able to make a rapid recovery. The economy drastically changed in some respects. One of the most striking changes was the rise of metalworking industries. This was made necessary when the Soviets required that a large part of the reparations be paid in metal goods. Several small steel plants and other metalworking establishments were built to meet this demand, and they continue to operate to the present day. The output includes a variety of machines, other secondary metal products, and basic metals. Metal goods now have about the same value as forest products in Finland's export trade.

ICELAND: AFFLUENCE SHAPED FROM BLEAK SURROUNDINGS

Iceland is a fairly large, mountainous island in the Atlantic Ocean just south of the Arctic Circle. Its rugged surface shows the effects of intense glaciation and vulcanism. Some upland glaciers and many active volcanoes and hot springs remain. The vegetation consists mostly of tundra, with considerable grass in some coastal areas and valleys. Trees are few, their growth discouraged by summer temperatures averaging about 52°F (11°C) or below as well as by the prevalence of strong winds. The cool summers are also a great handicap to agriculture. Mineral resources are almost nonexistent.

Despite the deficiencies of its environment, however, Iceland has been continuously inhabited at least since the ninth century and is now the home of a progressive and democratic republic with a population of about 265,000. Practically the entire population lives in coastal settlements (Fig. 5.6), with the

FIGURE 5.6
A residential section in Akureyri, Iceland's second largest city. Above the spic-and-span houses of this north-coast city towers one wall of the long fjord on which Akureyri is located.   *Len Kaufman*

largest concentration in and near the capital, Reykjavik, which has a metropolitan population of just over 100,000. Because of the proximity of the relatively warm North Atlantic Drift, the coasts of Iceland have winter temperatures that are mild for the latitude. Reykjavik has an average January temperature that is practically the same as New York City's.

The economy of Iceland depends basically on fishing, farming, and the processing of their products. Agriculture focuses on cattle and sheep raising and a limited production of potatoes and hardy vegetables. The real backbone of the economy is fishing, which supplied 76 percent of all exports by value in 1996. Manufacturing consists mostly of food processing, encouraged somewhat in recent times by the development of a small part of the island's considerable hydroelectric potential. Hydropower supplies 94 percent of Iceland's electricity.

Definitions and Insights

HYDROELECTRICITY

In the generation of electricity, there is no means more environmentally benign and geographically sensitive than water power. Water that is ponded—or captured in dams—to develop **water head**, or the volume of water necessary to turn turbines, possesses the potential for **multiple use**. Once the water moves through a complex of turbines and dynamos—generating power with virtually no waste product of any sort—it returns to a stream system that exists downstream from the dam. In small systems, power is generated by diverting flowing streams into small generating units and then returning water directly to the stream system. Thus, water is a **flow resource** or a **renewable resource**.

In hydroelectric complexes that require the construction of dams, there is the inevitable loss of some settlement land as rivers are stopped in their flow, ponded to depths of sometimes scores or hundreds of feet, and then released through major turbine and penstock networks. This system of power generation works best where there is little or only modest silt carried in the water being utilized; a heavier silt load in the reservoir behind the dam reduces the water head obtainable at that site and thus shortens the life of the hydroelectric project. In the larger projects built around massive river systems, major flood control components are commonly designed into the project. This has great potential benefit to downstream farming and urban populations.

For centuries, Iceland was a colony of Denmark. In 1918, it became an independent country under the same king as Denmark and in 1944 declared itself a republic. Iceland has strategic importance because of its position along major sea and air routes across the North Atlantic; the country is also a member of the North Atlantic Treaty Organization (NATO).

Greenland, the Faeroes, and Svalbard

Denmark and Norway possess outlying islands of some significance. Greenland, earlier a colony of Denmark, became a Danish province in 1953. It is the world's largest island and lies off the coast of North America. Greenland and the Faeroe Islands—between Norway and Iceland—are considered integral, although self-governing, parts of Denmark, and their peoples have the rights of Danish citizens. Although Greenland has nearly one fourth the area of the United States, about 85 percent is covered by an ice cap, and the population, mainly distributed along the southern coast, amounts to only about 57,000. Nearly all of this population is of mixed Eskimo and Scandinavian descent. The principal means of livelihood are fishing, hunting, trapping, sheep grazing, and the mining of zinc and lead. Under the auspices of NATO, joint Danish-American air bases are maintained there, and the United States has a giant radar installation at Thule in the remote northwest.

The Faeroes are a group of treeless islands where some 48,000 people of Norwegian descent make a living by fishing and grazing sheep. Norway controls the island group of Svalbard, located in the Arctic Ocean and commonly known as Spitsbergen, which is also the name of its main island. Although largely covered by ice, the main island has the only substantial deposits of high-grade coal in Northern Europe. Mining is carried on by Norwegian and Russian companies.

5.2   Mediterranean Patterns and Problems

On the south, Europe is separated from Africa by the Mediterranean Sea, into which three large peninsulas extend (Fig. 5.7). To the west, south of the Pyrenees Mountains, is the Iberian Peninsula, unequally divided between Spain and Portugal. In the center, south of the Alps, is Italy and its southern offshoot, the island of Sicily. To the east, between the Adriatic and Black

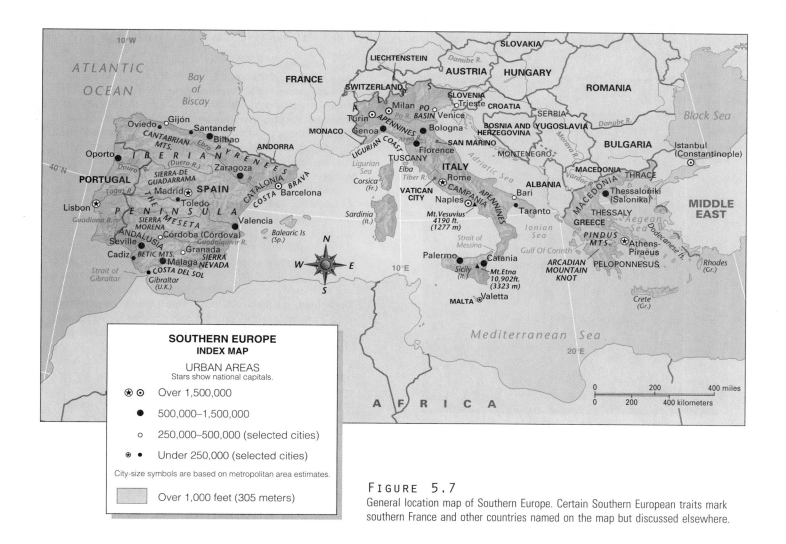

Figure 5.7

General location map of Southern Europe. Certain Southern European traits mark southern France and other countries named on the map but discussed elsewhere.

seas, is the Balkan Peninsula, from which the Greek subpeninsula extends still farther south between the Ionian and Aegean seas. The four main countries of Southern Europe, as considered in this text, are Portugal, Spain, Italy, and Greece. But Southern Europe also includes the small island country of Malta, the microstates of Vatican City (enclosed within Rome) and San Marino, and the British colony of Gibraltar. Three of the peninsular countries include islands in the Mediterranean, the largest of which are Italy's Sicily and Sardinia, the Greek island of Crete, and the Spanish Balearic Islands. Some Mediterranean islands lie outside these countries, the most notable being Corsica, which is a part of France, and Cyprus.

THE ECOLOGY OF MEDITERRANEAN AGRICULTURE

A distinctive natural characteristic of Southern Europe is its mediterranean (dry-summer subtropical) climate, which combines mild, rainy winters with hot, dry summers. Characteris-

FIGURE 5.8
A classic Mediterranean landscape in Tuscany, north central Italy, near Florence. The stone town of San Gimignano on the hilltop overlooks small farm homes and vegetation characteristics of the Mediterranean world. Note the closepacked rows of grapevines in the vineyard (deep green) to the immediate left of the stone houses in the center of the view. *Jay Fries/The Image Bank*

tics associated with this climate become increasingly pronounced toward the south. The northern extremities of both Italy and Spain have atypical climatic characteristics. Except for a strip along the Mediterranean Sea, northern Spain has a marine climate like that of northwestern Europe—cooler and wetter in summer than the typically Mediterranean areas—and the basin of the Po River in northern Italy is distinguished by a relatively wet summer and cold-month temperatures in the lowlands averaging just above freezing. Much of the high interior plateau of Spain, called the Meseta, is cut off from rain-bearing winds and has somewhat less precipitation and colder winters than is typical of the Mediterranean climate, although the seasonal regime of precipitation is characteristically Mediterranean.

Agriculture is based principally on crops that are naturally adapted to winter rainfall and summer drought (Fig. 5.8). Winter wheat is the most important crop. Barley, a more adaptable grain, tends to supplant wheat in some areas that are particularly dry or infertile.

Definitions and Insights

WINTER WHEAT

Although all agricultural patterns create a mosaic of crops and land-use patterns in **agricultural geography**, winter wheat deserves special note. This grain crop is generally planted at the end of the harvest of a summer crop. In September or as late as October, farmland is cleared of crops that were planted in spring, cultivated through the summer, and harvested before the first frost. Wheat is planted as fall temperatures drop, daylight shortens, and other farming activities generally diminish. Winter wheat is an unusually hardy grain, for it has to endure the snow cover and winter cold that gives it its name, and it tends to give a higher yield than does **spring wheat** which is planted in late spring and harvested in early fall. Tillers (grain heads) form on winter wheat, and when cold weather and snow come, the plant goes dormant. In the spring rains and/or snow melt, this wheat begins growing again, taking advantage of the moisture, the return of the longer hours of daylight, and the new warmth. Winter wheat is so attuned to this pattern that if this grain is planted in the spring and the seedlings do not go through a long cold period, the grain will not head.

There is a grand quality of timing to winter wheat. In the spring, before any other crops have grown long enough—or even been planted—to bring the farmer a harvest, winter wheat comes to fruition. Such timing produces a very popular and usable crop at the outset of the busy summer farming season. It is like getting an extra check on the twenty-fifth of the month when you are accustomed to being paid on the first of the month. Farmers in Europe and all around the world find the crop calendar of winter wheat to be significant in the overall farm economy, both because of yield and timing.

Other typical crops are olives, grapes, citrus, and vegetables. The olive tree and the grapevine have extensive root systems and certain other adaptations that allow them to survive the summer droughts. Olive oil is a major source of fat in the typical Mediterranean diet, and virtually all of the world supply is produced in countries that touch or lie near the Mediterranean Sea, or in a Mediterranean climate zone. The principal use of grapes is for wine, a standard household beverage in Southern Europe and a major export product. Where irrigation is lacking, vegetables are grown which will mature during the wetter winter or in the spring. The most important are several kinds of beans and peas. These are a source of protein in an area where meat animals make only a limited contribution to the food supply; feedstuffs are not available to fatten large numbers of animals, and parched summer pastures further inhibit the meat supply. Extensive areas too rough for cultivation are used for grazing, but the carrying capacity of such lands is relatively low. Sheep and goats, which can survive on a sparser pasturage than cattle, are favored. They are kept only partially for meat, and the total amount they supply is relatively small. In many places, grazing depends on transhumance—utilizing lowland pastures during the wetter winter and mountain pastures during the summer. In some areas, nonfood crops supplement the basic Mediterranean products. An example is tobacco, which is an export of Greece.

Mediterranean agriculture comes to its peak of intensity and productivity in areas where the land is irrigated. In such areas, relatively abundant and dependable moisture allows full exploitation of the subtropical temperatures. Fruits and vegetables, a large proportion of which are often destined for export, tend to supplement, and sometimes largely to displace, other types of products.

Although irrigation on a small scale is found in many parts of Southern Europe, a few irrigated areas stand out from the rest in size and importance. Among these are northern Portugal, the Mediterranean coast of Spain, the northern coast of Sicily, and the Italian coastal areas near Naples. The largest of all, the plain of the Po River in northern Italy, uses irrigation to supplement year-round rainfall. In northern Portugal, irrigated corn (maize) replaces wheat as the major grain, and some cattle are raised on irrigated meadows. Grapes are the other agricultural mainstay. In the Mediterranean coastal regions of Spain, irrigation makes possible the development of extensive orchards. Oranges are the most important product, with growth concentrated around the city of Valencia, which has given its name to a type of orange. Lemons are particularly important on the island of Sicily. The district around Naples, known as Campania, is more intensively farmed, productive, and densely populated than any other agricultural area of comparable size in Southern Europe. Irrigation water applied to exceptionally fertile soils formed of volcanic debris from Mount Vesuvius supports a remarkable variety of production that includes almost every crop grown in Southern Europe. Despite the intensive cultivation and productivity of this area, however, the overcrowded population on its tiny farms is notably poor.

Definitions and Insights

IRRIGATION

The act of **irrigation**—the watering of land by artificial methods—is one of the most significant aspects of the agricultural landscape. The Chinese talk about **"teaching water"** when they devise systems of irrigation. In Europe—and especially in Southern Europe in the mediterranean climate zone—**hydraulic control** is the factor that has taken farming from a marginal economic activity to a more prosperous one. Elements central to the irrigation process include **canals, dams, ponds, pumps, gravity flow, laterals, wells,** and **evaporation.** The process is thousands of years old and is at the center of being able to increase a region's arable land. In the Mediterranean world irrigation is particularly significant because of the **summer drought** climate pattern. There is good sunlight and often reasonably fertile soils in this climate zone, but there is seldom rainfall in the high sun season. When irrigation is introduced and sustained, little-used lands are turned into rich agricultural zones. There must be a network of dams, canals, ponds, and conduits that moves the water, and power for pumping if the system is not gravity flow. There must also be enough water available to flush the salts from the soil, because, in summer irrigation, there is the potential for excessive evaporation which can leave salts on the soil's surface. Only the addition of quantities of water in excess of crop needs can dissolve such salts and maintain a healthy crop environment. Wherever farmers have invested labor, capital, and ingenuity in designing effective irrigation systems, they have realized agricultural benefits, but with increased labor and capital requirements. Yet, all across the farming world, farmers make the choice to "teach water" to work for them.

RELIEF AND POPULATION DISTRIBUTION

In terrain and population distribution, as well as in climate and agriculture, the countries of Southern Europe present various points of similarity. Rugged terrain predominates in all four major countries. Plains tend to be small, to face the sea, and to be separated from each other by mountainous territory. Population distribution corresponds in a general way to topography, with the lowland plains being densely populated and the mountainous areas much less so, although a number of rough areas attain surprisingly high densities.

On the Iberian Peninsula, the greater part of the land consists of a plateau—the Meseta—with a surface lying between 2000 and 3000 feet (c. 600–900 m) in elevation. The plateau surface is interrupted by deep river valleys and ranges of mountains. Population density is generally restricted, mainly by lack of rainfall, to between 25 and 100 per square mile (c. 10–40 per sq km), compared to the European average of 290 per square

FIGURE 5.9
Rugged mountains in northern Italy where the Alps merge with the Apennines. Genoa, Italy's main seaport, is relatively near. The landscape exhibits a spectacular display of quarries producing the famous Carrara marble.   *Pierre Boulat/Woodfin Camp & Associates*

mile (*c.* 120 per sq km). The plateau edges are mostly steep and rugged. The Pyrenees and the Cantabrian mountains border the Meseta on the north, and the Betic Mountains, culminating in the Sierra Nevada, border it on the south. Most of the population of Iberia is distributed peripherally on discontinuous coastal lowlands.

In Italy, the Alps and the Apennines are the principal mountain ranges (Fig. 5.9). Northern Italy includes the greater part of the southern slopes of the Alps. The Apennines, lower but often rugged, form the north-south trending backbone of the peninsula, extending from their junction with the southwestern end of the Alps to the toe of the Italian boot, appearing again across the Strait of Messina in Sicily, where Mount Etna, a volcanic cone, reaches 10,902 feet (3323 m). Near Naples, Mount Vesuvius, at 4190 feet (1277 m), is one of the world's most famous volcanoes. West of the Apennines, most of the land between Florence on the north and Naples on the south is occupied by a tangled mass of lower hills and mountains, often of volcanic origin. Both Sicily and Sardinia, the two largest Italian islands, are predominantly mountainous or hilly.

Parts of the Italian highlands have population densities of more than 200 per square mile (*c.* 75 per sq km), but even these areas are much less densely populated than most Italian lowlands. The largest lowland, the Po Plain, has about two fifths of the entire Italian population, with nonmetropolitan densities frequently reaching 500–700 per square mile (*c.* 190–310 per sq km). Italy overall has a population density of nearly 500 per square mile, more than twice the density of any other large Mediterranean nation. Some other lowland areas with extremely high densities include the narrow Ligurian Coast centering on Genoa, the plain of the Arno River inland to Pisa and Florence, the lower valley of the Tiber River including Rome, and Campania around Naples.

In Greece, most of the peninsula north of the Gulf of Corinth is occupied by the Pindus Mountains and the ranges that branch from it. Extensions of these ranges form islands in the Ionian and Aegean seas. Greece south of the Gulf of Corinth, commonly known as the Peloponnesus, is composed mainly of the Arcadian mountain knot. Along the coasts of Greece, many small lowlands face the sea between mountain spurs and contain the majority of the people. A particularly famous lowland, although far from the largest, is the Attica Plain, still dominated as in ancient times by Athens and its seaport, Piraeus.

HISTORICAL CONTRASTS IN WEALTH AND POWER

Each of the four main countries of Southern Europe has played a major role at some period or periods in Western or world history. Centuries before the Christian era, Greek artists, architects, authors, philosophers, and scholars produced many of the ideas and works that laid the foundation for Western civilization. Between about 600 and 300 B.C., the sailors, traders, warriors, and colonists of the Greek city-states spread their culture throughout the Mediterranean area. A second period of Greek power and influence occurred in the Middle Ages, when Constantinople (modern Istanbul, Turkey) was the capital of the large Byzantine Empire. This Greek empire developed and diffused the Eastern Orthodox branch of Christianity.

Italy's main period of preeminence occupied approximately five centuries, during which the Roman Empire ruled over the whole Mediterranean basin and some lands beyond. Within the Empire, Christianity began and spread, and by the time the Empire collapsed, in the fifth century A.D., Rome had become the seat of the Popes and the Catholic Church. During the later Middle Ages, a number of Italian cities—notably Venice, Genoa, Milan, Bologna, and Florence—became powerful independent states. Venice and Genoa controlled maritime empires within the Mediterranean, and all the cities profited as traders between the eastern Mediterranean and Europe north of the Alps. After unification in the 19th century, Italy attempted to emulate past glories, but the empire that it acquired in northern Africa—the eastern Mediterranean and Ethiopia—was lost when Italy was defeated in World War II.

This excerpt from Louis L'Amour's *The Walking Drum* is presented here to give you a sense of how big a role trade and markets played in European life almost a thousand years ago. Although set in France, the images and the lore captured in this scene spill over into the rest of Mediterranean Europe and other regions as well. As you read these paragraphs, think about the pathways that had to be traveled to bring all of these goods and people together in this market setting. Think about the role of change agent played by the vendors and the soldiers who traveled with the merchant caravans from the distant source points for many of these goods

> The fair at Provins was one of the largest in France during the twelfth century. There was a fair in May, but the most important was that in September. Now the unseasonably cold, wet weather had disappeared, and the days were warm and sunny.
>
> Long sheds without walls covered the display of goods. Silks, woolens, armor, weapons, leather goods, hides, pottery, furs, and every conceivable object or style of goods could be found there.
>
> Around the outer edge of the market where the great merchants had their displays were the peasants, each with some small thing for sale. Grain, hides, vegetables, fruit, goats, pigs, and chickens, as well as handicrafts of various kinds.
>
> Always there was entertainment, for the fairs attracted magicians, troupes of acrobats, fire-eaters, sword-swallowers, jugglers, and mountebanks of every kind and description.
>
> The merchants usually bought and sold by the gross; hence, they were called *grossers*, a word that eventually came to be spelled *grocer.* Dealing in smaller amounts allowed too little chance for profit, and too great a quantity risked being left with odds and ends of merchandise. The White Company had come from Spain with silk and added woolens from Flanders. Our preferred trade was for lace, easy to transport and valued wherever we might go.
>
> Merchants were looked upon with disdain by the nobles, but they were jealous of the increasing wealth and power of such men [as the merchants in this chapter], or a dozen of others among us.
>
> The wealth of nobles came from loot or ransoms gained in war or the sale of produce from land worked by serfs, and there were times when this amounted to very little. The merchants, however, nearly always found a market for their goods.
>
> At the Provins fair there were all manner of men and costumes: Franks, Goths, Saxons, Englanders, Normans, Lombards, Moors, Armenians, Jews, and Greeks. Although this trade was less than a century old, changes were coming into being. Some merchants were finding it profitable to settle down in a desirable location and import their goods from the nearest seaport or buy from the caravans.
>
> Artisans had for some time been moving away from the castles and settling in towns to sell their goods to whoever passed. Cobblers, weavers, coppers, and armorers had begun to set up shops rather than doing piecework on order. The merchant-adventurers were merely distributors of such goods. . . .

The Church looked upon the merchants with disfavor, for trade was considered a form of usury, and every form of speculation considered a sin. Moreover, they were suspicious of the far-traveling merchants as purveyors of freethinking.

Change was in the air, but to the merchant to whom change was usual, any kind of permanence seemed unlikely. The doubts and superstitions of the peasants and nobles seemed childish to these men who had wandered far and seen much, exposed to many ideas and ways of living. Yet often the merchant who found a good market kept the information for his own use, bewailing his experience and telling of the dangers en route, anything to keep others from finding his market or his sources of cheap raw material.[1]

The main period of Spanish power and influence began in 1492, when a centuries-long struggle—the Reconquista—ended with Spain's defeat of the Kingdom of Granada, the last part of Iberia held by the Muslim Moors (Fig. 5.10). In that

FIGURE 5.10

A doorway to La Mezquita, or the Great Mosque of Cordoba, Spain. This Moorish temple was built in the 8th century and turned into a Christian cathedral in 1236. It exists today in southern Spain as one of the most visually stunning examples of the blending of Moorish and Christian sacred space.   *Kit Salter*

[1] L'Amour, Louis. *The Walking Drum.* Toronto: Bantam Books, 1984. Pp. 256–258.

same year, Columbus, a Genoese Italian navigator in the service of Spain, landed in the Americas in search of a sea route to India and East Asia. For a century thereafter, Spain was the greatest power in Europe and probably in the world. It rapidly built an empire in areas as widespread as Italy, the Low Countries, Latin America, Africa, and the Philippines. These possessions were then lost at various times during a long, slow decline of Spanish power over more than three centuries. The Canary Islands, off the Atlantic coast of Morocco, are still Spanish-held and governed as an integral part of Spain.

Portugal also played a part in expelling the Moors from Iberia and in the 15th century took the lead in seeking a sea route around Africa to the Orient. The first Portuguese expedition to succeed in the voyage, headed by Vasco da Gama, returned from India in 1499. For the better part of a century thereafter, Portugal dominated European trade with the East and built an empire there and across the Atlantic in Brazil. Then there was a rapid decline in Portuguese fortunes, although they controlled Brazil until the 19th century, when it gained independence, and they also held large African territories until the 1970s.

GEOGRAPHIC FACTORS INFLUENCING SOUTHERN EUROPE'S UNEVEN DEVELOPMENT

Today, Southern Europe is, as it has long been, one of the relatively poor parts of Europe, along with East Central Europe (see Table 3.1). The main factor contributing to Southern European poverty seems to be a lagging development of industry. Lack of industrial employment has left too many people in crowded small-farm areas where output per person (although not always per acre) is low. Lagging industrial development, in turn, seems to have been related to unfavorable social and resource conditions during the 19th and early 20th centuries, when many European countries were industrializing rapidly. Among the inimical social conditions in Southern Europe were

1. *Trade deficiency.* The absence of large-scale trade and therefore of capital accumulations to invest in manufacturing created a steady lack of development capital.
2. *Low educational levels.* The preponderance of poor and uneducated populations offered little in the way of skilled labor or dynamic markets, and in combination with the trade deficiency, worked to slow economic growth.
3. *Land tenure patterns.* The control of land and often government by wealthy landowners whose interests lay in maintaining the agrarian societies that they dominated meant that there was little drive for a significant change in the way that society and the economy were organized.
4. *Ineffectual government assistance.* The prevalence of unstable, undemocratic, and ineffective governments that were usually little oriented toward economic development served as a continuing impediment to real economic growth.

5. *Resource deficiencies.* Resource deficiencies included a lack of good coal and iron ore deposits, except in northern Spain; deficient wood supplies due to dryness and centuries of deforestation; and lack of reliable water power in the areas of mediterranean climate, where streams dry up in summer.
6. *Transportation difficulties.* The rugged topography of much of the region made construction of adequate rail and road systems difficult and expensive.

A variety of these factors worked in different combinations as major impediments to economic development in the region. Only three sizable industrial areas developed in Southern Europe before World War II. The most important was in the Po Plain of northern Italy. Here, imported coal and raw materials, hydroelectric power from the surrounding mountains, and cheap labor formed the basis for an area specializing in textiles but including some heavy industry, automobile production, an aircraft industry, and engineering. A second area very similar in its foundations and specialties, but developed on a much smaller scale, came into existence in the northeastern corner of Spain, in and around Barcelona (population: 4.1 million). A third area, focused on heavy industry, developed in the coal- and iron-mining section of northern coastal Spain. All three areas lie near the foot of mountains with humid conditions that provide the basis for hydroelectric power development.

ECONOMIC PROGRESS SINCE WORLD WAR II

In the later 20th century, the economies of Southern Europe began to show a new dynamism, shaped by both industrial modernization and an expanding role of the European Union. Industrial and service activities led the way, and there was a sharp decline in the proportion of the population still employed in agriculture. A number of factors played a part in this revitalization:

1. *Transportation changes.* Improved transportation, particularly in ocean transport, made the importation of fuels and materials for manufacturing more economical.
2. *Energy shift.* Coal ceased to be the dominant source of power in Southern Europe, replaced by oil or, in Italy, by natural gas and oil. The countries in Southern Europe became major importers of oil, as did their industrial competitors. Thus, Southern Europe moved to a more even footing in securing power than when it had to depend on large imports of coal. In Italy, the power situation also improved with the discovery of large natural gas deposits, principally under the Po Plain, as well as development of an active geothermal program.
3. *Market expansion.* Market conditions for Southern European exports improved greatly in the decades after World War II, largely as a result of the rising prosperity of northwestern Europe and the expansion of European markets. These exports included goods such as subtropical fruits and

Figure 5.11
The small island of Mykonos in the Aegean islands of Greece is one of many famous tourist sites in the Mediterranean world. The light-colored architecture, the fishing boats, and the deforested slopes are characteristically Mediterranean. Vegetation on the slope in the background has turned brown in the midsummer drought. Both the island and its only town are named Mykonos.   *Toshi Chatelin*

vegetables, Italian-built automobiles, and, less obviously, labor. Streams of temporary migrants from Southern Europe found jobs in booming, labor-short industries to the north. Absence of the migrants relieved unemployment, and their remittances added purchasing power in their home areas.

4. *Capital infusion.* There were other infusions of foreign money into the Southern European countries. Each country received American economic and military aid, and favorable economic circumstances and government policies attracted a considerable amount of foreign investment in industrial and other facilities. Southern Europe also saw a tremendous boom in tourism. To increasingly prosperous Europeans and Americans, Southern Europe offers a combination of spectacular scenery (Fig. 5.11), historical depth exceeding that of even the other parts of Europe, a sunny subtropical climate, the sea, and the beaches. The ancient cities, now modern and bustling with activity around their monuments of the past, supplemented attractions in new or greatly expanded resort areas such as Spain's Costa Brava and Costa del Sol, the Balearic Islands, the Italian Riviera, and various other places.

THE PO BASIN: INDUSTRIAL SHOWCASE OF SOUTHERN EUROPE

The Po Basin is the economic heart of modern Italy and the most highly developed and productive part of Southern Europe. Comprising just under a fifth of Italy's area, it contains about two fifths of Italy's population (who are more prosperous by far

than the rest) and accounts for approximately one half of its agricultural production and two thirds of its industrial output.

A very rapid expansion and diversification of Po Basin industry, previously dominated by textiles, occurred after World War II. Prominent components of this expansion were (1) the growth of engineering and chemical industries, (2) the development of a sizable iron-and-steel industry, mainly in northern Italy but partly in peninsular Italy, and (3) a shift to domestic natural gas and imported oil as the predominant sources of power—with hydroelectric energy continuing to play an important role.

Many cities in the basin share in the region's industry, but the leading ones are Milan (population: 3.8 million), Turin (population: 1.6 million), and Genoa (population: 805,000). Milan is Italy's industrial capital, largest city, and main center of finance, business administration, and railway transportation. It is advantageously located between the port of Genoa and important passes, now shortened by railway tunnels, through the Alps to Switzerland and the North Sea countries. Turin is the leading center of Italy's automobile industry. It is particularly associated with FIAT SpA, which is by far the largest Italian auto manufacturer. The port of Genoa functions as a part of the Po Basin industrial complex, although it is not actually in the basin. It is reached from the Milan-Turin area by passes across the narrow but rugged northwestern end of the Apennines. The city dominates the overseas trade of the Po area and is Italy's leading seaport. Genoa is now much larger and economically more important than its old medieval commercial and political rival, Venice (population: 425,000), whose location on islands at the eastern edge of the Po Plain is relatively unfavorable for serving present-day industrial centers.

MAJOR MEDITERRANEAN URBAN CENTERS OUTSIDE THE PO BASIN

The basically nonindustrial character of the leading cities of Southern Europe indicates the generally low level of industrial development in this region, aside from northern Italy and the two smaller industrial areas of northeastern and northern Spain. The cities are mostly political capitals or ports serving especially productive agricultural areas. An exception, however, is Athens, the largest city in Greece by an overwhelming margin. With a metropolitan population of 3 million (including the port of Piraeus), amounting to three tenths of the population of Greece and growing rapidly, Athens has no national rival (see page 131). It is the capital, main port, and also the main industrial center. This development is based on imported fuels, and materials are used to manufacture products that are sold almost entirely in a domestic market whose main center is Athens itself.

In Italy, the largest cities outside of the northern industrial area are the capital, Rome (population: 3.2 million), and Naples (population: 2.9 million). Rome (Fig. 5.12) has a location that is roughly central within Italy. Rome is predominantly a govern-

FIGURE 5.12
Panorama of Rome. The tree-lined Tiber River winds through the view. The low skyline and uniform architecture are characteristic of older European cities. The gleaming white structure in the left foreground is the Victor Emmanuel Monument, celebrating the 19th-century unification of Italy. The ancient Roman Colosseum is at the upper right. *Guido Alberto Rossi/The Image Bank*

mental, religious, and tourist center. Naples, located farther south, on the west coast of the peninsula, is the port for the populous and productive, but very poor, Campanian agricultural region described earlier. It is also the main urban center of one of Italy's major tourist regions, with attractions such as Mt. Vesuvius, the ruins of Pompeii, and the island of Capri in the vicinity.

In Spain, the three largest cities are the capital Madrid (population: 4.7 million) and the two Mediterranean ports of Barcelona (population: 4.1 million) and Valencia (population: 1.3 million). Madrid was deliberately chosen as the capital of Spain in the 16th century because of its location near the geographic center of the Iberian Peninsula, approximately equidistant from the various peripheral areas of dense population and of sometimes separatist political tendencies. Located in a poor region, Madrid has had little economic excuse for existence during most of its history. But its position as the capital has made it the center of the Spanish road and rail networks and thus has given it nodal business advantages. In Spain's recent economic boom, the city has begun to attract industry as well as tourists on a considerable scale.

Barcelona is Spain's major port and the center of a diversified industrial district. The main elements of Barcelona's industrialization have been (1) the early formation of a commercial class that supplied financing, (2) the availability of hydroelectric power, mainly from the Pyrenees, (3) cheap and sufficiently skilled labor, and (4) imported raw materials.

Valencia, Spain's third city in population (1.3 million), is the business center and port for an unusually productive section of coastal Spain. An extensive irrigation system in the area was developed by the Moors in the Middle Ages. Today, the density of population on irrigated land resembles that of Italy's Campania, but here, much of the agricultural effort goes into producing oranges, which are Spain's largest single agricultural export.

The two large cities of Portugal are both seaports on the lower courses of rivers that cross the Meseta and reach the Atlantic through Portugal. Lisbon (population: 2.3 million), on a magnificent natural harbor at the mouth of the Tagus River, is both the leading seaport of the country and the capital. The smaller city of Oporto (population: 1.6 million), at the mouth of the Douro River, is the regional capital of northern Portugal and the commercial center for trade in Portugal's famous port wine, which comes from terraced vineyards along the hills overlooking the Douro Valley.

ECONOMIC AND DEMOGRAPHIC TRENDS: THE SHIFT FROM AGRICULTURE TO INDUSTRY

Since World War II, Italy, Spain, Portugal, and Greece have all evolved from agricultural to industrial and service-oriented countries. Agricultural exports are still important, but manufactured exports are more valuable. Large tourist industries "marketing" sun, scenery, historic sites, and beaches supplement exports with an increasing economic significance.

As one goes farther south in Southern Europe, the countryside becomes progressively poorer in both environment and economy. Shortly after World War II, the Italian government

Regional Perspective

THE SHIFT FROM AGRICULTURE TO INDUSTRY

One of the most significant **demographic shifts** to take place anywhere in the world is the migration of rural youth to the city. In the excerpt from *The Walking Drum* that appears earlier in this chapter, you can see images of the sorts of city attractions and opportunities available to youth living in the countryside nearly one thousand years ago. When the market comes to town there are powerful changes that wash across the market square in the kinds of people, the sorts of goods, the entertainment, and the foods that appear from distant places. Such influences can work powerfully on the minds of underemployed youth. Because traditional inheritance laws tend to give land to first-born sons, second and third sons have little likelihood of ever gaining their own land.

To such youth, the idea of moving to the town or a city becomes very attractive. In that shift, there is the need not only to replace the labor function played by the **migrant** who leaves the farm, but also to fulfill the food, shelter, and social needs of the migrant once the move to the town or city is accomplished. In the **push-pull forces of migration** (the influences that drive immigrants away from the in-home places and attract them toward a new place, often a city), it is easy to see the sorts of images and realities that are at work. It must also be realized, however, that the pressure of such a demographic shift is a major factor in the capacity of an urban center to be productive and effective.

In the excerpt from Dickens' *Hard Times* that appeared in Chapter 4, just as in this chapter's images of the periodic market (a market that comes to a village only several times a month or week, and is not a permanent feature of the settlement), it can be seen that new opportunities are generally mixed blessings. There is an underside to the forces that pull these youth to town, but such a population shift plays a major role in the development of industrial resource bases as well as commercial settings. The conceptions of opportunity that pull these people off the farms are widely different, but the impact on the urban world is always powerful and costly—no matter the region or nation in which such a demographic shift occurs. These demographic shifts are the beginnings of the process of urbanization that so characterizes this century.

---

began efforts to develop the southern part of its peninsula. Agricultural development, land reform, and the introduction of subsidized industries were among the methods used to stimulate development. But the differential between north and south remains and is widely perceived in Italy as a major national problem. This was accentuated by the 1996 efforts of the Po Valley region to draw attention to the possibility of secession and the creation of a new country, to be called "Padania," made up of the northern third of the Italian peninsula. This region is home to most of Italy's heavy industry and considerable agricultural and resource wealth. Given the history of political change in Europe, even seemingly facetious suggestions like this have to be given consideration.

5.3   Legacies of Empire and Current Ambitions in East Central Europe

As defined in this text, East Central Europe consists of countries that are attempting to recover from more than four decades of Communist rule. They are Poland, the Czech Republic, Slovakia, Hungary, Romania, Bulgaria, the successor states of the former Yugoslavia, and Albania, occupying eastern Europe between the Baltic, Adriatic, and Black seas (Fig. 5.13). Prior to German reunification in 1990, an adjacent country, the former East Germany, also was Communist-controlled, and it will be brought into the present discussion where appropriate. Many of these countries (including East Germany) became parts of the former Soviet empire during World War II, when Soviet troops entered them in pursuit of retreating German armies in 1944 and 1945. Exceptions were Albania and the former Yugoslavia, where national Communist resistance forces took power on their own as German and Italian power collapsed. In the others, Soviet occupation forces supported local Communists, and all had Communist "satellite" governments subservient to the Soviets by 1948.

Following their takeover by Communism, the nations of East Central Europe (and East Germany) were reshaped along Communist lines, with one-party dictatorial governments; planning and direction of national economies by organs of the state; abolition of private ownership (with some exceptions) in the fields of manufacturing, mining, transportation, commerce, and services; abolition of independent trade unions; and varying degrees of socialization of agriculture. Attempts by East Germany in 1953, Hungary in 1956, and Czechoslovakia in 1968 to break away from Soviet control were crushed by the use of military force.

EAST CENTRAL EUROPE
INDEX MAP

URBAN AREAS
Stars show national capitals

⊛ ⊙   Over 3,500,000

✪ ●   1,000,000–3,500,000

★ ○   500,000–1,000,000

☆ ○   300,000–500,000

⊛ •   Selected smaller places

The independent successor nations of former Yugoslavia are underlined.

City-size symbols are based on metropolitan area estimates.

Mountainous areas

Major industrial concentrations

Main territorial changes since 1938

FIGURE 5.13

Index map of East Central Europe. The "mountainous areas" are broadly generalized. For identification of cities shown by letter in the major industrial concentrations, see Figure 5.1, p. 124. City-size symbols are based on metropolitan area estimates. Names (underline) and capitals of the independent successor states of former Yugoslavia are shown (see text). Serbia and Montenegro are separate republics that make up all that is left of federal Yugoslavia. Belgrade is the capital of both Serbia and Yugoslavia. Former Czechoslovakia (see inset) has split into the two independent countries known officially as the Czech Republic and Slovakia. Their capitals are Prague and Bratislava, respectively.

In 1989 and 1990, Soviet backing for the region's Communist order was abruptly withdrawn because of the inability of the Russian government to continue financial support, and also because of rapidly expanding calls for greater autonomy by the former satellite nations. Several Communist regimes (including East Germany's) quickly collapsed and were replaced by democratically elected non-Communist governments. However, "reformed" and renamed Communist parties and ex-Communist officials still play a significant role in the region. In late 1995, a communist was elected president of Poland, and the Communists gained more than 20 percent of the common vote—the largest of any party—in a parliamentary election in Russia. Governments are struggling to build democracies with capitalist economies out of the economic wreckage of frequently nonproductive, nonprofitable, noncompetitive, state-owned enterprises left as a Communist legacy.

POLITICAL GEOGRAPHY OF THE "SHATTER BELT"

The extension of Soviet power into East Central Europe at the end of World War II was in keeping with the history of the region. In the Middle Ages, several peoples in the region—the Poles, Czechs, Magyars, Bulgarians, and Serbs—enjoyed political independence for long periods and at times controlled extensive territories outside their homelands. Their situation then deteriorated as stronger powers—Germans, Austrians, Ottoman Turks, and Russians—pushed into East Central Europe and carved out empires. These empires frequently collided, and the local peoples were caught in numerous wars that devastated great areas, often resulted in a change of authority, and sometimes brought about large transfers of population from one area to another.

Definitions and Insights

SHATTER BELT

In regions characterized by frequent or even continual shifting of boundaries due to warfare, rapidly changing roles of authority, or frequent friction between two major nations, there tends to be a fluidity in the alignment and independence of the adjacent smaller nations. The result is a **shatter belt**. Historically the term has referred to the East Central European nations from Poland in the north to Albania in the south. In broader use, however, this term connotes any region that experiences continual demographic shifts, through ongoing migrations, efforts at **ethnic cleansing**—the often bloody removal of ethnic populations from a state or region controlled by different ethnic groups—and continual internecine warfare and border disputes. This pattern of geopolitical fracturing and splintering is at the base of enormous instability,

tension, local enmity, and, very often, local warfare. In the case of the former Yugoslavia, the shatter belt characteristic has a history of leading to much broader wars, and it carries that same potential even today.

ORIGINS OF THE PRESENT PATTERN OF COUNTRIES

The present pattern of countries in East Central Europe resulted from the disintegration of the Ottoman Turkish, Austro-Hungarian, German, and Tsarist Russian empires in the 19th and early 20th centuries. World War I hastened this process, in which Germany, Austria-Hungary, and Turkey were on the losing side. Russia, originally allied with the victors, suffered disastrous defeats and then withdrew from the war following the Bolshevik Revolution of 1917. It was not represented at the Paris Peace Conference of 1919, where the victorious Western powers rearranged parts of the political map of East Central Europe in an attempt to satisfy the aspirations of its various nationalities. Poland, which had been partitioned among three empires in the 1700s, was reconstituted as an independent country. Czechoslovakia—the homeland of the closely related Czech and Slovak peoples partitioned in 1993 by its own peaceful preference—was in 1919 carved out of the Austro-Hungarian Empire as an entirely new country. Hungary, greatly reduced in size, was severed from Austria. The Kingdom of Serbia, which had won independence from Turkey in the 19th century, was joined with the small independent state of Montenegro and several regions taken from Austria-Hungary to form the new Kingdom of the Serbs, Croats, and Slovenes, later known as Yugoslavia. Romania, which had been independent of Turkish control since the mid-19th century, was enlarged by territories taken from Austria-Hungary, Russia, and Bulgaria. Bulgaria had already achieved independence from Turkey in the second half of the 19th century, as had Albania immediately before the outbreak of World War I.

This longer view of regional historical geography reminds us that the civil war of the 1990s fought in the former Yugoslavia has long been characteristic of this shatter belt of Eastern Europe.

Definitions and Insights

IRREDENTISM

As a geographic term—meaning "unredeemed" in Italian—irredentism is daily becoming more common in the media. It refers to the partitioning of national boundaries to allow minority populations to be settled in their perceived homelands, or to the extending of national boundaries to reclaim territory and population that earlier belonged to the country seeking the change in borders. During the Cold War Period (1945–1989), the blanket of Soviet authority over Eastern Europe meant that traditional enmities were generally kept in check, holding traditional homeland battles to a minimum. To be continually assigned roles in the global confrontation between the Communists and the Free World meant that there was little geographic latitude for exercising calls for the return of homelands. However, after 1989, a "new world order" did slowly become apparent. Instead of the hoped-for peace and greater flourishing of human rights, the disappearance of Soviet authority allowed the region to become a powerful vortex of micronationalism. This process ripped apart the cultural fabric of different ethnic groups, religious groups, and language groups that had, in the past half century, experienced a reasonable blend of living traditions in a common landscape. When a government, or a nationalistic group, calls for the return of lands that lie in some other nation's or group's geopolitical realm, irredentism and the seeds for everything from disruptive propaganda campaigns to bloody civil wars emerge. These patterns are, tragically, more representative of the so-called new world order than are concerns for human and environmental rights.

There have been large population transfers in the region since the beginning of World War II. During that war and the immediate postwar years, transfers involving millions of Germans, Poles, Hungarians, Italians, and others "simplified" the ethnic distribution patterns but at enormous human cost. People often were uprooted without notice, losing all their possessions, and were dumped as refugees in a so-called "homeland" that many had never seen. The prewar populations of Germans in Poland and Czechoslovakia were expelled at the end of World War II and forced into East and West Germany. Many died in this process of forced migration. Millions of Jews were systematically killed throughout East Central Europe during the German occupation in World War II. Ethnic minorities now constitute only 1 percent of the population in East Central Europe's largest country, Poland, which transferred most of its German population to Germany and whose territory containing Lithuanians, Russians, and Ukrainians was taken away.

A number of countries still have sizable minorities—for example, Hungarian minorities in the Slovak Republic, Vojvodina in Serbia, and the mountainous area called Transylvania in Romania. However, while Communist control put a quieting blanket over most of East Central Europe, its collapse beginning in 1989 was accompanied by a rapid reemergence of some old ethnic quarrels. For example, quarrels among diverse ethnic elements have fractured the former federal state of Yugoslavia (see Fig. 5.14).

THE SLAVIC REALM OF EUROPE

The expulsion of Germans from East Central Europe and the extermination or flight of more than six million Jews in total during the period of Nazi control in the 1930s and 40s intensified the Slavic character of the region. Practically all of Europe's Slavic people live there, and they form a large majority within the region.

The Major Slavic Groups

The Slavic peoples are often grouped into three large divisions: (1) East Slavs of the Former Soviet Region, (2) West Slavs, including Poles, Czechs, and Slovaks, and (3) South Slavs, including Serbs, Croats, Slovenes, Bulgarians, and Macedonians. Another group is the Romanians.

Although the various Slavic peoples speak related languages, such languages may not be mutually intelligible. For example, the Polish language is not easily understood by a Czech, although Poles and Czechs are customarily grouped as West Slavs. Other significant differences also exist. Serbian and Croatian, for example, are essentially one language in spoken form, but the Serbs, like the Bulgarians and most of the East Slavs, use the Cyrillic alphabet (based on the Greek language), whereas the Croats use the Latin alphabet, as do the Slovenes and the West Slavs. The Romanian language is classed with the Romance languages derived from Latin; however, the contemporary Romanian language contains many Slavic words and expressions. Romanians are generally classified as non-Slavs and are an additional example of the ethnic variety of Eastern Europe.

A religious division exists between the West Slavs, Croats, and Slovenes, who are Roman Catholics for the most part, and the East Slavs, Romanians, Bulgarians, Serbs, and Macedonians, who adhere principally to various branches of the Orthodox Christian Church.

Non-Slavic Peoples

The principal non-Slavic elements in East Central Europe are the Hungarians and the Albanians. The Hungarians, also known as Magyars, are the descendants of nomads from the east who settled in Hungary in the ninth century. They are distantly related to the Finns and speak a language entirely distinct from the Slavic languages. Roman Catholicism is the dominant religious faith in Hungary, although substantial Calvinist and Lutheran groups have long existed there. The Albanians speak an ancient language that is not related to the Slavic languages except in a very distant sense. Excluding Turkey, Albania is the only European state in which Muslims form a majority of the population, although Slavic Muslims (Moslems) are the largest ethnic group in Bosnia and Herzegovina (in the former Yu-

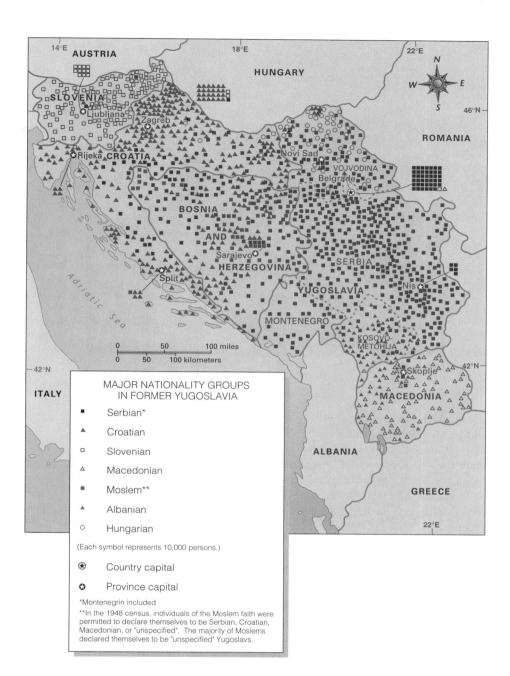

FIGURE 5.14

This map has been redrawn from a map in George W. Hoffman and Fred Warner Neal, *Yugoslavia and the New Communism* (New York, Twentieth Century Fund, 1962), p. 30, by permission of the authors and the Twentieth Century Fund. Names in the largest type denote the six republics that constituted the former Socialist Federal Republic of Yugoslavia. The Vojvodina and Kosovo-Metohija areas are former "autonomous provinces" brought under tight control by Serbia in 1988. The future status of the political divisions and boundaries within former Yugoslavia continued to be very uncertain in mid-1997.

goslavia). Serbia, Macedonia, Montenegro, and Bulgaria have many Albanian or Turkish Muslims. These groups are legacies from former Turkish rule.

PHYSICAL GEOGRAPHY: A CHECKERBOARD OF HABITATS

A glance at Figure 5.13 shows that East Central Europe forms a kind of crude checkerboard of mountains and plains. The *mountains*—located in the center and south—come in a wide range of elevations and degrees of ruggedness. The *plains*—concentrated in the north and in the center along the Danube—exhibit varying degrees of flatness, fertility, warmth, and humidity, and agriculturally they range from some of Europe's better farming areas to some of the poorest (Fig. 5.15). A handful of Europe's largest rivers carry the drainage of the region to the Black Sea, the Baltic Sea, or the North Sea.

The Northern Plain

Most of Poland lies in a large northern plain between the Carpathian and Sudeten mountains on the south and the Baltic Sea on the north. The plain is a segment of the North European Plain, which extends westward from Poland into Germany and

FIGURE 5.15

In the early 1990s, Poland had begun to utilize Western farm technology. This harvesting of winter wheat in the Poznán region of west central Poland shows teams of combine drivers working as a unit to bring in the harvest.   *East Foto*

eastward into the Ukraine in the Former Soviet Region. Central and northern Poland exhibit land that is rather sandy and infertile, with many swamps, marshes, and lakes. At the south, the plain rises to low uplands. Here, where Poland touches Europe's loess belt, are the country's most fertile soils.

Poland's largest rivers, the Vistula (Polish: *Wisla*) and Oder (Polish: *Odra*), wind across the northern plain from Upper Silesia to the Baltic. The Oder is the main internal waterway, carrying some barge traffic between the important Upper Silesian industrial area (with which it is connected by a canal) and the Baltic. Poland's main seaports—Gdansk (formerly Danzig), Gdynia, and Szczecin (formerly *Stettin*)—are along or near the lower courses of these rivers. Gdansk and Gdynia form a metropolitan area of about 900,000 people, and Szczecin has about 450,000.

The Central Mountain Zone

A central mountain zone is formed by the Carpathian Mountains and lower ranges farther west along the Czech frontiers. The Carpathians, considerably lower and less rugged than the Alps, extend in a giant arc for about a thousand miles (*c.* 1600 km) from Slovakia and southern Poland to south central Romania. West of the Carpathians, lower mountains enclose the hilly and highly industrialized Bohemian basin in the Czech Republic. On the north, the Sudeten Mountains and Ore Mountains separate this republic from Poland and the former East Germany. Between these ranges, at the Saxon Gate, the valley of the Elbe River provides a lowland connection and a navigable waterway from Bohemia to the industrial region of Saxony in Germany, and, farther north, to the German seaport of Hamburg and the North Sea. To the southwest, the Bohemian forest occu-

pies the frontier zone between Bohemia and former West Germany.

Between mountain-rimmed Bohemia and the Carpathians of Slovakia, a busy passageway is provided by the lowland corridor of Moravia in the Czech Republic. Through this corridor run major routes connecting Vienna and the Danube valley with the plains of Poland. Near the Polish frontier, the corridor narrows at the Moravian Gate between the Sudeten Mountains and the Carpathians. Just beyond this gateway, East Central Europe's most important concentration of coal mines and iron and steel plants developed in the Upper Silesian–Moravian coal field of Poland and the Czech Republic.

The Danubian Plains

Two major lowlands, bordered by mountains and drained by the Danube River and its tributaries, compose the Danubian plains. One of these, the Great Hungarian Plain, occupies two thirds of Hungary and smaller adjoining portions of Romania, Serbia, and Croatia. It is very level in most places and contains much poorly drained land near its rivers. The second major lowland is composed of the plains of Walachia and Moldavia in Romania, together with the northern fringe of Bulgaria. The Danubian plains contain the most fertile large agricultural regions in East Central Europe.

The Danube River, which supplies a vital navigable water connection between these lowlands and the outside world, rises in the Black Forest of southwestern Germany and follows a winding course of some 1750 miles (*c.* 2800 km) to the Black Sea. Upstream from Vienna, the river is swift and hard to navigate, although large barges go upstream as far as Regensburg, Germany. Below Vienna, the Danube flows leisurely across the

Hungarian and Serbian plains past Budapest (Fig. 5.16) and Belgrade. In the border zone between Serbia and Romania, the river follows a series of gorges through a belt of mountains about 80 miles (*c.* 130 km) wide. The easternmost gorge is the Iron Gate gorge. Beyond it, the Danube forms the boundary between the plains of southern Romania and northern Bulgaria. The river then turns northward into Romania and enters the Black Sea through a marshy delta. Traffic on the Danube is modest compared with the tonnage carried by barges on the Rhine or by road and rail in East Central Europe. The river flows mostly through agrarian areas, and heavy industries—the most important generators of barge traffic—are found in only a few places along its banks.

The Southern Mountain Zone

East Central Europe's southern mountain zone occupies the greater part of the Balkan Peninsula. Bulgaria, the Yugoslav successor states, and Albania, which share this zone, are very mountainous, although important lowlands exist in some places. In Bulgaria, the principal mountains are the Balkans, extending east-west across the center of the country, and the Rhodope Mountains in the southwest. These are rugged mountains that attain heights of over 9000 feet (*c.* 2800 m) in a few places. Between the Rhodope Mountains and the Balkan Mountains is the productive valley of the Maritsa River, constituting, together with the adjoining Sofia Basin, the economic core of Bulgaria. North of the Balkans, upland plains, covered with loess and cut by deep river valleys, slope to the Danube.

In southern Serbia and neighboring areas to the west and south, a tangled mass of hills and mountains constitutes a major

barrier to travel. Through this difficult region, a historic lowland passage connecting the Danube valley with the Aegean Sea follows the trough of the Morava and Vardar rivers. At the Aegean end of the passage is the Greek seaport of Salonika (Thessaloniki). The Morava-Vardar corridor is linked with the Maritsa Valley by an east-west route that leads through the high basin in which Sofia, the capital of Bulgaria, is located. Along the rugged, island-fringed Dalmatian Coast, the Dinaric Alps rise steeply from the Adriatic Sea. The principal ranges run parallel to the coast and are crossed by only a few significant passes.

The Varied Range of Climates

In most of East Central Europe, winters are colder and summers are warmer than in western Europe. In the Danubian plains, the humid continental long-summer climate is comparable to that of the corn and soybean region (Corn Belt) in the United States' Midwest, but the plains of Poland have a less favorable humid continental short-summer climate, comparable to that of the American Great Lakes region. The most exceptional climatic area is the Dalmatian Coast. Here, temperatures are subtropical and precipitation, concentrated in the winter, reaches 180 inches (about 260 cm) per year on some slopes facing the Adriatic. This is Europe's greatest precipitation. The climate of the Dalmatian Coast is classed as mediterranean (dry-summer subtropical). Together with spectacular scenery, this environmental attribute has fostered a sizable tourist industry.

CHANGING PATTERNS OF RURAL LIFE

Prior to World War II, large sections of East Central Europe were little affected by the industrial and urban modes of life that swept over northwestern Europe. In all countries, industries grew considerably between the two world wars, but they were concentrated in a handful of large urban districts or national capitals. The industries that existed often were financed and controlled from western Europe, giving industrial life a markedly "colonial" quality. Bohemia and Moravia were exceptions to the common pattern; as the industrial heart of the Austro-Hungarian Empire, these Czech lands developed diversified and efficient industries that were able to continue successfully when the empire disintegrated at the end of World War I. Another exception was the heavy industrial area in Upper Silesia, then held by Germany.

Although industry and urbanism have grown markedly since World War II, large areas in East Central Europe are still strongly rural, and most rural people still support themselves wholly or primarily by agriculture. On the whole, their efforts do not provide a very good livelihood. Great numbers of rural homes now have electricity, and overall levels of rural welfare are probably improved over what they were before World War

FIGURE 5.16

The Danube River at Budapest. Government buildings line the waterfront beyond the Chain Bridge. The domed structure is Hungary's national Parliament. The view was taken from Buda on the higher west bank, looking toward Pest across the river.   *H. Todd Stradford, Jr.*

II, but they still are well below those of urban places and far below those of rural areas in the countries bordering the North Sea.

Land Reform Before the Communist Era

The discrepancy in rural standards of living between East Central Europe and the North Sea lands has endured throughout modern times. For centuries prior to World War I, and, in some areas, up to World War II, the countryside of Poland, Slovakia, Hungary, and Romania was dominated by estates, and the history of estates often was one of exploitation or neglect of the laborers or tenants. Bulgaria, Serbia, Montenegro, and Albania were mainly lands of peasant farmers, characterized by small landholdings, but weak systems of education, brigandage, feuds, governmental exploitation, inadequate landholdings, poor transportation, and, in mountainous areas, poor soil severely handicapped this agrarian population. Most of these peasants were among the most poverty-stricken in Europe.

Where manufacturing developed—attracting surplus population from the land and providing urban markets for farm produce—the scene grew more prosperous. The height of rural prosperity was reached in Bohemia and Moravia, the industrial heart of the Austro-Hungarian Empire. But in most of East Central Europe, population accumulated on the land, the amount of land per person became smaller, strips of the fragmented small farms were subdivided and resubdivided among heirs, and the situation steadily deteriorated.

In the 19th and early 20th centuries, some of the pressure was relieved by the emigration of millions of East Central Europeans, principally to the United States. Those who remained clamored for land, and the result was a series of land-reform programs in various countries. The main intent was to break up large holdings and sell the land to small farmers. Redistribution of land was most successful, and benefits the greatest, in the relatively well-developed Czech provinces of Bohemia and Moravia.

PATTERNS OF ERRATIC AGRICULTURAL DEVELOPMENT UNDER COMMUNISM

Following World War II, the new Communist governments liquidated the remaining large private holdings in East Central Europe, and programs of collectivized agriculture on the Soviet model were introduced. Some farmland was placed in large state-owned farms on which the workers were paid wages, but most land was organized into collective farms owned and worked jointly by peasant families who shared the proceeds after operating expenses of the collective had been met. **Collectivization** met with strong resistance and in former Yugoslavia and Poland was discontinued in the 1950s. Today, all but a mi-

nor share of the cultivated land in Poland is privately owned. The remaining countries are pursuing programs to reprivatize their farmland, most of which was still in collective ownership in mid-1996. Conversion to private farming after four decades of collectivization presents huge obstacles, and progress is slow, even though images of this transformation continue to fuel hopes of the region's small farmers.

Despite recent attempts to diversify and intensify agriculture, farming in East Central Europe from a production standpoint remains primarily a crop-growing enterprise based on corn and wheat. These are the leading crops in the region from Hungary and Romania south. In the Czech and Slovak republics and Poland, with their cooler climates, corn is unimportant, but wheat is a major crop as far north as southern Poland's belt of loess soils. Most of Poland lies north of the loess belt and has relatively poor sandy soils; here, rye, beets, and potatoes become the main crops. Livestock raising is a prominent secondary part of agriculture throughout East Central Europe. The region's meat supply is much less abundant than in western Europe and comes primarily from pigs. The more southerly areas contain poor uplands that pasture millions of sheep, the main source of meat in Bulgaria, Albania, and parts of former Yugoslavia.

URBAN AND INDUSTRIAL DEVELOPMENT IN EAST CENTRAL EUROPE

East Central Europe has few cities of major size, although small regional centers, often old and historic, are widespread (Fig. 5.17). Only two areas—the Upper Silesian–Moravian coal

FIGURE 5.17
Eger, in Hungary, was founded about ten centuries ago and is one of the many European cities that were more prominent in the past than they are today. Older low-rise buildings lining narrow streets contrast with standardized high-rise structures dating from the Communist era.   *H. Todd Stradford, Jr.*

field of Poland and the Czech Republic, and the industrialized Bohemian basin—exhibit a closely knit web of urban places. Even the agglomeration of mining and metallurgical centers in Upper Silesia–Moravia—a Ruhr in miniature—contains only about 3.5 million people. The Polish part of Upper Silesia–Moravia is somewhat larger than Poland's capital, Warsaw (population: 2.4 million). In all the other countries, the largest cities are national capitals. None of the capitals, including Warsaw, ranks in Europe's top ten cities in metropolitan population. The largest, Budapest, has about 2.6 million people; Bucharest has 2.3 million; Prague, Sofia, and Belgrade each have between 1 and 1.7 million; Zagreb, Skopje, and Sarajevo each have over 500,000; and both Ljubljana and Tirane have 250,000 to 500,000. (The figures are estimates compiled prior to the civil wars that began in the former Yugoslavia in 1991.) In every instance, the capital is its country's leading center of diversified industries and serves also as a primate city.

INDUSTRIAL TRENDS IN THE COMMUNIST ERA

The ascension of Communism in East Central Europe following World War II inaugurated a new industrial and urban era in this region. With minor exceptions, the existing industries were taken out of private hands, and national economic plans were developed. Communist planners did not aim at a balanced development of all types of industry. Instead, they stressed the types that were deemed most essential to industrial development as a whole. Investment was channeled heavily into a few fields: mining, iron and steel, machinery, chemicals, construction materials, and electric power. These were favored at the expense of consumer-type industries and agriculture.

The planned expansion of mining and industry in East Central Europe under Communism produced notable increases in the total output of minerals, manufactured goods, and power. But it also produced inefficient, overmanned industries ineffective in their later competition for the world markets of the post-Communist period. The central planning agencies of the Communist governments maintained a rigid control over individual industries. Plant managers were directed to produce certain goods in quantities determined by state governmental planners, and the success of a plant was judged by its ability to meet production targets rather than its ability to sell its products competitively and at a profit. Political reliability and conformity to the central plan were qualities much desired in plant managers. This system resulted in shoddy goods that were often in short supply or in mountainous oversupply because production was not being driven or keyed by buyer demand and satisfaction.

A key element in East Central Europe's industrialization in the Communist era was its heavy dependence on the Soviets for vital minerals. These included iron ore, coal, oil, natural gas, and other minerals, which were imported in great quantities and

FIGURE 5.18
This 1989 anti-Communist rally of shipyard workers in Gdansk, Poland, reflects the surge of protest that rapidly overthrew the Communist order in most of East Central Europe and in former East Germany, as well. The meeting was addressed by former United States President George Bush and "Solidarity" leader Lech Walesa.   *Chick Harrity/U.S. News and World Report*

over long distances by rail or pipeline from Soviet mines and wells. In return, the Soviet Union was a large importer of East Central Europe's industrial products.

During the 1970s and 1980s, however, the region's trade relationship with the Soviet Union weakened. Trade and financial relations with Western countries, companies, and banks expanded rapidly. With relatively little to export, the East Central European countries borrowed massively from Western governments and banks to pay for imports from the West. This course was followed particularly by Poland and Romania. New industrial plants and equipment acquired in this way were supposed to be paid for by goods that would be produced for export to the West by industries that had learned efficient production and marketing techniques from the West. But then the Western economies began to slump, lowering the demand for imports. This led to a situation by the 1980s in which large debts to Western governments and banks needed to be repaid if countries were to maintain any credit at all. However, with the help of mismanagement by Communist bureaucrats, exports with which to pay were not being produced or could not be sold. The result was a severe impediment to economic expansion, falling standards of living as the governments squeezed out the needed money from their people, and, in Poland, social action through the formation of an independent trade union called **Solidarity** which paralyzed the country and threatened to upset Communist dominance (Fig. 5.18). These conditions, along with the USSR's own growing distress, set the stage for the withdrawal of Soviet control and de-Communization in 1989 and 1990.

FIGURE 5.19

One of the most popular and widespread signatures of a U.S. presence abroad is the nearly universal Marlboro Man. This Moscow scene, with the ten-story-high image of the Big Sky country cowboy associated with Marlboro cigarettes, is another sign of the rapid expansion and flow of Western, particularly American, cultural icons and influences into Europe and, in this case, beyond and into Russia. There are parallels to this image and others of numerous United States exports all through Europe, and more recently, East Central Europe as well.   *Swersey/Gamma Liaison*

THE POLITICAL REVOLUTIONS OF 1989 AND THE RECESSION OF COMMUNISM

As late as the early summer of 1989, the former East Germany and the East Central European countries seemed firmly in the control of totalitarian Communist governments. Then, in the late summer and autumn, the Communist order began to crumble. Public demands for freedom, democracy, and a better life gathered momentum in country after country. Communist dictators who had ruled for many years were forced out, and reformist governments took charge. This liberalizing process continued into the 1990s. By mid-1991, democratic multiparty elections had been held in all countries.

Definitions and Insights

THE GEOGRAPHY OF ECONOMIC TRANSFORMATION: EASTERN EUROPE

In a process that has had bold implications for the political as well as the economic landscape of Eastern Europe, the whole geography of economic activity underwent profound change in the early 1990s. Ten highlights of the economic restructuring of the region included:

1. *privatization*—the privatizing of state-owned enterprises;
2. *increasing the efficiency of state-run enterprises* by al-
lowing noncompetitive enterprises to fail and removing diseconomic support of poor performers;
3. *encouraging new private enterprises* to develop, in many cases with foreign capital and management playing a role in the growth of these new units;
4. *developing new institutions* required by a market-oriented economy (banks, insurance companies, stock exchanges, accounting firms, and so on);
5. *fostering joint enterprises* between state-owned firms and foreign firms;
6. *ending price controls* and allowing prices to reflect competition in the market—a shift that has been particularly difficult for the firms that paid little attention to markets, the need for their product, or quality;
7. *eliminating bureaucratic restrictions* on the private sector;
8. *making currencies internationally convertible* to increase trade and thus increase competition;
9. *converting farmland* from state and collective ownership to private ownership; and
10. *expanding access to international communications media* of all sorts to help local entrepreneurs learn from foreign examples—such as encouraging travel of potential foreign investors to plant sites to assess the prospects for success in joint ventures.

This economic transformation has been highly varied in its success in East Central Europe. The period from 1989 to the mid-1990s saw only spasmodic economic progress in the region. Economic hardship, uncertainty, and uneasiness were widespread. Loans by the International Monetary Fund (IMF) and other outside sources were helpful but modest in scale. Unemployment was high in all countries. Freeing of prices as a move toward a market economy often resulted in more and better goods in stores but at prices few local customers could afford. The increased prices created great hardship for pensioners and others living on low fixed incomes. Meanwhile, a high proportion of all industries remained state-owned despite various efforts to dispose of them to private owners. Such privatization was slowed not only by the decrepitude of many properties but also by resistance to change on the part of workers and managers whose jobs, security, and power were at stake. Similarly, the privatization of state-owned farmland moved very slowly. Most prospective farmers were reluctant to exchange the relative security of employment on collectivized farms for the potential hazards of private farming. The uncertainty generated by all these changes was further promoted by the success of Communist party candidates in the region in elections in the mid-1990s.

Hungary in particular has been a relatively bright spot economically. Its move toward free-market capitalism began early and has been aided by Western and Japanese investments larger than those received by any other East Central European country in recent decades. A relatively small but growing class of "newly rich" private entrepreneurs has emerged, often as participants in joint business ventures with foreign companies. This new "business bourgeoisie" contrasts with a far larger class of "newly poor" persons living near or below the poverty line.

The economically diversified Czech Republic, where a long tradition of economic accomplishment was undercut by Communism, is also enjoying more than usual success in breaking out of the Communist mold, although, like Hungary, it is still a poor area by broader European standards. Economic advance is less marked in adjacent Slovakia. In fact, the economic disparity between the two areas was an important factor leading to the dissolution of the Czech and Slovak Federal Republic (Czechoslovakia) on January 1, 1993, and the creation of the new independent countries of the Czech Republic and the Slovak Republic. Many Czechs felt that less industrialized and less prosperous Slovakia was a drag on the federal economy; Slovaks, in turn, resented "dictation" by the Czech majority. Slovakia also has a potential problem of political separatism within the minority community of about 600,000 ethnic Hungarians.

Poland undertook a much-publicized "shock therapy" by abruptly freeing prices in the hope of moving rapidly to a free-market economy; but despite Western encouragement and modest financial support, the country remains mired in serious inflation, unemployment, faltering production, and political discontent. Prospects seem even bleaker for Bulgaria, Romania, and Albania, in which former Communists remain influential. Conditions in Albania became so desperate in 1991 that many thousands fled the country by ship to Italy as economic refugees. This process was repeated in early 1997.

Political Splintering and Ethnic Warfare in the Former Yugoslavia

Economic progress in Yugoslavia was disrupted by ethnic warfare following dissolution of the Yugoslav federal state in 1991. Previously, under the Communist regime instituted by Marshal Josip Broz Tito during World War II, Yugoslavia had been organized into six Socialist People's Republics: Serbia (population: 85 percent Serb in 1981); Croatia (Croats 75 percent,

PROBLEM LANDSCAPE

BOSNIA'S ALTERNATE REALITY

In the outline in an article in the January, 1997 *Harper's Magazine*, the ways in which some form of peace can be achieved in Bosnia, even temporarily, will be threatened almost immediately by a continuing Serb drive for secession. In June, 1991, Slovenia and Croatia declared their independence from Yugoslavia. That political move to segment the already patchwork nation of Yugoslavia into smaller, more homogeneous political units is simply a renewed attempt toward "balkanization," or the generally "bloody" breakup of larger political units into smaller ethnic units. As soon as Croatia made its declaration in 1991, the Croatian Serbs took over one third of the new country and began "ethnic cleansing," the forced emigration—or killing—of peoples different from the ethnic group in power. As a result of this act, the tensions between the Serb peoples and the Muslim peoples of the former Yugoslavia have become a major focus of the world press.

In Bosnia, U.S. troops have been deployed since late 1995, as part of the Day-ton Accords signed by the presidents of Bosnia, Croatia, and Serbia on behalf of Serbia and the Bosnian Serb Republic. The Accords and the positioning in this region of 60,000 NATO troops has brought a slow-down to the military's attempts to continue ethnic cleansing, and in a sense, political efforts have replaced military efforts. The NATO presence has initiated some effort to reunite ethnically diverse populations, to bring Muslims back to their homes in cities that now are controlled by Serbs, and to stimulate government cooperation at all levels in this effort.

The "alternate reality" described in *Harper's* chronicles the dogmatic persistence in Serb plans for ethnic separatism. It also contends that if the Bosnian Serb leaders are able to achieve a Serb secession, the Croat leadership—also made up of peoples who declared their independence from Yugoslavia in 1991—will follow the same path. This will mean that the alleged gains of the Dayton Peace Accords will in all likelihood explode in a return of the warfare associated with this balkanization policy.

Early this century, Otto von Bismark was asked what would cause World War I, and he replied, "Some damn foolish thing in the Balkans." In 1914, the assassination of Archduke Francis Ferdinand of Austria-Hungary by a Serbian nationalist in Sarajevo did, in fact, act as the primary catalyst for World War I. The assassination was part of an effort to shift the political pattern of this region.

The same forces of ethnic cleansing and assassination continue to break up and realign the struggling populations of the Balkan peninsula today, and it is this part of central and eastern Europe that is breaking up into ever smaller and smaller ethnic, religious, and political groups—all generally demanding autonomy, political independence, and widespread political recognition.

Similar tensions (of varying intensities) dot the map in India, Sri Lanka, Canada, central Africa, and parts of Latin America.

1997. Bosnia's alternate reality. *Harper's Magazine*, 294:1760 [January]. PP. 19–20.

Serbs 12 percent); Slovenia (Slovenes 91 percent); Bosnia and Herzegovina (also often spelled "Hercegovina": Slavic Muslims 40 percent, Serbs 32 percent, Croats 18 percent); Montenegro (Montenegrins 69 percent, Slavic Muslims 13 percent, Albanians 6 percent); and Macedonia (Macedonians 67 percent, Albanians 20 percent, Turks 5 percent). In addition, two "autonomous provinces" were created: Kosovo (Albanians 77 percent, Serbs 13 percent) and Vojvodina (Serbs 54 percent, Hungarians 19 percent).

In 1988, Serbia took direct control of Kosovo and Vojvodina as part of Serbia's push (under a renamed Communist President, Slobodan Milosevic) for greater influence within federal Yugoslavia. This push was challenged by all the other republics except Montenegro; they demanded a looser federation with more autonomy for each republic, and in 1991 they all declared independence. Slovenia, Croatia, and Bosnia and Herzegovina were admitted to the United Nations in 1992. The secessions left federal Yugoslavia a rump state composed of Serbia and Montenegro. The Dayton Accords of 1995 attempted to institute a means of establishing a more viable series of ethnic and political units in the lands of the former Yugoslavia.

QUESTIONS

1. Name six countries touched by the Danube River.

2. Name the new political units formed from the breakup of the Former Yugoslavia.

3. Name the wettest place on the Balkan Peninsula and explain the geographic features that gives it that trait.

4. Name four national capitals located in the mediterranean climate zone.

5. Name four distinct environments located on the Scandinavian Peninsula.

The Former Soviet Region

2

6

A Geographic Profile of the Former Soviet Region

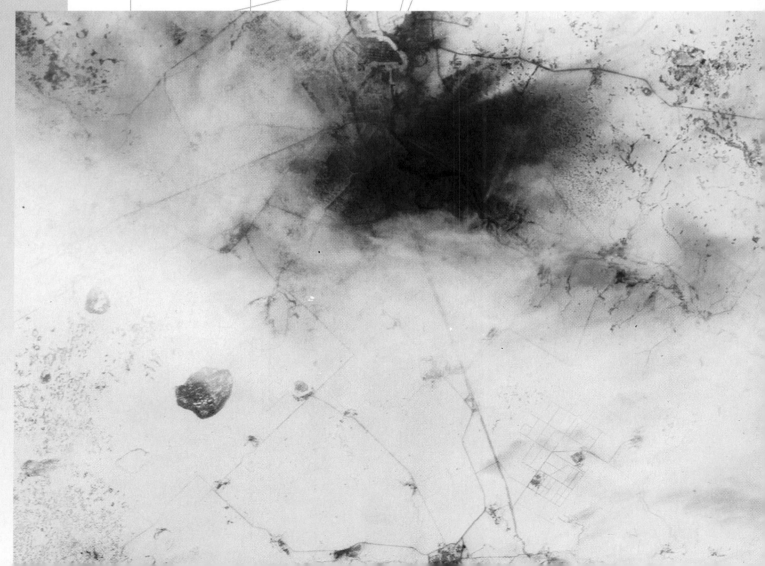

From 1917 to 1991, the huge area now known as the Former Soviet Region comprised a single Communist-controlled country called (from 1922) the Union of Soviet Socialist Republics (USSR). After World War II, the government based in the Kremlin fortress in Moscow exercised control over Communist "satellite" countries in eastern Europe and maintained strong influence in other Marxist states, particularly in LDCs not aligned with the capitalist West. For four decades, the Cold War between the Soviet bloc of nations and the Western bloc led by the United States dominated world politics. The perspectives and actions of the Soviet Union became major determinants of world events.

Suddenly, in 1991, the country split into 15 independent nations (see Table 6.1 and Definitions and Insights, page 155). Twelve of the units are associated loosely in the Commonwealth of Independent States (CIS), but tendencies toward further political fragmentation and decentralization are still great. Without the USSR's Red Army to impose order, long-simmering ethnic conflicts have boiled over. Political and economic re-development is under way, but the future of this huge territory is uncertain.

This chapter offers geographic and historical perspectives on a major world region in which momentous changes are taking place. Placing the 15 former republics of the Soviet Union, now 15 independent countries, in a single region is somewhat awkward but still necessary in the late 1990s. Although the Soviet government failed in its vision of creating a single, unified "Soviet people" from the hundreds of ethnic groups living within the boundaries of the USSR, it did succeed in transforming the human geography of the vast Soviet empire. For many years to come, the 15 countries of that crumbled empire will be attempting to sever or reestablish ties with the imperial hub of Moscow. For its part, the Russian government in Moscow will try vigorously to maintain influence over the former republics. Russia may even attempt to reexert control over some of the recently independent nations, especially those vital to the economic and political security of the Russian Federation.

◄ This U.S. Space Shuttle photograph of the Siberian city of Troitsk on the Uy River shows one of the environmental problems troubling the Former Soviet Region. Troitsk is one of three large industrial cities on the east side of the Urals where uncontrolled pollution from steel mills creates blighted landscapes and human health problems. Black soot is ejected from the steel mills onto the city and its surrounding snowclad landscape. Studies indicate that because of mills like these, only one percent of the children in nearby Magnitogorsk are healthy. *NASA*

TABLE 6.1

Independent Successor States of the Former Soviet Union: Area, Population, and Ethnic Data

Country	Land Area (Thousand sq mi)	Land Area (Thousand sq km)	Population Mid-1996 estimates (millions)	Ethnic Percentages Titular nationality[a]	Ethnic Percentages Russian	Ethnic Percentages Other
Slavic States and Moldova						
Russia (Russian Federation)	6593	17,078	147.7	85	85	Tartar 4 Ukrainian 2
Ukraine	233	604	51.1	73	22	Belarusian 1 Jewish 1
Belarus (former Belorussia)	80	208	10.3	78	14	Ukrainian 3 Jewish 1
Moldova (former Moldavia)	13	34	4.3	65	13	Ukrainian 14 Gagauz 4
Transcaucasia						
Georgia	27	70	5.4	70	6	Armenian 8 Azerbaijani 6 Ossetian 3 Greek 2 Abkhazian 2
Armenia	12	30	3.8	93	NA	Azerbaijani 3 other 4
Azerbaijan	33	87	7.6	83	6	Armenian 6 Lezgin 2
Central Asia						
Kazakstan	1049	2717	16.5	41	37	Ukrainian 5 German 5 Uzbek 2 Tatar 2
Uzbekistan	173	447	23.2	73	8	Tajik 5 Tatar 2
Turkmenistan	189	488	4.6	73	10	Uzbek 9 Kazak 2
Kyrgyzstan	77	199	4.6	52	22	Uzbek 13 Ukrainian 3
Tajikistan	55	143	5.9	64	7	Uzbek 24 Tatar 1
Baltic States						
Estonia	17	45	1.5	62	30	Ukrainian 3 Belarusian 2
Latvia	25	65	2.5	58	31	Belarusian 4 Ukrainian 3
Lithuania	25	65	3.7	81	9	Polish 7 Belarusian 2
Totals	8600	22,275	292.7	—	51	Ukrainian 16 Uzbek 6

[a]The titular nationality is the ethnic gorup for which the state is named.

Sources of Data: 1996 *World Population Data Sheet* (Population Reference Bureau); 1996 *Britannica Book of the Year; The New Cosmopolitan World Atlas* (Rand McNally, 1994).

Definitions and Insights

AREAL NAMES OF THE FORMER SOVIET REGION

Geographic study of the Former Soviet Region requires an understanding of important areal names. The region corresponds in extent with the former Union of Soviet Socialist Republics, often shortened to Soviet Union or USSR. This state came into existence in 1922 following the overthrow of the last Romanov tsar in the Russian Revolution of 1917 and the subsequent civil war. Prerevolutionary Russia is often called Old Russia, Tsarist Russia, Imperial Russia, or the Russian Empire.

During the Communist era from 1917 to 1991, Westerners often called the country Soviet Russia. In the past, the name Russia has been used loosely to refer to the entire country either before or after the Revolution, but it is now properly restricted to the huge Russian Federation, which was the largest of the 15 Soviet Socialist Republics (Union Republics or SSRs) that made up the Soviet Union. The full name of the Russian Federation during the Communist period was Russian Soviet Federated Socialist Republic (RSFSR). The loosely aligned Commonwealth of Independent States (CIS) was formed by 12 of the 15 former Union Republics late in 1991. Estonia, Latvia, and Lithuania are not members of this organization.

The area west of the Ural Mountains and north of the Caucasus Mountains has been known historically as European Russia. Similarly, the Caucasus and the area east of the Urals has been called Asiatic Russia, Soviet Asia, or the eastern regions. Here, Transcaucasia is the region in and south of the Caucasus Mountains (see Fig. 7.1, page 176); Siberia is a general name for the area between the Urals and the Pacific; and Central Asia is a general name for the arid area occupied by five states with large Muslim populations immediately east and north of the Caspian Sea.

6.1 Area and Population

With an area (including inland waters) of 8.6 million square miles (22.3 million sq km), the Former Soviet Region is smaller than Africa but larger in area than any other major world region recognized in this text (see Table 1.2, page 115). However, so much of it is sparsely populated that its estimated population of 293 million in mid-1996 ranked it only sixth among the eight world regions. The average population density of 34 per square mile (13 per sq km) is only half that of the United States. Great stretches of economically unproductive terrain separate many outlying populated areas from other population centers.

The eventful geopolitical history of the Former Soviet Region has given this region land frontiers with 12 countries in Eurasia (Fig. 6.1). Between the Black Sea and the Pacific, the region directly borders Turkey, Iran, Afghanistan, China, Mongolia, and North Korea. Pakistan and India also lie close by. In the Pacific Ocean, narrow water passages separate the Russian-held islands of Sakhalin and the Kurils from Japan. If India, Pakistan, and Japan are considered, the Asian part of the Former Soviet Region is a near neighbor of about half the world's people.

In the west, the region has frontiers with Romania, Hungary, Slovakia, Poland, Finland, and Norway. These differ greatly from the sparsely populated, arid, and mountainous frontiers in Asia. On the frontier between the Black and Baltic seas, international boundaries pass through well-populated lowlands that have long been disputed territory between the Russian state and other countries. During World War II (1939–1945), the Soviet Union expanded its national territory westward in several areas (see Chapters 5 and 7).

Stretching nearly halfway around the globe in northern Eurasia, the Former Soviet Region has formidable problems associated with climate, terrain, and distance. Most of it is handicapped economically by excessive cold, infertile soils, marshy terrain, aridity, and ruggedness. Natural conditions are more similar to those of Canada than to those of the United States. Interaction with a complex and demanding environment was a major theme in Russian and Soviet expansion and development. Nature provided large assets but also posed great problems.

6.2 The Natural Environment

THE ROLE OF THE CLIMATIC AND BIOTIC ENVIRONMENT

Russian and Soviet expansion and development went forward in a harsh climatic setting (Fig. 2.6). Severe winter cold, short growing seasons, drought, and desiccating summer winds that shrivel crops in the steppes are major handicaps. Associated with these rigorous conditions are significant advantages in the form of tillable soils, the world's largest forests, natural pastures for livestock, and a diverse wild fauna. However, many of these resources are only marginally useful because of unfavorable climatic factors.

Most parts of the Former Soviet Region display continental climatic influences characterized by long cold winters, warm to hot summers, and low to moderate precipitation. The severe winters are the result of the northerly continental location coupled with mountain barriers on the south and east. Four fifths of the total area is farther north than any point in the conterminous

Definitions and Insights

THE USSR AND OTHER LAND EMPIRES

Some geographers have characterized the Soviet Union as a **land empire**. Rather than establishing its colonies overseas as did imperial powers such as Spain and Great Britain, Russia founded its colonies in its own vast continental hinterland. Many of the colonized peoples had little in common with the Russian ethnic majority which ruled from faraway Moscow. Like former colonies of overseas empires such as Great Britain, non-Russian regions of the Soviet Union's periphery were drawn into a relationship of economic dependency upon the imperial Russian center. The Central Asian republics, for example, followed Moscow's demands to grow cotton, which was shipped to Moscow to be manufactured into expensive clothing marketed in Central Asia and elsewhere in the empire. This pattern contributed to the growth of the USSR's economy but often inhibited local development. Other entities characterized historically as land empires include the United States, China, Brazil, and the Ottoman, Moghul, Aztec, and Inca Empires.

THE COMMUNIST ECONOMIC SYSTEM

The Communist economic system that played such a central role in Soviet life following the Bolshevik Revolution was an attempt to put into practice the economic and social ideas of the 19th-century German philosopher Karl Marx (who actually spent most of his adult life in England and is buried in London). According to Marx, the central theme of modern history is a struggle between the capitalist class (bourgeoisie) and the industrial working class (proletariat). He forecast that exploitation of workers by greedy capitalists would lead the workers to revolt, overthrow the capitalists, and turn over ownership and management of the means of production to new workers' states. In the classless societies of these states, there would be social harmony and justice, with little need for formal government.

Marx's utopian vision has not materialized anywhere, but it did promote revolutions and provide guidelines for communist political systems in many countries. The ideas of Marx as Lenin interpreted and implemented them ("Marxism-Leninism") provided the philosophical basis for the Soviet Union's centrally planned "command economy"—an economy that is now being restructured with great difficulty. This economy developed as a mixture of advanced technology and an economic bureaucracy inherited from Tsarist Russia. Beginning in 1928, the Soviet command economy operated according to Five-Year Plans that prescribed goals of production for the nation—for example:

types and quantities of minerals, manufactured goods, and agricultural commodities to be produced; factories, transportation links, and dams to be built or improved; and locations of new residential areas to house industrial workers. The Bolsheviks did not develop this state-controlled economic machine from scratch. A huge state economy had existed under the tsars, gaining impetus as early as the reign of Peter the Great and continuing to develop up to World War I. The Bolsheviks implemented further dominance by the state in stages. The revolutionary leaders had essentially two aims: They meant to abolish the old aristocratic and capitalist institutions of Tsarist Russia and to develop a strong socialist state able to stand on an equal footing with the major industrial nations of the West. For the first decade, they consolidated their hold on the country and put a limited part of their program into effect. Large-scale industry, banking, and foreign trade were nationalized, while the New Economic Policy (NEP), announced in 1921, permitted some private trading, together with private ownership of small industries and agricultural land. This compromise policy was the result of a near breakdown in the newly instituted Communist economy during the difficult period of civil war and foreign intervention following the Revolution. However, it proved to be only a temporary expedient.

Lenin, the chief architect of the Revolution, died in 1924. It was mainly his successor, Josef Stalin, who forged the Soviet system. Abandoning the privatization of the New Economic Policy, he and his long, despotic regime embarked on full-scale Communism and used terror to sweep aside real and suspected opposition. Stalin's government launched a ruthless drive for comprehensive planning, forced socialization of the economy, and massive industrialization. The motivation was partly ideological and partly defensive. Insisting that because the world socialist revolution had not yet occurred, and because the USSR was surrounded by hostile capitalistic states such as Germany, Stalin resolved to "build socialism in one country." After his death in 1953, Stalin's successors abolished and modified many rigidities and cruelties of the Stalin era and were able to raise the Soviet standard of living somewhat. Such changes, however, did not revolutionize the country's basic structure and ideology, and economic conditions gradually worsened.

Under the Communists, an agency in Moscow called Gosplan (Committee for State Planning) formulated national plans for the Soviet Union. Once approved at the highest level, the plans were transmitted downward through a huge administrative bureaucracy (apparat) until they finally reached the operating level of individual factories, farms, and other enterprises. This unwieldy process generated inefficiency in a number of ways:

1. Fear of offending superiors made persons at lower levels reluctant to suggest ways to improve efficiency.
2. It was hard for planners in Moscow to manage an area larger than North America as though it were one gigantic corporation. Trying to coordinate such a huge and diverse body of

enterprises, materials, labor, and consumer demand from one central point was too great a task.

3. In freer economies, the market—the desires and abilities of consumers and businesses to make and buy things—largely determines what will be produced. The Soviet planning bureaucracy had no free market to guide it, so often goods were produced that people would not buy, or goods were not produced even though people would have liked to buy them.

4. Although Soviet decision-makers knew that more free enterprise might result in more efficient operation of plants and farms, they did not widely grant such freedoms. They were afraid of the uncertainties of free markets and unwilling to surrender their decision-making powers.

5. Gosplan stated production targets in quantitative rather than qualitative terms, often resulting in shoddy workmanship and unsalable goods. Enterprise managers often lobbied to keep their production targets as low as possible.

Communist planners changed the country's spatial organization, altered the interaction of people and nature, and added many new elements to the cultural landscape. They constructed new cities and transport links, enlarged older cities, carried out a massive expansion of mining and manufacturing, increased the amount of cultivated land, and reorganized the countryside by collectivized agriculture. These are still significant features on the landscapes of the Former Soviet Region.

The Soviet Union emphasized some grandiose economic projects. Such enterprises harnessed energies and resources of the whole country to achieve specific objectives. The government called on the people to sacrifice in order to make the country strong and provide a better life in the future; the term "Hero Project" was meant to incite enthusiasm. A few examples include the construction of large tractor plants in the 1920s and 1930s at Kharkov in Ukraine, Stalingrad on the Volga, and Chelyabinsk in the Urals; the "Virgin and Idle Lands" program of the 1950s to expand grain acreage east of the Volga; and the "Project of the Century," the construction of the Baikal-Amur Mainline (BAM) Railroad during the 1970s and 1980s, to provide an alternate link to the Pacific north of the Trans-Siberian Railroad in the Far East. These accomplishments were impressive, but they failed to give the Soviet people the higher standard of living they desired.

COLLECTIVIZED AGRICULTURE

Despite current attempts to privatize, the Former Soviet Region's economies continue to rely heavily on the socialized industries and farms inherited from the Communist system. But agriculture's future success will require that a legacy of past mismanagement be reversed.

Farmers and farming had troubled careers under the Soviets. Between 1929 and 1933, about two thirds of all peasant households in the Soviet Union were collectivized, and the more prosperous private farmers, known as *kulaks,* were killed,

exiled, sent to labor camps, or forced to starve. Collectivization was virtually complete by 1940. People resisted this program fiercely. Peasants and nomads slaughtered millions of livestock and burned crops to avoid turning them over to the "socialized sector." Government reprisals followed, including wholesale imprisonments and executions, together with confiscation of food at gunpoint (often including the peasants' own food reserves and seed). Famines took millions of lives. Soviet leaders disregarded these costs and reorganized the countryside into two types of large farm units (Fig. 6.9): the **collective farm** (*kolkhoz*) and the factory-type **state farm** (*sovkhoz*).

As originally conceived, the system of collectivized agriculture was to result in the following major advantages:

1. The old arrangement of small, fragmented individual holdings separated by uncultivated boundary strips would be replaced by larger fields incorporating the boundary strips, increasing the amount of cultivated land and promoting mechanized farming.

2. Increased mechanization would release surplus farm labor for employment in factories and mines, promoting industrialization and creating the large urban working class looked to as the principal support for the Communist system.

3. Mechanization, improved methods of farming, and reclamation of new land under state supervision would result in greater overall production.

4. Increased production, plus easier collection of surpluses from a greatly reduced number of farm units, would result in larger and more dependable food supplies for growing urban populations and greater tax revenues to use in building industry.

5. Liquidation of individual peasant farming would remove the most important capitalist element still remaining in the USSR.

FIGURE 6.9
A scene on a collective farm in Moldova. Advanced technology keeps company with holdovers from Old Russia in the economies of the Former Soviet Region.   *Sovfoto/Eastfoto*

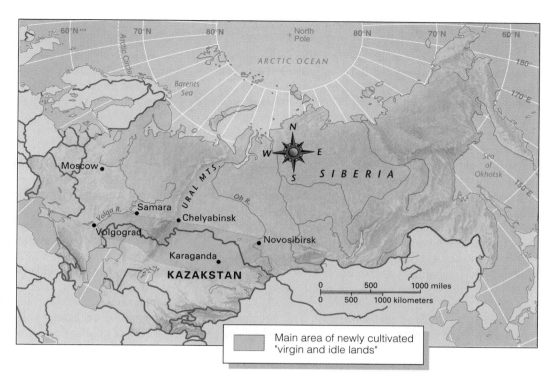

FIGURE 6.10
The area shown in green encloses most of the new lands brought under cultivation in the period 1954–1957. Modern international boundaries are shown.

6. Consolidation of individual farmsteads and villages into fewer but larger and sometimes planned communities on the sovkhozes and kolkhozes would permit the government to more efficiently and cheaply administer the rural population, monitor and indoctrinate the rural population, and provide services including education, health care, and electricity to a greater share of the rural population.

Soviet agriculture experienced many setbacks in the early phases of collectivization, and overall production was slow to expand. After World War II, and particularly after Stalin's death in 1953, important changes were made in Soviet farming in the hope of increasing productivity. The system of agricultural procurement, pricing, and taxation was revamped to induce farmers to produce more on collectively farmed land. By merging adjoining units, the number of collective farms was reduced. Some collectives were absorbed by or converted to state farms. Beginning in 1958, the collectives were allowed to own and operate their own tractors and farm machinery instead of contracting for them from machine-tractor stations. Restrictions on the use of personal plots and privately owned livestock were reduced.

An important facet of the drive to increase the national supply of farm products and diversify the diet to include more animal products was a sizable enlargement of cultivated area. In 1954, the Soviet Union instituted a program to increase the amount of grain (mainly spring wheat and spring barley) by bringing tens of millions of acres of **virgin and idle lands** or **"new lands"** into production in the steppes of northern Kazakstan and adjoining sections of western Siberia and the Volga region (Fig. 6.10). Hundreds of new state farms were organized. Between 1954 and 1960, the cultivated area of the Soviet Union was enlarged by over 90 million acres (36 million hectares), with most of the increase taking place in 1954–1956 in Kazakstan, Siberia, and the Volga region. Despite bad weather in some years, with crop failures and low production, the Virgin and Idle Lands Scheme added to Soviet grain output and moved the center of grain farming eastward. This lessened the impact of bad weather in a given year, as a poor winter wheat crop in the west could be offset by a good spring wheat crop in the east, and vice versa. However, there have been problems in maintaining satisfactory production on a long-term basis. The "new lands" have low precipitation and require careful management to conserve soil moisture and prevent wind erosion. Some of the most marginal land is no longer productive.

THE SUPERPOWER

The principal target of Soviet national planning after 1928 was a large increase in industrial output, with emphasis on heavy machinery and other capital goods, minerals, electric power, better transportation, and military hardware. The drive to industrialize had far more success than the planned expansion of agriculture. Masses of peasants were converted into factory workers, new industrial centers were created, and old ones were enlarged. The increase in urban population was phenomenal: In 1926 only 18 percent of the Soviet Union's population lived in cities, but by 1996 the average figure for 15 countries of the Former Soviet Region had risen to an estimated 56 percent (the world average is 43 percent; in the United States it is 75 percent).

The prodigious drive to remake the USSR industrially provided the base for a large expansion of railways and ocean shipping, the mass production of millions of new apartments, space flight, and a huge military machine. The Soviet Union became the military power strong enough to survive Germany's onslaught in World War II. After the war, the Soviets maintained large armed forces and accumulated a huge arsenal of conventional and nuclear weapons in the "arms race" against the world's only other superpower, the United States. At the height of the Cold War, 15 to 20 percent of the country's GNP was dedicated to the military (in contrast to less than 10 percent in the U.S.), representing a considerable diversion of investment away from the country's overall economic development. While Russian arms sales abroad continue to grow today, economic reformers are grappling with the chore of converting a large share of the military economy to production for civilian use (Fig. 6.11).

To meet the needs of its people, the USSR became a collectivized welfare state, with the government providing guaranteed employment, low-cost housing, free education and medical care, and old-age pensions. Social services were often minimal but in some sectors were quite successful; the literacy rate, for example, rose from 40 percent in 1926 to 98.5 percent in 1959. However, military-industrial superpower status was generally achieved at the expense of Soviet consumers, whose needs were slighted in favor of heavy metallurgy and the manufacture of machinery, power-generating and transportation equipment, and industrial chemicals. Lines of consumers waited in stores to purchase scarce items of clothing and everyday conveniences. The exasperation of shoppers confronted by long lines and empty shelves was an important factor generating dissatisfaction with the economic system and leading to demands that the system be redeveloped. Shortages of food and industrial products available to urban consumers were caused in considerable part by producers withholding goods in the hope of higher prices and by large-scale bartering of food from state farms and collectives in exchange for needed goods from large industrial enterprises.

6.5   The Second Russian Revolution

Internal freedoms and prosperity did not accompany geopolitical power in the Soviet Union. Near paralysis gripped the flow of goods and services in the late 1980s, when large demonstrations and strikes underscored public anger at a political and economic system that was sliding rapidly downhill. The Communist system came under open challenge on the grounds that it stifled democracy, failed to provide a good living for most people, and thwarted the ambitions of the country's many ethnic groups for a greater voice in running their affairs. The outpouring of dissent was unprecedented in Soviet history.

Worsening economic conditions led to official calls for fundamental reform in the mid-1980s. The revamping of the economic system became an urgent priority in the regime of Mikhail Gorbachev that began in 1985. Gorbachev initiated new policies of *glasnost* ("openness") and *perestroika* ("restructuring") to facilitate a more democratic political system, more freedom of expression, and a more productive economy with a market orientation. The government was reorganized, and the Communist Party's control over political and economic affairs eventually ended.

At the same time, there were increasing problems of disunity as the different republics and the ethnic groups organized by the Soviet system into nominally "autonomous" units took advantage of new freedoms to resurrect old quarrels and demand that their units be given more autonomy. Fighting among ethnic groups erupted in several republics. Meanwhile, in all of the republics, declarations of sovereignty and in some cases outright independence challenged the authority of the central government. In 1991, the USSR and most other nations recognized the independence of the three Baltic republics of Estonia, Latvia, and Lithuania. The process of the empire's disintegration had begun.

At the center of the crumbling empire, having failed to reverse the downward slide of the economy, Gorbachev faced growing sentiment to scrap the command system and move as rapidly as possible to a market-oriented economy. Early in 1991, strikes in coal fields vital to the national fuel supply compounded the country's economic troubles. Laboring in some of the world's most blighted industrial districts, the miners called for decentralization of political and economic control as a means of achieving a better life.

The miners' cause was championed by Boris Yeltsin, a leading advocate of rapid movement toward a market-oriented

FIGURE 6.11
The RONIS factory in the Russian city of Rostov-On-Don is one of Europe's largest producers of video cassettes and compact disks.   *Sovfoto/Eastfoto*

economy and greater control by individual republics over their own resources, taxation, and affairs. Yeltsin's political stature grew when, in June 1991, the citizens of Russia chose him president of the giant Russian Federation in the Soviet Union's first open democratic election. Gorbachev, himself elected president of the USSR by a Congress of People's Deputies rather than by the people as a whole, opposed Yeltsin's proposals for radical reform. He advocated more gradual movement toward free-market orientation within the state-controlled planned economy and insisted on the need for unity and continued political centralization within the Soviet state. Even Gorbachev's gradualist approach was too radical for a third body of opinion, held by hardliners in the Communist Party—the government bureaucracy, the military, and the police—who called for strong measures to enforce greater efficiency within the existing system.

Matters came to a head in August 1991, when an attempted coup by hardliners failed. Yeltsin gained enhanced stature from his defiant opposition to the coup. Gorbachev resigned his position as head of the Communist Party and began to work with Yeltsin to reconstruct the political and economic order. His efforts to preserve the Union and a modified form of the Communist economic system failed. His program of *perestroika* was supposed to rebuild a sense of individual involvement in the Russian economy, but Gorbachev seesawed on his reform program and ended up fighting the forces of liberalization he had unleashed. During the autumn of 1991, the Communist Party was disbanded and the individual republics seized its property. On December 25, 1991, Gorbachev resigned the presidency, and the national parliament formally voted the Soviet Union out of existence on the following day. A powerful empire had quickly and quietly faded away, to be replaced by 15 independent countries.

6.6   Yeltsin's Russia

Despite heightened public expectations for better life in Russia, criticism of the triumphant President Yeltsin and his policy of economic reform toward freer markets increased. Russia's parliament attempted to slow the pace of reform. In September 1993, acting under a new draft constitution, Yeltsin dissolved the parliament and set a December date for a referendum on the draft constitution and election of a new two-chamber parliament: a lower house, or State Duma, of 450 deputies, and an upper house, or Federation Council, composed of representatives from 88 subdivisions of the Russian Federation. Then, in October 1993, Russian army units quickly and violently suppressed an armed uprising in Moscow against the president by members of the old parliament and their supporters. In December, Russia's electorate approved the draft constitution (which gave the president broad powers), but the parliamentary elections did not yield a clear majority favoring rapid market reforms. The new legislature was generally hostile to Yeltsin, and was dominated by Communists and ultranationalists like Vladimir Zhirinovsky. Only about half of the Russian electorate voted in the 1993 elections. Those who stayed away and those who voted for Zhirinovsky did so in part to express anger at poverty, insecurity, crime, and mismanagement in Russia.

In the 1995 parliamentary elections, public dissatisfaction with the new Russia gave once-disgraced Communists nearly a quarter of the legislature's seats. Now the largest party in Russia's parliament, the Communists prepared to capture the prize of the Russian presidency by offering Gennadi Zyuganov as their candidate in the June 1996 election. Yeltsin challenged Zyuganov with the campaign slogan "Russia cannot afford another 1917!" Boris Yeltsin's prospects of winning appeared dim in advance of the election, especially because most Russians expressed dissatisfaction with the state of the economy. However, the majority preferred an uncertain future under Yeltsin rather than a return to the Communist past, and they elected him president for a second term.

Early in his first term, President Yeltsin had introduced a program of rapid economic reform—known widely as economic "shock therapy"—designed to replace the Communist system with a free-market economy. It removed price controls and encouraged privatization of businesses. Products and services from the private sector had already been indispensable during the Communist era, but now they were to be the centerpiece of the economy. Attempts to expand private sector goods and services are a prominent feature of current economic reform, and most of the former Soviet republics are in the process of "privatizing" their economies. In Russia the process has been swift; by the time the first phase of the privatization process officially ended in 1994, two thirds of Russian industry was in private hands and 40 million Russians owned shares in newly privatized firms. Progress in some industrial sectors and in some newly independent countries has been slow, as there are many obstacles to overturning state ownerships.

Results of "shock therapy" have been mixed. Former employees of the Communist Party resent the loss of their jobs and privileges. Printing of money to meet state obligations has caused inflation. Poor people living on fixed incomes are struggling to survive high prices for basic necessities. A new consumer-oriented society is developing in Russia, but along class lines. There is growing unemployment and homelessness, and a widening gap between rich and poor. Access to choice goods and services is not forbidden to ordinary citizens as it was during the Communist era, but prices are so high that most people cannot afford to pay them.

One of the major components of Russia's economic geography is an **underground economy**, also known as the "countereconomy" or "second economy" (or economy *na levo*—"on the left"). Widespread barter has resulted from the declining value of the ruble (Fig. 6.12). Many people resort to selling personal possessions to buy high-priced food and other necessities. Rampant black-marketeering has developed since prices were decontrolled. Much of this traffic is still officially illegal, but

FIGURE 6.12
Free enterprise thrives throughout the Former Soviet Region. Even the clothing of the once-mighty Soviet army is up for sale. *Sovfoto/Eastfoto*

there is a general tendency to overlook such transactions because they are so essential to the economy. Russia's new private entrepreneurs include a large criminal "mafia" that preys on government, business, and individuals, raising public fears and prompting some people to call for a return to the stability and security of the Soviet state.

With its economic problems and a variety of related troubles, Russia has been described as a "misdeveloped" country. Unlike any other industrialized nation, it is now experiencing a surging death rate, which rose to 15 per thousand by 1993, or a startling 20 percent increase over the previous year's figure. And life expectancy is falling; in 1996 it was just 57 years for men (versus 72 for men in the U.S., for example). Men in such LDCs as Bolivia, India, and Indonesia live longer than those in Russia. One projection puts the life expectancy for the Russian man at just 53 in the year 2005. The birth rate also is falling, from 14 per thousand in 1992 to 9 per thousand in 1996—an astonishing 36 percent decrease in just four years. In 1989, an average of 2.17 children were born to a Russian woman; by 1996, the number had fallen to 1.4.

With 800,000 more people dying than being born in 1996, Russia is the first industrialized country in history to be experiencing a sharp decrease in its population for reasons other than war, famine, or disease. The reasons for the falling population since 1991 include alcohol abuse; environmental pollution; a high rate of abortion (two for every live birth) and attendant surgical procedures that leave many women unable to bear children; a lack of antibiotics and other medical amenities; and, among many people, a lack of personal hope for the future.

However, by 1997, the outlook for Russia's economic future was generally brighter than it had been in several years. It appeared that Russia's market-oriented reforms had begun to

bear fruit. The economy was growing at a rate of 2 percent per year, following declines since the end of Communism in 1991. It remains to be seen whether this growth will translate into better prospects for the great majority of Russians or benefit only a select few.

6.7 Russia and the Wider World

With the collapse of the Soviet Union, the Russian Federation took over the property of the former government within the Federation's borders (including Moscow's Kremlin; see Fig. 6.13). Russia also took custody of the international functions of the USSR, including its seat at the United Nations. In early December 1991, the republics of Russia, Ukraine, and Belarus (formerly Belorussia) formed a loose political and economic organization called the Commonwealth of Independent States (CIS). Except for Estonia, Latvia, and Lithuania, the other republics eventually joined the CIS, in which Russia took a strong leadership role. CIS headquarters are in Minsk, Belarus. Disputes erupted, notably between Russia and Ukraine, over such questions as ownership of property inherited from the defunct USSR, particularly the Black Sea naval fleet based in Ukraine's Crimea region, and disposition and control of nuclear weapons.

Having persuaded the West of his commitment to economic reform, Yeltsin was rewarded with Western aid. However, more aid has been promised than delivered. In the wake of a scandal in which a United States Central Intelligence Agency employee named Aldrich Ames was discovered to have yielded

FIGURE 6.13
Moscow's Red Square, seat of power for the USSR and now the Russian Federation. Contrasting with the austere Lenin Mausoleum, which backs up to the Kremlin at right, are the whimsical onion domes of the 16th-century St. Basil's Cathedral on the center horizon. *Sovfoto/Eastfoto*

vital secrets to Soviet and then Russian agents, and in view of the Russian elections that brought antireformers, ultranationalists, and Communists to office, many U.S. lawmakers urged suspension of aid to Russia. For their part, many Russians perceived Yeltsin to be "selling out" to the West and resented their growing dependence upon the West. There is evidence that Russia is growing weary of the loss of superpower status and is reasserting its power abroad, particularly in the former Soviet republics which Russians call the "near abroad" (see "Regional Perspective" below).

Regional Perspective

RUSSIA AND THE "NEAR ABROAD"

Since the Soviet Union dissolved in 1991, many important links between Russia and the other successor states have persisted. Economic, ethnic, and strategic links are especially crucial. For example, one major economic link is the vital flow of Russian oil and gas to other states such as Ukraine, Belarus, and the Baltic states, which are highly dependent on this supply of energy (see Chapter 7, page 178). Ethnicity links approximately 25 million Russians who resided in the 14 smaller states at the time of independence. Although a considerable number have migrated into Russia since 1991, many have had difficulties finding housing and employment, and the great majority still live in the other countries. In certain areas, notably eastern Ukraine and Crimea, many ethnic Russians are causing political instability because they want to secede and form their own state or join Russia. Russians in the 14 non-Russian nations complain that governments and peoples discriminate against them.

In response, the Russian government has said that it has a right and a duty to protect Russian minorities in the other countries. Russia's perspective on the other 14 former members of the USSR is that they constitute the country's "**near abroad**," in which Russia's special interests and influence must be recognized. Some observers contend that Russia's new activism in the "near abroad" is an effort to reestablish the countries of the former socialist camp as buffers between Russia and the "far abroad."

In its national security interest, Russia forcefully asserts claims to a special sphere of influence in the near abroad. Along the outer frontiers of the cordon of successor states, Russia maintains a chain of military bases where, in 1997, an estimated 100,000 Russian soldiers were stationed. Within the near abroad, notably in Moldova, Tajikistan, Georgia, Armenia and Azerbaijan, Russian troops have engaged frequently as "peacekeepers" in local ethnic conflicts. There are fears in the West and among Russia's neighbors that "peacekeeping" is merely a euphemism for a restoration of Russian imperial rule. Some observers contend that Russia wants access to important resources and economic assets in its former republics, such as uranium in Tajikistan, aviation plants in Georgia, military plants in Moldova, and the Black Sea coast and naval fleet in Ukraine's Crimea. Noting Moscow's reluctance to recall Russian troops from the near abroad, some analysts fear that Russia's reasoning may be that Moscow's empire will again be expanded, and it is not worth withdrawing troops and dismantling bases only to redeploy and rebuild them later.

Within Russia there is a body of public opinion favoring reassertion of Russian control over its former empire, but the strength of this feeling is unknown. Many Western analysts anticipate that Russia will attempt increasingly to "Finlandize" its near abroad, referring to the Soviet Union's historic influence over its Nordic neighbor. This would involve Russia's insistence on authority over the foreign policy and security policy of such nations as Ukraine, Estonia, Latvia, and Lithuania. With the recent election of pro-Russian governments in Ukraine and Belarus, it looks increasingly likely that there will emerge a Slavic political and economic *troika* consisting of Russia, Ukraine, and Belarus, possibly dominated by Russia.

In sum, during the years immediately following the USSR's breakup, the Former Soviet Region consisted broadly of two units, Russia and Russia's "near abroad." New poles are now emerging. A significant reintegration around Russia may occur in the Slavic nations, while most of the Central Asian countries may gravitate toward neighboring Islamic nations such as Turkey or Iran (see Chapter 7, page 202).

7

FRAGMENTATION AND REDEVELOPMENT IN THE FORMER SOVIET REGION

Like other major regions of the world, the Former Soviet Region has a very uneven spatial structure of population and development. Nearly three fourths of the people, and an even larger share of the cities, industries, and cultivated land of this immense region, are packed into a triangular **Slavic Coreland** composing roughly one fifth of the region's total area (see Fig. 7.1). Lying mainly west of the Ural Mountains but narrowing eastward into Asia, the triangle is also called the **Fertile Triangle** or **Agricultural Triangle**. The rest of the Former Soviet Region lies mostly in Asia and consists of land where settlement is extremely spotty, handicapped by environments that are nonagricultural except in very limited areas. Most of the non-Slavic population of the region lives there (see Fig. 7.2), but many immigrant Slavs reside there too, generally in cities. These areas outside the coreland are a storehouse of minerals, timber, and waterpower, and governments have made major investments to develop these resources. This has resulted in a series of discrete production nodes widely separated by taiga, tundra, mountains, deserts, wetlands, and frozen seas. The nodes are scattered through the Caucasus region, Central Asia, the Far East, and the Northern Lands.

7.1 Peoples and Nations of the Slavic Coreland

Slavic peoples are the dominant ethnic groups in most of the Former Soviet Region, in both numbers and political and economic power. The major groups are the Russians (whose cultural geography is described in Chapter 6), Ukrainians, and Belarusians (Byelorussians). Most of this Slavic population lives in the triangular coreland extending from the Black and Baltic seas to the neighborhood of Novosibirsk in Siberia. The greatest part of the coreland is in Russia, but Belarus, Moldova, Ukraine, and northern Kazakstan also fall within it. The triangle has about one half the area and more than four fifths the population of the United States. Lowlands predominate — the only mountains are the Urals, a small segment of the Carpathians, and a minor range in the Crimean Peninsula. The original vegetation was mixed coniferous and deciduous forest in the northern part and steppe in the south. Moscow (population of city proper, 8.8 million; metropolitan area, 13.15 million) and St. Petersburg (formerly Leningrad; population of city proper, 4.47 million; metropolitan area, 5.53 million) are by far the largest cities. Moscow, capital of Russia before 1713 and of the Soviet

◀ Satellite image of Moscow, former capital of the vast Soviet Empire and now the Russian Federation. In the twentieth century, growth in this city on the banks of the Moscow River was designed to produce the ideal "socialist city" with decentralized, self-sufficient neighborhoods. Parks and agricultural areas outside the ring road around Moscow appear red in this false color image. © 1991 CNES, Spot Image Corporation/Photo Researchers, Inc.

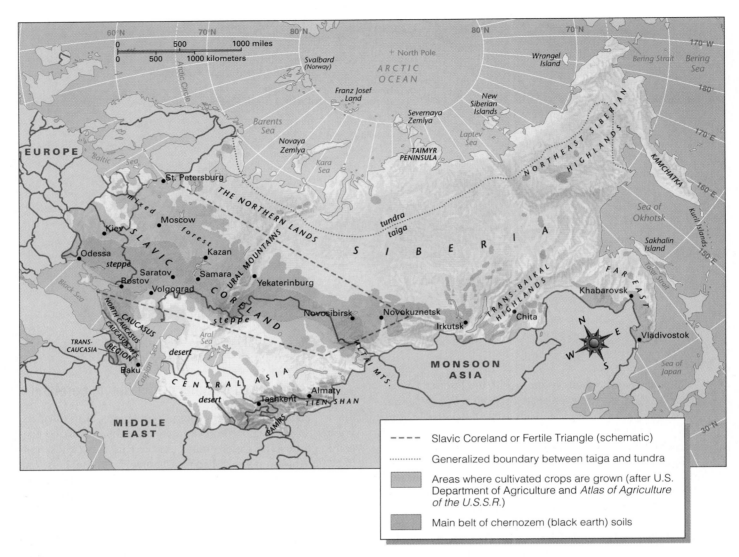

FIGURE 7.1

Major regional divisions of the Former Soviet Region as discussed in this text.

Union from 1918 to 1991, remains the most important manufacturing city, transportation hub, and cultural, educational, and scientific center in the Former Soviet Region (see satellite image, page 174, and Fig. 6.13). It is the political capital of the Russian Federation.

UKRAINE

Russians are by far the largest ethnic group in the coreland, followed by Ukrainians, most of whom live in the recently independent state of Ukraine. Although closely related to the Russians in language and culture, Ukrainians are a distinct national group. The name Ukraine translates as "at the border" or "borderland." In this area, armies of the Russian tsars fought for centuries against nomadic steppe peoples, Poles, Lithuanians, and Turks before the Russian Empire absorbed Ukraine in the 18th century. For three centuries Ukrainians were subordinate to Moscow, first as part of the Russian Empire and then as a Soviet republic. Ukraine's industrial and agricultural assets were always vital to the Soviet Union; the Bolshevik leader Lenin once declared, "If we lose the Ukraine, we lose our head." Since achieving independence from Moscow in 1991, Ukraine has restored the Ukrainian language to its educational system. However, a majority of Ukrainians (57 percent) polled in 1995 wanted Russian to be a second official language, along with Ukrainian.

Today, Ukraine is one of the most densely populated and productive areas in the Former Soviet Region. With about 51

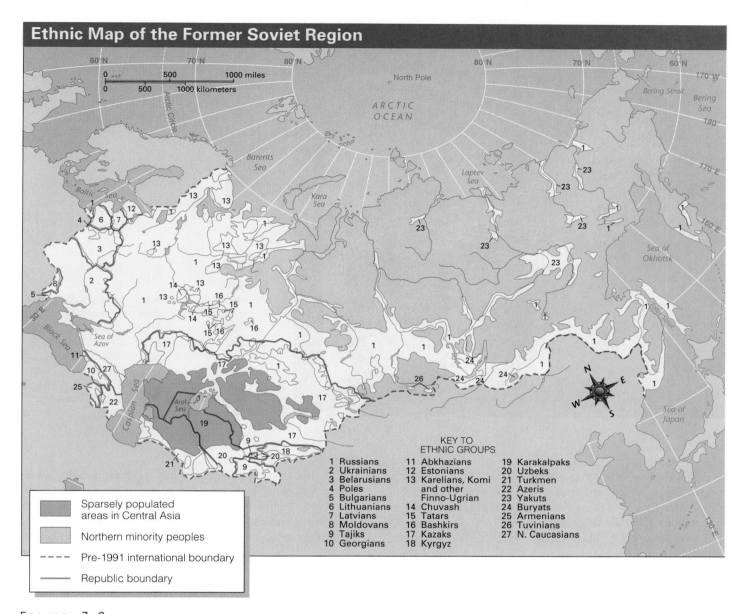

Ethnic Map of the Former Soviet Region

KEY TO
ETHNIC GROUPS

1 Russians	11 Abkhazians	19 Karakalpaks
2 Ukrainians	12 Estonians	20 Uzbeks
3 Belarusians	13 Karelians, Komi	21 Turkmen
4 Poles	and other	22 Azeris
5 Bulgarians	Finno-Ugrian	23 Yakuts
6 Lithuanians	14 Chuvash	24 Buryats
7 Latvians	15 Tatars	25 Armenians
8 Moldovans	16 Bashkirs	26 Tuvinians
9 Tajiks	17 Kazaks	27 N. Caucasians
10 Georgians	18 Kyrgyz	

Sparsely populated areas in Central Asia

Northern minority peoples

- - - - Pre-1991 international boundary

Republic boundary

FIGURE 7.2

The Soviet Union had difficulties trying to hold together such a vast collection of ethnic groups.
The Russian Federation faces many of the same challenges.

million people as of 1996, the republic is not far behind such countries as the United Kingdom and France in population.

Ukraine lies partly in the forest zone and partly in the steppe. On the border between these biomes is the historic city of Kiev (population of city proper, 2.64 million; metropolitan area, 3.25 million), Ukraine's political capital and a major industrial and transportation center (Fig. 7.3). The country is an important producer of wheat, barley, livestock, vegetable oil (mainly from sunflower seeds), beet sugar, and many other products, grown mainly in the steppe region south of Kiev. North of Kiev is Chernobyl, where an explosion at a nuclear

power station in 1986 rendered parts of northern Ukraine and adjoining Belarus incapable of safe agricultural production for years to come.

The Ukraine has large deposits of coal, iron, manganese, salt, and natural gas which contribute to heavy industry (see Fig. 7.5). The coal comes from the Donets Basin (Donbas) coal field near the industrial center of Donetsk (population of city proper, 1.12 million; metropolitan area, 2.13 million), where most of Ukraine's iron and steel plants are also located. Iron ore comes by rail from western Ukraine in the vicinity of Krivoy Rog (population of city proper, 713,000), the Former Soviet

FIGURE 7.3
The heart of Kiev, third largest metropolis of the Former Soviet Region, capital of Ukraine, and mother city of the Ukrainian and Russian branches of Orthodox Christianity. This view features the St. Sophia Cathedral (foreground) and Bogdan Khmelnitsky Square, named for a 17th-century Ukrainian nationalist leader. Kiev was devastated by the military actions of World War II, and massive reconstruction of buildings took place after the war. *Tass/Sovfoto*

Region's most important iron-mining district. Most of the ore is used in Ukrainian iron and steel plants, but some is exported to East Central European countries and Russia. This area is also an important center of chemical manufacturing, in large part due to an abundant supply of blast furnace wastes, coke-oven gases, salt deposits, natural gas piped from wells in Ukraine and the Caucasus, and oil piped from the Caucasus. Surrounding the inner core of mining and heavy-metallurgical districts in Ukraine is an outer ring of large industrial cities that carry on metal fabricating and other types of manufacturing. These cities include the Ukrainian capital of Kiev on the Dnieper River, the machine-building and rail center of Kharkov (population of city proper, 1.62 million; metropolitan area, 2.05 million) about 250 miles (400 km) east of Kiev, and the seaports and diversified industrial centers of Odessa (population of city proper, 1.10 million; metropolitan area, 1.19 million) on the Black Sea.

Since independence, Ukraine has taken important steps on the world stage. In 1994, the Ukrainian parliament voted overwhelmingly to become a nuclear-free nation. By electing to sign the Nuclear Nonproliferation Treaty, Ukraine committed itself to disposing of the 1800 nuclear warheads which had made it the world's third largest nuclear weapons power. The United States rewarded this move with a much-increased foreign aid package to Ukraine, which now is the third largest recipient of U.S. economic assistance, after Israel and Egypt.

At the same time, Ukraine has improved relations with Russia, which had been soured since the collapse of the USSR. The 1994 election of President Leonid Kuchma, who favors economic union and closer political ties with Russia, may reverse some of the problems that followed Ukraine's efforts to distance itself from Russia following independence. The Ukrainian economy had faltered, especially because the country remained extremely dependent upon Russia for oil, gasoline, natural gas, and uranium for its nuclear power reactors. With independence, Ukraine had to pay world market prices for these sources rather than the low prices the Soviets previously subsidized.

Another source of trouble with Russia has been the Crimean Peninsula, which Russia's Catherine the Great annexed in 1783 but which Soviet leader Nikita Khrushchev returned to Ukraine in 1954. In 1991, Ukraine gave the Crimea special status as an autonomous republic. Crimeans then elected their own president, a pro-Russian politician who advocated the Crimea's reunification with Russia. (Ethnic Russians make up 70 percent of Crimea's population and about 22 percent of Ukraine's total population.) Other residents want complete independence for Crimea. The region's future is uncertain, but as long as the government in Kiev is aligned with Moscow, tension over the Crimea may be diminished. The peninsula does enjoy an accommodating climate—sheltered by the Yaila

Mountains, the Crimea's south coast has the relatively mild temperatures and minimal summer rainfall associated with the mediterranean or dry-summer subtropical climate—and is a picturesque area of orchards, vineyards, and resorts, of which Yalta is the most famous.

Polls conducted in 1995 revealed a wide range of opinion on the course Ukrainians want for their country. Fully one third favored the re-creation of the Soviet Union. Another 31 percent wanted Ukraine to emulate the West. Only 24 percent believed the Ukraine should establish its own nonaligned pathway.

BELARUS

The much smaller state of Belarus (formerly called Belorussia and "White Russia") adjoins Ukraine on the north. It developed slowly in the past, mainly because of a lack of fuels other than peat. Recently, however, oil and gas pipelines from other areas such as Siberia have provided raw material for new oil-refining and petrochemical industries. Some oil production has begun in Belarus itself, and a large potash deposit provides fertilizer. Minsk (population of city proper, 1.63 million; metropolitan area, 1.69 million), the political capital and main industrial city, is also the headquarters of the USSR's successor organization, the Commonwealth of Independent States (CIS). Belarus is too cool, damp, and infertile to be prime agricultural country, although it has a substantial output of small grains, hay, potatoes, flax, and livestock products. Southern Belarus contains the greater part of the Pripyat (Pripet) Marshes, large sections of which have been drained for agriculture.

Like Ukraine, Belarus has agreed to eliminate the nuclear weapons it possessed since the breakup of the Soviet Union. Also like Ukraine, Belarus is reestablishing close ties with Moscow. In 1994, Belarus and Russia agreed to unify their monetary systems. In the agreement, Belarus gave up sovereignty over its currency and banking system in exchange for preferential access to Russian energy and other resources. In a 1995 referendum, an overwhelming majority of Belorussians voted for closer economic integration with Russia and for the adoption of Russian as their official language. In early 1996, Belarus and Russia agreed to form a "union state" which would more closely link the economies, political systems, and cultures of these predominantly Slavic countries.

MOLDOVA

Tiny Moldova (formerly Moldavia), made up largely of territory that the Soviet Union took from Romania in 1940, adjoins Ukraine at the southwest. Moldavians are a people with many Slavic cultural characteristics who speak a dialect of Rumanian. The area, however, has many problems of ethnic antagonism and political separatism within its borders. Ukrainians and Russians in the eastern part have been agitating violently to form a separate state, and a minority of nationalistic Moldavians, who claim that Moldavians are Romanians, want Moldova to join Romania.

Industry has increased markedly in Moldova since World War II, especially around the capital of Chisinau (population of city proper, 754,000). Mainly a fertile black-earth steppe upland, however, Moldova is primarily agricultural. An important specialty is the growing of vegetables and fruits, especially grapes for winemaking.

THE BALTIC STATES AND KALININGRAD

Estonia, Latvia, and Lithuania, often called the Baltic states, lie between Belarus and the Baltic Sea. On the basis of ethnicity, these countries are technically outside the Slavic Coreland, but they do share many environmental, economic, and historical traits with the other countries of the coreland. The Latvians and Lithuanians have borrowed culturally from Slavic neighbors but are not Slavs. The Estonians, related to the Finns, are also non-Slavic. The Estonians, Latvians, and Lithuanians have tried zealously to safeguard their national identities against Russian encroachment. They were part of the Russian Empire before World War I but became independent following the Russian Revolution. The Soviet Union reabsorbed them in 1940. Today, they are again independent states, but they are still linked closely to the remainder of the former Soviet Union in various ways. However, these nations have refused to become members of the Commonwealth of Independent States.

In these countries, dairy farms alternate with forests in a hilly landscape. Mineral resources are scarce; the most notable are deposits of oil shale in Estonia, mined for use as fuel in electric power plants. Like Belarus, the Baltic states benefit from natural gas and petroleum brought from fields in the Former Soviet Region located hundreds and even thousands of miles away. Expansion of industry after World War II helped make the area one of the more prosperous and technically advanced parts of the former USSR. Its largest city and manufacturing center is the port of Riga (population of city proper, 804,000) in Latvia.

About 85 percent of Latvia's non-Latvians are Russian speaking—mostly retired Soviet military officers and their offspring. While independent Lithuania and Estonia established procedures early on to grant citizenship to ethnic Russians, Latvia delayed taking such measures until late 1994 and gave in only under pressure from Russia and the West. The Latvian government feared that enfranchised Russians would destabilize the country in future elections or even try to reannex Latvia to Russia.

The presence of Russian troops in the Baltic countries was a serious foreign relations problem after the countries gained independence. With independence, only Lithuania (capital: Vilnius; population of city proper, 584,000) successfully negotiated the immediate withdrawal of Russian troops. A large number of these Russian forces resettled in neighboring Kaliningrad, an anomalous Russian enclave two countries away from Russia. As the northern half of the former German East Prussia, the territory was transferred to the USSR at the end of World War II to become the Kaliningrad Oblast. Soviet authorities expelled nearly all of Kaliningrad's Germans to Germany, and Russians now make up more than three fourths of the population. Kaliningrad remains a massive military and naval establishment, with defense workers and their dependents and retired military families comprising most of the population of about one million.

Unwanted Russian troops remained in Latvia and Estonia until late 1994. Before then, Russia had argued that it could not pull out its troops because Latvians and Estonians were violating the rights of 300,000 ethnic Russians in Estonia and a large minority in Latvia. Since independence, Estonia (capital: Tallinn; population of city proper, 443,000; see Fig. 7.4) has developed close economic ties with Finland, which quickly replaced Russia as the country's dominant trading partner.

7.2   Agriculture in the Coreland

The Fertile Triangle is the agricultural core of the Former Soviet Region and by far the leading area of production for all the major crops except cotton and for all the major types of livestock. There are two main agricultural zones within the Triangle: a black-soil zone in the southern steppes, and a nonblack-soil zone corresponding roughly to the region of mixed forest. The black-soil zone includes chernozem soils and associated chestnut and other grassland soils. Wheat does well on these soils, making the region one of the major wheat-growing areas of the world. The black-soil zone is also the principal producing area for sugar beets, sunflowers, hemp (grown mainly for oil), barley, and corn. Irrigation waters reach the area from the Dnieper, Don, Volga, and other rivers. In the **nonblack-soil zone**, with its cooler and more humid climate and poorer soils, rye has been traditionally the main grain crop for human consumption. However, increased liming and fertilization of soils, plus new frost-resistant and quick-ripening varieties of wheat have recently promoted more wheat cultivation. Agriculture in this zone also emphasizes dairy production, potatoes, oats, and hay.

Some countries of the Former Soviet Region, notably Russia and Ukraine, are global-scale producers of such farm commodities as wheat, barley, oats, rye, potatoes, sugar beets, flax, sunflower seeds, cotton, milk, butter, and mutton. The overall high output, however, does not reflect high agricultural productivity per unit of land and labor. The independent nations of the Former Soviet Region will be struggling for many years to overcome the agricultural inefficiencies of the state-operated and collective farms that long dominated the agricultural sector. On the state-operated farms, workers receive cash wages in the same manner as industrial workers. Bonuses for extra performance are paid. Workers on collective farms receive shares of the income after obligations of the collective have been met. As on the state farms, there are bonuses for superior output. These methods, however, fail to provide enough incentives for highly productive agriculture. There is still a strong tendency for farm machinery to stand idle because of improper maintenance; since no individual owns the machine, the incentive to repair it is diminished. There are shortages of spare parts. Poor storage, transportation, and distribution facilities, plus wholesale pil-

FIGURE 7.4
A downtown festival in Tallinn showcasing the unique culture of Estonia.
Sovfoto

fering, cause alarming losses after the harvest; as much as 40 percent of the grain grown in Russia, Ukraine, and Kazakstan in 1995 rotted in the fields. Younger people are deserting the farms for urban work, often living with the family in the countryside but commuting to their city jobs. With the shortage of younger male workers, women, children, and the elderly are doing a large share of the farm work.

In an effort to reverse such disturbing trends, all of the countries of the Former Soviet Region are promoting land reform by privatizing collective and state farms and developing more independent or "peasant" farming. Privatization of the collective and state farms has occurred rapidly in Russia, Armenia, and the Baltic states. After 1991, the Russian government held about 30 percent of the country's farms in reserve as state farms (sovkhozes) and allowed the remaining 70 percent to be privatized. By 1996, nearly all of the latter had been privatized. These farms have been reorganized in different ways: some are now joint-stock companies issuing shares to their members or to outsiders; some operate as peasant farming cooperatives; and some have been divided into separate independent farms.

Convinced that individuals who have a greater stake in the land will contribute to a more productive agriculture, Russian legislators are attempting to strengthen the rights and rewards of individual farmers. The importance of personal incentive in agriculture is clear; in 1995, personal plots and livestock provided an estimated one third of all meat and one half of the vegetables produced in Russia. This helped the country to weather its worst grain harvest in 30 years.

7.3   Industries of the Russian Coreland

The Soviet Union's emphasis on heavy industry and armaments required the use of huge quantities of minerals, most of which existed in the Soviet Union in adequate quantities. From the beginning of its Five-Year Plans, the Soviet Union stressed the need for heavy industry, which it developed in three principal areas of mineral abundance: Ukraine, the Urals, and the Kuznetsk Basin (Fig. 7.5). One of the most problematic legacies of the breakup of the Soviet Union is that production of industrial commodities is not evenly spread across the region. The main production of each one tends to come from only a few areas, such as textiles from the Moscow region; steel from Ukraine, the Urals, and areas immediately south and north of Moscow; automobiles from the Volga and Moscow areas; and raw cotton from Central Asia. The manufacture of machinery is widespread, but a given city specializes only in particular kinds of machinery, such as grain harvesters at Rostov-on-Don and textile machinery near Moscow.

This regional specialization, which the Former Soviet Region inherited from Communist planning, has made continued cooperation essential among the now-independent states. For example, a single factory in Belarus produces all the electric motors required by certain types of appliance factories throughout the Former Soviet Region. For the short term at least, it is in the interest of the producers and the consumers in the former republics to maintain trading relations.

INDUSTRIES OF THE MOSCOW AND ST. PETERSBURG REGIONS

The industrialized area surrounding Moscow is known as the Central Industrial Region, Old Industrial Region, or Moscow-Tula-Nizhniy Novgorod Region. These names indicate important characteristics of the region. It has a central location physically within the populous western plains and is functionally the most important area of the Former Soviet Region. It lies at the center of rail and air networks reaching Transcaucasia, Central Asia, and the Pacific, and is connected by river and canal to the Baltic, White, Azov, Black, and Caspian seas (Moscow calls itself the "Port of Five Seas").

This region ranks with Ukraine as one of the two most important industrial areas in the Former Soviet Region. South of Moscow lies the metallurgical and machine-building center of Tula (population of city proper, 535,000), and to the east is the diversified industrial center of Nizhniy Novgorod (formerly Gorkiy; population of city proper, 1.4 million; metropolitan area, 2.5 million) on the Volga. Textile milling, mainly using imported U.S. cotton and Russian flax, was the earliest form of large-scale manufacturing to be developed in the Moscow region. The area still has the Former Soviet Region's main concentration of textile plants (cottons, woolens, linens, and synthetics), with Moscow and Ivanovo the leading centers, but the breakup of the Soviet Union threatens to ruin the area's textile industry. The price of Central Asian cotton quadrupled with independence, forcing mills to close and consumers to turn to cheap clothing imports. Other light and heavy industries around Moscow have fared better. Metal-fabricating industries emphasizing skill and precision are the most important. Specialized chemicals such as pharmaceuticals are also made here.

The Moscow region achieved industrial eminence despite a notable lack of mineral resources. However, the area's well-developed railway connections, partly a product of political centralization, provide good facilities for an inflow of minerals, materials, and foods, and for a return outflow of finished products to all parts of the Former Soviet Region. A significant development in recent decades has been the construction of pipelines to bring petroleum, its products, and natural gas from fields located in several different areas. Piped oil and gas are now major fuels in the region and provide the basis for large petrochemical industries.

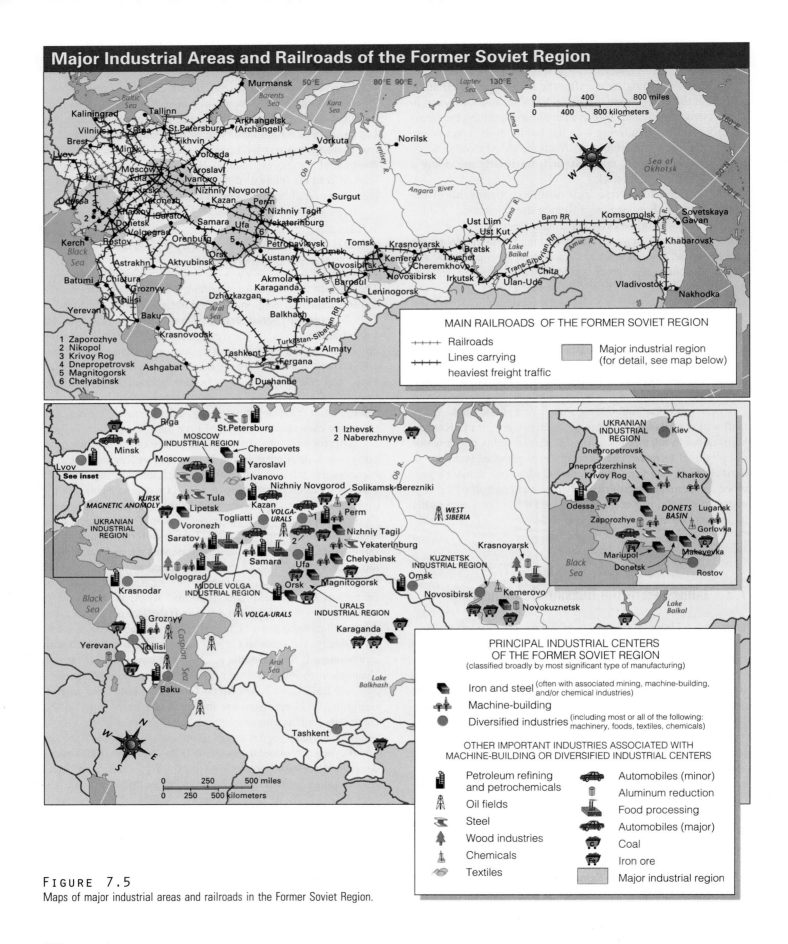

Major Industrial Areas and Railroads of the Former Soviet Region

50°E 80°E 90°E 130°E

Murmansk
Kaliningrad
Tallinn
Vilnius Riga
St.Petersburg
Brest Minsk Tikhvin
Lvov Vologda
Kiev Moscow Yaroslavl
Tula Ivanovo
Kursk Nizhniy Novgorod
Odessa Voronezh Kazan Perm
Kharkov Saratov Nizhniy Tagil
Donetsk Samara Ufa Yekaterinburg
Kerch Volgograd
Rostov Orenburg
Astrakhn Orsk Aktyubinsk Kustanay
Batumi Chiatura
Groznyy
Tbilisi
Yerevan Baku
Krasnovodsk

Arkhangelsk (Archangel)
Vorkuta
Norilsk
Surgut
Ust Llim
Ust Kut Bam RR Komsomolsk Sovetskaya Gavan
Tomsk Krasnoyarsk Bratsk Tayshet
Petropavlovsk Kemerovo Khabarovsk
Omsk Cheremkhovo Irkutsk Chita
Novosibirsk Novosibirsk Ulan-Ude Vladivostok
Akmola Barnaul Leninogorsk Nakhodka
Karaganda Semipalatinsk
Dzhezkazgan
Balkhash
Tashkent Almaty
Ashgabat Fergana
Dushanbe

1 Zaporozhye
2 Nikopol
3 Krivoy Rog
4 Dnepropetrovsk
5 Magnitogorsk
6 Chelyabinsk

Baltic Sea, Barents Sea, Kara Sea, Laptev Sea, Yenisey R., Ob R., Lena R., Angara River, Lake Baikal, Trans-Siberian RR, Amur R., Sea of Okhotsk, Black Sea, Aral Sea, Irtish R., Turkestan-Siberian RR

0 400 800 miles
0 400 800 kilometers

MAIN RAILROADS OF THE FORMER SOVIET REGION

++++ Railroads

++++ Lines carrying heaviest freight traffic

Major industrial region (for detail, see map below)

Riga St.Petersburg
Moscow Industrial Region Cherepovets
Moscow Yaroslavl
Ivanovo
Tula Nizhniy Novgorod Solikamsk-Berezniki
Lipetsk Kazan Perm
Togliatti Volga-Urals Nizhniy Tagil
Voronezh Yekaterinburg
Saratov Chelyabinsk
Samara Ufa
Volgograd Orsk Magnitogorsk
Middle Volga Industrial Region Urals Industrial Region
Krasnodar Volga-Urals
Groznyy Karaganda
Yerevan Tbilisi
Baku

Minsk
Lvov See inset
Kursk Magnetic Anomaly
Ukranian Industrial Region

1 Izhevsk
2 Naberezhnyye

Ob R.
West Siberia
Kuznetsk Industrial Region
Krasnoyarsk
Omsk Kemerovo
Novosibirsk Novokuznetsk

Black Sea, Caspian Sea, Aral Sea, Lake Balkhash, Lake Baikal

Tashkent

UKRANIAN INDUSTRIAL REGION
Kiev
Dnepropetrovsk
Dneprodzerzhinsk
Krivoy Rog Kharkov
Odessa Donets Basin Lugansk
Zaporozhye Gorlovka
Mariupol Makeyevka
Donetsk Rostov
Black Sea

PRINCIPAL INDUSTRIAL CENTERS OF THE FORMER SOVIET REGION
(classified broadly by most significant type of manufacturing)

Iron and steel (often with associated mining, machine-building, and/or chemical industries)

Machine-building

Diversified industries (including most or all of the following: machinery, foods, textiles, chemicals)

OTHER IMPORTANT INDUSTRIES ASSOCIATED WITH MACHINE-BUILDING OR DIVERSIFIED INDUSTRIAL CENTERS

Petroleum refining and petrochemicals

Oil fields

Steel

Wood industries

Chemicals

Textiles

Automobiles (minor)

Aluminum reduction

Food processing

Automobiles (major)

Coal

Iron ore

Major industrial region

0 250 500 miles
0 250 500 kilometers

FIGURE 7.5

Maps of major industrial areas and railroads in the Former Soviet Region.

182

FIGURE 7.6
St. Isaac's Cathedral at the heart of St. Petersburg, established on the Baltic Sea at the mouth of the Neva River to serve as Russia's "Window on the West." *Sovfoto/Eastfoto*

The St. Petersburg area (see Fig. 7.6) is not as great an industrial center as the Moscow region, Ukraine, or the Urals, but is still a significant presence in industrial development. As in the case of Moscow, local minerals have not formed the basis of this industrialization. Instead, metals from outside the area provide material for the metal-fabricating industries that are the leaders in St. Petersburg's diversified industrial structure. Supported by university and technological-institute research workers, the city's highly skilled labor force played an extremely significant role in early Soviet industrialization. They pioneered the development of many complex industrial products, such as power-generating equipment and synthetic rubber, and supplied groups of experienced workers and technicians to establish new industries in other areas. Industrial innovation now centers in Moscow, but St. Petersburg is still preeminent in some specialized industries such as the making of large turbines for hydroelectric stations.

INDUSTRIES OF THE URALS

The Ural Mountains contain an extraordinarily varied collection of useful minerals. Although deficient in coking coal, this highly mineralized area has valuable deposits of iron, copper, nickel, chromium, manganese, bauxite, asbestos, magnesium, potash, industrial salt, and other minerals. Sizable amounts of low-grade bituminous coal, plus lignite and some anthracite, are mined. The important Volga-Urals oil fields lie partly in the western foothills of the Urals, and a major gas field lies at the southern end of the mountains.

The former Communist regime fostered the development of the Urals as an industrial region well removed from the exposed western frontier of the Soviet Union. The major industrial activities are: heavy metallurgy, emphasizing iron and steel and the smelting of nonferrous ores; the manufacture of heavy chemicals, based on some of the world's largest deposits of potassium and magnesium salt; and the manufacture of machinery and other metal-fabricating activities.

The Soviets modernized and expanded tsarist-era metallurgical and machine-building plants in the Urals and constructed several immense new plants. Their most famous and spectacular creation was the Soviet Union's largest iron-and-steel center at Magnitogorsk (population of city proper, 440,000) in the southern Urals. Located near a reserve of exceptionally high-grade iron ore, this place was not even a village prior to 1931. Construction of a huge plant and a city to house the workers began that year. Other giant Soviet-built iron-and-steel mills are scattered through the Urals at Chelyabinsk, Nizhniy Tagil, and Orsk.

The years that saw the creation of Magnitogorsk also saw a large expansion of coal mining in the Kuznetsk Basin (Kuzbas), located in southern Siberia more than 1000 miles (*c.* 1600 km) to the east of Magnitogorsk. A railway shuttle developed, with Kuznetsk coal and coke moving to Magnitogorsk and other industrial centers in the Urals, and Urals iron ore (primarily from Magnitogorsk) moving to a new iron-and-steel plant in the Kuznetsk Basin. Thus was created the famous Urals-Kuznetsk Combine (*Combinat* in Russian). Each end of the Combine, however, soon became partially independent of the other as coal was exploited closer to the Urals and iron closer to the Kuznetsk Basin. Much iron ore now reaches the Urals from mines that have been developed in a huge ore formation between the Moscow region and Ukraine called the Kursk Magnetic Anomaly, or KMA (so named because of its disturbing effect on compass needles).

Yekaterinburg (formerly Sverdlovsk; population of city proper, 1.38 million; metropolitan area, 1.62 million), located at the eastern edge of the Ural Mountains, is the largest city of the Urals and the region's preeminent economic, cultural, and transportation center. The second most important center is Chelyabinsk (population of city proper, 1.15 million; metropolitan area, 1.33 million), located 120 miles (193 km) to the south. This grimy steel town was infamous in Old Russia as the point of departure for exiles sent to Siberia. Today, areas in the vicinity are highly polluted with radioactive waste from Soviet nuclear operations. In western Siberia, between the Urals and the large industrial and trading center of Omsk (population of city proper, 1.17 million; metropolitan area, 1.19 million), rail lines from Yekaterinburg and Chelyabinsk join to form the Trans-Siberian Railroad, the main artery linking the Far East with the coreland (see railroad map, Fig. 7.5). Omsk is a major metropolitan base for Siberian oil and gas development. It is also Siberia's most important center of oil refining and petrochemical manufacturing.

INDUSTRIES OF THE KUZNETSK REGION

From Omsk, the Trans-Siberian Railroad leads eastward to the Kuznetsk industrial region, the most important concentration of manufacturing east of the Urals. The principal localizing factor for industry here is the enormous reserve of coal mentioned above. The manufacture of iron and steel is also a major industrial activity of the Kuznetsk region, the main center being Novokuznetsk (population of city proper, 600,000). The industry draws its iron ore from various sources in the Asian sector of the Former Soviet Region. The Kuznetsk Basin produces steel primarily for use by fabricating industries in Siberia, but sizable quantities move to factories west of the Urals. Chemical manufacturing, based in large measure on byproducts of coke ovens, also is important.

The largest urban center of the Kuznetsk region is Novosibirsk, a diversified industrial, trading, and transportation center located on the Ob River at the junction of the Trans-Siberian and Turkestan-Siberian (Turk-Sib) railroads. Sometimes called the "Chicago of Siberia," Novosibirsk (population of city proper, 1.45 million; metropolitan area, 1.60 million) has developed from a town of a few thousand at the turn of the century. Hundreds of factories produce mining, power-generating, and agricultural machinery, tractors, machine tools, and a wide range of other products.

The city of Akademgorodok ("Academy Town" or "Science Town"; population of city proper, 100,000), Siberia's main center of scientific research, is an outlying satellite community of Novosibirsk. The city was established in 1957 as a parklike haven where some of the Soviet Union's greatest scientific minds could research and develop both civilian and military industrial innovations. Reflecting a general crisis in the post-Soviet military-industrial establishment, Akademgorodok's funding has shrunk dramatically, forcing many prominent scientists to seek menial jobs. Internationally there are fears that underpaid Russian nuclear scientists will sell their knowhow or materials to governments or terrorist organizations seeking to develop atomic weapons. There have already been hundreds of thefts of radioactive substances at nuclear and industrial institutions in the former USSR. The destination points of these materials remain largely unknown. Former Soviet nuclear experts have applied for work or are already working in Iran, Iraq, Algeria, India, Libya and Brazil.

Numerous industrial cities outside the main industrial concentrations are scattered through the Slavic Coreland. The most notable are cities along the Volga from Kazan southward, in the region Russians know as the Povolzhye. The largest are Samara (formerly Kuybyshev; population of city proper, 1.26 million; metropolitan area, 1.51 million), Volgograd (formerly Stalingrad; population of city proper, 1.01 million; metropolitan area, 1.36 million), Saratov (population of city proper, 911,100; metropolitan area, 1.16 million; heart of the Volga German

Autonomous Republic until World War II), and Kazan (population of city proper, 1.11 million; metropolitan area, 1.17 million). These four cities in the middle Volga industrial region have diversified machinery, chemical, and food-processing plants. Kazan also has a major university whose alumni include Vladimir Lenin and Leo Tolstoy. Maxim Gorki wrote about Kazan in *My Universities*; down along the Volga wharves, he found "a whirling world where men's instincts were coarse and their greed was naked and unashamed."

Soon after its founding in 1596 as a frontier post against the Tatars, Samara became the financial capital of the Volga wheat trade. Its strategic importance grew after 1914 when the greatest railway in Russia at the time, the Moscow-Orenburg line, was laid across the greatest river in Russia here. Under the Soviets, Samara was one of the military-industrial "closed cities" (closed to foreigners because of security concerns) whose industries Stalin commanded to be hastily built as a defensive measure against German attack in World War II. Like Nizhniy Novgorod upriver on the Volga, the city is undergoing a process known officially as "conversion," that is, the retooling of military enterprises for civilian-oriented manufacturing (Fig. 7.7).

In recent decades, the Volga cities have seen a marked upsurge of industrial activity, particularly the construction of dams and hydroelectric stations along the Volga River and its large tributary, the Kama. A second major development has been the rapid rise of the Volga region to leadership in the Former Soviet Region's automobile industry. Togliatti (population of city proper, 630,000) on the Volga, is by far the Former Soviet Region's leading center for the production of passenger cars. These come from the Volga Automobile Plant, which was built and equipped for the Soviet government by Italy's Fiat

FIGURE 7.7
The port of Saratov, on the lower Volga River. The Volga cities have traditionally had diversified industries, including the military production which is now being scaled back in the process of conversion to civilian production. *Wolfgang Kaehler/Gamma Liaison*

Company; the city itself takes its incongruous name from a former leader of Italy's Communist Party.

Large-scale exploitation of petroleum in the nearby Volga-Urals fields has contributed to the industrial rise of the Volga cities. Prior to the opening of fields in western Siberia during the 1970s, the Soviet Union's most important area of oil production was the Volga-Urals fields. Stretching from the Volga River to the western foothills of the Ural Mountains and containing many separate deposits of oil, these fields produced about two thirds of the USSR's oil in the middle 1960s. By the late 1980s, this figure dropped to well under one third of the countrywide total. Important petrochemical industries in many parts of the Former Soviet Region use oil and natural gas from these and other sources to manufacture synthetic rubber, artificial fibers, nitrogenous fertilizers, plastics, and other products. In the Volga-Urals region, Samara and Ufa (population of city proper, 1.10 million; metropolitan area, 1.12 million) are especially well-known oil refining and petrochemical centers, but all of the larger cities along the Volga and Kama rivers (such as Perm on the Kama: population of city proper, 1.11 million; metropolitan area, 1.18 million), as well as some smaller cities in the oil fields, have a share of these industries.

FIGURE 7.8
Rail yard on the electrified Trans-Siberian Railroad at the busy port of Nakhodka on the Pacific. The view shows passenger cars, timber for export, and a trainload of containers. Much container traffic moves between Pacific countries and European countries via this Trans-Siberian "bridge." *Tass/Sovfoto*

7.4   The Russian Far East

The Far East is the mountainous Pacific edge of Russia. Most of it is a thinly populated wilderness in which the only settlements are fishing ports, lumber and mining camps, and the villages and camps of aboriginal peoples. Port functions, fisheries, and forest industries provide the main support for most Far Eastern communities, and the output of coal, oil, and a few other minerals is small. Most of the Russians and Ukrainians who make up the majority of its people live in a narrow strip of lowland behind the coastal mountains in the southern part of the region. This lowland, drained by the Amur River and its tributary the Ussuri, is the region's main axis of industry, agriculture, transportation, and urban development. Several small to medium sized cities form a north-south line along two important arteries of transportation: the Trans-Siberian Railroad and the lower Amur River. At the south on the Sea of Japan is the port of Vladivostok (population of city proper, 648,000), kept open throughout the winter by icebreakers. About 50 miles (80 km) east of the city, the main commercial seaport area of the Far East has developed at Nakhodka (population of city proper, 165,000; see Fig. 7.8) and nearby Vostochnyy (East Port). Both ports are nearly ice-free.

The coastal regions of the Far East, from Vladivostok northward to the mouth of the Amur River, have a humid continental climate with monsoonal seasonality. As in nearby parts of Korea and China, most of the annual rainfall results from moist onshore winds of the summer monsoon. In contrast, the cold outflowing winds of the winter monsoon produce little precipitation. The average annual precipitation is 20 to 30 inches (c. 50–75 cm), of which three fourths or more falls from April through September.

North of Vladivostok lies a small district that is the most important center of the Far East's meager agriculture, producing cereals, soybeans, sugar beets, and milk for Far Eastern consumption. The Far East is far from self-sufficient in food and consumer goods; large shipments from the Russian coreland and from overseas supplement local production.

The diversified industrial and transportation center of Khabarovsk (population of city proper, 601,000) is located at the confluence of the Amur and Ussuri rivers, where the main line of the Trans-Siberian Railroad turns south to Vladivostok and Nakhodka and the Amur River turns north to the Sea of Okhotsk.

Before World War II, the Soviet Union and Japan held the northern and southern halves, respectively, of the large island of Sakhalin, which today has important forest and fishing industries and some coal, petroleum, and natural gas reserves. At the end of the war, the USSR annexed southern Sakhalin and the Kuril Islands and repatriated the Japanese population. Russian control of these former Japanese territories continues to be a major problem in relations between the two countries (see Chapter 14).

The Kurils are small volcanic islands that screen the Sea of Okhotsk from the Pacific. Fishing is the main economic activity. At the north, the Kurils approach the mountainous peninsula

FIGURE 7.9
Kamchatka is a land of fire and ice. The 11,364 feet (3464 meter) Koryak volcano towers over the city of Petropavlovsk   *Sovfoto/Eastfoto*

of Kamchatka (total population is 450,000, with 320,000 living in two cities). Like the islands, Kamchatka is located on the Pacific Ring of Fire—the geologically active perimeter of the Pacific Ocean—and has 23 active volcanoes (Fig. 7.9). Soviet authorities protected Kamchatka for decades as a military area, and no significant development activities occurred there. Kamchatka is now one of the last great wilderness areas on Earth, resembling the U.S. Pacific Northwest landscape of a century ago. A struggle is now underway between those who want to develop Kamchatka's gold and oil resources and those who wish to set the land aside in national parks and reserves where ecotourism would be the only significant source of revenue.

7.5   The Northern Lands of Russia

North and east of the Slavic Triangle and west of (and partially including) the Pacific littoral lie enormous stretches of coniferous forest (taiga) and tundra extending from the Finnish and

Norwegian borders to the Pacific. These outlying wilderness areas in Russia may for convenience be designated the Northern Lands, although parts of the Siberian taiga extend to Russia's southern border.

These difficult lands form one of the world's most sparsely populated large regions. The climate largely prevents ordinary types of agriculture, but hardy vegetables, potatoes, hay, and barley are grown in scattered localities, and there is some dairy farming. A few primary activities, including logging, mining, reindeer herding, fishing, hunting, and trapping support most people. Significant towns and cities are limited to a small number of sawmilling, mining, transportation, and industrial centers, generally along the Arctic coast to the west of the Urals, along the major rivers, and along the Trans-Siberian Railroad between the Slavic Triangle and Khabarovsk.

The oldest city on the Arctic coast is Arkhangelsk (Archangel; population of city proper, 416,000), located on the Northern Dvina River inland from the White Sea. Tsar Ivan the Terrible established the city in 1584 for the purpose of opening seaborne trade with England. Today, the city is the most important sawmilling and lumber-shipping center of the Former Soviet Region. Despite its restricted navigation season from late spring to late autumn, it is one of the more important seaports.

Another Arctic port, Murmansk (population of city proper, 468,000; see Fig. 7.10) is located on a fjord along the north shore of the Kola Peninsula west of Arkhangelsk, and is the headquarters for important fishing trawler fleets that operate in the Barents Sea and North Atlantic. Murmansk also has a major naval base and cargo port, and is home port to the icebreakers that escort cargo vessels and carry Western tourists to the North Pole and other High Arctic destinations. It is connected to the coreland by rail. The harbor is open to shipping all year, thanks to the warming influence of the North Atlantic Drift, an exten-

FIGURE 7.10
The ice-free port of Murmansk is situated on a fjord at the northern end of the Kola Peninsula.   *Sovfoto/Eastfoto*

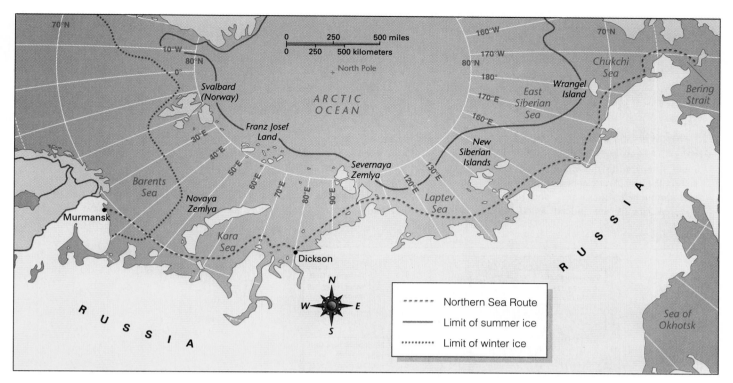

FIGURE 7.11
The Northern Sea Route.

sion of the Gulf Stream. Murmansk and Arkhangelsk played a vital role in World War II, continuing to receive supplies by sea from the Soviet Union's Western allies after Nazi forces captured and closed off the other ports of the western Soviet Union.

Murmansk and Arkhangelsk are western termini of the Northern Sea Route (NSR), a waterway the Soviets developed to provide a connection with the Far East via the Arctic Ocean (Fig. 7.11). Navigation along the whole length of the route is possible for only up to four months per year, despite the use of powerful icebreakers (including nine nuclear-powered vessels) to lead convoys of ships (Fig. 7.12). Areas along the route provide such cargoes as the timber of Igarka and the metals and ores of Norilsk. Some supplies are shipped north on Siberia's rivers from cities along the Trans-Siberian Railroad and then loaded onto ships plying the Northern Sea Route for delivery to settlements along the Arctic coast. The Northern Sea Route is now open to international transit shipping and, if not for the capricious sea ice, would be an attractive route. The distance between Hong Kong and all European ports north of London is shorter via the Northern Sea Route than through the Suez Canal. Russia is anxious to develop the Northern Sea Route more for its own exports, since Russian exports by land to western Europe must now pass over the territories of Ukraine, Belarus, and the Baltic countries.

The railroad that connects Murmansk with the coreland serves important mining districts in the interior of the Kola

Peninsula and logging areas in Karelia, which adjoins Finland. The Kola Peninsula is a diversified mining area, producing nickel, copper, iron, aluminum, and other metals, as well as phosphate for fertilizer. The peninsula and Karelia are physi-

FIGURE 7.12
The Russian nuclear-powered icebreaker *Yamal* at the North Pole in August, 1996. The former Soviet icebreaker fleet still escorts commercial ships on the Northern Sea Route, but during the summer carries Western tourists such as these to the High Arctic.   *Photo by J.J. Hobbs*

PROBLEM LANDSCAPE

RUSSIANS DESCRIBE EXTENSIVE DUMPING OF NUCLEAR WASTE

On the map, the Russian islands of Novaya Zemlya ("New Land") separating the Barents and Kara seas in the high Arctic appear remote, wild, and untouched. These ice- and tundra-covered islands are indeed an isolated wilderness, which is precisely why Soviet authorities perceived them as prized grounds for nuclear weapons testing and nuclear waste dumping.

Soon after the breakup of the Soviet Union in 1991, a former Soviet radiation engineer stepped forward with disturbing information: Soviet authorities had ordered the dumping of highly radioactive materials off the coast of Novaya Zemlya for the past 30 years (Fig. 7.A). From the 1950s through 1991, the Soviet navy and ice-

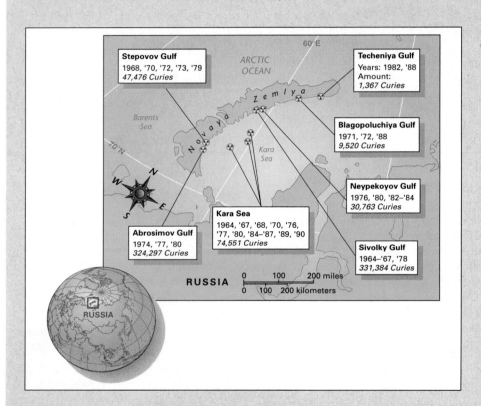

Stepovov Gulf
1968, '70, '72, '73, '79
47,476 Curies

Techeniya Gulf
Years: 1982, '88
Amount:
1,367 Curies

Blagopoluchiya Gulf
1971, '72, '88
9,520 Curies

Neypekoyov Gulf
1976, '80, '82–'84
30,763 Curies

Kara Sea
1964, '67, '68, '70, '76,
'77, '80, '84–'87, '89, '90
74,551 Curies

Abrosimov Gulf
1974, '77, '80
324,297 Curies

Sivolky Gulf
1964–'67, '78
331,384 Curies

ARCTIC OCEAN

Barents Sea

Kara Sea

Novaya Zemlya

RUSSIA

0   100   200 miles
0   100   200 kilometers

FIGURE 7.A
Sites around Novaya Zemlya used as Soviet nuclear-waste dumping grounds between 1964 and 1990.

cally an extension of Fennoscandia, located on the same ancient, glacially scoured, granitic shield that underlies most of the Scandinavian peninsula and Finland. Karelia, with its thousands of lakes, short and swift streams, extensive softwood forests, timber industries, and hydroelectric power stations, bears a close resemblance to nearby parts of Finland (Fig. 7.13). In fact, Finland annexed some of this area during World War II.

Important energy resources, including good coking coal, petroleum, and natural gas, are present in the basin of the Pechora River, east of Arkhangelsk. The principal coal-mining center is Vorkuta (population of city proper, 116,000), infamous for the labor camp where great numbers of political prisoners died during the Stalin regime. Coal from the Pechora fields moves by rail to St. Petersburg and to a large iron-and-steel plant at Cherepovets north of Moscow. The plant processes iron

breaker fleet secretly and illegally used the shallow waters off the eastern shore of No-vaya Zemlya as a nuclear dumping ground. At least twelve nuclear reactors removed from submarines and warships—six of which still contained their highly radioac-tive nuclear fuel—were sent to the shallow seabed. Some of these individual reactors contain roughly seven times the nuclear material contained in the Chernobyl reactor which exploded in 1986. Between 1964 and 1990, 11,000 to 17,000 containers of solid radioactive waste were also dumped here at shallow depths between 200 and 1000 feet. Soviet seamen reportedly cut holes in those "sealed" containers which otherwise would not have sunk. This dumping occurred even though the Soviet Union joined other na-tions in a 1972 treaty which allowed only low-level radioactive waste to be dumped at sea, and then only at depths greater than 12,000 feet.

The danger posed by the nuclear mate-rial is not limited to the islands' shores but extends through the entire ecosystem of Russia's Arctic seas, and possibly beyond. Contamination can begin at the base of the food chain (to the phytoplankton which "bloom" in the area each spring) and pass upward to higher trophic levels, where fish, seals, walrus, polar bears, and people feed. Alaskan authorities are worried, and Nor-wegians are particularly concerned. Nor-way's Prime Minister Gro Brundtland described the dumping as a "security risk to people and to the natural biology of north-ern waters." Western European and North American markets for Norwegian and Russian fish, which now represent signifi-cant exports for these countries, may refuse purchases if radiation levels rise.

Due to the long half-life of the nuclear materials, the dangers posed by 30 years of haphazard dumping near Novaya Zemlya may last thousands of years. The threat can be reduced greatly by retrieving the waste and disposing of it in terrestrial sites deep underground, which are considered safer. (Today, the Russian military supports a pro-posal to convert some of the shafts used in 1950s atomic weapons tests to permanent radioactive waste repositories.) The major obstacle to this step is the extreme costli-ness of such an operation—estimated to be in the hundreds of billions of dollars—which is well beyond Russia's economic means. An arms-agreement obligation for Russia to decommission about 80 nuclear submarines based in Murmansk is also costly and, because of the radioactive mate-rials involved, potentially dangerous. Con-fessing and confronting the nuclear legacy of the Soviet Union, Russia has asked the United States and other nations to help pay for an environment with fewer nuclear haz-ards. The United States, in turn, is asking Russia not to export its nuclear power tech-nologies to Iran and other countries where the West fears they may be used to develop weapons.

The predicament of Novaya Zemlya and the future of the vast wilderness in Rus-sia's Far East reflect important questions about environment and development all across the Former Soviet Region. Like the United States and other industrialized coun-tries, the Soviet Union experienced prob-lems of environmental pollution caused by rapid economic growth. Russia and the other countries must now deal with the So-viet legacy of decades of environmental ne-glect. Environmental cleanup is hindered by the massive cost of repairing past dam-age and a reluctance to bear the expenses of stringent enforcement of pollution-abate-ment measures. Some Russian sources esti-mate that 80 percent of all industrial enterprises in Russia would go bankrupt if forced to comply with environmental laws. A desperate search for hard currency and widespread corruption have led to other en-vironmental problems, including the wide-scale selling off of natural resources like timber, and to poaching of animals. With the breakup of the USSR, many environ-mentalists had predicted more protection for nature. They have been disappointed.

William J. Broad, "Russians Describe Extensive Dump-ing of Nuclear Waste." *The New York Times,* April 27, 1993: A1.

---

ore concentrates from the Kola Peninsula and ore from the Kursk Magnetic Anomaly to make steel that is used mainly in the industries of St. Petersburg. In 1994, U.S. and Norwegian oil companies formed a new company to explore for and possi-bly develop oil for export from the Pechora region.

East of the Urals, the swampy plain of western Siberia north of the Trans-Siberian Railroad has seen a surge of petro-leum and natural gas production since the 1960s. The West Siberian fields are the largest producers in the Former Soviet Region. Extraction and shipment of oil and gas take place there under frightful difficulties caused by severe winters, per-mafrost, and swampy terrain. Huge amounts of steel pipe (Fig. 7.14) and pumping equipment have been required to connect the remote wells with markets in the coreland and Europe.

Still farther east, the town of Igarka, located about 425 miles (*c.* 680 km) inland on the deep Yenisey River, is an

FIGURE 7.13
The village of Varzuga near Murmansk on the Kola Peninsula. The wooden pitched-roof homes, coniferous trees and a landscape heavily scoured by glaciation are characteristic of this region.   *Tass/Sovfoto/Eastfoto*

important sawmilling center accessible to ocean shipping during the summer and autumn navigation season. Northeast of Igarka lies Norilsk (population of city proper, 175,000), the most northerly mining center of its size on the globe. Rich ores yielding nickel, copper, platinum, and cobalt have justified the region's large investment in this Arctic city, including a railway connecting the mines with the port of Dudinka on the Yenisey downstream from Igarka.

In central and eastern Siberia, the taiga extends southward beyond the Trans-Siberian Railroad. A vast but sparsely settled hinterland is served by a few cities spaced at wide intervals along the railroad: Krasnoyarsk (population of city proper, 912,000) on the Yenisey River, Irkutsk (population of city

proper, 626,000) on the Angara tributary of the Yenisey near Lake Baikal, and others. From east of Krasnoyarsk, a branch line of the Trans-Siberian leads eastward to the Lena River. Years of construction in the 1970s and 1980s under very difficult conditions resulted in continuation of this line to the Pacific under the name Baikal-Amur Mainline, or BAM (Fig. 7.15). The chief purposes of the new railroad are to open important mineralized areas east of Lake Baikal and to lessen the strategic vulnerability caused by the close proximity of the Trans-Siberian Railroad to the Chinese frontier. The Pacific terminus is located on the Gulf of Tatary, which connects the Sea of Okhotsk with the Sea of Japan (see Fig. 6.1, page 157; and Fig.7.5).

The BAM railway passes through Bratsk (population of city proper, 255,000), site of a huge dam and hydroelectric power station on the Angara River. Lake Baikal, which the Angara drains, lies in a mountain-rimmed rift valley and is the deepest body of fresh water in the world at more than a mile deep in places. Lake Baikal is at the center of an environmental controversy over the estimated 8.8 million cubic feet of waste water which pour into it daily from two large paper and pulp mills on its shores. The waste jeopardizes Baikal's unique natural history as the lake contains an estimated 1800 endemic plant and animal species (see page 192).

Siberia's development is based in large part on hydroelectric plants located along the major rivers. The largest hydropower stations are two on the Yenisey (at Krasnoyarsk and in the Sayan Mountains) and two on the Yenisey's large tributary, the Angara (at Bratsk and Ust Ilim). These power stations and dams are among the largest in the world. They supply inexpensive electricity, consumed mainly in mechanized industries that use power voraciously. These include cellulose plants that process Siberian timber and aluminum plants making use of aluminum-bearing material (alumina) derived from ores in Siberia, the Urals, and other areas and imported from abroad via the Black Sea.

FIGURE 7.14
Laying an oil pipeline in Russia's Northern Lands. Immense quantities of steel have been required for thousands of miles of oil and gas pipes that have been laid in Russia and other parts of the Former Soviet Region since World War II.   *Tass/Sovfoto*

FIGURE 7.15
Linkup of tracklayers from east and west on the Baikal-Amur (BAM) railroad, September 19, 1984. The two ends of the line met in the mountainous taiga between Lake Baikal and the Pacific. The laying of a symbolic "golden rail" marked the completion of 10 years of work under difficult climatic and terrain conditions.   *Novosti/Sovfoto*

Scattered throughout central and eastern Siberia are important gold-mining centers, mainly in the basins of the Kolyma and Aldan rivers and in other areas in northeastern Siberia. (In Stalin's time, the Kolyma mines, located in one of the world's harshest climates, developed an evil reputation as death camps for great numbers of political prisoners forced to work there.) Large coal deposits are mined along the Trans-Siberian Railroad at various points between Irkutsk and the Kuznetsk Basin. Huge coal reserves in many areas between the Yenisey and the Pacific have scarcely begun to be exploited. There were discoveries of large deposits of diamonds in the eastern part of the Central Siberian Uplands in the 1950s, and Russia rapidly became a major world producer and exporter of both gem diamonds and industrial diamonds.

Aside from irregularly distributed centers of logging, mining, transportation, and industry peopled mostly by Slavs, the Northern Lands are home mainly to non-Slavic peoples who make a living by reindeer herding, trapping, fishing, and, in the more favored areas of the taiga, by cattle raising and precarious forms of cultivation (Fig. 7.16). The domesticated reindeer is especially valuable to the tundra peoples, providing meat, milk, hides for clothing and tents, and draft power. The Yakuts, a Turkic people living in the basin of the Lena River, are among the most prominent of the non-Slavic ethnic groups in the Northern Lands. Their political unit, Sakha (formerly Yakutia), is larger in area than any former Union Republic except the Russian Federation, within which it lies. Sakha has only about 1 million people (1996 estimate) inhabiting an area of 1.2 million square miles (3.1 million sq km). The capital city of Yakutsk (population of city proper; 187,000) on the middle Lena River has road connections with the BAM and Trans-Siberian railroads and with the Sea of Okhotsk. Riverboats on the broad and deep

Lena provide a connection with the Northern Sea Route and the southern rail system during the warm season. Sakha will figure prominently in future mineral and timber development with its large reserves of diamonds, natural gas, coking coal, other minerals, and timber.

FIGURE 7.16
Yakut reindeer herders in the northeast Siberian region of Sakha. *Sovfoto/Eastfoto*

Landscape in Literature

BAIKAL

Valentin Rasputin, a native and still a resident of Irkutsk in Siberia, is one of Russia's leading writers and most outspoken environmental activists. He is associated with a Russian literary tradition called "rural prose" or "village prose," which presents a realistic view of the often difficult lives of people in Russia's vast hinterland. In this passage, Rasputin argues that Lake Baikal (Fig. 7.B) is a spiritual resource—a marked contrast to the economic cornucopia that most Soviet (and now Russian) planners have perceived it to be.

"The sacred sea," "the sacred lake," "the sacred water"—that is what native inhabitants have called Baikal from the beginning of time. So have Russians, who had already arrived on its shores by the seventeenth century, as well as travelers from abroad, admiring its majestic, supernatural mystery and beauty. The reverence for Baikal held by uncivilized people and also by those considered enlightened for their time was equally complete and captivating, even though it touched mainly the mystical feelings in the one and the aesthetic and scientific impulses in the other. The sight of Baikal would dumbfound them every time because it did not fit their conceptions either of spirit or of matter: Baikal was located where something like that should have been impossible, it was not the sort of thing that should have been possible here or anywhere else, and it did not have the same effect on the soul that "indifferent" nature usually does. This was something uncommon, special and "wrought by God."

Baikal was measured and studied in due course, even, in recent years, with the aid of deep-sea instruments. It acquired definite dimensions and became subject to comparison, alternatively likened to Lake Tanganyika and to the Caspian Sea. They've cal-

FIGURE 7.B
Fishermen at work in Lake Baikal, Earth's deepest lake.   *Tass/Sovfoto*

culated that it holds one fifth of all fresh water on our planet, they've explained its origin, and they've conjectured as to how species of plants, animals and fish existing nowhere else could originate here and how species found only in other parts of the world many thousands of miles away managed to end up here. Baikal is not so simple that it could be deprived of its mystery and enigma that easily, but based on its physical properties it has, nevertheless, been assigned a fitting place alongside other great wonders that have already been discovered and described, as well it should. And it stands alongside them solely because Baikal itself, alive, majestic, and not created by human hands, not comparable to anything and not repeated anywhere, is aware of its own primordial place and its own life force.

Nature as the sole creator of everything still has its favorites, those for which it expends special effort in construction, to which it adds finishing touches with a special zeal, and which it endows with special power. Baikal, without a doubt, is one of these. Not for nothing is it called the pearl of Siberia. We will not discuss its natural resources at present, for that is a separate issue. Baikal is renowned and sacred for a different reason—for its miraculous, life-giving force and for its spirit, which is a spirit not of olden times, of the past, as with many things today, but of the present, a spirit not subject to time and transformations, a spirit of age-old grandeur and power preserved intact, of irresistible ordeals and inborn will.

From "Baikal," by Valentin Rasputin. In *Siberia on Fire: Stories and Essays by Valentin Rasputin.* DeKalb: Northern Illinois University Press, 1989: 188–189.

7.6   The Future of the Russian Federation

Russia's internal republic of Sakha is the largest in area of many autonomies (nationality-based republics and lesser units) in the Russian Federation. They are successors to the autonomous ASSRs, autonomous oblasts, and autonomous okrugs of the Soviet era (see inset map, Fig. 6.1). Scattered autonomies are located in other independent successor states of the former USSR, but the majority are in the Russian Federation, where they occupy over two fifths of the total area and represent somewhat under one fifth of the total population.

Nearly half of the population in Russia's autonomies is made up of ethnic Russians, who actually compose a majority in many units. Employment in new industries and mining attracted many of these Russians during the Communist period. Titular nationalities of the autonomies are ethnically diverse (see Fig. 7.2): some, such as the Karelians, Mordvins, and the Komi, are Finnic (related to the Finns and Magyars); others are Turkic (Tatars, Bashkirs, and Yakuts, for example), Mongol (Buryats near Lake Baikal, for example), and members of many other ethnic groups. Russian expansion brought these peoples into the Russian Empire at different times over a period of several centuries, and the Communist rulers of the USSR organized them into "autonomous" units ranked on a nationalities ladder with the former SSRs (Union Republics) at the top. Some of the lesser units eventually climbed the ladder to become Union Republics.

The largest areal units form a nearly solid band stretching across the northern part of the Northern Lands from Karelia, bordering Finland, to the Bering Strait. The Karelian, Komi, and Sakha (Yakut) republics are in this group. This band also includes several large but thinly inhabited units of aboriginal peoples that the Soviets designated as "autonomous okrugs." Pastoralism generally predominates over agriculture in the autonomies. Large cities are few; the only "million cities" are Kazan in Tatarstan, Ufa in Bashkiria, and Perm in Komi-Permyak. Manufacturing is poorly developed except in a handful of cities, such as Naberezhnyye Chelny in Tatarstan, where Russia's largest truck plant is located. However, minerals and mining are of major importance in some units. Oil and gas abound in the Volga-Urals fields in Tatarstan and Bashkiria, and there are high-grade coal deposits at Vorkuta in the Komi Republic. The diamonds mined in Sakha represent one fourth of the world's production.

Such resources have great potential significance in the geopolitical realm. Many of the former Soviet ASSRs, recalling Moscow's long standing promises of self-rule for them, have issued declarations of sovereignty asserting their right to greater self-direction of their internal affairs and greater control over their own resources. Moscow has rewarded some of them. The Sakha Republic, for example, in 1992 won the right to keep 45 percent of hard currency earnings from foreign sales of Sakha diamonds, compared with only a small fraction in the Soviet era. In turn, Sakha must now pay for the government subsidies which Moscow previously paid to local industries. Without credits from Moscow, Sakha no longer pays taxes to Moscow. Salaries and other indices of living standards in the Sakha Republic have risen since this agreement was implemented.

Such regional demands for greater self-rule threaten the unity of the Russian Federation. Eighteen out of 20 of Russia's republics (consisting of the 16 Soviet-era ASSRs plus 4 autonomous regions upgraded to republic status) signed a 1992 Federation Treaty which grants them considerable autonomy. The treaty calls for the devolution of power centralized in Moscow, and for more cooperation between regional and federal governments. Each republic of the Russian Federation is legally entitled to have its own budget, tax laws and other legislation, and foreign and domestic economic partnerships. But many republics are complaining that Moscow is not honoring the treaty, and they are looking for regional solutions to their economic and other problems. Karelia, for example, is pressing for closer ties with neighboring Finland. Several industrial areas have joined in the "Siberian Agreement," which may create a new center of power mostly independent of Moscow. The Sverdlovsk Region has already declared itself the "Urals Republic." The Vladivostok area wants to become the "Maritime Republic." Kaliningrad, Arkhangelsk, Irkutsk, and Krasnoyarsk are discussing similar declarations.

Chechnya (then joined with Ingushetia) and Tatarstan did not sign the 1992 Russian Federation Treaty and insisted on independence from Moscow. Oil-rich Tatarstan now has its own constitution, parliament, flag, and official language, and has since signed the treaty. Other autonomous regions will be watching to see what happens in Tatarstan and Chechnya. The oil-rich, Turkic-speaking Bashkirs and the Chuvash might push for more independence. This could begin a process that would virtually cut Russia in half. Russia's alarmed government is trying to keep the lid on the republics to prevent what happened to the USSR from happening to Russia itself. Chechnya is of particular concern.

Chechnya (population: 1.2 million), although officially designated as a republic within the Russian Federation, first declared its independence from post-Soviet Moscow in November 1991. The Chechens, who are Sunni Muslims, have periodically resisted Russian rule, and Moscow has punished them. Chechen defiance of Russian rule actually dates to the 18th century, when Tsarina Catherine II began expanding the Russian empire toward the Caucasus. It took Moscow nearly a hundred years to achieve control of Chechnya. Stalin exiled what he perceived as the hostile and subversive Chechens to Siberia and Kazakstan, and fully 30 percent of the Chechen people died. In 1957 the Soviet government "rehabilitated" the Chechens and allowed them to return to their ancestral homeland, where many entered disputes with Russians and other outsiders who had settled there.

Russian troops attempted but failed to seize control over Chechnya soon after its 1991 declaration of independence.

They then withdrew to form a blockade around Chechnya's borders. President Yeltsin chose not to send troops into the region until late 1994 when, many analysts argue, his hold on political power grew so weak that he saw invading Chechnya as an opportunity to precipitate a crisis that would allow him to cancel Russian elections. Russia also feared losing strategic hold on the oil pipelines crossing Chechnya and on Chechnya's own plentiful oil supplies. Russian troops failed in an attempted invasion on December 11; they then regrouped to take the capital city of Groznyy (population: 401,000 prior to the fighting in 1994–1995). Skilled Chechen fighters took a heavy toll on Russian lives on New Year's Eve. Russian forces withdrew to bombard Groznyy throughout January 1995. At a cost of an estimated 30,000–80,000 lives, Russian troops succeeded in exerting physical control over most of Chechnya, but rebel forces retained control in many remote villages.

The conflict reignited late in 1995 and burned through the politically-charged period leading up to the Russian presidential election of 1996. Russia's involvement in Chechnya was very unpopular among the Russian people. It revealed deep divisions in Russian political and military leadership and greatly weakened support for Boris Yeltsin. In April 1996, two months before the elections, Yeltsin announced a cease-fire and peace plan for Chechnya that most observers saw as a ploy to increase his chances of reelection. The war began again after Yeltsin's victory at the polls, and Chechen forces recaptured Groznyy in August. The prospect of further Russian losses led President Yeltsin's security chief, Alexander Lebed, to negotiate a peace agreement with the Chechens in September. The pact required an immediate withdrawal of Russian forces from Chechnya and deferred the question of Chechnya's permanent political status for five years—beyond the expiration of Yeltsin's second term in office. Before the end of the five year period, the Chechen people are to hold a referendum on the republic's future. In 1997 they elected a president they believe will be the father of their nation. As Moscow continues to insist that Chechnya is part of the Russian Federation, and Chechens continue to demand complete independence for the country they would call Ichkeria, the future of Chechnya and the Russian Federation remain uncertain. Thus far, none of the autonomies that have declared independence from Russia has actually secured it. Moscow and the world watch nervously.

7.7 The Caucasus

The far southern Caucasus region borders Russia between the Black and Caspian seas. It includes the rugged Caucasus Mountains, a fringe of foothills and level steppes to the north, and the area to the south known as Transcaucasia. The Greater Caucasus Range forms almost a solid wall from the Black Sea to the Caspian (Fig. 7.17). It is similar in age and character to the Alps but is much higher, with the highest point in Europe at Mt. El-

FIGURE 7.17
Mt. Kazbek (15,812 feet, 5033 meters) in the Greater Caucasus Range towers over a small settlement in northern Georgia.   *Sovfoto*

brus (5642 m, 18,510 feet), on the Russia/Georgia border. In southern Transcaucasia is the mountainous, volcanic Armenian Plateau. Between the Greater Caucasus Range and the Armenian Plateau are subtropical valleys and coastal plains where the majority of people in Transcaucasia live. Russians and Ukrainians predominate in the North Caucasus, but non-Slavic groups form a large majority in Transcaucasia (see Table 6.1).

The Caucasian isthmus between the Black and Caspian seas has been an important north-south passageway for thousands of years, and the population includes many different peoples who have migrated into this region at various times. At least 25 or 30 nationalities are distinguished, mostly small and confined to mountain areas that became their refuges in past times. In addition to Russians and Ukrainians, most of whom live north of the Greater Caucasus Range, the most important nationalities are the Georgians (Fig. 7.18), Armenians, and Azerbaijanis, each represented by an independent country. Throughout history, all have stubbornly maintained their cultures in the face of pressure by stronger intruders.

These nationalities have ethnic characteristics and cultural traditions that are primarily Asian and Mediterranean in origin. Their religions differ. The Azerbaijanis (also known as Azeri Turks) are Muslims. The Georgians belong to one of the Eastern Orthodox churches. The Armenian Church is a very ancient, independent Christian body; Armenia became the world's first Christian country late in the second century A.D. (Fig. 7.19). There has been a history of animosity between the Armenians and Azeri Turks (including a war between them in 1905), growing out of the Turks' persecution of Armenians in the Ottoman Empire prior to and during World War I. Turkey and Azerbaijan still have not accepted responsibility for the horrors of the Armenian Genocide, in which an estimated 1.5 million Armenians died between 1915 and 1918.

Historical animosity recently flared into massive violence between Armenians and Azeris over the question of Nagorno-Karabakh, a predominantly Armenian enclave within Azerbaijan and governed by Azerbaijan but claimed by Armenia. From 1992 to 1994, fighting over Nagorno-Karabakh took thousands of lives and created nearly a million refugees. Large numbers of Armenians fled from Azerbaijan (although not from Nagorno-Karabakh) to Armenia as refugees. There was also a refugee flight of Azeris from Armenia to Azerbaijan. Tens of thousands of Azeris also fled into Iran. Iranian Azeris, who make up a third of Iran's population, feel that Iran has leaned too far toward Armenia as a de facto ally against Turkish influence in the Caucasus and Central Asia.

Russia negotiated a cease-fire between Armenia and Azerbaijan in March 1994, which has held until now, but the countries are still technically at war. Armenians are angry that Moscow did not restore to Armenia the enclave of Nagorno-Karabakh, which the Bolsheviks had put under Azeri control. But the Armenians also embrace Russia as a strategic ally against their historic enemy, the Turks, who virtually surround them.

In Georgia, there also have been recent episodes of ethnic and political violence, particularly between the Georgians and the primarily Muslim South Ossetians, who would like to free themselves from Georgian control and establish an Ossetian nation. The North Ossetians, whom they wish to join, live adjacent to them in Russia.

FIGURE 7.19
The Holy Echmiadzin Cathedral in Echmiadzin, Armenia. With foundations dating to 301 A.D., this is reputed to be the oldest Christian church in the world.   *Bruce Brander/Photo Researchers, Inc.*

FIGURE 7.18
Ethnic Georgians selling produce in a market at Tbilisi, the main city and capital of Georgia.   *Katrinka Ebbe*

Another Muslim people, the Abkhazians, who make up 1.8 percent of Georgia's population, also seek freedom from the Georgians. In September 1993, Abkhazian separatists captured the Georgian Black Sea port of Sukhumi. Russian forces initially aided them, both to regain access to Black Sea resorts and to take revenge on Georgian President Eduard Shevardnadze, the former Soviet foreign minister whom many Russians hold partly responsible for the breakup of the USSR. Russia was also putting pressure on Georgia to rejoin the CIS. Shevardnadze agreed in October 1993 that Georgia would join the CIS, and agreed to allow Russian bases on Georgian soil and Russian troops on the border with Turkey. Russian tanks then drove back the Abkhazian rebels fighting the Georgian government and secured rail lines linking Russia with Georgia. The status of Abkhazia is not yet settled, but technically it is still part of Georgia.

A humid subtropical climate prevails in coastal lowlands and valleys of Georgia, south of the high Caucasus Mountains. Mild winters and warm to hot summers combine with the heaviest precipitation to be found anywhere in the Former Soviet Region—50 to 100 inches (*c.* 125–250 cm) a year. The lowland area bordering the Black Sea in Georgia is the most densely populated part of Transcaucasia. Such specialty crops as tea, tung oil, tobacco, silk, and wine, together with some citrus fruits, grow there on a marginal basis. There is too much freezing weather for the citrus industry to flourish; "subtropical" areas in Transcaucasia have many frosts. The lowlands of eastern Transcaucasia, bordering the Caspian Sea, receive so little precipitation in most places that they are classed as subtropical steppe.

The Caspian Sea, which is actually the world's largest freshwater lake in area, is an important fishery, especially for

Regional Perspective

OIL IN THE FORMER SOVIET REGION

The Soviet Union was the world's largest oil producer in the late 1980s. Then came the collapse of the Communist empire in 1991, when Russia and 14 newly-independent countries scrambled to consolidate economic assets and political interests. With the Soviet Union's oil production and transport infrastructure straddling several of the new countries, and cooperation among these countries lacking, oil production plummeted. By 1995, the Former Soviet Region's total output had declined 40 percent from its 1988 high. Oil exports remain a priority, however; Russian oil and gas exports provide about 80 percent of the country's hard currency revenues.

The Soviet-era oil refinery and pipeline system still links Russia with Kazakstan, Azerbaijan, Ukraine, Belarus and other countries. The needs of these countries to buy and sell oil could provide a strong impetus to hold the Commonwealth of Independent States together or to create new political and economic blocs. However, fossil fuels are such precious commodities—and control over them so enhances national power—that they tend to promote division rather than unity between states. Since the breakup of the Soviet Union, Russia has periodically shut off the pipelines supplying natural gas to Ukraine,

Belarus, and Moldova to obtain political concessions from those countries.

Azerbaijan's oil situation epitomizes the economic prospects and obstacles facing the Former Soviet Region. The Azeri resource is large; there are huge reserves of oil and gas under the bed of the Caspian Sea. Planners study Azerbaijan's oil with a difficult question in mind: What is the best way to get this oil to market? They are considering a variety of export pipeline routes (Fig. 7.C), each of which has a different set of geopolitical, technical, and ecological problems. Since the Caspian is essentially a large lake, Azerbaijan is effectively landlocked, and Azeri oil must cross the territories of other nations to reach market. These territories, however, are embroiled in wars and political unrest left in the wake of the collapse of the USSR. Russia wants to retain strong influence over Azerbaijan in part by insisting that Azeri oil reach export terminals on the Black Sea by passing through Russian territory, as it now does, through a pipeline to the Russian port of Novorosslysk. Turkey opposes this strongly, on the grounds that oil-laden ships navigating the narrow Bosporus and Dardanelles straits imperil Turkey's environment. In response, Iran has offered a swap with Azerbaijan: Iran would import Azeri oil for Iran's domestic needs and would ex-

port equivalent amounts of its own oil from established ports on the Persian Gulf.

Azerbaijan's leaders, however, want more ties with the West (which is generally hostile to Iran) and independence from Moscow. The United States and its Western allies anticipate that Azerbaijan and Kazakstan will become major counterweights to the volatile oil-producing states of the Middle East, and they want to be able to obtain Caspian Sea oil without relying on Russia's or Iran's goodwill. Azerbaijan, therefore, in 1995 signed an agreement with a U.S.-led Western oil consortium to export oil through a combination of new and old pipelines to the Georgian port of Batumi on the Black Sea. To avoid a direct confrontation with Russia over this sensitive issue, some of the initial production will pass through the existing Russian pipeline to Novorosslysk.

This compromise applies only to the first stage of production, through 1997. A permanent route has yet to be established, with both Turkey and Russia vying for the pipeline to pass through their territories. Russia insists that the environmental hazard of navigating the Turkish straits can be averted by offloading oil at the Bulgarian Black Sea port of Burgas, shipping it overland through a new pipeline to the Greek Aegean Sea port of Alexandroupolis, and

---

its sturgeon, which supply over 90 percent of the world's caviar (sturgeon eggs). When only the USSR and Iran bordered the Caspian, this resource was managed sustainably. Now, however, the recently independent countries bordering the Caspian are exerting tremendous pressure on this resource in an attempt to boost their market shares in competition with Russia and Iran. The result has been a rapid drop in the sturgeon population and soaring prices of caviar on the world market.

Oil and gas are the most valuable minerals of the Caucasus region. The main oil fields are along and underneath the

Caspian Sea around Azerbaijan's capital Baku (population of city proper, 1.08 million; metropolitan area, 2.02 million) and north of the mountains at Groznyy in Chechnya. Dry and windswept Baku, the largest city of the Caucasus region, was the leading center of petroleum production in Old Russia, and at the dawn of the 20th century was the world's leading center of oil production. Under the Soviets it continued its leadership until it was decisively surpassed in the 1950s by the Volga-Urals fields and again in the 1970s by the West Siberian fields. However, Baku oil continues to be prized for its high quality—it is low in sulfur and can be refined into high-octane gasoline

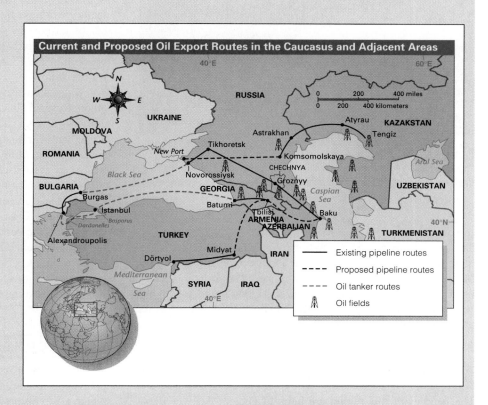

Current and Proposed Oil Export Routes in the Caucasus and Adjacent Areas

Existing pipeline routes
Proposed pipeline routes
Oil tanker routes
Oil fields

FIGURE 7.C
Numerous physical and political obstacles stand in the way of oil exports from the Caspian region.

reloading it there. The Turkish route would be far from the straits, skirting hostile Armenia to run across Anatolia to the Mediterranean port of Ceyhan. The Western powers prefer this route. So do the oil-producing countries, if only because the Russian route would lead through the breakaway republic of Chechnya—perhaps exposing the pipeline to a greater risk of sabotage. Proponents of the Turkish route also have security risks in mind: the pipeline would take a detour to skirt the volatile Kurdish-dominated region of southeastern Turkey.

and superior lubricating oils. Projected petroleum development is likely to once again establish the Caucasus as a vital world region.

Diverse manufacturing industries have developed in Transcaucasia on a small scale. Most factories are located in and near the political capitals: Baku; Tbilisi (population of city proper, 1.28 million; metropolitan area, 1.46 million) in Georgia; and Yerevan (population of city proper, 1.20 million; metropolitan area, 1.32 million) in Armenia. Oil-refining and petrochemical industries are significant in the areas that produce oil and gas. Despite these industries, the Caucasus region

vies with Ukraine's Crimea as a tourist destination. The main resorts are along the Black Sea coast and in the mountains of Georgia.

7.8 The Central Asian States

Across the Caspian Sea from the Caucasus region lie large deserts and dry grasslands, bounded on the south and east by high mountains. Like Transcaucasia, this Central Asian region

FIGURE 7.20
The elders of this community in Tajikistan are about to partake in the traditional meal celebrating Novruz, the Muslim New Year. *Sovfoto*

has a variety of non-Slavic nationalities who outnumber Slavs in all the countries. Those with the largest populations are four Muslim peoples speaking closely-related Turkic languages—the Uzbeks, Kazaks, Kyrgyz, and Turkmen—and a Muslim people of Iranian origins, the Tajiks (Fig. 7.20). Each has its own independent nation (see Table 6.1).

DIVISION AND CONQUEST

The peoples of this region are heirs of ancient oasis civilizations and of the diverse cultures of nomadic peoples. Small-scale social and political units such as clans, tribes, and petty autocracies were associated with particular oases. Larger empires controlled some of these at various times. Conquerors have repeatedly possessed this area, which lies on the famed Silk Road, an ancient route across Asia from China to the Mediterranean. Among them were Alexander the Great, the Arabs who brought the Muslim faith in the eighth century, the Mongols, and the Turks. By the time of the Russian conquests in the 19th century, Turkestan (as it was then known) was contested between feuding, tradition-bound Muslim khanates with political ties to China, Persia, Ottoman Turkey, and British India. Cultural and political fragmentation continued up to the Russian Revolution and into the period of civil war.

Nationalist identities and elites existed in Central Asia, but nations did not exist as large communities bound together by common interests and allegiance. During World War I, Central Asian peoples revolted against the tsarist government, which had begun to draft the Muslims for menial labor at the front. This generated widespread disorder and bloodshed until the Bolshevik government gained control in the early 1920s and organized the area into nationality-based units. The Central Asian

"nationalities" are themselves to a large degree artificial creations of the Bolsheviks.

Most of the 20th century saw a large incursion of Russians and Ukrainians into Central Asia, mainly into the cities. They included political dissidents banished by the Communists, administrative and managerial personnel, engineers, technicians, factory workers, and, in the north of Kazakstan, a sizable number of farmers in the "new lands" wheat region. Most of the Slavic newcomers live apart from the local Muslims. With the breakup of the Russian-dominated Soviet Union, their destiny is uncertain.

Kazakstan provides a good example of the dilemma facing ethnic Russians in Central Asia. At the time it became independent in 1991, Kazakstan's population of 17 million was roughly 40 percent Kazak and 39 percent Russian, with the rest a mix of nationalities. Many ethnic Kazaks who believe they were treated as second-class citizens until independence have been seeking to rectify their status. They are increasingly predominant in government and business. Kazak is now the official language and Russian is the language of "interethnic communication." By the end of 1995, the non-Kazak residents were told they had to decide if they wanted Kazak citizenship. With rising ethnic tensions between 1991 and 1996, about one million Russians, along with a half million ethnic Germans, emigrated from Kazakstan, and the exodus is continuing.

ENVIRONMENT AND AGRICULTURE

The five Central Asian states are composed predominantly of plains and low uplands, except for Tajikistan and Kyrgyzstan, which are extremely mountainous and contain the highest summits of the Former Soviet Region in the Pamir and Tien Shan ranges (see Fig. 6.6). The Kazak Upland in east central Kazakstan is a hilly area with occasional ranges of low mountains. Central Asia is almost entirely a region of interior drainage. Only the waters of the Irtysh, a tributary of the Ob, reach the ocean; all the other streams either drain to enclosed lakes and seas or gradually lose water and disappear in the Central Asian deserts.

Most people live in irrigated valleys at the base of the southern mountains (Fig. 7.21). There, most soils (many formed of loess) are fertile, the growing season is long, and rivers flowing from mountains provide irrigation water. The principal rivers in the heart of the region are the Amu Darya (Oxus) and Syr Darya, both of which empty into the enclosed Aral Sea. Their tributaries and other rivers provide a large share of the region's irrigation waters, which numerous canals carry.

Most of the larger irrigated districts are in Uzbekistan. Especially important are the fertile Fergana Valley on the upper Syr Darya, almost enclosed by high mountains and the oases around Tashkent (capital of Uzbekistan and Central Asia's largest city; population of city proper, 2.11 million; metropolitan area, 2.33 million), Samarkand (population of city proper,

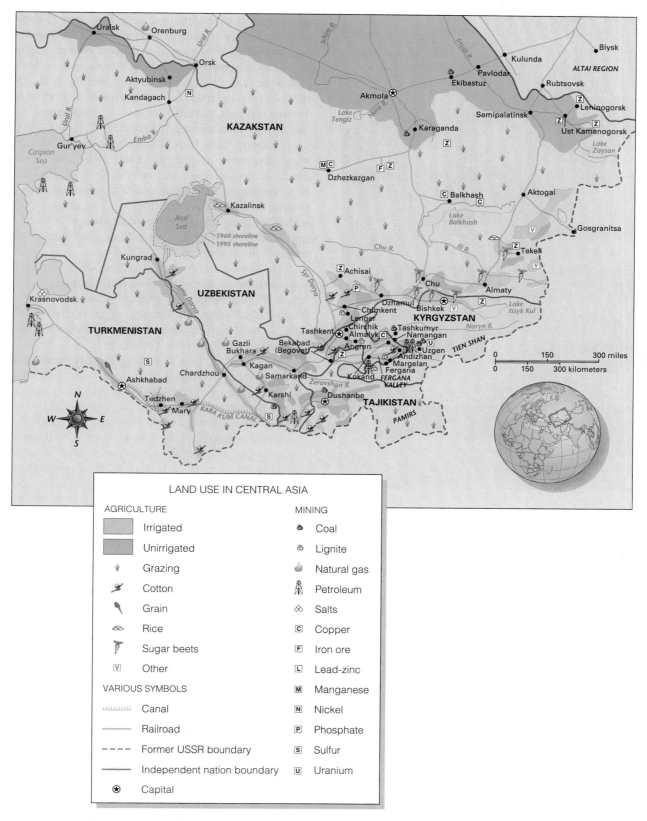

FIGURE 7.21

Agricultural lands and some important mining areas in the Central Asian countries.

FIGURE 7.22
A 16th-century madrasa (Islamic school) on the Registan Square in the heart of the storied ancient city of Samarkand, Uzbekistan.
Karen Sherlock/Tony Stone Images

366,000), and Bukhara (population of city proper, 224,000). These old cities became powerful political centers on caravan routes connecting southwestern Asia and the Mediterranean basin with eastern Asia. Samarkand is the successor to many earlier cities that occupied the site on loess hills beside the Zeravshan River. It was the capital of Tamerlane's empire, which encompassed much of Asia, and, like Bukhara today, has numerous superb examples of Islamic architecture (Fig. 7.22). Streams fed by melting snow and ice in some of the world's highest mountains have sustained these places through their long histories.

Cotton is the region's major crop. This area grows over nine tenths of the Former Soviet Region's output. Production supplies textile mills in the Moscow region, Tashkent, and other cities in far-flung locations, including many outside the Former Soviet Region. However, a significant environmental price was paid as a result of cotton's development here. Driven by the need for foreign exchange from export sales, Communist planners pushed an expansion of irrigated cotton production to the point that the rivers and the Aral Sea became highly polluted with agricultural chemicals draining from the fields. Diversion of water from the rivers caused the Aral Sea to shrink rapidly in volume and area, virtually destroying a vast natural ecosystem and an important regional fishery (Figs. 7.22 and 7.23).

Mulberry trees to feed silkworms are grown around the margins of many irrigated fields and along irrigation canals in the oases. Irrigated rice and other grains, sugar beets, vegetables, vineyards, and orchards of temperate fruits are also important components of this region's production. Commercial orchard farming is prominent around Almaty ("City of Apples"; population of city proper, 1.16 million; metropolitan area, 1.19 million), the former capital of Kazakstan.

Most of Central Asia outside the oases is too dry for cultivation. However, ethnic Russians and Ukrainians practice non-irrigated grain farming in northern Kazakstan, which extends into the black-earth belt. Huge acreages, primarily on chestnut soils, were planted in wheat under the "virgin and idle lands" scheme of 1954–1956 (see Chapter 6).

Many of the region's people previously were pastoral nomads who grazed their herds on the natural forage of the steppes and in mountain pastures in the Altai and Tien Shan ranges. Over the centuries, there was a slow drift away from nomadism. The Soviet government accelerated this process by collectivizing the remaining nomads—often by harsh measures in the face of strong resistance—and settling them in permanent villages. Livestock raising in the process became a form of ranching, but herdsmen must still accompany the grazing animals from one range to another. Sheep and cattle are the principal livestock animals, followed by goats, horses, donkeys, and camels.

FORCES OF MODERNIZATION

In recent decades, parts of Central Asia have become increasingly important in the production of minerals, particularly natural gas, coal, oil, iron ore, nonferrous metals, and ferroalloys. Kazakstan has the greatest natural resource wealth, and political stability under a nominally-democratic government since independence has made it the largest target for Western investment in Central Asia. Large reserves of bituminous coal are mined at Karaganda (population: 614,000) in central Kazakstan. The Karaganda coal basin has become a very important producer (although much less so than the Donets and Kuznetsk basins), and Karaganda itself has experienced spectacular growth. Much of the coal is shipped to metallurgical works in the Urals.

Despite Kazakstan's coal wealth, equipment and management problems have resulted in declining production in the country since the breakup of the Soviet Union. In 1996, the country's new capital was chosen to be established at Akmola (formerly Tselinograd; population: 277,000), northwest of Karaganda, as part of a larger plan to redevelop Kazakstan's industrial heartland.

Coal from Karaganda and other Central Asian locations, as well as from the Kuznetsk Basin, provides fuel for large-scale metallurgical production in the region. Of primary importance is the smelting of nonferrous metals, especially copper, lead, and zinc. The largest copper reserves and the most important copper mines of the Former Soviet Region are in central Kazakstan, midway between Karaganda and the Aral Sea. Complex ores yielding zinc and lead are mined in the Altai Mountain foothills of eastern Kazakstan. In 1995, prospecting began in Kazakstan in what may be the world's third largest gold deposit. This rich and diversified mineralized region of Central Asia also holds impressive reserves of chrome and nickel in the Ural foothills of northern Kazakstan, natural gas near the Amu Darya, petroleum in fields bordering the Caspian Sea, and large iron ore deposits in northern Kazakstan.

Petroleum and natural gas will be the most important resources in the near future of Central Asia. Kazakstan has the largest endowment and is working with Russian, Omani, and U.S. oil companies on arrangements to build a pipeline from the Caspian Sea oil fields around Tengiz to a new port north of the Russian Black Sea port of Novorosslysk (see Fig. 7.C). This would allow Kazakstan to boost its oil exports greatly. Turkmenistan (capital: Ashkhabad; population, 398,000), which also has frontage on the Caspian Sea, also has significant fossil fuel reserves. Its huge natural gas reserves are the fourth largest in the world; some observers have dubbed this small but potentially rich country the "new Kuwait."

Turkmenistan still has a Communist government. Ethnic Russians comprise about 10 percent of the population and Uzbeks another 10 percent. Natural gas exports by pipeline to western Europe are the country's main source of hard currency. The pipeline runs through Kazakstan, Russia, and Ukraine, and many analysts have advised Turkmenistan to de-

velop a new and more secure route through Iran and Turkey to the West.

Uzbekistan, Central Asia's most populous country (23.2 million), does not have access to the Caspian oil fields. The country is largely agricultural, with cotton exports the most important source of hard currency. Like other LDCs, Uzbekistan faces the challenge of transforming itself from an exporter of raw materials to an exporter of manufactured goods. The country is also trying to reverse the ecological damage done by years of Soviet emphasis on cotton production heedless of the damages done by pesticides and overirrigation. Since becoming independent, Uzbekistan has dramatically cut the area planted in cotton and increased production of food crops.

Kyrgyzstan, with its capital at Bishkek (population of city proper, 616,000), is, like the other Central Asian countries, the successor to a republic which Stalin carved artificially out of Russian Turkestan. Searching for identity in the post-Soviet period, the people and government of Kyrgyzstan have resurrected the oral epics which tell the story of Manas, a great warrior who, a thousand years ago, fought off the enemies of the Kyrgyz nomadic tribes and unified the tribes for the first time. As it recognizes these ancient roots, Kyrgyzstan is modernizing. Most of the Soviet-era state-owned businesses are now privatized, and in 1995 investors opened the country's first stock exchange.

CENTRAL ASIA AND THE WIDER WORLD

Central Asia's future is of outstanding international importance. In an attempt to create a common market and integrate their economies more effectively, the five countries formed a regional organization called the "United States of Central Asia" in 1993. Although primarily intended to meet regional economic needs, the group may become preoccupied with Central Asia's external relations. The region borders revolutionary Iran,

FIGURE 7.23
Rusting vessels on former sea bottom testify to the shrinkage of Central Asia's Aral Sea. The major cause of the catastrophic water loss is diversion of the Amu Darya and Syr Darya for irrigation of cotton. *G. Pinkhassov/Magnum*

war-torn Afghanistan, and the great power of China. Central Asia's population of over 50 million is growing rapidly. Many kinship relations exist between the Central Asian Muslims and their neighbors on the other side of the international frontiers. Along with internal ethnic conflicts and dissatisfaction over living conditions, these international affiliations may make the region a major political problem area.

Both Iran and Turkey are vying for increased influence in the region. Support for Iran is largely lacking, in part because the majority of Central Asians practice the Sunni Islam religion rather than the Shi'ite Islam religion of Iran. Ethnicity is also important; the four Turkic states are inclined to orient with Turkey, with which they have already established cultural ties such as educational exchanges and shared media. Turkey also has extended economic assistance and established many small business ventures and large construction contracts in Central Asia. Turkmenistan has good relations with both Turkey and Iran and has recently completed a rail link with Iran. Turkmenistan also tries to maintain good ties with Russia.

The region is also becoming a focus for rivalry between the historical enemies Turkey and Russia. Russians are now concerned that Turkey is winning too much influence in the new Muslim states of Central Asia. For their part, many Turks believe Russia will try to take over the Transcaucasus region again and then turn to Central Asia. Some Turks uphold a dream of Pan-Turkism, uniting all the Turkic peoples of Asia from Istanbul to the Sakha Republic in Russia's Siberia region—a prospect that Moscow does not like.

Kazakstan is developing strong ties with the West, and, like Ukraine and Belarus, has agreed to eliminate its nuclear weapons. In return, Kazakstan wants the West to help guarantee the security of its borders with Russia, China, the Caspian Sea, and three other Central Asian nations. The country also favors close coordination with Russia, and in 1996 these two countries and Kyrgyzstan signed an agreement calling for the creation of a common market to ensure the free flow of goods, services, and economic capital between them and to promote coordinated industrial and agricultural policies.

International concern about the region's stability focuses on Tajikistan (capital: Dushanbe; population, 595,000). Russian diplomats have a "domino theory" about the region stemming from a fear that Islamic fundamentalism might spread from Tajikistan to neighboring Central Asian countries. Tajikistan is still a Communist country, but its government has the backing of Washington and Moscow because they see the alternative as a radical Islamic government. At issue is whether or not, with support from Afghan guerrillas and Middle Eastern activists, Tajikistan's Islamists can gain control of the country and spread their influence into neighboring states like resource-rich Kazakstan. In 1992, a coalition of prodemocracy and Islamic rebels, comprised largely of members of the Garm and Badakhshan ethnic groups who have little economic and political power in the country, lost a bloody civil war against the Russian-backed government. Many refugees fled into neighboring Afghanistan, where resistance to the Communist government in Tajikistan is based. Islamic Iran also supports the rebels.

As of early 1997, Russia still had 25,000 troops in Tajikistan to back the government there as a shield against the Islamist movement. Russia also wants Tajikistan to be a buffer so that political violence does not spread into neighboring former republics on Russia's borders. President Yeltsin declared that Tajikistan's borders "are effectively Russia's"—its main barrier against the infiltration from Afghanistan of Islamic militancy, revolution, guns, and drugs. Despite Tajikistan's formal independence, in fact it is now a "client state" of Russia. Moscow currently funds 70 percent of the Tajik national budget. Ethnic Russians make up 12 percent of the country's population and Uzbeks another 23 percent. Secular Turkey also fears the specter of a rising tide of fundamentalism originating in Tajikistan and neighboring Afghanistan as forces of both tradition and change advance into Central Asia from the heart of the Middle East.

QUESTIONS

Page numbers refer to Saunders College Publishing 1996 Special Edition of Rand McNally's Atlas of World Geography.

1. On the map of Europe on pp. 140–141, identify:
    (a) The "port of five seas."
    (b) A wilderness region, now opening to tourism, with many volcanoes and thermal features such as geysers.

2. On the map of Europe on p. 114–115, identify:
    (a) A region belonging to Ukraine, but inhabited by a majority of ethnic Russians.
    (b) Members of the Former Soviet Region that have agreed to give up nuclear weapons.
    (c) The world's greatest source of caviar.

The Middle East

3

Crossroads
Cradle of Civilization
Culture Hearth
Center and Target of Empire Builders
Birthplace of Jews, Arabs, Turks, Persians,
Kurds, and Berbers
Heart of the Dry World
Irrigation-Dependent
Center of World Oil Reserves
War-Torn Region
Home to Interdependent Farmers, Pastoral
Nomads, and Urbanites

8

A Geographic Profile of the Middle East

What and where is the Middle East? The term itself is Eurocentric, created by Europeans and their American allies who placed themselves in the figurative center of the world. They began to use the term prior to the outbreak of World War I, when the "Near East" referred to the territories of the Ottoman Empire in the Eastern Mediterranean region, the "East" to India, and the "Far East" to China, Japan, and the western Pacific rim. With "Middle East" they designated as a separate region the countries around the Persian Gulf (known to Arabs as the "Arabian Gulf," and in this text as the "Persian/Arabian Gulf" and simply "The Gulf"). Gradually, the perceived boundaries of the region grew. Today, depending upon what source you consult, you might find that the Middle East includes only the countries clustered around the Arabian Peninsula, or (as in this text) spans a vast 6000 miles (9700 km) west to east from Morocco in northwest Africa to Afghanistan in central Asia, and 3000 miles (4800 km) north to south from Turkey, on Europe's southeastern corner, to Sudan, which adjoins East Africa (Figs. 8.1 and 8.2). Thus defined, the region incorporates 22 countries and the disputed Western Sahara region (Table 8.1), occupying 5.93 million square miles (15.36 million sq km) and inhabited by about 405 million people as of mid-1996.

Even the region's occupants themselves now use the term "Middle East." It is a fitting designation for their location and their cultures, for they are literally in the middle. The region is a physical crossroads, where the continents of Africa, Asia, and Europe meet and the waters of the Mediterranean Sea and the Indian Ocean mingle. Its peoples—Arab, Jew, Persian, Turk, Kurd, Berber, and others—express in their cultures and ethnicities the coming together of these diverse influences. Occupying as they do this strategic location, the nations of the Middle East have throughout history been unwilling hosts to occupiers and empires originating far beyond their borders. They have also bestowed upon humankind a rich legacy which includes the ancient civilizations of Egypt and Mesopotamia and the world's three great monotheistic faiths: Judaism, Christianity, and Islam. People outside the region tend to forget about such contributions as they associate the Middle East with military conflict and terrorism. The region may be on the eve of a new era of peace and reconciliation. The following two chapters attempt to make the complex, often destructive, but now hopeful events in the Middle East more intelligible by presenting the geographic context within which they occur.

◀ A false-color satellite image of the heart of the Middle East. The huge Arabian Peninsula dominates the image. The rugged mountains of Turkey and Iran are prominent below the Black Sea and Caspian Sea. The thin irrigated ribbons of the Nile, Tigris, and Euphrates rivers are clearly visible. At the bottom, the Horn of Africa points at the island of Socotra. *Geospace/Science Photo Library*

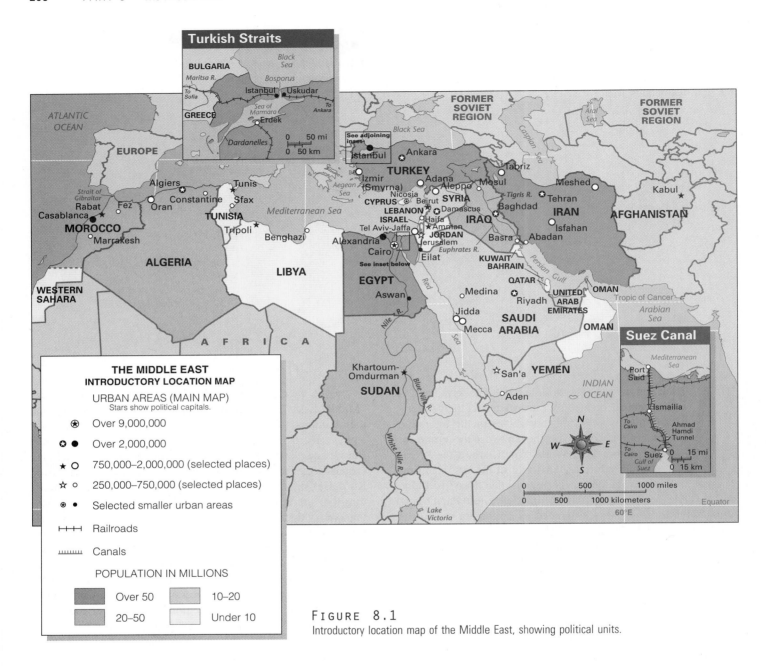

FIGURE 8.1
Introductory location map of the Middle East, showing political units.

FIGURE 8.2
The Middle East compared in latitude and area with the conterminous United States.

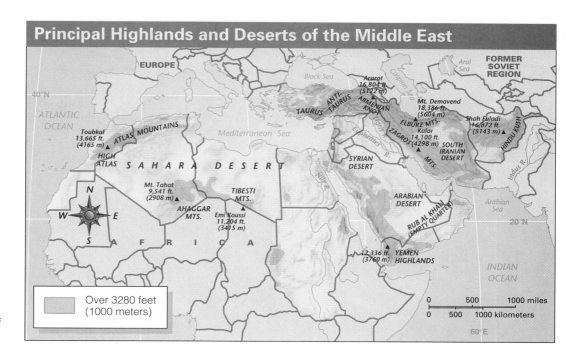

Principal Highlands and Deserts of the Middle East

FIGURE 8.3
Principal highlands and deserts of
the Middle East.

8.1   Environmental Setting

The margins of the Middle East are occupied mainly by oceans, seas, high mountains, and deserts: to the west lies the Atlantic Ocean; to the south, the Sahara Desert and the highlands of East Africa; to the north, the Mediterranean, Black, and Caspian seas, together with mountains and deserts lining the southern land frontiers of the Former Soviet Region; and to the east, the Hindu Kush mountains on the Afghanistan-Pakistan frontier and the Baluchistan desert straddling Iran and Pakistan. The land is composed mainly of arid or semiarid plains and plateaus, together with considerable areas of rugged mountains and isolated "seas" of sand.

HEAT AND ARIDITY

Aridity dominates the Middle East (see Fig. 8.3; the world precipitation map in Fig. 2.1; the climate map in Fig. 2.6; and the world vegetation map in Fig. 2.7). Most of the region is part of what geographers call the Dry World—a belt of deserts and dry grasslands extending across Africa and Asia from the Atlantic Ocean nearly to the Pacific. At least three fourths of the Middle East has an average yearly precipitation of less than 10 inches (25 cm), an amount too small for most types of unirrigated agriculture under the prevailing temperature conditions. Sometimes, however, localized cloudbursts release moisture that allows plants, animals, and even small populations of people—the Bedouins, Tuaregs, and other pastoral nomads—to live in the desert. Even the vast Sahara, the world's largest desert, which outsiders perceive as utterly desolate and barren, supports a surprising diversity and abundance of life. Plants, ani-

mals and even people have developed strategies of **drought avoidance** and **drought endurance** to live in this harsh biome. Plants either avoid drought by completing their life cycle quickly wherever rain has fallen, or endure drought by using their extensive root systems, small leaves, and other adaptations to take advantage of subsurface or atmospheric moisture. Animals endure drought by calling upon extraordinary physical abilities—a camel can sweat away a third of its body weight and still live, for example—or avoiding the worst conditions by being active only at night, or by migrating from one moist place to another. (Migration to avoid drought is the strategy pastoral nomads use.) Populations of people, plants, and animals are all but nonexistent in the region's vast sand seas, including the Empty Quarter of the Arabian Peninsula and the Great Sand Sea of western Egypt (Fig. 8.4).

FIGURE 8.4
Sand "seas" cover large areas in Saudi Arabia, Iran, and parts of the Sahara Desert in North Africa. This dune field is in central Saudi Arabia.
Ray Ellis/Photo Researchers, Inc.

TABLE 8.1   The Middle East: Basic Data

Political Unit	Area (Thousand sq mi)	Area (Thousand sq km)	Estimated Population (millions, mid-1996)	Estimated Annual Rate of Natural Increase (% 1996)
Arab League States				
Egypt	386.7	1001.4	63.7	2.2
Sudan	967.5	2505.8	28.9	3.0
Tunisia	63.2	163.5	9.2	1.7
Algeria	919.6	2381.7	29.0	2.4
Morocco	172.4	446.4	27.6	2.2
Lebanon	4.0	10.4	3.8	2.0
Syria	71.5	185.1	15.6	3.7
Jordan	35.5	91.9	4.2	2.6
Iraq	167.9	434.8	21.4	3.7
Saudi Arabia	830.0	2149.7	19.4	3.2
Kuwait	6.9	17.8	1.8	2.3
Bahrain	0.3	0.7	0.6	2.6
Qatar	4.2	11.0	0.7	1.6
United Arab Emirates	32.3	83.6	1.9	1.9
Yemen	203.9	527.9	14.7	3.2
Oman	82.0	212.5	2.3	4.9
Totals	3947.9	10,224.4	244.8	2.3 (mean)
Other Units				
Libya	679.4	1759.5	5.4	3.7
Turkey	301.4	780.6	63.9	1.6
Cyprus	3.6	9.2	0.7	0.9
Iran	636.3	1648.0	63.1	2.9
Afghanistan	250.0	647.5	21.5	2.8
Israel	8.0	20.8	5.8	1.5
Western Sahara	103.0	266.8	0.2	2.8
Totals	1981.7	5132.4	160.6	2.3 (mean)
Grand Totals	5929.6	15,356.8	405.4	2.3

Figures for Egypt, Syria, Jordan, and Israel are based on the extent of territory controlled by each of these countries prior to the 1967 Arab-Israeli War. Not shown in the table are five small enclaves still held by Spain along the Mediterranean coast of Morocco. Figures for the politically divided island of Cyprus pertain to the entire island. Figures for the Western Sahara pertain to the extent of territory prior to the Moroccan occupation.

Middle Eastern climates exhibit the comparatively large diurnal (daily) and seasonal ranges of temperature that are characteristic of dry lands. Summers in the lowlands are very hot almost everywhere. The hottest shade temperature ever recorded on Earth, 136°F (58°C), occurred in Libya in 1922. Many places regularly experience daily maximum temperatures over 100°F (38°C) for weeks at a time. Day after day a baking sun assails the parched land from a cloudless sky. Night brings some relief, when the lack of cloud cover and low humidity allows heat to radiate back into space. Human settlements located near the sand seas often experience the unpleasant combination of high temperatures and hot, sand-laden winds, creating the sandstorms known locally by such names as *simuum* ("poison") and *scirocco*. Only in mountainous sections and in some places near the sea do higher elevations or sea breezes temper the intense midsummer heat. The population of Alexandria explodes in summer as Egyptians flee from Cairo and other hot inland locations. In Saudi Arabia, the government relocates from Riyadh to the highland summer capital of Taif to escape the heat.

Estimated Population Density	Estimated Population Density	Infant Mortality Rate	Urban Population (%)	Cultivated Land (% of total area)	Per Capita GNP ($US: 1994)
Per sq mi	Per sq km				
155	55.8	62	44	2.8	710
29.1	11.2	80	27	5.5	300 (1992)
140.4	54.1	43	60	31.7	1800
30.4	11.7	55	50	3.3	1690
152.3	58.8	57	47	22.2	1150
761.8	294.1	28	86	29.9	4360
200.2	77.3	44	51	32.1	1170 (1991)
121.9	47.1	34	78	4.5	1390
121.7	47.0	67	70	12.5	710 (1991)
20.7	8.0	24	79	1.7	7240
254.8	94.9	12	96	0.3	19,040
2160.4	834.0	19	88	2.9	7500
131.3	50.1	11	91	0.6	14,540
68	26.3	23	82	0.5	21,420
63.6	24.6	83	23	2.8	280
18.3	7.1	24	12	0.3	5200
62	24	42 (mean)	49.7	5.6	5529 (mean)
8	3.1	63	85	1.2	6510
207.8	80.2	47	63	35.8	2450
194	76	9	53	47	10,380
97.2	37.4	57	58	11.1	1940 (1993)
72	27.8	163	18	12.4	220 (1988)
673.9	260.2	6.9	90	20.7	14,410
2	1	152	NA	0	NA
81	31	71 (mean)	56.6	11.1	5130 (mean)
68	26	50.6 (mean)	52.4	7.4	5407 (mean)

MOISTURE FROM SEAS AND MOUNTAINS

Lower winter temperatures bring relief from the summer heat, and the more favored places receive enough precipitation to grow winter wheat or barley and some other cool-season crops. In general, Middle Eastern winters may be characterized as cool to mild. But, very cold winters and slight-to-moderate snowfalls are experienced in the high interior basins and plateaus of Iran, Afghanistan, and Turkey. These locales gener-

ally have a steppe climate. Only in the southernmost reaches of the region, such as Sudan's upper Nile Basin, do temperatures remain consistently high throughout the year. A savanna climate and biome prevail there.

Most areas bordering the Mediterranean Sea have 15 to 40 inches of precipitation a year, falling almost exclusively in winter, while the summer is dry and warm—a typical Mediterranean climate pattern. Throughout history, people without access to perennial streams have stored this moisture to make it available later for growing those crops that require the higher

orographic or elevation-induced precipitation, the mountains tend to receive much more rainfall than surrounding lowland areas.

There are three principal mountainous regions of the Middle East (see Fig. 8.3). In northwestern Africa between the Mediterranean Sea and the Sahara Desert, the Atlas Mountains of Morocco, Algeria, and Tunisia reach over 13,000 feet (3965 m) in elevation. Mountains also rise on both sides of the Red Sea, with peaks up to 12,336 feet (3760 m) in Yemen. A larger area of mountains, including the highest peaks in the Middle East, stretches across Turkey, Iran, and Afghanistan. On the eastern border with Pakistan, the Hindu Kush range has peaks of over 25,000 feet (7600 m). The loftiest and best-known

FIGURE 8.5
"The Treasury," a temple carved in red sandstone, probably in the first century A.D., by the Nabateans at their capital of Petra in southern Jordan.
Bettina Cirone/Photo Researcher, Inc.

temperatures of the summer months. The Nabateans, for example, who were contemporaries of the Romans in what is now Jordan, had a sophisticated network of limestone cisterns and irrigation channels (Fig. 8.5). Rainfall sufficient for unirrigated summer cropping (dry farming) is concentrated in areas along the southern and northern margins of the region. The Black Sea slope of Turkey's Pontic Mountains is lush and moist in the summer, and tea grows well there (Fig. 8.6). In the southwestern Arabian Peninsula a monsoonal climate brings summer rainfall and autumn harvests to Yemen and Oman, probably accounting for the Roman name for the area: *Arabia Felix*, or "Happy Arabia" (Fig. 8.7).

Mountainous areas in the Middle East, like the river valleys and the margins of the Mediterranean, play a vital role in supporting human populations and national economies. Due to

FIGURE 8.6
Rainfall is heavy on the Black Sea side of Turkey's Pontic Mountains. Note that roofs are pitched to shed the rainfall. In the right center, the close-cut crop is tea. An almond tree is in bloom. A mosque dominates this Muslim village. *C. Adam Woolfitt/Woodfin Camp & Associates*

FIGURE 8.7
A typical high mountain landscape of Yemen, around the village of Menakha. Multistoried residences help conserve precious agricultural land. The terraces are uncultivated in this winter scene.   *George Holton/Photo Researchers, Inc.*

WHERE FORESTS STOOD

Extensive forests existed in early historical times in the Middle East, particularly in these mountainous areas, but overcutting and overgrazing have almost eliminated them. Since the dawn of civilization in this area, at around 3000 B.C., people have cut timber for construction and fuel faster than nature could replace it. Egyptian King Tutankhamen's funerary shrines and Solomon's Temple in Jerusalem were built of cedar of Lebanon. So prized has this wood been through the millennia that only a few isolated groves of cedar remain in Lebanon. Described in ancient times as "an oasis of green with running creeks" and "a vast forest whose branches hide the sky," Lebanon is now largely barren (Fig. 8.9). Lumber is still harvested commercially in a few mountain areas such as the Atlas region of Morocco and Algeria, the Taurus Mountains of Turkey, and the Elburz Mountains of Iran, but the total supply falls far short of demand.

mountain ranges in Turkey are the Taurus and Anti-Taurus, and in Iran, the Elburz and Zagros mountains. These chains radiate outward from the rugged Armenian Knot in the tangled border country where Turkey, Iran, and the Former Soviet Region meet. One of the world's most dramatic and culturally important mountains, Mount Ararat, is an extinct, glacier-covered volcano of 16,804 feet (5122 m) towering over the border region between Turkey and Armenia (Fig. 8.8). Many Biblical scholars think the ark of Noah lies high on the mountain, for in the book of Genesis this boat was said to have come to rest "in the mountains of Ararat."

MINERAL WEALTH AND SHORTAGES

A shortage of mineral resources, especially those useful for industrialization, also handicaps many parts of the Middle East. Good deposits of coal are rare, except in Turkey. The region is rich in petroleum and natural gas, but the largest deposits are confined to a few countries bordering the Persian/Arabian Gulf (Saudi Arabia, Iraq, Kuwait, Iran, United Arab Emirates, Oman, Qatar), plus Libya, Algeria, and Egypt in North Africa. By coincidence, the countries rich in oil tend to have relatively small populations, while the most populous nations have few

FIGURE 8.8
The mighty volcano of Mt. Ararat, rising to 16,804 ft (5122 m) to dominate the frontier regions of Turkey, Armenia, and Iran. At the bottom center, a small village is situated to take advantage of water runoff from the slopes above.   *Ray Ellis/Photo Researchers, Inc.*

FIGURE 8.9
One of the twelve groves of cedars remaining in Lebanon. People of the Mediterranean Basin harvested this valuable resource for about five thousand years, nearly depleting it. This symbol of Lebanon now has complete protection.   *Leonard Wolfe/Photo Researchers, Inc.*

oil reserves (Iran is an exception). Although scattered deposits of metals occur, only a few are important on a global scale. Large salt deposits are common, and phosphate rock, useful as a chemical and fertilizer material, is mined commercially in Morocco, Tunisia, and Jordan. Israel and Jordan extract potash, another chemical and fertilizer material, from the briny waters of the Dead Sea.

Aside from oil and gas, which currently account for all but a tiny percentage of Middle Eastern mineral production by value, the outlook for mineral extraction is poor. The region is generally handicapped by an inadequate natural resource base. Thus, despite the very high national GNP and per capita GNP that characterize some of the Middle Eastern nations, most countries in the region are clearly recognizable as LDCs.

8.2   World Importance of the Middle East

Throughout history, the sparsely populated deserts and mountains of the Middle East, separating the humid lands of Europe, Africa, and Asia, have been a hindrance to overland travel between those regions. Nevertheless, circulation of people, goods, and ideas has taken place along certain favorable routes, and the scattered population centers of the region have had a history of vigorous interaction with the outside world and with each other.

Cultures of this region have made many fundamental contributions to humanity. Many of the plants and animals upon which the world's agriculture is based were first domesticated in the Middle East between 5000 and 10,000 years ago, in the course of the Agricultural Revolution. The list includes wheat,

Definitions and Insights

THE FERTILE CRESCENT

The Fertile Crescent is an arc-shaped area of relatively high precipitation stretching from Iran's Zagros Mountains westward through northern Iraq, southern Turkey, Syria, Lebanon, Israel, and western Jordan (Fig. 8.10). Since about 10,000 years ago, relatively abundant rainfall (over 12 inches or 30 cm annually) provided excellent habitat for diverse plant and animal life. This resource wealth attracted human settlers, whose growing populations may have led by necessity to the earliest known domestications of plants and ani-

mals (see Chapter Two, page 37). The productive seasonal agriculture supported by good water supplies and soils allowed Neolithic (New Stone Age) cultures to flourish throughout the Fertile Crescent until about 3000 B.C. After that time, far more productive irrigated agriculture—largely outside the Fertile Crescent and along the river valleys in Mesopotamia and Egypt, made those regions the main centers of cultural development.

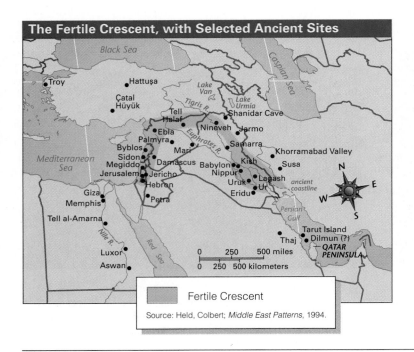

The Fertile Crescent, with Selected Ancient Sites

Source: Held, Colbert; *Middle East Patterns*, 1994.

FIGURE 8.10
The Fertile Crescent was a focal region for the domestication of plants and animals, leading to the emergence of urban cultures.

barley, sheep, goats, cattle, and pigs, whose wild ancestors people processed, manipulated, and bred until their physical makeup and behavior changed to suit human needs. The interaction between people and the wild plants and animals they eventually domesticated took place mainly in the **Fertile Crescent** stretching from Israel to western Iran (see page 212).

By about 6000 years ago, people sought higher yields by irrigating crops in the rich but often dry soils of the Tigris, Euphrates, and Nile River valleys. Their efforts produced the enormous crop surpluses that allowed **civilization**—a cultural complex based on an urban way of life—to emerge in Mesopotamia (literally, "the land between the rivers" Tigris and Euphrates) and Egypt. Accomplishments in science, technology, art, architecture, language, and other areas diffused outward from these centers of civilization. Egypt and Mesopotamia are thus among the world's great **culture hearths** (see Chapter Two, page 38).

Most of the Old World's great empires have included portions of the Middle East. Some were indigenous, including the Egyptian, Babylonian, Phoenician, and Assyrian empires. Foreign emperors and kings of Greece, Rome, and Christian Byzantium ruled the Middle East before the coming of Islam. Following the death of Islam's Prophet Muhammad in Arabia in 632, a great Islamic Empire arose and grew with such vigor that just a century later it stretched from Spain to Central Asia, with Damascus serving as its capital and cultural center. The most recent Islamic Empire collapsed in the early 20th century when World War I brought an end to the Ottoman Empire that had been based in Istanbul beginning in 1453. Foreigners once again asserted control over the region after the war, when victorious Great Britain and France divided the remains of the Ottoman Empire between them, laying the groundwork for what is known today as the "Arab-Israeli Conflict (see Chapter Nine)." After World War II, these great powers withdrew from the region and left newly independent nations with the task of reconciling often irreconcilable problems, such as those created when Britain and France made conflicting promises to both Arabs and Jews and drew national boundaries that bore little relationship to the distributions of local ethnic and religious groups.

Since World War II, several international crises have been precipitated by events in the Middle East. Although the colonial age has ended, strong outside powers heavily dependent upon Middle Eastern oil for their current and future industrial needs have competed for political influence and supplied their proxy armies. In strategic terms, the area is thus known as a shatter belt—a large, strategically-located region composed of conflicting states caught between the conflicting interests of great powers. For example, in the 1967 and 1973 Arab-Israeli wars, Soviet-backed Syrian forces fought U.S.-backed Israeli troops.

The United States has always maintained a precarious relationship with the key players in the Middle East arena. On the one hand, the U.S. has pledged unwavering support for Israel, but on the other it has courted Israel's traditional enemies such as oil-rich Saudi Arabia. That Arab kingdom and its neighbors around the Persian/Arabian Gulf possess more than 60 percent of the world's proven oil reserves and are thus vital to the long-term economic security of the Western industrial powers and Japan. The United States and its Western allies made it clear they would not tolerate any disruption of access to this supply when Iraqi troops directed by Iraqi president Saddam Hussein occupied Kuwait on August 2, 1990. U.S. president George Bush drew a "line in the sand," proclaiming "we cannot permit a resource so vital to be dominated by one so ruthless—and we won't."

The region's strategic crossroads location also often has made it a cauldron of conflict. From very early times, overland caravan routes, including the famous "Silk Road," crossed the Middle East with highly-prized commodities traded between Europe and Asia. The security of these routes was vital, and countries on either end could not tolerate threats to them. Then, in 1869, one of the world's most important waterways opened in the Middle East. Slicing 107 miles (172 km) through the narrow Isthmus of Suez, the British and French-owned Suez Canal linked the Mediterranean Sea with the Indian Ocean, saving cargo, military, and passenger ships a journey of many thousands of miles around the southern tip of Africa (see Fig. 8.1). Egyptian president Gamal Abdel Nasser's nationalization of the canal in 1956 led immediately to a British, French, and Israeli invasion of Egypt and a conflict known as the "Suez Crisis" and the "1956 Arab-Israeli War."

Because of its crossroads location, the Middle East has many important airports serving intercontinental routes, most notably in Cairo, Tel Aviv, and the United Arab Emirates cities of Dubai and Abu Dhabi. At the same time, Middle East infrastructure suffers a general lack of long-distance rail lines and highways.

8.3   The Heartland of Islam

The Middle East is so varied in its ways of living that generalizations about the culture of the entire region can be misleading. Nevertheless, a surprising amount of similarity can be discerned in modes of life prevailing in places as far apart as Morocco and Afghanistan. Because of Islam's powerful influence not merely as a set of religious practices but as a total way of life, it is possible to distinguish an Islamic Middle Eastern culture that is characteristic of almost the entire region. In the Middle East, only Israel, within its pre-1967 borders, and Cyprus have majority non-Muslim populations. (A Muslim is a person who practices the faith of Islam.) Christians are minorities in all nations of the Middle East except Cyprus.

There is a tendency for outsiders to associate Islam and the Middle East exclusively with Arabs (people whose native language is Arabic). However, while most of the region's occupants are Arabs, it is important to note that two of the region's most populous states—Turkish Turkey and Persian Iran—are mostly non-Arab. The languages of Turkish (in the Altaic language family) and Persian (in the Indo-European language family) are entirely unrelated to Arabic. About 20 million Kurds—a people living in Turkey, Iraq, Iran, and Syria—speak Kurdish (in the Indo-European language family). Many people in North Africa speak Berber (in the Afro-Asiatic language family). It is also important to recognize that although political circumstances made them enemies in the 20th century, Arabs and Jews lived in peace for centuries and share many cultural traits. For example, both speak Semitic languages (in the Afro-Asiatic language family) and recognize Abraham as their patriarch. Finally, it is important to note that the world's most populous Islamic nation, Indonesia, is neither Arab nor Middle Eastern.

ROOTS OF THE FAITH

An understanding of the religious tenets, culture, and diffusion of Islam is vital for appreciating the cultural geography of the Middle East. Islam is a monotheistic faith built upon the foundations of the region's earliest monotheistic faith, Judaism, and its offspring, Christianity. Indeed, Muslims call Jews and Christians "People of the Book," and their faith obliges them to be tolerant of these special peoples. Muslims believe that their Prophet Muhammad was the very last in a series of prophets who brought the Word to humankind. Thus they perceive the Bible as incomplete but not entirely wrong—Jews and Christians merely missed receiving the entire message. Muslims disagree with the Christian concept of the divine trinity, and regard Christ as a prophet rather than as God.

Muhammad was born in 570 A.D. to a poor family in the western Arabian (now Saudi Arabia) city of Mecca. Located on an important north-south caravan route linking the frankincense-producing area of southern Arabia (now Yemen and Oman) with markets in Palestine (now Israel) and Syria, Mecca was a prosperous city at the time. It was also a pilgrimage destination, because more than 300 deities were venerated in a shrine there called the Ka'aba (the Cube) (Fig. 8.11). Muhammad married into a wealthy family and worked in the caravan trade. Muslim tradition holds that when he was about 40 years old, Muhammad was contemplating in a cave outside Mecca when the Angel Gabriel appeared to him and ordered him to repeat the words of God which the angel would recite to him. Over the next 22 years, the prophet related these words of God (Allah) to scribes who wrote them down as the Qur'an (or Koran), the Holy Book of Islam.

During this time, Muhammad began preaching the new message, "There is no god but God," which the polytheistic people of Mecca viewed as heresy. As much of their income depended upon pilgrimage traffic to the Ka'aba, they also viewed Muhammad and his small band of followers as an economic threat. They forced the Muslims to flee from Mecca and take refuge in Yathrib (modern Medina), where a largely Jewish population had invited them to settle. There were subsequent skirmishes between the Meccans and Muslims, but in 630 the

FIGURE 8.11
The black-shrouded shrine known as the Ka'aba in Mecca's Great Mosque is the object toward which all Muslims face when they pray, and is the centerpiece of the pilgrimage to Mecca required of all able Muslims.
Nabeel Turner/Tony Stone Images

Muslims prevailed and peacefully occupied Mecca. The Muslims destroyed the idols enshrined in the Ka'aba, which became a pilgrimage center for their one God.

THE DIFFUSION OF ISLAM

After Muhammad's death in 632, Arabian armies carried the new faith far, and quickly. The two decaying empires that then prevailed in the Middle East—the Byzantine or Eastern Roman Empire, based in Constantinople (now Istanbul), and the Sassanian Empire based in Persia (now Iran) and adjacent Mesopotamia (now Iraq)—put up only limited military resistance to the Muslim armies before capitulating. Local inhabitants generally welcomed the new faith, in part because administrators of the previous empires had not treated them well, while the Muslims promised tolerance. Soon the Syrian city of Damascus became the center of a Muslim empire. Baghdad assumed this role in 750 A.D.

Arab science and civilization flourished in the Baghdad immortalized in the legends of the Thousand and One Nights. There were important accomplishments and discoveries in mathematics, astronomy, and geography. Scholars translated the Greek and Roman classics, and, if not for their efforts, many of these works would never have survived to become part of the modern European legacy. It was an age of exploration, when Arab merchants and voyagers visited China and the remote lands of southern Africa. Many important discoveries by the Arab geographers were recorded in Arabic, a language unfamiliar to contemporary Europeans, and had to be rediscovered centuries later by the Portuguese and Spaniards.

Definitions and Insights

Shi'ite and Sunni Muslims

A serious schism developed very early within Islam, and it persists today, continuing to have an important impact on relations among the Muslim nations and between some Muslim countries and the West. The split developed because the Prophet Muhammad had named no successor to take his place as the leader (caliph) of all Muslims. Some of his followers argued that the person with the strongest leadership skills and greatest piety was best qualified to assume this role. These followers became known as **Sunni,** or orthodox, Muslims. Others argued that only direct descendants of Muhammad, specifically through descent from his cousin and son-in-law Ali, could qualify as leaders. They became known as **Shi'a,** or Shi'ite, Muslims. The military forces of the two camps engaged in battle south of Baghdad at Karbala in 680 A.D. and in the encounter, Sunni troops caught and brutally murdered Hussein, a son of Ali. Thereafter, the rift was deep and permanent. The martyrdom of Hussein became an important symbol for Shi'ites, who still today regard themselves as oppressed peoples struggling against cruel tyrants—including some Sunni Muslims.

Today only two Muslim countries, Iran and Iraq, have majority Shi'ite populations. Significant minority populations of Shi'ites are in Syria, Lebanon, Yemen, and the Arab states of the Persian/Arabian Gulf. Shi'ite Iran challenged the United States in an infamous hostage ordeal in 1980 that began after Muslim clerics assumed control of Iran's government. After Israel's invasion of Lebanon in 1982, Lebanese Shi'ite Muslims sympathetic to Iran abducted several U.S. and other Western citizens and killed hundreds more. The Shi'ite Muslim government of Iran continues to be an outspoken critic of Saudi Arabia (a largely Sunni nation), causing constant apprehension on the part of the oil kingdom's Western allies. The West's fear of a Shi'ite rebellion leading to the emergence of a new Shi'ite state in southern Iraq contributed to the allies' decision to halt their assault on Iraq after Iraqi forces withdrew from Kuwait in 1991.

The Five Pillars of Islam

Despite their differences, Sunni and Shi'ite Muslims are united in support of the five fundamental precepts, or **pillars,** of Islam. The first of these is the profession of faith: "There is no god but God, and Muhammad is His Messenger." This expression is often on the lips of the devout Muslim, both in prayer and as a prelude to everyday activities. The second pillar is prayer, required five times daily at prescribed intervals. Two of these prayers mark dawn and sunset. Business comes to a halt as the faithful prostrate themselves before God. Muslims may pray anywhere, but wherever they are they must turn toward Mecca.

There also is a congregational prayer at noon on Friday, the Muslim sabbath.

The third pillar is almsgiving. In earlier times, Muslims were required to give a fixed proportion of their income as charity, similar to the concept of the tithe in the early Christian church. Today the donations are voluntary. Even Muslims of very modest means give what they can to the more needy.

The fourth pillar is fasting during Ramadan, the ninth month of the Muslim lunar calendar. During Ramadan, Muslims are required to abstain from food, liquids, smoking, and sexual activity from sunrise to sunset. The lunar month of Ramadan occurs earlier each year in the solar calendar and thus periodically falls in summer. In the torrid Middle East, that timing imposes special hardships on the faithful, who even if they are performing manual labor must resist the urge to drink water during the long, hot days.

The final pillar is the pilgrimage (*hajj*) to Mecca, Islam's holiest city. Every Muslim who is physically and financially capable is required to make the journey once in his or her lifetime. A lesser pilgrimage may be performed at any time, but the prescribed season is the twelfth month of the Muslim calendar. Those days witness one of Earth's greatest annual migrations, as about two million Muslims from all over the world converge on Mecca (Fig. 8.11). Hosting these throngs is an obligation the government of Saudi Arabia fulfills proudly and at considerable expense, but in recent years with some trepidation because Shi'ite militants have sometimes demonstrated at Mecca's holy places. Many pilgrims also visit the nearby city of Medina, where Muhammad is buried. Most Muslims regard the *hajj* as one of the most significant events of their lifetimes. All are required to wear simple seamless garments, and, for a few days, the barriers separating groups by income, ethnicity, and nationality are broken. They return home with the new stature and title of "hajj" but also with humility and renewed devotion.

Interpretations of the Faith

While all Muslims share the five pillars and other tenets, they vary widely in other cultural practices related to their faith, depending upon what country they live in, whether they are from the desert, village or city, and how much education and income they have. The governments and associated clerical authorities in Saudi Arabia and Iran insist upon strict application of Islamic law (*shari'a*) to civil life; in effect, there is no separation between church and state. The Qur'an does not state that women are required to wear veils, but it does urge them to be modest, and it portrays their roles as different from those of men. Clerics in Saudi Arabia insist that women wear floor-length, long-sleeved black robes and black veils in public, that they not travel unaccompanied by a male member of their families, and that they not drive cars. In Egypt, by contrast, Muslim women are free to appear in public unveiled if they choose. However, in most Muslim countries, conservative ideas about the role of

ISLAMIC FUNDAMENTALISM

A wave of Islamic "fundamentalism" has recently swept the Islamic world. Arguing that "Islam is the solution," Islamists (as they are more accurately known) reject what they view as the materialism and moral corruption of Western countries, and the political and military support these countries lend to Israel. Both Sunni and Shi'ite Muslims have advanced a wide range of Islamic movements, notably in Iran, Lebanon, Egypt, Afghanistan, Sudan, and Algeria.

Although nominally religious, the more radical of these movements have political ambitions. In 1993, followers of the radical Egyptian cleric Shaykh Umar Abdel-Rahman bombed New York City's World Trade Center as a protest against American support of Israel and Egypt's pro-Western government. And in an attempt to destabilize and replace Egypt's government, Islamists of the al-Jami'at al-Islamiyya movement attacked and killed foreign tourists in Egypt in the early 1990s. In the 1980s, members of the pro-Iranian Hizbullah, or "Party of God," in Lebanon kidnapped foreigners as bargaining chips for releasing comrades jailed in other Middle Eastern countries. Within Israel and the autonomous Palestinian territories of the West Bank and Gaza Strip, Palestinian members of HAMAS (an Arabic acronym for the Islamic Resistance Movement) have carried out terrorist attacks on Israeli civilians and soldiers in an effort to derail implementation of the peace agreements reached between the Israeli government and the Palestine Liberation Organization (PLO). In Algeria, years of bloodshed have followed the government's annulment of 1991 election results which would have given the Islamic Salvation Front (FIS) majority control in the parliament. Muslim sympathizers have carried the battle to France, bombing civilian targets in protest against the French government's support for the Algerian regime.

In all of these situations, a tiny minority of Muslims carried out terrorist actions which the great majority condemned. Mainstream Islamic movements have distinguished themselves through public service to the needy, and through encouragement of stronger moral and family values. For many people however, "Muslim" and "terrorist" have become synonymous—an erroneous association which can be overcome in part by careful study of the complex Middle East.

women are still very strong: They should be modest, retiring, good mothers, and keepers of the home. The Qur'an portrays women as equal to men in the sight of God, and, in principle, Islamic teachings guarantee the right of women to hold and inherit property.

Most Muslim women argue that what others often see as "backward" cultural practices are in fact very progressive. For example, their modest dress compels men to evaluate them on the basis of their character and performance, not their attractiveness. Segregation of the sexes in the classroom makes it easier for both women and men to develop their confidence and skills—a finding only now being applied in some U.S. classrooms. Sexual assault is rare. A married woman retains her maiden name. These apparent advantages can be weighed against the drawbacks that women are generally subordinate to men in public affairs and have fewer opportunities for education and for work outside the home.

8.4 The Middle Eastern Ecological Trilogy: Villager, Pastoral Nomad, and Urbanite

In the 1960s, the American geographer Paul English developed a useful model for understanding relationships between the three ancient ways of life that still prevail in the Middle East today: villager, pastoral nomad, and urbanite (Fig. 8.12). Villagers are the subsistence farmers of rural agricultural areas; pastoral nomads are the desert peoples who migrate with their livestock, following patterns of rainfall and vegetation; and urbanites are the inhabitants of the large towns and cities. Describing each of these as a component of the **Middle Eastern ecological trilogy,** English explained how each of them has an important, usually mutually beneficial, pattern of interaction with the other two.

The peasant farmers of Middle Eastern villages (the villagers) represent the cornerstone of the trilogy. They grow the staple crops such as wheat and barley which feed both the city-dweller and the pastoral nomad of the desert. Neither urbanite nor nomad could live without these. The village also unwillingly provides the city with tax revenue, soldiers, and workers. And before the mid-20th century, villages provided pastoral nomads with plunder, as the desert-dwellers raided their settlements and caravan supply lines. Generally, however, the exchange is beneficial: The nomads provide villagers with livestock products including live animals, meat, milk, cheese, hides, and wool, and with desert herbs and medicines, while educated and progressive urbanites provide technological innovations, religious instruction, and cultural amenities.

There is little direct interaction between urbanites and pastoral nomads, although some manufactured goods such as clothing travel from city to desert, and some desert-grown folk medicines pass from desert to city. Historically the exchange has been violent, as urban-based governments have sought to control the movements and military capabilities of the elusive and sometimes hostile nomads. Pastoral nomads once plundered rich caravans plying the major overland trade routes of the Middle

East. Governments did not tolerate such activities and often cracked down hard on those nomads they were able to catch.

In the 1970s, Paul English wrote an article marking the "passing of the ecological trilogy." He noted that cities were encroaching on villages, villagers were migrating into cities and giving some neighborhoods a rural aspect, and pastoral nomads were settling down—thus, the trilogy no longer existed. In reality, while the makeup and interactions of its parts have changed somewhat, the trilogy model is still valid and useful as an introduction to the major lifeways of the Middle East. It is especially significant that a given man or woman in the Middle East strongly identifies himself or herself as either a villager, a pastoral nomad, or an urbanite. This perception of self has an important bearing on how these people of very different backgrounds interact, even when they live in close proximity. Urban officials may work in rural village areas, but they remain at heart and in their perspectives city people, and usually live apart from farmers. Similarly, extended families of pastoral nomads may settle down and become farmers, but continue to identify themselves by affiliation with the nomadic tribe. Many continue to harvest desert resources on a seasonal basis, and retain marriage and other ties with desert-dwelling kinsmen.

THE VILLAGE WAY OF LIFE

Agricultural villagers historically represented by far the majority in Middle Eastern populations; only within recent decades have urbanites begun to outnumber them. In this generally dry environment, the villages are located near a reliable water source, with cultivable land near by. They tend to be composed of closely related family groups, with the land of the village often owned by an absentee landlord. Most often, the villagers live in closely spaced flat-roofed houses made of adobe, mud brick, or concrete blocks. Production and consumption focus on a staple grain. As land for growing fodder is often in short supply, villagers keep only a limited number of sheep and goats, and rely in part on nomads for pastoral produce. Residents of a given village usually share common ties of kinship, religion, ritual, and custom, and the changing demands of agricultural seasons regulate their patterns of activity.

Village life has been increasingly exposed to outside influences since the mid-18th century. Contacts with European colonialism brought significant economic changes, including the introduction of cash crops and modern facilities to ship them. Improved and expanded irrigation, financed initially with capital from the West, brought more land under cultivation. Recent agents of change have been the countries' own government doctors, government teachers, and land reform officers. Products of modern technology such as radios, television, sewing machines, and motorcycles have modified old patterns of living. The young and more ambitious have been drawn to urban areas. Improved roads and communications in turn have carried urban influences to villages, prompting villagers to become more integrated into national societies.

FIGURE 8.12

Faces of the Middle Eastern ecological trilogy: (a) villagers at a market in Luxor, Egypt; (b) Ma'aza Bedouin nomads in the Eastern Desert, Egypt; (c) and Israeli shoppers at the urban market of Makaneh Yehudah.
(a) Manly Kazmers, © 1996/Dembinsky Photo Associates; (b) Joseph Hobbs; (c) Richard Schiell, © 1996/Dembinsky Photo Associates

THE PASTORAL NOMADIC WAY OF LIFE

Pastoral nomadism emerged as an offshoot of the village agricultural way of life not long after plants and animals were first domesticated in the Middle East. Rainfall and the wild fodder it produces, although scattered, are sufficient resources to support small, mobile groups of people who migrate with their sheep, goats, and camels (and in some locales, cattle) to take advantage of this changing resource base. In mountainous areas they follow a pattern of **vertical migration** or **transhumance**, moving with their flocks from lowland winter to highland summer pastures. In the flatter expanses which comprise most of the region, the nomads have a pattern of **horizontal migration** over much larger areas where rainfall is typically far less reliable than in the mountains. In addition to selling or trading livestock in order to obtain foods, tea, sugar, clothing, and other essentials from settled communities, pastoral nomads also hunt, gather, work for wages, and, where possible, grow crops. Their many-faceted livelihood has been described as a strategy of "risk minimization," based on the exploitation of multiple resources so that some will support them if others fail.

Although renowned in Middle Eastern legends and in popular Western films like *Lawrence of Arabia*, pastoral nomads have been described as "more glamorous than numerous." It is still impossible to obtain adequate census figures on the number still living in the deserts of the Middle East, though estimates range from 5 to 13 million. The late 20th century has witnessed the rapid and progressive settling down, or **sedentarization,** of the nomads—a process attributed to a variety of reasons. In some cases prolonged drought virtually eliminated the resource base on which the nomads depended. Traditionally they were able to migrate far enough to find new pastures, but modern national boundaries now prohibit such movements. Some have returned with the rains to their desert homelands, while others have chosen to remain as farmers or wage laborers in villages and towns. On the Arabian Peninsula, the prosperity and technological changes prompted by oil revenues made rapid inroads into the material culture—and then the livelihood preferences—of the desert people; many preferred the comforts of settled life. Some governments, notably those of Israel and prerevolutionary Iran—unable to count, tax, conscript, and control a sizable migrant population—compelled nomads to settle.

Pastoral nomads of the Middle East identify themselves primarily not by their nationality but by their tribe. The major ethnic groups from which these tribes draw are the Arabic-speaking Bedouins of the Arabian Peninsula and adjacent lands; the Berber and Tuareg of North Africa; the Kababish and Bisharin of Sudan; the Yoruk and Kurds of Turkey; the Qashai and Bakhtiari of Iran; and the Pashtun of Afghanistan. Members of a tribe claim common descent from a single male ancestor who lived countless generations ago; their kinship organization is thus called a **patrilineal descent system.** It is also a **segmentary kinship system,** so called because there are smaller subsections of the tribe, known as clans and lineages, which are functionally important in daily life. Members of most closely related families comprising the lineage, for example, share livestock, wells, trees, and other resources. Both the larger clans, made up of numerous lineages, and the tribes possess territories. Members of a clan or tribe typically allow members of another clan or tribe to use the resources within its territory on the basis of "usufruct," or nondestructive mutual use.

Although some detractors have depicted pastoral nomads as the "fathers" rather than "sons" of the desert, blaming them for wanton destruction of game animals and vegetation, there are numerous examples of pastoral nomadic groups who have developed indigenous and very effective systems of resource conservation. Most of these practices depend upon the kinship groups of family, lineage, clan, and tribe to assume responsibility for protecting plants and animals.

THE URBAN WAY OF LIFE

The city was the final component to emerge in the ecological trilogy, beginning in about 4000 B.C. in Mesopotamia and 3000 B.C. in Egypt. Unlike the villages they resembled in many ways, the early cities were distinguished by their larger populations (more than 5000 people), use of written languages, and presence of monumental temples and other ceremonial centers. The early Mesopotamian city and, after the seventh century, the classic Islamic city, called the **medina,** had several structural elements in common (Fig. 8.13). The medina was characterized by a high surrounding wall built for defensive purposes. The city center was dominated by the congregational mosque and often an attached administrative and educational complex. Although Islam is often characterized as a faith of the desert, religious life has always focused in, and been diffused from, Middle Eastern cities. The importance of the city's congregational mosque in religious and everyday life is often emphasized by its large size and outstanding artistic execution.

A large commercial zone, known as a *bazaar* in Persian and *suq* in Arabic, and recognizable as the ancestor of the modern shopping mall, typically adjoined the ceremonial and administrative heart of the city. Merchants and craftsmen of different commodities occupied separate spatial areas within this complex, and visitors to an old medina today can still expect to find streets where spices, carpets, gold, silver, traditional medicines, and other goods are sold exclusively. Smaller clusters of shops and workshops were located at the city gates.

Residential areas were differentiated as quarters not by income group but by ethnicity; the medina of Jerusalem, for example, still has distinct Jewish, Arab, Armenian, and non-Armenian Christian quarters. Homes tended to face inward toward a quiet central courtyard, buffering the occupants from the noise and bustle of the street. The narrow, winding streets of the medina were intended for foot traffic and small animal-drawn carts, not for large motorized vehicles, which accounts for the traffic jams in some old Middle Eastern cities today and for the wholesale destruction of the medina in others.

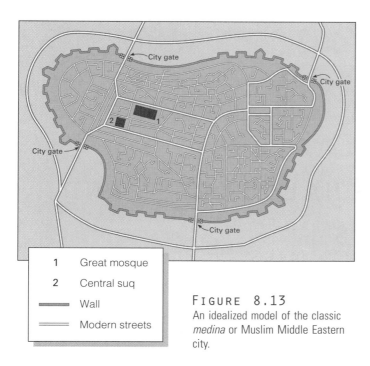

1	Great mosque
2	Central suq
———	Wall
=====	Modern streets

FIGURE 8.13
An idealized model of the classic *medina* or Muslim Middle Eastern city.

Those medinas which survive today are gently-decaying vestiges of a forgotten urban pattern. Periods of European colonialism and subsequent nationalism changed the face and orientation of the Middle Eastern city. During the colonial age, resident Europeans preferred to live in more spacious settings at the outer edges of the city, and later the national elite followed this pattern. In recent times, independent governments have followed Western building styles, with broad traffic arteries cutting through the old quarters and large central squares near government buildings. This opening up of the cityscape has scattered commercial activity along the wide avenues, thus diluting the prime importance of the central bazaar as the focus of trade.

Rural-urban migration and the city's own internal growth contribute to a rapid rate of urbanization which puts enormous pressure on services in the poorer Middle Eastern nations. Governments often build high-rise public housing to accommodate the growing population, contributing to a cycle in which the urban poor move into the new dwellings only to leave their old quarters as a vacuum to draw in still more rural migrants. In Cairo, millions of former villagers now live in the "City of the Dead," an extraordinary urban landscape composed of multistory dwellings erected above graves—a last resort for the poor who have no other place to go (Fig. 8.14). The overwhelmingly largest, or primate, city, so characteristic of Middle Eastern capitals, thus grows at the expense of the smaller city. Much of the rural-urban migration and subsequent urban gridlock and squalor could probably be avoided if governments invested more in the development of villages and smaller cities. The oil-rich countries with relatively small populations generally enjoy an urban standard of living equaling that of affluent Western countries. Modern industrial cities such as Jubail in Saudi Arabia, and others founded on oil wealth, were built virtually overnight, providing fascinating contrast to the region's colorful, complex ancient cities (see Fig. 8.15).

FIGURE 8.14
Cairo's City of the Dead. The domes mark tombs of Egypt's 15th-century ruling class. The twin-pillared structures in the center foreground are individual graves from the 20th century. Until 1967, these individual graves and the monumental tombs were the only features on this landscape on the eastern edge of Cairo. But population growth and refugee movements in the last three decades have led to massive settlement of this cemetery. Single- and multiple-family dwellings now cover most of the graves; the living dwell with the dead. *Marc Bernheim/Woodfin Camp & Associates*

FIGURE 8.15
The port of Jubail in Saudi Arabia's Eastern Province is an entirely modern, meticulously planned city, founded upon the kingdom's oil wealth. *Robert Azzi/Woodfin Camp & Associates*

9

THE MIDDLE EASTERN COUNTRIES: CHALLENGES AND OPPORTUNITIES

Conflict in the Middle East has been one of the most persistent and dangerous problems in global affairs in the late 20th century. This chapter introduces the key players and issues in conflict and discusses the central features of land and life in the individual countries of the Middle East. Hostilities in this region generally focus on questions of who owns land and water and are of great interest to geographers. The ongoing peace process represents an opportunity to resolve these issues and provide peace "dividends"—not just to the region's inhabitants but to the international community. The consequences of a breakdown in the peace process would likewise affect the wider world. The challenges posed in the Middle East in the late 1990s are enormous, and so are the opportunities.

9.1 The Arab World

The Arab world, stretching east-west from the Indian Ocean to Morocco and southward from the Mediterranean to the fringes of tropical Africa, is the major component of the Middle East. The peoples who inhabit it are ethnically and culturally diverse, but an overwhelming majority speak dialects of Arabic and are therefore known as Arabs. Originally, the Arabs were inhabitants of the Arabian peninsula, but conquest after their conversion to Islam spread their language and associated Islamic culture widely.

Today, most Arabs live in countries that belong to the League of Arab States, commonly known as the Arab League. This body, formed in 1945, is composed of Algeria, Bahrain, Djibouti, Egypt, Iraq, Jordan, Kuwait, Lebanon, Mauritania, Morocco, Oman, the Palestine Liberation Organization (PLO), Qatar, Saudi Arabia, Somalia, Sudan, Syria, Tunisia, the United Arab Emirates (UAE), and Yemen; Libya withdrew its membership in 1994. Quarrels between member states have periodically hindered their ability to function as a group. For example, Egypt was suspended from Arab League membership for a decade after it signed a peace treaty with Israel in 1979. Iraq's attack on its fellow league member Kuwait has effectively alienated it from the organization since 1990.

Nevertheless, a strong sense of Arab unity generally counteracts such divisiveness. This sense of Arab community has several important roots—namely, a common language, cultural heritage, and majority belief in Islam, coupled with pride in past achievements and the realization of a need for united efforts to improve social and economic conditions throughout the Arab world. In recent decades, regional unity has been en-

◀ A photograph looking northeastward from the U.S. Space Shuttle *Atlantis* in April, 1991. In the foreground is the Nile Valley. The Gulf of Suez and Gulf of Aqaba frame the triangle of Egypt's Sinai Peninsula. The political boundary between Egypt and Israel is clearly visible; agricultural development on the Israeli side gives Israel a shade of green. The rift valley follows the Gulf of Aqaba northward to the Dead Sea and beyond, up the Jordan River. *NASA/Mark Martin/Photo Researchers, Inc.*

hanced by opposition to perceived economic and political intervention by foreigners—a feeling most sharply focused on the state of Israel, which was founded under foreign auspices on a territory inhabited mostly by Arabs.

9.2 Israel, Jordan, Lebanon, and Syria: Uneasy Neighbors

THE PROMISED LAND

Depending upon one's perspective, the Jewish connection with the geographical region known as Palestine—essentially the area now composed of Israel, the West Bank, and the Gaza Strip—is most significant on a time scale of about 4000 years or about 100 years. According to the Bible, around 2000 B.C. God commanded Abraham and his kinspeople, known as Hebrews (later as Jews), to leave their home in what is now southern Iraq and settle in Canaan. God told Abraham that this land of Canaan—geographical Palestine—would belong to the Hebrews after a long period of persecution. The Bible says that the Hebrews did settle in Canaan, until famine struck that land. At the command of Abraham's grandson Jacob, the Hebrews—known then as Israelites—relocated to Egypt, where grain was plentiful. That began the long sojourn of the Israelites in Egypt, which, according to the Bible, ended in about 1200 B.C. when Moses led them out (the Exodus).

According to Jewish history, the prophecy of Abraham was first fulfilled when the Israelites settled once again in their "promised land" of Canaan. The Jewish king Saul unified the twelve tribes who were Jacob's descendants into the first united Kingdom of Israel in about 1020 B.C. In about 950 B.C. in Jerusalem—the capital of a kingdom enlarged by Saul's successor David—King Solomon built Judaism's First Temple. He located it atop a great rock known to the Jews as Even HaShetiyah, the "Foundation Stone," plucked from beneath the throne of God to become the center of the world and the core from which all the world was created.

The united Kingdom of Israel lasted only about 200 years before splitting into the states of Israel and Judah. Empires based in Mesopotamia destroyed these States: The Assyrians attacked Israel in 721 B.C. and the Babylonians sacked Judah in 586 B.C. The Babylonians destroyed the Temple and exiled the Jewish people to Mesopotamia, where they remained until conquering Persians allowed them to return to their homeland. In about 520 B.C. the Jews who returned to Judah, the land from which they take their name, rebuilt the Temple (the "Second Temple") on its original site. A succession of foreign empires came to rule the Jews and Arabs of Palestine: Persian, Macedonian, Ptolemaic, Seleucid, and, around the time of Christ, Roman. Herod, the Jewish king who ruled under Roman authority and was a contemporary of Christ, greatly enlarged the Temple complex.

THE JEWISH DIASPORA

The Jews of Palestine revolted against Roman rule three times between 64 and 135 A.D. The Romans quashed these rebellions in a series of famous sieges, including those of Masada and Jerusalem. The Romans destroyed the Second Temple, and a third has never been built. All that remains of the Second Temple complex is a portion of the surrounding wall built by Herod. Today this Western Wall, known to non-Jews as the "Wailing Wall," is the most sacred site in the world accessible to Jews (Figure 9.1). Religious tradition prohibits Jews from ascending the Temple Mount above—the area where the Temple actually stood. After the temple's destruction, that site was occupied by a Roman temple, and then in 691 by the Muslim shrine called the Dome of the Rock, which still stands today.

The victorious Romans scattered the defeated Jews to the far corners of the Roman world. Thus began the Jewish exile or **diaspora**. In their exile the Jews never forgot their attachment to the promised land. The Passover prayer ends with the words "Next year in Jerusalem!" In Europe, where their numbers were greatest, Jews were subjected to systematic discrimination and persecution, and were forbidden to own land or engage in a number of professions. Known as **anti-Semitism**, the hatred of Jews developed deep roots in Europe. This sentiment in part grew out of the fact that Christian Europeans were prohibited from practicing usury, or moneylending. They assigned this role to Jews but then resented paying debts to the despised moneylenders.

FIGURE 9.1
Sacred places in Judaism and Islam. The Western Wall (with shrubs on it) is part of the retaining wall which supported Judaism's holiest site, the Temple in Jerusalem. Near the bottom-left corner of the wall, a man is entering the ancient tunnel whose other end was opened in 1996. The tunnel opening sparked widespread unrest among Muslim Palestinians because one of Islam's holiest sites is the Dome of the Rock, whose golden dome rises above the Western Wall. The Dome of the Rock is widely believed to occupy the site where the Jewish Temple's Holy of Holies stood, and where the Ark of the Covenant rested for many centuries. *Mike Barlow © 1992/Dembinsky Photo Associates*

In the 1930s anti-Semitism became state policy in Germany under the Nazis, led by Adolf Hitler. Many German Jews—including Albert Einstein—fled to the United States, and others emigrated to Palestine in support of the **Zionist movement**, which aimed at establishing a Jewish homeland in Palestine with Zion (a synonym for Jerusalem) as its capital. Most Jews were not as fortunate as the emigrants. Within the boundaries of the Nazi empire that dominated most of continental Europe during World War II, Hitler's regime executed its "final solution" to the Jewish "problem." Nazi Germans and their allies killed an estimated six million Jews, along with other "inferior" minorities, including Gypsies and homosexuals. It was this **Holocaust** that prompted the victorious allies of World War II, from their powerful position in the newly formed United Nations, to create a permanent homeland for the Jewish people in Palestine.

THE BIRTH OF ISRAEL

The modern state of Israel was carved from lands whose fate had been undetermined since the end of World War I. The Ottoman Empire, based in what is now Turkey, had ruled Palestine and surrounding lands in the eastern Mediterranean since the 16th century. After the British and French defeated the

Turks in World War I and destroyed their empire, they divided the region between them. The British received the mandate for Palestine, Transjordan (modern Jordan), and Iraq, and the French the mandate for Syria (now Syria and Lebanon). British administrators of Palestine made conflicting promises to Jews and Arabs, on the one hand promoting Jewish immigration to Palestine with an eye to the eventual establishment of a Jewish state there, and on the other imposing limits on this immigration and implying that any Jewish state would not be established in the heart of predominantly-Arab Palestine. Placing themselves in a no-win position, the British ultimately decided to withdraw from Palestine in 1948 and leave the young United Nations with the task of determining its future.

The United Nations Partition Plan of 1948 implemented a "two state solution" to the problem of Palestine. It established an Arab state and a Jewish state. The plan was deeply flawed. The states' territories were long, narrow, and almost fragmented, giving each side a sense of vulnerability and insecurity (Figure 9.2). War broke out immediately between newborn Israel and the armies of the neighboring Arab countries of Transjordan, Egypt, Iraq, Syria, and Lebanon. The smaller but better-organized and more highly motivated Israeli army defeated the Arab armies, and Israel acquired its "pre-1967" borders (see Figure 9.2). Israel gained control of about 77 percent of the territory of the Palestine mandate, including the coastal

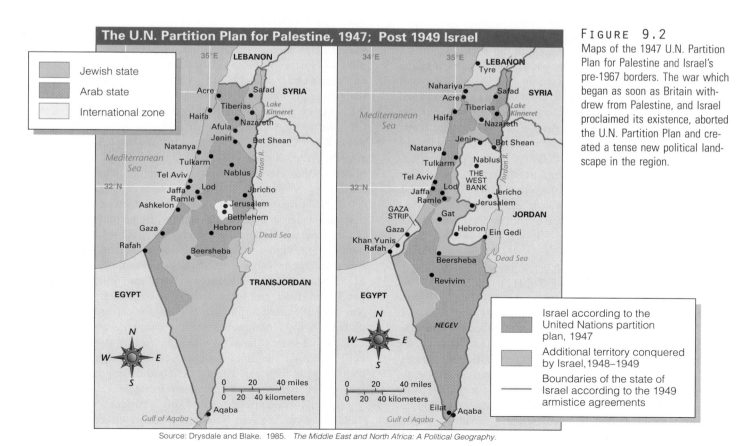

FIGURE 9.2

Maps of the 1947 U.N. Partition Plan for Palestine and Israel's pre-1967 borders. The war which began as soon as Britain withdrew from Palestine, and Israel proclaimed its existence, aborted the U.N. Partition Plan and created a tense new political landscape in the region.

Source: Drysdale and Blake. 1985. *The Middle East and North Africa: A Political Geography.*

areas, the northern hill country of Galilee, the dry and thinly populated southern triangle called the Negev, the western portion of the city of Jerusalem in the Judean hills, and a corridor leading to the city from the coastal plain. Egypt occupied the Gaza Strip, a piece of land on the Mediterranean shore adjacent to Egypt's Sinai peninsula. Inhabited mostly by Palestinian Arabs, the Gaza Strip was part of the independent Palestine the U.N. had envisioned. Transjordan occupied the West Bank—the predominantly-Arab hilly region of central Palestine on the west side of the Jordan River—and the entire old city of Jerusalem, including the Western Wall and Temple Mount. The Palestinian Arab state envisioned in the U.N. Partition plan was thus stillborn.

In keeping with national legislation known as the **Law of Return**, the Jewish state of Israel has always granted citizenship to any Jew who wishes to live there. Immigrants from Europe (Ashkenazi Jews) and from the Middle East (Sephardic Jews) have populated Israel since its founding. New waves of Jewish immigration followed the dissolution of the Soviet Union and a change of government in Ethiopia, where an ancient Jewish group called Falashas lived in isolation for thousands of years. Since 1985, Israel has accommodated about 500,000 Russian Jews and 30,000 Ethiopian Jews. Members of both groups often complain that other Israelis discriminate against them. In 1996, Falasha Jews rioted when they learned that Israeli hospitals had been routinely throwing away blood the immigrants donated because of a reported high rate of HIV infection among them.

In 1946, Jews made up 31 percent of the population of the mandate of Palestine. In 1996, they made up 81 percent of the population of 5.8 million living within Israel's pre-1967 borders. Palestinian Arabs with and without Israeli citizenship, and a variety of other ethnic groups, comprise the balance of the population. Another 2.3 million people live in the West Bank, Gaza Strip, and Golan Heights—areas occupied by Israel in the wars of 1967 and 1973. About 145,000 of those are Jewish settlers, and most of the rest are Palestinian Arabs (in the West Bank and Gaza Strip) and Druze and other Arabs (in the Golan Heights). Another 110,000 Jewish settlers live in East Jerusalem, which is also part of the territories Israel conquered in 1967.

A GEOGRAPHIC SKETCH OF ISRAEL

Aided by large amounts of outside capital, primarily from the United States, the Israelis have developed their republic as a modern Westernized state. Still far from self-sufficient economically, however, Israel has difficulty in financing both a large military budget and extensive social programs, including those which service the recent immigrants from Russia and Ethiopia.

Despite many obstacles, one of Israel's greatest achievements since independence has been the expansion and intensification of agriculture (see photograph, page 220, and Fig. 9.3).

Both cultivated land and irrigation in Israel proper have greatly increased. Production concentrates on citrus fruits (including the famous Jaffa oranges) which provide export revenue, and on dairy, beef, and poultry (for eggs and meat) products, as well as flowers, vegetables, and animal feedstuffs. Israel's intensive, mechanized agriculture, oriented to nearby urban markets, resembles in many ways the agriculture of densely populated areas in western Europe. Collectivized settlements called *kibbutzim* (singular, *kibbutz*) are a distinctively Israeli feature on the agricultural, and cultural, landscape. Many of these lie near the frontiers and have defensive as well as agricultural and industrial functions. Far more numerous and important in Israel's agricultural economy are other types of villages, including the small farmers' cooperatives called *moshavim* (singular, *moshav*) and villages of private farmers.

Agricultural development is concentrated in the northern half of the country, with its heavier rainfall (concentrated in the winter half-year in this Mediterranean climate) and more ample supplies of surface and underground water for irrigation. Annual precipitation averages 20 inches (50 cm) or more in most sections north of the city of Tel Aviv and in the interior hills where Jerusalem lies. Due to the rain shadow effect, very little rain falls in the deep rift valley of the Jordan River. From Tel Aviv southward to Beersheba, annual rainfall decreases to about 8 inches (20 cm), and in the central and southern Negev it drops to 4 inches or less. In the northern half of the country, irrigation water from varied sources (streams, lakes, springs, wells, artificial reservoirs to catch runoff, and reclaimed sewage water) helps intensify agriculture and overcome the summer drought. Surface water is scarce in the semiarid and arid south, and where underground water is found, it is often too saline for most crops. To expand agriculture along the coastal plain and in the northern Negev, Israel transfers large quantities of water from the north by the aqueduct and pipeline network of the National Water Carrier. The largest source for this network is Lake Kinneret (Lake Tiberias or Sea of Galilee), which is fed by the upper Jordan River (see also Regional Perspectives, page 226).

Israel's main hope for continued support of its growing population lies in expanding industry, trade, and services, including tourism. Industry has increased greatly since independence, although—rather like Japan—little of Israel's industrialization can be based on its own mineral resources. Metal-bearing ores and mineral fuels are scarce, although potash, bromine, and other materials extracted from the Dead Sea have provided an important basis for expanded chemical manufacturing. Immigrants and the country's own advanced educational system have provided the technical and scientific skills required for diverse industries, including many high-technology enterprises. Israel's diamond-cutting industry is its most internationally well-known enterprise.

Israel's proclaimed capital of Jerusalem (population: 429,000, excluding East Jerusalem in occupied territory) is the country's largest city. Tel Aviv-Jaffa (population: 327,000, city proper; 1.74 million, metropolitan area) on the Mediterranean

FIGURE 9.3

Israel and its neighbors, with the occupied territories. Most highland areas are predominantly mountains except in Israel and Jordan, where hills predominate. International boundaries show the status prior to the 1967 Arab-Israeli War.

coast is Israel's greatest industrial center. The leading seaport and center of heavy industry (including steel milling and oil refining) is Haifa (population: 226,000) in northern Israel. The ancient town of Beersheba (population: 111,000) is the northern gateway, administrative center, and main industrial center of the Negev frontier region. To the south lies Israel's small Red Sea port and resort town of Eilat. Thanks to successful peace accords with Egypt in 1979 and Jordan in 1994, open doors now link Eilat with the neighboring resort towns of Taba in the Egyptian Sinai and Aqaba in Jordan. Tourism is booming, and planners envision a "Red Sea Riviera" that will grow on the three countries' Gulf of Aqaba shores to meet the demands of European sun-worshippers.

THE ARAB-ISRAELI CONFLICT

Israel's relations with the Arab world have always cast a cloud on this young country. Central issues—all of which have geographic underpinnings—include (1) Israel's right to exist, (2) Israel's possession of occupied territories taken from Arab countries in warfare, (3) the rights of the Arab Palestinians who fled as refugees in wartime or were overrun by Israeli occupy-

ing forces, (4) the question of who should control Jerusalem, and (5) the question of Jordan River water appropriation. Poor relations between Israel and its Arab neighbors have exploded into full-scale war on five occasions since Israel's declaration of independence on May 14, 1948. These conflicts have profoundly affected the geography of the Middle East and have had a marked impact on international economies and political relations. In order to understand these effects and the issues of the current Middle East peace process begun in the 1990s, it is essential to be familiar with these five wars (including the 1948 war, described above), particularly as they have rearranged the boundaries of nations and territories.

The 1956 War.   Israeli forces invaded Egypt's Sinai peninsula shortly before a British and French invasion of the Suez Canal Zone in the autumn of 1956. Egypt's President Gamal Abdel Nasser had nationalized the Suez Canal earlier in the year, ostensibly as a means of paying for the construction of the Aswan High Dam. This antagonized the British and French, who had built and still owned the Canal, and viewed it as a vital asset to their national economic interests. At the same time, Israel wanted to destroy Soviet-made arms which Egypt had positioned in the Sinai, and to open the Gulf of Aqaba to Israeli

WATER IN THE MIDDLE EAST

In the arid Middle East, where most water is available either from rivers or underground aquifers that cross national boundaries, control over water is an especially difficult and potentially explosive issue. The ongoing peace process promises to settle some of the more contentious disputes. In its September 1995 accord with the PLO, for example, Israel pledged to increase Palestinian access to the fresh-water aquifers underneath the West Bank, which supply about 40 percent of Israel's water. Exact details still need to be negotiated. Jordan and Israel similarly promised in their 1994 peace treaty to reach an agreement on sharing waters from the Jordan River (which forms a portion of their common border) and its tributary, the Yarmuk River. They are discussing a joint venture to build the "Dead-Red Canal," which would connect the Gulf of Aqaba with the Dead Sea. The gravity flow of seawater to the Dead Sea would spin turbines and run generators to produce electricity to be shared between the two nations. Still to be addressed is the sharing of water between Israel and Syria. In occupying Syria's Golan Heights, Israel controls an important watershed and some of the northern bank of the Yarmuk River on the border with Jordan. For many years, Israel indicated it would never allow Syria and Jordan to construct the "Unity Dam" which would store waters of the Yarmuk River to be shared between those countries. Israel thus implied it would bomb the dam rather than allow it to deprive Israel of Jordan River water.

As a "downstream" riverine, or **riparian**, nation, completely dependent upon water originating outside the country, Egypt considers relations with upstream countries vital to its long-term national security. Egypt and Sudan are signatories of a Nile Waters Agreement that apportions water between them, but Egypt does not have similar agreements with other countries. Ethiopia's diversion of water from vital Nile headstreams is a potential future source of friction with Egypt and Sudan, but at present Ethiopia has little money to finance the dams, canals, and pumps that large-scale diversions would require.

Turkey is an "upstream" state and is the source of four fifths of Syria's water and two thirds of Iraq's. These downstream neighbors are distraught by the diminished flow and quality of water resulting from Turkey's comprehensive Southeast Anatolia Project. When completed, the project is expected to reduce Syria's share of the Euphrates waters by 40 percent and Iraq's by 60 percent. Also increasing the likelihood of serious future tension is a history of strained relations among Turkey, Syria, and Iraq, accompanied by the fact that no commonly accepted body-of-water law governs the allocation of water in such international situations.

---

shipping. Pressure from both the United States and the Soviet Union quickly forced all of the invaders to withdraw from Egypt. It was a sensational victory for the firebrand nationalist Nasser. Israel won the right of navigation through the Gulf of Aqaba. Great Britain and France were humiliated.

The 1967 War ("The Six Day War"). This struggle was precipitated in part when, by positioning arms at the Strait of Tiran **chokepoint** (see Fig. 9.4 and Definitions and Insights, page 226), Egypt again closed the Gulf of Aqaba to Israeli shipping. President Nasser and his Arab allies took several other belligerent but nonviolent steps toward a war they were ill-prepared to fight. Israel elected to make a preemptive strike on its Arab neighbors, virtually destroying the Egyptian and Syrian air forces on the ground. Israel gave Jordan's King Hussein an opportunity to stay out of the conflict. However, Jordan went to war and quickly lost the entire West Bank and the historic Old City of Jerusalem. The entire nation of Israel was transfixed by the news that Jewish soldiers were praying at the Western Wall. The Israeli army (Israeli Defense Forces, or IDF) also seized the Gaza Strip, Egypt's Sinai Peninsula (thus closing the Suez Canal), and the strategic Golan Heights section of Syria overlooking Israel's Huleh Valley, in which the upper Jordan River flows. Israel had tripled its territory in six days of fighting (see Fig. 9.3).

Definitions and Insights

CHOKEPOINTS

The Strait of Tiran at the mouth of the Gulf of Aqaba is one of the Middle East's many chokepoints, or strategic narrow passageways on land or sea that may be easily closed off by use of force—or even the threat of force. Other notable chokepoints in the area are the Strait of Gibraltar at the mouth of the Mediterranean Sea; the Dardanelles and Bosporus straits (together known as the Turkish Straits; see Fig. 9.4) linking the Black Sea and Aegean Sea; the Bab el-Mandeb at the mouth of the Red Sea; the Strait of Hormuz at the mouth of the Persian/Arabian Gulf; and the Suez Canal. Notable events in military history and the formation of foreign policy in the Middle East focus on these strategic places. For example, Iran's plans to station Chinese-made silkworm missiles on the Strait of Hormuz and thus threaten international oil shipments led to a new level of U.S. involvement late in the Iran-Iraq War of 1980–88; and Egyptian closure of the Strait of Tiran chokepoint to Israeli shipping in 1967 helped precipitate the Six Day War.

The 1973 War.   On October 6, 1973, the Jewish holy day of Yom Kippur, Egypt and Syria launched a surprise attack on Israel with the hope of rearranging the humiliating, stalemated political map of the Middle East. Egypt's army initially showed surprising strength, penetrating deep into Sinai and overturning Israel's image of invulnerability. Israeli troops soon surrounded Egypt's army, but in the ensuing disengagement talks Egypt won back the eastern side of the Suez Canal and by 1975 was able to reopen it to commercial traffic. Syrian forces won back a small sliver of the Israeli-occupied Golan Heights.

The United States supported Israel during the conflict. Fearing a move by Syria's ally the Soviet Union, President Richard M. Nixon during the 1973 War placed U.S. forces on an alert status that presumed a possible nuclear exchange. Led by Saudi Arabia, furious Arab allies of Egypt and Syria imposed an **oil embargo**, halting petroleum sales to the U.S. for eight months, beginning in October 1973, and precipitating the country's **energy crisis**, when oil supplies plummeted and prices soared.

In 1979, Israel returned Sinai to Egypt under the U.S.-sponsored Camp David Accords, and Egypt recognized Israel's rights as a sovereign state. The Gaza Strip, West Bank, and Golan Heights continued to be Israeli-held, and came to be known as the "Occupied Territories."

The 1982 War.   This conflict was one facet of a larger war in Lebanon involving both civil strife and international intervention. Lebanon had become the main stronghold of the Palestine Liberation Organization (PLO), which used the country as a base for guerrilla operations against Israel. Under the leadership of Yasir Arafat, the PLO had been established in 1964 with the ambition of establishing an independent Palestinian state either in the occupied territories or in Israel itself. In June 1982, Israel invaded Lebanon with the avowed intention of smashing the PLO and establishing security for the northern Israeli frontier. Syria, whose forces had previously occupied Lebanon's Bekaa Valley on an alleged peacekeeping mission for the Arab League, resisted the Israeli advance and gave support to the

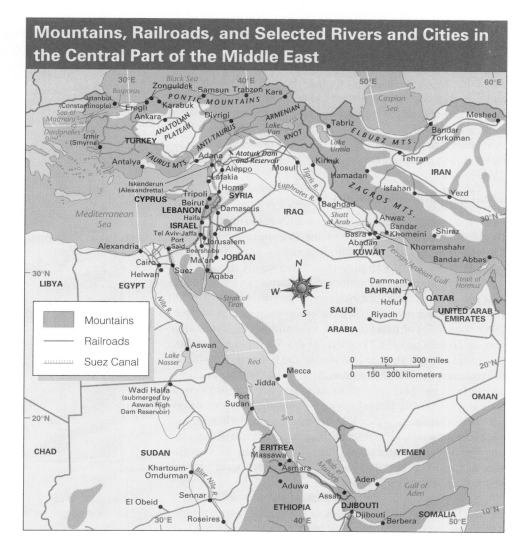

Mountains, Railroads, and Selected Rivers and Cities in the Central Part of the Middle East

FIGURE 9.4

Mountains, railroads, and selected rivers and cities in the central part of the Middle East. Mountainous areas are generalized.

PLO. Israeli troops routed and inflicted heavy casualties on both the Syrians and the PLO.

The Israeli push continued northward to the suburbs and vicinity of Beirut, Lebanon's capital, culminating in a ferocious aerial and artillery bombardment of the city itself that ended only when U.S. President Ronald Reagan exerted pressure on Israel to cease fire. Under supervision by U.S. and other international peacekeeping troops, the PLO withdrew from Beirut to Tunisia. After the multinational troops withdrew, alleged Muslim assailants killed Lebanon's Christian President Bashir Gemayal. Angry Christian forces then massacred hundreds of Palestinian and other civilians in refugee camps south of Beirut. U.S. and other multinational forces were again deployed to Beirut, and American ships used firepower to assist the new Christian president in widening his area of authority.

Lebanese Shi'ite Muslim guerrillas, seeking to avenge U.S. attacks on Lebanon and its support of Israel, soon bombed American and French targets in Beirut, killing over 200 U.S. Marines in a barracks near Beirut's airport. The multinational forces withdrew, but Syrian troops remained in Lebanon and allowed Iranian forces and Iran's Shi'ite Muslim allies in Lebanon to establish a stronghold in the Bekaa Valley town of Baalbek. Subjected to persistent guerrilla attacks, Israeli troops withdrew from all of Lebanon except a self-declared "security zone" in the south, inhabited by both hostile Shi'ite Muslims and by Israel's de facto Christian Lebanese allies. Israel continues to exert control over this zone and engages in regular skirmishes with Hizbullah, a Shi'ite Muslim organization sympathetic to Iran and hostile to the peace process between Israel and its Arab neighbors.

The Palestinian Question

The creation of Israel, the subsequent wars, and the Camp David Accords left the future of the Palestinians unresolved. The Palestinians are Arabs who historically have lived in the geographic region of Palestine (which is roughly the area now made up of Israel and the occupied territories) and who recognize themselves as a distinct ethnic group. About 700,000 Palestinians are now Israeli citizens residing within the pre-1967 boundaries of Israel. Another five million Palestinians live outside Israel's pre-1967 borders, in the West Bank, the Gaza Strip, and abroad. Prior to and during the fighting of 1948, approximately 800,000 Palestinian Arabs were forced, or chose, to flee from the new state of Israel to neighboring Arab countries (Fig. 9.5). The United Nations established refugee camps for these displaced persons in Jordan, Egypt, Lebanon, and Syria. Little was done to resettle them in permanent homes, and both the Arab governments and the refugees themselves continued to insist on the return of the refugees to Israel and the restoration of their properties there.

The refugee situation became more complex as a result of the 1967 war, in which Israel took over the areas where most of the camps were located. Many persons in the camps again took

flight, and Palestinians fleeing villages and towns in the newly-occupied territories joined them as refugees (see Fig. 9.5). Some later returned to their homes, but an estimated 116,000 either did not attempt to return or were denied permission by Israel authorities to do so.

After 1967, and particularly after the election of its conservative Likud Party to power in 1977, Israel moved to strengthen its grip on the occupied territories through security measures and the government-sponsored establishment of new Jewish settlements, most of which are spread through the West Bank (Fig. 9.6). These are generally not frontier farming settlements but communities inhabited by relatively prosperous middle-class Jews who commute to jobs in Israel. They are not built as temporary encampments but as permanent fixtures meant to create what have been described as "facts on the ground"—an Israeli presence so entrenched that its withdrawal would be almost inconceivable (Fig. 9.7). Some of the motivation for this settlement is ideological, as the West Bank is composed of his-

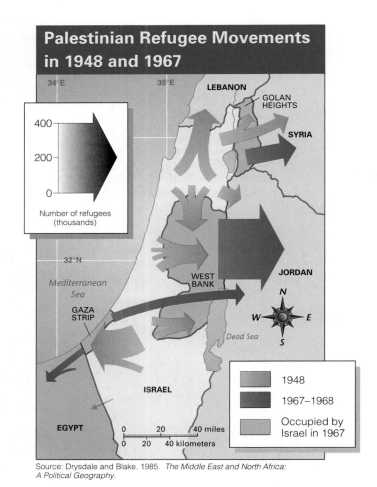

Source: Drysdale and Blake. 1985. *The Middle East and North Africa: A Political Geography.*

F I G U R E   9 . 5

Palestinian refugee movements in 1948 and 1967. Many who fled in the first conflict relocated again in the second.

FIGURE 9.6

Israeli and Palestinian areas of control in the West Bank in 1997.

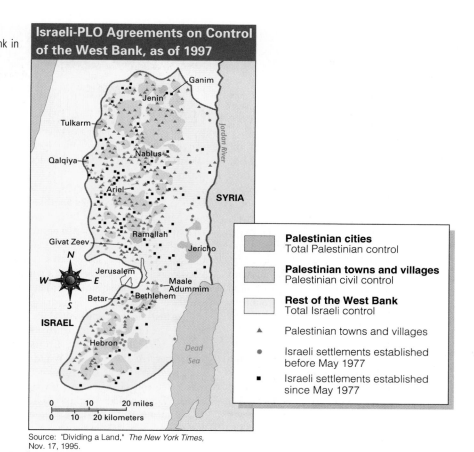

Source: "Dividing a Land," *The New York Times,* Nov. 17, 1995.

torical Judea and Samaria, which many devout Jews regard as part of Israel's historic and inalienable homeland.

The proliferation of new settlements, which persisted until Israel's liberal Labor Party came to office in 1992, increased the bitterness between the Arab world and Israel and drove a wedge into the Jewish population of Israel itself. Many Israelis strongly opposed further settlement in the occupied territories and annexation of these territories to the Jewish state, since, if the territories were annexed, Israel would become a country only approximately 60 percent Jewish; with an Arab minority

FIGURE 9.7

A Jewish settlement in the West Bank, near Jerusalem. The pitched roofs indicate that these dwellings have piped-in supplies of water. Most Palestinian homes in the West Bank lack plumbing and so have flat roofs on which water can be collected and stored. In the foreground is a Palestinian Arab shepherd with his flock.   *R. Norman Matheny/Christian Science Monitor*

PROBLEM LANDSCAPE

WITH A HANDSHAKE, NETANYAHU AND ARAFAT REVIVE PEACE TALKS

President Anwar Sadat of Egypt set the precedent in 1979: Arabs could make peace with Israel. Egypt's Arab neighbors scorned this populous and influential Arab country for a decade for its peacemaking, and Sadat was assassinated because of it. However, in the process, Israel restored the Sinai to Egyptian control and the United States rewarded Egypt handsomely with about $2 billion in aid annually. Along with about $3 billion awarded Israel annually, these countries have been the two largest recipients of U.S. foreign aid.

Israel and its remaining Arab adversaries stand to gain much more through the implementation of a comprehensive framework for peace. The United States initiated a sequence of events early in the 1990s that makes that prospect more likely than ever before. After Iraq invaded Kuwait in 1990, the U.S. and its European allies needed to build a firm coalition of Arab military and political will to oppose Iraq. Backed also by a sympathetic Soviet Union, the U.S. obtained this support in part by promising to pursue Middle East peace aggressively once Iraq was defeated. The presidents of both countries (George Bush in the U.S. and Mikhail Gorbachev of the Soviet Union) kept their pledges, and, in Madrid in October 1991, they sponsored the historic opening of talks between Israel and four Middle Eastern delegations: Egyptian, Syrian, Lebanese, and—because Israel did not recognize the PLO—a joint Jordanian-Palestinian delegation that excluded PLO members.

Little progress was made in the on-again, off-again peace talks until the Labor Party in Israel, under Prime Minister Yitzhak Rabin, came to power in 1992. Throughout the summer of 1993, Rabin's ministers undertook secret talks in Oslo, Norway, with members of the PLO. In early September came the historic announcement that Israel recognized the legitimacy of the PLO as the sole representative of the Palestinians. The PLO in turn recognized Israel's right to exist and renounced terrorism. On September 13, 1993, on the White House lawn, U.S. president Bill Clinton orchestrated a historic handshake between PLO leader Arafat and Israeli prime minister Rabin. The leaders signed an agreement known as the "Gaza-Jericho" Accord, which established a framework for peace with the following major points:

1. A five-year interim period of limited self-rule (autonomy) for Palestinians in the occupied territories, beginning in 1993 in the Gaza Strip and West Bank town of Jericho. Self-rule was to cover education, culture, health, taxation, and tourism. By the third year (1996), negotiations were to begin on the permanent status of the territories.
2. Withdrawal of Israeli troops from Palestinian areas in the Gaza strip, Jericho, and Palestinian towns in the West Bank.
3. Elections of a Palestinian Authority to oversee self-rule.
4. Formation of a Palestinian police force.
5. Continued Israeli protection of Jewish settlers in the Gaza Strip and West Bank.
6. Discussion of the possible repatriation of Palestinian refugees from the 1967 war.
7. Discussion of the status of Jerusalem, claimed by both Israelis and Palestinians as their "eternal" capital.

Israeli forces did withdraw from Jericho and from Palestinian areas of the Gaza Strip. Yasir Arafat took up residence in the Gaza Strip—another historic, previously unimaginable event. However, Israeli and Palestinian public approval of the peace process have periodically waxed and waned since September 1993. Arafat's PLO has been forced to deal with a series of violent attacks by the radical Palestinian organization HAMAS on Jewish civilians and soldiers. While these terrorist attacks caused many Israelis to reject the peace process, subsequent PLO detention and prosecution of Palestinian extremists also hurt Palestinian support for the PLO, which many Palestinians now see as an agent for Israeli interests. Many Palestinians turned against the peace process after February 1994, when a right-wing Israeli extremist opposed to Israel's negotiations with the Palestinians shot and killed Muslim Pales-

that has a higher birth rate than that of the Jews, Jews could eventually become the minority population in their own country.

In 1987, Palestinians instigated a popular uprising in the occupied territories known as the Intifada (Arabic for "the shaking"). Initially they relied on rocks and bottles to engage Israeli troops, who answered with rubber and metal bullets. International media coverage of a Palestinian "David" fighting an Israeli "Goliath" did much damage to Israel's image abroad, and many Israelis came to question the justice and importance to security of Israel's continued occupation. Popular opinion and the new government began to consider the previously unthinkable: giving the Palestinians at least some control over land and internal affairs in the occupied territories.

tinian worshipers in the Tomb of the Patriarchs, the holy place in the West Bank town of Hebron where the prophet Abraham is said to be buried.

Nevertheless, the peace process remained on track. In September 1995, Israel and the PLO signed another agreement to implement the withdrawal of Israeli troops from Palestinian cities in the West Bank. Palestinians then acquired self-rule over about 30 percent of the West Bank, with the rest, including all Jewish settlements, remaining in Israeli hands (see Fig. 9.6). The agreement reaffirmed that the final status of the territories would be negotiated in 1996.

Many of the approximately 100,000 Jewish settlers living in the West Bank (where about one million Palestinians also live) are vehemently opposed to the peace process. Regarding Jewish presence in the area as a God-given right, they fear the peace process will result in an independent Palestinian state and the forced withdrawal of Israeli Jews. On November 4, 1995, an Israeli Jew who held this view assassinated Prime Minister Yitzhak Rabin following a peace rally in Tel Aviv. The assassin proclaimed that his purpose was to destroy the peace process.

The short-term impact of Rabin's murder was a surge in popular Israeli support for his legacy of the peace process. This support was quickly reversed by a series of bloody suicide bombings carried out by HAMAS militants against civilian targets in Israel during early 1996. These attacks contributed to the defeat of Rabin's successor, Shimon Peres, and the victory of the conservative Likud Party's Benjamin Netanyahu, in Israel's 1996 general election.

As a candidate for prime minister, Netanyahu had vowed to step up Jewish settlement in the West Bank and to slow the pace of the peace process in the interest of Israel's security. He also promised that he would not meet with Palestinian leader Yasir Arafat, whom he labelled a terrorist. Under much international pressure to fulfill Israel's treaty obligations to its Arab neighbors, however, Prime Minister Netanyahu did hold face-to-face talks with Arafat late in 1996, and the men breathed new life into the flagging peace process. Soon afterward, Israeli authorities opened a new entrance to an ancient tunnel near the Western Wall and the Dome of the Rock. Some Palestinian Muslims perceived this action as an incursion against their sacred space, and mounted demonstrations against Israel. These quickly escalated into widespread unrest, with armed Palestinian police and unarmed Palestinian civilians alike engaging Israeli troops. Casualties were high on both sides, and the disturbances threatened to bring down the delicate peace process. However, early in 1997, Israel and the PLO signed an agreement providing for further Israeli troop withdrawal from Palestinian areas in the West Bank, including the city of Hebron.

The peace process has both progressed and faltered on other fronts as well. In October 1994, following the PLO's lead (and again under President Clinton's eye), Jordan signed a treaty with Israel. It called for an end to hostility between the two nations, establishment of full diplomatic relations, the opening of exchanges in trade, tourism, science, and culture, and an end to economic boycotts. Meanwhile, Syria and Israel have failed to reach a peace agreement because of a deadlock on the issues of recognition of Israel and return of occupied land: Israel refuses to return the Golan Heights to Syria until Syria recognizes the sovereignty and legitimacy of Israel, and Syria refuses to recognize the sovereignty and legitimacy of Israel until Israel returns the Golan Heights. Syrian President Hafez al-Assad may be waiting to see how successful the implementation of Israel's existing peace treaties is before concluding his own with Israel. And because Syria is the dominant power in Lebanon, with tens of thousands of troops stationed in that tiny, politically-weak country, peace between Israel and Lebanon is improbable unless Israel and Syria first come to terms.

Peace in Lebanon appeared to be an unlikely prospect following a 1996 Israeli military offensive, dubbed "Operation Grapes of Wrath," against the country. Israel vowed to quell Katyusha rocket attacks on northern Israel from Lebanon by the Syrian and Iranian-backed Hizbullah. Weeks of Israeli artillery and aerial bombardment of targets in Lebanon produced large civilian casualties but few apparent military successes for Israel. The "1996 Border War" seemed to harden Syria's resolve to remain as Israel's last frontline opponent.

Serge Schmemann, *The New York Times*, September 5, 1996: p. A1.

JORDAN, BETWEEN IRAQ AND ISRAEL

Jordan (1996 population: 4.2 million), bordering Israel on the east, is mostly desert and semidesert. The main agricultural areas and settlement centers, including the largest city and capital, Amman (population: 936,000, city proper; 1.63 million, metropolitan area), are in the northwest. Prior to the 1967 war with Israel, northwestern Jordan included two upland areas—the West Bank, or Palestine Hills, on the west, and the East Bank, or Transjordanian Plateau, on the east—separated by the deep rift valley occupied by the Jordan River, Lake Kinneret, and the Dead Sea (see Fig. 9.3). The valley is the deepest depression on the Earth's land surface, lying about 600 feet

(183 m) below sea level at Lake Kinneret and nearly 1300 feet (400 m) below sea level at the Dead Sea (the lowest point on Earth). Although annual precipitation in the valley increases northward to an average of more than 12 inches (30 cm) in the vicinity of Lake Kinneret, most of the valley bottom lies in deep rain shadow and receives less than 4 inches a year. The bordering uplands, rising to 2000 feet (610 m) above sea level and more and lying in the path of moisture-bearing winds from the Mediterranean, receive precipitation during the cool season that averages 20 inches or more annually within two north-south strips. It was within these belts, one in the West Bank (occupied by Israel in 1967) and the other in the hilly East Bank, that most of Jordan's pre-1967 population resided. The largest numbers were villagers supporting themselves by cultivating winter wheat and barley, vegetables, and fruits, especially olives and grapes.

Jordan is a poor country. Today only about 5 percent of Jordan is cultivated, although much larger areas are used to graze sheep and goats. Rainfall is so scarce that only small areas can be cultivated without irrigation, and water available for irrigation is very limited. Some food has to be imported, although irrigated vegetables (especially tomatoes), together with olives, citrus fruits, and bananas, provide substantial exports. Only limited manufacturing is carried on in Jordan, and the contrast with the large and technologically advanced industries of neighboring Israel is extreme. Natural resources are limited, although rock phosphate and potash extracted from the Dead Sea provide major exports. They are shipped through Jordan's small Red Sea port of Aqaba, located on the Gulf of Aqaba and adjoining Eilat in Israel. Aqaba also handles much transit traffic, mostly by truck, for Iraq and was especially important to Iraq during its war against Kuwait and the Western coalition.

There is an enormous potential for growth in Jordan's tourism industry, especially now that the country has established peaceful relations with Israel. Traditionally constrained within their tiny borders, Israelis are now enjoying the freedom of traveling in neighboring Arab states and enjoying such sites as Jordan's Nabatean city of Petra (see Fig. 8.5) and Egypt's Sinai beaches.

Jordan's economic and political situation has always been precarious. King Hussein, supported by the native Bedouins, has withstood various crises resulting from the entry of great numbers of Palestinians, including the PLO. For a time, Jordan was the chief operating base of the PLO, but in 1970 and 1971, the king expelled the armed Palestinian forces, who relocated to Lebanon. Palestinians still make up about 60 percent of Jordan's population, not including those in the Israeli-controlled West Bank that Jordan annexed in 1950 and that Israel took in the 1967 war.

LEBANON AND SYRIA

Israel's immediate neighbors on the north are Lebanon and Syria (see Fig. 9.3). Their physical and climatic features are similar to those of Israel and Jordan. As in Israel, narrow coastal plains backed by highlands front the sea. The highlands of Lebanon and Syria are loftier than those of Israel and Jordan, however; Mount Hermon and the Lebanon and Anti-Lebanon ranges are high mountains, reaching above 10,000 feet (c. 3050 m) in the Lebanon range. Cyclonic precipitation brought by air masses off the Mediterranean is supplemented by orographic rain and snow induced by the mountain chains that parallel the coast and block the path of the prevailing westerly winds.

Lebanon and Syria as a result have agricultural patterns that conform to the mediterranean climatic pattern of winter rain and summer drought. The coastal plains and seaward-facing mountain slopes are settled by villagers growing Mediterranean crops such as winter grains, vegetables, and drought-resistant trees and vines. Precipitation decreases and pastoralism increases toward the east. The Anti-Lebanon range and Mount Hermon are drier than the Lebanon range, which screens out moisture they would otherwise receive. These mountains, and the eastern slopes of the Lebanon range as well, are used primarily for grazing.

The Bekaa Valley, drained in the south by the Litani River, lies between the Lebanon and Anti-Lebanon ranges and is a northward continuation of the Jordan rift valley and the African Great Rift Valley. Baalbek, its chief city and the site of an important ancient Roman temple, became a Syrian military stronghold in 1976 and a center of pro-Iranian Hizbullah activities against Israel after 1982. The political and military volatility of the Bekaa was conducive to the emergence of a flourishing drug industry—with both hashish (from marijuana plants) and opium (from poppies) production—from the mid-1970s until the early 1990s (Fig. 9.8).

A semiarid zone covers central and northern Syria east of the mountains. In southeastern Syria, it trends gradually to the Syrian Desert, most of which lies in Iraq and Jordan. The semi-

FIGURE 9.8
Cultivation of marijuana for the production of hashish in Lebanon's Bekaa Valley. Weak governments or corrupt political systems provide fertile ground for drug production in many world regions.   *Pascal Maitre/Matrix*

arid region, much of which has good soils, is Syria's main producer of cotton, wheat, and barley. Cotton is grown as an irrigated summer crop, whereas wheat and barley, primarily nonirrigated, are winter crops.

Syria is more industrialized than Lebanon, but neither country compares with Israel industrially. Textile milling, agricultural processing, and other light industries predominate. Both countries are deficient in minerals, although Syria has an oil field of moderate importance in the east.

Three large cities of long historic importance are in Syria and Lebanon. Damascus (population: 1.3 million, city proper; 2 million, metropolitan area), Syria's capital, is in a large irrigated district in the southwest at the eastern foot of the Anti-Lebanon range. Its Umayyad Mosque, originally a Byzantine cathedral, is one of the great monuments of the Middle East. The main city of northern Syria, Aleppo (population: 1.27 million, city proper; 1.34 million, metropolitan area), is an ancient caravan center located in the "Syrian Saddle" between the Euphrates valley and the Mediterranean. And the old Phoenician city of Beirut (population: 509,000, city proper; 1.68 million, metropolitan area), on the Mediterranean, is Lebanon's capital, largest city, and main seaport. Equipped with modern harbor facilities and an important international airport, Beirut before 1975 was one of the busiest centers of transportation, commerce, finance, and tourism in the Middle East. Indeed, for both its prosperity and its natural beauty, pre-1975 Lebanon had earned the name "Switzerland of the Middle East." Then a precarious political balance broke down and the country was plunged into civil war. Large parts of the capital city were devastated, and only now is Beirut rising from the ashes in an ambitious, expensive reconstruction program.

The origins of the civil war of 1975–1976 were rooted in Lebanon's mixture of religious communities that entered at various times and became strongly localized in particular areas. Seventeen separate ethnic and religious groups, known as "confessions," are recognized officially (Fig. 9.9). Maronite Christians prevail in the Lebanon range north of Beirut, but close to the northern border with Syria Sunni Muslims predominate. Shi'ite Muslims are the majority in the Bekaa Valley and along Lebanon's southern border with Israel. The Shouf Mountains, a portion of the Lebanon range south of Beirut, are home to the Druze, who practice an ancient and distinct form of Shi'ite Islam. Greek Orthodox, Greek Catholic, and nearly a dozen other Christian sects are also present in Lebanon. Beirut itself is divided (literally, in the war years of the 1970s and 1980s, by the "Green Line") between predominantly Muslim West and Christian East. Warfare between the communities broke out from time to time, and foreign powers that gained influence with one or another group heightened the antagonisms.

What the French carved out of greater Syria and put together as Lebanon in 1920 was a collection of small geographic areas, each dominated by one or a few minorities. A 1932 census revealed a majority of Christians in Lebanon, and the National Pact, or constitution, drafted for independent Lebanon in 1946 distributed power according to those census figures. Mar-

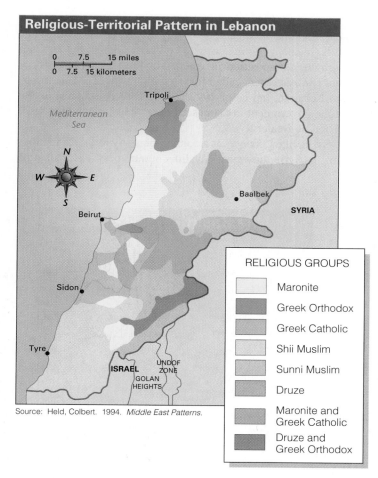

FIGURE 9.9
Generalized map of the distribution of religious groups in Lebanon.

onite Christians held a majority of seats in Parliament, followed by Sunni Muslims and Shi'ite Muslims. Top leadership positions were allocated by a formula still followed today, with the president a Maronite Christian, the prime minister a Sunni Muslim, and the speaker of Parliament a Shi'ite Muslim.

There was never an effort to unify the disparate geographic and ethnic enclaves of this mountainous land. Meanwhile, Muslim populations grew. Eventually they came to outnumber the Christians, who were unwilling to relinquish their political and economic privileges. Pushed from Jordan, PLO guerrillas established a virtual "state within a state" in the southern slums of Beirut and in Shi'ite-controlled areas of southern Lebanon from which they launched attacks on Israel.

The volatile mix exploded in 1975, with the basic formula of Muslims versus Christians, but complicated by shifting alliances with outside powers, including Israel and Syria. By the end of the war in 1976, as many as 100,000 people, most of them civilians, had been killed. Lebanon effectively had become a number of geographically-distinct microstates or fiefdoms—a politically unstable situation that set the stage for the next war in 1982.

Currently, Lebanon seems to be on a steady course of reconciliation and reconstruction. A peace accord known as the Taif Agreement has redistributed power to reflect the new demographic composition in which Muslims outnumber Christians. The agreement called for the withdrawal of Syrian troops from Lebanon, but as of 1997 they remained in place, a sign that the government in Damascus never has accepted the 1920 division of greater Syria.

Syria, like Lebanon, is a country beset by religious factionalism. In Syria, however, Muslims outnumber Christians nine to one. In 1970, a coup overthrew the ruling Sunni Muslim establishment and brought to power Hafez al-Assad, a member of the Alawite sect of Shi'ite Islam. Under his authoritarian rule as president, dominance of the government quickly passed to the Alawites. The police and army crushed opposition within the other communities. President Assad is a skilled politician who plays Middle Eastern affairs cautiously. His regime survived the economic and political setbacks accompanying the downfall of its benefactor the Soviet Union and has slowly adopted a more accommodating relationship with the U.S. and the West.

9.3   Egypt: The Gift of the Nile

At the southwest, Israel borders Egypt (officially the Arab Republic of Egypt), the most populous Arab country. This ancient land, strategically situated at Africa's northeastern corner between the Mediterranean and Red seas, is utterly dependent on a single river. Not only does the Nile supply the water that enables 64 million Egyptians to exist, but it created the alluvial flood plain and triangular delta on which more than 95 percent of all Egyptians live (see map, Fig. 9.10). The Greek historian and geographer Herodotus called Egypt "an acquired country—the gift of the Nile." The river's bountiful waters and fertile silt helped Egypt become a culture hearth that produced remarkable achievements in technology and cultural expression over a period of almost 3000 years until about 500 B.C., when a succession of foreign empires came to rule the land.

The Nile valley may appropriately be described as a "river oasis," for stark, almost waterless desert borders this lush ribbon (Fig. 9.11). Only 3 percent of Egypt is cultivated, and nearly all this land lies along and is watered by the great river. The conversion of the original papyrus marshes and other wetlands along the Nile to the thickly settled, irrigated landscape of today is a process that has been unfolding for more than fifty centuries. It is difficult to imagine that, in the time of the pharaohs, crocodiles and hippopotami swam in the Nile and game animals typical of East Africa roamed the nearby plateaus.

EGYPTIAN AGRICULTURE

Ancient Egyptian civilization was based on a system of **basin irrigation**, in which the people captured and stored in built-up embankments the flood waters which spread over the Nile floodplain each September (see Fig. 9.12 and Landscapes in Literature, page 237). Egyptian farmers allowed the water to stand for several weeks in the basins, where it supersaturated the soil and left a beneficial deposit of fertile volcanic silt washed down by the Blue Nile from Ethiopia. They drained the excess water back into the Nile and planted their winter crops in the muddy fields. They were able to harvest one crop a year with this method. They were able to produce a second or third

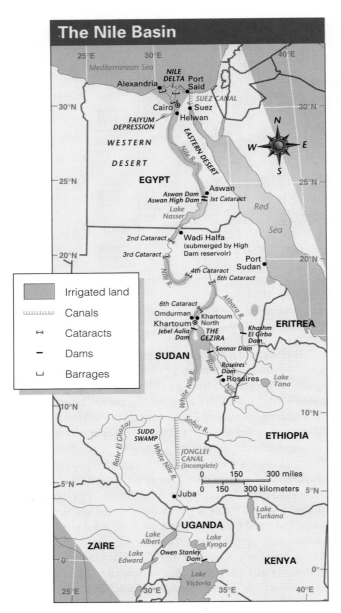

FIGURE 9.10

The Nile Basin. The widths of the irrigated strips along the Nile are exaggerated.

FIGURE 9.11
The life-giving ribbon of the Nile River near Aswan in Upper Egypt. The sailboat is called a felucca.   *Porterfield/Chickering*

crop on limited areas of land using water-lifting devices such as a weighted lever or *shaduf* (a bucket on a counterweighted pole, manually operated to raise water from canal to field) and an Archimedes' screw (a cylinder containing a screw, also manually turned to raise the water).

However, achieving this so-called **perennial irrigation** on a vast scale required the construction of barrages and dams—

innovations of the much later 19th and 20th centuries (Fig. 9.10). It was during this time—beginning late in the 19th century—that French and British colonial occupiers of Egypt, anxious to boost Egypt's exports of cotton (a summer crop demanding perennial irrigation), began Egypt's conversion from basin to perennial irrigation by constructing a number of barrages, or low barriers designed to raise the level of the river high enough that the water flows by gravity into irrigation canals. The barrages permit a constant flow of water to irrigated land even in periods of low water levels. They are not designed to store large amounts of water—a function now performed by two dams in Upper Egypt (the old Aswan Dam, completed in 1902, and the huge Aswan High Dam, completed in 1970) and several others along the Nile and its tributaries in Sudan and Uganda.

The enormous Aswan High Dam (Fig. 9.13) is located about five miles (eight km) upstream from the older Aswan Dam, on the stretch of the Nile known as the First Cataract (the Nile's many cataracts are areas where the valley narrows and rapids form in the river). Its reservoir, Lake Nasser, stretches for more than 300 miles (480 km) and reaches into northern Sudan.

Like all giant hydrological schemes, the Aswan High Dam has generated benefits and consequences, and its construction was controversial. On the plus side, by storing water in years of high flood for use in years of low flood, the High Dam provides a perennial water supply to Egypt's preexisting 6 million irrigated acres (2.4 million hectares), thus extending the cultivation period. Egypt weathered a two-year drought in the late 1980s, thanks to the excess waters stored in Lake Nasser. The dam made possible the reclamation of considerable amounts of additional farmland, particularly on the desert fringe adjoining the western Nile delta. Navigation of the Nile downstream from

FIGURE 9.12
The Pyramids of Giza, tombs of Egyptian pharaohs of the mid-third millennium B.C. The annual autumn flooding of the Nile brought productive regularity to Egyptian agriculture and supported a prosperous nation.
C. Christian Michaels/FPG International

FIGURE 9.13
Egypt's controversial High Aswan Dam. The Lake Nasser reservoir begins on the right. In the center rear is the dam's hydroelectric station, which generates much-needed and pollution-free electricity for Egypt's industries.
Deni McIntyre/Photo Researchers, Inc.

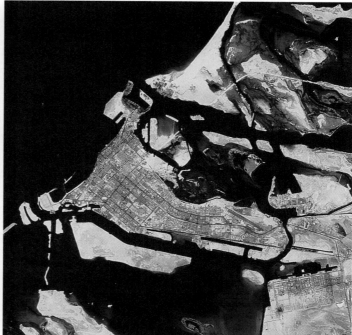

A   B

FIGURE 9.18

These extraordinary before-and-after satellite images reveal how oil wealth has transformed
some Middle Eastern landscapes. The first shows Abu Dhabi, capital of the UAE, in 1972. The
city occupies much of the natural island in the center. The second image of the same area,
recorded in 1984, reveals the outcome of huge investments, including humanmade islands and
deepwater port facilities.   *NRSC Ltd./Science Photo Laboratory*

piracy. Today the UAE is capitalizing on its central location between Asia, Africa, and Europe with the construction of sophisticated air and seaport facilities. Like nearby Abu Dhabi's
facilities, Dubai's new seaport of Jebel Ali is a colossal excava-
tion easily visible to astronauts and satellites in space (Fig.
9.18). Unlike its more conservative neighbors, the UAE has
promoted itself as a destination for sun-worshipping international tourists and shoppers. Russians comprise a substantial
proportion of the foreign visitors.

THE SOUTHERN MARGINS
OF ARABIA

Known to the ancient Romans as Arabia Felix, or "Happy Arabia," because of their verdant landscapes, Yemen (population:
14.7 million) and Oman (population: 2.3 million) occupy, respectively, the mountainous southwestern and southeastern corners of the Arabian peninsula.

The present Republic of Yemen was created in 1990 by the
union of the former North Yemen and South Yemen. Its political capital and largest city is San'a (population: 450,000, city
proper), located in the former North Yemen. Until the 1960s,

North Yemen was an autocratic monarchy, extremely isolated
and little known. In the mid-1960s, a republican government
supported by Egypt and opposed by Saudi Arabia took control
of large sections, at a cost of much bloodshed.

The former South Yemen, known until 1967 as Aden, a
British colony and protectorate, lies along the Indian Ocean between the former North Yemen and Oman. A Marxist state supported by the Soviet Union and other East-bloc nations, South
Yemen had hostile relations with neighboring countries and the
Western world. It served as an East-bloc foothold on the Arabian
Peninsula near the strategic chokepoint of Bab el-Mandeb connecting the Red Sea with the Indian Ocean. The main city, Aden
(population: 350,000, city proper), is located on a fine harbor
near the straits.

Civil war erupted between the north and the south in May
1994, with the south proclaiming its secession from the four-
year-old union. Northern forces besieged Aden and overran most
of the south. Today, an uneasy union holds the two Yemens together in a single official nation.

Yemen receives the heaviest rainfall in the Arabian Peninsula, enjoying a summer rainfall maximum derived from the
same monsoonal wind system that brings summer rain to the
Horn of Africa and the Indian subcontinent. Its main subsistence

FIGURE 9.19
The frankincense tree (*Boswellia sacra*) of Oman. The prized frankincense is the tree's gum, which in ancient times was harvested, like rubber, by cutting and "bleeding" the tree.   *Michel Viard/Matrix*

crops are spring-sown millet and sorghums. Most of the country's best agricultural land consists of terraces and is dedicated to the cultivation of qat (*Cathya edulis*), a stimulant shrub chewed by virtually all adults in the country on a daily basis (see Fig. 8.7). Most of the country's people live in the rainier highlands, while the coastal plain is extremely arid and sparsely populated.

Yemen's meager agricultural output and lack of industrial production has been compensated for during recent decades by earnings sent home by Yemeni workers in the oil fields and services of the wealthy Gulf countries. The support of Yemen's government for Saddam Hussein's regime in Iraq during the 1991 Gulf War prompted Saudi Arabia to expel millions of these expatriate Yemeni workers, leading to a precipitous decline in Yemen's economy. Oil production began in Yemen itself only in the 1980s. The fields are located in a contested border area between Yemen and Saudi Arabia and subjected periodically to raids and kidnappings by groups opposed to Yemen's government, so production has been slow to increase.

Most of the Sultanate of Oman lies along the Indian Ocean immediately west of the Strait of Hormuz entrance to the Persian/Arabian Gulf, with a small detached part bordering the Strait and the Gulf. As in Yemen, the coastal region is very arid (although verdant in irrigated spots), and most people live in interior highlands where mountains induce a fair amount of rain. For about 2000 years, beginning around 1000 B.C., Oman was the world's only producer of frankincense, a prized commodity which made the region a hub of important oceanic and continental trade routes (Figure 9.19). Now, crude oil from fields in the interior and piped to the coast furnishes practically all of Oman's exports.

The country's strategic position on the Strait of Hormuz—through which much of the oil produced by the Gulf States is transported—has made it an object of concern in industrial nations, particularly Great Britain, with which it has historic ties, and the United States. These Western powers have rights of access to base facilities in Oman in the event of any military emergency that threatens the continued flow of oil from the Gulf region. Allied forces used the bases after Oman joined the United Nations coalition against Iraq in the Gulf War of 1991.

KUWAIT AND THE GULF WAR OF 1991

With 9 percent of the world's proven oil reserves within its borders, tiny Kuwait (6900 sq mi, 17,800 sq km—about the size of the U.S. state of New Jersey) has the world's third largest store of proven petroleum resources (after Saudi Arabia and Iraq). Associated with the oil is much natural gas, which Kuwait uses for fuel within the country and also liquifies and makes into petrochemicals for export. Oil production is greatest in the Burgan field south of Kuwait City. Oil revenues have changed Kuwait from an impoverished country of boatbuilders, sailors, small traders, and pearl fishermen to an extraordinarily prosperous welfare state. Generous benefits have flowed to Kuwaiti citizens but not to the Asian and other non-Kuwaiti expatriates drawn by employment opportunities to this emirate. (By 1990, foreign workers actually outnumbered Kuwaiti citizens in the country's total population of 2,143,000.)

In August 1990, Kuwait took center stage in world affairs when invading military forces at the command of Iraqi President Saddam Hussein overran the country. Claiming that Kuwait was historically part of Iraq and calling it Iraq's "nineteenth province," the Baghdad regime attempted to annex Kuwait. Saudi Arabia and other Gulf oil states seemed in danger of invasion. Strenuous worldwide opposition to Iraq's threats and aggression resulted in a sweeping embargo on Iraqi foreign trade imposed by the Security Council of the United Nations. Led by the United States, a coalition of 28 U.N. members (including several Arab states) staged a rapid buildup of armed forces in the Gulf region.

Beginning in January 1991, the coalition's crushing air war drove the Iraqi air force from the skies, massively damaged Iraq's infrastructure, and hammered the Iraqi forces in Kuwait and adjacent southern Iraq with intensive bombardment. In the ensuing ground war, the Iraqis were driven from Kuwait within 100 hours. Overall, an estimated 110,000 Iraqi soldiers and tens of thousands of Iraqi civilians were killed, while the technologically superior coalition forces suffered only 340 combat deaths, of which 148 were American.

During the conflict, which came to be known as the Gulf War, the Iraqis also made war on the environment, setting fire to hundreds of oil wells in Kuwait (Fig. 9.20) and allowing damaged wells to discharge huge amounts of crude oil into the Gulf. Ecological damage was severe.

A formal cease-fire came on April 6, 1991, with harsh conditions imposed on Iraq by the U.N. Security Council. Particularly important was the abolition of Iraqi chemical, biological, and nuclear "weapons of mass destruction" capabilities. Iraq agreed to pay reparations for the damage its forces had caused, and it renounced its claims to Kuwait. To guarantee the boundary between Iraq and Kuwait, the United Nations sent a multinational U.N. force to police a security zone along the border. A rapid withdrawal of coalition forces from the war zone then took place. Today, U.N.-sponsored economic sanctions against Iraq continue, with shortages and food and medical supplies bringing great hardship to the country's population. Only in late 1996 did the U.N. agree to allow Iraq to sell limited quantities of oil abroad, with the proceeds limited to purchasing food and humanitarian supplies.

After the Gulf War cease-fire, Saddam Hussein's surviving forces mercilessly put down rebellions by Shi'ite Arabs in southern Iraq and by Kurds (members of a distinct non-Arab ethnic group) in northern Iraq. Great numbers of terror-stricken Kurds fled to the mountainous borderland where Iraq, Turkey,

FIGURE 9.20
Iraqi troops set more than 600 oil wells ablaze before retreating from Kuwait in 1991. This image shows some of the fires on April 28, 1991. Kuwait City occupies much of the peninsula in the top center. *Earth Observation Satellite Company, Lanham, Maryland/Science Photo Library*

Iran, and Syria meet. An international effort to aid them slowly gained momentum, hindered by rugged terrain, lack of roads, and the sheer number of people requiring food, water, medicine, and protection from Saddam Hussein's soldiers. To deter Iraq's military and provide a safe haven for the Kurds in the north and the Shi'ites in the south, the United Nations established "no-fly zones" north of the 36th parallel and south of the 32nd parallel, where Iraqi aircraft were prohibited from operating. Following these actions, and with U.N. monitoring, repair of the devastation throughout Kuwait and Iraq began, and large numbers of refugee Kurds moved back to their home areas.

Then, late in 1996, rival Kurdish factions in northern Iraq battled each other for control of key towns and villages. One faction called for and received military assistance from Iraq's army, and was soon victorious. Iraq's military employed only ground forces and therefore technically did not violate the northern "no-fly zone" order. However, the United States perceived the Iraqi incursion as an unacceptable threat to regional stability and U.S. national interests. In September, the Clinton administration launched a series of missile strikes against military targets in southern Iraq, and extended the southern "no-fly zone" northward from the 32nd to the 33rd parallel, touching the southern suburbs of Baghdad.

After the Gulf War, the emir of Kuwait returned to his country from Saudi Arabia, where he had sought refuge from the Iraqi invasion. He faced a restive Kuwaiti population intent on securing more democracy and personal freedoms than had been available under the emir's authoritarian rule in the past. Meanwhile, Kuwait continues to live nervously in Iraq's shadow. It has excavated a trench on its border with Iraq to slow any invasion, and has continued to practice war games with U.S. military troops in anticipation of future incursions by its northern neighbor.

IRAQ, TROUBLED OIL STATE OF MESOPOTAMIA

At the eastern end of the Fertile Crescent (see Definition and Insights, p. 212), the Republic of Iraq occupies one of the world's most famous centers of early civilization and imperial power. This riverine land of Mesopotamia has had a complex geopolitical history, having been a seat of empires, a target of conquerors, and, in the 20th century, a focus of oil development and political and military contention involving many other nations.

Interaction with the wider world began very early. The ancient Babylonians and Assyrians (occupying lands later known as Iraq) overcame neighboring peoples and built a succession of empires reaching from the Persian/Arabian Gulf to the Mediterranean; the earliest such empire, the Akkadian, dates to 2350 B.C. Irrigated land on the Tigris-Euphrates plain gave empire-builders a dependable food base, and the Fertile Crescent corridor between mountains to the north and deserts to the south was a passageway for military movement and trade. Much later, during the centuries of Arab ascendancy after the death of the Prophet Muhammad, Baghdad on the Tigris became the capital of an imposing Arab and Muslim empire.

Over thousands of years, such periods of imperial triumph alternated with times of internal disorder and weakness when outsiders such as the Persians and the Romans subjugated Iraq. In the 13th century Mongol invaders devastated the area, which subsequently became a part of the Turkish Ottoman Empire. British forces defeated the Turks in Iraq in World War I, and after the war Great Britain took over the country as a mandated territory under the League of Nations until it gained its independence in 1932.

Iraq's government was a pro-Western monarchy from 1923 to 1958, when a revolution engineered by army officers made the country a republic. Over the next two decades, Iraq's political instability was reflected in many coups and attempted coups as various factions struggled for power. By the middle 1970s, Iraq had become an authoritarian one-party state ruled by President Saddam Hussein in the name of the Arab Baath Socialist Party. Aspiring to leadership in the Arab community, this Sunni Muslim–based regime asserted its dedication to Arab solidarity, opposition to Israel, freedom from Western imperialism, and strong internal development for a self-dependent, secularized, and socialistic Iraq. In 1972, the Iraqi government signed a treaty of friendship with the Soviet Union, and Iraq received large Soviet arms shipments and equipment and technology for many new factories. After 1973, increasing oil revenues underwrote new social programs, large construction projects, and the creation of many new businesses under state, private, and joint ownership. New high-rise buildings, busy expressways, and an active commercial and social life gave metropolitan luster to ancient Baghdad.

In 1980, perceiving internal weakness in the neighboring fledgling Islamic Republic of Iran, as well as an opportunity to boost his stature in the Arab world and a means of securing land and mineral resources on the Iranian side of the Shatt al-Arab waterway ("Shore of the Arabs," formed by the confluence of the Tigris and Euphrates Rivers), President Hussein launched an invasion of his more populous neighbor. Thus began an eight-year war that proved enormously costly to both sides, with an estimated 500,000 people killed or wounded. Iran mounted a far better defense than Iraq had anticipated, expending its greater manpower in bloody "human wave" assaults to counter Iraq's greater firepower from tanks, artillery, machine guns, and warplanes.

Although the fighting was localized mainly along the international border, many noncombatant nations became involved in various ways. For example, Saudi Arabia and other wealthy Arab Gulf states provided large-scale financing and arms to Iraq because they feared the destabilizing effects of an Iranian victory, particularly what other Arab territories Iran might overrun, or what Arab populations might become receptive to Iran's revolutionary message that political Islam should replace secu-

lar governments and monarchies. The United States and the Soviet Union pursued their respective geopolitical interests in ways that generally benefitted Iraq, although neither of them declared formal support for either of the warring countries. Turkey, sharing a border with each, greatly expanded its trade with both. France sold Iraq sophisticated missiles and aircraft. A complex web of geopolitical relationships thus made the Iran-Iraq War a storm center in the political world. The war ended in stalemate in 1988, with neither side gaining significant territory or other assets.

Iraq occupies a broad, irrigated plain drained by the Tigris and Euphrates rivers (Fig. 9.4), together with fringing highlands in the north and deserts in the west. In recent decades, Iraq's government has expanded irrigation on the plain, partly through the construction of dams on the Tigris and Euphrates and their tributaries for multiple uses of flood control, hydroelectric development, and irrigation storage. An estimated 70 percent of Iraq's 21 million people—about 60 percent of whom are Shi'ite Muslims—are urban dwellers—a much higher percentage than in most Middle Eastern countries. Most of the rest are villagers depending on agriculture for a living. Small bands of nomadic Bedouins still raise camels, sheep, and goats in the western deserts. Some of the Kurdish tribesmen in the mountain foothills of the north are seminomadic graziers, although most Iraqi Kurds are sedentary farmers.

A Sunni Muslim people of Indo-European origins, the Kurds are by far the largest of several non-Arab minorities in Iraq. They have revolted against the central authority on numerous occasions, most recently during the Gulf War of 1991. The Kurdish question has international implications, because the area occupied by the Kurds also extends into Iran, Turkey, Syria, and the Former Soviet Region. European powers promised an independent state of Kurdistan at the end of World War I, but the governments concerned never took steps to create it. A book about the Kurds is appropriately entitled *No Friends But the Mountains*.[1]

Iraq has a generous endowment of oil, with reserves estimated at about 10 percent of the world's total. As was typical of the early oil industry in the Middle East, Great Britain and other Western nations produced and marketed Iraqi petroleum. After 1972, when Iraq nationalized its oil, and the 1973 energy crisis, the country was able to use its surging oil profits to modernize what had been a very traditional and poor economy. But, as was mentioned earlier, Iraq's military ventures after 1980 squandered the nation's wealth.

Baghdad (population: 3.8 million, city proper; 5.5 million, metropolitan area), Iraq's political capital, largest city, and main industrial center, is located strategically on the Tigris near its closest approach to the Euphrates River. Like other metropolises in the LDCs, Baghdad has received a huge influx of villagers seeking a better life. The Iraqi preference for single-family homes finds expression in an endless spread of

---

[1]Bullock, John. 1993. *No Friends But The Mountains*. London: Oxford University Press.

low-rise houses with garden plots. This mass of housing stretches outward for many miles from the cluster of office buildings and Western-style hotels in the downtown core of the city. The country invested heavily in rapid reconstruction of its capital city after the 1980–88 war.

Basra (population: 313,000, city proper; 900,000, metropolitan area) is Iraq's main seaport, though it is located upriver on the Shatt al-Arab. It suffered extensive destruction in the Iran-Iraq War and again in the Gulf War. Located in the predominantly Shi'ite southern portion of Iraq, Basra has long been the center of smoldering rebellion against the repressive Sunni-dominated government in Baghdad. One of the great en-

vironmental and cultural tragedies of the late 20th century has been Baghdad's systematic draining after 1991 of the marshes adjoining the southern floodplains of the Tigris and Euphrates rivers. Perceiving the ancient Shi'ite culture of the Maadan, or "Marsh Arabs," inhabiting the wetlands as enemies of the state, the Iraqi military built massive embankments and canals to drain the marshes. Authorities in Baghdad told international observers this was an "agricultural reclamation scheme," but its real purpose was to put an end to the agriculture and fisheries of the local people and to force them to flee into neighboring Iran and into settlements within Iraq where they could be controlled (Figure 9.21).

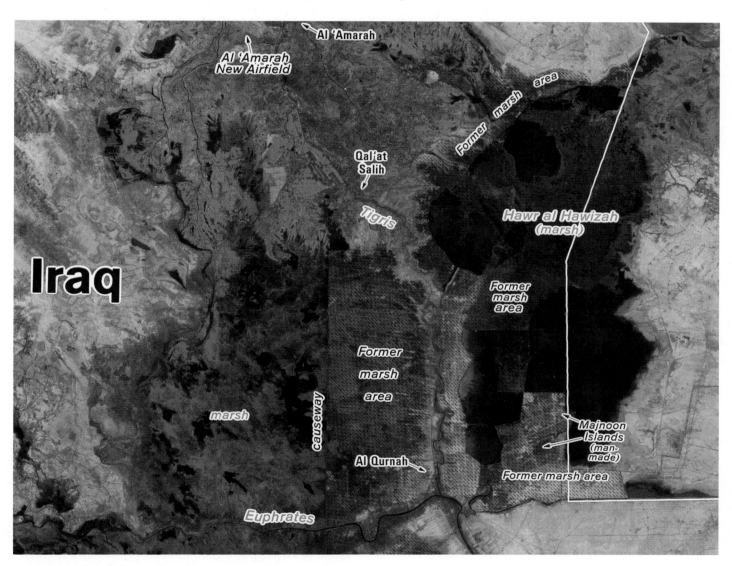

FIGURE 9.21
Southeastern Iraq in August 1992, after the draining of the marshes near the Tigris River was well underway. In order to drive out Shi'ite inhabitants, Iraqi engineers erected causeways to divert water away from and drain the wetland areas labelled former marsh areas here. The destruction of the southern marshes widened after this satellite image was taken. *Central Intelligence Agency,* "Iraq: A Map Folio."

OIL AND UPHEAVAL IN IRAN

Iran, formerly Persia, was the earliest Middle Eastern country to produce oil in large quantities. The Anglo-Persian Oil Company, now the British Petroleum Company (BP), began commercial production in 1912. The main oil fields form a line along the foothills of the Zagros Mountains in southwestern Iran. British-built pipelines linked the large refinery at Abadan with the inland oil fields. Abadan, which was heavily damaged by Iraqi shelling in the 1980–88 Iran-Iraq War, lies on the Shatt al-Arab waterway about 50 miles (80 km) from the Gulf. It is accessible only to small tankers, so nearly all exports of crude oil from Iran pass through a newer terminal at Kharg Island (see Figure 9.16), which allows supertankers to load in deep water.

Revenues from the oil industry (fully nationalized in 1973) are crucially important to Iran, a country where a dry and rugged habitat makes it hard to provide a good living for its rapidly growing population of 63 million (as of 1996). Arid plateaus and basins, bordered by high, rugged mountains, are Iran's characteristic landforms. Encircling mountain ranges prevent moisture-bearing air from reaching the interior, creating a classic rain-shadow desert. Only an estimated 11 percent of Iran is cultivated. Enough rain falls in the northwest during the winter half-year to permit unirrigated cropping (dry farming), but rainfall sufficient for intensive agriculture is largely confined to a densely populated lowland strip between the high, volcanic Elburz Mountains and the Caspian Sea. The western half of this lowland receives more than 40 inches (c. 100 cm) annually, including both summer and winter rainfall, and produces a variety of subtropical crops such as wet rice, cotton, oranges, tobacco, silk, and tea.

In the other regions of Iran, agriculture depends mainly or exclusively on irrigation. Many irrigated districts lie on gently sloping alluvial fans (triangular deposits of stream-laid sediments) at the foot of the Elburz and Zagros mountains; the capital, Tehran (Fig. 9.22; population: 6 million, city proper; 7.5 million, metropolitan area) is in one of these areas south of the Elburz. *Qanats* (underground sloping tunnels) furnish the water supply for a large share of the country's irrigated acreage. However, they are expensive to build and maintain, and an increasing part of Iran's water supply during recent times has come from stream diversions or from deep wells equipped with diesel pumps. Wheat is the most widespread and important crop, as it generally is in Middle Eastern areas having winter precipitation and summer drought. Cattle are raised in areas where there is enough moisture to grow forage for them, but drier areas support sheep and goats. Many livestock belong to seminomadic mountain peoples—the Qashqai (Kashgai), Baktiari, Lurs, Kurds, and others—whose independent ways have long been a source of friction between these tribesmen and the central government. Before the Islamic revolution in Iran in 1979, it was government policy to sedentarize Iran's nomads so that they did not threaten the state.

Prior to World War I, Iran was an exceedingly poor and undeveloped country, in marked contrast to its imperial grandeur when Persian kings ruled a great empire from Persepolis. In the 1920s, a military officer of peasant origins, Reza Khan, seized control of the government and began a program to modernize Iran and free it from foreign domination. He had himself crowned as Reza Shah Pahlavi, the founder of the new Pahlavi dynasty. Influenced by the modernizing efforts of Mustafa Kemal Ataturk in neighboring Turkey, the new Shah (king) introduced social and economic reforms. During World War II, Reza Shah displayed such pro-German leanings that British and Russian troops invaded his country in 1941 to secure a "back door" through which Allied supplies could reach the Russian front. Reza Shah was forced to abdicate, and in 1941 his young son took the throne as Mohammed Reza Shah Pahlavi.

Following the war, the new Shah continued and expanded the program of modernization and Westernization begun by his father. Progress was relatively slow until the 1970s, when Iran played a leading role in raising world oil prices. Mounting oil revenues underwrote explosive industrial, urban, and social development. New port facilities, hundreds of new factories, and surfaced highways were built. Extraction of coal, metals, and other minerals was stepped up. The new industrial development centered in Tehran, but Isfahan (population: 987,000, city proper) received a large steel mill, and Tabriz (population:

FIGURE 9.22
The Elburz Mountains rise dramatically behind Tehran, Iran's capital.
Bernard Silberstein/FPG International

971,000, city proper) in the northwest became a center of ma-chine-tool production to support the nation's industries.

In the countryside, the Shah's government attempted to up-grade agricultural productivity through land reform and the in-troduction of better methods, tools, seeds, fertilizers and animal varieties. Iran made major efforts to increase the supply of irri-gation water and electricity by building large storage dams and hydropower stations along mountain rivers fed by melting snows. The largest river-control scheme involved the Karun River and smaller rivers in the southern oil-bearing region called Khuzistan, which adjoins Iraq, where there continues to be an emphasis on such cash crops as sugarcane, sugar beets, and citrus fruits. Under the Shah, reforms in the agricultural sector were largely unrewarding, and poverty-stricken families from the countryside poured into Tehran and other cities. From this devoutly Muslim group of new urbanites came much of the support for the revolution that ousted the Shah in 1978–1979.

Oil money funded Iran's development under the Pahlavi dynasty, but the vast sums the regime spent on the military un-dercut the benefits to Iran's people. Arms came primarily from the United States, which allied itself with Iran as one of the "Twin Pillars" of American interests in the oil-rich Gulf region (the other pillar was Saudi Arabia). The 1978–79 revolution that overthrew the Shah and abruptly took Iran out of the Amer-ican orbit shattered these arrangements. The country became an Islamic republic governed by Shi'ite clerics, who included a handful of revered and powerful ayatollahs ("signs of God") at the head of the religious establishment, and an estimated 180,000 priests called mullahs. These religious leaders con-tinue to supervise all aspects of Iranian life and perform many functions allotted to civil servants in most countries. They base their authority on Shi'ite interpretations of the Islamic faith. (About 93 percent of Iran's population is Shi'ite.)

The central figure among the revolutionaries who over-threw the Shah was the Ayatollah Ruhollah Khomeini, an el-derly critic of the regime who was forced into exile by the Shah in 1964. Living in Iraq until 1978, when President Hussein ex-pelled him, and then in Paris, Khomeini sent repeated messages to Iran's Shi'ites that helped spark the revolution. During 1978, the country experienced many riots and strikes, which eventu-ally caused the Shah to flee the country and abdicate the throne. (After hospitalization in the United States and life in exile in Panama and Egypt, the Shah died in Cairo in 1981 and is buried there.) The Ayatollah Khomeini returned in triumph to Tehran in 1979, where he soon was able to take control of the govern-ment. Meanwhile, his followers held 52 U.S. diplomatic per-sonnel as hostages in Tehran for 444 days—an enormously publicized event that helped defeat U.S. President Jimmy Carter in his bid to be reelected in 1980.

The revolution was the product of overwhelming opposi-tion to the Shah and his regime among most elements of Iranian society. Opponents ranged from Communists to Westernized liberals to Shi'ite clerics and their followers. Grievances against the Shah included widespread corruption within the rul-ing circles, whereby the Shah, his numerous relatives, members of the government, and court favorites enriched themselves. Many people objected to the huge sums spent on arms, the ex-aggerated pomp of the Shah's court, the heavyhanded control of dissent by the secret police SAVAK, and the Shah's harassment of the influential merchants in the bazaars. Landowners (in-cluding the mullahs) whose holdings had been diminished or amalgamated by the Shah's land reforms were resentful. But, the driving force of the revolution was a furious tide of Shi'ite religious sentiment directed against modernization, Westerniza-tion, Western imperialism, communism, the exploitation of un-derprivileged people in Iran and elsewhere, and the institution of monarchy, which Shi'ite beliefs regard as illegitimate.

What has emerged since the revolution is a theocratic state, guided by edicts of the Ayatollah Khomeini (who died in 1989) and his successors, who have undertaken to create a society governed by Islamic law (shari'a). Political and religious dissi-dence has been crushed. Under the new regime, "polluting" in-fluences of Western thought and media have been purged. In contrast to their status during the Shah's reign, when they were encouraged to have stronger public roles, women in postrevolu-tionary Iran have lost many freedoms. While the government continues to be run by the Ayatollah's adherents, Western ana-lysts alternately perceive it to be either developing a more ac-commodating attitude toward the international community, or exporting ever more dangerous notions of revolutionary Islam to currently pro-Western countries in the Middle East.

9.8   Turkey, Where East Meets West

Once a Muslim state at the center of a disintegrating empire, Turkey transformed itself in the 20th century into a secular na-tional unit with fewer ties to traditional Islam and with a marked Western orientation. The Turks who organized the Ot-toman Empire beginning in the 14th century were pastoral no-mads from interior Asia (major Turkic elements still exist in the Former Soviet Region and China), and, from the 16th to the 19th century, their empire was an important power, centering on the large Anatolian peninsula between the Mediterranean, Aegean, and Black seas which forms most of the present Re-public of Turkey (see Fig. 9.4).

Modern Turkey was created from the wreckage of the old empire after World War I. Its founder, Mustafa Kemal Ataturk, was determined to Westernize the country, raise its standard of living, and make it a strong and respected national state. He in-augurated social and political reforms designed to break the hold of traditional Islam and open the way for modernization and Turkish nationalism. Islam had been the state religion un-der the Ottoman Empire, but Ataturk separated church and state, and to this day Turkey is the only Muslim Middle Eastern country to officially cleave them. Wearing the red cap called the *fez*, an important symbolic act under the Ottoman caliphs, was prohibited, and state-supported secular schools replaced the religious schools that had monopolized education. To facilitate public education and remove further traces of Muslim dominance, Latin characters replaced the Arabic script of the Qur'an. Slavery and polygamy were outlawed, and women were given full citizenship. Legal codes based on those of Western nations replaced Islamic law, and forms of democratic representative government were instituted, although Ataturk ruled in dictatorial fashion. Turkey has continued to have trouble establishing a fully democratic system, and there have been frequent periods of military rule. Despite decades of secularization, the current democratically-elected government is a coalition comprising avowedly secular politicians and Islamist leaders, including the prime minister, who wish to see Turkey reestablished on Islamic foundations.

From a physical standpoint, Turkey is composed of two units: the Anatolian Plateau and associated mountains occupying the interior of the country, and the coastal regions of hills, mountains, valleys, and small plains bordering the Black, Aegean, and Mediterranean seas. The Anatolian Plateau, an area of wheat and barley fields and grazing lands, is bordered on the north by the Pontic Mountains and on the south by the Taurus Mountains. The annual precipitation, concentrated in the winter half-year, is barely sufficient for grain. The plateau ranges from 2000 to 6000 feet (*c.* 600–1800 m), and is highest in the east where it adjoins the high mountains of the Armenian Knot. Its surface is rolling and windswept, hot and dry in summer and cold and snowy in winter, with a natural vegetation of short steppe grasses and shrubs. Production of cereals (especially barley, wheat, and corn) and livestock (especially sheep, goats, and cattle), employing both traditional and mechanized means, prevails in Anatolia.

Coastal plains and valleys along and near the Aegean Sea, Sea of Marmara, and Black Sea are generally Turkey's most densely populated and productive areas. Here are grown such cash crops as hazelnuts, tobacco, grapes for sultana raisins, and figs. Irrigated sugar beets are important in the small European section of Turkey known as Thrace. The Black Sea coastlands differ from the rest of Turkey in that they experience summer as well as winter rain and have much greater precipitation than other parts of the country (see Fig. 8.6). Turkey in general has a physical presence quite unlike the popular image of the Middle East as a desert region—it is, in fact, the only country in the region that has no desert, and cultivated land occupies about a third of Turkey's land area.

Relative poverty compared with the nearby European countries is a striking characteristic of present-day Turkey. There remains a very strong rural and agricultural component in Turkish life, with about two fifths of the population still classified as rural. However, the country has embarked on a course of change that promises to modernize its agriculture, expand industry, and raise its general standard of living. Early in the 1990s Turkey began an impressive agriculture scheme called the Greater Anatolia Project (GAP) or the Southeast Anatolia Project (SEAP). Its aim is to convert the semiarid southeast quarter of the country into the "Breadbasket of the Middle East." The centerpiece of this massive irrigation project is the Ataturk Dam, whose sluice gates closed down on the waters of the Euphrates River in 1991. About 20 other dams on the Turkish Euphrates and Tigris are also part of the project, which will provide hydroelectricity as well as water. The GAP is controversial, particularly because it is being implemented in the part of the country where a restive Kurdish population is seeking recognition, autonomy, and, among some factions, independence from Turkey. In addition, Turkey's Tigris and Euphrates waters flow downstream into neighboring states, raising serious questions about downstream water allocation and quality (see Regional Perspective, page 226).

The value of output from manufacturing in Turkey is already greater than from agriculture. Turkey's predominant industries are characteristic of LDCs: textiles, agricultural processing, cement manufacturing, simple metal industries, and assembly of vehicles from imported components. The country is still very dependent on imports for much of its machinery as well as many other types of manufactured goods. An iron and steel mill in the northern interior, based on Turkish coal and iron ore, is the largest heavy-industrial establishment. Turkey does produce some oil and has valuable deposits of chromium and other metallic and nonmetallic minerals, and its transportation facilities are more adequate than those of most Middle Eastern countries. This rapidly growing nation of 64 million (as of 1996) is a member of the North Atlantic Treaty Organization (NATO) and an associate member of the European Union. Turkey is thus a kind of "in-between" country: economically, it is well below the level of MDCs but above most of the world's LDCs; culturally, it is between traditional Islamic and secular European ways of living.

Istanbul (population: 6.6 million, city proper; 7.6 million, metropolitan area), formerly Constantinople and Byzantium, is Turkey's main metropolis, industrial center, and port (Fig. 9.23). One of the world's most historic and cosmopolitan cities, Istanbul was for many centuries the capital of the Eastern Roman (Byzantine) Empire. It became the capital of the Ottoman Empire when it fell to the Turks in 1453. However, the capital of the Turkish Republic was established in 1923 at the more centrally located and more purely Turkish city of Ankara (population: 2.7 million) on the Anatolian Plateau. Both Istanbul and Ankara have shantytowns that house migrants from impoverished rural Turkey. The same is true of Turkey's third largest city, the seaport of Izmir (population: 1.8 million, city proper; 1.9 million, metropolitan area) on the Aegean Sea.

Istanbul is located at the southern entrance to the Bosporus Strait, the northernmost of the three water passages (Dardanelles, Sea of Marmara, and Bosporus) that connect the Mediterranean and Black seas and are known as the Turkish Straits (see inset map, Fig. 8.1, page 206). The straits have long

FIGURE 9.23

The double-deck Karakoy (Galata) Bridge, which floats on huge pontoons, spans the Golden Horn inlet from the Bosporus in the heart of Istanbul, Turkey. In the background, minarets of historic mosques tower above the old walled city from which the Byzantine Empire and then the Ottoman Turkish Empire were ruled. Ferries such as the one in the foreground interconnect points along the Golden Horn and the nearby Bosporus. Road linkage between Istanbul and the main part of Turkey east of the Bosporus is provided by the high Bosporus Bridge, completed in 1973 and not shown in this view.   *R. Michaud/Woodfin Camp & Associates*

been a focus of contention between Turkey and Russia. In recent years, relations between the Turks and their neighbors to the north have been relatively tranquil, although the Turks, who have fought many wars against the Russians, maintain a high level of military preparedness within the NATO alliance. Some of Turkey's military posture has also been shaped by fragile relations with its NATO neighbor, Greece. The main questions recently at issue have been the status of Cyprus and conflicting claims to the oil and gas resources of the Aegean seabed.

9.9   Ethnic Separation in Cyprus

The large Mediterranean island of Cyprus, located near southeastern Turkey (see Fig. 9.4), came under British control in 1878 after centuries of Ottoman Turkish rule. In 1960, it gained independence as the Republic of Cyprus. An all-important problem on the island is the division between the Greek Cypriots, who are Greek Orthodox Christians comprising about three fourths of the estimated population of 700,000, and the Turkish Cypriots, who are Muslims comprising about one fourth. Agitation by the Greek majority for union with Greece (*enosis*) was prominent after World War II and led in the 1950s to widespread terrorism and guerrilla warfare by Greek Cypriots against the occupying British. Violence also erupted between Greek advocates of *enosis* and the Turkish Cypriots, who greatly feared a transfer from British to Greek sovereignty.

In 1974, a major national crisis erupted when a short-lived coup by Greek Cypriots, led mainly by military officers from Greece, temporarily overthrew President Makarios, who had followed a conciliatory policy toward the Turkish minority. Turkey then launched a military invasion that overran the northern part of the island. Cyprus was soon partitioned between the Turkish north and the Greek south. A buffer zone (the "Attila Line") sealed off the two sectors from each other, and even the main city of Nicosia was divided. A separate government was established in the north, and in 1983 an independent "Turkish Republic of Northern Cyprus" was proclaimed. Only Turkey has recognized this state. Meanwhile, the internationally recognized Republic of Cyprus functions in the Greek-Cypriot sector, which comprises somewhat more than three fifths of the island's land. Both republics have their capitals in Nicosia (population: 167,000, city proper), which was the capital before partition.

Prior to the partitioning of Cyprus, the north had dominated the economy, but since then the north has had severe economic difficulties while Greek Cyprus has prospered. The Turkish sector was seriously weakened during the 1974 crisis by the flight of an estimated 200,000 Greek Cypriots to the south as refugees. There was a return flow of Turkish Cypriots entering the north, and thousands of immigrants from Turkey settled in the north, but this immigration has not compensated for the almost total loss of Greek entrepreneurs, farmers, skilled workers, and consumers. Most outside nations refused to trade directly with Turkish Cyprus after the invasion, and the trade and economic assistance offered by an economically weak Turkey were insufficient to provide much momentum. The economically depressed north remains tied to the struggling economy of the Turkish mainland and is exceedingly dependent on aid from Turkey.

The Greek sector, by contrast, was able to make effective use of economic aid from Greece, Britain, the United States, and the United Nations. A construction program provided new housing, business buildings, roads, and port facilities. Tourism based on both beach and mountain resorts was greatly expanded. An efficient telecommunications system gave the south new links to all parts of the world; hundreds of new businesses were attracted to the south by such factors as favorable tax policies, modern facilities, dependable overseas communications, an educated and reasonably-priced labor force, the relative security provided by the island location, a government friendly to foreign business, and amenities provided by the tourism industry. A duty-free zone allows goods to be landed and transshipped without payment of customs duties.

9.10   Rugged, Strategic, Devastated Afghanistan

High and rugged mountains dominate Afghanistan, the only landlocked nation in the Middle East (Fig. 9.24). It has limited resources, poor internal transportation, and little foreign trade. Afghanistan is one of the poorest of the world's LDCs, with an annual per capita GNP of just $220 (as of 1988—latest figures available). Historically, it has occupied an important strategic location between India and the Middle East. Major caravan routes crossed it, and a string of empire-builders sought control of its passes. Through most of the 20th century, the highland country was remote from the main currents of world affairs.

FIGURE 9.25

This makeshift camp for Afghan refugees in the desert near Peshawar, Pakistan, is typical of the thousands of camps for displaced persons that still exist over the world. Many millions of persons have fled or been forcibly consigned to such places in the 20th century as a result of warfare, repression, ethnic conflict, natural disasters, disease, and famine. Some camps with more permanent facilities than those shown here have existed for many years.   *GAAL/Gamma Liaison*

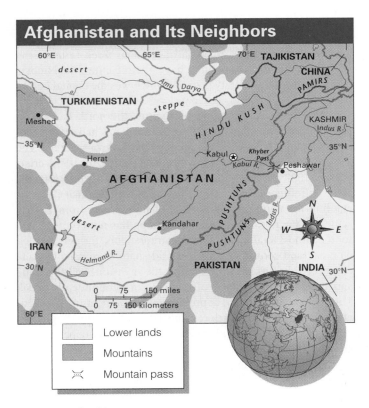

FIGURE 9.24

Afghanistan in its regional setting.

After the Islamic revolution in Iran in 1979, however, Afghanistan's location next to that oil-rich country made it once again the target of foreign interests. The Soviet military intervention of 1979 and the ensuing devastation catapulted the country into world prominence.

In 1973, the Afghan monarchy had been overthrown by a military coup, and a communist government oriented to the Soviet Union took power. Widespread rebellions followed, and in 1979 the Afghan government called on the USSR for assistance. In a surprise move, the Soviets responded with sizable military force, and a long internal war began. Widespread killing and maiming of civilians, sowing of land mines over vast areas, destruction of villages, burning of crop fields, killing of livestock, and pollution or destruction of irrigation systems by Soviet ground and air forces caused several million Afghan refugees to flee into neighboring Pakistan and Iran (Fig. 9.25). Meanwhile, arms from various foreign sources filtered into the hands of the *mujahadiin*, the anti-Soviet rebel bands who kept up resistance in the face of heavy odds. The United States was one of the powers supporting the rebels, and in that sense waged a proxy war against the Soviet Union in Afghanistan. The USSR's motives for its invasion of Afghanistan may have included a desire to prevent by force the spread of Iranian-style Islamic fundamentalism into this country (which is 74 percent Sunni Muslim and 25 percent Shi'ite Muslim) or into the Central Asian Soviet republics deemed vital to the superpower's security. For its part, the United States warned the Soviet Union that it would not tolerate further Soviet expansionism.

Only the Soviet Union's desire under Mikhail Gorbachev to put a more humanitarian face on the communist giant, and the realization that the USSR could not establish a lasting presence in Afghanistan and win its "Vietnam War," led to a withdrawal of Soviet troops in 1988 and 1989. It is estimated that of the 15.5 million people who lived in Afghanistan when it was invaded in 1979, one million died, two million were displaced from their homes to other places within the country, and six million fled as refugees into Pakistan and Iran. As of 1997, fewer than half of these refugees had returned.

The *mujahadiin* forces succeeded in overthrowing the communist government of Afghanistan in 1992, but since then rival factions among the formerly-united rebels have been engaged in civil warfare. By late 1996 one of the factions, the Taliban (backed by Saudi Arabia and Pakistan), controlled two thirds of the country's provinces, including the capital of Kabul. The Taliban immediately imposed a strict code of Islamic law in the regions under its control, and gained international notoriety for its austere administration, particularly for removing all women from the country's work force. The scars of war will last a very long time in this country. When peace does come, Afghanistan, like Kuwait, will have to confront another problem: clearing the millions of land mines sown across the landscape in wartime.

Definitions and Insights

THE LASTING LEGACY OF LAND MINES

Land mines are antitank and antipersonnel explosives, usually hidden in the ground, designed to kill or severely wound the enemy. There are many kinds: Soviet troops planted at least nine varieties in Afghanistan, including seismic mines, which are triggered by vibrations created by passing horses or people; mines with devices that pop up and explode when a person approaches; "butterfly" mines meant to maim rather than kill; and explosives disguised as toys, cigarette packs, and pens.

There are an estimated 10 million land mines in Afghanistan, or 40 per square mile (15 per sq km). Even after the fighting ends, reconstruction of this agricultural country cannot begin in earnest until the mines are cleared. The obstacles are enormous in rugged Afghanistan. Mines typically remain active for decades after they are planted, and tend to last longer in arid places. The mountainous terrain prevents systematic clearance of the explosives. Many mines were scattered indiscriminately from the air and will be difficult to locate. Heavy rains such as those which fell on Afghanistan in spring 1996 move the mines downhill and away from known locations. And it is expensive to clear mines; Kuwait spent $800 million to rid its landscape of them after the Gulf War.

So far, Western countries have donated $12 million toward the expense of clearing Afghanistan's land mines; Russia has contributed nothing.

Land mines are a global problem. Up to 110 million mines contaminate the land in 64 countries, and they kill or maim more than 25,000 civilians each year. The numbers and concentrations in some countries are staggering. The leader in numbers is Egypt, with 23 million mines; in density of mines, Bosnia-Herzegovina leads with 152 per square mile (58 per sq km). While 100,000 mines are cleared each year, two to five million more are planted.

In 1994, the United Nations General Assembly passed a resolution that would ban antipersonnel land mines, but too many countries around the world regard these as legitimate weapons to make this prospect likely. Costing as little as two dollars each, they are very affordable weapons for cash-strapped warring LDCs (and so have been dubbed the "Saturday-night specials of warfare"). Some countries such as Belgium and Austria have acted unilaterally to stop production and export of mines, hoping to inspire other nations to follow their example. In the meantime, these hidden horrors continue to be cleared, as one de-mining specialist put it, "one arm and one leg at a time."

Afghanistan's population is estimated at about 22 million, only about one third the size of neighboring Iran's. Most people live in irrigated valleys around the fringes of a mass of high mountains occupying a large part of the country. The country's second most populous area is the foothills and steppes on the northern side of the central mountains. Most of its inhabitants live in oases forming an east-west belt along the base of the mountains. Northern Afghanistan borders the Former Soviet Region, and millions of people on the Afghan side are related to peoples of the now independent Central Asian states.

The most heavily populated section is the southeast, particularly the fertile valley of the Kabul River, where the capital and largest city, Kabul (population: 1.4 million, city proper; elevation: 6200 ft/1890 m) is located. Most of the inhabitants of the southeast are Pushtuns (also known as Pashtuns or Pathans). Their language, Pushtu, is related to Farsi (Persian), which is the main language of administration and commerce in Afghanistan. The Pushtuns are the largest and most influential of the numerous ethnic groups that make up the Afghan state. The independent-minded tribal Pushtun people have always been slow to recognize the authority of central governments. In the days of Great Britain's Indian Empire, the area was rife with warfare among tribes, tribal raids on British-controlled areas, and British punitive expeditions against the tribes. In those days, Peshawar—on the Indian (now Pakistan) side of the Khyber Pass into Afghanistan—became a noted garrison town for British and Indian troops.

Since Britain's withdrawal from India in 1947, the border region has continued to be a source of friction between

Afghanistan and the new state of Pakistan. The friction has been partially due to border incidents but has also grown out of proposals that the Pushtun-inhabited areas of Pakistan be incorporated in a separate state ("Pushtunistan"), either independent or affiliated with Afghanistan. Pakistan has firmly opposed such proposals.

Southwestern Afghanistan is composed of a large desert basin, part of which extends into Iran. An area of interior drainage, it receives the waters of the Helmand River (see Fig. 9.24) and several others originating in Afghanistan's central mountains. Extensive irrigation works supported a large population here in ancient times, but Mongolian invaders largely destroyed the water system in the 13th and 14th centuries. Prior to 1979 the Afghan government, through its Helmand valley irrigation and power project, was reclaiming and resettling land by providing irrigation water from storage and flood-control reservoirs on the Helmand and other rivers.

The only minerals currently extracted are minor quantities of natural gas, coal, and a few others; Afghanistan is overwhelmingly a rural agricultural and pastoral country. The land exhibits a wide range of climatic conditions corresponding to differences in altitude, but is generally so mountainous and arid that only an estimated 12 percent is cultivated. Enough rain falls in the main populated areas during the winter half-year to permit dry farming of winter grains. A variety of cultivated crops, fruits, and nuts are important locally. Livestock raising on a seminomadic and nomadic basis is widespread. In general, Afghanistan's agriculture bears many of the customary Middle Eastern earmarks: traditional methods, simple tools, limited fertilizer, and low yields.

QUESTIONS

Refer to the Atlas of World Geography Special Edition as well as this chapter to answer the following questions.

1. *On the map of Asia on pp. 138–139, identify:*
   Four "chokepoints."

   The city where Judaism's First and Second Temples stood, and which both Israel and the Palestinian claim as their "eternal" capital (in the inset on p. 139).

2. *On the map of Northern Africa on pp. 128–129, identify:*

   A strife-torn country characterized by a largely Arab Muslim north and black Christian and animist south.

   The country where an Islamic party won national elections in 1992, only to have the government reject the results, setting off waves of domestic violence and increased emigration to France.

3. *On the map of the Middle East on pp. 144–145, identify:*

   The region where the national government is systematically draining wetlands in an effort to control and drive out a rebellious population.

Monsoon Asia

4

Ancient Cultures with Very Early Centers of Civilization
Diverse Cultures
Geopolitically Fragmented and Unstable
Postcolonial but Important Colonial Legacies Remain
Superpopulous with Half the World's People Unevenly Distributed
Developing and Generally Poor with Japan a Major Exception
Villagers Predominate but Urban Population is Huge and Growing
Physically Compact Continent Rimmed by Large Peninsulas and Islands
Topographically Broken
Climatically Diverse
Plains-Oriented, River-Oriented, Irrigation-Dependent
Contrasts and Conflicts Among Traditional Values System
and Dynamic Modernism

10 A GEOGRAPHIC PROFILE OF MONSOON ASIA

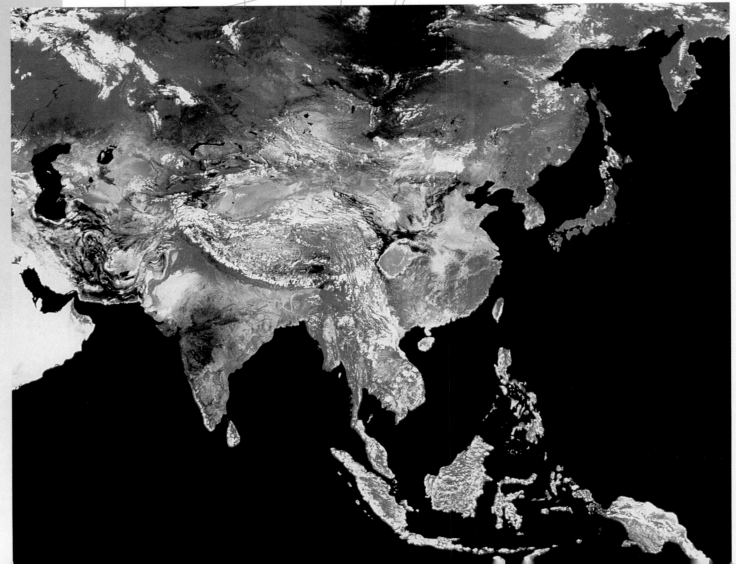

Asia extends, in a broad, sweeping crescent, from Pakistan and the Indus River through an arc of mountains, rivers, plains, islands, and seas to the northern island of Japan. In this landmass of more than 8 million square miles (20,761,000 sq km) lives more than half of the world's population (Fig. 10.1). The cultural makeup of these 3.3 billion people is as diverse and expressive as are the physical environments they inhabit. The world's highest mountain peaks, some of the longest rivers, and Earth's most highly transformed and densely settled river plains are the setting for some of our oldest civilizations and our most modern developing economies. Just as this region has played a major role in societal development over the past millennia, so, too, will it be a major force in defining the world of the 21st century.

"The Orient" is the term traditionally used to refer to the countries occupying the southeastern quarter of Eurasia. The term "orient" comes from the Latin meaning "to rise," or "to face the east." "Occident" means "to set" (as in the sun—in the west). The point from which this daily orbit of the sun was described early on was Western Europe. Hence, the Orient and the Occident early became European terms for spatial regional identification. In this text we use the term Monsoon Asia rather than the Orient for this region because of the significant role the seasonal shift of wind patterns plays all the way from Pakistan in the west to Japan in the east (see Fig. 10.1).

Monsoon Asia encompasses the following regions: East Asia, which embraces Japan, North and South Korea, China, Taiwan, Hong Kong, Macao, and countless nearshore islands; South Asia, which includes Pakistan, India, Sri Lanka, Bangladesh, and the mountain nations of Bhutan, Nepal, and offshore islands; Southeast Asia, which is the term used for the dominant peninsula jutting out from the southeast corner of the Asian continent and includes the countries of Burma (Myanmar), Laos, Cambodia, Vietnam, Malaysia, and Singapore; and the island world that rings this peninsula, which includes the countries of Indonesia, the Philippines, Brunei, and East Timor.

This crescent of Monsoon Asia has been home to some of the most important cultural developments of humankind—in landscape transformation, settlement patterns, religion, and political innovations. It is not the magnitude of population alone that makes understanding Monsoon Asia central to our study of world regional geography; it is also the pivotal role of past and present Asian cultural innovations that makes this region a major building block in the process of better understanding the nature of the world today.

◄ This composite satellite image is from the National Oceanic and Atmospheric Administration (NOAA) and shows the full landscape of Monsoon Asia with its great mountain highlands and plateaus, its desert expanses, ranges of borderland mountains, verdant river valleys, and the plains and river systems with dense accumulations of farming and urban populations. Image colors have red denoting vegetation; grey, semi-desert; and white represents desert sands. Mottled red and white in Southeast Asia is tropical rain forest. *Earth Satellite Corporation/Science Photo Library*

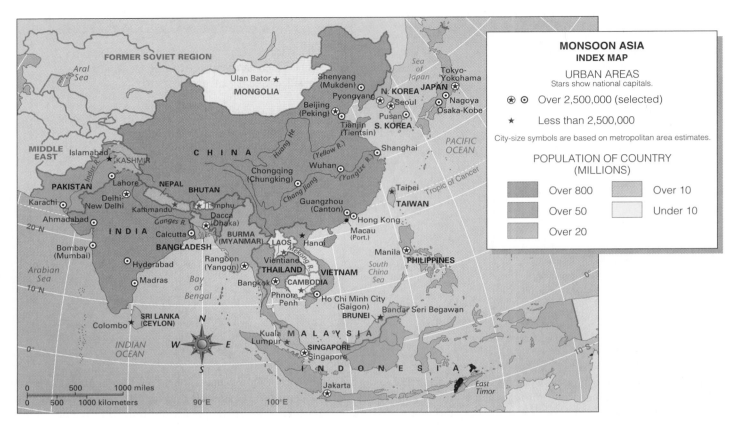

FIGURE 10.1

Introductory location map of Monsoon Asia, showing political units as of early 1997. East Timor is a former Portuguese colony occupied by Indonesia. Indonesia's claim is not recognized by the United Nations.

10.1 Definition and Basic Magnitudes

Table 10.1 portrays the demographic and geographic magnitude of Monsoon Asia. Within it lie the major subregions of the Indian Subcontinent, Southeast Asia, the Chinese Realm, and Japan and Korea. There is profound geographic variety in this region that encompasses less than one-eighth of the land surface of the earth but which is home to approximately 60 percent of the world's population. The birthrate ranges from a 0.32 annual rate of increase in Japan to nearly 3.0 in parts of Southeast Asia. In Hong Kong, Macao, and Singapore, some of the world's highest urban densities are found, while in Mongolia and Nepal, population densities are extremely low. Singapore is one of the few nations to claim a 100 percent urban population, while 92 percent of the Nepalese population is still rural and nonurban. Only 2 percent of Bhutan is arable, while farmers in India and Bangladesh both cultivate more than 50 percent of their land.

In gaining a feeling for the diversity of this region, it is important to see it today as a collection of distinctive approaches to achieving economic growth. This challenge is set against a backdrop of widely varying distinctive histories, particularly in the development of technology, farming, literature, and religion. At the same time, images in the West of prosperous Japanese and Hong Kong tourists festooned with cameras and sophisticated electronic gadgetry must be tempered with the reality of an Asian peasantry still working the land with traditional tools and only modest evidence of recent change in agricultural technology. Monsoon Asia, to a greater extent than other regions chronicled in our text, is a restless amalgam of tradition and innovation, rural industriousness and urban experimentation, and societal continuity and tense assimilation of Western popular culture. The region is in many ways an evocative microcosm of the restless forces and elements of contemporary world regional geography.

10.2 Varied Physiographic Settings

The setting in which the Asian drama is being enacted is a complex intermingling of many topographies (Fig. 10.2). However,

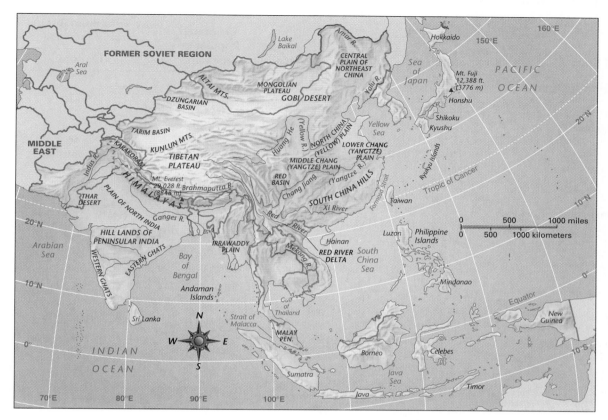

FIGURE 10.2
Major landforms and water features of Monsoon Asia.

a certain broad order appears if the surface features are conceived of as three concentric arcs, or crescents, of land: an inner arc of high mountains, plateaus, and basins; a middle arc of lower mountains, hill lands, river plains, and shallow basins; and an outer arc of islands and seas.

THE INNER HIGHLAND

The inner highland is composed of the world's highest mountain ranges, interspersed with plateaus and basins. At the south of the continent, the great wall of the Himalaya, Karakoram, and Hindu Kush mountains overlooks the north of the Indian subcontinent. At the north, the Altai, Tien Shan, Pamir, and other towering ranges separate this region of Asia from the countries of the former Soviet Union. Between these mountain walls lie the sparsely inhabited Tibetan Plateau, at over 15,000 feet (c. 4500 m) in average elevation, and the dry, thinly populated basins and plateaus of Xinjiang (Sinkiang) and Mongolia (Fig. 10.3).

FIGURE 10.3
Towering dunes and irrigated oasis fields have resulted from natural and human processes at work in the arid region called Xinjiang, western China.   *H. Todd Stradford, Jr.*

TABLE 10.1   Monsoon Asia: Basic Data

Political Unit	Area (thousand/sq mi)	Area (thousand/sq km)	Estimated Population (millions)	Estimated Annual Rate of Increase (%) (natural)
Indian Subcontinent				
India	1269	3287.6	936.5	1.77
Pakistan	310.4	803.9	131.5	2.98
Bangladesh	55.6	144	128.1	2.32
Nepal	54.4	140.8	21.6	2.44
Bhutan	18.1	47	1.8	2.34
Sri Lanka	25.3	65.6	18.3	1.24
Total	1732.8	4488.9	1237.8	1.96
Southeast Asia				
Burma (Myanmar)	261.9	678.5	45.1	1.84
Thailand	198.4	514	60.3	1.24
Vietnam	127.2	329.6	74.4	1.87
Cambodia	69.9	181	10.6	2.83
Laos	91.4	236.8	4.8	2.84
Malaysia	127.3	329.8	19.8	2.24
Singapore	0.245	0.634	2.9	1.06
Brunei	2.24	5.8	0.3	2.08
Indonesia	741.1	1919.4	203.6	1.56
Philippines	115.8	300	73.3	2.35
Total	1735.485	4495.534	495.1	1.77
Chinese Realm				
China (PRC)	3704.4	9597	1203.1	1.04
Hong Kong	0.4	1	5.5	0.6
Macao	0.007	0.02	0.5	1.03
Taiwan	13.9	36	21.5	0.96
Mongolia	604.2	1565	2.5	2.58
Total	4322.907	11,199.02	1233.1	1.04
Japan and Korea				
Japan	145.8	377.8	125.5	0.32
North Korea	46.5	120.5	23.5	1.78
South Korea	38	98.5	45.6	0.95
Total	230.3	596.8	194.6	0.64
Summary Total	8021.492	20,780.254	3160.6	1.49

Sources: The World Almanac and Book of Facts 1996, Statesman's Yearbook 1995–1996, Britannica Book of the Year 1997 and The World Factbook 1995.

RIVER PLAINS AND HILL LANDS

The area between the inner highland and the sea is principally occupied by river flood plains and deltas bordered and separated by hills and relatively low mountains. Major components are:

(1) the immense alluvial plain of northern India, built up through ages of meandering and deposition by the Indus, Ganges, and Brahmaputra rivers;

(2) the uplands of peninsular India, geologically an ancient plateau but largely hilly in aspect;

(3) the plains of the Irrawaddy, Chao Praya (Menam), Mekong, and Red rivers in peninsular Southeast Asia, together with bordering hills and mountains;

(4) the uplands and densely settled small alluvial plains of southern China;

(5) the broad alluvial plains along the middle and lower Chang Jiang (Yangtze River) in central China and the mountain-girt Red Basin on the upper Chang Jiang;

(6) the large delta plain of the Huang He (Yellow River) and its tributaries in North China, backed by loess-covered hilly uplands; and

(7) the broad central plain of Northeast China (Manchuria), almost completely enclosed by mountains.

Estimated Population Density		Infant Mortality Rate	Urban Population (%)	Arable Land (% of total area)	Per Capita GNP ($U.S.)
(sq mi)	(sq km)				
738	285	76.3	26	55	1360
424	164	99.5	32	23	1930
2304	890	104.6	17	67	1040
397	153	81.2	10	17	1060
98	38	118.6	13	2	700
724	280	21.3	22	16	3190
714	276	81	25	47	1408
171	66	61.6	25	15	930
303	117	35.7	19	34	5970
585	226	44.6	21	22	1140
150	58	109.6	13	16	630
52	20	99.2	19	4	850
155	60	24.7	51	3	8650
11,834	4569	5.7	100	4	19,940
132	51	24.7	90	1	16,000
275	106	65	31	8	3090
632	244	49.6	49	26	2310
285	110	55.1	24	14	3090
324	125	52.1	28	10	2500
13,805	5330	5.8	100	7	24,530
79,465	30681	5.4	100	0	10,000
1549	598	5.6	75	24	12,070
5	2	41.8	55	1	1800
285	110	51	29	9	5682
860	332	4.3	77	13	20,200
505	195	26.8	61	18	920
1198	463	20.9	74	21	11,270
845	326	10.9	74	14	15,779
394	152	60.9	29	18	4224

OFFSHORE ISLANDS AND SEAS

Offshore, a fringe of thousands of islands, mostly grouped in great archipelagoes, borders the mainland. On these islands, high interior mountains with many volcanic peaks are flanked by coastal plains where most of the people live. Three major archipelagoes incorporate most of the islands—the East Indies, the Philippines, and Japan. Sri Lanka, Taiwan, and Hainan are large, densely populated islands outside these archipelagoes.

Between the archipelagoes and the mainland lie the China Seas, and, to the north, the Sea of Japan. At the southwest, the Indian peninsula projects southward between two immense arms of the Indian Ocean—the Bay of Bengal and the Arabian Sea.

10.3 Climate and Vegetation

The climatic pattern of the region, like the physiographic, is one of almost endless variety. Two unifying elements, however, are present throughout those parts of Asia inhabited by considerable numbers of people. These are (1) the dominance of warm climates, and (2) a monsoonal regime of precipitation.

TEMPERATURE AND PRECIPITATION

In the most populated parts of Monsoon Asia, temperatures are tropical or subtropical. The principal exceptions exist in northern sections of China, Korea, and Japan. Here, summers are warm to hot in the lowlands, but the growing season is relatively short and winters are cold. The arid, sparsely populated basins and plateaus of Xinjiang and Mongolia also have warm summers and sometimes bitterly cold winters. The higher mountain areas and Tibetan Plateau have undifferentiated highland climates varying with the altitude, aspect, and latitude. Permanent snow fields and glaciers occur at the higher elevations.

Annual precipitation varies from near zero in parts of Xinjiang to more than 400 inches (1000 cm) in parts of the Khasi Hills of northeastern India. A monsoon climate, or at least a climate with monsoonal tendencies, prevails nearly everywhere in the populous middle arc of plains and hills all across Asia, and in many parts of the islands as well.

Definitions and Insights

MONSOON

A monsoon is a current of air blowing steadily from a given direction for several weeks or months at a time. In Asia, the monsoon is characterized by a summer monsoon blowing from the sea to the land and bringing with it high humidity, moist air, and generally predictable seasonal rains. The winter monsoon reverses these patterns as wind blows from the cold continental interior toward the sea. However, because there is little moisture in the source areas over land for the winter monsoon, there is little rain. A monsoon climate, then, is generally characterized by spring and summer precipitation and a long dry season in the low Sun (winter) cycle. Japan, however, experiences a relatively unusual pattern of winter precipitation related to the monsoon wind shifts. The winter winds blow eastward across the Sea of Japan and drop great quantities of very wet snow on the coast of western Honshu.

The monsoon has a dominant role in the cultures of Asia because agricultural success is tied so closely to the arrival of the spring and summer rains. Thus, the monsoon shapes not only the nature of farming, but literature and the personality of place as well.

TYPES OF CLIMATE

Monsoon Asia is characterized for the most part by a warm, well-watered climate. There are seven main types of Asian climate customarily recognized in Asian climatic classifications:

(1) tropical rain forest, (2) tropical savanna, (3) humid subtropical, (4) humid continental, (5) steppe, (6) desert, and (7) undifferentiated highland.

Most rainy tropical climates (rain forest and savanna) are generally found between the equator and the Tropics of Cancer or Capricorn, with major regions closer to the equator. Consequently, high temperatures are experienced throughout the year in the lowlands. The year-round growing season offers the maximum possibilities for agriculture from the standpoint of temperature. However, the high temperatures and heavy rains promote rapid leaching of mineral nutrients and destruction of organic matter, with the result that most soils are relatively infertile despite the surprisingly thick cover of trees and grasses they often support. A brief description of the Asian climate zones follows.

Tropical rain forest.   This climate zone is typically found within 5 or 10 degrees of the equator. Precipitation is spread throughout the year so that each month has considerable rain. The amount of precipitation is at least 30 to 40 inches (c. 75–100 cm), often 100 inches (c. 250 cm). Average temperatures vary only slightly from month to month; Singapore, for example, exhibits a difference of only 3°F between the warmest and coolest months. Monotonous heat prevails year round, although some relief is afforded by a drop of 10 to 25°F (6–14°C) in the temperature at night, and sea breezes refresh coastal areas. In such climates as this, it is said that "nighttime is the winter" of the region.

The tropical rain forest produced by the climatic conditions just described is characteristically a dense forest of large broadleaf evergreen trees, mostly hardwoods, from 50 to 200 feet (c. 15–60 m) in height, forming an almost continuous canopy of foliage. The trees are often entangled in a mass of vines, and dense undergrowth occurs wherever sufficient light can penetrate to the ground. The apparent vitality of this forest growth comes largely from the nutrient base created by the mass of **detritus**, or dead plant and animal remains lying at the surface.

Tropical rain forest climate and vegetation are characteristic of most parts of the East Indies, the Philippines, and the Malay Peninsula. Rain forest vegetation is also found in certain other areas (generally along coasts) that experience a dry season but in which the precipitation of the rainy season is sufficiently heavy to promote a thick growth of trees.

Tropical savanna.   The savanna climate, like the tropical rain forest climate, is characterized by high temperatures year round but is customarily found in areas farther from the equator, and the average temperatures vary somewhat more from month to month. However, the most important difference between the two climates is that the savanna has a well-defined dry season, lasting in some areas as long as six or eight months each year, creating a problem for agriculture.

The main Asian areas of tropical savanna climate occur in southern and central India, the greater part of the Indochinese peninsula, and eastern Java and the smaller islands to the east. In Asia, the characteristic natural vegetation associated with this climate is a deciduous forest of trees smaller than those of the tropical rain forest. Tall, coarse grasses, like bamboo—a very common vegetation form in African and Latin American savannas—are found only in limited areas, and even they are thought to have been produced by repeated burning of forests.

Humid subtropical. This climate zone occurs in southern China, the southern half of Japan, much of northern India, and a number of other countries. It is characterized by warm to hot summers, mild or cool winters with some frost, and a frost-free season lasting 200 days or longer. The annual precipitation of 30 to 50 inches (*c.* 75–125 cm) or more is fairly well distributed throughout the year, although monsoonal tendencies produce a dry season in some areas. The natural vegetation is a mixture of evergreen hardwoods, deciduous hardwoods, and conifers.

Humid continental. This climate characterizes the northern part of eastern China, most of Korea, and northern Japan. It is marked by warm to hot summers, cold winters with considerable snow, a frost-free period of 100 to 200 days, and less precipitation than the humid subtropical climate. Most areas experience a dry season in winter. The predominant natural vegetation is a mixture of broadleaf deciduous trees and conifers, although prairie grasses are thought to have formed the original cover in parts of northern China.

Steppe and desert. These climates, whose characteristics have been previously described, are found in Xinjiang and Mongolia, and in parts of western India and Pakistan.

Undifferentiated highland. The undifferentiated highland climate is most extensive in the Tibetan Highlands and adjoining mountain areas. It is made up of a broad range of montane microclimates that provide little base for plant growth and vary from place to place according to altitude, aspect, and latitude.

10.4 Cultural Landscapes and Signatures of the Past

A cultural landscape is sometimes described as a **palimpsest**—a term which describes an artist's canvas that has been sketched on, erased, and sketched on again, but with markings left from the earlier drawings. The landscape, in the same sense, has tangible and visible evidences of earlier human efforts to settle, to farm, and to create landscapes that satisfy human needs (and desires). Even though such needs change over time from cultural diffusion, the markings of earlier efforts stand in mute testimony to the variety of ways people interact with their landscapes.

Definitions and Insights

LANDSCAPE SIGNATURE

In learning to "read" a cultural landscape, there are features that have such distinctive associations with place that they serve to mark that place in a specific and unique way. The Golden Gate Bridge and the Empire State Building serve to identify San Francisco and New York City respectively; and the Eiffel Tower and Great Pyramids do the same for Paris and Cairo. The Great Wall and the Taj Mahal in China and India are two universally known features of human engineering. These features are called **landscape signatures**. They are generally features created by human modification of the environment, although sometimes physical features may be so distinctive—Niagara Falls, for example—as to play the same role. Landscape signatures are useful to know as a means of assigning landscape and regional identity to specific places.

EUROPEAN SIGNATURES OF COLONIALISM

Like much of the modern world, Asia is an area where many traditional geographic patterns were created or reshaped by Western colonialism. Today, the region is postcolonial and made up of many independent nations, some of which contain areas in which very early civilizations developed. Pride in this ancient heritage, coupled with resentment of indignities and repression suffered under colonial administrations, contributed to rising nationalism and, consequently, the end of Western colonial control after World War II. But the geographic heritage from the colonial past remains pervasive, and some of its major aspects are summarized here as essential background for the discussions that follow.

European penetration of this part of Asia began at the end of the 15th century. The early explorers (from Portugal, Spain, and the Netherlands) gradually extended political control over some islands and limited areas near the coast. In the 18th and 19th centuries, the pace of colonization and economic control quickened, and large areas came under European sway. By the end of the 19th century, Great Britain was supreme in India, Burma (Myanmar), Ceylon (now Sri Lanka), Malaya, and northern Borneo; the Netherlands possessed most of the East Indies (Indonesia); and France had acquired Indochina (now Vietnam, Laos, and Cambodia) and small holdings around the coast of India (Fig. 10.4). China, although retaining a semblance of territorial integrity, was forced to yield possession of strategic Hong Kong to Britain in the 1842 Treaty of Nanking and to grant special trading concessions and extraterritorial rights to various European nations and the United States

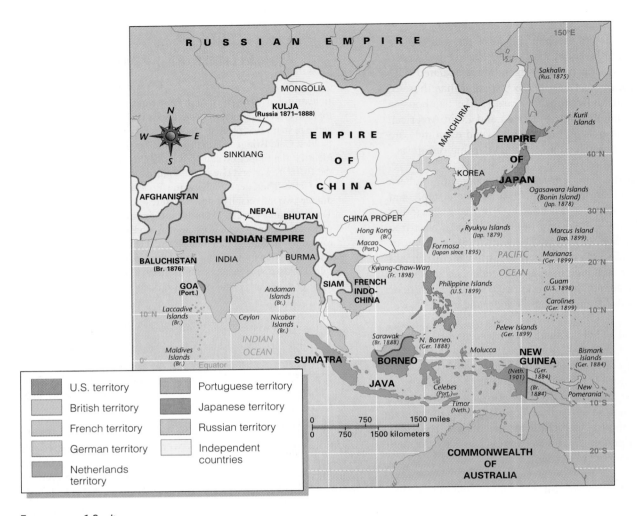

FIGURE 10.4

This map shows the general distribution of colonial realms—as well as independent countries—in Monsoon Asia at the beginning of the 20th century. The British influence is seen as the most dominant and the one that continued to the middle of the century, and was finally eclipsed almost totally with the 1997 return of Hong Kong to China. China had small regions controlled by the Germans, Russia, Japan, and the British but never fell under full control of any colonizing power. It was about this time that the United States also began to play a role in the development of some Pacific territories, with the Philippines and Hawaii as the most significant. After "The World in 1900" in R.R. Palmer. *Atlas of World History.* Chicago: Rand McNally 1957. 169–171.

through the latter half of the 19th century. In 1898, the Philippines, held by Spain since the 1500s, came into American control.

The few Asian countries to escape domination by the Western powers during the colonial age were (1) Thailand, which formed a buffer between British and French colonial spheres in Southeast Asia, (2) Japan, which withdrew into almost complete seclusion in the mid-17th century but which emerged in the later 19th century as the first modern, industrialized Asian nation and soon acquired a colonial empire of its own, and (3) Korea, which also followed a policy of isolation from foreign influences until 1876, when a trade treaty was forced on it by Japan.

During the age of European expansion, Asia constituted an extraordinarily rich colonial area. Western nations extracted vast quantities of tropical agricultural commodities and, in turn, found large markets for their manufactured goods. Westerners also found a fertile field for investment in plantations, factories, mines, transportation, communication, and electric-power facilities. Some of the region's most important cities—for example, Shanghai in China, Calcutta in India, and Singapore—were developed largely by Western capital as seaports serving Western colonial enterprise.

Western dominance of Asia ended in the 20th century by circumstances including (1) the weakening effect of the conflicts among the Western nations in the two world wars, (2) the rise of Japan to great-power status and its successful, although temporary, military challenge to the West in the early stages of World War II, and (3) the rise of anticolonial movements in areas subject to European control. After World War II, virtually all colonial possessions in Asia gained independence, except for Hong Kong (returned to China in mid-1997) and Macao (slated for transfer from Portuguese to Chinese sovereignty in 1999). Largely as a consequence of these changes, the Asia of the 20th century has been a region marked by revolution, war, and considerable turmoil.

RELIGION

Beyond temple architecture and sacred groves, religion influences societal organization, patterns of settlement and mobility, and even a people's inclination or disinclination to adopt foreign cultural practices. For these reasons religion assumes a profound role in our efforts to read the Asian landscape.

The variety of religions in Monsoon Asia is broad. Hinduism is dominant in India, although many other religions are practiced there as well. The Islamic faith prevails in Pakistan, Bangladesh, most parts of the East Indies, parts of the southern Philippines, parts of western China, and among the native Malays of the Malay Peninsula. Various forms of Buddhism are dominant in Burma, Thailand, Cambodia, Laos, Tibet, and Mongolia. The people of Sri Lanka and Nepal are uneasily divided between Buddhism and Hinduism. The religious patterns of the Chinese and Vietnamese are somewhat more difficult to describe, even if effects of communist rule are disregarded. Among them, Buddhism, Confucianism, and Taoism have all exerted an important influence, often in the same household. The same general situation has prevailed in both Koreas. In Japan, religious affiliations often overlap, with an estimated 84 percent of the Japanese population sharing the strongly nation-

alistic religions of Shintoism and Buddhism. The Philippines, with a large Roman Catholic majority, is the only Christian nation in Asia, although Christian groups are found in various other areas.

The religions just named are indigenous to Asia except for Christianity and Islam, both of which originated in the Middle East. Veneration of ancestors is prominent, especially among Chinese, Vietnamese, and Koreans. Often an elaborate ritual for everyday living is followed. Asian hill tribes—particularly in insular Southeast Asia—are largely animists, believing that natural processes and objects possess souls. Because of their cultural significance and widespread adoption in Asia, however, Buddhism and Confucianism are worthy of some additional consideration. (Hinduism also plays a major role in the religious geography of Monsoon Asia, and it is discussed in some detail in Chapter 11.)

Buddhism

Buddhism had its origin in India, and was diffused to China from Central Asia by the first century A.D. along the major Silk Road trade routes to Korea by the fourth century, and from there to Japan by 552. It was proclaimed the state religion of Japan in 593. It continues to be a significant religious influence there today. It reached Tibet in the middle of the eighth century, and its arrival was followed by decades of contest there between Chinese and Indian Buddhists (Fig. 10.5). The Chinese were defeated in this struggle and expelled from Tibet at the end of the eighth century.

In Japan, contemporary evolution of this religion has focused upon the *Soka Gakkai, the Value Creation Society* of the Nichiren sect of Buddhism. In this Japanese context, the belief system now promotes the use of mass media and effective organization to expand material benefit and contemporary happi-

Regional Perspective

THE MING VOYAGES

We often think of the Age of Discovery as simply the chronicle of what the West did after Columbus launched his maritime effort to get to Asia by sea in 1492. The map of the world might have a very different look if the Ming Voyages—from an earlier age of Chinese discovery—had been dealt with in a different way by the Chinese and other Asian nations. From 1405–1433 a Chinese Muslim admiral, Zheng He (Cheng Ho), commis-

sioned by the Ming Court in China, led seven long-range maritime expeditions. He sailed from a port in southeast China to large overseas Chinese communities in Indonesia and then on to India, the Arabian Sea, and even to Mombasa on the East African coast. There were 63 oceangoing junks (a Chinese sailing vessel) in Zheng He's fleet, the largest of which—at 440 feet long and 180 feet wide, with a crew of more than 400 men, was six times the size of the Niña, the Pinta, or the Santa Maria. The

Chinese certainly had the maritime power and the manpower to colonize new lands in the early 15th century, and were it not for the death of the sponsoring Chinese emperor in 1433 and the general Chinese inclination to stay focused on the world of East Asia, the Earth might have had a very different Age of Discovery. After 1433, these voyages ended when the new Ming Court prohibited any Chinese ships from leaving traditional coastal shipping lanes.

FIGURE 10.5
Buddhist monks in their saffron robes, Bangkok, Thailand. Buddhism is the main religion in several Southeast Asian countries, and also plays a role in other countries in Monsoon Asia. *Lorina Ebbe*

ness for its followers. While it has evolved in many distinct ways since its inception, the Nichiren sect has evolved the furthest from the initial abandonment of material goods and the self-discovery and pilgrimages of Siddhartha Gautama in fifth-century-B.C. India.

Confucianism

Confucianism is another religious belief system that has a continuing presence in Asia. While more fully a belief system associated with the presence of Chinese populations (both inside and outside of China proper), the broad spatial distribution of Chinese outside Asia means that it appears as a cultural influence in many countries. Confucianism comes from the writings of Confucius (*c.* 551–479 B.C.), collected primarily in his *Analects*. Mencius (*c.* 371–288 B.C.) later became a major force in the widespread diffusion of this belief system throughout China. The major tenets of Confucianism are embodied in a system of ethical precepts for the proper management of soci-

ety. Confucius never proclaimed his beliefs to constitute a religion; he was simply attempting to create a social contract between different classes central to Chinese government and society. He viewed his philosophy as secular, and its diffusion to Korea, Japan, and Vietnam has been part of the cultural innovation and baggage taken abroad by a steady stream of Chinese migrants to those places.

Confucianism has ceremonial activities that take place around ancestral shrines and Buddhist temples on the birthday of Confucius. The most significant role of this religion has been to provide a continuing link with the main stem of Chinese civilization for the billions of Chinese born and raised in China or who have migrated to other Asian nations.

10.5 Distinctive Contemporary Landscape Signatures

The current Asian landscape and culture are universally expressed in the three features highlighted here—garden agriculture, education, and use of water—for they have an influence on social and visual patterns all across this crescent of Asia.

GARDEN AGRICULTURE

Garden agriculture has a long tradition in Monsoon Asia. Because of the huge population in virtually all Asian nations, a shortage of land has always faced farmers. Rural families have traditionally relied on additional children as a source of inexpensive and significant labor, in the belief that more hands could do more farming. Whether you are traveling in the coastal lowlands of southeast China, the uplands of the island of Luzon in the Philippines, the hill lands of Burma, or the broad plain of northern India, there is a world of small garden plots that ring the cities and extend out to towns and villages as well. Although farm mechanization in the past three decades has introduced significant change in Asian agriculture, there continues to be the rich landscape signature of delicate, small, and generally privately worked garden plots.

EDUCATION

In Monsoon Asia, families value education. In China, literacy is the most frequently sought way out of the traditional pattern of rural poverty. If a young man or woman could pass local and then regional exams, the gateway to government administration was opened. Once through that doorway a life path in China could lead almost anywhere. This same dedication to education has existed in India, and the combined influences of the Chinese and Indian migrants who settled Southeast Asia have carried this value to that region as well.

One benefit of this stress on education has been that, as new technology has been introduced from the West in the post–World War II era, Asian populations have been especially effective in utilizing it in national development. The Japanese began this transformation in the 1860s when they decided to adopt and modify elements of military and industrial patterns in Europe and the United States. The pattern has continued with powerful success in China, South Korea, Taiwan, Singapore, and Hong Kong, just to name some of the most dynamic economies of the current era. Education has been central to the success with which economic development has been fomented in Asia. The technology associated with this development has shown up in transformation of both rural as well as urban landscapes.

"TEACHING WATER"

The Chinese describe irrigation practices as the process of "teaching water" how to do what the farmer needs it to do to make agriculture more productive (Fig. 10.6). In the landscapes of Asia, water has been "taught" to flow in canals, to pond in artificial reservoirs, to warm in slow moving sheets atop terraced rice fields, to flow in and out of different levels of farmland with gravity as the major force of movement, and to be the center of attention in many agricultural ceremonies. This talent of the Asian farmer and engineer to transform the earth so that water can be better controlled, or better "taught," comes from the realization that if labor is abundant and precipitation relatively scarce or unpredictable, useful landscape modifications allowing irrigation can be designed and created by manual labor. The benefit of such hydraulic control has traditionally led to higher

FIGURE 10.6
Terraced fields of irrigated rice in a characteristic landscape of rural Japan. Such a system of fields may have numerous individual owners, necessitating cooperative effort to keep the terraces in repair and the channels open so that water can move slowly by gravity from one field to the next.   *Brian Brake/Photoresearcher, Inc.*

FIGURE 10.7
In the design of the lobby of the Shanghai Hilton Hotel careful attention has been given to artistic motifs that reflect Chinese culture while plants and open space are mixed to assure comfortable surroundings for social as well as business conversations. There are even young musicians who play above the restaurant, creating an environment that seems wonderfully tranquil in comparison with the crowds and activity outside the hotel. Hotels such as these are big draws for tourists from all over the world.   *Kit Salter*

yields, more certainty of crop success, and labor opportunities for marginally unemployed farm populations.

Tourism has also grown enormously in Asia (Fig. 10.7). In Hong Kong, Thailand, South Korea, and Japan, for example, tourist income has increased five- to seven-fold in the past decade. The draw to touring Asia has long been fueled by the classic and traditional landscape treasures of art, architecture, and shrines and gardens. The upswing in these recent years has been stimulated by the construction of world class hotels in major cities all across Asia, and the continuing skill the region has in extending services and welcome to foreign visitors. This inflow of tourist dollars has not only been an important source of investment capital for the countries, but tourist traffic has led to the upkeep and significant refurbishing of traditional landscape features—such as temples, sacred burial sites, and monumental structures—that might have otherwise been given less attention in the current rush to modernize and, in many cases, Westernize.

10.6   Population, Settlement, and Economy

Approximately three fifths of the world's people live in Monsoon Asia (see Table 10.1). They range from small tribal groups with locally distinct cultures to major culture groups, like the Chinese and the Indians. Mongoloid peoples form a majority in China, Japan, Korea, Burma, Thailand, Cambodia, Vietnam, and Laos. But the majority of the people in India, although darker skinned than Europeans, are considered in many classification systems to belong to the Caucasian race; and similar peoples form a majority of the native inhabitants of the Malay Peninsula, the East Indies, and the Philippines.

DISTRIBUTION OF POPULATION

The densest Asian populations are found on river and coastal plains, although surprisingly high densities occur in some hilly or mountainous areas (see world population map in Chapter 2 and specific national population densities in Table 10.1). Higher mountains, steppes, deserts, and some areas of tropical rain forest are very sparsely inhabited.

Most countries have experienced large increases in population during recent centuries, especially since the beginning of the 19th century, and most of the additional people have accumulated in areas that were already the most crowded. Food production levels remain precarious, although such factors as increased grain supplies from new high-yielding varieties, the ability of Thailand and Burma to continue to export rice, and the availability of surplus grain from overseas countries are generally making it possible to maintain adequate levels of nutrition. Through irrigation, farmland expansion, and increasing use of chemical fertilizers and new seed stock, Asian agriculture has grown to be much more productive than it has been historically. How long this can continue in the face of continuing population increases—and, just as importantly, in the face of changing diet patterns in China that are moving away from rice—remains to be seen. While Japan has lowered its rate of population increase to below that of the United States and Canada, and China has considerably reduced its growth rate in the past decade by stringent birth-control measures built around a limit of one child per couple, most Asian countries continue to have high fertility rates characteristic of the "developing" world.

DOMINANCE OF VILLAGE SETTLEMENT

Although an estimated 900 million people in Monsoon Asia live in urban settlements, the main unit of Asian settlement is the village (Fig. 10.8). About two thirds of the region's people are residents of an estimated 1.9 million villages. Highly urbanized Japan, Singapore, Hong Kong, Macao, and Brunei do not fit this pattern, nor, in lesser measure, do Taiwan, South Korea, and Mongolia. But in all the other countries, the typical inhabitant is a villager. The Asian village is essentially a grouping of farm homes, although some villages house other occupational groups, such as miners or fishermen. Clusters of houses bunched tightly together are typical, and cheap and simple structures—often made of local clays and other building materials—are characteristic. Piped water and indoor plumbing

FIGURE 10.8

This winter scene from northern Japan serves to modify the general images that so often show Japan only in scenes of Tokyo traffic and crowding. This village scene from Tsumago in northern Honshu reflects the traditional architecture and street scale that is still representative for most of rural and small town Japan, particularly in areas distant from the environs of Tokyo.   *Art Wolfe/Tony Stone Images*

continue to be somewhat exceptional in village homes, although the availability of electricity has been steadily expanding. Details of village life vary according to culture and place; for instance, in Hindu villages there are segregated quarters for different castes, even though the caste system has been officially outlawed.

As in all corners of the agricultural world, the original siting of villages was closely adapted to natural conditions. For example, in flood plains the villages are slightly elevated—located on natural levees (raised river banks built up by deposition of sediments during floods), dikes, or raised mounds. Early villages in Indonesia were often built in defensible mountain sites, although the Dutch colonial administration gradually required the building of villages along main roads and trails in the lowlands to make it easier to exercise control, collect taxes, and draft soldiers or laborers for road work or other projects.

RURAL-TO-URBAN POPULATION SHIFT

Even though the village is central to Asian demography, this era of unprecedentedly rapid cultural diffusion and change since World War II has rapidly accelerated migration. Rural peoples have been leaving the farming life and going to the cities at a pace never before experienced. In the late 1940s, approximately 85 percent of China's population lived in the countryside. Although there were massive cities such as Shanghai, Beijing, and Guangzhou (Canton), the great bulk of the Chinese population was still deeply involved in rural activity and living in relatively small villages.

FIGURE 10.9
This street scene from Kathmandu, Nepal, shows the market elements that are attractive to local Nepalese and tourists alike. Even with the steady push toward greater urbanism by rural people in Monsoon Asia, there continues to be a great vitality evident in the market towns and large villages of this world. *Demetrio Carrasco/Tony Stone Images*

However, in the 1990s, China has undergone a demographic shift that has taken it to nearly 30 percent urban, in large part due to economic growth in urban (especially coastal) China. India also defined nearly 30 percent of its population as urban in 1996. This demographic shift has meant enormous problems for city management as the rural migrants seek space, jobs, services, food, and goods (Fig. 10.9). The magnitude of these demographic shifts in China and India is so great that such moves can mean the transfer of hundreds of millions of people within a decade in those two Asian countries alone. Those left behind in the countryside, consequently, have been prompted to introduce more agricultural mechanization in order to fill the labor gaps left by these migrating youth. At the same time, cities have been energetically absorbing this labor force into everything from small sweat-shop activities to major industrial enterprises.

The lure of the city—so powerfully expressed through the nearly universal availability of television and movies—has never played its siren song so successfully before. The whole demography of Monsoon Asia is shifting from a traditional rural past to an evolving urban present.

MEANS OF LIVELIHOOD

Japan was the first Asian country to develop modern types of manufacturing and modern cities on a large scale (Fig. 10.10). For more than a century it has been a major industrial power. But China and India also have important and dynamically expanding industrial bases. These two nations are the largest in the world in population, and they are much better supplied with mineral resources than Japan. They also have cheaper labor.

But Japan's labor force is more skilled, and Japan has shifted increasingly to types of industry requiring skilled labor and advanced technology. Not only does it lead Asia in high value added manufacturing, but it is far and away the leader in the provision of services in finance and other activities essential to steady economic growth. The remaining Asian countries present a very mixed picture in their development of manufacturing and services. Hong Kong and Singapore are outstanding in proportion to their size. South Korea and Taiwan also have a relatively high development. At the other end of the scale are many countries with very little modern industry and few services.

In Monsoon Asia as a whole, agriculture remains the major source of livelihood. In most countries—Japan being a conspicuous exception—the majority of the people are farmers. Two major types of agriculture—plantation agriculture and

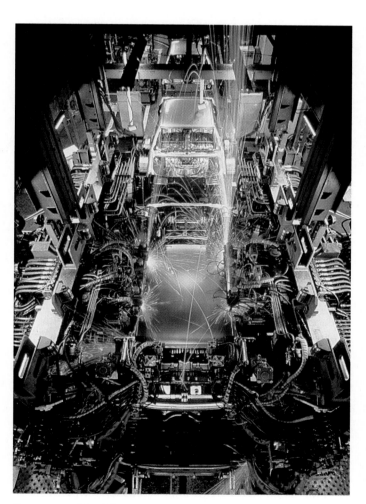

FIGURE 10.10
This high tech manufacturing conveyor belt in a Honda plant in Japan represents the sophistication that Japan has brought to its manufacturing. The spot welding robot is called "General Welder" by the employees who keep the line running smoothly. *Tadanori Saito/PPS/Photo Researchers*

shifting cultivation—are discussed in the chapter on Southeast Asia, the subregion in which these forms of farming are the most prominent. In the steppes and deserts, nomadic or seminomadic herding and oasis farming are practiced. Over large sections of Asia most farmers make a living by cultivation of small rain-fed or irrigated plots worked by family labor. In communist-held areas—particularly in China—the past four decades have seen a variety of experiments in agricultural organization. The current pattern has largely reverted to family farming operations with markets controlled—at least in part—by local and provincial governments, or, increasingly, the free market system has expanded. Although there has been some farm mechanization in Asia, the general practice (outside Japan) continues to be characterized by the steady input of large amounts of very arduous hand labor. Production is often of a semisubsistence character. This type of agriculture, which is often referred to as intensive subsistence agriculture, is built around the growing of cereals. Where natural conditions are not suitable for irrigated rice, such grains as wheat, barley, soybeans, millet, sorghums, or corn (maize) are raised. However, irrigated rice yields the largest amount of food per unit of area where conditions are favorable for its growth, and this crop is the agricultural mainstay in the areas inhabited by a large majority of Asian farmers. Because of the role of rice in the traditional Asian diet, it is the crop of choice, other things being equal.

Making the "Land Say Rice"

For most people in the West, bread is the staff of life. But for several billions in Asia, rice is the basic source of carbohydrate intake. (Protein is provided mostly by peas, beans, or fish instead of the meat, milk, and cheese of the West.) Irrigated rice ("wet rice") is grown in the tropics or subtropics of all the world regions discussed in this text, but the great majority of the world's supply comes from the monsoonal lands of eastern and southern Asia (see photograph, page 269). Generally, it is still produced primarily by hand labor or animal power, although mechanization plays an increasingly large role through small tractors and hand cultivators. Adequate water from streams, wells, springs, or ponds is a crucial factor, as the grain is produced in fields that are submerged under several inches of slowly moving fresh water. It is customary to start the seedlings in small seed beds and then transplant them to the flooded paddy fields (see Fig. 10.6). Wet rice is grown on various types of soil and under a variety of environmental conditions. One condition, however, must always be present: an impervious subsoil that retards the downward seepage of water. This capacity of the soil to hold water is more significant than the natural soil fertility because nutrients for plant growth in irrigated rice are introduced through the steadily moving water that surrounds the rice plant until the last weeks before harvest. Ideal conditions are often found on alluvial flood plains where fertile silt has been deposited on an impermeable layer of clay.

In many parts of Asia, relatively level farming plots have been created by terracing steep hillsides. This traditional act of landscape transformation is part of a more than two-thousand-year-old tradition of "making the land say rice." Such scenes of terraces marching up mountain flanks, water glistening in the spring and summer sunshine, serve as major landscape signatures of Asian farming. The creation of such terraces, however, demands not only abundant and cheap labor but also basic engineering skill in leveling land, constructing and maintaining stone or earthen terrace walls, and controlling water above, around, and throughout such landscapes.

The Green Revolution

As a step toward eliminating hunger by using science to increase rice yields, the International Rice Research Institute (IRRI) was founded in the Philippines in 1962. The institute is one facet of a worldwide research effort involving use of new grains in association with a modification of traditional farming practices. Governments also have invested increased capital in building better roads and bridges to enable farmers to get their surpluses to markets, and to bring new farming tools and materials to the countryside more efficiently (see Fig. 2.12 in Chapter 2).

Notable success has been achieved in breeding the new high-yielding varieties of seed stock, and there has been a large upsurge of production in certain areas where the new strains have been widely introduced. Peasant farmers have changed their **crop calendars** to accommodate an increasing dependence on chemical fertilizers as well. The entire effort has come to be known as the **Green Revolution**. Whether the term is too optimistic remains to be seen, as the "revolution" is still in an early stage.

For Asian farmers to capitalize fully on the Green Revolution, they must overcome many obstacles. For example, success requires inputs of capital that are often beyond the present means or inclination of peasant farmers, landlords, and governments to provide. Such expenditures are needed for water-supply facilities (for example, the tubewells that have burgeoned by hundreds of thousands in the Indo-Gangetic Plain of the northern Indian subcontinent), chemical fertilizers, and chemicals to control weeds, pests, and diseases. And as agriculture becomes more mechanized, considerable increases in the costs of machinery, fertilizers, and fuel have to be borne by farmers who have, in many cases, had little experience with the cash economy needed to achieve a positive return on their investment. Governments have to improve transportation so that the heavy inputs of fertilizer required by the new seed varieties can be delivered in a timely fashion. Not only must there be these associated infrastructural changes to support this "revolution," but the crop calendar of the farmer becomes much less forgiving as many of the newest seed grains demand more precise water, fertilizer, and cultivation requirements than traditional grains. Grain-storage facilities, now subject to plundering by rats, have to be improved.

Overcoming the financial obstacles is rendered more difficult by the widespread system of share tenancy. If a farmer is a share tenant, he generally has no security of tenure on the land, and thus he cannot be sure that money he invests in the Green Revolution will actually benefit him in the future. He may not wish to assume any additional risk, even if credit on reasonable terms is available. Even if he remains on his holding, the landlord may take up to half of the increased crop while bearing little or none of the additional expense. Landlords, in their turn, may be content to collect their customary rents without expending the additional capital necessary in this new mode of farming, or they may endeavor to turn tenants off the land in order to create larger spatial units that they themselves can farm more profitably with machinery and hired labor. In fact, landowners with large holdings often become the chief beneficiaries of the new technology, with many smaller farmers becoming a class of landless workers hired for low wages on a seasonal basis.

Other problems associated with the Green Revolution include damage to ecosystems by large infusions of agricultural chemicals, and the economic dislocations that result when rice-importing countries become more self-sufficient, thus causing hardships for rice exporters. The Asian experiment with the Green Revolution provides yet additional evidence of the widespread ramifications that are set in motion when any major technological shift—either agricultural or industrial—is introduced into societies. Indian or Filipino farmers, for example, who had developed patterns of reasonable self-sufficiency through traditional agricultural practices, can find their lives turned upside down as they try to adopt some of the standards of the Green Revolution.

The scale of culture change that must accompany the pattern shifts in a farmer's involvement in the Green Revolution may seem acceptable to an urbanite—whether from Asia or the world beyond—but to the traditional farmer, such changes are not only financially costly but serve to rupture long-held traditions. As you explore the various nations and landscapes of Monsoon Asia, keep in mind the impact the global flow of information, fads, and images has on human ambitions for an improved livelihood and lifestyle. The ways in which peoples of Monsoon Asia have responded to these pulsating influences of change and growth are significant to your understanding of world regional geography.

11

COMPLEX AND POPULOUS SOUTH ASIA

A triangular peninsula that thrusts southward a thousand miles (*c.* 1600 km) from the main mass of Asia splits the northern Indian Ocean into the Bay of Bengal and the Arabian Sea. The peninsula is bordered in the north by the alluvial plain of the Indus and Ganges (Ganga) rivers, north of which rise the highest mountains on Earth. The entire unit—peninsula, plain, and fringing mountains—is often called the Indian subcontinent. It contains the five countries of India, Bangladesh, Pakistan, Nepal, and Bhutan (Fig. 11.1), in an area a little more than one half the size of the 48 conterminous United States (Fig. 11.2). India outranks all of the other countries in both area and population. Off the southern tip of India, across the narrow Palk Strait, lies the island nation of Sri Lanka, which shares many physical and cultural traits with the subcontinent.

Mountains enclose the subcontinent on its landward borders. Its northern boundary lies in the Himalaya Mountains and the Karakoram Range. Nepal and Bhutan, on the southern flank of the Himalaya (Fig. 11.3), are small, rugged, and remote. They are buffers between India and China, which have engaged in sometimes violent border disputes since the early 1950s.

From each end of this massive wall, lower ranges trend southward to the sea. Until 1947, the entire area, except the small Himalayan states and Sri Lanka, was referred to as "India." It was for well over a century the most important unit in the British colonial empire—the "jewel in the crown." In 1947 it gained freedom, but in the process became divided along religious lines into two countries, the secular but predominantly Hindu nation of India and the Muslim nation of Pakistan. Pakistan had two parts, West Pakistan and East Pakistan, separated by Indian territory. In 1971, an Indian-supported revolt in East Pakistan led to the birth there of the new independent country of Bangladesh.

India is one of the most important nations that have gained independence since World War II. It is the world's largest democracy, a demographic giant in which 950 million people (as of mid-1996), or nearly one sixth of the human population, lives. Its area of 1.3 million square miles (3.3 million sq km) is exceeded by that of only a few other countries. Victory in the 1971 war with Pakistan confirmed India as the leading power in south Asia. India's detonation of an atomic bomb in 1974, and Pakistan's subsequent development of nuclear weapons, added a frightening new dimension to the conflict between these rivals. Pakistan and Bangladesh are much smaller than India in area and population, but both are among the world's ten largest countries in population. Altogether, about 1.2 billion people, more than one out of every five on Earth, lived in the subcontinent and Sri Lanka in 1996.

◄ Young women at a folk festival in the Indian city of Chandigarh. The cultural fabric of South Asia is extremely rich. *Frederica Georgia/Photo Researchers*

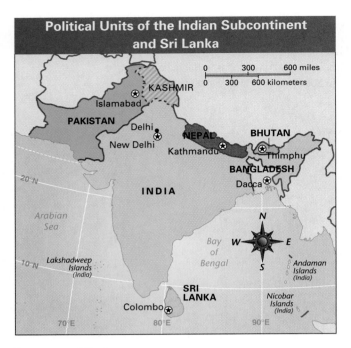

FIGURE 11.1

Political units of the Indian subcontinent and Sri Lanka. The future status of Kashmir, disputed between India and Pakistan, with some parts occupied by China, remains unresolved.

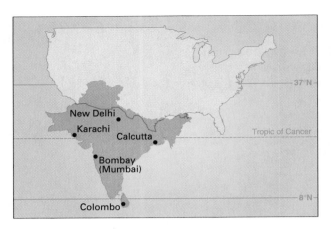

FIGURE 11.2

The Indian subcontinent and Sri Lanka compared in latitude and area with the conterminous United States.

11.1 The Cultural Foundation

The Indian subcontinent is one of the world's culture hearths. Among the animals first domesticated in the region were zebu cattle, now an important livestock component in many world regions. Before 3000 B.C., some of the world's earliest cities (notably Mohenjo Daro) developed on the banks of the Indus River in what is now Pakistan. Their inhabitants were probably the ancestors of the Dravidians, a dark-skinned people whose stronghold is now southern India.

Between 2000 and 1000 B.C. came the Aryans (or Indo-Aryans), tribal cultivators and pastoral nomads from the inner Asian steppes. They were the source of many ethnic and cultural attributes that are dominant in the subcontinent today, notably the Caucasoid racial majority, the numerically dominant Aryan languages such as Hindi—most of which were derived from the Sanskrit used by the conquerors—and Hinduism, which developed from the beliefs and customs of the Aryans. In the eighth century, Arabs brought the Islamic religion into India from the west, to what is now Pakistan. Waves of Muslim conquerors who came through mountain passes from Afghanistan and central Asia brought a wider spread of Islam in later centuries. The last great invasion, by a Mongol-Turkish dynasty called the Moguls (Mughals), began in the early 16th century, soon after the first Europeans came to India by sea in the Age of Discovery. Before it succumbed to rebellious local Hindus and

Sikhs and to the British Empire, the Mogul empire ruled nearly all of the subcontinent from its capital at Delhi. Some of the region's most spectacular artistic and scientific achievements date to the peak of Mogul power in the 16th and 17th centuries, including the Red Fort in Delhi and the tomb of Mogul emperor Shah Jahan and empress Mumtaz Mahal (the world-famous Taj Mahal) in Agra (Fig. 11.4). Muslims now make up about 12 percent of India's population—seemingly a small number except that it equals more than 100 million people, ranking India fourth among the world's countries with the largest Muslim populations, behind Indonesia, Pakistan, and Bangladesh.

British colonialism left an indelible mark on the subcontinent's cultural landscape. To serve the needs of their empire, the British developed the great port cities of Karachi, Bombay, Madras, and Calcutta, and India's current inland capital of New Delhi. The existence of English as the *de facto* national languages of India and Pakistan, and the presence in Great Britain today of large numbers of immigrants from the subcontinent, bear witness to the close relations between these countries.

This simple sketch of Dravidian, Aryan, Arab, and British influences can only begin to shed light on the extreme social diversity of the Indian subcontinent. There is an enormous range of ethnic groups, social hierarchies, languages, and religions among regions of the subcontinent, and even within single settlements. The subcontinent is the most culturally complex area of its size on Earth.

11.2 Regions and Resources

Physical conditions are extremely important to the overwhelmingly rural, resource-based economies of the Indian subcontinent and Sri Lanka. With relatively little foreign aid reaching

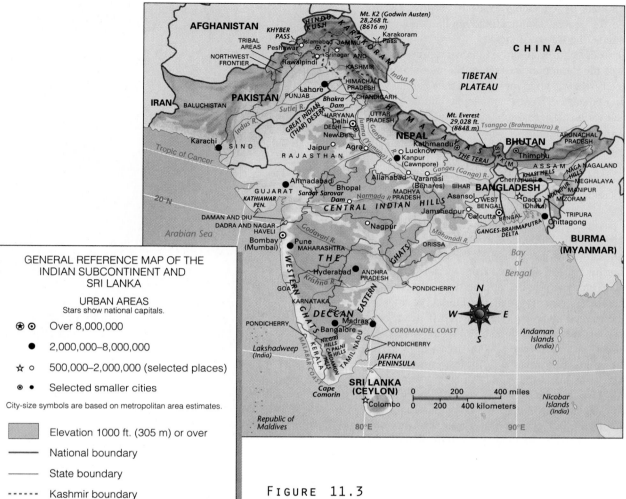

GENERAL REFERENCE MAP OF THE INDIAN SUBCONTINENT AND SRI LANKA

URBAN AREAS
Stars show national capitals.

⊛ ⊙   Over 8,000,000

●   2,000,000–8,000,000

☆ ○   500,000–2,000,000 (selected places)

⊛ ●   Selected smaller cities

City-size symbols are based on metropolitan area estimates.

▨ Elevation 1000 ft. (305 m) or over

——— National boundary

——— State boundary

------ Kashmir boundary

----- Cease-fire line

FIGURE 11.3

General reference map of the Indian subcontinent and Sri Lanka. Note how mountainous areas frame the northwestern, northern and northeastern portions of this region, while ocean waters set the boundaries in the south.

FIGURE 11.4

For Indians and foreigners alike, the Taj Mahal on the bank of the Jumna river symbolizes the cultural achievements and natural beauty of the Indian subcontinent.   *Glen Allison/Tony Stone Images*

the region, people's survival depends quite directly on local agricultural resources, and future industrialization will have to draw largely on available natural resources. Fortunately, the subcontinent has some generous endowments of natural resources. It also has colossal natural hazards, however, including floods, droughts, landslides, and earthquakes.

The subcontinent can be roughly subdivided into three natural areas: the outer mountain wall, the northern plain, and peninsular India.

THE OUTER MOUNTAIN WALL

A series of parallel mountain ranges forms an inverted "U" around the north of the subcontinent. In the west, the mountains extend northeastward from the Arabian Sea to Kashmir (see Fig. 11.3). From here, the longest section of the wall trends southeast and then eastward to the subcontinent's northeast corner in the Indian state of Arunachal Pradesh. Finally, a short eastern leg of the "U" reaches southward along the border with Burma (Myanmar) to the Bay of Bengal.

Pakistan's borders with Iran and Afghanistan traverse the western section of the mountain wall. Here the mountains are rugged almost everywhere, and in most places they extend well into the bordering countries. Desert and steppe climates reach upward to high elevations. These mountains provide only a few crossing places, the most famous of which is the Khyber Pass through the Hindu Kush mountains on the border between Pakistan and Afghanistan. Such passes played a fateful role for thousands of years as gateways into India for conquerors from the outer world, including the Aryans and the Arabs.

The towering northern segment of the mountain wall has been much less passable. Across the north of the subcontinent, the Himalaya extend about 1500 miles (c. 2400 km) from Kashmir to the northeastern corner of India. Paralleling the Himalaya on the northwest, and separated from them by the deep gorge of the upper Indus River, lies another exceedingly high range called the Karakoram. Beyond the Himalaya and the Karakoram is the rugged and very high Tibetan Plateau, controlled by China. The two ranges contain 92 of the about 100 world peaks whose elevations are 24,000 feet (7315 m) or higher, culminating in the Earth's two loftiest summits: Mt. Everest (29,028 ft/8848 m) in the Himalaya on the border between Nepal and China, and Mt. K2, or Godwin Austen (28,268 ft/8616 m), in the Karakoram within the Pakistani-controlled part of Kashmir (Fig. 11.5). Geographical surveys recently quieted a brewing debate over whether K2 might actually be higher than Everest: The Himalayan peak is the highest. Both great ranges, however, are young, dynamic, and still growing, so the two peaks may still be in competition. Although most passes through these ranges are higher than any peak in the Alps of Europe, a trickle of trade and cultural exchange has crossed these mountains for millennia. However, neither commerce nor military forces have ever crossed this barrier on any large scale.

The kingdoms of Nepal (capital Kathmandu; population 423,000) and Bhutan (capital, Thimpu; population 12,000) are situated on this mountain wall. Their landlocked locations and difficult topographies certainly contribute to their slow economic and social development. Both countries have productive agricultural regions in the "middle ranges," the bands of foothills between the vast lowland of the subcontinent and the towering Himalayan peaks. In Nepal the capacity of this productive area has been saturated, and poor peasants "marginalized" by rapid population growth have sought new ground on steeper slopes, in nearby Bhutan, and in Nepal's lowland region of the Terai. Once a malaria-plagued region of dense forest and savanna, the Terai helps serve as a "safety valve" for Nepal's overpopulation. Control of malaria began in the 1950s and people began migrating in to clear and cultivate the land. Nepal's Himalayan region is a popular destination for tourists. An estimated 120,000 trekkers visit annually, bound especially for the Everest region in the east and the Annapurna area of the west.

FIGURE 11.5
The jagged summit of Mount K2 in the Karakoram Range, a part of the Indian subcontinent's outer wall of high mountains. The view is from a climbers' camp on a glacial moraine at an elevation of nearly 15,000 feet (c. 4600 m).   *Galen Rowell/Mountain Light Photography*

They bring in much-needed foreign currency to this poor country, but also create problems by littering and inducing social change in traditional cultures. Bhutan has been far more careful in opening its doors to tourism, placing restrictions on the numbers permitted in and on the routes they may follow.

The mountains along and near the subcontinent's border with Burma (Myanmar) are much lower than the Himalayan ranges but are also nearly impenetrable. They are rugged, cloaked with dense vegetation, and soaked with rain. Cherrapunji in the Khasi Hills of India has an average annual rainfall of 432 inches (1097 cm), of which more than nine tenths falls in the summer half-year. It ranks with a station on Kauai in the Hawaiian Islands as one of the two spots with the greatest annual rainfall ever recorded on Earth.

THE NORTHERN PLAIN

Just inside the outer mountain wall lies the subcontinent's northern plain. This large alluvial expanse contains the core areas of the three major countries. In the west, the plain is split between Pakistan and India. The Indus River and its tributaries and distributaries cross the portion lying in Pakistan. The climate here is desert and steppe; the dry region straddling the two countries is known as the Thar Desert or Great Indian Desert. Irrigation water from the rivers is essential to the many millions of people living here.

Definitions and Insights

"Upstream" and "Downstream" Countries

The region of the subcontinent called the Punjab, meaning "Five Waters" (five Indus River tributaries), is split between two countries, with its rivers flowing through Indian territory before entering Pakistan (see Fig. 11.3). In geographical and geopolitical terms, India is an "upstream country" and Pakistan a "downstream country" where Punjab waters are concerned. There are no internationally accepted laws regarding water-sharing between upstream and downstream states, and the upstream country often exercises its geographical advantage to siphon off what the downstream country views as too much water. Riparian (river-owning) countries often come close to, or actually go to, war over the problem. In many cases, formal treaties have averted what could have become major international conflicts (see also Chapter 9, page 226).

After 1947, relations between India and newly-independent Pakistan eroded in large part because of the disputed waters of the Indus and the five Punjab rivers. For many years the parties could not agree on how much of the water India could divert and how much should remain for Pakistan. The countries finally reached a satisfactory resolution in the 1960 Indus Waters Treaty. The agreement allocates the water of the three eastern rivers of the Punjab to India, which in return allows unrestricted and undiminished flow of the two others, and the Indus River itself, into Pakistan.

East of the Punjab lies the part of the northern plain traversed by the Ganges (Ganga) River and its tributaries, especially the Jumna (Yamuna). The western portion of this region is known as the Doab, "the land between the two rivers." The northern plain is India's core region, containing more than two fifths of the country's population as well as its capital, numerous other major cities, and many places of great significance in the religions of Hinduism, Buddhism, and Jainism.

This region of India is relatively well-watered, especially toward the east, where an agriculture based primarily on irrigated rice supports population densities that are unusually high even for the subcontinent. At the narrow western end of the Ganges-Jumna section of the plain, the old fortified capital of Delhi arose as a bastion against invaders from beyond the western mountains (Fig. 11.6). The present (and adjoining) capital of New Delhi was founded after 1912 as the capital of British-controlled India. Today, the metropolitan population of Delhi-New Delhi is an estimated 8.4 million (7.2 million, city proper),

making it India's third largest metropolis after Bombay and Calcutta.

The northeasternmost part of the northern plain, crossed by the Brahmaputra River, lies in the Indian state of Assam. Assam has relatively good agricultural conditions and by Indian standards is lightly-populated, with only around 750 people per square mile (290 per sq km). From the viewpoint of people living in crowded Bangladesh, some parts of India such as Assam are lands of opportunity. In the 1980s, native inhabitants of Assam defending their traditional lands slaughtered thousands of illegal immigrants from Bangladesh. The total of such immigrants in Assam was already estimated at four million, and immigration continues despite ongoing violence.

The southeastern part of the northern plain lies in the region of Bengal, which is essentially the huge delta formed by

FIGURE 11.6
The Mogul Emperor Shah Jahan established Delhi as India's capital early in the seventeenth century. His monumental Red Fort still dominates this old city. Adjacent New Delhi is India's capital today. *Frederica Georgia/ Photo Researchers*

FIGURE 11.7

The port city of Calcutta in West Bengal, like the other great cities of India, is home to millions of poor people who have migrated from rural areas in search of a better life.   *Paolo Koch/Photo Researchers*

the combined flow of the Ganges and Brahmaputra. Bengal, which is divided between India and Bangladesh, is extremely densely populated. Another magnet for migrants from Bangladesh and from rural India is the English-founded port of Calcutta in the Indian state of West Bengal (Fig. 11.7). A city of 11.02 million in its metropolitan area (4.4 million, city proper), Calcutta has a worldwide reputation for teeming slums and dire poverty.

The Bangladesh portion of the delta could yield many million more emigrants as it ranks, along with the Indonesian island of Java, as one of the two most crowded agricultural areas on Earth. Its problems increase sharply during the many periods when floods strike (Fig. 11.8). Tropical cyclones from the Bay of Bengal cause some of the flooding. Deforestation upstream on the steep slopes of the Himalaya has increased water runoff and sediment load in the Ganges, also contributing to Bangladesh's flood problems. Bangladesh must also keep a wary eye on the potential for the sea level to rise if the world's temperatures increase according to many "greenhouse effect" models.

PENINSULAR INDIA

The southern peninsular portion of the subcontinent, entirely within India, consists mainly of a large volcanic plateau called the Deccan. It is relatively low, generally less than 2000 feet (c. 600 m) above sea level. Its rivers run in valleys cut well below the plateau surface. This topographic problem, together with the rivers' seasonal flow regime, makes it difficult to use their waters for irrigation of the upland surface without large inputs of capital and technology. Wells and the rain-catchment ponds called tanks (Fig. 11.9) are used, but water is in shorter supply

and population density is much lower than in the better-watered parts of the northern plain.

Hills and mountains outline the edges of the Deccan. In the north, belts of hills (the "Central Indian Hills" in Fig. 11.3) separate the plateau from the northern plain. On both east and west the plateau is edged by ranges of low mountains called Ghats ("steps"), which overlook low-lying alluvial plains along the coasts of the Bay of Bengal and the Arabian Sea. The low and discontinuous eastern mountains are known as the Eastern Ghats. The western mountains, which are higher and which present a relatively continuous west-facing escarpment, are the Western Ghats (Figure 11.10). The two ranges merge near the southern tip of the peninsula in clusters known locally as the Nilgiri, Palni, and Cardamon "hills." They are really rather high mountains, with some peaks rising above 8000 feet (c. 2400 m). The colonizing British founded several "hill stations" in them for recreation and administration during summer months, when the climate on the subcontinent's lowlands is stifling. They are now popular tourist destinations for Indians during that season.

The coastal plains between the Ghats and the sea are occupied by ribbons of extremely dense population, with the density reaching a peak in the far south along the Malabar Coast of western India and with a slightly lower density on the corresponding Coromandel coast in the east. The largest metropolis of the far south, Madras (recently officially renamed Chennois; population: 3.84 million, city proper; 5.42 million, metropolitan area) is a seaport on the Coromandel coast. The main me-

FIGURE 11.8

Bangladesh is both a low-lying coastal land and a downstream country of several large watersheds. As a result of storms blowing in from the Bay of Bengal and of accelerated runoff from rainfall and snowmelt on deforested mountains upstream, it is subjected to devastating floods. This is an aerial view following the cyclone (hurricane) which struck Bangladesh on April 29, 1991, leaving more than 120,000 people dead and inundating most of the country's precious farmlands.   *D. Aubert/SYGMA*

FIGURE 11.9
Tanks (artificial ponds in which to store water) are devices that help communities in the Indian subcontinent survive the dry season. They are especially numerous in the Deccan plateau of peninsular India. This one is in northwestern India. Some water buffalo, used both as draft animals and for milking, can be seen bathing in the pond. The animals must retreat to water in the dry season, as their skins are vulnerable to dry heat.

tropolises of the interior Deccan are Hyderabad (population: 3.06 million, city proper; 4.34 million, metropolitan area) and Bangalore (population: 2.66 million, city proper; 4.13 million, metropolitan area).

SRI LANKA

Sri Lanka (population: 18.4 million), formerly called Ceylon, is a tropical island country. It is a land apart; despite its proximity to the Indian subcontinent, some geographers place it in the Southeast Asian realm. Its main physical and cultural character-

FIGURE 11.10
Southwest India's mountains, known as the Western Ghats, have been widely cleared of native forest by logging and agricultural expansion. Indian elephants, which are indigenous to this part of the country, are still prized for their labors in the timber industry. This bull elephant on the Onengara Plantation in Kerala State is pulling a 3000-lb. log.   *Ted Wood*

istics are linked to the subcontinent, however. Its name conveys its physical beauty: Sri Lanka means "Resplendent Isle" in the native Sinhalese language, and medieval Arabs knew it as "Serendip," the island of serendipity (Fig. 11.11). The island consists of a coastal plain surrounding a knot of mountains and hill lands (see Fig. 11.3). Most people live either in the wetter southwestern portion of the plain, in the south-central hilly areas, and in the drier Jaffna Peninsula of the north. Coconuts and rice are the major crops of the low southwestern coast and Jaffna Peninsula, while tea and rubber plantations dominate the economy of the uplands. Colombo (population: 612,000, city proper; 2.05 million, metropolitan area), in the southwest, is the capital, chief port, and only large city.

Sri Lanka's economy is highly commercialized. Three cash crops—rubber, coconuts, and the world-famous Ceylon tea—occupy the greater part of the agricultural land and supply about 30 percent of export earnings. Clothing manufacture in the Colombo area has grown into an export industry. Expansion of rice production has been so successful that near self-sufficiency has replaced the need for major rice imports. This development has been helped by declining rates of population growth. However, with its limited land area, Sri Lanka has a high population density of 714 per square mile (276 per sq km), just below that of India.

External influences on this attractive island have diversified its population and sown seeds of unrest (see page 293). Centuries of recurrent invasion from India were followed by Portuguese domination in the early 16th century, Dutch in the 17th, and British from 1795 until the country was granted independence in 1948 as a member of the British Commonwealth. This eventful history, plus longstanding commercial importance of the sea route around southern Asia, has given Sri Lanka a polyglot population which includes Burghers (descendants of Portuguese and Dutch settlers) and Arabs.

The two major ethnic groups, distinguished from each other by language and religion, are the predominantly Buddhist

FIGURE 11.11
Verdant mountains and well-kept plantations are typical of the beautiful island country of Sri Lanka. The close-cropped portion of the hill in the center is a tea field. Note the electric lines that have reached the village here. *James Strachan/Tony Stone Images*

Sinhalese, making up about 82 percent of the population, and the Hindu Tamils, constituting about 9 percent. The light-skinned Sinhalese are an Indo-Aryan people who settled in Sri Lanka about 2000 years ago. The dark-skinned Tamils, whose main area of settlement is the Jaffna Peninsula and adjoining areas in the north, are descendants of early invaders and more recent imported laborers from southern India.

11.3 Climate and Water Supply

Climatic conditions in the subcontinent and Sri Lanka vary between remarkable extremes, from Himalayan ice and snow fields to the year-round tropical heat of peninsular India, and from some of the world's driest climates to some of the wettest.

Climatic and biotic types in the subcontinent include undifferentiated highland climates in the northern mountains; desert and steppe in Pakistan and adjacent India; humid subtropical climate in the northern plains; tropical savanna in the peninsula, except for a patch of tropical steppe in the rain shadow of the Western Ghats; and rain forest on the seaward slopes of the Western Ghats and the coastal plain at their base, and in parts of the Ganges-Brahmaputra Delta and the eastern mountains near Burma (see world climate map, Fig. 2.6, page 31).

Heat is nearly constant except in the mountains. The tropical peninsula is hot all year. The subtropical north has stifling heat before the "break" of the wet monsoon, and warm conditions even in the winter. Some of the warmest temperatures on the globe occur in the plains of Pakistan and northwestern India. Delhi, for instance, averages 94°F (34°C) in both May and June. The highest temperatures over most of the subcontinent occur in May and June, just before monsoonal rainfall brings relief.

Another near-constant is the extreme seasonality of rainfall, offering a stark contrast between a short, wet summer and arid or semiarid conditions the rest of the year. The "humid" parts of the subcontinent are actually rather dry for most of the year, while floods are often a threat or a reality during the short rainy season, which corresponds with a season of snow melt along upper river courses in the Himalaya. These striking conditions are caused by seasonally-reversing winds known as monsoons—a wet monsoon in the summer and a generally dry monsoon in the winter.

The **southwest** or **wet monsoon** is at its height from June to September, and most parts of the subcontinent receive the bulk of their annual rainfall during those months. There are two main arms of this monsoon. One arm, approaching from the west off the Arabian Sea, strikes the Western Ghats and produces heavy rainfall on these mountains and the coastal plain. The amount of rain diminishes sharply in the interior Deccan rain-shadow region to the east of the mountains. Here, the annual precipitation over a large area is barely sufficient for dry farming and in some years is so low that serious crop failures occur.

The second major arm of the wet monsoon, approaching from the Bay of Bengal, brings moderate amounts of rain to the eastern coastal areas of the peninsula and heavy precipitation to the Ganges-Brahmaputra Delta region and northeastern India. Wet monsoon winds pass up the Ganges Valley to drop moisture that diminishes in quantity from east to west. Both arms of the monsoon bring some rain to Pakistan, but the total is so small that semiarid or desert conditions prevail in most areas.

The **dry monsoon** of the winter half-year is also called the **northeast monsoon**, because it often blows from the northeast over the peninsula. It also frequently blows from the northwest over much of the northern plain. Blowing mainly from land to sea, this monsoon brings dry weather to most parts of the subcontinent, with occasional light rains in areas outside the tropics. An exception occurs in the far south of the peninsula. Here,

Landscape in Literature

THE MONSOON

The torrential monsoonal rains are a blessing for the subcontinent (Fig. 11.A). Without them, agriculture would be impossible and the summer heat unbearable. But when they come too early or too late, or drop too little or too much water, suffering can result. In her novel *Nectar in a Sieve*,[1] Kamala Markandaya depicted the monsoon season in rural India. It ended when the rain stopped, and when her hungry family fished and drained their terraced fields in preparation for rice planting.

Nature is like a wild animal that you have trained to work for you. So long as you are vigilant and walk warily with thought and care, so long will it give you its aid; but look away for an instant, be heedless or forgetful, and it has you by the throat.

That year the monsoon broke early with an evil intensity such as none could remember before. It rained so hard, so long and so incessantly that the thought of a period of no rain provoked a mild wonder. It was as if nothing had ever been but rain, and the water pitilessly found every hole in the thatched roof to come in, dripping onto the already damp floor. If we had not built on high ground the very walls would have melted in that moisture. I brought out as many pots and pans as I had and we laid them about to catch the drips, but soon there were more leaks than we had vessels. Fortunately, I had laid in a stock of firewood, and the few sticks that remained served at least to cook our rice, and while the fire burnt, hissing at the water in the wood, we huddled round trying to get dry. At first the children were cheerful enough—they had not known such things before, and the lakes and rivulets that formed outside gave them endless delight; but Nathan and I watched with heavy hearts while the waters rose and rose and the tender green of the paddy field sank under and was lost.

The paddy was completely destroyed; there would be no rice until the next harvesting. Meanwhile, we lived on what remained of our salted fish, roots and leaves, the fruit of the prickly pear, and on the plantains from our tree. At last the time came for the rice terraces to be drained and got ready for the next sowing. Nathan told me of it with cheer in his voice and I told the children, pleasurably, for the fields were full of fish that would feed us for many a day. Then we waited, spirits lifting, eyes sparkling, bellies painful with anticipation.

At last the day. Nathan went to break the dams and I with him and with me our children, sunken-eyed, noisy as they had not been for many days at the thought of the feast, carrying nets and baskets. First one hole, then another, no bigger than a finger's width, until the water eroded the sides and the outlets grew large enough for two fists to go through. Against them we held our nets, feet firm and braced in the mud while the water rushed away, and the fish came tumbling into them. When the water was all gone, there they were caught in the meshes and among the paddy, shoals of them leaping madly, wet and silver and good to look upon. We gathered them with flying fingers and greedy hearts and bore them away in triumph, with a glow at least as bright as the sun on those shining scales. Then we came and gathered up what remained of the paddy and took it away to thresh and winnow.

Late that night we were still at work, cleaning the fish, hulling the rice, separating the grain from the husk. When we had done, the rice yield was meagre—no more than two measures—all that was left of the year's harvest and the year's labor.

We ate, finding it difficult to believe we did so. The good food lay rich, if uneasy, in our starved bellies. Already the children were looking better, and at the sight of their faces, still pinched but content, a great weight lifted from me. Today we would eat and tomorrow, and for many weeks while the grain lasted. Then there was the fish, cleaned, dried and salted away, and before that was gone we should earn some more money; I would plant more vegetables . . . such dreams, delightful, orderly, satisfying, but of the stuff of dreams, wraithlike. And sleep, such sleep . . . deep and sweet and sound as I had not known for many nights; it claimed me even as I sat amid the rice husks and fish scales and drying salt.

FIGURE 11.A

The torrential monsoon rains are a regular and welcome feature of land and life in the Indian subcontinent. *Paolo Koch/Photo Researchers*

[1] Kamala Markandaya, *Nectar in a Sieve* (New York: John Day, 1954): 57–65.

the heaviest rainfall of the year falls along the eastern coast and in adjacent uplands during the four-month period from October through January. In addition to widespread drought, the winter monsoon brings cooler weather to the subcontinent, especially the north.

With precipitation so concentrated within a short part of the year, and with the extreme pressure of so many people on so little land, it is vital to extend the crop season by irrigation. People in the subcontinent therefore expend an enormous amount of labor to get additional water onto the land. Modern technology does the job in places where huge dams impound rivers and networks of canals distribute water from these reservoirs.

The main locales for such large-scale development are in the steppe and desert areas of the Punjab and the lower Indus Valley (Sind Province) in Pakistan. Since the 19th century, when the British began irrigation works in these areas, rapid colonization has changed dry lands from sparsely populated to densely populated areas. Both the Indian and Pakistani parts of the Punjab still normally produce surpluses of wheat. East and West Punjab and Pakistan's Sind Province produce surplus rice, including the Basmati variety, which is an important export.

A massive irrigation and hydroelectric power project is now underway in the north-central part of peninsular India, where the Sardar Sarovar Dam is being constructed on the Narmada River (see Fig. 11.3). Largely because it is expected to displace 100,000 tribal and rural people and inundate 32,000 acres of forest and 28,000 acres of cropland, the dam was opposed by environmentalists and human rights advocates both in India and abroad. The World Bank withdrew its financial support of the project in 1993 in response to these protests, but, asserting that its benefits will greatly outweigh its costs, India has pressed ahead with the dam's construction. It is the centerpiece of the massive Narmada Valley Development Project, where 30 dams will ultimately displace about one million people.

Tubewells are common where large-scale irrigation systems are absent. These are deep wells in which electric or diesel-powered engines raise water. Farmers also rely on "tanks," artificial ponds that fill in the wet season and serve as reservoirs in the dry season. There are hundreds of thousands of them in India (see Fig. 11.9). A huge labor input is characteristic of some forms of irrigation, where people and livestock use simple tools such as leather sacks, pots, and waterwheels to lift water from wells and tanks to channels and fields. Even if water can be impounded in tanks or lifted from shallow wells sunk to groundwater, its availability depends ultimately on the summer monsoon rains. Years may pass when rainfall is so scarce that tanks, wells, and some rivers dry up.

11.4   Food and Population

The major staple food crops of the subcontinent, along with population densities, correlate spatially with the amount of water available. Rice is the basis of life in the wetter areas and,

with its high caloric yields per acre, is associated with the highest population densities. Rice dominates in the delta area of Bengal (in both India and Bangladesh), the adjacent lower Ganges valley, and the coastal plains of the peninsula. Irrigated wheat is the staple crop and food in the drier upper Ganges valley of India and the dry Punjab of India and Pakistan. Here, the caloric yield and population densities are intermediate. Unirrigated sorghums and millets are dominant over most of the Deccan plateau and in other areas where low rainfall cannot be supplemented much by irrigation. Caloric yields from these crops are low, and so are population densities, although the "low" densities in some places are more than 200 per square mile (c. 80 per sq km). Maize is an important grain in the lower regions of the Himalaya. Rice is the preferred food almost everywhere, however, and where enough water is available there are patches of irrigated rice.

A host of minor crops supplements the staple grains. People grow many kinds of fruits and vegetables, protein-rich peas and beans, and oilseeds such as peanuts. Barley, sugarcane, coconuts, bananas, cashews, tobacco, and spices are important in certain areas. The overall agricultural economy is so large that some "minor" crops represent important shares in world production. Sugarcane, for instance, has a subordinate role in Indian agriculture, but India is the world's leading producer. Almost all of this crop is grown for consumption within India.

INDUSTRIAL AND EXPORT CROPS

Cotton, jute, tea, and rice are the main crops grown in the subcontinent for industrial use and export. Irrigated cotton in the Punjab and lower Indus Valley makes Pakistan a cotton producer exceeded among the world's states only by China and the United States. Textiles, yarn, and raw cotton contribute 60 per-

FIGURE 11.12
The Bengal region of both Bangladesh and India is the world's leading source area for jute, the raw material of burlap. These men are rowing a cargo of freshly-harvested and bundled jute near Dacca, Bangladesh.
Tim Gibson/Envision

cent of Pakistan's exports. In India, cotton is grown mainly in the interior Deccan on soils that hold moisture unusually well and produce a crop without irrigation. India's large textile industry absorbs most of the production.

The Ganges-Brahmaputra Delta is the world's greatest producing area for jute, the principal material for burlap (Fig. 11.12). The delta lies in the former province of Bengal, which was partitioned between India and (East) Pakistan when British rule ended. Most of the jute-growing areas lay on the Pakistani side (now Bangladesh), while the jute mills, which had developed in the Calcutta area, went to India. Since independence, India has increased raw jute production on its side of the border, and Bangladesh has developed jute mills of its own.

The canelike jute plant, grown in several feet of water, requires a long growing season, high temperatures, and much hand labor. It is well-suited to the delta, which is tropical, seasonally-inundated, and extremely densely settled. The plant became the commercial mainstay in the 19th century, after Scottish entrepreneurs developed the necessary technology.

Of the four major commercial crops in the subcontinent, only tea is exported on a sizable scale. India is the world's greatest exporter of tea. Unlike cotton, jute, and most other crops of the subcontinent, tea is principally a plantation crop. The plantations were developed, and are still largely owned, by British interests. Production is greatest in the northeastern state of Assam, with a secondary center in the mountainous far south of peninsular India. Coffee, rubber, and coconuts are minor plantation crops, and India exports products made from coir (coconut fiber).

Opium is a major illegal export crop of Pakistan. Cultivation is restricted to the rugged mountain region of the Northwest Frontier Province, inhabited mainly by Pushtun tribespeople, along the frontier with Afghanistan. Processing of the raw opium into heroin is also a significant industry in the region and in adjoining areas of Afghanistan. The government of Pakistan is under pressure from the United States and other heroin-consuming nations to crack down on the illegal trade.

Definitions and Insights

THE SACRED COW

Many attitudes, beliefs and practices associated with cattle make up the world-famous but often poorly understood "sacred cow" concept of India (Fig. 11.B). There are nearly 200 million cattle in India, representing about one third of the world total and the largest concentration of domesticated animals anywhere on Earth. India's dominant religion of Hinduism forbids the slaughter of cows but allows male cattle and both male and female water buffalo to be killed. Reverence for the cow is well-founded. Indians favor cow's milk, ghee (clarified butter), and yoghurt over dairy products from water buffalo. They value cows as producers of male offspring, which serve as India's principal draft animal. Both cows and bullocks provide dung, an almost universal fuel and fertilizer in rural India.

FIGURE 11.B
The human city-dwellers of India accommodate both wild and domestic animals, particularly cattle.   *Mike Barlow/Dembinsky Photo Associates*

Reverence for the cow in particular and cattle in general pervades the Hindu religion and mythology. The bull Nandi is associated with the Hindu god Shiva, and so people allow bulls to freely roam the streets of Indian cities. As a symbol of fertility cows are associated with (but not worshipped as) several deities. The mother of all cows, Surabhi, was one of the treasures churned from the cosmic ocean. Hindus honor cows at several special festivals. They use cow's milk in temple rituals. They believe the "five products of the cow"—milk, curds, ghee, urine, and dung—have unique magical and medicinal properties, particularly when combined. All over India there are *goshalas*, or "old folks' homes," for aged and infirm cattle.

Remarkably, the subcontinent countries of Pakistan, Bangladesh, and India are major exporters of leather and leather goods. In India the leather industry is mainly in the hands of Muslims, who do not share Hinduism's prohibitions against killing or eating cattle. Muslims are forbidden to eat pork, and, therefore, pigs, a major food resource in many developing countries, are of little importance here. They are eaten mainly by Christians, very low-caste Hindus (some of whom are pig breeders), and tribal peoples.

11.5 Industry, from Bhopal to Booming Bangalore

In the partition following independence from Britain, India received almost all of the subcontinent's modern industries and most of its mineral wealth and energy supplies. Pakistan, including the present Bangladesh, was left with a much smaller industrial infrastructure and fewer natural resources. In global terms, the subcontinent has a small but growing industrial base. Handicraft workers still exceed factory workers by many millions, and handicrafts supply many everyday needs for the population.

IS AGRICULTURAL PROGRESS KEEPING PACE WITH POPULATION GROWTH?

Poverty and marginal human health are already problems in the Indian subcontinent and Sri Lanka, and the prospect of continued high population growth raises the question of whether growth in food supplies can avert an eventual Malthusian crisis. These countries are among the poorest in the world, with 1994 per capita GNP estimates of $310 for India, Pakistan $440, Bangladesh $230, Nepal $200, Bhutan $400, and Sri Lanka $640. The region fares poorly in measures of social welfare too. Infant mortality rates are high. Average life expectancy is estimated at 59 years in India, 61 in Pakistan, 57 in Bangladesh, 55 in Nepal, 51 in Bhutan, and 73 in Sri Lanka. By comparison, in nearby China the average person may expect to live 70 years, and in the United States 76 years (by 1996 figures).

These averages conceal some very wide variations among social groups and regions. In India, for example, a few people are very wealthy, and there is an emerging middle class of as many as 200 million people. Most people, however, are even poorer than the averages indicate (Fig. 11.C). The prosperous minority is surrounded by a sea of overwhelming poverty. India and its neighbors are countries where "people overpopulation" is a problem: A great many people must rely on limited resources, particularly land. With 2304 people per square mile (890 per sq km), Bangladesh is the most densely populated country in the subcontinent, far outpacing its nearest contenders, India and Sri Lanka. The average rural dweller in Bangladesh ekes out a living from less than one thirtieth the crop land supporting the average person in rural United States.

Such crowding is a relatively recent phenomenon in the region, resulting from the declining death rates that accompanied the spread of modern public health and medical techniques to the subcontinent. The British began this process when they colonized the subcontinent in the 18th and early 19th centuries. Population growth has accelerated rapidly since independence, as birth rates have remained relatively high while death rates have continued to fall. Given increases in urbanization and industrialization, and dividends from India's strenuous family planning programs, future population growth in the region is expected to slow. The population base is already so vast, however, that even modest growth will add huge numbers. The issue of population growth is especially critical for Pakistan, which has never mounted an effective family planning program, mainly because of opposition from Muslim religious authorities who regard birth control as intervention against God's will.

Agricultural output in South Asia has been increasing rapidly since independence. Most notably, despite its huge and rapidly growing population and irregular distribution of food supplies, India has managed to remain self-sufficient in food production, and has large strategic reserves of staple grains. The successes of agricul-

FIGURE 11.C
Large numbers of people in South Asia are poor and homeless, but widespread famine is unknown. *Dave Watts/Tom Stack & Associates*

India's leading factory industries are cotton textiles, jute, and iron and steel. Modern cotton mills are concentrated mainly in the Bombay (recently renamed Mumbai) metropolitan area (population: 9.9 million, city proper; 12.6 million, metropolitan area) and neighboring areas along and near the west coast. Of basic importance to cotton manufacturing are the large home market, cheap labor, hydroelectric power from stations in the Western Ghats, and the large domestic production of cotton. As a major world producer of cotton goods, India supplies cloth to enormous numbers of its impoverished citizens and is regularly a leading exporter of textiles and clothing.

The Bengal is an important industrial region. India's jute industry is concentrated in and near Calcutta. Jute is only a minor export for India, but in Bangladesh raw jute and jute prod-

ture in the subcontinent have been due mainly to increased use of artificial fertilizers, introduction of new high-yield varieties of wheat and rice associated with the Green Revolution (see Chapter 10, page 272), application of more labor from the growing rural population, increased irrigation, the spread of education, and the development of government extension institutions to aid farmers. At the time of independence, fertilizer use was very limited and artificial fertilizers were unknown, unavailable, or unaffordable. For example, the dried manure of the world's largest cattle population was (and still is) used as fuel for cooking. This source is necessary because of the fuel-wood shortages caused by deforestation, and because of the shortages and high costs of fossil fuels. But diverting animal manure to fuel removes an important potential source of natural fertilizers from the land.

A large increase in the use of artificial fertilizers began in South Asia after the mid-1960s, especially in conjunction with the use of new strains of wheat and rice in the Green Revolution. The new developments took root first in the Punjab area of India and Pakistan and are continuing to spread from there. They have generally been accompanied by increasing irrigation, both from wells and from large new reservoir and canal systems. Pumping by electric motors and gasoline engines is essential to much of this new irrigation. Agricultural development thus has required the spread of rural electrification and of fuel distribution facilities.

Social conditions and services in rural areas are improving in some parts of South Asia, notably in India, suggesting that development is progressing and that agricultural production has so far been sufficient to

keep up with population growth—and that a Malthusian catastrophe long-predicted for the region is not imminent. In India, the literacy rate among people over age 15 grew from 16 percent in 1947 to 53 percent in 1991. Education, including government extension, increasingly is reaching the villagers (Fig. 11.D), influencing their agricultural practices as well as their health and family planning.

One of the challenges confronting India is raising the status of women. Despite a 1961 ban on dowries (the money and gifts given by a bride's parents to the groom), the practice is continuing. So is the rate of the killing of brides who do not provide enough dowry. The burden placed on the bride's family has also prompted parents to abort

female fetuses, which they now can often detect with ultrasound.

Technology is thus a mixed blessing in this traditional society. Generally it plays a constructive role, since where face-to-face efforts are not possible, radio and television can encourage positive change in the most remote villages. India now has its own space program, and has put several communications satellites into orbit. Such accomplishments, with the possibility for much further improvement, indicate that there is a process of development unfolding in South Asia. The challenge will be for technology and human resources to keep agricultural growth apace with or ahead of population growth.

FIGURE 11.D

Education and agricultural production are advancing in South Asia despite great odds. These are Sikh schoolchildren in the Punjab region of India. *Joseph A. Astroth, Jr.*

ucts are major exports. India's iron and steel industry is concentrated in the Bengal and in adjacent parts of the northeastern portion of the peninsular uplands. The most valuable assemblage of mineral resources in India—including iron ore, coal, manganese, chromium, and tungsten—is found in and near this area. These resources have led to the rise of small- to medium-sized industrial and mining cities in a region that in other respects is one of the least developed parts of the country. The largest center is Jamshedpur (population: 670,000, city proper; 850,000, metropolitan area), 150 miles (*c.* 240 km) west of Calcutta.

India has made expansion of the iron and steel industry a major objective in its industrialization. Output has increased but is still inadequate for India's needs and represents only

FIGURE 11.13
Bombay (Mumbai) produces the action and romance films which are the staples of Indian cinema.   *Glen Allison/Tony Stone Images*

about 3 percent of world total. Except for coal of coking quality, which will probably have to be supplemented by imports, resources are adequate for expanded production in the future. The country has large reserves of noncoking coal (very important in India's overall energy economy), iron ore, and manganese.

Until quite recently India has been intent on keeping out most foreign manufactures, and has sought industrial self-sufficiency. With this policy India's industries have expanded rapidly and its output of manufactured goods has become diversified and sophisticated, but inefficiency in some sectors has led recently to a greater openness to imports. Smaller industries such as engineering and chemicals are growing at a faster rate than older sectors like the textile industries.

Today, there are very few items among the great variety of machines used in a modern society that are not produced somewhere in India. There are several automobile and truck assembly plants in the country. Motorcycles are manufactured and bicycles, produced in many small factories, have become a significant export. More recent additions to the array of manufacturing include tractors, bulldozers, data-processing equipment, silicon chips for electronics, and computer software. Bangalore is known as India's "silicon valley," and its space-age character presents a startling contrast to the largely rural and timeless image that many outsiders have of India. Bombay is India's "Wall Street" and also its "Hollywood," the center of a prolific and accomplished film industry (and the world's largest) that produces movies of international stature as well as the regular fare which a huge domestic market consumes passionately (Fig. 11.13).

India produces a wide range of chemical products, from vitamins to fertilizers. Inadequate factory safety measures set the stage for a horrendous accident in the chemical industry on December 2, 1984. In the north-central city of Bhopal, the cooling system failed in a tank containing highly toxic gas used in the manufacture of pesticides at a Union Carbide factory. The ensuing explosion killed nearly 15,000 people and seriously injured more than 200,000, the majority of whom were poor people living in the plant's shadow. The multinational corporation paid a $470 million dollar settlement for an accident that cost India an estimated $4.1 billion dollars in economic damages alone.

Pakistan and Bangladesh are still far behind India in total industrial output. However, Pakistan has made rapid progress in some areas since independence. Its principal success has been the development of a cotton-textile industry. Pakistan's main industrial centers are its two largest metropolitan areas: the seaport of Karachi (population: 4.9 million, city proper; 5.3 million, metropolitan area) at the western edge of the Indus River delta and the cultural center of Lahore (population: 2.7 million, city proper; 3.03 million, metropolitan area; see Fig. 11.14) in the Punjab. The capital city of Islamabad, the "City of Islam" (population: 204,000, city proper) is a very modern metropolis mainly of administrative significance. In Bangladesh, industrial development centers in Dacca (Dhaka; population: 3.64 million, city proper; 6.54 million, metropolitan area), the capital and largest city, and Chittagong (population: 1.57 million, city proper; 2.34 million, metropolitan area), the main seaport.

Inadequate energy resources are likely to be a major problem confronting further industrialization in the subcontinent. The major domestic sources of commercial energy are India's coal deposits and the hydropower generated in all the countries but Bangladesh. All countries in the subcontinent are heavily dependent on imported energy, especially oil, and would probably be more so if their levels of industrial development and consumption were higher. India has modest but rapidly growing oil output from fields in Assam, Gujarat, and offshore of Bom-

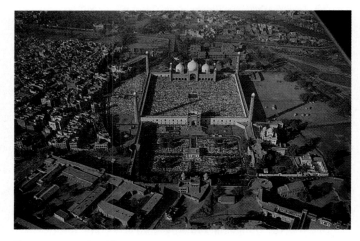

FIGURE 11.14
Lahore's seventeenth century Badshahi Mosque, built by the Mogul Emperor Aurangzeb, is thronged with worshippers on Fridays.   *Franke Keating/Photo Researchers*

PROBLEM LANDSCAPE

BATTLE OVER KASHMIR SEEMS ONLY TO WORSEN

"Kashmir is a Paradise on Earth," wrote Samsar Koul in the introduction to his book *Beautiful Valleys of Kashmir and Ladakh.** Until the late 1980s the stunning snow-crowned peaks and flower-laden valleys of this western Himalayan region lured millions of Indian tourists and ranks of Western trekkers each year. Now Kashmir (called Jammu and Kashmir in India) is famous for the blood spilled in a seemingly-intractable and especially violent conflict between supporters of Indian and Pakistani claims to the region.

Before independence, Kashmir was a princely state administered by a Hindu maharajah. Perhaps three fourths of Kashmir's estimated population of 10 million is Muslim, which is the basis of Pakistan's claim to the territory. But, under the partition arrangements, the ruler of each princely state was to have the right to join either India or Pakistan, as he chose. Kashmir's Hindu ruler chose India, which is the legal basis of India's claim.

After partition, fighting between India and Pakistan led to a cease-fire line leaving eastern Kashmir, with most of the state's population, in India, and the more rugged western Kashmir in Pakistan. Beginning in the mid-1950s, China pressed its own claims to remote northern mountain areas of Kashmir, and occupied some of the border territories. Pakistan ceded part of the occupied territory to China and established friendly relations with its giant northern neighbor, while India rejected China's claims. Until now, India's recurrent small-scale military actions have failed to dislodge Chinese forces. India now holds about 55 percent of the old state of Kashmir, Pakistan 30 percent, and China 15 percent.

Beginning with the conflict that accompanied partition and independence, three wars between India and Pakistan have effected little change in Kashmir. In 1965, conflict began in Kashmir, spread to the Punjab, and escalated to a brief but indecisive full-scale war involving tanks, air-borne forces, and widespread air raids. Renewed hostilities in Kashmir in 1971, as part of the war in which India supported the revolt of Bangladesh from Pakistan, again did not alter the political landscape of Kashmir. The Siachen Glacier at about 20,000 feet (2800 m) in the Karokoram earned the title of "world's highest battlefield."

In 1989, Kashmir's Muslim majority escalated the campaign for secession from India, and the strife has continued ever since. International observers have accused both sides of gross human rights violations. In 1995, the separatists adopted a new technique of kidnapping foreign trekkers, demanding freedom for jailed comrades in exchange for their release. They beheaded a Norwegian hiker when their demands were not met, and continued to take and hold more hostages.

*Mysore, Frelia: Wesley Press, 1963. John F. Burns, "Battle Over Kashmir Seems Only to Worsen." *The New York Times*, July 9, 1995: A3.

bay. Natural gas from domestic wells is especially important to Pakistan, and smaller quantities are in India and Bangladesh. India derives about two percent of its electricity from nuclear power plants.

11.6 Social and Political Complexities

On a physical map, the Indian subcontinent looks like a discrete unit, marked off from the rest of the world by its mountain borders and seacoasts. But the social complexity of this area is so great that it has never been unified politically except during a relatively brief period in the 19th century and the first half of the 20th century, when the outside power of Great Britain imposed this unity. Even then, a variety of political units under indigenous rulers retained varying degrees of autonomy, although all of these "princely states" were ultimately under British control.

As soon as British power withdrew, the divisive force of conflicting social groups asserted itself, splitting the subcontinent into Muslim and Hindu countries. Within a quarter-century, social and geographic divisiveness split Pakistan into two countries. All countries of the subcontinent and Sri Lanka have serious problems in internal or international relations, and sometimes both.

RELIGION AND THE PROBLEMS OF SECTARIANISM

The major social divisions within the subcontinent are in religion and language. Religious differences are particularly troubling (Fig. 11.15). The most serious division has been between the two major religious groups, the Hindus and the Muslims. Indigenous Hinduism was the dominant religion at the time Islam made its appearance. It continues to have the largest number of adherents, with an estimated 83 percent of the population

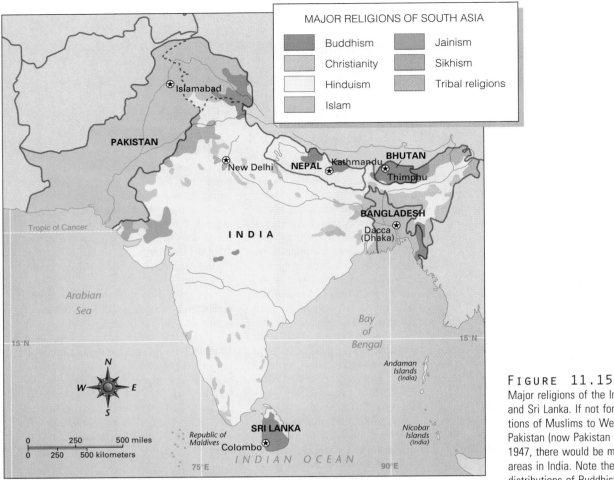

Source: Stansfield and Zimolzak. 1982. *Global Perspectives: A World Regional Geography.*

FIGURE 11.15

Major religions of the Indian subcontinent and Sri Lanka. If not for the massive migrations of Muslims to West Pakistan and East Pakistan (now Pakistan and Bangladesh) in 1947, there would be many more Muslim areas in India. Note the markedly localized distributions of Buddhists, Jains and Sikhs.

of India, 11 percent of Bangladesh's population, and 2 percent of Pakistan's.

The simple faith of Islam is described in Chapter 8. It is much more difficult to characterize the complex and regionally varied faith of Hinduism. It lacks a definite creed or theology. It is very absorptive, encompassing an unlimited pantheon of deities and an infinite range of types of permissible worship. Briefly, it has the following notable elements distinguishing it from the region's other religions: Most Hindus recognize the social hierarcy of the **caste system**, deferring authority to the highest (Brahmin) caste. They practice rituals to honor one or another of five principal deities, attending temples to worship them in the form of sanctified icons in which the gods' presence reside. They believe in reincarnation and the transmigration of souls. They are supposed to be tolerant of other religions and ideas. They participate in folk festivals to commemorate legendary heroes and gods. To earn religious merit and to struggle toward liberation from the bondage of repeated death and rebirth, they make pilgrimages to sacred mountains and rivers.

The Ganges is a particularly sacred river to Hindus, who believe it springs from the matted hair of the god Shiva, and who make pilgrimage to its city of Varanasi (Benares) in the state of Uttar Pradesh (Fig. 11.16). Many elderly people go to die in this city and to be cremated where their ashes may be thrown into the holy waters. The Indian government is now attempting to clean up the Ganges, polluted in part by incompletely cremated corpses; many of the faithful poor cannot afford to buy the fuel needed for thorough immolation. Officials in Varanasi have released scavenging water turtles to dispose of the cadavers.

Seldom in history have two large groups with such differing beliefs lived in such close association with each other as have Hindus and Muslims in the subcontinent. Islam holds to an uncompromising monotheism and prescribes uniformity in religious beliefs and practices. Hinduism is monotheistic for some believers but polytheistic for others, and asserts that a variety of religious observances is consistent with the differing natures and social roles of humans. The exuberant and noisy celebrations of the Hindu faith are a striking contrast to the austere ceremonies of Islam. Islam has a mission to convert others to the true religion, while most Hindus regard proselytizing as essentially useless and wrong. Islam's concept of the essential equality of all believers is a total contrast to the inequalities of the caste system endorsed in Hinduism. Islam's use of the cow for food is anathema to Hinduism.

Definitions and Insights

THE CASTE SYSTEM

Serious problems of religious division exist not only between Hindus and minority groups in India, but also within Hinduism. A fundamental feature of Hinduism has been the division of its adherents into the most elaborate caste system ever known. Traditionally, every Hindu is born into a particular caste. Caste membership is inherited and cannot be changed. Particular castes are associated with particular religious emphases, and their members are expected to follow traditional caste occupations. With certain exceptions, marriage outside the caste is forbidden, and meals may be taken only with fellow caste members or those from higher castes. Castes form a hierarchy that determines a person's social rank, with the Brahmin caste at the top (comprising about 5 percent of all Hindus) and three others (Kshatriya, Vaisya, and Sudra) below.

At the bottom of the social ladder are "the untouchables" (about 26 percent of all Hindus). They are not part of the caste system but are literally "outcast" because, according to Hindu belief, they are not twice-born. The untouchables are so-called because they traditionally performed the worst jobs, such as handling of corpses and garbage, and therefore their touch would defile caste Hindus.

Forces of modernization are challenging this ancient, rigid social system. Brahmin privileges are being increasingly challenged and in most areas have been restricted by law. Indian leaders of high caste have championed the untouchables' cause for both moral and practical reasons. Indian law now forbids recognition of untouchables as a separate social group. Disintegration of the caste system has been especially rapid in the cities, where it has been hastened by the close intermingling of different castes in factories, public eating places, and public transportation. Confronting the fact that upper castes account for less than one fifth of India's population but command more than half of the best government jobs, the Indian government has instituted an affirmative action program to allocate more jobs to members of lower castes.

Practical differences have sometimes added to doctrinal and cultural differences to produce outright conflict between the groups. During the British occupation, the formerly subordinate Hindus came to dominate the civil service and most businesses. Many Muslims feared the results of being incorporated into a state with a Hindu majority, and their demands for political separation led to the creation of the two independent states (India and Pakistan) from the British colony.

Immediately preceding and following partition in 1947, violence broke out between the two peoples on a huge scale, and hundreds of thousands of lives were lost in wholesale massacres. Mass migrations between the two countries involved more than 15 million people. In Pakistan today the Mohajirs, or "Migrants"—the families of the Muslims who fled India in 1947—are still struggling for integration in the society. Indigenous Muslim Pakistanis periodically carry out attacks of terror against the Muslim "newcomers," especially in Karachi and other cities in the southern province of Sind where they are most numerous. In 1971, the events that created Bangladesh—revolt, repression, and Indian intervention against Pakistan—again brought huge casualties and millions of refugees.

Pakistan also has internal problems centering on religious divisions, mainly between majority Sunni Muslims (70 percent of the country's population) and Shi'ite Muslims (30 percent). Shi'ite Muslims opposed the government's efforts in the 1980s to establish an avowedly Islamic state, and there was violent conflict between members of the two sects. Pakistani factions who preferred a Western-style secular democracy also rioted in protest against the Islamization of the government. Authorities responded by cracking down on such opposition, but the underlying antagonisms remain.

Sectarian violence (known in the region as "communal" violence) between Hindus and Muslims is a constant threat to India's social and political fabric. For example, a 16th century mosque in the city of Ayodhya in Uttar Pradesh was sacred to

FIGURE 11.16
The Hindu faithful pray and bathe on the banks of the sacred Ganges River in Varanasi. *Joel Simon/Tony Stone Images*

FIGURE 11.17
The Golden Temple (established in 1577, rebuilt in 1764) in Amritsar is the holiest shrine of the Sikh religion.   *Paolo Koch/Photo Researchers*

both faiths. Hindus revered it as the birthplace of the god-king Ram, but it was also a Muslim place of worship. Backed by Hindu nationalists in the provincial government, a mob of about 250,000 Hindu fundamentalists demolished the mosque in December 1992 with the intention of building a Hindu temple on the site. The ensuing communal violence between Hindus and Muslims throughout India, notably including the cosmopolitan and usually tolerant urbanites of Bombay, left thousands dead in the weeks which followed.

In addition to the Muslims who make up 11 percent of the population in India, there are other significant religious minorities in that country. About 17 million Sikhs are concentrated mainly in the unusually prosperous state of Punjab. Their faith emerged in the 15th century as a religious reform movement intent on narrowing the differences between Punjab's Hindus and Muslims. It is now India's fourth largest religion, after Hinduism, Islam, and Christianity. Sikh men are recognizable by their turbans, which contain their never-shorn hair, and by a bracelet called a "bangle" worn on the right arm. They regard men and women as equals, and disavow the caste system recognized by Hindus.

In India's Punjab, where Sikhs make up 60 percent of the population and Hindus 36 percent, the 1980s and early 1990s were violent years during which about 20,000 people died in armed clashes. Sikh factions intent on establishing their own homeland of Khalistan—a movement prompted in part by New Delhi's plans to divert water from the Punjab—challenged Indian authority and turned Amritsar's Golden Temple, the holiest site in the Sikh faith, into their military stronghold (Fig. 11.17). In a controversial move to quell the revolt, Prime Minister Indira Gandhi ordered troops to storm the Golden Temple in June 1984. The resulting deaths and desecration led directly to Mrs. Gandhi's assassination by her own Sikh bodyguards later that year. Indian authorities accused Pakistan of arming the Sikh militants.

There are about 22 million Christians in India, most of whom live in the south of the peninsula, and Buddhists, Jains,

Parsis, and members of a variety of tribal religions make up the numerous smaller remaining religious minorities. Most of the estimated 7 million Buddhists are recent converts from among India's lowest castes. Numbering perhaps 200,000 and concentrated mainly in Bombay, the famously entrepreneurial Parsis have attained wealth and economic power far out of proportion to their number. Their religion is the ancient pre-Islamic Persian faith of Zoroastrianism, known mainly for its reverence of fire.

India's 5 million Jains also have influence beyond what their numbers suggest, as they control a significant share of India's business. Their faith, founded upon the teachings of Vardhamana Mahavira (a contemporary of Buddha, *c.* 540–468 B.C.), is renowned for its respect for geographical features and animal life. Jains believe that souls are in people, plants, animals, and nonliving natural entities such as rocks and rivers. Jainism has taken the principle of nonviolence (*ahimsa*, also present in Hinduism) to mean they should not even harm microbes. Jain worshippers therefore often wear masks to prevent inhalation of microscopic organisms, and Jains cannot be farmers because they would destroy plant life and living organisms in the soil (Fig. 11.18).

Religious and political troubles are also rife in the mountain country of Bhutan. The country is a monarchy ruled from the capital of Thimphu by a Buddhist king of the Drukpa tribe, to which most of the government ministers also belong. The government forbids political parties, and is being challenged by a number of outlawed organizations including the Bhutan People's Party (BPP). The BPP represents ethnic Nepalis, who are Hindu and who for decades have been migrating out of their own overcrowded Himalayan kingdom to work the productive soils of Bhutan. Ostensibly as a means of promoting national unity, Bhutan's king has outlawed the use of the Nepali language in Bhutan's schools, and insisted that all inhabitants of Bhutan wear the national dress and hairstyle of the Drukpa tribe. In the early 1990s violent clashes ensued between ethnic Nepalis and Drukpas.

FIGURE 11.18
Jain women in prayer at the Ranakpur Temple, in India's Rajasthan region.
Nicholas DeVore/Tony Stone Images

Nepal itself has suffered violent consequences of one-party rule. The country was an absolute monarchy until 1991, when prodemocracy riots and demonstrations forced King Birendra to accept a new role as constitutional monarch and permit elections for a national parliament.

In Sri Lanka, internal violence reflects antagonisms between ethnic groups and discontent with economic and political conditions, especially among the Tamils. Since the mid-1980s, thousands of people have been killed and many more rendered homeless as the Tamil majority in the north has fought for independence from Sri Lanka's overall majority Sinhalese government. Guerrilla warfare by the organization of Tamil separatists called the Tamil Tigers in 1987 led to military actions by Indian troops invited in to help restore order. They withdrew in 1990, but the troubles have continued. The 1991 assassination in southern India of Indian prime minister Rajiv Gandhi, son of the slain Indira Gandhi, has been linked to Indian sympathizers of the Tamil separatist movement in Sri Lanka. In December 1995, Sri Lankan government troops succeeded in retaking the Tamil's geographical stronghold of the Jaffna Peninsula. This failed to end the conflict, however, as a devastating Tamil car bombing of downtown Colombo just a month later illustrated.

PROBLEMS OF LANGUAGE

Language is another strongly divisive factor in India and Pakistan. The languages of the subcontinent fall into two principal families—in the north, the Aryan languages of the Indo-European language family, and in roughly the southern third of peninsular India, the Dravidian languages of the Dravidian language family. The various languages within each of these groups are fairly closely related to one another. Languages of major importance, each with at least tens of millions of speakers, include four languages of the Dravidian group and eight Aryan languages. Hundreds of other languages and dialects are also spoken. Many people speak two or even three languages, but overall literacy rates are low and communication is often a problem.

Since British colonial times, English has been the *lingua franca* and the de facto official language of the subcontinent. However, only a very small percentage of the population, mainly the educated, are literate in English. This fact, along with nationalist sentiment, have prompted calls for indigenous languages to be adopted as official languages. After independence and partition, disputes arose in each country over which language should be chosen. India decided on Hindi, an Aryan language spoken by the largest linguistic group, and made it the country's national language in 1965. This government decision resulted in strong protests, especially in the Dravidian south. The result has been a proliferation of official languages to satisfy huge populations; India now has 15, all of which appear on the country's paper currency. Bloody rioting between speakers of different languages has occurred at times in areas of India where such peoples are mixed, and demands by language groups have been major factors in reshaping the boundaries of a number of India's internal political units. In Pakistan, Urdu, a language similar to Hindi but written in Perso-Arabic script, is the official language. As in India, English is widely used in business and government. In Bangladesh, Bengali is the predominant and official language.

The subcontinent's accomplishments in recent decades have been remarkable. India has lowered its birth rate and achieved considerable growth in agriculture and industry. More uniquely among the world's LDCs, it has done so while maintaining, despite all its internal divisions and frictions, a Western-style representative democracy. But sectarian violence between Hindus and Muslims, a growing Hindu nationalist movement which threatens India's remarkable diversity, Sikh desire for autonomy, and the problem of Kashmir pose long-term difficulties for this South Asian giant. Pakistan and Nepal must confront the problems of rapid population growth, deforestation, and land degradation where the balance between people and resources is already precarious. Bangladesh faces the huge challenge of dealing with the environmental consequences of deforestation and repeated flooding assaults from the Bay of Bengal. Sri Lanka's future may be brightened either by a forcible reunification of the country, or by the government's willingness to tolerate a de facto partition of the country into Sinhalese and Tamil realms, so long as the Tamils do not push for independence. All of these nations struggle with the chronic problem of poverty, which ultimately is the source of many of the region's most severe environmental and political dilemmas.

QUESTIONS

Page numbers refer to *Saunders College Publishing 1996 Special Edition of Rand McNally's Atlas of World Geography*.

On the map of Southwestern Asia on page 143, identify:

The world's highest mountain.

A country experiencing civil war between a mainly Hindu Tamil north and a mainly Buddhist Sinhalese south.

The two cities known, respectively, as the "Wall Street" and the "Silicon Valley" of India.

A region, disputed between India and Pakistan, with a predominantly Muslim population.

The low-lying flood-prone country known formerly as East Pakistan.

12

THE TIGERS AND TRIBULATIONS OF SOUTHEAST ASIA

outheast Asia, from Burma (Myanmar) in the west to the Philippines in the east, is a fragmented region of peninsulas, islands, and intervening seas. East of India and south of China, between the Bay of Bengal and the South China Sea, the large Indochinese peninsula projects southward from the continental mass of Asia (Fig. 12.1). From it, the long, narrow Malay Peninsula extends another 900 miles (*c.* 1450 km) toward the equator. Ringing the south and east of this continental projection are thousands of islands, among which Sumatra, Java, Borneo, Celebes (Sulawesi), Mindanao, and Luzon are outstanding in size. Another large island, New Guinea, east of Celebes, is culturally a part of the Melanesian archipelagoes of the Pacific World, but its western half, held by Indonesia, is politically part of Southeast Asia.

Southeast Asia is composed politically of 10 states: Burma, Thailand, Laos, Cambodia, Vietnam, Malaysia, Singapore, Indonesia, Brunei, and the Philippines. With the exception of Thailand, which was never a colony, all of these states have become independent from their colonial powers since 1946. These ten nations together occupy a land area less than one half that of China, and are fragmented both politically and topographically. There is also a high degree of cultural fragmentation, which has often led to contention and warfare between differing cultural groups. Outside intervention has generally complicated and worsened local discord, producing enormous suffering in the region and long-lasting traumatic effects among the foreign soldiers who fought losing battles against the determined inhabitants of Southeast Asia.

12.1 A Region of Diverse Cultural Influences

Southeast Asia is one of the world's culture hearths, having contributed domesticated plants (including rice) and animals (including chickens) and the achievements of civilizations to a wider world. Cultural innovations and accomplishments have been diffused from several centers which themselves were built upon many influences reaching Southeast Asia from abroad. Indian and Chinese traits were especially strong in shaping the region's cultural geography. Hinduism came from India, and Hinayana Buddhism from India through Ceylon (Sri Lanka). The Chinese cultural influences of Confucianism, Taoism, and Mahayana Buddhism were particularly strong in Vietnam. China periodically demanded and received tribute from states in Burma (now Myanmar), Sumatra, and Java. The faith of Islam came from the west in the 14th and 15th centuries and took permanent root; Southeast Asia is the eastern margin of the

◀ A floating village in Vietnam's Mekong Delta. The Mekong River is a vital artery in mainland Southeast Asia, where rivers and canals lend an amphibious quality to land and life. *L. Rebman/Photo Researchers*

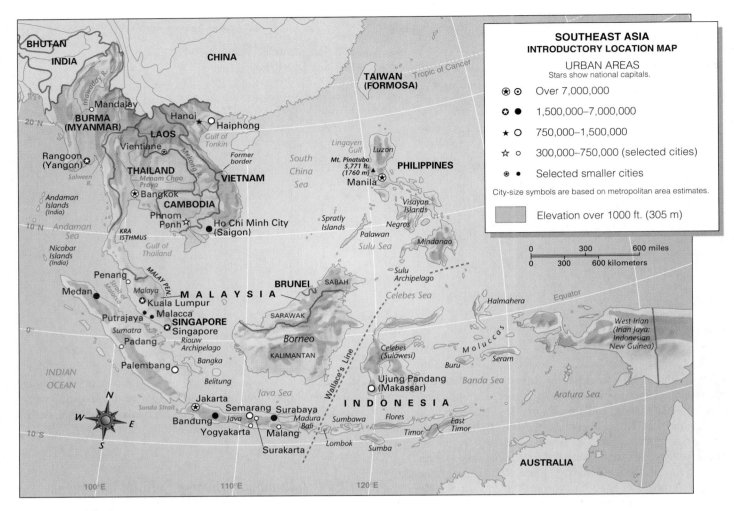

FIGURE 12.1
Introductory reference map of Southeast Asia. Indonesian military forces occupy East Timor, but the United Nations does not recognize Indonesia's claim to this former Portuguese colony.

world of Islam, and Indonesia has more Muslims than any other country in the world today.

The historic and contemporary monumental architectures of Southeast Asia reflect these diverse origins. In the ninth century, the advanced Sailendra culture of the Indonesian island of Java built Borobudur, the world's largest Buddhist *stupa* (shrine) and still the largest monument in the Southern Hemisphere (Figure 12.2a). In the 12th century, a Hindu king built the extraordinary complex of Angkor Wat as the capital of the Khmer empire based in Cambodia. The first independent Thai kingdom emerged in the 13th century, and in the 14th century Thai kings ruling from the great city of Ayuthaya (near modern Bangkok) extinguished the Khmer empire and extended their own control into the Malay Peninsula and Burma. The world's largest mosque stands in modern Brunei, a tiny oil-rich country

on the northern tip of the island of Borneo, and the international culture of capitalism is behind the ongoing construction of the world's tallest buildings (1462 feet; 446 m)—the Petronas towers in the Malaysian city of Kuala Lumpur (Figure 12.2b).

12.2   Area, Population, and Environment

It is approximately 4000 miles (*c.* 6400 km) from western Burma to central New Guinea and 2500 miles (*c.* 4000 km) from northern Burma to southern Indonesia (Figs. 12.1 and 12.3). Despite these great distances, the total land area of Southeast Asia is only about 1.8 million square miles (4.6 mil-

FIGURE 12.2A
The ninth-century Borobudur Buddhist temple on the island of Java is a symbol of the rich cultural history of Southeast Asia.   *F. Stuart Westmorland/Photo Researchers*

FIGURE 12.2B
Modernity and prosperity characterize some nations in Southeast Asia today. The twin Petronas Towers in Kuala Lumpur are the world's tallest buildings.   *Robin Moyer/Gamma Liaison*

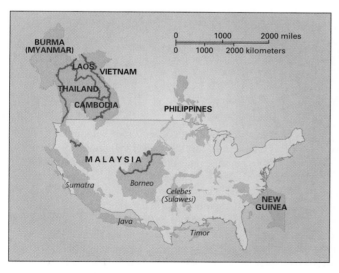

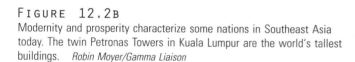

FIGURE 12.3
Southeast Asia compared in area (but not latitude) with the conterminous United States.

lion sq km), or approximately half the size of the United States (including Alaska). The estimated population of the region was 495 million in mid-1996, giving it an overall density of 285 per square mile (slightly less than 110 per sq km)—an average that embraces very great extremes within the region (see Table 10.1, page 263). The average population density of Southeast Asia is high compared with that of much of the world but low compared with most other areas on the seaward margins of east Asia. For example, although Southeast Asia has nearly four times the overall density of the United States, its density is well under half that of the Indian subcontinent and China proper (the humid eastern part of China south of the Great Wall).

Southeast Asia has been less populous than India and China proper throughout history. The entire region probably contained only about 10 million people around the year 1800 and lacked widespread dense populations. Environmental difficulties probably had much to do with the historical slowness of population growth in Southeast Asia. By land the region is relatively isolated, since the northern Indochinese Peninsula is an area of high and rugged mountains. Over many centuries, the ancestors of most of the present inhabitants entered the area as recurrent thin trickles of population, crossing this mountain barrier as refugees driven from previous homelands to the north.

There are formidable environmental challenges within the region itself. Southeast Asia is truly tropical, with continuous heat in the lowlands, torrential rains, a prolific vegetation difficult to clear and keep cleared, soils that are generally leached and poor, and a high incidence of disease. A four- to six-month dry season on the Indochinese peninsula and on scattered smaller areas in the islands has unfavorable agricultural effects, compounded by the high evaporation rates produced by tropical heat. Despite the lush vegetation, most soils have little fertility. The nutrients of the natural ecosystem are largely retained in the natural vegetation itself; when people clear the natural vegetation and plant crops, soil quality deteriorates rapidly. In mountainous areas, deforested slopes erode rapidly; therefore, during the rainy season, rivers carry enormous volumes of mud and silt.

More spectacular environmental difficulties relate to the location of part of the region on the Pacific "ring of fire" (see Chapter 14), subjecting especially Indonesia and the Philippines to earthquakes and volcanic eruptions. Violent wind-and-rain storms known locally as typhoons (known elsewhere as hurricanes and cyclones) often strike coastal Vietnam and the northern Philippines, usually during the summer.

As is typical in LDCs throughout the world, the recent introduction of modern medical and other technologies into the region have reduced the effects of these environmental difficulties and have lowered death rates, producing a tremendous population increase in the last century and a half. This increase, which accelerated after World War II, is now slowing but is still rapid. Between 1973 and mid-1996, population grew from an estimated 316 million to 495.1 million, a 20-year increase of 179 million people, or 57 percent. People overpopulation is generally not yet the problem that it is in India or China, but the current regional growth rate of 1.8 percent means the population will double in 39 years—which would create the prospect of people overpopulation. During recent times, the race between population growth and food supply has seen food supply winning in some countries but losing to population growth in others. From the 1970s through the mid-1990s, large increases in food production significantly outpaced population growth in Malaysia and Indonesia. At the other end of the scale were Cambodia, Laos, and Vietnam, in which war, repression, and inefficient communalization of agriculture thwarted significant progress.

Southeast Asia as a whole is among the world's poorer regions. As of 1994, the per capita gross national product (GNP) of 6 of its 10 countries was below the average for the world's LDCs. In contrast, Malaysia and Thailand were above average for LDCs, and because of their high rates of industrial productivity and economic growth, they are often described as "Asian Tigers," or at least "tiger cubs" (along with Singapore, Hong Kong, South Korea, and Taiwan). In fact, the highly successful commercial-industrial city-state of Singapore has a per capita GNP that places it in the category of MDCs (see Table 10.1, pages 262–263).

12.3 The Economic Pattern

Despite rapid urbanization in recent years, the majority of the people in Southeast Asia are still farmers, and manufacturing is less developed than in Japan, China, and India. Mining contributes significantly to the incomes of a few countries.

Regional Perspective

ASEAN

ASEAN—the Association of Southeast Asian Nations—is a group which consisted until recently only of noncommunist Southeast Asian states. Founded in 1967, the organization included Singapore, Malaysia, Thailand, Indonesia, Brunei, and the Philippines. ASEAN's former adversary, Vietnam, joined in 1995 with the promise of strengthening regional ties and adopting the more open economic system the Vietnamese call "doi moi," or renovation. Since 1977, one of ASEAN's stated major policy objectives has been to rid Southeast Asia of superpower influences and to create a "zone of peace, freedom, and neutrality." With the withdrawal of United States troops from large bases in the Philippines in 1991, however, ASEAN members have grown concerned that they lack a security "umbrella" and that China or possibly even Japan might become more aggressive in the region. ASEAN is not a military pact, but is becoming more concerned with military security as the perceived power of these northern neighbors grows. Several ASEAN members, notably Indonesia, Malaysia, and Singapore, have strengthened ties with the U.S. military to deter other powers from attempts to dominate the region. Cambodia, Laos, and Burma are likely to join ASEAN soon. The organization thus promises to become a large trading bloc and a substantial regional counterweight to China.

FIGURE 12.4
A farmer plowing and hoeing a rice field in Bali, Indonesia, in preparation for planting. Rice seedlings grown in a seedbed will be transplanted by hand to the flooded field. The soil must have a creamy consistency when the young plants are set out. Water buffaloes (upper left) are the main work animals used in Indonesian rice farming. *CARE*

SHIFTING CULTIVATION

Permanent agriculture has been difficult to establish in most parts of this mountainous tropical realm. Much land is used only for a shifting, subsistence form of cultivation, in which a migratory farmer clears and uses fields for a few years and then allows them to revert to secondary growth vegetation while he moves on to clear a new patch. Since about 15 years are necessary for an abandoned field to be restored to forest that can again be cleared for farming, only a small proportion of the land can be cultivated at any one time. This extensive form of land use can support only sparse populations, with farmers often growing crops using only simple digging sticks and little additional cultivation. Unirrigated rice is the principal crop and corn, beans, and root crops such as yams and cassava are also grown. Agriculture is supplemented by hunting and gathering in forested areas. These migratory subsistence farmers often differ ethnically from adjacent settled populations, having historically been driven to refuge in the backcountry by stronger invaders.

SEDENTARY AGRICULTURE

Most of Southeast Asia's people live, often in extremely dense clusters, in scattered areas of permanent sedentary agriculture. These core regions stand in striking contrast to the relatively empty spaces of the adjoining districts. Superior soil fertility in these areas appears to have been the main locational factor in most instances. In Southeast Asia, soils of better-than-average fertility have ordinarily resulted from one or more of the following factors:

1. the presence of lava and volcanic ash of proper chemical composition to weather into fertile soils;

2. the accumulation and periodic renewal of plant nutrients washed from upstream areas and deposited on river flood plains and deltas;
3. a slowing down of the leaching process in areas having a distinct dry season and relatively low rainfall;
4. occurrence in some areas of uplifted coral platforms (relics of once-submerged coral reefs) that weather into a superior soil.

A few areas of Southeast Asia exhibit fairly dense populations without any corresponding soil superiority. Such areas are ordinarily characterized by plantation agriculture based on crops adapted to poor soils, such as rubber (see below).

The typical inhabitant of the areas of permanent sedentary agriculture is a subsistence farmer whose main crop is wet rice, grown with the aid of natural flooding or by irrigation (Fig. 12.4). In some drier areas, unirrigated millet or corn replaces rice. The major crop is supplemented by secondary crops that can be grown on land not suited for rice and other grains. Prominent secondary food crops include coconuts, yams, cassava, beans, and garden vegetables. Some farmers grow a secondary cash crop such as tobacco, coffee, or rubber. In addition, the harvesting of fish is important both along the coasts and along inland streams and lakes. Because of their importance in the food supply, many fish are harvested from artificial fish ponds or from flooded rice fields, in which they are grown as a supplementary "crop."

PLANTATION AGRICULTURE

Subsistence agriculture in Southeast Asia exists alongside tropical plantation cash crops grown for export. The region is one of the world's major supply areas for such crops. This highly commercial type of production is a legacy of Western colonialism which began in the 16th century. Until the 19th century, however, the newcomers were interested mainly in the region's location on the route to China. They were generally content to control patches of land along the coasts and to trade for some goods produced by local people. During the 19th century, the Industrial Revolution in Europe and North America greatly en-

larged the demand for tropical products, and as a result, Europeans extended their control over almost all of Southeast Asia. European colonists used their capital and knowledge, along with indigenous and imported labor, to bring about a rapid increase in production for export. The colonizers gave emphasis to certain local commodities, such as copra (coconut meat) and spices, and introduced a number of entirely new crops.

Definitions and Insights

RUBBER

The most notable cash crop introduced into Southeast Asia was rubber, in the 1870s. The Englishman Sir Joseph Priestley had given the name "rubber" to the latex of the *Hevea brasiliensis* tree in 1770, when he discovered he could use it to rub errors off the written page. Extensive exploitation of this tree in its native Brazil gave rise to the "rubber boom" of the late 19th century, when Brazilian rubber traders were so wealthy that they used bank notes to light cigars and sent their shirts to be laundered in Europe. In 1876 an Englishman named Henry Wickham smuggled 70,000 rubber seeds out of Brazil. De-

lighted botanists at London's Kew Gardens cultivated them and shipped them on to Ceylon (Sri Lanka) and Southeast Asia for experimental commercial planting. Rubber trees proved exceptionally well-adapted to the climate and soils of South and later Southeast Asia (Fig. 12.A). Back in Brazil, however, a fungus known as leaf blight wiped out the rubber plantations, and by 1910, bats and lizards came to inhabit the mansions of Brazil's rubber barons. Fortunately, the Kew Gardens stock was free of the fungus; by 1940, 90 percent of the world's natural rubber came from Asian plantations.

FIGURE 12.A
The rubber tree has been more commercially successful in Southeast Asia than in its native Brazil.   *Thomas D. W. Friedman/Photo Researchers*

The usual method of introducing commercial production was to establish large estates or plantations managed by Europeans but worked by indigenous labor or labor imported from other parts of Monsoon Asia (see Chapter 2, page 42). Development of these enterprises was aided by a favorable climate, the abundance of land, and the availability of cheap transportation by water. The major difficulty was the recruitment of adequate labor from a population already fully engaged in food production and not poor enough to be drawn away from village life into a wage labor economy. The solution in many areas was large-scale importation of contract labor from India and China. These massive migrations further complicated an already complex ethnic mixture, and account in part for the unusual distribution of ethnic groups in the region today.

Plantation activity came to have widespread repercussions for the economic life of local inhabitants. Many of them learned by example and entered commercial production on a small scale on their own. Today, small landowners command an important share of the export production of most "plantation" crops in Southeast Asia.

Many types of plantation cash crops have been produced in Southeast Asia. Wide fluctuations have occurred over the years in the crops grown, the centers of production, the amounts exported, and the prosperity of the producers. Factors such as fluctuating world demands, regional and world competition, changing political conditions, and the occasional ravages of plant diseases have caused these shifts. The major plantation cash crops now exported from Southeast Asia are:

1. **Rubber.**   Over four fifths of the world's natural rubber is produced in Southeast Asia, primarily in Malaysia, Indonesia, and Thailand. However, synthetic rubber, most of which is made from petroleum, now supplies the greater part of the world's rubber consumption.
2. **Oil Palm and Coconut Palm Products.**   Palm products consist primarily of palm oil, coconut oil, and copra (dried coconut meat from which oil is pressed). Malaysia is the world leader in palm oil production, and the Philippines and Indonesia dominate the world output of coconut palm products.
3. **Tea.**   Indonesia is the region's leading producer of tea, and is the world's fourth largest.

Many nations in Southeast Asia produce other export crops. These include coffee from Indonesia, cane sugar from the Philippines and Indonesia, pineapples (of which the Philippines and Thailand have become the world's leading producers and exporters), and many others. However, the countries of Southeast Asia are generally coming to depend less heavily on their agricultural exports as mining and manufacturing activities develop.

COMMERCIAL RICE FARMING

One significant impact of Western colonialism on the economy of Southeast Asia was the stimulation of commercial rice farming in areas that had formerly been unproductive. The development of plantation agriculture, mining, and trade in Southeast Asia provided a market for rice by creating a large class of people who worked for wages and had to buy—rather than produce—their food. During the same period, Western economic, medical, and sanitary innovations helped decrease death rates and bring about an enormous increase in population and a growing demand for food. Western technology facilitated the bulk processing and movement of rice and aided in the development of drainage, irrigation, and flood-control facilities needed to produce it on a large scale in previously undeveloped areas. As a consequence, an Asian pioneer movement into areas capable of expanded rice production took place.

Three of these areas gained prominence in commercial rice growing—namely, the deltas of the Irrawaddy, Chao Praya, and Mekong rivers, located in Burma, Thailand, and Vietnam and Cambodia respectively. Almost impenetrable wetlands and uncontrolled floods had kept these deltas thinly settled, but incentives and methods for settlement have turned them into densely populated areas within the past century. The farms there are larger than most in east Asia, and the surpluses of rice produced on them until the 1960s represented nearly one third of the world's total rice exports. War and economic dislocation then ended the ability of Vietnam and, in the 1970s, Cambodia, to generate this surplus production. Expansion of production in Thailand partially replaced the loss of the Mekong Delta as a surplus rice-producing area, and also in Thailand, corn (maize) became a major food export. In the mid-1990s, Thailand, Burma, and the resurgent Vietnam had become the region's leading rice exporters. Since most Southeastern Asian countries are also major grain importers, there is a substantial regional market for rice.

AGRICULTURAL GROWTH AND DEFORESTATION

Near areas of dense settlement in Southeast Asia are large areas still relatively undeveloped, despite the fact that they apparently are capable of supporting large populations. Clearing of forests, draining of swamps, and construction of irrigation systems proceeds when population growth demands such expansion. Many environmentalists view the attendant destruction of tropical rain forest here as an international ecological problem. Southeast Asia's forests were already largely destroyed by the early 1990s; for example, only 15 percent of Thailand's forests remained at that time, along with 14–20 percent of the Philippines', and 16–19 percent of Vietnam's. Deforestation has moved into Malaysia, where 47 percent of the land area is still forested, and to Indonesia, where 54 percent is forested. Malaysia is now the world's largest exporter of tropical hardwoods, and, at current rates of deforestation, it will have logged all of its principal forest reserves in Sarawak on the island of Borneo by the year 2005.

Although Southeast Asia's tropical forests are smaller in total area than those of central Africa and the Amazon Basin, they are being destroyed at a much faster rate, most significantly by commercial logging for Japanese markets rather than by subsistence farmers. Environmentalists fear the irretrievable loss of plant and animal species and the potential contributions to global warming that deforestation in this region may cause. In 1995, both Malaysia and Indonesia moved to slow the destruction, with Indonesia taking the aggressive step of banning the use of fire to clear forests.

Many of the species that find habitats in these forests are **endemic** (found nowhere else on Earth). Indonesia contains 10 percent of the world's tropical rain forests and, after Brazil, is known as the world's second most important "megadiversity" country—with about 11 percent of all plant species, 12 percent of all mammal species, and 17 percent of all bird species within its borders. The Southeast Asian region is also particularly significant in biogeographical terms because of the so-called "Wallace's Line"—named for its discoverer, English naturalist Alfred Russel Wallace—which divides it (see Fig. 12.1). Nowhere else on Earth is there such a striking local change in the composition of plant and animal species in such a small area on either side of this "divide." East of the line (separating the Indonesian islands of Bali and Lombok, for example), marsupials are the predominant mammals, while west of the line, placental mammals prevail. Similarly, bird populations are remarkably different on either side of the line.

PRODUCTION AND RESERVES OF MINERALS

Petroleum is the most important mineral resource in Southeast Asia, although the region's reserves and production are not impressive on a world scale. Seven countries—Indonesia, Malaysia, Brunei, Burma, Thailand, Vietnam, and the Philippines—produce oil. Their combined output is about 5 percent of world oil production, with Indonesia the leading producer. Indonesia's main oil fields are in Sumatra and Indonesian Borneo (known as Kalimantan).

Although it does not loom large on the world scene, Southeast Asian oil is very important to the countries that own and produce it. Oil and associated natural gas provide nearly all the exports of Brunei and make that tiny Muslim country one of the world's wealthiest nations as measured by per capita GNP (Fig. 12.5). Oil, gas, and refined products supply almost one fourth of Indonesia's exports and about one tenth of Malaysia's. Singapore profits as a processor of oil. The principal customer for Southeast Asian oil and natural gas (shipped in liquid form) is Japan, which is the leading trade partner of several Southeast Asian countries.

Considerable mineral wealth other than oil and gas exists in Southeast Asia. A wide variety of metal-bearing ores are extracted in many locations, and many unexploited reserves remain. Tin is the most important ore currently mined, with about one fifth of world output from Malaysia and another fifth from Thailand and Indonesia combined. The Philippines produces

FIGURE 12.5
Brunei is a tiny, prosperous Muslim nation. This mosque of Omar Ali Saifuddin was designed by an Italian architect and cost five million dollars to build.   *Alan Evrard/Photo Researchers*

the greatest variety of metals—notably copper, chromite, nickel, silver, and gold—in significant amounts. The most serious mineral deficiency in the region is the near absence of high-grade coal. However, there are sizable hydropower potentials that could supply additional energy.

12.4   The Countries

Although the countries of Southeast Asia have many broad similarities, each has its own distinctive qualities, exhibiting a different combination of environmental features, indigenous and immigrant peoples, economic activities, and culture traits. Many of the characteristics and problems of these countries are the outcome of European imperialism, but in no two countries have the results of colonialism been the same. Following are portraits of the major countries.

BURMA (MYANMAR)

Burma (renamed Myanmar) by its military government in 1989; 1996 population: 46 million) is centered in the basin of the Irrawaddy River and includes surrounding uplands and mountains. Within the basin are two distinct areas of dense population: the Dry Zone, around and south of Mandalay (population: about 533,000, city proper), and the Irrawaddy Delta, which includes the capital and major seaport of Rangoon (Yangon); population: 2.7 million, city proper; 2.8 million, metropolitan area) (Fig. 12.6). The Dry Zone has been the historical nucleus of the country. The annual rainfall of this area—34 inches (87 cm) at Mandalay—is exceptionally low for Southeast Asia, and there is a dry season of about six months. The people are supported by mixed subsistence and commercial farming, with millet, rice, and cotton as major crops. During the past century, the Dry Zone has been surpassed in population by the Delta, where a commercial rice-farming economy now provides one of the country's two leading exports (forest products is the other). The Irrawaddy forms a major artery of transportation uniting the two core areas.

The indigenous Burman people, most of whom live in the Irrawaddy Basin, number an estimated 69 percent of the country's population. A minority of around 1 million Indians also lived here prior to World War II. Between the end of World War II and 1970, however, most of these Indians were ejected from this area through expropriations of property and other oppressive measures resulting from the nationalization of the economy, including land ownership. The Shan Plateau of eastern Burma is inhabited by the ethnic Shans to the north and ethnic Karens to the south and in the delta of the Salween River. They

account, respectively, for about 9 and 7 percent of the country's population. And a variety of hill-dwelling tribes inhabit the Arakan Mountains of the west and the northern highlands. About 90 percent of Burma's citizens are Buddhists.

Great Britain conquered Burma in three wars between 1824 and 1885. It became an independent republic outside the British Commonwealth in 1948. There has been almost constant civil war since independence. At one time no fewer than eight different rebellions were in progress. Both communism and ethnic separatism have supplied major motivations for rebel forces drawn principally from the Karens, the Shans, a number of less numerous hill peoples, and Burmese Muslims who fear the creation of a pure Buddhist state. The economy has been badly damaged by the fighting, and for years was mismanaged under a form of rigid state control promoted as "The Burmese Way to Socialism." Burma became the poorest noncommunist country in Southeast Asia, but, since the late 1980s, it has begun to abandon socialism in favor of a free-market economy and has enjoyed considerable economic growth. China is Burma's main trading partner and military ally.

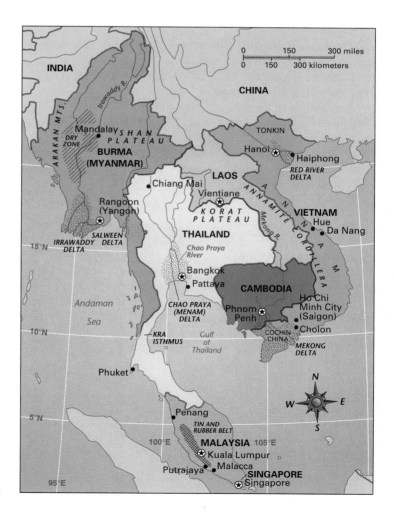

Figure 12.6

Index map of the Southeast Asian mainland. Stars show political capitals and the independent state of Singapore. Area symbols show river deltas, the Dry Zone of Burma (Myanmar), and the Tin and Rubber Belt of Malaysia.

The repressive military government that seized power in Burma in 1988 has yielded little to popular pressure for democratization. In 1995 the government did release the opposition leader and Nobel Peace Prize laureate Aung San Suu Kyi, who had been under house arrest since leading her anti-military political party to victory in the 1990 elections (the results of which the military government nullified). It is uncertain whether the release of Ms. Suu Kyi, popularly known as "the Lady" and as the daughter of the founder of modern Burma, signals a new beginning for democracy in the troubled country.

THAILAND

Thailand (formerly Siam; 1996 population: 61 million) is located in the delta of the Chao Praya River, known to Thais as the Menam ("The River"). Annual floods of this river irrigate the rice that is the most important crop in this Southeast Asian land of relative abundance and progress. Thailand is now the world's largest exporter of rice.

On the lower Chao Praya is Bangkok (population: 5.62 million, city proper; 7.06 million, metropolitan area), the capital, main port, and only large city. It is a notoriously crowded and polluted city; air pollution is responsible for as many as 400 deaths yearly. But it is also a city of stunning architectural marvels, particularly its Buddhist temples. Bangkok's network of canals, thronged with both commercial and residential activities, give it a distinctively amphibiouslike character; especially renowned is its "floating market" (Fig. 12.7).

Areas outside the delta are more sparsely populated and include the mountainous territories in the west and north—inhabited mainly by ethnic Karens and a variety of mountain tribes—and the dry Korat Plateau to the east, populated by the Thai, related Laotian, and Cambodian peoples. To the south in

FIGURE 12.8

Trained elephants have long played an important role in Southeast Asian ceremonies, traditional warfare, and everyday transportation and work. Those in the photo are participants in a ceremony held annually in eastern Thailand to commemorate the use of elephants in past wars.   *Gary E. Johnson*

the Kra Isthmus, a part of the Malay Peninsula, live more than 2 million Malays. Most Malays are Muslims, but an estimated 95 percent of the people of Thailand are Buddhists whose faith includes elements borrowed from Hinduism and local spiritualism. Spectacular Buddhist temples are a prime attraction for one of Southeast Asia's largest tourist industries, as are ceremonies celebrating Thailand's monarchy and other historical traditions (Fig. 12.8).

Governed today as a democracy in which military influence is strong and in which the monarchy has largely symbolic functions, Thailand was the only Southeast Asian country to preserve its independence throughout the period of colonialism. This peculiar status was due to then Siam's position as a buffer between British and French colonial spheres. A number of border territories were lost, however, and the country has exhibited irredentist (separatist) tendencies for many years (see Definitions and Insights, Chapter 5, page 41). Thailand's attention recently has focused more on combating insurgencies within its own outlying territories than on attempting to regain lost territories.

The major ethnic minority of Thailand is Chinese. An estimated 12 percent of the population, or 7 million ethnic Chinese, reside in the country and are claimed as citizens by both Thailand and China. They control much of the country's business, a situation that has roused ill feeling on the part of the Thais and makes the loyalty of the Chinese a matter of importance. Thailand has enjoyed more internal tranquility than most of Southeast Asia, but since the late 1960s, guerrilla insurrections motivated both by communist aims and by ethnic separatism have persisted in the northeast and south. In the late 1970s, the situation was further complicated by a mass movement of refugees from adjoining Cambodia. In 1993, Malay guerrillas

FIGURE 12.7

The Floating Market in Bangkok, Thailand, typifies the amphibious nature of life in Southeast Asia. Water transportation is vitally important in the river deltas and islands where most Southeast Asians live.   *Lovina Ebbe*

seeking independence for Thailand's Muslim south burned schools and ambushed government troops in that area.

In spite of conflict and refugee problems, Thailand has one of the strongest economies in Southeast Asia. From 1993–95, the economy grew a robust 8 percent annually. The country is shifting away from its traditional exports of garments and textiles in favor of "medium-tech" industries such as computer assembly. Gem cutting and seafood production are also important industries. And through the mid-1990s, tourism was Thailand's leading source of foreign exchange.

Thailand's assets and attractions are diverse. A liberal social climate has given rise to the unique phenomenon of sex tourism, with package tours catering to an international clientele, particularly Japanese men. However, the spread of HIV among prostitutes in the red-light districts of Bangkok and the once-popular beach resorts of Pattaya and Phuket (see Fig. 12.6) now threaten both the country's public health and its lucrative tourist trade. Thailand's government has responded with an aggressive and increasingly successful anti-AIDS public awareness campaign, which has resulted in a decrease in the number of new HIV infections every year since 1991.

Other Southeast Asian countries lag far behind Thailand in attacking the AIDS scourge. The government of the Philippines has been unsuccessful in mounting an anti-AIDS campaign because of opposition from the country's powerful and anticontraception Catholic clergy. Church officials have denounced the government's anti-AIDS program as "intrinsically evil," and have set boxes of condoms on fire at antigovernment demonstrations. Similarly, in the conservative and largely Islamic nations of Indonesia and Malaysia, Muslim clerics denounce the governments' anti-AIDS campaigns as efforts to encourage promiscuity.

Definitions and Insights
THE "GOLDEN TRIANGLE"

The "Golden Triangle" is the remote region where Laos, Thailand, and Burma meet. None of these countries exercises successful control over its frontier in this area. As in the Northwest Frontier Province of Pakistan, the absence of a strong central government presence and the existence of ideal growing conditions have made the Golden Triangle one of the world's greatest centers of opium poppy cultivation (Fig. 12.B). "China White," or Asian-grown heroin, originates here and is commanding a growing share of the global heroin market. In 1983, only about 3 percent of the heroin used in the United States came from the Golden Triangle. By 1995, more than half did. Nearly all of the region's heroin refining is done in Burma. Remote laboratories there produced about two thirds of the world's total heroin output.

About 100 tons of heroin leave the Golden Triangle each year, bound for markets in the U.S., Australia, and Western Europe. International authorities succeed in stopping only an estimated 10–20 percent of that flow. In recent years, with new heroin-producing areas cropping up in Mexico and elsewhere, the price of heroin to the consumer has dropped and its quality has risen. In the United States these trends are noticeable in the growing appeal of heroin among middle and upper income groups, in the use of inhaled and other previously-rare forms of heroin consumption, and by the increasing numbers of deaths due to heroin overdose. The problem is also growing in the source region: Both Thailand and Laos have high rates of heroin addiction and in the early 1990s began experiencing an epidemic of AIDS due to both intravenous heroin use and associated prostitution. In 1994 an estimated 150,000–450,000 of Burma's 43 million people were infected with HIV, and in Thailand, an estimated 800,000 of 57 million people were HIV-positive.

FIGURE 12.B
A 15-year-old girl of one of the hill tribes in Thailand's Golden Triangle region lances a closed poppy (*Papaver somniferum*) so that the opium can be collected later in the day. Most of the raw opium will be processed into heroin.   *Steven L. Raymer/National Geographic Image*

VIETNAM, CAMBODIA, AND LAOS

Vietnam, Cambodia, and Laos are the three states that emerged from the former French colony of Indochina. Vietnam is the largest of the three and is geographically the most complex. Three major regions have long been recognized in Vietnam: Tonkin, the northern region, consists of the very densely populated delta of the Red River and a surrounding frame of sparsely populated mountains; Annam, the central region, includes the sparsely populated Annamite Cordillera and numerous small, densely populated pockets of lowland along its seaward edge; and Cochin China, the southern region, lies mainly in the delta of the Mekong River and has a fairly high population density. Almost nine tenths of Vietnam's people are ethnic Vietnamese. Vietnamese are related closely to the Chinese in many cultural respects, including language and the shared religious elements of Confucianism, Buddhism, and ancestor worship. A Catholic minority reflects the French occupation, and there are minor native religions.

Many minority groups complicate the region's ethnic makeup. Lao people form a minority in the mountains of northern Vietnam, and Cambodians are a minority in southern Vietnam. In the 1980s, Vietnam systematically settled hundreds of thousands of Vietnamese in the eastern part of conquered Cambodia. In the mid-1970s, there were perhaps 5 million people belonging to a variety of mountain tribes in Vietnam and Laos. Some of them remain, but the number is again uncertain, as many were allied with the United States during the Vietnam War and have since then been subjected to devastating attacks by Vietnamese forces.

Until the mid-1970s, there was a large Chinese minority in the major cities of the three countries, numbering perhaps 2 million altogether and dominating much of the commerce of the region. How many remain is unknown, as they were victimized by repressions that created millions of Indochinese refugees after 1975.

Cambodia and Laos occupy interior areas except for Cambodia's short coast on the Gulf of Thailand. They are very different countries physically, and a particular culture dominates each. Cambodia is mostly plains along and to the west of the lower Mekong River, although it has mountainous fringes to the northeast and southwest. Almost all the 11 million people of Cambodia are Cambodians, also known as Khmer people. Their language is Khmer, and their predominant religion is Buddhism.

Laos is extremely mountainous, sparsely populated, and landlocked. It has 68 identified ethnic groups, the largest group of which is the ethnic Lao, who are linguistically and culturally related to the Thais, and whose dominant religion is Buddhism. Their largest numbers are in the lowlands, especially in and near the capital, Vientiane. Most of the ethnic minorities of Laos live in the mountain regions, including the country's second largest group, the Hmong.

The Vietnam War

France conquered Indochina between 1858 and 1907. The French first extinguished the Vietnamese Empire, which covered approximately the territory of today's Vietnam, and defeated the Chinese whom the Vietnamese called upon for help. Later, the French took Laos (in 1893) and western Cambodia (in 1907) from Thailand. France's administration of Indochina, centered in Hanoi (population: 906,000, city proper; 1.28 million, metropolitan area) and Saigon (formally Ho Chi Minh City, but still known popularly as Saigon; population: 4.32 million, city proper), left a considerable cultural imprint. Its major economic success lay in opening the lower Mekong River area to commercial rice production, thus converting both Vietnam and Cambodia into important exporters of rice. Japanese forces overran French Indochina in 1941, initiating five decades of warfare in the area. In World War II, a communist-led resistance movement was formed to carry on guerrilla warfare against the Japanese. When French forces attempted to reoccupy the area after the end of World War II, these forces fought to expel them. After eight years of warfare, the Vietnamese, with Chinese material support, destroyed a French army in 1954 at Dien Bien Phu in the mountains of Tonkin, and France withdrew to end its century of colonial occupation.

Four states came into existence with France's departure. Laos and Cambodia became independent noncommunist states. North Vietnam, where resistance to France had centered, emerged as an independent communist state. South Vietnam gained independence as a noncommunist state that received many anticommunist refugees from the north. Vietnamese Catholics were prominent among those who fled southward. This partition of Vietnam into two states (see Fig. 12.1) was largely the work of United States power and diplomacy.

From the outset, South Vietnam was a client state of the United States, while North Vietnam became allied first with communist China and then with the Soviet Union. Meanwhile, warfare in Vietnam continued, eventually to a different conclusion. Between 1954 and 1965, an insurrectionist communist force supported by North Vietnam and known as the Viet Cong achieved increasing successes in South Vietnam in an attempt to reunify the country on its own terms. A similar situation existed in Laos. Increasing intervention by the United States in South Vietnam escalated from a small scale to a large scale in 1965 and eventually involved the full-scale commitment of half a million U.S. military personnel (Fig. 12.9a). North Vietnam responded by committing regular army forces against American and South Vietnamese troops. The United States never invaded North Vietnam, although American air strikes there were devastating. American bombs, napalm, and defoliants such as Agent Orange caused enormous damage to the natural and agricultural systems of Vietnam, destroying an estimated 5.4 million acres of forest and farmland in an effort described in military parlance as "denying the countryside to the enemy."

The U.S. avoided leading an invasion of the North in part because of the risks posed by Chinese and Soviet support of the

FIGURE 12.9A
U.S. troops in combat in a Vietnamese village in 1972. The war inflicted enormous suffering on all sides.   *Johner Vie Retro Guerre Du Vietnam/Gamma Liaison*

A

FIGURE 12.9B
Vietnam opened its doors to capitalism and Western goods in the mid-1990s.   *Noboru Komine/Photo Researchers*

B

North. In limiting the theater of ground warfare however, the United States found itself unable to expel from the South a North Vietnamese army that was determined, skillfully commanded, increasingly better equipped by its allies, accomplished in guerilla tactics, and willing to bear heavy losses. About one million Vietnamese died in the conflict, and about 58,000 U.S. soldiers and support staff perished. Almost all American forces were withdrawn in 1973 from the costly war, which was extremely divisive at home. There are still 2238 Americans listed as "missing in action" in Indochina.

North Vietnam completed its conquest of South Vietnam in 1975. Saigon was renamed Ho Chi Minh City, after the original leader of Vietnam's communist party. Vietnam entered a period of relative internal peace; however, this unusual condition did not preclude border conflicts with Cambodia and Vietnam's 1979 conquest of that country, continued warfare against Cambodian resistance groups, warfare against resistance groups in Laos, the attempted extermination of hill tribes that had been friendly to the United States, and a brief border war with China to resist Chinese aggression in 1979. Within its borders, reunited Vietnam unleashed a repression strong enough to cause a massive outpouring of refugees. This situation intensified, especially for the country's ethnic Chinese, after China invaded Vietnam in 1979.

Definitions and Insights

THE BOAT PEOPLE

The desperation of many people in Vietnam and its neighboring wartorn countries created the phenomenon of the "boat people"—hundreds of thousands of refugees who fled onto the open ocean on large rafts, with chances of survival variously estimated at 40 to 70 percent. The total refugee flow from Vietnam since 1975 has been estimated at about 800,000.

The United States has provided permanent homes for more than half of those who have been resettled, with France, Canada, and Australia also receiving sizable numbers. Large numbers of Vietnamese and Laotian refugees occupy camps in China near its borders with those countries. Boat people have also filled refugee camps in Hong Kong, Malaysia, Indonesia, Thailand, and the Philippines (Fig. 12.C).

FIGURE 12.C
Warfare and poverty in Vietnam prompted an exodus of boat people. While some have been accommodated in new lands, others have languished in refugee camps like this one in Hong Kong.   *Alan Evrard/Photo Researchers*

The United States and other Western governments have withdrawn the welcome mat for the refugees, and the Asian countries are already undertaking or are considering measures to repatriate the Vietnamese boat people within their borders to Vietnam. There are widespread fears that, in the wake of China's acquisition of Hong Kong in 1997, China will forcibly repatriate its population of boat people to Vietnam.

Vietnam Today

Vietnam is now restoring its wartorn landscape through large-scale reforestation, agricultural reclamation, and nature conservation programs, aided by international organizations such as the World Wildlife Fund. Animal species new to science have been discovered in the Vu Quang Nature Reserve in the rugged Annamite Cordillera region on the border with Laos. These include two hoofed mammals—the Vu Quang ox and the giant muntjac, a new species of barking deer. Such natural treasures, along with Vietnam's tigers, rhinoceroses, and elephants may help lure international ecotourists and provide much-needed revenue.

Tourism is already booming, especially in the south, augmenting a generally positive picture of Vietnam's economy (Fig. 12.9b). The government halted collectivized farming in 1988, turning control of land over to small farmers under 20-year lease agreements. The results have been impressive; Vietnam since 1988 has become a major exporter of rice. In addition to joining ASEAN, Vietnam in 1995 restored full diplomatic relations with the United States, and U.S. businesses began scrambling for consumers in this promising market of 72 million people. The dual personalities of north and south are being perpetuated in the new economic climate, with most investment and infrastructural improvement focused in the south. Along with a growing gap between rich and poor people in the country, this trend could slow overall development for the country.

Like that of many LDCs, however, Vietnam's progress is also threatened by the specter of uncontrolled population growth. Since the war with the United States ended in 1975, Vietnam's population has increased by more then 60 percent. It will probably reach 168 million by 2025, a staggering figure in view of the country's limited size and resource base. The government in Hanoi has mounted a family planning campaign in an effort to reduce birth rates, but so far most Vietnamese have ignored it and continue to prefer large families.

Cambodia's Killing Fields

Cambodia escaped the kinds of ills known to Vietnam until relatively late, but then suffered even worse catastrophes. After five years of fighting in the country, a communist insurgency overcame the American-backed military government in 1975. Known as the Khmer Rouge and led by a man with the *nom de guerre* of Pol Pot, this communist organization ruled the country for four years with a savagery almost unparalleled. In 1973, the population had been estimated at 7.5 million, but by 1979 it was estimated at only about 5 million. Mass murders between 1975 and 1979 eliminated nearly a third of the population. Some intended victims managed to escape as refugees from Cambodia's "killing fields" to adjoining Thailand. The Cambodian communists' stated policy was to build a "new kind of socialism" by eradicating the educated and the rich, emptying the cities, and breaking up the family. The population of Phnom Penh, the capital and by far the largest city, was expelled on a death march to the countryside. The city's population dropped from over 1 million in 1975 to an estimated 40–100,000 in 1976. By 1996, it had recovered to an estimated 300,000.

Landscape in Literature

THE POETRY OF VIETNAM

Vietnamese poetry was born out of Chinese influences more than a thousand years ago. Over time, it developed a distinct Vietnamese flavor. Although an educated elite, Vietnamese poets (in contrast to their Chinese counterparts) interacted freely with and borrowed heavily from the oral traditions of peasants. The result is a unique poetry that has firm roots in the land and life of Vietnam. Here are a few selections that offer insight into the world view, landscape appreciation, and wisdom of Vietnam's people from early times until the beginning of the colonial age.

"Reply to a Northerner Who Asked About Annam's Customs," by Ho Quy Ly (1336–?). (Note: Annam was the name given to Vietnam by the T'ang court in 679 A.D. when it was a dependency of China. A Northerner is a person from China.)

> You asked about things of Annam.
> Jade bottles brim with fresh-brewed
>   wine.
> Gold knives carve up delicious fish.
> Each year, for two or three full
>   months,
> Spring gardens burst with peach and
>   plum.

"A Song of Wild Freedom," by Tran Quoc Tang (1252–1313). (Note: The song describes a Buddhist monk who carried all of his belongings in a homespun bag at the end of a stick. The author likens fleeting time to the glimpse one catches of a white colt galloping past a crack in the window. According to Buddhism, the four elements are earth, water, fire, and air.)

> Behold the sky and earth—how vast,
>   how vast!
> Rid of all chains, lean on your staff
> and roam.
> Climb higher, higher—mountains
>   harbor clouds.
> Dive deeper, deeper—oceans hoard
>   the flood.
> When hungry, eat a feast of
>   scentwood smoke.
> When sleepy, make your bed in
>   Neverland.
> For music, tootle on a holeless flute.
> For silence, burn the incense of
>   Release.
> If wearied, journey toward the realm
>   of Bliss.
> If thirsty, drink the nectar of No Care.
> In Van-nien Village, hum the K'ao-
>   p'an ode.
> On Cuu-khuc River, sing the Ts'ang-
>   lang song.
> Row up Ts'ao Creek and bow to
>   Master Lu.
> Scale Mount Shih-t'ou and consort
>   with Old P'ang.
> Your joy? That kept inside a
>   homespun bag.
> Your madness? That of him who
>   spread the word.
> Tut, tut! What's wealth or rank? A
>   floating cloud.
> Oho! What's time? A gallop past a
>   crack.
> Officialdom is rigged with traps—so
>   what?
> The fickle world blows hot and
>   cold—how now?
> Keep your shirt dry in water, deep or
>   shallow.
> When needed, do your part; when
>   useless, hide.
> Free all four elements, don't tie them
>   up.
> Bustle and fret no more—your
>   karma's done.
> You've met your heart's desire—
>   you've found your place.
> Let no constraint or duress cripple
>   life.

"Thinking of Home at Dusk," by Lady Thanh Quan (19th century):

> Sunset—the sky's all blurred with
>   shades of dusk.
> A conch blares somewhere, blends
>   with faint post drum.
> Back at fair havens, fishermen rest
>   their oars.
> Bound for lone hamlets, herdboys tap
>   on horns.
> Winds sweep through woods—birds'
>   wings wear out in flight.
> Dew falls on roads—the travelers'
>   steps make haste.
> One stays at home, the other roams
>   the world.
> To whom confide what chills or
>   warms my heart?

"Looking Far Ahead," by Tran Te Xuong (1870–1907). (Note: The poem reflects the author's despair over the present and future of his country under French domination.)

> Through five night watches I've
>   stayed wide awake.
> I'm looking far ahead—I feel a jolt.
> To ancient wisdom people have gone
>   blind.
> The human race may vanish, all
>   wiped out.
> Hills, dug for wealth, will someday
>   crumble down.
> Seas, opened up for trade, in time will
>   tilt.
> The earth, once hollowed out, will
>   fade away.
> Folks say a comet's started whirling
>   round.

In 1979, the communist/nationalist imperialism of Vietnam came to dominate Cambodia's political landscape. There had been escalating border conflicts between the two countries, and in 1979 Vietnamese troops quickly conquered the greater part of Cambodia and installed a puppet communist government. Then the Khmer Rouge, supported by China (Vietnam had by then become a close ally of China's rival, the Soviet Union), began a guerrilla resistance to the Vietnamese. Two noncommunist Cambodian guerrilla forces also took the field. These three resistance forces often quarreled among themselves, but all maintained the conflict with the Cambodian/Vietnamese army of the Phnom Penh government, which periodically pursued them into Thailand and engaged Thai border military units. At first, the Vietnamese rejected international food aid for Cambodia and followed a policy of systematically starving the country, apparently to weaken resistance. Later they attempted to promote recovery under their puppet government. Peace negotiations among Cambodia's contending factions led, in 1991, to an agreement resulting in the restoration of democratic government under United Nations supervision. A new constitution with new leaders took effect in 1993, and Cambodia's former king Norodom Sihanouk returned to the throne. The Khmer Rouge participated in democratic reform at first, but then withdrew from the process and remain as a potential threat to stability in Cambodia.

Even in this relatively peaceful time, the people of Cambodia must deal with a persistent scourge of war: land mines. Sown by hostile forces during years of warfare, the antipersonnel explosives remain active indefinitely. More than 35,000 Cambodians have been maimed by mines, and 200–300 more casualties occur each month. People sow new mines, imported mainly from China and Singapore, to protect their property in Cambodia.

The Lao Way

Laotians also suffer from ordnance dating to the Vietnam war. Between 1964 and 1973, U.S. warplanes dropped more than two million tons of bombs—more than the U.S. dropped on Germany in World War II—on the Laotian frontier with Vietnam in an effort to disrupt communist supply lines on the nearby Ho Chi Minh trail and to prevent communist troops from entering Laotian cities. Today millions of unexploded American cluster bombs remain in the region. Smaller than a tennis ball, the cluster bomb contains more than 100 steel ball bearings. Many curious Laotian children who play with them are killed or dismembered. About half of the territory of Laos may contain unexploded ordnance, according to United Nations studies, creating a serious deterrent to farming in a country with inadequate food supplies.

The administrative capital of Laos is Vientiane (with just 125,000 people), located on the Mekong River where it borders Thailand (Fig. 12.10). In 1975, when Laos came under Vietnamese occupation, Vientiane replaced the royal capital of Luang Prabang ("Royal Holy Image"), located upriver on the Mekong, when the Pathet Lao organization overthrew the Laotian monarchy and established a new "People's Democratic Republic." The communist dictatorship now rules Laos in a benign fashion, encouraging a market economy and foreign investment. The government calls its conservative approach to economic reform and social change "the Lao Way." To slow the potential flood of foreign world views and materials into the country, Laos has built only one bridge to Thailand across the

F I G U R E 12.10

Lan Xang Avenue is the main street of the Laotian capital of Vientiane, on the Mekong River. *Noboru Komine/Photo Researchers*

Mekong River, and has resisted appeals for the construction of a second bridge.

Landlocked Laos is now one of the world's least developed countries, ranking 133 out of the 173 countries on the United Nations Human Development Index. About 85 percent of its people are peasant farmers. Exports are limited to timber, hydroelectric power (to Thailand), and garments and motorcycles assembled by cheap Laotian labor from imported materials. Like Vietnam and Cambodia, Laos suffers from the **brain drain** prompted by years of warfare, political turmoil, and underdevelopment—conditions which caused the best-educated and most talented of the country's inhabitants to flee. After the 1975 takeover, 343,000 Laotians, mostly members of the Hmong tribe, fled to neighboring Thailand. Many Hmong refugees subsequently emigrated to the United States.

Given the conditions of warfare, political turmoil, oppression, and inefficient economic systems, it is not surprising that the three Indochinese countries are, with Burma, the poorest in Southeast Asia. By 1997 the outlook seemed to be improving as tensions with the outer world relaxed, privatization of communist economies increased, the supply of goods and services improved, considerable foreign aid and investment flowed in, and the number of returning refugees—voluntary and forced—exceeded outmigrants.

MALAYSIA AND SINGAPORE

The small island of Singapore (Fig. 12.11), narrowly separated from the southern tip of the mountainous Malay Peninsula, lies at the eastern end of the Strait of Malacca, which is the major passageway through which sea traffic funnels between the Indian Ocean and the South China Sea. For centuries, European sea powers contested for control of this passageway. British control over the strait became continuous from 1824 and it was exercised from the port of Singapore, founded on the southern side of the island five years earlier. Under the British, Singapore, with its relatively central position among the islands and peninsulas of Southeast Asia, developed into not only a major naval base but also the region's major **entrepôt**. Singapore ranks with Rotterdam as the world's largest ports in tonnage of goods shipped.

From Singapore, the British gradually extended their political hold over the adjacent southern end of the Malay Peninsula. This expansion gave Britain control over the part of the peninsula which is now included, along with northwestern Borneo (except Brunei), in the independent country of Malaysia. This southern end of the peninsula is known as Malaya, Peninsular Malaysia, or West Malaysia. The Borneo section is composed of the units (former colonies) of Sarawak and Sabah, which together are known as East Malaysia.

During the British period, Malaya developed a highly commercialized and relatively prosperous economy. Tin and rubber became the major commercial products, with oil palms and coconuts of secondary importance. Chinese and British companies pioneered the tin-mining industry in the late 19th century. The building of railroads gave access to a line of rich tin deposits along the western foothills of the mountains. These rail lines also provided transportation for rubber when new electrical and automotive industries stimulated a greatly increased demand for rubber during the early 20th century. A densely populated belt of tin mines and rubber plantations (see Fig. 12.6) developed in the foothills between Malacca and the hinterland of Penang (George Town; population: 248,000, city proper; 500,000, metropolitan area). Within this belt, the inland city of Kuala Lumpur (population: 919,600, city proper; 1.48

FIGURE 12.11
Singapore, located at a crossroads where the Pacific and Indian Oceans meet, exhibits new skyscrapers of a rising business center, older buildings once occupied by the former British colonial administration (foreground), and a spacious harbor (background).   *Ron McMillan/Gamma Liaison*

million, metropolitan area) became the leading commercial center and capital of the country (see Fig. 12.2b). The country's future capital, Putrajaya, is under construction about 25 miles (40 km) south of Kuala Lumpur.

The development of a commercial economy gave Malaya an ethnically mixed population. The native Muslim Malays played only a minor role in this development, often preferring to remain subsistence rice farmers in small coastal deltas. Chinese and Indian immigrants and their descendants became the principal farmers, wage workers, and businessmen of the tin and rubber belt. So heavy was the immigration that 30 percent of Malaysia's population today is Chinese and 8 percent is Indian. The growth of Singapore involved even heavier immigration from overseas, so that over three fourths of Singapore's present population of 3 million is Chinese.

Ethnic antagonisms between Malays and Chinese have been fundamental in shaping the political geography of Malaysia and Singapore. Malaya accepted independence in 1957 only on the condition that Singapore not be included. This was to ensure that the Malays would have a majority in the new state. In 1963, however, the Federation of Malaysia was formed, composed of Malaya, Singapore, and the former British possessions of Sarawak and Sabah in sparsely populated northern Borneo. The non-Chinese majorities in Borneo were counted on to counterbalance the admission of Singapore's Chinese. This experiment in union lasted only until 1965, when Singapore was expelled from the federation and left to go its own way as an independent state.

Malaysia and Singapore are among the more economically successful of the world's formerly colonial states that have received independence since World War II. Singapore has become one of the outstanding centers of manufacturing, finance, and trade along the Pacific Rim. Its industries include oil refining, machine building, and many others. The tourists it attracts each year far outnumber its resident population. It has now reached a level of income higher than that of many of the poorer countries in Europe. Its prosperity is accompanied by a strict sense of propriety. For example, chewing gum is banned and graffiti artists are flogged.

Malaysia's dependence on exports of rubber and tin has greatly lessened, although it still leads the world as an exporter of both commodities. A more diversified export pattern now includes such items as crude oil (from northern Borneo), timber, palm oil, and electronic components. Malaysia is the world's largest exporter of semiconductors and refrigerators, and has recently begun manufacturing automobiles for export. Malaysia's annual per capita income is far below Singapore's but is quite high among the LDCs and exceeds that of any other state in Southeast Asia except Brunei and Singapore (see Table 10.1, pages 262–263). Throughout the country the slogan "Wawasan 2000" ("Vision 2000") is emblazoned on buses, billboards, and pamphlets; it refers to the year by which national economic planners intend Malaysia to be a fully developed country.

INDONESIA

Indonesia is by far the largest and most populous country of the region (1996 population: 201 million), and is the fourth most populous country on Earth. Its 13,600 islands comprise over two fifths of Southeast Asia's land area and contain about two fifths of its population. The large population of Indonesia results from an enormous concentration of people on the island of Java. About 130 million people lived there in mid-1996, representing nearly two thirds of Indonesia's population and about one fourth that of all Southeast Asia. The island's population density is about 2500 per square mile (970 per sq km). In contrast, the remainder of Indonesia, which is about 13 times larger than Java in land area, has an average density of only about 115 per square mile (*c.* 45 per sq km).

This extraordinary concentration of population on one island is owed in part to the superior fertility of Java's soils, the best of which have been derived from materials poured out of its many volcanic peaks. However, it also has cultural and historical roots, particularly resulting from the concentration of Dutch colonial activities in Java. Other islands of Indonesia (with the notable exception of Borneo) have areas of fertile volcanic soil, but such areas, although generally more densely populated than adjoining nonvolcanic areas, seldom attain the extremely high population densities found on Java. The economic development of some islands has been handicapped by their unfriendly coastlines, along which coral reefs, cliffs, and wetlands create difficulties of access.

The Dutch East India Company, after lengthy hostilities with Portuguese and English rivals and with indigenous states, secured effective control of most of Java in the 18th century. Large sections of the remaining islands, however, were not brought under colonial control until the 19th century, and only in 1904 was the conquest of northern Sumatra completed. The Netherlands undertook strenuous efforts to exploit the natural wealth of Java with the introduction of the **Culture System** in 1830. Under this system, Dutch colonizers required Javanese farmers to contribute land and labor for the production of export crops under Dutch supervision. From the Dutch point of view, this harsh system was successful. The Culture System was abolished in 1870, but indigenous commercial agriculture continued to expand along with plantation production to support increasing numbers of people. The introduction of the Culture System thus appears to have set off the enormous increase in Java's population, which is more than 20 times larger than it was a century and a half ago. Development of the other main islands began later and has been less intensive. Nevertheless, the eastern coastal plain of Sumatra, inland from the great fringing swamp, has now surpassed Java in agricultural exports.

Java's concentration of people and production, and Sumatra's importance in export production, are evident in the distribution of Indonesia's larger cities. Of 12 cities having estimated metropolitan populations of more than 500,000, 8 are on Java. These include four "million-cities," headed by the country's capital and main seaport, and the largest metropolis in Southeast Asia, Jakarta (population: 8.23 million, city proper; 10.2 million, metropolitan area). Java's second largest city is Surabaya (population: 2.47 million, city proper). Of the remaining Indonesian cities with more than half a million people, three are in Sumatra and one (Ujung Pandang) is in Celebes (Sulawesi). In all 12 cities, and many others as well, a minority of Chinese immigrants and their descendants form a commercial class that has frequently been the target of resentment by other Indonesians.

The Indonesian state has had a turbulent career. Increasing nationalism during the Japanese occupation of World War II led to a bitter struggle for independence from the Netherlands following the war. This struggle finally succeeded in 1949. Another confrontation with the Netherlands in 1962 brought western New Guinea (West Irian) from Dutch control into the Indonesian state. Being Papuan rather than Indonesian in culture, West Irian had not originally been included in Indonesia. However, a United Nations plebiscite in 1969 ratified Indonesian control, and relatively friendly relations have been reestablished with the Netherlands.

The presence of some 300 different ethnic groups, combined with physical fragmentation and economic problems, have made it difficult to attain peace, order, and unity in Indonesia. Although 87 percent of the population is at least nominally Muslim (Indonesia has a larger total population of Muslims than any other nation in the world), and despite the presence of an official national language (Bahasa Indonesian), great cultural diversity is reflected in the fact that over 200 languages and dialects are in use. Various groups in the outer islands have resented the dominance of the Javanese, and such animosities have escalated at times to armed insurrections. For example, the government has been carrying on a long struggle against an independence movement led by the Catholics of East Timor—a former Portuguese possession that Indonesia occupied after the collapse of the Portuguese colonial empire in the 1970s.

In 1965 ("The Year of Living Dangerously"), an attempted communist coup against the longstanding regime of Indonesian President Sukarno was unsuccessful and resulted in the massacre of about 300,000 communists and their supporters by the Indonesian army and by Islamic and nationalist groups. It also brought to power a government in which army influence was predominant and which still controls the country, under the leadership of the president, General Suharto. International human rights organizations regularly accuse the Suharto regime of widespread abuse, but in the mid-1990s the now pro-Western and procapitalist government showed signs of greater tolerance of opposition. The main political opposition is composed of Muslim fundamentalists who want to restructure the state and society according to Islamic principles.

Indonesia has made striking economic gains in recent years but remains very poor. The gains have centered in the oil industry (in the hands of the state firm Pertamina) and in food production. Oil fields located mainly in Sumatra and Kalimantan produce 2 to 3 percent of the world's oil, and their production

PROBLEM LANDSCAPE

DANGEROUS ISLES: EVEN BARREN ROCKS ATTRACT CONFLICT

About 60 islands make up the Spratly Island chain, which lies in the South China Sea between Vietnam and the Philippines (see Figure 12.1). They are an idyllic tourist destination, where divers can hire luxury boats to explore the coral reefs and palm-lined beaches of remote atolls. However, the islands are much more significant for their strategic location between the Pacific and Indian oceans. During World War II, Japan used the islands as a base for attacking the Philippines and Southeast Asia. Still more significantly, as much as one trillion dollars in oil and gas may lie beneath the seabed around the Spratlys.

Not surprisingly, many nations covet control of the Spratlys. Six nations claim some or all of the islands: China, Vietnam, Taiwan, Malaysia, Brunei, and the Philippines. During the Cold War, the competing claimants felt it was too hazardous to push their claims on the islands. As the Cold War drew to a close, however, the situation became more volatile. In 1988 the Chinese Navy invaded seven of the islands occupied by Vietnam, killing about 70 Vietnamese soldiers. The other powers placed soldiers, airstrips, and ships on the islands. In 1992, China again landed troops in the islands and began exploring for oil in a section of the seabed claimed by Vietnam. In 1995,

China moved to expand its territorial claims on islands claimed already by the Philippines. Indonesia, which has no claims on the Spratlys, is sponsoring unofficial workshops on joint efforts in oil exploration among the six claimants in an effort to defuse the emerging crisis. There are fears that as the countries' petroleum needs grow, they will seek more aggressively to gain control of the Spratlys. An incident in these remote islands could trigger a much wider and more serious conflict in Asia.

Richard C. Hottelet, "Dangerous Isles: Even Barren Rocks Attract Conflict." *The Christian Science Monitor*, March 7, 1996: 18.

has been expanding. Crude oil and liquefied natural gas account for about 25 percent of Indonesia's exports and have tied its economy very closely to that of Japan, which is the main market for these products. Indonesia is a member of OPEC, the Organization of Petroleum Exporting Countries.

FIGURE 12.12
The cultural arts of Bali's Hindu people are world-renowned, making this Indonesian island a popular tourist destination.   *George Hunter/Tony Stone Images*

Tourism also brings welcome foreign exchange earnings to Indonesia. The country's principal attraction by far is the legendary island of Bali, home of Indonesia's only remnant of the India-born Hindu culture that permeated the islands from the 4th to 16th centuries (Fig. 12.12). Bali is world-famous for its unique culture and forms of art and dance.

Indonesia is still an important world source of various tropical agricultural products, and their production and export are still of major consequence to many Indonesians. However, such commodities no longer dominate the country's commerce as they have through nearly all of Indonesia's history. In addition to oil and gas, wood products and clothing make important contributions to the export trade. There has also been a remarkable increase in rice production (Fig. 12.4). The rapid increase in population by 59 percent, or 74 million people, in the 21 years from 1975 to 1996 nevertheless has had the effect of keeping Indonesia poor. The government lacks an aggressive family planning program, but is promoting emigration from Java and other densely-populated islands to some of the sparsely inhabited outer islands of the vast Indonesian archipelago.

THE PHILIPPINES

The Philippine archipelago includes over 7000 generally mountainous islands. The two largest islands, Luzon and Mindanao—almost equal in size—account for two thirds of the total area. Most of the mountainous districts are inhabited by a sparse population of shifting cultivators. In northern Luzon, however, the Igorot tribes have developed a spectacular system of wet-rice cultivation on terraced mountainsides.

Most of the population of the Republic of the Philippines is concentrated in three areas (see Fig. 12.1):

1. *The Visayan Islands in the center of the archipelago.* Here, soils derived from volcanic materials and uplifted corals support an intensive agriculture based on rice and corn. Negros Island is a major center of plantation sugar production.

2. *The plains extending from south of Manila to the Lingayen Gulf and north along the west coast of Luzon.* Rice is the main food crop here, and sugarcane, much of it produced on small native farms, is the main cash crop.

3. *The southeastern peninsula of Luzon.* Subsistence rice and commercial coconut production are basic here except in the extreme southern part, where large plantations utilize volcanic soils to produce abaca or Manila hemp.

The Philippines were a Spanish colonial possession governed from Mexico from the late 16th century until 1898. Spanish missionary activity in the Philippines succeeded in creating the only Christian nation in Asia. With the exception of the Japanese occupation of 1942–1944, the Philippines were controlled by the United States from 1898 until 1946, when independence was granted.

The country's Spanish legacy is still important. Ever since its founding in 1571, the Spanish-oriented capital of Manila (population: 1.59 million, city proper; 9.65 million, metropolitan area) has been the major metropolis and only large city of the islands. The society created by Spain was composed of a small upper class of Hispanicized Filipino landowners and a great mass of landless peasants. Problems created by this maldistribution of agricultural land have remained as a major source of difficulty for the Philippines. Discontented peasants gave much support to a communist-led revolt after World War II. It was eventually suppressed, in large part through granting land (generally on sparsely populated Mindanao) to surrendered rebels. Revolt flared up again in the 1960s, however, and continues on a reduced scale in the 1990s.

American involvement began with the suppression of a Philippine independence movement but eventually ended with Philippine independence, along with economic development and tutelage in democracy. Preferential treatment in the U.S. market stimulated the growth of major Philippine export industries—coconut products, sugarcane, and abaca. Education in English grew so that even now English is spoken by about 45 percent of the people and serves as a bridge between many of the diverse linguistic groups of the population. After independence, Tagalog, one of the most widely used indigenous languages, became the principal base for the official national language of Pilipino, spoken by 55 percent of the populace. English also remains an official language. Both Tagalog and English are second languages for the majority of Spanish speakers.

The Philippine economy has expanded considerably since independence. Until the 1970s, the most striking growth was in agricultural exports such as coconut products, sugar, bananas, and pineapples, and in copper and other metal exports. More recently, there has been growth in manufacturing and food production. Labor-intensive manufacturing, such as electronic devices and clothing, now account for about 40 percent of all exports. Japanese and American capital, together with American agricultural science and technology, have been very important in Philippine economic development. The United States and Japan buy more than half of all Philippine exports.

Despite encouraging progress in recent decades, the Philippine republic remains poor and potentially explosive politically, and the country is not yet in the class of the "Asian Tigers." Economic expansion has been hard-pressed to stay ahead of the 70 percent growth in population during the 21 years between mid-1975 and mid-1996. The predominant Roman Catholic culture (Catholics make up 83 percent of the population, with 11 percent adhering to other Christian denominations) encourages large families, and the Philippines has a high annual population growth rate of 2.1 percent (while food production has been increasing at only about 1 percent annually in the 1990s). Church officials have labeled the government's current promotion of family planning as "demographic imperialism" masterminded by the United States.

Compounding the problem, land and wealth continue to be extremely unevenly distributed, as Philippine society continues to be dominated by just a few hundred wealthy families. On the island of Mindanao, rebels of the Moro Islamic Liberation Front have been fighting the government in an effort to establish a separate Islamic state. Their rebellion is rooted in economic as well as political grounds with claims that Muslims have not benefitted from the region's economic growth. Other defiant rebel populations became more moderate and entered negotiations with the government after 1986, when the country's dictator, Ferdinand Marcos, was ousted in a popular uprising and replaced by a democratically-elected president, Corazón Aquino. Marcos and his wife, Imelda, had looted the country of billions of dollars over a period of twenty years and headed a corrupt and abusive regime virtually ignorant of the needs of the ordinary Filipino.

The United States, which has maintained friendly ties with its former colony and long depended on important military bases there, has watched with much concern the country's struggle to resolve its difficult problems. During the 1980s many Filipino legislators strongly resisted the prospect of renewing the leases under which the U.S. held two military bases on Luzon, Subic Bay and Clark Field. Then, in 1991, the eruption of volcanic Mount Pinatubo put an end to the debate over the fate of these two American outposts (Fig. 12.13); volcanic ash forced their closure. Pinatubo also ejected enough ash high into the Earth's atmosphere to cool temperatures globally over the following several years, at least temporarily confounding efforts to track the trend in global warming due to the greenhouse effect.

FIGURE 12.13

After lying dormant for six centuries, Mount Pinatubo in the Philippines exploded into life in 1991 with a violence that caused over 350 deaths and enormous property damage. Both the Philippines and Indonesia lie in the volcanic and earthquake-prone ring of fire around the Pacific rim.   *Alberto Garcia/SABA*

QUESTIONS

Page numbers refer to *Saunders College Publishing 1996 Special Edition of Rand McNally's Atlas of World Geography.*

On the map of Southeastern Asia on pages 148–149, identify:

A small island group, possibly rich in fossil fuels, contested by six countries.

The Golden Triangle region, where much of the world's opium and heroin supplies are produced.

The capital city that was nearly depopulated during the terrifying years when the Khmer Rouge were in power.

One of the world's two largest ports in terms of tonnage of goods shipped.

The island containing nearly one-quarter of the entire population of Southeast Asia.

13

CHINA: AMBITIOUS BLENDING OF PAST AND FUTURE

So often images of China are built almost entirely around population issues. While there is real dimension to having the world's largest population—as China does—the real drama of China relates to the ways in which its people have made their landscape accommodate such populations. From the earliest dynasties dating back to the second millennium B.C.—and the monumental construction of China's Great Wall in the third century B.C.— to the present, the story of China has been one of steady manipulation of a not particularly fertile landscape. China is a nation that has maintained a distinctly Chinese perspective for nearly 4000 years, despite major foreign infusion through Mongol, Manchu, and Western control of all or part of the country.

Outsiders, like the young engineer depicted in the excerpt from John Hersey's *Single Pebble* below, have long had a fascination with China. We have sent missionaries, businesspeople, young scholars hopeful of understanding this giant of the East, and—now—countless tourists. There is a grandeur to China that we have long appreciated, and our interaction with it runs the gamut from the engineer seeking the right place for a massive dam on the Chang Jiang (Yangtze River) as is depicted in the literature selection, page 322, to a South Carolina textile businesswoman seeking a source for new shirts she wants to produce offshore and sell in the global market to the hundreds of students involved in language study and academic interchange. China has long been on the horizon of interest and opportunity for people of the West, and especially for merchants of the United States.

China currently engages in active industrial and economic development in an effort to pull away from its past as a monumental agricultural nation. At the same time, the need to feed its population and raise fiber for its enormous textile base forces it to stay attentive to growth in the rural sector as well. With the reclaiming of Hong Kong in 1997 after a century and a half of British control, and with relations with the United States ever uncertain, China faces a broad array of demands on its social and economic fabric. Not only is China continually contested by the forces of nature, such as the annual rise and fall of the Chang Jiang, but it must deal with the insurgencies that have played an ongoing role in fomenting instability in this enormous nation. China continues to face forces both in nature and in humankind that seem to contest the country's efforts to gain

◀ This night scene of Shanghai's Nanjing Road radiates the energy of this dominant city in China. It has been the entry port for foreign cultures for the past century and a half, and continues today as a center of Chinese flirtation with many aspects of world culture. Shanghai, more than Beijing or Guangzhou, has a night life that gives a special personality to this city. This is one of the busiest shopping streets in Shanghai, and a favorite place to see and be seen for the Chinese who are rapidly becoming both urban and urbane. *Jeff Greenberg/Photo Researchers*

stability and independent economic development. The unfolding Chinese drama represents an effort to blend strengths of the past with realities of the present.

13.1   Area and Population

China has the world's largest total population, estimated at 1.21 billion (excluding Taiwan) in 1997—over one fifth of the world's people—and increasing by 13 million per year. Population is a very serious matter for an industrializing but underdeveloped country whose area of about 3.7 million square miles (9.6 million sq km) is only slightly larger than that of the United States (Fig. 13.1) but whose inhabitants outnumber the United States population nearly 5 to 1. This population lives in a state officially called the People's Republic of China (PRC), which plays a major role in world affairs (Fig. 13.2). It is, nonetheless, still relatively poor and largely agricultural despite major industrial development and economic growth under the Communist regime that came to power in 1949 (see Table 10.1).

Over 800 million Chinese live in rural settlements and are largely engaged in and supported by agriculture. This accumulation of peasant humanity must wrest food for itself and for the ever expanding urban population from a landscape where nearly nine tenths of the land is not in cultivation because of steep slopes, dryness, short growing seasons, and/or technological inadequacies. China's arable area, concentrated in the plains and river valleys of humid China (see page 321), is about half the area of the United States (see Fig. 10.2), but through **multiple-cropping** practices (growing successive crops on a given field throughout the year), the sown acreage may actually be larger than the area farmed in the U.S. China's arable area provides, on the average, about one third acre of arable land per person to the agricultural population, and only one fourth of an acre per person to the total population. This per capita average equals approximately one eighth of the U.S. average.

FIGURE 13.1
China compared in latitude and area with the conterminous United States.

POPULATION POLICY

China's success in reducing population growth in the past decade has been one of the global success stories in population management (Fig. 13.3), though not without controversy and serious social impact. The drive for smaller families—begun seriously in the late 1970s—was motivated by the fact that after the initial recovery from war in the 1950s, there was scarcely any progress in China's per capita food output for two decades. Increases in agricultural output were largely matched by rapid population growth, while the Mao Zedong regime made little or no effort to curb the birth rate (Figure 13.4).

Definitions and Insights

MAO ZEDONG

Much of the history of China in this century has been shaped by the influence and leadership of Mao Zedong (Mao Tsetung). He was one of the founders of the Chinese Communist Party (CCP) and in the 1920s was actively involved in efforts to organize and improve life for China's peasants. In 1927 the CCP split from the Kuomingtang (Nationalists) and China began a bloody civil war that was to last until 1949. Mao was the major leader in the 1934–35 **Long March** (a 6000-mile flight on foot of nearly 100,000 communists being pursued by Nationalist soldiers; see page 326) and worked steadily to have China's military fight the Japanese rather than fight among themselves during the Second Sino-Japanese War (1937–1945). With the defeat of Japan in World War II, the CCP and the Nationalists returned to their civil war and, in 1949, two million Chinese Nationalists under Chiang Kai-shek fled to Taiwan and Mao and the CCP claimed victory. From then until his death in 1976, Mao led the People's Republic of China through a demanding and often counterproductive series of experiments in social organization in the Chinese countryside: In 1957 he initiated the Great Leap Forward and the Chinese Commune movement, and in the mid-1960s he promoted the Cultural Revolution, all the while making efforts to expand and increase the productivity of China's industries as well. By the time he died, China had regained a global prominence that Mao felt it had been denied for a century and a half because of Western colonialism and interference in China's domestic affairs. Two years after Mao's death, Deng Xiaoping gained the role of major authority in China, and, for the first time since the 1949 revolution, China opened its doors to the West and began to tailor its economic efforts, devoting more attention to personal incentives for agricultural and industrial success. With this came a new attention to the production of consumer goods and a relaxation of Maoist migration control. Nevertheless, even though many of Mao's favorite programs have been disbanded, he continues to be a figurehead of significance in China.

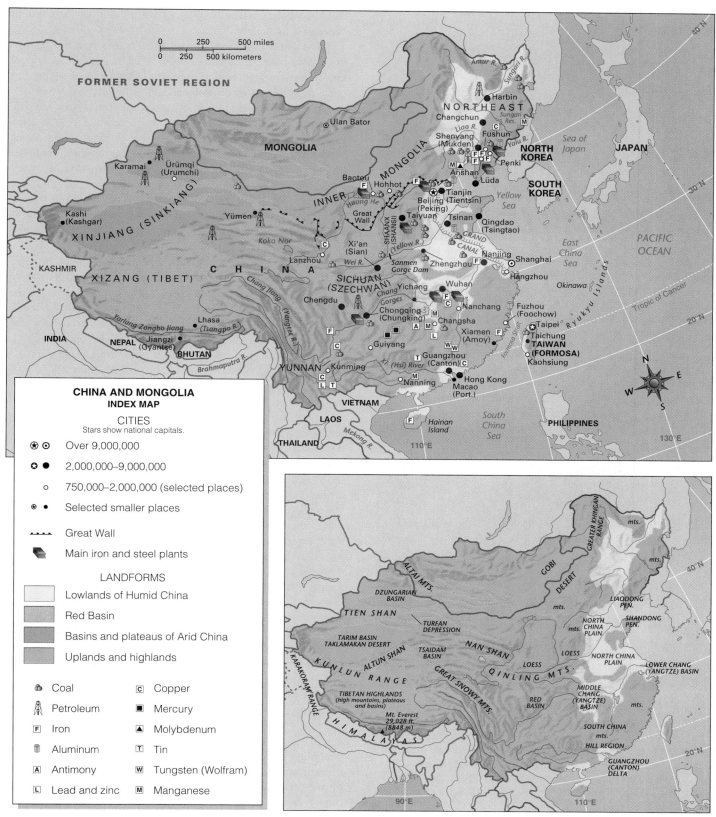

FORMER SOVIET REGION

MONGOLIA

NORTHEAST

Ulan Bator

Harbin
Changchun
Shenyang (Mukden)
Fushun
Penki
Anshan
Lüda
NORTH KOREA
SOUTH KOREA
JAPAN
Sea of Japan

Karamai
Ürümqi (Urumchi)
INNER MONGOLIA

Baotou
Hohhot
Huang He
Beijing (Peking)
Tianjin (Tientsin)
Yellow Sea
Qingdao (Tsingtao)
Tsinan
Taiyuan
SHAANX (SHANSI)
GRAND CANAL

Kashi (Kashgar)
XINJIANG (SINKIANG)
Koko Nor
Yümen

Lanzhou
Xi'an (Sian)
Wei R.
Zhengzhou
Nanjing
Shanghai
East China Sea
PACIFIC OCEAN

KASHMIR
XIZANG (TIBET)   C H I N A
Sanmen Gorge Dam
Yichang
Wuhan
Hangzhou
Okinawa
Ryukyu Islands
Tropic of Cancer

SICHUAN (SZECHWAN)
Chang Jiang
Chang Gorges
Nanchang
Fuzhou (Foochow)
Chengdu
Chongqing (Chungking)
Changsha
Xiamen (Amoy)
Taipei
Taichung
TAIWAN (FORMOSA)
Kaohsiung

Yarlung Zangbo Jiang
Lhasa (Tsangpo R.)
Jiangzi (Gyantse)
INDIA
NEPAL
BHUTAN
Brahmaputra R.
YUNNAN
Guiyang
Guangzhou (Canton)
Hong Kong
Macao (Port.)

Kunming
Xi (Hsi) River
Nanning
VIETNAM
LAOS
THAILAND
Mekong R.
Hainan Island
South China Sea
PHILIPPINES

Amur R.
Sungari R.
Liao R.
Yalu R.
Sungari Res.
Great Wall

40°N
30°N
20°N
110°E
130°E

CHINA AND MONGOLIA
INDEX MAP

CITIES
Stars show national capitals.

⊛ ⊙   Over 9,000,000

✪ ●   2,000,000–9,000,000

○   750,000–2,000,000 (selected places)

⊛ •   Selected smaller places

‥‥‥   Great Wall

◤   Main iron and steel plants

LANDFORMS
Lowlands of Humid China
Red Basin
Basins and plateaus of Arid China
Uplands and highlands

Coal
Petroleum
F Iron
Aluminum
A Antimony
L Lead and zinc

C Copper
Mercury
A Molybdenum
T Tin
W Tungsten (Wolfram)
M Manganese

GREATER KHINGAN RANGE
ALTAI MTS.
GOBI DESERT
DZUNGARIAN BASIN
TIEN SHAN
TURFAN DEPRESSION
LIAODONG PEN.
NORTH CHINA PLAIN
SHANDONG PEN.
TARIM BASIN
TAKLAMAKAN DESERT
TSAIDAM BASIN
NAN SHAN
LOESS
NORTH CHINA PLAIN
LOWER CHANG (YANGTZE) BASIN
KUNLUN ALTUN SHAN RANGE
QINLING MTS.
KARAKORAM RANGE
GREAT SNOWY MTS.
RED BASIN
MIDDLE CHANG (YANGTZE) BASIN
TIBETAN HIGHLANDS (high mountains, plateaus and basins)
Mt. Everest 29,028 ft. (8848 m)
HIMALAYAS
SOUTH CHINA mts.
SOUTH CHINA HILL REGION
GUANGZHOU (CANTON) DELTA

mts.

40°N
20°N
90°E
110°E

FIGURE 13.2
General location map of the Chinese realm.

FIGURE 13.3
The Chinese government has been promoting "one couple, one child" for over a decade in an effort to slow down the annual population increase. This poster exhorts young couples to observe this constraint with the phrase "For the sake of a prosperous today and a beautiful tomorrow," limit your family to one child.   *Forrest Anderson/Gamma Liaison*

In the early years of the revolution, Mao felt that every new pair of hands born to China could be a productive addition to the country's economic ambitions. Then, after Mao's death in 1976, the regime instituted probably the most stringent program of birth control in the world. This has culminated in the "one

FIGURE 13.4
This 10-feet-high painting of Chairman Mao Zedong (Mao Tse-tung) maintains its position atop the central gateway to the massive complex of The Forbidden City in Beijing. Mao appears to look out on Tiananmen Square, nearly 100 acres of open space cleared by the removal of makeshift housing in the years following the victory of the Communists over the Kuomintang in 1949. It was in this square in June, 1989 that China defined its disinclination to allow democracy to grow by the massacre of unarmed students who were protesting governmental restrictions on public assembly.   *C.L. Salter*

child campaign," which aims to limit the number of children per married couple to one.

The government continually exhorts its young families to limit family size to a single child. It takes note of individual family birthing patterns—especially in the cities— and maintains surveillance through local authorities. It dispenses free birth-control devices, free sterilization operations, free hospital care in delivery, free medical care for the child, free education for the child, an extra month's salary each year for the parents, and other favors and preferences to induce compliance with the one-child-per-family norm. Those who violate the norm are subjected to constant social and political pressure, denial of the privileges accorded one-child families, pay cuts, and fines. Women who become pregnant without permission are pressured to have a free state-supplied abortion, with a paid vacation provided.

Under such programs, China's birth rate declined by 1995 to 1.03 percent, the lowest in any major LDC, and its rate of natural population increase dropped to about 60 percent of the average for other LDCs. But the situation remains critical. China's population is so large that even the relatively low birthrate achieved by 1995 still meant a net increase of approximately one million new mouths a month, or a gain of approximately the urban population of New York City annually! This is in a country where the death rate has also been drastically lowered by better food availability, better medicine, and improved and more readily available public health facilities. The regime hopes to decrease childbearing to the extent that population in the 21st century will stabilize at 1.3 billion people, but some population experts are projecting a population of 1.4 billion by the year 2010, or 1.5 billion by 2025. In the early 1990s, government officials were reporting widespread disregard of the "one couple, one child" policy—particularly in the countryside, as peasants became wealthier and hence more ready to pay the fines imposed for violations of the national population policy. Urban populations, always more tightly controlled by working and living units called *danweis*, have been more responsive to the continual exhortation for couples to have only one child. At the same time, however, the rapid growth of the urban population in China has also led to decreasing birth rates in the cities because of the increasing importance of material goods and lifestyle patterns that do not necessarily revolve around parenting.

It is, moreover, important to realize that China's relative success with the diminution of the national birthrate has significant social costs as well as economic benefits. For a nation that has had a long tradition of large families—particularly in the countryside—this shift to a single child has generated a great deal of pain and social dislocation. Parents who have taken care of their parents and grandparents increasingly anticipate a future of uncertain residence in their own old age. If the contemporary family has only one child and that child is a girl, for example, there is the fear that she will move away to her husband's city when she is married. This will leave her parents both distant and without the "social security system" that is traditional. The birth of a girl child in a nation so demanding in its "one couple, one child" policy has led to female infanticides as couples continue to try for a male child.

There is also a new class of child called "the little emperor," who is a single child, most often male, much loved, and often spoiled. Historically there were usually many siblings to diffuse parental care across the lives of a number of children. Now with few siblings and an improvement in the standard of living, China is fearful of the power and consequence of having so many children growing up without the social benefit of extended family and sibling interaction.

These unexpected social outcomes of a successful family planning campaign server to remind people that virtually every change in one social domain has the capacity to send ripples of influence across all or at least many other areas of social concern. Population planning is particularly likely to have such varied impacts.

13.2 Natural Environment and Major Landscape Elements

China is vast in size but, like all extremely large countries, contains a great deal of unproductive land. In China, this land lies mainly in the western interior half of the country (see Fig. 10.2). Here are huge outlying areas principally composed of high mountains and plateaus, together with arid or semiarid plains, where rainfall is generally insufficient for agriculture. This dry, sparsely settled country probably contains about 5 percent of China's population—mostly various peoples other than the **Han Chinese**, particularly outside the scattered urban areas of this region. The term Han Chinese is commonly used to designate the dominant ethnolinguistic group in China; the term Han is derived from the first great dynasty of China (206 B.C. to 220 A.D.). There are more than 50 non-Han Chinese ethnolinguistic groups—nearly all of them racially of Mongoloid stock—who number nearly a hundred million, or about 8 percent of the Chinese population. They live primarily in the western and southwestern parts of China, and are the major populations in the Autonomous Regions (Fig. 13.5).

The arid interior area is in marked contrast to the better-watered, more densely settled eastern core of the country. Thus, China is divided into a western half, or **Arid China**, and an eastern core region, or **Humid China**, in which the country's population, developed resources, and productive capacity are heavily concentrated. A rough boundary between the two major divisions is an arc drawn from the northeastern corner of India to the northern tip of China's Northeast. This line corresponds in a general way to the stretch of land whose average annual rainfall is 20 inches (c. 50 cm), with Arid China to the west and Humid China to the east.

FIGURE 13.5
The Special Economic Zones (SEZs) of southeast China were organized in the late 1970s to showcase China's ability to deal with the trading, manufacturing, and urban demands of the Western world. This map shows their concentration along China's southeast coast, the traditional area of China's main interaction with trading nations from the West. These SEZs today have become some of the most rapidly growing urban centers in all of China.

ARID CHINA

The principal regions of Arid China are the Tibetan Highlands, Xinjiang (Sinkiang), and Inner Mongolia. All three are identified historically with non-Han populations, although now—because of four decades of politically promoted domestic migration away from the east coast toward the borders with Russia—Han Chinese comprise the great majority in Inner Mongolia and may become a majority in Xinjiang.

Tibetan Highlands

The very thinly inhabited Tibetan Highlands (see Population Map, page 47) occupy about one fourth of China's area. Included are Tibet itself (in Pinyin: Xizang; see Definitions and Insights below) and fringes of adjoining provinces.

Definitions and Insights

PINYIN SPELLINGS

The Chinese government uses a system of official place-name spellings known as "Pinyin." It superseded older spellings from the traditional Wade-Giles transliteration system, so that "Peking," for example, became "Beijing," "Canton" became "Guangzhou," and "Yangtze Kiang" became "Chang Jiang." A few letters in the Pinyin scheme have sounds substantially different from the way these letters are customarily pronounced in English: "c" in Pinyin is pronounced as though it were "ts"; "q" is "ch"; "x" is "sh"; "z" is "dz"; and "zh" is "j." The system is based on the pronunciations of Chinese characters in the standard form of Chinese called Mandarin. Because of the widespread adoption of Pinyin spellings in Western news media, maps, and other published materials, it has been decided to cite the Chinese place names in this text in their Pinyin forms. But in most instances, the familiar older spelling such as Peking or Canton is given in parentheses the first time or two that the word is used. After all, Pinyin was only adopted in 1958, and all but a very small fraction of the material in Western libraries uses the older spellings. In the case of some place names such as Shanghai, the newer and older spellings are the same. To make the reading less cumbersome, "Tibet" has been retained, although this region is called Xizang in Pinyin.

Most of this vast area is a very high, barren, and mountainous plateau averaging nearly three miles (c. 5000 m) in elevation. To the northwest, high basins of internal drainage are common, some containing large salt lakes. To the southeast, the plateau is cut into ridge and canyon country by the upper courses of great rivers such as the Tsangpo (the Brahmaputra of India), the Mekong, and China's own Chang Jiang (Yangtze Kiang or Yangtze River) and Huang He (Hwang Ho or Yellow River). It is along the Chang Jiang's flow through the eastern edge of this plateau—in movement toward the well-settled middle Chang Jiang basins of central China—that the Chinese have begun to focus their plans to capture and control the power of this greatest river of China, which travels more than 3,400 miles in its descent from the Tibetan Highlands to its delta in the Shanghai area in the east. The Three Gorges Project is in the same area that is chronicled in John Hersey's *Single Pebble* literature selection that follows.

The following two passages come from a short novel about an American engineer who travels through the gorges on the Yangtze River in China in the 1920s. He is looking for the best geologic location for a proposed dam across this largest river in China.*

> I became an engineer. I found my way into hydraulics, and not many years along, while still a youthful dam surveyor, I was chosen by the big contracting firm for which I worked to go to China and study the river called by the Chinese "the Great," the Yangtze, to see whether it would make sense for my company to try to sell the Chinese government a vast power project in the river's famous gorges.
>
> This was half my life ago, in the century's and my early twenties; the century and I were both young and sure of ourselves then.
>
> I spent a year preparing myself for the trip. I applied myself to spoken Mandarin Chinese and got a fair fluency in it. I read all I could find on the Yangtze; I learned of its mad rise and fall, of the floods it loosed each year, killing unnumbered people and ruining widespread crops; of its fierce rapids and beautiful gorges, and of its endless, patient traffic of hundreds of junks towed upstream and rowed down by human motive power.
>
> Even after my studies, though, I could scarcely visualize this storied, treacherous river, and being an ambitious young engineer I could only think of it as an enormous sinew, a long strip of raw, naked, cruel power waiting to be tamed. I had much yet to learn.
>
> I took passage on a steamer to Shanghai, and after an impatient month in that transplanted Western city I was able to talk my way onto a British gunboat, the *Firefly*, which was going upriver as far as Ichang, at the gate of the gorges, on a patrol such as British ships were then allowed by treaty to make on certain Chinese rivers on behalf of British business interests in the interior.
>
> The thousand miles from Shanghai to Ichang were long. The landscape was flat; the river was enormously wide and sluggish. Where was the Yangtze's brutal power? I was let down. We made no stops, and everything on board was British and regular, and I witnessed a riverbank China but did not feel it.
>
> We arrived at length in Ichang. I went immediately, as I had been told to do, to our consul in that city, and because bandits and revolutionaries were said to be harassing the few flat-bottomed steamboats then trading in the gorges above Ichang, he urged me to go upriver not by steamer but by junk, as he thought I would travel unnoticed that way and would have more leisure for my study. And so with his help I arranged a passage with a thin, gaunt junk owner whose Chinese I could understand quite well, for he was a Szechuan man, from Wanhsien, and the Szechuanese dialect is not too far from pure Mandarin. . . .

*John Hersey, *A Single Pebble* (New York: Alfred A. Knopf, 1956): 3–5, 136–138.

Later in the novel, the engineer is on the junk passing through one of the most difficult places in the gorges, being hauled by trackers who are on shore. Harnessed to the junk by long, taut woven bamboo ropes, the Chinese men are pulling it upstream.

> This seemed to me the most terrible place on the whole river. Men working with chisels had cut out of the steep cliff a running rectangle of rock to make this path. It was scooped out of the flat face of the mountain, which was too perpendicular to permit an ordinary ledge being formed. The path had a ceiling, an inner wall, and a floor of solid rock; all it had for outer wall was peril. . . .
>
> Where it began, the path was about thirty feet above the surface of the water, so that from the deck of the junk we looked up at it, more than half the height of our mast. Su-ling told me that in winter, at low water, at what the river men call "zero," the path would be more than sixty feet above the surface, while late in spring . . . it would be nearly as much, or more, under the surface. When the melted snows of many mountains of Tibet course toward the sea, and when, riding the crests of those thaws, the run-off of spring rains that have fallen on half a million square miles of Chinese hills flow too down the Great River, its power becomes unimaginable, even to a hopeful young hydraulic engineer.
>
> These were the very cliffs again which, in its record year, the river had climbed in a short time two hundred and seventy-five feet.

Lower elevations, warmer temperatures, and greater precipitation in parts of the southeastern plateau result in some extensive grasslands and stands of conifers. Around the edges of the Tibetan Highlands are huge mountain ranges: the Himalayas on the south; the Karakoram and Kunlun Shan to the north and northwest; and the Qiling Shan and others to the east.

The people of the Highlands, numbering approximately 5 million (2.3 million in the political area called the Tibetan Autonomous Region) are predominantly sedentary farmers and animal herders (Fig. 13.6). Agricultural land is extremely limited and found only at elevations lower than about 13,000 feet (*c.* 4300 m), along the valleys of the Tsangpo and other major streams and their tributaries. A few hardy crops, especially barley and root crops, are basic to agriculture in this restrictive environment. In grasslands at higher elevations, a nomadic minority graze their flocks of yaks, sheep, and goats. Yaks are particularly important not only as durable beasts of burden in these high elevations, but also as providers of meat, milk and butter, leather, and hair traditionally woven into cloth.

The Tibetan Highlands are under Chinese political control, and Lhasa, the capital, now has more than 500,000 Han Chinese residents. This province's population—beyond the capital—is comprised predominantly of Tibetans, who are distinguished by their own language and by their adherence to Lamaism, the Tibetan variant of Buddhism. Prior to the Communist era, China had sometimes exerted loose control over Tibet, but Chinese authority vanished with the overthrow of the Manchus in 1911. From 1912 until the Chinese Communists' conquest of 1951, Tibet existed as an independent state, with its capital and main religious center at Lhasa. After the Chinese completed roads to

FIGURE 13.6

Yaks continue to play both a ceremonial and utilitarian role in the farm lands of Tibet. They do better in the high plateaus of this region than any of the traditional lowland farm animals. The banners are a reminder of the traditionally significant role these animals have played in this land of harsh farming conditions. Even though China has made some effort to introduce small scale farm machinery to its agriculture, traditional patterns still persist—especially in the distant upland regions. *D.E. Cox/Tony Stone Images*

Tibet in late 1954, an increase in restrictive measures by the Communist government contributed to the rise of guerrilla warfare. This culminated in a large-scale Tibetan revolt in 1959 and the flight of the Dalai Lama (the spiritual and political head of Lamaism) and many other refugees to India. Subsequently, the Chinese drove most of the monks from their monasteries, expropriated the large monastic landholdings, and made institutional changes similar to those designed to implement socialism and prohibit organized religion in the rest of China. Tibetan resistance to Chinese rule and aggressive repressions by the Chinese continue.

Xinjiang (Sinkiang)

Xinjiang adjoins the Tibetan Highlands on the north. It has an area of roughly 635,000 square miles (*c.* 1.6 million sq km) and a population of 15.7 million (as of 1995). It consists of two great basins—the Tarim Basin to the south and the Dzungarian Basin to the north (see Fig. 10. 2). These basins are separated by the lofty Tien Shan range. The Tarim Basin is rimmed to the south by the mountains bordering Tibet, and the Dzungarian Basin is enclosed on the north by the Altai Shan and other ranges along the southern border of the former Soviet Union and Mongolia. Both basins are arid or semiarid. The Tarim Basin is particularly dry, since it is almost completely enclosed by high mountains that block off rain-bearing winds. This basin includes the Taklamakan Desert—which, translated, means "Once you get in, you'll never get out"—perhaps the driest region in Asia. The basin varies in altitude from 2000 to 6000 feet

FIGURE 13.7
This is a scene from Dunhuang in the arid interior of China in western Gansu Province. This oasis setting was a major stop on the Silk Road that has connected northern China with the Mediterranean region for more than 2000 years. Dunhuang has now become an important tourist destination because of the discovery of scores of caves with delicate religious paintings on their walls.   *C.L. Salter*

(about 600–1800 m) above sea level; the smaller adjoining Turfan Depression drops to 928 feet (283 m) below sea level. The Dzungarian Basin averages about 1000 feet (305 m) in elevation, is more open than the Tarim Basin, and has somewhat more rain, although not enough for much agriculture.

The great majority of Xinjiang's population is concentrated in oases (see World Population Map, page 47), located mainly at points around the edges of the basins where streams from the mountains enter the basin floors. For many centuries, these oases were stations on caravan routes crossing central Asia from Humid China toward the Middle East and Europe on the historic Silk Road (Fig. 13.7). Under the People's Republic, the ancient routes have been superseded by modern transportation. A railroad connects Lanzhou (Lanchow; population: 1.5 million), a major industrial center and supply base in northwestern China proper, with Urumqi (Urumchi; population: 1.2 million), the capital of Xinjiang; several major roads link the oasis cities, and several major airfields have been built in the region.

The transport links are key elements in a drive to expand the economic significance of this remote part of China and bring it under firmer political control. Early in the Communist era, demobilized soldiers and urban youths from eastern China were organized into quasi-military production and construction divisions to push expansion of irrigated land on state-operated farms in this arid landscape. Now cotton is a major crop. The region also has deposits of coal, petroleum, iron ore, and other minerals, and a considerable expansion of mining and manufacturing has occurred under the auspices of the People's Republic.

This region has been valuable both for its isolation and its resources. It was in Xinjiang that the Chinese exploded their first nuclear bomb in 1964. It continues to be the most mineral-rich province in China and has the potential for considerable political as well as economic activity in the decades to come.

Development in Xinjiang serves Chinese political purposes by tying this outlying area more closely to China's core. The population was for centuries predominantly Muslim and Turkic—groups who have often been restive under Chinese rule. In addition, some frontier regions have been contested by both China and Russia in the past. However, Chinese control is probably tighter now than ever before, strengthened by the new economic development that has brought in so many Han Chinese immigrants that Xinjiang may now have a majority of Chinese in its rapidly growing population.

Inner Mongolia

North and northwest of the Great Wall, rolling uplands, barren mountains, and lifeless basins stretch into the arid interior of Asia. Here lie slightly more than a million square miles (c. 2.6 million sq km) of dry terrain that is divided about equally between Inner Mongolia, the area nearest the Great Wall, and the independent country called Mongolia. An overwhelming majority of Inner Mongolia's present population (22.6 million in 1996) is Han Chinese. This population is concentrated in irrigated areas along and near the great bend of the Huang He. By far the greater amount of territory of Inner Mongolia is still the habitat of a very sparse population of Mongol herdsmen. Although they make use of some areas of grassy steppe, much of their territory includes the notably barren, gravel-strewn Gobi Desert—a harsh and desolate country of widely spaced springs and meager pasturage with less than 5 inches (13 cm) of precipitation per year. It is from these severe landscapes that the Mongols emerged to control China and an expanse of land that ranged from the Korean peninsula in the east to the margins of Poland in the west in the 13th century. The desert landscape displays the shadows of an extraordinary history of Asia, through remnants its of the Chinese Great Wall, caravan oases and routes, and temples hidden in desert caves at the end (or the beginning) of the long treks to and from distant markets.

The traditional picture of nomadic Mongol tribesmen herding their flocks of sheep and goats and using camels and horses for riding and pack purposes is fast disappearing. The economic core of Inner Mongolia today is composed of the agricultural areas near the Huang. The irrigated areas, which have been expanded by the People's Republic, produce mainly oats and spring wheat, while unirrigated fields are used mainly for drought-resistant millet and kaoliang—a sorghum grain used for feed and liquor. The main city of the area is Baotou (Paotow; population: 1.2 million), an expanding industrial center on the Huang. One of China's major iron and steel works was opened at Baotou with Soviet help in the 1950s, using nearby resources of coal and iron ore.

HUMID CHINA

Humid China, sometimes referred to as Eastern or **Monsoon China**—the core region of the country—includes the densely settled parts of China south of the Great Wall and the Northeast. The ancient provinces south of the Wall are often referred to by outsiders as **"China proper,"** although the Chinese do not employ the term. China proper includes two major divisions, North China and South China, which differ from each other in various physical, economic, and cultural respects.

Although each area exhibits variety from place to place, South China may be characterized in general as subtropical, humid, and hilly or mountainous, with irrigated rice as the main crop (Fig. 13.8). North China, on the other hand, is continental and subhumid, has larger stretches of level land, and depends mainly on nonirrigated grain crops other than rice. Prior to collectivized agriculture, farms in the North averaged about twice the size of those in the South; in the latter region, however, double-cropping practices and the dominance of irrigated rice, with its higher yields, compensated for the smaller size of farms. Oxen and other draft animals such as mules and donkeys are common in the North, but water buffaloes, with their ability to withstand heat and work in the muddy rice fields, predominate in the South. The North Chinese are typically taller, heavier, more purely Mongoloid, and lighter complexioned, and speak mainly Mandarin (also called North Chinese) or one of its variants. The South Chinese tend to be shorter and somewhat darker and speak a variety of mutually unintelligible Chinese

FIGURE 13.8
The treadmill used here to pump water from an irrigation canal to the fields is one example of the arduous manual labor still applied in massive amounts to Chinese agriculture. China has learned the value of labor intensive agriculture partly because of its need to support—and utilize—such a large population.   *Paolo Koch/Photo Researchers*

languages or dialects that are different from Mandarin. The use of Mandarin Chinese is currently expanding, notwithstanding a dynamic independence being expressed by southeast China as Guangzhou (Canton) develops more rapidly and with more prosperity than any other region in the country—and whose predominant language is Cantonese.

North China and South China are worlds of major cultural and physical differentiation. As in the United States, there is a strong sense of regional identity that radiates from both of these worlds.

North China

Affected mightily by the monsoon wind patterns that characterize so much of this Asian region, North China has hot, humid, and rain-filled summers and correspondingly cold and dry winters. When the winter monsoon is pouring out from interior Asia, long, dark, cold winter days dominate North China. Because of the historic disinclination to have interior heating in homes in this region, winters are particularly bitter as people attempt to break the cold with just the heat from the cooking stove or, traditionally, the heated bed called the *kang*.

Although the region receives between 15–21 inches of precipitation annually, it is plagued with relatively high variability from year to year (Figure 13.9). Such a rainfall pattern has led to a history of droughts and floods, reminding one again of the burden of being a densely-settled area of low precipitation and high variability. The droughts that have swept across North China have been instrumental in the initiation of the massive Chang Jiang Water Transfer Project, which is designed to possibly bring surplus water from the Three Gorges region of the Chang Jiang through aqueducts to the heavily farmed but water-deficient North China Plain.

Floods in North China almost all relate to the continual problem of managing the Huang He (Yellow River). It is around this 2500 mile long river and the Wei He tributary (east of Xian) that Chinese civilization was founded. The Huang drops down from its source area on the Tibetan Plateau and courses for more than a thousand miles through the loess soils of North China. These airborne soils blown from the Gobi Desert to the west and northwest are very poorly structured and they collapse easily into the river system. The river carries loess as silt to the Yellow Sea in the east. It has been this physical process that has built up the broad, fertile, and relatively level North China Plain. Chinese recorded history chronicles 26 major channel changes and different discharge points to the Yellow Sea in the past 1500 years.

Floods are particularly significant on the Huang He (Yellow), for the enormous quantity of silt that is deposited along the final 500–600 miles of the river's flow (from the Tai Hang Mountains on the western edge of the North China Plain) has elevated the actual stream channel of the river *above* the surrounding landscape. This phenomenon is called a **superjacent stream**. In the case of the Huang He, the river flows within a

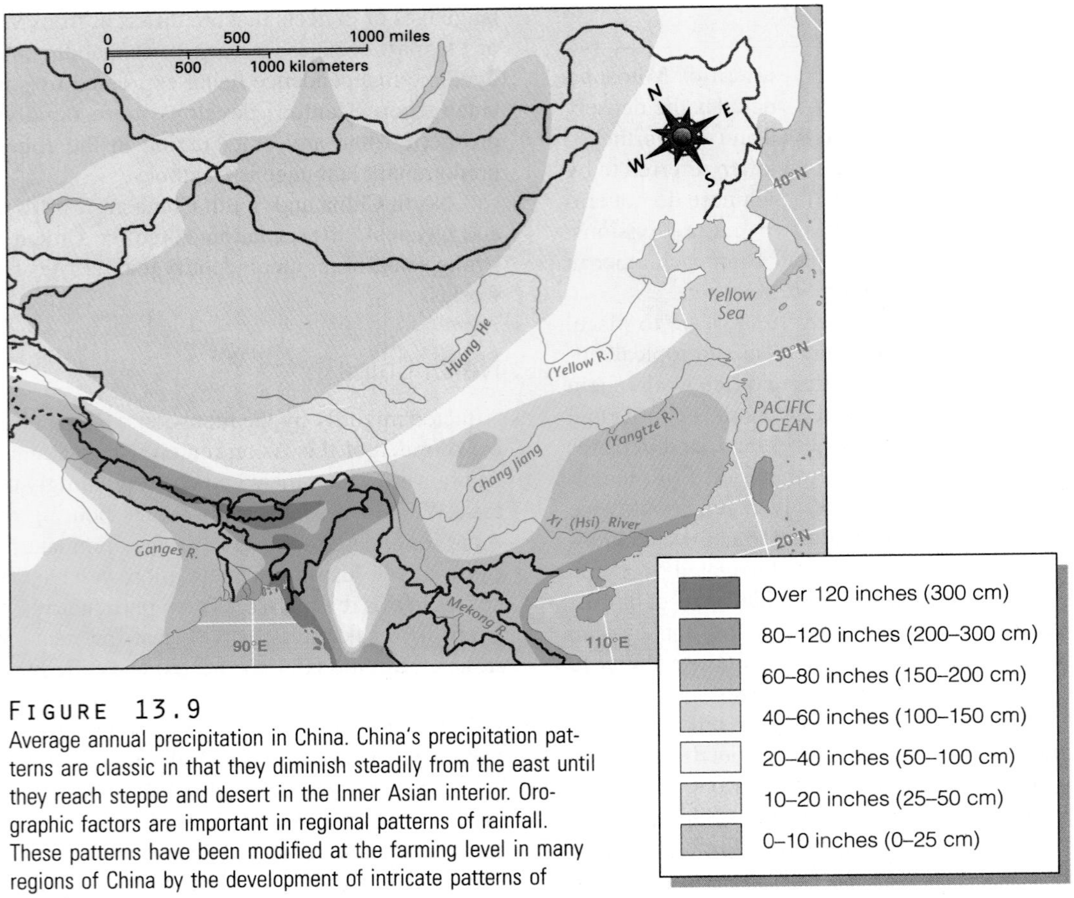
0 500 1000 miles
0 500 1000 kilometers
N
E
W
S
40°N
Yellow Sea
30°N
PACIFIC OCEAN
20°N
Huang He
(Yellow R.)
Chang Jiang
(Yangtze R.)
Xi (Hsi) River
Ganges R.
Mekong R.
90°E
110°E
Over 120 inches (300 cm)
80–120 inches (200–300 cm)
60–80 inches (150–200 cm)
40–60 inches (100–150 cm)
20–40 inches (50–100 cm)
10–20 inches (25–50 cm)
0–10 inches (0–25 cm)

FIGURE 13.9
Average annual precipitation in China. China's precipitation patterns are classic in that they diminish steadily from the east until they reach steppe and desert in the Inner Asian interior. Orographic factors are important in regional patterns of rainfall. These patterns have been modified at the farming level in many regions of China by the development of intricate patterns of irrigation.

critical diking system all the way from the Tai Hang Mountains to the current delta north of the Shandong peninsula (Fig. 13.10). When the dikes break during times of flood, the river pours out of the channel and down to the lower, densely settled farm lands and cities of the North China Plain. Even after the dike is repaired and the river has dropped to more manageable levels, the great flood of water that issued forth from the elevated stream bed has nowhere to go. There is no stream system by which it can easily flow eastward into the Bo Hai Sea. Such widespread ponding and floods have been associated with the Huang He and the North China Plain for centuries.

The major crops in North China are winter wheat, millet, and kaoliang, with some land given over to summer rice crops and corn and a wide variety of kitchen vegetables. North China has exported wheat to South China since the seventh century A.D. Historically much trade has been carried along the Grand Canal, a civil engineering project of the Sui Dynasty (581–618 A.D.). This 1050 mile (1700 km) long inland waterway linked Hangzhou in the Chang Jiang delta region to just southwest of Beijing. Wheat and coal were barged southward and rice was carried north. Although the Grand Canal has never attained a status equal to the image of the Great Wall, it has come much closer to achieving its initial design goals than has the Great Wall.

In North China the uplands region lies in the Tai Hang Mountains to the west of the North China Plain. These mountains are, in many areas, covered with loess soils that are easily worked and naturally fertile. Historically, the countless small villages here have remained isolated, with their inhabitants moving through very small known worlds. In 1935, the famous 6,000 mile Long March was terminated in Yan'an in Shaanxi Province in North China, with Mao Zedong confident he and his 6000 remaining troops were finally secure in the isolation of these uplands. During the early decades of the Communist Revolution era (1950s–1970s) villages in this arid and mountainous region made valiant efforts to gain self-sufficiency in grain production and farm activity.

In a cultural sense, the North Chinese have long felt that they were born in and of the real China because all of the early dynasties were centered either around the Huang He and its tributaries (especially the Wei He in Shaanxi Province) or on the North China Plain. The people of the north tend to be physically distinctive, have more wheat in their diet, traditionally speak Mandarin Chinese as opposed to other dialects, and generally follow the current government's edicts more closely than do the Southerners.

South China

The Qinling Mountains, at approximately 10,000 feet (3048 m), have historically served as the accepted dividing line between North and South China. These mountains run east to west and divide stream drainage between the Huang He to the north and the Chang Jiang to the south. The agricultural dominance of South China comes from the 3400-mile-long (*c.* 5440 km) Chang Jiang that flows from its origin in the Tibetan Plateau to the delta just north of Shanghai on the east. The more than 700,000 square miles (1,813,000 sq km) of waters from both monsoon and orographic rains that are drained by this largest of Chinese rivers give it a magnitude unmatched by any other in China, and by few rivers in the world. Historically, the Chang Jiang has served as a major east-west transit corridor for China. It can accommodate freighters that draw ten feet all the way from the East China Sea to Wuhan in the central basin. From Wuhan the river can be traveled by smaller freighters to Yichang. At that point there has historically been a classic **break-of-bulk** point for further passage through the Chang Jiang Gorges: Freight and passengers are off-loaded and put on flat-bottomed junks and sampans which were drawn upstream by human trackers who pulled and rowed the junks through the Three Gorges (in exactly the style and difficulty described in Hersey's *Single Pebble* on page 323) to Wanxian in Sichuan Province.

The two major basins of the Chang Jiang east of the gorges are broad, densely settled areas with agricultural populations, small towns, and some expanding industrial cities (Wuhan is the most significant of them). The Chang Jiang has long presented the Chinese with difficult settlement decisions, however. Because of the fertile soils and the irrigation benefits associated with residence in the great river flood plains, these lowlands have long been settled and farmed. Such geographic benefits of soil and river transportation, however, always have another side to them—just as parallel environments do in the United States and other nations with heavily settled flood plains. The Chang Jiang Water Transfer Project is the latest in Chinese efforts to harness the enormous strength of this waterway and turn it into a hydroelectric power source and an irrigation and settlement resource, not only for the middle and eastern basins of the Chang Jiang, but also possibly for the North China Plain.

The Red Basin—at the western end of the passage through the gorges of the Chang Jiang—contains one of the largest concentrations of population in China, estimated at somewhat more than 120 million people. Nature has provided little level land in the basin, but rice and other crops are grown on enormous numbers of small fields in narrow ribbons of valley land and on terraced hillsides. The Red Basin served as the center of the Chinese government during World War II because of its isolation and distance from Japanese bombers. It continues today to seem almost like an independent nation because of its environmental resources, distance from central government, and history of relative self-sufficiency.

South of the Chang Jiang basins is the real cultural heartland of South China: the delta region at the mouth of the Xijiang in the Guangdong lowlands. Its central features are the city of Guangzhou (Canton) and the numerous small, densely farmed and settled lowlands on the Xijiang and its tributaries and **distributaries** (the smaller channels by which a river takes its silt and water out to sea). This whole area was not brought

FIGURE 13.10
The Huang He (Yellow River), long known as "China's Sorrow," is subject to huge floods and is confined by massive dikes such as the one in this photo. In many places the bed of the river's stream channel lies at a higher elevation than the surrounding plain so widespread disaster results if rising water breaches or laps over the man-made barriers. *Lowell Georgia/Photo Researchers*

under Chinese control until the Qin and Han dynasties (*c.* 200 B.C.), or nearly 1000 years after China proper had developed into a viable political state.

The coastal zone between the deltas of the Chang and the Xi is so difficult to penetrate that it was not annexed to China until the third century A.D. The mountains rise to between 1500 and 2500 feet (*c.* 460 and 760 m), although some peaks rise to 6000 feet (*c.* 1830 m). The rivers are short, swift, and unnavigable, and each basin constitutes a unit isolated except on the east. The size of the towns at the rivers' mouths is limited by the productivity of the basins they serve. Farming is restricted to small and scattered valley lands and adjacent mountain flanks. The coast is dotted with fishing hamlets. This is the only section of China in which the people have taken much interest in seafaring. The difficulty in traveling overland between the basins on the Chang Jiang and the southern river communities of the Xijiang is such that until just several decades ago the only option for travel to Yunnan province in the southwest, for example, was to boat down the Chang Jiang to the East China Sea, sail south to the mouth of the Xijiang, and go upstream toward the desired destination. New roads have changed this now.

The other dimension of South China is the mountainous landscape that lies west of the Guangzhou lowlands. These hills—mostly in Yunnan and Kweizhou provinces—are part of a deeply dissected 6000-foot-high limestone plateau. It is here that some of the world's most dramatic **karst** (limestone hills with caves and sinkholes) landscape features are found (Fig. 13.11). The area is also home to many of China's small minority populations who continue to sustain themselves with an intensive self-sufficiency in hillside farming, one aspect of which, because of the isolation of the region and the distance from the seat of national authority, has been active cultivation of the opium poppy.

China's Northeast

The Northeast—better known to the West as Manchuria—is the part of Humid China lying north and east of the Great Wall. It is made up of the three provinces of Liaoning, Jilin, and Heilungjiang. It has an area of about 310,000 square miles (*c.* 800,000 sq km) and a population of approximately 105 million (1996). To the west are the Mongolian steppes and deserts; to the east, the area is separated from the Sea of Japan by the Korean peninsula and by Russian territory.

The Manchu emperors of Qing Dynasty China (1644–1911) kept their homeland relatively closed to the Chinese until the very end of their rule. The political instability of East Asia in the early 20th century led the Manchus to encourage Chinese to migrate to the relatively open lands of the Northeast. Millions of Chinese farmers moved to these provinces, pioneered its agricultural frontier, and gave it an overwhelming majority of ethnic Chinese at the outset of the Sino-Japanese War in 1937. During the 20th century, this region became a subject of contention and an area of conflict among China, Russia,

FIGURE 13.11
This is a classic "karst landscape." This geographic term—meaning a limestone landscape with numerous caves and sinks and sharply sided mountains and hills—is Serbo-Croat in its origin, but this region in China is one of the most dramatic examples of these physical landscape features anywhere in the world. The flatlands are rich in lime but the broken topography caused by the sinks and depressions presents problems to farmers and settlements. In southwest China, karst landscapes have been important as tourist draws, especially in the region shown here in Guilin, Guangxi-Zhuang Autonomous Region. *C.L. Salter*

and Japan, all of which have controlled it at one time or another. Russian interest has focused on the Chinese Eastern Railway—a shortcut across the northern two provinces between Vladivostok and the Trans-Siberian Railroad east of Lake Baikal—and on the naval and port facilities of Lüda, which the Russians have used at various times.

Japan has long been interested in developing the Northeast's impressive industrial resources and its potential for surplus food production. After the Russo-Japanese War of 1904–1905, fought mainly in Liaoning, Japan became increasingly active in the area and took over actual control in 1931 with the establishment of the puppet state of Manchukuo. In 1945, during the final days of World War II, Soviet forces took it from the Japanese. Eventually, the Chinese Communists incorporated the Northeast into the People's Republic, beginning this region's role as the center of broadly expanded industrial development. Although significant factory plant and equipment was uprooted and shipped back to industrial centers in the former USSR by the Russians at the very end of World War II, the industrial and transportation infrastructure created in the 1930s and 1940s during the Japanese occupation were important assets for the PRC in its economic development of this region.

It was the northward movement of the United Nations troops on the Korean peninsula in 1950 that brought the Chinese army into the Korean Conflict. The North Korean army was being pushed back nearly to the Yalu River—the border between China and North Korea—when the Chinese decided to step in and stop this advance on their newly developing industrial heartland in Liaoning. The Communists' geopolitical and military decision to enter the war on North Korea's side to help push back the U.N. troops from their approach to the Yalu River border is not surprising when you see the industrial importance this region played for revolutionary China.

The Northeast consists of a broad, rolling central plain surrounded by a frame of mountains that seldom rise higher than 6000 feet (a little over 1800 m). The mountains on the east and north contain much valuable hardwood and softwood timber. The central plain, oriented northeast-southwest, is approximately 600 miles long (966 km) by 200 to 400 miles wide. The soils are very fertile, and the summer rainfall is generally sufficient for the crops that are grown, although spring drought is often troublesome. Conditions are not favorable for irrigated rice, and the winters are generally too severe for winter grains. However, soybeans, kaoliang, millet, corn, and spring-sown wheat do well during the relatively short but warm frost-free season of 150 to 180 days. Compared to the farms of South China, farmlands in the Northeast are relatively larger and have lower population densities. There is a much greater possibility for future mechanization of agriculture here than in the parts of Humid China south of the Great Wall.

The industries of the region are centered mainly in or near the largest city, Shenyang, in Liaoning. The largest center of iron and steel production is Anshan (population: 1.4 million, city proper), south of Shenyang. The industry is based on deposits of iron ore in a belt that crosses the southern part of the region. Substantial deposits of coal also exist, although the total reserves are far smaller than those of China south of the Great Wall, and only a small fraction of Manchurian coal is suitable for coking. The Daqing oil fields are located northwest of Shenyang in Heilongjiang Province. These fields have been developed into one of China's richest oil production centers since their discovery and energetic development in the 1960s.

13.3   Continuity and Revolution in China

Scholars estimate that the Chinese had already become a major political entity with the Shang Dynasty (*c.*1523–1027 B.C.), well before the beginning of the Christian era in the West. At this early time, Chinese culture had already developed an agricultural system able to support large numbers of people on small areas of intensely worked arable land. During the Shang Dynasty the great alluvial plain of the Huang He in North China already was occupied by Chinese peasants with a garden type of agriculture based on the hoe, irrigated crops, varied fertilizers including night soil (human wastes), intensive use of flat alluvial lands, terracing of fertile hillsides to create flat land, a high degree of control over water, private ownership of land, use of family labor, and siting of villages on less fertile land. Little attempt was made to farm land not fertile enough to yield

a living to families equipped only with hoes. From the Huang He basin, this system spread throughout the humid eastern part of the country by a process of diffusion that included migrations and conquests over a period of many centuries.

Definitions and Insights
Night Soil

Night soil is the term used to define the human waste collected from both rural and urban households in China. The culture pattern of making use of this often neglected resource through its collection and storage in outdoor pits has given China an historically productive agriculture, especially in the market gardens that surround urban centers. The collection and use of this organic resource also promoted frugality and recycling of other organic resources, helping to support Chinese household self-sufficiency. It has only been in the last three decades that chemical fertilizers have begun to reduce the relative utility of night soil (Fig.13.A).The use of chemical fertilizers, however, requires considerable expenditure by Chinese farm households. It also introduces a problem in urban waste management that the Chinese have not traditionally had to deal with.

FIGURE 13.A
The early morning collection and transportation of 'night soil' from the houses in Chinese cities and towns continues to be a symbol of traditional household economy. This organic resource is carried by barge to outlying rural farming areas and there it is used as an enrichment for the growing of urban foodstuffs. As China modernizes there is, however, increasing pressure to adopt municipal sewage systems that simply remove, treat, and then dispose of urban household wastes.   *C.L. Salter*

In East Asia, China has historically been as dominant politically as it has been culturally, but this political dominance has often alternated with political disunity and weakness. Unification of the Chinese people into one empire was first accomplished in the third century B.C.; after that time, periods of strong central government, generally occurring soon after the beginning of a new dynasty, alternated with periods of weak central government and internal warfare. During periods of unity and strong central power, China generally extended its authority widely, both by direct conquest and by accepting the fealty and tribute of lesser powers around it. Under the Han Dynasty (206 B.C.–220 A.D.) for example, Chinese armies pushed far westward across central Asia and seem to have barely missed contact with the Romans in the vicinity of the Caspian Sea. Under the Yuan (Mongol) Dynasty (1280–1367), the Chinese empire controlled Burma, Indochina, Korea, Manchuria, and Tibet. Under the Manchu (Qing) Dynasty (1644–1911), the Chinese held large areas that eventually were added to the Asian part of the Russian empire. But strong dynasties tended to weaken with time, and then internal disorder, partition, and civil war generally followed. During such periods of weakness, parts of China often were conquered by nomadic invaders (such as Mongols and Manchus) from the north and northwest. Nevertheless, through all the rise and fall of dynasties, a distinctive Chinese civilization persisted and intensified.

AN EMERGING WORLD POWER

Present day China has important roots in the distant past and its history reflects the interaction with the West associated with centuries of trade and influential cultural connections.

The 1997 return of Hong Kong to China brings to a close more than 150 years of British control and development of this major port city in southeast China. The island of Hong Kong was granted to England in the 1842 Treaty of Nanking. This concluded a war called The Opium War because of the role opium had played in fomenting a Chinese need for British trade. From the late 18th century until the 1839 opening of hostilities, China had been a poor trading partner, claiming little interest in English goods. The British, on the other hand, were eager to get and trade ceramics, textiles, art work, and exotic food stuffs from China. It was finally opium that developed a trade need that secured an informal trading relationship between these two nations. China has long disavowed this era, and its associated treaty and loss of Hong Kong. This period of colonial control is approximately coincident with steady civil unrest, the collapse of the Qing Dynasty, disorder, a costly civil war and civil disorder ultimately culminating in the Tiananmen massacre in 1989.

To China the return of the prosperous entrepôt of Hong Kong seems only just after having to endure more than a century and a half of being subordinate to the British administration in this rich port city. In the dark era of British control of the island port, and with the repeated efforts of other nations to extract economic and political concessions, China built up a bitter resentment of the West. Such a resentment was intensified by China's prior history as the dominant power in Asia through its great achievements in technology, culture, religion, and numerous other fields. Pride in culture, love of the past, and disinterest in the outside world contributed powerfully to China's failure to keep pace with Western development and power in the 19th century, and violation of Chinese pride, as well as territorial encroachment by outside powers, helped foster a powerful nationalism during the 20th century. The figure most central to this explosion of Chinese nationalism was Communist Party Chairman Mao Zedong.

These circumstances took on added significance with the emergence of mainland China as a military power under Communist rule. With the explosion of its first nuclear device in October 1964 and successful nuclear tests afterward, China gained a new importance in world affairs. However, even before their first nuclear explosion, the Chinese Communists had already demonstrated their military prowess—in the Korean War (1950–1953) and in a short-lived border war with India in 1962. The haltingly successful efforts of communist planners to build an industrial society have added to the country's power potential. Chinese foreign policy not only impinges on the many countries that border China but is a very important factor in world affairs generally. The PRC's military war games in the Taiwan Strait during Taiwan's spring 1996 elections and the decision to augment their national investment in defense by 13% after the death of Deng Xiaoping in March, 1997 further demonstrated China's demand for world notice in the military as well as political and demographic spheres.

CHINA'S PUSH TO INDUSTRIALIZE

Mass production of industrial goods began in China around the beginning of the 20th century. Prior to that time, Imperial China's handicraft industries produced high-quality goods that attracted Western traders in spite of China's official disinterest in such trade. To this day, handicraft and shop-scale industries continue to supply many of China's simpler needs.

The early factory industries were largely foreign-owned and were attracted to China by the huge market for cheap goods, by cheap labor, and sometimes by certain natural resources. As in so many countries, the first large-scale factory industry to develop was the manufacture of cotton textiles. Japanese and other foreign firms, together with some native Chinese companies, developed a major cotton industry by the 1930s. This industry, and others, developed in such coastal cities as Shanghai and Tianjin. Meanwhile, the Japanese used the minerals of the Northeast (in their puppet state of Manchukuo) to make that area the first center of heavy industry in China.

In their efforts to industrialize, China has directed national economic development in four distinct national drives:

Communization and repair of wartime damage, *1949–1952*. The industrial structure, which had been badly hurt by invasion and civil war from 1937 to 1949, was brought back to prewar production levels under the new system of state ownership.

Regional Perspective

THE SILK ROAD

In the early Han Dynasty (from 206 B.C. to about 8 A.D.) the Chinese began to effect a steady exchange of goods for items from the Mediterranean trading ports. From China went silk (not raised in the West until the sixth century), and to China came wool, gold, silver, and glass. The overland route for this exchange—called the Silk Road—was approximately 5000 miles long, going from

X'ian (the early Chinese capital of Chang'an) to Damascus along the Great Wall and the Taklamakan Desert, over the Pamir Mountains, and through Baghdad. From there goods were shipped to various European ports. It was this route that played a role in the diffusion of Buddhism into China in the first century A.D. Almost no one ever followed the whole route, for travel was dangerous and every caravan load had to pass numerous check points and

pay considerable tolls. Venetian explorer **Marco Polo** is sometimes described as the first person to travel the complete route, and it was, in part, the power of his description of this grand overland route of trade, diffusion, and cultural exchange that made Christopher Columbus so determined to find a sea route to replace it. The route is now being considered as a possible base for a United Nations trans-Asian international highway (Fig. 13.B).

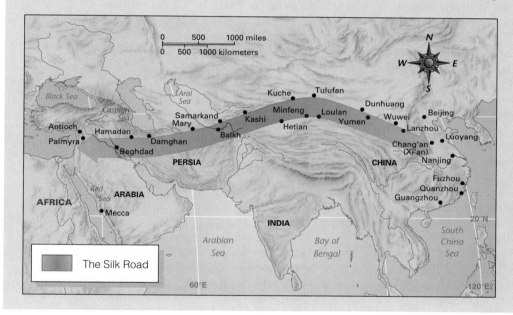

FIGURE 13.B

The Silk Road (sometimes called The Silk Route) is the nearly 5000 mile link that connected China with European markets and civilization several centuries before the time of Christ. This series of partial trails and routes was almost never traveled by a single person or caravan, but rather was a trade route for many caravan groups each crossing specific terrain. Silks, paper, artgoods, and spices went west from China while gold and silver bullion, glass goods and metal ware were major goods going east. The Silk Road has been a major diffusion route of religions, language, and customs as well.

This was done even while China was expending considerable capital and manpower in its involvement in the Korean conflict.

Heavy-industry development using the Soviet pattern, *1952–1960.* Aided by the Soviet Union, Chinese planners gave priority to coal, oil, iron and steel, electricity, cement, certain machine-building industries, and armaments. China's resources for these industries are considerable, and the country's centrally **planned economy**—a system in which major economic activity is initiated by government planners—achieved major successes in developing them, as has been the case in many communist countries. There was also a spatial component to this drive, for the Russians advised the Chinese to diminish the coastal concentration of industrial plants. This led to expansion into new industrial landscapes at Wuhan on the Chang Jiang, at Baotou in Inner Mongolia, and even into Inner Asia

near the [then] Soviet border in Xinjiang. Pre–Communist era coal production had peaked at 66 million metric tons per year under Japanese military control in 1942, but by 1958 China's output was 270 million tons and the country ranked third in the world, after the USSR and the United States. Subsequently, output increased to 1116 million tons in 1992 (with the U.S. at 947 million tons), making China the world's largest producer. Meanwhile, a maximum pre–Communist era steel production of 900,000 tons per year was multiplied by a factor of 9, to 8 million tons, by 1958. By 1992, steel output had reached 81 million tons, placing China ahead of Russia (58 million) for third position in the world, but behind Japan (98 million) and the United States (93 million).

Progress in heavy industry has been built on a rich resource base. China has coal resources estimated at about one tenth of the world total. They are widely spread, but the major producing fields lie under the North China Plain and its bordering hills and in Liaoning Province in the Northeast. Iron-ore resources are less abundant, but there are workable deposits in a number of locations, the largest deposits also being in Liaoning. The southern region of the Northeast has been the country's leading heavy-industrial region since modern industries were developed there by the Japanese, but widespread coal and ore deposits have facilitated Chinese Communist development or expansion of these industries at other scattered centers in China proper, in Inner Mongolia, and along the Chang Jiang. China's large production of many alloys and a number of other metals have further facilitated development. The discovery and development of major new oil fields in the Northeast, Gansu, Sichuan, and in the western reaches of Xinjiang province has significantly enriched China's resource base in the past three decades. China has gone from being a petroleum importer just four decades ago to now serving as a major Asian exporter of petroleum products.

China's entire resource history has been and continues to be troubled by the lack of a national transportation system that allows easy and efficient movement of heavy commodities from extraction or production sources to refining centers or, more importantly, to market or export facilities.

Near-isolation under late Maoism *1960–1978*. This phase of Communist China's industrial development began with the country's rift with the Soviet Union in 1958–1960 and continued until new economic policies were introduced after the death of Mao in the late 1970s and early 1980s. Progress continued in heavy industry, but the country's general industrial advance slowed in pace, technical development, quality of output, and modernization by (a) near isolation from more advanced industrial countries, (b) the "Cultural Revolution" (see below) begun by Mao in 1966 and carried on nearly until his death, and (c) the customary difficulties that nonmarket, centrally planned economies on the Soviet model have had in developing consumer-oriented industries and advancing technology.

The Cultural Revolution (1966–1972) was Mao's attempt to maintain an ongoing "permanent revolution" in China. Politically the movement was driven by Mao in his effort to throw a newly established urban and intellectual elite off balance by promoting purges of virtually anyone who had offended Communist Party functionaries or local youth who donned the red armbands of the youthful Red Guards. The Red Guards were loosed on the populace to harass those who could be accused of putting learning, skill, expertise, or personal matters above revolutionary enthusiasm or Marxist goals and principles. Government oppression in this campaign is thought to have led to a million assassinations, executions, or suicides, and to have to provided lifetime jobs for those deemed most loyal to the Maoist revolution, regardless of their job performance. Reforms being initiated in the late 1970s began the process of removal of such people from often unproductive positions.

Geographically the Cultural Revolution was a revolution in mobility as well. For the first time in the lives of many Chinese it was suddenly acceptable to travel randomly about the country, taking advantage of free rides on buses, trains, and trucks for the purpose of "making revolution." This meant that Chinese who had lived within known worlds of very small scale were now making journeys all across China in an effort to root out the libraries and lifestyles of people who had wandered, or who were accused of wandering, from the purist revolutionary route that Chairman Mao had outlined.

The impact of the Deng Xiaoping reform era, *1978–1997*. After a short period of internal confusion after the 1976 death of Mao Zedong, another Long March veteran—Deng Xiaoping—ascended to supreme power in China. Under his leadership, China introduced the **Four Modernizations** that focused the nation's energy on accelerated development of agriculture, industry, science and technology, and national defense (Fig. 13.12). Since the late 1980s, China has done a complete about-face in its interaction with the West.

Definitions and Insights

THE CHINESE COMMUNE EXPERIMENT

Soon after China's declaration of success in the Korean War (1953), Chairman Mao began to push for a more rapid change from individual village farming to more organized farming units. Initially China experimented with agricultural cooperatives in the early 1950s. These expanded the farming base from family farms to numerous village families working land as a unit. In 1957, Mao felt that China needed to make a more dramatic move toward communal farming. With the establishment of the first rural communes in that year, Mao enlarged the scale of all agricultural units. Whereas cooperatives had nested hundreds of farmers within a decision-making unit, the commune brought thousands and sometimes tens of thousands of farmers together with crop and marketing decisions all made by people's committees and commune managers. Mao worked to eliminate all private plots so that all farming energy would be focused on communal lands. Wages came from work points gained for labor given to communal effort, but increasingly there was dissatisfaction felt because of the belief that more work points were given to politically opportunist workers than to bona fide agrarian workers. Production began to suffer, private plots crept back into the landscape, and by the end of the 1970s, farm management had reverted largely to household levels, with the introduction of the **Household Responsibility System** in 1978 replacing the Maoist agricultural communes. In the swing toward free-market experimentation in the 1980s, the communes lost almost all direct involvement in Chinese agricultural production. The experiment that was so closely tied to the Marxist goals of Chairman Mao was quietly allowed to lapse with little fanfare. As market forces began to reward the most productive farmers in the late 1980s and 1990s, the commune experiment became another segment of the difficult history of the Chinese Revolution and Mao Zedong.

FIGURE 13.12
View of a market in Shanghai. The signs in Chinese urge support for China's "Four Modernizations": the modernization of industry, agriculture, science and technology, and national defense. This drive to modernize continues to be a major feature of present-day China.   *Jesse Wheeler*

The combination of these four major shifts in governmental control since 1978 have launched China on a pathway of economic change that has brought significant benefit to rural and city folk alike. For example, before the Deng Xiaoping reforms, China had 6 washing machines for every 100 families. The number now is 86. Before Deng, there was one refrigerator per hundred households; now fully half of the Chinese households have one. Not only has the life of the average Chinese been transformed, but light industrial capacity in China is expanding at an unprecedented rate to satisfy both domestic markets and expanding export needs.

A great variety of manufactured goods are produced in China, including some relatively sophisticated products such as those needed to maintain the country's nuclear arsenal. But China's trade pattern is still that of a country in the early stages of industrialization. It emphasizes export of labor-intensive manufactured goods along with resources or related products such as textiles and clothing, oil, and oil products. The scattered oil and gas deposits in various parts of the country now account for approximately 5 percent of world oil production, which is enough to supply the country's own low per capita usage and also provide a major export. The enormous potential of both the Chinese market and China's labor force has made foreign investors endure high office space costs, a complex and largely unyielding bureaucracy, and wide governmental swings in political attitudes toward joint ventures. The country's main trading partners and its primary sources of foreign investment and technology are Hong Kong, Japan, and the United States, with Hong Kong playing the role of an effective middleman between China and the outside world.

13.4   Landscape Legacies from the Maoist Revolution

Politically, China, like virtually all nations, has alternated between periods of strength and periods of weakness. But, despite these fluctuations, it has continued to be the home of the dominant culture of East Asia since very ancient times. Surrounding countries and regions show evidence of the long-standing influence of Chinese culture on their languages and writing, religion, agriculture, arts, crafts, institutions, ways of thinking, and histories. Since the 18th century, however, the balance has swung, and China itself has come under the powerful and sometimes disruptive influence of Western culture, although China never was totally immune to outside influences.

China's cultural ascendancy before its Western era was due in part to the creativity of Chinese culture, but it was also because of the fact that the Chinese have been an especially numerous people since prehistory. Creativity is attested to by many early pioneering Chinese inventions and discoveries, such as gunpowder, paper, the wheelbarrow, the magnetic compass, and their system of writing. It is also attested to by the organizational and engineering skills that were necessary for the building of the Great Wall (much of which was completed before the end of the first millennium B.C.) and the subsequent construction of the Grand Canal to link the Chang Jiang valley with the populations and resources of the North China Plain. There have been, as well, a number of geographical efforts in resource management and development that have been particularly significant in postrevolution China:

Landscape modification and intensification of existing agricultural techniques.   China has attacked the problem of agricultural production by intensifying the use of the existing agricultural base in order to raise yields per unit of land. Irrigation and flood-control measures, when applied to land already farmed, may be seen as one phase of this effort. Another approach lies in the application of scientific knowledge and experimentation to develop better seed, equipment, and farming techniques, together with the development of administrative and propaganda machinery to spread their use. A third approach lies in increased fertilization. Chinese farmers have long fertilized their fields with animal and human manures, pond mud, plant residues and ashes, oilseed cakes, and green manure crops. Although these traditional fertilizers are still by far the most important source of plant nutrients, the current regime also stresses the use of chemical fertilizers. China's relatively small chemical-fertilizer industry has been rapidly expanded by the purchase of entire plants from abroad. However, the amounts used are still relatively low.

Land reclamation.   One thing that has often impressed observers in China is the small use that is made of low-yielding second-rate land even in places where population is excessively crowded on adjacent good land. For instance, in South China, narrow valleys often teem with an overcrowded and struggling population, but broader adjacent uplands are little used. Apparently the Chinese farmer and his family, unequipped with machinery, cannot till a sufficient amount of second-rate land by hand labor to make a living.

However, in the early 1950s, just after the communists achieved control at the end of their civil war, their statements placed considerable stress on land reclamation and increasing the amount of land available for potential cultivation. Small gains in the amount of cultivated land were made through reclamation of poorly drained land and extension of irrigation into deserts. Since the 1950s, however, there has actually been a slight decrease in the amount of land under cultivation, owing largely to the expansion of the urban base in China. This decrease in farmland suggests the difficulties of the land-reclamation approach, and in recent years the Chinese have scaled down their estimates of reclaimable land. Potential areas of significant size are claimed for the Northeast, Xinjiang, and scattered areas elsewhere, but these areas are agriculturally marginal and too costly to bring into production under present conditions, even though there was major effort in this process in the 1960s and 1970s.

Mechanization, crop breeding, and control of diseases and pests. Other methods to increase farm output are not yet of first importance but may be crucial in the future. Mechanization has developed slowly. The use of tractors is growing, but primarily in the Northeast and in the west where more extensive farm fields are located. Thus, there has been progress compared to the situation existing in 1949, but, by and large, the rural scene still is dominated by human and animal power and simple hand tools.

Definitions and Insights

DAZHAI: A MODEL LANDSCAPE

During the decades when Mao Zedong was in control of the Chinese government, he once claimed that "there are no unproductive regions; there are only unproductive people." During the 1960s, after the cooperative support of the USSR had ended and China was largely dependent upon its own resources and ingenuity, Mao launched a national campaign promoting self-reliance. In 1964, a small village in the mountains of Shanxi Province in North China was elevated to the role of China's first **"model landscape."** The village was Dazhai (Tachai) and Mao's dictum was "In agriculture, learn from Dazhai!" For more than a decade, the stories of Dazhai's terrace construction, water control, farming experiments, and village growth became daily news in China's press. Literally millions of peasant agriculturalists from all over China came to the village to learn from the peasant folk who had remade their landscape (with no government aid, it was claimed) into a much more productive one. These visitors made plans themselves for replicating the Dazhai villagers' landscape transformation. It was not until after Mao's death in 1976 that officials began to realize that not all of China could shape their village landscapes in parallel fashion (Fig. 13.C). There were too many environmental niches for China to have only one model landscape to guide this era of self-reliance and expansion of the country's arable base. Dazhai remains an extraordinary example of a whole village becoming a national "folk hero," playing a role in and shaping national agrarian policies in unprecedented fashion.

FIGURE 13.C

In 1963 Chairman Mao proclaimed "In Agriculture, Learn from Dazhai!" In this call for national emulation of the work being done by Dazhai village—with its population of fewer than 500 people in the mountains just west of the North China Plain—he established China's first national model for landscape transformation. For more than a decade, agricultural leaders from all over China came to the Shanxi Province mountain village to see their terraces, irrigation works, cave-dwellings, and hear of their dedication to remaking their difficult landscape into a more productive unit. By the time of Mao's death in 1976, China had begun to argue that no single village could serve as a model for a country as environmentally diverse as China, but for a while visitors from all parts of China and other countries came to witness these patterns of bold Dazhai landscape transformation. *C.L. Salter*

Progress also has been reported in crop breeding and in the control of insects and plant diseases, but little definitive information is available. The lack of trained agricultural scientists and technicians has been a hindrance. There is evidence that insecticides and herbicides are coming into wider use. Pest control was greatly publicized during the early years of the People's Republic, with great campaigns to eradicate rats, mice, flies, sparrows, and other pests. Observers were amazed at the sight of Chinese peasants fanatically hunting down pests, propelled by the exhortation of the government. A major objective of these campaigns was to reduce food losses caused by the pests, which have been credited with the destruction of many millions of tons of grain each year. Antipest campaigns against flies, mosquitoes, and rats have had, in addition, an obvious public health benefit. The evidence suggests that these programs have had at least a limited success. They also were politically important as they demonstrated the government's capacity to marshal enormous numbers of people in a wide variety of national campaigns—many of which were focused upon modification of the environment.

Projects to control and conserve water and to expand the irrigated acreage have been actively promoted. Floods and droughts have always periodically afflicted Chinese crop production, especially in North China. In the past, they have caused millions of deaths from famine. Thus, a major objective is to prevent water from running off in floods and to conserve it in reservoirs for times when it is needed. Three specific types of projects have been and are being carried out to achieve this: (1) Construction of large numbers of small-scale dams, ponds, canals, and dikes by local communities; (2) Building of large-scale dams and associated structures to control larger streams, several projects of which are complete or under way; (3) Reforestation, which also relates closely to water control. Only 14 percent of China is forested, and large areas are without significant vegetative cover. Most forest growth in China proper has long since been removed, primarily by peasants for use as fuel. The lack of vegetative cover on nonfarm lands causes a rapid runoff of much of the rain that falls, thus raising stream levels and increasing flood hazard as well as effecting serious erosion and the siltation of reservoirs and crop fields. The Chinese Communists have stressed large-scale projects and have greatly increased the amount of standing timber.

Definitions and Insights

The Chang Jiang Water Transfer Project

The Great River—the Chang Jiang—has long been central to a Chinese sense of place. This river drains more than 700,000 square miles and has seasonal shifts in river depth of more than 100 feet in the gorges that lie between Sichuan Province in the west and the broad agricultural basins of central China

to the east. The current Chang Jiang Water Transfer Project—called the Three Gorges Dam for short—is the latest and largest effort made by the Chinese to control the enormous power and potential of that river. In 1919, Sun Yat-sen (sometimes called the George Washington of China) suggested that a major dam be built to help control the Chang Jiang. (It was probably that call that stimulated John Hersey's novel *Single Pebble*, excerpts from which begin on pages 322–23.) One goal of the contemporary project is to build a series of dams so strong and high that the flow of river water through the heavily populated provinces east of the gorges can be evened out, reducing or eliminating flood threats to a region that has suffered major inundations for thousands of years. The dam would be 610 feet high, 1.3 miles long (2.1 km) and the reservoir would extend upstream for 385 miles (620 km). The boldness and scale of this landscape modification project makes it akin to the building of the Great Wall and the Grand Canal in the eyes of the Chinese. There has been much disapproval—both domestic and foreign—of the plan because of its potential for environmental disturbance to a whole network of water and plant ecosystems that rely upon the Chang Jiang. The World Bank, after expending $8.7 million for a feasibility study, decided not to fund any part of the project. In May 1996, the U.S. Export-Import Bank also refused to provide any funding for the megadam. The project is also disfavored because of the number of villages that will be drowned by the reservoir that will develop behind this tallest of all China dams. It is estimated that completion of the project will displace 1.4 million people, submerge 13 cities, including the lower half of Wanxien, hundreds of villages and 115,000 acres of farmland. The fact that the project is being carried forward is yet another example of the strength of China's wish to be independent, even in a world of global economic, political, and environmental interdependence. The Three Gorges Dam project is a political, environmental, and economic statement further declaring China's wish to reshape its present, and to create a more productive future—with electric power being a major agent in such transformation.

A major objective of water conservancy measures has been to increase the amount of land under irrigation and thereby increase output. China claims to have increased the amount of arable land under irrigation by 200 percent since 1949. It appears that a considerable increase has been achieved, that many of the existing irrigation works have been improved, and that use of power equipment has increased the efficiency of irrigation in some areas. However, in arid areas there continues to be a problem with salinization because of the inability to gain enough water to not only nourish the crops but to "wash the salts" from the soil as well.

Improvement of transportation.   To integrate China's widespread resources and widely scattered producing and consuming centers into a functioning whole, the government has placed

FIGURE 14.4
Wood is a vital resource in Japan. This wooden home is in Shirakawa, north of Tokyo on the island of Honshu.   *Jean Kugler/FPG International*

much smaller population, the nation's heavily forested mountainsides supplied the needed wood, but if today's demand for forest products were met with domestic reserves, the country would be rapidly deforested. The varied local environments of Japan support a wide range of valuable trees, mainly broadleaf types toward the south and needleleaf conifers toward the north. Many mountainsides are quilted with rectangular plots of different kinds of trees planted in rows. Most first-growth timber is long since gone, but some virgin stands are still protected in preserves. Forests are important in Japanese culture, and this affluent country does not wish to lose them. With its enormous demands for wood products, Japan therefore is the world's largest importer of wood, mostly hardwoods from the tropical rain forests of southeast Asia.

Kerosene stoves have largely replaced the charcoal used formerly for household cooking and heating, but Japanese people still use wood for many purposes (Fig. 14.4). Single-family dwellings continue to be built largely of wood, which provides more safety from earthquake shock waves than more rigid construction. However, wood makes the dwellings more vulnerable to the fires that typically sweep through Japanese cities after earthquakes rupture gas lines. Paper made from wood is used to cover the sliding partitions between rooms in homes and apartments. Japan's large publishing industry requires a great deal of paper, and other industrial and handicraft uses are large.

MINERALS, WATER, AGRICULTURE, AND FISHERIES

Small deposits of many different minerals, notably coal, iron ore, sulfur, silver, zinc, copper, tungsten, and manganese, supported early industrialization in Japan. The supplies of all of these minerals are inadequate for Japanese needs today, so, like most other raw materials, they must be imported. Japan's mountain streams provided much of the power for early industrialization, but they are so swift, shallow, and rocky that they are of little use for navigation. Most of the water used for irrigation in Japan comes from these streams, which emerge from the mountains, divide into distributaries, and then cross the plains in beds elevated above the level of the cropland. These elevated beds make it possible to get much of the water to the land by gravity.

Only about one eighth of this small country's area is arable; Japan is not one of the world's major crop-growing nations. Most cultivable land is in mountain basins and in small plains along the coast. Terraced fields on mountainsides augment this level land. Even on flat land, great numbers of Japan's fields are terraced for the growing of irrigated rice, the country's basic food and most important crop (Fig. 14.5). Irrigated rice is nearly everywhere on arable land, even in most parts of the northern island of Hokkaido. Production of this grain is heavily subsidized by the Japanese government and promoted by a powerful farm lobby, which aims at rice self-sufficiency, with other foods being imported as needed. The country's farmers usually achieve a net export of rice by intensive use of scarce cropland. However, a scarcity of rice throughout east Asia in the early 1990s, and United States pressure on Japan to reduce its massive trade surplus with the U.S., prompted Japan in 1995 to open its rice market to foreign imports for the first time. Rice

FIGURE 14.5
Hanging rice on racks to dry in the island of Kyushu, Japan.   *Todd Stradford, Jr.*

consumption per person has nevertheless been declining in recent years as the Japanese diet has become more diverse.

Mild winters and ample moisture permit double cropping—the growing of two crops a year on the same field—on most irrigated land from central Honshu south. Rice is grown in summer, and wheat, barley, or some other winter crop is planted after the rice harvest. Overall, about one third of Japan's irrigated rice fields are sown to a second crop, and more than half of the unirrigated fields are double cropped. **Intertillage**—the growing of two or more crops simultaneously in alternate rows—is also common. Dry farming also produces a variety of crops: sugar beets in Hokkaido; all sorts of temperate zone fruits and vegetables; potatoes, peas, and beans, including soybeans; and, in the south, tea, citrus fruits, sugarcane, tobacco, peanuts, and mulberry trees to feed silkworms.

Changes in Japan's agricultural sector have accompanied a surge in urban-industrial employment and general affluence in recent decades. Farmers have been moving to towns and the number of agricultural workers has dropped dramatically (from about 25 percent of the labor force in 1962 to 7 percent in 1997). A huge increase in farm mechanization has made it possible to till the land with far less human labor. Millions of small tractors and other types of mechanized equipment are now used on Japan's tiny farms. With this technology and heavy applications of fertilizers (both chemical and processed from human waste), plus improved crop varieties, Japanese agriculture has achieved some of the world's highest yields per unit of land.

The island people of Japan view the ocean as an important resource and promising frontier. The country has an ambitious "inner space" program of deep-sea exploration to learn how to warn of earthquakes and tsunamis and to mine the sea's living and mineral riches. A variety of food fish are present in the waters surrounding Japan, and Japanese fishermen range widely

through the world's major ocean fishing grounds. Japan is the leading nation in ocean fisheries and the Japanese enjoy the world's highest per capita consumption of fish, including *sashimi* or raw fish, and other marine foods (Fig. 14.6). The Japanese have also historically enjoyed whale meat, but pressure from international environmentalists compelled Japan in 1985 to support a worldwide moratorium on commercial whaling. Early in the 1990s, along with Iceland and Norway, Japan lodged reservations against the ban and has since resumed small-scale whaling of minke whales for what Japan insists are "scientific research" purposes. According to Japanese sources, only a few people consume the whale meat after the animals are studied.

In recent years the Japanese have begun to eat less fish and rice and to adopt a more Western diet. While sea fish continue to be the main source of animal protein in the Japanese diet, increased consumption of meat, milk, and eggs has been reflected in a growing number of cattle, pigs, and poultry on Japanese farms. By 1993, livestock surpassed rice as the country's most valuable agricultural product. The animals require heavy imports of feed, and the increasing variety in people's tastes mean many more food imports. The United States is Japan's largest supplier of imported feedstuffs and food.

JAPAN'S CORE AREA

Most of Japan's economic activities and a large majority of its people are packed into a corridor about 700 miles (*c.* 1100 km) long. This megalopolis extends from Tokyo, Yokohama, and the surrounding Kanto Plain on the island of Honshu in the east, through northern Shikoku, to northern Kyushu in the west (see Fig. 14.1). This highly urbanized core area contains more people than live in the Boston-to-Washington "megalopolis" on the

FIGURE 14.6
Ocean fish are a major element in the Japanese diet. This photo shows an open-air fish market in Tokyo.
Yamaguchi/Gamma Liaison

United States' east coast. All of Japan's greatest cities and many lesser ones have developed here, on and near harbors along the Pacific and the Inland Sea. They are sited typically on small, agriculturally productive alluvial plains between the mountains and the sea.

The largest urban agglomeration in the world—Tokyo-Yokohama and suburbs (population: 30.3 million by one estimate; other figures differ according to how the metropolitan area is defined)—has developed on the Kanto Plain, Japan's most extensive lowland. Geographers regard Tokyo-Yokohama as one of the world's three truly "global cities," rivaled in size and significance only by New York City and London (Table 14.1). Originally, Tokyo grew as Japan's political capital and Yokohama as the area's main seaport. This urban complex functions today as Japan's national capital, one of its two main seaport areas (the other is Osaka-Kobe), its leading industrial center, and the commercial center for northern Japan (Fig. 14.7). The Japanese government is contemplating a relocation of the capital to an inland site less prone to tsunamis and earthquakes.

Beyond Tokyo, in the entire northern half of Japan, there is only one major city, Sapporo (population: 1.67 million, city proper; 1.9 million, metropolitan area), on Hokkaido. About 200 miles (*c.* 320 km) west of Tokyo lies Japan's second largest

FIGURE 14.7

Tokyo's famous shopping street, called the Ginza, is reserved for pedestrians on Sunday. The view suggests the opulence of present-day Japan. *Low Jones/The Image Bank*

TABLE 14.1	**Leading Cities of the World, by Metropolitan Area Population (1995)**	
	Tokyo-Yokohama, Japan	30,300,000
	New York, New York	18,087,251
	Sao Paulo, Brazil	16,925,000
	Osaka, Japan	16,900,000
	Seoul, South Korea	15,850,000
	Los Angeles, California	14,531,529
	Mexico City, Mexico	14,100,000
	Moscow, Russia	13,150,000
	Bombay, India	12,596,243
	London, England	11,100,000
	Rio de Janeiro, Brazil	11,050,000
	Calcutta, India	11,021,918
	Buenos Aires, Argentina	11,000,000
	Paris, France	10,275,000
	Jakarta, Indonesia	10,200,000
	Manila, Philippines	9,650,000
	Cairo, Egypt	9,300,000
	Shanghai, China	9,300,000
	Delhi, India	8,419,084
	Chicago, Illinois	8,065,633
	Istanbul, Turkey	7,550,000
	Teheran, Iran	7,500,000
	Beijing, China	7,320,000
	Bangkok, Thailand	7,060,000

Source: *Goode's World Atlas*, 1995.

urban agglomeration, in the Kinki District at the head of the Inland Sea. Here, three cities—Osaka, Kobe, and Kyoto—together with their suburbs form a metropolis of nearly 20 million people; this complex is also among the world's 10 most populous urban areas (see Table 14.1).

The inland city of Kyoto was the country's capital from the eighth century A.D until 1869 and is now preserved as a shrine city where modern industrial disfigurement is prohibited (Fig. 14.8). Some U.S. military planners wanted to target Kyoto with a nuclear weapon in August 1945, but the opinion prevailed that it should be avoided because of its historical importance. Kyoto continues to be the destination of millions of pilgrims and tourists annually.

Osaka grew mainly as an industrial center, and Kobe originally developed as the district's deepwater port. Both cities now combine port and industrial functions. Another huge metropolis, Nagoya (population: 2.15 million, city proper; 4.8 million, metropolitan area), is situated directly between Tokyo and Osaka, on the Nobi Plain at the head of a bay. It is both a major industrial center and a seaport.

West of the Kinki District, smaller metropolitan cities spot the coreland. The largest are metropolises of over 1.5 million people: the resurrected Hiroshima on the Honshu side of the Inland Sea, Kitakyushu (a collective name for several cities) on the Strait of Shimonoseki between Honshu and Kyushu, and Fukuoka on Kyushu. Hiroshima was destroyed on August 6, 1945, when the United States dropped the first atomic bomb ever used in warfare, detonating it directly over the center of the city. With no topographic barriers to deter the effects of the explosion, the city was obliterated, and an estimated 140,000 died by the end of 1945, when radiation sickness had taken its greatest toll. Kitakyushu and Fukuoka developed as the original centers of heavy industry in Japan, located in 19th-century fashion

on and near the country's major coal deposits. They are still industrial centers, but the region's coal production has declined.

Among the several smaller cities of northern Kyushu is Nagasaki, which, on August 9, 1945, was hit by a second, and final, U.S. nuclear bomb—Nagasaki was chosen during the flight mission after clouds obscured the primary target. The bomb detonated over the suburbs rather than the city center, and hills helped to diminish the explosion's impact, but even so, an estimated 70,000 died by year's end. The bombing of Nagasaki effectively brought World War II to its end.

14.2 Historical Background

Japan's eventful history has seen successive periods of isolationism and expansionism, and of economic and military accomplishment and defeat. Japan today is near the top of most indicators of prosperity in the MDCs. The remarkable success story of Japan can best be appreciated by considering its turbulent history and the difficult home environment in which the Japanese have always lived.

EARLY JAPAN

The Japanese are descended from a number of primarily Mongoloid peoples who reached Japan from other parts of eastern Asia at various times in the distant past. An earlier non-Mongoloid people, the Ainu, were driven into outlying areas where some still live, principally in Hokkaido. The oldest surviving Japanese written records date from the eighth century A.D. These depict a society strongly influenced by China, often through cultural and ethnic traits reaching Japan by way of Korea. A distinct Japanese culture gradually evolved, and today it is linguistically and in many other ways different from its Chinese and Korean antecedents. Japanese legendary traditions extend back to the reign of Jimmu, the first emperor, whose accession is ascribed to the year 660 B.C., but which probably took place centuries later. These traditions are important in the Japanese worldview, which says that the islands and their emperor have a divine origin. Jimmu was said to have descended from the Sun Goddess, and the Japanese have known their homeland as the "Land of the Gods." A Japanese concept of Japan as unique and invincible developed early on, shaped by traditions such as that of the *kamikaze*, the "divine wind" that repelled a Mongol attack on Japan in the 13th century.

The early emperors gradually extended control over their island realm. A society organized into warring clans emerged. By about the 12th century, powerful military leaders called *shoguns*, who actually controlled the country, diminished the emperor's role. Meanwhile, the provinces were ruled by nobles, or *daimyo*, whose power rested on the military prowess of their retainers, the *samurai*. This structure resembled the European feudal system of medieval times.

EARLY CONTACTS WITH EUROPE

Adventurous Europeans of the Age of Discovery reached Japan in the early 16th century. The Portuguese came in the 1540s, and people from other European nations followed. Most of these early arrivals were merchants and Roman Catholic missionaries. Japanese administrators allowed the merchants to set up trading establishments and open commerce, and allowed the missionaries to preach freely. There were an estimated 300,000 Japanese Christians by the year 1600.

These contacts were short-lived. Japan had entered a period when a series of military leaders imposed central authority on the disorderly feudal structure. After winning the battle of Seikigahara in 1603 with arms purchased from the Portuguese, warlord Ieyasu Tokugawa proclaimed himself "Nihon Koku Taikun," Tycoon of All Japan, and became the first shogun to

Regional Perspective

Nuclear Power and Nuclear Weapons in the Western Pacific

In 1995, on the fiftieth anniversary of the bombing of Hiroshima and Nagasaki, Americans and Japanese did much soul-searching about the use of nuclear weapons. The surviving decision makers generally continued to insist that the use of these weapons spared many thousands of lives, both American and Japanese, that would have been lost if the U.S. had instead undertaken an invasion of the Japanese homeland. For their part, most Japanese continue to condemn the decision, and it remains official policy that Japan will never develop or use atomic weapons.

However, Japan has come to rely on nuclear power for about 33 percent of the country's electricity needs. Hydropower still supplies 12 percent, but, with fossil fuels almost absent, the critical lack of energy has compelled the nation to develop the world's most energy-efficient economy and to adopt a technology to which many Japanese are averse. Japan imports large quantities of plutonium for use in its nuclear power industry, and the long-distance ocean shipment from Europe of this very hazardous material has contributed to Japan's poor reputation in environmental affairs. International protests against the plutonium shipments caused Japan in 1994 to postpone the construction of a series of nuclear breeder reactors, which use and create recyclable plutonium.

Japan continues to worry about the potential nuclear threat from three adversaries, all of which possess or have had programs to develop nuclear weapons: Russia, China, and North Korea. The Japanese feel that the West dismissed such potential threats from Russia too readily when the USSR dissolved. Japan still has territorial disputes with Russia, particularly involving the four Kuril islands of Kunashiri, Etorofu, Shikotan, and Habomai, which the government of Josef Stalin seized at the end of World War II and which are just off the coast of northeast Hokkaido (see Fig. 14.9 and Fig. 6.1). Japan and China also have a territorial dispute in the East China Sea, and if, as suspected, oil is discovered there, relations between the two countries could deteriorate. China continued to test nuclear weapons through 1996, adding to the tension in Japan. Finally, Japan fears reunification in Korea, which, as a Japanese colony from 1910–1945, has a particular historic dislike of Japan. There is speculation that Japan may do the unthinkable—develop nuclear weapons—to counter the perceived threat of North Korea's nuclear weapons program.

The world also continues to look nervously at the troubled relations between North and South Korea and between North Korea and the West. A crisis flared in the spring of 1994 when heavily-armed North Korea refused to permit full inspection of its nuclear facilities by the International Atomic Energy Agency. The country was suspected of separating plutonium that could be used in making nuclear bombs. Some analysts believed that North Korea had already manufactured one or two nuclear weapons. In 1993, the North Koreans had successfully tested a new medium-range ballistic missile capable of carrying a nuclear warhead to most areas of Japan. Unofficial diplomatic talks between former U.S. President Carter and North Korean leader Kim Il Sung helped scale down the tension level. The U.S., Japan, and South Korea agreed to assist North Korea in the construction of two nuclear power plants in exchange for a freeze on North Korea's nuclear weapons program.

---

rule the entire country. During the era of the **Tokugawa Shogunate,** from 1600 to 1868, the Tokugawa family acquired absolute power and shaped Japan according to its will.

To maintain power and stability, the early Tokugawa shoguns wanted to eliminate all disturbing social influences, including foreign traders and missionaries. They feared that the missionaries were the forerunners of attempted conquest by Europeans, especially the Spanish, who held the Philippines. The shoguns drove the traders out, and nearly eliminated Christianity in persecutions during the early 17th century. After 1641, a few Dutch traders were the only Westerners allowed in Japan. Their Japanese hosts segregated them on a small island in the harbor of Nagasaki, even supplying them with a brothel so that they would not be tempted to venture out and pollute Japan's ethnic integrity. Under Tokugawa rule, Japan settled into two centuries of isolation, peace, and stagnation.

Westernization and Expansion

When foreigners made a serious attempt to reopen the country to trade two centuries later, they would not be thwarted by a Japan that had fallen far behind in technology. Visits in 1853 and 1854 by American naval squadrons of "Black Ships" under Commodore Matthew Perry (who wanted to establish refueling stations in Japan for American whaling ships) resulted in treaties opening Japan to trade with the United States; this was "gunboat diplomacy." The major European powers were soon able to obtain similar privileges. Some of the great feudal authorities in southwestern Japan opposed accommodation with the West. United States, British, French, and Dutch ships responded in 1863 and 1864 by bombarding coastal areas under the control of these authorities. These events so weakened the faltering prestige of the Tokugawa Shogunate that a rebellion overthrew the ruling shogun in 1868. The revolutionary leaders restored the sovereignty of the emperor, who took the name Meiji, or "Enlightened Rule." The revolution of 1868 is known as the **Meiji Restoration**.

The men who came to power in 1868 aimed at a complete transformation of Japan's society and economy. They perceived that if Japan were not to fall under the control of Western nations, its military impotence would have to be remedied. They saw that this would require a reconstruction of the Japanese economy and of many aspects of the social order. They approached these tasks with energy and intelligence. They abolished feudalism, but only after a bloody revolt in 1877. The Meiji leaders cemented the power of a strong central government that would remodel and modernize the country while resisting foreign encroachment. Under this style of government, which lasted until 1945, democracy was instituted but was strictly limited, so that Japan was generally ruled by small groups of powerful men manipulating the machinery of government and the prestige of the emperor. Military leaders were very prominent in this power structure.

The new government pressed its people to learn and apply the knowledge and techniques that Western countries had accumulated during the centuries of Japan's isolation. Foreign scholars were brought to Japan, and Japanese students were sent abroad in large numbers. Japan adopted a constitution modeled after that of imperial Germany. The legal system was reformed to be more in line with Western systems. The govern-

ment used its financial power, which it obtained from oppressive land taxes, to foster industry. New developments included railroads, telegraph lines, a merchant marine, light and heavy industries, banks, and other financial institutions. Wherever private interests lacked the capital for economic development, the government provided subsidies to companies or built and operated plants until private concerns could acquire them. Rapid urbanization accompanied this process of rapid industrialization.

So spectacular were the results of the Meiji Restoration that within 40 years Japan had become the first Asian nation in modern times to attain the status of a world power. Outsiders often spoke of the Japanese in derogatory terms as mere imitators. The Japanese, however, knew which elements of Western technology they wanted, and they adapted them successfully to Japanese needs.

THE PREWAR EMPIRE

Japan emerged as an imperial power after abandoning the isolationism of the Tokugawa period. With only brief interruptions, the country pursued an expansionist policy between the early 1870s and World War II. By 1941 Japan controlled one of the world's most imposing empires. The extent and dates of the empire's acquisitions are shown in Figure 14.9. Between 1875 and 1879, Japan absorbed the strategic outlying archipelagoes of the Kuril, Bonin, and Ryukyu Islands. Japan won its first acquisitions on the Asian mainland by defeating China in the Sino-

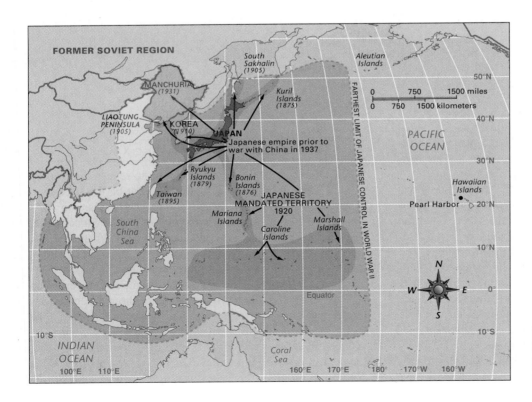

FIGURE 14.9

Map showing overseas areas held by Japan prior to 1937 and the line of maximum Japanese advance in World War II.

Japanese War of 1894–1895, which broke out over disputes in Korea. In the humiliating 1895 treaty of Shimonoseki, China ceded the island of Taiwan (Formosa) and Manchuria's Liaotung peninsula to Japan, and recognized the independence of Korea. Feeling threatened by Japan's emerging strength so close to its vital port of Vladivostok, Russia demanded and won a Japanese withdrawal from the Liaotung peninsula, and leased the strategic peninsula from China. Japan opened a war against Russia in 1904 by attacking the peninsula's main settlement of Port Arthur. Russia lost the ensuing battle of Mukden, which involved more than half a million soldiers, making it history's largest battle to that date. Russia also lost the Russo-Japanese War of 1904, representing the first defeat of a European power by a non-European one. This victory restored the Liaotung peninsula to Japan, gave Japan control over southern Sakhalin Island, and established Korea as a Japanese protectorate until Japan annexed it in 1910.

As a result of World War I, the Caroline, Mariana, and Marshall Islands were transferred from defeated Germany to Japan as a mandated territory. Encroachments on China in Manchuria followed in the 1930s, and an all-out attack that overran the most populous parts of China began in 1937. Japanese troops committed many atrocities in the assault on China; the most notorious was the "Rape of Nanking" (now Nanjing) in 1937, when the invaders took control of China's temporary capital, killing an estimated 300,000 Chinese civilians and soldiers, and raping about 20,000 Chinese women. In 1940, Japan seized French Indochina after Germany crushed France, and after Japan's carrier-based air attack crippled the American Pacific Fleet at Pearl Harbor, Hawaii, on December 7, 1941, Japanese forces rapidly overran Southeast Asia as far west as Burma, much of New Guinea, and many smaller Pacific islands.

The motives for Japanese expansionism were mixed. They included a perception of national superiority and "manifest destiny," a desire for security and great-power recognition, the desire of military leaders to aggrandize themselves and gain control of the Japanese government, and a desire on the part of industrialists in Japan to gain sources of raw materials (such as Manchurian and Korean coal) and markets for Japanese industries. Although the methods of gaining them were often illegitimate, the desire for materials and markets was solidly based on need. Expansion of industry and population on an inadequate base of domestic natural resources had made Japan dependent on sales of industrial products outside the homeland. Such sales provided, as they do now, the principal funds with which the country purchased the imported foods, fuels, and materials that it required.

Prewar Japan turned to military solutions for several problems. The main difficulties were economic, as worldwide depression hurt the trade-dependent Japanese economy. Japan chose to create a Japanese-controlled Asian realm which would insulate the Japanese economy from the vicissitudes of the world economy and elevate the nation to the rank of a leading world power. In 1938, Japan therefore proclaimed its "New Order in East Asia" and in 1940 and 1941 this widened into the "Greater East Asia Co-Prosperity Sphere," a euphemism for Japanese political and economic control over China and Southeast Asia.

Japan's militaristic and colonial enterprises proved to be disastrous at home and abroad. Japanese soldiers inflicted great sufferings on civilians throughout the western Pacific. Even now nations are demanding formal apologies from Japan. In August 1995, on the fiftieth anniversary of Japan's defeat in World War II, Japan's Prime Minister Tomiichi Murayama did issue a formal general apology for his nation's role in the war.

August 1945 indeed found Japan completely defeated. Its overseas territories, acquired during nearly 70 years of successful imperialism, were lost. Its great cities were in ruin. In addition to 1.8 million deaths in its armed services, Japan had suffered some 8 million civilian casualties from American bombing of the home islands, and most of its major cities were over half destroyed. Tokyo's population had fallen from nearly 7 million people to about 3 million, most of whom were living in shacks. In 1950, five years after the beginning of the U.S. military occupation of Japan, national production still stood at only about one third of its 1931 level, and annual per capita income was $32. One study showed that the average urban worker was subsisting on 1600 calories per day, rather than the approximately 2000 needed for an active adult life.

14.3   Japan's Postwar "Miracle"

Without benefit of colonies or spheres of influence, Japan has become an economic superpower since World War II. The nation's explosive economic growth after its defeat has been one of the most startling and significant developments of the late 20th century; it is widely known as the Japanese "Miracle."

Japan's postwar relations with the United States were focused mainly on the American military occupation of the country. The occupation of 1945 to 1952 instituted a series of significant reforms, several of which contributed to Japan's subsequent economic successes. The divine status of the emperor was officially abolished; indeed, when Emperor Hirohito announced Japan's surrender on the radio, it was the first time the stunned Japanese public had ever heard the voice of this mythic figure. Some U.S. decision makers wanted him to be tried as a war criminal, but more moderate voices who recognized the constructive role of the emperor in Japanese society prevailed. General Douglas MacArthur cautioned the American president to keep him in power for the sake of postwar stability: "He is the symbol which unites all Japanese," MacArthur argued. "Destroy him and the nation will disintegrate."

A new U.S.–written constitution made Japan a constitutional monarchy with an elective parliamentary government.

With the rejection of divine monarchy, the strongly nationalistic Shinto faith lost its status as Japan's official religion, although worship at Shinto shrines was allowed to continue; with Buddhism, it still dominates Japan's spiritual life. There was extensive land reform in the countryside to do away with a near-feudal landlordism. Some major Japanese companies were broken up as a means of reducing their monopolistic hold on the economy. Women were enfranchised. A democratic trade union movement was established. The formerly dominant military officer corps was purged. Japan was forbidden to rearm, except for small "self-defense" forces.

When the occupation ended in 1952, leaving U.S. bases in Japan but returning control of Japanese affairs to the Japanese, Japan was placed under American military protection. The U.S. military umbrella over Japan remains controversial today. Many Japanese regard it as an outmoded vestige of colonialism and the war. And in 1995, the trial and conviction of three U.S. servicemen accused of raping a local child in Okinawa (where three quarters of the U.S. bases and more than half of the U.S. troops in Japan are stationed) intensified Japanese public attention to this issue. Public protests and diplomatic appeals led the United States early in 1996 to agree to return control of a major airbase in Okinawa to Japan, and to relocate some U.S. troops from Okinawa to mainland Japan. On behalf of the American people, U.S. President Clinton also publicly apologized for the rape. In a subsequent referendum, the people of Okinawa voted overwhelmingly in favor of a further U.S. military withdrawal from the island.

EXPLAINING THE MIRACLE

Observers of Japan cite different reasons for the country's economic success. Proponents of dependency theory (see Chapter 2, page 42) argue that Japan, never having been colonized, escaped many of the debilitating relationships with Western powers that crippled many potentially wealthy countries. Others emphasize Japan's association with the U.S. following World War II. Some of the postwar U.S.–imposed reforms worked well economically, and relations between the two countries since the war have also generally stimulated Japan's industrial productivity. Without large military expenditures, much of Japan's capital was freed to invest in economic development.

Rapid recovery from the postwar depths began when the U.S. called on Japanese production to support American forces in the Korean War of 1950–1953. By 1954, Japanese steel production was the largest in the country's history. The country's textile and clothing industries supplied American military needs, then expanded exports so rapidly that United States industries were asking for protection by 1956. Military procurement for American forces continued to support the Japanese recovery through the late 1950s, and U.S. economic aid continued to flow to Japan. The U.S. permitted many Japanese products to have free access to the United States market for long periods, and allowed Japan to continue to protect its economy from imports. Japanese firms had relatively free access to American technology, which they frequently improved on, introducing the improvements into production ahead of American firms.

By the 1970s, Japan had become an industrial giant with a GNP far exceeding that of any country except the United States and the Soviet Union. The U.S. and European nations found themselves at a strong disadvantage against Japanese competitors in many industries. They busily sought ways to match this competition or protect themselves from it without doing too much damage to their exporter firms, consumers, and overall trade relations.

By the 1980s the United States and Japan were at odds over Japan's massive trade surplus with the U.S. The issue brought the countries to the brink of a trade war in 1995, but Japanese concessions averted the conflict.

Another explanation of Japan's economic rise stresses the low cost and high quality of Japan's work force (Fig. 14.10). During the 1950s and 1960s, when the Japanese economy had remarkable growth rates averaging 10 percent a year, Japanese labor was very inexpensive compared with that of the United States and major European industrial countries. Not only were wages low, but the quality and productiveness of the workforce were high. However, during the 1970s and early 1980s this advantage declined as strikes by Japanese workers won large wage increases. By the mid-1980s, some analysts were pointing out that differential labor costs were no longer crucial in much of the competition between Japanese and Western firms. This situation continues in the late 1990s.

FIGURE 14.10
Japan is one of the world's leaders in high technology. The man in the cradle is at work in a high-voltage laboratory. *Alan Levenson/Tony Stone Images*

Definitions and Insights

CAPITAL GOODS IN JAPAN'S ECONOMY

One reason for Japan's phenomenal economic growth rates of about 10 percent per year in the 1950s and 1960s was the country's emphasis on producing **capital goods** rather than **consumer goods.** Capital goods are goods that are used to produce other goods, and consumer goods are goods that individuals use for their own short-term needs and wants. After World War II, Japanese decision makers insisted that industrial production be geared to producing capital goods such as machinery, equipment, and robotics that would manufacture the automobiles, ships, electronics, and other products which Japan would export. Japan invested as much as 30 percent of its domestic output in such capital goods. By contrast, after World War II the United States invested only about 10 percent in capital goods and instead dedicated much of its domestic output to producing consumer goods such as televisions, homes, and clothing for the enjoyment of American citizens.

The emphasis on capital versus consumer goods may be understood as an economic choice between quality of life today and quality of life tomorrow. With greater emphasis on consumer goods, the United States has emphasized production for present consumption, sacrificing production for future generations and settling for modest annual economic growth. With more emphasis on capital goods, Japan has emphasized austerity for today's consumer in favor of future benefits, favoring high annual economic growth. The economic growth patterns of the two countries demonstrate the results: From 1960 to 1993, Japan's domestic output grew at about 6 percent per year, while in the U.S. it expanded at about 3 percent.

Observers often point to several unique features of Japanese management and employment to explain the country's success. One is recruitment through an extremely challenging (some say brutal) educational system in which technical training has strong emphasis. A rigorous and stressful testing system controls admission to higher education. As indicators of Japan's educational emphasis on production rather than service, the country trains many more electrical engineers than does the U.S., while the U.S. has about 20 times more lawyers than Japan. Executives and supervisory personnel are more likely to have been trained in science and technology in Japan than in the U.S., and research and development budgets are higher and long-range planning more common in Japanese than in American companies.

Japanese management strategies emphasize benevolence toward employees, encouragement of employee loyalty, and participation of workers in decision making. About 20 percent of Japanese workers enjoy guarantees of lifetime employment in their firms, and large Japanese companies are generally active in providing housing and recreational facilities for their employees. Compared with Western companies, Japanese firms generally pay more attention to employee suggestions and give employees more information about important matters affecting the company. The Japanese style of decision making seeks widespread agreement before actions are implemented. Some Japanese companies operating plants in the United States have apparently achieved good results in productivity and employee satisfaction by using such participatory methods with American workers. Also, the range of salaries between employees and managers is much smaller in Japan than in the U.S., which may contribute to better morale among Japanese employees.

One essential factor in Japan's postwar economic growth was a high level of investment in new and generally very efficient industrial plants. The very low level of military expenditure and the ability of companies to operate at low profit margins helped boost this investment. Investment capital has also been freed by government policies which cut expenditures on amenities and services such as roads, antipollution measures, parks, housing, and even higher education. Japanese industrial cities have grown explosively and have also become very highly polluted areas of dense and inadequate housing, with few public amenities and snarled transportation. The money "saved" has been available for more direct investment in industrial growth, but the costs to Japanese society have been high.

Also crucial to the high level of industrial investment has been a remarkably high rate of savings by the Japanese people, who have many incentives not to spend all of their income. Employees commonly receive only about two thirds of their earnings each year as regular wages or salary, with the other third in the form of one or two lump-sum payments. Most Japanese workers prefer this because it automatically and at least temporarily saves one third of their income. In addition, buying on credit is less common in Japan than in the United States. There are few mechanisms for lending, so most Japanese consumers must save money to purchase a large-ticket item. Workers must also save to take care of their children and themselves because Japanese higher education is expensive, and because the country's welfare system, pensions, and social security are quite inadequate. Mandatory retirement, the end of the "lifetime" employment Japanese companies offer, is generally at 55, but social security payments do not begin until age 60; retirees can expect to exist largely on savings for five years.

Some analysts cite elements of Japan's political culture to explain the country's economic successes. One political party, the Liberal Democratic Party, has been repeatedly reelected to power since the time Japan regained its sovereignty in 1952. It is conservative and strongly business-oriented and business-connected. It has promoted Japanese exports with financial policies that have tended to keep the yen (the Japanese unit of currency)—and thus Japanese goods—inexpensive. The party has also cooperated very closely with Japanese business in the development of new products and new industries.

Some analysts believe that the country's postwar economic miracle grew from an intense spirit of achievement and enterprise among the Japanese. Notably, many Japanese attribute this industrious spirit to Japan's geography as a resource-poor island nation. In order to overcome the constraints nature has placed on them, the Japanese people feel they must work harder. This rationale, plus Japan's strategic location on the Pacific Rim, led the vice minister for international affairs at Japan's Trade Ministry to explain at a 1992 press conference that part of the reason for Japan's chronic trade surplus with other countries was "geography."

Despite all of these factors in Japan's economic favor, the Japanese "miracle" did not last. The peak of Japan's postwar success came in the 1980s. A powerful economic boom led to speculative rises in stock prices and land prices. At the end of the 1980s, the total value of all land in Japan was four times greater than that of the United States. The Tokyo Stock Exchange was the world's largest, based on the market value of Japanese shares. Then the "bubble economy" burst, and real estate and stock prices fell by more than 50 percent. Japan had near-zero economic growth through the first six years of the 1990s, and the Japanese industries whose growth had seemed unstoppable experienced an increasing loss of market share to U.S. and European producers.

14.4 Japanese Industry

The gradual evolution of industry over more than three centuries which characterized Europe and the United States has been compressed into little more than a century in Japan. Early iron and steel development was concentrated in northeast Kyushu, on and near the country's most significant coal resources. The product went largely into building a railway system and both commercial and naval shipping. Japanese coal, supplemented increasingly by imports, powered the country's factories, railways, and ships. Hydroelectricity was developed on many mountain streams and became an important element in the energy economy.

Petrochemicals and other oil-based industries reflect its great dependency upon imports. Almost devoid of petroleum reserves, the country obtains almost all of its oil and gas as imports, mostly from the Persian/Arabian Gulf region. Japan therefore has carefully avoided alignment with the West in most Middle Eastern disputes, and only after much pressure from the U.S. and its Western European allies did Japan contribute financially to the multinational Desert Shield/Desert Storm operations which ousted Iraqi troops from Kuwait in 1991.

The phenomenal advance of industry since the early 1950s has been marked by a series of overlapping booms, along with

some declines, in various sectors. Cotton textiles and clothing, then based on inexpensive labor, led the way in the early 1950s. (Japanese labor is no longer inexpensive, and these industries are now declining.) In the later 1950s, boom periods in iron and steel, shipbuilding, and chemicals (especially fertilizers) set in. Japan now has the world's largest iron and steel industry, and some of the world's largest steel plants occupy coastal locations in the Japanese core area. With Kyushu's small coal reserves now inadequate, Japan imports most of its coal from the United States and Australia. Most of the iron ore also comes from Australia. Relatively low transport costs for these materials are made possible by the oversized ocean carriers that Japan builds; Japan is still the world's leading shipbuilder (Fig. 14.11).

In the 1960s, electronics, cameras and optical equipment, petrochemicals, synthetic fibers, and automobiles (notably those produced by Mazda, Honda, Toyota, Mitsubishi, and Nissan) became boom industries.

In the 1970s and 1980s, Japan became deeply involved with the manufacture of computers and robots. Supported by the government, Japanese companies attempted to take over the world industry leadership from the United States' computer companies, and by the early 1980s, Japan established a clear lead over all other countries in the robotization of industry.

However, for the first time in the postwar era, Japanese industry in the 1990s is not characterized by world supremacy in any single category of manufacturing. While Japanese manufacturing evolved with exceptional speed from relatively simple industries based on domestic natural resources and inexpensive labor to highly sophisticated and futuristic industries based on a highly-skilled and educated workforce, Japanese industry must now look for new directions in which to establish leadership. As always, in order to overcome the severe limitations imposed by a small, resource-poor island, the Japanese must live by their wits.

FIGURE 14.11

The port of Tokyo. The island nation of Japan thrives on seaborne trade.
Karen Kasmauski/Matrix

14.5   The Social Landscape

Japan has one of the most homogeneous populations on Earth. Historically, except for a longstanding community of ethnic Koreans (who have often suffered discrimination in Japan), few non-Japanese have settled in the country. Only with the country's booming prosperity and growing labor shortages of the 1980s did Japan open its doors to a few unskilled and low-skilled immigrants from such countries as Pakistan, Bangladesh, Thailand, Peru, and Brazil. This shortage of racial and ethnic diversity has had mixed results for Japan. Many observers believe that it has helped the country achieve a sense of unity of purpose, allowing the Japanese to persist through periods of adversity—especially the postwar years of reconstruction. However, the Japanese have also earned a reputation for intolerance of ethnic minorities.

Until recently, Japan enjoyed the status of being one of the world's safest, most secure, and most crime-free nations. The Japanese often derided countries such as the United States for rampant crime, violence, homelessness, and other social problems. The recession of the 1990s, however, resulted in growing joblessness and homelessness in Japan, and began to tarnish the country's self-satisfied image. The country's official unemployment rate of 3.2 percent was challenged by a government survey indicating that the actual unemployment rate approached 9 percent in 1995. And the country's sense of internal cohesion suffered a great blow in 1995 when a cult called Aum Shinrikyo carried out nerve gas attacks in Japanese cities as part of an apparent plan to precipitate war in the country.

The legendary Japanese work ethic has had its advantages and drawbacks: It has helped the Japanese create a prosperous country, but it has also created a nation of workaholics beset with the same problems—stress, suicide, depression, and alcoholism—experienced by workers in the world's other MDCs. Women have not achieved parity with men in the workplace, and they complain increasingly of discrimination and sexism. College graduates who are not hired during the annual recruiting season face difficult obstacles to entering the job market. And for those it affects, the practice of lifetime employment makes it difficult to change jobs. One Japanese critic declared that workers in his country "are owned like pets by their companies and housing is offered as the equivalent of dog food."

Crowding and overdevelopment, accompanied by a striking lack of amenities, are inevitable facts of life in Japan. Officially designated parks comprise 14 percent of Japan's land area, but many of these areas are heavily developed with roads, houses, golf courses, and resorts. There has been enormous growth in the popularity of winter sports destinations in northern Honshu and Hokkaido, and of hot springs and mineral bath resorts throughout the country. Reflecting dissatisfaction with too much growth and development, the conservation ethic is already strong and growing in Japan. However, it is difficult for the Japanese to truly "get away from it all" at home. Actual

FIGURE 14.12
A bullet train glides through the Ginza district of Tokyo. Japan has an efficient, high-speed rail network.   *Pete Seaward/Tony Stone Images*

wilderness areas comprise only 0.015 percent of Japan's land area.

Daily life for the Japanese is devoid of many of the benefits associated with the MDCs. Apartments and homes are generally very small. Only about half of the country's homes are connected to modern sewage systems. Smog from automobile exhaust and industrial smokestacks is a persistent health hazard. The ever-present traffic jams are known locally as traffic "wars." Japan's much-vaunted rail system, including its high speed *hikari*, or "bullet" trains, which carry traffic between northern Honshu and Kyushu at speeds averaging 106 miles (170 km) per hour, is chronically overburdened with passengers (Fig. 14.12). Summarizing these living conditions, the Japanese politician Ichiro Ozawa recently described Japan as having "an ostensibly high-income society with a meager lifestyle."

Japan faces a troubled period as it must deal with the aftershocks of earthquakes, internal dissent, and economic recession. Most analysts expect the country to recover, perhaps turning current adversity into opportunity as it has in the past. The country's highly-educated and skilled workforce is a major resource. Although natural assets are lacking, Japan's favorable location should also assist in its recovery. The western Pacific region is expected to witness dynamic economic growth through the first decade of the 21st century, and Japan is well-positioned to benefit from trade with the countries of this region. Japan continues to be a critical member of the international community; even during recession it has maintained its status as the world's largest donor of foreign aid.

Landscape in Literature

BOXCAR OF CHRYSANTHEMUMS

In this story by Enchi Fumiko,* the narrator attempts to return home to Karuizawa after visiting friends in the nearby town of Ueda, northwest of Tokyo on the island of Honshu. It is too late to catch the express train, so she boards a local train. The experience brings her close to the retreating rural side of Japanese life, and she recognizes a link between rural and urban Japan. She also has a brush with chrysanthemums, the flowers that are the symbol of Japan's royal family (the emperor occupies the "Chrysanthemum Throne").

Well, even if this is a local I'll be in Karuizawa in two hours or so for sure, I said to myself as I turned to look out the window. The moon was nearly full and silhouetted in dark blue the low mountains beyond the fields beside the tracks. The plants in the rice paddies were ripe, so of course I couldn't see the moon reflecting on the water in the fields. Plastic bird rattles here and there glittered strangely as they reflected the moonlight. The clear dark blue of the sky and the coolness of the evening air stealing into the deserted car made me realize keenly that I was in the mountain region of Shinshu, where fall comes early.

I was thinking that I wanted to get home quickly when the train shuddered to a stop at a small station. We had been moving for not more than ten minutes. I couldn't complain about the stop since the train wasn't an express, but it didn't start up for a long time even though no one was getting on or off. It finally moved but then stopped again at the next station, took its time and wouldn't start again, as it had done before. . . .

When the train stopped for the fourth time at a fairly large station, I got off to ease my irritation. It would probably stop for ten minutes or so, and even if it started suddenly I thought I could easily jump on such a slow-moving train.

A few passengers got off. Some long packages wrapped up in straw matting were piled up near one of the back cars, and the station attendants were loading them into the car as if they were in no big hurry. The packages were all about the same size, bulging at the center like fish wrapped in reed mats, but the station attendants were lifting them carefully in both hands, as if they were handling something valuable, and loading them into the soot-covered car.

I was watching the scene and wondering what the packages were when I suddenly noticed a moist, plant-like smell floating in from somewhere . . . the packages were being loaded one after another, and I realized that the fragrance in the air was coming from them.

"Oh, that one over there! Those are our mums!" the old man yelled suddenly, extending his arms as if he were swimming toward the package the station attendant was about to load on the train. The old man's little nose was twitching like a dog's. "That smell . . . It's the Shiratama mum. . . ."

I turned around. A middle-aged man in a gray jacket was sitting across the aisle from me. His face was dark and wrinkled from the sun, but he didn't look unpleasant. I realized that he hadn't been on the train before we stopped at the last station. . .

I said, "Were those chrysanthemums they were loading in that boxcar? That old man was smelling them, wasn't he?"

"Yes, those were the chrysanthemums they grow in their garden. Mums are the only thing the old man cares much about. When they send some off he comes with his wife to watch, whether it's late at night or early in the morning."

"Then all of the packages are chrysanthemums?"

"That's right. Most of the flowers that go to Tokyo at this time of year are. Lots of farmers around here grow flowers, but they're all sold in Tokyo, so it's a big deal. Not only flowers, either. The people who work in the mountains around here collect tree roots, branches and other stuff, put prices on them and send them to Tokyo. Once they get them to market they can sell them, I guess, because money is always sent back. Tokyo's a good customer that lets the landowners around here earn money that way. . . ."

Now I recalled fondly that I had been riding on a freight train full of chrysanthemums. In those dark, soot-covered cars hundreds and thousands of beautiful flowers were sleeping, in different shades of white, yellow, red, and purple, and in different shapes. Their fragrance was sealed in the cars. Tomorrow they would be in the Tokyo flower market and sold to florists who would display them in front of their shops.

* From Enchi Fumiko, "Boxcar of Chrysanthemums," *This Kind of Woman: Ten Stories by Japanese Women Writers,* ed. Yukiko Tanaka and Elizabeth Hanson (Stanford: Stanford University Press, 1976, 72–86).

14.6   Divided Korea

Japan is closest to the Asian mainland in the extreme west, where the peninsula of Korea lies only about 110 miles (177 km) away (see Fig. 14.1). Korea's political history during the 20th century, first as a colony of Japan and then as a divided land composed of the separate countries of North and South Korea, has tended to obscure its distinctive culture and contributions to the world. Although the Koreans have been influenced by Chinese culture, and to a lesser extent by Japanese culture, they are ethnically and linguistically a separate people.

LIABILITIES OF KOREA'S LOCATION

The two Koreas occupy an unfortunate geographical location. These small countries are near larger and more powerful neighbors (China and Russia) that have frequently been at odds with one another and with the Koreans. North Korea adjoins China along a frontier that follows the Yalu and Tumen rivers, faces Japan across the Korea Strait, and, in the extreme northeast, borders Russia for a short distance. For many centuries, the Korean peninsula has served as a bridge between Japan and the Asian mainland. From an early time, both China and Japan have been interested in controlling this bridge, and Korea was often a subject or vassal state of one or the other. However, from the late 7th century to the 20th century, Korea was a unified state, sometimes invaded and forced to pay tribute, but never destroyed as a political entity. The decline of Chinese power in the 19th century was accompanied by the rise of modern Japan, whose influence grew in Korea. From 1905 until 1945 Korea was firmly under Japanese control; in 1910 it was formally annexed to the Japanese Empire. The legacy of hostility and occupation continues to cast a shadow over relations between the Koreas and Japan, and disputes periodically emerge between them over such issues as control of islands and fishing rights (see Definitions and Insights, below).

Definitions and Insights

JAPAN, KOREA, AND THE LAW OF THE SEA

Some 90 miles (145 km) between the shores of Japan and South Korea in the Sea of Japan lie two small, inhospitable islands known to Japanese as the Takeshima Islands and to Koreans as the Tokdo Islands (see Fig 14.1). South Korea has actually controlled the islands since 1956, but with only one Korean couple and some Coast Guard personnel living on the islands, Japan has periodically asserted its right to them. In 1996 they became the focus of a dispute between Japan and Korea, not because of any riches they contain, but because of a 1970s United Nations treaty known as the **Convention on the Law of the Sea,** which would permit their sovereign power to have greater access to surrounding marine resources.

The Law of the Sea was initiated in an effort to apportion ocean resources as equitably as possible and to avoid precisely the kind of conflict now brewing between Japan and South Korea. The treaty gives a coastal nation mineral rights to its own continental shelf, a territorial water limit of 12 miles off shore, and the right to establish an exclusive economic zone (EEZ) of up to 200 miles offshore (in which, for example, only fishing boats of that country may fish). The power that controls offshore islands such as the Takeshima/ Tokdo can extend the area of its exclusive economic zone even further.

By early 1996, 85 nations had ratified the treaty. Korea ratified the treaty late in 1995. Japan was preparing to ratify in 1996 when news reached Tokyo that South Korea had plans to build a wharf on the islands. To avoid provoking Japan, North Korea, and China, South Korea avoided declaring an exclusive economic zone off its waters, which would include the islands. But fears that Korea may build facilities and station more people on the islands as a step toward establishing an EEZ—and thereby excluding Japanese fisherman from the area—caused Japan to restate its claim to the islands in 1996. South Korean officials answered with military exercises near the islands, and South Korean civilians staged loud demonstrations outside Japan's embassy in Seoul. The issue may have to be settled in the same international legal arena in which it originated.

Japan lost Korea at the end of World War II. In accordance with the victorious allies' agreements, Soviet forces occupied Korea north of the 38th parallel and United States forces occupied Korea south of that line. Korea was to have become a unified and independent country, but the occupying powers could not reach agreement on the formation of a government. The occupying powers therefore set up separate governments: the democratic Republic of South Korea in the south, under the auspices of the United Nations, and the People's Republic of Korea in the north, a Chinese communist satellite. The occupying powers withdrew most of their forces in 1948 and 1949.

North Korea attacked South Korea in 1950. United Nations units, made up mostly of U.S. forces, entered the peninsula to repel the communist advance. Late in 1950, when the North Koreans had been driven back almost to the Manchurian border, China entered the war and drove the United Nations forces south. A stalemate then developed just north of the 38th parallel until an armistice was arranged in 1953. The border between the two Koreas, often the most tense boundary on earth, still follows this armistice line. The 151-mile-long (243 km), 2.5-mile-wide (4 km) demilitarized zone is a virtual no-man's-land of mines, barbed wire, tank traps, and underground tunnels

FIGURE 14.13
A United Nations observer peers across the Demilitarized Zone to North Korea. This is one of the most tense borders on Earth.   *Mary Beth Camp/Matrix*

(Fig. 14.13). The world's largest concentration of hostile troops faces off on either side of it. Remarkably, with the near absence of human activity, rare birds like the Manchurian crane and other endangered animal species have taken refuge in this narrow strip.

Few lands have ever been more devastated than Korea after several years of warfare covering the length and breadth of the peninsula. The tragedy was all the greater in that it is not a poor land by nature. During their period of control, the Japanese developed transportation, agriculture, and industry based on a sizable reserve of mineral and power resources. The Korean people received few benefits, however, because the increased production was put mainly to Japanese uses. The division between north and south handicapped the economy after 1945, and after 1950 the enormous physical destruction of the Korean War set the countries back again. The physical scars of the conflict have been erased, however, and the economies of both the North and South Korean states have been expanded with the help of considerable outside aid.

A state of cold war and periodic border incidents have nevertheless persisted between North and South Korea. The two countries were preparing for an unprecedented summit meeting when the North Korean leader died suddenly in July 1994, rendering future meeting prospects uncertain. Relations between the two countries thawed encouragingly in 1995 when the South sent emergency stocks of rice to the North, but they soon iced over again in 1996, when a North Korean spy submarine

ran around in South Korea. North Korea apologized for the incident, and began diplomatic talks with South Korea, the U.S., and China in 1997. The border between the two Koreas is unlikely to be the tripwire for a third World War, as was feared in the days of the U.S.–Soviet Union Cold War. Any incident along the border could, however, plunge the two Koreas into war and lead inevitably to the intervention of United States troops, beginning with the 37,000 American soldiers now stationed there, and possibly even to the intervention of Chinese forces. Some analysts fear that Korea may yet become a nuclear battlefield.

CONTRASTS BETWEEN THE TWO KOREAS

The Korean armistice line divides one people into two very different countries (see Fig. 14.14a, b and Table 10.1, pages 262–263). North Korea has an area of about 46,500 square miles (120,500 sq km) inhabited by 24 million people in 1996, with a density approaching 500 per square mile (*c.* 185 per sq km). In 1996, South Korea had 45 million people in only 38,000 square miles, averaging somewhat more than 1150 per square mile (*c.* 455 per sq km). South Korea is a republic that has fluctuated between attempts at democracy and a repressive military dictatorship. It has a capitalist economy heavily dependent on relationships with the United States and Japan. North Korea is a rigid and very tightly controlled communist state which vacillated in its principal ties between China and the Soviet Union (Fig. 14.14). With the collapse of the USSR, China became North Korea's main ally and benefactor. From World War II until his death in 1994, the dictator Kim Il Sung—whom North Koreans called "The Great Leader"—governed the nation. His son, Kim Jong Il, is now in power and to date has not changed the country's militaristic character. In 1995, North Korea spent a staggering 26 percent of its gross domestic product on its military, compared with South Korea's expenditure of only 3.4 percent, China's spending at 2 percent, and Japan's at 1 percent.

There are marked physical contrasts between the two countries. Although both are predominantly mountainous or hilly, North Korea is the more rugged. Relatively level lowland is found mainly along the western side of the peninsula in both countries, but South Korea also has some extensive lowland areas in the southeast. In both countries, mountains rise toward the eastern side of the peninsula and drop abruptly into the Sea of Japan with little or no coastal plain. They are highest and most rugged in northeastern North Korea adjoining Manchuria.

North Korea has a humid continental long-summer climate, with hot summers and quite cold winters. South Korea is mostly a humid subtropical area with much shorter and milder winters. Both are strongly monsoonal, with precipitation concentrated in the summer.

(a)

(b)

FIGURE 14.14

The two Koreas are like night and day. Affluent shoppers enjoy the fruits of free enterprise in South Korea, while military security preoccupies austere, Stalinist North Korea.   (a) *Portfield/Chickering/Photo Researchers* (b) *Mary Beth Camp/Matrix*

Rice is the staple food in both countries. Rice-growing conditions are best in South Korea and diminish northward. Unlike the North, South Korea is able to double-crop irrigated rice fields by growing a dry-field winter crop such as barley after the rice has been harvested. In the North, the main supplementary crop is corn, which must be grown in the summer and therefore cannot occupy the same fields as rice.

Another physical contrast between the two Koreas is that most of the nonagricultural natural resources are in North Korea. All of the peninsula's major resources—coal, iron ore, some lesser metallic ores, considerable hydropower potential, and forests—are more abundant in the North than in the South. North Korea was thus originally the more industrialized state, featuring a typical communist emphasis on mining and heavy industry, together with hydroelectric production and timber products. However, while the North remained mired in this stage of development, the South took over industrial leadership in the 1970s and 1980s by very rapidly developing a dynamic and diversified capitalist industrial economy. In an effort to begin to close the gap, North Korea opened the Rajin-Sonbong Free Economic and Trade Zone in the mid-1990s. Centered at the extreme northwest corner of North Korea around the twin cities of Rajin and Sonbong (combined population: 140,000),

and ringed by barbed wire, the 288-square-mile (720 sq km) trade zone is intended to attract foreign investment and industries that will capitalize on this site's favorable location near the borders of Russia and China.

In the mid-1990s, South Korea's main exports were electrical and electronic equipment and appliances, automobiles, textiles, shoes, iron and steel (from a huge plant on the southeast coast north of Pusan), and ships from a major shipbuilding industry. Large investments from Japan and the United States, as well as access to markets in those countries, were fundamental to this explosive development. Also important were the skills of South Koreans receiving higher education abroad and at home (the country has the world's highest number of Ph.D.'s per capita), inexpensive and increasingly skilled Korean labor, and vigorous backing of industrialization by strong and sometimes ruthless governments.

South Korea is one of Asia's success stories, an "Asian Tiger" with a rocketing economic growth rate of 9.9 percent (as of 1995) and industries that have been causing concern to competitors in Japan and the U.S.; for example, Samsung electronics has established a growing market share in an industry dominated by Japanese firms. South Korea is also poised for especially strong growth in the automobile (notably Hyundai cars and trucks), semiconductor, information processing, telecommunications, and nuclear energy industries (South Korea is the world's third largest producer of nuclear energy, after France and Japan).

The burgeoning cities of South Korea are visible evidence of economic growth. The capital city of Seoul (Fig. 14.14b) has reached a metropolitan area population of 15.85 million (10.63 million, city proper), ranking it among the top five most populous cities in the world (see Table 14.1). The port of Pusan has 3.8 million people (metropolitan area) and the inland industrial center of Taegu more than 2 million. By contrast, the only large city in North Korea is the capital, Pyongyang, with a metropolitan population of 2.36 million.

The Pacific World

5

15 A GEOGRAPHIC PROFILE OF THE PACIFIC WORLD

Covering fully one third of the Earth's surface, the Pacific World is mostly water. The Pacific itself, the world's largest ocean, is bigger than all the Earth's continents and islands combined. Before World War II, the Western world spawned legends about this ocean and its islands as a kind of utopia. Some islands still do have an idyllic quality for European and American visitors. However, there has long been trouble in this paradise. On many islands, traders, whaling crews, labor agents, and other opportunists exploited the indigenous peoples, reducing their numbers and disrupting their cultures. The military battles of World War II shattered whatever idyllic quality remained in many islands. Yet today, the Pacific mystique, which has been perpetuated in the works of many noted writers, forms part of the allure for the massive development of tourism in such places as the U.S. state of Hawaii, the island of Tahiti, and other islands of the South Pacific.

The Pacific World region (often called Oceania) encompasses Australia, New Zealand, and the islands of the mid-Pacific lying mostly between the Tropics. The Pacific islands nearer the mainland of east Asia, the Former Soviet Region, and the Americas are excluded here on the basis of their close ties with the adjoining continents. Large areas of the eastern and northern Pacific that contain few islands are also largely discounted. Because Australia and New Zealand are sufficiently different from the tropical island realms to the north, they could be considered a separate world region, but they are included here in the Pacific World because of their strong political and economic interests in the tropical islands, their similar insular character, and the ethnic affiliations of their original inhabitants with the peoples of those islands.

15.1 Melanesia, Micronesia, and Polynesia

The Pacific islands are commonly divided into three principal regions: Melanesia, Micronesia, and Polynesia (Fig. 15.1). The islands of **Melanesia** (Greek: "black islands"), bordering Australia on the northeast, are relatively large. New Guinea, the

◀ Earth's largest ocean, the vast Pacific. This is a composite of several thousand satellite images.
Tom Van Sant/Geosphere Project, Santa Monica/Science Photo Library

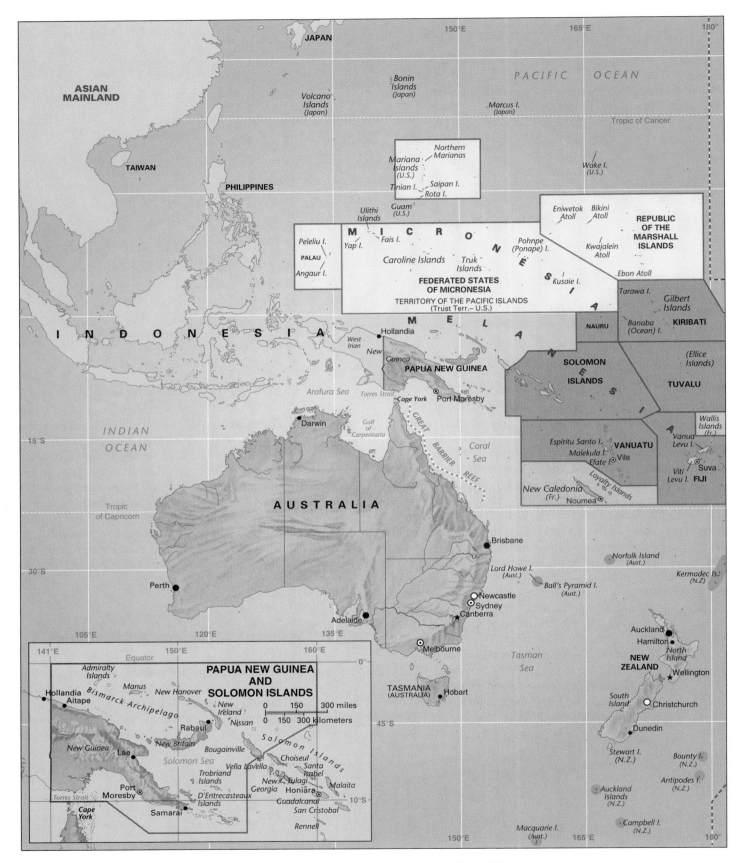

FIGURE 15.1   General reference map of the Pacific World. In some cases, the tints for political affiliation do not show precise political boundaries. Note the peculiar jog in the international date line near Kiribati. In a bid to attract more tourists by being the first country in the world to greet the new millenium, Kiribati unilaterally decided in 1997 to shift the date line eastward by more than 2000 miles.

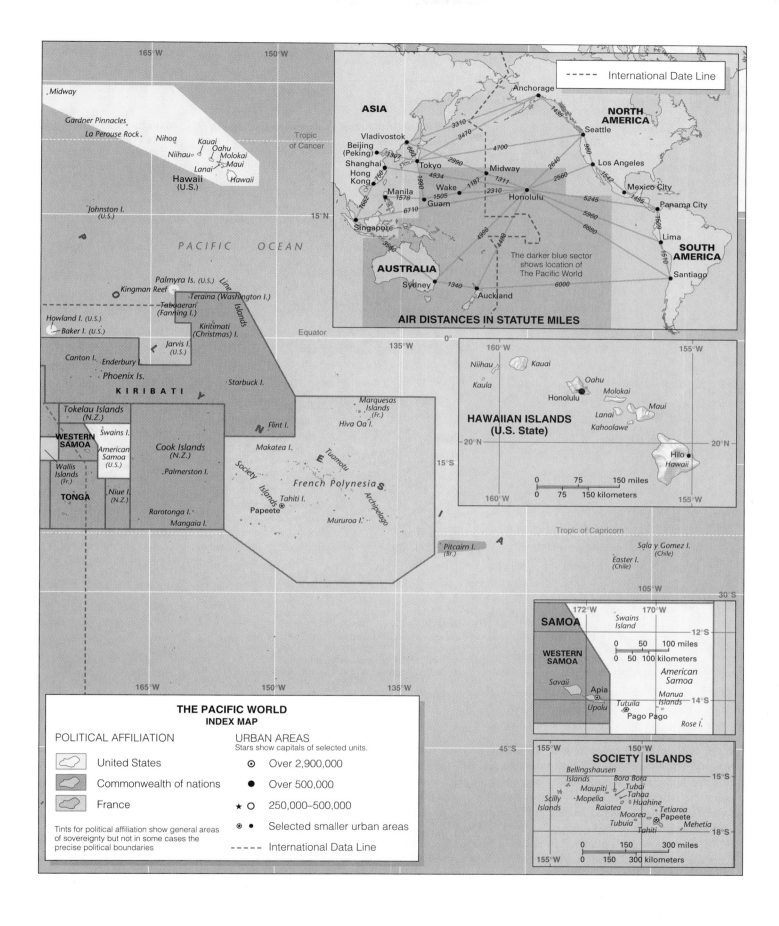

International Date Line -----

AIR DISTANCES IN STATUTE MILES

ASIA

NORTH AMERICA

Anchorage

Vladivostok
Beijing (Peking)
Shanghai
Hong Kong
Tokyo
Manila
Wake
Guam
Singapore

Seattle
Los Angeles
Midway
Honolulu
Mexico City
Panama City
Lima
SOUTH AMERICA
Santiago

3310
3470
4700
1490
660
2990
2640
960
750
4934
1990
1311
2560
1540
1505
1187
2310
5245
1495
1602
1578
6710
5960
8880
1509
3560
4996
1468
6000
1340

The darker blue sector shows location of The Pacific World

AUSTRALIA
Sydney
Auckland

Midway

Gardner Pinnacles
La Perouse Rock
Nihoa
Kauai
Niihau
Oahu
Molokai
Lanai
Maui
Hawaii (U.S.)
Hawaii

Tropic of Cancer

165°W   150°W   15°N

Johnston I. (U.S.)

PACIFIC OCEAN

Palmyra Is. (U.S.)
Kingman Reef
Teraina (Washington I.)
Tabuaeran (Fanning I.)
Howland I. (U.S.)
Baker I. (U.S.)
Kiritimati (Christmas) I.
Jarvis I. (U.S.)
Canton I.   Enderbury I.
Phoenix Is.
Starbuck I.
K I R I B A T I
Line Islands

Equator   135°W   0°

Marquesas Islands (Fr.)
Hiva Oa I.

Tokelau Islands (N.Z.)
Swains I.
WESTERN SAMOA
American Samoa (U.S.)
Flint I.
Makatea I.
Cook Islands (N.Z.)
Palmerston I.
Wallis Islands (Fr.)
Niue I. (N.Z.)
TONGA
Rarotonga I.
Mangaia I.
Society Islands
Tahiti I.
Papeete
Tuamotu
French Polynesia
Archipelago
Mururoa I.

15°S

Tropic of Capricorn

Pitcairn I. (Br.)

Sala y Gomez I. (Chile)
Easter I. (Chile)

105°W   30°S

HAWAIIAN ISLANDS (U.S. State)

160°W   155°W
Niihau   *Kauai*
Kaula
Honolulu   *Oahu*   *Molokai*
Lanai   *Maui*
Kahoolawe
20°N
Hilo   *Hawaii*

| 0 | 75 | 150 miles |
| 0 | 75 | 150 kilometers |

160°W   155°W   20°N

SAMOA

172°W   170°W
Swains Island
12°S
WESTERN SAMOA
Savaii
Upolu   Apia
Tutuila   Pago Pago
American Samoa
Manua Islands
Rose I.
14°S

| 0 | 50 | 100 miles |
| 0 | 50 | 100 kilometers |

SOCIETY ISLANDS

155°W   150°W
Bellingshausen Islands
Bora Bora
Maupiti   *Tubai*
Mopelia   *Tahaa*
Scilly Islands   *Raiatea*   *Huahine*
Moorea   *Tetiaroa*
Tubuia   Papeete
Tahiti   *Mehetia*
15°S
18°S

| 0 | 150 | 300 miles |
| 0 | 150 | 300 kilometers |

155°W   150°W

165°W   150°W   135°W   45°S

THE PACIFIC WORLD
INDEX MAP

POLITICAL AFFILIATION

⬡ United States

⬡ Commonwealth of nations

⬡ France

Tints for political affiliation show general areas of sovereignty but not in some cases the precise political boundaries

URBAN AREAS
Stars show capitals of selected units.

⊙ Over 2,900,000

● Over 500,000

★ ○ 250,000–500,000

⊛ ● Selected smaller urban areas

----- International Data Line

TABLE 15.1

Summary of Main Pacific Island Groups

Island Realms	Main Island Groups	Well-Known Individual Islands or Atolls	Countries Exercising Political Control	Main Exports of Political Units (1994)
Melanesia		New Guinea	Papua New Guinea	Crude oil 26%; gold 26%; timber 18%
			Indonesia (Irian Jaya province)	
	Bismarck Archipelago	New Britain, New Ireland	Papua New Guinea	
	Solomon Islands	Guadalcanal	Solomon Islands	Timber products 56%; fish products 20%; palm oil products 8%
		Bougainville	Papua New Guinea	
	Vanuatu (former New Hebrides)	Espiritu Santo	Vanuatu	Copra 26%; beef and veal 17%; timber 10%; re-exports 22%
	Fiji Islands	Viti Levu, Vanua Levu	Fiji	Sugar 38%; gold 10%; fish 6%; timber 5%
		New Caledonia	France (overseas territory; also includes the Loyalty Islands and other small islands)	Nickel 90% (1990)
Micronesia		Nauru	Nauru	Phosphate 100%
	Caroline Islands	Kusaei, Pohnpei (Ponape), Palau (Belau), Truk, Yap	United States (U.N. Trust Territory of the Pacific Islands); Federated States of Micronesia[a]	Coconut products
	Mariana Islands	Guam	United States (self-governing territory)	Clothing and other manufactured items
		Saipan, Tinian	Northern Marianas[a]	Coconut products
	Marshall Islands	Bikini, Eniwetok	Republic of the Marshall Islands	Chilled and frozen fish 68%; coconut oil 9%; pet fish 1%

[a]The United States Trust Territory of the Pacific Islands has developed into (1) the Republic of the Marshall Islands and the Federated States of Micronesia (in the Carolines), where "compacts of free association" with the United States have replaced the original trusteeship agreements; (2) the Northern Marianas, a self-governing commonwealth forming a political union with the United States and whose people are U.S. citizens; and (3) Palau (Belau), where a compact of free association with the United States is being negotiated. Source of data on exports: mainly *Britannica Book of the Year,* 1996.

largest, is about 1500 miles long (*c.* 2400 km) and 400 miles across (*c.* 650 km) at the broadest point. In general, these islands are hot, damp, mountainous, and carpeted with dense vegetation. **Micronesia** (Greek: "tiny islands") includes thousands of small and scattered islands in the central and western Pacific, mostly north of the equator. Before World War II, Japan held Micronesia under a mandate from the League of Nations, except for the Gilbert Islands, which were a British possession, and Guam, a United States possession. After the war, the former Japanese islands became the Trust Territory of the Pacific Is-

lands, administered by the United States (see footnote, Table 15.1). **Polynesia** (Greek: "many islands") occupies a greater expanse of ocean than does either Melanesia or Micronesia. It is shaped like a rough triangle whose corners are at New Zealand, the Hawaiian Islands, and remote Easter Island.

Each of these three island regions contains a number of distinct island groups. Table 15.1 lists the major groups, together with some noteworthy individual islands, the nations exercising political control, and the main exports.

Diverse peoples inhabit the huge Pacific realm. Four dis-

TABLE 15.1

Summary of Main Pacific Island Groups *(Continued)*

Island Realms	Main Island Groups	Well-Known Individual Islands or Atolls	Countries Exercising Political Control	Main Exports of Political Units (1992)
	Kiribati	(Republic of Kiribati includes Tarawa and the rest of the Gilbert Islands, plus the Phoenix Islands, most of the Line Islands, and Ocean Island)	Kiribati	Copra 67%; fish 11%
Polynesia	Hawaiian Islands	Oahu, Hawaii, Maui, Kauai	United States (U.S. state)	Sugar; pineapples
	Tuvalu (former Ellice Islands)		Tuvalu	Clothing 30%; copra 22% (1990)
	Tonga Islands		Tonga	Squash 49%; vanilla beans 15%; root crops 3%
	Samoa Islands		Western Samoa	Taro 58%; coconut cream 21%; beer 8%
			United States (American Samoa is a non–self-governing U.S. territory)	Canned tuna
	Marquesas Islands Tuamotu Archipelago Society Islands	(The Marquesas, Tuamotus, and Societies form French Polynesia, governed from Tahiti in the Society Islands)	France (overseas territory)	Re-exports 58%; black cultured pearls 34% (data are for French Polynesia as a whole) (1990)
	Cook Islands	Rarotonga	New Zealand (self-governing territory)	Clothing

tinct geographical races call the region home. The dark-skinned but non-Negroid Australian Aborigines are indigenous to Australia. The Melanesians, who are also dark-skinned, inhabit New Guinea and islands to the east. The dark-skinned Micronesians, who live west of Polynesia and north of Melanesia, probably originated from the mixing of Melanesians and Southeast Asians. The lighter-skinned Polynesians are relative latecomers to the region; they descended mainly from a culture which originated in the interior of Southeast Asia about 5000 years ago, and then advanced southward into the Malay Peninsula. From there, over a period of several thousand years, people of this "Malayo-Polynesian" culture embarked on voyages to Madagascar, New Zealand, Easter Island, and finally Hawaii, "island-hopping" all the way and accomplishing extraordinary feats of navigation from their outrigger canoes.

Hundreds of distinct cultures emerged from the complex origins of the Pacific peoples. New Guinea and the Solomon Islands alone are home to 800–900 different languages, making them—on a per capita basis—the most linguistically diverse region in the world.

Definitions and Insights

DEFORESTATION AND THE DECLINE OF EASTER ISLAND

"Easter Island is Earth writ small," wrote traveller Jared Diamond. Recent archeological and paleobotanical studies suggest that the civilization that built the island's famous monolithic stone statues destroyed itself through overpopulation and abuse of natural resources (Fig. 15.A).

When the first human colonists from eastern Polynesia reached remote Easter Island in about 400 A.D., they found the island cloaked in subtropical forest. Plant foods, especially from the Easter Island palm, and animal foods, notably porpoises and seabirds, were abundant. The human population grew rapidly in this prolific habitat. The complex, stratified society that emerged on the island grew to an estimated 7000 to 20,000 people between 1200 and 1500 A.D., when most of the famous statues were built. Apparently in association with their religious beliefs, Easter Islanders erected more than 200 statues, some weighing up to 82 tons and reaching 33 feet (10 m) in height, on gigantic stone platforms. At least 700 more statues were abandoned in their quarry sites and along roads leading to their would-be destinations, "as if the carvers and moving crews had thrown down their tools and walked off the job," Diamond observed.

"Its wasted appearance could give no other impression than of a singular poverty and barrenness," the Dutch explorer Jacob Roggeveen wrote of the island on the day he discovered it—Easter Sunday, 1722. Not a single tree stood on the island. The depauperate landscape bears testimony to the fate of the energetic culture which built the great statues. By 800 A.D., people were already exerting considerable pressure on the island's forests for fuel, construction, and ceremonial needs. By 1400, people and the rats they introduced to the island caused the local extinction of the valuable Easter Island palm. Con-

tinued deforestation to make room for garden plots and to supply wood to build canoes and to transport and erect the giant statues probably eliminated all of the island's forests by the 15th century. By then people had hunted to extinction many terrestrial animal species and could no longer hunt porpoises because they lacked the wood needed to build seagoing canoes. Crop yields declined because deforestation led to widespread soil erosion. There is evidence that in the ensuing shortages, people turned on each other as a source of food. By about 1700 A.D., their population began a precipitous decline to only 10–25 percent of the number who once lived on this isolated Eden.

15.2   Land and Life on the Islands

The landscapes, seascapes, and cultural geographies of the almost countless islands scattered across the tropical Pacific vary widely. However, there are generally two types of islands—"high" and "low" (Fig. 15.2a, b). Most high islands are the result of volcanic eruptions, but high continental islands (islands attached to continents before sea level changes and tectonic activities isolated them) are much larger and more complex in origin. The low islands are made of coral, a material composed of the skeletons and living bodies of small marine organisms that inhabit tropical seas. These very different settings offer quite different opportunities for human livelihoods.

HIGH ISLANDS

Some of the volcanic high islands of the Pacific comprise **island chains.** These are formed by oceanic crust sliding over a stationary "**hot spot**" in the Earth's mantle where molten magma is relatively close to the crust. As the crust slides over the hot spot, magma rises through the crust to form new volcanic islands. The Hawaiian Island chain is an excellent example. The big island of Hawaii is now situated over the hot spot and is still active. It is one of the chain's youngest members. The Pacific Plate of the Earth's crust is moving northwestward here; therefore, older volcanic islands which were born over the hot spot have moved northwestward where they have become inactive. Still older islands farther northwest in the chain have submerged to become **seamounts,** or underwater volcanic mountains.

Many of the high volcanic islands are spectacularly scenic (Fig. 15.2a). Steep slopes predominate, and some islands have peaks thousands of feet high. However, there are great variations in elevation, slope, soil, rainfall, and plant life from island to island and even on a given island. Among islands of this type, the Hawaiian, Samoan, and Society groups, all located in Polynesia, are especially well known. The main cities and seaports

FIGURE 15.A

The Polynesian people who erected these *moai* statues between 1200 and 1500 A.D. deforested most of Easter Island—in part to supply levers and rollers for the statues—and thus precipitated the demise of their civilization.   *George Holton/Photo Researchers*

(a)

(b)

FIGURE 15.2

Characteristic high and low islands of the Pacific. The high island of Bora Bora (Fig. 15.2a) in French Polynesia is an atoll in the making, roughly in the middle stage of the sequence pictured in Figure 15.4. Clouds are forming over the higher parts of the island. The photograph was taken at low tide; note the extensive sandbars. In the foreground, note the airstrip which has facilitated tourist access to this lovely, world-famous island. With less diverse attractions and resources, low islands (Fig. 15.2b) are most associated with the coconut civilizations of the Pacific. Low islands are extremely vulnerable to the dangers which would be posed by global warming.   *(a) Tony Stone Images/Paul Chesley, (b) FPG International/Karl & Jill Wallin*

in the tropical Pacific islands are in the volcanic high islands and the continental islands. Valuable minerals are scarce, but the soil is generally fertile. Varied physical conditions in the volcanic islands allow a diversity of tropical crops. The islands' original peoples depended mainly on starchy foods such as breadfruit, plantain, taro, sweet potatoes, and yams. Pigs and sea fish supplied animal protein. Europeans, North Americans, and east Asians subsequently introduced other crops and animals, including arrowroot and cassava; bananas; tropical fruits such as mangoes, pineapples, papayas, and citrus; coffee and cacao; sugarcane; and cattle, goats, and poultry.

The newcomers established sugar plantations in some islands, most prominently in the Hawaiian Islands, Fiji Islands (Fig. 15.3), and Saipan in the Marianas. They also introduced laborers from outside the region, because the local peoples generally proved too few in number or refused to work in the cane fields. Successive infusions of Chinese, Portuguese, Japanese, and Filipinos brought a polyglot character to Hawaii. British plantation owners imported mainly Hindu Indian laborers to Fiji, where indigenous Fijians now outnumber ethnic Indians only slightly. Ethnic tensions after independence (1970) led to a military takeover of the government by native Fijians in 1987. In Saipan, Japanese interests using Japanese labor developed the island's sugar plantations. The sugar industry did not survive the devastating U.S. military invasion of Saipan in World War II, but sugar is still important in Hawaii and Fiji.

Some high islands to the north and northeast of Australia are continental islands, including New Guinea; New Britain

and New Ireland in the Bismarcks; New Caledonia; Bougainville and smaller islands in the Solomons; the two main islands of Fiji; and a few others. Mountains over 16,000 feet (4877 m) high rise in New Guinea, and lower but still imposing mountains exist in various other islands. Valuable minerals are extracted in a few of these islands, including the nickel of the

FIGURE 15.3

Sugarcane is a major commercial crop in a few high islands of the Pacific World. The small locomotive in the view is hauling a load of cane on a plantation in the Fiji Islands.   *William A. Noble*

French-held island of New Caledonia and the copper of Bougainville in Papua New Guinea.

Mineral wealth has sometimes brought unrest to these islands, notably New Caledonia and Bougainville. Separatists demanding independence for nickel-rich New Caledonia clashed with French police in 1988. An accord signed after the confrontation guarantees a 1998 referendum for New Caledonia's people, who may then choose to become independent or remain under French rule. In 1988 a crisis also shook Bougainville, a former colony of Australia which Australia insisted in 1975 should become part of newly-independent Papua New Guinea. Angry landowners calling themselves the "Black Rambos" destroyed mining equipment and power lines to force the closure of the world's third largest copper mine. They were generally younger landholders who were receiving none of the royalties and compensation payments made by New Guinean and Australian mining interests to an older generation of the island's inhabitants. Residents of Bougainville concerned about the ruinous environmental effects of copper mining and disturbed that most mine laborers were imported from mainland Papua New Guinea, with little mining revenue remaining on the island, joined the opposition. The growing crisis drove world copper prices to record high levels. Opposition to the Australian-backed mining interests of Papua New Guinea in Bougainville has grown since 1988 into a full-fledged independence movement led by the Bougainville Revolutionary Army (BRA). Late in 1989, war broke out when the BRA leader declared independence from Papua New Guinea. Most Bougainvilleans want to be united again with the Solomon Islands, whose people they are culturally and ethnically related to. Papua New Guinea has responded with an economic blockade around Bougainville to prevent supplies from reaching the rebels, who salvage World War II era weapons and ammunition to fight superior troops.

LOW ISLANDS

The low islands are generally smaller than the volcanic high islands and lack the resources to support dense populations. These islands generally pose several natural hazards to human habitation. Their low elevation above sea level provides little defense against huge storm waves and tsunamis, the waves generated by earthquakes. The lime-rich soils are often so dry and infertile that trees will not grow, and a shortage of drinking water often prevents permanent settlement. The low islands are typically fringed by the waving coconut palms that are a mainstay of life and the trademark of the "South Sea isles." The islands of the republic of Kiribati are typical of the picturesque low islands idealized by Hollywood and travel brochures (Fig. 15.2b).

The low islands are made of coral and usually form an irregular ring around a lagoon. Such an island is called an atoll.

Generally the coral ring is broken into many pieces, separated by channels leading into the lagoon, but the whole circular group is commonly considered one island. Charles Darwin devised a still widely-accepted explanation for the three-stage formation of atolls (Fig. 15.4). First, coral builds a fringing reef around a volcanic island. Then, as the island slowly sinks, the

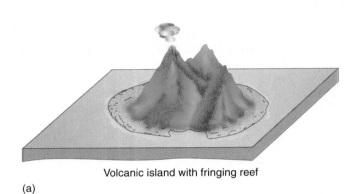

Volcanic island with fringing reef

(a)

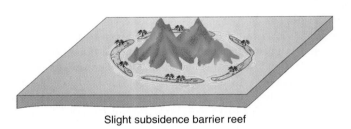

Slight subsidence barrier reef

(b)

An atoll

(c)

FIGURE 15.4

Charles Darwin's explanation of the development of an atoll. First (top), a fringing coral reef is attached to the volcanic island's shore. Then, as the island subsides (center), a barrier reef forms. With continued subsidence, the coral builds upward, and the volcanic center of the island finally becomes completely submerged, forming an atoll (bottom). *After Gabler et al. 1987, p. 562, Essentials of Physical Geography.*

coral reef builds upward and forms a barrier reef separated by a lagoon from the shore. Finally, the volcanic island sinks out of sight and a lagoon occupies the former land area whose outline is reflected in the roughly oval form of the atoll.

THE VULNERABLE ISLAND ECOSYSTEM

Island ecosystems in the Pacific region, like those across the globe, are typically inhabited by **endemic** plant and animal species—species found nowhere else in the world. They result from a process in which ancestral species colonize the islands from distant continents, and, over a long period of isolation and successful adaptation to the new environments, evolve to become new species. Gigantic and flightless animals, like the moa bird of New Zealand and the giant tortoises of the Galapagos and Seychelles islands, are typical of endemic island species.

Generally developing in the absence of natural predators and inhabiting relatively small areas, island species have proven to be especially vulnerable to the activities of humankind. People purposely or accidentally introduce new plant and animal species (called **exotic** species) which often prey upon or overtake the endemic species. Habitat destruction and deliberate hunting also lead to extinction—for example, of New Zealand's moas and the dodo birds of the Indian Ocean island of Mauritius, now "dead as dodos." Indigenous inhabitants of New Zealand set fires to hunt the giant, ostrichlike moa. The local people had already killed off most of the birds by the time the Maori people colonized New Zealand around 1350 A.D., and by 1800 the moa was extinct.

From one end of the Pacific World to the other, there is heightened concern about human-induced environmental changes. The U.S. Hawaiian Islands have the distinction of being known to ecologists as the "extinction capital" of the world because of the irreversible impacts that exotic species, population growth and development have on indigenous wildlife. Environmentalists are also concerned about the impact of commercial logging on the island of New Guinea. About 22,000 plant species grow there, fully 90 percent of them endemic. Ecologists consider nearby New Caledonia to be one of the world's biodiversity "hot spots" because of ongoing human impact on its unique flora and fauna (see Chapter 2, page 33).

Human-induced extinctions complement a long list of natural hazards to island species. For example, island animals are particularly prone to natural catastrophes such as volcanoes. Most animals find it difficult or impossible to evacuate in the event of an eruption, and the island's distance from other lands may hinder or prevent recolonization. So, while island habitats have a higher rate of endemism than similar climates on mainlands, environmental difficulties often cause the number of different kinds of species (the species diversity) to be lower.

15.3 Subjugation, Independence, and Development

Europeans began to visit the Pacific islands early in their Age of Exploration. Spanish and Portuguese voyagers were followed by Dutch, English, French, American, and German explorers. Many famous names are connected with Pacific exploration, including Magellan, Tasman, Bougainville, La Perouse, and Cook. By the end of the 18th century, Europeans knew virtually all of the important islands. For a long period, the European governments exercised only nominal control. European whaling crews, sandalwood traders, indentured labor contractors, and other adventurers abused the indigenous peoples. On island after island, European penetration decimated the islanders and disrupted their cultures. The intruders introduced venereal and other infectious diseases, alcohol, opium, forced labor, and firearms, which greatly increased the slaughter in tribal wars. Four-and-a-half centuries of turbulent Western influence in the Pacific islands do not reflect well on the outsiders.

In recent times, plantation agriculture, mining, fishing, military facilities and activities, and an expansion of tourism have attracted Western personnel and capital to the islands. These activities are widespread except for mining, which is confined largely to New Caledonia, Papua New Guinea, and Nauru (see Table 15.1). The new surge of Western interest has accompanied a steady process of decolonization. Since the end of World War II, the United States, Britain, Australia, and New Zealand have abandoned most of their colonies in the region. Only France has insisted on holding onto all of its colonies.

The Pacific islands are now midstream between colonialism and independence. Once entirely colonial, the region is a mixture of units still affiliated politically with outside countries and states that have become fully independent (see Table 15.1). Evolution of the islands away from colonial status began with New Zealand's granting of independence to Western Samoa in 1962. As of 1997, nine fully independent states had emerged, with populations ranging from 4.3 million in Papua New Guinea—which is exceptionally large in population and area for this region—to 7500 on tiny Nauru island. The "typical" Pacific island country has about 100,000 to 150,000 people in an area of 250 to 1000 square miles (*c*. 650–2600 sq km), consists of a number of islands, is quite poor economically, is an ex-colony of Britain, New Zealand, or Australia, and depends heavily on foreign economic aid.

In some cases colonial powers have delayed independence to would-be island nations. They explain that such territories are too small and isolated, or are not economically viable, for independence. Some islands remain dependent because they confer unique military or economic advantages on the governing power. For example, Guam and American Samoa are useful to the United States for military purposes, and French Polynesia has been the locale for French atomic testing (see Regional Perspectives, page 374).

Regional Perspective

FRENCH NUCLEAR WEAPONS TESTING IN THE SOUTH PACIFIC

"The French fail to appreciate that the end of the Cold War really created in most people's minds a sense that the nuclear age was over, or at least was winding down; that we'd put behind us the nuclear arms race, we'd put behind us the nuclear balance of terror," declared Australian Foreign Minister Gareth Evans in July 1995. "What this French decision did was really start to open all those cans all over again."

Evans spoke in condemnation of France's decision that June to resume underground testing of nuclear weapons on the Mururoa ("Place of Deep Secrets") atoll in French Polynesia after a three-year hiatus. France insisted that a series of 1995 and 1996 tests would be the last before the country signed the Comprehensive Test Ban Treaty in 1996.

Crew members of *Rainbow Warrior 2*, the flagship of the environmentalist organization Greenpeace, confronted French Navy ships near Mururoa in July 1995. French commandos used tear gas to overwhelm the crew and seize the ship. Greenpeace managed to televise the event, inflicting enormous public relations damage on France. It was the continuation of a long-running feud between Greenpeace and France, which had first come to a head in 1985 when French agents attacked and sank the original *Rainbow Warrior* at anchor in New Zealand's port of Auckland as it was preparing to sail to the Mururoa test site. The revelation that the French government was behind the attack was a serious embarrassment for France and boosted antinuclear sentiment throughout the region.

Throughout the Pacific there were strong negative reactions to the 1995 weapons tests. Strongly antinuclear New Zealand protested most loudly. In Australia, customers boycotted imported French goods and local French restaurants, and postmen refused to deliver mail to the French embassy and consulates. Antinuclear protestors firebombed the French consulate in Perth. Australia joined New Zealand in filing a case at the World Court in The Hague aimed at halting French nuclear tests. Worldwide, informal boycotts diminished sales of French wines. In French Polynesia itself, activists for independence from France used the issue to highlight international attention to their demands. Over several days in September 1995, hundreds of anti-French demonstrators in Tahiti rampaged through French Polynesia's capital city of Papeete, looting and setting parts of the city on fire and causing $11 million worth of damage to the island's international airport, a vital touristic hub.

Throughout the controversy, French authorities insisted that the tests and the atoll itself were safe. About 1500 French military personnel live on Mururoa, subsisting largely on supplies flown in from Tahiti but also on fish caught in local waters. French spokesmen swam for press photographers in Mururoa's lagoon to demonstrate the safety of the water above the test site, which was located deep in the basalt basement rock where the blasts had been detonated since 1981.

Despite the nuclear tests, the majority of inhabitants of French Polynesia have long favored continued French rule. Tourism and other local businesses contribute only about 25 percent of the territory's revenue; the remainder comes from French economic assistance, largely in the sectors supporting the nuclear testing program. Many locals, however, whose parents and grandparents gave up subsistence fishing and farming for jobs in the military and its service establishments, are worried about their future. France did sign the test ban treaty and ceased its nuclear tests in 1996—raising fears about the potential economic impact if France withdraws its investments from the region.

France is not the only nuclear power to conduct controversial weapons tests in the Pacific. Throughout the 1950s the United States used the Bikini atoll in the Marshall Islands as one of its chief testing grounds (the island's fame at the time led French designers to name a new two-piece bathing suit after it). United States government authorities relocated the indigenous inhabitants of Bikini to another island, which the Bikinians have never accepted as their new home. But returning to Bikini is a hazardous prospect given its long legacy of radioactive soil. One detonation on Bikini had the power of a thousand Hiroshima bombs and produced radioactive dust that fell downwind on the Rongelap atoll, where children played in the dust "as though it were snow," one observer wrote. The Marshallese today suffer a legacy of cancers and deformities linked to these weapons tests.

While Bikini Atoll has not yet been declared safe for permanent habitation, in 1997 it was declared open as a tourist destination. The island's lack of human inhabitants for more than 50 years has had the remarkable effect of rendering Bikini a true marine wilderness. Many more tourists are sure to come to marvel at Bikini's natural treasures, and the Bikinians may one day return home to profit from the boom.

Another characteristic of the Pacific islands is a general lack of industrial development, combined in many cases with high population densities. The poverty typical of LDCs prevails in the region. The major exception is tiny Nauru, whose earnings from phosphate—the product of thousands of years of accumulation of seabird excrement, or **guano**—have provided a high average income. Nauru's economy depends totally on annual exports of 2 million tons of phosphate and on revenues earned from overseas investments of profits from phosphate sales (the tallest building in Melbourne, Australia, is a product

FIGURE 15.5
Phosphate mining has stripped most of the soil and vegetation cover from the island of Nauru, whose inhabitants are now looking for a new home. In the foreground is a plant for processing the raw phosphate into exportable fertilizer; behind is a bare stockpile from which the phosphate has been mined. *William E. Ferguson/William E. Ferguson Photography*

of Nauruan investments and is known affectionately as "Birdshit Tower"). This resource is expected to be depleted completely by the year 2000, and because the mining of phosphate has stripped the island of soil and vegetation, the people of Nauru are now looking for a new island home (Fig. 15.5).

Nauru is an extreme example of how, like LDCs elsewhere, the Pacific nations tend to rely on a few primary products for major exports. Before the intrusion of outsiders, the island economies were based heavily on subsistence agriculture, gathering, and fishing, with coconuts a major element in the food supply. Many countries still rely heavily on modern commercialized versions of the coconut and fishing economy. But, along with this commercial reliance goes a continuing dependence on coconuts and fish as major elements in a widespread pattern of subsistence activities. Indeed, the coconut is so important in these islands that they have been said to have a "coconut civilization" (Fig. 15.6). The coconut provides both food and drink for the islanders, and the dried meat, known as copra, is the only significant export from innumerable islands. The husks and shells of the nuts have many uses, as do the trunks and leaves of the coconut palms. People make baskets and thatching from the leaves, for example, and use timber from the trunk for building and furniture making.

Many of the island political units listed in Table 15.1 rely mainly on exports other than coconut products or fish. These units fall into four major categories:

1. *Countries and colonies in which mineral extraction has replaced subsistence economies.* New Caledonia, Nauru, and Bougainville are outstanding examples. Another important producer is Papua New Guinea, where petroleum, gold, and copper provided 67 percent of the country's exports in 1994.

2. *Colonies whose economies are supported and shaped by military activities*, such as Guam, American Samoa, and French Polynesia.

3. *Fiji*, which developed under British rule as a source of plantation sugar using labor largely imported from India. Sugar is still the largest export (see Fig. 15.3), but Fiji also exports gold, fish, and timber. Another Pacific locale where sugar is a valuable source of income is the U.S. state of Hawaii.

4. *The Cook Islands*, a self-governing territory in free association with New Zealand and which is developing an industrial economy. Labor-intensive clothing manufacturing is the leading export.

In sum, the Pacific islands' economic picture is one of nonindustrial economies (1) exporting products that were important in their precolonial subsistence activities; (2) exporting plantation crops introduced by Westerners; (3) exporting minerals desired by industrial nations; and (4) deriving income from activities connected with the military needs of occupying powers. The most dynamic new element is the rapid growth of tourism as a basic industry, with the U.S. and Japan the main sources of visitors.

FIGURE 15.6
Coconuts are vital in the diets, material cultures and economies of many Pacific islanders. This boy is a Western Samoan. *Nik Wheeler/Westlight*

16

PROSPEROUS, REMOTE
AUSTRALIA AND NEW ZEALAND

Australia and New Zealand are unique. Far removed from Europe, their dominant cultures are nevertheless European. In the underdeveloped Pacific World, they are prosperous. Distinctive plants and animals inhabit the often odd landscapes of these islands, stirring a sense of wonder among indigenous people, European settlers, and foreign tourists alike. Writer Alan Moorehead described the reactions of early European visitors: "Everything was the wrong way about. Midwinter fell in July, and in January, summer was at its height. In the bush there were giant birds that never flew and queer, antediluvian animals that hopped instead of walked or sat munching mutely in the trees. Even the constellations in the sky were upside down."

16.1 Peoples and Populations

There is a basic kinship between these two unusual countries, deriving from similarities in population, cultural heritage, political problems and orientation, type of economy, and location. Australia and New Zealand are among the world's minority of prosperous countries. Australia's per capita gross national prod-

uct of $18,000 is comparable to that of Britain, although well below that of the United States and the most prosperous European countries. A World Bank study published in 1995 ranked Australia as the world's wealthiest country, based on its natural resource assets divided by the total population. New Zealand is less affluent, but it is still quite prosperous compared with most nations of the world. In both countries there are relatively few people to spread the prosperity among; Australia's 18.3 million people plus New Zealand's 3.6 million (as of 1996) amount to only about two thirds of the people living in California.

Both countries trace their prosperity to the wholesale transplanting of a culture—from the industrializing Great Britain—to the remote Pacific beginning in the late 18th century. Australia (established originally as a penal colony for British convicts) and New Zealand are products of British colonization and strongly reflect the British heritage in the ethnic composition and culture of their majority population. The Australians and New Zealanders speak English, live under British-style parliamentary forms of government, acknowledge the British sovereign as their own, and attend schools patterned after those of Britain. Only very recent decades of immigration from continental Europe and Asia have given Australia a more cosmopolitan population.

◀ Australian Aborigines believe that the landscapes and living things of the world were created by ancestral beings during the "Dreamtime." This is Aboriginal rock art depicting a Wandjina, a creator-spirit of the Kimberley region, believed to be responsible for storms and other events in the sky. *Photo by Joseph J. Hobbs*

PROBLEM LANDSCAPE

AUSTRALIA GRANTS ABORIGINES RIGHT TO CLAIM NATIVE TITLE

As in the United States, newcomers to Australia forged a new nation by dispossessing the ancient inhabitants of the land. And as in the United States, in Australia there is a long history of white racism, discrimination, and abuse against blacks, who in Australia's case are the original inhabitants. Aborigines were not even counted in the national census until 1967. The most disadvantaged group in Australian society, Aborigines suffer from high infant mortality, low life-expectancy, and high unemployment relative to non-Aboriginal populations. Like Native Americans, a disproportionately large share of Aborigines fall prey to the economic and social costs of alcoholism. Statistics indicate they are 29 times more likely than non-Aborigines to end up in jail. Once in jail, they are more likely than non-Aborigines to commit suicide or be killed by guards.

In 1992, the Australian government announced a five-year, $113 million program to improve the justice system for native Australians. It aimed at placing Aborigines in Aboriginal-run "bail hostels" rather than white jails, educating white police about Aboriginal culture, and treating alcoholism and drug abuse among Aborigines.

One of the largest issues of contention between Australia's indigenous people and the white majority is land rights. As in the United States, European newcomers pushed the native people off productive ranching, farming, and mining lands into special reservations on inferior lands. The Europeans used a legal doctrine called *terra nullius*, meaning "the land was unoccupied," to lay claim to the continent. Under the 1976 Land Rights Law, Aborigines were permitted to seek title to vacant state land ("Crown land") that they could prove a historical relationship to, but few succeeded. Following new legislation in 1992, the government began to return titles on a parcel-by-parcel basis—generally in very marginal lands—to Aborigines who had argued their rights successfully. However, most Aborigines were unhappy with the terms of the returned titles, which allowed native title to coexist with but not supersede the established crown title. The government retained mineral rights to the land, and could therefore lease native land to mining companies.

Late in 1993, Aboriginal leaders expressed their objections to the legislation in a declaration known as the Eva Valley statement, which made four demands:

- Aborigines should have veto rights over mining and pastoral leases on native-title land.
- Mining and pastoral leases cannot extinguish native title to the land.
- There should be no access to developers without Aboriginal permission.
- The federal government, not the states, should control native title issues.

Prime Minister Paul Keating rejected the Eva Valley statement, arguing that the government had already made enough concessions. Aboriginal leaders took their case to the United Nations in Geneva, where they argued that existing legislation to validate mineral leases on native land was a breach of international human rights conventions which Australia had signed.

Late in 1993, after the longest senate debate in the country's history, Foreign Minister Gareth Evans succeeded in pushing the Native Title bill through Australia's parliament. The new legislation addressed the Aborigines' major objections, providing the following concessions:

- Aborigines now have the right to claim land leased to mining concerns once the lease expires.
- Aborigines now have the right to negotiate with mineral leaseholders over development of their land. However, they do not have the right of veto they asked for.

Although they are fully independent nations, loyalty to Britain has long been an outstanding characteristic of Australia and New Zealand. The countries still belong to the Commonwealth of Nations. In both world wars, the two countries immediately came to the support of Britain and lost large numbers of men on battlefields far from home. In World War II, United States forces helped to frustrate a threatened Japanese assault on their homelands. (Since that time, the two countries have sought closer relations with the U.S., and British power has declined sharply.) Their small populations and remote insular locations account for the two countries' seeking ties with strong allies, especially naval powers such as Great Britain and the United States.

Both Australia and New Zealand have minorities of indigenous inhabitants. The native Australian people are known as "Aborigines." Colonizing whites slaughtered many of their an-

- Where their native title has been extinguished, Aborigines are entitled to compensation paid by the government.
- With the help of a Land Acquisitions Fund, impoverished Aborigines can buy land to which they have proven native title.

The Native Title bill is likely to be of great consequence in South and Western Australia, where there is much vacant Crown land and many Aborigines able to file claims on it. Those states are also heavily dependent upon mining. Members of Australia's Mining Industry council are unhappy with the new legislation. They worry that their mine leases will run out before the minerals do, which will require them to negotiate with the Aboriginal titleholders.

In the Northern Territory, which has its own land-rights legislation, vast tracts of land have already been given back to Aborigines. Aborigines there make up 24 percent of the population, and now control 32 percent of the land. Aborigines also hold title to Australia's two greatest national parks, Uluru (Ayers Rock) (Fig. 16.A) and Kakadu, both in the Northern Territory. In an arrangement which has become a model worldwide for management of national parks where indigenous people reside, the Aboriginal owners of the reserves have leased them back to the Australian government park system, which comanages them with the Aborigines. Tourism programs in Uluru and Kakadu now highlight the cultural resources of the parks, in addition to their natural wonders. At both parks visitors can enjoy interpretive natural history walks that also emphasize Aboriginal culture and which are conducted by the land's oldest and most knowledgeable inhabitants.

Although there has been remarkable progress in recent years in this area and others involving the Aborigines, the issue of rights to land has not been resolved to the satisfaction of all. "The handover of Ayers Rock is a turning point in Australia's race relations," declared Charles Perkins, then the country's top Aboriginal civil servant. "It's a recognition that Aboriginal people were the original owners of this country." But Aboriginal activist Galarrwuy Yunupingu declared, "I'd only call it a victory if the governor-general came and gave us title to the whole of Australia, and we leased *that* back to the Commonwealth."

Catherine Foster, "Australia Grants Aborigines Right to Claim Native Title," *The Christian Science Monitor*, December 23, 1993, p. 3.

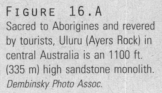

FIGURE 16.A
Sacred to Aborigines and revered by tourists, Uluru (Ayers Rock) in central Australia is an 1100 ft. (335 m) high sandstone monolith.
Dembinsky Photo Assoc.

cestors and drove the majority into marginal areas of the continent (see Problem Landscape on this page). The Aboriginal population now numbers about 260,000. Living mainly in the tropical north, this minority exists mostly on the fringes of white society. Some carry on a more traditional way of life in the dry Australian wilderness.

In New Zealand, the dominant indigenous group is the Maori, a Polynesian people concentrated mainly on North Island. Whites broke the Maori hold on the land in a bloody war between 1860 and 1870. The Maori, who in 1900 seemed destined for extinction, have rebounded to make up almost 10 percent of New Zealand's population. Although their socioeconomic standing is still depressed, their situation is much better than that of the Australian Aborigines. The Maori are being increasingly integrated, both socially and racially, with New Zealand's white population. At the same time, a

Landscape in Literature

ABORIGINAL VIEWS OF EARTH

"The Aborigine clings to his native soil with every fiber of his being," wrote the ethnographer Carl Strehlow. "Mountains and creeks and springs and water holes are to the Aborigine not merely interesting or beautiful scenic features. They are the handiwork of ancestors from whom he himself has descended. The whole country is his living, age-old family tree."

According to Aborigines, Earth and all things on it were created by the "dreams" of humankind's ancestors during the period of the "Dreamtime." These ancestral beings sang out the names of things, literally "singing the world into existence." The Aborigines call the paths which the beings followed the "Footprints of the Ancestors" or the "Way of the Law"; whites know them as "Songlines." In a ritual journey called "Walkabout," the Aboriginal boy on the verge of adulthood follows the pathway of his creator-ancestors.

In this passage, Jack Sullivan, whose mother was Aboriginal and father European, tells the anthropologist Bruce Shaw some accounts of the creation of people, places, and animals. Sullivan, who was born in 1901, "came over to the white side" in his late twenties and took on the life of a white cattle drover, moving in search of

work between the cattle stations of northwest Australia. The accounts suggest that, like many Aborigines today, Sullivan has forgotten many of the details of the Dreamtime stories passed down orally through the generations. However, the words Sullivan spoke into Shaw's tape recorder convey the essential Aboriginal senses of wonder and reverence for places and living things on the landscape. They also carry a note of sadness that people today are forgetting how places came to be.

I don't know how the first lot of blackfellers [Aborigines] were bred from the earlier generation. There are different words about how the world began, the way things went on before humans like you and I came out. It was the same as they reckon among the white men before we came out. There was some sort of man who may have been taller or shorter, different, until a big earthquake sort of washed up and wiped them all out. A big water came over and they were all drowned. White and black tell you the same story, just like you hear in another country or see in the paper where a volcano bursts up or there is an earthquake. Perhaps the land gets broken and the water comes up there.

A lot of whites used to tell me this too. I suppose you have heard of or seen those things in the science papers or books? An old Chinaman once told me it happened like that in China, that they dug up this 12 foot man who lived before the Chinamen, together with a lot of early day jars full of gold that had been planted. They took up a little bit of an acre and fiddled around digging up everything, and when they opened the lid there was gold. Well those people all died out and then we came out all over this world. It happened around Jerusalem too. China was a very old country and they reckon that Jerusalem was too. And they reckon with that China Wall now, how was that done before the machines? They reckon they were big blooming stones and how did they carry them there? Just like in Jerusalem, they reckon there was a cold chisel made from copper to cut the iron like a wood chisel, and that there was a mosquito net made from gold. There you are. That was before all the machines. How they made that nobody knows.

I heard once from Duncan and Mandi that in this Hidden Valley they had sort of an early day turnout fur-

Maori cultural and political movement is resisting the loss of traditional culture which this integration may cause.

16.2   The Australian Environment

Australia is truly a "world apart." Located in Earth's Southern Hemisphere (see Fig. 16.1) where landmasses are few and far between, it is both an island and a continent. Its natural wonders and curiosities have provided endless inspiration—and sometimes consternation—to its inhabitants and to travelers and visiting writers. On arriving in Botany Bay—now Sydney Harbor—in 1788, British major Robert Ross declared, "I do not

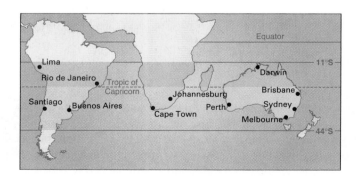

FIGURE 16.1
Australia and New Zealand compared in latitude and area with southern Africa and South America

ther up where nobody should go, right in the centre where there was a sort of round cliff and a big long tree. I don't know whether it was a tree or a stone but it was a big long thing standing right in the centre up that creek where it finished. That was where everything must have rushed in the early days, beginning too late and they all ended up there. The first lot were birds and goannas [monitor lizards] and other animals. Those animals were human beings too before the present black-fellers, according to the old people who used to show us the caves.

Old Waddi told me a good tale once when we were sitting in my camp about all the times when they were travelling around. He talks man to man sensibly about all turnouts right from the jump to the finish. You would get a lot of information from him about these olden days. He knows this country inside out although he never went much over Auvergne. His country was only from Keep River and Newry this side of Auvergne to Carlton, Ningbing and Ivanhoe. That was his beat. He can show you the places, tell you where this has been and where that goes and where this goes, all the years and cor-

ners about the Law [the Aboriginal scheme of customary behavior] and corroborees [ceremonial dances; see Fig. 16.B] that went through. And he can figure out where there was a blackfeller in the early days who may have been a turtle turned out of rock down the Keep River from Newry somewhere. That was where they all came in and finished up after a bit of a row according to the big corroboree. Some turned out of birds and some into rocks and trees. [Waddi pointed a place out to me once where] a little mob of ducks had been and ended up in stone streaking along the river. And other places he said were turtles,

birds, fish and crocodiles all water-logged; then the water sort of dried off and left them. It was one of those places they went to every few years to have a corroboree and sing so that all the animals will come out and run over the land. That little place where they finished was where the birds went up from the rocks and shells. Nobody ever mentioned it before Waddi said it. Some fellers knew but forgot it. Waddi still had it in his mind.

From Jack Sullivan, *Banggaiyerri: The Story of Jack Sullivan as Told to Bruce Shaw.* (Canberra: Australian Institute of Aboriginal Studies, 1983), 139—141.

FIGURE 16.B

Australian Aborigines performing a corroboree (ceremonial dance). Note the man wearing a cross; many Aborigines are Christians.
Penny Tweedie/Tony Stone Images

scruple to pronounce that in the whole world there is not a worse country than what we have yet seen of this. All that is contiguous to us is so very barren and forbidding that it may with truth be said here that nature is reversed; and if not so, she is nearly worn out."[1] Mark Twain marveled at the "Land Down Under" in his book *Following the Equator*: "To my mind the exterior aspects and character of Australia are fascinating things to look at and think about. They are so strange, so weird, so new, so uncommonplace, such a startling and interesting contrast to the other sections of the planet."[2]

[1]Quoted in: Evans, Howard Ensign and Evans, Mary Alice. Australia: *A Natural History,* Washington, D.C.: Smithsonian Institute Press 1983. p.12.

[2]Twain, Mark. *Following the Equator: A Journey around the World.* Hartford, Conn.: American Publishing Company, 1897.

Including the offshore island of Tasmania, Australia has an area of nearly 3 million square miles (7.8 million sq km), approximately equal to the area of the contiguous United States. However, most of the continent is very sparsely populated, and in total population Australia is a relatively small country. The sparseness of Australia's population and its concentration into a comparatively small part of the total land area are closely related to the continent's physical characteristics, among which aridity (accentuated by high temperatures and rapid evaporation) and low average elevation are outstanding. Australia is the world's oldest continent. This has given the elements time to wear down the surface of the land, so that it is also the world's flattest continent. On the basis of climate and relief, Australia has four major natural regions: the humid eastern highlands, the

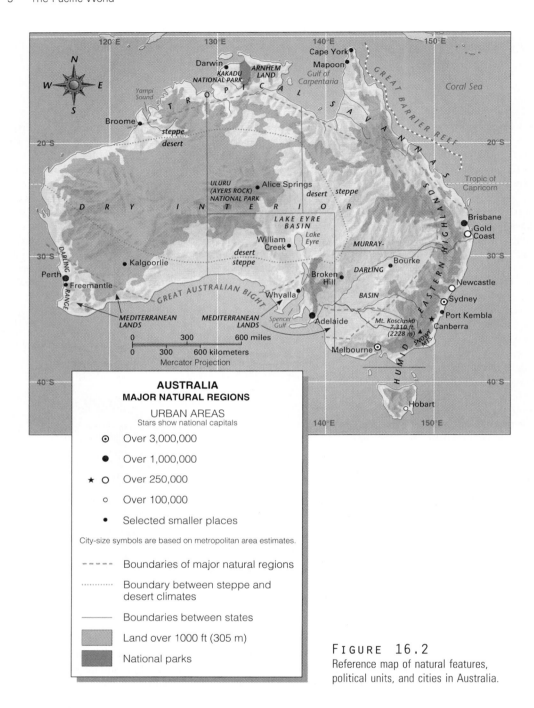

FIGURE 16.2
Reference map of natural features, political units, and cities in Australia.

Within the map:

AUSTRALIA
MAJOR NATURAL REGIONS

URBAN AREAS
Stars show national capitals

⊙   Over 3,000,000

●   Over 1,000,000

★ ○   Over 250,000

○   Over 100,000

•   Selected smaller places

City-size symbols are based on metropolitan area estimates.

- - - - -   Boundaries of major natural regions

··········   Boundary between steppe and desert climates

———   Boundaries between states

    Land over 1000 ft (305 m)

    National parks

tropical savannas of northern Australia, the "Mediterranean" lands of southwestern and southern Australia, and the dry interior (see Fig. 2.6, page 3, and Fig. 16.2).

THE HUMID EASTERN HIGHLANDS

Australia's only major highlands extend along the east coast from just north of the Tropic of Capricorn to southern Tasmania in a belt 100 to 250 miles wide (161–410 km). Although com-

plex in form and often rugged, these highlands seldom reach elevations of 3000 feet (914 m). Their highest summit and the highest point in Australia is Mount Kosciusko, which rises only 7310 feet (2228 m). The highlands and the narrow and fragmented coastal plains at their base constitute the only part of Australia that does not experience a marked period of drought each year. However, although onshore winds from the Pacific bring rain each month, the strong relief reduces the amount of agricultural land in this most favored of Australia's climatic areas. South of the city of Sydney and at higher elevations to the

FIGURE 16.3
Large parts of Tasmania are rugged and wild. This is a winter scene in the alpine habitat of Mount Field National Park. *Martin Withers/Dembinsky Photo Associates*

north, the climate is marine west coast, despite the location. North of Sydney higher summer temperatures change the classification to humid subtropical, while still farther north, beyond approximately the parallel of 20°S, hotter temperatures and greater seasonality of rain cause subhumid conditions.

Tasmania is a rugged, beautiful island off Australia's southeast coast (Fig. 16.3). About 7 percent of the land is dedicated to national parks and other protected areas. The state's western region is heavily forested and dotted with lakes. Numerous rivers cut rapid, short courses to the sea, representing potential sources of hydroelectric power. At this latitude—roughly comparable to that of the U.S. state of Wyoming—snow sometimes falls on the high peaks even in the height of summer (November–January).

TROPICAL SAVANNAS OF NORTHERN AUSTRALIA

Northern Australia, from near Broome on the Indian Ocean to the coast of the Coral Sea, receives heavy rainfall during the season locals call "the wet"—the Southern Hemisphere summer season. The coastal zone is subjected in summer to hurricanes, known as cyclones in Australia; on Christmas Eve in 1974, a powerful typhoon flattened the city of Darwin (which has since been rebuilt). After such deluges, northern Australia experiences almost complete drought during the winter, which is six months or more of the year. This highly seasonal distribution of rainfall is the result of monsoonal winds that blow onshore during the summer and offshore during the winter. The seasonality of the rainfall, combined with the tropical heat of the area, has produced a savanna vegetation of coarse grasses with scattered trees and patches of woodland (Figure 16.4). The long season of drought, the poverty of the soils, and the lack of

highlands which would nourish large perennial streams all combine to reduce agricultural possibilities. The alluvial and volcanic soils that support large populations in some tropical areas are almost completely absent in northern Australia.

"MEDITERRANEAN" LANDS OF THE SOUTH AND SOUTHWEST

The southwestern corner of Australia and the lands around Spencer Gulf have a mediterranean or dry-summer subtropical type of climate, with subtropical temperatures, winter rain, and summer drought. In winter, the Southern Hemisphere belt of the westerly winds shifts far enough north to bring precipitation to these districts, while in summer this belt lies offshore to the south and the land is dry. Crops introduced from the mediterranean-climate lands of Europe generally do well in these parts of Australia, but the shortage of highlands to catch moisture and supply irrigation water to the lowlands limits agricultural possibilities.

THE DRY INTERIOR

The huge interior of Australia is desert, surrounded by a broad fringe of semiarid grassland (steppe) that is transitional to the more humid areas around the edges of the continent. This is the "outback" that has lent so much to Australian life and lore. Locals know it as "the bush," the "back of beyond," and the "never-never." Altogether, the interior desert and steppe cover more than one half of the continent and extend to the coast in the northwest and along the Great Australian Bight in the south. This tremendous area of arid and semiarid land is too far south to get much rain from the summer monsoon, is too far north to

FIGURE 16.4
Kakadu National Park in Australia's Northern Territory. This gentle landscape receives torrential rains during "the wet," and is home to Aboriginal people and a rich variety of wildlife. *Martin Withers/Dembinsky Photo Associates*

Since then, the rate of immigration has slowed, but immigration continues and has been opened considerably to skilled non-whites. Half of the immigrants to Australia in 1995 came from Asian countries. An expanding manufacturing sector has provided many of the jobs for new Australians. Meanwhile, the country's very efficient agricultural and mining sectors provide the basic support for a relatively affluent population employed mostly in urban services and manufacturing.

Tourism has recently become an important industry in Australia, particularly since the worldwide acclaim of the 1986 film *Crocodile Dundee*. (Crocodiles themselves occupy an important place in Australia's tourist business. In Kakadu National Park and other sections of northern and western Australia, many tours promise visitors a closeup look at "salties," the great saltwater crocodiles that are ancient and often fatal neighbors of people in the region.) Australia's attractions are sufficiently diverse to draw many different styles of tourists to various destinations on the continent. The Great Barrier Reef,

located off the northeast coast, is one of the world's most prized scuba-diving areas. Hotels and other services on nearby Queensland beaches of the "Gold Coast" cater to those looking for an idyllic, tropical, restful holiday. The geological features of Uluru (Ayers Rock) and the nearby Olgas in the country's center are on the itinerary of most international tourists. For many visitors, Sydney is as lovely a port city as San Francisco and Venice. Adventurous tourists raft the whitewaters and hike the mountains of Tasmania.

16.4   New Zealand, Pastoral and Urban

Located more than 1000 miles (*c.* 1600 km) southeast of Australia, New Zealand consists of two large islands, North Island and South Island (separated by Cook Strait) and a number of smaller islands (Fig. 16.11). North Island is smaller than

FIGURE 16.11
Index map of New Zealand.

South Island but contains the majority of the population, which numbered about 3.6 million in mid-1996. Since the total area is approximately 104,000 square miles (269,000 sq km), the population density averages just 35 per square mile (14 per sq km). New Zealand is thus a sparsely populated country, although not so much so as Australia.

ENVIRONMENT AND RURAL LIVELIHOOD

Much of New Zealand is rugged. The Southern Alps, often described as one of the world's most spectacular mountain ranges, dominate South Island (Fig. 16.12). These mountains rise above 5000 feet (*c.* 1500 m), with many glaciated summits over 10,000 feet, and in the southwest are deeply cut by coastal fjords. The mountains of North Island are less imposing and extensive, but many peaks exceed 5000 feet. New Zealand's highlands lie in the Southern Hemisphere belt of westerly winds and receive abundant precipitation. While lowlands generally receive more than 30 inches (76 cm), distributed fairly evenly throughout the year, highlands often receive more than 130 inches (330 cm). Precipitation drops to less than 20 inches (51 cm) in small areas of rain shadow east of the Southern Alps. Temperatures typical of the middle latitudes are moderated by a pervasive maritime influence. The result is a marine west coast climate, with warm-month temperatures generally averaging 60° to 70°F (16° to 21°C) and cool-month temperatures of 40° to 50°F (4° to 10°C). Highland temperatures are more severe, and a few glaciers flow on both islands.

Rugged terrain and heavy precipitation in the Southern Alps and the mountainous core of North Island explain an almost total absence of people from one third of New Zealand. About one fifth of the country is completely unproductive, except for the appeal of mountains to an expanding tourist industry. New Zealand's attractions for visitors include golfing, fishing, hunting, hiking, and camping in spectacular mountainous national parks, and geyser watching around the Rotorua thermal area of North Island.

Well-populated areas are restricted to fringing lowlands around the periphery of North Island and along the drier east and south coasts of South Island. The climate of New Zealand's lowlands is ideal for growing grass and raising livestock. More than one half of the country's total area is in pastures and meadows that support a major sheep and cattle industry (Fig. 16.13). Pastoral industries contribute greatly to the country's export trade. Meat, wool, dairy products, and hides together account for about one half of what New Zealand sells abroad.

The earning power of these pastoral exports has given New Zealand a moderately high level of living (per capita GNP in 1994 of $13,190, comparable to that of Ireland). The United Kingdom was the main market until recently. When the U.K. joined the European Common Market in the 1970s, however, a new situation emerged because the Market (now the European Union) includes countries with surpluses competitive with New Zealand's. In response, New Zealand has tried to foster industrial growth and to expand trade with other partners. These efforts have not been entirely successful. The economy suffered during the 1970s and 1980s, and the country slipped downward in the ranks of the world's affluent countries. A sizable emigration took place, mainly to Australia. In addition to the economic troubles, the emigrants complained of boredom in an isolated, placid society in which tax policies made it difficult for almost anyone to make or keep a high income.

FIGURE 16.12
The Southern Alps tower over pastoral landscapes on New Zealand's South Island. *Paul Chesley/Tony Stone Images*

Dairy cattle on the lush pasture associated with New Zealand's cool and humid climate. The view is from a locality in the North Island not far from New Zealand's world-famous district of hot springs and geysers.
Thomas Brown

INDUSTRIAL AND URBAN DEVELOPMENT

In New Zealand, as in Australia, a high degree of urbanization and active attempts to develop manufacturing supplement the basic dependence on pastoral industries (Fig. 16.14). The two countries are similar in their conditions and purposes of urban and industrial development. New Zealand's resources for manufacturing do not equal those of Australia, but coal, iron, and other minerals are present in modest quantities. There is much potential for hydropower development. New Zealand's magnif-

icent natural forests enjoy excellent growing conditions, and production and exports of forest products have become increasingly important as the country attempts to reduce its dependence on sheep and dairy exports.

New Zealand is not as urbanized as Australia. The country in 1996 had only six urban areas with estimated metropolitan populations of more than 100,000: Auckland (population: 315,668, city proper; 855,571, metropolitan area) and the capital of Wellington (population: 150,301, city proper; 375,000, metropolitan area) at opposite ends of North Island; Napier-Hastings (population: 110,000, city proper) on the southeastern

FIGURE 16.14

About 85 percent of the people of New Zealand live in cities. This is Queenstown, on Lake Wakatipu in South Island. *Chad Ehlers/Tony Stone Images*

coast of North Island; Hamilton (population: 110,000, city proper) not far south of Auckland; and Christchurch (population: 335,000, city proper) and Dunedin (population: 115,000, city proper) on the drier east coast of South Island. Christchurch is the main urban center of the Canterbury Plains, which contain the largest concentration of cultivated land in New Zealand. About one half of the country's population lives in these six modest-sized metropolises. By comparison, the Sydney metropolitan area in Australia has approximately the same population as all of New Zealand.

Africa South of the Sahara

6

A Huge, Complex Mosaic of Environments, Peoples, Cultures, and Political Systems

Stereotyped and Misunderstood, with an Accomplished Past

A Plateau of Plateaus, with Several of the World's Great Rivers and Lakes

Mainly Tropical, but with a Midlatitude South and Frequent Drought Conditions

Wilderness Areas and Wildlife in Danger

Long Exploited and Colonized by Foreign Interests

A Supplier of Raw Materials (especially Minerals) and Cash Crops for Foreign Markets

Strongly Rural and Agricultural

South Africa is Most Developed Economically

Now Postcolonial, Politically Fragmented, and Unstable, with Emerging Democratic Systems

Unevenly Populated, with Low Overall Density but Rapid Growth

The World's Most Impoverished Region, with Poor Prospects

A Geographic Profile of Africa South of the Sahara

Geographers typically recognize that the continent of Africa has two major divisions: North Africa—the predominantly Arab and Berber realm of the continent—and culturally complex Africa south of the Sahara, often called sub-Saharan or Black Africa. This text regards North Africa as part of the greater Middle East (see Chapters 8 and 9). This chapter sketches the broad outlines of the geography of Africa south of the Sahara and Chapter 18 examines the major African subregions (Figure 17.1): West Africa, the Sahel, the Horn of Africa, East Africa, Southern Africa, West Central Africa, and the Indian Ocean islands. Area and population data are in Table 17.1.

The region of Africa south of the Sahara is culturally complex, physically beautiful, and problem-ridden. In presenting an introductory perspective, we focus on the region's diverse environments, peoples and modes of life, major population concentrations, European colonial legacies, and major current problems.

17.1   Area and Population

Africa south of the Sahara (including Madagascar and other Indian Ocean islands) is the largest in land area of all the major world regions discussed in this book. Its 8 million square miles (21.8 million sq km) make it more than twice the size of the United States. "People overpopulation" is apparent in many areas, and yet much of the region is sparsely populated. With a population of 568 million as of mid-1996, the region's average population density is substantially less than that of the United States. However, this density gap will close rapidly; the current rate of natural population increase in Africa south of the Sahara is 2.7 percent per year, or about four times that of the United States.

The relatively low population density of Africa south of the Sahara as a whole obscures the fact that a majority of this region's people live in a small number of densely populated areas that together occupy a small share of the region's total area. The

◀ Earth from space. *NASA*

TABLE 17.1

Africa: Basic Data

Political Unit	Area (Thousand sq mi)	Area (Thousand sq km)	Estimated Population (millions, mid-1996)	Estimated Annual Rate of Natural Increase (% 1996)
The Horn of Africa				
Ethiopia	437.8	1133.9	57.2	3.1
Eritrea	45.3	117.4	3.6	2.8
Somalia	246.2	637.7	9.5	3.2
Djibouti	8.5	22.0	0.6	2.2
Totals[a]	737.8	1911	70.9	2.8 (mean)
The Sahel				
Mauritania	397.9	1030.6	2.3	2.5
Senegal	75.7	196.1	8.5	2.7
Mali	478.8	1240.0	9.7	3.1
Niger	489.2	1267.0	9.5	3.4
Burkina Faso	105.9	274.1	10.6	2.8
Gambia	4.4	11.3	1.2	2.7
Cape Verde	1.6	4.0	0.4	1.8
Chad	495.8	1284.1	6.5	2.6
Totals	1618.3	5307.2	48.7	2.7 (mean)
West Africa				
Guinea	94.9	245.8	7.4	2.4
Ivory Coast (Côte d'Ivoire)	124.5	322.3	14.7	3.5
Togo	21.9	56.8	4.6	2.7
Benin	43.5	112.6	5.6	3.1
Sierra Leone	27.7	71.7	4.6	2.7
Ghana	92.1	238.4	18.0	3.0
Nigeria	356.7	923.8	103.9	3.1
Guinea-Bissau	13.9	36.1	1.1	2.1
Liberia	43.0	111.3	2.1	3.1
Totals	818.2	2118.8	162	2.9 (mean)
West Central Africa				
Cameroon	183.6	475.3	13.6	2.9
Gabon	103.3	267.6	1.2	1.5
Republic of the Congo (former Congo)	132.0	341.9	2.5	2.3
Central African Republic	240.5	622.7	3.3	2.5
Democratic Republic of Congo (former Zaire)	905.6	2345.5	46.5	3.2
Equatorial Guinea	10.8	28.0	0.4	2.6
São Tomé and Principe	0.4	1.0	0.1	2.6
Totals	1576.2	4082	67.6	2.5 (mean)
East Africa				
Kenya	225.0	582.7	28.2	2.7
Tanzania	364.9	945.1	29.1	3.0
Uganda	91.1	235.9	22.0	3.3
Rwanda	10.2	26.3	6.9	2.7
Burundi	10.7	27.8	5.9	3.0
Totals	702.7	1817.8	92.1	2.9 (mean)
Southern Africa				
Zambia	290.6	752.6	9.2	3.0
Malawi	45.7	118.5	9.5	3.0
Zimbabwe	150.9	390.8	11.5	2.5
Angola	481.4	1246.7	11.5	2.5
Mozambique	313.7	812.4	16.5	2.7
South Africa	473.3	1225.8	44.5	2.3
Namibia	318.1	824.0	1.6	2.7
Botswana	224.6	581.7	1.5	2.7
Lesotho	11.7	30.4	2.1	2.6
Swaziland	6.7	17.4	1.0	3.2
Totals	2316.7	6000.3	108.9	2.7 (mean)
Indian Ocean Islands				
Madagascar	226.7	587.0	15.2	3.2
Mauritius	0.8	2.0	1.1	1.3
Comoros	0.7	1.9	0.6	3.6
Mayotte	0.14	0.4	0.1	4.9 (1992)
Réunion	1.0	2.5	0.7	1.6
Seychelles	0.2	0.5	0.1	1.5
Totals	229.54	594.3	17.8	2.7 (mean)
Grand Totals	7999.44	21831.4	568	2.74

Sources: Population Reference Bureau, *World Population Data Sheet; Britannica Book of the Year 1996.*
[a]Area totals do not always add exactly because of disputed territories and estimates.

Estimated Population Density, 1996	Estimated Population Density, 1996	Infant Mortality Rate	Urban Population (%)	Cultivated Land (% of total area)	Per Capita GNP ($U.S.: 1994)
Per sq mi	Per sq km				
126	49	120	15	13	130
78	30	105	17	4	115
27	11	122	24	2	150
65	25	115	77	1000 acres (400 hectares only)	780
96	37	116	16	9	294 (mean)
6	2	101	39	0.2	480
109	42	68	43	12	610
19	7	106	26	2	250
20	7	123	15	3	230
97	38	94	15	13	300
335	130	90	26	18	360
252	97	65	44	11	910
13	5	122	22	3	190
30	9	96 (mean)	35	12	416 (mean)
71	27	139	29	3	510
115	44	88	46	12	510
189	73	143	35	45	320
124	48	86	36	17	370
163	63	143	35	8	150
179	69	66	36	19	430
268	103	87	16	36	280
77	30	140	22	12	240
62	24	113	44	4	498 (1990)
197	76	112 (mean)	24	23	368 (mean)
74	29	65	41	15	680
11	4	95	73	2	3550
20	8	109	58	1	640
13	5	97	39	3	370
49	19	108	29	4	220 (1991)
37	14	103	37	8	430
339	131	50.8	46	39	250
43	17	89.7 (mean)	34	5	877 (mean)
127	49	62	27	8	260
82	32	92	21	4	90
245	95	115	11	34	200
659	254	110	5	47	210
593	229	102	6	53	150
136	51	96 (mean)	18	11	182 (mean)
33	13	107	42	7	350
273	105	134	17	18	140
75	29	53	31	7	490
24	9	137	32	3	620 (1989)
58	22	148	33	4	80
88	34	46	57	11	3010
5	2	57	32	1	2030
7	3	41	46	1	2800
175	68	79	16	11	700
137	53	93	30	11	1160
47	18	89.5	42	5	1138 (mean)
65	25	93	26	5	230
1431	553	18.1	44	52	3180
758	293	80	29	45	510
722	278	NA	NA	NA	NA
680	263	8	73	19	3080 (1990)
426	165	12.9	50	16	6210
78	30	42.4 (mean)	29	5	1524 (mean)
71	26	90.9	32.6	9	8007

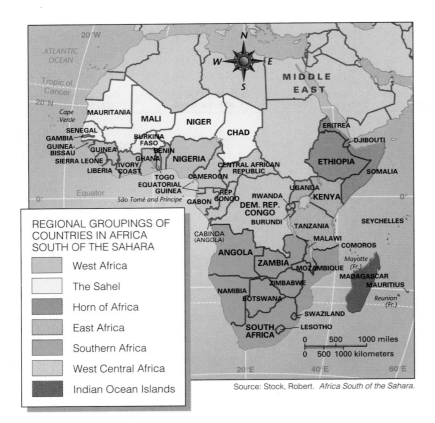

REGIONAL GROUPINGS OF
COUNTRIES IN AFRICA
SOUTH OF THE SAHARA

- West Africa
- The Sahel
- Horn of Africa
- East Africa
- Southern Africa
- West Central Africa
- Indian Ocean Islands

Source: Stock, Robert. *Africa South of the Sahara.*

FIGURE 17.1
Informal regional groupings of countries south of the Sahara.

principal areas of population are (1) the coastal belt bordering the Gulf of Guinea in West Africa, from the southern part of Africa's most populous country, Nigeria, westward to southern Ghana; (2) the savanna lands in the northern third of Nigeria; (3) the highlands of Ethiopia; (4) the highland region surrounding Lake Victoria in Kenya, Tanzania, Uganda, Rwanda, and Burundi; and (5) the eastern coast and parts of the high interior plateau (High Veld) of the Republic of South Africa (for population distribution, see Fig. 2.17, page 47; for place locations, see Fig. 17.2). Each of these population concentrations has a strong impact on the political and economic geography of Africa. Lesser population concentrations are scattered irregularly through the sparsely inhabited deserts, steppes, and grassy and forested expanses of tropical "bush" that make up most of Africa.

17.2 The African Environment

Africa south of the Sahara is both rich in natural resources and beset with environmental challenges that make economic development difficult. It is home to some of the world's greatest concentrations of wildlife and to some of the most degraded habitats.

LARGE PLATEAUS AND MAJOR RIVERS

Most of Africa is a vast plateau—actually a series of plateaus—rising to varying elevations (see inset, Fig. 17.3). The plateau surfaces are interrupted by prominent river systems such as the Nile, Niger, Congo, Zambezi, and Orange. Lowland plains form a narrow band around the coasts. Inland from the coast, escarpments mark the transition to the plateau surface, which typically lies at an elevation of more than 1000 feet (305 m). Near the Great Rift Valley in the horn of Africa and in southern and eastern Africa (see Definitions and Insights, page 403), the general elevation rises to 2000–3000 feet, with considerable areas at 5000 feet and higher (see inset, Fig. 17.3). The highest peaks and largest lakes of the continent are located in this highland belt. The loftiest summits lie within a 250-mile radius (*c*. 400 km) of Lake Victoria. They include Mount Kilimanjaro (19,340 feet/5895 m) and Mount Kirinyaga (Mount Kenya; Figure 17.4) (17,058 feet/5200 m), which are volcanic cones, and the Ruwenzori Range (16,763 feet/5109 m), a nonvolcanic massif produced by faulting. Lake Victoria, the largest lake in Africa, is surpassed in area among inland waters of the world only by the Caspian Sea and Lake Superior. There are several other large lakes in East Africa, including Lake Tanganyika and Lake Malawi.

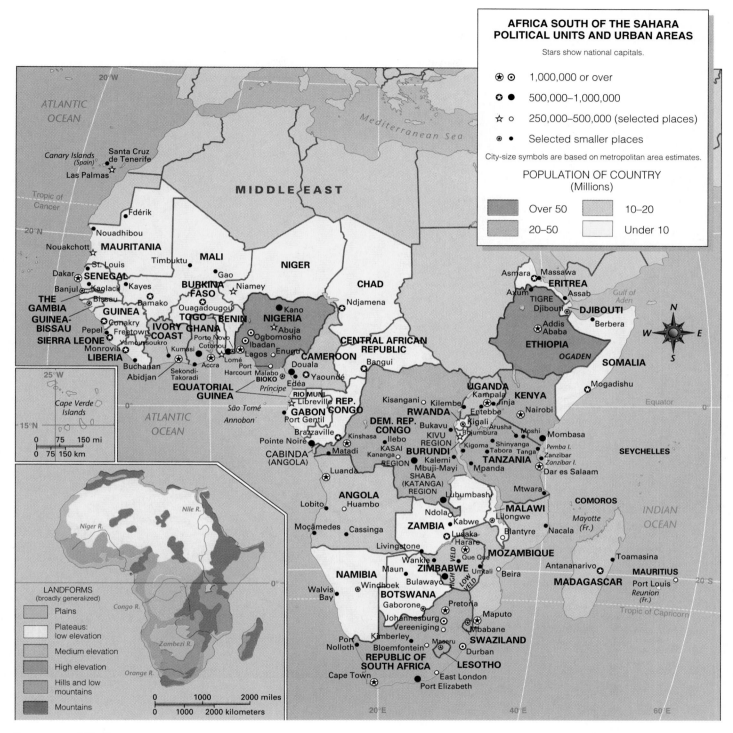

FIGURE 17.2
Reference map of countries, cities, and landforms of Africa south of the Sahara.

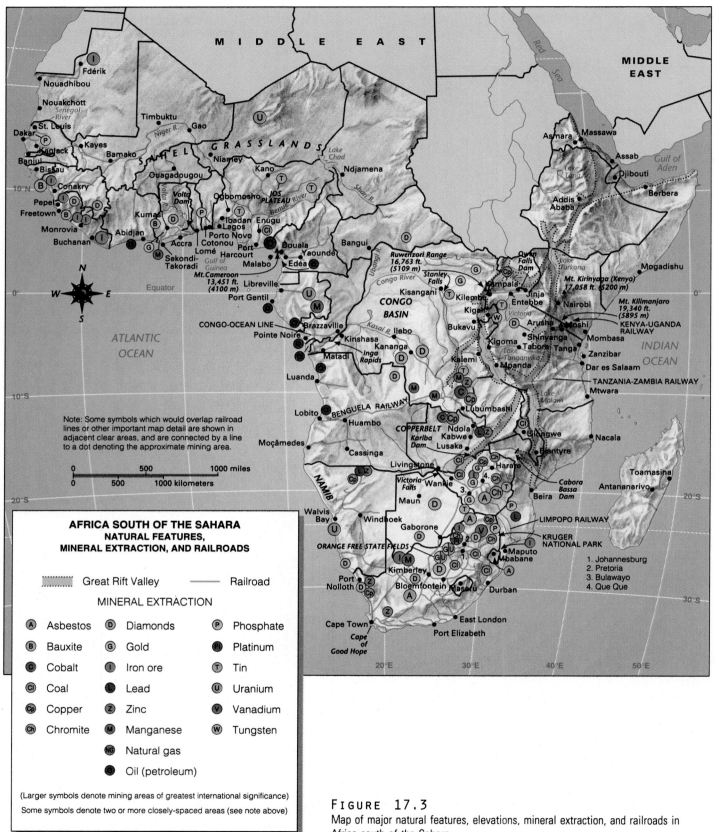

FIGURE 17.3

Map of major natural features, elevations, mineral extraction, and railroads in Africa south of the Sahara.

FIGURE 17.4
Kenya's Mount Kirinyaga (Mt. Kenya: 17,058 ft./5200 m), an extinct volcano, is Africa's second highest mountain.   *Baron Wolman/*
Tony Stone Images

Definitions and Insights

THE GREAT RIFT VALLEY

One of the most spectacular features of Africa's physical geography is the **Great Rift Valley**, a broad, steep-walled trough extending from the Zambezi valley (on the border between Zimbabwe and Zambia) northward to the Red Sea and the valley of the Jordan River in southwestern Asia (see Fig. 17.3). Its relationship to the tectonic movement of crustal plates is still poorly understood. However, most Earth scientists believe it marks the boundary of two crustal plates which are rifting, or tearing apart, causing a central block between two parallel fault lines to be displaced downward, creating a linear valley. This movement will eventually cut much of southern and eastern Africa away from the rest of the continent and allow sea water to fill the valley.

The Great Rift Valley actually has several branches. Lakes, rivers, seas, and gulfs already occupy much of it. It contains most of the larger lakes of Africa, although Lake Victoria, located in a depression between two of its principal arms, is an exception. Some, like the 4823-foot-deep (1470 m) Lake Tanganyika (the world's second deepest lake after Russia's Lake Baikal), are extremely deep. Most have no outlet. Volcanic activity associated with the Great Rift Valley has created Mount Kilimanjaro, Mount Kirinyaga, and some of the other great African peaks, along with lava flows, hot springs, and other thermal features. Faulting along the Great Rift Valley in Ethiopia, Kenya, and Tanzania has also exposed remains of the earliest known ancestors of *Homo sapiens*.

The physical structure of Africa has influenced the character of African rivers. The main rivers rise in interior uplands and descend by stages to the sea. At various points they descend abruptly, particularly at plateau escarpments, so that their courses are interrupted by rapids and waterfalls. These often block navigation a short distance inland. Low water at certain seasons and shallow and shifting delta channels also hinder navigation of many African rivers. Africa's discontinuous inland waterways are interconnected by rail and highways more than on any other continent.

Among the important rivers that have built deltas are the Niger, Zambezi, Limpopo, and Orange. In contrast, the Congo River has scoured a deep estuary 6 to 10 miles (10–16 km) wide, which ocean vessels can navigate to the seaport of Matadi in the Democratic Republic of Congo, about 85 miles (*c.* 135 km) inland. The Congo is used more for transportation than is any other African river.

The frequent falls and rapids of rivers in Africa south of the Sahara have a positive side: They represent a great potential source of hydroelectric energy. Large power stations exist on the Zambezi River at the Cabora Bassa Dam in Mozambique and at the Kariba Dam (Fig. 17.5), which Zimbabwe and Zambia share; at the Kainji Dam on the Niger River in Nigeria; and at the Akosombo Dam on the Volta River in Ghana. Many other power stations of varying size are scattered over the continent. However, only about 5 percent of Africa's hydropower potential has been realized (compared to 59 percent in North America, for example). Many of the best sites are remote from large markets for power. The largest single potential source of hydroelectricity in Africa, and possibly in the world, is the stretch of rapids on the Congo River between Kinshasa and Matadi. The Democratic Republic of Congo's Inga Project is developing this resource.

FIGURE 17.5
The Kariba Dam straddles the Zambezi River between Zimbabwe and Zambia and provides hydroelectric power to both countries.
Michael Busselle/Tony Stone Images

CLIMATE, VEGETATION AND MOISTURE

The equator bisects Africa, so most of the region lies within the low latitudes and has tropical climates. One of the most striking characteristics of Africa's climatic pattern is its symmetry or regularity. This is due mainly to the continent's position astride the equator, coupled with its generally level surface. (The broad pattern of climates is depicted in Figure 2.6 on page 30; see also the biomes map, Fig. 2.7, page 32). Areas of tropical-rain-forest climate center around the great rain forest of the Congo basin in central and western Africa. The forest merges gradually into a tropical savanna climate on the north, south, and east. This is the climatic and biotic zone which supports the famous large mammals of Africa. The savanna areas, in turn, trend into steppe and desert on the north and southwest. A broad belt of drought-prone tropical steppe and savanna bordering the Sahara Desert on the south is known as the Sahel (see Regional Perspective, Chapter 18). There is desert on the coasts of Eritrea, Djibouti, and Somalia in the Horn of Africa. In South Africa and Namibia, a coastal desert, the Namib, borders the Atlantic. The "Kalahari Desert," which lies inland from the Namib, is better described as steppe or semidesert than as true desert. Along the northwestern and southwestern fringes of the continent are relatively small but important areas of mediterranean climate, while eastern coastal sections and adjoining interior areas of South Africa have a humid subtropical climate. High elevations moderate the temperatures of extensive interior areas that lie within the realm of tropical savanna climate in the east and south of the continent.

Total precipitation on the African continent is very large but unevenly distributed. Some parts of Africa receive an overabundance of rain, while other areas have scarcely any. Even in the rainier parts of the continent, large areas have a long dry season, and wide fluctuations occur from year to year in the total amount of precipitation. As a result, one of the major needs in Africa is better control over water. In the typical village household, women carry water by hand from a stream or lake or a shallow (often polluted) well. Use of more small dams would help provide water storage throughout the year, especially in the seasonally rainy areas.

SOILS

Among Africa's most productive lands are alluvial soils on river plains. Other especially fertile soils are found in scattered areas where volcanic parent materials occur, particularly in parts of the East African and Ethiopian highlands. A third group of better-than-average soils are the grassland soils found in some areas of tropical steppe and tropical highland and in the midlatitude grasslands of the High Veld in South Africa. These soils are not entirely comparable to the highly fertile grassland soils of North America as they are more difficult to cultivate and lose their fertility more quickly under continuous cropping.

Soils of the deserts and regions of mediterranean climate are generally poor. True soils are absent over broad areas of desert. The same is true in the mediterranean areas, with the exception of some valleys which have fertile soils where materials transported from adjoining slopes have accumulated. In the tropical rain forests and savannas, reddish tropical soils generally dominate. These are characteristically infertile once the natural vegetation is removed, and can support only shifting agriculture (see discussion on page 413).

ANIMAL LIFE

Africa has the planet's most spectacular and numerous populations of large mammals. However, while film documentaries promote a perception outside Africa that the continent is a vast animal Eden, the reality is less positive. Human population growth, urbanization, and agricultural expansion are taking place in Africa, as elsewhere in the world, at the expense of wildlife. Hunting and competition with domesticated livestock also take their toll. Numbers of many species have now diminished to the point that they are protected by law. Such laws are difficult to enforce, and poaching on a large scale has devastated some species (see Regional Perspective on page 405). However, Africa is still home to some of the world's most extraordinary and successfully managed national parks, such as the Kruger National Park in South Africa and the Ngorogoro Crater National Park in Tanzania.

The tropical grasslands and open forests of Africa are the habitats of the large herbivorous animals—such as the elephant, buffalo, antelope, zebra, and giraffe—and also of carnivorous and scavenging animals, such as the lion, leopard, and hyena. The tropical rain forests have fewer of these "game" animals (as Africans call them); the most abundant species here are insects, birds, and monkeys, with the hippopotamus, the crocodile, and a great variety of fish present in the streams and rivers draining the forests and wetter savannas.

17.3 Cultures of the Continent

Many non-Africans are unaware of the achievements and contributions of the cultures of Africa south of the Sahara. The African continent was the original home of the human race. After about 5000 B.C., indigenous people were responsible for agricultural innovations in four culture hearths: the Ethiopian Plateau, the West African savanna, the West African forest, and the forest-savanna boundary of West Central Africa. Africans in these areas domesticated such important crops as millet, sorghum, yams, cowpeas, okra, watermelons, coffee, and cotton. From Africa, these diffused to agricultures in other world regions.

Civilizations and empires emerged in Ethiopia, West Africa, West Central Africa and Southern Africa. In the first century A.D., the Christian empire based in the northern

Regional Perspective

MANAGING THE GREAT HERBIVORES

Elephants and rhinoceroses are Africa's largest and most endangered herbivores. Their plight has accelerated in recent years, and so have local and international efforts to maintain their populations in the wild. The problem has compelled countries with very different wildlife resources to work together toward solutions.

In 1970, there were about two-and-a-half million African elephants living on the continent. Poaching and habitat destruction reduced their numbers to 1.8 million by 1978. Today (as of 1997), there are an estimated 350,000. The main reason for the sharp decline is that elephants have something people prize: ivory. Poaching for ivory was reducing African elephants at a rate of 10 percent annually when delegates of the 112 signatory nations of the Convention on International Trade in Endangered Species (CITES) met in 1989. The organization succeeded in passing a worldwide ban on the ivory trade, and since then the precipitous decline has halted.

CITES member states won the ivory ban over the strong objections of southern African states, led by Zimbabwe. While the East African nations of Uganda, Kenya, and Tanzania were suffering crashing populations, Zimbabwe was experiencing what it regarded as an elephant overpopulation problem. Zimbabwe had 5000 elephants in 1900. Today there are an estimated 77,000, and they are increasing at a rate of 4 percent annually. There are also healthy and growing elephant populations in Botswana, Malawi, Namibia, and South Africa. Before the worldwide ivory ban, these countries profited from the sustainable harvest and sale of elephant ivory, hides, and meat. At the 1989 CITES meeting, these countries argued that they should not be punished for

their success in protecting the great mammals. They appealed for an exemption from the ivory ban so that they could earn foreign export revenue from a sustainable yield of their elephant populations. The majority of CITES members rejected this appeal, arguing that any loophole in a complete ban would subject elephants everywhere to illegal poaching, resulting in the loss of the African elephant. The elephant-rich countries reluctantly supported this position. Meanwhile, these countries continue to cull (kill) "excess" elephants. Zimbabwean officials argue that their country can support only 45,000 animals. Meat from the cull of about 5000 elephants yearly goes to needy villagers and crocodile farms. This resource helped the country weather the drought and near-famine of 1992.

If elephant ivory is like gold, rhinoceros horns are like diamonds. Men in the Arabian peninsula nation of Yemen prize daggers with rhino horn handles. Although Western scientists deny the medicinal efficacy of powdered rhino horn, traditional medicine in East Asia (particularly in Taiwan, China, and Malaysia) makes wide use of it, including as an aphrodisiac. These demands, and the current black-market value of about $25,000 per horn, have led to a precipitous decline in population of black rhinoceroses in Africa. There were an estimated 65,000 black rhinos in Africa in 1982; poaching had reduced that number to about 2500 in 1997.

Zimbabwe is on the front line in the war to protect the black rhino and appears to be losing. As recently as 1984 there were as many as 2000 of the animals in the country. Having reduced the numbers in countries to the north, poachers turned to rhino-rich Zimbabwe. Since 1984 there has been a steady increase in the numbers of poachers crossing international boundaries

to kill rhinos in Zimbabwe. Most of them come across the Zambezi River from Zambia. Since per capita gross national product in Zambia is only about $350, the prospect of making hundreds or thousands of dollars in a night's work is irresistible to many. Even the order to Zimbabwean wildlife rangers to shoot poachers on sight, in effect since 1985, has not slowed the slaughter. Armed with automatic weapons, poachers have killed more than 1500 rhinos since 1984. In the same period, poachers killed four rangers, while rangers killed more than 150 poachers.

With fewer than 300 rhinos surviving in Zimbabwe, wildlife officials have turned to more desperate measures. In 1991 they began dehorning rhinos to make them unattractive to poachers. To do this, a marksman tranquilizes the animal and two assistants use a chain saw to remove the two horns. It is not a permanent solution; the horn grows back at a rate of three inches yearly, so each animal must be regularly re-dehorned. In 1995, wildlife authorities admitted that the program had failed. Poachers continue to kill the animals, perhaps out of spite, or because they cannot tell in the dark whether or not the prey has horns—or perhaps because they are after even a few inches of horn stump. Zimbabwean wildlife officials are now considering the possibility of opening a legal trade in rhino horns. They would raise rhino herds on state farms and regularly harvest their regrowing horns for sale. South Africa supports this idea of sustainable harvest. Like Zimbabwe, South Africa is sitting on a stockpile of tons of confiscated rhino horns, and would profit greatly from their legalized trade. In addition, with about 900 living animals, South Africa is the last stronghold of the black rhino and does not want to become the next frontline state in the rhinoceros war.

Ethiopian city of Axum controlled the ivory trade from Africa to Arabia. Ethiopian tradition holds that a shrine in Axum still contains the Biblical Ark of the Covenant and the tablets of the Ten Commandments, which disappeared from the Temple in Jerusalem in 586 B.C. Several Islamic empires, including the

Ghani, Mali, and Hausa states, emerged in West Africa between the 9th and 19th centuries. All of these agriculturally-based civilizations controlled major trade routes across the Sahara. They profited from the exchange of slaves, gold, and ostrich feathers for weapons, coins, and cloth from North Africa. Three kingdoms arose between the 14th and 18th centuries in what is the present area of southern Democratic Republic of Congo and northern Angola. These included the Kongo kingdom, which had productive agriculture and was the hub of an interregional trade network for food, metals, and salt. In what is now Zimbabwe, the Karanga kingdom of the 13th to 15th centuries built its capital city at the site known as Great Zimbabwe. Its skilled metalworkers mined and crafted gold, copper, and iron, and merchants traded these metals with faraway India and China.

Beginning in the 16th century, European colonialism began to overshadow and inhibit the evolution of indigenous African civilization. However, the artistic, technical, and entrepreneurial skills of the region's peoples continued to flourish. These traits are today part of a greater African culture which is poorly understood and often stereotyped in the wider world. Surprisingly, in this diverse and often fragmented continent, there are many shared cultural traits which together comprise what one observer describes as "Africanity." The geographer Robert Stock identifies the following eight constituent elements of the African identity:

1. A black skin color.
2. A unique conceptualization of the relationship between people and nature. Indigenous African religions emphasize that spiritual forces are manifested everywhere in the environment, in contrast with the introduced Christian and Muslim faiths, which tend to see nature as separate from God and people as apart from and superior to nature.
3. An identity tied closely to the land, with many people dependent on hunting, herding, and farming. This dependence on the land reinforces the sense of closeness to nature. Africans tend to treat the land as communal rather than individual property.
4. Emphasis on the arts, including sculpture, music, dance and storytelling, as essential to the expression of African identity.
5. A view of Africans as individuals making up links in a continuing "chain of life," in which reverence of ancestors and nurturing of children are virtues. Parents prefer to have many children to keep the chain growing, and strive to educate them in the traditions of the ethnic group.
6. Extended rather than nuclear families, with parents and their children living and interacting with grandparents, cousins, nieces, nephews, and other relatives.
7. Respect for wise and fair authority, with village elders, "big men," and tribal chiefs endowed with powers they are expected to wield to benefit the group.
8. A shared history of colonial occupation that contributes to a unified sense that in the past, Africans were humiliated and oppressed by outsiders.

Despite the common features of "Africanity," however, Africa south of the Sahara is culturally and ethnically diverse. African nations in this region vary greatly in their tribal or ethnic composition. (Politically and socially, Africans traditionally identified themselves by their tribe, recognizing members of the tribe as all those descended in kinship from a single tribal founder, or "eponym." Because "tribe" and "tribalism" have acquired connotations of primitive feuding between hostile rivals, many Africanists now prefer to use the terms "ethnic group" and "ethnicity" in their place.) Some, like Somalia, Lesotho, Swaziland, and Botswana, are very homogeneous, while Tanzania, Cameroon, and Nigeria have hundreds of ethnic groups. Conflict between tribes or ethnic groups is actually rather rare in Africa south of the Sahara, with the notable exceptions of the recent bloodshed between Hutus and Tutsi in Rwanda and Burundi.

There is also great linguistic diversity in Africa south of the Sahara. By one count, the peoples of this region speak more than 1000 languages. Most of the peoples of sub-Saharan Africa belong to one of four broad language groupings:

1. The Niger-Congo language family—the largest—which includes the many West African languages and the roughly 400 Bantu subfamily languages. Most of these are spoken south of the equator.
2. The Afro-Asiatic language family, including Semitic branch languages (such as Arabic and Amharic) and tongues of the Cushitic (such as Oromo and Somali) and Chadic (such as Hausa) branches. People living in the area adjoining the Sahara, from West Africa to the Horn of Africa, speak these languages. Even some of the Niger-Congo languages originating south of the Sahara, such as the Swahili (Kiswahili) tongue spoken widely in East Africa, have borrowed much from Arabic and other languages with roots elsewhere. The prominence of Arabic words in Swahili reflects a long history of Arab seafaring along the Indian Ocean coast of Africa; Swahili in fact means "coastal" in Arabic.
3. The Nilo-Saharan language family of the central Sahel region, the northern region of West Central Africa, and parts of East Africa.
4. The Khoisan languages of the western portion of southern Africa, which the Bushmen (San) and related peoples speak.

17.4   The Geographic Impact of Slavery

Until about a thousand years ago, the cultures of Africa south of the Saharan desert barrier remained largely unknown to the peoples north of the desert. Egyptians, Romans, and Arabs developed contacts with the northern fringes of sub-Saharan Africa, and some trade filtered across the Sahara, but to most outsiders the "Dark Continent" was a self-contained, tribalized land of mystery. Even at the opening of the 20th century, vast areas of interior tropical Africa were still little known to Westerners.

The tragic impetus for growing contact between Africa and the wider world was slavery (Fig. 17.6). Over a period of 12

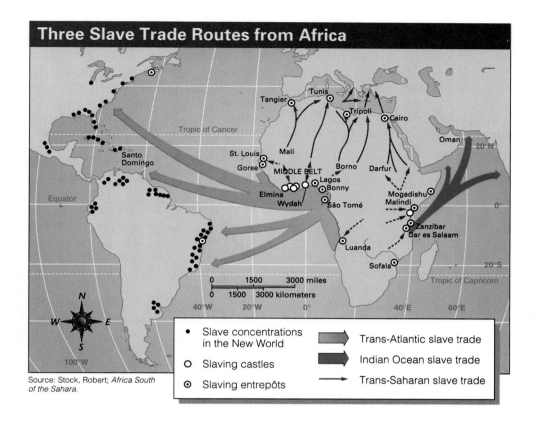

Three Slave Trade Routes from Africa

Source: Stock, Robert; *Africa South of the Sahara*.

- • Slave concentrations in the New World
- ○ Slaving castles
- ◉ Slaving entrepôts

➡ Trans-Atlantic slave trade
➡ Indian Ocean slave trade
→ Trans-Saharan slave trade

FIGURE 17.6
Slave export trade routes from Africa south of the Sahara.

centuries, as many as 25 million people from Africa south of the Sahara were forced to become slaves, to be exported as merchandise from their homelands. The trade began in the seventh century, with Arab merchants using trans-Saharan camel caravan routes to exchange guns, books, textiles, and beads from North Africa for slaves, gold, and ivory from Africa south of the Sahara. As many as two thirds of the estimated nine-and-a-half million slaves exported between the years 650 and 1900 along this route were young women who became concubines and household servants in North Africa and Turkey. Male slaves usually became soldiers or court attendants. From the 8th to 19th centuries, about five million more slaves were exported from East Africa to Arabia, Oman, Persia (modern Iran), India, and China. Again, most were women who became concubines and servants.

The notorious and lucrative traffic in slaves provided the main early motivation for European commerce along the African coasts, and it inaugurated the long era of European exploitation of Africa for profit and political advantage. The European-controlled slave trade was the largest by far. Between the 16th and 19th centuries, the capture, transport, and sale of slaves was the exclusive preoccupation of trade between the European world and West Africa. Portuguese and Spaniards began the trade in the 15th century, and a century later, English, Danish, Dutch, Swedish, and French slavers were active. The industry boomed with the development of plantations and

mines in the New World. Populations of Native Americans in North and Latin America were insufficient for these industries, so the Europeans turned to Africa as a source of labor (see Chapter 2, page 42). The peak of the trans-Atlantic slave trade was between 1700 and 1870, when about 80 percent of an estimated 10 million slaves made the crossing. In escape attempts, in transit, and in the famines and epidemics that followed slave raids, probably more than 10 million died.

Slaves were a prized commodity in the **triangular trade** linking West Africa with Europe and the Americas. European ships carried guns, alcohol, and manufactured goods to West Africa, exchanging them there for slaves. They then transported the slaves to the Americas, exchanging them for gold, silver, tobacco, sugar, and rum to be carried back to Europe. As "raw material" and as the labor working the mines and plantations of Latin America, the West Indies, and North America, slaves generated much of the wealth that made Europe prosperous and helped spark the Industrial Revolution.

While Europeans carried out the trade, their physical presence was limited to coastal shipping points. Africans were the intermediaries who actually raided inland communities to capture the slaves and assemble them at the coast for transit shipment. West African kingdoms initially acquired their own slaves in the course of waging local wars. As the demand for slaves grew, these kingdoms increasingly went to war for the sole purpose of capturing people for the trade. As the exports grew, so did the practice of Africans keeping African slaves. Even after Great Britain (in 1807) and the other European countries abolished slavery—finally bringing an end to the trans-Atlantic trade in 1870—slavery flourished within Africa. By the end of the century slaves made up half the populations of many African states. Slavery has not yet died out in the region; in Mauritania, some light-skinned Moors still enslave blacks, although the national government has outlawed this practice three times.

17.5   Colonialism

European penetration of the African interior began in 1850 with a series of journeys of exploration. Missionaries like David Livingstone, as well as traders, government officials, and now-famous adventurers and scientific explorers such as James Bruce, Richard Burton, and John Speke, undertook these expeditions. By 1881, when Africans ruled about 90 percent of the region, these exploits had revealed the main outlines of inner African geography, and the European powers began to scramble for colonial territory in the interior. By 1900, only Ethiopia and Liberia had not been colonized. Much of the carving up of

Africa took place at the Conference of Berlin in 1884 and 1885, when the French, British, Germans, Belgians, Portuguese, Italians, and Spanish established their respective spheres of influence in the region. Africa south of the Sahara became a patchwork of European colonies—a status it retained for more than half a century (Fig. 17.7)—and Europeans in these possessions were a privileged social and economic class.

At the outbreak of World War II in 1939, only three countries—South Africa, Egypt, and Liberia—were independent. The United Kingdom, France, Belgium, Italy, Portugal, and Spain controlled the rest. But after the war, mainly in the 1960s and 1970s, a sustained drive for independence changed Africa from a colonial region to one comprising more than one fourth of the world's independent states. This was a peaceful process in most instances, but bloodshed accompanied or followed independence in several countries, including Angola and Democratic Republic of Congo.

European colonization of Africa south of the Sahara produced many of the negative attributes of underdevelopment described in Chapter 2 (see pages 42-43). These included marginalization of subsistence farmers, notably those who colonial authorities—intent on cash crop production—displaced from quality soils to inferior land. In addition, European use of indigenous labor to build railways and roads often took a

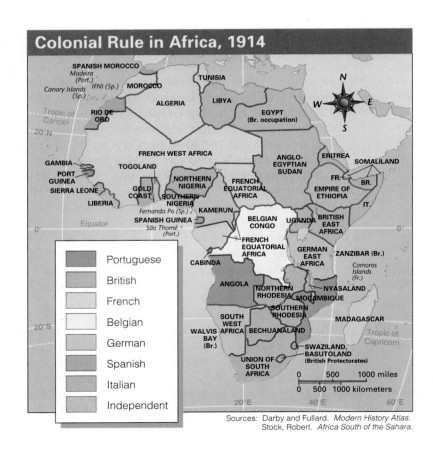

Sources: Darby and Fullard. *Modern History Atlas.*
Stock, Robert. *Africa South of the Sahara.*

FIGURE 17.7

Colonial rule in 1914. Germany lost its colonies after World War I.

high toll in human lives and disrupted countless families. The colonizers also often corrupted traditional systems of political organization to suit their needs, sowing seeds of dissent and interethnic conflict.

The European colonial enterprise did have some positive impact. The colonies, and the independent nations that succeeded them, were the beneficiaries of new cities and the transport links built with forced or cheap African labor; new medical and educational facilities (often developed through Christian missions); new crops and better agricultural techniques; employment and income provided by new mines and modern industries; new governmental institutions; and government-made maps useful for administration and planning. Such innovations were very helpful, but they were distributed unequally from one colony to another, and were inadequate for the needs of modern societies when independence came.

17.6   Shared Traits of African Countries Today

It is impossible in the scope of this book to examine in detail every nation in this complex region. The following generalizations provide an introductory overview, and subsequent sections in Chapter 18 deal with more specific regional and national issues. (The country of South Africa, which has a distinct historic and economic legacy, is an important exception to many of these generalizations, as discussed in Chapter 18.)

1. *Great poverty is characteristic.* Africa south of the Sahara is the most impoverished of the world's regions; 18 of the world's 20 poorest countries are there (see Table 17.1). The region's poverty-related problems include a high incidence of illiteracy, hunger, and disease; inadequate facilities for transportation and communication; and a lack of domestic capital to foster increased agricultural and industrial production. The average African eats 10 percent less than he or she did 20 years ago. Most African societies lack a middle class and the prospect of upward economic mobility. Instead, most are hierarchical, with only a very small (often tribally-based) elite controlling the lion's share of the nation's wealth (see Chapter 2, pages 39–41).

Despite these grim features, in recent decades there have been some changes for the better. Improved standards of health and literacy in many areas have resulted from the work of national and international governmental and nongovernmental agencies. The extension of roads, airways, and other transportation facilities has promoted the marketing of farm products, including perishable items, from formerly inaccessible areas. Stores and markets in both rural and urban areas stock a variety of manufactured goods from overseas and from African sources. Modern factories have been established in many urban centers, and improved agricultural techniques have been introduced in many areas. Such changes have affected some peoples and

FIGURE 17.8
The typical rural dwelling in Africa south of the Sahara is made of locally-available materials and, although having an earthen floor, is kept meticulously clean. This couple's four-room home is in the Kasai region of southern the Democratic Republic of Congo.   *Lovina Ebbe*

areas much more than others, and the total impact of change represents only a beginning in lifting the region from its current state of underdevelopment.

2. *Most people live in rural areas.* The rural areas of individual countries range generally between an estimated 65 and 85 percent; the most rural include Rwanda (95 percent), Burundi (94 percent), Uganda (89 percent), Ethiopia, Niger, and Burkina Faso (85 percent), and Nigeria and Lesotho (84 percent). (See Table 17.1.) Life in villages is the rule, although dispersed homes on individual farms are common in some areas. The most common rural home is a small hut made of sticks and mud, with a dirt floor, a thatched roof, and no electricity or plumbing (Fig. 17.8). In southwestern Nigeria, many farmers live in cities, from which they commute to their fields.

3. *Subsistence agriculture is the main occupation in nearly all countries.* The great majority of farmers in Africa south of the Sahara operate on a subsistence basis, producing so little surplus for sale that they have little cash income and few savings. Women do a large share of the farmwork—they produce 80 to 90 percent of Africa's food—in addition to household chores and the bearing and nurturing of children (Fig. 17.9). Mechanization is rare, fertilizers are expensive, and therefore crop yields are low. Storage facilities are often so poor that pests and bad weather ruin much of the harvest. Subsistence agriculture is discussed in more detail on page 413.

4. *Per capita food output in most countries has declined or has not increased since independence.* The world's most rapid rates of population increase, government policies unfavorable to agriculture, and damage from warfare and drought have been among the factors thwarting efforts to maintain self-sufficiency in food production. Substantial imports of food are now common in the majority of coun-

FIGURE 17.9
Women with headloads in the Shaba region of southeastern Democratic Republic of Congo. Much of tropical Africa's produce moves to local markets in this manner. In the background at the left is an oil palm. The climate is tropical savanna.   *Katrinka Ebbe*

tries. A traditional emphasis on large families threatens to nullify any gains in food production that could be achieved. As in most LDCs, African parents want to have large families for several reasons: so that they will have extra hands to work the fields; so that they will be looked after when they are old or sick; and, in the case of girls, so that they will receive the "bride wealth" a groom pays in a marriage settlement (see Chapter 2, page 45). Large families also convey status.

The Malthusian scenario seems constantly to loom large over Africa (see page 50). Analysts fear the consequences of what they call the "one percent gap": Since the 1960s, the population of Africa south of the Sahara has grown at a rate of about 3 percent annually, while food production in the region has grown at only about 2 percent per year. Providing an adequate domestic food supply has become increasingly difficult because of prolonged droughts, rapid population increases, and unwise land-use practices brought on by population pressure. Inadequate transportation and maintenance and the high cost of imported oil, fertilizer, food, and other necessities reduce food output. Governments have contributed to food shortages by favoring industry over agriculture and export crops over food crops (see Chapter 2, page 42). Many regimes have propelled food shortages into full-blown famines by investing in domestic or international warfare rather than in getting food to people.

5. *Drought is a persistent problem in most countries.* Several countries include deserts where rain falls infrequently, but annual dry seasons are the rule even in the more humid ma-

jority of Africa south of the Sahara. Although all droughts create problems, the most devastating effects in this heavily agricultural part of the world occur when the normal rainy season fails to come for a year or more. This condition has been very widespread in the region since the late 1960s. One area of repeated drought and human suffering is the Sahel, the east-west belt of dry grass and shrub lands south of the Sahara Desert (see Regional Perspective, Chapter 18, and Fig. 17.3, page 402). In the early 1990s, severe drought afflicted South Africa, Zimbabwe, and several neighboring countries in southern Africa.

6. *Lack of education hinders development.* In only about one third of the countries is more than one half the population reported to be literate. In Mauritania, literacy in adults (over age 15) was estimated at 38 percent in 1995; in Burkina Faso, that figure was 35 percent. In Burkina Faso, as in many countries of the region, the literacy rate for women is only about half that for men. Authorities on population and development concur that much higher literacy rates for women will be needed to begin to bring down the high population growth rates that underlie many problems (see Chapter 2, p. 53). Lack of schooling for women in much of Africa south of the Sahara is a formidable barrier to improved family health and the creation of a more skilled labor force. All countries except South Africa are short of skilled workers, particularly workers with administrative and managerial skills. In most countries, college graduates were few at the time of independence, and more schools of higher education are needed.

7. *Poor transportation hinders development* (Fig. 17.10). Since independence, the new nations have been able to

FIGURE 17.10
Many African river crossings have no bridges and it is necessary to use ferries, although some streams can be forded during the dry season. This photo shows vehicles waiting for the ferry at a river crossing in the Shaba region of southeastern Democratic Republic of Congo. Backup of traffic at these bottlenecks may take hours or even days to clear.   *Katrinka Ebbe*

build relatively few surfaced roads or railroads. Those left over from colonial days have often deteriorated. Overland transportation is both inadequate in extent and plagued by severe maintenance problems affecting both vehicles and routes. Some of the reasons include tropical heat, high humidity, dust, prevalence of unpaved roads, lack of lubricants and spare parts, low levels of technical skill, and governments lacking a sense of public service. Africa south of the Sahara critically needs a good international transportation network, coupled with a lowering of trade barriers, in order to create market opportunities on a vastly enlarged scale.

8. *There are serious public health problems.* A high incidence of disease and parasites affecting people, domesticated animals, and cultivated crops has been one of the main hindrances to African development. Insects carry many of the major diseases; mosquitoes, for instance, carry malaria, yellow fever, and dengue, or "breakbone fever." The tsetse fly carries sleeping sickness and nagana, a destructive disease affecting cattle and horses (see Definitions and Insights, page 414). Large numbers of Africans are afflicted by digestive diseases and parasites, including dysentery, typhoid and paratyphoid fever, bilharziasis (schistosomiasis), hookworm, and other types of intestinal worms. Contaminated water and other unsanitary conditions are largely responsible for such afflictions. Other diseases common in Africa include tuberculosis, filariasis, nutritional deficiency diseases, pneumonia, yaws, leprosy, influenza, trachoma, venereal diseases, and many fungoid diseases of the skin. Limited and localized outbreaks of the fatal Ebola virus in West Africa and West Central Africa in 1995 caused great concern in the international medical community.

Recently AIDS (acquired immunodeficiency syndrome), which apparently originated in Africa, has spread so extensively in Africa that it is a new and major plague. The World Health Organization estimates that Africa has almost two thirds of the world's AIDS cases. An estimated 9 percent of Uganda's entire population, and 30 percent of all Ugandan adults, carry the virus (HIV) which causes AIDS. About 30 percent of Rwanda's adults are also infected. Overall, by 1995 an estimated 10 million Africans had been diagnosed with AIDS, and two million had died.

Despite the many health problems, there has been much progress recently in combating diseases such as river blindness in Africa. Enough is known about the control of tropical diseases and parasites to greatly reduce their incidence if means are available for the technical knowledge to be fully applied. But in vast, poverty-stricken Africa, the need for medical assistance far outruns the money and personnel available for such assistance.

9. *The national economies of all countries except South Africa are underindustrialized and overly dependent on the export of a few primary products*, particularly minerals and cash crops. Dependency theorists often point to Africa as a prime example of how colonialism created lasting disadvantages for the colonized. Africa's place in the commercial world is mainly that of a producer of foods and raw materials for sale outside the region. In most nations, one or two products supply more than two fifths of all exports—for example, coffee and tea in Kenya (Fig. 17.11). Each country is therefore vulnerable to international oversupply of an export on which it is vitally dependent (see Chapter 2, pages 42–43). The value of imports generally exceeds that of exports in Africa south of the Sahara, with imports consisting overwhelmingly of manufactured goods, oil products, and/or food. Cash crop and mineral exports are described in more detail on page 414.

10. *Almost all countries are heavily in debt to foreign lenders.* Many billions of dollars in outside grants and loans during the postcolonial era have failed to eliminate poverty in the

FIGURE 17.11

Tea pickers at work in the central highlands of Kenya. Cash crops and other raw materials are typical exports of Africa south of the Sahara.
M.P. Kahl/Photo Researchers

region. Instead, the debts which African governments owe to international lenders such as the International Monetary Fund (IMF), the World Bank, and private banks often compound economic woes and contribute to destructive resource uses (see Chapter 2, pages 42–43).

Much of the indebtedness dates to the late 1970s, when large amounts of loan capital became available to lending institutions in the form of "petrodollars" recycled from Middle Eastern oil states. Nations undertook many costly development projects with borrowed money. Western financiers, planners, and contractors gave optimistic assessments of the benefits to be expected from such projects, and African leaders were ready to accept loans as a way to reap quick benefits from newly won independence. Now, however, many countries are having great difficulty in meeting even the interest payments on their debts. The IMF and other lenders have resisted requests of African debtor nations that these institutions reschedule interest payments and advance new loans. As a condition of further support, the lenders often demand that African debtors put their finances in better order by such measures as revaluating national currencies downward to make exports more attractive, increasing the prices paid to farmers for their products, and reducing corruption, especially among government officials and urban elites. Although many African debtor nations have attempted to conform to such demands, internal political factors slow progress. Loss of economic privileges angers urban elites on whom governments depend for support. Relaxation of price controls on food to stimulate production is often met by rioting among city-dwellers who have tight budgets already allocated for other costs of living.

11. *Economic and humanitarian assistance to the region has slowed since the end of the Cold War.* During the Cold War years, the Soviet Union and United States and their respective allies extended aid generously to many African nations to boost their competing interests. The ending of the Cold War has constricted these aid pipelines; the U.S., for example, cut its economic assistance to Africa south of the Sahara by 30 percent between 1985 and 1992. The bitter experience of the United States in Somalia in the early 1990s also has helped to suppress the West's appetite to extend humanitarian assistance. After 1984 and 1985, when television images of famine-wracked Ethiopia prompted American and Western European citizens to give generously, popular sympathy and support during a series of subsequent African emergencies waned. This lack of apparent public and official interest in crises abroad is known as **donor fatigue**.

12. *Authoritarian governments have been the rule since independence, but progress toward democracy is now widespread.* True democracy has been slow to take root in the region. Military governments and one-party states (often dominated by a single ethnic group) have been common. In the mid-1990s, military and other special interests seized power, overturned the results of popular elections, or stifled opposition in the Central African Republic, Nigeria, Kenya, Ghana, Cameroon, Ivory Coast, and elsewhere. Some postcolonial regimes have been guilty of atrocities, and human rights violations by governments have been commonplace. In contrast, free and fair elections have brought new hope and better human rights records to Mali, Benin, Tanzania, Zambia, Mozambique, Malawi, and South Africa. Namibia, Botswana, and Zimbabwe have also dealt effectively with many of the challenges of democratization and development.

13. *There is serious political instability in many countries.* This is often based on tribal rivalries and antagonisms of long standing, although many of what appear to be tribally-based differences are actually related to issues of economic class and political representation, as between the Tutsis and Hutus in Rwanda and Burundi (see Chapter 18). Coups, failed coup attempts, political murders, armed rebellions, and civil wars have punctuated the political histories of many countries since independence. There has been large-scale slaughter, often including massacres of unarmed civilians, in a long list of countries including Liberia, Sierra Leone, Nigeria, Ethiopia, Uganda, Mozambique, Angola, Rwanda, Burundi, Democratic Republic of Congo, and Zimbabwe. Many other countries have experienced serious civil disorder. Such conflicts and crises have drained national treasuries, discouraged foreign investment, and sidetracked progress toward nationhood.

14. *A diverse array of political, economic, and social ideas from the West, the Communist bloc, the Muslim world, and Africa south of the Sahara itself has influenced governments since independence.* Western democratic ideas have rubbed shoulders with African ideas of chieftainship, communist ideas of one-party dictatorship, and Muslim ideas of theocratic government. Capitalism has coexisted with socialism, and individualism with collectivism. Different governments have adopted different combinations of such ideas. Various forms of socialism have been particularly widespread, resulting in centrally planned national economies, tight state control over wages and prices, state ownership of major economic enterprises, and attempts at collectivization of agriculture and rural life. Such measures have produced disappointing economic results in most countries, although some socialistic governments have been able to expand their social services. The poor economic performance of socialism has sparked a widespread recent trend toward freer enterprise.

15. *Although formal political colonialism has vanished, most countries still have important links with the colonial powers that formerly controlled them,* and many foreign corporations that operated in colonial days still maintain an important presence (see Chapter 2, page 42). A good example is the Democratic Republic of Congo, where all mineral

deposits and production have been nationalized, but the same Belgian interests that monopolized the mining industry under the colonial regime still carries on mining for the Democratic Republic of Congo government. France, but not Britain, has a long history of postindependence intervention in the political and military affairs of its former African colonies. France is the only ex–colonial power to keep troops in Africa (with the highest numbers in Djibouti, Gabon, and the Central African Republic). France also takes steps to ensure that most of its former colonies trade almost exclusively with France, and it in turn supports national currencies with the French treasury.

17.7   Agricultural and Mineral Wealth

Because most people of the region are subsistence farmers or livestock herders, and as the region's export economy is heavily dependent upon cash cropping and mineral production, these activities are described here in more detail.

SUBSISTENCE AGRICULTURE

Over half of the people of Africa south of the Sahara depend directly on agriculture or pastoralism for a livelihood. In most areas, cultivators and their families grow a large share of the crops and livestock for their own or for local sale; this is subsistence agriculture. In the steppe of the northern Sahel, both rainfall and cultivation are scarce and precarious. The savanna of the southern Sahel, with its greater and more dependable rainfall, is a major area of rainfed cropping despite its long dry season. Unirrigated millet, sorghum, corn (maize), and peanuts are major subsistence crops in the savanna. In the tropical savannas south of the equator, corn is a major subsistence crop in most areas (see Definitions and Insights, page 416), with manioc and millet also widely grown. Corn, manioc, bananas, and yams are major food crops grown in the rain forests for subsistence and local sale. Oil palms provide household cooking oil in these areas.

Many African farmers practice land rotation between crop and fallow years. This practice is commonly known as **shifting cultivation** (Fig. 17.12). Most tropical soils lose their natural fertility quickly when they are cropped, and after two or three years must be rested for several years (often 10 to 15 years and more) before they will again produce a crop. During the fallow period, the land reverts to wild vegetation. The need for fallowing can be lessened or eliminated by fertilization, but most African farmers cannot afford to buy chemical fertilizer, animal manure is generally not available, and there is no tradition of systematic fertilization with human wastes as in China and Japan. Some garden plots adjoining huts (known as the "women's land" because they are cultivated by the housewives) are kept more continuously productive by applications of ashes, house sweepings, and goat, chicken, and human manures. Such

FIGURE 17.12

These remote villages on a ridgetop dirt road grow crops under a system of shifting cultivation. The light-colored patches on the slope at the lower right represent fields currently used. The locale is the Kasai region in southern the Democratic Republic of Congo. Few of the world's people are quite so poor, isolated, and bereft of conveniences as the African families in this view.   *Jesse H. Wheeler, Jr.*

plots grow vegetables, melons, and bananas for household use, whereas the fields tilled in shifting cultivation generally grow staple crops such as corn, millet, and manioc.

People of the village generally work together to clear fresh land and harvest crops. Fire is a major tool in clearing, and the resultant ashes provide temporary fertilization. Farmers cultivate with hoes, and leave large stumps to be left to decay in the fields.

The system of shifting agriculture is widespread in the world's tropics. Although it is not a very productive system, it traditionally has minimized soil erosion because most of the land is not in cultivation at any given time. It does provide a bare living for people too poor to afford fertilizer or farm machinery. Unfortunately, there has been a recent widespread trend in Africa for farmers to reduce the length of the fallow period in the cycle of shifting cultivation, or abandon it altogether. This is a direct result of a surging growth in population: With more mouths to feed, in many places there is simply not enough land to provide the "luxury" of fallow. The tragic consequence is that short-term overuse of the land is eroding the soil and rendering it less able to support agriculture in the long run. When the best available lands are exhausted, farmers are often "marginalized" to places unsuitable for farming, such as semiarid lands or slopes that are too steep to till without the consequence of disastrous erosion (see Chapter 2, page 43). These farmers often drive pastoralists from traditional grazing lands. Confined to smaller areas in which to browse and graze, the nomads' cattle, sheep, and goats often overgraze vegetation and compact the soil. Such dilemmas of land use are the root of chronic proximity to famine in many countries of Africa south of the Sahara. Unwise political leadership is a common and dangerous addition to this volatile mix.

THE IMPORTANCE OF LIVESTOCK

Many peoples in sub-Saharan Africa are pastoral, particularly in the enormous tropical grasslands both north and south of the equator. Herding of sheep and hardy breeds of cattle is particularly important in the Sahel. Although cattle raising is widespread through the savannas, cattle are largely ruled out over extensive sections both north and south of the equator by the disease called nagana, which is carried by the tsetse flies that also transmit sleeping sickness to humans (see Definitions and Insights, below). In Africa's tropical rain forests, tsetse flies are even more prevalent and few cattle are raised, but goats and poultry are common (as they are also in tsetse-frequented savanna areas).

Definitions and Insights

AFRICA'S GREATEST CONSERVATIONIST

Not much larger than the common housefly, the tsetse fly of sub-Saharan Africa packs a wallop. This insect carries two diseases, both known as trypanosomiasis, which are extremely debilitating to people and their domesticated animals. People contract "sleeping sickness" from the fly's bite, while cattle contract nagana. Since the 1950s there have been widespread efforts to eradicate tsetse flies so that people can grow crops and herd animals in fly-infested wilderness areas. Where the efforts have been successful, people have cleared, cultivated, and grazed the land. The results are mixed. While people have been able to feed growing populations in the process of opening up these lands, they have also eliminated important wild resources and in many cases caused erosion, desertification, and salinization of the land. Where the tsetse fly has been eliminated, so has the wilderness. The diminutive tsetse fly thus may be characterized as a **keystone species**— one which affects many other organisms in an ecosystem. The loss of a keystone species—in this case, a fly that keeps out humans and cattle—can have a series of destructive impacts throughout the ecosystem. For its role in maintaining wilderness in Africa south of the Sahara, some wildlife experts know the tsetse fly as "Africa's Greatest Conservationist."

Most Africans who live by tilling the soil also keep some animals, even if only goats and poultry. Among African peoples such as the Maasai of Kenya and Tanzania and the Tutsi (Watusi) of Rwanda and Burundi, livestock not only contribute to daily diet but are an indispensable part of customary social, cultural, and economic arrangements (see Landscapes in Literature, Chapter 18). Cattle are particularly important in this regard, although sheep and goats also play a role. The Maasai are probably the most famous African example of close dependence on cattle (Fig. 17.13). They milk and carefully bleed the animals for each day's food, and tend them with great care. The Maasai give a name to each animal, and herds play a central role in the main Maasai social and economic events through the year.

Zebu cattle represent status and wealth in Madagascar, and livestock owners tend to want higher numbers of the animals, rather than better quality stock, to enhance their standing. The resulting population of zebu on the island is about 10 million. Their forage needs have grave consequences for Madagascar's rain forests and other wild habitats, however. People clear the forests and repeatedly set fire to the cleared lands to provide a flush of green pasture for their livestock, causing a rapid retreat of the island's natural vegetation (Fig. 17.14).

As in Madagascar, cattle represent wealth in many societies of Africa south of the Sahara, and people tend to value them for their numbers rather than their quality. They have traditionally constituted the "bride wealth" that changes hands in the marital arrangements of African tribal societies. While men do not "buy" wives, it is a universal custom for the bridegroom to make a large gift to the bride's family in advance of the marriage; men customarily pay this "bride price" (or "bride wealth") in cattle or other livestock, although they may also pay it in cash or merchandise. The gift signifies that the groom's intentions are serious and that he has financial prospects for supporting the bride. It also compensates the bride's father for the loss of her labor in the household and the fields. Should the marriage break up on account of the wife's misdeeds, the husband is entitled to the return of the bride wealth.

CASH CROPS

The proportion of crops grown for export to overseas destinations and for sale in African urban centers has risen significantly in recent times. This has been partly a result of taxation and other governmental pressures. It has also been due to people's desire for cash with which to purchase food and manufactured goods such as hardware, utensils, bicycles, radios, and clothing. Governments also tend to favor such crops as a means of gaining foreign exchange with which to buy foreign technology, industrial equipment, arms, and consumption items for the elite.

Most export crops are grown on small farms, with some also grown on plantations and estates. Large plantations and estates have never become established to the degree they have in Latin America and Southeast Asia. Political and economic pressures forced many of them out of business during the period of transition from European colonialism to independence. Governments of new independent states have nationalized some white-owned plantations, and many whites were obliged to sell their land to Africans. Now operated by Africans, many of these units provide a very important source of tax revenue and for-

FIGURE 17.13
Cattle are central to the traditional livelihood and cultural system of the Maasai of southern Kenya.   *Renee Lynn/ Tony Stone Images*

eign exchange. (One exception is South Africa, where the most important producers and exporters are the ethnic Europeans, who raise livestock, grains, fruit, and sugarcane. But throughout tropical Africa as a whole, there is a growing trend in export production from small, black-owned farms.) The most valuable export crops are coffee, cacao, cotton, peanuts, and oil palm products. Crops of lesser significance, but very important in some regions, include sisal (grown for its fibers), pyrethrum (used in insecticides), tea, tobacco, rubber, pineapples, bananas, cloves, vanilla, cane sugar, and cashew nuts.

FIGURE 17.14
Zebu cattle are extremely important in the cultures and household economies of rural Madagascar. There are about 10 million cattle in this country of 15 million people.   *Art Wolfe/Tony Stone Images*

Definitions and Insights
CROP AND LIVESTOCK INTRODUCTIONS AS THE "CURSE OF AFRICA"

Many of the important food and export crops of Africa are not native to the continent. For example, corn (maize), manioc, peanuts, cacao, tobacco, and sweet potatoes were introduced from the New World. Cattle, sheep, chickens, and other domesticated animals now characteristic of Africa south of the Sahara also originated outside the region. Such nonnative species are known as **exotic species**, or introduced species.

Some exotic species such as manioc and chickens have adapted well to the environments and human needs in Africa south of the Sahara. Others are more problematic. Known in the human diet as "mealie-meal," corn has become the major staple food for southern Africans. However, the people of southern Africa came to regard corn as more of a scourge than a blessing during the severe regional drought of 1992. Introduced from relatively well-watered Central America, corn has flourished in African regions as long as rains or irrigation water have been sufficient. But the crop failed altogether during the 1992 drought. The situation was particularly critical in Zimbabwe, where the government had just sold its entire stock of corn reserves in an effort to repay international debts. Zimbabwe suddenly became a large importer of corn. Many in the country proclaimed corn to be "the curse of Africa," lamenting that, with its high water requirements, it should never have been allowed to become the principal crop of this drought-prone region. Agricultural specialists argued that native sorghum and cowpeas are much better adapted to African conditions and should be cultivated as the major foods of the future.

There are also growing questions about whether the cow is an appropriate livestock species for Africa. The traditional approach to improving cattle pastoralism in sub-Saharan Africa, and to boosting livestock exports, has focused on the need to develop new strains of grasses that can feed cattle and other nonindigenous strains of livestock, since the native grasses of African savannas suit the wild herbivores but generally not the domesticated livestock. Hence, cattle frequently overgraze their own rangelands and cause their erosion.

A new approach turns this problem around by asking whether it might be more appropriate to commercially develop the indigenous animals already suited to the native fodder and soil conditions. Kenya, Zimbabwe, South Africa, and other countries are now maintaining "game ranches" where antelopes—such as eland, impala, sable, waterbuck, and wildebeest—along with zebras, warthogs, ostriches, crocodiles and even pythons, are bred in semicaptive and captive conditions to be slaughtered for their meat, hides, and other useful properties (Fig. 17.A). Safari hunting and tourism pro-

vide supplemental revenue in some of these operations. So far, these enterprises have demonstrated that indigenous large herbivores produce as much or more protein per area than domestic livestock under the same conditions. In commercial terms, the systems that have incorporated both wild and domestic stock have proven to be more profitable than systems employing only domestic animals. The World Wildlife Fund and other proponents of this "multispecies" system argue that it is an economically and ecologically sustainable option of land use in Africa. They emphasize that ranchers who keep both wild and domesticated animals will be better able economically to weather periods of drought, and will require a much lower capital input than they would expend on cattle or other domesticated livestock alone.

Having convinced many ranchers of the economic value of wildlife, the government of Zimbabwe has begun to enlist peasant farmers in protecting, and thus benefiting from, game animals. In an innovative and so far successful program known as CAMPFIRE (Communal Area Management Program for Indigenous Resources), the government has turned over management of wild animals to local councils, which are permitted to sell quotas of hunting licenses for elephant and other game. Profits from safari hunting return to the local communities to fund a wide range of development projects.

FIGURE 17.A
Oryx antelopes maintained in a semidomesticated state on the Galana game ranch in Kenya. Husbandry of indigenous wildlife may prove to be a viable alternative to keeping cattle and goats, which have many harmful impacts on soil and vegetation in Africa south of the Sahara.   *Sven-Olof Lindbald/Photo Researchers*

MINERALS

The export of minerals has had a particularly strong impact on the physical and social geography of Africa south of the Sahara. The three primary source areas are: South Africa and Namibia, the Democratic Republic of Congo–Zambia–Zimbabwe region, and West Africa, especially the areas near the Atlantic Ocean (see Fig. 17.3). Notable mineral exports from these regions include precious metals and precious stones, ferroalloys, copper, phosphate, uranium, petroleum, and high-grade iron ore, all destined principally for Europe and the United States.

Large multinational corporations, financed initially by investors in Europe or America, do most of the mining in Africa. Mining has attracted far more investment capital to Africa than any other economic activity; money is invested directly in the mines, and a great many of the transportation lines, port facilities, power stations, housing and commercial areas, manufacturing plants, and other elements in the continent's infrastructure have been developed primarily to serve the needs of the mining industry.

Great numbers of workers in the mines are temporary migrants from rural areas, often hundreds of miles away (Fig. 17.15). The recruitment of migrant workers has had important cultural and public health effects. Millions of Africans who work for mining companies have come into contact with Western ideas as well as those of other ethnic groups, and have carried these back to their villages. The growing numbers of radios and televisions appearing in homes across the continent, even in remote villages, also convey an unprecedented wealth of information about the wider world (Fig. 17.16). Unfortunately,

the phenomenon of migrant labor has also helped spread HIV, the virus that causes AIDS. Since most miners are single or married men who spend extended periods away from their wives or girlfriends, it is common that, in their loneliness, they exchange the virus with prostitutes and then return home to their villages, where they spread it still further. The resolution of this dilemma is one of the key challenges facing the huge resource-rich, problem-ridden region of Africa south of the Sahara.

FIGURE 17.16
There is no electricity in this village in Niger, but students receive some education and the entire community obtains news and entertainment from a television powered by photovoltaic cells. *John Chaisson/Gamma Liaison*

18

THE ASSETS AND AFFLICTIONS OF COUNTRIES SOUTH OF THE SAHARA

The countries of Africa south of the Sahara are diverse in size and geographic personality. In this chapter, sketches of the major subregions and their countries—and more in-depth portraits of Ethiopia, the Democratic Republic of Congo, and South Africa—illustrate some of the region's characteristic and unique challenges.

18.1   The Horn of Africa

East of the Sudan section of the Nile basin, a great volcanic plateau rises steeply from the desert. This highland and adjacent areas occupy the greater part of the "Horn of Africa"—named for its prominent protrusion from the continent into the Indian Ocean—and include the countries of Ethiopia, Eritrea, Somalia, and Djibouti (see Fig. 17.1, page 400). Much of the area lies at elevations above 10,000 feet (3050 m), and one peak in Ethiopia reaches 15,158 feet (4620 m). The highland, where the Blue Nile, Atbara, and other Nile tributaries rise, receives its rainfall during the summer half of the year. Temperature conditions vary from tropical to temperate as altitude increases. Crops ranging from bananas, coffee, and dates, to oranges, figs, and temperate fruits, to cereals can be produced without irrigation. Large expanses of upland provide pasture for a variety of livestock, but primarily cattle and sheep.

EUROPEAN IMPERIALISM

European powers seized coastal strips of the Horn of Africa in the latter 19th century (see Fig. 17.8, page 409). Britain was first, with British Somaliland in 1882; then France annexed French Somaliland in 1884; and, finally, Italy asserted dominance over Italian Somaliland and Eritrea in 1889. These areas along the Suez–Red Sea route have long had strategic importance, although their economic importance has been negligible.

Italy attempted to extend its domain from Eritrea over the much more attractive and potentially valuable land of Ethiopia in 1896, but Ethiopian tribesmen annihilated the Italian forces at Aduwa. Forty years later, in 1936, a second attempt was successful. Hopes of developing the country's potential wealth and of using Ethiopia as a location in which to resettle the surplus Italian population were frustrated in World War II, when British Commonwealth forces (primarily South Africans) defeated the Italian forces throughout the region. Ethiopia was restored to independence, and Eritrea was federated with it as an autonomous unit in 1952. Ethiopia incorporated Eritrea as a province in 1962.

Following World War II, Italian Somaliland was returned to Italian control to be administered as the Trust Territory of Somalia under the United Nations, pending independence in 1960 as a republic. The present Somali Democratic Republic, or Somalia, includes both the former Italian territory and former

◄ The future of Africa south of the Sahara lies with its children; 46 percent of the region's population is under age 15. This boy is from the Democratic Republic of Congo. *Sally Mayman/Tony Stone Images*

British Somaliland, which chose to unite with Somalia. The people of much smaller French Somaliland remained separate and eventually became independent as the Republic of Djibouti, with its capital at the seaport of the same name. Influential elements among the Somali peoples have advocated the formation of a "greater Somalia," which would include the three Somaliland units plus substantial parts of Ethiopia and possibly some of Kenya. Somali claims to Ethiopian areas rest primarily on seasonal use of these lands by Somali pastoralists who migrate with their livestock across the international boundary. The Ethiopian government opposes such proposals to annex parts of its territory and has advanced counterclaims to parts of Somalia. There has been much fighting between regular military forces of the two countries.

THE PEOPLES OF ETHIOPIA

Ethiopia is inhabited by an estimated 57 million people of diverse ethnic and cultural origins. About 45 percent, including the often politically dominant Amhara peoples, adhere to the Coptic Christian faith, an ancient branch of Christianity that came to Ethiopia in the fourth century from Egypt, where there is still a sizable Coptic minority. The entire area has had important cultural and historical links with Egypt, the Fertile Crescent, and Arabia. The Ethiopian monarchy even based its origins and legitimacy on the union of the Biblical King Solomon and the Queen of Sheba, who tradition holds gave birth to the first Ethiopian emperor, Menelik. Until a Marxist coup brought an end to the emperorship in the 1970s, Ethiopia's rulers were always Christian. Ethiopia has many outstanding Christian artistic and architectural treasures, including the 11 churches of Lalibela, carved from solid rock in the 12th and 13th centuries (Fig. 18.1).

FIGURE 18.1

The churches of Lalibela, carved from volcanic rock in the 12th and 13th centuries, are among Ethiopia's many Christian cultural treasures.
Lawrence Manning/Tony Stone Images

The rest of Ethiopia's people are divided among those of the Muslim faith (who make up about 40 percent of the population), those practicing a variety of indigenous religions, and some Protestant, Evangelical, and Roman Catholic Christians. The great majority of the population is poor and illiterate. East of the mountain mass of Ethiopia, lower plateaus and coastal plains descend to the Red Sea, the Gulf of Aden, and the Indian Ocean. Extreme heat and aridity prevail at these lower levels, and nomadic and seminomadic Muslim tribesmen make a living by herding camels, goats, and sheep. Dwellers of scattered oases carry on a precarious agriculture. The arid lowland sections have many characteristics more typical of the Middle East than other areas of Africa south of the Sahara.

PROBLEMS OF DEVELOPMENT IN ETHIOPIA AND ERITREA

In 1974, the aging emperor Haile Selassie, who had held Ethiopia's population together in a loose political union, was deposed by a civilian and military revolt against the country's feudal order. Causes of the revolt included widespread drought and famine during the early 1970s, separatist dissension in some areas, and general discontent with the country's social and economic backwardness. A Marxist dictatorship, strongly oriented to the former Soviet Union, emerged. Thousands of middle-class intellectuals were executed or jailed. Privately-owned land and industries were nationalized, and collectivization of agriculture began. American influence, which had been strong under Haile Selassie, was eliminated. Munitions and advisors from the Soviet Union and other East-bloc nations poured in to equip and train one of Africa's largest armies, and the Soviets gained access to base facilities in Ethiopian ports. Such foreign interventions contributed to the strife that has hindered development in the region.

Ethiopia has substantial resources, possibly including significant mineral wealth. However, its resource base remains largely undeveloped, although with foreign assistance the government has been able to develop a limited number of factories and a few dams to harness some of the country's large hydroelectric potential. A poor transportation system—whose best feature is the 3000 miles (c. 4800 km) of good roads built by the Italians during their occupation in the 1930s—is a major factor in keeping the country isolated and underdeveloped. The French-built railroad from Djibouti to Ethiopia's capital Addis Ababa (population: 1.9 million, city proper; 2 million, metropolitan area) lacks feeder lines and has never been as successful as hoped. Many roads into the Ethiopian heartland are rough tracks, although better connections have gradually been developing. Air services—especially Ethiopia's remarkable fleet of DC-3s more than 50 years old but kept in top condition—help somewhat in overcoming the deficiencies in ground transportation. However, the high and rugged part of Ethiopia is badly handicapped by inaccessibility. A good share of its population

FIGURE 18.2
Much of Ethiopia's landscape is a rugged volcanic plateau. In this view, highlands rise over pasturelands and small cultivated fields in Ethiopia's northern Tigre province.   *Herman Emmet/Photo Researchers*

lives in villages on uplands cut off from each other and from the outside world by precipitous chasms (Fig. 18.2). These formidable valleys were cut by streams that carried huge amounts of fertile volcanic silt to the Nile and thus provided material for the famous nourishing floods in Egypt (see Chapter 9, pages 234–236).

Ethiopia has been further troubled in recent times by political and military turmoil in the former Ethiopian province of Eritrea, the adjacent parts of northern Ethiopia, and in the borderland between Ethiopia and Somalia. In Eritrea, both Muslims and Christians (each roughly 50 percent of the population) resented their political subjugation to Ethiopia's Christian Amhara majority. Terrorism and guerrilla actions escalated into civil war in the early 1960s. The Ethiopian army made a major effort to stamp out the revolts of two Eritrean "liberation fronts" (one Marxist and the other not), but resistance continued. At the same time there was also a rebellion in Tigre Province adjoining Eritrea. In 1991, these combined rebel forces captured Addis Ababa and took over the Ethiopian government. This allowed the rebels in Eritrea to establish an autonomous regime, and Eritrea became independent in May 1993 following an overwhelmingly favorable vote in a referendum. Today, many geographers point to Eritrea as a model for sustainable development in Africa south of the Sahara; since independence, the country has reduced its foreign debt, cut its population growth rate, and taken other promising steps toward eliminating the burdens of underdevelopment.

With Eritrea's independence, Ethiopia suddenly found itself a landlocked country. As a result, Eritrea's small Red Sea ports of Massawa and Assab now handle the bulk of Ethiopia's foreign trade. Both Eritrean ports have highway connections to Addis Ababa, and Massawa has a short rail connection to adjacent highlands (see Fig. 17.3). The railroad serves Eritrea's main city and manufacturing center, Asmara (population: 358,100, city proper), located in highlands at an elevation of more than 7000 feet (2130 m).

In addition to the devastation wrought by war, droughts centering in its war-torn north have stricken Ethiopia since the late 1960s. For years there has been little or no rain in areas inhabited by millions of people. Hunger has beset both people and livestock, to the point that desperate farmers have fed their families with grain needed for seed and have sold their oxen needed for plowing. Ethiopian villagers are especially vulnerable to drought because of their heavy dependence on rainfed agriculture. The volcanic soils are fertile, but the streams are entrenched so far below the upland fields that little irrigation is possible.

In 1984 and 1985, drought in Ethiopia reached the point of widespread catastrophe. An estimated one million people perished. Great numbers of Ethiopians left their villages and gathered in makeshift refugee camps in the hope of being fed. Lack of drinking water and outbreaks of disease compounded the basic problem of insufficient grain and other food. In the late summer of 1984, hundreds of thousands of penniless refugees fled to camps in neighboring Sudan. Sudan was beset by droughts and famines of its own but was at least more accessible to outside relief agencies than was highland Ethiopia. Within Ethiopia, relief agencies encountered great difficulty in getting food to the hungry because of such factors as the limited capacity of the available ports, competition for use of the ports by general export, import, arms shipment and traffic, the insufficient quantity and quality of trucks and servicing, the lack of road access to starving villagers, and government bombing and strafing of relief trucks in rebellious areas. However, the massive and sustained relief effort helped reduce the impacts of the country's worst drought.

Recent economic statistics for Ethiopia depict a country that is still desperately underdeveloped. Coffee, often picked from wild trees, made up 67 percent of the scarce exports in 1992–93. Manufacturing is largely confined to simple textiles and processed food. Addis Ababa and Eritrea account for the bulk of the area's manufacturing. Wood is in very short supply as a result of deforestation; in 1950, 40 percent of the surface of Ethiopia was forested, but today less than 1 percent is. Programs of reforestation and afforestation are critically needed.

HOMOGENEOUS, WAR-TORN SOMALIA

Between Ethiopia and the Indian Ocean lies the poor, lowland country of Somalia. Some 99.8 percent of the people here are Sunni Muslims and 98.3 percent are ethnic Somalis—surely one of the most homogeneous populations in the world.

Moisture is so scarce that only 2 percent of the country is cultivated. Livestock raising on a nomadic and seminomadic basis is the main support for Somalia's simple economy; live animals (goats, sheep, camels, cattle), shipped mainly to Saudi Arabia, Yemen, and the United Arab Emirates, are the main exports, along with fish products and bananas. Food processing is the chief industry, but there is almost no mineral production. Somalia once received aid from the former East bloc but lost that resource when the bloc transferred its military aid to Ethiopia. Somalia then granted the United States access to base facilities at Mogadishu (population, about 1 million), the capital, located in former Italian Somaliland, and at the much smaller city of Berbera in former British Somaliland.

Somalia was badly mauled in a rebellion that overthrew its government in early 1991. Fighting between the new rulers and resistance forces continued in 1992. The opposing forces in this destructive civil war were regional clan groups of the Somali people (the country has six major clans and numerous subclans). Contending forces inflicted great damage on Mogadishu, from which large numbers of people fled to the countryside as refugees. Massive famine threatened the devastated country in mid-1992. International efforts to provide relief were hindered by banditry in a country where law and order had disintegrated.

United States military forces intervened in late 1992 with "Operation Restore Hope," a mission to help protect relief workers and aid shipments. However, American policy makers decided to use military force to neutralize the power of one of the country's leading "warlords," Mohamed Farrah Aidid. In a gruesome sequence of events, 18 U.S. servicemen stranded in Aidid's territory in Mogadishu were killed in a single day in 1993. American public outcry at the tragedy brought an end to the American mission in 1994. The U.S. then handed over the main responsibility for aiding Somalia to United Nations forces.

While considerable progress has since been made in salvaging Somalia, still unresolved is the future status of northern Somalia, where local political and clan leaders declared independence in 1990 for a new country to be called Somaliland. Once part of British Somaliland, this area escaped the destruction that ravaged the south in the last stages of Somalia's civil war.

18.2   The Sahel

The Sahel region extends eastward from the Cape Verde Islands to the Atlantic shore nations of Mauritania, Senegal, and The Gambia and inland to Mali, Burkina Faso ("Land of the Upright Men"; formerly Upper Volta), Niger, and Chad (see Fig. 17.1). In colonial times, The Gambia was a British dependency. Mauritania, Senegal, Mali, Burkina Faso, Niger, and Chad were French. Cape Verde was Portuguese. Cape Verde's dry volcanic islands lie well out in the Atlantic off Senegal, but in population, culture, economy, and historical relationships the islands are so akin to the adjacent mainland that they fit easily into a Sahelian context. Ancient Mali, with its legendary port of Timbuktu on the Niger River, was an important trading intermediary between the Guinea Coast and Islamic North Africa before seafaring Europeans took over this trade.

Climatically, the region ranges from desert (in parts of the Sahara) in the north through belts of tropical steppe and dry savanna in the south. The name Sahel in Arabic means "coast" or "shore," referring to the region as a front on the great desert "sea" of the Sahara. Since the late 1960s the area has been subjected to severe droughts, which, in combination with increased human pressure on resources, has prompted a process of desertification (see Regional Perspective, pages 424–425).

This area of Africa has seen many dramatic changes in climate and vegetation over its long history. There is abundant evidence, particularly in the form of prehistoric rock drawings, that much of the now extremely arid northern Sahel and Sahara region was a grassy, well-watered savanna approximately ten thousand years ago. Many of the large mammals now associated with East Africa frequented the region, along with bands of pastoralists who kept large cattle herds. Now, even though under much less favorable conditions, the raising of sheep, goats, camels, and cattle, combined where possible with subsistence farming, continues to be the major livelihood for peoples of the Sahel.

One of the most prominent features of the Sahel's landscape is Lake Chad, located north of landlocked Chad's capital, N'Djamena (population: 529,555, city proper). This huge lake is situated in a vast open plain. It has no outlets, and yet it is not very salty, because groundwater carries most of the salts away from the lake. Lake Chad is very shallow, with maximum depths between 7 and 23 feet (2–7 m). Characteristic of shallow lakes in arid lands, its surface area increases dramatically with rainy periods and shrinks in times of drought. Recent decades of drought have seen its area diminish substantially, but prolonged rains would quickly enlarge it. The waters of Lake Chad are very important economically for fish and irrigated agriculture.

Mining, manufacturing, and urbanization are generally insignificant in this subsistence-oriented region of Africa. Mauritania exports iron ore and Senegal exports phosphates. Niger exports uranium for France's nuclear power program. The largest metropolises are Senegal's main port and capital of Dakar (population: 1.5 million, city proper), and Bamako (population: 658,275, city proper), the capital and main city of Mali (Fig. 18.3). Dakar was formerly the capital of the immense group of eight colonial territories known as French West Africa. It is the commercial outlet for much of the peanut belt in the Sahelian and West African savanna, and is a considerable industrial center by African standards. Dakar also is a tourist and resort center with fine beaches and an exotic blend of French and Senegalese culture.

FIGURE 18.3
The faith of Islam unites most of the disparate peoples of the Sahel. This is the mosque and market at Djenne, near a tributary of the Niger River in southern Mali.   *Explorer/Photo Researchers*

The Sahel is also the border region between the predominantly Arab and Berber North Africa and the predominantly black Africa south of the Sahara. Relations between the ethnic groups are good in most countries, but there are problems in some. In Mauritania, for example, the government—led by the country's majority Arabic-speaking Moors—began expelling black Mauritanians across the border into Senegal in 1995. Analysts fear the Mauritanian government aims to "cleanse" the country of all of its 40 percent black population, and warned in 1996 that a race war was imminent.

18.3   West Africa

AREA, POPULATION, AND PHYSICAL ENVIRONMENT

West Africa, here defined as extending from Guinea-Bissau eastward to Nigeria (see Fig. 17.1), is larger than one might realize from looking at the map. The nine political units in the region make up more than 8.18 million square miles (about 21.18 million sq km), or nearly one fourth of the area of the United States. Distances are correspondingly great. It is about 600 miles (*c.* 1000 km) from the coast of Nigeria to the country's northern border. The distance along the savanna belt between Kano in northern Nigeria to the Atlantic shore of Guinea-Bissau is about 1700 miles (*c.* 2700 km). Inadequate transportation compounds the isolating effects of such distances.

The region is less impressive in population totals. In 1996, the nine countries had an estimated total population of 162 million, or 103 million fewer than that of the United States. Four

countries had under 5 million people, and only three—Ivory Coast, with 15 million; Ghana, with 18 million; and Nigeria, with 104 million—exceeded 10 million. The most striking single aspect of West Africa's population distribution is that about two thirds of the region's population resides within one country, Nigeria (which also has 18 percent of the population of the entire region of Africa south of the Sahara). This unusual concentration of people, along with advanced commercialization and urbanization, existed even before European contact. It may have been related to the existence of powerful kingdoms that were able to impose a degree of internal order and security, and to the region's agricultural productivity and resource wealth.

Environmental contrasts within West Africa are extreme. Climatically, the region ranges from belts of tropical steppe, dry savanna, and wetter savanna, to some areas of tropical rain forest along the southern and southeastern coasts (see Figs. 2.6 and 2.7). Most of the countries' most populous areas are in the zone of tropical savanna climate. Temperatures are tropical throughout West Africa. Along the southern coast, temperatures in the coolest month of the year average in the high 70s Fahrenheit (mid-20s Centigrade) and in the hottest month average in the high 80s (low 30s Centigrade). Most uplands are not high enough to bring significant reductions in temperatures.

West Africa exhibits some topographical variety (see Figs. 17.2 and 17.3). To the north lies an east-west belt of uplands, from interior Guinea-Bissau to eastern Nigeria. This area includes hills, low mountains, and plateaus in various stages of dissection. Along the immediate coast is a narrow belt of coastal plain, widening in some sections, such as the Niger Delta, and extending inland along the lower courses of major rivers. Parts of the belt are mangrove swamp, and its straight reaches of coast and barrier sandbars provide few good natural harbors.

ETHNIC AND POLITICAL COMPLEXITY

West Africa was an advanced part of Africa south of the Sahara when the European Age of Discovery began. A series of strong pre-European kingdoms and empires developed there, both in the forest belt of the south and in the grasslands of the north. In colonial times it became a French and British realm, except for a small Portuguese dependency and independent Liberia. Of the present countries, four—The Gambia, Sierra Leone, Ghana, and Nigeria—were British colonies; four—Guinea, Ivory Coast (officially Côte d'Ivoire), Togo, and Benin (formerly Dahomey)—were French; and Guinea-Bissau was Portuguese. Liberia is an exceptional country, becoming Africa's first republic after 1822, when 5000 freed American slaves sailed there and settled with support from the U.S. treasury. Problems between Americo-Liberians and the indigenous people have continued ever since, reaching a full-scale civil war in 1989. In 1995, a new coalition government brought the promise of peace

Regional Perspective

DROUGHT AND DESERTIFICATION IN THE SAHEL

The Sahel region, like much of Africa south of the Sahara, experiences periodic drought. This is a naturally-occurring climatic event in which rain fails to fall over an area for an extended period, often years. When rain does return, the arid and semiarid ecosystems of the Sahel come to life with a profusion of flowering plants, insects, and herbivores like gazelles, whose populations climb when foods are abundant. These ecosystems are resilient, meaning they are able to recover from the stress of drought and have mechanisms to cope with a natural cycle that includes periods of dryness and rain.

Desertification is the destruction of that resilience and the biological potential of arid and semiarid ecosystems. It is an unnatural, human-induced condition that has afflicted the Sahel severely since the late 1960s, and, as such, it is different from the natural phenomenon of drought (Fig. 18.A). However, the recent history of the Sahel suggests that drought can be the catalyst which initiates the process of desertification. During several years of good rains in the 1950s and early 1960s, the Fulani (Fulbe) and other Sahelian pastoralists allowed their herds of cattle and goats to grow very large, since the animals represent wealth and capital and people naturally wanted their numbers to increase beyond the immediate subsistence needs of their families. However, due to declining death rates, the numbers and sizes of families keeping livestock were also very large. Thus, the unprecedented numbers of livestock built up during the rainy years made the Sahelian ecosystem vulnerable to the impact of drought on an unprecedented scale.

Drought struck the region in 1968, persisting through 1973. Annual plants failed to grow, so the large herds of cattle and goats turned to acacia trees and other perennial sources of fodder. They ate all of the palatable vegetation.

This destruction was a critical problem, because plants play an important role in maintaining soil integrity, helping to intercept moisture and funnel it downward through the root system. Plant litter and decomposer organisms working around the plant contribute to soil fertility and help stabilize the soil. With hungry livestock in the region eating the plants, the landscape changed. No longer anchored and replenished, good soils eroded. "Junk" plants like Sodom apple (*Calotropis*) replaced palatable plants, and an almost impermeable surface formed on the land. This degraded ecosystem lost its resilience, so that even when rains returned, vegetation did not recover.

Meteorologists studying the Sahel drought discovered another important link between precipitation and the removal of vegetation. They noticed that when people and livestock reduced the vegetative cover, they increased the **albedo,** or the amount of the sun's energy reflected by the ground. Fewer plants meant more solar energy deflected back into the atmosphere, which then reduced precipitation locally. This connection is known as the **Charney Effect,** named for the scientist who documented it. It was a tragic sequence: people responding to drought actually perpetuated further drought conditions.

The 1968–1973 drought devastated the great herds of the Sahel. An estimated three and one-half million cattle died. Two million pastoral nomads lost at least half of their herds, and many lost as much as 90 percent. Farmers dependent on unirrigated cropping were also affected. Harvests were less than half of the usual crop for 15 million farmers, and for many there was no harvest. The Niger and Senegal rivers dried up completely, depriving irrigated croplands in Niger, Mali, Burkina Faso, Senegal, and Mauritania. The cost in human

and reconciliation to Liberia, but in early 1996 the war resumed with a new ferocity.

In West Africa today, the English, French, and Portuguese languages continue in widespread use, are taught in the schools, and are official languages in the respective countries. European languages are a highly useful means of communication in this region, where hundreds of indigenous languages and dialects, often unrelated, are spoken. Economic, political, and cultural relationships between West African countries and the respective European powers that formerly controlled them continue to be close. France, for instance, supplies economic and military aid to its former colonies in this and other parts of "Francophone" (French-speaking) Africa. France further cultivates good relations with its former dependencies by inviting their leaders to summit conferences to discuss matters of common interest. Similarly, Britain maintains ties with "Anglophone" Africa through meetings of the Commonwealth of Nations to which all its former colonies in Africa south of the Sahara belong.

As is often the case in Africa, the governments of the region must deal with serious problems growing out of the ethnic complexity of national populations. European governments contributed initially to some present troubles by drawing arbitrary boundary lines around colonial units without proper concern for ethnicity. As a result, the typical West African country today, like most countries elsewhere in Africa south of the Sahara, is a collection of ethnic groups that have little sense of

lives was estimated at 100,000 to 250,000. Environmental refugees poured into towns like Nouakchatt, the capital of Mauritania, which were unprepared to deal with such a large and sudden influx.

The Sahelian crisis of 1968–1973 resulted in an international effort to combat desertification using recommendations developed by the United Nations in 1977. The U.N. report, however, soon became a case study in how *not* to deal with an environmental crisis in the developing world. Most of the recommendations were universal, high-technology solutions that proved impossible to implement in the villages and degraded pastures of the Sahel. Following the recommended guidelines, many international agencies attempted to relieve the suffering of Sahelian pastoralists. For example, they dug deep wells to water livestock, but failed to anticipate that these wells would act like magnets for great numbers of people and animals. While there was plenty to drink, the animals starved to death when they decimated what vegetation remained around the new water supplies.

Since the disappointment of the 1977 U.N. plan, and with the lessons relief organizations learned as a result, there has been greater focus on sustainable development, with its local rather than universal solu-tions. An example is the technique of "rainwater harvesting," in which local villagers use local stones and tools to construct short rock barriers that follow elevation contours. Water striking these barriers backs up to saturate and conserve the soil, allowing more productive farming. The challenge to find such solutions is a continuous one; since 1968, drought has established an almost permanent presence in the Sahel.

FIGURE 18.A

Desertification in progress is shown graphically in this photo from Mali, in the Sahel. As growing human populations intensify the impact of naturally-occurring drought, the Sahara creeps southward.   *Steve McCurry/Magnum*

identification with the national unit. The region is inhabited by peoples who speak a large variety of languages and dialects, and who in some cases have a history of suspicion and hostility toward one another. A variety of religious faiths also coexist here. Many people hold traditional animistic beliefs, in which they associate spiritualism with a variety of natural features and ancestral beings, rather than with a single deity. The Muslim faith is dominant in the north, and over centuries has spread southward. Protestant and Roman Catholic missionaries converted many West Africans to Christianity.

Only modest numbers of Europeans have ever lived in West Africa, primarily because of their perceptions of an inhospitable climate, fears of disease, the historical opposition of slave-trading interests to other forms of European enterprise, the opposition of well-organized African kingdoms to encroachment by white settlers, and policies by European colonial powers that limited European land ownership. Almost all Europeans in West Africa have been of a governmental, professional, managerial, commercial, or technocratic class, and the great majority have lived in the political capitals and scattered other cities. The presence of an entrenched European settler class owning large blocks of the best farmland has therefore not been a circumstance contributing to internal political troubles of West African countries in the age of independence, as it was in Kenya, Zimbabwe, and other countries outside of West Africa.

AGRICULTURE AND MINING

The West African countries exhibit the customary economic pattern of the world's LDCs. They depend heavily on subsistence and near-subsistence crop farming and livestock grazing, along with a few agricultural and/or mineral export specialties. In the wetter south, subsistence agriculture relies mainly on root crops such as manioc and yams, and on maize, the oil palm, and, in some areas, irrigated rice. In the drier, seasonally rainy grasslands of the north (in northern Nigeria and northern Ghana, for example), nomadic and seminomadic herding of cattle and goats is important, along with subsistence grain crops of millet and sorghum.

The major export specialties of West African agriculture are cacao, coffee, and oil-palm products in the wetter, forested south, and peanuts and cotton in the drier north. Ivory Coast leads the world in cacao exports. Nigeria was once the world leader in exports of palm oil and peanuts, but these exports largely vanished when Nigerian agriculture was allowed to deteriorate during the oil boom of recent decades. There is a marked correlation between the distribution of these resources and the distribution of Nigeria's major ethnic groups. The main oil-palm belt of Nigeria is in the denser rain forest of the southeast, inhabited by the Ibo people. The main cacao belt is in the lighter forests of the southwest, inhabited by the Yoruba people. The peanut belt is in the open grasslands of the north, inhabited by the Hausa and Fulani peoples.

One peculiarity of West African commercial agriculture is the almost complete dominance of African small landholder production. With a few exceptions, foreign owned and managed

PROBLEM LANDSCAPE

"NIGERIA EXECUTES CRITIC OF REGIME; NATIONS PROTEST"

Oil is the economic lifeblood of populous Nigeria. Until the early 1990s it helped to make Nigeria one of the more prosperous and enviable countries of Africa south of the Sahara. But after 1993, when the military government of General Sani Abacha annulled democratic elections and imprisoned the apparent victor, corruption, human rights abuses, and mismanagement of the country's resource wealth made Nigeria an outcast in the world community.

Nevertheless, the multinational Royal Dutch/Shell Group continues to produce (as of 1996) about half of this OPEC member's output of two million barrels of oil per day. A small portion of that production comes from wells drilled in a 350-square-mile section of the Niger Delta region inhabited by 500,000 people of the Ogoni tribal group. Shell has generated about $30 million in revenue from oil drilled since 1958 on Ogoni lands in this region, known as the Rivers State. According to their spokesmen, the Ogoni have derived few benefits and have suffered much from oil development in their homeland. They complain that for more than 30 years, oil spills have tainted

their croplands and water, destroying their crops and fisheries, while the flaring off of natural gas has polluted their air and caused acid rain. They note that despite the enormous revenue generated from oil drilled on their land, little money has returned to the area. Most people live in palm-roofed mud huts. Half of the delta region lacks adequate roads, water supplies, and electricity. Schools have few books, and clinics have few medical supplies.

In an attempt to improve their plight, Ogoni activists founded the Movement for the Survival of the Ogoni People (MOSOP) in 1990. Thirty-two Ogoni chiefs and elders issued an "Ogoni Bill of Rights" in which they declared the right to a safe environment and more federal support of their people. One of the movement's leaders, a popular author, playwright, and television producer named Ken Saro-Wiwa, also called for self-determination for the Ogoni. Saro-Wiwa wrote that he despaired of his country because his government's leaders had "hearts of stone and the brains of millipedes; because Shell is a multinational company with the ability to crush whomever it wishes; and because the petro-

leum resources of the Ogoni serve everyone's greed." After an antigovernment rally by hundreds of thousands of Ogoni people in January 1993, government police razed 27 villages, killing about 2000 Ogonis and displacing a further 80,000. Shell has suspended oil production in the region since that violence occurred.

In another violent incident, Ogoni activists killed four founding members of MOSOP in May 1994. According to family members and supporters of the slain men, Saro-Wiwa incited their murders by stating that they "deserved to die" for not taking a more active position against the government and Shell, and by denouncing their efforts to settle a dispute with the neighboring Andoni people. Government authorities arrested Saro-Wiwa and thirteen of his associates, charging them with responsibility for the murders. In his opening defense statement, Saro-Wiwa said "the crime of the Ogoni people is that they had the temerity to ask for their rights from both the Government of Nigeria and from Shell. What the criminal partners of this alliance are trying to tell the world by this arraignment is that the black man, even the best of them, is

plantations never achieved a notable foothold, although foreign traders and companies have stimulated commercialization of small-farm agriculture. West African agriculture has had a long succession of difficulties in recent years. These have included extended droughts, bushfires, plant and animal diseases and pests, low prices on world markets, ill-advised government policies, migration of younger workers to cities, and high expenditures for imported food and oil-derived fertilizer.

The other main export industry of West Africa is mining (see Fig. 17.3, page 402). This region has contributed resources to supply expanding industrial needs worldwide, although its total mineral output to date has been modest on a world scale. For many West African countries, minerals have become very important as a source of export earnings, government finance, and funds for economic development. The most spectacular and

significant mining development has been Nigeria's emergence as a producer and major exporter of oil (see Problem Landscape, page 426). Commercial oil development began here in the 1950s. By 1983, crude oil constituted 96 percent of Nigeria's exports by value, and in 1996 it still made up 90 percent of the country's foreign export earnings. Extraction thus far has taken place mainly in the mangrove-forested Niger Delta and offshore in the adjacent Gulf of Guinea.

Other prominent mineral products of West Africa include iron ore, bauxite, diamonds (Fig. 18.4), phosphate, and uranium. Although the region supplies only a tiny fraction of the world's iron ore, its export is very important to Liberia. The shipments, which in recent years have been largely suspended because of political instability in Liberia, consist largely of high-grade ore. The main bauxite resources lie near the

---

no better than a criminal." For its part, the government accused Saro-Wiwa of supporting violent means against both the regime and oil interests. On October 31, 1995, the government court found Saro-Wiwa and eight other defendants guilty, and ordered their execution.

An international outcry ensued. Saro-Wiwa had already earned an international reputation as an environmentalist (he received the 1995 Goldman Environmental Prize) and as a human rights activist (he was a 1995 nominee for the Nobel Peace Prize). Environmental and human rights groups mounted an international media and diplomatic campaign to win him clemency. These defenders claimed that Saro-Wiwa was framed and did not receive a fair trial. Under scrutiny for its Nigerian operations, Shell admitted responsibility for frequent oil spills (40 percent of the company's spills worldwide had occurred in Nigeria), but argued that MOSOP had exaggerated the environmental impact of the company's operations. Shell resisted international pressure to use its influence with the Nigerian government to secure clemency for the men, arguing, "It is not for a commercial organization like Shell to interfere in the legal processes of a sovereign state such as Nigeria." The company also accused Mr.

Saro-Wiwa of supporting violence. It later withdrew that accusation, but expressed no regret at the death sentence. The company finally wrote to the Nigerian leader asking that Saro-Wiwa's life be spared on humanitarian grounds.

On November 10, 1995, the government hanged Ken Saro-Wiwa and his eight codefendants. Many analysts regarded the executions as General Abacha's effort to deter actions by other more serious rivals, especially within the army. As Nigeria is a thinly-glued collection of 250 ethnic groups, Abacha may also have seen the Ogoni struggle for minority rights as a Pandora's box that would lead to the breakup of the nation. Redistribution of oil revenue also would have diminished his income and that of the country's other ruling senior military officers. The government waged its own public relations war after the executions, claiming that it had taken action to return 13 percent of oil proceeds to regions where oil is produced to stimulate development (a figure now officially stated to be 3 percent). The government also promised to hand over power to an elected government in 1998.

The international community weighed its options in the months following the executions. The United States, Britain, and

other countries recalled their ambassadors to Nigeria. The World Bank announced that it would not extend a loan of $100 million for a project to develop liquefied natural gas (which, ironically, would have helped eliminate one of Saro-Wiwa's main concerns, the problem of pollution caused by the flaring-off of natural gas). At the urging of South African president Nelson Mandela, the 52-member organization of the British Commonwealth suspended Nigeria from membership. Mandela pushed the U.S. and Britain for even stronger measures to punish the Nigerian government, including an embargo of Nigerian oil. That action would certainly hurt the country, which exports 45 percent of its oil to the U.S. and 40 percent to Europe. Fearing the prospects of instability, civil war, and refugee migrations that such an action might touch off in an already unstable region, however, the U.S. and other major members of the international community declined to embargo Nigeria's oil or take any additional steps against its government.

Howard J. French, "Nigeria Executes Critic of Regime; Nations Protest." *The New York Times*, November 11, 1995: p. A1.

FIGURE 18.4
Small-scale diamond mining in Ivory Coast (Côte d'Ivoire), West Africa. The miners will screen for diamonds the small piles of dirt they have extracted by hand from shallow holes. The scrubby vegetation, a secondary growth that is typical of a deforested area, is common in West African tropical rain forest areas.   *Manaud/Elbaz/Matrix*

southwestern bulge of the West African coast; Guinea, by far the main producer and exporter, derives the bulk of its export revenue from bauxite and alumina (the second-stage product between bauxite and aluminum). Togo exports phosphate. Gold, bauxite, and manganese are important to Ghana, as are bauxite and diamonds to Sierra Leone. However, much of Sierra Leone's potential wealth from diamonds is lost to smugglers, many reportedly from the entrepreneurial class of Syrians and Lebanese who settled in the country decades ago. The diamond purge and antigovernment guerrilla fighting since 1991 have contributed to a steady decline recently in Sierra Leone's standard of living.

URBANIZATION

The world's large cities are associated mostly with trade, centralized administration, and large-scale manufacturing. West Africa's principal cities are also national capitals and seaports, but, as the region has only minor manufacturing, its cities have developed primarily in its areas of major commercial agriculture. In nearly every country in the region, the capital is a primate city (see Chapter 4, page 102), far larger than any other city.

A total of only 33 percent of West Africa's inhabitants live in urban areas (as of 1996), but the urban population is increasing very rapidly. Nigeria has the most impressive urban development. Of the 11 largest metropolitan cities in West Africa, three are clustered in the Yoruba-dominated southwestern part of Nigeria, in and near the densely populated belt of commercial cacao production.

The largest of these three—and by far the largest in West Africa—is Lagos (population: 1.2 million, city proper; 3.8 million, metropolitan area). Its development surged in the early 20th century when the British colonial administration chose Lagos as the ocean terminus for a trans-Nigerian railway linking the Gulf of Guinea with the north. Harbor facilities able to handle large modern ships had to be developed. Lagos is now West Africa's leading cargo port and industrial center. It is a sprawl of shanties on dirt streets surrounding a modest knot of modern high-rise buildings and older low-rise structures in the commercial, governmental, and residential core near the harbor. During the oil boom of recent decades, migrants from all parts of Nigeria (and many from outside the country) flooded into the city. With its severe traffic congestion, rampant crime, inflated prices, and inadequate public facilities, Lagos gives the impression of urbanism out of control. It has long been Nigeria's political capital, but by 1992 most governmental functions were moved to a new capital at Abuja in the interior.

About 50 miles (80 km) north of Lagos on the railway lies Nigeria's second largest city, Ibadan (population: 1.14 million, city proper), in the heart of the cacao belt. Ibadan is a city of traders and craftsmen but also includes a surprising number of farmers. Like other cities in southwestern Nigeria, Ibadan has the character of both a city and a huge village; many farmers living in these cities commute 10 to 20 miles or more to their fields outside the city.

About 50 miles north of Ibadan is Nigeria's third largest city, Ogbomosho (population: 660,000, city proper). Still further north, close to the border with Niger, is Kano (population: 699,900, city proper). Kano is the most important commercial and administrative center of Nigeria's north and has long been a major West African focus of Islamic culture.

Immediately west of Nigeria on the Guinea Coast lies Cotonou (population: 533,212, city proper), the main city and port of Benin. Further west, in southern Ghana—another agricultural area with a high degree of commercialization—are two more of the 11 largest West African metropolises. The larger of the two, Accra (population: 949,113, city proper; 1.39 million, metropolitan area), is Ghana's national capital. Until recently it was the main seaport, despite the lack of deep-water harbor facilities and the consequent need to transfer cargo from ship to shore through the surf. A modern deep-water port has been developed at Tema a short distance east of Accra, and that city is now the country's main port and industrial center. Well inland from Accra lies the country's second largest city, Kumasi (population: 385,200, city proper), in the heart of Ghana's cacao belt, which was once the world's greatest cacao producing area.

Today, the world's leading cacao producer and exporter is Ivory Coast (Côte d'Ivoire), which contains yet another of West Africa's 11 largest cities, Abidjan (population: 1.93 million, city proper). Abidjan, which has a considerable Western veneer in its buildings and businesses, is Ivory Coast's main seaport. Although the inland city of Yamoussoukro is the official capital, Abidjan is the de facto capital of a country which has vowed to

FIGURE 18.5
Only about one third of the people of Africa south of the Sahara live in cities, but with considerable rural to urban migration, that number is growing. This is very modern Abidjan, the leading city of Ivory Coast (Côte d'Ivoire), with St. Paul's Cathedral in the foreground.   *Charles O. Cecil/ Visuals Unlimited*

achieve economic prosperity comparable to that of the "Asian tiger" countries (Fig. 18.5). Also on the Guinea Coast are three more seaport capitals: Monrovia (population: 1 million, city proper, including numerous war refugees) in Liberia; Freetown (population: 469,776, city proper) in Sierra Leone, on West Africa's finest natural harbor; and Conakry (population: 800,000, city proper) in Guinea.

TRANSPORTATION

Inadequate transportation is a problem for West Africa. Good natural harbors are scarce due to the abundance of offshore sandbars and the tendency of river mouths to be choked with silt. It has often been necessary to transfer goods by tender between ship and shore, and these countries incur large expenses as they attempt to provide modern port conditions and facilities. Inland from the coast lies the formidable problem of adequate connection between the ports and the interior. Rapids, together with annual seasons of low water, limit the utility of rivers for transport. Some rivers do carry traffic, notably the Niger and its major tributary the Benue, the Senegal River, and the Gambia River. The Gambia River is the main transport artery for The Gambia—a remarkable country nearly surrounded by Senegal and consisting merely of a 20-mile-wide strip of savanna grassland and woodland extending inland for 300 miles (*c*. 500 km) along either side of the river (Fig. 18.6).

Railways are much more important than rivers in the transport structure of West Africa but are few and widely spaced. Instead of forming a network, the rail pattern comprises a series of individual fingers that extend inland from various ports but which usually do not connect with other rail lines (see Fig. 17.3, page 402). The region's road network is dense in well-populated areas, but most roads are unpaved or unreliable in bad weather. Much of the region's area and many of its people are still remote from any good highway. Nevertheless, truck traffic is sizable. Not surprisingly, a very large part of the limited long-distance passenger travel within the region moves by air. If much of West Africa is to be integrated into the commercial world, however, either air freight must become much more economical than it is now, or road and rail facilities must be greatly expanded.

FIGURE 18.6
An aerial view of gallery forest along the Gambia River in Niokolo National Park, Senegal.   *Norman Myers/Bruce Coleman Inc.*

18.4   West Central Africa

The subregion of West Central Africa is flanked by Cameroon and the Central African Republic to the north and the Democratic Republic of Congo (formerly Zaire) to the south (see Fig. 17.1, page 400). In colonial days, France held three of the present units—Central African Republic, Republic of the Congo, and Gabon—in the federation of colonies called French Equatorial Africa. Cameroon consisted of two trust territories administered by France and Britain. Belgium held the Democratic Republic of Congo, Spain held Equatorial Guinea, and Portugal held the islands of São Tomé and Principe.

A narrow band along the immediate coast of West Central Africa is composed of plains below 500 feet (*c.* 150 m) in elevation. As in West Africa, the coast is fairly straight and has few natural harbors. Shallow lagoons behind coastal sandbars are common, and mangrove swamps—with their impenetrable tangle of stiltlike tree roots—fringe many lagoons and delta channels. Most of the region is composed of plateaus with surfaces that are undulating, rolling, or hilly. A conspicuous exception to the generally uneven terrain is a broad area of flat land in the inner Congo basin, once the bed of an immense lake. The greater part of West Central Africa lies within the drainage basin of the Congo River. The heart of the region has a tropical rain forest climate, grading into tropical savanna to the north and south (see Fig. 2.7, page 32). Only scattered parts of West Central Africa, primarily along the eastern and western margins, are mountainous. The most prominent mountain ranges lie in the eastern part of the Democratic Republic of Congo near the Great Rift Valley and in the west of Cameroon along and near the border with Nigeria. Vulcanism has been important in mountain-building in both instances.

Hundreds of ethnic groups speaking many languages and dialects live in West Central Africa. The majority speak Bantu languages of the Niger-Congo family, while Nilo-Saharan language speakers predominate in the drier areas of grassland north of the rain forest. Small bands of Pygmies (Twa) live in the rain forest of the Congo basin. Within West Central Africa as a whole, population densities are low and populated areas are few and far between. Rough terrain, thick forests, wetlands, and areas with tsetse flies tend to isolate clusters of people from each other. The savanna lands and highlands are generally more densely populated than the rain forests. Most countries have an overall density well below that of Africa south of the Sahara as a whole (see Table 17.1, pages 398–399).

Subsistence agriculture supports people in most areas. Production of export crops is small and takes place mainly in areas of rain forest, although some cotton is exported from savanna areas. Southern Cameroon is an important cash crop area, where coffee and cacao are the main export crops, supplemented by bananas, oil-palm products, natural rubber, and tea. These products come from African small farms and plantations, including plantations in western Cameroon that were developed in a period of German control before World War I and now are

government operated. They occupy an area of volcanic soil at the foot of Mt. Cameroon (13,451 ft/4100 m), a volcano that is still intermittently active (Fig. 18.7). The mountain is notable for its extraordinarily heavy rainfall, averaging around 400 inches (1000 cm) a year.

The island of Bioko is notable for its cacao exports. Bioko and the adjacent mainland territory of Río Muni, formerly held by the Spanish, received independence in 1968 as the Republic of Equatorial Guinea. Cacao production in Bioko developed on European-owned plantations, but the newly independent country's leaders expelled the plantation owners.

Exports of minerals, principally oil, aid the financing of some countries—most notably Gabon, Republic of the Congo, and Cameroon, where oil revenues have created per capita GNPs that are exceptionally high for Africa south of the Sahara (see Table 17.1, pages 398–399). Cameroon in 1996 was engaged in an ongoing oil-related dispute with Nigeria over control of the Bakassi Peninsula, which juts into the Gulf of Guinea near the border between the two countries. Control of the peninsula will give the dispute's victor ownership of promising offshore oil reserves.

Cities are few in these countries. Most of the larger ones are political capitals and/or seaports, including Republic of the Congo's capital city of Brazzaville (population: 693,712, city proper) and its seaport of Pointe Noire (population: 576,206, city proper). Brazzaville, located directly across the Congo River from the much larger capital city of Kinshasa in the Democratic Republic of Congo, was once the administrative center of French Equatorial Africa. Despite Republic of the Congo's huge offshore oil wealth, years of political mismanagement have perpetuated underdevelopment. There is not yet a paved road between Brazzaville and Pointe Noire. The 320-mile (515-km) Congo-Océan Railway does connect the cities. Another pair of cities connected by rail is Cameroon's seaport of Douala (population: 810,000, city proper) and the country's

FIGURE 18.7
The volcanic soils of northwest Cameroon support lush pastures and productive farms.   *Jane Thomas/Visuals Unlimited*

inland capital, Yaoundé (population: 649,000, city proper). But, aside from a handful of rail fingers reaching inland from seaports, West Central Africa is largely devoid of railways, and roads only poorly serve most areas. Isolation and poor transport are major limiting factors for economic development.

THE DEMOCRATIC REPUBLIC OF CONGO: A CASE STUDY IN COLONIALISM AND ITS LEGACY

The Democratic Republic of Congo is Africa's second largest country in area (after Sudan), but its estimated population in 1996 was only 47 million, or well under one half that of the continent's most populous country, Nigeria. Nearly half of the Democratic Republic of Congo is an area of tropical rain forest in which the population is still extremely sparse. The national population comprises some 250 ethno-linguistic groups speaking more than 75 languages. The Democratic Republic of Congo has a colonial history and faces modern challenges which are representative of the region of West Central Africa, and so merits in-depth discussion (see Fig. 18.8).

Colonial Background

Belgian penetration of the Congo basin began in the last quarter of the 19th century, following exploration by an American, Henry M. Stanley. During this early period, the colony was virtually a personal possession of the Belgian King Léopold II, whose agents ransacked it ruthlessly for wild rubber, ivory, and other tropical products gathered by Africans. This lawless era inspired Joseph Conrad's famous novel *Heart of Darkness* (1902), in which the trader Kurtz, at the point of death, evokes the ravaged Congo Region with the cry "The horror! The horror!" In the 1890s, colonial authorities built a railroad to connect the port of Matadi with the capital of Léopoldville (now Kinshasa), thus avoiding a stretch of rapids on the Congo River and providing access to the interior. In 1908, the Belgian government formally annexed the greater part of the Congo basin, creating the colony of Belgian Congo.

A Belgian-appointed governor-general administered the Belgian Congo colony. From the start of Belgian administration, economic life was dominated by large corporations given concessions to do business in specific geographic areas. In many cases, government and private concerns joined in economic enterprises. Neither whites nor Africans were allowed to vote. Belgium took the position that the first requirement for the colony was a sound economic base and an acceptable standard of living for both whites and Africans, with political development to come later.

Belgian authorities tried to draw Africans more fully into general economic life than was the case in most African colonial possessions. Rather than being confined to unskilled labor, Africans were brought into occupations requiring considerable skill. They learned to operate trains, riverboats, steam shovels,

FIGURE 18.8
Fishermen working shallows along the Congo River, the Democratic Republic of Congo's lifeline.   *George Holton/Photo Researchers*

bulldozers, and electric furnaces; to work as carpenters, masons, telegraphers, and typists; and to serve as postal clerks, nurses, elementary school teachers, and pastors of African Christian churches. The Belgians emphasized elementary education, and a larger proportion of the children attended elementary schools (operated mainly by Roman Catholic and Protestant missions) than was true in most African colonies. But Belgian policies barred all but a tiny handful from receiving a college education; almost none received training as doctors, lawyers, or engineers. The lack of well-educated African leaders trained in management techniques proved disastrous when independence came.

Independence

Pressure for independence built up rapidly in 1959, and Belgium yielded to the demands of Congolese leaders. In 1960, three months before adjacent French Equatorial Africa became the independent "People's Republic of the Congo," the Congo colony became the independent Republic of the Congo. (In 1971 it took the name Zaire, meaning "River." Following the overthrow of Zaire's government in 1997, the country was renamed "Democratic Republic of the Congo," and is now generally referred to simply as "Congo." The neighboring People's Republic of the Congo is now generally referred to as "Congo Republic.") The new country quickly faced major crises that its government was unable to handle. These included a large-scale mutiny within the Belgian-trained Congolese army, regional separatist movements and rebellions—notably a declaration of independence by the mineral-rich Katanga (now Shaba) Province in the southeast—and ethnic warfare, especially in the diamond-mining area of Kasai Province.

With Republic of the Congo disintegrating into chaos, Belgian troops intervened to protect and evacuate Belgian nationals. Shortly afterward, the United Nations dispatched an emergency force to restore order and assist the government in Katanga. After intense fighting between United Nations and Katanga forces, the secession movement in Katanga ended in 1963. However, new rebellions and political crises centering in the Republic of the Congo's remote east followed the withdrawal of U.N. military units in 1964. Government forces eventually contained the rebels with the aid of white mercenaries, who themselves subsequently became disaffected and for a while exerted independent control over sizable areas in the east. This episode ended with the withdrawal of the mercenaries from the country under guarantees of safe passage.

The country's economy then entered a difficult era that continues today. Major problems have included governmental corruption and mismanagement, squandering of funds on ill-advised projects, large withdrawals of foreign aid, a decline in the world price of copper (the country's chief export), the high cost of imported oil products after 1973, and civil war beginning in 1997 (see below).

Economic and Urban Geography Today

Like many other states in Africa south of the Sahara, the Democratic Republic of Congo has only a modest development of manufacturing plants, producing mainly simpler types of goods. Typical commodities are cotton textiles, shoes, bricks, cement, wood products, processed foods, cigarettes, some chemicals, and simple metal products. The country has been important in the world economy mainly as an exporter of minerals and tropical agricultural products. A large proportion of the minerals are smelted or concentrated in the Democratic Republic of Congo and then are further refined at plants in Belgium and in other overseas industrial countries. Among the leading mineral exports, copper, cobalt, and other metals come principally from the Shaba region and diamonds from the adjoining Kasai region (see Fig. 17.3). Copper is by far the leading export. The Shaba region is also a major world producer of cobalt, an alloy with vital high-technology uses (for example, in jet aircraft engines, to resist heat and abrasion). Diamonds, primarily industrial but including some gemstones, are secured in the Kasai region to the east and west of Kananga (population: 393,030, city proper).

Most nonmineral exports of the Democratic Republic of Congo, including palm products, rubber, some coffee, and wood, come from areas of tropical rain forest near the Congo River. Coffee is grown in many different parts of the Democratic Republic of Congo. The variety known as *arabica*, grown in the eastern highlands, brings a particularly high price. Corporate plantation agriculture is still important, but foreign ownership of agricultural land is prohibited.

Matadi is the Democratic Republic of Congo's only important seaport. Rapids block navigation on the Congo just upstream from Matadi, and goods destined for Kinshasa must be transshipped by rail or truck about 230 miles (370 km). Located on the Congo River at the lower end of another stretch of navigable water about 1000 miles (c. 1600 km) long, Kinshasa (population: 3 million, city proper) is the Democratic Republic of Congo's capital, largest city, and principal manufacturing center. Another well-known city is Lubumbashi (population: 851,381, city proper), formerly Elisabethville, in the copper-mining region of the southeast. One of the Democratic Republic of Congo's major arteries of transportation connects these two cities. It is composed of a railway from Lubumbashi to Ilebo on the lower Kasai River (see Fig. 17.3) and a navigable water connection from Ilebo to Kinshasa via the Kasai River and the Congo River. The Democratic Republic of Congo formerly used the Benguela Railroad across Angola to export copper through the Angolan port of Lobito, but this route to the sea has been largely useless for many years because of Angola's civil war. Deterioration of roads handicaps travel within the country.

The Democratic Republic of Congo is so deeply mired in poverty that its per capita GNP is only about $220, a remarkably low figure for a country with such an abundance of resources. Only a handful of the world's countries are poorer. At present there is little optimism that the Democratic Republic of Congo's economy will soon improve. Billions of dollars derived from mineral sales or foreign loans had been spent on grandiose "white elephant" projects and pocketed by political leaders and their families. A heavy foreign debt has had to be rescheduled because of the Democratic Republic of Congo's difficulty in paying even the interest. The country's longstanding autocratic ruler, Mobutu Sese Seko, acquired a massive fortune during his reign of more than 30 years. Political opposition and public discontent with the 80 percent unemployment rate and staggering rates of inflation led to riots and unrest between 1991 and 1993. Production of copper and cobalt plummeted, and by 1997 the country's economy appeared to be on the brink of collapse.

Civil war rocked the Democratic Republic of Congo in 1997. Late in 1996 and early in 1997, Zairean army troops attempted to demilitarize and expel a perceived weak force of indigenous Tutsis living in the country's far eastern region. The assault backfired when these Tutsis, supported by neighboring Tutsi-dominated governments in Rwanda and Burundi, and by other sympathetic countries in the region, took up yet more arms and initiated fierce attacks on government forces. The Tutsi counterattack quickly widened into an antigovernment revolt comprised of many ethnic factions and led by a non-Tutsi guerrilla named Laurent Kabila. In spring 1997, Kabila's forces pushed westward virtually unopposed through the huge country and in May occupied Kinshasa with little loss of life. President Mobutu fled Zaire on the eve of Kabila's triumph. Kabila then declared himself president of the country he now called the Democratic Republic of Congo, and promised that political reforms would live up to the country's new name. Foreign companies rushed to sign mining contracts with the fledgling government, and throughout Congo there was hope that decades of neglect and poverty could be reversed in this resource-rich giant of West Central Africa.

18.5   East Africa

East Africa comprises three former British dependencies—Kenya, Tanzania, and Uganda—and two former Belgian dependencies—Rwanda and Burundi—that lie between the Congo basin and the Indian Ocean (see Fig. 17.1, page 400). Kenya and Tanzania front the Indian Ocean, but Uganda, Rwanda, and Burundi are landlocked. Prior to their independence, the three British units varied in political status. Uganda was a protectorate administered with African interests paramount. Kenya was a colony and protectorate in which a white settler class had great political influence. Kenya and Uganda are now republics.

Tanganyika (formerly German East Africa, now Tanzania) was a United Nations trust territory administered by Britain. Now a part of Tanzania, the islands of Zanzibar and Pemba, which lie north of Dar es Salaam and some 25 to 40 miles (40–64 km) off the coast, were once a British-protected Arab sultanate. The city of Zanzibar was the major political and slave-trading center, and **entrepôt** of an Arab empire in East Africa. While Arabs formerly controlled political and economic life, the majority of the population was, and is, African. Africans have been in control since 1964, when a revolution overthrew the ruling sultan. In 1964, the political union of Tanganyika and Zanzibar was formed with the name "United Republic of Tanzania." Zanzibar and Pemba have a parliament and president separate from Tanzania's.

The small but densely populated highland countries of Rwanda and Burundi were also once a part of German East Africa. When Germany lost its colonies at the end of World War I, the League of Nations placed the two units under Belgian administration as the mandated territory of Ruanda-Urundi. This area later became a Belgian trust territory under the United Nations and then received independence in 1962 as two countries.

AREA, POPULATION, AND PHYSICAL ENVIRONMENT

Although a coastal lowland fringed by coral reefs and mangrove swamps occupies the eastern margins of Kenya and mainland Tanzania, the terrain of the five East African countries is composed mainly of plateaus and mountains. The plateaus lie generally at elevations of 3000 to 6000 feet (c. 900–1800 m) but rise higher in some places, and in much of eastern and northern Kenya, they descend to low plains. Individual plateau surfaces vary in elevation, topography, climate, and vegetation. Often, the land opens out broadly into undulating or rolling country, punctuated by remnant hills or mountains formed of older resistant rock. All five countries include sections of the region's Great Rift Valley. International frontiers follow the floor of the Western Rift Valley for long distances and divide lakes Tanganyika, Malawi, and Albert and smaller lakes among different countries. The more discontinuous Eastern Rift Valley crosses the heart of Tanzania and Kenya. Lake Victoria, lying in a shallow downwarp between the two major rifts, is divided among Uganda, Kenya, and Tanzania. None of these three countries is predominantly mountainous, but all have mountains in some areas, most commonly in close proximity to the rift valleys.

Vulcanism, often on a massive scale, accompanied formation of these valleys in many places. The two highest mountains in Africa, Mt. Kilimanjaro (19,340 ft/5895 m; see Fig. 18.10) and Mt. Kirinyaga (formerly Mt. Kenya; 17,058 ft/5200 m), are extinct volcanoes. Both are majestic peaks crowned by permanent snow and ice and visible for great distances across the surrounding plains. In some places, most notably in southwestern Kenya, erosion has sculpted lava flows into hills and mountains. East Africa's most productive agricultural districts, including many that contribute to the export trade, tend to be localized in these volcanic areas of unusually fertile soils (Fig. 18.9). Vulcanism was also important in building the fertile highlands of Rwanda and Burundi. The Virguna Mountains along the Rwanda-Uganda border include many active volcanos. One of the region's most active volcanos is the prominent Mt. Elgon (14,176 ft/4321 m) on the Uganda-Kenya border. On the Uganda–Democratic Republic of Congo border is Margherita Peak (16,762 ft/5109 m), part of a nonvolcanic range known as the Ruwenzori, which was commonly identified with the "Mountains of the Moon" depicted by early Greek geographers on their maps of interior Africa.

The climate of most of East Africa is classified broadly as tropical savanna, a category embracing a wide range of temperature and moisture conditions. Altitude has a strong effect on temperatures, and some agricultural districts, even on the

FIGURE 18.9

East Africa is blessed with extensive areas of volcanic soil where commercial and subsistence crops thrive. The main crop in this region near Kigali, Rwanda, is banana.   *Walt Anderson/Visuals Unlimited*

equator, are so tempered by elevation that midlatitude crops like wheat, apples, and strawberries do well. Kenyan tourist brochures boast accurately that Nairobi (elevation: 5500 ft/1676 m) has a springlike climate year-round. There are great variations in the amount, effectiveness, and dependability of precipitation and in the length and time of occurrence of the dry season (or seasons). Over much of East Africa, moisture is too scarce for a truly productive nonirrigated agriculture. Long droughts occur, even in what would normally be the rainy season. Relatively little of the present crop acreage is irrigated. Adequate moisture and good soil are linked only in the scattered areas that support the bulk of East Africa's population.

East Africa's natural vegetation is composed largely of deciduous woodlands interspersed with grasslands. Forest growth of commercial value is scarce. Softwoods are in short supply, but there are several ongoing government afforestation programs. The most luxuriant forests grow on rainy mountain slopes and along the northern coastal plain of Lake Victoria. The deforestation, cultivation, and overgrazing of steep slopes in the mountainous countries of Rwanda and Burundi and in southwestern Uganda have produced severe erosion, lessening the ability of the land to yield more than bare subsistence to the dense and rapidly increasing population.

In Uganda, Kenya, and Tanzania, park savanna—composed of grasses and scattered flat-topped acacia trees—stretches over broad areas and is the habitat for some of Africa's largest populations of large mammals. Great herds of herbivores and their predators and scavengers migrate seasonally across these countries to follow the changing availability of food and water. The countries manage numerous parks and wildlife reserves, and in Kenya and Tanzania these are responsible for an enormous share of the nations' hard currency revenues. Among the most famous are the Serengeti Plains and Ngorogoro Crater of Tanzania, and the Masai Mara and Amboseli National Parks of Kenya (Fig. 18.10).

Most parts of East Africa are not thickly populated, yet this is one of the world's fastest growing areas of population growth. The five countries make up some 703,000 square miles (1.82 million sq km), roughly the size of the U.S. state of Texas, with a total population in 1996 of 92 million (about one third the population of the United States). The populations of the five East African countries were increasing at rapid rates as of 1996: Kenya at 2.7 percent natural increase per year, Tanzania 3.0 percent, Uganda 3.3 percent; Rwanda 2.7 percent; and Burundi 3.0 percent. The estimated average density of population in 1996 for East Africa as a whole was 341 per square mile (132 per sq km). The population tends to be heavily concentrated in a few areas of dense settlement that support hundreds of people per square mile. The three largest and most important areas with a greater-than-average density are (1) in a belt along the northern, southern, and eastern shores of Lake Victoria, (2) in south central Kenya around and north of Nairobi, and (3) in most of Rwanda and Burundi.

FIGURE 18.10
The world's greatest concentrations of large mammals are in East and Southern Africa. This is Amboseli National Park in southern Kenya, with the Kilimanjaro volcano rising to 19,340 ft. (5895 m) on the border with Tanzania.   *Daryl Balfour/Tony Stone Images*

ECONOMIC AND URBAN GEOGRAPHY

Most of East Africa's subsistence agriculturalists are cultivators, but some are pastoralists. Nearly the entire range of subsistence and export crops typical of Africa south of the Sahara is to be found in the region. For the area as a whole, the main subsistence crops are maize, millets, sorghums, sweet potatoes, plantains, beans, and manioc. (Manioc is a particularly valuable plant in areas prone to crop failure because the roots can be left in the ground for several years as a reserve food supply.) Some cultivators depend exclusively on crops for subsistence, and others also keep livestock. Cattle are the most important livestock animal (see Landscape in Literature, page 436).

The most valuable export crop is coffee, but tea is also important in Kenya. Kenya and Tanzania also produce sisal for its valuable fiber. The islands of Zanzibar and Pemba rely mainly on millions of clove trees (primarily on Pemba) that provide the bulk of the world's supply of cloves. A recent fall in world clove prices has hurt the islands' economies, however, and in 1995, they announced plans to transform Zanzibar into a free economic zone modeled after Singapore and Hong Kong.

Prior to the new era of political independence, East Africa was home to some 300,000 Asians (Indians, Pakistanis, and Goans, from the former Portuguese colony of Goa on the west coast of India), 100,000 Europeans, and 60,000 Arabs (the figures are estimates for 1958). Most Asians had commercial, industrial, and clerical occupations, although some owned estates producing export crops. They dominated retail trade and owned factories. Some were wealthy, and practically all were more prosperous than most Africans. Arabs were primarily a mercantile class. Europeans were a professional, managerial, and administrative class. In the "White Highlands" of southwestern Kenya, about 4000 European families owned and operated estates that were worked by African laborers to produce export crops and cattle. Since independence, many Europeans have left, and Africans occupy much of the farmland formerly on European estates.

The most productive parts of East Africa are bound together by railways that form a connected system leading inland from the seaports of Mombasa in Kenya and Dar es Salaam and Tanga in Tanzania (see Fig. 17.3). Mombasa (population: 465,000, city proper) is the most important seaport in East Africa. Situated on an island connected with the mainland by causeways, the city has two harbors: a picturesque old harbor still frequented by Arab sailing vessels (*dhows*) and other small ships, and a deep-water harbor that has modern facilities for handling large vessels (Fig. 18.11).

From Mombasa, the main line of the Kenya-Uganda Railway leads inland to Kenya's capital, Nairobi (population: 1.5 million, city proper), the largest city and most important industrial center in East Africa. It is also the busiest crossroads of international air traffic and the main outfitting and departure point for safaris into East Africa's world-renowned national parks. Nairobi was founded early in the 20th century as a construction camp on the railway, which continues westward to Kampala (population: 773,463, city proper), the capital and largest city of Uganda. Dar es Salaam ("House of Peace," population: 1.1 million, city proper) is the capital, main city, main port, and main industrial center of Tanzania, with rail connections to Lake Tanganyika, Lake Victoria, and Zambia. Rwanda's capital is in the highlands at Kigali (population: 237,782, city proper), and Burundi's capital is on the northernmost point of Lake Tanganyika at Bujumbura (population: 236,334, city proper). Only about 6 percent of the populations of Rwanda and Burundi is urban.

In comparison with many other African countries, the East African nations lack mineral reserves and mineral production. Manufacturing is in an early stage of development, limited mainly to agricultural processing, some textile milling, and the making of simple consumer items. Both Mombasa and Dar es Salaam have refineries that process imported crude oil. There is a large cotton-textile mill at Jinja, Uganda, that operates with hydroelectricity from a station at the nearby Owen Stanley Dam on the Nile. The Tana River scheme in Kenya produces hydropower and supplies irrigation waters.

SOCIAL AND POLITICAL ISSUES

The five East African countries face many social and geopolitical dilemmas. None has had much internal peace, order, or prosperity, and none has had easy relations with its neighbors. Ethnic rivalries and conflicts have beset all of them.

Aside from small minorities of Asians, Europeans, and Arabs, the population of East Africa is composed of a large number of African tribes; Tanzania alone, for example, has about 120 tribal groups. Among the large tribes are the Kikuyu and Luo of Kenya, the Baganda of Uganda, the Sukuma of Tanzania, the pastoral Maasai of Kenya and Tanzania, and the Tutsi and Hutu of Rwanda and Burundi. Most East African peoples speak Bantu tongues of the Niger-Congo language family, but some in northern Uganda, southern and western Kenya, and northern Tanzania speak Nilo-Saharan languages. Swahili, a Bantu language drawing heavily on Arabic for vocabulary, is a widespread lingua franca (and is the national language of Tanzania), as is English. Religions in all five countries include animist, Muslim, and Christian elements. The Muslim religion reflects a long history of Arab commercial enterprise in the region, including slave trading.

Since independence, African governments have driven the Asian minority from its commercially predominant position. In Uganda, where Asians were the backbone of commerce and trade, the country's former leader Idi Amin abruptly expelled them from the country as refugees in the 1970s. Some have recently begun to return.

FIGURE 18.11
Mombasa's new harbor of Kipevu on Kenya's Indian Ocean coast.   *J. C. Carton/Bruce Coleman Inc.*

Landscape in Literature

CATTLE IN TUTSI CULTURE

Cattle are central to the lives of many pastoral and mixed agricultural-pastoral peoples in Africa south of the Sahara (see Figs. 17.13 and 17.14). Numerous ethnic groups hold the animals in great esteem and are highly dependent on them economically. This passage offers insight into the importance of cattle to the Tutsi (Mututsi) of highland Burundi (Urundi). There is also brief reference to the traditional rivalry between the Tutsi and the Hutu (Bahutu), which exploded into slaughter in the mid-1990s. This selection is from a traditional genre of Tutsi oral literature known as the *amazina*, which praises warriors, cattle, hunters, and hunting dogs.

No one has travelled in Urundi without understanding how much the cow means to the country. The cow is loved for its beauty and for its usefulness and for the prestige it brings. Its beauty is not extraordinary. But what a man loves he considers beautiful.

The cow is loved for its great use-

fulness. Nothing that comes from it is without value. The Murundi [a person of Burundi] has need of every part of the cow, from its hoof to the tip of its horn.

The poor man arrives at the slaughter-house and if he fails to get meat he at least goes off with a cow's hoof.

The hunter or fighter who comes to the slaughter-ground remembers the bow and the string stretching it and takes away the cow's tendons.

The maker of bark-cloth sees the horns, remembers the board on which he beats out the cloth and so takes away the points of the horns (to beat the cloth with on the board).

A man who is longing for meat and indeed anyone who is hungry for meat comes to the slaughter-ground; when he sees a piece of meat he thinks of what he possesses and uses some of it. It does not matter whether it be grain or metal (with which to barter) so long as he can buy some meat.

So the cow is of worth to its owner and to his neighbours. Nothing of the cow goes to waste, not even its skin. I have said "not even," but really the skin is of no small value to people.

Before the Europeans came and taught us a different manner of dressing, the cow's skin was worn in the same way as the beaten-out bark-cloth. The cow was carefully skinned, the skin nailed out to dry, then scraped and beaten. The butter was rubbed on and when the skin was soft enough it was put on. Poor people did not trouble with all this preparation. They simply put on the skin without its being made soft. Because of this they were called indibati (rustlers), the reason being that when they walked the skin began to make the noise: ndibati, ndibati (rustle, rustle).

But the value of the dead cow is not as much as that of a living one. Although, as an individual, I have spoken of the value of a dead cow, I would not have dared to do so if I had been with a group of people, espe-

---

Episodes of bloodshed have punctuated East Africa's history. From 1905 to 1907, when mainland Tanzania was German-controlled, German authorities put down a great revolt against colonial control, costing 75,000 African lives. Immediately prior to its independence, Kenya experienced a bloody guerrilla rebellion led by a secret society called the Mau Mau within the large Kikuyu tribe. Antagonism directed against British colonialism and European ways in Kenya cost about 11,000 African and 95 white settler lives over the period 1951–1956.

In postcolonial Uganda, a savage orgy of murder, rape, torture, and looting during the notorious regimes of Idi Amin and Milton Obote dwarfed such preindependence turmoil. Both leaders were Ugandan northerners—Amin belonging to a relatively minor Muslim tribe, the Kakwa, and Obote a member of the predominantly Christian Langi tribe. From independence in 1962 until 1971, Obote governed Uganda dictatorially as prime minister. His supporters initially were made up of a coalition of northerners and the southern Baganda, Uganda's largest tribe

and the one that the British colonial government had favored for recruitment into the civil service. But the Baganda did not support the Obote government for long, and Obote was forced to rely on his northern kinsmen to maintain his power. In 1971, their allegiance proved insufficient to withstand a military coup by the army commander, Amin. At the outset, soldiers that Amin had recruited from Sudan, Zaire, and his own Muslim tribe massacred Obote's northern soldiers in their barracks. Bands of soldiers then terrorized the country until Amin was ousted in 1979.

In 1981, Obote returned to power, and undisciplined soldiers began a new reign of terror. It lasted until 1985, when an army coup exiled Obote. The ravaged country finally had a respite in 1986 when Yoweri Museveni, the leader of a southern guerrilla force that had developed in opposition to Obote, took over the government. The formidable task of rebuilding began, beset by great difficulties because of the collapsed economy and the continuing cultural and political fragmentation of Uganda among some 40 tribes, mostly unable to communicate

cially a group of those who possess cows. Are any of you unaware that a Mututsi looks after his cow, brings it up so that it gives birth to many calves and when it gets old he cannot bring himself to eat it, but instead exchanges it for a less-valuable cow? Have you never seen him weeping and wailing when a cow dies in the enclosure or through falling into a pit?

Think what the cow means to him. He calls it his friend and his companion. And is the cow not both food and health to him? That well-filled-out body, those fat children, are they not the result of the milk and the butter of the cow? Those fields with good crops, are they not the result of the compost provided by cow manure and the grass that cows have slept on? That strong and prosperous homestead, is it not dependent on the cow—its very offspring?

Would you dare say to me: "The cow does not mean all that much to a Mututsi. To lose a cow is not such a terrible catastrophe; it does not mean

deep poverty to the Mututsi or irrepressible loss"? I ask you, since the cow really controls him and his family, how could he love his family and dislike the cow since the cow provides for them? Let him love the cow for he has good reason so to do.

But this is not the only reason why he loves the cow. He loves it for the prestige it gives him. A kraal [enclosure] where cows come home at night is not a lonely place. A kraal where cows come home at night is well established, it is respected, it is the kraal of a true man.

In the evening he brings home his cows, the smoke goes up from the fire and the neighbours say: "So and so has brought home his cows." In the morning he lights the fire again, the cows low and the neighbours say: "So and so is leading his cows out to pasture." Listen to that lowing of the cow and to its breathing at night, and tell me if such a kraal where there are cows is compared to a kraal where only crickets chirp at night? Isn't this what made a Mututsi laugh when he

came across some Bahutu who had no cows? He found them talking together, smiling and laughing merrily. He withdrew from them and said to his friends: "These Bahutu are not even sad! They have no cows and yet they can find something to laugh about!"

When you find him among his fellows, listen to the way that which is dearest to his heart comes out in his conversation. When he wants to express astonishment he talks of a man who has given him a cow! If he wants to praise you he wishes you to have the company of this benefactor! When he is happy he swears by such and such a cow! If he hates you he expresses the hope that all your cows will be taken from you, that you may have no milk, no cows, and that you may drink water only. Only God could conceive of a worse curse than that!

From Monsignor Ntuyahaga, "Here Come the Cows," *A Selection of African Prose*, ed. W. W. Whiteley. (Oxford: Clarendon, 1964), 142–145.

with each other in their native tongues and bearing heightened suspicions and animosities from the recent disastrous decades. Sporadic internal violence, border conflicts, and economic problems continued into the 1990s, although tensions had eased and reconstruction was proceeding in 1997.

For a time, an East African Community existed, within which Kenya, Tanzania, and Uganda cooperated in economic matters. This union broke apart in 1977. International borders within the area have been closed for long periods. Meanwhile, the countries have followed divergent economic philosophies. Kenya has embraced many capitalist ideas and has attracted some Western investment. At the same time, highly centralized control has been evident in certain aspects of the Kenyan economy, such as agricultural marketing. Although much formerly white-owned farmland has been redistributed among African small landholders (with compensation paid to the whites), there are still several hundred whites whose large holdings, worked by African laborers, make a vital contribution to Kenya's export trade in tea and other farm products. Economically, Kenya is a

relatively successful country of Africa south of the Sahara, although with a per capita GNP of only $260 (in 1994) it is certainly not prosperous. Recent political unrest and rampant crime in Nairobi has threatened the vital business of international tourism in Kenya, which contributes 40 percent of the country's hard currency revenue. In addition, since 1990, the international community has accused President Daniel Arap Moi's government of corruption and human rights abuses against political and tribal opponents. In 1992, Britain, the U.S., and other donor countries threatened to cut off aid to Kenya unless Moi implemented democratic reforms. He appeared to do so, but Moi's victory in elections late that year was regarded widely as fraudulent. Limited international economic assistance has resumed, however, in part to encourage greater democratization in advance of upcoming elections.

Discrepancies between the haves and have-nots are much greater in Kenya than in neighboring Tanzania, where former president Julius Nyerere instituted an order based on the principle of African socialism, or *ujamaa*. Intended mainly to create self-sufficiency and redistribute land, the system did improve social services but also led the country's economy to near collapse. The effort to eradicate rural poverty and increase harvests through the collectivization of scattered farming communities met with little success.

Like many states in Africa south of the Sahara, Tanzania has squandered money on some "white elephant" projects. One was a multimillion-dollar paper mill that had to be shut down because operating costs were so high that the paper was cheaper to import than to produce.

In the mid-1990s, Rwanda and Burundi were engaged in disputes that had tragic consequences. About 85 percent of the population in Burundi and 90 percent in Rwanda is composed of the Hutu (Bahutu), and most of the remainder of the two populations is Tutsi (Watusi). These are not separate tribal or ethnic groups; the peoples speak the same language, share a common culture, and often intermarry. The distinction between Hutu and Tutsi is rather one of socioeconomic class. Historically, the Tutsi were a ruling class who dominated the Hutu majority. Their power and influence were measured especially by the vast numbers of cattle the Tutsi owned (see Landscape in Literature, page 436). Tutsi who lost cattle and became poor came to be identified as Hutu, and Hutu who acquired cattle and wealth often became Tutsi.

Colonial rule in East Africa attempted to polarize these groups, thereby increasing antagonism between them. German and Belgian administrators differentiated between them exclusively on the basis of cattle ownership: anyone with fewer than 10 animals was Hutu, and anyone with more was Tutsi. The

colonists replaced all Hutu chiefs with Tutsi chiefs. These leaders carried out colonial policies that often imposed forced labor and heavy taxation on the Hutus. Education and other privileges were reserved almost exclusively for the Tutsi. Europeans mythologized the Tutsis as "black Caucasian" conquerors from Ethiopia who were a naturally superior, aristocratic race whose role was to rule the peasant Hutus.

Ferocious violence between these two peoples has marked Rwanda and Burundi intermittently since the states became independent in the early 1960s. After independence, the majority Hutus came to dominate the governments of both Rwanda and Burundi. In 1994, the death of Rwanda's Hutu president in a plane crash (in which Burundi's president also died) sparked civil war in Rwanda between the Hutu-dominated government and Tutsi rebels of the Rwandan Patriotic Front (the RPF, based in Uganda). Hutu government paramilitary troops systematically massacred Tutsi civilians with automatic weapons, grenades, machetes, and nail-studded clubs. Women, children, orphans, and hospital patients were not spared. An estimated 500,000 died. International media tracked these horrors, but outside nations did nothing to stop the fighting.

Well organized and motivated Tutsi rebels seized control of most of the country and took power in the capital, Kigali, in 1994. That precipitated a flood of two million Hutu refugees mainly into neighboring Zaire (now the Democratic Republic of Congo) (Fig. 18.12). Although the Tutsi-dominated government promised to work with Hutus to build a new multiethnic democracy, Hutu insurgents, based mainly in eastern Zaire, continued to carry out attacks against Tutsi targets. These Hutu

FIGURE 18.12
Warfare and drought in post-colonial Africa have driven millions of Africans from their homes as refugees. This harrowing scene shows terrorized Rwandans fleeing into Tanzania in 1994 to escape the savage massacres of civilians in Rwanda's civil war. *Scott Daniel Peterson/Gamma Liaison*

operations all but ended in 1997 when the mostly Tutsi forces of Laurent Kabila's rebel alliance established dominance in the region. Kabila's troops incited a reverse flow of Hutu refugees back into Rwanda, as the international community watched anxiously to see if the victorious Tutsis would carry out revenge killings against Hutu civilians or militia.

A parallel situation existed in Burundi in 1996, when the Tutsi-dominated army overthrew the country's Hutu president. The new Tutsi president vowed to end the ethnic bloodshed that had cost 150,000 lives since October 1993, when Tutsis murdered Burundi's first elected Hutu president, sparking large-scale Hutu retaliatory slaughter of Tutsis. But the civil war instead worsened. Even under strong pressure from Hutu rebels, Burundi's Tutsi-dominated government is unlikely to settle for anything less than control of the army and half of the government ministries. Tutsis argue that only such strength would protect them from more bloodshed in the event that a Hutu president comes to power. For their part, Hutu opponents to the regime argue that they are fighting for democracy. They want an integrated army and a constitutional government like the one that existed in 1993, when the first Hutu president was elected.

18.6   Southern Africa

In the southern part of Africa south of the Sahara are five countries—Angola, Mozambique, Zimbabwe, Zambia, and Malawi—that share the basin of the Zambezi River and have long had important relations with each other. The first two were former Portuguese colonies and the latter three were formerly British. Still farther south are South Africa and four countries whose geographical problems are closely linked to South Africa: the three former British dependencies of Botswana, Swaziland, and Lesotho, and the former German colony of Namibia. These ten states are grouped here as the countries of southern Africa (see Fig. 17.1, page 400). As the region's dominant power, South Africa is described here in the greatest detail; the others are depicted in groupings of countries.

SOUTH AFRICA

South Africa was one of the world's most controversial countries until its new beginning in 1994. Organized from four British-controlled units in 1910 as the Union of South Africa, it was for five decades a self-governing constitutional monarchy under the British crown. In 1961, following a close majority vote by the ruling white population, it became a republic outside the British Commonwealth.

The economic power of South Africa (1996 population: 45 million) is now comparable to that of the rest of Africa com-

bined. It accounts for about one third of Africa's output of manufactured goods and one third of the electricity generated on the entire continent. Its diversified industries include iron and steel, machinery, metal manufacturing, weapons, automobiles, tractors, textiles, chemicals, processed foods, and others. The United Kingdom and other Western countries have invested heavily in South Africa, and many thousands of jobs in those countries are directly linked to trade with it. South Africa's military strength, which probably includes nuclear weapons (the country will not disclose this information), contributes to its global significance. It is strategically important for its frontages on both the Atlantic and Indian oceans, its possession of Africa's finest transport network (vitally important to many African countries that trade with and through South Africa), and its diversified mineral wealth.

South Africa's physical environment has played a major role in its unusual course of development (Fig. 18.13). Lying almost entirely in middle latitudes, it has annual temperatures generally between 60° and 70°F (16°–21°C). It is a cooler land than most of Africa south of the Sahara. In combination with its natural resource wealth, this climate made the area very attractive for European settlement; it was one of the "neo-Europes" of the colonial era (see Chapter 2). Consequently, South Africa is the main center of European settlement and advanced economic development in an underdeveloped continent that Europeans dominated until very recently.

Visitors from western Europe and North America will find in South Africa most of the institutions and facilities to which they are accustomed. However, the Europeanized cultural landscape does not reflect the majority culture of this unusual country. Whites represent only about 13 percent of the total population, and blacks outnumber them more than five to one. There are two other large racial groups: the so-called "coloreds," of mixed origin, and the Asians, most of whom are Indians (Table 18.1). Many peoples make up the black African majority. The Zulu and Xhosa tribes, both concentrated in hilly sections of eastern South Africa near the Indian Ocean, are the most populous of the nine officially recognized tribal groups.

A huge economic gulf separates South Africa's impoverished black Africans from about 6 million whites of European

TABLE 18.1	**Population Elements of South Africa**		
Racial Group	**Total Number (1996 estimates, in millions)**	**Percent of Total Population**	
Black African	33.9	76.1	
European	5.7	12.8	
Colored	3.8	8.5	
Asian	1.2	2.6	

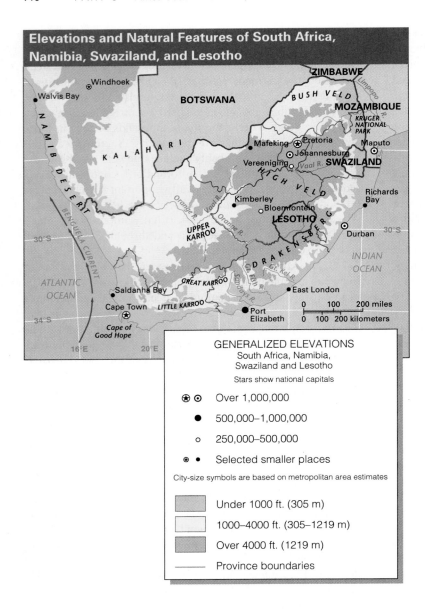

Elevations and Natural Features of South Africa, Namibia, Swaziland, and Lesotho

GENERALIZED ELEVATIONS
South Africa, Namibia,
Swaziland and Lesotho

Stars show national capitals

⊛ ⊙ Over 1,000,000

● 500,000–1,000,000

○ 250,000–500,000

⊛ • Selected smaller places

City-size symbols are based on metropolitan area estimates

Under 1000 ft. (305 m)

1000–4000 ft. (305–1219 m)

Over 4000 ft. (1219 m)

——— Province boundaries

FIGURE 18.13
Major natural features of South Africa and surroundings.

descent who dominate the economy and enjoy a lifestyle comparable to that of people in western Europe and North America. Because of the poverty of the non-European majority, South Africa's overall per capita GNP of $3010 is typical of such LDCs as Brazil and Turkey. For unskilled and semiskilled labor with which to operate their mines, factories, farms, and services, the whites of South Africa have always relied mainly on low-paid black workers in most of the country and colored workers in the Western Cape Province. The white-dominated economy could not operate without them, and the Africans in turn are extremely dependent on white payrolls. While the races are interlocked economically, however, relations between them have historically been very poor and have been structured to benefit the whites.

How South Africa's Racial Situation Developed

The British-Afrikaner Division.   South Africa's white population is split between the British South Africans and the Afrikaners, or Boers (Dutch: "farmers"). The Afrikaners speak Afrikaans, a derivative of Dutch, as their preferred language. They outnumber the British approximately three to two. The Afrikaners are the descendants of Dutch, French Huguenot, and German settlers who began coming to South Africa more than three centuries ago. The Dutch East India Company established the earliest permanent settlement at Cape Town in 1652 as a way station to provide water, fresh vegetables, meat, and repairs for company vessels plying the Cape of Good Hope route to the Far East. Although the company did not intend to annex large areas of land, agricultural settlements slowly expanded in valleys near Cape Town, and a pastoral frontier society developed farther inland. Its impact on indigenous Africans was catastrophic. White settlers killed, drove out, and, through the unintentional introduction of smallpox, decimated most of the original Khoi and San (Bushman) inhabitants. They put the survivors to work as servants and slaves. When these proved to be insufficient in number, the colonists brought slaves from West Africa, Madagascar, East Africa, Malaya, India, and Ceylon (now Sri Lanka).

Great Britain occupied the Cape Colony in 1806, during the Napoleonic Wars. Friction developed between many Boers and the British authorities, who imposed tighter administrative and legal controls than the Boers were accustomed to. The British abolished slavery throughout their empire in 1833, contributing to the anti-British sentiments of the slave-holding Boers. Boer discontent resulted in the Great Trek, a series of northward migrations by which groups of Boers, primarily from the eastern part of the Cape Colony, sought to find new interior grazing lands and establish new political units beyond British reach. After some earlier exploratory expeditions, the main trek by horse and ox-wagon began in 1836. It resulted in the founding of Natal, the Orange Free State, and the Transvaal as Boer republics. Britain annexed Natal in 1845, but recognized Boer sovereignty in the Transvaal and the Orange Free State in 1852 and 1854 respectively.

While Boer disaffection with Britain was building, British settlers were coming to South Africa in increasing numbers. The area's natural resource wealth helped shape subsequent events. The British colonies and Boer republics might have developed peaceably side by side if diamonds had not been discovered in the Orange Free State in 1867 and gold in the Transvaal in 1886. These discoveries set off a rush to these republics of prospectors and other fortune hunters and entrepreneurs from outside the region—particularly from Britain. Ill feeling led to the Anglo-Boer War in 1899. British troops defeated the Boer forces decisively, ending the war in 1902 but leaving behind a reservoir of animosity that exists to this day.

The Union of South Africa was established in 1910 as a self-governing dominion under the British crown. Dutch became an official language on par with English (this was later altered to specify Afrikaans rather than Dutch). Pretoria in the Transvaal became the administrative capital as a concession to Boer sentiment. However, the national parliament now meets at Cape Town and the supreme court sits at Bloemfontein (population: 250,000, city proper) in the Orange Free State.

Since 1910, and particularly since the Nationalist (now National) Party took power in 1948, Afrikaners have dominated the political life of South Africa by virtue of their greater numbers and cohesion. Racial segregation characterized South African life from 1652 onward, but it was first systematized after 1948 under a comprehensive body of national laws supporting the official governmental policy of "separate development"

of the races" (subsequently called "multinational development" and commonly known as *apartheid;* see Definitions and Insights, page 441). The new laws imposed racially-based restrictions and prohibitions on the entire population, but they weighed most heavily on black Africans. They ruled out significant sharing of power by black people in a unified South Africa. The apartheid system drew furious criticism from many other nations. Most of the world community ostracized South Africa, and many countries, including the United States, imposed economic sanctions against it.

Definitions and Insights

APARTHEID

The official South African policy of "separate development of the races," known best by its Afrikaner name, **apartheid,** was the world's most anachronistic vestige of colonialism and formalized racism in the late 20th century. The apartheid laws mandated that all persons be classified and registered by race. Marriage and sexual relations between whites and nonwhites were prohibited. Segregated educational facilities at all levels existed for whites, coloreds (mixed-blood people), Asians, and black Africans. Public facilities were segregated by race. Political rights were denied to nonwhites. With limited exceptions, black Africans obtained citizenship only in tribal "homelands." They could not vote or hold office in "white" South Africa and had no representation in the white-controlled national parliament. Coloreds and Asians had no parliamentary representation until 1984, when a new, white-adopted constitution granted them some representation. A system of permits known as the "pass laws," applying mainly to the black population, restricted travel within South Africa. Law and custom imposed job discrimination in favor of whites. The broad powers of South Africa's minister of justice and police forces repressed dissent. Accused blacks had few rights. Authorities often broke up protest gatherings on the grounds that they did not have a proper permit or represented a threat to public order. The most publicized such incident happened in 1960 at Sharpeville near Johannesburg, where police killed 69 blacks protesting the "pass laws." A long period of reconciliation awaits the races in the new South Africa.

The Homelands Scheme.   The white government declared that racial peace and justice would exist only if the races were separated territorially and socially so that each race could develop within its allotted niche. There was to be no such thing as a unified South African nationality. Instead, the country was to be a collection of "nationalities" living side by side. The Nationalist Party policy of apartheid culminated in the scheme to establish tribal states or homelands which were nominally intended to achieve "independence." These prescribed racial areas for

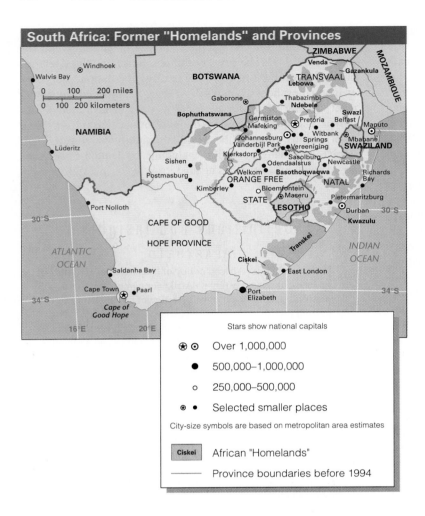

South Africa: Former "Homelands" and Provinces

Stars show national capitals

⊛ ⊙ Over 1,000,000

● 500,000–1,000,000

○ 250,000–500,000

⊛ ● Selected smaller places

City-size symbols are based on metropolitan area estimates

| Ciskei | African "Homelands" |

——— Province boundaries before 1994

FIGURE 18.14

South Africa's former provinces and former African "homelands," which were abolished in 1994. The homelands have been merged into South Africa's nine new provinces, shown in Figure 18.16.

homes, business, and private lands included numerous "homelands" (also known as "native reserves," "Bantustans," and "national states") and the "independent republics" of Transkei, Ciskei, Bophuthatswana, and Venda (see Fig. 18.14).

The ten South African homelands were territorial units reserved for Africans and run by elected African governments. Organized essentially on a tribal basis, they collectively incorporated hundreds of separate tracts of land that began to be set apart as African reserves soon after the Union of South Africa was formed. South Africa's laws permitted the government to classify a black South African as a legal resident of one of the homelands, even though he might never have seen it and had no sense of affiliation with the tribe on which the homeland was based. Whites ejected several million Africans not wanted in "white" South Africa from their homes and transferred them to homelands. They removed greater numbers of blacks from urban areas and white farms and dumped them in poverty-stricken and crowded areas unequipped to handle new influxes of people. Millions of homeland residents continued to work in areas of South Africa outside their homelands. The government regarded them as temporary visitors, to be permitted in "white" South Africa only in such numbers as the white economy needed. Black breadwinners often migrated for long periods to jobs outside their homelands while the rest of the family stayed behind. The magnitude of destruction of black family life that this system caused in South Africa is beyond calculation.

The New South Africa

Black unrest became so widespread and violent in 1984–1986 that the government declared a state of emergency. Blacks killed many other blacks whom they perceived to be collaborating with the whites. An intensive government crackdown resulted in the wholesale banning of political activity by antiapartheid groups, intensified restrictions on media, and the jailing of thousands of black Africans perceived as protest leaders. The government lifted the state of emergency in 1990, but massive violence continued. Much of the fighting was between two large, tribally-based rival factions, the Xhosa-dominated African National Congress (ANC), led by Nelson Mandela, and the Zulu-dominated Inkatha Freedom Party (IFP), led by Chief Mangosuthu Buthelezi.

A complete official turnabout on the issue of apartheid resolved the crisis. It began in 1989 after Frederik W. de Klerk, an insider in the Afrikaner political establishment, became president. Urged on by white South African business leaders, anti-apartheid activists, the powerful force of South African and international political opinion, the impacts of international economic boycotts against South Africa, and his own stated convictions concerning the injustices and unworkability of apartheid, de Klerk launched a broad-scale program to repeal the apartheid laws and put South Africa on the road to revolutionary governmental changes under a new constitution. Four years of intensive negotiation involving all interested parties resulted in an agreement to hold a countrywide, all-race election to create an interim government that would hold office for not more than five years. In April 1994, voters turned out in massive numbers to elect a new national parliament, which then chose Nelson Mandela as president. Only 4 years earlier, the white government had freed Mandela after a 27-year period of political imprisonment. Mandela's African National Congress received 62.6 percent of the votes in the nation as a whole, followed by the Nationalist party (20.4 percent) and the Inkatha Freedom Party (10.5 percent).

This historic election ended white European political control in the last bastion of European colonialism on the African continent. The apartheid laws themselves became null and void (Fig. 18.15). The bizarre homeland units were abolished and merged into South Africa's nine new provinces (see Fig. 18.16). South Africa now faces the chore of reducing the friction be-tween the ruling ANC and the IFP that still leads to bloodshed between the factions in the Kwazulu/Natal region. Writing a permanent constitution will be a major parliamentary task, to be completed by 1999.

South African Regions

The Transvaal Provinces and the Orange Free State.   The interior of South Africa is a plateau with a general elevation of 3000 to 6000 feet (*c.* 900–1800 m; see Fig. 18.13). Two river systems, the Orange (including its tributary, the Vaal) and the Limpopo, drain most of it. At the extreme east is the High Veld, originally a grassland similar to the midwestern prairies of the United States. The northeastern province of Northern Transvaal is an area of woodlands and savanna grasses known as the "Bush Veld."

South Africa is the world's largest producer of gold (Fig. 18.17). The gold-mining industry, which supplies the country's most important export, is almost entirely confined to the High Veld. The greater part of the gold mined thus far has come from the Witwatersrand (formerly Rand; see Fig. 18.16), an area in the province of Gauteng (formerly Pretoria/Witwatersrand/Vaal [PWV]). Johannesburg, South Africa's largest city (population: 712,507, city proper; 4 million, metropolitan area), plus the adjacent impoverished and overcrowded black township of Soweto, and nearby Pretoria (population: 525,583, city proper; 1.1 million, metropolitan area) are located in this province. Immense amounts of capital from British and other sources are invested in the mines (the world's deepest), which are operated by large corporations and employ African workers recruited from South Africa and other African countries.

FIGURE 18.15
Racial segregation in South Africa remains a fact of life in many ways, but the system is being breached on both governmental and private levels. This snapshot shows black and white faces intermingled at a private school for girls in Johannesburg. The children are attending morning religious services. *J. Kyle Keener/Philadelphia Inquirer/Matrix*

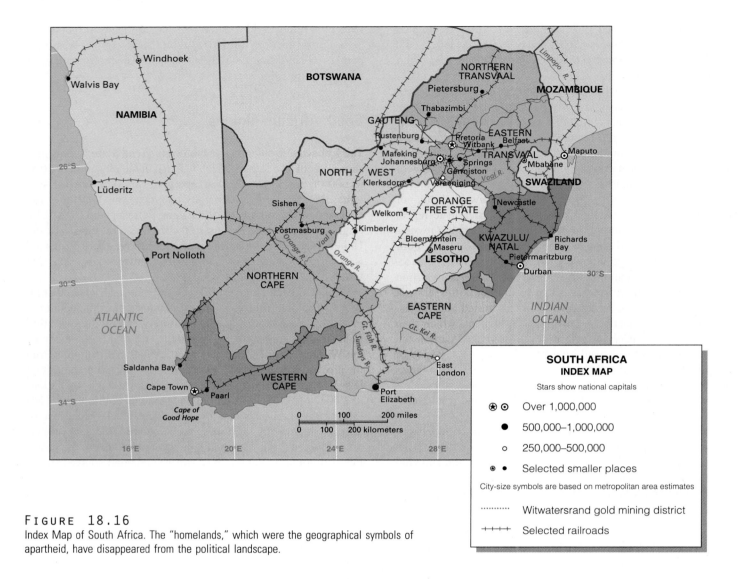

FIGURE 18.16
Index Map of South Africa. The "homelands," which were the geographical symbols of apartheid, have disappeared from the political landscape.

SOUTH AFRICA INDEX MAP

Stars show national capitals

⊛ ⊙   Over 1,000,000

●   500,000–1,000,000

○   250,000–500,000

⊛ ●   Selected smaller places

City-size symbols are based on metropolitan area estimates

·········   Witwatersrand gold mining district

+++++   Selected railroads

Recovery of uranium from mine tailings and as a byproduct of current gold mining has made South Africa the largest uranium producer on the continent. And since 1867, the country has been an important producer of diamonds from alluvial deposits along the Vaal and Orange River valleys and from kimberlite "pipes" (circular rock formations of volcanic origin). Kimberley (population: 155,000, city proper) in the Orange Free State is the administrative center for the diamond industry.

The mining industry in South Africa requires large quantities of electricity. Most of this is supplied by generating plants powered by bituminous coal, which is present in vast quantities near the surface on the interior plateau. These coal deposits, which are the largest in Africa, are immensely important to South Africa, as the country has no petroleum or natural gas output and has only a modest hydroelectric potential. Coal is a major export, and it has facilitated development of Africa's leading iron and steel industry, concentrated at Pretoria and around Johannesburg. Iron ore comes by rail from large deposits in the northeast and central parts of the country.

The Transvaal provinces and adjacent areas also compose one of the world's most vital mineral-exporting regions, with exports including copper, asbestos, chromium, platinum, vanadium, antimony, phosphate, and other minerals.

The High Veld is the country's leading area of crop and livestock production. The principal farm animals of the High Veld are beef cattle, dairy cattle, and sheep. Its main crops are corn (maize) and wheat. Much of the corn production becomes livestock feed, and surpluses become exports, although corn is also a major item of diet for the country's black population. Much of South Africa is too dry for nonirrigated farming. The High Veld gets about 20 to 30 inches (c. 50 to 75 cm) of rain in an average year, but the total amount varies considerably from year to year, and drought often occurs during what would normally be the rainy season. Fortunately, air masses from the Indian Ocean bring most of the precipitation the High Veld receives during the summers, when it is most needed for crops. In 1992, South Africa suffered from a regional drought and had to import large quantities of food and livestock feed.

Many dams have been constructed on South Africa's rivers to conserve water and provide hydroelectricity. In the early 1960s, the country inaugurated its most important water-control scheme, the Orange River Project. Water stored behind Orange River and Vaal River dams meets nearby urban and industrial needs, and a major share irrigates fields and orchards. Tunnels through mountain ranges also transfer water to southward-flowing rivers that serve expanded irrigated areas and other users in the Cape provinces.

KwaZulu/Natal.   KwaZulu/Natal occupies hilly terrain between the Indian Ocean and the Drakensberg, a mountainous escarpment at the edge of the interior plateau. The leading commercial crop of the area is sugarcane, grown mainly on large, white-owned farms worked by African laborers, including those drawn from the Zulu tribe—the region's most populous ethnic group. Historically, however, most workers in the cane fields were indentured laborers brought from India beginning in the 1860s. The majority chose to remain in South Africa when their terms of indenture ended, and some Indians came as free immigrants. Laws prohibited new immigration after 1913 and restricted internal migration of Indians from one province to another. Currently, about 85 percent of South Africa's Indian population is in KwaZulu/Natal, mainly in and near Durban (population: 715,669, city proper; 1.7 million, metropolitan area), KwaZulu/Natal's largest city and main industrial center and port. Most Indians are employed in commercial, industrial,

and service occupations. Like the Europeans, Africans, and coloreds, the Indians of South Africa do not form a unified cultural group. They are divided among Hindu, Muslim, and Christian elements and they speak a variety of Indian languages in addition to English and Afrikaans.

Cape Provinces.   Except for a mountainous fringe along the southern and southeastern coasts, nearly all of the Cape provinces area is semidesert and desert. Sections adjoining KwaZulu/Natal share the humid subtropical climate of that area. The vicinity of the Western Cape Province has a mediterranean climate, being the only African area south of the Sahara where this climate type occurs (see Fig. 18.18). Average temperatures are similar to those of the same climate in southern California. The early settlers established vineyards within the area of mediterranean climate. Today the area's grapes and wines sell both in South Africa and abroad. The province also produces a variety of other fruits, including citrus fruits (mostly oranges), deciduous fruits, and pineapples. Sheep ranching is a major occupation in the semiarid interior. The arid western coast, swept by the cold, northward-moving Benguela Current, supports Africa's most important fisheries (see Fig. 18.13). Upwelling cool water brings up nutrients from the ocean floor and provides a fertile habitat for plankton, the basic food of fish. These waters harbor many kinds of fish, with anchovies the most abundant and commercially important. Most of the catch is processed into fishmeal and fish oil for South African uses.

Like KwaZulu/Natal, the Cape provinces have a distinctive racial group—South Africa's coloreds. Nearly nine tenths live in the provinces, primarily in and near Cape Town (population:

FIGURE 18.17
Johannesburg, the "City of Gold," is surrounded by huge dumps of waste material from former gold-mining operations. One of these dumps occupies the center of the view. Note the automobile expressway at the bottom of the view. In the distance lie residential areas occupied by whites. *Jesse H. Wheeler, Jr.*

FIGURE 18.18
A characteristic rural landscape in the mediterranean climate of South Africa's Western Cape
Province. The middle distance is dominated by fields of ripe winter wheat at the base of one of
the Cape Ranges.   *Jesse H. Wheeler, Jr.*

854,616, city proper; 1.9 million, metropolitan area), the country's second largest city and one of its four main ports. This group originated in the early days of white settlement as a product of sexual relations between Europeans (Dutch East India Company employees, settlers, and sailors) and non-Europeans, including slaves, Khoi (Hottentot), Malagasy, West African, and various south Asian peoples. The resulting mixed-blood people vary in appearance from persons with pronounced African features to others who are physically indistinguishable from Europeans. About nine tenths speak Afrikaans as a customary language, and most of the remainder speak English. Like great numbers of other South Africans, many are bilingual. Culturally, they are much closer to Europeans than to Africans, and many with light skins have successfully "passed" into the European community.

Most coloreds work as domestic servants, factory workers, and farm laborers, and perform other types of unskilled and semiskilled labor. Many are employed in fishing. There is a small but growing professional and white-collar class. Coloreds have always had a higher social standing and greater political and economic rights than black South Africans, although they have ranked considerably below the Europeans in these respects and far below them in level of living. Among coloreds there is a social stratification, generally on the basis of lightness of skin. One of South Africa's outstanding cultural characteristics is that not only is the population divided among four major racial groups, but within each group there is further splintering.

South Africa's Future

Because of an extraordinary web of uncertainty and potential mischances, there are many questions about the future of South Africa. Grounds exist for both optimism and pessimism. Certainly the country's present mood seems upbeat, helped along by a growing willingness of whites to forsake racist attitudes; friendly gestures by foreign governments, including removal of sanctions; positive attitudes by potential investors; resumption of membership in the United Nations; and the existence of South Africa's rich resources and well-developed infrastructure (see Fig. 18.19). But a long and difficult road lies ahead as South Africa's new government tries to keep peace, redress past wrongs, and narrow the gap between economically privileged whites and underprivileged nonwhites, particularly the millions of desperately poor black South Africans in urban shantytowns and the former homelands. Some analysts fear that a growing exodus of whites—those who fear or will not accept the new realities—will drain the country of wealth and expertise when both are desperately needed.

NAMIBIA

Namibia (population: 1.6 million) has a long history of association with neighboring South Africa. South African forces overran this former German colony during World War I, and after the war the League of Nations mandated it to South Africa under the name of South West Africa. The terms of the mandate called for administration of the territory as an integral part of South Africa. After World War II, the United Nations repeatedly called upon South Africa to place the territory under the trusteeship system, but the South African government refused to comply.

Independence in Namibia was linked to events in neighboring Angola. South Africa had long supported an Angolan guerrilla movement opposed to Angola's Cuban-backed Marxist government, and South African military forces actively intervened in the Angolan struggle in the mid-1970s. Eventually South African forces withdrew but continued to attack Angolan bases of SWAPO (South West Africa People's Organization), a guerrilla group fighting for independence. In 1988, South Africa agreed to grant independence to Namibia, but only if the would-be Namibian government under SWAPO leadership could convince its main source of support, the government of Angola, to expel Cuban troops from Angola. But without resolution, protracted hostilities proved costly to South Africa economically and politically. They also eroded the morale of the young South Africans drafted for the fighting. Such burdens eventually led the South African government to change course and grant independence to Namibia on March 21, 1990. SWAPO is now the major political party, but its Marxist leanings have not prevented the country from promoting capitalist investment.

The Namib Desert extends the full length of Namibia in the coastal areas. Inland is a broad belt of semiarid plateau country, merging at the east with the Kalahari semidesert of Botswana

FIGURE 18.20
A surprising abundance and diversity of wild animals and plants are adapted to the harsh conditions of the Namib Desert in Namibia. This is an oryx antelope known as the gemsbok.   *Peter Lamberti/Tony Stone Images*

(Fig. 18.20). With its scattered livestock forage, the Kalahari section is peopled mainly by Bantu-speaking herders and by remnants of the Khoi and San (Bushman) populations. Some of the San are among the world's last remaining hunters and gatherers (see Fig. 2.11, p. 36), but the majority have given up the nomadic way of life in recent decades and have settled as farmers and wage laborers. The roughly 80,000 whites in Namibia live mostly on the interior plateau. Windhoek (population: 125,000, city proper), the capital, is located there. Walvis Bay, which South Africa ceded to Namibia in 1994, has a deepwater harbor and is the port for Namibia's seaborne trade.

Namibia's economy is tied closely to that of South Africa. South African, British, and U.S. mining companies export Namibian diamonds and various metals. Namibia has established an economic zone in the Atlantic stretching 200 miles offshore, in which fishing and fish processing based at Walvis Bay have become important industries. Large wildlife populations responding to dramatic fluctuations in wet and dry conditions make Etosha National Park an attraction for many international tourists who represent an important source of hard currency to Namibia. The young country is attempting to provide conditions favorable both to foreign investors and to Namibia's white community of business entrepreneurs and ranchers.

FIGURE 18.19
An entrepreneurial class is steadily emerging in South Africa's nonwhite communities. This is a black businessman in Cape Town.   *David Barritt/Gamma Liaison*

BOTSWANA, LESOTHO, AND SWAZILAND

Three states bordering South Africa—Botswana (formerly Bechuanaland), Lesotho (formerly Basutoland), and Swaziland—were protectorates ("High Commission Territories") of

Britain until the 1960s. Botswana and Lesotho gained independence in 1966, and Swaziland, the last absolute monarchy in southern Africa, became independent in 1968. Lesotho, a mountainous country containing the highest summits in the Drakensberg escarpment, is surrounded by South Africa, and Swaziland is nearly so, although it has a short frontier with Mozambique. Botswana, most of which is semidesert with wetlands in the north, is bordered on three sides by South Africa and Namibia and on the fourth side by Zimbabwe. With a wealth of wildlife attractions in the Okavango Delta and adjacent national parks, Botswana is a growing destination for international nature-loving tourists.

Of the three, Swaziland has the most varied resources and the most diversified economy. Western Swaziland has mountains along the Drakensberg escarpment that receive heavy rainfall. Rivers that rise there and flow to the Indian Ocean supply irrigation projects in the lower and drier east. Mountain slopes planted with conifers form the basis for wood pulp and lumber industries. Mining is an important source of revenue, as is tourism—which features gambling casinos—and sugar production. Europeans using Swazi labor have developed most of the larger enterprises. Race relations have been better than in many African countries, and the economic benefits to Swaziland have been considerable.

Mining companies have been prospecting actively in Botswana and Lesotho; diamond yields in 1992 (latest figure available) amounted to 79 percent of the total value of exports from Botswana. The black populations of these two countries raise livestock and a little grain, largely on a subsistence basis, and some Europeans own cattle ranches in eastern Botswana. Large numbers of African workers from all three countries migrate to jobs in South Africa, primarily in mining. All three countries are highly dependent on South Africa for trade and many kinds of services, and South Africa has established a customs union with the three and with Namibia.

ANGOLA AND MOZAMBIQUE

Portugal played an active role as an imperial power in southern Africa from a very early time during the age of European colonial expansion. Portugal was, in fact, the earliest colonial power to build an African empire. The epic voyage of Vasco da Gama to India in 1497–1499 by the Cape route was the culmination of several decades of Portuguese exploration along Africa's western coasts. During the 16th century, Portugal controlled an extensive series of strong points and trading stations along both the Atlantic and Indian Ocean coasts of the continent. Later, as stronger powers outdistanced this earliest empire builder, Portugal managed to retain footholds along both coasts. By 1964, Portugal's principal African possessions, Angola (Portuguese West Africa) and Mozambique (Portuguese East Africa), were the most populous European colonies still remaining in the world. In both units, Portugal fielded large military forces to

combat guerrilla opposition. In 1974, Portugal's resistance came to a breaking point, and an army coup installed a regime in Lisbon that granted the two colonies independence.

The Portuguese did little to develop their African territories until after World War II. A poor country itself, Portugal did not have the resources to undertake development on a large scale. It did receive aid from Great Britain, Belgium, and South Africa to build railways and harbor facilities servicing mineral-rich areas those countries controlled. After the war, Portugal did more to improve its territories. Aiding this effort was income from Angolan mineral wealth: diamonds, iron ore, and, in the years immediately before independence, an increasing output of oil.

Angola and Mozambique were attractive enough to white settlers that a policy of government-assisted immigration increased the European population substantially. By 1974, about 450,000 Europeans, mainly Portuguese, lived in Angola and perhaps 200,000 more lived in Mozambique. They nearly monopolized the skilled occupations and controlled most of the wealth. During the disorders accompanying independence, most of these settlers fled to Portugal, South Africa, and Zimbabwe (then Rhodesia). Guerrilla warfare then wiped out earlier economic gains and wrecked social services in Angola and Mozambique. Both countries face difficult problems of rebuilding in the late 1990s.

Angola is composed physically of a narrow coastal lowland and expanses of plateau in the interior. The climate is tropical savanna, with a fringe of tropical steppe in the west. The steppe grades into coastal desert at the southwest. Prior to independence, Angola had important linkages with Zaire (now the Democratic Republic of Congo), and Zambia by virtue of a rail line, the Benguela Railway, connecting Angola's seaport of Lobito with the Shaba mining region of Zaire and the adjacent Copperbelt of Zambia. During many years of civil war in Angola, guerrilla forces repeatedly interdicted the railway and rendered it inoperative for long periods. Zaire turned to its own port of Matadi and to South African ports for most overseas traffic. From its unfortunate landlocked position, Zambia has relied mainly on South African ports and the inefficient "Tazara" railway across Tanzania to the congested port of Dar es Salaam. Chinese aid developed the Tazara line, completed in 1976, but after the Chinese withdrew, the line lapsed into mismanagement and neglect.

When the Portuguese withdrew from Angola in 1975, a Marxist-Leninist liberation movement supported by the Soviet Union and Cuba took over the central government in the seaport capital and largest city, Luanda (population: 1.46 million, city proper; Fig. 18.21). But its authority was not recognized by other liberation movements, one of which—the National Union for the Total Independence of Angola (UNITA)—came to control large parts of the country. South Africa and the United States supplied aid to this anti-Marxist movement headed by Jonas Savimbi, a member of Angola's largest tribe. Large quantities of Soviet arms, thousands of Cuban troops, and many Soviet and East German military advisors supported the

FIGURE 18.21

Luanda, the capital and largest city of Portugal's former colony of Angola, exhibits the common pattern of cities in Africa south of the Sahara: a modest cluster of high-rise buildings in the city center is surrounded by a sea of small one-story houses, often on dirt streets and generally roofed with metal sheeting. *Sarah Errington/Hutchison Library*

government in Luanda. Cuban contingents guarded vital oil fields, particularly in Angola's coastal exclave of Cabinda north of the Congo River mouth.

In 1988, representatives of Angola, Cuba, and South Africa signed an American-mediated agreement under which Cuba and South Africa would withdraw militarily from Angola, and South Africa would grant independence to Namibia. In May 1991, Angola's warring factions signed a cease-fire accord, but it fell apart, and warfare continued until a peace agreement allowing UNITA's participation in all levels of government was reached in 1994. UNITA continues to seek some official control over Angola's promising diamond-producing area.

Cabinda is the country's main oil source, and other fields are scattered along the Angolan coast between the Congo mouth and Lobito. Crude oil and modest amounts of refined products make up nine tenths of Angola's exports, and by 1996 Angolan oil comprised 7 percent of U.S. oil imports. In addition to oil and its products, Angola exports small amounts of diamonds, coffee, and other items. Diamonds are concentrated in the northeast, where UNITA rebels mined them for years to finance their antigovernment operations. Coffee was once Angola's main export, but its production fell as a consequence of warfare in the countryside and the flight from Angola of the European farmers who grew most of the crop. Most Africans in Angola supported themselves precariously during the civil war by subsistence farming and cattle raising.

Portugal's other ex-colony, Mozambique, has the distinction of being the world's poorest country. It consists of a coastal lowland rising to interior uplands and plateaus. Most interior areas are low in elevation. The country has a tropical savanna climate, with a natural vegetation composed primarily of grasslands with scattered trees. Marshy and mangrove vegetation is conspicuous along the coast. The lower Zambezi River divides Mozambique into two fairly equal parts. There are two main port cities: Maputo (population: 1.07 million, city proper), the capital, in the extreme south, and Beira (population: 298,847, city proper), located about 125 miles (200 km) south of the Zambezi mouth. Under Portuguese rule, much of Mozambique's commercial economy was closely linked with the economies of nearby states, and the two ports had rail connections with adjoining landlocked countries (see Fig 17.3, page 402) and handled much international transit trade as well as overseas trade for Mozambique itself. British capital played a major role in financing port facilities at Maputo and Beira, as well as rail lines from the ports to the interior.

This transportation net focusing on the Mozambican ports suffered severely from many years of guerrilla warfare in the country, which first involved African insurgencies against the Portuguese and then warfare by African dissidents against the Marxist government that took over from the Portuguese in Maputo. The highly destructive civil war—costing one million lives—between the Maputo government and the South African-backed rebel organization called RENAMO (Mozambique National Resistance) dragged on until October 1992 when a peace accord was signed and the United Nations sent a military peacekeeping force to Mozambique. Elections took place peacefully in October 1994, and a large share of the five million people displaced by the war began returning to their villages. Zimbabwe recalled its troops sent to Mozambique during the war to guard rail, highway, and oil pipeline connections to the port of Beira. Since 1989, Mozambique's government has been shifting away from Marxism-Leninism in favor of a more market-oriented economy and greater democracy.

Under the Portuguese, the European population of Mozambique was mainly urban, although there were several settlement schemes for European farmers and also some European-controlled plantation agriculture. Since independence, Mozambique's African farmers, beset by civil war and years of extreme drought, have produced little more than bare subsistence. There has been some export production in recent years of such crops as cotton, cashew nuts, tea, and sugar, but imports of farm products have far exceeded farm exports. Shrimp are the largest export. Meanwhile, international initiatives have been under way to redevelop the damaged port of Beira and upgrade the transportation facilities in the "Beira Corridor" to Zimbabwe.

ZIMBABWE, ZAMBIA, AND MALAWI

Prior to the post–World War II movement for African independence, the countries of Zimbabwe, Zambia, and Malawi (then known as Southern Rhodesia, Northern Rhodesia, and Nyasaland) were British colonies. In 1953, the three became linked politically in the Federation of Rhodesia and Nyasaland, or the Central African Federation. It had its own parliament and prime minister but did not achieve full independence. The racial policies and attitudes of Southern Rhodesia's white-controlled government created serious strains within it, and these helped bring about dissolution of the Federation in 1963. Northern Rhodesia then achieved independence as the Republic of Zambia, and Nyasaland became the independent Republic of Malawi.

But the government of Southern Rhodesia, the main area of European settlement in the Federation, was unable to come to an agreement with Britain concerning Britain's demand for the political participation of the colony's African majority. In 1965, the government of Southern Rhodesia issued a Unilateral Declaration of Independence (UDI), and in 1970 it took a further step in separation from Britain by declaring itself to be Rhodesia, a republic that would no longer recognize the symbolic sovereignty of the British crown. Britain and the Commonwealth responded with a trade embargo. Negotiators for Britain and the Rhodesian government unsuccessfully attempted to arrive at a formula to allow political rights and participation for the 95 percent of Rhodesia's population that were black, and thus to achieve reconciliation and world recognition of Rhodesia's independence. Although the economic sanctions were ineffective, pressure on the white government increased in the mid-1970s with black guerrilla warfare against the government and the end of white (Portuguese) control in adjacent Mozambique. Independence for Rhodesia (renamed Zimbabwe) as a parliamentary state within the Commonwealth of Nations finally came in 1980.

Most whites left Zimbabwe at that time, but about 100,000 remained. The whites have been participating in Zimbabwe's political life, setting an early example to South Africa of an effective transition to majority black rule. The future course of relations between the country's two largest ethnic groups—the majority Shona and the Ndebele—remains uncertain. Many Ndebele people want autonomy for the Ndebele stronghold in the south known as Matabeleland.

Zimbabwe, Zambia, and Malawi occupy highlands with a tropical savanna climate. Over nine tenths of the annual rainfall comes in the six months from November to April. The natural vegetation is primarily open woodland, although tall-grass savanna predominates on the High Veld of Zimbabwe. The Zimbabwean High Veld, lying generally at 4000 feet (c. 1200 m) and higher, forms a broad divide between the drainage basins of the Zambezi and Limpopo rivers. Composed of a band of territory about 50 miles (80 km) broad by 400 miles (c. 650 km) long, oriented southwest-northeast across the center of Zimbabwe, the High Veld has been the main area of European settlement in the three countries. Most of Zimbabwe's whites live on it. They include about 4500 farmers whose large holdings, worked by black laborers, produce most of the country's agricultural surplus to feed urban populations and to export. Tobacco grown on these white-owned farms is an important export. Other agricultural exports include cotton, cane sugar, corn and roses.

The main axis of economic development in Zimbabwe lies along the railway connecting the largest city and capital, Harare (formerly Salisbury; population: 681,000, city proper; 890,000, metropolitan area), in the northeast with the second largest city, Bulawayo (population: 620,936, city proper), in the southwest. Along and near the axis are the principal areas of European farming and mines producing gold, nickel, asbestos, copper, chromium, and other minerals, primarily for export. This axis also corresponds with an extraordinary geological feature called the Great Dike, an intrusion of younger, mineral-rich ma-

FIGURE 18.23

On November 16, 1855, the Scottish missionary Dr. David Livingstone became the first non-African to view a waterfall known to the Africans as Mosi Oa Tunya, "the Smoke that Thunders." "The falls are singularly formed," Livingstone wrote. "They are simply the whole mass of the Zambezi waters rushing into a fissure or rent made right across the bed of the river. The falls had never been seen before by European eyes, but scenes so lovely must have been gazed upon by angels in their flight." Victoria Falls—named by Livingstone for the Queen of England—were soon dubbed one of the "Seven Natural Wonders of the World." The Zambezi River here forms the border between Zambia (on the right) and Zimbabwe. *Gregory G. Dimijian/Photo Researchers*

FIGURE 18.22

Zimbabwe's Great Dike, seen from space. About 300 miles (500 km) long and 6 miles (10 km) wide, the mineral-rich Great Dike was formed about 2.5 billion years ago when molten lava filled and widened a fracture in the earth's crust. Near the center of this image, note how faulting has offset the dike. *NASA/Science Photo Library*

terial into older rock. Ranging in width from 2 to 7 miles (3 to 11 km), the Great Dike crosses the entire nation and is readily visible to astronauts in space as a giant stripe on the landscape (Fig. 18.22).

Iron ore mined at Que Que, about midway between Harare and Bulawayo, is used there to produce steel in southern Africa's only iron and steel plant. The Wankie coal field in western Zimbabwe supplies coal to smelt the ore.

Zimbabwe's manufacturing industries are more important, both in value of output and in diversity of production, than those of any other African nation south of the Sahara except Nigeria and South Africa. Despite this production, the country's exports are dominated overwhelmingly by unprocessed minerals, tobacco, cotton, and corn, and many manufactured goods must still be imported. Tourism is of growing importance today, thanks to such attractions as the ruins of Great Zimbabwe, a wealth of animal life and other natural marvels, and good air and road links.

Below the spectacular Victoria Falls (regarded traditionally as one of the seven "Natural Wonders of the World," and another significant international tourist destination today; Fig. 18.23), the middle course of the Zambezi River separates Zambia from Zimbabwe. Hydropower from the Kariba Dam (completed in 1962; see Fig. 17.5, p. 403) on this stretch of the Zambezi flows to both countries. Zambia is much poorer than its southern neighbor. Much of it is wilderness with a sparse human population. As in Zimbabwe, the population lives mainly on a subsistence basis, but in Zambia, agriculture and cattle-keeping are less productive than in Zimbabwe.

Zambia's economy depends heavily on copper exported from several large mines developed with British and American capital. An urban area has developed at each mine, and the mining area, known as the Copperbelt, consists of a series of separate population nodes strung close to Zambia's frontier with the adjacent copper-mining Shaba region of southeastern Democratic Republic of Congo. The Copperbelt, which has the majority of Zambia's manufacturing plants and copper mines, is largely an urban region and contains a high proportion of the country's small European population. Copper, most of which is shipped in refined form, composed 86 percent of Zambia's exports in 1993 (latest figure available). Other mineral production includes cobalt, a byproduct of copper mining; zinc and lead mined along the railway between the Copperbelt and Zambia's capital and largest city, Lusaka (population: 982,362, city proper); and coal production from extreme southern Zambia.

Malawi is a long, narrow, densely populated country along the western and southern margins of Lake Malawi. Its population is almost entirely black African and is even more rural and agricultural than that of most African states. Tobacco is the main item in Malawi's export trade, with tea and cane sugar also of some importance. There is little mineral wealth, and manufacturing industries are meager. The country has long supplied many migrant laborers to mines in Zimbabwe and South Africa, and their earnings are an important adjunct to the internal economy of this poor country.

18.7 Indian Ocean Islands

The islands and island groups off the Indian Ocean coast of Africa are unique in their cultures and natural histories. Madagascar, the Comoro Islands, Réunion, Mauritius, and the Seychelles exhibit African, Asian, Arab, European, and even Polynesian ethnic and cultural influences (Fig. 18.24). As island ecosystems, they are homes to many endemic plant and animal species.

Madagascar and the smaller Comoro Islands are former colonies of France, while Réunion remains an overseas possession of France. Mauritius became independent of Britain in 1968 as a constitutional monarchy under the British crown. The Seychelles, which gained independence in 1976, is a republic within the Commonwealth of Nations.

Madagascar (in French, "La Grande Ile") is the fourth largest island in the world, nearly 1000 miles (c. 1600 km) long and about 350 miles (c. 560 km) wide. It lies off the southeast coast of Africa and has geological formations similar to those of the African mainland. Its distinctive flora and fauna include

FIGURE 18.24
The 15 million people of Madagascar have a unique culture, with roots in both Africa and the Malayo-Polynesian realm. This is a scene in Antsirabe, near the center of the island.   *Hubertus Kanus/Photo Researchers*

most of the world's lemur and chameleon species (see Definitions and Insights, page 453). Some of Madagascar's early human inhabitants migrated from the southwestern Pacific region, bringing with them the cultivation of irrigated rice and other culture traits. There was also a later influx of Africans from the mainland. The official language is Malagasy, a Malayo-Polynesian tongue that has acquired many French, Arabic, and Swahili words.

The east coast of Madagascar rises steeply from the Indian Ocean to heights of over 6000 feet (1829 m). Because the island lies in the path of trade winds blowing across the Indian Ocean, the east side receives the heaviest rain and has a natural vegetation of tropical rain forest. The remainder of the island has a vegetation consisting primarily of savanna grasses with scattered woody growth. An eastern coastal strip—low, flat, and sandy—is backed by hills and then by escarpments of the central highlands. Paddy rice is the principal food crop of the coastal zone, and coffee, vanilla, cloves, and sugar are grown for export. The most densely populated part of the island is the central highlands, where the economy is a combination of rice growing in valleys and cattle raising on higher lands. Antananarivo (formerly Tananarive; population: 1.25 million, city proper), the capital and largest city, is located in the central highlands. It is connected by rail with Toamasina (formerly Tamatave; 127,441, city proper), the main seaport, located on the east coast.

Madagascar has experienced a much smaller degree of economic development than such large tropical islands as Java, Taiwan, and Sri Lanka. Deep economic troubles and political dissension have marked its recent history under a socialist government. Unwise decision making has made Madagascar fertile ground in the 1990s for international drug-money launderers, loan sharks, and con artists who have purloined much of this poor nation's financial assets.

Definitions and Insights

MADAGASCAR AND THE THEORY OF ISLAND BIOGEOGRAPHY

In a recent issue of the journal *Scientific American*, Harvard University entomologist Edward Wilson and Stanford University biologist Paul Ehrlich asserted that if tropical rain forests continue to be cut down at the present rate, a quarter of all of the plant and animal species on Earth will become extinct within the next 50 years.* They based their estimate on a model which correlates habitat area with the number of species living in the habitat. This **theory of island biogeography** emerged from observations of island ecosystems in the West Indies. The theory is that the number of species found on an individual island corresponds with the island's area, with a tenfold increase in area normally resulting in a doubling of the number of species. If island "A," for example, is 10 square miles in area and has 50 species, 100-square-mile island "B" may be expected to support 100 species.

What makes the theory useful in projecting species losses is the inverse of this equation: a tenfold diminution in area will result in a halving of the number of species. One-square-mile island "C" therefore can be expected to hold just 25 species. In applying the model, ecologists treat habitat areas as if they were islands. So, if people cut down 90 percent of the tropical rain forest of the Amazon Basin, for example, the theory of island biogeography suggests they would eliminate half of the species of that ecosystem. Scientists caution that the theory is only a tool meant to help in making rough estimates; the actual number of species lost with habitat removal may be higher or lower.

As a rough guideline, the theory of island biogeography is useful in projecting and attempting to slow the rate of extinction in the world's biodiversity "hotspots," such as Madagascar. Over 90 percent of Madagascar's plant and animal species are endemic, occurring nowhere else on Earth. Extinction of species was well underway soon after people arrived on the island; the giant, flightless elephant bird (*Aepyornis*) was among the early casualties. But human activities, particularly the clearing of forest to grow rice and provide pasture for zebu cattle, are eliminating habitat areas on the island at a more rapid rate than ever before. Scientists are anxious to learn whether some of Madagascar's remaining plants might be useful in fighting diseases such as AIDS and cancer. Already Madagascar's rosy periwinkle has yielded compounds

*Edward O. Wilson, 1989. "Threats to Biodiversity." *Scientific American* 261(3): 60–66.

FIGURE 18.B
Although people have lived in Madagascar for less than 2,000 years, it is now Earth's most eroded country. Since arriving on the island, people have cleared and burned its forest to cultivate crops and provide pastures for zebu cattle. As heavy rains strip the topsoil, people create new fields and pastures from the forest. They use fire to stimulate the growth of new livestock fodder. Near the bottom center, note where erosion has begun to collapse the abandoned terraces. Rice has been planted in the paddies below the terraces.   *David Austen/Tony Stone Images*

effective against Hodgekin's disease and lymphocytic leukemia. Other species like the rosy periwinkle could become extinct before their useful properties even become known.

How urgent is the task to study and attempt to protect plant and animal species in Madagascar? Scientists turn to the theory of island biogeography for an answer. Although people have lived on Madagascar for less than 2000 years, they have succeeded in removing 90 percent of the island's forest and setting a stage for some of the most ruinous erosion seen anywhere on Earth (Fig. 18.B). The theory of island biogeography suggests that in the process, they have caused the extinction of roughly half of the island's species. With Madagascar's human population expected to double in 22 years, and with pressure on the island's remaining wild habitats expected to increase accordingly, the task of conservation is extremely urgent.

The Comoro Islands, located about midway between northern Madagascar and northern Mozambique, are of volcanic origin. They exhibit a complex mixture of African, Arab, Malayan, and European influences. Aside from the island of Mayotte, the Comoro group forms the Federal Islamic Republic of the Comoros (short form: Comoros), in which Islam is the official religion and both Arabic and French are official languages. By its own choice, Mayotte, which has a Christian majority, remains a dependency of France. France has a defense treaty with the Comoros, which it invoked in 1995 to put down a coup led by a French mercenary.

The coelacanth, a primitive fish thought to have become extinct soon after the end of the Mesozoic Era (the age of dinosaurs), has been discovered alive in Comoro Islands waters since 1938. Limited fishing, subsistence agriculture, and exports of vanilla provide livelihoods to the 600,000 inhabitants of the these islands.

Réunion, east of Madagascar, is the tropical island home to 700,000 people, mainly of French and African origin. The island was uninhabited prior to the coming of the French in the 17th century. Elevations reach 10,000 feet (3048 m). The volcanic soils are fertile, and tropical crops in great variety are grown. Cane sugar from plantations supply 75 percent of all exports by value, with rum (derived from cane) supplying an additional 2 percent. Réunion also has a significant international tourist industry.

Sugar is the mainstay of the larger island of Mauritius, which lies nearby. Sugarcane, grown mainly on plantations, occupies most of the cultivated fields on this volcanic landscape. Mauritius has a larger proportion of lowland area than Réunion. Its population is composed mainly of descendants of Indians brought in by the British to work the cane fields, and also includes descendants of 18th-century French planters, and some blacks, Chinese, and racial mixtures. The island has developed a clothing industry which supplied more export value than sugar in 1993 (latest figures available; clothing was 55 percent of all exports; sugar, 25 percent). International tourism also produces valuable revenues ($300 million in 1993) for its 1.1 million people.

The unusual core of the Seychelles island group is composed of granites, which typically are found only on continental mainlands. The exceptional physical beauty of the islands, and the renowned friendliness of its people of French, Mauritian, and African descent, have made the Seychelles a famous tourist destination among those wealthy enough to afford travel to this remote area (Fig. 18.25). The tourist boom began when an airport was built in 1971.

Many food requirements have to be imported to the islands. The country's largest exports consist of petroleum products, although the islands have no crude petroleum and no refining industry; the products are simply imported and then re-exported. Fish products are the only other important exports. The Aldabra Atoll, far to the south of the main group of islands, is home to a population of giant tortoises and other endemic animals.

FIGURE 18.25
La Digue Island in the Seychelles. In the center, note the granite boulders which are very unusual on an oceanic island. They add to the natural beauty which has made this island group a prized tourist destination.
Chad Ehlers/Tony Stone Images

As a group, the Seychelles are a fitting symbol of the plight and potential of the entire region of Africa south of the Sahara. They are culturally diverse, have an abundance of some resources, and utterly lack other vital assets. Outsiders are drawn to the often imaginary Eden the Seychelles and Africa evoke, but would do well to make a stronger effort to understand the even more marvelous facts of land and life in the region.

Latin America

7

Comprises a Wide Latitudinal and Longitudinal Spread

Environmentally Diverse

Politically Fragmented, with Many States and a Few Colonial Remnants

Unevenly Populated, Ethnically Varied

Strong Native-Indian and African Associations

Many Geographic Residues from Past Eras

An "Emerging" Area, but Relatively Urbanized

Tourism Increasingly Important

Economically Fragile, with Oversized National Debts, Much Poverty, and Excessive Dependence on Nonindustrial Exports

Rich in Resources but Underindustrialized

Agriculturally Diverse with Extreme Contrasts in Productivity

Poor Regional Unity

Politically Volatile, but with Increasing Stability and Democracy

"In-Between" Prosperity: Still Poor, but Better Off than Most of Africa and Asia

Geographic Profile of Latin America

The land portion of the Western Hemisphere south and southeast of the United States is commonly known as Latin America (Fig. 19.1). The name reflects the importance of culture traits inherited primarily from the Latin European colonizing nations of Spain and Portugal. Spanish is the major language, while Portuguese is the language only of the region's largest country, Brazil. Dialects of French are spoken in Haiti and in France's dependencies of French Guiana, Guadeloupe, and Martinique. Some small units in the Caribbean Sea or on the nearby mainland do not exhibit Latin culture except in minor ways. They mainly speak English or Dutch, and some are still dependencies of the United Kingdom, the United States, or the Netherlands (Table 19.1). Their presence in an area of original Spanish control reflects early colonial struggles among European nations for territories suitable for sugarcane plantations. Important precolonial influences in some Latin American countries are manifested by native Indian languages, cultural practices, and dialects. But Latin influences are predominant today in the region as a whole and are shown not only in Latin-derived (Romance) languages but also

in the dominance of the Roman Catholicism that arrived with the first ships from Latin Europe (Fig. 19.1). Actually, language and religion are merely the most obvious cultural importations, which also included such major elements as land tenure arrangements, governmental practices, legal systems, social structures, and economic systems.[1]

It should not be assumed that European cultures have been transplanted to Latin America without essential modification or that a uniform culture prevails today in even the most strongly Latinized parts of the region. Underneath a veneer of sameness—promoted by the widespread use of the Spanish language—important distinctions exist between regions, between countries, and within countries. Among the many contributing factors are diverse pre-European and colonial experiences, different resource bases, divergent political systems, and

[1]The authors wish to thank Brian K. Long for his valuable assistance in the initial writing of Chapters 19 and 20 in the 1995 edition. These chapters also contain some material written originally by Dr. Richard S. Thoman but updated here and presented in a revised format.

◀ Rio de Janeiro occupies a spectacular setting of bays, peninsulas, islands, low mountains, and world-famous beaches serving a large international tourist clientele. Coastal cities and mountain highlands are now ambitiously seeking the tourist trade because of the foreign exchange it captures for the domestic economy. *Brissaud-Figaro/Gamma Liaison*

TABLE 19.1

Latin America: Basic Data

Political Unit	Area Thousand sq mi	Area Thousand sq km	Estimated Population (millions)	Estimated Annual Rate of Natural Increase
Caribbean America				
Mexico	761.6	1972.5	94	2.2
Guatemala	42	108.9	11	2.9
Belize	8.9	23	0.2	3.3
El Salvador	8.1	21	5.9	2.6
Honduras	43.3	1121.1	5.5	2.8
Nicaragua	50	129.5	4.2	2.7
Costa Rica	19.7	51.1	3.4	2.2
Panama	30.2	78.2	2.7	1.8
Guyana	83	215	0.8	1.8
Suriname	63	163.3	0.4	1.6
French Guiana	35	91	0.2	2.06
Cuba	42.8	110.9	10.9	0.8
Dominican Republic	18.8	48.7	7.5	2.3
Haiti	10.7	27.8	6.5	2.3
Jamaica	4.2	11	2.6	1.8
Trinidad and Tobago	2	5.1	1.3	1.2
Bahamas	5.4	13.9	0.3	1.34
Barbados	0.2	0.4	0.3	0.5
Antigua and Barbuda	0.2	0.4	0.1	1.2
Dominica	0.3	0.8	0.1	1.33
Grenada	0.1	0.3	0.1	2.4
St. Christopher and Nevis	0.1	0.3	0.04	1.3
St. Lucia	0.2	0.6	0.2	2.0
St. Vincent and the Grenadines	0.1	0.3	0.1	1.8
Puerto Rico	3.5	9.1	3.8	1.0
U.S. Virgin Islands	0.2	0.35	0.1	1.33
British Virgin Islands	0.06	0.15	0.01	1.42
Cayman Islands	0.1	0.3	0.03	0.98
Montserrat	0.04	0.1	0.01	0.57
Anguilla	0.04	0.1	0.01	1.61
Turks and Caicos Islands	0.2	0.4	0.01	0.83
Guadeloupe	0.7	1.8	0.4	1.2
Martinique	0.4	1.1	0.4	0.9
Netherlands Antilles	0.37	0.96	0.2	1.3
Total	1235.51	4209.46	163.31	1.97
South America				
Andean Countries				
Colombia	439.7	1138.9	36.2	2.1
Venezuela	352.1	912.1	21	2.1
Ecuador	109.5	283.6	10.9	2.3
Peru	496.2	1285.2	24.1	2.1
Bolivia	424.2	1098.6	7.9	2.6
Total	1821.7	4718.4	100.1	1.89
Brazil	3286.5	8512	160.7	1.7
Southern Midlatitude Countries				
Argentina	1068.3	2766.9	34.3	1.2
Chile	292.3	757	14.2	1.6
Uruguay	68	176.2	3.2	0.8
Paraguay	157	406.7	5.4	2.8
Total	1585.6	4106.8	57.1	1.33
Summary Total	7929.31	21,546.66	481.21	1.63

Sources: *The World Almanac* and *Book of Facts 1996, Statesman's Yearbook 1995–1996, Britannica Book of the Year 1996, 1997,* and *The World Factbook 1995.*

Estimated Population Density, 1996	Estimated Population Density, 1996	Infant Mortality Rate	Urban Population (%)	Arable Land (% of total area)	Per Capita GNP ($U.S.)
Per sq mi	Per sq km				
123	48	26	71	12	4010
262	101	52.2	39	12	1190
24	9	34.7	48	2	2550
723	279	38.9	45	27	1480
126	49	43.4	47	14	580
84	33	50.3	63	9	330
173	67	10.3	44	6	2380
89	34	15.8	55	6	2670
8.7	3	47.7	33	3	530
7	3	30.2	49	0	870
4	2	15.1	—	0	6000
256	99	8.1	74	23	—
399	154	49.5	61	23	1320
610	236	107.5	32	20	220
607	234	16.1	53	19	1420
642	248	18.5	65	14	3740
48	18	24.3	84	1	11,790
1544	596	19.2	38	77	6530
384	148	17.8	31	18	6970
285	110	9.9	61	9	2830
720	278	12.1	32	15	2620
395	152	19.4	42	22	4760
652	252	20.5	48	8	3450
894	345	17.2	25	38	2120
1085	419	12.8	73	8	7000
715	276	12.5	—	15	—
225	89	19.33	—	20	—
331	128	8.4	—	0	—
330	127	11.69	—	20	—
202	78	17.3	—	—	—
84	32	12.6	—	2	—
586	226	8.5	48	18	—
930	359	7.3	81	10	—
549	212	9	92	8	—
132	51	31.5	63	11	—
82	32	26.9	67	4	1620
60	23	26.5	84	3	2760
99	38	37.7	59	6	1310
49	19	52.1	70	3	1819
19	7	70.6	58	3	770
55	21	31.9	64	3	4927
49	19	57.2	76	7	3370
32	12	28.8	87	9	8060
48	19	14.3	85	7	3560
47	18	16.3	90	8	4650
34	13	24.1	50	20	1570
36	14	24	83	10	7225
61	23	39.3	70	7	5721

FIGURE 19.1
Roman Catholicism is the leading religion in all Latin American countries except some relatively small units colonized by countries other than Spain or Portugal. The old Cathedral of Lima, Peru is one of the more majestic among great numbers of Catholic buildings spread through Latin America from Mexico and Cuba to Argentina and Chile. *Guilio J. Barbero*

4. diverse governmental philosophies and structures, characterized by relatively frequent shifts; and
5. countries that have distinctive personalities despite enough overall similarity and cohesion to fit together reasonably well in an identifiable major world region.

19.1 Definition and Basic Magnitudes

With a land area of slightly more than 7.9 million square miles (over 20.5 million sq km), Latin America is surpassed in size by Africa, the Former Soviet Region, North America (including Greenland), and Asia (see Table 1.2). However, its maximum latitudinal extent of more than 85 degrees, or nearly 5900 miles (*c.* 9500 km), is greater than that of any other major world region, and its maximum east-west measurement, amounting to more than 82 degrees of longitude, is also impressive. Yet Latin America is not so large as these figures might suggest, for its two main parts are offset from each other. The northern part, known as Caribbean or Middle America—which includes Mexico, Central America, and the islands of the Caribbean—trends sharply northwest from the north-south orientation of the continent of South America. The latter is thrust much farther into the Atlantic Ocean than is the Caribbean realm or Latin America's northern neighbor, North America. In fact, the meridian of 80°W, which intersects the west coast of South America in Ecuador and Peru, passes through Pittsburgh, Pennsylvania. Brazil lies less than 2000 miles (under 3200 km) west of Africa (see Fig. 19.2, inset).

differential rates of development. The result is that Latin America presents many faces to the world, including:

1. extraordinary variety in environmental settings;
2. population groups of great diversity;
3. local economies that run the scale from technologically advanced to extremely underdeveloped;

	Leading Exports of Selected Latin American Countries Reported in 1996	
TABLE 19.2		
Country	**Commodity**	**Percent of Total Exports**
Venezuela	Crude petroleum and petroleum products	78
Cuba	Sugar	63
Mexico	Metallic products, machinery, equipment	58
Chile	Industrial products	44
Ecuador	Crude petroleum	42
Panama	Bananas	40
Colombia	Forestry and fisheries products	32
Belize	Sugar	32
El Salvador	Coffee	31
Costa Rica	Bananas	30
Dominican Republic	Ferronickel	28
Honduras	Bananas	28
Paraguay	Soybean products	27
Guatemala	Coffee	21
United States	Machinery and transport (for comparison)	49

Britannica Book of the Year 1996. Encyclopaedia Britannica, Inc., 1996.

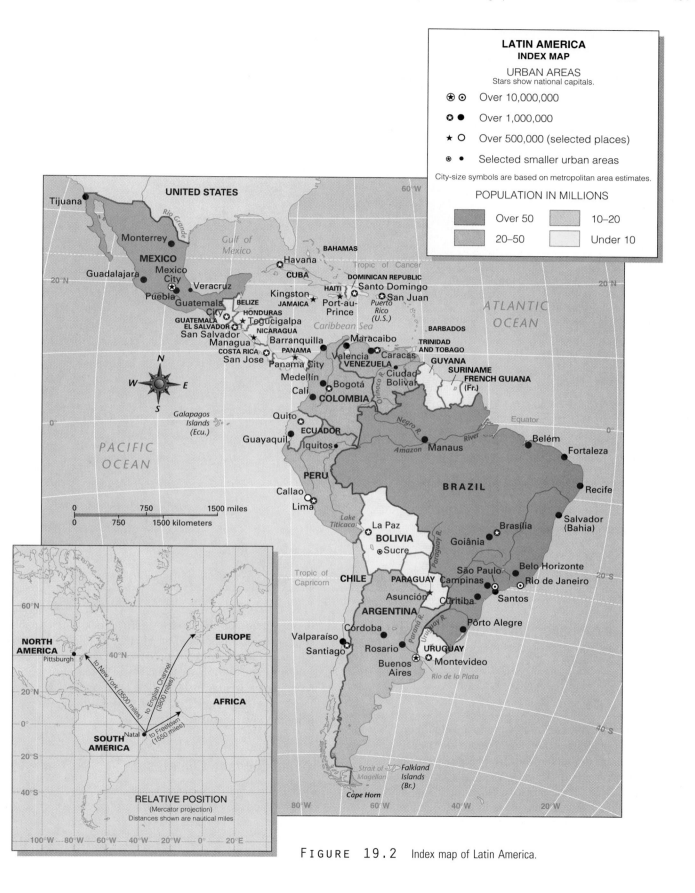

LATIN AMERICA
INDEX MAP

URBAN AREAS
Stars show national capitals.

⊛ ⊙   Over 10,000,000

✪ ●   Over 1,000,000

★ ○   Over 500,000 (selected places)

⊛ •   Selected smaller urban areas

City-size symbols are based on metropolitan area estimates.

POPULATION IN MILLIONS

	Over 50		10–20
	20–50		Under 10

FIGURE 19.2   Index map of Latin America.

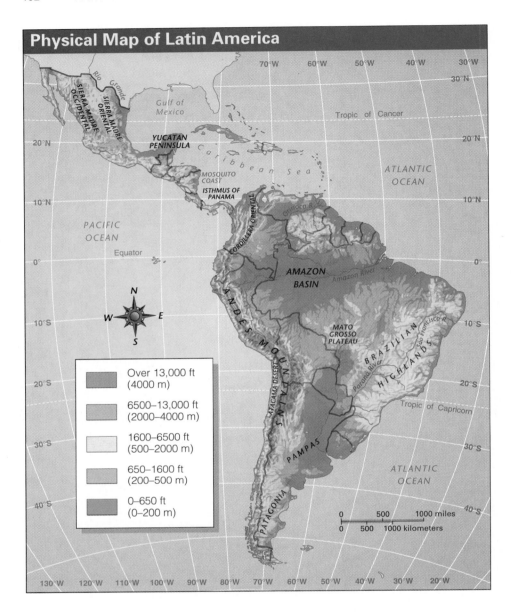

Physical Map of Latin America

Over 13,000 ft
(4000 m)

6500–13,000 ft
(2000–4000 m)

1600–6500 ft
(500–2000 m)

650–1600 ft
(200–500 m)

0–650 ft
(0–200 m)

FIGURE 19.3
A map of physical features in Latin America.

19.2   Physical Geography: A Great Diversity of Habitats

TOPOGRAPHIC AND CLIMATIC VARIATIONS

Latin America is characterized by pronounced differences in elevation and topography from one area to another (Fig. 19.3). Low-lying plains drained by the Orinoco, Amazon, and Paraná-Paraguay river systems dominate the north and central part of South America and separate older, lower highlands in the east from the rugged Andes of the west. In Mexico, a high interior plateau broken into many basins lies between north-south trending arms of the mountains called the Sierra Madre. High mountains within Latin America—largely contained within the Sierra Madre and the Andes—form a nearly continuous line from northern Mexico and the southern United States to Tierra del Fuego at the southern tip of South America. Most of the smaller islands of the West Indies are volcanic mountains, although some islands made of limestone or coral are lower and flatter. The largest islands have a more diverse topography, including low mountains.

Humid Lowland Climates

Most Latin American lowlands are characterized by humid climates, which may be subdivided into "types" according to precipitation and temperature characteristics. Each type is associated with a dominant form of natural vegetation or ecosystem. The tropical rain forest climate, with its heavy

FIGURE 19.4
A map of climate patterns in Latin America.

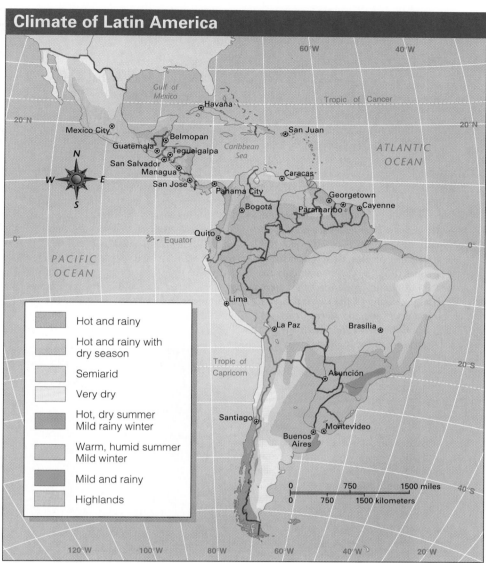

Climate of Latin America

Hot and rainy

Hot and rainy with
dry season

Semiarid

Very dry

Hot, dry summer
Mild rainy winter

Warm, humid summer
Mild winter

Mild and rainy

Highlands

Source: Rand McNally. *Atlas of World Geography.*

year-round rainfall, monotonous heat and humidity, and associated superabundant vegetation dominated by large broadleaf evergreen trees, lies primarily along or very near the equator, although segments extend to the margin of the tropics in both the Northern and Southern hemispheres. The largest block lies in the basin of the Amazon River system. Additional areas are found in southeastern Brazil, eastern Panama, the western coastal plain of Colombia, on the Caribbean side of Central America and southern Mexico, and along the eastern (windward) shores of some Caribbean islands, particularly Hispaniola and Puerto Rico (Fig. 19.4).

On either side of the principal region of tropical rain forest climate, the tropical savanna climate extends to the vicinity of the Tropic of Capricorn in the Southern Hemisphere and, more discontinuously, to the Tropic of Cancer in the Northern

Hemisphere. In this climate zone, the average annual precipitation decreases and becomes more seasonal as one approaches the poles. The mean temperature decreases and the broadleaf evergreen trees of the rain forest grade into tall savanna grasses, woodlands, or deciduous forests that lose their leaves in the dry season.

Still farther poleward in the eastern portion of South America lies a sizable area of humid subtropical climate, which has cool winters unknown to the tropical zones. Its Northern Hemisphere counterpart is north of the Mexican border in the southeastern United States. In Latin America, this climate is associated principally with a natural vegetation of prairie grasses in the humid pampas of Argentina, Uruguay, and extreme southern Brazil. On the Pacific side of South America, a small strip of mediterranean or dry-summer subtropical climate

Regional Perspective

ALTITUDINAL ZONATION IN LATIN AMERICA

One of the most significant features of Latin America's climatic pattern with respect to the distribution of economic activity and population is a series of highland climates arranged into zones by altitude. This zonation results from the fact that air temperature decreases with altitude, at a normal rate of approximately 3.6°F (1.7°C) per 1000 feet (304.8 m) of elevation. At least four major zones (Fig. 19.A) are commonly recognized in Latin America: the *tierra caliente* (hot country), the *tierra templada* (cool country), and the *tierra fría* (cold country, and the *tierra helada* (frost country). At the foot of the highlands, the *tierra caliente* is a zone embracing the tropical rain forest and tropical savanna climates discussed earlier. The zone reaches upward to approximately 2500–3000 feet above sea level at or near the equator and to slightly lower elevations in parts of Mexico and other areas near the margins of the tropics. In this hot, wet environment are grown such crops as rice, sugarcane, bananas, and cacao, often on a plantation basis. Populations of black Africans, brought to the New World as slaves, are concentrated in many of the *tierra caliente* zones.

The *tierra caliente* merges almost imperceptibly into the *tierra templada*. Although sugarcane, cacao, bananas, oranges, and other lowland products reach their respective uppermost limits at some point in this higher level, the *tierra templada* is most notably the zone of the coffee tree. In the *tierra templada*, coffee can be grown with relative ease (although by no means are all the soils suitable); at lower altitudes the crop encounters difficulties because of excessive heat and/or moisture. The upper limits of this zone—approximately 6000 feet (1800 m) above sea level—tend also to be the upper limit of European-induced plantation agriculture in Latin America. In its distribution, the *tierra templada* flanks the rugged western mountain ranges and, in addition, is the uppermost climate in the lower uplands and highlands to the east. Thickly inhabited sections occupy large areas in southeastern Brazil, Colombia, Central America, and Mexico. Although broadleaf evergreen trees characterize the moister, hotter parts of this zone, coniferous evergreens replace them to some degree toward the zone's poleward margins. In such places as the highlands of Brazil or Venezuela where there is less moisture, scrub forest or savanna grasses appear—the latter generally requiring more water.

In brief, the *tierra templada* is a prominent zone of European-influenced settlement and of commercial agriculture. Six metropolises exceeding 2 million in population—São Paulo, Belo Horizonte, Caracas, Medellín, Cali, and Guadalajara— are in this zone, while two others—Mexico City and Bogotá—lie just above it. Four smaller cities that are national capitals and the largest cities in their respective countries are

Altitudinal Zonation in Latin America

Snowline

Treeline

—15000 ft. (4600 m)
TIERRA HELADA
highland grains and tubers, sheep, guinea pigs, llama, alpaca, vicuña

—12000 ft. (3600 m)

TIERRA FRÍA
wheat, barley, maize, cool weather vegetables, apples, pears, dairying, short horn cattle

—6000 ft. (1800 m)

TIERRA TEMPLADA
coffee, maize, warm weather vegetables, cut flowers, short horn cattle

—3000 ft. (900 m)

TIERRA CALIENTE
sugar cane, tropical fruits, lowland tubers, maize, rice, poultry, pigs, zebu cattle

—Sea level

Source: Clawson, *Latin America and the Caribbean.*

FIGURE 19.A

A map of altitudinal zonation in Latin America. Because of the profound differences in relief in the landscapes of Latin America, specific patterns of land use have evolved in relation to differing altitudinal zones. This map shows their elevations and general land use patterns.

in central Chile is similar to that in southern California. To the south in Chile is a strip of marine west coast climate, which occupies the lower slopes of bleak, rainy, windswept, glaciated, and essentially uninhabited mountains at the southern end of the Andes.

Dry Climates and the Factors that Produce Them

The humid climates of Latin America have a more or less orderly and repetitive spatial arrangement. One may expect to find similar climates in generally similar positions on all major

located in the *tierra templada*: Guatemala City, San Salvador, Tegucigalpa, and San Jose. Others, like Rio de Janeiro, which are situated at lower elevations, have close ties with predominantly residential or resort towns in these cooler temperatures.

The *tierra fría* (at 6,000–10,000 ft) can be distinguished from the other zones by the related facts that it often experiences frost and is often the habitat of a native Indian economy with a strong subsistence component. European colonization in Latin America has driven some native Indian settlement upslope and into the *tierra fría* zone, although some major populations—the Inca of Peru for example—had already selected upland locations for their settlement before the arrival of Columbus. These upland settlements are most extensive in

Ecuador, Peru, and Bolivia, and this type of economy is also very evident in Colombia, Guatemala, and southern Mexico. The *tierra fría* is comprised of high plateaus, basins, valleys, and mountain slopes within the great mountain chain that extends from northern Mexico to Cape Horn. By far the largest areas are in the Andes, although areas in Mexico are sizable. The upper limit of the zone is generally placed at about 10,000 feet for locations near the equator and at lower elevations toward either pole. This line is usually drawn on the basis of two generalized phenomena: (1) the upper limit of agriculture, as represented by such hardy crops as potatoes and barley, and (2) the upper limit of natural tree growth.

Another zone lies above the other three and consists of the alpine meadows,

sometimes called paramos, along with still higher barren rocks and permanent fields of snow and ice. This zone is known as the *tierra helada.* It has some grains and animals (llama, alpaca, sheep) but is largely above the mountain flanks that are central to upland Indian settlement and agriculture. In the highlands of Latin America, the *tierra fría* tends to be a last retreat and a major home of the indigenous peoples, except in Guatemala and Mexico (see Fig. 19.B) and is characterized by small settlements and what Europeans and Americans might consider rather primitive ways of life. However, certain valuable minerals like tin and copper are located here, attracting modern types of large-scale mining enterprise into the *tierra fría* and, the *tierra helada* of Bolivia and Peru.

FIGURE 19.B
Andean Indians at a market in the town of Saquisili, Ecuador, located about a two-hour bus ride from Quito. The derby hats are customarily worn by Indians in a number of Andean countries. *Melanie S. Freeman/ Christian Science Monitor*

land masses of the world. But the region's dry climates—desert and steppe—are often due, at least in part, to local circumstances such as the presence of high landforms that block moisture-bearing air masses or the presence of offshore cool ocean water. In northern Mexico, such climates are partly associated

with the global pattern of semipermanent zones of high pressure which create arid conditions in many areas of the world along the tropics of Cancer and Capricorn. However, the aridity is also due in part to the rain-shadow effect caused by the presence of high mountain ranges on either side of the Mexican plateau. In Argentina, the extreme height and continuity of the Andean mountain wall are largely responsible for the aridity of large areas, particularly the southern region called Patagonia. Here, the Andes block the path of the prevailing westerly winds, creating heavy orographic precipitation on the Chilean side of the border but leaving Patagonia in rain shadow. But in the west coast tropics and subtropics of South America, the Atacama and Peruvian deserts cannot be explained so simply. Here, shifting winds that parallel the coast, cold offshore currents, and other complexities, as well as the Andes Mountains, are important conditions that combine to create one of the world's driest areas. However, the mountains serve to restrict this area of desert to the coastal strip. Arid and semiarid conditions also are experienced along the northernmost coastal regions of Colombia and Venezuela, and in the Sertao of northeastern Brazil.

Mountains play such a major role in the development of climate patterns and overall demographic patterns of Latin America that a whole pattern of regionally unique zonation has been developed for the region. It is built around **altitudinal zonation** (see Regional Perspective, pages 464–465) and is characterized by four dominant regional zones: the *tierra caliente, tierra templada, tierra fría,* and *tierra helada.*

19.3 Population Geography: Uneven Spacing, Diverse Stocks, Rapid Growth

In mid-1996, the world was estimated to contain 5.7 billion people, of whom the Latin Americans represented some 480 million, or almost 8.4 percent of the world's population. Because Latin America is a large and diverse region with well over twice the land area of the United States, it is not surprising that its population is unevenly distributed and exhibits enormous variations in density. There is, however, some overall order in the fact that most of Latin America's people are packed into two major geographic alignments. The larger of these two areas—called the rim—is a discontinuous ring around the margins of South America. The second—a relative highland—extends along a volcanic belt from central Mexico southward into Central America. The two alignments are outlined reasonably well by the pattern of population shown in Figure 2.15.

MAJOR POPULATED AREAS

The South American "rim" of clustered population contains roughly two thirds of Latin America's people, with the majority living in and around cities located on or very near the coast. Two major segments of the rim can be discerned. One—much

the larger in population and area—extends along the eastern margin of the continent from the mouth of the Amazon River in Brazil southward to the humid pampa (subtropical grassland) around Buenos Aires, Argentina. The second segment, located partly on the coast and partly in high valleys and plateaus of the adjacent Andes Mountains, stretches around the north end and down the west side of South America. This crescent begins in the vicinity of Caracas, Venezuela, on the Atlantic coast and arches around to the vicinity of Santiago, Chile, on the Pacific coast. This second segment is more fragmented than the first, as it is broken in many places by steep Andean slopes or coastal desert. For example, a strip of hot, rainy coast lies between the Amazon mouth and Caracas in which the thinly populated interior of South America extends to the ocean. In the far south, the population rim is again broken in the rugged, rainswept southern Andes and the dry lands of Argentina's Patagonia that lie in the Andean rain shadow on the Atlantic side of the continent. These latter territories have only a scanty human population, although millions of sheep graze in Patagonia and adjacent Tierra del Fuego.

Before the European conquest, it is thought that few native Indians lived along the Latin American coasts. Major population centers were located in the Andes, and the coasts were only lightly inhabited by migratory native peoples. The high intermontane areas within the Andes already exhibited a pattern of settled communities, ruled by the Inca Empire and approxi-

FIGURE 19.5
The old city of Cuzco in the Peruvian Andes, shown here, was the capital of the Inca Empire when the Spanish conquerors came to Latin America. The city, called "The City of the Sun" by the Incas, is located in a basin at an altitude of about 11,000 feet (c. 3400 m). *Guillo J. Barbero*

FIGURE 19.6

A landscape in central Mexico between Mexico City and Guadalajara. The corn fields and conifers betoken a relatively high altitude; the cinder cone of an extinct volcano is being quarried for road-building material.   *Jesse H. Wheeler, Jr.*

mately parallel to contemporary demographic patterns (Fig. 19.5). But when the Europeans began their quest and settlement of South America in the 1490s, the coasts, only lightly settled by the indigenous peoples, lay open to whatever settlement pattern the newcomers might devise. Because the Europeans approached from the sea at many separate points, the development of ports as bases for subsequent penetration of the interior was the first major settlement task. Many early ports eventually grew into sizable cities, and a few became major metropolitan centers. In some instances—notably Lima, Caracas, and Santiago—the main city developed a bit inland but retained a close connection with a smaller coastal city that was the actual port. Around the ports, agricultural districts developed, spread, and shipped an increasing volume of trade products overseas. The ships that took these goods away returned with more from Europe. They also carried passengers, including government officials, in both directions. Slave ships brought the Africans who provided the principal labor on European-owned plantations, primarily in the tropical lowlands. Populations along or near the coasts multiplied and are still doing so today.

Meanwhile, the ever powerful lure of gold and silver stimulated the penetration of the Andes and the Brazilian Highlands. After looting the stores of precious metals that had been accumulated by the indigenous peoples for ceremonial use and trade, the newcomers opened mines, or in many places took over old mines, and urban centers arose to service the mines. Some highland cities gained a wider importance as centers for new ranching or plantation areas. Such functions still endure, although great numbers of older mines have closed, and mining today is far more diversified and less dependent on gold and silver than was true in the early colonial age. A few highland cities—of which the largest is Bogotá in Colombia—have grown into sizable metropolises. These places are separated from the seaports by difficult terrain that required feats of considerable engineering skill before satisfactory transportation links were developed. Some of the main seaports and highland cities became important centers of colonial government as well as economic nodes, and several are national capitals today, such as La Paz in Bolivia and Quito in Ecuador. Rapid development of manufacturing industries and explosive population growth have characterized such cities in recent times.

The second major alignment of populated areas in Latin America lies on the mainland of Caribbean or Middle America to the northwest of the South American "rim." Composed of the majority of Middle America's people, it extends along an axis of volcanic land that dominates central Mexico and from there reaches southeastward through southern Mexico and along the Pacific side of Central America to Costa Rica. This belt is characterized by good soils, rainfall adequate for crops but not excessive, and enough elevation in most places to moderate the tropical heat but still permit the raising of many tropical crops such as coffee. The belt of high population density that extends across central Mexico from the Gulf of Mexico to the Pacific Ocean was already Mexico's center of population when the Spanish conquerors came. This was the principal domain of the large and technologically advanced Aztec Empire, which, together with its vassal states, composed much of Middle America's population at that time. Then, as today, the soils of this area's high volcanic plateaus and flat-floored basins provided the foundation for a productive agriculture based primarily on maize (Fig. 19.6). Spanish occupation did not alter the dominance of the area within Mexico, and roughly half of Mexico's 94 million people (1996) now live in, around, or between the region's two principal cities of Mexico City and Guadalajara.

Definitions and Insights

TEOTIHUACAN: THE FIRST GREAT AMERICAN METROPOLIS

The vast city of Teotihuacan, located in the valley of Mexico, was the first true urban center in the Western Hemisphere. Its construction is thought to have begun approximately 2000 years ago, and by 500 A.D. it was as large as London was in 1500 A.D. Located some 30 miles from Mexico City, it covered about eight square miles. The city was laid out on a grid that was aligned to mesh with stars and constellations that were central to the time keeping of the Teotihuacanos. As of yet, the origins, language, and culture of these people are not fully understood. The city may have had 200,000 residents at the time of its zenith as a ceremonial center in 500 A.D. There were some 2000 apartment compounds, a twenty-story Pyramid of the Sun, a Pyramid of the Moon, and a 39-acre civic and religious complex called the Great Compound or the Citadel (Ciudadela). As described by National Geographic Society anthropologist George Stuart, "The whole is a masterpiece of architectural and natural harmony."[2] It still stands as a monument to native American engineering skill and cultural intensity.

In Pacific Central America also, the basic pattern of population distribution already was established when the Spanish took over. Today, the majority of people live in highland environments, but considerable numbers live in lowlands as well. The ongoing vulcanism of the highlands periodically deposits ash into the highland basins, and this weathers into fertile soils that grow maize, beans, and squash for local food and the coffee that is Central America's largest export. Between the highlands and the Pacific Ocean lies a coastal strip of seasonally dry lowlands where cattle and cotton are grown largely for an export market. By contrast, the Atlantic side of Central America is lower, hotter, rainier, and far less populous than the Pacific side. Here, on large corporate-owned plantations, are grown most of the bananas that are Central America's agricultural trademark and second most valuable export after coffee.

VARIATIONS IN POPULATION DENSITY

Average population densities for entire countries conceal the fact that in nearly every mainland country there is a basic spatial configuration consisting of a well-defined population core (or cores) with an outlying sparsely populated hinterland.

[2]George E. Stuart, "The Timeless Vision of Teotihuacan," *National Geographic,* Vol. 188, No. 6 (December 1995): 2–35.

Densities within the cores are often greater than anything to be found within areas of comparable size in the United States. On the other hand, most mainland countries have a larger proportion of sparsely populated and relatively unorganized terrain than is true of the United States. Figures on population density for the greater part of Latin America are extraordinarily low, averaging under 2 persons per square mile over approximately half of the entire region. This pulls down Latin America's average density—and that of most major Latin American countries—to levels somewhat below the average density for the United States.

The pattern of core versus "outback" (to use an Australian term) can be seen in Brazil and Argentina. Most Brazilians, for example, live along or very near the eastern seaboard south of Recife, a port city in northeastern Brazil, whereas large areas in the interior are still thinly populated. In Argentina, approximately three fourths of the population is clustered in Buenos Aires or the adjacent humid pampa—an area containing a little over one fifth of Argentina's total land. The less densely settled regions in Latin America are generally quite distinctive from the core. While there is a steady migration from the outback to the core, there is very little reverse migration of the core population going to the outback, with its relatively more demanding and less convenient setting for people who have grown used to the environmental and urban amenities of the core region. In Latin America, population patterns reflect this demographic imbalance in Brazil, Argentina, and Mexico as well as in the Andean countries (see Fig. 2.15 for world dot population map).

Not every country displays the pattern of core versus outback. A very different picture of population density and distribution is evident in El Salvador and Costa Rica—both of which are composed of thickly settled volcanic land—and in most Caribbean islands (see Table 19.1). Because the combined population of the islands in the Caribbean is considerably less than that of central Mexico alone, they do not contain a concentration of people comparable in scale to either of the two major Latin American concentrations discussed earlier. However, they do stand out for exceptionally heavy densities, which have been created over a long period by the accumulation of expanding population on territories of very restricted size. Most often the crowding is aggravated by high rates of natural population increase and/or the prevalence of steep slopes that restrict possibilities for further farming or settlement; this is also true of parts of El Salvador and Costa Rica.

ETHNIC DIVERSITY

Although most Latin Americans speak Spanish or Portuguese and embrace some form of Roman Catholicism, the region's inhabitants have highly varied ancestries—indigenous, black African, white European, Asian, or mixtures. And while Europeans, Africans, and Asians came from various parts of their respective continents, the native peoples represented a multiplicity of nations from an even larger base in both the Northern and Southern hemispheres. Despite the overall dominance in Latin America of culture traits derived from Europe, only three of this region's 35 nations—Argentina, Uruguay, and Costa Rica—have preserved white European racial strains on a large scale with little admixture by Indians or blacks. This is due in part to the fact that these nations received relatively large numbers of European immigrants during the late 19th and early 20th centuries. Scattered districts in other countries also are predominantly European or profoundly native Indian.

Native Indians (see Fig. 19.B) compose an estimated 40 to 55 percent of the population in the highland nations of Guatemala, Bolivia, Ecuador, and Peru. Various sources estimate that 40 to 82 percent of the population in highland nations are composed of native Indians. Ecuador at 60 percent and Bolivia at 82 percent are the highest. Native Indians are a significant minority overall in Mexico at 25–30 percent, but comprise over 50 percent in some southern states of the country.

Native Indians are also a major population element in the basin of the Amazon River and in Panama, where scattered lowland Indians maintain their aboriginal cultures. However, the outer world is pushing in on them despite some efforts on the part of governments to protect their ways of life. Over the centuries after Columbus, the encounters of lowland Indians with Europeans generally had catastrophic results for the aborigines, who often resisted exploitation fiercely but were no match for the Europeans' diseases (especially smallpox), guns, liquor, and relentless hunger for more land. The more numerous highland Indians, who were more advanced, more distant, and better organized than the lowlanders, were somewhat more successful in accommodating European encroachment and exploitation but still suffered enormous population losses at European hands. Disease was the most active killer.

Black Latin Americans of relatively unmixed African descent are found in the greatest numbers on the Caribbean islands and along hot, wet Atlantic coastal lowlands in Middle and South America. These are the areas to which African slaves were brought during the colonial period, primarily as a source of labor for sugar plantations. Slavery was gradually abolished during the 19th century, although not until the 1880s in Brazil and Cuba. By that time, slavery had generated large fortunes for many owners of plantations and slave ships and had introduced African peoples to the region who have made cultural contributions that are today a varied, colorful, and important part of Latin American civilization.

The ethnic diversity of Latin America is further enhanced by relatively small concentrations of Asians in scattered places. For example, about one half of the population of Guyana and Trinidad and Tobago is made up of people whose ancestors came from the Indian subcontinent, and one half of Suriname's population is Indian or Indonesian. There are nearly one million Japanese in Brazil—the largest concentration outside Japan, including the United States. These Asians came mainly as contract laborers for sugar plantations in the 19th century after the end of slavery, and many stayed after their contracts expired.

Latin America has escaped many of the racial tensions that grip much of the world. This results, in part, from the fact that the majority of Latin Americans have a mixed racial composition. Most of the region exhibits a primary mixture of Spanish and native Indians, resulting in a heterogeneous group known as *mestizos*.

RAPID POPULATION GROWTH AND EXPANSIVE URBANIZATION

Irrespective of ethnic composition and type of culture, most Latin American countries today find themselves in the second stage of what has been called the **demographic transition.** Startling rates of urban and metropolitan growth have been a major Latin American characteristic during recent times. Recent estimates (1996) give a figure of 70 percent urban for the population of the region as a whole, compared to a world average of 43 percent. A major element in the rapid growth of Latin American cities is a huge and continuing migration from countrysides that often are so overpopulated and poor as to be no better than rural slums.

Definitions and Insights

HUMAN MIGRATION: THE OFTEN IRRATIONAL POWER OF PUSH-PULL FACTORS

In any consideration of human migration, there is a discussion of the environmental and social factors that cause someone to leave an area (the **push factor**) and the oftentimes corresponding factor that attracts (the **pull factor**) someone to a specific locale. Most often economic factors are given the most importance in assessing migration flows, particularly those from the countryside to the city. However, very often a young man or woman—and this is particularly true in Latin America—will leave a position of steady underemployment in the rural sector and go to the city, even if there is no certainty about employment there at all. Images of the city are often filled with "the good life" and mythic opportunities for economic gain. As a result, these urban centers become terribly overcrowded with unskilled, generally uneducated rural youth who have left their very simple homes in the countryside, hoping to gain money with which they can help support rural family members. Very often this critical migration ends with economic failure, but the pressure to hide such lack of success is so strong that there is very little reverse migration, or return to the countryside. Even where there is active industrial growth and employment opportunity—such as in Mexico City and São Paulo, Brazil (two of the four largest cities in the world)—there are many more migrants and applicants than factory or service-sector positions. As a result, the slums grow with a steadiness that challenges every urban planner and municipal administrator.

Even in rural areas that are better off, cities are viewed by the ambitious young as places to get ahead rapidly and enjoy exciting amenities while doing so. Deeply rural and tradition-bound villages offer little competition to the bright lights and swirling action of the metropolis. But a great many rural dwellers flee to the cities out of sheer desperation induced by drought, exhausted land, depressed farm prices, runaway inflation, chronic unemployment, guerrilla warfare, or other ills that beset the countrysides of many developing parts of the world in our time. In Latin America, the urban explosion is especially problematic in the major urban centers in the rimland on the margins of the agricultural lowlands in both South and Central America. Caring for the huge influxes of people has strained the services of Latin American cities beyond their limits. As a result, every large city has big slums, which often take the form of ramshackle shantytowns on the urban periphery. Sometimes they are built on hillsides overlooking the city, as at Rio de Janeiro, where some of the world's more unsightly housing (Fig. 19.7)—called *favelas* in Brazil—commands one of the world's more imposing views. Such shantytowns, which are practically universal in the world's less-developed countries, are full of underemployed and ill-fed people, who still may prefer their present plight to the deplorable conditions of the outback and depressed rural landscapes from which they came.

19.4 "Emerging" or "Developing" Nations: Characteristics and Frustrations

Latin America is an "emerging" or "developing" region that does not yet provide a good living for most of its people. It is not the worst off of the major world regions; in fact, its overall per capita GNP is almost four times that of Africa (see Table 1.2). But compared with North America or Europe, the Latin American region as a whole is quite poor. This tends to be masked by the glitter of the great metropolises, with their forests of new skyscrapers—but even these places are rimmed by wretched shantytowns, and further out lie depressed countrysides. In an attempt to overcome their economic deficiencies and constrain popular discontent, many Latin American governments have borrowed heavily from the international banking community, including leading banks in the United States. By the mid-1990s, unpaid loans had reached staggering proportions, and the Latin American debtor nations were having great difficulty in mustering even the annual interest payments, let alone surplus capital to pay toward the principal of these troubling loans—a situation so often associated with the process of "emerging" or "developing."

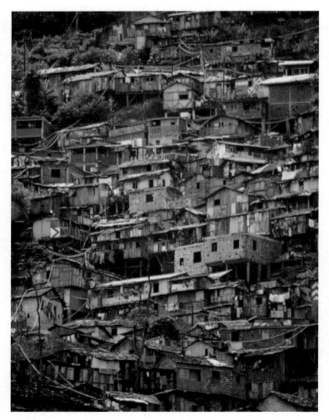

FIGURE 19.7
Many of Latin America's poorest people live in ramshackle housing that clings to nearly vertical slopes. The shantytown in this photo overlooks scenic Rio de Janeiro; the locale is the suburb of Santa Marta. *Stephanie Maze/Woodfin Camp & Associates*

TWO SYSTEMS OF AGRICULTURE: LATIFUNDIA AND MINIFUNDIA

Although the total number of Latin Americans employed in agriculture has not declined very much in recent decades, there has been a sizable percentage decline in agriculture as a component of both employment and value of product in national economies. The actual percentages vary a great deal from country to country. Dependence on agricultural exports has also dropped as national economies have become more diversified, but in many countries more than half of all export revenue is still derived from products of agricultural origin (Table 19.2).

Farms in Latin America are often divided into two major classes by size and system of production. Large estates with a strong commercial orientation are **latifundia** (singular: latifundio). These estates, whether called *haciendas,* plantations, or some other name, are owned by families or corporations. Some have been in the same family ownership literally for centuries. The desire to own land as a form of wealth and symbol of prestige and power has always been a strong characteristic of Latin American societies. Huge tracts were granted by Spanish and

FIGURE 19.8

Sugarcane on a Jamaican plantation. Occupying a basin rimmed by hills, this family-owned plantation is one of the few remaining commercial sugarcane producers in Jamaica. It bears continuing witness to the early role of cane cultivation on slave plantations as a stimulus to colonial settlement of the West Indies and many areas of the Latin American mainland. The plantation has diversified into cattle raising and citrus production; the darker green in the photo at the base of the slope marks a grove of orange trees. *Jesse H. Wheeler, Jr.*

Portuguese sovereigns to members of the military nobility (**conquistadors**) who led the way in exploration and conquest. Some of this land has been reallocated to small farmers by government action from time to time—the *ejido* (see Definitions and Insights, page 472) in Mexico is a good example—but in many countries a very large share of the land is still in the hands of a small, wealthy, landowning class.

Agricultural production on the large estates characteristically involves the employment of large numbers of landless, often illiterate, workers. Profits benefit primarily the owners, whereas the workers are often underpaid and only seasonally employed. Today, many owners have business or professional occupations in cities, leaving the actual direction of their estates to managers. In such cases, an owner and his family may reside exclusively in the city, or they may live part of the time on the estate. There are great variations in the care and efficiency with which estates are managed and kept up. Some are flourishing enterprises, whereas others are drained of wealth by absentee owners and held primarily for the prestige they confer. These large landholdings have been declining in number, but they still produce a sizable share of the coffee, sugar (Fig. 19.8), cotton, livestock products, and other Latin American farm commodities that enter world markets. A special variant of the large estate is the single-crop commercial plantation owned by a company or a syndicate. The capital to establish these enterprises has come principally from the United States or Europe. Bananas are a characteristic product.

Minifundia are smaller holdings with a strong subsistence component. The people who farm them generally lack the capital to purchase large and fertile properties, and hence they are relegated to marginal plots, often farmed on a sharecropping basis. Individuals who do own land are frequently burdened by

indebtedness, the foreclosure of farm loans, and the fragmented nature of farms, which are becoming smaller and smaller in size as they are subdivided through inheritance and the reversal of promised governmental programs of land distribution. Such farms produce food primarily for family use but also for the local market. The crops most commonly raised are maize (Fig. 19.9), beans, and squash, although many other crops are locally important, especially in the different climatic zones of the high-

FIGURE 19.9

Maize (corn) is a major crop all over Latin America. In this photo, a government drug inspector in Mexico is about to burn the harvested corn, which contains concealed heroin. Many ingenious means are used to smuggle illegal drugs into the United States. *Andy King/Sygma*

land regions. Although these small farmers make up the bulk of Latin America's agricultural labor force, food has to be imported into many areas, and many of the people are poorly nourished. Productivity is low because of the marginal quality of the land and the generally rudimentary agricultural techniques. There is little capital with which to buy machinery, fertilizers, and improved strains of seeds. Soil erosion and soil depletion are making serious inroads in many areas.

Definitions and Insights

LAND REFORM EFFORTS

The historical geography of Latin America is filled with attempts—sometimes very serious and bloody, sometimes less painful governmental innovations—to achieve a better balance between those who work the land and those who own the land. The main attention has been given to breaking up existing properties or bringing vacant land (whether owned by the public, by private individuals, or by the Catholic church) into cultivation, usually by small farmers. Some new farms have been structured as communal holdings, reflecting indigenous traditions or 20th century revolutionary plans of agrarian reform. In Cuba, the communist government that took control in 1959 placed the land from expropriated estates in large farms owned and operated by the state. Workers on these farms are paid wages. The Nicaraguan Sandinistas implemented land reform as part of their revolutionary efforts in the 1980s.

The details and success of land reform schemes have varied sharply from one Latin American country to another. Such schemes can help relieve poverty in the countryside, but they do little for the majority of Latin America's poor, who live in cities. In Mexico, there was great fanfare at the introduction of the *ejido* (communally farmed land in Mexican Indian villages) early in the 20th century. It has been so effective that *ejidos* currently account for 50 percent of the cultivated land in Mexico and produce nearly 70 percent of the beans, rice, and corn of this nation. The *ejido* has been the most successful land reform program attempted in Latin America, including Cuba. Puerto Rico has had public battles over "pan, tierra, y libertad" (bread, land, and liberty) because of the difficulty in getting real distribution of farmland to the farming people, and the January 1994 insurrection in Chiapas, Mexico, had land distribution at its heart. Virtually every nation in Latin America can point to events, martyrs, and programs birthed in protest or blood in an effort to achieve more equitable land ownership. Nevertheless, the great majority of the arable land in Latin America continues to be owned by wealthy farming families, agribusiness operations (both domestically and foreign-owned), and the church. There is a world of small farmers, but they seldom own as much as 20 percent of a nation's arable land.

MINERALS AND MINING: SPOTTY DISTRIBUTION AND UNEQUAL BENEFITS

Latin America is a large-scale producer of a small number of key minerals that are very significant to the outside nations where they re-marketed. Only a handful of Latin American nations gain large revenues from such exports; in most countries, mineral production is relatively minor. Even in the countries that do have a large value of mineral output—notably Mexico, Venezuela, Chile, Ecuador, and Brazil—much of the profit appears to be dissipated in the form of showy buildings, corruption, ill-advised development schemes, and enrichment of the upper classes and foreign investors. Benefits to the broader population have tended to be rather minimal. Nonetheless, significant infrastructural development, including many new highways, power stations, water systems, schools, hospitals, and employment opportunities, has been made possible by mineral revenues. Among Latin America's most important known mineral resources are petroleum, iron ore, bauxite, copper, tin, silver, lead, zinc, and sulfur. However, local use of these minerals for industrialization has been retarded by a lack of good coal, especially the coking coal that is very important in steel production, although Brazil has realized a major improvement in its production and now is the top automobile producer in Latin America.

All but a small proportion of Latin America's petroleum is extracted in the Caribbean Sea–Gulf of Mexico area, particularly the central and southern Gulf coast of Mexico and northern Venezuela (see Table 19.2). Other oil fields in Latin America are widely scattered, with the principal ones located along or near the Atlantic coast in Brazil, Argentina (in Patagonia), and Colombia, or in sedimentary lowlands along the eastern flanks of the Andes in every country from Trinidad and Tobago to Chile. Natural gas is extracted in many areas that produce oil, but Latin American production is not yet of major world consequence. Mexico, Venezuela, and Argentina are the largest producers.

Latin America is a major world area in the production of metal-bearing ores. Most of the extracted ore is shipped to overseas consumers in raw or concentrated form, although the region has scattered iron and steel plants, as well as smelters of nonferrous ores, which process metals for use in Latin American industries or for export (Fig. 19.10). Deposits of high-grade iron ore in the eastern highlands of Brazil and Venezuela are the largest known in the Western Hemisphere and are among the largest in the world; the two countries are Latin America's main producers and exporters of ore. Brazil is the main producer of iron and steel by far, with Mexico second. Most of the region's production of bauxite—the major source for aluminum—comes from Jamaica, Brazil, Suriname, and Guyana. The deposits in these countries are located relatively near the sea and are of critical importance to the industrial economies of the United States and Canada. Huge unexploited bauxite deposits in Venezuela are a major resource for the future.

FIGURE 19.10

Mineral wealth in the Andes was a major stimulus of the early European interest in control and, later, settlement in the New World countries that had these resources. Tin continues to be one of the minerals still extracted from the Andean uplands. The mining operations also continue to rely on hand labor for much of their productivity. This photo is from a Bolivian tin mine.

Elizabeth Harris/Tony Stone Images

Regional Perspective

MEXICO, THE UNITED STATES, AND NAFTA: THE BURDENS OF "EMERGING"

Mexico's history since the middle of the 19th century has been powerfully shaped by the proximity and economic magnitude of the United States. In 1993, when the North American Free Trade Agreement (NAFTA) was signed, Mexico anticipated an era of steady economic growth through the new trade relations stimulated by this pact with the United States and Canada. In 1994, Mexico realized some of these hopes with steadily increasing exports to the U.S., but the assassination of presidential candidate Luis Donaldo Colosio in March of that year sent shock waves through the international community. Mexico was instantly set back to the image of a politically unstable country. Because the entire process of "emerging" as an economy is based upon a nation's capacity to provide a solid political base for foreign investment, this descent into political instability has proved very costly.

By the end of 1994, a run on the peso had pulled the country into economic chaos, and it was not until the United States cobbled together an international aid package ($20 billion from the U.S., $17.8 billion from the International Monetary Fund) that the Mexican currency was stabilized. However, to gain these external dollars, Mexican president Ernesto Zedillo Ponce de Leon had to inaugurate economic reforms that led to enormous domestic dissatisfaction (see Chapter 20. This process further fomented support for the rebellion in the southern state of Chiapas that had begun on New Year's Day, 1994. By 1996, the swirl of heavy foreign debt, aggressive demands for domestic budget restraints, United States pressure for stronger police action against drug trafficking, and the northward flow of undocumented Mexicans crossing into the U.S. put Mexico in the center of troubles flowing from all directions.

The process of attempting to plan for the most productive use of domestic resources (including human labor) to respond to the potential of foreign markets and associated capital investment has proven to be a monumental dilemma for not only Mexico, but for Brazil, Argentina, and most other Latin American nations faced with heavy international debt and controls designed by the international monetary community.

Low-grade but comparatively abundant copper deposits are located in the Atacama Desert of northern Chile and in the arid and semiarid sections of Mexico. Additional reserves are found in the Andes, especially in Peru and Chile. Chile is overwhelmingly the largest copper producer in Latin America and ranks number one in the world. Most of Latin America's known

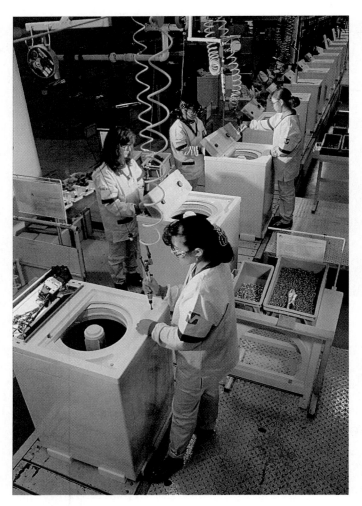

FIGURE 19.11

The *maquiladora* movement stimulated by NAFTA—that is, the legal capacity to assemble units in Mexico with Mexican wages and export them without tariffs to the United States market—has been a great boon to manufacturing plants in Mexico. This assembly line of Whirlpool washers is a good example of this economic change. There are also very high-tech car assembly production lines in operation through the *maquiladora* innovation. *Gabriel Covian/The Image Bank*

reserves of tin are in Bolivia or Brazil, and those countries produce most of the region's output. The silver of Mexico, Peru, and Bolivia—sought from the very beginning of Spanish colonization—is found principally in mountains or rough plateau country. Mexico and Peru are by far the largest Latin American producers of silver and of lead and zinc as well. Peru's deposits of all three minerals are in the Andes, and Mexico's are mainly in the dry northern and north-central sections of the country. In recent decades, Mexico has become a significant source of native sulfur. The main fields border the Gulf of Mexico, as do the very important adjacent fields of the United States. Sulfur is used primarily for the manufacture of sulfuric acid, which has many important industrial uses.

THE INCREASING IMPORTANCE OF MANUFACTURING AND SERVICES

Although they are still very spotty in distribution and often lacking in complexity and sophistication, manufacturing industries in Latin America are making an increasingly important contribution to national economies. Most parts of the region are still a long way from full industrialization, but certain districts have an impressive array of factories, including some of the most modern types. Because the industrial revolution has come to Latin America very late, the region does not face large problems of industrial obsolescence and is able to bypass the centuries of development that took place in older industrial countries by importing modern technology and equipment. There is a regionwide push to attract industry from abroad with offers of tax exemptions, cheap labor, and other inducements. The development of the *maquiladora* (assembly operations in Latin America that combine components manufactured from other countries with local labor) movement and NAFTA (see Regional Perspective, page 470) have done a great deal to try to keep Latin American labor occupied in locally run factories (Fig. 19.11).

By far the most numerous manufacturing establishments in Latin America are household enterprises or small factories that employ fewer than a dozen workers and sell their products mainly in home markets. Larger operations are almost always located in the larger cities and often are branch plants of overseas companies. The Latin American countries that have the largest output of factory goods are Brazil, Mexico, and Argentina. The industrial scene is dominated by a handful of large metropolises, mostly notably São Paulo, Mexico City, and Buenos Aires. Brazil actually ranks in the top ten of the world's nations in production of manufactured goods, although its per capita ranking would be much lower than that.

The proportion of Latin Americans in service employment has risen sharply during the past quarter century as people from the countryside have flooded into the cities. Expanding government and business bureaucracies, the increasing affluence of urban upper and middle classes, an increase in tourism, and a general increase in the facilities required for even a minimal servicing of hordes of new urban dwellers have created many millions of new service jobs. Even at general pay scales far below those of western Europe or North America, such jobs have given new opportunity to desperately poor migrants from overcrowded rural areas, and they have helped accommodate the explosive natural increase of population within the cities themselves.

In many Caribbean islands and scattered places on the mainland, tourism particularly has become a major economic asset (Table 19.3 and Fig. 19.12). The recent development of major resort complexes in Mexico, the Caribbean islands, Costa Rica, parts of Brazil, and other Latin American countries has expanded tourism as a major player in the generation of

TABLE 19.3	International Tourism to Selected Latin American Destinations, 1992			
Country	**Total No. Tourists (000s)**	**Percentage of Tourists by origin (%)**		**Total Tourist Receipts ($U.S.) (000,000s)**
		Americas	Europe	
Argentina	3030.9	83.0	12.1	3090
Aruba	541.7	89.6	9.8	443
Bahamas	1398.9	89.1	8.7	1244
Barbados	385.4	59.0	40.0	463
Brazil	1474.9	72.4	23.2	1307
Chile	1283.3	89.8	8.0	706
Colombia	1075.9	92.3	7.1	705
Costa Rica	610.6	83.5	14.8	431
Cuba	460.6	49.1	47.5	382
Dominican Republic	1523.8	n.a.	n.a.	1054
Ecuador	403.2	80.2	16.9	192
El Salvador	314.5	92.4	6.6	49
Guatemala	541.0	77.5	19.6	243
Jamaica	909.0	77.1	20.1	858
Martinique	320.7	18.5	80.8	282
Mexico	17,271.0	97.0	2.1	5997
Panama	311.4	90.2	6.3	207
Puerto Rico	2639.8	70.2	n.a.	1511
Uruguay	1801.7	83.2	2.7	381
U.S. Virgin Islands	390.0	94.9	3.3	792
Venezuela	433.5	54.9	43.2	432

Adapted from David L. Clawson, *Latin America and the Caribbean: Lands and Peoples.* Dubuque, Iowa: Wm. C. Brown, Publishers, 1997: 281.

FIGURE 19.12
Almost nothing is as important as landscape elements in the selection of tourist destinations. This major hotel in Nassau in the Bahamas features natural-looking water displays, lush vegetation and flowering trees, and comfortable accommodations—all with English as the spoken language. The economic importance of becoming a major tourist destination means very much to the economies of many Latin American countries.   *Timothy O'Keefe/ Tom Stach & Associates*

foreign exchange. Cancún, Mexico, for example, was recently described in a tourist brochure as "Cancún—Mexico's American Resort . . . Cancún may seem more American than Mexican." The region's proximity to the relatively wealthy and mobile North American and Canadian populations, the expansion and improvement of regional air transport, and the increasing inclination of the North Americans to spend money for tourism have all brought considerable disposable income to the countries and coastal areas that have made these landscape investments. Passenger cruise ships now stop at ports of call in the Antilles, on the Pacific coast of Mexico as well as at Cancun, and at Santa Marta and Cartagena in Columbia. The transiting of the Panama Canal continues to be important as a tourist activity, although the canal is now too narrow for the largest of the transoceanic cargo ships.

Cruise companies are also increasing the number of sailing routes along the east coast of South America with stops in the Amazon Basin and in coastal ports of Brazil, Uruguay, and Argentina. San Juan, Puerto Rico, is now a major port of embarkation for tourist cruises destined for the eastern Antilles, with passengers regularly being flown from U.S. and Canadian cities to San Juan for boarding and embarkation. Countries in which tourist dollars amount to more than 20 percent of the nation's foreign exchange include Argentina, Dominican Republic, and the Bahamas.

19.5 Political Geography: Problems of Fragmentation and Instability

Latin America is organized into political units that vary tremendously in size and political status. Aside from some smaller countries (mainly Caribbean island states) that gained independence after World War II, Latin America generally broke away from European control in the 19th century. The combined population of Latin America's still-dependent units is only about 5 million (as of 1994) and is found primarily in units still affiliated politically with the United States, France, the Netherlands, or the United Kingdom. All of these—except the United Kingdom's Falkland Islands colony—are located in the Caribbean islands or, in the case of French Guiana, on the nearby mainland of South America. The different units vary in their relationships to the overseas nations with which they are linked. Most of them, however, have much latitude in handling their internal affairs.

Today, the fragmented political order of Latin America stands in marked contrast to the two massive political units—the United States and Canada—that occupy a territory of comparable area in North America. One can visualize the complexities that would have resulted had the individual states of the U.S. and the provinces of Canada become separate nations divided from each other by international boundaries. In the United States and the more populous parts of Canada, most state or provincial boundaries run through areas where settlement is continuous on both sides of the boundary and the boundary line is crisscrossed by numerous highways and railroads. People in great numbers and goods in great quantities move back and forth across these boundaries with general ease every day. But in Latin America most mainland borders run through outlying areas where there are few people, few roads, and almost no railroads. All over the Latin American mainland, the core regions of the different countries tend to be well insulated from each other by distance, unpopulated wilderness, and a dearth of international land routes. From a functional standpoint, the national population nodes and their capitals might almost be in separate continents. Each country tends to go its own way, communicating with the outside world by air or sea and maintaining its most crucial relationships with nations outside of Latin America. Any united effort to develop Latin America's resources and alleviate

Definitions and Insights

INSURRECTIONS AS A GOVERNING NORM

Guatemala had two military coups in eighteen months during the early 1980s; Grenada's Marxist government experienced a leftist-oriented military coup in 1983 that was ended by United States military invasion; in early 1989 a military coup in Paraguay overthrew a strongman who had held the office of president, with dictatorial powers, for many years; a military coup ousted a democratically elected president in Haiti in 1991 (a promised U.S. military action in 1995 returned him to power); a democratically elected president in Peru dissolved the national parliament and took dictatorial control of the government in early 1992 for the announced purpose of dealing more efficiently with guerrilla warfare and other national problems; and serious guerrilla warfare was in progress in a number of countries between 1994 and 1996. The insurrection in the Mexican state of Chiapas on New Year's Day in 1994 continues to keep Mexico's government in the throes of political uncertainty even as it tries to stabilize the nation's economic systems. The historical patterns of domestic unrest leading to armed insurrections make for both images and realities of governmental uncertainty. The political forces that foment such instability relate to traditions of land tenure, the economic gap between the wealthy and the poor, and real and perceived foreign control of the best agricultural lands. The guerrilla rebellion as a Latin American response to economic difficulty and political frustration is a part of regional reality in much of this region's history.

poverty through economic development remains rather minimal, despite increasing efforts in recent years to promote greater cooperation and integration in both the political and economic spheres.

The United States has played a major role in the last century of Latin American development. By the beginning of World War I, the U.S. had invested $8.5 billion of private capital in this region, with one third in Argentina, one third in Mexico, and one quarter in Brazil. By 1914, the United States had been instrumental in the design and construction of the Panama Canal, and played a political role in breaking away critical Panamanian territory from Columbia—an act that seemed essential to achieve initiation of the construction of the canal at the time. By the same time, the U.S. had invaded Cuba and been central to political change in Mexico. By the 1990s, U.S. troops, or "military advisors," had been in Santa Domingo, Panama, Guatemala, Mexico, Haiti, Nicaragua, Honduras, El Salvador, Grenada, and very nearly a second time in Cuba in the 1960s.

Parallel to the record of United States military involvement in Latin America has been its ongoing centrality to regional economic development. The U.S. has been generally the largest export market for Latin American commodities and goods, and has served as the largest supplier of imports. Other continuing and significant U.S.–Latin American interaction includes combined efforts in dealing with the illicit drug trade that serves as the major source for foreign marijuana and cocaine in the United States. There has also been the steady migration—both legal and illegal—of young males streaming north in the hope of expanded labor opportunities in American cities and farms. As distinct and independent as Latin American nations want to be, an inexorable influence continues to be worked on the region by the United States. That influence—as is so often the case in relationships shaped by spatial proximity—has both negative and positive impacts on Latin American development.

In the diversity of Latin America all of the elements for steady development and democratic reform are present. In many of the 35 countries in this region, there are historical periods of success in both of those realms. At the same time, the region has known great frustration as single-crop or single-mineral economies have been depressed by rapid and deep drops in prices for these goods, leaving the Latin American export country in major trouble because of a too-narrow base for economic stability. Populations are varied in stock, but not so dense as to be an overall burden on the country—except in the cases of overcrowded urban centers that have become migration destinations for many rural youth seeking new, urban opportunity. The recent inflow of tourist dollars and the steady efforts to slow population growth, even within the context of a Roman Catholic region, are taken by many as signs of optimism for planners and citizens alike. It is with keen anticipation that the peoples and governments of Latin America will observe the continuing impact of NAFTA during these next years, trying to assess both economic and environmental effects of these significant attempts at regional cooperation.

20

THE DIVERSITY OF LATIN AMERICA

The themes in this chapter's literature example (page 492) introduce us to a full set of geographic realities that characterize Latin America. The small nation in the story by Carmen Naranjo has, as a neighbor, a country that is large and untrustworthy. It is plagued by too much foreign debt, controlled by international agencies that have lent it money and now seek to control its economy, and has a government that takes care of itself but seems to be able to do little for the mass of the population, which suffers from poverty, hunger, and lack of any hope of changing this. It also has a natural resource that it sees as tiresome and worthy of sale for financial benefit. The solution that the president comes upon—after holding an international beauty contest for "Miss Underdeveloped"—is the sale of its country's abundant rain.

These characteristics are given in a harsh hyperbole, but all of these themes play across the pages of the daily press in Latin America. The United States has long had an interest in the exploitation of the resources of Latin America, and has had financial involvement that has demonstrated its capacity to turn such resources into capital. But such benefit to the Latin American nations often comes with a growing debt structure. The United States has also been a major force in the various international agencies that monitor debt loads and attempt to change government and management patterns so that large loans can be repaid on schedule. Such arrangements have traditionally created feelings of dependency and irritation among Latin American nations. The story of the debt and poverty that led to the selling of the rain spins out exactly such feelings. While the details of such a situation change over time and differ from country to country, these elements exist as part of Latin American reality. Geographically, there is great variety in environments and human geography; economically, there is a troublesome frequency in the existence of financial and political problems akin to those in Naranjo's story.

Consider the breadth of national and regional geographic variety in Latin America. Mexico and Argentina, for example, speak Spanish and espouse Roman Catholicism but in many other respects are extremely different. The influence of indigenous people is strong in Mexico but scanty in Argentina; Mexico's core region in volcanic tropical highlands around Mexico

◄ Slash and burn agriculture is one of the most powerful agents of landscape change in Latin America. Major forest regions are often a long distance from direct governmental protection. A family can turn forest richness into a year or two of varied crops and then the infertility of the soils drives the family on to other forest reserves. In this process small populations are given traditional returns on family labor, but primary forest reserves are often lost to unproductive volunteer grasses and failed land use. The tension over this land use is further intensified if ranchers with machinery come in and clear-cut the forests in an effort to create rangeland for beef exports to North America.   *Paul Edmonson/ Tony Stone Images*

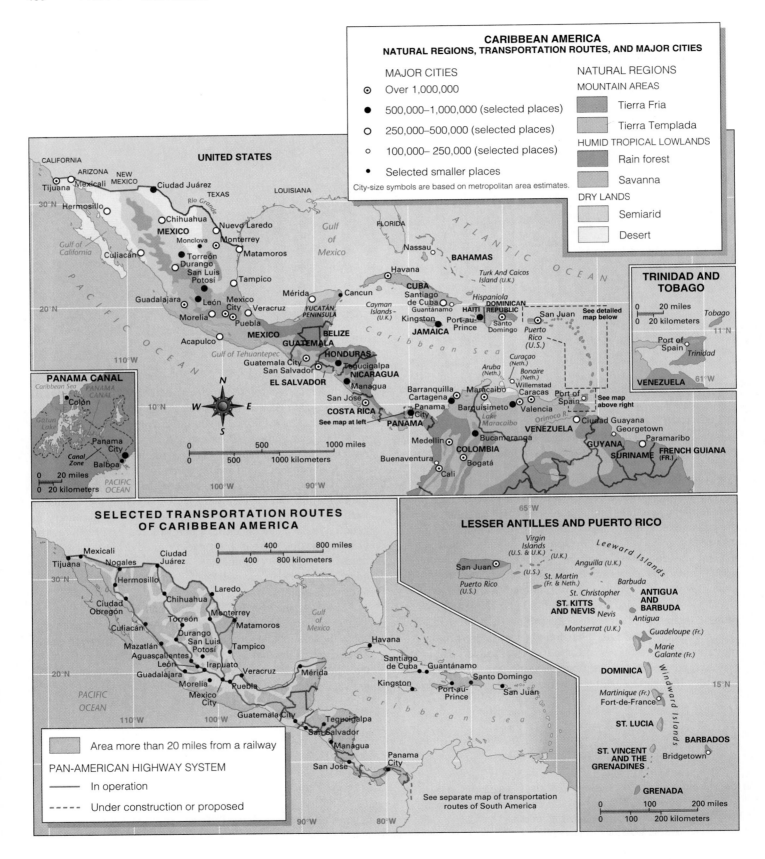

CARIBBEAN AMERICA
NATURAL REGIONS, TRANSPORTATION ROUTES, AND MAJOR CITIES

MAJOR CITIES

⊙ Over 1,000,000

● 500,000–1,000,000 (selected places)

○ 250,000–500,000 (selected places)

○ 100,000– 250,000 (selected places)

• Selected smaller places

City-size symbols are based on metropolitan area estimates.

NATURAL REGIONS

MOUNTAIN AREAS

Tierra Fria

Tierra Templada

HUMID TROPICAL LOWLANDS

Rain forest

Savanna

DRY LANDS

Semiarid

Desert

TRINIDAD AND TOBAGO

0   20 miles

0   20 kilometers

Tobago

11°N

Port of Spain   Trinidad

VENEZUELA   61°W

PANAMA CANAL

Caribbean Sea

PANAMA CANAL

Colón

Gatún Lake

Panama City

Canal Zone

Balboa

0   20 miles

0   20 kilometers

PACIFIC OCEAN

UNITED STATES

CALIFORNIA

ARIZONA   NEW MEXICO

Tijuana   Mexicali

Ciudad Juárez   TEXAS   LOUISIANA

30°N   Hermosillo   Rio Grande

Chihuahua

MEXICO   Nuevo Laredo   FLORIDA   Nassau   BAHAMAS

Monclova   Monterrey

Gulf of California   Matamoros

Culiacán   Torreón   Havana

Durango   Tampico   CUBA   Turk And Caicos Island (U.K.)

San Luis Potosí   Santiago de Cuba   Hispaniola   DOMINICAN REPUBLIC

Guadalajara   Mérida   Cancun   Guantánamo   HAITI

León   Mexico City   Cayman Islands (U.K.)   Kingston   Port-au-Prince   San Juan   See detailed map below

Morelia   Veracruz   JAMAICA   Santo Domingo   Puerto Rico (U.S.)

Puebla   YUCATÁN PENINSULA   Caribbean Sea

Acapulco   MEXICO   BELIZE   Curaçao (Neth.)

GUATEMALA   Aruba (Neth.)   Bonaire (Neth.)   See map above right

Gulf of Tehuantepec   HONDURAS   Willemstad

Guatemala City   Tegucigalpa   Barranquilla   Maracaibo   Caracas   Port of Spain

San Salvador   NICARAGUA   Cartagena   Valencia   See map above right

EL SALVADOR   Managua   Barquisimeto

San José   Panama City   Ciudad Guayana   Georgetown

COSTA RICA   See map at left   VENEZUELA   Paramaribo

PANAMA   Bucamaranga   GUYANA   SURINAME   FRENCH GUIANA (FR.)

Medellin   Lake Maracaibo   Orinoco R.

COLOMBIA   Buenaventura   Bogatá

Cali

0   500   1000 miles

0   500   1000 kilometers

100°W   90°W

SELECTED TRANSPORTATION ROUTES OF CARIBBEAN AMERICA

0   400   800 miles

0   400   800 kilometers

Mexicali   Ciudad Juárez

Tijuana   Nogales

Hermosillo   Laredo

Ciudad Obregón   Chihuahua

Culiacán   Monterrey

Torreón   Matamoros

Mazatlán   Durango   San Luis Potosí   Tampico   Gulf of Mexico

Aguascalientes   Irapuato

León   Veracruz   Havana   Santiago de Cuba   Guantánamo

Guadalajara   Mérida

Morelia   Puebla   Kingston   Santo Domingo

Mexico City   Port-au-Prince   San Juan

PACIFIC OCEAN   Guatemala City   Caribbean Sea

Tegucigalpa

San Salvador

Managua   Panama City

San José

See separate map of transportation routes of South America

Area more than 20 miles from a railway

PAN-AMERICAN HIGHWAY SYSTEM

──── In operation

- - - - Under construction or proposed

110°W   100°W   90°W   80°W

LESSER ANTILLES AND PUERTO RICO

65°W

Virgin Islands (U.S. & U.K.)   Leeward Islands

San Juan   (U.K.)   Anguilla (U.K.)

Puerto Rico (U.S.)   (U.S.)   St. Martin (Fr. & Neth.)   Barbuda

St. Christopher   ANTIGUA AND BARBUDA

ST. KITTS AND NEVIS   Nevis   Antigua

Montserrat (U.K.)   Guadeloupe (Fr.)

Marie Galante (Fr.)

DOMINICA   15°N

Martinique (Fr.)   Windward Islands

Fort-de-France

ST. LUCIA   BARBADOS

ST. VINCENT AND THE GRENADINES   Bridgetown

GRENADA

0   100   200 miles

0   100   200 kilometers

City contrasts with the expansive midlatitude lowland plains of Argentina's core around Buenos Aires. Bolivia has both influences expressed in its urban centers.

Not only do the Latin American countries differ widely among themselves, but generally there is strong internal diversity within each country. Some countries have great potential for future development; others are more limited. In this chapter, the strongly regionalized character of Latin America is surveyed through four regional groups of countries: (1) the northern realm of **Caribbean America,** (2) the **Andean countries,** (3) giant **Brazil,** and (4) the **countries of the southern midlatitudes.** (For statistical data, see Table 19.1; also see Figs. 19.1, 20.1, 20.2, and 20.3.)

20.1   The Intricate Mosaic of Caribbean America

Caribbean America includes (1) Mexico, by far the largest country in area and population, (2) the small Central American states of Guatemala, El Salvador, Belize, Honduras, Nicaragua, Costa Rica, and Panama, and (3) the numerous islands in the Caribbean Sea or near it. Of the many island political units, the largest are **Cuba** (which encompasses the largest island), **Haiti and the Dominican Republic** (which share the second largest), and **Jamaica.** The entire area presents a patchwork of physical features, races, cultures, political systems, population densities, and pursuits. Most of the political units are independent, but a few are still dependencies of outside nations. We begin with the more prominent unit, Mexico.

MEXICO: AN UNRECOGNIZED GIANT

The federal republic of Mexico, officially the *Estados Unidos Mexicanos* (United Mexican States), is by far the largest, most complex, and most influential country in Caribbean America. Compared to most of the world's countries, Mexico is a spatial giant, but its large size tends to go unrecognized, largely because it lies in the shadow of the United States. Its capital, Mex-

ico City, stands now as the world's largest city in population (Fig. 20.4). Triangular in shape and situated next to the compact bulk of the conterminous United States, Mexico looks comparatively small on a world map, but its area of 762,000 square miles (*c.* 2 million sq km) is nearly eight times that of the United Kingdom, and its elongated territory would stretch from the state of Washington to Florida. The country's huge population (an estimated 94.8 million in 1996; expected to reach 120 million by the year 2010) makes Mexico by the far the largest nation in which Spanish is the main language. Mexico is also a big country economically, ranking quite high among the world's nations in annual output of many different commodities, such as metals, oil, gas, sulfur, sugar, coffee, corn, citrus fruit, cacao, and cattle.

Mexico's remarkable diversity of mineral and agricultural production is made possible by great environmental variety. Metal-bearing ores exist in many places, mined particularly in mountainous terrain of northern and central Mexico. Iron ore from scattered locations supports a fairly sizable iron and steel industry, which also can draw on substantial coal deposits, some suitable for coking. The coastal lowlands along the Gulf of Mexico share the oil, natural gas, and sulfur deposits that are a marked feature of adjacent coastal Texas and Louisiana. In agriculture, great variety is achieved despite a mountainous environment short of tillable land. Only one eighth of the country is cultivated; good crop land exists mainly (1) on the floors and lower slopes of mountain basins with volcanic soils, or (2) on scattered and generally small alluvial plains. Agricultural habitats vary according to elevation, rock types, physiographic history, and atmospheric influences. Mountains too steep to cultivate without excessive erosion dominate the terrain in most areas, but they are flanked by plains, plateaus, basins, and foothills that can be tilled if enough water is available.

A little over one half of Mexico lies north of the Tropic of Cancer and is an area dominated by desert or steppe climates with hot summers and winters that are warm to cool depending on altitude. South of the Tropic, where about four fifths of all Mexicans live, seasonal temperatures vary less, but altitudinal differences create conditions ranging from high temperatures in the *tierra caliente* of lowlands along the Gulf of Mexico, through moderate heat in the *tierra templada* of low highlands,

◀ FIGURE 20.1

General reference maps of Caribbean America. Curaçao and Bonaire comprise the Netherlands Antilles; Aruba is a separate country. Antigua and Barbuda form one political unit, as do St. Kitts and Nevis, and St. Vincent and the Grenadines. Major highways in the transportation inset are supplemented and interconnected by other surfaced highways not shown. Places shown by lettered symbol in the inset but not named on the main map include Ciudad Obregon and Manizales on the western route, and Aguascalientes, Irapuato, and San Francisco del Rincon (not lettered) on the central route between Durango and Mexico City.

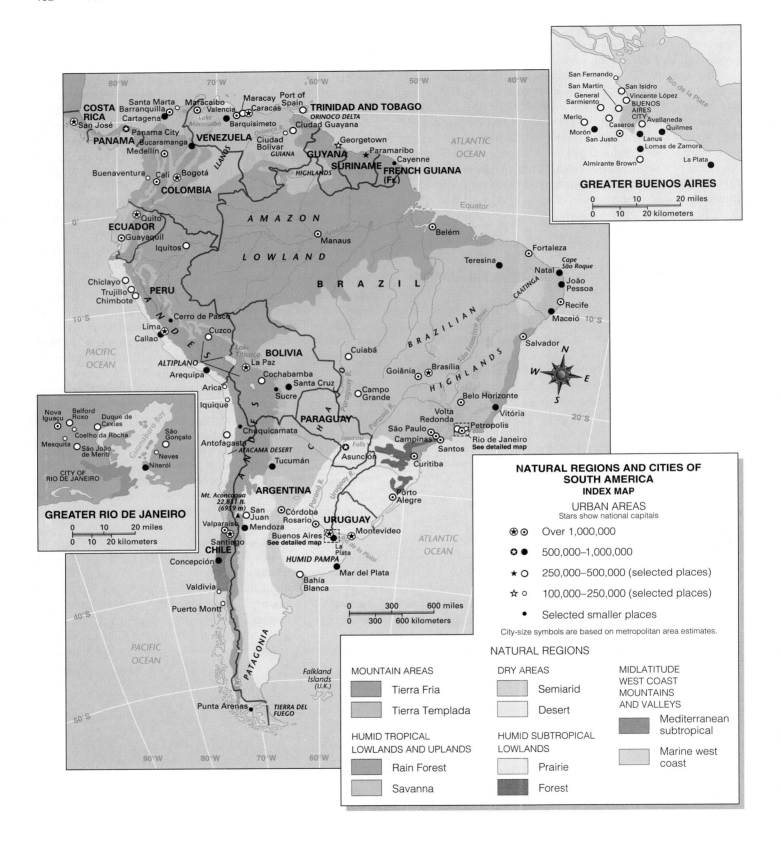

GREATER BUENOS AIRES

San Fernando
San Martin
General Sarmiento
San Isidro
Vincente López
BUENOS AIRES CITY
Merlo
Avellaneda
Morón
Caseros
Quilmes
San Justo
Lanus
Lomas de Zamora
Almirante Brown
La Plata

| 0 | 10 | 20 miles |
| 0 | 10 | 20 kilometers |

GREATER RIO DE JANEIRO

Nova Iguaçu
Belford Roxo
Duque de Caxias
Mesquita
Coelho da Rocha
São Gonçalo
São João de Meriti
Neves
Niterói
CITY OF RIO DE JANEIRO
Guanabara Bay

| 0 | 10 | 20 miles |
| 0 | 10 | 20 kilometers |

COSTA RICA
San José
PANAMA
Panama City
Santa Marta
Barranquilla
Cartagena
Maracaibo
Lake Maracaibo
Valencia
Maracay
Caracas
Barquisimeto
VENEZUELA
Bucaramanga
Medellín
LLANOS
Orinoco R.
Ciudad Bolivar
Port of Spain
TRINIDAD AND TOBAGO
ORINOCO DELTA
Ciudad Guayana
GUIANA
HIGHLANDS
GUYANA
Georgetown
Paramaribo
SURINAME
Cayenne
FRENCH GUIANA (Fr.)
Buenaventura
Cali
Bogotá
COLOMBIA
Quito
ECUADOR
Guayaquil
Iquitos
AMAZON LOWLAND
Amazon R.
Manaus
BRAZIL
Belém
ATLANTIC OCEAN
Equator
Fortaleza
Teresina
Natal
Cape São Roque
João Pessoa
Recife
Maceió
Chiclayo
Trujillo
Chimbote
PERU
Cerro de Pasco
Lima
Callao
Cuzco
Lake Titicaca
ALTIPLANO
La Paz
Arequipa
Cochabamba
BOLIVIA
Santa Cruz
Sucre
Arica
Iquique
Cuiabá
Goiânia
Brasília
BRAZILIAN
São Francisco River
HIGHLANDS
CAATINGA
Salvador
Campo Grande
Paraguay R.
Belo Horizonte
Volta Redonda
Vitória
Petropolis
Rio de Janeiro
See detailed map
Antofagasta
Chuquicamata
ATACAMA DESERT
PARAGUAY
Asunción
Iguassu Falls
Paraná R.
São Paulo
Campinas
Santos
Curitiba
Pôrto Alegre
Uruguay R.
Tucumán
Mt. Aconcagua
22,831 ft.
(6919 m)
San Juan
Mendoza
Valparaíso
Santiago
CHILE
Concepción
ARGENTINA
Córdoba
Rosario
URUGUAY
Montevideo
Buenos Aires
See detailed map
La Plata
Rio de la Plata
HUMID PAMPA
Mar del Plata
Bahía Blanca
ATLANTIC OCEAN
Valdivia
Puerto Montt
PACIFIC OCEAN
PATAGONIA
Falkland Islands (U.K.)
Punta Arenas
TIERRA DEL FUEGO

PACIFIC OCEAN

NATURAL REGIONS AND CITIES OF SOUTH AMERICA
INDEX MAP

URBAN AREAS
Stars show national capitals

⊛ ⊙ Over 1,000,000

✪ ● 500,000–1,000,000

★ ○ 250,000–500,000 (selected places)

☆ ○ 100,000–250,000 (selected places)

● Selected smaller places

City-size symbols are based on metropolitan area estimates.

| 0 | 300 | 600 miles |
| 0 | 300 | 600 kilometers |

NATURAL REGIONS

MOUNTAIN AREAS
Tierra Fría
Tierra Templada

HUMID TROPICAL LOWLANDS AND UPLANDS
Rain Forest
Savanna

DRY AREAS
Semiarid
Desert

HUMID SUBTROPICAL LOWLANDS
Prairie
Forest

MIDLATITUDE WEST COAST MOUNTAINS AND VALLEYS
Mediterranean subtropical
Marine west coast

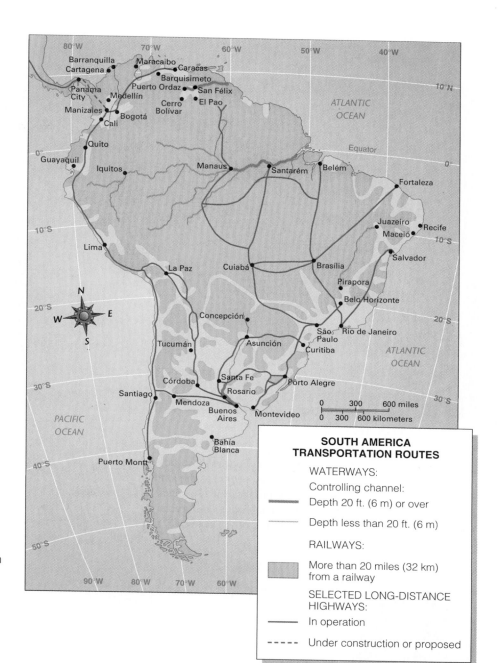

SOUTH AMERICA TRANSPORTATION ROUTES

WATERWAYS:

Controlling channel:

———— Depth 20 ft. (6 m) or over

·········· Depth less than 20 ft. (6 m)

RAILWAYS:

▨ More than 20 miles (32 km) from a railway

SELECTED LONG-DISTANCE HIGHWAYS:

——— In operation

----- Under construction or proposed

◀ FIGURE 20.3

The basic skein of major highways in the Pan-American Highway System is shown, together with new long-distance roads reaching into the deep interior of Brazil.

to cool temperatures in the *tierra fría* of the highlands (see Chapter 19, page 464). Nearly all areas inside the tropics are humid enough to support rainfed crops, although irrigation is often used to increase the diversity of crops and/or the intensity and productivity of agriculture. Large areas have a dry sea-son in the low-sun period when irrigation becomes a necessity for cropping.

On the hemispheric and world stage, Mexico speaks with a respected voice, although the country's relative poverty and its great economic dependence on the United States make its

◀ FIGURE 20.2

General reference maps of natural areas and major cities in South America. Note the clearly defined arrangement of cities around the rim of the continent and the enormous empty expanses in the interior. See also the world landforms map on which the Brazilian Highlands are shown as "plateaus." Chuquicamata in Chile is that country's most famous copper-mining settlement.

FIGURE 20.4
A view of downtown Mexico City.
The building with the prominent
archway (upper center) is the
Monument to the Revolution.
Portfeild Chickering/
Photo Researchers, Inc.

influence less potent than might otherwise be true of so large a nation. Mexico has over one third as many people as the United States but in 1996 had a gross national product only 3 percent as great. Economic relations with the United States are crucial to Mexico because approximately two thirds of the country's for-eign trade is with the United States, and the latter nation is over-whelmingly the main source of outside capital. The 1994 North American Free Trade Agreement (NAFTA) between the United States, Canada, and Mexico has stimulated considerable eco-nomic activity on both sides of the Mexico–United States border.

Regional Perspective

MAQUILADORAS

One of the most significant economic innovations that links Mexico and the United States are the fabrication centers in which components from Mexico, the United States, and other countries are assembled for sale and export. These plants are called **maquiladoras**. Begun in 1965, they expanded by 70 percent in the decade from 1970 to 1980, versus a growth rate of only 40 percent in Mexico overall. The number of border maquiladora plants has more than doubled in the past decade. They are generally located just south of the border in Mexico, where there is relatively inexpensive Mexican labor and easy transport of partially assembled goods from north of the border to these **free port** facilities. (Only 5 percent of the components used in the manufacturing at these plants are Mexican, while the rest are elements brought from the U.S. or other plants.) Such plants were important even before the 1994 initiation of NAFTA but have become particularly important now that tariff barriers have been further lowered. The savings in labor are what have made the maquiladoras enormously successful in the border region, where access to good truck and rail transport facilities have helped them to thrive. United States opponents of this aspect of NAFTA activity claim that environmental conditions are not watched as closely as in the U.S., and that manufacturers simply transfer low-wage and environmentally detrimental processing to Mexico. In Mexico since 1994, the economic importance of the maquiladoras has surpassed that of petroleum. In 1994–1995, maquiladoras produced more than 14 percent worth of export trade for Mexico, concentrated in such border cities as Nuevo Laredo and Matamoros—which combined accommodate more than 2000 maquiladoras.

NORTH AMERICAN FREE TRADE AGREEMENT (NAFTA) I

The origins of NAFTA are varied, but three motivations have been central to this multinational agreement among the United States, Canada, and Mexico: (1) a wish by all three to promote Mexican economic growth and development; (2) hope for expanded export markets for the U.S. and Canada; and (3) a desire to decrease the flow of illegal migrants from Mexico to the U.S. through the expansion of Mexican manufacturing activity. The earlier success among the European nations in their creation of the European Economic Community (EEC) led the NAFTA countries to assume that lower tariff barriers in Latin America and North America would enable the affected economies to expand into markets for which they were best suited. On the American side of the ledger, there had been a decline in U.S. textile and shoe industries. Because of the relatively low wages in Mexico, NAFTA markets seem attractive. Before the December 1994 collapse of the Mexican economy, there was a strong upswing in export of U.S. and Canadian consumer goods to Mexico, especially electronics and automobiles. On January 1, 1994, the day that NAFTA became official, an Indian uprising in Mexico's southern state of Chiapas drew attention to the fact that the Mexican government had population segments who were against NAFTA. These rebels expressed their fears that whatever economic benefits were going to come from NAFTA were probably not going to find their way to the impoverished Indian populations. And although there has been much expansion in the exports from the **maquiladora** factories (see Regional Perspective, page 484) just south of the U.S. border, only about 5 percent of the components of these plants (plus the labor) are Mexican. There also continue to be tensions over potential environmental deterioration associated with NAFTA, the loss of jobs in American trucking and manufacturing, and American demands on the Mexican government for economic reform. The entire experiment with NAFTA innovations continues to provoke ongoing concern on all sides of the issues, and in all three countries.

There is inevitably a touchiness in relations between Mexico and the United States due to (1) the disparity between them in wealth, power, and cultural influence, (2) the historic fact that Mexico lost more than one half its national territory to the United States in the 19th century, and (3) Mexicans' national pride, derived from great cultural longevity. Mexico's aboriginal Indian civilizations were more highly developed than those within the territory of the present United States, and in the early 17th century, when the first tiny English settlements of the future United States were struggling for survival at the edge of a vast wilderness, the court of the Spanish Viceroy in Mexico City was an opulent and far-reaching center of imperial power.

Geographic Signatures from Indian, Spanish, and Mexican Eras

Mexico's human geography exhibits many elements from three major eras: the *Indian era* prior to the Spanish conquest of the early 16th century, the *Spanish colonial era* to the early 19th century, and the *Mexican era* following Mexican independence in 1821. In the Indian era, the country was the home of Indians speaking hundreds of languages or distinct dialects and possessing cultures more advanced than any others in the Americas except those within the Inca Empire in the Andes of South America. Great builders in stone, the aboriginal peoples constructed the mighty Pyramids near Mexico City (Fig. 20.5) and the huge temples and palaces of the Mayan culture in Yucatán. Today, richly varied Indian cultures persist and the Indian racial component in the population is pronounced: The greater part of the people are *mestizos* with mixed Spanish and Indian blood. The present settlement pattern has been powerfully influenced by the distribution of Indians at the time the Spanish came. The conquerors sought out concentrations of Indians as laborers and potential Christian converts. Hence, the locations of Spanish-inspired settlement tended to conform to the population pattern already established.

The Spanish colonial era lasted nearly four centuries. Following the overthrow of the Aztec Empire by Hernando Cortez (Hernán Cortés) in 1521, the Spanish Crown established the Viceroyalty of New Spain, ruled from Mexico City and eventually encompassing not only the area now included in Mexico but also most of Central America and about one quarter of the territory now occupied by the 48 conterminous American states. A Hispanic pattern of life, still highly evident today, was established by Spanish administrators, fortune hunters, settlers, and Catholic priests. Racial, ethnic, and cultural residues of their activities are apparent in recent estimates that *mestizos* comprise 60 percent of Mexico's population while Caucasians (nearly all of Spanish ancestry) comprise 9 percent. Spanish is spoken as a first language by 93 percent of the populace, and religious affiliation in Mexico is 90 percent Roman Catholic. Other strong Spanish influences are found in architectural styles and urban layouts; towns and cities characteristically have a Spanish-inspired rectangular grid of streets surrounding a *plaza* at the town center. Cattle ranching begun by the Spanish also endures as a way of life in large sections of Mexico; the Spanish introduced not only cattle, but also wheat, sugarcane, sheep, horses, donkeys, and mules. The consuming Spanish hunger for precious metals greatly expanded mining, particularly of silver. Many of Mexico's larger cities were founded as silver-mining camps or regional service centers for such mining areas.

Under the absolutist government of imperial Spain, a small Spanish aristocracy monopolized power, wealth, prestige, and education. Mexico was governed by a hierarchy of officials appointed by the Crown, and a large share of the best land was acquired by aristocratic Spanish owners of the large estates called

FIGURE 20.5

The Pyramid of the Sun (shown in photo) and the Pyramid of the Moon are prominent pre-Aztec structures rising in the midst of maize and maguey fields near Mexico City. They are sited on broad avenues floored by stone and lined with ruined stone temples and priests' quarters. The entire assemblage, known as Teotihuacan, is what remains of a city that may have housed 200,000 people prior to the city's destruction by fire some 14 centuries ago.   *Jesse H. Wheeler, Jr.*

haciendas. Meanwhile, the Church, holding official status under devoutly Catholic Spain, spread its influence, structures, and extensive landholdings everywhere. Innumerable towns and cities of today originated essentially as Catholic missions in the midst of clustered Indian populations.

When the Mexican era began in the early 19th century, Mexico was overwhelmingly rural, illiterate, village-centered, Church-oriented, and poor. Huge landholdings belonging to a few thousand wealthy families or the Church enclosed nearly all the best farmlands. In the capital, an educated elite governed and exploited the country with little benefit to the Indian and *mestizo* farmers in the countryside. The bloody and chaotic Mexican Revolution of 1810–1821 ended Spanish control and instituted a short-lived Mexican Empire that expired in 1823 and was followed by a federal republic in 1824. Then came a long period of political instability, with personalities of many persuasions contending for power. During its first three decades of independence, Mexico had some 50 governments, none of which had resources enough to develop the country economically or power enough to exact more than limited tribute and token allegiance from powerful generals who governed local areas with the support of large landowners. Meanwhile, from the 1830s through the 1840s huge blocks of the nation's territory were lost, including the Central American part of former New Spain, Texas, California, and areas now comprising New Mexico, Arizona, and parts of other states.

During the 1850s and 1860s, a dynamic Indian reformist, Benito Juárez, attained power for a time, but the country then settled into a long period of despotic personal rule by President Porfirio Díaz, which began in 1876 and did not end until 1910. Democratic processes were largely suspended and the rural population sank deeper into poverty even as heavy foreign investments were made in railroads, the oil industry, metal mining, coal mining, and manufacturing. Political discontent over national disunity and weakness, internal oppression and corruption, foreign exploitation, poverty, and a dynamic need for land reform then led to the Mexican Revolution of 1910–1920, which ousted Díaz and subsequently turned into a destructive struggle among personal armies of regional leaders. In 1917, a new constitution was enacted under which land reform was an urgent priority. Since 1910, huge acreages have been expropriated from *haciendas* by the government and redistributed or restored to landless farmers and farm workers, most particularly in a program that provided *ejidos,* or small farming units, for village farmers. Large *haciendas* still exist, especially in the dry north, but they contain only a very small proportion of Mexico's cultivated land.

Agriculture is still a highly important source of livelihood (Fig. 20.6), although recent estimates report Mexico's population to be 17 percent urban. Corn is by far the most important crop. Mexico is almost certainly the area where corn was first domesticated and from which it spread to become the premier crop of the New World and many overseas areas. Although not an export of any importance, it is central to the Mexican diet. Beans, also an inheritance from the Indian past, are another universal Mexican crop and food.

Today, the total value of output from factories and mines is several times that of output from farms. All of Mexico's economic production needs to be seen in a regional context, however, as the output of particular commodities tends to be rather sharply localized within particular areas.

Regional Geography of Mexico

Mexico is strongly regionalized. Although innumerable areas have distinctive personalities, space restricts us here to regions of the broadest scale. We focus on (1) **Central Mexico**—the country's core region; (2) **Northern Mexico**—dry, mountainous, and increasingly intertwined with the United States; (3) the **Gulf Tropics**—composed of hot, wet coastal lowlands along the Gulf of Mexico, together with an inland fringe of *tierra templada* in low highlands; and (4) the **Pacific Tropics**—a poorly developed southern outlier of humid mountains and narrow coastal plains.

Central Mexico: Volcanic Highland Core Region.

It is very common for a Latin American country to have a sharply defined core region where the national life is centered. Mexico is a classic example of such a cultural and demographic pattern. About one half of the population inhabits contiguous highland basins clustered along an axis from Guadalajara (population: 2.4 million) at the northwest to Puebla (population: 1.3 million) at the southeast. Toward the eastern end of the axis is Mexico City (population: 15 million; other estimates for a broader metropolitan territory range upward to 20 million or more). The basins comprise Mexico's core region, which is often referred to as Central Mexico or the Central Plateau. The basin floors vary in elevation from about 5000 feet (1524 m) to 9000 feet (2727 m), with each basin separated from its neighbors by a hilly or mountainous rim. Centers of basins contain flat land, which may be swampy. The swampiness has led populations to cluster on higher sloping land, and population pressure has induced cultivation of steep slopes, causing much soil erosion. Most areas are too high and cool for coffee or the other crops of the *tierra templada*. The prevailing *tierra fría* environment dictates growth of crops that are less subject to frost damage, such as corn, sorghums, wheat, and potatoes. In addition, cattle (both dairy and beef) and poultry are raised. Agriculture still relies heavily on human labor and animal draft power. A productive agriculture has been maintained here for millennia, based on

FIGURE 20.6

Sugarcane is one of the most productive of the *tierra caliente* crops. This scene in Veracruz state in Mexico illustrates the continuing role of hand labor in the cane harvest.   *Byron Augustin/Tom Stack & Assoc.*

fertile and durable soils derived from lava and ash ejected by the many volcanoes.

Mexico City was a major urban and political focus long before its conquest by Cortez in 1521. Near the city are impressive pyramids and other monuments built by Indian precursors of the Aztecs. But Indian political power reached its apex under the Aztec Empire in the centuries immediately before the Spanish came. The Aztecs, notorious for their insatiable practice of ritual human sacrifice, came to the Basin of Mexico from the north and built their capital on an island in a shallow salt lake, with causeways to the shore (Fig. 20.7; see also Definitions and Insights below).

Definitions and Insights

URBAN SUBSIDENCE

The decision of the Aztecs to build their most significant ritual and demographic center on an island in the middle of what was then Lake Texcoco is a good example of simple origins confounding complex later histories. Paris, for example, was initially settled on an island in the Seine River with no anticipation of the size of the urban center that would grow out of that insular origin. The initial urban center built in the mid-14th century in the Lake Texcoco site was the Aztec commercial and sacred center, **Tenochtitlan.** The city was lined with canals and dike constructions to control flooding and protect water resources for the city. It was here that Cortez finally defeated the Aztecs in 1521. Mexico City was later founded in this same general locale. Most of the lake subsequently was drained, and, for the past four and a half centuries, city structures have been constructed on the drained land. The result has been a gradual **subsidence,** or settling, of buildings (still in progress today) as the lake sediments contract. In recent times, this problem has been aggravated by overpumping of ground water to support an urban settlement of more than 20 million people. But despite the unstable subsurface, including the fact that the city is underlain by a geologic fault, Mexico City now has imposing modern office towers with foundations designed to withstand both subsidence and earthquake shocks. This combination of two such significant environmental hazards underlying what is generally considered to be the largest city in the world reminds us that initial settlement decisions seldom anticipate the demographic, political, or cultural magnitude a ritual center can achieve as time passes.

Further environmental hazards in Mexico City are posed by severe pollution from automobile exhaust and factory chimneys. It is often characterized as having the world's worst air quality for a major urban center. Temperature inversions in the confined atmosphere of the basin keep pollutants from dissipating and regularly create some of the world's most notorious smogs. Disposal of household wastes from millions of shanties

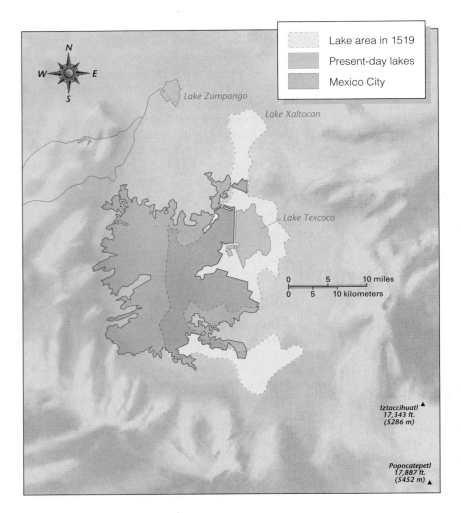

Legend:
- Lake area in 1519
- Present-day lakes
- Mexico City

Lake Zumpango

Lake Xaltocan

Lake Texcoco

0   5   10 miles
0   5   10 kilometers

Iztaccihuatl
17,343 ft.
(5286 m)

Popocatepetl
17,887 ft.
(5452 m)

FIGURE 20.7

This map of ongoing land subsidence in Mexico City highlights an environmental problem of major proportions. Having selected a lake bed for the area of initial urban settlement here, city dwellers have been plagued with periodic subsidence for centuries. The enormous city growth and associated expanded withdrawal of water from below the surface of the city, and the ever heavier urban burden placed upon this landscape, have led to this steady and perplexing problem.

that have no indoor plumbing is another major problem. But despite such vexations and discomforts, there is a relentless flow of migrants to the Basin of Mexico. Unemployment is high, but many newcomers soon find low-wage service or industrial jobs, and others subsist with the aid of relatives until they can find work.

Explosive urbanism has outrun the efforts of Mexico City's planners to cope with it. One mayor of the city commented that administering this monstrous sprawl was like "repairing an airplane in flight." Nonetheless, the city does have some spectacular planning achievements to its credit—not least of which are a modern subway system over 60 miles (100 km) long and a system to pump water to the city through the encircling mountain walls of the Basin of Mexico.

Amid the many memorials to Mexico City's dramatic past, the modern city carries on its functions as the nation's leading business and industrial center. It is estimated that approximately one third of all manufacturing employees in Mexico work in thousands of factories in metropolitan Mexico City. Most factories are small, and manufacturing is devoted largely

to miscellaneous consumer items, although some plants carry on heavier manufacturing such as metallurgy, oil refining, and the making of heavy chemicals.

Northern Mexico: Localized Development in a Dry and Rugged Outback. North of the Tropic of Cancer, Mexico is characterized by ruggedness, aridity, an extensive ranching economy supporting a generally sparse population, and some widely separated spots of more intense activity based principally on irrigated agriculture, mining, heavy industry, and diversified border industries adjacent to the United States. Most of Northern Mexico is above 2000 feet (610 m) in elevation, but there are lowland strips along the Gulf of Mexico, the Gulf of California, and the Pacific. The climate is desert or steppe, with scanty xerophytic vegetation except where elevations are high enough to yield more rainfall. In the latter areas, some uplands are carpeted with steppe grasses, and mountains rising still higher have productive coniferous forests. Two major north-south ranges of high mountains—the Sierra Madre Occidental in the west and the Sierra Madre Oriental in the east—tower above the surrounding areas. Surface transport across them is poorly developed. Most forest land is in the Sierra Madre Occidental. This is Mexico's main source of timber, and lumbering is important in some places. Within Northern Mexico as a whole, ranching is the most widespread source of livelihood, although the total number of people supported by it is small.

In addition to being Mexico's main area of livestock ranching, Northern Mexico also provides the principal output of metals, coal, and products of irrigated agriculture. Iron ore, zinc, lead, copper, manganese, silver, and gold are the main metals produced. The ores, which often contain more than one metal, are smelted with coal mined in Northern Mexico. Some coal is made into coke and used for the manufacture of iron and steel at a number of places, particularly around the city of Monterrey (population: 2.1 million), which is Mexico's leading center of heavy industry and Northern Mexico's most important business and transportation hub.

Irrigated agriculture in the dry north has expanded greatly since World War II with the building of many multipurpose dams to store the water of rivers originating in the Sierra Madre and emptying into the Gulf of California or the Rio Grande. New or expanded irrigated districts now are widespread, with the main clusters located in the northwest and along or near the Rio Grande. Some large and important districts immediately along the United States border have been in existence for a long time, including one along the lower Rio Grande Valley directly opposite a larger district on the U.S. side, and the Mexican Mexicali Valley, or the continuation of California's Imperial Valley north of the Gulf of California. Cotton is a major product of Northern Mexico's irrigated agriculture, and varied fruits and vegetables for shipment to U.S. markets also are grown. In some irrigated areas, the Mexican government has helped develop agricultural colonies of small landowning farmers who

have migrated from more crowded parts of the country. Throughout Northern Mexico, the new or expanded oases stand out as islands of agricultural productivity amid large areas that are either nonagricultural or used for ranching or nonirrigated grain farming yielding low returns per unit of land.

Since World War II, there has been a large upsurge of manufacturing in widely separated Mexican cities immediately adjacent to U.S. cities along the international boundary. The largest are Tijuana (population:750,000, city proper; probably well over 1 million, metropolitan area) adjacent to San Diego, California, and Ciudad Juárez (population: 800,000, city proper) across the Rio Grande from El Paso, Texas. In "free zones" within these and smaller cities, U.S. firms have established manufacturing plants that now employ large numbers of Mexican low-wage workers. The companies are allowed to move components into Mexico duty-free and ship finished products back to the United States, where duty is charged only on the value added to each item by labor cost and the cost of any components not shipped from the United States. The factories in the "free zones" are only one instance of the close intertwining of Mexicans and Americans along the border. Large numbers of Mexicans cross the boundary each day to work or shop in the United States, and there is a return flow of huge numbers of American shoppers and tourists.

The Mexican border cities are collecting points for Mexican and Central American migrants who attempt to enter the United States illegally and often succeed despite the efforts of the U.S. Border Patrol to apprehend them (Fig. 20.8). The tension in this zone has also grown because of the smuggling of drugs into the U.S. through the long border. A U.S. immigration law in 1986 gave amnesty to illegal migrants who could prove residence in the United States for a stated time, but the law prescribed tougher measures to try to halt the continuing illegal

FIGURE 20.8
Due to the disparity in income and employment opportunities between the United States and Mexico, the long border between the two countries sees endless attempts by Mexicans and Central Americans to enter the United States illegally. In this photo, the point of entry is a hole in the fence along the Rio Grande at El Paso, Texas. *Alon Reininger/Contact Press Images*

flow. It remains to be seen how well such measures can succeed along this bicultural border when great poverty lies closely adjacent to affluence, and strong demand for low-wage Mexican labor exists on the U.S. side. This tension came to a head in California during the election in 1994, when Proposition 187—prohibiting the state from providing any welfare or medical support to undocumented aliens (the great majority of whom in California come from Latin America)—brought national attention to the steadily increasing flow of illegal migrants to the United States.

The Gulf Tropics: Oil and Gas, Plantations, Maya, Tourism.

On the Gulf of Mexico side, Mexico has tropical lowlands that have a much lower population density than Central Mexico but which are very important to the national economy. Their most valuable product is oil, discovered in huge quantities in the mid-1970s. Just as in coastal Texas and Louisiana, some fields are onshore and others lie underneath the Gulf. The Mexican oil industry is a government monopoly (carried on by a government corporation, *Petróleos Mexicanos*, or PEMEX), and the United States is its largest foreign customer. The bonanza oil strikes in the 1970s seemed so promising that the government borrowed heavily from foreign banks to finance both oil development and a wave of other economic and welfare schemes. Dependence on oil income grew to the point that crude oil represented 67 percent of all Mexican exports in value in 1983. Subsequently, however, a worldwide oil glut caused prices to skid, and by 1990, crude oil accounted for only 33 percent of exports. Economic reforms in Mexico during the 1980s reduced inflation and spurred growth, but by 1994 the country's external debt was nearly $128 billion and a sizable share of export revenue was being spent on debt service. The 1994 collapse of the Mexican peso came in large part because of the combination of massive foreign debt and diminishing oil revenues due to sinking prices. Nonetheless, recent restructuring of the economy has created a more hopeful outlook for the future.

In addition to petroleum, natural gas is produced in the Gulf lowlands and is piped to customers in Mexico and the United States. Construction of a pipeline to the United States was widely criticized in Mexico because of sensitivity concerning the country's large and growing economic dependence on its large neighbor. There is much resentment over Mexico's role as a supplier of natural resources to feed the voracious U.S. industrial economy. But the country's need for foreign exchange has sidetracked questions of national sensibility in favor of exports to the United States, not only of oil and gas but also of native sulfur from deposits in the Gulf Tropics and various minerals from the dry north.

The Gulf Tropics are Mexico's main producer of tropical plantation crops, which contribute in a relatively minor way to the export trade. Cacao, sugarcane, and rubber are leading export crops in the lowlands, and coffee exports come from a strip of *tierra templada* along the border between the lowlands and Mexico's Central Plateau.

The greater part of Mexico's Gulf Tropics lies in the large Yucatán Peninsula. In pre-Hispanic times, the Maya people developed a notable civilization there and in adjacent parts of Mexico, Guatemala, and Belize. Its existence was unknown until ruined cities overgrown by jungle were found in the 19th century. Recent translations of Mayan writing and ethnographic studies of their descendants reveal an enduring culture with an elaborate past. Why they abandoned their cities is a mystery.

The Pacific Tropics: Mountainous Southwestern Outlier.

South of the Tropic of Cancer, rugged mountains with a humid climate lie between the densely populated highland basins of Central Mexico and the Pacific Ocean. This difficult terrain is far more thinly populated than Central Mexico, has a high incidence of relatively unmixed Indians living in traditional ways, and has relatively little economic production. There are two exceptions to the comparative underdevelopment. The coastal resort of Acapulco is on a spectacular bay about 190 miles (c. 300 km) south of Mexico City (Fig. 20.9). This port city of the cliff-diver images continues to play an important role in attracting tourist dollars. There is also a major new iron and steel complex about 150 miles up the coast from Acapulco. Its operation is based on local deposits of iron ore and is intended to stimulate a more general economic growth within this region.

F I G U R E   20.9
The resort hotels that have grown up around the early resort city of Acapulco play a strong role in the expanding importance of tourist dollars in the Mexican—and Caribbean—economy. The fact that the mountains push right up to the edge of the sea, and that water, beaches, warm climate, and proximity to reasonably convenient airports all occur here, has helped this Mexican city and area grow rapidly. *Mark Lewis/Gamma Liaison*

CENTRAL AMERICA: FRAGMENTED DEVELOPMENT IN AN ISTHMIAN BELT OF RELATIVELY WEAK STATES

Between Mexico and South America, the North American continent tapers southward through an isthmian belt of small and poor countries known collectively as Central America. Five countries—Guatemala, Honduras, Nicaragua, Costa Rica, and Panama—have seacoasts on both the Pacific and the Atlantic oceans (Caribbean Sea) (see map, Fig. 20.1), while the other two have coasts on one ocean only: El Salvador on the Pacific and Belize on the Atlantic. If the seven countries formed one political unit, they would have an area only about one fourth that of Mexico (see Table 19.1) and a population only one ninth that of the United States. Geographical fragmentation is a major characteristic of Central America which manifests itself in the fragmented pattern of its political units. Five of the present countries—Guatemala, El Salvador, Honduras, Nicaragua, and Costa Rica—originally were governed together as a part of New Spain, in a unit called the Captaincy-General of Guatemala. Composed of many small settlement nodes isolated from each other by mountains, empty backlands, and poor transportation, the unit never established strong geopolitical cohesion. Separate feelings of nationality became strong enough to fracture overall unity after the end of Spanish imperial rule in 1821. The area gained independence from Mexico in 1825 as the Central American Federation, but in 1838–1839 this loose association segmented into the five republics of today. In recent decades, there have been serious international tensions affecting the five-country area, associated in part with antigovernment guerrilla warfare in various countries. However, by the mid-1990s, tensions had lessened, and the area appeared to be entering a new era of greater political stability.

In Guatemala, El Salvador, Honduras, and Nicaragua, society is composed of groups that differ sharply in wealth, social standing, ethnicity, culture, and opinion. Wealthy owners of large estates have formed a social aristocracy since the first land grants were made in early colonial times by the Spanish Crown. Members of this group tend to be of relatively unmixed Spanish descent, although the great majority have some Indian blood. They have, in general, monopolized wealth, defended the status quo, and exercised great political power. Occupying a middle position in society are small land-holding farmers, largely *mestizos*, who own and work their own land. At the bottom of the scale are millions of landless *mestizo* tenant farmers and hired workers, and the many native peoples (mainly in Guatemala) who live essentially outside of white and *mestizo* society. In the early 1990s, urban dwellers formed a majority in Nicaragua (62 percent of its population) but were in the minority in the other three nations. In each country, the national capital is overwhelmingly the largest city and serves not only as the demographic hub, but also as the economic and political center of the nation.

Costa Rica has had a rather different historical development from the other four countries of the old Central American Federation. It was the area farthest away from the center of government in Guatemala and was left largely to its own devices in developing its economy and polity. Spanish settlers early eliminated the smaller numbers of indigenous people there, and the ethnic pattern that developed was dominated overwhelmingly by whites of Spanish descent. By one estimate, more than 85 percent of the present Costa Rican population is white, whereas the same source shows an overwhelming predominance of *mestizos* and/or Indians in El Salvador, Honduras, Nicaragua, and Guatemala. No precious metals of significance were found and no large landed aristocracy developed. The basic population of small land-owning European immigrant farmers or their descendants managed to create one of Latin America's most democratic societies despite some episodes of dictatorial rule.

Panama and Belize were not part of the Central American Federation, and they achieved independence much later. Panama was governed from Bogotá as a part of Colombia until 1903, when it broke away as an independent republic. The United States was deeply involved in the political maneuvering that led to Panamanian independence, and in 1904 the United States was granted control over the Panama Canal Zone through which the famous transoceanic Canal was opened in 1914. In 1978, after long agitation by Panama for return of the Canal Zone and control over the canal, the United States ratified a treaty under which Panama would receive full control in stages, ending in 1999. Meanwhile, in 1989, a U.S. military invasion of Panama ousted its president, Manuel Noriega, who was captured, brought to the United States, and subsequently convicted of drug-smuggling charges by a United States court.

Belize, on the Caribbean side of Central America, was held by Great Britain during the colonial age under the name of British Honduras. Remote, poor, undeveloped, and nearly uninhabited, the colony was used largely as a source of timber, with African slaves brought in to do the woodcutting. It was also claimed by Guatemala as part of its national territory. Not until 1981 did Belize become independent. It is now a parliamentary state governed by a prime minister and bicameral legislature; the British monarch, represented by a governor-general, is the chief of state. English is the official language and **ecotourism** (carefully planned tours to environmentally sensitive areas at relatively expensive rates), is steadily more important as Cuba's political distance from the United States continues to spawn new tourist destinations in and on the margins of Caribbean Latin America.

The Central American states are among the poorest in Latin America. Mineral wealth is generally lacking, although Guatemala has some oil production and there is scattered mining of metals in all countries—including silver mining at a few sites that have been operational since early colonial times.

Commercial agriculture producing coffee, bananas, cotton, sugarcane, beef, and a miscellany of other export commodities is very patchy in distribution and not very impressive in overall output, although it does provide the greater part of Central American exports. All the countries except overcrowded El Salvador still have pioneer zones where new agricultural settlement is taking place. There is some manufacturing in the few large cities, but the products are limited largely to simple consumer goods for domestic markets. There are a few plants assembling foreign-made cars, but such bits and pieces of productive enterprise yield scanty returns in relation to the needs of the impoverished population, and the situation is worsened by the fact that a large share of the profit from such enterprises makes its way into very few pockets.

Physical Belts and Their Activities

The distribution of settlement—and the accompanying economic life—in Central America strongly relates to three parallel physical belts: the **volcanic highlands,** the **Caribbean lowlands,** and the **Pacific lowlands.** All Central American countries except Belize share the volcanic highlands that reach south from Mexico. Here lie the core regions, including the political capitals, of Guatemala, El Salvador, Honduras, Nicaragua, and Costa Rica. The only cities with over a million people in their metropolitan areas are San Salvador (1.5 million), Guatemala City (2 million), and San José (1.4 million)—all national capitals. Tegucigalpa (Honduras) and Managua (Nicaragua) are between 500,000 and 1 million in population. Dotted with majestic volcanic cones and scenic lakes, the highlands are the most densely settled large section of Central America. Climatically, they are classed largely as *tierra templada,* and most of Central America's coffee is grown there. Increasing population pressure has resulted in deforestation to the point that only a minor fraction of the land is still forested. There has been a considerable development of hydroelectricity along highland rivers.

As in the literature selection below, there is popular sentiment in this region that the government has too much concern for governmental cabinets and high offices and too little for the needs of the common people. Central America has historically had revolutionary movements that embody these fears, and the recent histories of Nicaragua and El Salvador have demonstrated this potential for popular uprising, and for the United States becoming involved at one level or another in such unrest. In 1979, for example, the governing Somoza family—with a long history of U.S. support—was thrown out by the Sandinista revolutionaries in Nicaragua. After a very contentious eleven-year rule characterized by continual battles with the Contras, (Nicaraguan military rebels given support by the U.S.), the Sandinistas were replaced by an elected president, and, by the early 1990s, a reasonably stable government was established with some of the Sandinista reforms institutionalized. Nearly one third of the countries in Central and South America have simi-

lar histories of U.S. military, economic, or political intervention in the governing of Latin American countries.

The hot and rainy Caribbean lowlands touch every Central American country except El Salvador. This coastal strip of tropical rain forest still contains large wooded areas and is quite thinly and irregularly settled, although before the European conquest, it supported in Guatemala and Belize the same advanced Mayan civilization that left such impressive stone structures in Mexico's Yucatán Peninsula. Throughout Spanish colonial times and far into the postcolonial era, economic activity in these disease-infested lowlands was confined to subsistence hunting, fishing, gathering, and shifting cultivation, plus occasional commercial activity such as woodcutting or the growing of sugarcane on scattered plantations. A few settlements were established as bases for British pirate ships preying on Spanish ships of commerce. Some black African slaves were brought in, providing the initial basis for Central America's relatively small black and *mulatto* populations. In the late 19th century, some black workers migrated there from Jamaica to work on the banana plantations being developed by U.S. fruit companies to serve the American market. At the turn of the century, several companies were consolidated into the United Fruit Company (now part of United Brands). An up-and-down history of banana growing saw many plantations eventually shift to the Pacific side of Central America because of devastating plant diseases on the Caribbean side. Today, exports of bananas, grown in part on the Caribbean lowlands and in part on the Pacific lowlands, continue to be very important in the economies of Honduras, Panama, and Costa Rica (Fig. 20.10). The Caribbean lowlands represent the principal area in Central America where agricultural settlement is expanding on a pioneer basis. Settlers generally come from overcrowded highlands and are often aided by government colonization schemes featuring grants of land and the building of roads to get products to market and bring in supplies.

Author Carmen Naranjo sets the stage for a satire entitled "And We Sold the Rain" by describing a very poor Latin American country. It is plagued by the classic problems of too little food, too many people, and a crippling gap between poor and rich and city and countryside. In an effort to take the population's mind off of hunger and these problems, the country's president decides to sell a natural resource—the rain that pours down on this impoverished place continually, causing everything to flood and people to grow weary of the wetness. He sells the rain to a Middle Eastern king of a country called the Emirate of the Emirs. This excerpt begins just after the president has explained to his people how the resources gained from the sale of the rain will let his Latin American country be stronger, independent, and more prosperous.

> The people smiled. A little less rain would be agreeable to everyone, and the best part was not having to deal with the six fat cows [the International Monetary Fund, the World Bank, the Agency for International Development, the Embassy, the International Development Bank, and perhaps the European Economic

Commission], who were more than a little oppressive. Moreover, one couldn't count on those cows really being fat, since accepting them meant increasing all kinds of taxes, especially those on consumer goods, lifting import restrictions, spreading one's legs completely open to the transnationals, paying the interest, which was now a little higher, and amortizing the debt that was increasingly at a rate only comparable to the spread of an epidemic. And as if this were not enough, it would be necessary to structure the cabinet in a certain way. . . .

The president added with demented glee, his face garlanded in sappy smiles, that French technicians, those guardians of European meritocracy, would build the rain funnels and the aqueduct, [and provide a] guarantee of honesty, efficiency, and effective transfer of technology.

By then we had already sold, to our great disadvantage, the tuna, the dolphins, and the thermal dome, along with the forests and all Indian artifacts. Also our talent, dignity, sovereignty, and the right to traffic in anything and everything illicit.

F IGURE   20.10
One of the most productive uses of the *tierra caliente* zone is the raising of bananas. In this photo an inspector in Costa Rica is looking at a major shipment heading for the U.S. market.   *Martin Rogers/Tony Stone Images*

The first funnel was located on the Atlantic coast, which in a few months looked worse than the dry Pacific. The first payment from the emir arrived—in dollars!—and the country celebrated with a week's vacation. A little more effort was needed. Another funnel was added in the north and one more in the south. Both zones immediately dried up like raisins. The checks did not arrive. What happened? The IMF garnisheed them for interest payments. Another effort: a funnel was installed in the center of the country, where formerly it had rained and rained. It now stopped raining forever, which paralyzed brains, altered behavior, changed the climate, defoliated the corn, destroyed the coffee, poisoned aromas, devastated canefields, desiccated palm trees, ruined orchards, razed truck gardens, and narrowed faces, making people look and act like rats, ants, and cockroaches, the only animals left alive in large numbers.

To remember what we once had been, people circulated photographs of an enormous oasis with great plantations, parks, and animal sanctuaries full of butterflies and flocks of birds, at the bottom of which was printed, "Come and visit us. The Emirate of Emirs is a paradise."

The first one to attempt it was a good swimmer who took the precaution of carrying food and medicine. Then a whole family left, then whole villages, large and small. The population dropped considerably. One fine day there was nobody left, with the exception of the president and his cabinet. Everyone else, even the deputies, followed the rest by opening the cover of the aqueduct and floating all the way to the cover at the other end, doorway to the Emirate of the Emirs.

In that country we were second-class citizens, something we were already accustomed to. We lived in a ghetto. We got work because we knew about coffee, sugar cane, cotton, fruit trees, and truck gardens. In a short time we were happy and felt as if these things too were ours, or at the very least, that the rain still belonged to us.

A few years passed; the price of oil began to plunge and plunge. The emir asked for a loan, then another, then many; eventually he had to beg and beg for money to service the loans. The story sounds all too familiar. Now the IMF has taken possession of the aqueducts. They have cut off the water because of a default in payments and because the sultan had the bright idea of receiving as a guest of honor a representative of that country that is a neighbor of ours.[1]

The relatively narrow Pacific lowlands, classed climatically as tropical savanna, are shared by all Central American countries except Belize. They were long the habitat of sparse and unproductive cattle ranching, but recently this pastoral pursuit has taken on a new dynamism, emphasizing better breeds and pastures. Beef destined for market in the United States is loaded into refrigerated trailers, which then travel by all-weather highway to a Guatemalan port on the Caribbean. From here, the trailers move by ferry ships to Miami, Florida.

In addition to their beef output, the Pacific lowlands grow irrigated cotton for the Japanese market. Various small fishing ports exist, as in the Caribbean lowlands. In both instances,

[1] From Carmen Naranjo, "And We Sold the Rain," trans. by Jo Anne Engelbert, *Worlds of Fiction.* eds. Roberta Rubenstein and Charles R. Lawson. (New York: Macmillan Publishing Co., 1993) 948–952.

Regional Perspective

THE PANAMA CANAL

The Panama Canal, providing an interocean passage through the narrow Isthmus of Panama, is extremely important to Panama and to the broader shipping world as well. It was through the political and nearly military involvement of the United States in the early 20th century that Panama declared its independence from Colombia, followed by hasty recognition by the U.S. government. The canal's construction led to an enormous increase in the importance of Central America, the Caribbean Sea, the west coast of the United States, and Panama itself. More than one half of Panama's national population lives close to the canal, and an overwhelmingly large share of the national income is generated in this central core area stretching for about 40 miles (64 km) from Panama City on the Pacific to Colón on the Atlantic. The presence of the canal, combined with the security of United States military protection, has stimulated much international banking and other business activity in Panama. It is fortunate that such opportunities exist, for Panama is not blessed with many resources other than its geographic position. Although sovereignty over the canal and the Canal Zone will transfer to Panama in 1999, the U.S. military presence is still very much in evidence. However, now that the 110-foot lock size is too narrow to accommodate the world's largest cargo ships—generally enormous oil carriers—the canal is losing some of its strategic importance. There continue to be discussions about lock enlargement or the construction of an entirely new canal. For the 20th century, however, the canal changed the transport patterns of centuries of Pacific and Atlantic shipping.

shrimp is by far the most valuable product. There is no seaport of much importance in Pacific Central America except Panama's capital, Panama City (population: 800,000), and its significance is much heightened by the presence of the Panama Canal (Fig. 20.11).

The eastern section of Panama adjoining Colombia contains a roadless stretch where the continuity of the Pan-American Highway System is broken, in what is called the Darien Gap (see Fig. 19.1). Thick rain forests, mountains, and swamps have presented serious obstacles to the bridging of this last remaining gap in a surfaced highway linkage from Canada and the United States to Argentina, Chile, and southern Brazil.

THE HETEROGENEOUS CARIBBEAN ISLANDS

The islands of Caribbean America—commonly known as the West Indies—are very diverse. They encompass a wide range of sizes and physical types, great racial and cultural variations, and a notable variety of political arrangements and economic mainstays. Cuba, the largest island, is primarily lowland, although low mountains exist at the eastern and western ends of the island. The next largest islands—Hispaniola, Jamaica, and Puerto Rico—are steeply mountainous or hilly. Trinidad, just off South America, has a low mountain range in the north (a continuation of the Andes) and level to hilly areas elsewhere. Most of the remaining islands, all comparatively small, fall into two broad physical types: (1) low, flat limestone islands rimmed by coral reefs, including the Bahamas (northeast of Cuba) and a few others; and (2) volcanic islands, such as the Virgin Islands, Leeward Islands, and Windward Islands,

stretching along the eastern margin of the Caribbean from Puerto Rico toward Trinidad (see Fig. 20.1). Each of the volcanic islands consists of one or more volcanic cones (most of which are extinct or inactive), with limited amounts of cultivable land on lower slopes and small plains. Barbados, east of

FIGURE 20.11

When the Panama Canal was completed in 1914, it was thought to have been built much too large for the shipping that might move through the Isthmus of Panama. This photo of the Miraflores Locks in the canal shows how close these container ships come to the edges of the concrete lock channels. The truly massive contemporary oil tankers are both too wide and too long for the locks of the canal system now. Even with such inadequacies, the Panama Canal continues to play a major role in shipping patterns from East Asia to the Atlantic seaboard and Europe. Panama City lies in the background.   *Will & Den McIntyre/Photo Researchers*

the Windward Islands, is a limestone island with moderately elevated, rolling surfaces.

The islands are warm all year, although extremely hot weather is uncommon. Precipitation, however, varies notably, with windward slopes of mountains often receiving very heavy precipitation; leeward slopes and very low islands may have rainfall so scanty as to create semiaridity. The natural vegetation varies from luxuriant forests in wet areas to sparse woodland in dry areas. Most areas have two rainy seasons and two dry seasons a year. Hurricanes, approaching from the east and then curving northward, are a scourge of the northern islands in late summer and autumn.

Plantation agriculture and tourism are the economic activities for which these islands are best known. Sugarcane (Fig. 19.8), the crop around which the island economies were originally built, is still the main export crop in Cuba, Barbados, and a few other places. Other commercial crops include coffee, bananas, tobacco, cacao, spices, citrus fruits, and coconuts. Production comes from large plantations or estates worked by tenant farmers or hired laborers, from small farms worked by their owners, or, in Cuba, from state-owned farms. Mineral production is absent or unimportant on most islands, the most conspicuous exceptions being Jamaica (bauxite), Trinidad (petroleum), and Cuba (metals). Manufacturing has made sizable gains during recent times, and it now provides the largest exports from some islands. Tourism is a major source of revenue in many islands, most notably the Bahamas. Most tourists come from the United States (see Table 19.3).

Emigration has long been an active option for this region's inhabitants, most of whom are economically underprivileged. Millions of West Indians now live in the United States, Great Britain, Canada, France, or the Netherlands—mainly in cities. The money they send back to relatives is a major source of island income.

The Greater Antilles

The four largest Caribbean islands are together called the **Greater Antilles;** the smaller islands are known as the **Lesser Antilles.** Of the five present political units in the Greater Antilles, Cuba, the Dominican Republic, and Puerto Rico were Spanish colonies, Haiti was French, and Jamaica was British. This diversity of colonial origins is only one of many ways in which these units have distinct geographic personalities.

By any standard, Cuba is the most important Caribbean island unit. Centrality of location, large size, a favorable environment for growing sugarcane, some useful minerals, and proximity to the United States are important factors contributing to Cuba's leading role in the Caribbean. Cuba lies only 90 miles (145 km) across the Strait of Florida from the United States. At the western end of the island, Havana developed on a fine harbor as a major colonial Spanish stronghold. The island would stretch from New York to Chicago if superimposed on the United States and is nearly as large in area as the other

Caribbean islands combined. Gentle relief, adequate rainfall, warm temperatures, and fertile soils create very favorable agricultural conditions. Agriculture has always been the core of the Cuban economy, and sugarcane has been the dominant crop. Cane sugar still accounted for the largest share of the Cuban exports by value in 1995. The 1959 Communist revolution, led by Fidel Castro, shifted the ownership of most Cuban land to the state. About 90 percent of all farm land is now under state control. However, as in most Communist regimes, state-controlled agriculture collapsed in Cuba and a 35-year United States trade embargo with Cuba has made it difficult to find effective replacement exports for the island.

In 1963, a great international crisis erupted when the Soviet Union began to build facilities in Cuba that would have accommodated nuclear missiles capable of striking Washington, New York, and other cities of the eastern United States. The presence of the Soviet missile silos and launchers was discovered through the use of **aerial photo interpretation and remote sensing,** geographic research tools of increasing utility. Although the missiles were ultimately withdrawn under intense pressure from the U.S. government, led by its president, John F. Kennedy, Cuba has remained a storm center of political life in the Western Hemisphere. One of the curious geographic outcomes of the tense geopolitics in the Cold War era in the 1960s had Cuba shipping its sugar to mainland China, while Taiwan (then a close ally of the U.S.) shipped its sugar to the United States.

Aid to Cuba from the former Soviet Union fell sharply when the USSR ended its support for the Soviet bloc in 1989, and the USSR and the communist party collapsed in 1991. This profound Soviet change has placed new strains on an economy that already was in serious trouble. Castro's efforts in the mid-1990s to have the 1960 U.S. economic sanctions removed were brought to American attention by the 1994 flight of tens of thousands of boat people from Cuba (and thousands more from Haiti at the same time) (Fig. 20.12). These men, women, and children launched themselves from the northern and western shores of their island hoping either to reach the Florida peninsula or to be picked up by the U.S. Coast Guard, and ultimately to be granted asylum in the U.S. Castro argued that if the U.S. relaxed its sanctions, the Cuban economy would be able to gain enough strength to make flight to the United States relatively less attractive for Cubans. There has been modest change in U.S. immigration regulations for Cubans, but the sanctions have not been relaxed significantly. Castro's decision to shoot down two Cuban exiles who were piloting planes engaged in ritual badgering of Cuba in February 1996 brought the full political force of the Cuban refugees in Florida to bear on the U.S., and a strongly anti-Castro posture returned to Washington, D.C. The 1996 passage of the Helms-Burton legislation permits U.S. citizens to sue foreign nationals whose firms are using land or factories expropriated from U.S. businesses by Castro's government. This legal development has turned North American and European nations against the United States as they claim

FIGURE 20.12

In the mid-1990s, a pattern began of makeshift rafts setting sail from Cuba, and later Haiti, for the Florida peninsula. Often overcrowded and never prepared for bad weather or rough water, these flotillas were more interested in being intercepted by the U.S. Coast Guard than actually crossing the nearly 100 miles of open sea that lies between Cuba and Florida. Once the United States changed its initial policy of bringing the so-called "boat people" to the U.S., the flow of rafters decreased. These men have come ashore on the Grand Caymans, having launched themselves from the southern shore of Cuba. Their raft lies just offshore. *Susan Blanchet Kessel/Dembinsky Photo Assoc.*

such policies infringe on their sovereignty. With the mainland of the United States barely 90 miles distant from Cuba, the political and economic significance of this island continues to be a major thorn in U.S.–Latin American relations.

Metalliferous ores (primarily nickel) are of some importance in Cuba's export trade. All mineral resources were nationalized there in 1960. The country's manufacturing centers on industries that process food and Cuban-grown tobacco and that manufacture cotton and synthetic textiles, cement, chemical fertilizers, pharmaceuticals, and automobile tires. Havana (Habana; population: 2.2 million) is Cuba's main industrial center, as well as its political capital, largest city, and main seaport. Even with its economic troubles, Cuba's government, aided in a massive way by the former Soviet Union, has had some success in raising the adequacy of medical care, housing, and education over the years, particularly in comparison to the levels of such services during the twenty-five years of the Batista regime that Castro replaced in 1959.

The mountainous island of Hispaniola is shared uneasily by the French-speaking Republic of Haiti and the Spanish-speaking Dominican Republic. In Haiti, black population forms an overwhelming majority, although a small *mulatto* elite (perhaps 1 percent of the population) controls a major share of the wealth. Population density is extreme, urbanization is scanty, and poverty is rampant. The agricultural economy, largely of a subsistence character, reflects little influence from the former French colonial regime, which was driven out at the beginning of the 19th century. The republic, independent since 1804, has a West African flavor in its landscape, crops, and other cultural expressions. Deforestation, erosion, and poor transportation are major problems. Aside from subsistence farming, Haiti's small economy concentrates on light manufacturing (aided by low-wage labor), coffee growing, and tourism. Many Haitians have emigrated to the United States, where more than 600,000 are estimated to reside; thousands tried to gain asylum—with the

associated economic potentials—in the United States by fleeing from Haiti in open boats in 1994 and 1995 until this exodus ended with the reinstatement of deposed president Aristide—made possible by the arrival in Haiti of U.S. Marines in October 1994.

The Dominican Republic bears chiefly a Spanish cultural heritage. Santo Domingo, its capital, is the oldest European city established (*c*.1496) in the New World, possessing the first European church, school, and hospital as well. An independent country since 1844, the republic has a history of internal war, foreign intervention (including periods of occupation by U.S. military forces), and misrule, although recent peaceful transfers of power point toward greater governmental stability. The increasingly diversified economy relies on tourism, mining, oil refining, and agricultural products such as sugar, meat, coffee, tobacco, and cacao. Maquiladora export production is playing an increased role in the republic's economy—a pattern developed in the NAFTA world of the Mexico–U.S. borderlands. The Dominican Republic also produces baseballs for the U.S. market, and has sent some of the best major league baseball players to the U.S. as well. Urbanization is more pronounced than in Haiti, and the capital of Santo Domingo (population: 2.5 million, city proper) is considerably larger than Haiti's capital of Port-au-Prince (metropolitan population: 1.3 million).

Puerto Rico, smallest and easternmost of the Greater Antilles, is a mountainous island with coastal strips of lowland. The island was ceded to the United States by Spain after the latter's defeat in the Spanish-American War of 1898, and was administered by Congress as a federal territory until 1952. In that year, it became a self-governing commonwealth voluntarily associated with the United States. Puerto Rico's pre–World War II economy was agriculturally based, with a strong emphasis on sugar. There were accompanying wide disparities within the populace in income and living conditions. After the war, a determined effort by Puerto Ricans to raise their living standard brought great changes. Hydroelectric resources were harnessed, a land-classification program provided a basis for sound agricultural planning, and a thriving development of manufacturing industries and tourism began, financed in large measure by capital from the United States. A complex package of tax incentives was organized to attract American businesses and industries, and a steady migration stream sent Puerto Ricans to the U.S.—primarily to the metropolitan New York area—also providing capital for the island in the form of family remittances.

Today, agriculture has been far outstripped in product value by manufacturing. Petrochemicals, pharmaceuticals, electrical equipment, food products, clothing, and textiles are among the major lines. Sizable copper and nickel deposits have recently been found. Although Puerto Ricans have chosen thus far to remain a commonwealth, there is strong pressure within the island for statehood, and a small minority have expressed a desire for total independence. San Juan (population: 1.8 million) is the capital and main city and port.

Jamaica, independent from Great Britain since 1962, is located in the Caribbean Sea about 100 miles (*c.* 160 km) west of Haiti and 90 miles (*c.* 145 km) south of Cuba. Its population of 2.6 million (as of 1996) is largely descended from black African slaves transported there by the Spanish and (after 1655) the British. When slavery was abolished in 1838, most blacks left the sugar plantations on the coastal lowlands and migrated to the mountainous interior. Bauxite, tourism, and agriculture have constituted Jamaica's economic base for decades. Since the production and export of aluminum-bearing bauxite ore and alumina began in 1952, the island has been one of the world leaders in their output and export. It has also been an important focus of the West Indian tourist trade for many years, basing its tourism on distinctive cultural and historical attractions, a pleasant climate, scenic seaside and mountain landscapes, many miles of beaches, and extensive development of resorts. Agricultural production, long centered on sugarcane, has diversified into fruits, vegetables, rice, flowers, and cattle. Today, few sugar plantations remain. Illegal export of marijuana from small mountain plots has been common. Labor-oriented light industries such as textiles, frozen foods, and electronics are being strongly pushed. The Jamaican government has tried to foster economic diversity in order to offset a falling world demand for bauxite and sugar, but unemployment and poverty remain chronic problems. The capital city of Kingston (population: 820,000) had a notorious reputation in colonial times as a base for piracy. It lies on the southeastern coast, whereas the main tourist developments are in the north and northwest.

The Eastern Caribbean Islands: Common Threads but Differing Circumstances

Over 3 million people live on the eastern Caribbean islands scattered in a 1400-mile (2250-km) arc from the U.S. and British Virgin Islands east of Puerto Rico to the vicinity of Venezuela. This screen of small islands defining the eastern edge of the Caribbean Sea contains a variety of political entities. The Netherlands Antilles and Aruba are Dutch dependencies evolving toward greater self-government. Guadeloupe and Martinique are French overseas departments, and Anguilla and Montserrat are still British dependencies. The British, however, have withdrawn from Caribbean America in a massive way, starting with independence for Jamaica and Trinidad and Tobago in 1962. Since then, independence from British rule in the eastern Caribbean islands has been gained by Barbados (1966) and six smaller units. The British also granted independence to Bermuda in 1968, to the Bahamas in 1973, and to the mainland unit of Belize in 1981, but still retain their colonies of the British Virgin Islands, Cayman Islands, and Turks and Caicos Islands. All their former colonies have retained membership in the Commonwealth of Nations.

The economies of the eastern Caribbean islands suffer from many serious problems, including unstable agricultural exports, limited natural resources, a constantly unfavorable trade balance, embryonic industries lacking economies of scale, and chronic unemployment. A shared trait of the islands is their history of European colonialism and plantation slavery. Such generalizations contribute to some understanding of the islands but give little hint of the individuality these small units present when viewed in detail. One example—Trinidad and Tobago—is discussed here.

Trinidad and Tobago is the largest, richest in natural resources, and most ethnically varied of the island units in the eastern Caribbean. Trinidad, by far the larger of the two islands, has offshore oil deposits in the south. Yet Trinidad refines more oil than it pumps. Imported crude oil from various sources is refined and then shipped out, mainly to the United States. Some 43 percent of the population is black, most of whom live mainly in the urban areas and are often employed in the oil industry. Another 36 percent are the descendants of immigrants from the Indian subcontinent—a population stream that developed for indentured workers after the abolition of slavery. This element lives primarily in the intensively cultivated countryside. About 16 percent are classified as "mixed," and there are small numbers of Chinese, Madeirans, Syrians, European Jews, Venezuelans, and others. This ethnic complexity of Trinidad is reflected in religion, architecture, language, diet, social class, dress, and politics.

The Tourist-Dependent Bahamas

The Commonwealth of the Bahamas includes nearly 700 islands, only 22 of which are inhabited. The majority of the population (280,000 in 1996; 85 percent black, 15 percent white and mixed) lives on New Providence Island, where Nassau, the capital city (population: 175,000), is located. After World War II, tourism grew so rapidly that today it is the most important component in the national economy. Visitors come mainly from the United States. International banking and business activities also generate important revenues. Reexports of crude petroleum dominate the export trade.

20.2   The Andean Countries: Highland and Lowland Contrasts in the South American Indigenous Coreland

The South American countries of Colombia, Venezuela, Ecuador, Peru, and Bolivia—all within the tropics and traversed by the Andes—are grouped in this text as the Andean countries (see Figs. 20.2 and 20.3). Common interests among them were recognized in 1991, when the five countries agreed to create a new free-trade zone. A few outstanding traits shared by these countries are summarized here.

1. *Environmental zonation.* This region's zonation, both vertical and horizontal, is pronounced. The natural setting of each country features local environments that offer extreme contrasts in environmental setting and often are very isolated by natural barriers. There are, in addition, demographic patterns that relate to these zones as well (see Chapter 19, page 466).

2. *Fragmented patterns.* There is widespread fragmentation in patterns of settlement and economic development because of the broken topography that characterizes the Andean countries. This is particularly notable in Peru.

3. *Populous highlands and sparsely settled frontiers.* Every country contains both well-populated highlands and an extensive pioneer fringe of lightly populated interior lowlands or low uplands awaiting development. Some interior areas already yield minerals (notably oil) on an export basis.

4. *Dynamic coastal lowlands.* Every country except Bolivia has an economically productive coastal lowlands. Bolivia originally had a Pacific coastal zone but lost it to Chile during a 19th-century war, thus becoming a landlocked state—as is adjacent Paraguay.

5. *Significant indigenous populations.* All countries had relatively large Indian populations in the mountains, valleys, and basins at the time of the 16th-century Spanish conquest, and the old settlement areas are still major zones of dense population and strong political influence today, especially in the highlands.

6. *Significant native Indian landscape signatures.* The Incas of Peru were one of the most creative and productive pre-Columbian (pre-1492) cultures in Latin America. At the time of the Spanish conquest, the Incas had authority over South America from Quito, Ecuador, south for nearly 2000 miles to the Rio Maule in Chile. From their monumental ceremonial center in Machu Picchu, the Incas engineered a system of roads, bridges, and settlements. Their power was centered in two cities—Cuzco in the south and Quito in the north. They used geographic and topographic surveys and extended military control throughout this empire from 1200 until the arrival of Spain's **Francisco Pizarro** in 1532. In the building of roads, bridges, and other structures they demonstrated extraordinary skill, even though they had neither paper nor a writing system. (Records were kept by knotted ropes.) When Pizarro landed he was warmly greeted by an Inca leader, Atahualpa. The Spanish conquistador welcomed the Incan dignitary into his camp and then killed him, launching the destruction of the empire that had ruled a great deal of the western side of the Andes for more than three centuries. Of all the riches Pizarro gained from this conquest, perhaps the most important is the potato, a tuber domesticated by predecessors of the Inca and diffused widely to the Old World during the Age of Discovery.

7. *Dominant Hispanic traditions.* Each one of the Andean countries maintains strong Hispanic traditions but at the same time has an important heritage from Andean Indians. The Spanish language is official in all five countries (although Indian languages also are official in Bolivia and Peru), and over nine tenths of all religious adherents in every country embrace Roman Catholicism. These badges of Iberian culture were superimposed by the Spanish on advanced Indian cultures within the Inca Empire. Today, Bolivia, Peru, and Ecuador have Indian majorities, and cultural survivals from pre-Spanish times contribute to a keen sense of indigenous identity in each. In all the rest of Latin America, only Guatemala has an indigenous majority, although Indian strains are very pronounced in many countries with important *mestizo* elements. The latter is also true of Colombia and Venezuela, and even where native Indians amount to only 1 or 2 percent of the total but *mestizos* form the largest ethnic element.

8. *Endemic poverty.* All the countries in this region are relatively poor, although oil wealth makes Venezuela considerably better off than the others. However, considerable capital flows into many of these countries through the cultivation, processing, and marketing of illegal drugs—especially cocaine—but it is difficult to track this money and accurately discuss its broader significance (see Problem Landscape Box, page 500).

9. *Primate cities.* In most countries except Ecuador and Colombia, the national capital towers above all other cities in size and importance, reflecting characteristics of primate cities (see Chapter 4, page 102).

COLOMBIA: LARGE RESOURCES AND PROBLEMS IN A DIFFICULT TERRAIN

Within its wide expanse of the Andes, Colombia includes many populous valleys and basins, some in the *tierra fría* and others in the *tierra templada*. The great majority of Colombia's *mestizos* and whites are scattered among these upland settlement clusters, which tend to be separated by sparsely settled mountain country with extremely difficult terrain. Lower valleys and basins, along with some coastal districts, developed plantation agriculture employing slave labor. Consequently, 14 percent of the country's present population is *mulatto* and 4 percent is black.

The Highland Metropolises: Bogotá, Medellín, Cali

Colombia's inland towns and cities have tended to grow as local centers of productive mountain basins. The three largest—Bogotá, Medellín, and Cali—are diverse in geographic character. The largest Andean center in all of South America is Bogotá (population: 5.1 million), Colombia's capital. It lies in the *tierra fría* at an elevation of about 8700 feet (*c.* 2650 m—much higher than Denver, Colorado). In this region, the Spanish seized the land, put the Indians to work on estates, and improved agricultural productivity by introducing new crops and animals—notably wheat, barley, cattle, sheep, and horses. However, exploitation and disease reduced Indian numbers so drastically that the population took a long time to recover. The Spaniards were not able to develop an export agriculture, as night coolness and frosts precluded such crops as sugarcane and coffee, and early transport to the coast via the Magdalena River and its valley was extremely difficult. Slopes were so formidable that railroad connection with the Caribbean was not achieved until 1961. Modern governmental functions and the provision of air, rail, and highway connections have led to explosive urban growth in recent decades. The Bogotá area has even found a large agricultural export market in flowers grown for air shipment to the United States.

Medellín (population: 2.1 million) lies in the *tierra templada* at about 5000 feet (1524 m) in a tributary valley above the Cauca River. The area lacked a dense Indian population to provide estate labor, was difficult to reach, and consequently remained for centuries a region of shifting subsistence cultivation emphasizing corn, beans, sugarcane, and bananas. The introduction of coffee and the provision of railroad connections changed this situation in the early 20th century. Excellent coffee could be produced under the cover of tall trees on the steep slopes, and the Medellín region became a major producer and exporter. Local enterprise and capital developed the city itself into a considerable textile-manufacturing center, and it grew from about 100,000 in population in the 1920s to its present large size.

Cali (population: 1.7 million), located in the *tierra templada* some 3000 feet (915 m) above sea level on a terrace overlooking the flood plain of the Cauca River. Cali had easier connections than did Medellín with the outside world via the Pacific port of Buenaventura. The Spanish developed sugarcane plantations in the Cali region and also produced tobacco, cacao, and beef cattle. Initial Indian labor was followed by that of African slaves. From the late 19th century onward, dams on mountain streams provided hydropower for industrial development. This asset, along with cheap labor, attracted foreign manufacturing firms, and Cali grew at a rate comparable to that of Medellín.

Colombian Coasts and Interior Lowlands

The Caribbean and Pacific coasts of Colombia contrast sharply with each other in environment and development. The Pacific coastal strip has tropical rain forest climate and vegetation. In the north, mountains rise steeply from the sea, but there is a marked coastal plain farther south. The area is thinly populated, with a large proportion of the population of African ancestry. The main settlement is Buenaventura (population: 135,000), an important port but not a large city. Its principal advantage is that it can be reached from Cali via a low Andean pass. The Caribbean coastal area is much more populous. Three ports have long competed for the trade of the upland areas lying to the south on either side of the Magdalena and Cauca rivers. Barranquilla (population: 1.2 million) and Cartagena (population: 725,000) are the larger and more important cities in terms of legitimate trade. But the somewhat smaller city of Santa Marta (population: *c.* 400,000), known as a beach resort as well as a port, may well be more important in trade today. It is reputed to be the control center for Colombia's huge illegal export trade in cocaine and marijuana.

Most of the Caribbean lowland section of Colombia is a large alluvial plain in which agricultural development has been handicapped by floods. Nevertheless, the region has a history of ranching and banana and sugarcane production. The Colombian government is now fostering settlement by small farmers who grow corn and rice.

Colombia's interior lowlands east of the Andes form a resource-rich but sparsely populated region that makes little contribution to the national economy as yet. It is a mixture of plains and hill country, with savanna grasslands in the north and rain forest in the south. The grasslands are used for ranching and the forest for shifting cultivation. Considerable mineral wealth awaits exploitation.

Colombian Potentials

Colombia has the resources for notable economic development in the future. Energy resources include: a large hydropower potential; coal fields in the Andes, the Caribbean coastal area, and the interior lowlands; oil in the Magdalena Valley as well as in newly discovered and developing fields in the interior; and large supplies of natural gas from the area near Lake Maracaibo. Iron ore from the Andes already supplies a steel mill near Bogotá, and large Andean reserves of nickel are under development. At least one half of the country is forested, with most of the forest made up of mixed hardwood species in the tropical rain forest. Because of the intermixture of many different kinds of trees, this type of growth has relatively low utility for forest industries other than the making of plywood. Of the foregoing resources, only oil yet plays much of a part in the country's export economy. In 1996, petroleum and its products represented approximately 15 percent of all legal Colombian exports by

THE GEOGRAPHY OF COCA IN LATIN AMERICA

One of the classic responses of upland farmers to the difficulty and cost of transporting agricultural commodities to the markets in the adjacent lowlands has been to turn grain into liquor. The "white lightning" that is such a part of the mountain lore of the Appalachian region in the United States was a farmer's response to how little money he made carrying bags of grain to the markets compared to the money he could earn selling illegal or untaxed liquor. The **coca trade** of Latin America is being driven, in part, by the same economic dynamics. In Latin America illegal liquor is not the farmer's response to this geographic isolation from grain markets; it is coca, the common name for the shrub *Erythroxylon* or *E. coca*.

This shrub is found in the upland regions of the Andes of South America, as well as in the mountain regions of Caribbean Middle America. Before the Spanish came to Latin America, coca was limited in its use. It was chewed as a stimulant only by the upper classes of the Inca. The Spanish, however, saw that if it was chewed by the "indios"

who were forced into heavy mining labor in the extraction of mountain gold and silver ores, they could endure longer hours and carry heavier loads. The chewing of coca also reduced hunger pangs, diminishing the realization of the inadequate diet so characteristic of the upland Indians during the Spanish demands for mine labor.

Today approximately 90 percent of the world's cocaine comes from Colombia, Peru, Brazil, and Bolivia. The mountain uplands of these four countries provide good cover, an accommodating environment, and considerable independence from the control of the central government. This native plant can be brought to harvest in under a year and a half, whereas upland fruit trees will take four years to produce a harvest. An acre of coca shrubs will generate between $1000–$1300 an acre in cash yield—approximately four to five times the return on any other mountain cash crop. In Bolivia from 1960 to 1985, the return on raising coca increased an average of 11 percent annually. Given the modest return on upland farming and the associated costs of

moving bulk commodities to lowland market centers, it is clear why coca has become the upland crop of choice for many subsistence farmers—and many export-oriented mountain farmers as well.

The cultural pattern that has developed around the raising of coca is the involvement of lowland drug-traffickers, who contract to receive, process, transport, and sell cocaine from the coca raised by the upland farmers. This trafficking has caused whole networks of transporters, judges, police, customs agents, pilots, and **mules** (the people used to carry the most valuable drugs). The scale of this coca trade is such that in 1996 a major drug kingpin, Juan Garcia Abrego, was arrested by Mexican authorities and extradited to Houston for trial. Howard LaFranchi pointed out that it is alleged that Abrego was masterminding a trade that brought "100 tons of cocaine annually to United States markets over the past ten years, earning him annual revenues of more than $20 billion."

One researcher on the role of coca in the economy of Latin America pointed out

value, with coffee representing 24 percent. However, illegal cocaine and marijuana may well have been the largest category of exports.

VENEZUELA: SHIFTING GEOGRAPHIC PATTERNS OF DEVELOPMENT

Venezuela's population is concentrated mainly in Andean valleys and basins not far from the Caribbean coast, as it was when the Spanish arrived in the 1500s. The highlands contain the main city and capital, Caracas (population: 4 million). At an elevation of 3300 feet, Caracas has a *tierra templada* climate. Around Caracas, the Spanish found a relatively dense Indian population to draw on for forced labor, as well as some gold to mine and climatic conditions amenable to commercial production of sugarcane and, later, coffee. Labor needs associated with

sugarcane led to the introduction of African slaves, but Venezuelan plantation development was relatively small. Approximately one tenth of the present population is classified as black. Until the middle of the 20th century, Caracas was a fairly small city and its valley was still an important and highly productive area of farming. Now, the city fills and overflows the valley, and Venezuelan agriculture is now more generally associated with other highland valleys and basins.

Discovery of oil resources in the coastal area around and under the large Caribbean inlet called Lake Maracaibo have made Venezuela a relatively fortunate Latin American country in recent decades. It was one of the most influential charter nations in the 1961 founding of the Organization of Petroleum Exporting Countries (OPEC). Oil revenues have provided a per capita income that is fairly high for Latin America (see Table 19.1), although Venezuela is far from being a rich nation. The oil-based income has fueled rapid urbanization, attracted foreign immigration, and financed industrialization. However, in

that "the coca industry . . . in less-developed countries . . . is labour intensive, decentralized, growth-pole oriented, cottage-industry promoting, and foreign-exchange earning. If the coca industry were completely licit . . . it could be the final answer to rural development. . . ."[1] Another author pointed out, ". . . the cocaine industry is expanding swiftly to other Latin American republics—Ecuador, Panama, Venezuela, Brazil, Argentina and Guatemala—at a speed linked, ironically, to the intensity of eradication in the Andean countries."[2]

Coca represents a truly dramatic problem. The engine that drives the economic development of much of Latin America is the United States market, and for coca, this is its largest market. But coca is also the crop and production activity that the U.S. most reviles. Even while informal market mechanisms work to promote expansion of *E. coca* in the U.S., the government, the military, and the media all make major efforts to force the upland farmers and their associated traffickers to replace this crop with cocoa, cotton, sugarcane, or any other upland crop that could be accommodated by the microgeographies of the *tierra templada* or the *tierra fría* where the coca plants are now raised. Environmentalists claim that besides the more than 171,000 hectares of coca cultivated in Bolivia and Peru, "producers of illicit drugs have deforested as much as 1 million hectares of fragile jungle forest lands in the Amazon region."[3]

Like so many agricultural patterns that hold sway over the peasant farmer, the success of this cropping pattern is dependent upon a world beyond the borders and beyond the uplands of the farmer in this problem landscape. The farmer's well-being is not only threatened by a drop in the world price (street price) of his crop, but it is also put under stress by local efforts to eradicate the raising of the crop—through everything from military encroachment to governmental efforts to interest the farmers in alternative crops. Such programs are almost always funded by external (most often U.S.) sources, so the peasant farmers never know when such support for alternative farming and processing will collapse.

To see the global linkage that ties two very desperate parts of the world together, consider what happens when the New York and Los Angeles police departments effect a massive sweep to jail drug sellers in their cities and change the flow of cocaine on the streets of the U.S. for even a few months. The market forces shift back up the connecting linkages that go from urban drug dealer to smuggler to Peruvian or Bolivian or Colombian trafficker to the processor in the mountain highland, to, finally, the peasant farmer who raises the actual crop. In a sense, like all farmers, his family's success depends on forces over which he has absolutely no influence. While we think of this problem landscape as a major factor in our own urban and suburban instability because of drug abuse, the Latin American upland farmer sees it as a problem that is just as real and just as disturbing to his patterns of life and economy.

[1] Howard LaFranchi, "Mexico Polishes Images with US," *The Christian Science Monitor*. January 17, 1996:7.

[2] Mario de Franco and Ricardo Godoy, "The Economic Consequences of Cocaine Production in Bolivia: Historical, Local, and Macroeconomic Perspectives," *Journal of Latin American Studies*. Vol. 24, No. 2 (May 1992): 375–406.

[3] U.S. State Department Dispatch, "Fact Sheet: Coca Production and the Environment." Vol. 3, No. 9 (1992):165.

recent years the country's income has suffered because of a decline in world oil prices.

Thus, the highlands have supported Venezuela in the past, and the coastal lowland is crucial to the present. But the much larger share of the country that lies inland from the Andes is a major hope for the future. This sparsely inhabited section is mostly tropical savanna in the Guiana highlands and in lowlands (often swampy) along the Orinoco River and its tributaries. For centuries, remote interior Venezuela has been Indian country (in the highlands) or poor and very sparsely settled ranching country (in the lowlands), but development is now quickening. Government-aided agricultural settlement projects are occupying parts of the lowlands, and major oil and gas reserves are beginning to be exploited. In the Guiana highlands after World War II, enormous iron-ore reserves south of Ciudad Bolívar and Ciudad Guayana began to be mined for export by American steel companies, and huge bauxite reserves in the highlands continue to await exploitation. Venezuela has pushed the development of a major industrial center at Ciudad Guayana (population: over 300,000, city proper). Ocean freighters on the Orinoco carry export products away and bring in foreign coal for industrial fuel. Major steel, aluminum, oil-refining, and petrochemical industries are in operation, with further diversification planned.

Angel Falls (3212 ft) is the world's highest waterfall and is part of the reason that Venezuela has dedicated nearly one third of its area to protected parks and reserves.

ECUADOR: ANDEAN TENACITY, DYNAMIC COAST, OIL-RICH AMAZON

In Ecuador, the Andes form two roughly parallel north-south ranges. Between them lie basins floored with volcanic ash. These basins are mostly in the *tierra fría*, at elevations between

7000 and 9500 feet (*c.* 2100–2900 m). Quito (population: 1.3 million), Ecuador's capital and second largest city, is located almost on the equator, high up on the rim of one of the northern basins. Its elevation of 9200 feet (2800 m) yields daytime temperatures averaging 55°F (*c.* 13°C) every month of the year, so that daily fluctuations between warm daytimes and cold nights are the principal temperature variations.

The first Spanish invaders came to Ecuador from Peru in the 1530s. They found dense Indian populations in Ecuador's Andean basins. The conquerors appropriated the land and labor of the Indians and built Spanish colonial towns, often on the sites of previous Indian towns. But they found no great mineral wealth here, and both climate and isolation worked against the development of an export-oriented commercial agriculture. Even today, these basins have an agriculture with a strong subsistence component and sales mainly in local markets. Potatoes, grains, dairy cattle, and sheep are the main crop and livestock emphases. The isolated Ecuadorian Andes attracted relatively few Spaniards, except to Quito, and did not require African slave labor. Consequently, the mountain population has remained predominantly pure Indian, although with some *mestizo* and white admixture. The isolation has recently decreased somewhat with the growth of a certain amount of industry in Quito and lesser Andean cities.

Definitions and Insights

THE "IRISH" POTATO

One of the most significant crops to diffuse from the Peruvian uplands to Europe and Asia was the *Solanum tuberosum*, or the white potato. It was initially raised by the Indians in the Andes and the Incas in Peru and taken to the Old World in around the 1570s. For two centuries the potato slowly took hold in Britain, and by the middle of the 19th century, it was particularly dominant in Ireland. A potato blight in Ireland in the 1840s led to an enormous decline in productivity and millions of Irish fled their country for the east coast of the United States. The tuber should truly be called the Peruvian potato rather than the Irish potato, but it got its name to differentiate it from the sweet potato that followed approximately the same diffusion path. The Irish potato became important because it could be grown in sandy soils, in a wide variety of environmental conditions, in high or low moisture, and did reasonably well when kept in the ground until needed. In terms of yield, the lowly potato gives a higher caloric yield per unit of land than wheat, rice, or corn. It continues today to be a major crop because of its great flexibility in growth setting—potatoes are raised in all 50 U.S. states—and because of the range of foods and drink that can be coaxed from the tuber.

A drastic shift of Ecuador's economy and population toward the coast has taken place in the 20th century. Through the 19th century, approximately nine tenths of the people still lived in the highlands, but that proportion now has been reduced to slightly under one half. The growth zone near the coast is a mixture of plains, hills, and low mountains, with a climate varying from rain forest in the north, through savanna along most of the lowland, to steppe in the south. The main focus of development has been the port city of Guayaquil (population: 1.6 million), located at the mouth of the Guayas River, which drains a large alluvial plain where bananas for export (Ecuador is the world's leading banana exporter) and rice for domestic consumption are grown. On adjoining slopes, coffee is planted for export, and along the Pacific there is a considerable fishing industry. Anchovy harvests have long been a feature of the coastal fishing industry. The Galapagos Islands, located in the Pacific Ocean off the western coast of Ecuador, have been claimed by Ecuador since 1832 and currently are significant in the Latin American promotion of ecotourism—the careful touring of concerned (but well-paying) groups through delicate and often endangered environments. The population of the coastal region is mainly *mestizo* and white.

In the 1960s, major oil finds in the little-populated transAndean interior began to be exploited. The area lies under a tropical rain forest climate in the upper reaches of Amazon drainage. A several-hundred-mile pipeline across the Andes was constructed from the oil fields to a small port on the Pacific, and Ecuador became an oil-exporting country. In 1991, oil and its products provided about 40 percent of the country's foreignexchange earnings. Other major mineral resources are not known to exist. Possible future mineral discoveries in the interior rain forest add urgency to a border dispute with Peru, which controls considerable territory of this type claimed by Ecuador. In 1981, and again in 1994–1995, a bout of armed conflict between the two countries occurred along their undefined boundary in the mountains of southeastern Ecuador.

PERU: FRAGMENTED DEVELOPMENT IN AN ENVIRONMENT OF EXTREMES

Peru is a poor nation with an extraordinarily fragmented distribution of population, induced by an environment that exhibits extremes of ruggedness (in the Andes), dryness (along the coast), or wetness (in the trans-Andean Amazon lowland). More than one half of the population, predominantly pure Indian, lives in scattered basins and valleys in the Andes. Like the Andean Indians in other countries, these highlanders are descendants of an ancient civilization ruled in the three centuries before Spanish times by the Inca Empire, which originated around Cuzco (population: 260,000) in Peru's southern Andes (see Fig. 20.2). The highlands provide a notably poor environment for the development of surplus-producing agriculture. The mountain ranges are considerably higher on the average than those of Ecuador or Colombia, and cultivable valleys and basins tend to be small, densely populated, and in the *tierra fría*, where cool temperatures preclude such export crops as coffee, sugarcane, or cacao and have largely limited the options of farmers to potatoes, grains, and livestock raised for home consumption and sale in local markets. Farmed areas in central and southern sections of the Peruvian Andes tend to be so dry despite their elevation that high productivity is limited to irrigated areas. (Long before the Spanish invasion, Indian engineers were adept at constructing irrigation systems.) Today, considerable cash is earned as tourists travel to varied Inca sites, especially to the monumental city of Machu Picchu in the Andes Mountains (Fig. 20.13). At the same time, many farmers, as in other Andean countries, have turned to the illegal but profitable growing and sale of the coca leaves from which cocaine is extracted. Peru is reputed to be the world's largest supplier.

A narrow, discontinuous coastal lowland lies between the Peruvian Andes and the sea. Climatically, it is a desert of remarkable aridity associated with cold ocean waters along the shore. Air moving landward becomes chilled over these waters, and over the land is stable. Onshore, a relatively cool and heavy layer of surface air underlies warmer air. For this reason, the turbulence needed to generate precipitation is absent, although the surface air does produce heavy fogs and mists during the cooler part of the year which moderate temperatures somewhat. The great majority of rivers from the Andes carry so little water that they die out before reaching the sea. But irrigated agriculture has been practiced along these river valleys as they cross the desert coastal plains, at least intermittently, for millennia. In 1535, the Spanish founded Lima (population: 6.2 million) in an area of this type a few miles inland from a usable natural harbor. The city proved relatively well located for reaching the Indian communities and mineral deposits of the Andes, especially the Cerro de Pasco silver deposits which, for a time, led the world in silver production. Lima was designated the colonial capital of the viceroyalty of Peru and became one of the main cities of Spanish America. In the late 19th century, foreign companies became interested in the commercial agricultural possibilities of the Peruvian coastal oases. As a result, commercially oriented oases now spot the coast, producing varying combinations of irrigated cotton, rice, sugarcane, grapes, and olives. Small ports associated with the oases are generally fishing ports and also fish processing centers, as the cold ocean along the coast is exceptionally rich in marine life. But this resource provides an uncertain livelihood. For a few years in the 1960s and early 1970s, Peru led the world in volume of fish caught (largely anchovies for processing into fishmeal and oil), and the port of Chimbote (population: over 225,000, city proper) was the world's leading fishing port. But during the 1970s, a change in water temperature—associated with the weather phenomenon called El Niño (see Definitions and Insights, page 504)—plus overfishing, led to a drastic decline and widespread unemployment in the industry. However, by 1991, the fisheries had recovered to the point that Peru was the world's fourth nation in total fish catch (after China, the former Soviet Union, and Japan).

FIGURE 20.13
One of the most dramatic landscapes of South America is the remains of Machu Picchu—known also as the "Lost City of the Incas." This highland (8000 ft [c. 2400 m]) city and ceremonial center of the Incas is atop the Urubamba Valley in Peru and serves both as a major tourist destination and an archaeologic treasure of major consequence. It became known to the West in 1911.
Robert Fried/Tom Stack & Assoc.

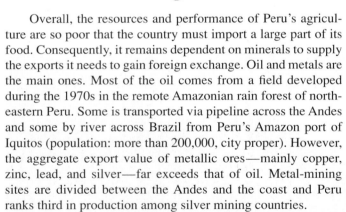

Definitions and Insights

EL NIÑO

"El Niño" is the popular name given to a weather condition that has made major news for decades. The name means "The [Christ] Child," referring to a shift in the temperature of ocean currents off the west coast of South America that occurs most often in December with the arrival of warm tropical waters spreading from west to east across the Pacific Ocean. Normal ocean currents give the west coast of South America a resident cold current (historically called the **Humboldt Current,** but now known as the **Peru Current**). This cold current is vital not only to the fishing industry, but also to temperature patterns on the adjacent lands of Latin America. However, once every four to six years, El Niño comes and changes everything. This thick layer of warmer water comes across the Pacific in the late summer and fall, often arriving off the coast of South America around Christmas. It caps the usually upwelling cold water associated with the Peru Current. Fishing is changed completely because of the shift in water temperatures. Regional air movement becomes unstable because the warm ocean surface of El Niño sets in motion a whole different set of storm and precipitation patterns than those associated with the normal cold Peru Current. The somewhat mythic power of El Niño is given credit for stimulating weather changes, droughts, hurricanes, tornados, and even modifying human behavior.

Overall, the resources and performance of Peru's agriculture are so poor that the country must import a large part of its food. Consequently, it remains dependent on minerals to supply the exports it needs to gain foreign exchange. Oil and metals are the main ones. Most of the oil comes from a field developed during the 1970s in the remote Amazonian rain forest of northeastern Peru. Some is transported via pipeline across the Andes and some by river across Brazil from Peru's Amazon port of Iquitos (population: more than 200,000, city proper). However, the aggregate export value of metallic ores—mainly copper, zinc, lead, and silver—far exceeds that of oil. Metal-mining sites are divided between the Andes and the coast and Peru ranks third in production among silver mining countries.

More than one half of Peru occupies interior Amazonian plains with a tropical rain forest climate. Fewer than 2 million people inhabit this large area. The main settlement, Iquitos, originated in the natural-rubber boom of the Amazon Basin during the 19th century and has tended to be more oriented to outside markets via the Amazon than to Andean and Pacific Peru. The Amazon River is now open to oceangoing vessels for 2200 miles (*c.* 3540 km) into the interior.

Economic development in Peru has been hampered by governmental instability. The country has gyrated wildly between civilian and military rule. Even the military governments have followed very divergent policies—sometimes leftist and sometimes rightist—with one government undoing the "reforms" of another. The 1980s saw widespread guerrilla activity by such movements as the "Shining Path"—which purportedly took its inspiration from Communist China's Mao Zedong (Mao Tse-tung). By 1992, this challenge to national stability had resulted in more than 5000 deaths. Despite the unrest, Peru held a democratic election in 1985, which brought about the first peaceful and democratic change of government in four decades. Then, in April 1992, the elected president Alberto Fujimori—the son of Japanese immigrants to Peru—and the armed forces dissolved the elected Congress in favor of rule by decree. The president claimed that the new powers would allow him to deal more effectively with economic problems, the illegal drug trade, and guerrilla warfare. By the mid-1990s, Peru had regained reasonable stability and was devoting itself to the timely payment of the interest costs on very large foreign debts. Fujimori was reelected president in 1995, even with his history of suspending certain civil rights to fight the rebels of the Shining Path movement. The 1996–1997 siege of the Tupac Amaru rebel group in Peru on the Japanese Embassy is a further example of ongoing tensions between classes in Peru.

LANDLOCKED BOLIVIA: ANDEAN CORE AND LOWLAND FRINGE

Bolivia has extreme Andean conditions and is by far the poorest Andean country. Like the other countries, it has a sparsely populated tropical lowland east of the mountains. But, unlike the others, it has no coastal lowland to supplement or surpass the population and production of its mountain basins and valleys. Agriculture still employs two fifths of Bolivia's people and is limited by extreme *tierra fría* conditions combined with aridity. La Paz (population: 1.1 million, city proper) is the main city and is the de facto capital (Sucre is the legal capital, but the supreme court is the only branch of the government actually located there). Lying at an elevation of about 12,000 feet (*c.* 3700 m), La Paz averages 45°F (8°C) in the daytime in its coolest month and only 53°F (12°C) in its warmest month. Aridity is a further problem; the city averages only 22 inches (56 cm) of precipitation per year, with 5 months essentially dry. A more rural cluster of people lives at a comparable elevation around Lake Titicaca on the Peruvian border. At 12,507 feet, this lake is the highest navigable lake in the world and boasts regular steamer service among lakeside ports high in the Andes. Smaller settled clusters are scattered through Andean Bolivia, mainly along the easternmost of the two main Andean ranges and the *Altiplano*—a very high basin between the two towering ranges. At such high altitudes, export-oriented commercial agriculture has not been possible. Not only is the climate unfavorable, but products would have to reach the sea via 13,000-foot (*c.* 4000 m) passes to the Pacific coast or perhaps a long route to the Rio de la Plata. Accordingly, the infusion of Spanish blood was relatively light, and the country has remained predominantly (over 80 percent) Indian.

Known as "upper Peru" by the Spanish, Bolivia was settled by Europeans partly for the land and labor of its Indians but in good part for its minerals. In the late 16th century, the Bolivian mine of Potosi was producing about one half of the world's silver output. Since the late 19th century, tin has been the mineral most sought in the Bolivian Andes, although exports of tin have now become far less valuable than exports of natural gas from Bolivia's eastern savanna lowlands. Since the 1920s, oil and gas discoveries have been made around the old lowland frontier town of Santa Cruz (population: 700,000, city proper). Andean Indians, Europeans, and Japanese have settled around Santa Cruz in growing agricultural areas. Farther north, where the lowlands have a tropical rain forest climate and where intermediate Andean slopes lie in the *tierra templada*, other pioneer agricultural settlements have been attracting immigrants from the highlands who come to pan for gold, raise cattle (meat is flown to La Paz), cut wood, and grow coffee, sugar, and, illegally, the coca plant. Nevertheless, even with the historic richness of the Potosi mine and the newer lowland settlements, Bolivia today is among the poorest nations in all of Latin America.

20.3   Brazil: Development Eras in an Emerging Tropical Power

Brazil is such a huge and important country that it is treated here as a major subdivision of Latin America. It has an area of about 3.3 million square miles (8.5 million sq km) and a population estimated at 161 million as of 1996. Although smaller in area than the entire United States, Brazil is larger than the 48 conterminous states (Fig. 20.14). This giant country had only about 61 percent as many people as the United States in 1995, but its annual rate of natural increase is so much higher (1.22 percent as opposed to 0.7 percent in the U.S.) that the gap between the two countries will narrow rapidly to the end of the century and beyond.

Rapid population growth is an important facet of the recent dynamism that is giving Brazil an increasingly important role in hemispheric and world affairs. The dynamism is also dramatically expressed in such related phenomena as (1) explosive urbanism that has made metropolitan São Paulo (population: 17 million) and Rio de Janeiro (population: 11 million) two of the world's largest cities; (2) the rise of manufacturing to the point that manufactured exports now exceed agricultural exports in value; and (3) the diversification of export agriculture to such a degree that the traditional overdependence on coffee has disappeared (see Table 19.2).

By providing an expanding domestic market and labor force, rapid population increase can make possible economies of scale and an accelerating economic development in Brazil.

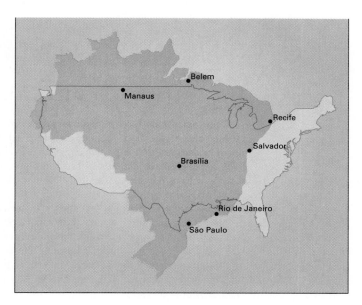

FIGURE 20.14
Brazil compared in area with the conterminous United States.

But today's burgeoning Brazilian population also poses serious problems for the nation. Brazil is hard-pressed to keep economic output ahead of population increase and to keep education and other social services at adequate levels. The country now has a sizable complement of millionaires, a substantial and growing middle class, and large numbers of technically skilled workers, but the benefits from development have been distributed so unevenly that many millions still live in gross poverty. And Brazil remains deeply mired in a huge foreign debt, owed by a government whose ambitions have outrun its finances in recent times. However, some portents are favorable: Brazil does have abundant natural resources of many kinds, many friends among the world's advanced economic powers, and a remarkable degree of internal harmony among the diverse racial, ethnic, and cultural elements that make up the Brazilian population.

ENVIRONMENTAL REGIONS AND THEIR UTILIZATION

Brazilian development must contend with many problems associated with humid tropical environments. The equator crosses Brazil near the country's northern border, and the Tropic of Capricorn lies just south of Rio de Janeiro and São Paulo. Only a relatively small southern projection extends into midlatitude subtropics comparable to those of adjoining sections of Uruguay and Argentina. Within the Brazilian tropics, the climate is classed as tropical rain forest or tropical savanna nearly everywhere, although relatively small areas of *tierra templada* exist in some highlands, and there is one section of tropical

Regional Perspective

THE TROPICAL RAIN FOREST

The tropical rain forest is one of the best known landscapes in the world. Even if people have not experienced it directly—or have not seen movies, TV shows, or magazine articles about its fragile lushness—they tend to have images in their mind of this important ecosystem. This particular system's importance is based in both perception and reality. In perception, the fact that more than 3,500,000 square miles of rainforest existed just two years ago leads people to feel a false sense of security and less concern about the threats to this particular biome. It is true that there is an abundance of animal and plant life in these environments that is unmatched in any other landscape. It is estimated that there are approximately 2000 different plant species in one square mile.

While less than 300 may be trees, there are more than 1500 other species, and it is in that reservoir of abundant life that scientists feel they may find life-saving plants. The reality, however, is that this ecosystem is under "attack."

The forces that are causing the elimination of the rain forest are classic. On the one hand, **swidden operators** come into these unmonitored forests and burn away forest cover for a year or two of **milpa** (slash and burn) farming. When the land loses its modest fertility, the **milpero** moves on, going deeper into the forest. On the other hand, ranchers armed with major earth-moving and forest-destroying equipment come into the forest to tear out tree stock and open the land for cattle ranching, often for the export of beef to the United States. Although the initial scale of each of these landscape

changes is vastly different, the end result of either process is the removal of a forest cover that cannot be regenerated. The loss is not just to the plants, but to the myriad fauna that also call the tropical rain forest their home. The removal of this plant cover also reduces the size of the natural "lung" that these rich forests represent. Geographer Michael Goulding has studied this painful transformation for nearly two decades and says that ". . . [the government has] no effective policy to prevent ranchers from transforming complex [Amazonian tropical rain forest] ecosystems into pasture. That the magnificent biodiversity of the flooded forests could be swept away is a chilling thought, even under the tropical sun."[1]

[1] Michael Goulding, "Flooded Forests of the Amazon," *Scientific American*. Vol. 268, No. 3 (March 1993): 114–120.

---

steppe climate near the Atlantic in the northeast. The overwhelming predominance of tropical rain forest and tropical savanna makes it hard to keep the soil fertile and in place, poses severe maintenance and disease problems, creates general health and sanitation problems, and makes water supply difficult in the huge areas that have a dry season.

On the opposite side of the coin, Brazil is not handicapped to any marked degree by excessively high or rugged terrain. Nowhere does it reach the Andes. Mountains exist, but they are limited in extent and elevation. This leaves a country made up largely of relatively low uplands or extensive lowlands, the latter found primarily in the drainage basin of the Amazon River or in strips along the Atlantic coast.

The physiographic diversity of Brazil is very great when viewed in detail, but its major components can be subsumed under three landform divisions: the Brazilian Highlands, the Atlantic Coastal Lowlands, and the Amazon River Lowland. The country also includes a part of the Guiana Highlands that extends across the border from Venezuela, but this region of hills and low mountains is sparsely populated and has comparatively little importance in Brazil's present development.

The Brazilian Highlands

Of very large importance, however, is the area called the Brazilian Highlands. This is a triangular upland that extends from about 200 miles (*c.* 325 km) south of the lower Amazon River

in the north to the border with Uruguay in the south. On the east, it is generally separated from the Atlantic Ocean by a narrow coastal plain, though it reaches the sea in places. Westward, the Highlands stretch into parts of Paraguay and Bolivia. From here, the northern edge runs raggedly northeastward toward the Atlantic, steadily approaching the Amazon River but never reaching it. The topography is mostly one of hills and river valleys, interspersed with tablelands and scattered ranges of low mountains. Elevations are highest toward the Atlantic. From Salvador to Pôrto Alegre, the Highlands descend very steeply to the coastal plain or the sea. One half of Brazil's metropolitan cities of a million or more population are seaports spotted along this coastal lowland: in the north are Fortaleza (population: 2 million), Salvador (population: 2.5 million), and Recife (population: 2.9 million); in the south lie Rio de Janeiro, Santos (population: 1.2 million), and Pôrto Alegre (population: 3 million). Two of Brazil's three historic capitals are among them: Salvador (formerly Bahia), the colonial capital from the 16th century to 1763, and Rio de Janeiro (Fig. 20.15), which was made the colonial capital in 1763 and then, with independence in 1822, the national capital, until 1960. In that year, the capital was moved to the new inland city of Brasília. These locational shifts were motivated by changing economic and political alignments within the country, although the shift to Brasília is more symbolic of the country's hopes for future inland development than it is of the locational realities of present development.

The descent in some places is a single formidable slope; in others, a series of steps. Throughout its length this dropoff is called the **Great Escarpment.** It has presented major obstacles to penetration of the interior from the coast. For the most part, the escarpment is only 2000 feet (610 m) or so in height, but a central section from north of Rio de Janeiro nearly to Pôrto Alegre forms the seaward edge of the highest and most mountainous section of Brazil, in which there are many peaks between 3000 and 9000 feet. Many short rivers come down the escarpment directly toward the Atlantic, but all except one of the major streams drain down the long slope of the Highlands westward toward the Paraná. The one important exception is the São Francisco River, which flows northward from this elevated area and then turns east through the Great Escarpment to reach the sea north of Salvador.

Another exceptional section of the Brazilian Highlands occurs in the northeast, where climatic maps show an interior area of semiarid steppe climate near Brazil's eastern tip. Actually, the steppe area is only a shade drier than a larger surrounding area in which the tropical savanna climate is so dry as to be almost steppe. The whole area, both climatic steppe and the marginal tropical savanna around it, is sometimes called the **Caatinga,** a word referring to its natural vegetation of sparse and stunted xerophytic forest. Wet enough to attract settlement at an early date, it has been a zone of recurrent disaster because of wide annual fluctuations in rainfall. Years of drought periodically witness agricultural failure, hunger, and mass emigration. The whole region called the northeast is considered the poorest region of Brazil.

The Atlantic Coastal Lowlands

The narrow Atlantic Coastal Lowlands lie between the edge of the Brazilian Highlands and the Atlantic in most places, with the continuity broken occasionally by seaward extensions of the Highlands. Climatically, the northern part of the discontinuous ribbon of plain is tropical savanna, the central part is tropical rain forest, and the southern part is humid subtropical. This was the first part of Brazil to be settled by Europeans, and it has been in agricultural use for a changing variety of crops since the 1500s.

The Amazon River Lowland

The other major lowland area of Brazil is radically different from the coastal lowlands in development and size. This Amazon Lowland reaches the Atlantic along a coast that extends some distance on either side of the Amazon's mouth. From this coastal foothold, it extends inland between the Guiana Highlands and the northern edge of the Brazilian Highlands, gradually widening toward the west. Far in the interior, beyond Manaus (population: over 1 million, city proper), the Lowland widens abruptly and extends across the frontiers of Bolivia, Peru, and Colombia to the Andes.

The Amazon Lowland is composed generally of rolling or undulating plains except for the broad flood plains of the Amazon and its many large tributaries (Fig. 20.16). The flood plains are often many miles wide and lie below the general plains level. They are nearly flat, but most of the Lowland lies slightly higher, is more uneven, and is not subject to annual flooding. Tropical rain forest climate and vegetation prevail nearly everywhere, although southern fringes are classed as tropical savanna. Manaus and the ocean port of Belém (population: 1.4 million) are great exceptions to the level of development in the rest of the Lowland; most of the Amazon region is still very undeveloped and sparsely populated (see Fig. 2.15, page 46). However, in recent decades the Brazilians have opened large

F I G U R E   2 0 . 1 5
A dramatic photo of one of the most frequently visited cities in Latin America, Rio de Janeiro, Brazil.   *Brissaud-Figaro/Gamma Liaison*

FIGURE 20.16
This aerial shot of the lower reaches of the Amazon River in Brazil gives a good feeling for the immensity of this river, the world's largest by volume. It drains more than 2 million square miles and moves fresh water scores of miles into the Atlantic at its mouth.   *Jacques Jangoux/Photo Researchers*

areas by building trans-Amazon highways to augment river and air transportation. The result has been a considerable influx of people into Amazonia. These new settlers live, generally, on a semisubsistence basis, employing various forms and combinations of ranching, agriculture, mining, lumbering, and fishing. Many ecologists fear that the push into the Amazon will eventually destroy the trees and the irreplaceable ecology of the rain forest. The number of living species of plants and animals reaches its maximum here (for example, 600 types of palms, 80,000 plant species, 30,000,000 insect species, and 500 species of fish), and destruction of the forest would cause staggering losses to the world's biodiversity.

ERAS OF BRAZILIAN DEVELOPMENT

In the 1494 **Treaty of Tordesillas,** Spain and Portugal agreed on a line of demarcation along a meridian which, as it turned out, intersected the South American coast just south of the mouth of the Amazon. Because of this, the eastward bulge of the continent was allocated to Portugal. The Portuguese subsequently expanded beyond this line into the rest of what is now Brazil, with international boundaries drawn in remote and little-populated regions east of the Andes where Portuguese and Spanish penetration met.

The Sugarcane Era

The colonial foundations of Brazil were laid in the 16th century. A Portuguese fleet, well off-course on a voyage around Africa, discovered the Brazilian coast in 1500, and a sporadic trade with local Indians for "Brazilwood" (a native source of dye) began. Settlement beyond the first small trading posts was promoted by Portugal in the 1530s in order to combat French and Dutch incursions. Meanwhile, the coastal settlements became increasingly valuable as they rapidly developed exports of cane sugar to Europe and brought shiploads of slaves from Africa to labor in the cane fields. Slave traders in search of Indian slaves penetrated the interior to some degree, and Catholic missions were founded among the Indians. Unfortunately, the missions often brought uncontrollable disease epidemics and slaving expeditions in their wake.

Expansion of coastal sugar plantations and increasing penetration of the interior dominated 17th-century Brazil. The colony became the world's leading source of sugar. Although some sugarcane was grown as far south as the vicinity of the Tropic of Capricorn, the main plantation areas developed farther north in more thoroughly tropical areas that also lay closer to Europe. Meanwhile, missionaries penetrated well up the Amazon, and a series of wide-ranging expeditions, based largely in São Paulo, took place in search of Indians to enslave or quick wealth through mineral discoveries. In the late 1600s, sizable gold finds were made in the Brazilian Highlands some hundreds of miles to the north and northwest of São Paulo in what is now Minas Gerais state.

The Gold and Diamond Era

During the 18th century, Brazil's center of economic gravity shifted southward, in response to the expansion of gold and diamond mining in the highlands north and northwest of São Paulo and Rio de Janeiro. By the mid-1700s, Brazil was producing nearly one half the world's gold. In the latter part of the century this output declined, but by that time much of the highland around the present city of Belo Horizonte (founded in 1896 as a state capital) had been settled, along with areas near São Paulo and spots in the far interior. Despite the difficulty of traversing the Great Escarpment between Rio de Janeiro's magnificent harbor and the developing interior, the city became the port and economic focus for the newly settled territories. Gold and diamonds were ideal high-value commodities to move by pack animals over poor roads and trails to the port.

Besides the attraction of mineral wealth, a factor contributing to the rise of the new areas was the economic distress of the sugar coast to the north. This was a result of increasing development of competing sugar production in other New World colonies (notably in the West Indies). One solution was relocation of plantation owners with their slaves to the newly developing areas in Brazil. Another was movement into the Caatinga (Sertão) area inland from the sugar coast. There, a ranching economy began, augmented by cotton growing in the next century. These 18th-century events set a pattern of economic hardship for this northeastern part of Brazil that has never been broken: (1) continued dependence on hard-pressed coastal sugar and inland cotton production, (2) poverty of unusual severity even for Latin America, (3) recurrent disasters brought by drought, and sometimes by flood, in the climatically unreliable interior, and (4) flows of population to other parts of Brazil. Such flows turn into strong currents in the worst times.

Rubber and Coffee Booms

Nineteenth-century Brazil was changed by rubber and coffee booms, by much-intensified settlement of the southeastern section, by the beginnings of modern industry, and by extension of the ranching frontier far into the Brazilian Highlands. The country underwent these changes as a colony until 1822, as an independent empire with a strongly federal structure until 1889, as a military dictatorship until 1894, and finally as a civilian-ruled republic. Its political history at this time was turbulent, and not until 1888 did the country emancipate its large slave population.

Among the major economic developments of the 19th century, the boom in wild-rubber gathering in the Amazon Basin had the most ephemeral impact. This followed the discovery of vulcanization—making rubber a valuable material—by the American Charles Goodyear in 1839. At that time, Amazonia was the only home of the rubber tree (*Hevea brasiliensis*). A "rush" up the rivers followed, and thousands of widely scattered settlers began to tap the rubber trees and send latex downstream. Manaus grew as the interior hub of this traffic and Belém as its port. But then some seeds of the rubber tree were smuggled to England, and seedlings transplanted to Southeast Asia became the basis for rubber plantations. By 1920 the Amazon production, which depended on tapping scattered wild trees, was practically dead.

Coffee was the preeminent boom product of Brazil during the 19th century. It had long been produced in Brazil's northeast in small quantities, and by the late 1700s was an important crop around and inland from Rio de Janeiro. During the 1800s, production spread in the area around São Paulo and then increased explosively in output during the last decades of the century. This growth was related to the growing world market for coffee, the emergence of expanded and faster ocean transport, nearly ideal climatic and soil conditions in this part of the Brazilian Highlands, and the arrival of railway transport to move the product over land. A spectacular and critical transport development was the completion by British interests in 1867 of a railway linking the coffee-collecting city of São Paulo with the ocean port of Santos at the foot of the Great Escarpment. The coffee frontier crossed the Paraná River to the west and spread widely to the north and south of São Paulo.

Production doubled and redoubled. In the late 1800s, over a million immigrants a year from overseas poured into Brazil, mostly into the region near São Paulo, to work in the coffee industry. Still more millions arrived in the São Paulo region from other parts of Brazil itself. The frontier of ranching and pioneer agriculture moved farther to the interior ahead of the coffee-planting frontier. São Paulo began its rapid growth as the business center and the collecting and forwarding point for the coffee industry; it also began a dynamic career as an industrial center. These developments were facilitated by very early hydroelectric production in the vicinity.

Twentieth-Century Development to the End of World War II

Brazil's development during the 20th century was not spectacular until after World War II. The rubber and coffee booms of the 19th century both collapsed in the early years of the 20th century. Brazil became a minor factor in world rubber production. It continued to lead the world in coffee production and export, as it does today, but increasing international competition depressed coffee prices, and the industry struggled economically. The most positive developments of this period were:

1. The continued growth of early-stage industrialization in the São Paulo area;
2. Some continued expansion of the railway network;
3. The arrival, beginning in 1925, of Japanese immigrants who began to make important contributions to agriculture, especially truck farming;
4. Successful government attacks on two widespread debilitating diseases—malaria and yellow fever—during the 1930s; and
5. A major expansion of irrigated rice production in the Paraíba Valley between Rio de Janeiro and São Paulo.

Rice is a staple of the Brazilian diet, as are corn, beans, and manioc. Overall, however, the country struggled through this period without any major export boom. In 1930, at the beginning of the worldwide Great Depression, Brazil's economy practically collapsed, and in that year democratic government was replaced by a rightist dictatorship that lasted until 1945.

Quickened Development Since World War II

Since 1945, the pace of development in Brazil has generally been very rapid. However, fluctuations have occurred from time to time, including a downturn in the 1980s. Growth took place under elected governments between 1945 and 1964 but occurred most rapidly between 1964 and 1985 under a military dictatorship with technocratic leanings and great ambitions. In 1985, the dictatorship relinquished power to a new elected government, which then had to face the accumulated problems of stalled "take-off." Among these were a crushing national debt that the military regime had contracted to finance expansion, and widespread economic misery caused by governmental attempts to keep up with payments on the debt. This pair of constraints (the Latin American plight alluded to directly in the literature selection earlier in this chapter) on governmental stability plays a role all through the region. The International Monetary Fund has brokered a major proportion of the loans that have enabled some Latin American governments to stay in authority. Conditions for such loans generally include massive efforts to reduce governmental spending and control inflation, and monetary policies that support the local currency. Such policies are never popular, although the cash inflows that come from these IMF loans are always welcome.

The provision of a far more adequate internal transportation system has been an essential foundation of Brazil's growth in the last few decades. This has involved primarily the building of a network of long-distance paved highways and emphasis on trucks as movers of goods. Major urban centers have been connected, and additional highways have been pushed into sparsely populated and undeveloped areas. The new roads have attracted ribbons of settlement, have facilitated the flow of pioneer settlers to remote areas, and have furthered the economic development of such areas by connecting them to markets (for rubber, minerals, timber, fish, tourism, and some agricultural products) and sources of supply. But the single most spectacular instance of interior development has been the building of a new capital in a more interior and central location. Built as a totally planned city on empty savanna in the Brazilian Highlands, Brasília was occupied by the government in 1960. Its growth, accompanied by unwanted adjacent shantytowns, has been such that the new federal district of about 2300 square miles (6000 sq km) had an nearly 1.8 million people by 1995 (Fig. 20.17).

FIGURE 20.17
It is alleged that Oscar Neimeyer, the primary architect of Brasília, drew his first designs for this new capital city on the back of an envelope. Created in thinly settled savanna in 1956 and assigned the role of national capital in 1960, Brasília has not been completely successful in shifting the seat of national government from Rio de Janeiro into the interior. This is the Congress building, designed by Neimeyer.   *George Holton/Photo Researchers*

Definitions and Insights

GROWTH POLE

The concept of a **growth pole** in economic development is ordinarily associated with the promotion of a major industry that spawns numerous supporting, complementary industrial activities. In 1956, a decision was made to stimulate a growth pole in the creation of a totally new capital city—Brasília—for the country, to be built on a sparsely settled savanna landscape in central Brazil. The whole culture of the government in Rio de Janeiro—the traditional capital—was geared to the **environmental amenities** of the coastal location. In selecting central Brazil for the location of the new federal capital, the government was trying to find the most effective way to cause a significant demographic shift in Brazil, spurring migration toward the interior of the country. For more than a decade after it became the capital in 1960, Brasília was only modestly settled, and even today it has a metropolitan population of nearly 1,800,000 while Rio de Janeiro continues to grow with nearly 6,000,000; São Paulo—another coastal city—has more than 10,000,000. Even the relatively low Brasília population numbers decline on weekends when there is much travel back to Rio de Janeiro. Thus, new settlement plans to bring people away from the coasts into the interior of Brazil and other parts of South America have been very checkered in their outcomes.

AGRICULTURAL EXPANSION: SUGAR, CITRUS, SOYBEANS

Three major agricultural boosts in older-settled territories have contributed to the country's recent economic growth. A program of the government in recent years to modernize and reinvigorate the old sugar industry had considerable success, and Brazil is again a major sugar exporter. In 1962, a severe freeze in Florida gave Brazil an entering wedge into the world market for orange concentrate. Orange growing had been gaining momentum for some years in the Rio de Janeiro–Paraíba Valley–São Paulo region, but only for the domestic market. By 1968, Brazilian exports of orange concentrate surpassed those of the United States, which had been the leading supplier, and by the 1980s Brazil was overwhelmingly the world's largest exporter. Also in the 1960s, Brazil became a major soybean producer and exporter, with production centered in the subtropical southeast and in a new district in the tropical Brazilian Highlands. By the mid-1990s, soy products had become Brazil's biggest agricultural export commodity.

BRAZIL'S NEW INDUSTRIAL AGE

Even more significant has been the extremely rapid expansion of Brazilian manufacturing. In the 1980s, the value of manufactured goods exported from Brazil began for the first time to exceed the combined value of all other exports. The products involved today are not just the simple ones of early-stage industrialization, although textiles and shoes are still major items. Steel, machinery, automobiles and trucks, ships, chemicals, and plastics are all important exports, as are weapons—especially types developed by the Brazilian military during its armed suppression of urban guerrilla opposition during the 1960s and 1970s. Aircraft are also manufactured, although mainly for the extensive domestic market that one would expect to find in a country so large. An industrial core area has developed that accounts for the greater part of this manufacturing.

Metals represent the other main contribution of Brazil's resource wealth to Brazilian industry. The Brazilian Highlands are highly mineralized with ores yielding numerous metals, outstanding among which are iron, bauxite, manganese, tin, and tungsten. Known reserves give Brazil about one eighth of the world's iron ore, and the country is a leading producer and exporter. The first modern steel mill began production in 1946 at Volta Redonda, located in the Paraíba Valley between Rio de Janeiro and São Paulo. Others have been built in the industrial core zone since that time, and Brazil has become a considerable exporter of steel.

Brazil's greatest resource deficiency is in fossil fuels. Known supplies of coal are restricted to the extreme southeast and are very inadequate for the country's needs, although they do provide some coking coal. Domestic petroleum supplies are known thus far only in the northeast. Located offshore for the most part, they are the result of intense exploration after the drastic rise in world oil prices in 1973. But they supply only a little over half of Brazil's oil consumption (as of 1994), and crude oil is a major Brazilian import. This is true despite a relatively successful program to develop and use gasohol, whose 20 to 25 percent alcohol content is currently supplied by sugarcane grown in the northeast and in the area around São Paulo. Unless the problem of oil supply can be solved, the country's continuing development drive may be seriously hampered.

20.4 Countries of the Southern Midlatitudes

Four countries of southern South America—Argentina, Chile, Uruguay, and Paraguay—differ from all other Latin American countries in that they are essentially midlatitude rather than low-latitude in location and environment. Northern parts of all the countries except Uruguay do extend into tropical latitudes, but the core areas lie south of the Tropic of Capricorn and are climatically subtropical. Argentina barely extends into the tropics in the extreme north, centers in subtropical plains around Buenos Aires, and reaches far southward into latitudes equivalent to those of southern Canada. Chile also extends south to these latitudes. Uruguay's comparatively small territory is entirely midlatitude and subtropical, and most Paraguayans live in the half of their country that lies south of the Tropic of Capri-

corn. The core areas of the three more eastern countries—Argentina, Paraguay, and Uruguay—have the same humid subtropical climatic classification as the southeastern United States. Chile's midsection or core area has a mediterranean (dry-summer subtropical) climate like that of California.

Besides their atypical locations and climates, these four Latin American countries share certain other regional characteristics. Agriculturally, for instance, they compete more with other midlatitude countries (in both the Northern and Southern hemispheres) than with most other Latin American or other tropical countries. Argentina is an important export producer of such crops as wheat, corn, and soybeans, and thus competes with North American farmers. Both Uruguay and Argentina are sizable exporters of meat and other animal products. Paraguay exports soybeans and cotton. Chile imports more food than it exports, but it does export California-like fruits, vegetables, and wine, as does California. Because these countries never developed labor-intensive and export-oriented plantation agricultures in colonial times, they have a relatively small black population. Argentina and Uruguay have been major magnets for European immigration and have populations estimated to be 85 percent "pure" European in descent. Paraguay and Chile, in contrast, attracted fewer Europeans and are about 90 percent *mestizo*. Uruguay, Argentina, and Chile are among the most developed countries in Latin America. They have relatively low dependence on agricultural employment, with only 10 to 15 percent of their employed workers still in agriculture. Paraguay, with more than 45 percent of its labor force still on farms in the 1990s, is less developed and poorer than the other countries, partly because of its history of landlocked isolation from the outside world.

ARGENTINA: SHAPING GEOGRAPHY BY NATURE, IMMIGRATION, MARKETS, AND POLITICS

Argentina is a major Latin American country. Its area of nearly 1.1 million square miles (2.8 million sq km), second in Latin America only to that of Brazil, is about one third the area of the 48 conterminous American states. Argentina's estimated population of 35 million in mid-1996 ranked it far behind Brazil and Mexico within Latin America, but placed it in close proximity to Colombia (36.2 million) for fourth place.

Argentinian Agriculture

Agricultural products that Argentina exports through its capital city of Buenos Aires (population: 3 million) and lesser ports are sufficient to make the country one of the world's principal sources of surplus food. The leading agricultural exports are wheat, corn, soybeans, and beef. Agriculture is concentrated heavily in the country's core region, known as the humid pampa. ("Pampa" comes from an Indian word for plains.) This is the Argentine part of the area identified on the map of natural regions as humid subtropical lowland prairie (see Fig. 20.2). The environment is outstanding for agriculture. Level to gently rolling plains lie under a climate similar to that of the southeastern United States, featuring hot summers, cool to warm winters, and an annual precipitation averaging 20 to 50 inches (c. 50–130 cm). But the area is specially advantaged by soils that are more fertile than might be expected from the climatic conditions. The grasses of the original prairie provided an unusually high content of organic material (humus), just as they did in the prairie areas of North America. In addition, soil fertility has been enhanced by the material called loess, carried by the wind from drier areas to the west and deposited on the plains. The soils have a loose structure of fine particles, affording both a high concentration of plant nutrients and the maximum opportunity for plant roots to be nourished by feeding on individual soil particles. The local beef industry is fed from the grasses and grain that flourish in this setting, leading directly to the fact that the Argentine per capita beef consumption is the highest in the world.

From the 16th century to the late 1800s, the humid pampa was sparsely settled ranching country, and parts of it remained the domain of Indians. Then, in the late 19th and early 20th centuries, Argentina experienced a rapid and dramatic shift to commercial agriculture and urbanization. There was heavy European immigration, and population grew rapidly. Commercial contact with Great Britain contributed greatly to Argentina's new dynamism. At the time, the British market for imported food was increasing rapidly; refrigerated shipping came into use, allowing meat from distant sources such as Argentina to be imported; and a British-financed rail network (the most dense in Latin America) was built in Argentina to tap the countryside for trade. The network focused on Buenos Aires, which rapidly became a major port and a large city. British cattle breeds, favored by the tastes of British consumers, were introduced to Argentine ranches, or *estancias* (Fig. 20.18), along with alfalfa as a feed crop. Immigrant farmers assumed a major role in agriculture, generally as tenants on the large ranches that continued to dominate the pampa economically and socially. The immigrants came primarily from Italy and Spain but also from a number of other countries, giving present-day Argentina a complex mixture of Spanish-speaking citizens from various European backgrounds. Soon an export agriculture developed in which wheat, corn, and other crops supplemented and then surpassed the original beef exports.

Argentine areas outside the humid pampa are less environmentally favored, less productive, and less populous. The northwestern lowland just inside the tropics—also called El Gran Chaco—has a tropical savanna climate and supports cattle ranching and some cotton farming. Westward from the lower Paraná River and the southern part of the pampa, the humid plains give way to steppe (Fig. 20.19) and then desert, with production and population declining accordingly. Population in these western areas is concentrated in oases along the eastern foot of the Andes, where mountain streams emerging onto the eastward-sloping plains have been impounded for irrigation. Vineyards and, in the hotter north, sugarcane are special emphases in a varied irrigated agriculture. Córdoba (population: 1.2 million) is the largest oasis service center and the second largest city in all of Argentina. To the south of the pampa lie thinly populated deserts and steppes in the hill country of Patagonia, a bleak region of cold winters, cool summers, and incessant strong winds. Indian country until the early 20th

FIGURE 20.18
This scene, reminiscent of beef cattle and cowboys in the American West, shows a ranch in Argentina's humid subtropical climate. Argentine exports of meat and grain compete in world markets with comparable midlatitude products from farms and ranches in the United States. *Juan-Pablo Lira/The Image Bank*

FIGURE 20.19

Iguassu Falls, which Brazil shares with Argentina near the Paraguay border, is seen in this shot from the Argentine side. These falls symbolize the great energy potentials available to many Latin American countries in the form of falling water. Such potentials are vital to "emerging" nations that often lack adequate deposits of mineral fuels. Such resources are, however, also valuable as tourist draws in Latin America.

Robert Fried/Tom Stack Assoc.

century, Patagonia now has millions of sheep on large ranches but little other agricultural production, although oil production has been gaining importance on the south coast.

Urban and Industrial Argentina

Only about one tenth of Argentina's labor force is directly employed in agriculture and ranching. The great majority of Argentinians are city dwellers employed in an economy where service occupations are very important and manufacturing employs about twice as many people as does agriculture. Metropolitan Buenos Aires (and the surrounding province) has about

one third of the national population. It was settled in the late 16th century as a port on the broad Paraná estuary called the Rio de la Plata. But until Argentina became reoriented toward British and then world markets, the city remained a rather isolated outpost of a country centered on the north and the western oases, with relatively little trade by sea. In the 19th century, Buenos Aires first became the national capital and then underwent explosive growth as a seaport, rail center, manufacturing city, and receiver of immigrants.

The growth of industry in Argentina has conformed to a sequence often found in nations emerging from an underindustrialized state:

1. *Local markets.* Industrialization began with the government fixing tariffs to help small-scale industries supply consumer items for sale within the country, and to support plants processing agricultural exports (for example, meat-packing plants). This is called a policy of **import substitution** (encouragement of local firms to produce goods most frequently imported) and it restricts consumption of popular foreign goods through the imposition of high tariffs on goods from overseas.
2. *Wartime isolation.* Some consumer industries such as textile manufacturing and shoemaking were expanded with government support when the country was cut off from imported supplies during World War I and again during World War II.
3. *State-led industrialization.* After World War II, a deliberate state-led policy of industrialization was instituted. There have been lapses in this policy and fluctuations in the mechanisms of support, but considerable success has been achieved in the quantitative growth of manufacturing and diversification of industrial output. Imported coal to supply power has been replaced by oil produced within Argentina (mainly from fields on the Atlantic coast of Patagonia), together with natural gas from the archipelago of Tierra del Fuego and the Andean piedmont. In addition, there is considerable hydroelectric output, and some production of atomic power. Steel plants using imported ore and coal have been built near Rosario on the navigable Paraná River, and diversification has extended into the manufacture of chemicals, machinery, and automobiles.

However, the economic health of a relatively large urbanized service and industrial economy is open to considerable question in Argentina's case. Manufacturing has never become competitive on the world market, and the country's trade pattern is still dominated by agricultural exports and industrial imports. Manufacturing contributes considerably less to national production than might be expected in a country of Argentina's size and economic status, and recent rapid growth of the small-enterprise service sector of the economy reflects in part a decline in some industries, with resulting unemployment. And despite sizable industrial gains, Argentina has not become a prosperous country. In 1995, its per capita GNP of $7990 was lower than that of poor European countries such as Portugal ($10,190) or Greece ($8870), and was a little more than half that of relatively poor Spain ($13,120).

The Disastrous Impact of Argentine Politics

Political turmoil and government mismanagement of the Argentine economy have taken a heavy toll on the country. After initial disturbances following independence, Argentina became a constitutional democracy that was actually controlled by wealthy landowners. This situation continued for many decades until 1930, when a military coup toppled the constitutional government. Between 1930 and 1983, the country was controlled either by military *juntas* or by Juan Perón or his party, with brief intervals of semidemocratic government. Although there were variations, both the military generals and the Perónists were ultranationalist and semifascist. They distrusted democracy, communism, and free-enterprise capitalism, and believed in authoritarianism, violent repression of opposition (which itself was often violent), and state ownership and/or direction of major parts of the economy. They courted the support of a growing urban working class by providing it with jobs, pay, and benefits beyond what was justified by its productivity and output, and they paid for these things by inflating the currency and borrowing abroad. Inflation rates eventually reached hundreds of percents a year, lowering savings and leading to the flight of investment capital and an economy heavily focused on currency manipulation and speculation rather than on productive investment.

In 1991, a treaty to establish a free-trade zone was signed by Argentina, Brazil, Uruguay, and Paraguay. Many of Argentina's inefficient state-run enterprises have been sold to private owners during the past decade, and this program is continuing.

CHILE: ASSETS AND LIABILITIES OF DIVERSE REGIONS

Chile's peculiar elongated shape tends to conceal the country's sizable area, which is more than double that of Germany. From its border with Peru to its southern tip at Cape Horn, Chile stretches approximately 2600 miles (*c.* 4200 km), but in most places its width is only about 100–130 miles (*c.* 160–210 km). Given a favorable natural environment, Chile might not find its curious shape a serious obstacle to development, but the country has actually had to struggle with rugged terrain, extreme aridity, and cool wetness to such a degree that only one relatively small section in the center has any considerable population. The long interior boundary lies mostly near the high Andean crest, with the populous core areas of neighboring

countries lying some distance away on the other side of the mountains. Although the Andean barrier has by no means been impassable during Chile's four and a half centuries of colonial and independent existence, traffic across the mountains has been difficult and is still light.

The Middle Chilean Core Region

Chile's populous central region of mediterranean (dry-summer subtropical) climate occupies lowlands between the Andes and the Pacific from about 31 to 37 South latitude. Mild wet winters and hot dry summers characterize this area, as they do southern California at similar latitudes on the west coast of North America (Fig. 20.20). This strip of territory has always been the heart of Chile's core area. Topographically, it consists of lower slopes in the Andes, a hilly central valley, and low coastal mountains. The area was the southern outpost of the Inca Empire and was occupied in the mid-1500s by a small Spanish army that approached from the north. Members of the army made use of land grants from the King of Spain to institute a rather isolated ranching economy. Grants varied in size by the military rank of the grantees, and this fact, together with the conquered status of the Indians, created a society marked by great social and economic inequality. At one end of the scale was the aristocracy,

FIGURE 20.20
This grape harvest in the central valley of Colchagua, Chile, is evocative of the images of California. Both locales share a mediterranean climate, and vineyards and wine production play important economic roles in both regions. *Carl Frank/Photo Researchers*

composed of those holding enormous ranches. At the other end was a mass of landless cowboys and subsistence farmers. Men from all ranks of society married Indian women quite freely, sometimes several at a time. Thus was created the Chilean nation of today, over 90 percent of which is classified as *mestizo*.

The owners of great *haciendas* dominated the Chilean core area and the country itself well into the 20th century, and ranching remained the primary pursuit in middle Chile. Agriculture, both irrigated and nonirrigated, was carried on but was secondary in importance and generally supplemental to ranching. Wheat, vineyards, and feed crops became agricultural specialties. During the 1960s and early 1970s, population pressure and resulting political demands led to large-scale land reform that somewhat reduced the role of many landowners. Growing population pressure in the core region has contributed to rapid urbanization in recent decades. The national capital of Santiago, founded at the foot of the Andes in the 16th century, now has a metropolitan population of more than 4 million. Metropolitan Valparaíso, the core region's main port and coastal resort, has another million, and the port of Concepción has about 700,000. Altogether, the area of mediterranean climate is the home of about 75 percent of Chile's 14.2 million people. The population of the country was 85 percent urban by a recent estimate—exceeded among the Latin American republics only by that of Uruguay (90 percent) and Argentina (87 percent).

To both north and south, the core area contains transition zones where population density lessens until areas of very scanty population are reached. On the north, the population declines irregularly across a narrow band of steppe to the arid and almost empty wastes of the Atacama desert. In the south, the transition zone is much larger and more important economically. South of Concepción, rainfall increases, average temperatures decline, and the mediterranean type of climate gives way to a notably wet version of the marine west coast climate, whose natural vegetation is impressively dense forests. The southernmost part of the mediterranean climatic area and the adjacent northern fringe of the marine west coast area as far south as the small seaport of Puerto Montt have become part of Chile's core in the past century, forming a less populous transitional end of the core. The Araucanian Indians native to these forests fought off Inca expansion, and they were also able to check most Spanish expansion for three centuries. They were eventually assimilated into Chilean society rather than truly conquered. Their full-blooded Indian descendants are still numerous in this part of Chile. Also distinctive is a less numerous German element descended from a few thousand immigrants who settled on this wild frontier in the mid-19th century. Today, the German community is economically and culturally very important, even putting its stamp on architectural styles. Most people in the southern transition zone, however, are wholly or partly descendants of *mestizo* settlers who came from the core of Chile in the past century as population pressure increased. Agriculture in this green landscape of woods and pastures is de-

voted largely to the raising of beef cattle, dairy cattle, wheat, hay, deciduous fruits, and root crops.

Northern and Southern Chile

North of the core, Chile is desert, except for areas of greater precipitation high in the Andes. This part of Chile, the Atacama Desert, is one of the world's few utterly rainless areas. Mild temperatures, high relative humidity, and numerous fogs are the result of almost constant winds from the cool ocean current offshore. Chile gained most of the Atacama by defeating Bolivia and Peru in the War of the Pacific (1879–1883). Bolivia had previously included a corridor across the Atacama to the sea at the port of Antofogasta, and Peru owned the part of the desert farther north. The war was fought over control of mineral resources, primarily the Atacama's world monopoly on sodium nitrate, a material much in demand at that time for use in fertilizers and the manufacture of smokeless powder. Since then, other mineral resources of the Atacama and the adjacent Andes, principally copper, have eclipsed sodium nitrate in importance.

South of Puerto Montt, the marine west coast section of Chile continues to the country's southern tip in Tierra del Fuego, the cool island region of windswept sheep pastures and wooded mountains divided in ownership between Chile and Argentina. In this little-populated strip, the central valley and coastal mountains of areas to the north continue southward, but here the valley is submerged and the Pacific reaches the foot of the Andes along a rugged fjorded coastline. The coastal mountains project above sea level only in higher parts that form offshore islands. Extreme wetness, cool temperatures, violent storms, and dense mountain forests are characteristic. Little agriculture beyond some grazing of sheep has been possible, and population is very sparse. Some oil and gas are produced along or near the Strait of Magellan, which separates the South American mainland from Tierra del Fuego.

Geography and the Chilean Economy

Chile's prosperity is highly dependent on the world prices of minerals that it exports from the Atacama and nearby Andes. Copper is by far Chile's leading export (35 percent of all exports in 1994). Chile is also the world's leading producer and exporter of copper. However, a rather broad array of primary resources is allowing the country to diversify its exports and could form the basis for major industrialization. In recent years, for example, forestry and fishing have expanded rapidly, and paper and fishmeal have become important exports. Although its known oil and gas resources have thus far left it dependent on the world market for more than half of its hydrocarbon supplies, Chile actually has more than adequate power resources for major industrialization in the future. These include coal fields, especially near Concepción, where the country's iron and steel industry has developed; Andean water power potential, little developed as yet; uranium deposits to support an atomic power industry that began in the early 1980s; and the possibility of using the high winds of the southern region to generate electricity in the future. Relatively little industrialization has been accomplished up to the present, although enough has been done to employ somewhat under 20 percent of the labor force in manufacturing.

Like most Latin American countries, Chile must solve serious political problems if it is to achieve the development and prosperity its resources warrant. A period of democratic government led to the election of a Marxist administration in 1970 by a minority vote against badly divided opposition. This government's drastic changes in the political and economic structure of the country threatened Chile's democracy and created chaos in its economy. A military coup in 1973 then fastened a repressive rightist dictatorship on Chile. It undid many of the Marxist measures that had been instituted, and the country experienced considerable economic success in the new military regime's early years. This dictatorship achieved a degree of popularity with much of Chile's political right and center at the same time that it was violently repressing much of the left. Then the economy was buffeted by world recession and depressed copper prices in the late 1970s and early 1980s, and opposition to the regime expanded and intensified. Military rule was replaced by a new civilian government in 1990, but friction still exists between government and nongovernment factions.

URUGUAY: COSTLY DEPENDENCE ON AGRICULTURE IN AN URBANIZED WELFARE STATE

Uruguay, with an area the size of Missouri and bordered only by Argentina and Brazil, is a part of the humid pampa. Its independence from Argentina and Brazil resulted from its historic buffer position between them. In colonial times, there were repeated struggles between the Portuguese in Brazil and the Spanish to their south and west over possession of this territory flanking the mouth of the main river system in southern South America. After Argentine independence from Spain, these struggles continued at the same time that a movement for political separation grew in the territory north of the Rio de la Plata. Great Britain eventually intervened, fearing Portuguese expansion south of the river, and the two contenders signed a treaty in 1828 that recognized Uruguay as an independent buffer state between the two larger countries.

Uruguay is very similar physically to the adjacent core region of Argentina—composed largely of rolling plains, hilly in some sections, with a humid subtropical climate. The natural vegetation was tall-grass prairie, with ribbons of woodland in river valleys, when the Spaniards first entered Uruguay in the 16th century. Until the 19th century, this was remote ranching country exporting only hides, some salted beef, and mules. Distant mines in Brazil were the main markets. Development then quickened, largely as a result of British initiatives and capital. During the 19th century, sheep were introduced together with techniques for breeding better livestock. Barbed-wire fencing gave better control over pastures and breeding. The coming of refrigerated shipping aided the transportation of meat to overseas markets, and just after 1900, some relatively modern meat-packing plants were built. However, crop farming has not replaced pastoralism to the same degree as on the Argentine pampa; ranching provides Uruguay's basic support to a far greater extent than in Argentina.

Uruguay is especially known for its attempt to become one of the world's first welfare states. This effort began during a period of democratic government in the early 20th century and achieved such initial success that Uruguay gained an international reputation as an enlightened Latin American state. Although much progress was made in spreading the benefits of a relatively prosperous economy in a socially equitable way, the economy proved unable to support the system during the 1960s and 1970s. This period was characterized by violent internal dissension, protracted guerrilla opposition, military dictatorship (1973–1985), and falling standards of living. Such conditions resulted essentially from the growth of a relatively inefficient urban, industrial, and governmental superstructure to such a size that the agriculturally dependent economy could not generate surpluses adequate to finance it.

Montevideo is the only large city in the country, and its metropolitan population of 1.6 million includes about one half of all Uruguayans. The city has dominated the country almost from Montevideo's inception in the 17th century, as first a Brazilian (Portuguese) and then an Argentine (Spanish) fort. Its port facilities handle Uruguay's important waterborne trade. As the national capital, Montevideo directs an economy that developed such a high degree of government ownership that the majority of workers came to be employed directly or indirectly by the government. The city is the main center of Uruguay's manufacturing industries, which employed about one fifth of the country's labor force in the mid-1980s but which export little except textiles. Meanwhile, agriculture and ranching, with 11 percent of the labor force, supplies the country's meat-rich diet and the greater part of all Uruguayan exports. This structure began to fail in the 1960s as costs outran production, and the structure was devastated by the high oil prices imposed by oil exporters in the 1970s. Aside from hydropower, which supplies nearly all electricity, Uruguay has almost no natural resources to support industrialization and depends heavily on imports of fuels, raw materials, and manufactured goods, which must be purchased with the proceeds from agricultural and textile exports. In the early 1990s, Uruguay was attempting to cope with these problems and to preserve its recently restored democracy. Lower world oil prices were a factor easing the burdens of this conflict-racked country. An extensive program is underway to privatize government-owned industries.

DEVELOPMENT PROBLEMS AND PROSPECTS IN LANDLOCKED PARAGUAY

The landlocked and underdeveloped country of Paraguay has long been isolated from the main currents of world affairs. When a nation is landlocked, a wide range of geography-based conditions are set in motion. There is the continual difficulty in getting bulk goods to a coastal shipping lane, except perhaps by river or canal barge on a waterway controlled by an adjacent nation. All transit into the landlocked nation depends on the capacity and inclination of the nation's neighbors to build, maintain, and keep open cross-country road networks, mountain passes, bridges, and other elements of a surface transport system. Beyond the transportational difficulties and the political and economic costs that have to be paid to move goods, even people, from the landlocked nation to an ocean coast—even in this time of expanding air linkages—there is the psychological cost of such a geographic condition. The limits of the map on the margins of a country has been the spark of conflict for centuries. To be on the ocean or on major international rivers creates a very distinct view of one's place in the world.

The Spanish founded Paraguay's capital city of Asunción (current population: 750,000) in the 1530s at a site far enough north on the Paraguay River to promise security against the warlike Indians of the pampas to the south. In addition, high ground at the riverbank gave security from floods. Asunción became the base from which surrounding areas were occupied. But when independence came, Paraguay had no way to reach the sea except through Argentina or Brazil. The main route that developed reached down the Paraguay and Paraná rivers to Buenos Aires. But this route is much longer than it appears to be on maps, owing to meandering of the rivers, and navigation was hindered by such difficulties as shallow and shifting channels, fluctuations in water depth between seasons, and sandbars. Transport costs were so high that Paraguay could not develop an export-oriented commercial economy, and the country remained locked into a very predominantly self-contained economy based on the resources of its own territory.

Today, after some boundary shifts due to past military conflicts, Paraguay has about the area of California. The Paraguay River flows southward across it and divides it into distinctly different eastern and western sections. The eastern section is hill country mixed with plains that are prone to flooding. The climate is humid subtropical, which here has produced a vegetation mainly of forest. Temperature and rainfall conditions are similar to those of central Florida. It is in this eastern section, and especially in the hill country east and southeast of Asunción, that the great majority of Paraguayans have always lived. Cassava, corn, sugar, bananas, citrus fruits, and livestock are important in their semisubsistence agriculture.

West of the Paraguay River is the region known as the **Chaco** ("hunting ground"), which extends into Bolivia, Argentina, and Brazil. This is a very dry area, increasingly so toward the west. Precipitation averages 20 to 40 inches (c. 50–100 cm) annually, but high temperatures cause rapid evapotranspiration, and porous sandy soils absorb surface moisture. These conditions produce an open xerophytic forest in the east, which thins westward and then gives place to coarse grasses on a vast plain of flat alluvium where wide areas are flooded during the wet season. Paraguayans in the entire western region number only in the neighborhood of 100,000. A striking element in this sparse population is provided by scattered Mennonite colonies that aggregate approximately 20,000 people.

Paraguay's comparatively low level of development manifests itself in a number of ways, one of which is total population. Only an estimated 5.4 million people occupied its sizable territory in 1995. This small population results partly from the fact that Paraguay has been a country of emigration for most of its history and has never attracted any large immigration. The latter circumstance has given the country a racial mix containing few whites and a high proportion of Guaraní Indian blood in the 95 percent of the population classed officially as *mestizo*. Urbanization and industrialization have not gone far. Asunción is the only urban place with more than 150,000 people, and about 45 percent of the labor force is still employed directly in agriculture. Per capita income is far below that of Argentina or Uruguay.

In addition to isolation, with consequent lack of commercial opportunities, two other factors need to be mentioned as causes for Paraguay's low level of development. One is the almost total lack of mineral resources that has deterred immigration of capital and people. The other is one of the world's more unfortunate military histories. A Paraguayan attempt to force an outlet to the sea resulted in a war between 1865 and 1870 in which Paraguay was pitted against the combined forces of Argentina, Brazil, and Uruguay. This conflict devastated the country and drastically checked its population growth. About four fifths of the entire population perished or fled. A fairly bloody and economically exhausting conflict, the Chaco War, then was fought against Bolivia in the years 1932–1935. Although Paraguay "won" this war by gaining ownership of the disputed Chaco, Bolivia kept the area around Santa Cruz that has become important in oil and natural gas production.

Recently, the pace of Paraguay's development has quickened. Modern road and air links with the outside world and between parts of Paraguay itself have been forged. With this access to outside markets, exports of soybeans and cotton have grown rapidly and have come to dominate the country's trade. The future may well see larger and more dramatic changes due to three huge new hydroelectric installations along the Paraná River on the southeastern border of the country (Fig. 20.20). These were partly operative in 1996, with construction work continuing. Paraguay is exporting electricity to pay its share of the costs of these projects, which it undertook jointly with Argentina and Brazil. The new hydroelectric output will give Paraguay a far more favorable energy situation in the future than this fuel-deficient country has had in the past.

ANTARCTICA—FUTURE RESOURCE BASE?

Antarctica is the world's fifth largest continent, with an area of 5,500,000 square miles (14,245,000 sq km) lying south of the tip of South America—virtually filling the Antarctic Circle. It is the setting of enormous human dramas in exploration, bravery, and foolishness as people have crossed the continent with dog sleds, on skis, on foot, pulling their gear, in airplanes, and now in tour helicopters. The mystery of the place comes from its winter of darkness, with continual problems of "whiteouts" caused by light refraction on an extensive snow and ice surface covering some 95 percent of the continent. Mirages are common, and average temperatures during the "summer" months barely reach 0° F. Winter averages are the coldest in the world, with winter mean temperatures averaging −70° F (−57° C).

The whole continent is alive with ice formations created by the 2–10 inches(5–25 cm) of annual precipitation that then move toward the open seas of the South Atlantic. Glacial action of all sorts provides a continual breaking away of valley glaciers, ice streams, and tongues all crashing into the sea of the margins of the continent. There is virtually no human settlement beyond research teams that have constructed an array of semipermanent structures on the small areas of exposed land that lie near the outer edges of the continent, and on the island of Little America in the Ross Sea.

There has long been an imagery of considerable resource wealth lying beneath the cap of ice, sometimes thousands of feet thick. Although many nations now claim some rights to the continent, and support periodic settlements of research scholars, neither the resources nor the fauna are likely to lead to a significant exploitation of this enormous continental mass. It will continue to be, however, another frontier in the geography of the future.

QUESTIONS

Refer to the Saunders *Atlas of World Geography, Special Edition* as well as this chapter to answer the following questions.

1. What is the length of the Amazon River, and what quantity of fresh water does it pour into the Atlantic each second? What percent of the world's fresh water is in this river system, and what is the area of its drainage basin?

2. What are the dimensions of the rain forest in South America? Explain why the loss of this environment is so significant—both to the present and the future.

3. What and where are the Nazca Lines? What is thought to be their origin, and what role do they play today in the host country's economy?

4. What impact did of the Panama Canal have on shipping in the early 20th century? Where else might have been considered for a transisthmus canal in Central America?

5. What are the three fastest growing countries in population in Latin America? How would you characterize the population distribution patterns of Latin America?

North America

8

Geographic Profile of North America

National and Regional Characteristics
of Canada

The Geography of Diversity and Change
in the United States

21

GEOGRAPHIC PROFILE OF NORTH AMERICA

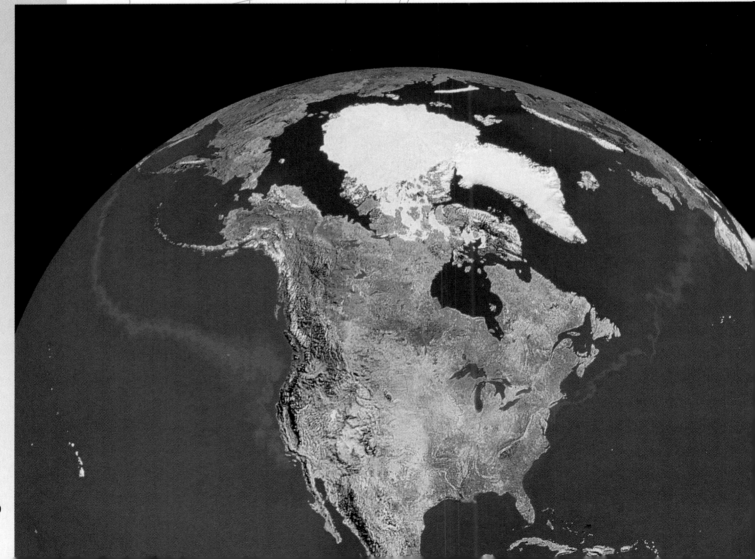

Nomenclature poses a continual challenge for geographers. Names that make good sense at one time often make much less sense at another time (did you ever wonder why the university in Evanston, Illinois, is called Northwestern?). The world region we explore in this chapter consists of two major countries—the United States of America and Canada. The term "Anglo-America" has sometimes been used as a regional title for this pair because of the pervasive and significant influences of Britain (the home of the Anglo-Saxons) on the development of the region. However, the past half-century has seen both a significant change in patterns of immigration and an increasing ethnic sensitivity about group identities. While the "Anglo" influences continue to be considerable in the U.S. and Canada, other powerful voices and presences are helping to shape current landscapes and culture. There is no term in current usage that embraces all of the ethnic and cultural diversity expressed in these two large countries and their distinct ethnic worlds within other worlds, so we turn to a more common and locational geographic term: North America.

North America, however, is also flawed as a collective term for these two countries. Often, Mexico is also included in the concept of North America—recall the North American Free Trade Agreement (NAFTA). Nevertheless, in this text we deal with Mexico as part of the world of Latin America because of similarities of language, culture, ethnic mixtures, and economic and cultural patterns among the nations that lie south of the southern border of the United States.

Another argument could be raised claiming that the names of the region ought to be derived from Native American groups (as so many U.S. state names are) because of their profound influence on the initial settlement of the region. However, in considering the 20th and 21st centuries, that argument loses power. Thus, we mark the lands north of Mexico as North America.

21.1 Definition and Basic Magnitudes

Possession of a large area assures a country of neither wealth nor power. But it does afford at least the possibility of finding and developing a wider variety of resources and, other things being equal, of supporting a larger population than might be expected in a small country (Table 21.1). The fact that there is no direct relationship between area on the one hand and wealth and power on the other may be seen by comparing the North Amer-

◄ North America is not the largest landmass in Earth's collection of continents, but it is fortunate to have one of the most productive agricultural regions of any landmass in the world. In this photograph, the broad, extensive plains regions spread from the central United States up into Canada, serving as a resource center for the food needs of the entire world. This satellite image also gives clear evidence of the various north-south mountain systems as well as the arctic reaches and major river systems of the continent. *World Sat International/Science Source/Photo Researchers*

TABLE 21.1

North America: Basic Data

Political Unit	Area		Estimated Population (millions)	Estimated Annual Rate of Natural Increase (%)
	(thousand sq mi)	(thousand sq km)		
Canada	3850.8	9976.1	29.0	0.63
United States	3617.8	9372.6	265.2	0.69
Total	7468.6	19,348.7	294.2	0.68

Sources: *The World Almanac and Book of Facts 1996, Statesman's Yearbook 1995–1996, Britannica Book of the Year 1996,* and *The World Factbook 1995.*

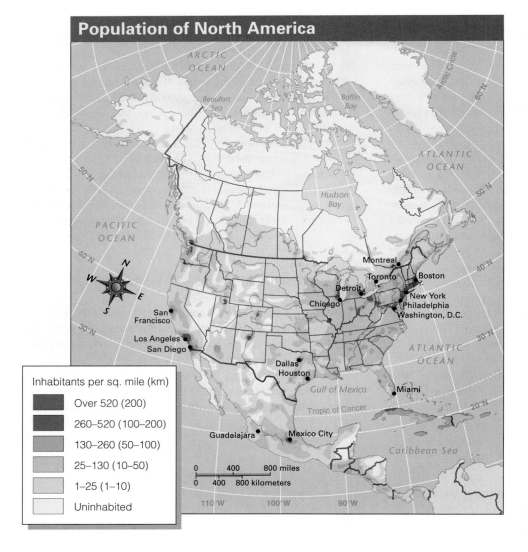

Population of North America

Inhabitants per sq. mile (km)

- Over 520 (200)
- 260–520 (100–200)
- 130–260 (50–100)
- 25–130 (10–50)
- 1–25 (1–10)
- Uninhabited

FIGURE 21.1

A population map of North America.

Estimated Population Density		Infant Mortality Rate	Urban Population (%)	Arable Land (% of total area)	Per Capita GNP ($U.S.)
(per sq mi)	(per sq km)				
8	3	6.8	77	5	22,760
73	28	7.88	75	20	25,850
39	15	7.8	75	12	25,550

ican countries themselves. Canada is slightly the larger in area (3.85 million sq mi/9.98 million sq km, compared with 3.62 million sq mi/9.37 million sq km for the United States), but is less wealthy and much less powerful on the global stage than the United States. In population, Canada had nearly 29.0 million in 1996, while the United States total was 265 million in that same year. Thus the geographical impact of size and population on either wealth or political power has yet to be fully determined (Fig. 21.1).

The large Arctic island of Greenland (population: about 57,000) is included in the concept of North America because of its spatial proximity to Canada (Fig. 21.2) even though it has mainly an Inuit and Danish culture and is a self-governing part of Denmark. This range of international influences on both the landscape and the historical past of this region is continued today by an active pattern of international trade.

21.2   Benefits and Characteristics of Diverse Environmental Settings

Even in a broad-scale view that suppresses much detail, the natural environments of North America are remarkably diversified. This diversity reflects a number of major circumstances and leads to numerous "microgeographies," or distinctive landscapes that have their own identities and use patterns.

MAJOR LANDFORM DIVISIONS

The major landform divisions are shown in Figure 21.3. The glacially scoured and thinly populated Canadian Shield is formed of ancient rocks that are rich in metal-bearing ores. The surface is generally rolling or hilly. Agriculture, handicapped by poor soils and a harsh climate, is limited and marginal. Water power, wood, iron, nickel, and uranium are major resources. The United States section bordering Lake Superior is called the Superior Upland.

The Arctic Coastal Plains occur in two widely separated sections, each of which is largely unpopulated wilderness: The Alaska section along the Arctic Ocean contains rich oil deposits, exploited on a major scale, but the section along Hudson Bay is economically insignificant. Mountainous Greenland is buried under permanent ice except for thinly inhabited coastal fringes of tundra. Fishing is the main economic activity there.

The Gulf–Atlantic Coastal Plain is low and relatively level, with soils that are generally sandy and relatively infertile. Agriculture involves many different specialties in scattered areas of intensive development. Pine or mixed oak-pine forests cover large areas. Swamps and marshes are abundant, and the coast is indented by river mouths and bays on which many prominent seaports are located. Stands of swamp hardwoods occur in poorly drained areas, and prairie grasslands were the original vegetation in certain coastal areas in Texas and southwestern Louisiana and in some inland areas underlain by limestone. Oil, gas, sulfur, and salt are major resources in the Gulf Coast section, where they provide the basis for major oil-refining and chemical industries.

The Piedmont, inland from the Gulf–Atlantic Coastal Plain, is formed of harder and more ancient rocks, is somewhat higher and more hilly, and has somewhat better soils—though now much depleted and eroded in the center and south because of a past emphasis on the growing of clean-tilled cotton, tobacco, and corn in a hilly area with heavy rains. Many cities, such as Richmond, Washington, D.C., and Baltimore, are spotted along the fall line between the two physiographic regions, where falls and rapids mark the head of navigation on many rivers and have long supplied water power (Fig. 21.4).

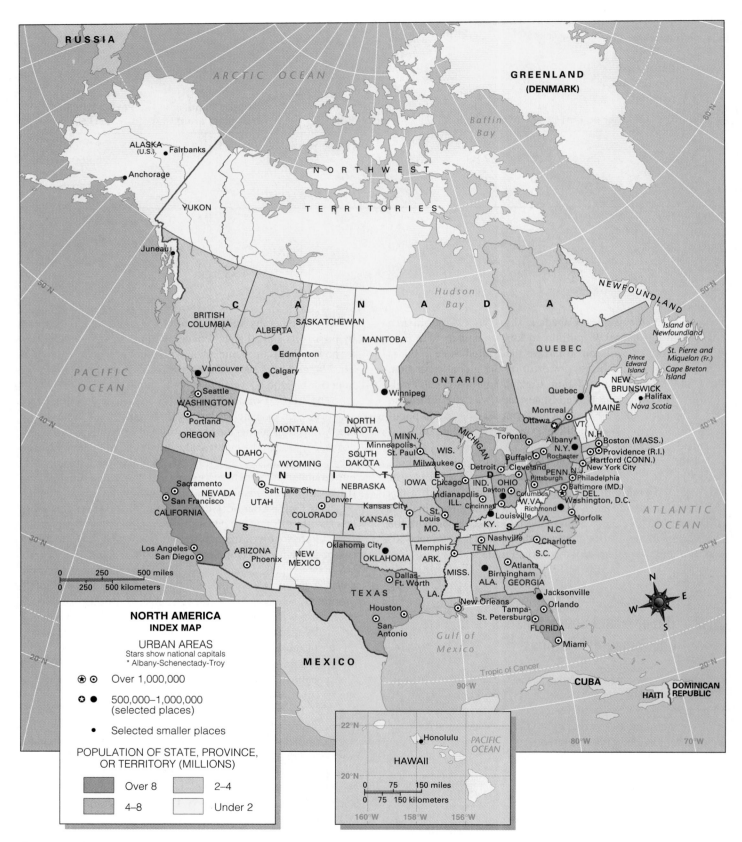

FIGURE 21.2

Political divisions and main cities of North America. City-size symbols conform to metropolitan area estimates given in this text.

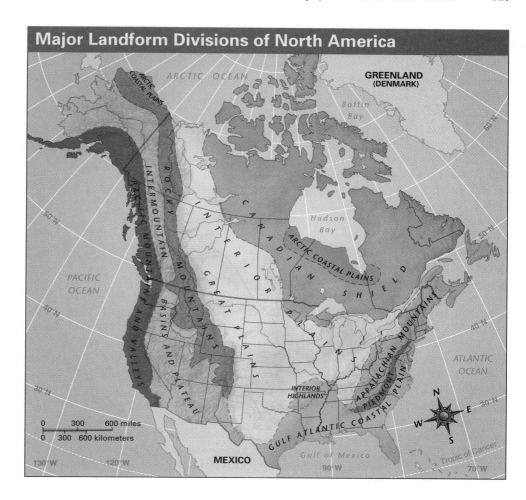

Major Landform Divisions of North America

FIGURE 21.3
Major landform divisions of North America.

The Appalachian Highlands form a complex system in which mountain ranges, ridges, isolated peaks, and rugged dissected plateau areas are interspersed with narrow valleys, lowland pockets, and rolling uplands. Elevations are relatively low, and most land is occupied by coniferous, deciduous, or mixed forests. Coal, water power, and wood are notable resources. The Appalachian Highlands have a surprisingly large population for a highland area, and their importance in the North American economy is significant.

The Interior Highlands have many physical similarities to the Appalachian Highlands, but they are lower and have only minor coal deposits. The Interior Highlands are divided by the Arkansas River valley into the dissected Ozark Plateau on the north and the east-west–oriented ridges and valleys of the low Ouachita Mountains on the south.

The Interior Plains lie mostly within the drainage basins of the Mississippi, St. Lawrence, Mackenzie, and Saskatchewan rivers. The western third of this region is called the Great Plains. Huge expanses of gently rolling or flat terrain exhibit some of the world's finest soils and most productive farms. The best soils developed under prairie or steppe grasslands, but some very good soils in the east developed under deciduous forest. Large quantities of beef, pork, corn, soybeans, and wheat

FIGURE 21.4
This photo of Richmond, Virginia, shows many of the geographic qualities that made fall-line cities take on early urban and industrial importance. Settlement began here in 1645 using the falls as a power source. The city, also the capital of Virginia, is now rebuilding the early wharfing and industrial hubs around the falls, and is continuing to renew the much later urban building stock. With its contemporary railway, highway, and Interstate networks weaving all through the city, Richmond serves as a strong example of the power of transportation to stimulate human settlement.
Andrea Pistolesi/The Image Bank

are produced. Coal, oil, gas, and potash are present in major quantities.

The Rocky Mountains, generally high and rugged, consist of many linear ranges enclosing valleys and basins. Easy passes are almost nonexistent. Ranching, mining (of both metals and fuels), and recreation are major activities. Many major rivers rise in these mountains, including the Columbia, Colorado, Missouri, and Rio Grande.

The Intermountain Basins and Plateaus consist generally of high plateaus trenched by canyons (in the Colorado Plateau or the volcanic Columbia-Snake Plateau sections) or—in the extensive Basin and Range Country of Nevada, Utah, and adjacent states—hundreds of linear ranges separated by basins of varying size. Shielded from moisture-bearing winds by high mountains on both the west and the east, the Intermountain area is this continent's largest expanse of dry land. Mining of copper, oil, gas, coal, uranium, and other minerals is prominent in scattered spots, and numerous major dams impound irrigation water and generate much hydroelectricity along the Columbia and Colorado rivers.

The Pacific Mountains and Valleys lie parallel to the Pacific shore and contain North America's highest mountains. In California, the Central Valley between the Sierra-Cascade mountain system on the east and the Coast Ranges on the west is a huge exhibit of productive irrigated agriculture on flat alluvial land. Farther north, the Willamette–Puget Sound Lowland forms another major valley between the Coast Ranges and the Cascade Range of Oregon and Washington. Extreme western Canada and southern Alaska are dominated by high mountains, with many spectacular glaciers. The economy of the Pacific Mountains and Valleys has grown rapidly on the basis of a major collection of natural resources: abundant forests, fisheries, rivers to produce hydroelectricity and supply irrigation water, oil and gas in California, gold that once drew settlers to California and British Columbia, and both climatic and scenic resources undergirding development in the California subtropics.

Major Climatic Regions

The North American climatic pattern exhibits both regional variance and largeness of scale. The United States includes a greater number of major climatic types than does any other country in the world, and even Canada is more varied than is commonly assumed. The variety of economic opportunities and possibilities afforded by this wide range of climates is one of the basic factors underlying the economic strength of this world region. The lack of any large producing area for tropical specialty crops, however, requires the import of such commodities as coffee, tea, cocoa, and bananas.

Each climate region is shown by name and color on Figure 21.5. The tundra climate, characterized by long, cold winters and brief, cool summers, has an associated vegetation of mosses, lichens, sedges, hardy grasses, and low bushes. The subarctic climate has long, cold winters and short, mild summers, with a natural vegetation of coniferous snow forest re-sembling the Russian taiga. Population is extremely sparse in the tundra and subarctic climates and is practically nonexistent in Greenland's ice-cap climate. Scattered groups of people engage in trapping, hunting, fishing, mining, logging, and military activities; many live largely on welfare. The humid continental climate with short summers is characterized by long and quite cold winters and short warm summers. Agriculturally, there is a heavy emphasis on dairy farming except in the extreme west, where spring wheat production is dominant. The humid continental climate with long summers has cold winters and hot summers; agriculturally, this belt, which includes the agricultural richness of the Midwest, is marked by an emphasis on dairy farming in the east and a corn-soybeans-cattle-hogs combination in the midwestern (interior) portion.

In the humid subtropical climate, winters are short and cool, although with cold snaps, and summers are quite long and hot. Agriculture is rather diverse from place to place, with some major specialized emphases in cattle, poultry, soybeans, tobacco, cotton, rice, peanuts, and a wide range of fruits and vegetables. Within each of the three humid climatic regions just described, the pattern of natural vegetation is complex, and each region has areas of coniferous evergreen softwoods, broadleaf deciduous hardwoods, mixed hardwoods and softwoods, and prairie grasses. Extreme southern Florida has a small area of tropical savanna climate, which is a major climate in adjacent Latin America. The state of Hawaii has a tropical rain forest climate. Along the Pacific shore of the United States and Canada, the narrow strip of marine west coast climate is associated with the barrier effect of high mountains near the sea, ocean waters offshore which are warm in winter and cool in summer relative to the land, and winds prevailingly from the west throughout the year. The mild, moist conditions have produced a magnificent growth of giant conifers (most notably the redwood and the Douglas fir) that provides the basis for the large-scale development of lumbering. Lush pastures support dairy farming, as they do in the corresponding climatic region in northwestern Europe.

The mediterranean or dry-summer subtropical climate of central and southern California, in which nearly all precipitation comes in the winter half-year, is associated with irrigated production of cattle feeds, fruits, cotton, and a great range of other crops. These crops, together with associated livestock, dairy, and poultry production, make California the leading U.S. state in total agricultural output. In the semiarid steppe climate, occupying an immense area between the Pacific littoral of the United States and the landward margins of the humid East and extending north into Canada, temperatures range from continental in the north to subtropical in the south. The natural vegetation of short grass, bunch grass, shrubs, and stunted trees supplies forage for cattle ranching, which is the predominant form of agriculture. Areas that are less dry are often used for wheat, and both wheat and other crops are grown in scattered irrigated areas, often associated with major rivers such as the Columbia, Snake, Arkansas, or Rio Grande. The desert climate of the U.S. Southwest is associated with scattered irrigated dis-

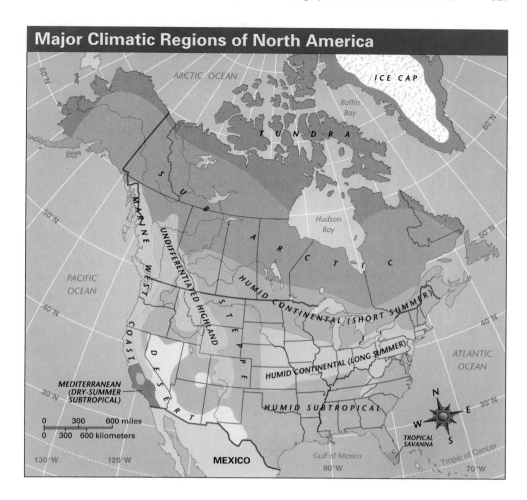

Major Climatic Regions of North America

FIGURE 21.5
Major climatic regions of North America.

tricts (oases) and settlements emphasizing mining, recreation, and retirement. High, rugged mountains, such as the Rockies and the Sierra Nevada, have undifferentiated highland climates varying with latitude, altitude, and exposure to moisture-bearing winds and to sun. More detailed discussion of climatic regions and major landform divisions, with many relevant maps and photographs, can be found in Chapters 22 and 23.

21.3   The Abundance of North America: A Major Regional Characteristic

Both the United States and Canada are wealthy nations, with Canada much the lesser in this respect. Canada's gross national product (GNP) is a little less than one tenth as great as that of the United States. On a per capita basis, Canada's is $22,760; in the U.S., it is $25,850. The output of goods and services in the United States is more than 80 percent of that of all Europe. In producing these goods and services, the U.S. consumes nearly 35 percent of the world's annual energy consumption. The ranking of the United States and Canada among the fifteen nations with the highest per capita GNPs can be seen in Table 21.2. The *relative* standing of the North American countries

TABLE 21.2	The World's Top Fifteen Countries in Per Capita GNP Values in U.S. Dollars	
Political Unit		**Values in U.S. Dollars**
1. United States		$25,850
2. Hong Kong*		24,530
3. Cayman Islands		23,000
4. Luxembourg		22,830
5. Canada		22,760
6. U.A.E.		22,480
7. Liechtenstein		22,300
8. Norway		22,170
9. Switzerland		22,080
10. Qatar		20,820
11. Australia		20,720
12. Japan		20,200
13. Singapore		19,940
14. Denmark		19,860
15. Sweden		18,580

*Hong Kong was a British Crown Colony until July 1, 1997, when it was returned to the People's Republic of China.

Data sources same as Table 21.1.

TABLE 21.3

North America's Share of the World's Known Wealth and Production

Item	Approximate Percent of World Total		
	U.S.	Canada	North America
Area	7	7	14
Population	4.6	0.5	5.1
Agricultural and Fishery Resources and Production			
Arable land	13	3	16
Meadow and permanent pasture	8	1	9
Production of:			
Wheat	12	5	17
Corn (grain)	46	1	47
Rice	1.4	0	1.4
Soybeans	52	1	53
Cotton	26	0	26
Milk	14	1	15
Number of:			
Cattle	11	1	12
Hogs	8	1	9
Fish (commercial catch)	6	1	7
Forest Resources and Production			
Area of forest/woodland	6	8	14
Wood cut	15	5	20
Sawn wood production	22	11	33
Wood pulp production	46	19	65
Minerals, Mining, and Manufacturing			
Coal reserves	24	1	25
Coal production	18	1	19
Coal exports	25	8	33

Sources: *World Almanac 1996, Statistical Abstract of the U.S. 1995, Statistical Yearbook 1995, Population Reference Bureau 1996, World Factbook 1995, Statesman's Yearbook 1995–1996.*

compared with other nations has been declining since World War II because of more rapid economic growth in some other countries.

The development of such a high degree of wealth and power in North America is a matter too complicated to be fully explained in the context of an introductory text and is a phenomenon subject to various interpretations. But it is clear that the United States and Canada possess, or have possessed, a number of specific assets on which they have been able to capitalize. Some of the more important ones include:

1. *Size.* Both countries are large in area and are among the five largest in the world.

2. *Internal unity.* Both possess a comparatively effective internal unity, although Canada has had some difficult problems in this respect in recent decades as Quebec continues to attempt to withdraw from Canada and gain independence.

There have also been major negotiations in both countries over Native American demands for expanded land rights and the right to their own self-governing territories.

3. *Resource wealth.* Both are outstandingly rich in natural resources and have well-developed capacities for making use of all levels of resource wealth.

4. *Size of populations.* The combined population of both countries is large, yet neither country is overpopulated—although there are images of urban densities and poverty, especially in the United States, that suggest possible problems.

5. *Role of technology.* Both countries developed mechanized economies rather early and under very favorable conditions. Recent innovations in robot technology have led to a continued replacement of human labor with machine labor in, for example, automobile manufacturing in both Canada and the United States. It is likely that such patterns of labor shifting

North America's Share of the World's Known Wealth and Production (Continued)

Item	Approximate Percent of World Total		
	U.S.	Canada	North America
Published proved oil reserves (tar sands and oil shale)	2	1	3
Crude oil production	12	2	14
Crude oil refinery capacity	20	3	23
Natural gas production	24	5	29
Natural gas reserves	4	3	7
Potential iron ore reserves	5	6	11
Iron ore mined	6	4	10
Pig iron produced	6	2	8
Steel produced	12	2	14
Aluminum produced	19	7	26
Copper reserves	17	5	22
Copper mined	19	8	27
Lead reserves	22	13	35
Lead mined	12	10	22
Zinc reserves	13	15	28
Zinc mined	7	18	25
Nickel mined	1	21	22
Gold mined	14	8	22
Phosphate rock mined	27	0	27
Potash produced	6	23	29
Native sulphur produced	21	0	21
Asbestos mined	1	18	19
Uranium reserves	9	12	21
Uranium produced	15	29	44
Electricity produced	24	6	30
Potential hydropower	4	3	7
Production of hydropower	17	16	33
Nuclear electricity produced	31	4	35
Sulfuric acid produced	26	3	29

will continue, even though labor unions do what they can to slow down the pace of such replacement because of the associated loss of employment.

6. *Neighborly relations.* Despite occasional irritations and quarrels, the relations between these North American nations are generally friendly and cooperative.

7. *Geographical diversity.* A remarkable range of environmental settings has both allowed and stimulated full use of human ingenuity in economic development.

The combination of these influences creates a regional base that has a rich history of economic development and relative abundance for its populations. This is not to say that North America is not without significant problems of poverty and imbalance in the distribution of the products of this abundance, but taken as a whole, North America is a region of continued strong potential for "the good life."

21.4 The Beneficial Resource Wealth of North America

The natural resources of the North American continent are outstanding in variety and abundance. In part, this simply reflects their large territorial base. But it is also probable that no other comparable area on Earth contains so much natural wealth. This abundance is largely due to the inventiveness of the region's peoples in the exploitation of nature—in the discovery and utilization of resources. However, the value of North American resources is also the product of scientific and technological advances in other lands. Coal, for example, became a major resource following discoveries in England before the superior coal resources of North America were known. Certain important factors affecting this region's world position in resources and production are summarized in Table 21.3.

Definitions and Insights
RESOURCES

One of the most important definitions you will learn in this text is: **"Resources are cultural appraisals."** This means that what we call "natural resources" are—with the exception of air, water, and space—actually earth products we define as having value for us. For example, before it was seen to be an energy source, coal was a troublesome rock because it broke up too easily to be useful for building. **Brea**—the term used in Southern California for the tarlike substance that used to seep to the surface in oil-rich lands in the Los Angeles basin—was the reason some land was relatively cheap in the first Los Angeles land boom of the 1880s. Brea lands were the cheapest land you could buy then. Such lands now, of course, are among the most expensive because of petroleum's resource value to us now. Brea has not changed, but society has appraised that sticky, black product—petroleum—to be a resource of considerable value.

Human appraisals change because of shifts in technology, in societal desires, and in availability. Coal, for example, is diminishing in its value as a resource not because its characteristics are changing, but rather because—as a society—we currently place a higher value on clean air. Since burning natural gas is less harmful to our atmosphere, the value of natural gas is increasing while the resource value of coal is diminishing. The societal forces that lead to changing human appraisals are difficult to predict but are continually at work. Every time you go to a flea market or a garage sale, for example, you are likely to see things that you may consider to be outdated junk; yet someone else has appraised those old 33-rpm records or that tube radio as having value. Resources are human appraisals and therein lies much of the dynamic of the economic and geographic tensions that define our current world.

FIGURE 21.6
The photo shows Douglas fir logs being loaded in the United States' Pacific Northwest for shipment to Japan. Sawmills and pulp-and-paper mills in Japan depend heavily on North American wood. This is only one instance of the dependence of Japan's industrial economy on U.S. and Canadian raw materials.   *Don Lamont/Matrix*

AGRICULTURAL AND FOREST RESOURCES

The United States probably ranks first among the world's nations in arable land of high quality. A much smaller proportion of Canada is arable, but it still has a total of more arable acreage than all but a handful of countries. Such abundance has helped North America become the main food-exporting region in the world. Canada and the United States also have abundant forest resources. Canadian forests are larger in extent than those of the United States but are not so varied. For example, Canada has no forests comparable to the redwood stands of California or the fast-growing pine forests of southeastern United States. Wood has been an abundant material in the development of both coun-

tries. Even now, the United States cuts more wood each year than any other country in the world. It both exports and imports large quantities, with the exports from private lands going largely to Japan (Fig. 21.6) and the imports coming overwhelmingly from Canada. Canada's timber industry is considerably smaller, although it is still one of the greatest in the world; the country's small population compared to the size of its forest resources allows it to be the world's greatest exporter of wood.

MINERAL RESOURCES

North America is also outstandingly rich in mineral resources. Such resources played a major role in the development of the region's economy, and mineral output is still huge. Canada is a major exporter of mineral products, largely to the United States. But the United States, although still a major producer, has also become a major importer. This situation reflects the partial depletion of some American resources, combined with growing foreign production. It also reflects the enormous demands and buying capacity of the U.S. economy.

Abundant energy resources have been basic to the rise of the North American economy and remain crucial to its operation and expansion. Coal was the largest source of energy in the 19th-century industrialization of the United States. Today, the country ranks with China and Russia as one of the three world leaders in magnitude of estimated coal reserves. In coal production, China and the United States are the world leaders, each producing about one fourth of the world output. The United States is the world's largest coal exporter by a wide margin. Canada is also an important coal exporter, with reserves and production that are small on a world scale but large in terms of Canada's needs.

Petroleum fueled much of the development of the United States in the 20th century, and for decades the country was the world's greatest producer, with reserves so large that the question of long-term adequacy received little thought. In 1994, the United States was still the largest producer of oil in the world (followed by Saudi Arabia and Russia), supplying approximately one eighth of world output. Even with that production level, the United States still imports approximately just under half of its own domestic needs. Canada's reserves and production were and are small in comparison but sufficient to be the basis of a continuing energy boom in its western province of Alberta.

Natural gas became a fuel of rapidly growing significance in the later 20th century. In 1994, the United States was the world's largest consumer of gas (ahead of Russia by a small margin) and was a major importer (from Mexico and Canada). It ranked second in the world in output (behind Russia, but far ahead of any other country), producing one quarter of the world's gas from 4 percent of known world reserves. Canada, with smaller but rapidly increasing proven reserves, produced about 5 percent of world output in 1994 and was a major exporter.

Potential water power in the United States and Canada is small on a world scale (see Table 21.3), but development has been so intense that the two countries rank with Russia as the world leaders in hydropower capacity and output (Fig. 21.7). Canada generates approximately 62 percent of its electricity from hydropower; the United States generates 9 percent. At the same time, nuclear power can draw on the large uranium resources in each country and is an important secondary source of electricity in both.

Nonfuel mineral resources in North America present a similar picture of abundance. Long exploitation on massive scales has somewhat depleted the American reserves, especially of the ores easiest to mine and richest in content, and even large mineral outputs within the United States are insufficient to supply the huge demands of the domestic economy. In many such cases, Canada has been able to assume the role of principal foreign supplier. A major example is iron ore. The United States originally had huge deposits of high-grade ore, principally in Minnesota's Mesabi Range and in lesser ore formations near western Lake Superior, but by the end of World War II this ore had been seriously depleted. New sources of high-grade ore then were developed in Quebec and Labrador, principally to supply the continuing American demand for ore of this quality. Meanwhile, the Lake Superior region somewhat revived its mining industry by extracting lower-grade ore, which is present in vast quantities but must be processed in concentrating plants before it can be shipped at a profit to iron and steel plants.

As Table 21.3 shows, there is major production of many other nonfuel minerals by both the United States and Canada. In fact, the mineral wealth of North America makes it easier to summarize gaps in the resource array than to describe the resources that are present. Major minerals in which both countries appear to be truly deficient include chromite, tin, diamonds, high-grade bauxite, and high-grade manganese ore.

FIGURE 21.7
North America's large-scale development of hydropower is symbolized by this photo of Glen Canyon Dam and Lake Powell, situated on the Colorado River between the states of Arizona (right) and Utah (left). Note the power station at the foot of the dam, the extreme desert environment, the escarpment in the background—separating two levels of the Colorado Plateau—and the mountains rising above the general surface of the plateau in the distance.
William Campbell/Time Magazine

21.5   Population Advantages

From the beginning of settlement until the 1960s and 1970s, North America was a region of rapidly growing population. By 1800, approximately two centuries after the first settlements, there were more than five million people in the United States and several hundred thousand in Canada. Growth since that time is summarized in Table 21.4. The population of this region long increased at a rate much more rapid than that of the world as a whole. As a result, one element in the mounting importance and power of the United States, where most of this growth occurred, was its possession of an increasing share of the world's population. By the 1980s, however, the population growth rates of both the United States and Canada were well below those of most of the world, except for Europe, Japan, and the former Soviet Union. Together, the United States and Canada now account for about 5 percent of the estimated world population and about 14 percent of the world's land area (excluding Antarctica). Most of North America's 295 million people (as of 1996) live in the eastern half of the region, from the St. Lawrence Lowlands and Great Lakes south and east. The estimated average population density of the two countries in 1996 was only 73 per square mile (28 per sq km) for the United States and 8 per square mile (3 per sq km) for Canada, with its tremendous expanse of sparsely settled northern lands. And by world standards, even the United States is far from overpopulated (see Fig. 21.1).

Definitions and Insights

POPULATION STATISTICS

One of the constants in the language of geography is the discussion of **population statistics.** These numbers are used to identify **density, growth, mortality, migration, location,** and **patterns of change.** There is value in these statistical indices, but it is important also to see the shortcomings in such data. Population density, for example, is the statistical product of dividing an area by the number of people living within that area. The United States has a population density of 73 people per square mile, or 28 per square kilometer. The Maldive Islands have a population density of 2000 per square mile, or 772 per square kilometer. And the overall population density of Denmark is 313 people per square mile (121 per sq km) while that of Mongolia is only 5 per square mile (2 per sq km). What do the differences in these numbers teach us? The answer is nothing, unless we also know something about the nature of urbanization, the environmental setting, and the potential for effective agriculture in each place. When you see a table that compares population densities among countries in a region—such as we have in this text—you will need to consider such additional information in order to understand the nature, significance, and utility of population statistics.

21.6   Unity within the United States and Canada

The welding of their large national territories into effectively functioning units represents a major accomplishment by the United States and Canada and a major source of their wealth and power. Aside from separatist aspirations in French Canada—which came very close to pulling away from Canada in late 1995—the two countries are not weakened by chronic ethnic/political separatism to the same degree as many other countries. Traditionally, most ethnic minorities in North America have been relatively small and/or geographically scattered, and they have been strongly Americanized or Canadianized. Since both nations have grown from patterns of immigration, they have developed a general openness to accommodation, assimilation, and acculturation (Fig. 21.8).

However, this openness has not precluded dissension among population groups. The original American Indian population was overwhelmed by European settlement and disease and thus was reduced to a politically weak and economically underprivileged minority segregated mainly in western and northern areas. And both Canada and the United States have one exceptionally large ethnic minority which has been the focus of serious problems of national unity. For example, about 26 percent of Canada's people are French in language and culture. These French Canadians are concentrated mostly in the lowlands along the St. Lawrence River, where they form the majority in the province of Quebec. Their ancestors in Quebec, some 60,000 in number, were left in British hands when France was expelled from Canada in 1763. They did not join the English-speaking colonists of the Atlantic seaboard in the American Revolution, and by their refusal to do so they laid the basis for the division of North America into two separate countries. Until the late 20th century, they increased rapidly in numbers, and they have clung tenaciously to their distinctive language and culture.

At times, major controversies arise between this group and other Canadians, and a provincial government with separatist aims was elected by Quebec's voters in the 1970s. However, the Quebec electorate then refused to approve its government's movement toward secession from Canada, and this possibility temporarily receded, only to flare up with renewed heat in the late 1980s. The Canadian government has attempted to preserve national unity through the recognition of French as a second official language and the provision of legal protection for French-Canadian institutions. In October 1995, the Québécois, by popular referendum, voted to remain a province within Canada. The vote, however, was 51–49 percent, so that the issue of Quebec separatism is still very much alive in Canada and will likely stay on the national agenda for some time to come.

Black people of African descent, one eighth of the United States population, are the principal minority in that country. Unlike the French Canadians, they are not largely confined spatially to only one part of the country, although they do predominate in a number of large cities, and particularly in the center of

TABLE 21.4	**Population Growth in the United States and Canada, 1850–1996**				
Year	U.S. Population (millions)	Canadian Population (millions)	Total (millions)	Total Increment (millions)	Total Increment (%)
1850	23.3	2.4	25.7	—	—
1900[a]	76.1	5.4	81.5	55.8	217
1950[a]	151.1	14.0	165.1	83.6	103
1960[a]	179.3	18.2	197.5	32.4	20
1970[a]	203.2	21.6	224.8	27.3	14
1980[a]	226.5	24.3	250.8	26.0	12
1996[a]	265.2	30.0	295.2	44.4	17

[a]Actual dates for Canada are 1901, 1951, 1961, 1971, 1981. The decennial census is taken 1 year later in Canada than in the United States. The table uses census data except for 1996 midyear estimates.

those cities. Historically, black Americans as a group have not been an active force toward sectionalism in the United States. Nevertheless, a conflict of attitudes toward black slavery was among the important factors that nearly split the United States into two nations in the American Civil War, and race relations are still a major source of friction in American politics. The primary setting for most of the race relations difficulties continues to be urban America. The disparity of levels of wealth and well-being is nowhere more visually and culturally evident than in

urban centers. There is a segment of the African-American population that has moved into the middle class and above, but a larger segment resides in the largely impoverished and troubled central cities, where problems in housing, jobs, crime, and racial isolation continue to tear at a traditional American sense of unity, or at least the hopes for a community of peoples within the diversity of the country.

Spanish-speaking people are a smaller ethnic minority in the United States than are African Americans, but their growth

FIGURE 21.8
In San Francisco—as in many of the major urban centers in North America—a racial sharing of public space has occurred in recreation and in work. This scene from a city park is instructive of such casual interaction among differing population groups. *C.L. Salter*

Regional Perspective

MOBILITY AND RECREATION IN NORTH AMERICA

One of the most dynamic regional images of North American culture is that of vast highway networks, motels and roadside commercial strips, and an abundance of motor homes, recreational vehicles, and cars loaded with children and gear—especially in the summer months. This pattern of seemingly continual motion has led to the emergence of abundant car rental facilities, expanding networks of interstate and limited-access highways, hurried weekend trips to not-so-nearby parks or second homes and cabins, and increasing expansion of government-maintained parks and recreational facilities. Large-scale Disney World kinds of recre-ation destinations are especially popular, just as is the network in both Canada and the United States of magnificent national parks created around landscape features of stunning beauty. But there is also a growing world of smaller regional, even municipal, green space that holds retail development at bay and provides environmental and recreational amenities for generally modest entrance fees and a generally accepted tax burden. These landscapes of leisure and the pattern of steadily expanding mobility that increases their accessibility are major components of the important culture changes of the past three decades. The construction of new stadiums, aquariums, convention cen-ters, riverboat casinos, auto racing tracks, golf courses, spas, artificial multiuse lakes, and theater complexes of all sizes, combined with a general governmental inclination to support such projects with tax breaks, demonstrates the economic benefits of such mobility and recreation. Add to those features the plethora of fast-food restaurants that explode upon the scene as soon as a new highway interchange is opened, and you can see the impact of our enormous fascination with staying in motion and spending considerable money on recreation, travel, and convenience.

rate is more rapid. A little more than 9 percent of the United States population is Hispanic. The majority are of Mexican origin and are strongly concentrated in California, Texas, New Mexico, Arizona, and Colorado. Other important Hispanic groups include people of Puerto Rican origin—most heavily represented in New York City and in the northeast—and people of Cuban origin, who live largely in southern Florida. Estimates vary as to how many million other Hispanic people, mainly Mexicans and Central Americans, are actually resident in the United States as illegal immigrants. Like African Americans, American Hispanics are not numerically or politically dominant in any state. They are, however, gaining in relative proportion in many of the large urban centers, especially in the southwest. Nationally, Asian Americans equal nearly 3 percent of the U.S. population. While initial settlement was concentrated in the distinctive "Chinatown" landscapes that so boldly gave this population (whether Chinese or not) its image in the 19th and early 20th century, Asian Americans have been steadily moving out of the central city and into white and mixed neighborhoods. The closest approach to minority dominance of a state is in multiethnic Hawaii, where Asian Americans have become very powerful politically (although without any hint of separatism). Among the many Asian-American groups in Hawaii, Japanese Americans are by far the most numerous. In the United States as a whole in the recent decades, there have been large increases in the Filipino, Korean, Viet-namese, and Cambodian populations, thereby further expanding the characteristics of Asian Americans in the North American landscape.

Even apart from ethnic antagonisms, however, regional conflicts are bound to occur in countries so large and varied. In both countries, democratic governments using federal systems have been able to mitigate such conflicts somewhat. Governmental responsiveness to majority opinion, combined with safeguards for minority and individual rights, has tended to prevent revolutionary pressures from building up. Division of powers between a central federal government and the 50 states or 10 Canadian provinces gives latitude for governmental expression of regional differences and for local solution of regional problems, thus preventing such differences and problems from becoming nationally disruptive. The effectiveness of their democratic federal systems in reducing political pressures has been reinforced by the two countries' relative prosperity, which has tended to allay political discontent. The result is that North America has been characterized thus far by great stability of government. The American Civil War is the only large-scale violent civil conflict that has ever occurred in either country, and neither government has ever been overthrown by force. Thus, aside from the Civil War and the recent rise of Quebec separatism, the two governments have not been obliged to devote large energies to preserving the state's existence or territorial integrity against internal stresses.

FIGURE 21.9
Transcontinental railroads have played an enormously important role in the development of the United States and Canada, and they continue to do so today. The photo shows containers being double-stacked on specially designed low-slung rail cars. They are often shipped for thousands of miles to and from seaports.
Michael L. Abramson

The governments of both countries have assiduously fostered national unity through the provision of adequate internal transportation. In both, transportation networks have had to be built not only over long distances but primarily against the "grain" of the land. Effective transport links between east and west have been the most imperative, but most of the mountains and valleys have a north-south trend, which has presented a series of obstacles to cross-country transport. The coasts of both countries were first tied together effectively by heavily subsidized transcontinental railroads (Fig. 21.9) completed during the latter half of the 19th century. Subsequently, national highway and air networks further enhanced national unity (Fig. 21.10).

FIGURE 21.10
One of the prominent landscape signatures of North America is a busy freeway. This photo of a Los Angeles freeway at dusk illustrates the fascination "Angelenos" have with auto-mobility. While this image perhaps makes sense at the rush hour, it is instructive to know that such traffic density in Southern California is common for nearly 15 hours a day. This dependence on the automobile and the independence it allows—even though it means very crowded roadways—is a major characteristic of urban North America. *Scott Robinson/ Tony Stone Images*

21.7 Mechanization and Productivity

The high productivity and national incomes of North American countries have come about essentially through the use of machines and mechanical energy on a lavish and ever-increasing scale during the past century and a half. From an early reliance on waterwheels to power simple machines, the United States increasingly exploited the power generated from coal by steam engines. Later, the total power used to drive machines increased enormously through use of oil, internal-combustion engines, and electricity. In achieving this unusually energy-abundant and mechanized economy, the United States was able to take advantage of a unique set of circumstances. For one thing, it had resources so abundant and varied that they attracted foreign capital and also made possible domestic accumulation of capital through large-scale and often wasteful exploitation. There was also a labor shortage, which attracted millions of immigrants as temporarily low-wage workers but also promoted higher wages and labor-saving mechanization in the long run. A relatively free and fluid society encouraged striving for advancement; large segments of the population subscribed with an almost religious fervor to ideals of hard work and economic success. Seldom did national energies have to be diverted excessively into defense and war.

Most of the foregoing American conditions were duplicated in Canada, and some of them in a number of other countries, but one American advantage was not: the existence in the United States of a large, unified, and growing internal market that allowed great economic organizations to specialize in the mechanized mass production of a few items, thus lowering the unit costs of production and further expanding the size both of the producing firms and of the market. In recent decades, however, this enormous American asset has been significantly eroded, along with some others, by the enormous expansion of the global market. Transportation and communication have become so cheap and rapid, and many artificial trade barriers have been so reduced, that the world market rather than the domestic market has become the essential one for major industries.

21.8 Cooperative Relations Between the United States and Canada

For many years after the American Revolution, the political division of North America between a group of British colonial possessions to the north and the independent United States to the south was accompanied by serious friction between the peoples and governments on either side of the boundary. This heritage of antagonism was a result of (1) the failure of the northern colonies to join the Revolution; (2) the use of those colonies as British bases during that war; (3) the large proportion of the northern population composed of Tory stock driven from U.S. homes during the Revolution; and (4) uncertainty and rivalry concerning who had ultimate control of the central and western reaches of the continent. The War of 1812 was fought largely as a United States effort to conquer Canada, though this intent was not specified at the time. Even after its failure, a series of border disputes occurred, and U.S. ambitions to possess this remaining British territory in North America were openly expressed; suggestions and threats of annexation were made in official quarters throughout the 19th century and into the 20th century.

In fact, Canada's emergence as a unified nation is in good part a result of American pressure. After the American Civil War, the military power of the United States took on a threatening aspect in Canadian and British eyes. Suggestions were made in some American quarters that Canadian territory would be a just recompense to the United States for British hostility to the Union during the Civil War. However, the British–North America Act, passed by the British Parliament in 1867, brought an independent Canada into existence by the combining of Nova Scotia, New Brunswick, and Canada into one Dominion under the name of Canada. With the Act, Great Britain sought to establish Canada as an independent nation capable of morally deterring U.S. conquest of the whole of North America. Although hostility between the United States and its northern neighbor did not immediately cease with the establishment of an independent Canada, relations have improved gradually, and the frontier between the two nations has ceased to be a source of insecurity. This frontier, stretching completely unfortified across a continent, has become more a symbol of friendship than of enmity.

The bases of Canadian-American friendship are cultural similarities, the material wealth of both nations, and the mutual need for and advantages of cooperation. A large volume of trade moves across the frontier and strengthens both countries economically and militarily. Each is a vital trading partner of the other, although Canada is much more dependent on the United States in this respect than is the United States on Canada. In 1996, for example, Canada supplied 19 percent of all United States imports by value and took 22 percent of all United States exports, making Canada the leading country in total trade with the United States. On the other hand, in 1996 the United States supplied 65 percent of Canada's imports and took 81 percent of its exports. The United States also supplies large quantities of capital to Canada, which has been an important factor in Canada's rapid economic development during recent decades but which also has been a frequent source of Canadian disquiet because of the amount of U.S. control over the Canadian economy it represents. However, Canadians in turn invest heavily in the United States. Except for Canadian exports of automobiles and auto parts to the United States, the main pattern of trade between the two countries is the exchange of Canadian raw and intermediate-state materials—primarily ores and metals, timber and newsprint, oil and natural gas—for American manufacturing. A free-trade pact to unite the two in one market was signed in 1988. In 1992, Canada joined the United States and Mexico in signing the North American Free Trade Agreement (NAFTA) that was enacted in 1994.

22

NATIONAL AND REGIONAL
CHARACTERISTICS OF CANADA

Writer Michael Ondaatje's novel about European migrants coming to Canada to find a new life in the early decades of the 20th century provides a window on the difficulties—as well as some of the humor—of this migration and transition. The struggles of adjusting to a new home, learning a new language, and finding an adequate job are all illustrated in these images of Nicholas on his move from the Balkans to Toronto.

Nicholas was twenty-five years old when war in the Balkans began. After his village was burned he left with three friends on horseback. They rode one day and a whole night and another day down to Trikala, carrying food and a sack of clothes. Then they jumped on a train that was bound for Athens. Nicholas had a fever, he was delirious, needing air in the thick smoky compartments, waiting to climb up on the roof. In Greece they bribed the captain of a boat a napoleon each to carry them over to Trieste. By now they all had fevers. They slept in the basement of a deserted factory, doing nothing, just trying to keep warm. . . .

Two of Nicholas' friends died on the trip. An Italian showed him how to drink blood in the animal pens to keep strong. It was a French boat called *La Siciliana*. He still remembered the name, remembered landing in Saint John and everyone thinking how primitive it looked. How primitive Canada was. They had to walk half a mile to the station where they were to be examined. They took whatever they needed from the sacks of the two who had died and walked towards Canada.

Their boat had been so filthy they were covered with lice. The steerage passengers put down their baggage by the outdoor taps near the toilets. They stripped naked and stood in front of their partners as if looking into a mirror. They began to remove the lice from each other and washed the dirt off with cold water and a cloth, working down the body. It was late November. They put on their clothes and went into the Customs sheds.

Nicholas had no passport, he could not speak a word of English. He had ten napoleons, which he showed them to explain he wouldn't be dependent. They let him through. He was in Upper America.

He took a train for Toronto, where there were many from his village; he would not be among strangers. But there was no work. So he took a train north to Copper Cliff, near Sudbury, and worked there in a Macedonian bakery. He was paid seven dollars a month with food and sleeping quarters. After six months he went to Sault Ste. Marie. He still could hardly speak English and decided to go to school, working nights in another Macedonian bakery. If he did not learn the language he would be lost.

The school was free. The children in the class were ten years old and he was twenty-six. He used to get up at two in the morning and make dough and bake until 8:30. At nine he would go to school. The teachers were all young ladies and were very good

---

◄ The landscapes of farming and particularly wheat cropping are of major importance in the heart of Canada. This scene from Thunder Bay on the northwest shore of Lake Superior shows the grain silos that are so common a signature along the rail lines and in large shipping facilities. Thunder Bay is an international port because of the connection to the Atlantic provided by the St. Lawrence Seaway and it is from here that a major portion of Canada's large annual grain export is launched.

Thomas Kitchin/Tom Stack & Assoc.

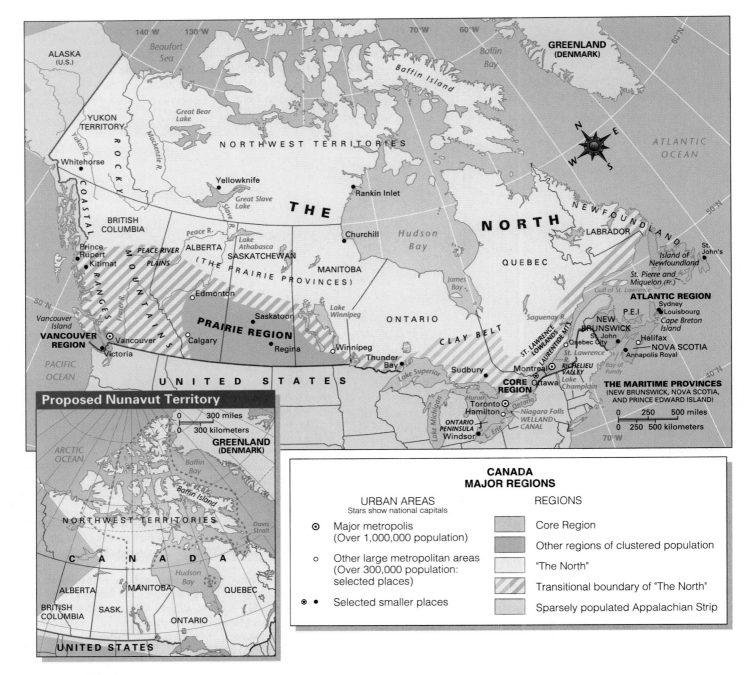

FIGURE 22.1

Major regions of Canada. In 1991 the Canadian government undertook to create a new Inuit (Eskimo)-controlled Nunavut Territory of land taken from the east and north of the Northwest Territories. In May 1993, Canada's prime minister signed the Nunavut Agreement, setting in motion the final steps to create Nunavut (Inuit: "our land") as a separate territory. The final transfer will take place in 1999.

people. During this time in Sault he had translation dreams—because of his fast and obsessive studying of English. In the dreams trees changed not just their names but their looks and characters. Men started answering in falsettos. Dogs spoke out fast to him as they passed him on the street.

When he returned to Toronto all he needed was a voice for all this language. Most immigrants learned their English from recorded songs or, until the talkies came, through mimicking actors on stage. It was a common habit to select one actor and follow him throughout his career, annoyed when he was given a small part, and seeing each of his plays as often as possible—sometimes as often as ten times during a run. Usually by the end of an east-end production at the Fox or Parrot Theaters the actors' speeches would be followed by growing echoes as Macedonians, Finns, and Greeks repeated the phrases after a half-second pause, trying to get the pronunciation right.

This infuriated the actors, especially when a line such as "Who put the stove in the living room, Kristin?"—which had originally brought the house down—was spoken simultaneously by at least seventy people and so tended to lose its spontaneity. When the matinee idol Wayne Burnett dropped dead during a performance, a Sicilian butcher took over, knowing his lines and his blocking meticulously, and money did not have to be refunded.

Certain actors were popular because they spoke slowly. Lethargic ballads, and a kind of blues where the first line of a verse is repeated three times, were in great demand. Sojourners walked out of their accent into regional American voices.[1]

As you learn of Canada and North America, keep in mind the demands of human migration, immigration, and being an outsider trying to find a new home and future in a world very different from one's own origins. Not only has this dynamic been important for the cities of Canada, but it also has shaped much of the urban, town, and even rural growth of North America.

Canada is a highly developed nation—affluent, industrialized, technologically advanced, and urbanized. But in some ways it is curiously unlike most developed countries. These differences are closely related to Canada's internal geography and location adjacent to the United States.

22.1   Canada: Highly Developed and Economically Unique

Canada's prosperity is derived in considerable part from an industrial output sufficiently great to place the country among the world's top 10 or 11 manufacturing nations. Products in which it ranks exceptionally high include newsprint, hydroelectricity, commercial motor vehicles, and aluminum. The country is very urbanized (Fig. 22.1), with 77 percent of its population classed as urban in 1996 (compared to 75 percent in the United States) and with about three tenths of its people living in the three largest metropolitan areas: Toronto (population: 3.9 million); Montreal (population: 3.1 million); and Vancouver (population: 1.6 million). These dynamic metropolises convey a positive image of an advanced and prosperous country. Impressively good national statistics on health, housing, and education provide further evidence of the same. However, the enormous clustering of the Canadian population up against its southern border further demonstrates the country's bittersweet relationship with its neighbor, the United States (see Fig. 21.2).

Along with its prosperity, Canada exhibits some striking departures from the expected patterns of developed countries. One

From Michael Ondaatje, *In the Skin of a Lion* (New York: Penguin Books, 1987): 45-47.

of these is an unusual pattern of trade. Despite a fair-sized export market for automobiles and parts and some other fabricated items, Canada is mainly an exporter of raw or semifinished materials, energy, and agricultural products, and is mainly an importer of manufactured goods and speciality food items. By contrast, most developed nations export mainly manufactured goods and import a mixture of manufactured and primary products.

A second departure from other developed nations is Canada's overwhelming dependence on trade with a single partner, the United States. In 1997, the U.S. took 80 percent of Canada's exports and supplied 67 percent of Canada's imports. No other developed nation comes near these percentages of trade with just one other country.

Still another major divergence from other developed countries lies in the degree to which the Canadian economy is financed and controlled from outside the country, and in this Canada again has an overwhelming dependence on the United States. Somewhat more than three quarters of all foreign investment in Canada is American. The extent of United States economic dominance in Canada has long been a sore point with many Canadians, who resent imputations that their country is an American economic "colony." Such irritations tend to be exacerbated by the degree to which American mass-produced culture has permeated Canada and the degree to which Canada is taken for granted in the United States. Few U.S. schools offer courses on Canada, and a high percentage of American students cannot even name Canada's capital. American news media give relatively scanty coverage to Canada.

There has been for some time a movement in Canada to achieve a more predominant position for Canadians within their own economy, to strengthen economic relations with industrial countries other than the U.S., and to heighten cultural self-determination. But there seems to be no way for Canada to pull away from the United States without risking its own prosperity. The country's economy is geared to a close interchange with the gigantic economy next door. In 1988, the two countries moved even closer when they signed a free-trade pact to abolish all tariffs on trade between them and to remove or lower many other restraints on trade. In 1994 this arrangement was expanded when the United States, Canada, and Mexico initiated the North American Free Trade Agreement (NAFTA), but in a process that was not without dissent (see Definitions and Insights, Chapter 20, page 485, and Fig. 22. 2).

22.2   Canada's Regional Structure of Diverse Landscapes

In addition to the problem of maintaining satisfactory relationships with the United States, Canada also has problems posed by its own regional structure. The country is divided between a thinly settled northern wilderness that occupies most of Canada's area and a narrow and discontinuous band of more

FIGURE 22.2
Street signs, business names, and traffic signs in French-speaking Canada reflect the duality of language in this region of early French settlement. Although Quebec voted not to secede by a very slim margin in 1995, the potential for a positive vote is always on the minds of the people of Quebec, and of the larger Canada as well. This is Rue Saint Denis in Montreal.   *Thomas Kitchin/Tom Stack & Assoc.*

large parts extend into seven of the ten provinces that stretch across southern Canada and make up the major political subdivisions of Canada's federal system. Even within the provinces, most of the land is lightly inhabited, and nearly all the people are concentrated in limited areas of relatively close and continuous settlement, separated by wedges of thinly populated terrain. These population clusters lie just north of the country's boundary with the 48 contiguous United States. An estimated 75 to 80 percent of Canada's people live within 100 miles (161 km) of the border. This has facilitated the development of strong ties between Canadian regions and adjoining American regions.

Canada's population clusters can be grouped into four main regions: (1) an *Atlantic region* of peninsulas and islands at Canada's eastern edge; (2) a culturally divided *core region* of maximum population and development along the lower Great Lakes and St. Lawrence River in Ontario and Quebec provinces; (3) a *Prairie region* in the interior plains between the Canadian Shield on the east and the Rocky Mountains on the west; and (4) a *Vancouver region* on or near the Pacific coast at Canada's southwestern corner. Each region is marked by a distinctive physical environment and cultural landscape (see photograph, page 538); each has a different ethnic mix and distinctive economic emphases. Each plays a different role within Canada and has its own set of relationships with the United States and other countries. All the regions have important links with the Canadian North. (For statistical data on Canadian areas, see Table 21.1.)

22.3   Atlantic Canada: Physically Fragmented, Ethnically Mixed, Economically Struggling

The easternmost of Canada's four main populated areas is characterized by a series of strips or nodes of population, largely along or very near the seacoast, in the provinces of Newfoundland, New Brunswick, Nova Scotia, and Prince Edward Island. The four provinces are commonly called the Atlantic Provinces, and three, excluding Newfoundland, have long been known as the Maritime Provinces. Their main populated areas are separated from the far more populous Canadian core region in Quebec and Ontario by areas of mountainous wilderness. Prince Edward Island, which is Canada's smallest but most densely populated province (Table 22.1), is a lowland occupied largely by a continuous mat of farms and small settlements, but the main populated areas of New Brunswick, Nova Scotia, and Newfoundland occupy valleys and coastal strips within a frame of lightly settled upland. As a province, Newfoundland includes the dependent territory of Labrador on the mainland, but in common usage the name Newfoundland is restricted to the large island opposite the mouth of the St. Lawrence. Formerly a self-governing Dominion, it was the last province to join

populous regions stretching from ocean to ocean in the extreme south (see Fig. 21.2). These southern regions are sharply different from each other, and are often in political conflict with each other and with the federal government at Ottawa.

About four fifths of Canada has cold tundra and subarctic climates; more than one half is in the glacially scoured and agriculturally sterile Canadian Shield (Fig. 22.3), and the part near the Pacific is dominated by rugged mountains. Such conditions create the huge expanse of nearly empty land that Canadians call "the North." More than one half of it lies in territories administered from the country's capital in Ottawa, Ontario, but

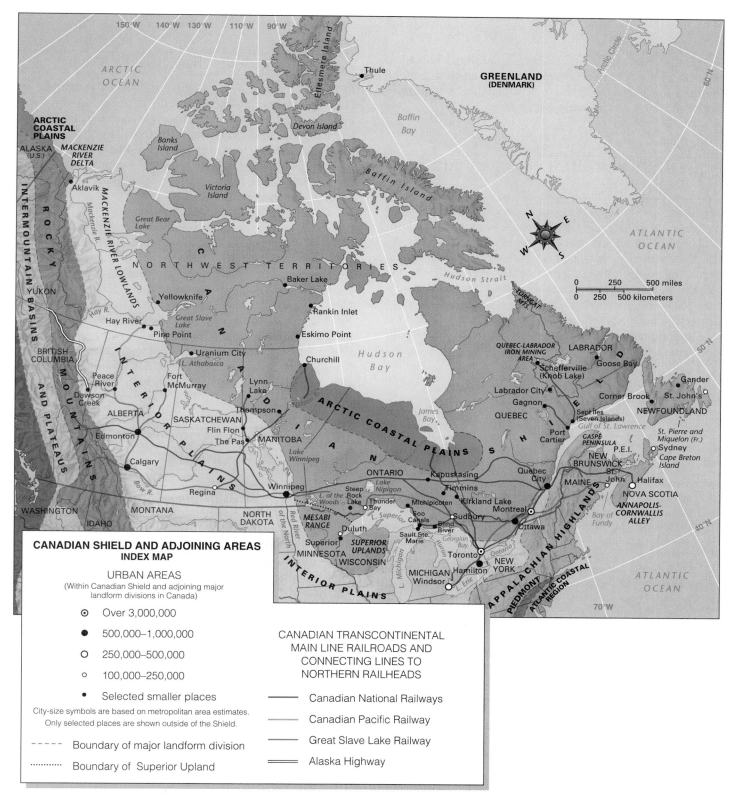

CANADIAN SHIELD AND ADJOINING AREAS
INDEX MAP

URBAN AREAS
(Within Canadian Shield and adjoining major landform divisions in Canada)

⊙ Over 3,000,000

● 500,000–1,000,000

○ 250,000–500,000

○ 100,000–250,000

● Selected smaller places

City-size symbols are based on metropolitan area estimates. Only selected places are shown outside of the Shield.

- - - - - Boundary of major landform division

·············· Boundary of Superior Upland

CANADIAN TRANSCONTINENTAL
MAIN LINE RAILROADS AND
CONNECTING LINES TO
NORTHERN RAILHEADS

——— Canadian National Railways

——— Canadian Pacific Railway

——— Great Slave Lake Railway

═══ Alaska Highway

FIGURE 22.3

General reference map of the Canadian Shield and adjoining areas. Only certain portions of Canada's Arctic islands share the rock formation of the Canadian Shield. However, the overall physical conditions of the islands are sufficiently similar to conditions in the Shield that it is convenient for the purposes of this text to regard the entire island group as an extension of the Shield.

Canada (in 1949). Bleak and nearly uninhabited Labrador, situated on the Canadian Shield, is very much a part of Canada's North. It is important to Newfoundland for revenues from iron mining and hydroelectric production.

A Centuries-Old Tradition of Fishing

The fishing industry has always been of outstanding importance in Canada's Atlantic region. Before the beginning of settlement in the early 17th century, and quite possibly before the discovery of the Americas by Columbus in 1492, European fishing fleets began operating in the waters—primarily in the "banks"—along and near these shores. The area lies relatively close to Europe, and its waters have always been exceptionally rich in fish, but it has suffered from steady overfishing and the near disappearance of the cod that have been the mainstay of this region's productivity.

Definitions and Insights
FISHING BANKS: GENESIS AND DECLINE

Elevated portions of the sea bottom known as "banks" are located off the Atlantic coast from near Cape Cod to the Grand Banks, which lie just off southeastern Newfoundland. The Grand Banks have been at the heart of the fishing economy in Canada's northeastern region for nearly five centuries. The shallowness of the ocean and the mixing of waters from the cold Labrador Current and the warm Gulf Stream foster a rich development of the tiny organisms called plankton, on which fish feed. Inshore fisheries supplement the catch from the banks. Although many kinds of fish and shellfish are caught, the early fleets fished mainly for cod, which were cured on land before making the trip to European markets. Both the British and the French established early fishing settlements in Newfoundland, but the French—who had 150 vessels fishing these banks in 1577—were driven out by the British in 1763. The island has remained strongly British in population, as its limited development has attracted very few other immigrants.

In 1977 the Canadian government began to enforce a 200-mile offshore jurisdiction which prohibited overt foreign competition in the fishing of the Grand Banks. This held off the decline somewhat, but the greater efficiency of fishing technology propelled this industry toward difficult times. In the 1990s, the overfishing of cod in the banks led to a serious decline in fishing productivity. In Newfoundland this decrease in yield has led to unemployment that runs higher than 20 percent, more than three times the overall Canadian level.

Agricultural Hardships

In general, Atlantic Canada provides hard environments for farming. The four provinces lie at the northern end of the Appalachian Highlands and have topographies dominated by hills and low mountains. Soils that are predominantly mediocre to poor exist under a cover of mixed forest. Many soils that once were farmed wore out quickly and were abandoned. In the 20th century, agriculture has been concentrated on relatively small patches of the best land.

Climate also handicaps agriculture. The Maritime Provinces have a humid continental short-summer climate, while Newfoundland has a subarctic climate. The entire region is humid and windy, with much cloud cover and fog. Strong gales are frequent in the winter. Summers are cool, and colder conditions occur as elevation increases; a good part of upland Newfoundland is tundra. Thus, most areas in the Maritimes and Newfoundland have always been agriculturally marginal, although some more favored lowlands with better soils—notably Prince Edward Island and the Annapolis-Cornwallis Valley of Nova Scotia—are exceptions. Potatoes, dairy products, and apples are major agricultural specialties in the Maritimes.

Geographic Profile of a Declining Region

During the first two thirds of the 19th century, the Maritime Provinces became a relatively prosperous area with a preindustrial economy. Local resources supported this development, including many harbors along indented coastlines, abundant fish for local consumption and export, timber for export and for use in building wooden sailing ships, and land—albeit not very good—for the expansion of an agriculture partly subsistence in character and partly serving local markets. The region carried on an extensive commerce with Great Britain and the West Indies. Then followed an era (still continuing) of relative economic decline within an industrializing Canada. The ports were so far from the developing interior of the continent that most shipping bypassed them in favor of the St. Lawrence ports or the Atlantic ports of the United States. Halifax, Nova Scotia, and St. John, New Brunswick, are the main ports of the region today, and—while each handles considerable traffic, with an emphasis on containers—neither is in a class with Montreal or the main U.S. ports.

In the meantime, fishing, forestry, and agriculture suffered various disabilities. Fishing did not prove to be an adequate basis for a prosperous modern economy (Fig. 22.4), primarily for the following reasons:

1. *Tariff barriers.* Tariff barriers were imposed by the United States against Canadian fish during the competitive efforts of Canada and the United States to preserve their own domestic markets.

TABLE 22.1

Canadian Provinces and Territories: Basic Data

Political Unit	Land Area (thousand/ sq mi)	Land Area (thousand/ sq km)	Estimated Population (thousands)	Estimated Population Density (sq mi)	Estimated Population Density (sq km)	Economic Production (% of Value) (farm receipts)	Economic Production (% of Value) (mineral output)	Economic Production (% of Value) (manufacturing)
Atlantic Provinces								
New Brunswick	27.8	72.1	752.5	26	10	1.34	2.5	2
Nova Scotia	20.4	52.8	928.4	46	18	1.46	1.52	1.69
Prince Edward Island	2.2	5.7	134.7	61	24	1.12	0.01	0.12
Newfoundland-Labrador	143.5	371.7	569.9	4	2	0.28	2.08	0.48
Total	193.9	502.3	2385.3	12	5	4.2	6.11	4.32
Core Provinces								
Quebec	523.9	1356.8	7260.9	14	5	16.16	7.43	25.67
Ontario	344.1	891.2	10,890.3	32	12	21.51	13.5	47.51
Total	867.9	2248.0	18,250.2	21	8	37.57	20.93	73.18
Prairie Provinces								
Manitoba	211.7	548.4	1126.1	5	2	10.34	3.21	2.15
Saskatchewan	220.3	570.7	1005.4	5	2	20.29	8.62	4.83
Alberta	248.8	644.4	2719.5	11	4	19.78	48.23	6.45
Total	680.9	1763.5	4581.1	7	3	51.11	60.06	13.43
Pacific Province								
British Columbia	359.0	929.7	3728.3	10	4	7.11	11	9.07
Territories								
Yukon	184.9	479.0	29.7	0.16	0.06	0.01	1	—
Northwest Territories	1271.4	3293.0	65.1	0.05	0.02	—	2	—
Total	1456.4	3772.0	94.9	0.07	0.02	0.01	3	—
Grand Total, Canada	3850.8	9976.1	28,434.5	8	3	100	100	100

Sources: Britannica Book of the Year 1997, The World Almanac and Book of Facts 1996, Stateman's Yearbook 1995-1996 and The World Factbook 1995.

Note: The dates of these data span from 1992 to 1996 due to varying times of report and acquisition.

2. *Global competition.* Competition from major world meat exporters and newly developed fishing areas in other parts of the world, particularly after the end of World War II, had grown to such size that it began to cripple the Canadian fishing economy.

3. *Foreign competition on the local scene.* The fishing conditions of the Grand Banks and other nearby waters of the Maritime Provinces were so productive that fishing fleets from other nations and regions—many possessing high-tech fishing operations—began to drive Canadian fishermen from their own waters.

4. *Disappearance of the resource.* The combination of expanding numbers of fishermen, more sophisticated and effective technology, and steady overfishing led to a major and continuing decline in the number of fish in this traditionally rich area.

Wood industries also suffered when forests became depleted by overcutting, ships began to be made of iron and steel, and competition set in from new areas of forest exploitation farther west. The position of agriculture in the Maritimes was undermined with the building of railroads, which facilitated

FIGURE 22.4

The fishing villages of Nova Scotia suggest a link with the past that is more a mark of marginal fishing than tourist delight in the contemporary context of Canada's Maritime Provinces. The small craft and family operations of this photo have a very difficult time competing in a world fishing economy that is increasingly high tech in its design and function.
Thomas Kitchin/Tom Stack & Assoc.

settlement of Canada's interior and thus brought cheaper farm products into the Atlantic region. Today, quite a few people in the Maritime Provinces still farm, but they do so largely on a small-scale, part-time basis while earning their living mainly from other employment.

As economic challenges have developed, the Atlantic region has attempted to meet them by developing industry. For example, many small cotton-textile factories were built during the 19th century. But these early industrial ventures largely failed when they were undercut by competition from mills in New England, central Canada, and Europe that were located closer to major markets. Sizable coal reserves existed near the town of Sydney on northern Cape Breton Island in Nova Scotia, and iron ore was present on Newfoundland's Bell Island. On the basis of these resources, a coal-mining industry and an iron-and-steel plant were developed at Sydney which flourished for some time, largely by supplying steel for Canadian railway construction. But eventually both the plant and the coal mines declined, with consequent economic depression in the Sydney area. The plant at Sydney was bought by the provincial government after World War II to forestall its closure; it continues to operate at a reduced scale and on a subsidized basis in order to provide employment. The local coal industry has been handicapped by both the increasing cost of extracting the coal, which is deep-lying, and the shrinking market for coal due to its widespread replacement by oil as a fuel. Some coal mines have closed.

Isolation from markets afflicts almost all major enterprises in the Atlantic region. The market within the region is small and divided into many small clusters of people separated by sparsely populated land. The largest city in the Atlantic provinces—Halifax, Nova Scotia—has a metropolitan population of only about 320,000, and only two other cities—St. John's, Newfoundland, and St. John, New Brunswick—surpass 125,000. The entire area has only 2.4 million people, or considerably less than metropolitan Toronto or Montreal. It is a region of out-migration and is the poorest of the four populated regions of Canada.

Currently, the basic economic activities that support Atlantic Canada's struggling economy are numerous but relatively small:

1. *Specialized agriculture.* Specialized commercial potato farming is somewhat successful in Prince Edward Island and the St. John's River Valley of eastern New Brunswick.
2. *Continued fishing.* Protected from foreign vessels since 1977 by the extension of Canadian control to waters 200 miles (322 km) offshore, commercial fishing continues on a small scale.
3. *Tourism.* Tourism, based on the region's scenic beauty, cool summers, and a sense of splendid isolation, is handicapped by some of those same qualities because of the region's distance from population centers.

4. *Forestry.* Pulp and paper manufacturing are making a modest comeback and undergoing steady growth because of the expansion of the international market for paper and wood products.
5. *Energy.* The export of electricity to adjacent provinces and the United States, especially from Labrador's large Churchill Falls hydropower installation and from New Brunswick, has been a steady factor in the modest expansion of economic health in the region.
6. *Minerals.* Atlantic Canada does possess a few large metal mines. Both coal and iron ore have a traditional importance in the region's mineral resources, but they pale compared to current anticipation of petroleum wealth. Offshore oil and gas exploration in the 1970s and 1980s found the Hibernia Field off southeastern Newfoundland. While it is thought that these fields might possess enormous reserves, the extraction is particularly difficult because of local weather conditions and ocean floor characteristics. Production has only moved ahead since the mid-1980s. This potential source of wealth for a region that has been traditionally the poorest in Canada has led to considerable intranational bickering over the ownership of the petroleum reserves.

Extremely important to the economy of the Atlantic region is a high level of subsidies from the federal government in the form of (1) pension and welfare payments, (2) the stationing of military forces in the region (Halifax is a major naval base), (3) the presence of many federal administrative offices in the region, (4) direct payments to provincial treasuries, and (5) funding for economic development projects and efforts to attract industry.

22.4   Canada's Culturally Divided Core: Ontario and Quebec

The core region of Canada, with about two-thirds of the nation's population, has developed along lakes Erie and Ontario and thence seaward along the St. Lawrence River. In this area, the Interior Plains of North America extend northeastward to the Atlantic. They are increasingly constricted seaward—on the north by the edge of the Canadian Shield and on the south by the Appalachian Highlands. The two provinces that divide the core region—Ontario and Quebec—both incorporate large and little-populated expanses of the Shield; Quebec has a strip of Appalachian country along its border with the United States. But most of the core region lies in lowlands bordering the Great Lakes or the St. Lawrence from the vicinity of Windsor, Ontario, to Quebec City, Quebec. The entire lowland area is often loosely termed the **St. Lawrence Lowlands,** although the Ontario peninsula between lakes Huron, Erie, and Ontario is frequently recognized as a separate section.

CONTRASTS BETWEEN THE ONTARIO AND QUEBEC LOWLANDS

The Ontario and Quebec sections of the core region show marked differences in types of agriculture and cultural features. The Ontario peninsula—strongly British to its roots—echoes the U.S. Midwest's cropbelts on a small scale, with corn and livestock production, dairy farming, and growth of various specialty crops such as tobacco. Along the St. Lawrence in Quebec, the French heritage is clearly evident in the landscape signatures of the French "long-lot" pattern of strip-shaped agricultural landholdings.

Definitions and Insights

LONG-LOTS IN FARMING SETTLEMENTS

One of the most distinctive French landscape signatures in the agricultural world is the **long-lot**. Long, narrow fields lie at right angles to ribbons of dense settlement along roads, where houses and buildings form elongated villages. In earlier times, the settlements closely hugged the banks of the St. Lawrence and some tributaries, but as settlement expanded, the long-lot pattern was repeated inland along roads. (This is a common feature of farmland patterns along the margins of the Mississippi River in Louisiana as well.) Like most landscape signatures and land tenure patterns, this unusual geometry had a very specific function. The French felt that there would be a more stable farming economy if the rural population held a variety of land types rather than just river floodplains or upland flanks. The long-lots also gave each farmer access to the river, and, given the early significance of river transport, this was extremely valuable. For these reasons, the settlement was built around a frontage on the river and good floodplain flat land. The farmland then extended away from the river, onto some slope land on the natural levees of the river, then to land behind the levees, and then to interior land. This long-lot system meant that in times of flood, no farmer lost all of his productive land. All the farmers of a given village suffered from flooding, but they also had higher land which provided a farming base during the periodic flooding. This pattern has traveled with the French to their colonial settlements all around the world, and it generally persists today.

In the Quebec landscape along the St. Lawrence, large Catholic churches are prominent landscape features. The Quebec lowlands lie farther north than the Ontario peninsula and have a harsher winter climate. Dairy farming is the predominant form of agriculture. Together, Ontario and Quebec account

FIGURE 22.5

This photo shows Quebec City, once the capital of France's North American empire and now the capital of Quebec Province. At the right is the Lower Town along the St. Lawrence River; at the left, the Upper Town. Grain elevators symbolize the city's seaport function. The hotel called the Chateau Frontenac with its copper sheathing is a reminder of the city's heritage from Imperial France. *Jesse H. Wheeler, Jr.*

for about two fifths of the value of products marketed from Canadian farms.

The core area contains more than 18 million people. Ontario's population is basically British in origin but has many minority ethnic groups as shown in the literature selection on page 539. By contrast, about four fifths of the people in Quebec are of French origin (Figs. 22.5, 22.6). This represents the biggest exception to the dominance of English culture in North America and is the only instance in which a national minority controls a provincial or state government.

Within the core, life focuses on Canada's two main cities: Toronto in Ontario (population: 4,338,400) and Montreal, the second largest French city in the world, in Quebec (population: 3,328,000). Other large cities of the core include Ottawa (population: 925,000), Quebec City (population: 695,000), and Hamilton (population: 641,000). Ottawa, the federal capital, is located just inside Ontario on the Ottawa River boundary between the two provinces. Quebec City, the original center of French administration in Canada, is the capital of Quebec Province. Hamilton, a Lake Ontario port, is the main center of Canada's steel industry, supplying steel for the automobile and other metal-fabricating industries of Ontario. A good part of the urban development in the core is associated with manufacturing. Three fourths of Canada's total manufacturing development is here, with somewhat more in Ontario than in Quebec. But almost three times as many people are employed in the service industry, reflecting the region's status as the business and political center of the country.

HISTORICAL EVOLUTION

The coreland began as an entry to the interior of North America, and this gateway function continues today. The French founded Quebec City in 1608 at the point where the St. Lawrence estuary leading to the Atlantic narrows sharply. Fortifications on a bold eminence allowed control of the river. From Quebec, fur traders, missionaries, and soldier-explorers soon discovered an extensive network of river and lake routes—with connecting portages—reaching as far as the Great Plains and the Gulf of Mexico. Montreal, founded later on an island in the St. Lawrence River, became the fur trade's forward post toward the interior Indian-dominated wilderness. Between Quebec City and Montreal, a thin line of settlement evolved along the St. Lawrence and formed the agricultural base for the colony. Population grew slowly in this northerly outpost where the winter was harsh and none but French Catholics were welcome. When finally conquered by Britain in 1759, French Canadians numbered only about 60,000.

A few British settlers came to Quebec after the conquest, and this immigration increased rapidly during and after the American Revolution, thus laying the foundations for the British segment of Canada's core. Many of the early English-speaking immigrants were Loyalist refugees from the newly independent United States, whose rebellion French Canada had refused to join. They were soon joined by more English settlers immigrating directly from Europe. These newcomers formed a sizable minority in French Quebec (Fig. 22.6), including a British commercial elite in Montreal that persists to this day, but they also settled west of the French on the Ontario peninsula. By 1791, the British government found it expedient to separate the two cultural areas into different political divisions—known at that time as Lower Canada (Quebec) and Upper Canada (Ontario), taking their directional terms from their positions on the St. Lawrence River.

THE RISE OF COMMERCE, INDUSTRY, AND URBANISM

Until Canada was formed as an independent country in 1867, Quebec and Ontario developed largely in a preindustrial fashion, with economies based primarily on agriculture and forest exploitation. With its larger expanses of good land and a somewhat milder climate, the Ontario peninsula became more agriculturally productive and prosperous than Quebec. It specialized in surplus production and export of wheat. On a superior natural harbor, the Lake Ontario port of Toronto developed and became the colonial (now the provincial) capital and Ontario's largest city. But the bulk of Canada's exports at this time moved eastward to meet ocean shipping, and this movement favored Montreal. Until about 1850, most shipping from the Atlantic went no farther upstream than Quebec City. But subsequent improvements along turbulent sections of the river enabled ocean ships to navigate safely as far inland as the major rapids (Lachine Rapids) at Montreal. Quebec City was bypassed and its growth began to slow, although it has continued to be an active ocean port. In the later 19th century, Montreal decisively surpassed its French-Canadian rival and became the leading port and largest city in Canada. But only recently has Quebec City been exceeded in trade by the Pacific port of Vancouver and surpassed in population by Toronto.

As Montreal's port traffic grew, the city's business interests facilitated it by building a pioneering railway network based in Montreal. This network made use of lowland routes that radiated from Montreal: the St. Lawrence Valley leading east to the Atlantic and west to the Great Lakes; the Ottawa River Valley along which Canada's transcontinental railroads found a passage west and north into the Shield; and the Richelieu Valley– Champlain Lowland leading south toward New York City.

Britain created an independent Canada in 1867 by confederating Quebec, Ontario, New Brunswick, and Nova Scotia. By this time, the core area was already dominant in agriculture and population. In the following century, it was transformed into today's urban-industrial region. In its evolution to industrial dominance, the coreland profited from a number of clear advantages over the rest of Canada: superior position with respect to transport and trade, the best access to significant resources, accessibility to the largest markets, and the most advantageous labor conditions.

FIGURE 22.6

An October 1995 referendum held in Quebec Province in Canada pitted the French-speaking citizens who wanted to achieve an independent Quebec against people of the same province who wanted to maintain the bilingual life style that had come to characterize this early 18th century home to French settlers. The move to secede failed by the narrowest of margins and the issue is bound to be revived again in the near future.
A. McInnis/Gamma Liaison

Definitions and Insights

THE ST. LAWRENCE SEAWAY PROJECT

Very important among the core region's advantages were its lowland and water connections between the Atlantic and the interior of North America. In eastern North America, the only other connection of this sort through the Appalachian barrier is the lowland route connecting New York City with Lake Erie via the Hudson River and Mohawk Valley. Rapids on the St. Lawrence River long barred ship traffic upstream from Montreal, but the lowland along the river facilitated railway construction, and the building of some 19th-century canals bypassed the rapids and allowed small ocean vessels to enter the Great Lakes. Finally, in the 1950s, the river itself was tamed by a series of dams and locks in the **St. Lawrence Seaway** project. Since then, good-sized ships have been able to reach the Great Lakes via the river. The Welland Canal, which bypasses Niagara Falls between lakes Ontario and Erie by means of a series of stairstepped locks, predated the Seaway. The canal admits shipping vessels to the four Great Lakes above the falls. Farther up the lakes, the "Soo" Locks and Canals enable ships to pass between Lake Huron and Lake Superior, and natural channels interconnect lakes Erie, Huron, and Michigan. The total length of the St. Lawrence Seaway project is 2342 miles (3769 km), and cities on the margins of the Great Lakes—both in Canada and the United States—now serve as international ports. The Seaway provides an eight-meter-deep waterway in the heart of North America, and it moves wheat, iron ore, petroleum, and durable goods from mid-April until approximately mid-December, with the assistance of ice breakers. Construction was begun in the 1950s, although its full complement of canals, locks, and dams include transportation components that had been built decades earlier. The project has changed commercial geography in the heartland of North America in many ways. It also, through unwanted events, has changed the ecology of the region as well, particularly in the case of the arrival of zebra mussels that have been diffused into these interior waters from the ocean and are now a major blight on shipping and machinery efficiency because of their skyrocketing numbers and persistence.

RESOURCE AVAILABILITY

In addition to the water routes that connect it to the world and to all the shores of the Great Lakes, the Canadian core area has had a number of other major resources to support industrialization:

1. *Agricultural resources.* Farmland good enough for a large commercial agriculture, especially in Ontario, has supplied materials for food processing industries and a market for farm equipment industries.

2. *Forest resources.* Forests, mainly in the adjacent southern edge of the Canadian Shield, once supplied lumber for export and now supply wood for a very large pulp and paper industry, mainly in Quebec along the St. Lawrence and its north bank tributaries.

3. *Energy generation.* Especially at the descent from the Shield to the plains, many high-volume, steeply falling rivers drive one of the world's larger concentrations of hydroelectric plants. Major hydroelectric plants are also present at Niagara Falls, along the St. Lawrence as part of the Seaway project, and, more recently, at far northern sites in the Shield. Electricity is cheap in the coreland and is exported to the United States. Coal requirements are met from nearby Appalachian fields in the United States, while oil comes from Atlantic or Middle Eastern ports by tanker up the St. Lawrence, and both oil and natural gas come by pipeline from western Canada.

4. *Bulk commodity transport.* Metallic minerals are varied and abundant in the Shield to the north and west of the core area. Among them are very large deposits of iron ore located near the western end of Lake Superior on both sides of the international border, while others, more recently developed, are located in northeastern Quebec and adjacent Labrador. A large development of metal processing and fabricating industries, especially in Ontario, uses these Shield minerals, which also are exported.

Through the combination of these resources, their processing, and their distribution to both domestic and foreign markets, the Canadian core region has established a solid base for continuing economic development and relative prosperity.

MARKET AND LABOR ADVANTAGES

Ever since industrialization began, the population of the core area has represented the main cluster of Canadian consumers and labor. A high-tariff policy to forestall American competition was adopted in Canada shortly after its confederation. While it fell short of shutting out foreign industrial products, which Canada imports in large quantities, it did favor the growth of a large variety of Canadian industries, mainly in the core region and often including branch plants of U.S. companies "jumping" the tariff wall. However, the policy has long aroused protest from other parts of Canada in which consumers preferred to buy cheaper American goods. It also has fostered many high-cost Canadian industries that lack international competitiveness. The exceptions to this latter phenomenon tend to be mainly industries making products desired in the United States.

Thus, the Canadian pulp-and-paper and mining industries, with the U.S. market available for their products, tend to be large and efficient. In 1965, the two countries negotiated a free-trade

PROBLEM LANDSCAPE

"THAT'S THAT, UNTIL QUEBEC TRIES AGAIN"

The blue-and-white banners waved, the chanteurs sang their throats out against les federastes, the balloons soared, the placards bounced like demented bedsprings. Then everything, slowly, went flat. The 8,000 mostly young French-speakers packed to the roof of Quebec City's ice hockey arena shifted from thunderous triumph to unease, alarm, quiet gloom, at last defiant despair. As a nail-bitten evening ended on October 30 [1995], Quebec, by one percentage point in a huge 94% turnout, had voted to stay with Canada and—as these separatists saw it—against itself.[1]

As noted in this excerpt from an article in *The Economist*, Quebec came very close to withdrawing from Canada in October 1995. In a referendum asking the Québécois to decide whether they wanted to continue in a provincial role—with special considerations—or become a sovereign and independent nation, just barely more than 50 percent of the voters decided to stay a part of Canada.

This issue of French separatism in Quebec has a long history. Quebec's government was French-controlled, whereas its economy was directed largely by an upper-class English minority. There was an unequal language status: French speakers who wanted to rise economically generally found it necessary to become fluent in English, and French Canadians who migrated to other provinces had to make adjustments to predominantly English-speaking environments; however, the English Canadians of Quebec could go about their lives with little or no knowledge of French. Quebec's large population gave it a large representation in Canada's Parliament, and the province produced some powerful national political leaders. At the same time, the rapid natural increase of the Quebec French seemed to assure the continuation of the French language and culture.

For the better part of two centuries, the consequences of this cultural dichotomy within Canada did not produce serious disturbances, but widespread and open French dissatisfaction arose after World War II. Its rise appears to have been related to marked

demographic changes in Canada and probably also to a sharply increased concern with the general plight of minorities in the Western world. Population growth in Quebec slowed drastically as the province became more urbanized and secular, until by the 1970s the rate of increase was only about one half the rate of Canada as a whole. Furthermore, most new immigrants bypassed the province in favor of Ontario or provinces farther west, and the immigrants who did settle in Quebec—many of whom had no French background—tended toward assimilation with the English minority. Quebec's population began to decline as a percentage of Canada's, and a perceived threat to the province's French culture and heritage intensified grievances of long standing.

The result was the rapid rise, in the 1960s and 1970s, of a separatist political party, the *Parti Québécois*, in Quebec. It advocated the secession from Canada of an independent Quebec that could use its sovereign powers to protect its cultural heritage. The party achieved 41 percent of the Quebec vote in 1976, giving it control of

---

agreement with respect to the production and import and export of automobiles. This sparked rapid expansion of the auto industry into southern Ontario. General Motors, Ford, and Chrysler all have plants there—some of which are in Detroit's Canadian satellite of Windsor (population: 260,000). However, metropolitan Toronto is the main center of the industry. More recently, the inauguration of NAFTA has begun to cause major industrial readjustments—generally characterized by the more labor intensive firms (clothing especially, moving to Mexican locales) in Canada's core region, just as it has in other parts of North America.

INDUSTRIAL DIFFERENCES BETWEEN ONTARIO AND QUEBEC

Canada's core region produces a wide range of manufactured goods, and for most of these products it accounts for more than three fourths of Canada's national output. Within the core, there are marked differences between the industrial emphases of Ontario and those of Quebec. During the formative decades of industrialization, Ontario had cheaper access to United States coal. This favored the development of iron-and-steel capacity and metal products industries there, and these types of indus-

the provincial government. It enacted measures such as requiring the use of French in business and requiring all newcomers to the province to be educated in French. In 1980, it held a provincial referendum asking the voters to authorize provincial negotiations with the federal government for independence. But 58 percent of Quebec's voters rejected this request, and in 1985, the separatists were again defeated at the polls. Thus, separatism receded temporarily, but by 1988 it had returned as a live issue. In that year, Quebec passed a law forbidding any use of English on outdoor commercial signs. Antagonistic feelings flared in English Canada and were reciprocated by the French. Then, in 1990, the so-called **Meech Lake Accord** was rejected. This accord was an effort by the government of Prime Minister Brian Mulroney to accommodate the growing frustration of the Quebec people. This proposal offered to revise the Canadian constitution to recognize Quebec as a "distinct society" within Canada, with the right to protect its distinctiveness; it also gave all the provinces, including Quebec, increased powers relative to the federal government. Proposed by the federal government in 1987, this agreement was greatly favored in Quebec. But it required

approval by all ten provinces, and it was rejected by Newfoundland and Manitoba. Intense and far-reaching constitutional debate thus continued in Canada in 1994, coming to a near climax in October 1995 when a plebescite took place in Quebec. In that election, the decision for Quebec to stay in the body politic of Canada was preferred by less than one-percent difference of the voting population—51 to 49—with an impressive 94 percent of the registered voters casting ballots.

The effects of these 40 years of controversy on Canada have been considerable. The federal government has made French an official language, along with English, all across Canada. It has guaranteed that French Canadians outside Quebec may have their children educated in French. And it has weakened its powers in a variety of ways by passing them on to the provinces—for when Quebec demands and receives concessions, other provinces demand concessions in parallel fashion. Such comparable demands have come especially from the western provinces, particularly oil-rich Alberta. Concessions by the federal government have resulted in a higher degree of provincial control over resources. But the federal government has not been

willing to concede that an independent Quebec would be treated economically as though it were still part of Canada. Hence, how Quebec would fare as an independent nation separating two sections of Canada from each other is a major question. Whether the remaining provinces could hold together as one nation is another question.

The issue of Quebec as a potentially independent political entity in the southeastern region of Canada—and as a new independent state within North America—is a very real one, and one that will be addressed again soon by the populations of both Quebec and other units of the Canadian population as well. Political units that have held together for centuries in some cases will find themselves exploring the possibilities—and the possible benefits and costs—of independence as nations within nations attempt to reorganize their political, social, and economic affiliations. Quebec is in the lead of this pack, but there are possible followers all across the globe.

[1]*The Economist.* November 4, 1995: 45.

tries are still more prominent in Ontario than in Quebec. Montreal also has industries focused on making transport equipment and other metal goods, but these are less typical of Quebec than are more labor-intensive industries such as apparel manufacturing. Such industries were established primarily to deal with a traditional surplus of labor in Quebec created by an unusually rapid natural increase of Quebec's Roman Catholic population. This labor surplus led to migration into other parts of Canada and the United States and also to the establishment of manufacturing industries that were in need of notably cheap labor. Textiles, shoes, and clothing are typical products. The differences

between the two provinces still persist, and they have resulted in lower average wages and incomes in Quebec, which has also experienced a steady decline in its population growth.

CULTURAL DIVERGENCE AND ITS CONSEQUENCES

The economic differences between Quebec and Ontario are a serious matter but probably less so than are the cultural differences. In Quebec, 82 percent of the people speak French as a preferred language and 11 percent speak English. This is the only province in which French speakers predominate and control the provincial government. On the other hand, Ontario, where 77 percent speak English as a first language and only 5 percent speak French, is the largest in population of the nine predominantly English-speaking provinces. Major political consequences have arisen from this situation (see Fig. 22.6). See the Problem Landscape box on page 550.

22.5   The Prairie Region: Oil and Wheat Triangle

Manitoba, Saskatchewan, and Alberta are identified as a group under the name "the Prairie Provinces." Isolation is one factor that has led to this grouping. Their populations are separated from other populous areas by hundreds of miles of very thinly inhabited territory. To the east, the Canadian Shield separates them from the populous parts of Ontario, and on the west, the Rocky Mountains and other highlands separate them from the Vancouver region. To the north lies subarctic wilderness, and to the south are lightly populated areas in Minnesota, North Dakota, and Montana.

The prairie environment that gives the provinces their regional name exists only in some southern sections of the three, and it is to these sections that we ascribe the name "Prairie region" in this text. The region is a triangle of natural grasslands extending to an apex about 300 miles (*c.* 500 km) north of the United States border. It lies within the Interior Plains between the Shield and the Rockies, with the southern base of the triangle resting on the United States boundary. On all sides except the south, the grasslands are bordered by vast reaches of forest. Most of the southern part of the triangle is a northward continuation of steppe grasslands from the Great Plains of the United States. However, toward the forest edges the soils are moister, and the original settlers found true prairies: taller grasses with scattered clumps and riverine strips of trees. It was to the prairies that settlers were most attracted, and these moister grasslands still form a more densely populated, arc-shaped band near the forest edges.

The Prairie region was settled late, but quite rapidly, in a few decades after 1890. Settlers found reasonably good land, with a climate of long and harsh winters, short cool summers, and marginal precipitation. Only hardy crops could be grown. Early subsistence farming was succeeded by specialization in spring wheat for export, and the Prairie region became the Canadian part of the North American Spring Wheat Belt (Fig. 22.7). The settlers were predominantly English-speaking people from eastern Canada and the United States, but they included notable minorities of French Canadians, Germans, and Ukrainians. The ethnic composition of the region still reflects these elements.

Although the economy has diversified, agriculture is still basic. The region produces about one half the value of farm products in Canada and sells over one half of the world's wheat exports. Agriculture has been diversified to include other hardy crops such as barley and rapeseed (for vegetable oil and fodder), together with raising of livestock.

Urbanization and nonagricultural industries have become increasingly important, especially in Alberta and Manitoba. The three main metropolises are Edmonton (population: 840,000) and Calgary (population: 755,000) in Alberta, and Winnipeg (population: 650,000) in Manitoba. These cities have all grown near the corners of the Prairie triangle where connections to the outside world are focused. Winnipeg developed where the rail lines through the Shield crossed the Red River and entered the Prairie region; Calgary settled near a pass over the Rocky Mountains that gave a route toward Vancouver; and Edmonton was established near another Rocky Mountain pass leading to the Pacific. Both Winnipeg and Edmonton have developed additional functions as metropolitan bases for huge sections of Canada's North. None of the three Prairie provinces has yet become a notable manufacturing center, but the richness of the environmental setting has turned this region into an important tourist destination for Canadians and Americans alike.

Minerals are the foundation of recent advances in the region's economy, with this region now producing more than 60% of Canadian mineral production. Coal deposits underlie both the Rocky Mountain foothills and the plains in Alberta, as well as western Saskatchewan. They have recently begun to provide a major export. Nickel (Fig. 22.8), copper, zinc, and other metals come from the Canadian Shield of Manitoba; the Saskatchewan Shield produces uranium and southern

FIGURE 22.7
The broad expanses of the Prairie landscapes in central western Canada are identified by the clusters of grain elevators, railroad lines, and box cars, and by wide open, very productive, spring wheat farmlands. This scene is from Nanton, Alberta, in a region that is a major source of wheat in world trade.   *Thomas Kitchin/Tom Stack & Assoc.*

Regional Perspective

RESOURCE SEPARATISM

Resource separatism occurs when, within the boundaries of a single nation, one region has a particularly rich resource base but sees the cash rewards for such good fortune, or regional productivity, go to the federal government and other parts of the country. In the province of Alberta, the presence of fossil fuels has led to the potential for tensions. On a world scale, Canada's oil output is still small, but it has led to explosive growth for Calgary and Edmonton, which are the oil industry's main business centers. Another result of the oil boom has been a

running quarrel between Alberta and the federal government, centering on resource control and especially on federal price control of oil piped from Alberta to eastern Canada. Alberta's economic dynamism and aggressive political stance seem likely to continue, as the province is much richer in potential energy than in present energy production. It is estimated that "tar sands" in northern Alberta may contain two to three times the energy equivalent of all the oil now thought to exist in the Middle East. Major extraction must await improved technology and/or a time when greater scarcity of oil yields higher prices.

The oil fields that enrich Alberta and Saskatchewan have their origins south of the Gulf of Mexico, course up northward through Texas (where they produce one sixth of the U.S. total oil output) and extend further north through the Prairie provinces. Alberta, particularly, is interested in gaining a greater percentage of the foreign exchange generated by the gas, tar sands, and crude oil of these beds. Resource-based domestic tension such as this reminds us that one can never tell exactly what outcomes crop up from the discovery of resource wealth.

Saskatchewan produces major quantities of potash. But it is petroleum and associated natural gas that have had the most economic impact on the Prairie region. The region's oil industry began near Calgary, Alberta, in the 1940s. As production has multiplied and new fields have been brought in, the industry has remained predominantly in Alberta, with a minor portion in Saskatchewan and a tiny share in southern Manitoba. An important product extracted from natural gas in Alberta is sulfur, of which Canada has become the world's largest exporter.

22.6   The Vancouver Region: Core of British Columbia

Canada's transcontinental belt of population clusters is anchored at the Pacific end by a small region centering on the seaport of Vancouver at the southwestern corner of British Columbia. The region contains no more than 5 or 10 percent of British Columbia's area but has over one half of the province's population. Metropolitan Vancouver has 1.8 million people, and another

FIGURE 22.8

A scene in the Manitoba subarctic. Wispy conifers of little commercial value characterize this northern section of the subarctic. Marshes and swamps are common in the cool, glaciated environment. The road at the left is made of waste material from two operating levels of an underground nickel mine nearby.
Jesse H. Wheeler, Jr.

FIGURE 22.9
A portion of the harbor at Vancouver, British Columbia, Canada's busiest seaport. The container port is in the center of the photo, which was taken from the city's central business district. The growth and expansion of this city in recent years reflects Canada's shifting trade connections with the nations of the Pacific Rim.   *Porterfield Chickering/Photo Researchers*

300,000 reside in the metropolitan area of the province's capital city, Victoria (population: 311,000). Vancouver is located on a superb natural harbor at the mouth of the Fraser River. This valley provides a natural route for railways and highways across the grain of rugged mountains and plateaus in Alberta and British Columbia (Fig. 22.9). Victoria is located at the southern end of Vancouver Island—the southernmost and largest of a chain of islands along Canada's Pacific coast.

Victoria was the leading commercial and industrial center in British Columbia before the completion of Canada's first transcontinental railroad, the Canadian Pacific, in 1886. Vancouver then became the country's main Pacific port, and it has recently become the largest seaport of the entire country. It has profited from continuing development and a buildup of foreign trade in British Columbia and the Prairie provinces, as well as Canada's increasing exports of resource commodities to Japan. Vancouver exports huge quantities of such bulk commodities as grain, wood in various forms, coal, sulfur, metals, potash, and asbestos. Sea trade is fundamental to Vancouver's economy, but the city is far more than just a seaport. Despite its peripheral location within British Columbia, it is the province's main industrial, financial, and corporate administrative center. As such, it has numerous links with production nodes scattered through a sparsely populated and mountainous province larger in area than Texas and California combined.

THE PROVINCE OF BRITISH COLUMBIA

British Columbia, along with Yukon Territory to its north and a narrow fringe of Alberta, is the high part of Canada. The fjorded Coastal Ranges along the Pacific and the Rockies of the interior are high and spectacular, and the Intermountain Plateaus between them are largely rugged country. Only the northeastern corner of the province is plains country. This is "Peace River Country," the extreme northern outpost of Prairie region grain and livestock agriculture. The province is also distinctive climatically within Canada by virtue of its coastal strip of marine

FIGURE 22.10
Timber continues to play a major role in the economy of Canada, especially in the province of British Columbia. This resource has found a vast market in the Pacific Rim countries and this photo shows logs on Slocan Lake in British Columbia, working their way toward a local sawmill. *Gordon Fisher/Tony Stone Images*

west coast climate. Here, moderate temperatures and heavy precipitation are characteristic. The moisture supports the dense coniferous forests that are one of the province's principal resources, and the mild temperatures, combined with spectacular scenery, attract tourism and retirees. This setting has been also important as a migration destination for Chinese who left Hong Kong prior to its return to China in mid-1997. The climate of the interior is subarctic in the north and varies according to elevation and exposure in the south. Like the Coastal Ranges, the interior is largely forested, although not so luxuriantly.

Scattered small communities exploit abundant natural resources at increasing scales from an increasing number of locations. Forest products come from both the interior and the coast (Fig. 22.10). Salmon fishing continues along the coast, as it has for a century. More important than forestry and fishing today is a boom in the mining of metals (copper, molybdenum, silver, lead, and zinc) and coal (Fig. 22.11). Older metal production sites in the south are being supplemented by operations farther and farther north. The driving factor is Japanese and U.S. demand, with development being supported by foreign capital, largely Japanese. The province is a storehouse of energy resources in the form of both coal and hydropower, with coal shipped to Japan and electricity exported to the United States. Near the coastal town of Kitimat, hydroelectricity is used to produce aluminum from imported alumina (the second-stage material from bauxite ore). Wages tend to be high in the small

and isolated production nodes scattered over British Columbia, but so is the cost of living, and cultural amenities are few. Most new residents of British Columbia settle in cosmopolitan Vancouver or in Victoria, known as the most "English" of Canadian cities.

22.7   The Canadian North: Wilderness and Resource Frontier

The North comprises most of Canada, but where its southern edge lies is arguable. It certainly includes that major section of Canada that is very sparsely populated because of harsh physical environments, but population density fades gradually northward. In this discussion, we use the southern edge of the subarctic climate as, for the most part, the southern edge of the North, but do not include the island of Newfoundland—with its Atlantic associations and somewhat denser population. This definition places large sections of all the provinces except the Maritimes in the North. In addition, we may justify inclusion in the North of parts of the Canadian Shield which project south of the subarctic climate into southeastern Manitoba, Ontario, and Quebec. These areas are so handicapped by poor soils as to be sparsely populated and basically Northern.

FIGURE 22.11
In Sparwood, British Columbia, coal mining is big business. This enormous truck has been designed as the most economical and efficient means of moving vast quantities of coal from the pit to the loading area for both domestic shipping and export.   *Linda Brown*

The southern fringes of the North, often called the "Near North," are spotted with widely scattered islands of development of considerable importance to Canada's economy and, through trade connections, to the economy of the United States. Many of these are mining settlements. The Shield here has a wide variety of metallic ores, although geological processes have left it lacking in fossil fuels. The mining settlements range in size from small hamlets through small metropolitan areas. The largest is Sudbury (metropolitan population: 158,000), located in Ontario north of Lake Huron. Particularly noteworthy among many metals produced in the North are nickel and copper around Sudbury, iron ore from Quebec and Labrador, and uranium mined just north of Lake Huron and in northern Saskatchewan.

Some Northern settlements manufacture pulp and paper or smelt metals. A scattering of towns producing wood pulp and paper is spread across the southern fringes of the North. Most are small, but in two places clusters of mills combine with other functions to produce fairly sizable agglomerations. A string of towns along the Saguenay River tributary of the St. Lawrence in Quebec has a combined population of more than 180,000 people, supported by pulp-and-paper plants and by the manufacture of aluminum from imported raw materials brought in by ship. The other agglomeration is at Thunder Bay, Ontario (population: 140,000), on the northwestern shore of Lake Superior. In addition to pulp and paper, Thunder Bay derives support from being at the Canadian head of Great Lakes navigation. Its port links the Prairie provinces with the Ontario-Quebec core area, the United States, and overseas points.

Water resources are of major importance in the southern fringe of the North. Hydroelectricity supports local mines and mills and is sent to southern Canada and the United States. Rivers and lakes are so characteristic of the Shield that much of the latter appears from the air to be an amphibious landscape. There is a particular cluster of hydroelectric plants where St. Lawrence tributaries from the north rush down through the Laurentide "Mountains"—the somewhat raised southern edge of the Shield in Quebec. But increasing demand, plus increasing ability to transmit electricity for long distances, has led to major developments much farther north. These are located near James Bay, in Labrador, and in Manitoba near Hudson Bay. A remarkably small number live in this vast area. With agricultural possibilities so limited, a large increase in population seems unlikely in the foreseeable future. Agricultural settlements are rare, and most of them are declining.

"TERRITORIAL" CANADA: BLEAK AND EMPTY FEDERAL DOMAIN

North of 60 degrees latitude lies the part of Canada not yet considered sufficiently developed and populous for provincial status. This bleak area—known as "Territorial Canada"—is under the direct administration of the Canadian federal government.

Two administrative territories have been established, and a third one is moving toward formation (see Fig. 22.1). In the northwest, Yukon Territory is cut off from the Pacific by the panhandle of Alaska, once claimed by Canada but acquired by the United States. It is mostly rugged plateau and mountain country. The much larger Northwest Territories actually lie east of Yukon Territory. They include a fringe of the Rockies, a section of the Interior Plains along which the Mackenzie River flows northward to the Arctic Ocean, a huge area of plains and hills in the Canadian Shield, and the many Arctic islands. Two huge lakes, Great Bear Lake and Great Slave Lake, both drained by the Mackenzie River, mark the western edge of the Shield. Yukon Territory and the western part of the Northwest Territories have mainly a subarctic climate and taiga vegetation, with tundra near the Arctic Ocean, but in the eastern part of Canada tundra extends far to the south (Fig. 22.12). The Nunavut Territory occupies eastern and northern areas withdrawn from the Northwest Territories.

The population of Territorial Canada is only about 96,000 (as of 1996). About one fifth of the inhabitants in each of the present territories are native Indians, and in the Northwest Territories about one third—some 15,000 to 20,000—are Inuit (Eskimos). Most whites are government, corporate, or church employees, often relatively transient. Indians and Inuit still carry on some hunting, gathering, and fishing from homes in small fixed settlements; practically all Inuit inhabit prefabricated government-built housing in widely separated villages along the Arctic Ocean or Hudson Bay (Fig. 22.13). Unemployment among both Indians and Inuit is very high, and a large proportion live essentially on welfare.

Europeans and white North Americans have been seeking resources from these far northern reaches since the 17th century, when England's Hudson's Bay Company first established

FIGURE 22.12
A classic view of the arctic tundra around Churchill, Manitoba. The ground is carpeted with lichens and grass, and only the spindliest of trees is to be seen. *Dominique Braud/Tom Stack & Assoc.*

FIGURE 22.13

Hunting and trapping are relatively rare occupations in the present world, although small numbers of people still gain some income from such activities. This photo shows hides of arctic wolves curing in the sun at the village of Eskimo Point on Hudson Bay in Canada's Northwest Territories. The month is August, but the Inuit woman is warmly dressed for this tundra climate. Today, Canada's Inuit live in prefabricated houses like the one seen here. Fuel oil for household heating and cooking comes in oil drums brought by ship. The hides came from a hunt by the woman's husband during the preceding winter. He pursued a pack of wolves on his snowmobile and then shot them when they could run no more.   *Jesse H. Wheeler, Jr.*

fur-trading posts on the shore of Hudson Bay. The chief 20th-century quest has been for minerals. Mines produce metals or, in one instance, asbestos. Exploration for minerals is active, focusing on oil and gas known to exist in quantity in the Mackenzie Valley and in the Beaufort Sea section of the Arctic Ocean near the Mackenzie delta. Besides income from the few mining enterprises, the territories also receive funding from federal government payrolls in the territorial capitals of Whitehorse (Yukon) and Yellowknife (Northwest Territories).

Energetic efforts to cope with the problems of the territorial North are being made. Isolation has been mitigated by the airplane and by telecommunications that now bring telephone service, radio, and even television to remote settlements. A network of schools and medical clinics blankets the area. Air transportation enables the seriously ill to be flown to the few hospitals in the region or to hospitals in cities such as Edmonton or Winnipeg that serve as metropolitan bases for both the provincial and the territorial North. Meanwhile, the increasingly assertive Indian and Inuit populations are pressing land claims to mineralized areas and pipeline routes and are demanding regulations to prevent environmental disruptions by mining companies.

Regional Perspective

NATIVE LAND CLAIMS

A great deal of the landscape and resources that make up North America has been gained from battles, trials, and land purchases from the native populations who were here before Europeans "discovered" this world and before they began steady settlement and development of these lands. This is particularly true in the North and Territorial Canada, and in the west and northwest of the United States. This reality has been the state of things for literally centuries, but significant changes began to take place in the 1970s, when the United States was keenly interested in exploiting the petroleum resources of the Northern Slope in Alaska. The Alaskan native populations had been in litigation with the U.S. government for years over failure to honor early treaties, and the Trans-Alaska pipeline suddenly gave a whole new importance to the litigation between the Alaskan Federation of Natives and the U.S. government. This led to the enactment of the Alaska Native Claims Settlement Act (ANSCA) which gave 44 million acres of land to approximately 50,000 native people in Alaska—in addition to a cash settlement of $1 billion.

This also led to the formation of 250 village corporations in Alaska, with a structure that provided a role for every participating native Alaskan in village dynamics. In Canada, there were a series of negotiations and settlements from the mid-1970s until the 1990s. Although details were distinct in each case, there was generally an effort to provide some cash for the extraction of resources—including the generation of hydroelectric power—and also to give the native peoples access to the management of traditional lands in varying degrees. As new resources are found, new patterns of tourism develop, and more legal precedents are set, it is certain that both Canada and the United States will have to negotiate further details on this process of settlement and land capture that began centuries ago.

23

THE GEOGRAPHY OF DIVERSITY AND CHANGE IN THE UNITED STATES

The historical geography of the United States reflects a continuing series of migrations. The country was peopled by immigrants from Europe and slaves from Africa, and then from all over the world. Those who came of their own volition were seeking some new environment, some new setting that would be safer, more productive, more satisfying. American history and geography are made up of the products of this restlessness.

Such steady movement means a continuing change in the country's cultural landscape. To the places peopled by new arrivals, there come new languages, new customs and costumes, new architecture, and new mixes of the cultural personalities created by these demographic blendings. To the landscapes left behind when migrants leave, there is also profound change. The nature and magnitude of such change depends in part on the factors that set a particular migration in motion. However, while different migrations have distinct sets of catalysts and migration paths, there are elements of similarity in all the migrations that the United States has known. For example, people move because industrial advances replace human labor, and as a result, people who worked suddenly have no work. Leaving behind their past and its worn landscape, they head west—most often the case in American history—and try to find a place to begin again, and anew. These dynamics have tagged along with migrants coming to the United States—indeed, to all migration target destinations—since humans began to explore and "hit the road" for new havens, new settings, and new beginnings. Our cultural landscape is covered with evidence of this human search for a better shelter for their ambitions.

◄ Office development shown here at Tysons Corner, Virginia, symbolizes the kind of sprawl that has revolutionized the peripheries of American cities in the age of expressways. Tysons Corner, which has only recently become an incorporated place, grew explosively in an area where several expressways intersect. Such landscapes as this are occurring in all corners of the United States, often described as "edge cities" and acting as powerful magnets for further out-migration from American inner cities.
John Troha/Black Star

Definitions and Insights

VERNACULAR REGION

Geography has given much energy to the concept of regions and regionalization. We have **formal regions**, which are defined by administrative factors such as state boundaries. We have **functional regions**, which take on their character by the economic or cultural activity taking place around a **node**, with mappable **interaction space**. (The area of primary distribution of the St. Louis *Post Dispatch* newspaper is such a region.) Both of these regions can be defined and limited by objective criteria. There is, however, a third type of region called a **vernacular region,** which exists in the minds of people and whose boundaries are much more difficult to define. Such a region, for example, might be "Southern California." There are many such areas in the United States because extreme physical and cultural variety is a major characteristic of the country. Although much cultural variety has been muted by nationwide standardization during the 20th century, the United States still has marked differences from place to place. This chapter tries to foster an understanding of such variety by analyzing broad-scale regions. For this purpose, the United States has been divided into four multistate regions—Northeast, South, Midwest, and West—with Alaska and Hawaii treated separately as distinctive outliers. State lines are used as boundaries of the major regions, with full recognition that such boundaries are arbitrary separations of cultural landscapes that tend to merge with each other. These four regions are made more functional by our use of state boundaries, but you will still find that people are ready to argue with you—and with our definitions—about the borders, and even the cultural or economic characteristics, of these four vernacular regions. That's good geography—to be able to argue about the utility of regional criteria. Enjoy the process, and see how the arguments help you understand the nature of regions.

23.1 The Northeast: Intensively Developed and Cosmopolitan

Of the four major regions, the Northeast is the most intensively developed, densely populated, ethnically diverse, and culturally intricate. It incorporates the nation's main centers of political and financial activity, and it is the area in which relationships with foreign nations are the most elaborate (e.g., through the United Nations and the capital at Washington, D.C.). Strong traditions of intellectual, cultural, scientific, technological, political, and business leadership persist there. This is the area where the country's political and economic systems had their main beginnings, and it historically has been the chief reception center for foreign immigrants.

The Northeast consists of six New England states (Maine, New Hampshire, Vermont, Massachusetts, Connecticut, and Rhode Island), plus five Middle Atlantic states (New York, New Jersey, Pennsylvania, Delaware, and Maryland), and the District of Columbia—the country's capital. Delaware and Maryland have important historical links with the states of the South, but their present character classifies them more appropriately as Middle Atlantic states.

As of 1995, about 53 million people, or 20 percent of the national population, lived within the Northeast, on 5 percent of the nation's area (Table 23.1). Hence, the regional population density is several times that of the South, Midwest, or West. But the density is much more extreme within a narrow and highly urbanized belt stretching about 500 miles (*c.* 800 km) along the Atlantic coast from metropolitan Boston (Massachusetts) through metropolitan Washington, D.C. The belt is often called the Northeastern Seaboard, Northeast Corridor, Boston-to-Washington Axis, "Boswash," or "Megalopolis" (see Fig. 23.1 and photograph, page 558). Here, seven main metropolitan areas contain about 40 million people: Boston (5.4 million), Providence (1.1 million), Hartford (1.2 million), New York (19.5 million), Philadelphia (5.9 million), Washington, D.C. (4.5 million), and Baltimore (2.4 million).[1] Additional population within this belt amounts to roughly 6 million, giving a total of about 46 million, or approximately 85 percent of the people in the Northeast. The leading cities of the belt make up the country's main centers of political decision-making, corporate headquarters, finance, trade (retailing), and services. An index of the centrality of this region is illustrated by the fact that 40 percent of all office space in the United States is located within 50 miles (80 km) of New York City. Manufacturing is relatively less important, but the belt still accounts for about one sixth of the nation's manufacturing. Congested trafficways and active intercity commercial and cultural linkages bind this mosaic of metropolitan areas together (Fig. 23.2).

THE NORTHEASTERN ENVIRONMENT

Most of the Northeast lies in the Appalachian Highlands, but relatively small sections lie in the Atlantic Coastal Plain, the Piedmont, or the Interior Plains (Fig. 23.3). Atlantic Coastal

---

[1]City populations in this chapter are rounded 1996 estimates for Metropolitan Statistical Areas (MSAs), Primary Metropolitan Statistical Areas (PMSAs), or Consolidated Metropolitan Statistical Areas (CMSAs), with the most inclusive figure being used (extrapolated from U.S. Bureau of the Census' *Statistical Abstract of the United States,* 1996). In Census Bureau terminology, MSA refers to an urban aggregate composed of a county, or two or more contiguous counties, meeting specified requirements as to metropolitan status. In New England, MSAs are defined by cities and towns rather than by counties. Designation of MSAs represents an attempt to give a realistic picture of functional urban aggregates, each of which often includes more than one incorporated urban place ("political city"), plus unincorporated urbanized areas and stretches of countryside where agriculture may be important. But even the countryside is closely bound to the services, markets, and employment afforded by the urban places. A PMSA is, in general, a larger aggregate, and a CMSA is the most inclusive unit of all. Figures cited in this chapter for the very largest cities, such as New York, Los Angeles, and a considerable list of others, are for CMSAs. Any of the three types of metropolitan units may include counties in more than one state.

TABLE 23.1	**United States Regions: Comparative Data**			
	Percent of National Totals			
	Northeast	South	Midwest	West
Total area	5	24	22	32
Total population	20	35	24	22
Land in farms	2	30	36	32
Cropland	3	17	63	12
Value of farm products sold	6	31	39	24
Value of livestock and products sold	8	34	41	18
Value of crops sold	5	26	38	30
Value added by manufacture shipments	19	32	31	17
Value of mineral output	3	60	11	20
Value of retail sales	22	32	25	22
Total personal income	25	30	23	22
Total metropolitan population	25	30	22	23

Source: U.S. Bureau of the Census, *Statistical Abstract of the United States,* 1996. National totals from which regional percentages are calculated include Alaska and Hawaii.

Plain areas include (1) Cape Cod in Massachusetts, (2) Long Island in New York state, (3) southern New Jersey, (4) the Delmarva Peninsula, which includes nearly all of Delaware plus the Eastern Shore of Maryland (east of Chesapeake Bay) and parts of Virginia, and (5) the Western Shore of Maryland inland to the boundary between Baltimore and Washington, D.C. The Plain is low and flat to gently rolling, with many sand dunes and marshes or swamps. Soils are sandy and range from very low to mediocre in fertility. The Plain is indented by broad and deep estuaries ("drowned" lower portions of rivers). The largest Northeastern seaports are on the estuaries of the Hudson River (New York); the lower Delaware River (Philadelphia); and the Patapsco River (Baltimore) near the head of Chesapeake Bay.

Between the Coastal Plain and the Appalachians, the northern part of the Piedmont extends across Maryland, Pennsylvania, and New Jersey to New York. The Piedmont is higher than the Coastal Plain but lower than the Appalachians. Where the old erosion-resistant igneous and metamorphic rocks of the Piedmont meet the younger, softer sedimentary rocks of the Coastal Plain, many streams descend abruptly in rapids or waterfalls, and the boundary between the two physical regions is known as the fall line. It is marked by a line of cities that have developed along it: Washington, D.C., Baltimore, Wilmington (Delaware), Philadelphia, Trenton (New Jersey), and New Jersey suburbs of New York City (see Fig. 23.1). Most of the Piedmont is rolling, with soils generally better than those of the Coastal Plain or Appalachians.

Aside from a narrow strip of the Interior Plains along the Great Lakes and St. Lawrence River (see Figs. 23.1 and 23.3), the rest of the Northeast lies in the Appalachian Highlands. But in New York, a narrow lowland corridor—the Hudson-Mohawk Trough—breaks the Appalachians into two quite different subdivisions. This Trough is composed of the Hudson Valley from New York City northward to Albany and the Mohawk Valley westward from Albany to the Lake Ontario Plain. Northeast and north of the Trough, the old hard-rock mountains of New England and northern New York form several ranges, characterized by rough terrain, cool summers, snowy winters, and poor soils. The Adirondack Mountains of northern New York rise as a roughly circular mass, surrounded by lowlands. Heavily forested and pocked by numerous lakes, these glaciated mountains reach over 5000 feet (1524 m) in elevation. They are bounded on the east by the [Lake] Champlain Lowland, which reaches northward into Canada as a continuation of the Hudson Valley. East of it the relatively low Green Mountains occupy most of Vermont and extend southward to become the Berkshire Hills of western Massachusetts and Connecticut. Farther east across the narrow valley of the upper Connecticut River, the White Mountains occupy northern New Hampshire and extend into Maine. Here, Mount Washington in New Hampshire rises to 6288 feet (1917 m), the highest elevation in the Northeast. The Adirondacks and the mountains of New England are major recreation areas for the Northeast's urban populations.

In New England, the hilly areas between the mountains and the sea, sometimes termed the New England Upland, are considered to be part of the Appalachians, but they are relatively low rather than mountainous. The original soils were persistently stony (caused by the deposition of stones by glaciation), but the early settlers were able to use them for subsistence agriculture once the stones and trees were laboriously cleared.

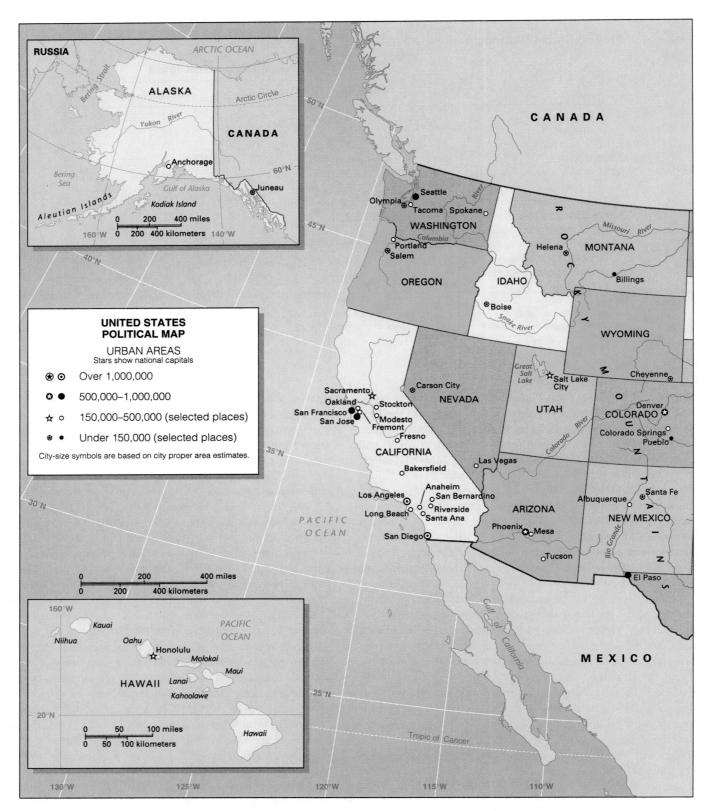

FIGURE 23.1
General political map of the United States.

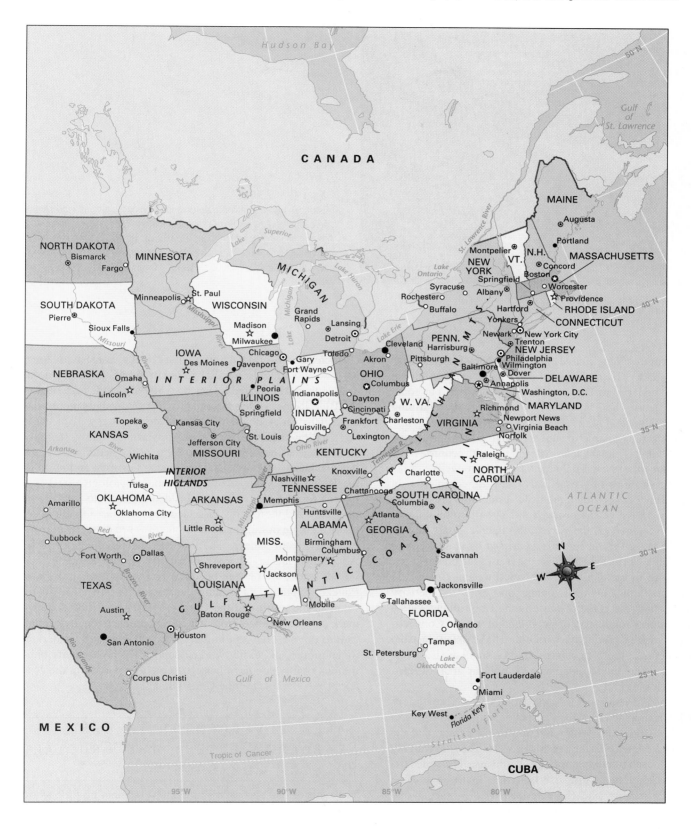

FIGURE 23.2
Rush-hour traffic in downtown Boston, Massachusetts, is representative of the congested Boston-to-Washington "Megalopolis" along the Atlantic shore of the Northeast. Downtown Boston is notorious for its winding, cluttered maze of streets, some of which originated as cow paths in colonial times. *Neal J. Manschel/Christian Science Monitor*

Boulders from the fields were used to build New England's famous stone fences. Today, many of these have been torn down to facilitate modern farming or urban development, but some derelict fences may still be seen in woodlands where farming has been abandoned (Fig. 23.4). The poor soils of the Upland proved increasingly unable to support a stable long-term commercial agriculture, and the late 18th and the 19th centuries saw a large movement of New Englanders from unproductive farms to Northeastern cities or—especially with the completion in 1825 of the Erie Canal—to more fertile agricultural areas to the west. This migration gave a strong New England cultural flavor to localities scattered all the way to the Pacific Coast. Figure 23.5 sets the stage for the patterns created in large part by migrants from the Northeast all across the rest of the country.

Embedded in the old hard-rock New England Upland are scattered, small lowlands with younger and softer sedimentary rocks and better soils. One of these is the Connecticut Valley Lowland, a north-south strip of good land along the Connecticut River in which a productive agriculture has existed since the first settlements in the 17th century. Today, small areas of surviving farmland are interspersed with the built-up areas of Hartford and other cities.

South of the Hudson-Mohawk Trough, the Northeastern Appalachians include three distinct sections: the **Blue Ridge** in the east, the **Ridge and Valley Section** in the center, and the **Appalachian Plateau** in the north and west. They lie roughly parallel, with each trending northeast-southwest. The series extends far beyond the Northeast into the South. The Blue Ridge—long, narrow, and characterized by old igneous and metamorphic rocks—has various local names. Often it forms the single ridge its name implies, but it broadens into many ridges in the South. West of it, the Ridge and Valley Section,

characterized by folded sedimentary rocks, consists of long, narrow, and roughly parallel ridges trending generally north and south and separated by narrow valleys. On the east, next to the Blue Ridge, valley floors are wider and ridges more scattered. This valley-dominated eastern strip is essentially a single large valley known as the Great Appalachian Valley. Its limestone floor has decomposed into some of the better soils of the Appalachians. The Appalachian Plateau lies west and north of the Ridge and Valley Section. Although geologically it is a plateau formed of horizontally bedded sedimentary rocks, it is so deeply and thoroughly dissected in most places that it is actually a tangle of hills and low mountains separated by narrow, twisting stream valleys. The northern part is often called the Allegheny Plateau, whereas, in parts of eastern Kentucky and farther south, the plateau becomes known as the Cumberland Plateau. Notable east-facing escarpments—the Allegheny Front in the north and the Cumberland Front in the south—mark the eastern edge, from which elevations gradually decline toward the west. Except in New York, the Plateau is underlain by enormous and easily worked deposits of high-quality bituminous coal which have been extremely important in the economic development of the United States.

EARLY DEVELOPMENT IN THE SEABOARD CITIES: THE CRUCIAL ROLE OF COMMERCE

The roots of the Northeast's urban development go far back in time. Boston was founded by English settlers and New Amsterdam (presently New York City) by the Dutch in the early 1600s, although the latter was subsequently annexed and renamed by

England in 1664. Philadelphia was founded by the English in the later 1600s and Baltimore in the early 18th century. All four cities were major urban centers by the time of the American Revolution (1775–1783), although by today's standards they were quite small.

Except for Washington, D.C., the largest metropolises of the Northeastern Seaboard gained their initial impetus as seaports, frequently located at river mouths. New York City was founded at the southern tip of Manhattan Island, on the great natural harbor of the Upper Bay (Fig. 23.6). It shipped farm produce from lands now occupied by built-up areas of metropolitan New York City and from estates up the Hudson River. Philadelphia was sited at the point where the Delaware River is joined by a western tributary, the Schuylkill. Philadelphia's local hinterland—the Pennsylvania Piedmont and Great Valley, plus parts of southern New Jersey—was productive enough to make the city a major exporter of wheat and helped it become the largest city (population: about 40,000 in 1776) in the thirteen colonies. Baltimore was founded on a Chesapeake Bay harbor at the mouth of the Patapsco River and shipped tobacco and other farm products from its local hinterland in the Piedmont and the Coastal Plain. Boston (Fig. 23.7) had a somewhat different early development since its New England hinterland produced little agricultural surplus for export. Instead, colonial New England placed its main emphasis on activities connected with the sea. Its forests yielded superior timber with which to build ships, and both timber and ships were exported, as were cod and other fish that were abundant along this coast. Whaling was important in the 18th and 19th centuries. New Englanders also developed a wide-ranging merchant fleet and trading firms with far-flung interests. Such activities characterized many ports, of which Boston was the largest.

Definitions and Insights

SITE SELECTION: THE NATIONAL CAPITAL

The selection of the site for a national capital is customarily invested with considerable care. The seat of government must blend economic and political centrality. It must also have the physical potential to support the nearly sacred image of a major capital because of the ceremony attached to its location. In the United States, the selection of the site of Washington, D.C., for the federal capital was not made until nearly two centuries after the founding of the initial Jamestown settlement in 1607. In 1800, a 69-square-mile block of land lying in near-swamp along the Potomac River at the junction of Maryland and Virginia was designated the site for the American capital city. At the time, the United States consisted of a narrow belt of states along the Atlantic. The Potomac location was on the border between the agrarian southern states—with their large slave populations—and the more diversified northern states, with fewer slaves and widespread abolitionist senti-

ments. The site also placed the capital on the fall line between seaboard and upcountry sectional interests. Many individual American states have their capitals similarly located between such sections.

In 1878, Washington, D.C., annexed the city of Georgetown, but it grew to be a large city primarily as a consequence of the post–World War II expansion of the federal government. Today, its metropolitan area encompasses both the District of Columbia and large suburban areas in Maryland and Virginia. It also possesses a monumental landscape that marks it clearly as a national center.

THE RACE FOR MIDWESTERN TRADE

As the present Midwest was settled, primarily between 1800 and 1860, its agricultural surpluses and need for manufactured goods greatly expanded the trade of the cities of New York, Philadelphia, and Baltimore. The ports engaged in a race to establish transport connections into the interior. New York City surpassed its rivals, largely because of its access to the Hudson-Mohawk Trough—the only continuous lowland passageway in the United States through the Appalachians (see Figs. 23.1 and 23.3). The Erie Canal was completed along this corridor in 1825. Connecting Lake Erie at Buffalo to the navigable Hudson River near Albany, the canal was a key link in an all-water route from the Great Lakes to New York City and it helped reduce transport costs between New York and the Midwest to a small fraction of their previous level. Settlement of the Midwest was stimulated, and a flood of trade was directed through the port of New York. In 1853, using the same route, New York interests became the first to connect a growing Chicago with an east coast port by rail. By the 1850s, New York had become the leading United States port and city by a wide margin. Although surpassed by New York, the competing ports of Baltimore and Philadelphia shared in Midwestern trade. Major railways were eventually pushed through the Appalachians into the Midwest from both ports. Boston was never a real competitor in this race, although it has remained the largest seaport in New England.

EARLY INDUSTRIAL EMPHASES ON THE NORTHEASTERN SEABOARD

Industrialization got its first big boost during the same period that the Northeast established connections to the interior. Favorable factors included accumulated capital from commercial profits; access to industrial techniques pioneered in Great Britain; skills from previous commercial operations and handicraft manufacturing; cheap labor supplied by immigrants (see

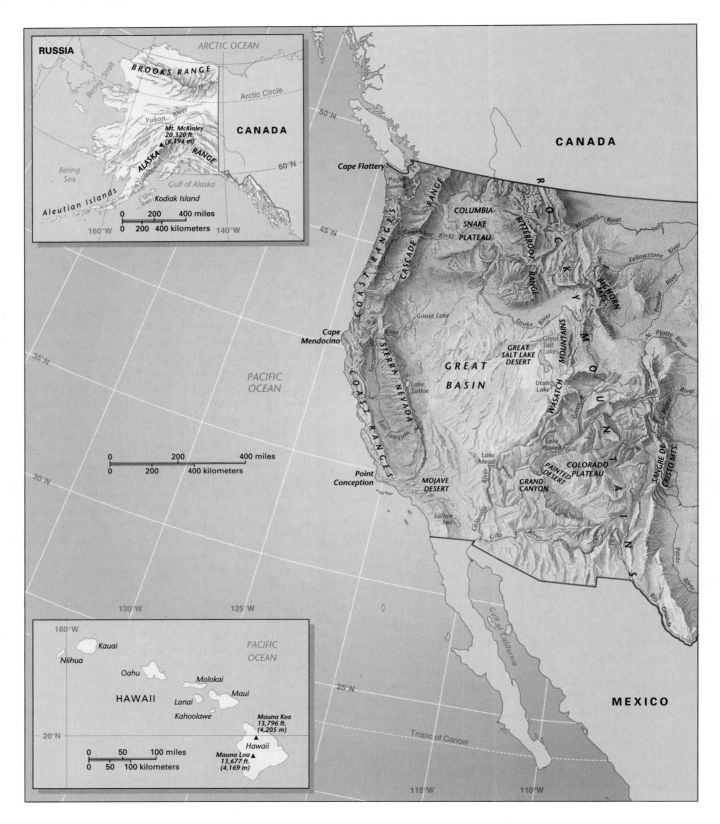

FIGURE 23.3

Major physiographic features of the United States.

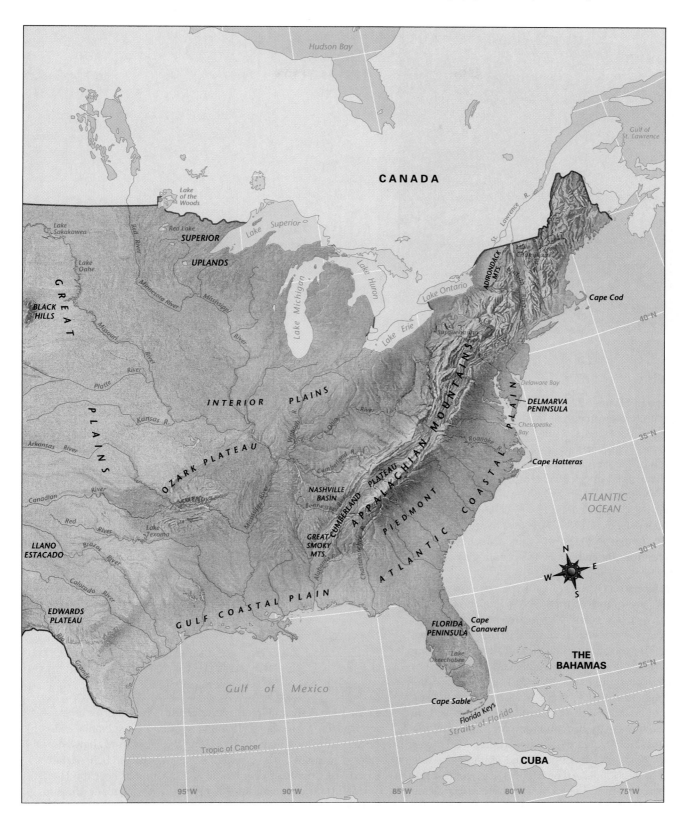

FIGURE 23.4
Originally, the farmland in New England had to be cleared of boulders left behind by continental glaciers. So much farmland has been abandoned that one often finds stone walls winding through patches of woodland where crops once were grown. Trying to interpret symbols of landscape change like this makes geography a kind of detective work, leading to a clearer understanding of what humans have done in their creation of cultural landscapes.   *Jesse H. Wheeler, Jr.*

literature selection in Chapter 22, page 539); and the presence of energy resources and raw materials that were more important in early industrialization than they are today. Also favorable was a United States high-tariff policy that restricted imports and secured much of the nation's market for its own emerging manufacturers. The first American factory, a water-powered textile plant using pirated British technology, was established by Samuel Slater on the Blackstone River at Pawtucket (near Providence), Rhode Island, in 1790, and the subsequent growth of factory industry in the Northeast was rapid. Much new technology soon came from inventions by the region's technically minded people.

The circumstances and results of industrialization were different in different parts of the Northeast. New England had suitable waterpower near its ports, with many small and swift streams to turn early waterwheels, but this region lacked coal for metalworking. It specialized in textiles, drawing wool from New England farms, cotton from the South, and abundant labor from recent European immigrants, especially the Irish. However, one New England state, Connecticut, took a different path. It expanded a colonial specialty in metal goods—based on small local ore deposits—by becoming a major manufacturer of machinery, guns, and hardware (which it still is). Mill towns sprang up across southern New England and up the coast into southern Maine. In the Middle Atlantic states, early industrialization also included textiles, but there was more emphasis on metals, clothing, and chemicals. The emphasis on metals reflected small local deposits of iron ore, especially in southeastern Pennsylvania. Then, in the second quarter of the 19th

century, the largest deposits of anthracite coal in the United States became a major source of power. This coal lay deep underneath the Pennsylvania Ridge and Valley Section between Harrisburg (population: 610,000) and Scranton–Wilkes-Barre (combined population: 637,000), but was made accessible despite the terrain by the determined building of canals and railroads and by large capital investment in the mines themselves. For many decades, anthracite was very significant industrially. Usable in blast furnaces without coking, it fueled large-scale iron and steelmaking in various eastern Pennsylvania cities. Iron and steel from the early plants contributed greatly to the first industrial surge in the Northeast by furnishing material for the manufacture of machinery, steam engines, railway equipment, and other metal goods.

THE RISE OF INDUSTRY IN THE INTERIOR NORTHEAST

During the early and middle 19th century, two interior areas—the Hudson-Mohawk corridor and western Pennsylvania—also began to develop industrially. Along the Hudson-Mohawk corridor, a string of cities emerged: Albany, Troy, and Schenectady (tricity metropolitan population: 875,000) on or near the Hudson; Buffalo (population: 1.2 million with adjacent Niagara Falls) at the Lake Erie end of the route; and Rochester (population: about 1.1 million) and Syracuse (population: 760,000) between them. Diverse industries developed: flour milling,

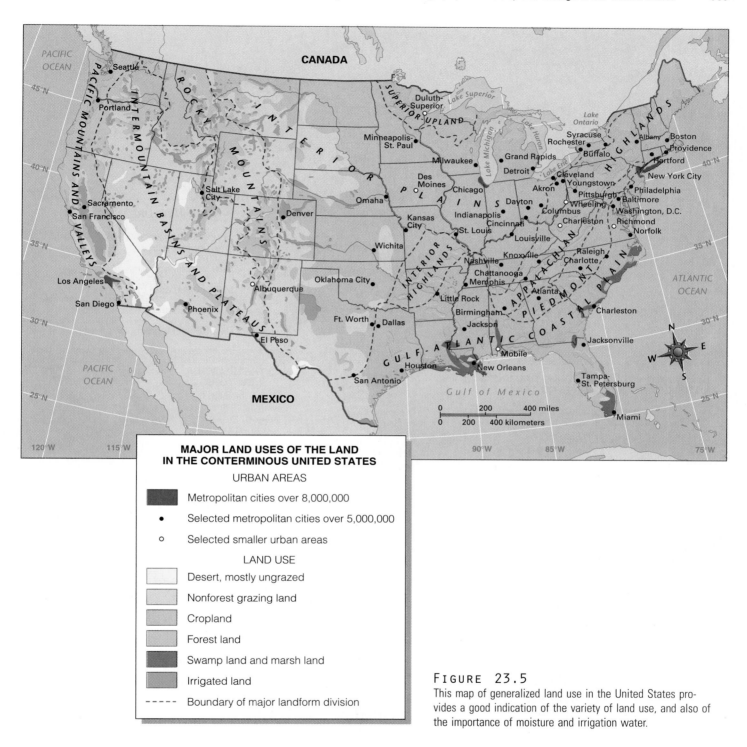

FIGURE 23.5

This map of generalized land use in the United States provides a good indication of the variety of land use, and also of the importance of moisture and irrigation water.

chemical and textile production, the Eastman Kodak photographic company in Rochester, the General Electric Company in Schenectady, major iron and steel manufacturing in Buffalo, and the manufacture of miscellaneous metal goods and precision instruments at various locations.

A second area of early development in the interior Northeast lay in the Appalachian Plateau of western Pennsylvania, especially in Pittsburgh (population: 2.4 million). In the mid-18th century, laborious land routes were pioneered from Baltimore and Philadelphia to the "Forks of the Ohio" where

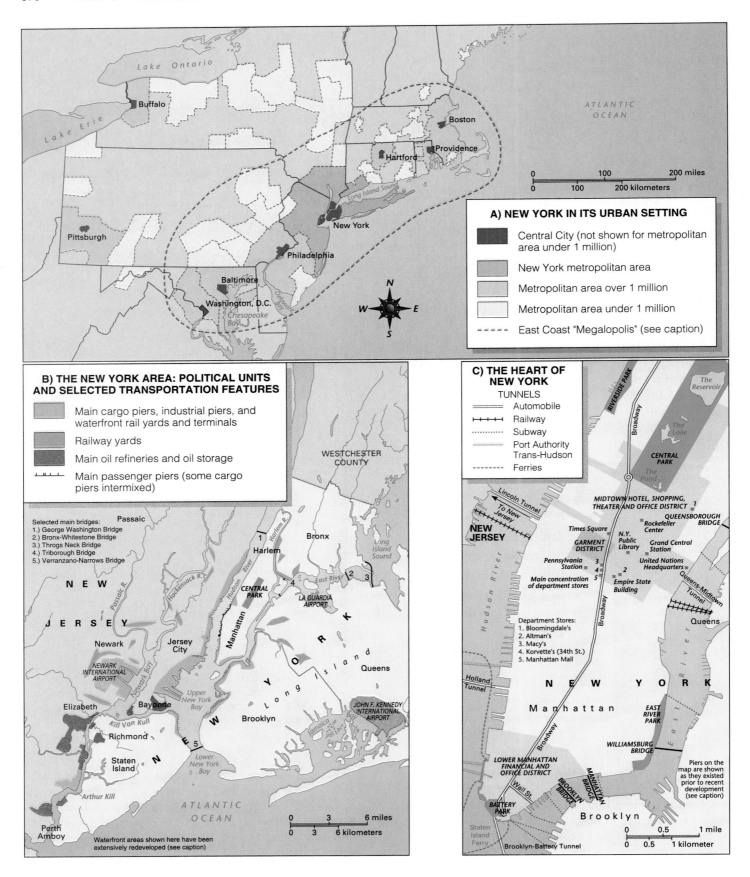

A) NEW YORK IN ITS URBAN SETTING

- Central City (not shown for metropolitan area under 1 million)
- New York metropolitan area
- Metropolitan area over 1 million
- Metropolitan area under 1 million
- East Coast "Megalopolis" (see caption)

Lake Ontario
Lake Erie
ATLANTIC OCEAN
Buffalo
Boston
Providence
Hartford
New York
Pittsburgh
Philadelphia
Baltimore
Washington, D.C.
Chesapeake Bay
Long Island Sound

B) THE NEW YORK AREA: POLITICAL UNITS AND SELECTED TRANSPORTATION FEATURES

- Main cargo piers, industrial piers, and waterfront rail yards and terminals
- Railway yards
- Main oil refineries and oil storage
- Main passenger piers (some cargo piers intermixed)

Selected main bridges:
1.) George Washington Bridge
2.) Bronx-Whitestone Bridge
3.) Throgs Neck Bridge
4.) Triborough Bridge
5.) Verrazano-Narrows Bridge

Passaic
Passaic R.
NEW JERSEY
Hackensack R.
Newark
Jersey City
NEWARK INTERNATIONAL AIRPORT
Elizabeth
Bayonne
Richmond
Staten Island
Arthur Kill
Perth Amboy
Kill Van Kull
Newark Bay
Upper New York Bay
Lower New York Bay
Hudson River
Manhattan
CENTRAL PARK
Harlem
Harlem R.
Bronx
WESTCHESTER COUNTY
Long Island Sound
East River
LA GUARDIA AIRPORT
NEW YORK
Long Island
Queens
Brooklyn
JOHN F. KENNEDY INTERNATIONAL AIRPORT
Jamaica Bay
ATLANTIC OCEAN

Waterfront areas shown here have been extensively redeveloped (see caption)

C) THE HEART OF NEW YORK

TUNNELS
- Automobile
- Railway
- Subway
- Port Authority Trans-Hudson
- Ferries

RIVERSIDE PARK
Broadway
The Reservoir
The Lake
CENTRAL PARK
The Pond
Lincoln Tunnel
To New Jersey
NEW JERSEY
MIDTOWN HOTEL, SHOPPING, THEATER AND OFFICE DISTRICT
Times Square
GARMENT DISTRICT
Rockefeller Center
N.Y. Public Library
QUEENSBOROUGH BRIDGE
Grand Central Station
Pennsylvania Station
United Nations Headquarters
Main concentration of department stores
Empire State Building
Hudson River
Holland Tunnel
Queens-Midtown Tunnel
Queens
NEW YORK
Manhattan
EAST RIVER PARK
Broadway
East River
WILLIAMSBURG BRIDGE
LOWER MANHATTAN FINANCIAL AND OFFICE DISTRICT
Wall St.
BATTERY PARK
BROOKLYN BRIDGE
MANHATTAN BRIDGE
Staten Island Ferry
Brooklyn-Battery Tunnel
Brooklyn

Department Stores:
1. Bloomingdale's
2. Altman's
3. Macy's
4. Korvette's (34th St.)
5. Manhattan Mall

Piers on the map are shown as they existed prior to recent development (see caption)

◀ FIGURE 23.6

New York stands at the center of a strip of highly urbanized land that was christened "Megalopolis" by the French geographer Jean Gottmann in 1961. The dashed line is not intended to
indicate a precise boundary but shows broadly the area included by Gottmann in his Megalopolis concept. Many waterfronts and other areas shown on the two maps of New York are being
redeveloped for new commercial, residential, and recreational uses. This is especially true of
the piers along both sides of the Hudson River. In the map "The Heart of New York," the shading for the "Midtown Hotel, Shopping, Theater, and Office District" covers the areas in which
most of Manhattan's well-known hotels, restaurants, department stores, specialty shops, theaters, concert halls, and museums are found. The Midtown area also includes numerous office
buildings, most notably the cluster of skyscrapers in Rockefeller Center. At the southern end of
the area near the main department stores is New York's Garment District with its large workforce and numerous workrooms, sweatshops, and showrooms crowded into a remarkably small
segment of this densely built-up city. The Lower Manhattan Financial and Office District
includes the imposing cluster of office skyscrapers in the Wall Street area and, farther north,
City Hall and a large assemblage of other government buildings occupied by city, county, state,
and federal offices. Note the complex of bridges, tunnels, and ferries that ties Manhattan Island
to the rest of metropolitan New York.

Pittsburgh developed. There, the Allegheny River from the
north and the Monongahela River from the south join to form
the Ohio River, a broad highway on which pioneers could float
west. Pittsburgh became the main center for outfitting settlers
and building riverboats, and it developed the needed ironworking based on Pennsylvania iron ore. As ironworking expanded,
rolling mills to shape iron in larger quantities were built, and
iron began to be brought by steamboats from furnaces in localities as far west as Missouri. Just before the Civil War, huge
high-grade iron ore deposits near Lake Superior in Wisconsin
and Minnesota were discovered and transported via the Great
Lakes to Pennsylvania steel centers. The ore-bearing formation
in Minnesota, called the Mesabi Range, eventually became the
main source for the dynamic growth of the iron and steel industry in the Northeast. The Pittsburgh area subsequently underwent a meteoric rise in iron and steel production. For many
years it was the world leader, although it eventually was surpassed by the Chicago area and various districts overseas.
Meanwhile, Appalachian coal mining spread into West Virginia, Kentucky, and Ohio, and the coal was shipped in huge
quantities to power industrial development in both the Northeast and Midwest (Fig. 23.8).

FIGURE 23.7

The city of Boston, Massachusetts, seen here from the Charles River, is
an appropriate symbol of the strong impact of English culture on North
America. The city is named for a town in eastern England, and the river is
named for an English king (Charles I). The golden dome of the State
House (capitol) symbolized the transfer of legal and governmental institutions from England. Boston played a key role in the Thirteen Colonies and
the American Revolution, and it has since been an important diffuser of
culture to North America.   *Grapes Michaud/Photo Researchers*

FIGURE 23.8

Barge loads of Appalachian coal in the Ohio Basin awaiting movement to
downriver industrial markets. Huge and varied coal resources have been a
major factor supporting industrialization of the Northeast and Midwest.
Nickelsburg/Gamma Liaison

The urban-industrial trends established in the Northeast during these earlier times have continued through the 20th century, during which time the region has experienced massive population growth and economic development. However, development in other parts of the United States has diminished the relative predominance of the Northeast, and in more recent decades the region has undergone economic stress, with slowing or even reversal of growth in many industries and localities.

THE PRIMACY OF NEW YORK CITY

The United States' largest city is centrally located within the Northeastern Seaboard's urban strip, midway between Boston and Washington, D.C. An enormous harbor, an active business enterprise, and a central location within the economy of the colonies and the young United States had already made New York City the country's leading seaport before access to the Hudson-Mohawk route gave it an even more decisive advantage. Superior access to the developing Midwest then moved New York rapidly to unquestioned leadership among American cities in size, commerce, and economic impact.

With some 7.3 million people in 1995, New York City proper has over twice the population of the next largest U.S. city, Los Angeles (population: 3.5 million, city proper); and the city's metropolitan area (population: 19.5 million) is considerably larger than the second largest metropolis, which is again Los Angeles (population: 15 million). New York is still one of the nation's leading seaports in container traffic and total value of waterborne commerce, and its three international airports (John F. Kennedy, La Guardia, and Newark) make the area a major world center of air transportation. The metropolitan area vies closely with the Los Angeles and Chicago metropolises as one of the country's three largest concentrations of manufacturing. It is by far the largest center of wholesale trade, publishing, advertising, broadcasting, and performing arts, and is a major center of retail trade and services. The United Nations headquarters confers international prestige. But perhaps the most impressive aspect of New York's stature relative to other U.S. metropolises lies in the city's leadership in finance and business management. It overshadows any other American center in funds controlled by banks and insurance companies, as well as in assets of industrial corporations headquartered there.

New York City centers on the less than 24 square miles of Manhattan Island (see Fig. 23.6). The island is narrowly separated from the southern mainland of New York state on the north by the Harlem River, from Long Island on the east by the East River, and from New Jersey on the west by the Hudson River. Manhattan is one of five "boroughs" of the city proper. The others are The Bronx on the mainland to the north, Brooklyn and Queens on the western end of Long Island, and Staten Island to the south across Upper New York Bay. In addition, the broader New York metropolitan area includes much of the population and industry of northern and central New Jersey, communities on Long Island east of Brooklyn and Queens, southernmost New York state just north and west of The Bronx, and southwestern Connecticut. Expressways and mass-transit lines converge on bridges, tunnels, and ferries that link the boroughs to each other and to northern New Jersey (see Fig. 23.6). Port activity is heavily concentrated in New Jersey along the Upper Bay and Newark Bay, although some traffic continues to be handled in various New York sections of the waterfront. The whole area is held together in part by an amazing 722 miles of subway system begun a century ago.

INDUSTRIAL EMPHASES OF THE NORTHEAST TODAY

The economic dominance of the Northeast to the whole of the U.S. economy is apparent in a brief look at some of its major specialties:

1. *Clothing design and manufacturing.* Clothing manufacturing is carried on in many locations but is concentrated in metropolitan New York City. The nation's greatest organizing center in this industry is Manhattan's Garment District (see Fig. 23.6). Aside from some high-fashion firms that still do their manufacturing on the premises—and numerous sweat shops that continue to job clothes out of the upper stories of tenement buildings and old professional buildings—the Garment District has become primarily a center for management, design, and merchandising, with a major share of the actual manufacturing decentralized to other locations within the metropolis, the nation, or abroad. The clothing industry in Manhattan was originally an outgrowth of New York's role as an importer of European-made cloth and clothing. It was favored by a large number of skilled immigrant workers in the later 19th century, among them Jewish tailors fleeing from persecution in Tsarist Russia.

2. *Iron and steel manufacturing.* Iron and steel production is still found in the Northeast, although employment and output have been drastically reduced in recent years because of competition from cheaper imported steel, large imports of products made from foreign rather than American steel, and the substitution of aluminum or plastics for steel in industrial processes, especially in automobile manufacturing.

3. *Chemical industries.* Chemical manufacturing is very diversified and widespread, and includes production of both industrial chemicals in bulk and a multiplicity of consumer products. The outlying areas of metropolitan New York City form the largest center, although the DuPont Company, which is the world's largest chemical manufacturing concern, has both its headquarters and a major share of its manufacturing in Wilmington, Delaware.

4. *Photographic film and equipment.* Photographic equipment

manufacturing is dominated by the world's largest photographic company, Eastman Kodak, which has both its headquarters and main plant in Rochester, New York. The Polaroid Corporation, based principally in the Boston area, is another major name in this industry.

5. *Electronics manufacturing*. Advanced electronics manufacturing, centering on computers, is widespread in the Northeast. This industry is a major specialty in eastern Massachusetts, being especially concentrated in industrial parks along Route 128 and other beltways that ring the Boston area. Graduates of the nearby Massachusetts Institute of Technology have been heavily involved in electronics development there. Some old textile towns such as Lowell and Lawrence on the Merrimack River have largely converted to electronics or other high-technology industries. They are examples of the widespread "recycling" and adaptive reuse of old mill towns in the Northeast. Eastern Massachusetts was once the nation's leading center of textile milling, but most of this industry has long since moved to the South, leaving behind substantial turn-of-the-century building stock that is now being adapted to high-tech industrial use.

6. *Electrical equipment manufacturing*. Concentrated in southern New England, New York, and Pennsylvania, electrical equipment manufacturing is dominated by the General Electric Company, headquartered at Stamford, Connecticut, and with major production facilities in Schenectady, New York. Westinghouse Electric, headquartered in Pittsburgh, is another major company with plants in the Northeast and other areas.

7. *Aircraft engines*. The manufacture of aircraft engines and helicopters is heavily concentrated near Hartford, Connecticut, where it is carried on by United Technologies Corporation, maker of the famous Pratt and Whitney engines and Sikorsky helicopters.

8. *Nuclear submarines*. The manufacture of Trident nuclear submarines in the New London, Connecticut, area continues New England's long shipbuilding tradition as well as Connecticut's long tradition of armaments manufacturing. It was at a factory built near New Haven, Connecticut, in 1798 to supply muskets under a federal contract that Eli Whitney originated the use of interchangeable product parts in the manufacture of firearms. This practice is central to modern mass production of innumerable types of goods.

9. *Publishing and printing*. Publishing and printing are industries in which the Northeast decisively overshadows the rest of the country. Most of the publishing industry is in the New York metropolis, although the Government Printing Office in Washington, D.C., is the world's largest publishing enterprise and printing plant. Many of the book and magazine printing phases of the industry have been decentralized to other parts of the country, but editorial and management functions remain concentrated in the Northeast.

Definitions and Insights

ADAPTIVE REUSE

One of the key factors in any effort to keep vitality in any urban place is the **adaptive reuse** of earlier building stock. This means taking abandoned commercial or factory buildings and converting them to some economic function that brings jobs, commercial activity, and people back to a once relatively productive area. This has been done with particular success in New England. In any downtown that has been losing customers and businesses for some years, the transformation of empty retail stores into speciality shops or trendy restaurants has become a commonplace example of adaptive reuse of older buildings. This process is important because (a) it gives new vitality to places and often well-designed buildings that once had economic and social centrality but subsequently lost it; (b) it helps create a sense of the value of the past as old landscapes are visited and utilized again; and (c) it maintains the structures in their historic context as landscape reference points or pivotal buildings. While adaptive reuse is most often considered in the context of old central business districts, the process has become increasingly appropriate in the resettlement of old residential neighborhoods as well. It serves as a bellwether for the revitalization of neighborhoods and helps in the fuller use of at least some sections of a graying, declining, and sometimes abandoned urban landscape.

This assemblage of manufacturing, planning, design, decision-making, and management capabilities has also prompted the establishment of tens of thousands of smaller, complementary, and ancillary industries that make their living by supplying materials and information to this concentration of dominant businesses. In the geography of commerce, there is great importance placed upon proximity, interaction, and timely supply of goods and information. New York City and the Northeast have historically been very effective in this union.

THE ROLE OF SPECIALIZED AGRICULTURE

In colonial times New England had only pockets of good land, concentrated around Boston and Providence and in the lowland along the Connecticut River, but hard toil enabled farm families to subsist and produce food for people engaged in fishing, trade, and handicraft manufacturing. In the 19th century, competition from the agriculturally favored Midwest after the 1825 completion of the Erie Canal drove many Northeastern farmers from the land. Others began shifting toward specialties adapted particularly to the Northeast's natural conditions and markets. The

FIGURE 23.9

Amish (Mennonite) farmer turning the soil with a horse-drawn plow in north central Ohio. Here, a major Amish concentration centers in Wayne and Holmes counties near Akron and Massillon. While Lancaster County, Pennsylvania, remains the most famous area of Amish settlement, there are many other settlements in scattered parts of the United States. Deeply traditional cultural practices persist in such areas, just as they do in Lancaster County. The soil developed under the broadleaf forest in the Ohio area is not first-rate, but it has been made highly productive by skilled Amish management.   *Fred Wilson*

main adaptation was a conversion from general grain and livestock farming to dairy farming. Distant areas with better land could not ship fluid milk to Northeastern markets in competition with Northeastern farmers. Dairy farming was also logical for New England especially, where poor soil and sloping land were more appropriate for pasture or hay than tilled crops. The region also has some very distinctive farm landscapes created by the Amish who have always had a distinctive attitude toward technology and farming (Fig. 23.9).

Other agricultural specialties also developed. The earliest was large-scale "truck farming," producing fresh summer vegetables and fruits to be hauled overnight to city markets (a tradi-

FIGURE 23.10

The Baltimore Inner Harbor development—called Harborplace—of James Rouse is one of the most successful reclamations of an old warehouse and wharf area in the country. The current uses of this initial 18th century harbor facility feature a two-story shopping mall with abundant outdoor restaurants and cafes, a new aquarium, relic brick factory power plant buildings converted into elegant townhouses and condominiums, and easy access to the new Camden Yards, home of the Baltimore Orioles. Rouse called such (re)developments "festival marketplaces." Berthing for local watercraft and tour ships that cruise the Chesapeake Bay are all part of this urban area that, thirty years ago, looked as though it would never be presentable to the public eye again, or economically useful. It now serves as one of Baltimore's most popular tourist destinations.   *Jerry Watcher/Photo Researchers*

tion that continues today). This developed mainly on the poor and sandy but easily tillable soils of the Coastal Plain, where high output value per acre justified high inputs of fertilizer. Another major recent specialty is broiler-chicken production, which has become the largest source of farm income in Maryland and Delaware. And in four extraordinarily urbanized and densely populated states—Massachusetts, Rhode Island, Connecticut, and New Jersey—greenhouse products have become the leading source of agricultural income. In the present-day Northeast, agriculture is far overshadowed by other ways of life and land uses, but it is nevertheless highly efficient. Delaware is the country's most productive state agriculturally per unit of total area, and Maryland, New Jersey, and Pennsylvania also gain significant state income from agriculture. The other states are less productive, with New Hampshire and Maine ranking very low. Generally, agricultural products are produced mostly on small farms, with a large input of capital and labor resulting in a large output per acre used. Such intensity is required by the high cost of land and is permitted by favorable market locations.

THE "OLD METROPOLITAN BELT" VERSUS THE "SUN BELT": INDUSTRIAL CHANGE AND SHIFT IN THE NORTHEAST

The Old Metropolitan Belt in this region contains the great majority of cities that were already prominent by the early 20th century. It stretches from the Northeastern Seaboard across the Midwest as far as Kansas City, Omaha, and Minneapolis-St. Paul, and is often called the Frost Belt, Snow Belt, or, at least in part, the Rust Belt. The Old Metropolitan Belt has recently been growing quite slowly in population and industrial output, with some areas actually declining in one or both of these. Most of the country's old "smokestack industries"—a term applied particularly to the steel and other heavy metallurgy industries—are in the Middle Atlantic states and eastern Midwest. Today, these industries are often characterized by outmoded buildings and equipment, depressed sales and profits, high unemployment, abandoned industrial buildings, and unattractive surroundings. By contrast, the Sun Belt in the South and West has been enjoying faster growth in population and jobs, including those in manufacturing. ("Sun Belt" is a very elastic term, but common usage and geographic logic both suggest an area encompassing most of the South, plus the West [in this text's regional system for the United States] at least as far north as Denver, Salt Lake City, and San Francisco.) The loss of manufacturing employment in the Old Metropolitan Belt, including the Northeast, reflects such factors as (1) obsolete plants being closed, (2) shifting of production to more profitable locations in other regions or overseas, (3) reduced employment in surviving plants reequipped with new labor-saving machinery, and (4) failure of new plants to locate in the old Metropolitan Belt in sufficient numbers to offset the job losses in existing industries.

Employment in service industries has been expanding quite rapidly in the Northeast, resulting in an expansion of total employment. But despite this, the region has been growing in population relatively slowly, with heavy outmigration from the Middle Atlantic region. The main exceptions to this slow growth are the peripheries of some metropolitan areas into which suburban and exurban development are expanding. Notable examples include the expansion into New Hampshire from Boston; into Connecticut and New Jersey from New York City; and into Maryland from Washington, D.C. The state of Vermont, whose population growth has recently been high, is not suburban, but much of this quiet and scenic state is exurban, attracting "refugees" from metropolitan areas, owners of second homes and vacation homes, and some diligent commuters, including those travelling by air. And despite the weight of economic and urban problems, the Northeast has recently been the scene of spectacular new development in the great metropolises, both in old decaying cores and around the peripheries such as the Baltimore Inner Harbor area (Fig. 23.10).

Definitions and Insights
GENTRIFICATION

Just as adaptive reuse has been a significant dynamic in the transformation of old, underused, or abandoned building stock to current economic use, **gentrification** is a process that has been giving whole neighborhoods and residential landscapes a new lease on life. This term denotes the upgrading of generally inner-city residential properties in a manner attractive to more affluent new occupants. Professional people, often with no children, have been active players in this process. However, gentrification has drawn criticism because it often displaces the poorer original residents, who then have difficulty in finding new lodgings that they can afford. The process usually takes the following course: In many urban landscapes, whole neighborhoods go through a descending spiral as single-family homes become rental units and then multifamily homes. This is often accompanied by the departure of the initial population and the replacement of that stock with immigrant peoples or minority peoples who find the central location and the relatively low cost of shelter to be the best deal they can get in an American city. These populations, with varying degrees of success, keep some life in the old residential stock while local stores and services often decline into very marginal conditions. The gentrification classically occurs when people with some capital and a willingness to invest **"sweat equity"** (personal funds and labor) begin to take an interest in the neighborhood because of its lower costs, architectural character, and proximity to the central city's professional, legal, and governmental offices and facilities. Like so many geographic processes that are embodied in a changing landscape, gentrification initiates a complex series of social and political changes as the landscape is remade.

23.2   The South: Old Traditions and New Directions

Various definitions of the South are possible, and none is entirely satisfactory (see Fig. 23.1). The region is discussed here as a block of 14 states: 5 along the Atlantic from Virginia to Florida; 4 along the Gulf of Mexico from Alabama to Texas; and the 5 interior states of West Virginia, Kentucky, Tennessee, Arkansas, and Oklahoma. Arguments could be made on various grounds for excluding some of these states. For example, West Virginia and Kentucky did not secede from the Union; and the western parts of Oklahoma and Texas lie in dry environments more typical of the West than the South. Missouri was a slave state and thus part of the South before the Civil War but today seems more typically Midwestern.

PHYSICAL ENVIRONMENTS AND THEIR UTILIZATION

Most of the South has a humid subtropical climate, with summers that are long, hot, and wet. January average temperatures range from the 30s (degrees Fahrenheit) along the northern fringe to the 50s near the Gulf Coast, and the 60s in southern Florida and the southern tip of Texas. More than 40 inches (c. 100 cm) average annual precipitation is characteristic, rising to over 50 inches in many Gulf Coast and Florida areas and more than 80 inches in parts of the Great Smoky Mountains. High humidity is characteristic, intensifying heat discomfort in summer and chilliness in winter. There are both advantages and disadvantages to these climatic conditions. They foster rapid and abundant forest growth, and forest-based industries are of major importance. Agriculturally, they provide long growing seasons and heat and moisture for a wide variety of crops. But insects and pests also flourish, and the heavy rainfall leaches out soil nutrients and causes severe erosion and flooding.

The Predominance of Plains

Most of the South is classified topographically as plains, but the plains form sections of three major landform divisions—the Gulf-Atlantic Coastal Plain, the Piedmont, and the Interior Plains (see Fig. 23.3). The Coastal Plain occupies the seaward margin from Virginia to the southern tip of Texas, including all of Florida, Mississippi, and Louisiana and portions of all the other Southern states except West Virginia. It is low in elevation, and large areas near the sea and along the Mississippi River are flat, but most inland sections have an irregular surface. Much marshy and swampy land occurs near the sea and in parts of the northward-reaching Mississippi Alluvial Plain (Mississippi Lowland). The Coastal Plain was originally forested (and the greater part still is), aside from isolated belts or patches of grassland such as the marsh grasses of the Ever-

glades in southern Florida, the "coastal prairies" along the Gulf of Mexico in Louisiana and Texas, and prairie grasslands in some inland areas (such as the Black Waxy Prairie of Texas) that are underlain by limestone. Forest growth is dominated by pines and oaks except for gums, cypresses, and other "swamp hardwoods" in poorly drained areas. Soils are generally sandy, heavily leached, and deficient in plant nutrients. The Coastal Plain has a long history of soil depletion and abandonment of worn-out cropland. Many former agricultural areas have reverted to forest. The most fertile soils are associated with river alluvium, chalky limestone bedrock, or organic material in drained swamps and marshes.

The Piedmont extends across central Virginia, the western Carolinas, northern Georgia, and into east-central Alabama. Its surface is mainly a rolling plain, with some hilly areas. The natural cover is mixed forest in which broadleaved hardwoods, especially oak, tend to predominate over pines. Large areas have reverted to forest from former agricultural use. Soils tend to be poor, and many have suffered serious damage through past cultivation of clean-tilled row crops (cotton, corn, and tobacco) on easily eroded slopes subject to frequent downpours. The region's fall line (often where early industry began), lies between the Piedmont and the Coastal Plain.

The Interior Plains section of the South occupies central and western Texas, plus all of Oklahoma except the easternmost part. In Texas, the Balcones Escarpment forms an abrupt boundary between the higher Interior Plains and the lower Coastal Plain. The eastern edge of the higher land is hilly, especially toward the south in the Edwards Plateau. Such cities as Dallas, Austin, and San Antonio lie on the Coastal Plain at the foot of this hilly belt. Westward, the plains rise in elevation and become more level. The elevated plains of western Texas and Oklahoma are not typically Southern. Their steppe and desert environments are more typical of the West, and irrigation, ranching, and dry farming of wheat and sorghums are important to the sparse population. But culturally, these areas are rather strongly Southern; in a large majority of counties the Southern Baptist Convention is the largest religious organization. Catholic Hispanics predominate in the far south of the Interior Plains, but a large majority of the Hispanics in Texas (and African Americans as well) live in the Coastal Plain.

Hills, Mountains, and Small Plains of the Upland South

Large parts of the South are hilly or mountainous. In the central and eastern South, two large embayments of rough country project into the South from the Northeast and Midwest. The larger of the two lies mainly in the Appalachian Highlands but includes some rough land west of the Appalachians known as the Unglaciated Southeastern Interior Plain (see Fig. 23.3). Farther west, the second large embayment is the southern part of the Interior Highlands. The three physical areas—Appalachians,

Unglaciated Southeastern Interior Plain, and Interior Highlands—are often termed the Upland South.

Southern areas within the Appalachian Highlands extend from western Virginia to eastern Kentucky and southward to northern Alabama. The Highlands include all the major physical subdivisions already distinguished in the Middle Atlantic states, with the total complex narrowing southward (see Figs. 23.1, 23.3). The Blue Ridge extends from the Potomac River to northern Georgia. Its highest section is called the Great Smoky Mountains. Here, Mount Mitchell in western North Carolina is the highest point in the eastern United States, at 6684 feet (2037 m). The Ridge and Valley Section, including the Great Appalachian Valley, runs parallel to the Blue Ridge and inland from it, from western Virginia into Alabama. The southern section of the Appalachian Plateau, known as the Cumberland Plateau in parts of Kentucky and areas to the south, is quite wide in the north, where it occupies eastern Kentucky, most of West Virginia, and a small section of Virginia. It narrows rapidly across eastern Tennessee and then widens to occupy much of northern Alabama. As in Pennsylvania, it is edged on its east by an abrupt escarpment, is mostly dissected into hills and low mountains, and is underlain by very large deposits of high-quality coal.

Central Kentucky, central Tennessee, and part of northern Alabama are in the Unglaciated Southeastern Interior Plain. This part of the Interior Plain, bounded by the Tennessee River on the west and south, was not subject to the glacial smoothing that occurred on the Plains farther north. Hence, a good part of this "plains" area is actually hill country, the result of dissection by streams over a long period of time. True plains are present but are not extensive. Two of them—the Kentucky Bluegrass Region around Lexington and the hill-studded Nashville Basin around and south from Nashville, Tennessee—are underlain by limestone that has weathered into good soils. Both areas are famous for their agricultural quality in a region of poor to mediocre soils and difficult slopes.

In the South, the Interior Highlands occupy northwestern Arkansas and most of the eastern margin of Oklahoma. From Arkansas, they extend northward into the southern part of the Midwestern state of Missouri. Composed of hill country, low mountains, and some areas of plains, these highlands are split by the broad east-west Arkansas Valley followed by the Arkansas River. North of the valley, the Boston Mountains form the southern edge of the extensive Ozark Plateau (Figure 23.11). South of the valley lie the Ouachita Mountains of west central Arkansas and adjacent Oklahoma. Most of the Ozark Plateau has been dissected into somewhat subdued hill country, but some parts of the Plateau, especially the Boston Mountains, attain mountainous character. The Ouachita Mountains consist of roughly parallel ridges and valleys running east and west.

The Southern Interior Highlands have not been rich in resources. Most soils are quite poor, with many very stony. Most of the area is still forested, but it has been heavily cut over, and the present timber is generally rather poor. Mineral deposits are varied but often not of sufficient value to justify present exploitation. The main mineral wealth extracted today from the Interior Highlands, primarily lead, comes from the section outside the South in Missouri. A number of large artificial lakes used for recreation are major resources in both Missouri and Arkansas.

COMPARISON WITH OTHER COUNTRIES AND MAJOR AMERICAN REGIONS

Like the other three major American regions discussed in this text, the South is sufficiently large, populous, and productive to be an impressive nation in its own right. With an area of 886,000 square miles (2.3 million sq km, or 24 percent of the United States) and a 1995 estimated population of 92.3 million (35 percent of the nation's population), the 14 Southern states could constitute a country ranked twelfth in the world in population and exceeded in area only by the United States and eleven other countries. The country would be, as the region now is, a major world producer of a long list of resource commodities, agricultural products, and manufactured goods. Such metropolises as Dallas–Fort Worth, Houston, Atlanta, and Miami would be noteworthy in any country (Fig. 23.12).

Regional patterns of the South in population density, urbanism, ethnicity, and income are distinct from those of the other major regions. Until recently, it was a region of farms, villages, and small cities, with low population density compared to that of the Northeast and Midwest. But for the past several decades, the South has been a region of rapid population growth, second only to the West, while its farm population has decreased drastically. It has now surpassed the Midwest in overall population density, although both are far behind the Northeast. Because recent Southern growth has been overwhelmingly urban, previous contrasts in this respect also are rapidly fading. The South has more small and medium-sized metropolitan areas than other regions of the country (because of recent rapid growth of many small cities). However, it has fewer centers of truly large size than might be expected from its population. The largest urban complex, Dallas–Fort Worth, has only 4.3 million people, and only it and Houston (population: 4.1 million), Miami (3.4 million), Atlanta (3.3 million), and Tampa–St. Petersburg (2.2 million) are among the more than 20 American metropolitan areas with over 2 million people. But these Southern urban complexes continue to grow very rapidly.

The population of the South has less ethnic and racial variety than that of the other regions. The overwhelming majority is comprised of white Protestants of British ancestry, African-American Protestants, or Mexican-American Catholics. Most striking is the heavy preponderance of white Protestants with Anglo-Saxon origins. The original European settlers along the

A

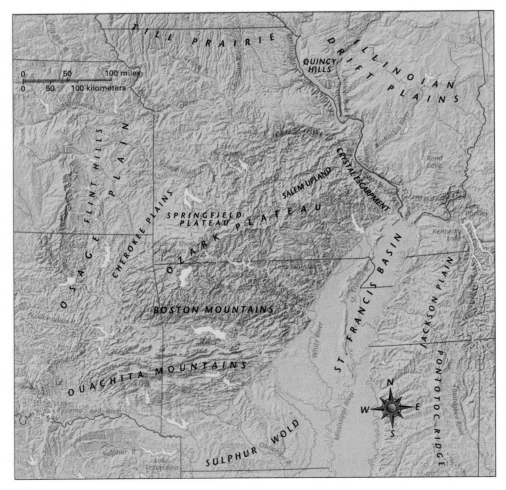

B

FIGURE 23.11

A scene in the southwestern edge of the Missouri Ozarks. (a) The water body lined with boathouses is an arm of one of the many artificial lakes in the Interior Highlands. (b) The subdued hilly topography evident on the map is characteristic of this physiographic area.   *Jesse H. Wheeler, Jr.*

FIGURE 23.12

This scene of the Dallas skyline at dusk does a good job of illustrating the drama of this Texas financial and commercial center. Dallas, in combination with Ft. Worth, exists as the major commercial and financial center for central Texas. Architects and engineers have given much attention to the clear light of this setting, draping glass over the surfaces of the tall buildings that often serve as icons or billboards of specific Texas firms. *James Blank/FPG International*

South have improved at the same time economic conditions of many Northeastern cities have deteriorated.

Other distinctive racial and ethnic elements in the South's population are less widely distributed. Especially notable are Mexican Americans, Cuban Americans, and Cajuns (of French descent). The 1990 census reported that the four states leading in percentage of population "of Hispanic origin" were New Mexico (38 percent), Texas (26 percent), California (26 percent), and Arizona (19 percent). In Texas, Hispanic people primarily of Mexican descent live mainly in a broad band of territory along the Gulf of Mexico and the Rio Grande, where they form a much higher proportion than the statewide 26 percent. Natural increase and immigration have rapidly expanded their numbers (in Texas and elsewhere) during recent decades.

South Florida has received intermittent immigration of another "Hispanic origin" group—Cuban Americans. Almost a million refugees have fled from communist Cuba since the revolution of 1959, and the majority of them or their descendants now live in metropolitan Miami. Cuban Americans, drawn largely from the middle and upper classes of precommunist Cuba, have shown rather rapid upward mobility economically. They also have been assimilating into the general population through marriage and rapid adoption of U.S. culture traits. Meanwhile, they have imprinted aspects of their own culture on south Florida. The country's Mexican American population, on the other hand, largely continues to be a group whose Hispanic identity remains strong. The largest non-Hispanic minority language group in the South are the French-descended Cajuns of southern Louisiana.

THE SOUTH'S EARLY AGRICULTURAL HERITAGE

Reliance on cash crops began early in the South. Jamestown, Virginia, just upstream on the James River from present-day Norfolk, Virginia, was founded in 1607 as the first enduring English settlement in what would become the United States. The area was found suitable for tobacco, a New World plant already in high demand in Europe. Tobacco cultivation is very labor-intensive, and enough workers could not be mustered from the colonists or the local Indians. The labor shortage was solved by importation of African slaves, beginning in 1619. Soon the Virginia "Tidewater" (the Coastal Plain adjacent to Chesapeake Bay) became the first large region of slave-plantation agriculture in the South. Production spread to adjacent Maryland and Delaware and inland onto the Piedmont. So prominent did the plantation aristocracy of Virginia become that all but one of the first five presidents of the United States were drawn from it.

In the late 17th century and 18th century, a second outstanding plantation area developed in Coastal Plain swamps inland from the port of Charleston, South Carolina. In this hot, wet environment, irrigated rice and indigo became the staple

Atlantic were British. While small groups of Spanish Catholics settled in Florida and Texas, and some French Catholics in Louisiana and elsewhere along the Gulf Coast, their numbers were insignificant compared with the tide from Great Britain.

Another way in which the South stands out ethnically from the other major regions is in its high proportion of African-American population. This is largely a result of the importation of African slaves for plantation labor in the pre–Civil War South. The five American states that rank highest in percentage of African Americans—from about one fourth to somewhat over one third—are Mississippi, Louisiana, South Carolina, Georgia, and Alabama, which together form a Coastal Plain belt. Migration out of the South, mostly in the last 50 years, has given some northern states and California large African American minorities. New York and California both have larger African-American populations than any Southern state, but the percentage is only 16 percent in New York and 7 percent in California. African-American populations outside the South are very concentrated in large metropolitan areas. Recently, a sizable reverse migration of African Americans from North to South has occurred as economic and social conditions in the

exports. Then, in the 1790s, technological innovation catapulted cotton into the lead as the main crop of the plantation economy. (Cotton previously had been an expensive luxury because of the amount of hand labor initially required to separate the cottonseeds from the fibers. Connecticut inventor Eli Whitney solved this problem by inventing the mechanical cotton gin.) Cotton became a cheap raw material for the new textile factories of the Industrial Revolution in England and New England. A long boom in cotton growing ensued until the Civil War broke out in 1861. Picking the crop was labor-intensive, and slavery expanded along with it.

Definitions and Insights

AGRICULTURAL INNOVATION

Eli Whitney may have done more to transform the agricultural landscape of the American South than any other single individual. He was a Yale graduate who was without work and visiting a friend on a cotton plantation in Georgia in 1792. There was a strong market for cotton in England to supply the burgeoning textile industries there, but American short-staple cotton—the most easily grown in the South—was difficult to extract from the seed. The labor costs of the hand separation were so great as to make cotton a losing proposition unless one could grow the more easily processed long-staple variety. Whitney spent ten days pondering this problem and then created a model machine that, in fact, effectively separated the short-staple cotton from the seed and produced a marketable product with unprecedented quickness and economy. He and his friend received a patent by 1794 and began production. While the cotton gin made fortunes for the many adopters of this agricultural innovation, Whitney and Miller made little money because of endless litigation with the ever-more wealthy plantation owners.

However, that is not the end of the story. In the manufacture of his gin, Whitney had realized the value of using interchangeable parts in the design and construction of his machine. He later used this method to revolutionize the manufacture of muskets in the last decade of the 18th century, and in 1801 he arranged a demonstration for president-elect Thomas Jefferson. He asked officials to pick random parts from piles of musket components and he then assembled a workable gun from these parts. The success of this innovation—directly derived from his work with the cotton gin—introduced the concept of mass production in the U.S.

By the Civil War, cotton plantations extended westward to eastern Texas and tobacco plantations to central Missouri. For the most part, the plantations that grew cotton were a feature of the southern parts of the Coastal Plain, where they tended to cluster in certain favored areas. The first great center of production was the offshore islands and adjacent mainland of Georgia and South Carolina. This Sea Island District continued to be an important producer until the crop was devastated by the boll weevil (a destructive insect) in the 20th century. By 1860, the black belt (so called for the color of its soils) of Alabama and Mississippi and the flood plain of the Mississippi River (Mississippi Lowland) had become the principal centers of production, although the crop had spread farther west into eastern Texas. The black belt is a crescent-shaped area extending east and west through central Alabama and then curving into northeastern Mississippi (see Fig. 23.1). Its soils, developed from limestone, were exceptionally fertile until they were worn out by excessive cropping in cotton, which, like tobacco, is notoriously hard on soils. The Mississippi Lowland proved more durable as a cotton producer and remains important today. This alluvial plain, some 600 miles long (966 km) from the mouth of the Ohio River to the Gulf of Mexico, and often over 50 miles (80 km) wide (see Fig. 23.3), is a remarkable feature in the physical geography of North America—comparable to the great flood plains and deltas that support dense populations in the Orient. It has deep and fertile alluvial soils, but many sections away from the river are swampy.

The dramatic expansion of the plantation economy that made the South the world's chief supplier of cotton was associated with a relative failure of other development. Nothing rivaling the industrialization and urbanization of the Northeast occurred in the South. Investment went mainly into land and slaves, and few cities of much size developed. Although slavery was ended by the Civil War, heavy dependence on cash crops was not. Dependence on cotton and tobacco continued, and both generally failed to yield their cultivators a good living. Cotton prices were adversely affected by increasing competition from foreign growers, and the South's own cropland gradually deteriorated. The depths of distress were reached in the Great Depression of the 1930s, when large numbers of cultivators and their families were forced to seek government relief in order to live.

ECONOMIC CHANGE SINCE THE GREAT DEPRESSION

Since the 1930s, the South has made great progress in narrowing the economic gap that set the region apart for so long. Many factors have contributed to this, but probably none has been more important than the end of dependence on hand-tilled and hand-picked cotton. The collapse of world trade in the Great Depression, with attendant record low prices for cotton, marked the beginning of the end of the Cotton Belt. Its demise was hastened by the introduction of a successful mechanical cotton-picker in the 1940s, which ended the need for concentrated labor in the cotton fields. Large and expensive, mechanical pickers could be used most advantageously by large farms on level land, producing high yields. Hence, their use favored a westward shift of cotton growing toward irrigated areas in the Texas High Plains, California, and Arizona. Cotton has continued to be important on the Mississippi Alluvial Plain, but relatively little cotton is grown east of there. Even on the Mississippi Plain, the farm economy depends heavily on corn, soybeans, and rice as well as cotton.

Disastrous unemployment might well have resulted from the mechanization of cotton production and the attendant abandonment of cotton over large areas. For many families, such unemployment did create great hardship, at least for a time. Many displaced agricultural workers, both black and white, joined a massive job-seeking migration to northern industrial cities that was already underway before the Great Depression. But the bulk of the South's redundant cotton workers eventually were absorbed by an expanding job market in the accelerating nonagricultural sector of the Southern economy. Many factors have contributed to this expansion: cheap labor, federal military expenditures and programs of economic development, intensified development of Southern natural resources, the introduction of air-conditioning, and enormous improvements in transportation with the creation of the Interstate Highway System and the expansion and improvement of airways and waterways. The continued migration of retirees to the South has also played a major role in U.S. demographic change and growth of the South.

AGRICULTURAL GEOGRAPHY OF THE SOUTH TODAY

What are the main features of the South's agricultural geography after a half-century of revolutionary changes in crop and livestock emphases, farm mechanization, farm markets, and the number of farmers? In detail the picture is intricate, but some generalizations are possible:

1. *In appearance and land use, the greater part of the South today is not agricultural.* Over large areas, less than half the land is in farms, and even where land is farmed, the amount of cropland is apt to be greatly exceeded by woodland and/or pasture.
2. *The present South is not outstandingly agricultural compared with the country as a whole.* Only about 2 percent of the national population still lives on farms, and the proportion in the South is very close to this—higher than in the Northeast or West, but much lower than in the Midwest.
3. *Southern farms tend to be small, low in production, and often part-time.* The prevalence of part-time operations and of small full-time farms means that output per farm over most of the South is far below the national average.
4. *Within the South as a whole, animal products decisively surpass crops in total farm sales.* Beef cattle, broiler chickens (Fig. 23.13), and dairy cattle are the main sources of animal products sold. Rising cattle production has been fostered by (1) the presence of much land that is too poor for crops but amenable to pasture, (2) improvements in pasture grasses and cattle breeds, (3) increased availability of chemicals for pest and disease control, and (4) greater emphasis on soy-

beans, sorghums, and citrus by-products for cattle feed. Soybeans have become the most widespread and important Southern crop. The leguminous soybean plant adds nitrogen to the soil and can be used for hay. But its greatest value lies in the beans that are pressed for soybean oil, with the residue (oilcake) being fed to cattle and poultry. Grain sorghums, which have low moisture requirements, have become a mainstay for cattle feeding in drier parts of Texas and Oklahoma. The greater part of Florida citrus fruits are now processed for frozen concentrate, and the residue is fed to cattle; the main market is the large ranch industry in southern Florida. Broiler chicken production has become a specialty of poorer agricultural areas, with major concentrations in the Interior Highlands of Arkansas and the Appalachians of Georgia and Alabama. Dairy farming is growing in importance as Southern urban markets grow.

5. *The historic staples of tobacco and cotton are still major elements in the agriculture of some parts of the South.* Most of the nation's tobacco is grown in the eastern and northern South. In 1996, it was the leading source of income in the agriculture of North and South Carolina, Kentucky, and Tennessee. Nearly all the South's remaining cotton is grown on large mechanized farms in the Mississippi Alluvial Plain or in west Texas.
6. *Many islandlike districts of intensive agriculture produce a variety of other specialties.* Often such districts represent the

FIGURE 23.13
Raising and processing broiler chickens has become a massive agricultural industry in the American South. The broilers in this photo will be processed at a plant in Arkansas operated by the food company ConAgra.
Mike Sprague

main producing areas of their kind in the United States. Some major specialties and areas include citrus fruits in the central part of the Florida peninsula; truck crops in the Florida peninsula; sugarcane in the Louisiana part of the Mississippi Lowland and near Lake Okeechobee in Florida (on a drained section of the northern Everglades); rice in the Mississippi Lowland and the prairies along the coasts of Louisiana and Texas; peanuts in southern Georgia and adjoining parts of Alabama and Florida; and race horses, bred particularly in the Kentucky Bluegrass around Lexington, where the horse farms are showpieces of conspicuous wealth.

INDUSTRIAL CHANGE IN THE SOUTH AND ITS GEOGRAPHIC OUTCOME

After the Civil War, Southern leaders began a drive to industrialize the region, and this effort is still in progress. The earliest large-scale industries to develop were cotton-textile and apparel factories using cheap labor from economically depressed countrysides. In the late 19th and early 20th centuries, this development gained regional importance, and from the 1930s to the 1950s it sapped the cotton-textile industry of New England and eventually gave national preeminence in the industry to the Southern Piedmont from Virginia to Alabama. The new factories were built in small towns and cities—many of which offered tax advantages and free land to attract mills. Piedmont mills sold cloth first to apparel factories in the Northeast, then to new Southern factories built at many locations in the Piedmont and Coastal Plain. In the 1940s, Southern industrialization accelerated and diversified. Low wages remained a powerful factor, but many other advantages developed, including new hydroelectric facilities built by the federal government in the Tennessee River Basin and elsewhere; federal improvements in river transportation and harbors; and joint federal and state construction of the Interstate Highway System, which gave the South cheap and fast new connections with markets elsewhere. Air-conditioning made Southern workplaces more comfortable and productive. When synthetic fabrics arose to challenge cottons and woolens, the South had abundant raw materials (wood and hydrocarbons) to make synthetic fibers. Unfortunately, the offshore movement of textile plants in the last decade has been most significant in the recent decline of the Southern textile industry.

An extraordinarily valuable regional store of energy resources—oil, natural gas, coal, and falling water—powered new Southern industries and provided large energy exports to markets outside the region. Increasing technical skills became available through a combination of on-the-job experience, improved research and teaching by Southern universities, and import of technology to such places as the Oak Ridge atomic research facility near Knoxville, Tennessee, and the space flight facilities of the National Aeronautics and Space Administration (NASA) at Huntsville, Alabama; Houston, Texas; and Cape Canaveral, Florida. As population increased and incomes grew, Southern regional markets became large enough to attract plants that could profit from nearness to those markets. By the early 1990s, the South was reaching a degree of industrialization generally commensurate with that of the entire nation (see Table 23.1); at least two Southern states—Texas and North Carolina—are now major industrial states.

THE MAJOR SOUTHERN INDUSTRIAL BELTS

The largest cluster of industrial development in the South lies along the Piedmont, with extensions into the adjacent Coastal Plain and Appalachians. Within this concentration, which may be termed the Eastern South Industrial Belt, textiles are the most prominent branch of manufacturing, but clothing, chemical products (including synthetic fibers), furniture, tobacco products, and machinery are also important. Much of this region's development, as well as that of adjacent areas in the South, has been fostered by the Tennessee Valley Authority (TVA). The TVA was founded in 1933 by the federal government during Franklin Roosevelt's New Deal era in order to promote economic rehabilitation of the deeply depressed Tennessee River Basin. It is a public corporation that built many dams, locks, and hydrostations on both the Tennessee and its tributaries. These installations controlled floods; allowed barge navigation as far upstream as Knoxville, Tennessee; generated hydroelectricity; and provided recreation at numerous reservoirs. The demand for electricity soon outran the capacity of the hydrostations, and many thermal plants were then built, fired by coal or using nuclear energy. The success of the TVA in helping industrialize several Southern states is a well-publicized facet of the industrial surge that has spread ever more widely across the South in recent decades. While in the North industrial development focused on the huge metropolises that arose in the railway age, in the South most development came later and took advantage of long-distance power transmission and high-speed truck transportation to spread widely into small cities and towns.

The second main belt of industrial development in the South lies along and near the Gulf Coast from Corpus Christi, Texas, to New Orleans and Baton Rouge, Louisiana. The chemical industry, including oil refining, is the dominant industrial sector within this West Gulf Coast Industrial Belt. It is based on deposits of oil, natural gas, sulfur, and salt. Texas and Louisiana produce nearly one third of the oil, over three fifths of the natural gas, and practically all the native sulfur output of the United States, with the Gulf Coast strip the leading area of production for all three. Machinery and equipment for oil and gas extraction, oil refining, and chemical manufacturing are also important products. Major refineries and petrochemical plants are clustered at several locations, most notably metropolitan Houston and Beaumont, Texas, and Baton Rouge, Louisiana.

Many other industries are present in the South, some of which are widespread while others are localized in a few places. Food processing, machinery, furniture, and pulp-and-paper industries are widely distributed, with food industries especially important in Florida and pulp-and-paper plants particularly prominent near the Gulf Coast from eastern Texas to Florida and along the Atlantic Coast in Georgia. (For raw material, the latter industry depends largely on the pine trees that grow rapidly in the wet subtropical climate.) Aircraft and aerospace industries are prominent in the Atlanta and Dallas–Fort Worth areas, and both auto assembly and the manufacture of auto components are expanding, especially in Tennessee and Kentucky (Fig. 23.14).

TEXAS-OKLAHOMA METROPOLITAN ZONE

The Texas-Oklahoma Metropolitan Zone contains the two largest urban clusters in the South—Dallas–Fort Worth and Houston—plus four other metropolises with populations of 750,000 or over: San Antonio (population: 1.4 million) and Austin (population: 964,000), Texas, and Oklahoma City (population: 1.07 million) and Tulsa (population: 745,000), Oklahoma. Houston is a seaport by virtue of the Houston Ship Channel, constructed inland to the city in the late 19th and early 20th centuries. The city has become a major port serving Texas and the southern Great Plains, in addition to being the principal control, supply, and processing center for the oil and gas fields and the chemical industry of the Gulf Coast. Dallas–Fort Worth, San Antonio, and Austin developed at or near the western edge of the Coastal Plain in a north-south strip of territory known as the Black Land Prairie, where unusually fertile limestone-derived soils supported a productive agriculture and towns and cities to service it (Fig. 23.15). Dallas–Fort Worth and San Antonio also serve large trading hinterlands stretching far to the west, within which there are no comparable competing centers. In the case of Dallas–Fort Worth, the same is true in other directions except for the competition of Houston to the south. Dallas–Fort Worth is the dominant business center for this huge region. San Antonio is a major military base and retirement center, and Austin is the site of the strongly funded University of Texas, a center for high-technology industries, and the seat of Texas state government. All these cities are involved in the oil and gas industry.

FLORIDA METROPOLITAN ZONE

A second area of exceptional metropolitan development is the Florida peninsula. Two Florida metropolitan areas here have estimated populations of more than 2 million each—Miami (3.4 million) and Tampa (2.2 million)—and three more have over 900,000 each: Orlando (1.4 million), Jacksonville (975,000), and West Palm Beach (955,000). The distribution of Florida's main metropolitan areas reflects the importance of amenities at-

FIGURE 23.14

Toyota has brought a considerable portion of its automobile production capacity to the United States, working especially on its models that have a major share in the domestic auto market. The majority of its Camry and Avalon production, for example, occurs in this Toyota plant in Georgetown, Kentucky. Such production centers are highly mechanized.   *Renato Rotolo/Gamma Liaison*

tractive to tourists and retirees. Proximity to the ocean is important, but winter warmth is even more crucial. Miami's marginally tropical winters have undoubtedly helped it to grow faster than Jacksonville in the extreme northeast. But Florida's cities are by no means totally devoted to vacation and retirement functions. Miami has become a major point of contact between the United States and Latin America in regard to air traffic, tourist cruises, banking, and the drug traffic. Tampa Bay affords a major harbor well placed for the trade of central Florida's citrus belt and phosphate mines. High-technology industries and regional corporate offices are increasingly attracted to Florida as the state's population grows and improved transportation and communication reduce the disadvantages of location in places remote from the Old Metropolitan Belt of the Northeast. For example, Orlando's extraordinarily rapid recent growth has been spurred by corporate office development as well as by the presence of Disney World—the world's largest tourist center and tourist attraction.

SOUTHERN PIEDMONT AND COASTAL VIRGINIA METROPOLITAN ZONE

The third major Southern concentration of large metropolises lies along the Piedmont, the fall line, and the nearby Virginia Coastal Plain. Piedmont and fall line metropolises include Richmond (population: 917,000), Virginia; Charlotte (popula-

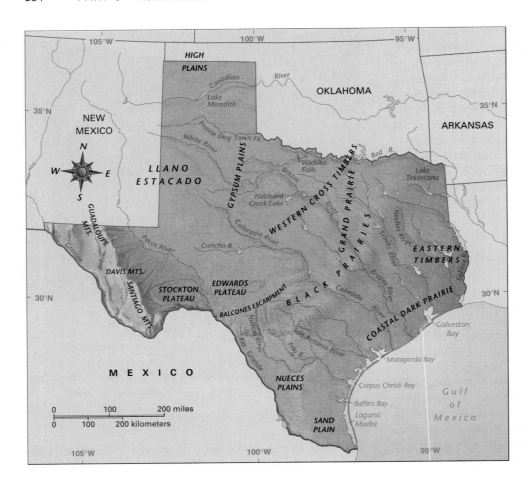

FIGURE 23.15
Physical map of Texas.

tion: 1.3 million), Greensboro (population: 1.1 million), and Raleigh (population: 965,000), North Carolina; and Greenville (population: 873,000), South Carolina. These metropolises have tended to form by the coalescing and integrating of cities located near each other, resulting in multiple-centered city clusters, although each is identified here by the name of its leading central city.

However, the two largest metropolises of this metropolitan zone differ in function from those just named. In the south, Atlanta (population: 3.3 million), Georgia, ranks fourth in metropolitan population within the entire South. It has diverse industries but stands out as a major inland transport center. Railroads and highways have generally avoided the Great Smoky Mountains, which lie just north of Atlanta. Instead, they skirt the southern end and converge on the city. Its ascendancy as a regional transport center today is seen in its airport, which is the fourth busiest in the United States as measured by volume of passenger traffic (after Chicago-O'Hare, Dallas–Fort Worth, and Los Angeles International). These transport facilities and its superior location have been exploited to make Atlanta the largest Southeastern business and governmental center.

In the north on the Coastal Plain is the seaport of Norfolk (population: 1.5 million), Virginia. Located on the spacious

Hampton Roads natural harbor at the mouth of the James River, Norfolk ships Appalachian coal by sea and is the main Atlantic base for the United States Navy.

METROPOLISES OF THE UPLAND SOUTH

The entire Upland South (Appalachians, Unglaciated Southeastern Interior Plain, and Southern Interior Highlands) contains only five metropolitan areas of over 500,000 people. Louisville (population: 981,000), Kentucky, began as a river port at the "Falls of the Ohio," where goods had to be transshipped around this break in transportation. It was also the closest point of contact on the Ohio River for the earliest settled and most productive agricultural area in the state—the Bluegrass Region. By the time a canal had been built to bypass the falls (in 1830), Louisville was a leading presence on the river, and it has continued to grow as a diversified regional center.

Two comparable centers, Nashville (population: 1.1 million) and Knoxville (population: 631,000), Tennessee, also began in locations associated with river transport and limestone soils. Nashville is located where the Cumberland River, a tributary of the Ohio, extends farthest south into the hilly but fertile Nashville Basin of central Tennessee. Knoxville is situated far up the Tennessee River from the Ohio in an area of limestone valleys within the Ridge and Valley Section of the Appalachians.

The same themes of transport advantages, relatively good soils, and growth as the major center for a considerable region are repeated in the case of Little Rock (population: 538,000), Arkansas. In this case, the exceptional land lies mainly in the Mississippi Alluvial Plain to the immediate east and secondarily in the Arkansas Valley to the west. The valley also provides a lowland route channeling traffic through the Interior Highlands, and the Arkansas River flows into the Mississippi.

The major exception, however, to the general pattern of urban location and development in the Upland South just described is Birmingham (population: 872,000), Alabama, which is the only large center of iron and steel industry in the South. Its development in the Ridge and Valley Section of the Appalachians was based on large deposits of both coal (from the nearby Cumberland Plateau) and iron ore.

OTHER SOUTHERN METROPOLISES

Other Southern cities with metropolitan populations of over 500,000 are New Orleans (population: 1.3 million) and Baton Rouge (population: 558,000), Louisiana; Memphis (population: 1.1 million), Tennessee; Charleston (population: 522,000), South Carolina; and El Paso (population: 665,000), Texas. The first three are Mississippi River ports. New Orleans was founded by the French just upstream from the head of the "bird's foot" delta where the Mississippi divides into separate channels flowing to the Gulf. In the riverboat era before the Civil War, New Orleans was New York's main competitor as a port. It then underwent relative decline as a port in the railway era but it remains a major seaport, especially for shipping grains and soybeans from the Midwest and the Mississippi Alluvial Plain. Somewhat upriver, Baton Rouge is accessible to medium-sized ocean tankers and is a major center of the Gulf Coast oil refining and petrochemical industries. Farther upriver, Memphis developed as a river port where the Mississippi flows adjacent to higher ground at the east side of its broad flood plain. The city's function as the main business center for the greater part of the Mississippi Alluvial Plain (and other nearby areas) dates back to antebellum days when a major sector of the "Cotton Kingdom" developed in "The Delta"—a regional name then in use for the large part of the Mississippi Alluvial Plain adjacent to Memphis in northwestern Mississippi. The remaining metropolises lie at opposite ends of the South. Charleston was the port and commercial capital of the first major plantation area outside of Virginia. It was the leading city of the South at the time of the American Revolution, then languished as the plantation economy moved westward, and has resumed rapid growth with the rise of the modern Southern economy.

23.3   The Midwest: Agricultural and Industrial Prodigy

The existence of a Midwest (Middle West) in the United States is widely recognized, but perceptions differ as to its extent (see Definitions and Insights, page 560). In this chapter, the term Midwest is applied to 12 states called the North Central states in federal publications. They are commonly subgrouped into the East North Central states of Ohio, Indiana, Illinois, Michigan, and Wisconsin and the West North Central states of Minnesota, Iowa, North Dakota, South Dakota, Nebraska, Kansas, and Missouri. The two subgroups are referred to here as the eastern Midwest and the western Midwest. The Mississippi River separates them except for part of the Wisconsin-Minnesota boundary (see Figs. 23.1 and 23.3). All states in the eastern Midwest border one or more of the Great Lakes, but only Minnesota does so in the western Midwest. In the eastern Midwest, Ohio, Indiana, and Illinois are bounded on the south by the Ohio River; in the western Midwest all states except Minnesota front on the Missouri River.

The use of the term "Midwest" for these 12 states is arbitrary to some degree but can be justified by spatial (mappable) characteristics distinguishing them as a group from the Northeast, South, and West. Among such characteristics are the Midwest's (1) interior "heartland" location; (2) distinctive plains environment; (3) important associations with major regional lakes and rivers; (4) ethnic patterns; (5) outstandingly productive combination of large-scale agriculture, industry, and transportation; and (6) rural landscapes famed for their rectangularity, symmetry, and prosperous appearance.

THE MIDWEST IN A WORLD AND NATIONAL COMPARISON

The Midwest is like the other major United States regions in that it has the size and production of a major nation. Encompassing about 770,000 square miles (c. 2 million sq km), the Midwest is surpassed in area only by the United States as a whole and 12 other countries. A total population of about 63 million is exceeded only by the populations of the United States and 13 other countries (as of mid-1996). In output of certain goods—for example, corn, soybeans, pork, and farm machinery—the Midwest is unsurpassed by any nation other than the entire United States. Within the United States, the Midwest is the largest regional producer of both agricultural commodities and manufactured goods, and it has an extraordinary web of transportation lines—centered on Chicago—that is crucially important in binding the nation's regions together. But perhaps the most notable geographic attribute of the region is its nearly solid occupancy by intensive economic development. The continuously productive character of its broad plains contrasts sharply with the islandlike character of development in the Northeast, South, and West.

THE WORLD'S MOST PRODUCTIVE PLAINS

By far the greater part of the Midwest is classified as plains (see Fig. 23.1). For the most part, smoother terrain in the Midwest has resulted from glacial deposition that buried a more irregular previous terrain under glacial debris (ground moraine or outwash). The flattest areas often are lacustrine plains—the floors of former lakes where glacial meltwater accumulated and sediments washed in and settled. The largest such area is the floor of old Lake Agassiz in Minnesota, the Dakotas, and adjacent Canada.

Rougher topography is most common near the margins of the Midwest. Only two small areas are mountainous: the Black Hills of South Dakota (and Wyoming) and a narrow strip of the Canadian Shield along Lake Superior in Minnesota. Other rough areas along the margins of the region include dissected parts of the Ozark Plateau of southern Missouri, the dissected Appalachian Plateau in eastern Ohio, the fantastically eroded South Dakota Badlands and other hilly parts of the western Dakotas, and the extensive Sand Hills of northern Nebraska. Away from the margins of the region, hilly areas occur in scattered places. One of these is the Unglaciated Hill Land of southeastern Wisconsin, which seems to have escaped the most recent (Wisconsin) glaciation.

MIDWESTERN AGRICULTURE AND ITS ENVIRONMENTAL SETTING

These vast Midwestern plains produce a greater output of foods and feeds than any other area of comparable size in the world. The Midwest commonly accounts for more than two fifths of all United States agricultural production by value. This output is unevenly spread among the Midwestern states, with Iowa leading. Within the nation, Iowa is the third-ranking state in value of farm products, exceeded only by much larger California with its irrigation-based agriculture, and by Texas. Ordinarily, five Midwestern states—Iowa, Nebraska, Illinois, Kansas, and Minnesota—are among the nation's top seven in total agricultural output, and nine states are among the top fifteen. The lowest-ranking Midwestern states are Michigan, handicapped by large areas of infertile sandy soil, and the Dakotas, handicapped by dryness, rough land toward the west, and a shorter frost-free season.

Within most of the Midwest, soils are exceptionally good. The best soils developed under grasslands, which supplied abundant humus. Such outstandingly fertile soils normally accompany steppe climates in the middle latitudes, where the fertility is often counterbalanced by precipitation so low as to be marginal for agriculture. However, the remarkable feature of the Midwest is that its soils—located in a more humid continental climatic area—still developed under a tall-grass prairie. Such climatic conditions normally produce a natural vegetation of forest, and trees grow well if they are planted. But in the early 19th century, the first settlers found large expanses of prairie, or a prairie-forest mixture, as far east as northwestern Indiana. From there, the prairie region extended west in a broadening triangle to the eastern parts of the Dakotas, Nebraska, and Kansas, and on into the South in Oklahoma and Texas. Of the Midwestern states, only Michigan and Ohio lacked considerable expanses of prairie. Thus, a very large part of the region had soils of a richness generally found only in semiarid areas but which in the Midwest were combined with humid conditions more favorable to most crops. The existence of the anomalous triangle of prairie bordered by forests both north and south can be accounted for in two main ways. One explanation holds that the prairie resulted from periodic drought conditions, discouraging tree growth in this normally humid area. The other explanation suggests that Native Americans may have burned the forest to create grazing range for game. Possibly both factors were important, and prairie fires set by lightning also may have helped perpetuate the prairie once it was established (grass is less susceptible than tree seedlings to lasting damage by fire). Whatever the causes, the occurrence of grassland soils in a humid area where forests might normally be expected provided an extraordinarily favorable agricultural environment.

Another natural condition contributing to the unusual excellence of many Midwestern soils is the widespread presence of loess deposits made of dust particles. These wind-laid soil materials are thought to have accumulated in glacial times when strong winds picked up the particles from dry surfaces where the ice had melted but where no protective mantle of vegetation had yet become established. Because most of the nutrients that nourish plants are contained in these finer soil particles, loess is associated with soils of exceptional fertility in various parts of the world.

In the eastern Midwest, the soils formed under natural broadleaf forests are not exceptionally fertile by nature, but neither are they poor (Fig. 23.1). The principal areas of poor soil in the Midwest lie in (1) the uplands of the Ozarks and the Appalachian Plateau; (2) sandy areas of needleleaf forest in northern Michigan, Minnesota, and Wisconsin; and (3) rougher western parts of the four westernmost states. After long use, the present quality of Midwestern soils depends heavily on how they have been handled. Where properly fertilized and protected against depletion and erosion, they have proved remarkably durable. Although unwise practices have caused much soil damage, huge expanses remain highly productive. Aided by science and technology, the soils of the Midwest as a whole have higher yields today than ever before.

The Legendary American Corn Belt

Corn is the single leading product of Midwestern farms, and the Corn Belt where this crop dominates the landscape has become an agricultural legend for its productivity. Sales of beef cattle in

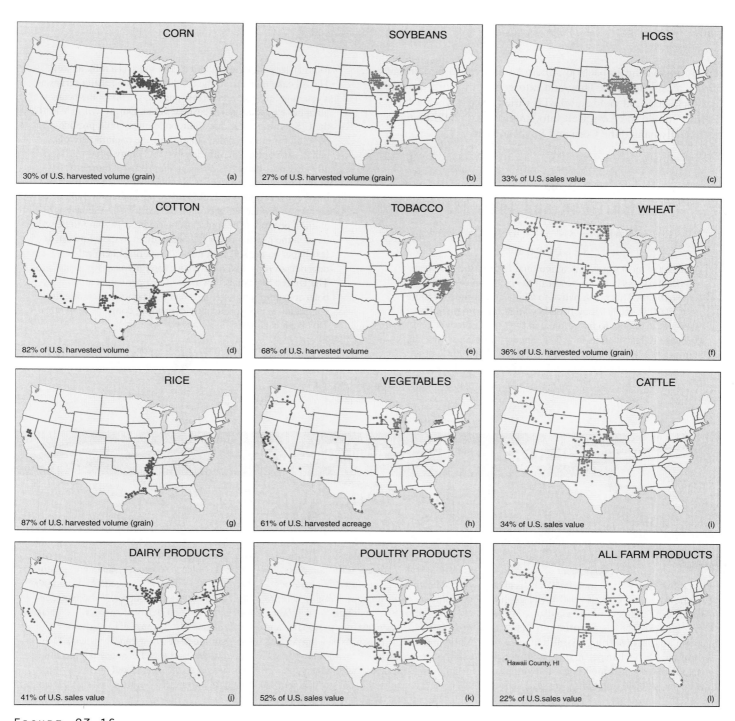

FIGURE 23.16

This panel of maps shows the 100 leading U.S. counties for selected crops and farm products. Each map also tells the reader what percent of the U.S. total production of this commodity comes from the aggregate of these 100 counties (except for rice, where only the top 50 counties are shown). *Cartography by Joseph H. Astroth, Jr., and Jennifer Nichols.*

the Midwest are actually greater than receipts from corn, but the cattle are raised primarily on corn. Over four fifths of American corn production and around two fifths of world production come from the Midwest. Highly favorable conditions for large-scale corn growing are provided by fertile soils, hot and wet summers, and huge areas of land level enough to permit mechanized operations. The corn plant itself and the methods of production have been greatly improved by extensive research.

Soybeans are another major element in Corn Belt agriculture (Fig. 23.16). Long cultivated in China, they have had a

587

rapidly increasing impact on American and world agriculture during recent decades. They add nitrogen to the soil, provide livestock feed, and yield an oil with varied uses. Soybeans have recently turned the Corn Belt into a region that could be more realistically named the Corn and Soybean Belt.

Definitions and Insights

SOYBEANS—AN UNPARALLELED RESOURCE

One of the most important agricultural finds in the Midwest this century has been the power of the bean—more exactly, the **soybean.** This plant, native to China and farmed for more than 5000 years, has long been known as a useful plant because of its ability to fix nitrogen in the growth process, and because it does well in as wide a range of farming settings as corn. The Chinese have revered the soybean because it provides high protein as a vegetable, can be ingested as *tofu*

(doufu), as a cooking oil, and can be used as a milk as well. The oil cake made from the residue from processing serves as a valuable fertilizer and a feed, and the greens of the bean plant have good organic fertilizer value. During World War II, the U.S. realized that this common legume could provide oil not only for our kitchens, but also for the manufacture of paints, soap, glycerin, printing ink, and many other traditionally chemical-based products. In the Midwest currently, soybean promotional groups are showing ways that the bean can be effectively added to fuel to lower dependence on imported petroleum products and expand the market for regional agricultural produce. The future of the soybean is going to be writ large in the heartland, for it is well-suited to the Midwest, is less harsh on the soil than many competing chemical-fertilizer-dependent crops, and has a major foreign market in east Asia. The United States has now moved to a per capita annual soybean consumption of greater than 250 pounds, meaning the continuance of a significantly expanding domestic market. This bean is here to stay.

PROBLEM LANDSCAPE

"SMALL-TOWN BLUES"

One of the most amazing population statistics of this century is this: Between 1900 and 1990, 80 percent of America's counties lost population. That means that eight out of ten counties had fewer people at the beginning of this decade than they did in 1900. In that same period our national population grew from 76 million to 250 million. Or, in harsher reality, that means that thousands and thousands of communities have been declining in population steadily since the beginning of the 20th century. Most of this depopulation has occurred in small towns of several thousand or, more likely, several hundred residents.

In many ways geography is at the very heart of this problem landscape. Small town settlements emerged as vital clusters of merchants and town folk located on some avenue of transportation—the highway, railway, river, or combination. These clusters were surrounded by the broad open

farmlands being cleared and farmed in the late 18th century and all through the 19th century. Dirt roads made moving farm produce to local markets, train stations, and riverside wharfs difficult, so there was a geographic wisdom for merchants and farmers alike to locate proximate to each other, and churches, services, professionals, and entertainment all began expansion around such places. The small towns that became county seats had a particular advantage, for they were the seat of local government as well. Particularly fortunate towns became home to food processing firms, shoe or textile factories, or any number of small scale industries that added a factory payroll to these centers.

Then, not long after the turn of this century, the geography of these settlements was changed. Farm market roads were black-topped. The larger highways that linked the small towns to each other—and all of them to regional urban centers—were improved.

The delivery truck and the private automobile began the space-time convergence that reduced the functional distance between places. Mileage stayed the same but the time required to cover it changed. Accordingly, personal decisions about work, shopping, entertainment, and personal services led to an acceptance of longer travel to gain access to a higher order of goods and services. These commuting and travel changes left the small town ever more isolated and the major urban centers ever more populated. The 20 percent of the U.S. counties that gained population between 1900 and 1990 were nearly all in large metropolitan urban clusters. America was voting for the larger city with its feet (and its cars) and with its life decisions about employment, residences, and shopping.

Why does this scenario represent a problem landscape? In one sense, it is no problem. These demographic shifts have been freely made by citizens following their employment, or their own best sense

Landscape Geometry of the Corn Belt and Its Origins

The Corn Belt is limited on the east and south by rougher country, on the north by cooler summers and poorer soils, and on the west by semiaridity. It extends over 800 miles (*c.* 1300 km) east to west and 300 to 600 miles (*c.* 500–1000 km) north to south, from western Ohio to eastern South Dakota, Nebraska, and Kansas. It is famous for its rectangular landscape (Fig. 23.17), created originally under federal laws requiring that federally owned land—in the public domain—be surveyed into townships six miles square, sections one mile square, and quarter sections one-half mile square. The survey lines ran north-south and east-west, and most roads came to follow section boundaries. This system, prescribed by the Confederation Congress under the Land Ordinance of 1785, imposed a uniformity on most of the United States that was absent in the earliest areas of settlement, where various local survey systems were employed. As settlement proceeded westward, the public domain was disposed of by sale or gift of rectangular plots to settlers. This dis-tinctive landscape pattern has persisted. Seen from the air, to-day's great sweep of Corn Belt rectangles, variegated in tone by use and season, is a compelling sight.

The Dairy and Wheat Belts: Adaptations to Restrictive Environments

The northern and western margins of Midwestern agriculture are not part of the Corn Belt. Wisconsin, most of Michigan, and much of Minnesota form the western sector of the Dairy Belt, whose eastern sector stretches across the northern Northeast and adjacent southern Canada. Corn is widely grown, but most of it is chopped in an immature state for silage. Much more of the land is kept in hay and pasture to feed dairy cattle. Wisconsin, the country's leading dairy state, produces around one sixth of the nation's fluid milk (see Fig. 23.16); and Wisconsin, Minnesota, and Michigan market about one quarter of the nation's fluid milk.

---

of how they want to live and where they want to locate. However, in terms of cultural development, the small town has a nearly sacred place in the heart of American national identity that is not shared by the large city, the inner city, the suburb, or even the isolated farmstead. It is known as "a place where you don't have to lock your door, and you can let your children come into downtown alone." It is the small town that seems to represent the American community, with a concern for people's well-being, for tradition, and for community self-reliance. The family farms that historically occupy the hinterland of these towns of several thousand or less are banner carriers of American tradition.

As farm families get into their sedans and drive through the increasingly empty streets of the small towns that lie between their farms and the regional mall or Wal-Mart that they are taking their farm dollars to, there is an uneasiness as the parents look at the closed stores on Main Street. These are the towns where they bought shoes, drank sodas, and went to the dentist several decades ago when they were children. They are witness to a demographic and economic change that is the product of innumerable individual, small economic decisions even while it produces a source of broad social uneasiness as more "Going Out of Business" signs go up.

This American problem landscape has two futures. On the positive side, increasing mobility gained through telecommunications (Internet communication, faxes, personal computers, and networking) has made it possible for more and more urbanites to play out much of their professional interaction in a very different setting. New England, for example, is witnessing a major increase in home purchases to metropolitan "refugees" coming to buy, retrofit electronically, and "cocoon" in small town settings. The same is happening in the Midwest as small towns take on new personality with speciality shops, tourist elegance, and at-tractive home prices for urban professionals.

The other side of this problem landscape is the continuing decline into abandonment of the small towns that cannot seem to muster architectural excitement, speciality events, or fairs, or that are simply too far away and too culturally distant from the dynamics of the urban environment. Our guess is that at least one third of the students who read these pages will find themselves at some point (or at several points) making a decision about whether to opt for the glitz of the city or the cozy traditions of the American small town.

This is a landscape that can be easily investigated. Go find a small-town General Store or cafe and see what it feels like to buy a soda, a pair of Carhartt chinos, or a few fishing lures in a place that remains a powerful part of America's past.

Hornik, Richard 1989. "Small-Town Blues." *Time Magazine*. March 27, 1989. 66–68.

FIGURE 23.17
This eastern Nebraska scene, featuring a beef-cattle feedlot and rectangular grain fields, mirrors the continuous productivity of America's agricultural heartland between the Appalachians and the Rockies. Planted clumps of trees mark farmsteads in an expanse once occupied by natural prairie.   *Grant Heilman/Grant Heilman Photography*

West of the Corn and Dairy Belts, lower precipitation focuses agriculture on wheat and ranching. Here, central and western Kansas form the heart of the Winter Wheat Belt, which extends into Oklahoma, Texas, and Colorado. The less productive Spring Wheat Belt centers in North Dakota, although it extends beyond there in all directions. Wheat farming shares the two wheat belts with ranching, and cattle are a more important source of farm income than wheat in many localities. Feed crops adapted to minimal moisture, such as sorghums in Kansas and barley in North Dakota, support livestock production (see Fig. 23.16).

In the Midwest as a whole, sales of animal products exceed sales of crops. Beef cattle are the largest source of income from animal products, but the Midwest occupies an even more impressive national position in the marketing of hogs (Fig. 23.16).

Local Specialties

Many local areas depend heavily on agricultural specialties that are less important in the total picture of Midwestern agriculture than corn, soybeans, wheat, beef, pork, or milk. For example, a fruit belt exists in Michigan along the shore of Lake Michigan. Climatic effects of the lake make the beginning and end of the growing season more reliable by reducing the chance of untimely freezes. North Dakota has become the nation's leading state in growing sunflowers, cultivated for the oil pressed from their seeds. This production comes from rich flatlands near the Red River in the former bed of glacial Lake Agassiz. The district extends into Minnesota and is also prominent in sugar-beet and potato growing.

The Risky Nature of Midwestern Agriculture

Despite its impressive resources and accomplishments, Midwestern agriculture is a risky enterprise. Natural hazards such as drought and plant diseases increase the normal business risk, but the main hazards are sharp fluctuations in world supply and demand and therefore in market prices. Adverse natural and economic conditions in varying combinations have forced waves of farmers out of business at various times in the past. Between 1930 and 1992, for example, the number of farms in the Midwest decreased from slightly over 2 million to approximately 777,000. Most of the land of defunct farms was bought or rented by surviving farmers who expanded their acreage. Hence, the amount of land in farms decreased by only 6 percent, and the mean size of farms increased from about 180 acres (73 ha) to about 442 acres (166 ha). Increased mechanization allowed farmers to work larger acreages but also increased their capital costs and hence increased their borrowing and their vulnerability to low market prices.

A notable feature of the American agricultural experience has been the inability of government subsidy programs to halt the process whereby some farms fail and others expand. Such programs, costing many billions of dollars and brought about by a mountain of legislation, have been a controversial facet of American economic and political life since the beginning of Roosevelt's New Deal in the 1930s. As U.S. farms become fewer, larger, and more highly capitalized (in the Midwest and elsewhere), corporate forms of farm ownership and enterprise are increasingly evident.

THE URBAN AND INDUSTRIAL MIDWEST

Across the Midwest, the level of industrial development declines from east to west. Thus, the eastern Midwest states are highly industrialized and, as of 1991, accounted for 22 percent of total United States manufacturing value added, whereas the West North Central states ("western Midwest") accounted for only 8 percent. The greater industrialization of the eastern Midwest reflects that section's closer proximity to the markets of the Northeast and Europe, its dense network of railways converging on Chicago, its access to both the Great Lakes and river transportation, and its access to high-quality Appalachian coal from Ohio, Pennsylvania, West Virginia, Illinois, and Kentucky.

Landscape in Literature

"THE HOUSES WERE LEFT VACANT ON THE LAND"

In *The Grapes of Wrath,* John Steinbeck created his most powerful novel of the American landscape and its people. Although he did not get his Nobel Prize for Literature until 1962, it is understood that it is this 1939 novel that really brought him to the world's attention. The following passage comes from the time when poor farmers of the Great Plains were finding that traditional patterns of farming and tenancy no longer worked. Stories about the potential for success in farming in California were rolling through the drought-stricken lands like wildfire. People pulled up the modest roots they had, bought old cars or trucks, and headed west. This passage creates strong images of the world left behind.

The houses were left vacant on the land, and the land was vacant because of this. Only the tractor sheds of corrugated iron, silver and gleaming, were alive; and they were alive with metal and gasoline and oil, the disks of the plows shining. The tractors had lights shining, for there is no day and night for a tractor and the disks turn the earth in the darkness and they glitter in the daylight. And when a horse stops work and goes into the barn there is a life and a vitality left, there is a breathing and a warmth, and the feet shift on the straw, and the jaws champ on the hay, and the ears and the eyes are alive. There is a warmth of life in the barn, and the heat and smell of life. But when the motor of a tractor stops, it is as dead as the ore it came from. The heat goes out of it like the living heat that leaves a corpse. Then the corrugated iron doors are closed and the tractor man drives home to town, perhaps twenty miles away, and he need not come back for weeks or months, for the tractor is dead. And this is easy and efficient. So easy that the wonder goes out of the land and the working of it, and with the wonder the deep understanding and the relation. And in the tractor man there grows the contempt that comes only to a stranger who has little understanding and no relation. For nitrates are not the land, nor phosphates; and the length of fiber in the cotton is not the land. Carbon is not a man, nor salt nor water nor calcium. He is all these, but he is much more, much more; and the land is so much more than its analysis. The man who is more than his chemistry, walking on the earth, turning his plow point for a stone, dropping his handles to slide over an outcropping, kneeling in the earth to eat his lunch; that man who is more than his elements knows the land is more than its analysis. But the machine man, driving a dead tractor on land he does not know and love, understands only chemistry; and he is contemptuous of the land and of himself. When the corrugated iron doors are shut, he goes home, and his home is not the land.

The doors of the empty houses swing open, and drifted back and forth in the wind. Bands of little boys came out from the towns to break the windows and pick over the debris, looking for treasures. And here's a knife with half the blade gone. That's a good thing. And—smells like rat died here. And look what Whitey wrote on the wall. He wrote that in the toilet in school, too, an' teacher made 'im wash it off.

When the folks first left, and the evening of the first day came, the hunting cats slouched in from the fields and mewed on the porch. And when no one came out, the cats crept through the open doors and walked mewing through the empty rooms. And then they went back to the fields and were wild cats from then on, hunting gophers and field mice, and sleeping in ditches in the daytime. When the night came, the bats, which had stopped at the doors for fear of light, swooped into the houses and sailed about through the empty rooms, and in a little while they stayed in dark room corners during the day, folded their wings high, and hung headdown among the rafters, and the smell of their droppings was in the empty houses.

And the mice moved in and stored weed seeds in corners, in boxes, in the backs of drawers and kitchens. And weasels came in to hunt the mice, and the brown owls flew shrieking in and out again.

Now there came a little shower. The weeds sprang up in front of a doorstep, where they had not been allowed, and grass grew up through the porch boards. The houses were vacant, and a vacant house falls quickly apart. Splits started up the sheathing from the rusted nails. A dust settled on the floors, and only mouse and weasel and cat tracks disturbed it.

On a night the wind loosened a shingle and flipped it to the ground. The next wind pried into the hole where the shingle had been, lifted off three, and the next, a dozen. The midday sun burned through the hole and threw a glaring spot on the floor. The wild cats crept in from the fields at night, but they did not mew at the doorstep any more. They moved like shadows of a cloud across the moon, into the rooms to hunt the mice. And on windy nights the doors banged, and the ragged curtains fluttered in the broken windows.

From John Steinbeck, 1939. *The Grapes of Wrath* (New York: Viking Press, 1939), 157–159.

have become the outstanding industry. This area is now commonly referred to as Silicon Valley, after the chips that carry microcircuits. The rise of Silicon Valley as a center of high technology is due in part to the attractiveness of the San Francisco area as a place to live and the nearby presence of two outstanding universities—the University of California at Berkeley near the industrial center of Oakland on the east side of the Bay, and Stanford University in nearby Palo Alto.

The Superproductive Agriculture of the Central Valley

About one tenth of the value of United States farm products comes from California. Largely produced in the Central Valley and supported by extensive irrigation systems, these products represent an output of amazing variety. The state's two leading crops by value are grapes—retailed principally in the form of wine—and cotton. A large majority of the grapevines in the United States are in the southern and central parts of the Central Valley or in the valleys of the Coast Ranges slightly north of San Francisco. California's cotton acreage, mainly in the southern half of the Central Valley, puts the state second only to Texas in cotton output (as of 1992).

Various fruits, vegetables, and nuts are often the top-ranking crops of particular localities. Almost the complete range of subtropical and temperate fruits and vegetables is produced somewhere in California's orchard and truck-farming districts. These districts are largely in the Central Valley but also are found in the Salinas Valley and other broad valleys of the Coast Ranges south from San Francisco, as well as in the irrigated desert of the Imperial Valley, which lies below sea level in California's southeast. The California citrus industry, largely shifted now from Southern California to the San Joaquin Valley, is second only to Florida's. And rice is the main crop of the Sacramento Valley.

But the two most valuable items sold from California farms are milk and beef cattle to supply markets within the state. Expanding local demand for basic foods by the large and increasing population tends to increase the proportion of farmland devoted to irrigated alfalfa and other feeds for cattle (Fig. 23.24). San Francisco is the economic capital for the agriculture of the Central Valley and nearby Coast Range valleys, and, like Los Angeles, it performs major financial and business functions for great areas in the West. But San Francisco's importance has not precluded the emergence of some fairly large cities as regional centers for parts of the Central Valley. Sacramento (population: 1.6 million), the state capital, occupies a central location in the Central Valley on the main rail and highway routes between San Francisco and the Donner Pass over the Sierra. Fresno (population: 835,000) is the dominant center for the San Joaquin Valley, and Bakersfield (population: 610,000) in the extreme south is both an agricultural service center and the operating base for oil fields in its area.

Seattle and Portland and Their Pacific Northwest Setting

The Puget Sound Lowland of Washington and the Willamette Valley of Oregon to its south are not irrigated oases. They have abundant precipitation and luxuriant natural forests, as do the adjoining Coast Ranges and Cascades. But these humid conditions end abruptly near the crest of the Cascades. Hence, the area is oasislike—a sharply limited area of abundant water and clustered population within the generally dry West. This population is centered in two port cities, each located on a navigable channel extending inland through the Coast Ranges. Seattle (population: 3.2 million), Washington, located on the eastern shore of Puget Sound, has access to the interior via passes over the Cascades. Portland (population: 2 million), Oregon, is on the navigable lower portion of the Columbia River where the

FIGURE 23.24
Sprinkler irrigation in the flat, rich Imperial Valley. Lettuce and other vegetables, cotton, alfalfa, hay, and feeder cattle are important products of the Valley.   *Steve Liss/Time Magazine*

Willamette tributary enters it, and has a natural route to the interior via the Columbia River Gorge through the Cascades. Outside these metropolises, the Lowland contains somewhat over a million other people.

Fishing, logging, and trade built the cities of the Lowland in the latter 19th and early 20th centuries. But it was agricultural land in the Willamette Valley that drew the first large contingents of settlers to "Oregon Country" in the 1840s. Agriculture is still of some importance, although it is handicapped by hilly terrain and leached soils. The leading products are milk and truck crops, and a local Oregon wine industry is expanding in importance. Fishing, primarily for salmon, is still carried on, but the Pacific Northwest fish catch is now much smaller than that of any one of the three leading states in this industry—Alaska, Massachusetts, and Louisiana (as of 1993). However, Oregon and Washington are still the leading states in production of lumber, and they still have one quarter of the sawtimber resources of the United States. The trees are mostly coniferous softwoods, and many are enormous in size, especially the famed Douglas fir. The timber industry, now under heavy attack by environmentalists, is still basic to the economies of both states.

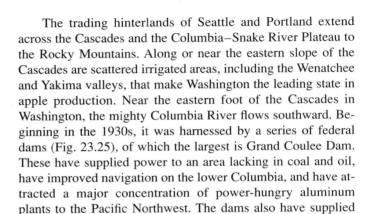

Definitions and Insights

LOCATIONAL ANOMALIES: THE AIRCRAFT INDUSTRY IN THE PACIFIC NORTHWEST

Ordinarily geography takes pride in being able to determine the locational factors that explain the site of a factory (presence of resources, labor, transportation), a city (water, agricultural land, networks of transportation and communication), or a suburban tract development (access to jobs, proximity to both city and more natural landscapes). There are, in essence, certain factors that must come into play to site any one of these features of the cultural landscape. In the aircraft industry, there is need for clear weather and a maximum number of days of sunlight for construction and for field-testing aircraft. Therefore, it is curious to note the geographical location of the world's largest aircraft producer—the Boeing Aircraft Company. Natural resources fail to account for this single leading industry of the Pacific Northwest, although the locally abundant Sitka spruce—the largest spruce species in the world—was an excellent wood for early aviation production. The later development of aluminum production in the region because of relatively inexpensive electricity was also significant. But the major initial draw was simply the geographic preference of founder William E. Boeing. The company began in Washington in 1917 (in the Seattle suburb of Everett), expanded vastly during World War II, and has since maintained its position as a world leader despite a series of drastic production and employment fluctuations. It was the Boeing "B" that one saw in the B-17, the B-29, and the B-52, and it is the same Boeing that has been deeply involved in modern jetliner and aerospace manufacturing. Since Boeing did not want to leave his native Pacific Northwest, his firm overcame various environmental difficulties and made the site and situation of the Seattle environs work very effectively for its needs.

The trading hinterlands of Seattle and Portland extend across the Cascades and the Columbia–Snake River Plateau to the Rocky Mountains. Along or near the eastern slope of the Cascades are scattered irrigated areas, including the Wenatchee and Yakima valleys, that make Washington the leading state in apple production. Near the eastern foot of the Cascades in Washington, the mighty Columbia River flows southward. Beginning in the 1930s, it was harnessed by a series of federal dams (Fig. 23.25), of which the largest is Grand Coulee Dam. These have supplied power to an area lacking in coal and oil, have improved navigation on the lower Columbia, and have attracted a major concentration of power-hungry aluminum plants to the Pacific Northwest. The dams also have supplied the water for a huge expansion of irrigation in the Columbia Plateau of both Washington and Oregon. However, they have also helped kill the area's significant salmon fishing industry.

In southeastern Washington, the Columbia Plateau rises high enough to have a steppe rather than a desert climate and is mantled by a deep hilly covering of fertile and water-retentive loess. This area, called the Palouse, grows unirrigated wheat, despite an annual precipitation averaging as low as 9 inches (23 cm) in some places. At the northeastern edge of the Palouse, Spokane (population: 400,000) serves as the local capital of the "Inland Empire" in the northern Columbia Plateau and adjacent Rockies.

Another major industry that has been sited in the Pacific Northwest because of owner environmental preferences is the Microsoft Corporation of Redmond, Washington. The firm was initially founded in Albuquerque, New Mexico, in 1975, but by 1979 the two founders moved the headquarters to the Seattle area. It has since grown to be the largest computer software company in the world, and is now a source of major payroll and export revenues for the Northwest and the United States.

Population Clusters in the Mountain States

The eight inland states in the Mountain West are drier than those in the Pacific Mountains and Valleys, lack the abundant irrigation water of California, and lack the advantage of ocean transportation. These handicaps have been so constricting that the eight states had only 15 million people in 1993. About half of these are in a handful of metropolitan areas in irrigated districts. Outside of them, the common pattern in a Mountain West state is that of a few hundred thousand people scattered across a huge area—on ranches and farms, on Indian reservations, or in small service centers and mining towns.

FIGURE 23.25
The Dalles Dam on the Columbia River permits salmon to migrate via the fish ladder in the foreground. Note the power-transmission lines in the distance. The dam, located on the border between Oregon and Washington, is one of 56 major dams in the Columbia River watershed. *Robert Harbison/Christian Science Monitor*

Phoenix and Tucson.   The largest population cluster in the Mountain West lies in the Basin and Range Country of southern Arizona. Here live about three quarters of Arizona's people, in the adjacent metropolitan areas of Phoenix (population: 2.4 million) and Tucson (population: 700,000). Water from the Salt and Colorado rivers (see Fig. 23.23), plus groundwater, makes it possible to support this development in a subtropical desert. Phoenix was originally the business center for an irrigated district along the Salt River. The district expanded after 1912 when the first federal irrigation dam in the West (Roosevelt Dam) was completed on the river. Today, cotton is the area's main crop, although hay to support dairy and beef cattle is important. Growth in both Phoenix and Tucson accelerated after the introduction of air-conditioning made them attractive retirement and business centers. Each is the site of a major state university, and Phoenix is the state capital. Tourism to both urban and nonurban destinations, the increasing income level of Sunbelt retirees, and copper mining are also basic to Arizona's economy.

The Colorado Piedmont.   Another important population cluster stretches along the Great Plains at the eastern foot of the Rocky Mountains. Within this area, called the Colorado Piedmont, metropolitan Denver contained 2.2 million people in 1995, and more than 800,000 lived in metropolitan Colorado Springs (452,000 in city proper) or smaller cities. These cities developed on streams emerging from the Rockies. The stream valleys provided routes into the mountains, where a series of mining booms began with a gold rush in 1859. One mineral rush followed another through the later 19th century, with various minerals being exploited at scattered sites. The Colorado Piedmont towns became supply and service bases for the mining communities, which often were short-lived. Irrigable land along Piedmont rivers afforded a farming base to provide food

for the mines, and agricultural servicing became an important function of the towns. In time, agriculture grew into a prosperous business with more stability than mining.

During the 20th century, the Piedmont centers have grown rapidly. Irrigated acreage has been expanded greatly, fostering a boom in irrigated feed to support some of the nation's largest cattle feedlots. The Colorado Rockies have also become an important producer of molybdenum, vanadium, and tungsten—comparatively rare alloys that are much in demand for high-technology uses. High energy prices in the 1970s led to expanded extraction of fuels from the state's huge coal reserves and scattered oil and gas deposits. And tourism in the Rockies has profited greatly from increasing American affluence, better transportation, and the rise in the popularity of skiing. Finally, Denver became a major center for the western regional offices of the federal government, and Colorado Springs became the site of the United States Air Force Academy and the nearest large city to the headquarters of the North American Air Defense Command, located in a redoubt inside a mountain.

The Wasatch Front (Salt Lake) Oasis.   Third in size among the population clusters of the interior West is the Wasatch Front, or Salt Lake, Oasis of Utah, composed of metropolitan Salt Lake City–Ogden (population: 1.2 million) and Provo (population: 291,000) to its south. It contains about three fourths of Utah's people, in a north-south strip between the Wasatch Range of the Rockies and the Great Salt Lake, with a southward extension along the foot of the Wasatch. In the rest of Utah, there are only about 400,000 people. The oasis was developed as a haven for people of the Mormon faith when they were expelled from the eastern United States in the 1840s. After a difficult trek westward, they began irrigation development based on water from the Wasatch Range, and they built at Salt Lake City the temple

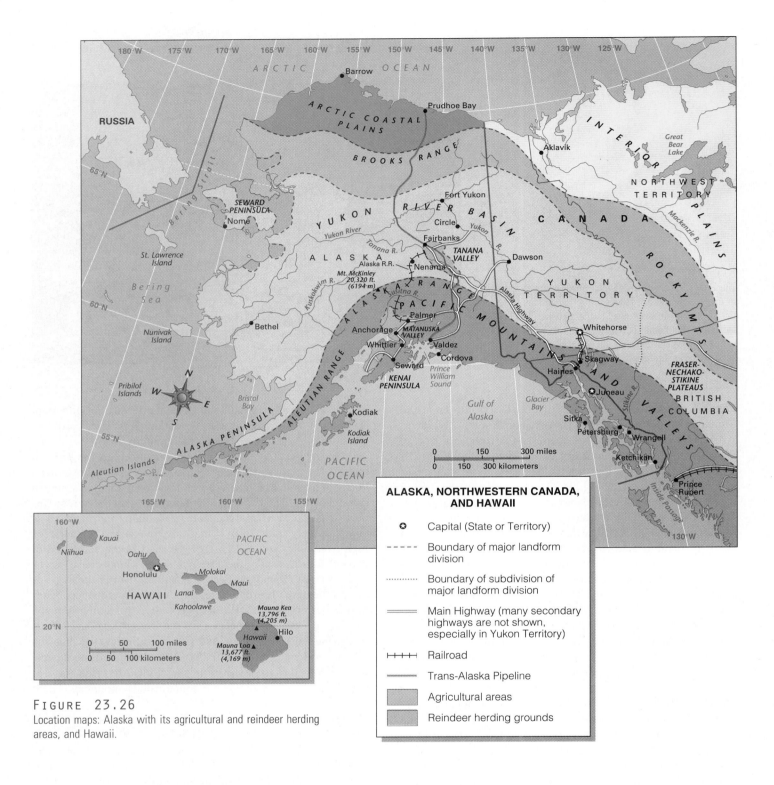

ALASKA, NORTHWESTERN CANADA, AND HAWAII

✪	Capital (State or Territory)
- - - -	Boundary of major landform division
··········	Boundary of subdivision of major landform division
═══	Main Highway (many secondary highways are not shown, especially in Yukon Territory)
⊢⊢⊢⊢	Railroad
──	Trans-Alaska Pipeline
▭	Agricultural areas
▭	Reindeer herding grounds

that is the symbolic center of the Mormon religion and culture. From this center, Mormon settlers have spread widely into all the surrounding states and California. Thus, Salt Lake City is a regional religious and cultural capital; in addition, with no large competing metropolis within hundreds of miles, the city is the business capital for a huge although sparsely populated area outside the Mormon core region at the foot of the Wasatch.

The Less Populous States of the Interior West. The other five states of the interior West are even less populous than Utah. Their aggregate population of about 6 million, in states that are very large in area, is less than that of the San Francisco metropolis. The people of New Mexico are largely strung along the north-south valley of the upper Rio Grande, where Albuquerque (population: 646,000) is the main city, although it competes economically with El Paso, Texas, which is located on the Rio Grande just south of New Mexico. New Mexico's economy is a typical Western mix of irrigated agriculture, ranching, mining, tourism, some forestry in the mountains, and federally funded defense contracting. Similarly, most of the population of Idaho is in oases strung along a river—in this case, the Snake River, as it crosses the volcanic Snake River Plains in the southern part of the state. Agricultural output is more substantial in Idaho, nationally known for its potatoes. But tourism and federal money yield less income than they do in New Mexico, and the largest city of the state, Boise, has only about 150,000 people in the city proper.

Except for a small section that lies in the Sierra Nevadas, the state of Nevada is desert, with very little irrigation. Metropolitan Las Vegas (population: 1.1 million) and Reno (population: 283,000) have more than four fifths of the state's population, and the state's laws allow gambling and related tourism to overshadow the usual Western mix of economic activities. Casinos and associated entertainment and convention business play a major role in the state's wealth. The current pace of population growth in the state is also a testimony, again, to the importance of air-conditioning.

In Montana and Wyoming, there is little irrigation and no warm-winter climate to encourage year-round tourism or to attract retirees. Economies based on farming, ranching, mining, and summer tourism support extremely sparse populations.

23.5   Alaska and Hawaii

In the middle of the 19th century, the United States began to acquire Pacific dependencies. Two of these, Alaska and Hawaii, were admitted to the Union as states in 1959. The United States acquired Alaska by purchase from Russia in 1867, and Hawaii was annexed in 1898. Acquisition of a Pacific empire was motivated partly by defense considerations, and these have continued to play an important role in both Alaska and Hawaii. Large military installations exist in both states, and defense expenditures are a major element in the economies of these states. In geography, these two most recent states are as different from each other as is possible, and in many ways they stand apart from the other 48 as well.

ALASKA'S DIFFICULT ENVIRONMENT

Alaska (from *Alyeska*, an Aleut word meaning "the great land") makes up about one sixth of the United States by area, but is so rugged, cold, and remote that its population was only 606,000 in 1994. The state's difficult environment can be conveniently assessed in four major physical areas (Fig. 23.26):

1. Alaska's *Arctic Coastal Plain* is an area of tundra along the Arctic Ocean north of the Brooks Range. Known as "the North Slope," it contains very large deposits of oil and natural gas.

2. The barren *Brooks Range* forms the northwestern end of the Rocky Mountains. Summit elevations range from about 4000 to 9000 feet (c. 1200–2800 m).

3. The *Yukon River Basin* lies between the Brooks Range on the north and the Alaska Range on the south. This is the Alaskan portion of the Intermountain Basins and Plateaus. The Yukon is a major river, navigable by riverboats for 1700 miles (c. 2700 km) from the Bering Sea to Whitehorse in Canada's Yukon Territory. In Alaska, the Basin is generally rolling or hilly, with some sizable areas of flat alluvium that often are swampy. Its subarctic climate is associated with coniferous forest that generally is thin and composed of relatively small trees. Fairbanks (population: 31,000, city proper) is interior Alaska's main town. The Alaska Highway connects it with the "Lower 48" (states) via western Canada, and both paved highways and the Alaska Railroad lead southward to Anchorage and other ports on the Gulf of Alaska. Fairbanks is a major service center for the Trans-Alaska Pipeline from the North Slope oil area to the port of Valdez in the south. The city is located on the Tanana River tributary of the Yukon and is associated with a small agricultural area in the Tanana Valley.

4. The *Pacific Mountains and Valleys* occupy southern Alaska, including the southeastern panhandle. The mainland ranges of the panhandle are an extension of the Cascade Range and the British Columbia Coastal Ranges, and the mountainous offshore islands are an extension of the Coast Ranges of the Pacific Northwest and the islands of British Columbia. The **Inside Passage** from Vancouver and Seattle lies between the islands and the mainland. On the mainland, some mountain peaks exceed 15,000 feet (4572 m). Fjorded valleys extend long arms of the sea inland. A very cool and wet marine west coast climate prevails, fostering coniferous forests that are a major resource, exploited primarily for the Japanese market. From Juneau (population: 27,000, city proper) northwestward, a majestic display of glaciers defines a long stretch of coast (Fig. 23.27). However, the highest mountains of Alaska do not lie along the coast but are well inland in the Alaska Range, where Mt. McKinley rises to 20,320 feet (6194 m) about 100 miles north of Anchorage. This is the highest peak in North America. Anchorage is the state's largest city (population: 258,000) and has more than two fifths of the state's total population. At the southwest, the Alaska Range merges into the lower mountains occupying the long and narrow Alaska Peninsula and the rough, almost treeless Aleutian Islands.

Immediately south of Anchorage, the Kenai Peninsula lies between Cook Inlet and the Gulf of Alaska, and still farther to the southwest is Alaska's largest island, Kodiak. These features mark the western end of the horseshoe of coastal mountains that curves across the north of the Gulf of Alaska from the southeastern panhandle. The climate of lower areas in this western coastal section is harsher than in the panhandle and is classified as relatively mild subarctic.

FROM FUR TRADE TO OIL AGE: ERAS OF ALASKAN DEVELOPMENT

After approximately two centuries of occupation by Russia and the United States, the population of Alaska was still only 73,000 in 1940. Eighteenth-century Russian fur traders were spread thinly along the southern and southeastern coasts. Evidence of this era persists in the many Russian place names. In 1812, the Russians established an agricultural colony on the California coast at Fort Ross north of San Francisco Bay. This effort to more adequately provision the Alaskan posts was abandoned in 1841, and then, in 1867, remote and unprofitable Alaska was sold to the United States. By that time, hunting had nearly wiped out the sea otters, whose expensive pelts had provided the main incentive for Russia's Alaskan fur trade. Only recently under American and Canadian governmental protection has a substantial beginning been made at replenishing these mammals. Under the Americans, there was a burst of settlement connected with gold rushes in the 1890s and for a few years thereafter, followed by a very slow increase connected largely with fishing and military installations.

Beginning with World War II, during which Japanese troops invaded the Aleutian Islands, growth became more rapid. Today, metropolitan Anchorage at the head of Cook Inlet

FIGURE 23.27
The spectacular river of ice is the Margerie Glacier at Glacier Bay, Alaska. In the foreground is the cruise ship *Noordam* of the Holland America Line. Such dramatic landscapes as this one have made Alaska grow in popularity as a tourist destination, bringing increasingly significant tourist dollars to the state.   *Holland America Line*

is the main population cluster. The rest of Alaska's people are erratically distributed over the state, but living mainly in or near Fairbanks, the state capital of Juneau in the panhandle, or other small and widely spaced towns and villages. Of the total population, about one sixth is composed of "Alaska Natives" (Inuit, or Eskimos; American Indians; or Aleuts). Long the victims of exploitation and neglect by whites, these peoples have recently received some redress in large land grants and cash payments

Regional Perspective

PACIFIC RING OF FIRE

If you were to chart the location of the approximately 600 active volcanoes on earth right now, you would see that well over three quarters of them would be located in a broken ring that begins in Tierra del Fuego at the southern tip of South America and arcs northward in a massive horseshoe along the west coast of South and North America, across the Aleutian Islands of Alaska, down through the island arcs of East Asia, the island chains of the southwest Pacific, and

terminates in North Island of New Zealand. This string of active volcanoes is known as the **Ring of Fire.** It outlines the approximate boundaries of the **lithospheric plates** upon which the surface of the earth lies. **Plate tectonics** is the monumental process that describes the slow but inexorable motion of these plates as they shift atop the **mantle** or underlying rock of the Earth's core. The volcanoes of the Ring of Fire are caused by the intersection of these massive plates and the subsequent **subduction** that occurs when one plate is drawn beneath an-

other. The **subduction zones** are areas of enormous friction and seismic instability. Volcanic eruptions at Mt. St. Helens in Washington in 1986 and Mt. Pinatubo in the Philippines in 1990, and the massive 1964 Alaska earthquake and the devastating 1995 Kobe earthquake all represent recent expressions of the hazards of settling on the Ring of Fire. They are hallmarks of the power and potential destruction that lie along these zones of plate intersection.

under the Alaska Native Claims Settlement Act passed by Congress in 1971.

The burst of growth since 1940 has been principally due to (1) increased military and governmental employment, (2) the rise of the fishing industry to major importance, with salmon by far the main product, (3) revenues accruing from the development of Arctic oil (discovered in 1965), (4) the rise of air transport as a major carrier of both people and freight, and (5) visits by increasing numbers of tourists, both domestic and foreign. Agriculture consists largely of high-cost dairy farming in a few restricted spots. Both food and other consumer goods come overwhelmingly from the "Lower 48" states.

HAWAII: AMERICANIZATION IN A POLYGLOT SOUTH SEAS SETTING

The state of Hawaii consists of eight main tropical islands (see Fig. 23.26, page 604), lying just south of the Tropic of Cancer and somewhat over 2000 miles (*c.* 3200 km) from California. These islands are all volcanic and largely mountainous, with a combined land area of about 6400 square miles (*c.* 17,000 sq km), which is somewhat greater than the combined areas of Connecticut and Rhode Island. The island of Hawaii is by far the largest and the only one to have active volcanic craters. The prevalent trade winds and mountains produce extreme precipitation contrasts within short distances. Windward slopes receive heavy precipitation throughout the year (rising to an annual average of 460 inches/1168 cm on Mt. Wai-aleale on the island of Kauai—the wettest spot on the globe), while nearby leeward and low-lying areas may receive as little as 10 to 20 inches (25–50 cm) annually. Some areas are so dry that growing crops requires irrigation. Temperatures are tropical except for colder spots on some mountaintops. Honolulu averages 81°F (27°C) in August and 73°F (23°C) in January; the average annual precipitation in the city is 23 inches (58 cm), with a relatively dry high-sun period ("summer") and a wetter low-sun period ("winter").

Hawaii was more populous than Alaska and was the site of a Polynesian monarchy when it was acquired by the United States. By 1994 its population was 1.2 million, of whom about 875,000 were concentrated in metropolitan Honolulu. The metropolis occupies the island of Oahu, with the city proper located near the southeastern corner of the island. Immigration from the Pacific Basin has given the state's population a unique mix by ethnic origin: one third white Caucasian, one quarter Japanese, somewhat under one fifth native Polynesian, 14 percent Filipino, 6 percent Chinese, and a small contingent of Koreans and Samoans. Although there has been ongoing tension between native islanders and the immigrants who come from the U.S. mainland and East Asia, much racial and ethnic mixing has taken place. In general, the population is thoroughly Americanized.

Hawaii's economy is based overwhelmingly on military expenditures, tourism, and commercial agriculture. The main defense installations, including the Pearl Harbor naval base, are on Oahu. The main agricultural products and exports—cane sugar and pineapples—have long been grown on large estates owned and administered by Caucasian families and corporate interests but worked mainly by non-Caucasians. Work on such estates has often been the first employment of new immigrants. Changing demographic patterns, land values, and the declining role of agriculture in Hawaii have all led to steadily changing landscapes in the islands. Agriculture now amounts to only about 1 percent of the annual gross state product. Sugar has lost importance while tourism has been the great growth industry of recent decades, propelled by the development of air transportation and the growing affluence of mainland America. Visitors from Japan and other Pacific rim countries are also common. The main attractions for tourists are the tropical warmth, beaches, spectacular scenery, exotic cultural mix, and highly developed resorts. However, some of the "South Seas" glamour of the islands has been eroded by the remorseless spread of commercial and residential development.

QUESTIONS

Refer to the *Saunders Atlas of World Geography Special Edition* as well as this chapter to answer the following questions.

1. What generalizations about the spatial patterns of poverty and unemployment in the United States can be made by reviewing maps in the atlas?

2. Forests cover what percent of North America? What percent of the continent is suitable for agriculture?

3. Name and rank the six rivers in North America that are among the 25 rivers with the greatest volume in the world.

Name and rank the two North American rivers that are among the world's top 15 rivers in terms of size of drainage basin.

4. Explain why Chicago, Illinois, is called "The Windy City." Explain why the correct explanation is perhaps surprising to a geography student.

5. Discuss the climate patterns to the east and west of the 100 degree meridian of longitude in North America, utilizing the atlas to explain the probable reasons for such significant pattern differences.

REFERENCES AND READINGS

Geography and related disciplines provide an overwhelming number and variety of references and readings appropriate to study of this text. The sampling that follows is weighted toward the period 1989–1992 (for newer readings, see pp. R-33 through R-42). In addition to books, it includes timely articles from periodicals. Citations are grouped under the nine part titles of the book, with subgroupings by topic and geographic area. For comparable earlier lists of references, see Wheeler and Kostbade, *World Regional Geography* (Saunders, 1990) and Wheeler, Kostbade, and Thoman, *Regional Geography of the World* (Holt, Rinehart & Winston, 1975 and 1969 editions).

The Geographer's Field of Vision

The citations under this heading pertain to the entire text or to Chapters 1–3.

General Bibliography

Current Geographical Publications (University of Wisconsin—Milwaukee Library for the American Geographical Society; 10 issues a year) provides the most useful running list of books, articles, and maps. A massive annotated list of 2903 books, atlases, and serials is contained in C. D. HARRIS, editor in chief, *A Geographical Bibliography for American Libraries* (Washington, DC: Association of American Geographers/National Geographic Society, 1985).

Encyclopedias, Gazetteers, Dictionaries

Geography students and teachers should not overlook the utility of the geographic material—often written by professional geographers—in standard encyclopedias such as the *Encyclopaedia Britannica* (1988). The one-volume *Columbia Encyclopedia* (Columbia[1], 1975), a marvel of succinct writing, contains so much essential information from all fields of knowledge that it is practically a library in itself. Among other things, it functions as a world gazetteer, containing entries for thousands of places. A far larger gazetteer is L. E. SELTZER, ed., *The Columbia–Lippincott Gazetteer of the World, With 1961 Supplement* (Columbia, 1962). *Webster's New Geographical Dictionary* (Merriam–Webster, 1984) is a world gazetteer describing 47,000 places. See also D. MUNRO *et al.*, eds., *Chambers World Gazetteer: An A–Z of Geographical Information* (Cambridge and W. R. Chambers, 1988). Dictionaries of geographical terms, concepts, and quotations include A. N. CLARK, *Longman Dictionary of Geography: Human and Physical* (Longman, 1985); T. P. HUBER *et al.*, *Dictionary of Concepts in*

Human Geography (Greenwood, 1988); B. GOODALL, *The Facts on File Dictionary of Human Geography* (Facts on File, 1987); R. J. JOHNSTON, ed., *The Dictionary of Human Geography*, 2nd ed. (Basil Blackwell, 1986); J. SMALL and M. WITHERICK, *A Modern Dictionary of Geography* (Edward Arnold, 1986); L. D. STAMP and A. N. CLARK, eds., *A Glossary of Geographical Terms*, 3rd ed. (Longman, 1979); J. O. WHEELER and F. M. SIBLEY, *Dictionary of Quotations in Geography* (Greenwood, 1986).

Atlases and World Histories

Goode's World Atlas, 18th ed. (Rand McNally; rev., 1991) is a versatile and relatively inexpensive all-purpose student atlas incorporating topical and regional maps, many kinds of statistical data, and a pronouncing index. Larger and more elaborate reference atlases include (among many) the *New Cosmopolitan World Atlas* (Rand McNally, 1992), *The Economist Atlas* (Economist Books, 1991), and *Webster's New Geographical Atlas, United States & The World* (Gallery Books, 1991); G. BARRACLOUGH, ed., *The Times Atlas of World History*, 3rd ed. (Times Books/Hammond, 1989) is a magnificently produced combination of multicolored maps with textual summaries. See also *The Harper Atlas of World History* (Harper & Row, 1987). The two historical atlases can be highly useful in providing context for all parts of the present text, as can the historical summaries and maps in such books as M. DOCKRILL, *Atlas of Twentieth Century World History* (Harper-Collins, 1991); J. A. GARRATY and P. GAY, eds., *The Columbia History of the World* (Columbia, 1990); W. H. McNEILL, *A World History*, 3rd ed. (Oxford, 1979).

Sources of Statistics

The annual statistical publications of the United Nations are basic, especially the *Statistical Yearbook, Demographic Yearbook, Yearbook of International Trade Statistics*, and *FAO Production Yearbook*. The annual *Population Data Sheet* published by the Population Reference Bureau, Washington, DC, offers an extraordinarily handy tabulation of social and economic data by country. Annual statistical profiles of countries and tables of comparative international data occupy many pages in the timely and informative *Britannica Book of the Year*; the *Statesman's Year-Book* also includes much statistical data, as do other yearbooks and almanacs. The annual *Geographical Digest* (George Philip) is both a convenient source of statistics and a record of significant geographical

[1]"Columbia" denotes Columbia University Press. Subsequent citations to university presses follow this model.

changes. The annual *Statistical Abstract of the United States* is a vast summation of all kinds of statistics, including some international comparisons by country. V. SHOWERS, *World Facts and Figures*, 3rd ed. (Wiley, 1989) is a general compilation of data concerning natural features, climate, cities, population, and other subjects. See also U. S. CENTRAL INTELLIGENCE AGENCY, The World Factbook 1991 (U.S. Government Printing Office), organized by political units, with maps in color.

Periodicals

The *Geographical Magazine*, written by professional geographers for a general readership, is especially valuable as an adjunct to this text. Illustrated with maps and photos in color, it carries large numbers of short and timely articles on geographical areas and topics. Many are cited in this reference list. Smaller numbers of articles of the same general kind are available in *Focus* (American Geographical Society: quarterly). The familiar *National Geographic* (National Geographic Society: monthly) is invaluable for maps, and its lavishly illustrated articles provide up-to-date information for general readers. Within the more strictly professional geographical journals in the English language (most of which are quarterlies), articles directly pertinent to this text are most likely to be found in *Geography, GeoJournal, Geographical Review, Journal of Geography, Annals of the Association of American Geographers* (cited below as "*Annals AAG*"), *Geographical Journal*, Institute of British Geographers *Transactions* (cited as "*Transactions IBG*"), *Geoforum*, and *Landscape*. Names of still other journals in geography and many other fields will become apparent in the lists of citations for specific areas and topics.

In our rapidly changing world, it is hard to keep abreast of day-to-day developments. Regular use of reliable geographically-oriented newspapers and magazines such as the *Christian Science Monitor, New York Times, Wall Street Journal*, and *Economist* can be very helpful. *Foreign Affairs* (quarterly) provides serious analyses of important world areas and situations; *Current History* (9 issues a year) regularly devotes entire issues to articles on major world areas such as Latin America or China; and the *Swiss Review of World Affairs* (monthly; published in English) provides fine journalistic coverage in articles that are exceptionally concise. Many well-written articles of geographic import also appear in such mass-circulation periodicals as *Technology Review, The Scientific American, Natural History, Fortune, The New Yorker, The Atlantic Monthly, The New Republic*, and the weekly newsmagazines.

Geography as a Discipline

The nature of geography as a field of study and action has been explored by professional geographers in a flood of articles, scholarly addresses, papers, and books over a period of many years. A good summation of the field's history is P. E. JAMES and G. J. MARTIN, *All Possible Worlds: A History of Geographical Ideas*, 2nd ed. (Wiley, 1981). Also, basic, although not easy reading for undergraduates, are R. HARTSHORNE, *The Nature of Geography: A Critical Survey of Current Thought in the Light of the Past, Annals AAG*, 29 (1939), 171–658; subsequently reprinted in book form by the AAG; and *Perspective on the Nature of Geography* (AAG, 1959). Other useful books and articles include ASSOCIATION OF AMERICAN GEOGRAPHERS, *Geography and International Knowledge* (1982); J. O. M. BROEK *et al., The Study and*

Teaching of Geography (Merrill, 1980); C. A. FISHER, "Whither Regional Geography?" *Geography,* 55 (1970), 373–389; G. L. GAILE and C. J. WILLMOTT, *Geography in America* (Merrill, 1989); E. D. GENOVESE and L. HOCHBERG, eds., *Geographic Perspectives in History* (Basil Blackwell, 1989); P. GOULD, *The Geographer at Work* (Routledge, 1985); P. HAGGETT, *Geography: A Modern Synthesis*, 3rd ed. (Harper & Row, 1983; a major introductory text); R. A. Harper, "Geography in General Education: The Need to Focus on the Geography of the Field," *Journal of Geography*, 81 (1982), 122–139; J. F. HART, "The Highest Form of the Geographer's Art," *Annals AAG*, 72 (1982), 1–29; A. HOLT–JENSEN, *Geography: History and Concepts: A Student's Guide,* 2nd ed. (Barnes & Noble Books, 1988); M. S. KENZER, ed., *On Becoming a Professional Geographer* (Merrill, 1989); R. KING, ed., *Geographical Futures* (Sheffield, England: Geographical Association, 1985); J. T. KOSTBADE, "A Brief for Regional Geography," *Journal of Geography* 64 (1965), 362–366, also, "The Regional Concept and Geographic Education," *ibid.*, 67 (1968), 6–12; G. S. LUDWIG *et al.; Directions in Geography: A Guide for Teachers* (Washington, DC: National Geographic Society); W. E. MALLORY and R. SIMPSON-HOUSLEY, eds., *Geography and Literature: A Meeting of the Disciplines* (Syracuse, 1987); B. MARSHALL, ed., *The Real World: Understanding the Modern World Through the New Geography* (Houghton Mifflin, 1991); D. W. MEINIG, ed., *On Geography: Selected Writings of Preston E. James* (Syracuse, 1971); R. L. MORRILL and J. M. DORMITZER, *The Spatial Order: An Introduction to Modern Geography* (Duxbury, 1979); R. MURPHEY, *The Scope of Geography*, 3rd ed. (Methuen, 1982); S. J. NATOLI and A. R. BOND, *Geography in Internalizing the Undergraduate Curriculum* (AAG, 1985); W. NORTON, "Human Geography and the Geographical Imagination," *Journal of Geography*, 88 (1989), 186–192; W. D. PATTISON, "The Four Traditions of Geography," *Journal of Geography,* 63 (1964), 211–216; C. L. (KIT) SALTER, "What Is the Essence of Geographic Literacy?" *Focus,* Summer 1990, 26–29; J. P. STOLTMAN, "Geography's Role in General Education in the United States," *GeoJournal*, 20 (1990), 7–14; YI-FU TUAN, "A View of Geography," *Geographical Review,* 81 (1991), 99–107; G. F. WHITE, "Geographers in a Perilously Changing World," *Annals AAG*, 75 (1985), 10–15.

Maps and Mapping

A few key references include W. F. ALLMAN, "A Sense of Where You Are" (geographic information systems and computer-generated maps), *U.S. News & World Report,* April 5, 1991, 58–60; "All Over the Map," *The Economist,* July 11, 1992, 83–84 (how name changes affect map and atlas publishing); G. BOULTON, "Relict Geography: Maps as Records," *Teaching Geography*, 16 (1991), 173–174; L. A. BROWN, *The Story of Maps* (Little, Brown, 1949); M. F. BURRILL, "The Wonderful World of Geographic Names: Things Learned and Things Yet to be Learned," *Names*, 39 (1991), (181–190; J. CAMPBELL, *Introductory Cartography*, 2nd ed. (Wm. C. Brown, 1991); D. CHURBUCK, "Geographics" (use of computerized mapping in business and government), *Forbes*, January 6, 1992, 262–267; P. GOULD and R. WHITE, *Mental Maps*, 2nd ed. (Allen & Unwin, 1986); D. GREENHOOD, *Mapping*, rev. ed. (Chicago, 1964); S. P. JESSUP and E. CARY, "Geographic Information Systems: What They Are, and How They Work," *Focus,* Summer 1989, 10–12; J. S. KEATE, Understanding

Maps (Longman, 1982); also, *Cartographic Design and Production*, 2nd ed. (Longman, 1989); P. LEWIS, *Maps and Statistics* (Wiley, 1977); M. MONMONIER, *How to Lie with Maps* (Chicago, 1991); also, *Maps with the News: The Development of American Journalistic Cartography* (Chicago, 1989); P. C. MUEHRCKE, *Map Use: Reading, Analyses, and Interpretation*, 2nd ed.; J. P. Publications, 1986); W. E. PHIPPS, "Cultural Commitments and World Maps," *Focus*, Summer 1991, 7–9; A. H. ROBINSON *et al.*, *Elements of Cartography*, 5th ed. (Wiley, 1984); M. M. THOMPSON, *Maps for America: Cartographic Products of the U. S. Geological Survey and Others* (Government Printing Office, 1979); N. J. W. THROWER, *Maps and Man: An Examination of Cartography in Relation to Culture and Civilization* (Prentice-Hall, 1972); E. R. TUFTE, *The Visual Display of Quantitative Information* (Cheshire, CT: Graphics Press, 1983); also, *Envisioning Information* (Graphics Press, 1990); J. N. WILFORD, JR., *The Mapmakers* (Knopf, 1981; a history of cartography for the general reader).

General Physical Geography

Multivolume series includes TIME–LIFE BOOKS, *Planet Earth* (1982–1984; 11 lavishly illustrated volumes for the general reader) and D. W. GOODALL, ed., *Ecosystems of the World* (Elsevier, 1977–   : to include 29 vols.; very technical). Some individual books and articles for general readers are D. ATTENBOROUGH, *The Living Planet: A Portrait of the Earth* (Collins/BBC, 1984); D. BRUNSDEN and J. DOORNKAMP, eds., *The Unquiet Landscape: Series from the Geographical Magazine* (Wiley, 1978); D. J. CRUMP, ed., *Our Awesome Earth: Its Mysteries and Its Splendors* (National Geographic Society, 1986); "Dynamic Earth," *Geographical Magazine*, vols. 15–17 (1983–1985); 15 articles by various authors); A. GOUDIE, *The Human Impact on the Natural Environment*, 3rd ed. (Basil Blackwell, 1990); R. H. GROVE, "Origins of Western Environmentalism," *Scientific American*, July 1992, 42–47; F. W. LANE, *The Violent Earth* (Croom Helm, 1986), a book about destructive natural phenomena; S. P. PARKER, editor in chief, *McGraw-Hill Encyclopedia of Environmental Science* (McGraw-Hill, 1980); "Planet of the Year: Endangered Earth" (a special issue), *Time*, January 2, 1989; M. G. RENNER, "War on Nature" (military-generated toxic waste), *World Watch*, May–June 1991, 18–25; R. REPETTO, "Accounting for Environmental Assets," *Scientific American*, June 1992, 94–100; J. SEAGER, ed., *The State of the Earth Atlas* (Simon & Schuster, 1990); "Sharing: A Survey of the Global Environment," *The Economist*, May 30, 1992, 24 pp.; I. G. SIMMONS, "Ingredients of a Green Geography," *Geography*, 75 (1990), 98–105; B. L. TURNER, II *et al.*, eds., *The Earth as Transformed by Human Action: Global and Regional Changes in the Biosphere over the Past 300 Years* (Cambridge, 1991); J. WEINER, *The Next One Hundred Years: Shaping the Fate of Our Living Earth* (Bantam Books, 1990).

Some dictionaries of physical geography and environment include A. S. GOUDIE, ed., *The Encyclopaedic Dictionary of Physical Geography* (Basil Blackwell, 1985); R. W. DURRENBERGER, *Dictionary of the Environmental Sciences* (National Press Books, 1973); S. E. STIEGELER, ed., *A Dictionary of Earth Sciences* (Rowman & Allanheld, 1983; J. B. WHITTOW, ed., *The Penguin Dictionary of Physical Geography* (Allen Lane/Penguin, 1984). Numerous references to the environment are contained in K. A. HAMMOND, G. MACINKO, and W. B. FAIRCHILD, eds.,

Sourcebook on the Environment: A Guide to the Literature (Chicago, 1978).

Among the many general textbooks in physical geography are R. E. GABLER *et al.*, *Essentials of Physical Geography*, 3rd ed. (Saunders College, 1987); A. GOUDIE, *The Nature of the Environment: An Advanced Physical Geography* (Basil Blackwell, 1989); T. L. McKNIGHT, *Physical Geography: A Landscape Appreciation* (Prentice-Hall, 1990); W. M. MARSH and J. DOZIER, *Landscape: An Introduction to Physical Geography* (Wiley, 1986); A. N. and A. H. STRAHLER, *Modern Physical Geography*, 4th ed. (Wiley, 1992).

Landforms, Climate, Biogeography, and Soils

References on landforms include: H. BLUME, *Colour Atlas of the Surface Forms of the Earth* (Harvard, 1992); G. H. DURY, *The Face of the Earth*, 5th ed. (Allen & Unwin, 1986), a concise introduction; A. K. LOBECK, *Geomorphology: An Introduction to the Study of Landforms* (McGraw-Hill, 1939), a classic book of timeless value, well illustrated with diagrams and photographs. R. E. MURPHY, "Landforms of the World," is a large map in color which appeared as Map Supplement 9, *Annals AAG*, 58, no. 1 (March 1968). See also the physiographic diagrams by A. K. LOBECK (available from Hammond, Inc.) and landform maps by E. RAISZ (available from Erwin Raisz, 130 Charles St., Boston, MA, 02114). For spectacular photos from space, see O. W. NICKS, ed., *This Island Earth* (Washington, DC: NASA, 1970); N. M. SHORT, *Mission to Earth: LANDSAT Views the World* (NASA, 1976); also, with R. W. BLAIR, JR., *Geomorphology from Space: A Global Overview of Regional Landforms* (NASA, 1986).

Some general references on climate include W. F. ALLMAN and B. WAGNER, "Climate and the Rise of Man," *U. S. News & World Report*, June 8, 1992, 61–67; R. A. BRYSON and T. J. MURRAY, *Climates of Hunger: Mankind and the World's Changing Weather* (Wisconsin, 1977); C. F. COOPER, "What Might Man-Induced Climatic Change Mean?" *Foreign Affairs*, 56 (1978), 500–520; J. R. GRIBBIN, ed., *Climatic Change* (Cambridge, 1978); M. HULME, "Global Warming," *Progress in Physical Geography*, 15, no. 3 (1991), 310–318; H. H. LAMB, *Climate, History and the Modern World* (Methuen, 1982); P. E. LYDOLPH, *The Climate of the Earth* (Rowman & Allanheld, 1985); W. H. MacLEISH, "Water, Water Everywhere, How Many Drops to Drink?" *World Monitor*, December 1990, 54–58; C. PARK, "Transfrontier Air Pollution: Some Geographical Issues," *Geography*, 76 (1991), 21–35; S. P. PARKER, editor in chief, *McGraw-Hill Encyclopedia of Ocean and Atmospheric Sciences* (McGraw-Hill, 1980); E. A. PEARCE and G. SMITH, *The Times Books World Weather Guide*, updated ed. (Times Books, 1990); C. TICKELL, "Human Effects of Climatic Change . . . ," *Geographical Journal*, 156 (1990), 325–329; G. T. TREWARTHA, *The Earth's Problem Climates*, 2nd ed., (Wisconsin, 1981).

On biogeography, soils, and diseases, see D. BARRACLOUGH, "The Earth Taken for Granted" (focus on soil), *Geographical Magazine*, March 1990, 36–38; E. M. BRIDGES, *World Soils*, 2nd ed. (Cambridge, 1978); also, "Soil: The Vital Skin of the Earth," *Geography*, 63 (1978), 354–361; A. S. COLLINSON, *Introduction to World Vegetation* (Allen & Unwin, 1977); S. R. EYRE, ed., *Vegetation and Soils: A World Picture*, new ed. (Edward Arnold, 1975); G. HARRISON, *Mosquitoes, Malaria and Man: A History of the Hostilities Since 1880* (Dutton, 1978); G. M.

HOWE, ed., *A World Geography of Human Diseases* (Academic, 1977); A. S. MATHER, "Global Trends in Forest Resources," *Geography*, 314 (1987), 1–15; I. G. SIMMONS, *Biogeography: Natural and Cultural* (Edward Arnold, 1979); D. STEILA, *The Geography of Soils: Formation, Distribution, and Management* (Prentice-Hall, 1976); J. TIVY, *Biogeography: A Study of Plants in the Ecosphere*, 2nd ed. (Longman, 1982).

The Tropics: Geography and Development

For general views, see M. HINTZ, *Living in the Tropics: A Cultural Geography* (Franklin Watts, 1987); B. W. HODDER, *Economic Development in the Tropics*, 3rd ed. (Methuen, 1980); D. H. JANZEN, "The Uncertain Future of the Tropics," *Natural History*, 81 (1972), 80–89; J. G. LOCKWOOD, *The Physical Geography of the Tropics: An Introduction* (Oxford, 1976); J. OLIVER, "A Study of Geographical Imprecision: The Tropics," *Australian Geographical Studies*, 17 (1979), 3–17.

Books and articles on tropical rain forests and savannas include B. N. FLOYD, "The Rain Forest and the Farmer: Observations and Recommendations," *GeoJournal*, 6 (1982), 433–442; N. GUPPY, "Tropical Deforestation: A Global View," *Foreign Affairs*, 62 (1984), 928–965; D. R. HARRIS, ed., *Human Ecology in Savanna Environments* (Academic, 1980); D. JUKOFSKY, "Problems and Progress in Tropical Forests," *American Forests*, July/August 1991, 48–51, 76–77; D. LEWIS, "You Can't See the Wood for the Trees" (tropical deforestation), *Geographical Magazine*, July 1990, 22–27; M. MARGOLIS, "Tomorrow's Trees" (world reforestation), *World Monitor*, November 1991, 34–40; D. C. MONEY, *Tropical Rainforests* (Evans Brothers, 1980); N. MYERS, *The Primary Source: Tropical Forests and Our Future* (Norton, 1984); A. NEWMAN, *Tropical Rainforest* (Facts on File, 1990); S. NORTCLIFF, "The Clearance of Tropical Rainforest," *Teaching Geography*, 12 (1987), 110–113; J. R. OSGOOD, "Tropical Timber: An Importer's Point of View," *Unasylva*, 31, no. 124 (1979), 26–28; J. PROCTOR, "Tropical Rain Forest," *Progress in Physical Geography*, 15 (1991), 291–303; P. W. RICHARDS, *The Tropical Rain Forest* (Cambridge, 1952, 1964); also, "The Tropical Rain Forest," *Scientific American*, December 1973, 58–67; P. A. STOTT, "Tropical Rain Forest in Recent Ecological Thought: The Reassessment of a Nonrenewable Resource," *Progress in Physical Geography*, 2 (1978), 80–98; T. C. WHITMORE, *An Introduction to Tropical Rain Forests* (Oxford, 1990); M. WILLIAMS, "Deforestation: Past and Present," *Progress in Human Geography*, 13 (1989), 176–208.

Arid, Polar, and Mountain Environments; Natural Hazards

A small selection of titles on arid lands and desertification includes T. BINNS, "Is Desertification a Myth?" *Geography*, 75 (1990), 106–113; A. T. GROVE, "Desertification," *Progress in Physical Geography*, 1 (1977), 296–310; R. L. HEATHCOATE, *The Arid Lands: Their Use and Abuse* (Longman, 1983); D. L. JOHNSON, ed., "The Human Face of Desertification," *Economic Geography*, 53 (1977), 317–432; P. PETROV, *Deserts of the World* (Wiley, 1976); A. WARREN, "Shifting Margins of the Desert," *Geographical Magazine*, 16 (1984), 457–462.

Selected references on polar and mountain environments include N. J. R. ALLEN, ed., *Human Impact of Mountains* (Rowman & Littlefield/Barnes & Noble, 1987); T. ARMSTRONG *et al.*,

The Circumpolar North (Methuen, 1978); F. BRUEMMER *et al.*, *The Arctic World* (Sierra Club, 1985); J. GERRARD, *Mountain Environments: An Examination of the Physical Geography of Mountains* (MIT Press, 1990); D. K. HAGLUND and M. H. HERMANSON, "Changing Resource Use in the Arctic," *Focus*, May-June 1983, 1–16; J. D. IVES and R. G. BARRY, eds., *Arctic and Alpine Environments* (Methuen, 1974); L. W. PRICE, *Mountains and Man: A Study of Process and Environment* (California, 1981); D. E. SUGDEN, *Arctic and Antarctic: A Modern Geographical Synthesis* (Barnes & Noble, 1982); P. B. STONE, ed., *The State of the World's Mountains: A Global Report* (Zed Books, 1992); J. H. ZUMBERGE, "Mineral Resources and Geopolitics in Antarctica," *American Scientist*, 67 (1979), 68–77.

On natural hazards, see E. A. BRYANT, *Natural Hazards* (Cambridge, 1991); I. BURTON, R. W. KATES, and G. F. WHITE, *The Environment as Hazard* (Oxford, 1978); J. E. BUTLER, *Natural Disasters* (Heinemann, 1976); J. CORNELL, *The Great International Disaster Book* (Scribner, 1976); M. T. FARVAR and J. P. MILTON, eds., *The Careless Technology: Ecology and International Development* (Natural History Press, 1972); R. WARD, *Floods: A Geographical Perspective* (Macmillan, 1978).

General Economic Geography, Economic Development, and Resources

R. J. BARNET, "The World's Resources," a three-part article, *The New Yorker*, March 24, March 31, April 7, 1980; B. E. COATES, "Our Unequal World" (world poverty), *Geography*, January 1992, 1–9; S. CORBRIDGE, "Third World Development," *Progress in Human Geography*, 15 (1991), 311–321; R. A. EASTERLIN, "Why Isn't the Whole World Developed?" *Journal of Economic History*, 41 (1981), 1–17; "Economic Development" (special issue), *Scientific American*, September 1980; D. K. FORBIS, *The Geography of Underdevelopment: A Critical Survey* (Croom Helm, 1984); M. FREEMAN, *Atlas of the World Economy* (Simon & Schuster, 1991); P. P. KARAN *et al.*, "Technological Hazards in the Third World," *Geographical Review*, 76 (1986), 195–208; M. KIDRON and R. SEGAL, *New State of the World Atlas*, 2nd ed. (Heinemann/Simon & Schuster, 1984); P. E. KNOX and J. AGNEW, *Geography of the World Economy* (Edward Arnold, 1989); J. MATTILA, "Innovation: A Neglected Locational Factor," *GeoJournal*, 9 (1984), 187–197; B. MITCHELL, *Geography and Resource Analysis*, 2nd ed. (Longman, 1989); P. J. TAYLOR, "Understanding Global Inequalities: A World-Systems Approach," *Geography*, 77 (1992), 10–21; D. WHEELER, *Human Resource Policies, Economic Growth, and Demographic Change in Developing Countries* (Oxford, 1984); WORLD BANK, *World Development Report 1992*: "Development and the Environment" (Oxford University Press for the World Bank, 1992); WORLD RESOURCES INSTITUTE, *World Resources* (Basic Books, from 1986); WORLDWATCH INSTITUTE, *Worldwatch Papers* (Washington, DC; a continuing series of paperbound reports on economic, social, political, and economic subjects); also, *State of the World* (Norton, annual), and World Watch, a bimonthly magazine on the environment.

Agriculture and Rural Life

I. R. BOWLER, "Agricultural Geography," *Progress in Human Geography*, 14 (1990, 569–578); B. CURREY and G. HUGO, eds., *Famine as a Geographical Phenomenon* (Reidel, 1984); L. T.

EVANS, "The Natural History of Crop Yields," *American Scientist*, 68 (1980), 388–397; "Food and Agriculture" (special issue), *Scientific American*, 235, no. 3 (September 1976); E. GRAHAM and I. FLOERING, eds., *The Modern Plantation in the Third World* (Croom Helm/St. Martin's, 1984); D. GRIGG, *An Introduction to Agricultural Geography* (Hutchinson, 1984); also, "World Patterns of Agricultural Output," *Geography*, 71 (1986), 240–245; D. Q. INNIS, "The Future of Traditional Agriculture," *Focus*, January-February 1980, 1–16; E. J. KAHN, JR., *The Staffs of Life* (Little, Brown, 1985; a survey of major food crops, written for the general reader); RESOURCES FOR THE FUTURE, "Feeding a Hungry World" (special issue), *Resources*, Spring 1984, 1–21; C. O. SAUER, *Agricultural Origins and Dispersals: The Domestication of Animals and Foodstuffs*, 2nd ed. (MIT, 1969); J. L. SIMON, "World Food Supplies," *Atlantic Monthly*, July 1981, 72–76; B. L. TURNER, II and S. B. BRUSH, *Comparative Farming Systems* (Guilford Press, 1987); J. R. TARRANT, "World Food Prospects for the 1990s," *Journal of Geography*, 89 (1990), 234–238; S. WHITMORE, "Agricultural Geography," *Progress in Human Geography*, 15 (1991), 303–310.

Urban, Industrial, and Transportation Geography

J. AGNEW, J. MERCER, and D. SOPHER, eds., *The City in Cultural Context* (Allen & Unwin, 1984); J. BIRD, *Centrality and Cities* (Routledge, 1977); S. D. BRUNN et al., *Cities of the World: World Regional Urban Development* (Harper & Row, 1983); K. HOGGART, "The Changing World of Corporate Control Centres," *Geography*, 76 (1991), 109–120; H. M. MAYER et al., *A Modern City: Its Geography* (National Council for Geographic Education, 1970; articles reprinted from the 1969 *Journal of Geography*); L. MUMFORD, *The City in History: Its Origins, Its Transformations, and Its Prospects* (Harcourt, Brace & World, 1961); R. J. ROSS and G. J. TELKAMP, eds., *Colonial Cities* (Martinus Nijhoff, 1985); J. R. SHORT, *An Introduction to Urban Geography* (Routledge, 1984); "The Beautiful and the Dammed" (demerits of large dams), *The Economist*, March 28, 1992, 93–97; TIME–LIFE BOOKS, *The World Cities* (1976–1978; volumes on individual cities); J. E. VANCE, JR., *Capturing the Horizon: The Historical Geography of Transportation* (Harper & Row, 1986); also, *This Scene of Man: The Role and Structure of the City in the Geography of Western Civilization* (Harper's College, 1977).

Population Geography

G. CHARLES, "Hobson's Choice for Indigenous Peoples: Stuck Between Cultural Extinction and Misery," *World Press Review*, September 1992, 26–28; PAUL and ANNE EHRLICH, *The Population Explosion* (Hutchinson, 1990); D. J. M. HOOSON, "The Distribution of Population as the Essential Geographical Expression," *Canadian Geographer*, 17 (1960), 10–20; H. JONES, *Population Geography*, 2nd ed. (Chapman, 1990); E. J. KAHN, JR., "The Indigenists," *The New Yorker*, August 31, 1981, 60–77 (survival of tribal peoples); N. KLIOT, "The Era of Homeless Man" (world survey of refugees), *Geography*, 72 (1987), 109–121; G. J. LEWIS, *Human Migration: A Geographical Perspective* (St. Martin's, 1982); *Population Bulletin* (4 issues a year), *Population Today* (11 issues a year), and other publications of the Population Reference Bureau, Washington, DC (issues of *Population Today* include useful one-page summaries of individual countries); J. A. McFALLS, JR., "Population: A Lively Introduction," *Population Bulletin*, October 1991, 1–43; C. TYLER, "The World's Manacled Millions" (increasing slave population), *Geographical Magazine*, January 1991, 30–35.

Cultural Geography

A. GETIS, J. GETIS, and J. FELLMANN, *Human Geography: Culture and Environment* (Macmillan, 1985); J. F. HART, *The Look of the Land* (Prentice-Hall, 1975); J. B. JACKSON, *Landscapes: Selected Writings of J. B. Jackson* (Massachusetts, 1970); W. A. D. JACKSON, *The Shaping of Our World: A Human and Cultural Geography* (Wiley, 1985); T. G. JORDAN and L. ROWNTREE, *The Human Mosaic: A Thematic Introduction to Cultural Geography*, 5th ed. (Harper & Row, 1990); J. LEIGHLY, ed., *Land and Life: A Selection from the Writings of Carl Ortwin Sauer* (California, 1963, reprinted, 1974); G. J. LEVINE, "On the Geography of Religion," *Transactions IBG*, n.s., 11 (1986), 428–440; D. W. MEINIG, ed., *The Interpretation of Ordinary Landscapes: Geographical Essays* (Oxford, 1979); "New Christendom: The Southern Cross," *The Economist*, December 24, 1988, 61–66 (a survey of world Christianity); D. E. SOPHER, *Geography of Religions* (Prentice-Hall, 1968); J. E. SPENCER and W. L. THOMAS, JR., *Introducing Cultural Geography*, 2nd ed. (Wiley, 1978); C. TYLER, "Speaking in Tongues" (the world's languages), *Geographical Magazine*, May 1991, 10–14; W. L. THOMAS, JR., ed., *Man's Role in Changing the Face of the Earth* (Chicago, 1956; a massive, impressive survey, with contributions by many notable authorities); H. J. VIOLA and C. MARGOLIS, eds., *Seeds of Change: A Quincentennial Commemoration* (Washington, DC: Smithsonian Institution Press, 1991); P. L. WAGNER and M. W. MIKESELL, eds., *Readings in Cultural Geography* (Chicago, 1962).

Political Geography; Ethnicity

F. ANSPRENGER, *The Dissolution of the Colonial Empires* (Routledge, 1989); G. ARNOLD, *Wars in the Third World Since 1945* (Cassell, 1991); T. ARKELL, "Environmental Determinism and the Balance of World Power," *Geographical Magazine*, November 1990, 18–22; "Defence in the 21st Century: Meet Your Unbrave New World," *The Economist*, September 5, 1992, 22 pp.; J. L. GADDIS, "Toward the Post–Cold War World," *Foreign Affairs*, Spring 1991, 102–122; C. HARRIS, "Power, Modernity, and Historical Geography," *Annals AAG*, 81 (1991), 671–683; R. HARTSHORNE, "The Functional Approach in Political Geography," *Annals AAG*, 40 (1950), 95–130; H. J. JOHNSON, *Dispelling the Myth of Globalization: The Case for Regionalization* (Praeger, 1991); J. KEEGAN, ed., *The Times Atlas of the Second World War* (Harper & Row, 1989); P. KENNEDY, *The Rise and Fall of the Great Powers: Economic Change and Military Conflict from 1500 to 2000* (Random House, 1987); D. LEWIS, "Conflict of Interests" (relation of indigenous peoples to national parks), *Geographical Magazine*, December 1990, 18–22; M. W. MIKESELL, "The Myth of the Nation-State," *Journal of Geography*, 82 (1983), 257–260; and, with A. B. MURPHY, "A Framework for Comparative Study of Minority-Group Aspirations," *Annals AAG*, 81 (1991), 581–604; R. E. H. MELLOR, *Nation, State and Territory: A Political Geography* (Routledge, 1989); J. S. NYE, JR., "What New World Order?" *Foreign Affairs*, Spring 1992, 83–96; J. O'LOUGHLIN, "Political Geography: Coping with Global Restructuring," *Progress in Human Geography*, 13 (1989),

412–426; also, "Political Geography: Attempting to Understand a Changing World Order," *ibid.*, 14 (1990), 420–437; also, "Political Geography: Returning to Basic Conceptions," *ibid.*, 15 (1991), 322–339; also, with H. VAN DER WUSTEN, "Political Geography of Panregions," *Geographical Review*, 80 (1990) 1–20; A. ROYLE, "The World's Remaining Colonies: Tugging at the Apron Strings," *Geographical Magazine*, June 1991, 14–16; P. M. SLOWE, *Geography and Political Power: The Geography of Nations and States* (Routledge, 1990); P. J. TAYLOR, *Political Geography: World-Economy, Nation-State and Locality* (Longman, 1989); "The Political Geography of the Post Cold War World" (a multiauthored symposium of short articles), *Professional Geographer*, 44 (1992), 1–29; "The World's Wars: Tribalism Revisited" (with map), *The Economist*, December 21/January 3, 1992, 45–46; U.S. DEPARTMENT OF STATE, "Status of the World's Nations," *Geographic Notes*, special issue, vol. 2, no. 1, Spring 1992; A. ZÄNKER, "Geopolitics: The Forgotten Dimension," *Swiss Review of World Affairs*, September 1992, 6–8.

Part 1: Europe

Europe: General Geographic Surveys and Analyses

S. ARLETT and J. SALLNOW, "European Centres of Dissent," *Geographical Magazine*, September 1989, 6–9; F. BRAUDEL, *Civilization and Capitalism 15th–18th Century*, 3 vols. (Collins/Harper & Row, 1981); D. BURTENSHAW *et al.*, *The European City: A Western Perspective* (Wiley, 1991); W. J. CAHNMAN, "Frontiers Between East and West in Europe," *Geographical Review*, 39 (1949), 605–624; A. CLARK, "America's Comeback" (in Europe) *World Monitor*, July 1992, 32–34; H. D. CLOUT *et al.*, *Western Europe: Geographical Perspectives*, 2nd ed. (Longman, 1989); N. COLCHESTER, "The European Community: Altered States," *The Economist*, July 11, 1992, 30 pp.; J. CUISENER, ed., *Europe as a Culture Area* (Mouton, 1979); "Divided They Stand: Europeans Step Back from Unity," *World Press Review*, August 1992, 9–14; "Europe's Immigrants: Strangers Inside the Gates," *The Economist*, February 15, 1992, 21–24; A. FONTAINE, "The Real Divisions of Europe," *Foreign Affairs*, 49 (1971), 302–314; C. R. FOSTER, ed., *Nations Without a State: Ethnic Minorities in Western Europe* (Praeger, 1980); R. GIBB, "The Channel Tunnel Rail Link: Implications for Regional Development," *Geography*, 77 (1992), 67–69; J. GOTTMANN, *A Geography of Europe*, 4th ed. (Holt Rinehart & Winston, 1969; dated, but still very valuable for regional description and historical detail); E. HADINGHAM, "Europe's Mystery People" (The Basques), *World Monitor*, September 1992, 34–42; G. W. HOFFMAN, ed., *Europe in the 1990s: A Geographic Analysis*, 6th ed. (Wiley, 1989); W. ILBERY, *Western Europe: A Systematic Human Geography*, 2nd ed. (Oxford, 1986); T. G. JORDAN, *The European Culture Area: A Systematic Geography* (Harper & Row, 1973); P. L. KNOX, *The Geography of Western Europe: A Socio-Economic Survey* (Rowman & Littlefield/Barnes & Noble, 1984); A. LAMBERT, "Fast Forward" (high-speed rail network), *Geographical Magazine*, March 1992, 11–14; P. LEBRET, "The Loess of West Europe," *GeoJournal*, 24 (1991), 151–156; E. LICHTENBERGER, "The Nature of European Urbanism," *Geoforum*, no. 4 (1970), 45–62; A. MacLEOD, "The Channel Tunnel," *Christian Science Monitor*, June 10, 1992, 10–11; R. J. McCALLA, "The Geographical Spread of Free Zones Associated with Ports," *Geoforum*, 21 (1990), 121–134; V. H. MALMSTRÖM, *Geography of Europe: A Regional Analysis* (Prentice-Hall, 1971); A. B. MURPHY, "The Emerging Europe of the 1990s," *Geographical Review*, 81 (1991), 1–17; K. NEBENZAHL, *Atlas of Columbus and the Great Discoveries* (Rand McNally, 1990); *Oxford Regional Economic Atlas: Western Europe* (Oxford, 1971; superb cartography; aging but still extremely useful); W. H. PARKER, ed., *Western Europe: Challenge and Change* (Guilford, 1991); P. REVZIN, "The Rocky Road to European Unity," *1992 World Book Year Book* 208–215; "Separatism: Western Europe Discovers Its 'Baltics,' " *World Press Review*, January 1992, 16–17; G. F. TREVERTON, "The New Europe," *Foreign Affairs*, 71 (1992), 94–112; D. TURNOCK, "Europe Reintegrates," *Geographical Magazine*, May 1990, 14–18; "Unity Jitters: How Does Europe Get There from Here?" *World Press Review*, August 1992, 9–14; O. WAEVER, "Three Competing Europes: German, French, Russian," *International Affairs*, 66 (1990), 477–493.

Europe: Historical Circumstances and Historical Geography

From an overwhelming number of possibilities, some selections are: F. ANSPRENGER, *The Dissolution of the Colonial Empires* (Routledge, 1989); J. P. V. D. BALSDON, *Romans and Aliens* (North Carolina, 1979); E. Barker *et al.*, eds., *The European Inheritance*, 3 vols. (Oxford, 1954); D. J. BOORSTIN, *The Discoverers* (Random House, 1983); M. E. CHAMBERLAIN, *Decolonisation: The Fall of the European Empires* (Basil Blackwell, 1985); H. C. DARBY, "The Face of Europe on the Eve of the Great Discoveries," in *The New Cambridge Modern History* (Cambridge, 1957), vol. 1, chap. 2; W. G. EAST, *An Historical Geography of Europe*, 5th ed. (Methuen/Dutton, 1966); D. K. FIELDHOUSE, *The Colonial Empires* (Dell, 1966); E. W. FOX, *The Emergence of the Modern European World: From the Seventeenth to the Twentieth Centuries* (Blackwell, 1991); E. A. GUTKIND, *International History of City Development*, 8 vols. (Free Press, 1964–1972); R. C. HARRIS, "The Simplification of Europe Overseas," *Annals AAG*, 67 (1977), 469–483; E. L. JONES, *The European Miracle: Environments, Economies and Geopolitics in the History of Europe and Asia*, 2nd ed. (Cambridge, 1986); W. LAQUER, *Europe in Our Time: A History 1945–1992* (Viking, 1992); W. R. MEAD, "The Discovery of Europe," *Geography*, 67 (1982), 193–202; R. C. MOWAT, *Decline and Renewal: Europe Ancient and Modern* (Oxford: New Cherwell Press, 1991); H. M. MUNRO, *The Nationalities of Europe and the Growth of National Ideologies* (Cambridge, 1945); J. H. PARRY, *The Age of Reconnaissance: Discovery, Exploration, and Settlement 1450–1650* (Praeger, 1969/California, 1981); also, *The Discovery of the Sea* (California, 1981); N. J. G. POUNDS, *An Historical Geography of Europe* (Cambridge, 1990); and, with S. S. BALL, "Core-Areas and the Development of the European States System," *Annals AAG*, 54 (1964), 24–40; W. P. WEBB, *The Great Frontier* (Houghton Mifflin, 1952); J. A. WILLIAMSON, "The Expansion of Europe," in J. Bowles, ed., *The Concise Encyclopedia of World History* (Hawthorn Books, 1958), chap. 11, pp. 263–278; E. R. WOLF, *Europe and the People Without History* (California, 1992); W. WOODRUFF, *Impact of Western Man: A Study of Europe's Role in the World Economy* (Macmillan, 1966); D. S. WHITTLESEY, *Environmental Foundations of European History* (Appleton-Century-Crofts, 1949).

British Isles: General Background and General Geography

Atlas of Britain and Northern Ireland (Oxford: Clarendon Press, 1963; an extraordinary array of large, detailed topical and regional maps in color; perhaps the finest national atlas in the world); R. BLYTHE, *Akenfield: Portrait of an English Village* (Pantheon, 1969); T. BAYLISS-SMITH and S. OWENS, eds., *Britain's Changing Environment: From the Air* (Cambridge, 1990); A. G. CHAMPION and A. R. TOWNSEND, *Contemporary Britain: A Geographical Perspective* (Edward Arnold, 1990); P. COONES and J. PATTEN, *The Penguin Guide to the Landscape of England and Wales* (Penguin, 1986); H. C. DARBY, "The Regional Geography of Thomas Hardy's Wessex," *Geographical Review*, 38 (1948), 426–443; K. DE BRES, "English Village Street Names," *Geographical Review*, 80 (1990), 56–67; J. FERGUSON-LEES and B. CAMPBELL, *Mountains and Moorlands* (Hodder & Stoughton, 1978); A. GOUDIE, *The Landforms of England and Wales* (Basil Blackwell, 1990); J. HARDISTY, *The British Seas: An Introduction to the Oceanography and Resources of the North-West European Continental Shelf* (Routledge, 1990); D. HERBERT, "Place and Society in Jane Austen's England," *Geography*, 76 (1991), 193–208 (extensive list of references); "Human Geography of Contemporary Britain," *Geographical Magazine*, 53–54 (1981–1982; 10 articles by various authors); R. J. JOHNSTON and V. GARDINER, eds., *The Changing Geography of the United Kingdom*, 2nd ed. (Routledge, 1991); D. LOWENTHAL and H. C. PRINCE, "The English Landscape," *Geographical Review*, 54 (1964), 309–346; also, "English Landscape Tastes," *ibid.* 55 (1965), 186–222; P. C. LUCAS, *Britain: An Aerial Close-up* (Crescent Books, 1984; spectacular color photos in an oversized format); G. MANLEY, *Climate and the British Scene* (Collins, 1952); "Map of Cherished Land," *Geographical Magazine*, 46, no. 1 (October 1973; a folded, multicolored separate map of Britain's scenic and recreational resources); J. B. MITCHELL, ed., *Great Britain: Geographical Essays* (Cambridge, 1962; a distinguished volume of regional studies by major authorities); J. J. NORWICH, ed., *Britain's Heritage* (Continuum, 1983; an oversized, lavishly illustrated survey touching many major topics); ORDNANCE SURVEY, *The Ordnance Survey National Atlas* (Hamlyn Publishing, 1986); D. C. D. POCOCK, "The Novelist's Image of the North," *Transactions IBG*, n.s., 4 (1979), 62–76; G. PRICE, *The Languages of Britain* (Edward Arnold, 1984); L. D. STAMP and S. H. BEAVER, *The British Isles*, 6th ed. (St. Martin's, 1972; a well-known text); J. A. STEERS, ed., *Field Studies in the British Isles* (Thomas Nelson, 1964); N. STEPHENS, ed., *Natural Landscapes of Britain From the Air* (Cambridge, 1990); J. W. WATSON and J. B. SISSONS, eds., *The British Isles: A Systematic Geography* (Thomas Nelson, 1964).

British Isles: Historical Background and Historical Geography

M. ASHLEY, *The People of England: A Short Social and Economic History* (Louisiana State, 1982); J. CAMERON, *To the Farthest Ends of the Earth: 150 Years of World Exploration by the Royal Geographical Society* (Dutton, 1980); A. J. CHRISTOPHER, *The British Empire at Its Zenith* (Croom Helm, 1988); H. C. DARBY, ed., *A New Historical Geography of England* (Cambridge, 1973); R. F. DELDERFIELD, *God Is an Englishman* (Simon &

Schuster, 1970; an extraordinarily geographic novel offering a panorama of mid-19th century England); W. G. EAST, *The Geography Behind History*, revised and enlarged ed. (Norton, 1967); "The Historical Geography of England from Earliest Settlement to the Victorian City," *Geographical Magazine*, 52–53 (1970–1971; 16 articles by various authors); W. G. HOSKINS, *The Making of the English Landscape*, (Hodder & Stoughton, 1988; reissued, with updating commentary); H. KEARNEY, *The British Isles: A History of Four Nations* (Cambridge, 1989); J. LANGTON, "The Industrial Revolution and the Regional Geography of England," *Transactions IBG*, n.s., 9 (1984), 145–167; B. LAPPING, *End of Empire* (St. Martin's, 1985); R. McCRUM, W. CRAN, and R. MacNEIL, *The Story of English* (Viking, 1986); J. G. WILLIAMSON, "Why Was Britain's Growth So Slow During the Industrial Revolution?" *Journal of Economic History*, 44 (1984), 687–712.

Great Britain: Economy, London

On British agriculture and fisheries, see D. GRIGGS, *English Agriculture: An Historical Perspective* (Oxford, 1989); B. ILBERY, "A Future for the Farms?" *Geographical Magazine*, 59 (1987), 249–254; D. SYMES, "UK Demersal Fisheries and the North Sea: Problems in Renewable Resource Management," *Geography*, 76 (1991), 131–142; J. A. TAYLOR, "The British Upland Environment and Its Management," *Geography*, 63 (1978), 338–353.

On British mining, industry, and transportation, see G. C. BAND, "Fifty Years of UK Offshore Oil and Gas," *Geographical Journal*, 157 (1991), 179–189; M. CHISHOLM and R. A. GIBB, "The Impact of the Channel Tunnel," *Geographical Journal*, 152 (1986), 314–353; J. P. COLE and K. DE BRES, "A New Channel Crossing: Light at the End of the Tunnel?" *Focus*, Winter 1988, 1–9, 28; P. DICKEN, "Japanese Industrial Investment in the UK," *Geography*, 75 (1990), 351–354; I. HARDILL, "The Recent Restructuring in the British Wool Textile Industry," *Geography*, 75 (1990), 203–210; D. HEAL, "Gas in a Competitive Market," *Geographical Magazine*, November 1991, 5–7; I. HEYWOOD and S. OPENSHAW, "Radioactive Britain" (nuclear energy), *Geographical Magazine*, June 1989, 30–32; "Industrial UK Up-to-Date II," *Geography*, 75 (1990), 348–378 (includes several articles updating the British industrial scene); M. SILL, "The Changing Face of Coal" (in Britain), *Geographical Magazine*, December 1990, 4–6.

On London, see R. CLAYTON, ed., *The Geography of Greater London* (George Philip, 1964); K. DE BRES, "Jam Yesterday, Jam Today and Certainly Jam Tomorrow: A Personal View of the Costs, Benefits and Congestions of London's New Orbital Road," *Focus*, Winter 1989, 12–16; J. M. Hall, *Metropolis Now: London and Its Region* (Cambridge: 1990); K. HOGGART and D. R. GREEN, eds., *London: A New Metropolitan Geography* (Edward Arnold, 1991); R. W. HORNER, "The Thames Barrier Project," *Geographical Journal*, 145 (1979), 242–253; B. LENON, "The Geography of the 'Big Bang': London's Office Building Boom," *Geographical Magazine*, 72 (1987), 56–59; "London's Docklands," *The Economist*, February 13, 1988, 67–69; R. J. C. MUNTON, "Green Belts: The End of an Era?" *Geography*, 71 (1986), 206–214; F. J. OSBORN and A. WHITTICK, *New Towns: Their Origins, Achievements and Progress*, 3rd ed. (Leonard Hill, 1977); S. PAGE, "The London Docklands: Redevelopment Schemes in the 1980s," *Geographical Magazine*, 72 (1987), 59–63; A. SAMPSON, "City of London: Bridge to the Past," *Geo*, November 1979, 38–56; D. P. SHAW, "A Barrier to Tame the

Thames," *Geographical Magazine* 55 (1983), 129–131; J. SHEP-HERD et al., "Londoners Are Moving Out," *Geographical Magazine*, 54 (1982), 669–673; D. THOMAS, "London's Green Belt: The Evolution of an Idea," *Geographical Journal*, 129 (1963), 14–24; E. ZWINGLE, "Docklands: London's New Frontier," *National Geographic*, July 1991, 32–50.

British Isles: Scotland, Wales, Ireland

On Scotland, see I. H. ADAMS, "One Hundred Years Later: Does Scotland Exist?" *Scottish Geographical Magazine*, 100 (1984), 113–122; J. A. AGNEW, "Place and Political Behaviour: The Geography of Scottish Nationalism," *Political Geography Quarterly*, 3 (1984), 191–206; "Land of the Scots," *Geographical Magazine*, 51–52 (1979; 6 articles by various authors); D. J. SINCLAIR, "Scottish Identity," *Geographical Magazine*, 50 (1978), 803–807; H. D. SMITH, "The Role of the Sea in the Political Geography of Scotland," *Scottish Geographical Magazine*, 100 (1984), 138–150; also, H. D. SMITH et al., "Scotland and Offshore Oil: The Developing Impact," *ibid.*, 92 (1976), 75–91; C. W. J. WITHERS, " 'The Image of the Land': Scotland's Geography Through Her Language and Literature," *Scottish Geographical Magazine*, 100 (1984), 81–95.

On Wales, see J. AITCHISON and H. CARTER, "Battle for a Language" (Wales), *Geographical Magazine*, March 1990, 44–46; F. V. EMERY, *Wales* (Longman, 1969; "The World's Landscapes"); J. MORRIS, *The Matter of Wales: Epic Views of a Small Country* (Oxford, 1984); E. THOMAS, *Wales* (Oxford, 1983); J. ZARING, "The Romantic Face of Wales," *Annals AAG*, 67 (1977), 397–418. On Ireland, see K. S. BOTTIGHEIMER, *Ireland and the Irish: A Short History* (Columbia, 1982); R. W. G. CARTER and A. J. PARKER, eds., *Ireland: Contemporary Perspectives on a Land and Its People* (Routledge, 1990); E. E. EVANS, *The Personality of Ireland: Habitat, Heritage and History* (Cambridge, 1973); also, "The Personality of Ulster," *Transactions IBG*, 51 (November 1970), 1–20; J. FULTON, *The Tragedy of Belief: Division, Politics, and Religion in Ireland* (Oxford, 1991); S. GRIMES, "Ireland: The Challenge of Development in the European Periphery," *Geography*, 77 (1992), 22–32; J. H. JOHNSON, "The Political Distinctiveness of Northern Ireland," *Geographical Review*, 52 (1962), 78–91; also, "The Context of Migration: The Example of Ireland in the Nineteenth Century," *Transactions IBG*, n.s., 15 (1990), 259–276; IRISH NATIONAL COMMITTEE FOR GEOGRAPHY, *Atlas of Ireland* (Dublin: Royal Irish Academy, 1979); K. A. MILLER, *Emigrants and Exiles: Ireland and the Irish Exodus to North America* (Oxford, 1985); J. K. MITCHELL, "Social Violence in Northern Ireland," *Geographical Review*, 69 (1979), 179–201; K. O'ROURKE, "Did the Great Irish Famine Matter?" *Journal of Economic History*, 51 (1991), 1–22; A. R. ORME, *Ireland* (Aldine, 1970; "The World's Landscapes"); D. G. PRINGLE, *One Island, Two Nations? A Political Geographical Analysis of the National Conflicts in Ireland* (Wiley, 1985); R. N. SALAMAN, *The History and Social Influence of the Potato* (Cambridge, 1949: discusses the causes of the 19th-century Irish emigration); S. WICHERT, *Northern Ireland Since 1945* (Longman, 1991); C. WOODHAM-SMITH, *The Great Hunger: Ireland 1845–1849* (Harper, 1963; a graphic account of the Irish famine of the 1840s, in the perspective of general relationships between Ireland and England).

France

J. BEAUJEU-GARNIER, *France* (Longman, 1975: "The World's Landscapes"); R. BERNSTEIN, *Fragile Glory: A Portrait of France and the French* (Knopf, 1990); F. BRAUDEL, *The Identity of France*, vol. 1, *History and Environment* (1990), vol. 2, *People and Production* (1990) (HarperCollins); H. D. CLOUT, "A New France?" *Geography*, 67 (1982), 244–250; I. EVANS, "Powerhouse in Decline" (Lorraine), *Geographical Magazine*, 59 (1987), 78–83; P. GALLIOU and M. JONES, *The Bretons* (Basil Blackwell, 1991: "The Peoples of Europe"); L. GALLOIS, "The Origin and Growth of Paris," *Geographical Review*, 13 (1923), 345–367; R. HARTSHORNE, "The Franco–German Boundary of 1871," *World Politics*, 2 (1950), 209–250; M. McDONALD, *We Are Not French! Language, Culture and Identity in Brittany* (Routledge, 1989); P. PINCHEMEL et al., *France: A Geographical, Social and Economic Survey* (Cambridge, 1986); I. SCARGILL, "Regional Inequality in France: Persistence and Change," *Geography*, 76 (1991), 343–357; also, "Studying French Regions in the National Curriculum," *Teaching Geography*, 17 (1992), 3–7; M. STETSON, "France's Tottering Nuclear Giant," *World Watch*, January-February 1991, 38–39; R. TIERSKY, "France in the New Europe," *Foreign Affairs*, Spring 1992, 131–146.

Germany

R. D. ASMUS, "A United Germany," *Foreign Affairs*, 69 (1990), 63–76; J. T. BERGNER, *The New Superpowers: Germany, Japan, the U.S. and the New World Order* (St. Martin's, 1991); C. BERTRAM, "The German Question," *Foreign Affairs*, 69 (1990), 45–62; B. BRYSON, "Main-Danube Canal Linking Europe's Waterways," *National Geographic*, August 1992, 2–31; J. P. and F. J. COLE, "The New Germany," *Focus*, Fall 1991, 1–6; C. ELLGER, "Berlin: Legacies of Division and Problems of Unification," *Geographical Journal*, 158 (1992), 40–46; D. HAMILTON, "A More European Germany, A More German Europe," *Journal of International Affairs*, 45, no. 1 (Summer 1991), 127–149; C. D. HARRIS, "Unification of Germany in 1990," *Geographical Review*, 81 (1991), 170–182; A. HULL and S. KENNY, "Ruhr Economy in Decline," *Geographical Magazine*, 55 (1983), 516–521; K. KAISER, "Germany's Unification," *Foreign Affairs*, 70, no. 1 (1991), 179–205; R. A. NENNEMAN, "Europe's Center of Gravity" (Germany), *World Monitor*, June 1992, 40–45; "Not as Grimm as It Looks: A Survey of Germany," *The Economist*, May 23, 1992, 18 pp.; E. POND, "Germany in the New Europe," *Foreign Affairs*, Spring 1992, 114–130; C. RAPAPORT, "Why Germany Will Lead Europe," *Fortune*, September 21, 1992, 149–158; G. RITTER and J. HADJU, "The East-West German Boundary," *Geographical Review*, 79 (1989), 326–344.

The Benelux Countries

J. C. BOYER, "Rotterdam," *Annales de Géographie*, 100 (1991), 101–115 (in French, with English abstract); R. DANIËLS, "Rotterdam, City and Harbour," *Cities*, 8 (1991), 283–291; M. DE SMIDT and E. WEVER, *An Industrial Geography of the Netherlands* (Routledge, 1990); A. K. DUTT and S. HEAL, "Delta Project Planning and Implementation in the Netherlands," *Journal of Geography*, 78 (1979), 131–141; J. FITZMAURICE, *The Politics of Belgium: Crisis and Compromise in a Plural Society*, 2nd ed. (C. Hurst, 1988); J. HOUSTON, "Sands of Time" (sand dunes), *Geographical Magazine*, March 1992, 1–5; *I.D.G. Bulletin*, issued (from 1973) by the Information and Documentation Centre for the

Geography of the Netherlands, Utrecht, Netherlands; J. I. ISRAEL, *Dutch Primacy in World Trade 1585–1740* (Oxford, 1989); A. M. LAMBERT, *The Making of the Dutch Landscape: A Historical Geography of the Netherlands*, 2nd ed. (Academic 1985); A. B. MURPHY, *The Regional Dynamics of Language Differentiation in Belgium: A Study in Cultural-Political Geography* (University of Chicago Geography Research Paper 227, 1988); E. SCHRIJVER, "How the Dutch Made Holland," *Geographical Magazine*, 51 (1979), 567–572; G. STEPHENSON, "Cultural Regionalism and the Unitary State Idea in Belgium," *Geographical Review*, 62 (1972), 501–523; *Tijdschrift voor Economische en Sociale Geografie*, (The Netherlands, special issue), 63, no. 3 (May-June 1972), 124–235; P. THOMAS, "Belgium's North-South Divide and the Walloon Regional Problem," *Geography*, 76 (1990), 36–50 (extensive references); K. VROOM, "Antwerp: A Modern City with a Significant Historic Heritage," *GeoJournal*, 24 (1991), 277–284; P. WAGRET, *Polderlands* (Barnes & Noble, 1972); G. G. WEIGEND, "Stages in the Development of the Ports of Rotterdam and Antwerp," *Geoforum*, 13 (1973), 5–15.

Switzerland and Austria

R. BERNHEIM, "Jubilee Time: Switzerland Revisited," *Swiss Review of World Affairs*, August 1991, 11–15; A. DIEM, "The Alps," *Geographical Magazine*, 56 (1984), 414–420; G. W. HOFF-MAN, "The Survival of an Independent Austria," *Geographical Review*, 41 (1951), 606–621; C. STADEL, "The Alps: Mountains in Transformation," *Focus*, January-February 1982, 1–16; B. WAL-LACH, "Geneva," *Focus*, Fall 1986, 10–13.

Countries of Northern Europe

J. COOK, "Norway: The New Kuwait," *Forbes*, January 6, 1992, 60–61; J. D. DAVIS, "International Oil: The Scandinavian Dimension," *Annals of the American Academy of Political and Social Science*, 512 (November 1990), 79–87; A. DE SHERBININ, "Iceland," *Population Today*, May 1990, 12; F. HALE, "Norway's Rendezvous with Modernity," *Current History*, 83 (1984), 173–177, 184; R. K. HELLE, "Reindeer Husbandry in Finland," *Geographical Journal*, 145 (1979), 254–264; A. HOLT-JENSEN, "Norway and the Sea: The Shifting Importance of Marine Resources Through Norwegian History," *GeoJournal*, 10 (1985), A. J. HOLT-JENSEN, "The Norwegian Oil Economy," *GeoJournal* (Supplementary Issue), 3 (1981), 81–92; B. S. JOHN, *Scandinavia: A New Geography* (Longman, 1984); R. D. LIEBOWITZ, "Finlandization: An Analysis of the Soviet Union's 'Domination' of Finland," *Political Geography Quarterly*, 2 (1983), 275–287; J. LUKACS, "Finland Vindicated," *Foreign Affairs*, Fall 1992, 50–63; W. R. MEAD, *An Historical Geography of Scandinavia* (Academic, 1981); also, "Finland in a Changing Europe," *Geographical Journal*, 157 (1991), 307–315; also, "Problems of Norden," *ibid.* 151 (1985), 1–10; also, "Recent Developments in Human Geography in Finland," *Progress in Human Geography*, 1 (1977), 361–375; also, "Sweden in Retrospect," *Geography*, 70 (1985), 36–44; K. SCHERMAN, *Daughter of Fire: A Portrait of Iceland* (Little, Brown, 1976); B. SCUDDER, "Energy Galore" (Iceland), *Geographical Magazine*, September 1990, 40–44; F. SINGLETON, *A Short History of Finland* (Cambridge, 1989); A. SÖMME, ed., *A Geography of Norden: Denmark, Finland, Iceland, Norway, Sweden* (Wiley, 1962; a major work by Scandinavian and Finnish geographers).

Countries of Southern Europe

M. BECKINSALE and R. BECKINSALE, *Southern Europe: The Mediterranean and Alpine Lands* (London, 1975); H. BRÜCKNER, "Man's Impact on the Evolution of the Physical Environment in the Mediterranean Region in Historical Times," *GeoJournal*, 13 (1986), 7–17; B. BRYSON, "The New World of Spain," *National Geographic*, April 1992, 2–33; R. COMMON, "The Study of Regional Geography in Southern Europe," *Scottish Geographical Magazine*, 98 (1982), 180–188; F. J. COSTA and A. G. NOBLE, "Evolving Planning Systems in Madrid, Rome, and Athens," *Geo-Journal*, 24 (1991), 293–303; P. P. COURTENAY, "Madrid: The Circumstances of Its Growth," *Geography*, 44 (1959), 22–34; P. DENNIS, *Gibraltar and Its People* (David & Charles, 1990); "Italy," *Geographical Magazine*, 59 (1987; articles by various authors in separate issues, from no. 4, April); J. GERLACH, "Tourism and Its Impact in Costa del Sol, Spain," *Focus*, Fall 1991, 7–11; R. KING, *The Industrial Geography of Italy* (Croom Helm/St. Martin's, 1985); also, "Southern Europe: Dependency or Development?" *Geography*, 67 (1982), 221–234; also, "The Three Italies: Recent Changes in the Regional Economic Geography of Italy," *Geographical Viewpoint* (Dublin, Ireland), 17 (1988–1989), 5–23; extensive list of references; also, "Italy: From Sick Man to Rich Man of Europe," *Geography*, 77 (1992), 153–169; M. H. LEVINE, "The Basques," *Natural History*, April 1967, 44–51; A. B. MOUNTJOY, *The Mezzogiorno*, 2nd ed. (Oxford, 1982; "Problem Regions of Europe"); J. NAYLON, "Industry on the Brink" (Spain), *Geographical Magazine*, 57 (1985), 436–440; also, "Modern Spain's Bitter Legacy," *Geographical Magazine*, 56 (1984), 570–574; also, "Ascent and Decline in the Spanish Regional System," *Geography*, 77 (1992), 46–62; E. C. SEMPLE, *The Geography of the Mediterranean World: Its Relation to Ancient History*, (Holt, 1931); "Settlement and Conflict in the Mediterranean World" (special issue), *Transactions IBG*, n.s., 3 (1978), 255–380; C. D. SMITH, *Western Mediterranean Europe: A Historical Geography of Italy, Spain and Southern France Since the Neolithic* (Academic, 1979); D. STANISLAWSKI, *The Individuality of Portugal: A Study in Historical-Political Geography* (Texas, 1959); P. WAGNER, "Wines, Grape Vines and Climate," *Scientific American*, June 1974, 106–115.

Countries of East Central Europe

U.S. CENTRAL INTELLIGENCE AGENCY, *The Former Yugoslavia: A Map Folio* (July, 1992); D. DANTA, "Romania: View from the Front," *Focus*, Summer 1991, 17–20; H. C. DARBY *et al.*, *A Short History of Yugoslavia from Early Times to 1966* (Cambridge, 1966; many useful maps); D. DODER, "Albania Opens the Door," *National Geographic*, July 1992, 66–93; H. F. FRENCH, "Eastern Europe's Clean Break with the Past" (environmental damage and cleanup), *World Watch*, March-April 1991, 21–27; C. GATI, "From Sarajevo to Sarajevo," *Foreign Affairs*, Fall 1992, 64–78; E. KRAFT, "Bosnia's Muslims as a Nationality," *Swiss Review of World Affairs*, August 1992, 24–25; D. LAWDAY, "A Troubled Past Clouds Yugoslavia's Future," *1992 World Book Year Book*, 472–475; W. Z. MICHALAK and R. A. GIBB, "A Debt to the West: Recent Developments in the International Financial Situation of East-Central Europe," *Professional Geographer*, 44 (1992), 260–271; W. B. MORGAN, "Economic Reform, the Free Market, and Agriculture in Poland," *Geographical Journal*, 158 (1992), 145–156, A. B. MURPHY, "Western Investment in

East-Central Europe: Emerging Patterns and Implications for State Stability," *Professional Geographer*, 44 (1992), 249–259; J. NEWHOUSE: "The Diplomatic Round: Dodging the Problem" (Yugoslavia), *The New Yorker*, August 24, 1992, 60–71; R. PETKOVIC, "The Lebanon of the Balkans?" (Yugoslavia), *World Press Review*, May 1991, 24–26; W. PFAFF, "Reflections: The Absence of Empire," *The New Yorker*, August 10, 1992, 59–69; J. J. PUTNAM, "Different Communism: Hungary's New Way," *National Geographic*, 163 (1983), 225–261; S. P. RAMET, "War in the Balkans," *Foreign Affairs*, Fall 1992, 79–98; D. S. RUGG, *Eastern Europe* (Longman, 1985, "The World's Landscapes"); J. RUPNIK, "Czech Off: Slovakia's Halfhearted Secession," *The New Republic*, August 17 & 24, 1992, 15–16; J. SACHS, "Building a Market Economy in Poland," *Scientific American*, March 1992, 34–40; and, with D. LIPTON, "Poland's Economic Reform," *Foreign Affairs*, Summer 1990, 47–66; G. SCHÖPFLIN, "The End of Communism in Eastern Europe," *International Affairs*, 66 (1990), 3–16; J. SMOLOWE, "Land of Slaughter" (Serbs in Bosnia-Hercegovina), *Time*, June 8, 1992, 32–36; also, "The Balkans: Why Do They Keep On Killing?" *ibid.*, May 11, 1992, 48–49; G. SCHWARZ, "Problems of Privatization in Eastern Europe," *Swiss Review of World Affairs*, August 1991, 4–6; D. SPOONER, ed., "The Changing Face of Eastern Europe and the Soviet Union," *Geography*, 75 (1990), 239–277; D. STATKOV, "Monarchists and Nationalists in Bulgaria," *Swiss Review of World Affairs*, August 1991, 8–10; C. THOMAS, "Yugoslavia: The Enduring Dilemmas," *Geography*, 75 (1990), 265–268; J. THOMPSON, "East Europe's Dark Dawn" (pollution), *National Geographic*, June 1991, 36–70; D. TURNOCK, "The Planning of Rural Settlement in Romania," *Geographical Journal*, 157 (1991), 251–264; also, "Postwar Studies on the Human Geography of Eastern Europe," *Progress in Human Geography*, 8 (1984), 315–346; also, "The Danube–Black Sea Canal and Its Impact on Southern Romania," *GeoJournal*, 12 (1986), 65–79; also, *Eastern Europe: An Economic and Political Geography* (Routledge, 1989); L. WESCHLER, "Deficit" (Poland), *The New Yorker*, May 11, 1992, 41–77.

Part 2: The Former Soviet Region

General and Political

A selection of textbooks and miscellaneous general references includes: J. H. BATER, *The Soviet Scene: A Geographical Perspective* (Edward Arnold, 1989); A. BROWN *et al.*, gen. eds., *Cambridge Encyclopedia of Russia and the Soviet Union* (Cambridge, 1982); M. J. BRADSHAW, "New Regional Geography, Foreign-Area Studies and Perestroika," *Area*, 22 (1990), 315–322 (extensive list of references on regional geography and Soviet studies); also, ed., *The Soviet Union: A New Regional Geography?* (Wiley, 1991); R. S. CLEM, "Russians and Others: Ethnic Tensions in the Soviet Union," *Focus*, September-October 1980, 1–16; J. P. COLE, *Geography of the Soviet Union* (Butterworth, 1984); J. DEWDNEY, "Population Change in the Soviet Union, 1979–1989," *Geography*, 75 (1990), 273–277; also, *USSR in Maps* (Holmes & Meier, 1982); P. DOSTAL and H. KNIPPENBERG, "The 'Russification' of Ethnic Minorities in the USSR," *Soviet Geography*, 20 (1979), 197–219; G. HAUSLADEN, "Themes for Teaching the Geography of the Soviet Union," *Journal of Geogra-*

phy, 90 (1991), 141–148; S. LOFFREDO and S. KALISH, "Russia," *Population Today*, June 1992, 12; P. E. LYDOLPH, *Geography of the U.S.S.R.*, 5th ed. (Elkhart Lake, WI: Misty Valley Publishing, 1990); also, *Geography of the U.S.S.R.*, 3rd ed. (Wiley, 1987; a detailed text organized by economic regions, whereas the 1990 edition is topically organized); W. H. PARKER, *The Soviet Union*, 2nd ed. Longman, 1983; "The World's Landscapes"); E. POND, *From the Yaroslavsky Station: Russia Perceived*, 3rd ed. (Universe, 1987; distinguished journalism); D. K. SHIPLER, *Russia: Broken Idols, Solemn Dreams* (Times Books, 1983; by a journalist); G. SMITH, "The Soviet Underclass," *Geographical Magazine*, May 1990, 4–7; H. SMITH, *The New Russians* (Random House, 1990; by a major journalist); L. SYMONS *et al.*, *The Soviet Union: A Systematic Geography* (Hodder & Stoughton, 1990); M. DE VILLIERS, *Down the Volga: A Journey Through Mother Russia in a Time of Troubles* (Viking, 1991; a travel account); M. ZARAEV, "An Insider's Guide to the New Russia," *World Monitor*, November 1991, 42–47.

For recent political changes, see "A Victory Gone Sour" (Russia, a year after the failed coup) *U.S. News & World Report*, August 24, 1992, 43–48; S. BIALER, "The Death of Soviet Communism," *Foreign Affairs*, Winter 1991/92, 166–181; M. EDWARDS, "Mother Russia on a New Course," *National Geographic*, February 1991, 2–37; M. I. GOLDMAN, "Three Days That Shook My World," *World Monitor*, October 1991, 30–33; R. G. KAISER, "Gorbachev: Triumph and Failure," *Foreign Affairs*, Spring 1991, 160–174; A. KARATNYCKY, "The Ukrainian Factor," *Foreign Affairs*, Summer 1992, 90–107; P. KENNEDY, "What Gorbachev Is Up Against: The Problem Isn't *In* the System—It *Is* the System," *Atlantic Monthly* (June 1987), 29–43; A. KOZYREV, "Russia: A Chance for Survival," *Foreign Affairs*, Spring 1992, 1–16; M. MANDELBAUM, "Coup de Grace: The End of the Soviet Union," *Foreign Affairs*, 71 (1992), 164–183; M. MILIVOJEVIĆ, "Crises of Nationalism," *Geographical Magazine*, November 1990, 23–26; J. SALLNOW, "What Price Perestroika?" *Geographical Magazine*, January 1990, 10–14; U. SCHMID, "Rude Awakening on the Neva" (St. Petersburg), *Swiss Review of World Affairs*, June 1992, 9–11; D. K. SIMES, "America and the Post-Soviet Republics" *Foreign Affairs*, Summer 1992, 73–89; T. SZULC, "The Great Soviet Exodus," *National Geographic*, February 1992, 40–65; "Westward No? Russia Looks at the World," *The Economist*, July 4, 1992, 19–24.

Historical Background

A. CATTANI, "Past and Present in the Crimea," *Swiss Review of World Affairs*, September 1992, 20–23; E. CRANKSHAW, *The Shadow of the Winter Palace: Russia's Drift to Revolution, 1825–1917* (Viking, 1976); "Doctor Zhivago (feature film; video; novel by Boris Pasternak); W. G. EAST, "The New Frontiers of the Soviet Union," *Foreign Affairs*, 29 (1951), 591–607; H. J. ELLISON, "Economic Modernization in Imperial Russia: Purposes and Achievements," *Journal of Economic History*, 25 (1965), 523–540; P. E. GARBUTT, "The Trans-Siberian Railway," *Journal of Transport History*, 6 (1954), 238–249; J. R. GIBSON, "Russian Expansion in Siberia and America," *Geographical Review*, 70 (1980), 127–136; C. J. HALPERIN, *Russia and the Golden Horde: The Mongol Impact on Medieval Russian History* (Indiana, 1985); S. J. LINZ, ed., *The Impact of World War II on the Soviet Union* (Rowman & Allanheld, 1985); S. MASSIE, *Land of the Firebird:*

The Beauty of Old Russia (Simon & Schuster, 1980); E. E. MEAD, ed., *Makers of Modern Strategy: Military Thought from Machiavelli to Hitler* (Princeton, 1943), chap. 14, "Lenin, Trotsky, Stalin: Soviet Concepts of War," pp. 322–364; A. S. MORRIS, "The Medieval Emergence of the Volga-Oka Region, *Annals AAG*, 61 (1971), 697–710; P. N. PAVLOV, "Fur Trade in the Economy of Siberia in the 17th Century," *Soviet Geography*, 27 (1986), 43–82; "Reds" (feature film; video); H. E. SALISBURY, *Black Night, White Snow: Russia's Revolutions 1905–1917* (Doubleday, 1978).

Cities; Economy; Resources; Environmental Damage

On cities, see R. A. FRENCH, "The Changing Russian Urban Landscape," *Geography*, 68 (1983), 236–244; G. M. LAPPO, "The City of Moscow and the Moscow Agglomeration," *GeoJournal*, Supplementary Issue, 1 (1980), 45–52; B. A. RUBLE, *Leningrad: Shaping a Soviet City* (California, 1990); T. SHABAD, "New Major Soviet Cities," *Soviet Geography*, 27 (1986), 59–65; D. J. B. SHAW, "Planning Leningrad," *Geographical Review*, 68 (1978) 183–200.

On the economy, see K. L. ADELMAN and N. R. AUGUSTINE, "Defense Conversion: Bulldozing the Management," *Foreign Affairs*, Spring 1992, 26–47; M. J. BRADSHAW, "Joint Ventures for Perestroika," *Geographical Magazine*, October 1990, 10–14; E. K. CROMLEY and P. R. CRAUMER, "Physician Supply in the Soviet Union, 1940–1985," *Geographical Review*, 80 (1990), 132–140; R. CONQUEST, *Harvest of Sorrow: Soviet Collectivization and the Terror-Famine* (Oxford, 1986); L. DIENES, *Soviet Asia: Economic Development and National Policy Choices* (Westview, 1987); N. C. FIELD, "Environmental Quality and Land Productivity: A Comparison of the Agricultural Land Base of the USSR and North America," *Canadian Geographer*, 12 (1968), 1–14; M. I. GOLDMAN, "Kapitalism" (in Russia), *World Monitor*, April 1992, 28–33; S. HEDLUND, *Crisis in Soviet Agriculture* (Croom Helm, 1984); W. H. PARKER, "The Soviet Motor Industry," *Soviet Studies*, 32 (1980), 515–541; B. RUMER, "Beating Swords into . . . Refrigerators?" *World Monitor*, January 1992, 36–40; M. J. SAGERS and T. SHABAD, *The Chemical Industry in the USSR: An Economic Geography* (Westview, 1990); M. J. SAGERS and T. MARAFFA, "Soviet Air-Passenger Transportation Network," *Geographical Review*, 80 (1990), 266–278; S. F. STARR, "Technology and Freedom in the Soviet Union," *Technology Review*, May-June 1984, 38–47; I. STEBELSKY, "Agriculture of the Soviet Union and Eastern Europe," *Journal of Geography*, 84 (1985), 264–272; J. WANNISKI, "The Future of Russian Capitalism," *Foreign Affairs*, Spring 1992, 17–25; A. YABLOKOV, "Disarming Soviet Industry," *World Press Review*, August 1992, 42.

On resources and environmental damage, see B. M. BARR and K. E. BRADEN, *The Disappearing Russian Forest: A Dilemma in Soviet Resource Management* (Rowman & Littlefield, 1988); (Environmental Problems in the Commonwealth of Independent States; articles about pollution), *U. S. News & World Report*, April 13, 1992, 40–51; M. FESHBACH and A. FRIENDLY, JR., *Ecocide in the USSR: Health and Nature Under Siege* (Basic Books, 1992); P. HOFHEINZ, "The New Soviet Threat: Pollution," *Fortune*, July 27, 1992, 110–114; G. B. KNECHT, "Russia: Low on Fuel," *Atlantic Monthly*, August 1992, 30–34; R. G. JENSEN, T. SHABAD, and A. W. WRIGHT, eds., *Soviet Natural Resources in the World Economy* (Chicago, 1983; massive, detailed, authoritative, with many maps); R. C. MORAIS *et al.*, "Blowout" (of pipelines), *Forbes*, July 20, 1992, 65–68; P. R. PRYDE, *Environmental Management in the Soviet Union* (Cambridge, 1991); M. SMITH, "A Tale of Death and Destruction" (environmental pollution), *Geographical Magazine*, March 1990, 11–14.

Caucasus Region and Central Asia

M. ATKIN, "The Survival of Islam in Soviet Tajikistan," *Middle East Journal*, 43 (1989), 605–618; L. R. BROWN, "The Aral Sea: Going, Going . . . ," *World Watch*, January-February 1991, 20–27; W. FIERMAN, ed., *Soviet Central Asia: The Failed Transformation* (Westview, 1991); V. M. KOTLYAKOV, "The Aral Sea Basin: A Critical Environmental Zone," *Environment*, January-February 1991, 4–9, 36–38; H. KRAMER, "Soviet Islam: Adjusting to an Atheistic State," *Swiss Review of World Affairs*, November 1979, 23–27; E. LEE, "Uzbekistan," *Population Today*, September 1992, 11; R. A. LEWIS, ed., *Geographic Perspectives on Soviet Central Asia* (Routledge, 1992); U. MEISTER, "Islam in Azerbaijan," *Swiss Review of World Affairs*, April 1987, 10–11; M. MILIVO-JEVIĆ "Soviet Central Asia Cuts Loose," *Geographical Magazine*, November 1990, 30–33; H. MYKELBOST, "Armenia and the Armenians," *Norsk Geografisk Tidsskrift*, 43 (1989), 135–154; P. FUHRMAN, "Follow the Ancient Silk Road" (Turkic relationships), *Forbes*, September 14, 1992, 392–400; M. B. OLCOTT, "Central Asia's Catapult to Independence," *Foreign Affairs*, Summer 1992, 108–130; A. ROXBURGH, "Georgia Fights for Nationhood," *National Geographic*, May 1992, 82–111; B. Z. RUMER, *Soviet Central Asia: A Tragic Experiment* (Unwin Hyman, 1989); "The Scramble for Central Asia: A Global Contest for Hearts, Minds, Money," *World Press Review*, July 1992, 9–14; D. SNEIDER, "The Soviet 'Ecocidal' Legacy" and "The Death of the Aral Sea: Case Study in USSR Ecology," *Christian Science Monitor*, June 11, 1992, 10–11; R. WRIGHT, "Islam, Democracy and the West" (includes a substantial section on Central Asia), *Foreign Affairs*, Summer 1992, 131–145.

Siberia; Far East; Northern Lands

D. BELT, "The World's Great Lake" (Baikal), *National Geographic*, June 1992, 2–39; R. CULLEN, "A Reporter at Large: Siberia," *The New Yorker*, July 27, 1992, 34–52; L. DIENES, "The Development of Siberian Regions: Economic Profiles, Income Flows and Strategies for Growth," *Soviet Geography*, 23 (1982), 205–244; also, "Siberia: Perestroyka and Economic Development," *ibid.*, 32 (1991), 445–457; V. A. DOIBAN *et al.*, "Economic Development in the Kola Region, USSR: An Overview," *Polar Record*, 28 (1992), 7–16; M. EDWARDS, "Siberia: In From the Cold," *National Geographic*, March 1990, 2–39; N. LOUIS, "Gulags and Goldmines" (mismanagement of Siberian resources), *Geographical Magazine*, February 1992, 28–34; E. B. MILLER, "The Trans-Siberian Landbridge, A New Trade Route Between Japan and Europe: Issues and Prospects," *Soviet Geography*, 19 (1978), 223–243; V. L. MOTE, "BAM, Boom, Bust: Analysis of a Railway's Past, Present, and Future," *Soviet Geography*, 31 (1990), 321–331; also, "A Visit to the Baikal-Amur Mainline and the New Amur-Yakutsk Rail Project," *ibid.* 26 (1985), 691–719; also, "Containerization and the Trans-Siberian Land Bridge, *Geographical Review*, 74 (1984), 304–314; A. RODGERS, ed., *The Soviet Far East: Geographical Perspectives on Development* (Routledge,

1990); P. ROSTANKOWSKI, "The Decline of Agriculture and the Rise of Extractive Industry in the Soviet North," *Polar Geography and Geology*, 7 (1983), 289–298; J. SALLNOW, "The Soviet Far East: A Report on Urban and Rural Settlement and Population Change, 1966–1980," *Soviet Geography*, 30 (1989), 670–683; P. SEIDLITZ, "Gas from Siberia," *Swiss Review of World Affairs*, June 1983, 14–19; P. VITEBSKY, "Perestroika Among the Reindeer Herders," *Geographical Magazine*, June 1989, 22–25; K. WARREN, Industrial Complexes in the Development of Siberia," *Geography*, 63 (1978), 167–178.

Part 3: The Middle East

Middle East as a Whole

For general and historical references, see *Aramco World* (bimonthly; a richly illustrated magazine for general readers published by Aramco Corporation; wide-ranging subject matter on the Arab and Islamic worlds); F. BARNABY, "The Nuclear Arsenal in the Middle East," *Technology Review*, May-June 1987, 27–34; P. BEAUMONT et al., *The Middle East: A Geographical Study*, 2nd ed. (Wiley, 1988); G. BLAKE et al., *The Cambridge Atlas of the Middle East and North Africa* (Cambridge, 1987); M. BRAWER, *Atlas of the Middle East* (Macmillan, 1989); S. B. COHEN, "Middle East Geopolitical Transformation: The Disappearance of a Shatterbelt," *Journal of Geography*, 91 (1992), 2–10; M. ELLIOTT, "Water Wars" and E. ANDERSON, "The Violence of Thirst," *Geographical Magazine*, May 1991, 28–34; D. FROMKIN, "How the Modern Middle East Map Came to be Drawn" (creation of new borders after the Ottoman Empire collapsed in 1918), *Smithsonian*, May 1991, 132–148; C. C. HELD, *Middle East Patterns: Places, Peoples, and Politics* (Westview, 1989); C. HELLIER, "Draining the Rivers Dry" (water supply in the Middle East), *Geographical Magazine*, July 1990, 32–35; M. INDYK, "Watershed in the Middle East," *Foreign Affairs*, 71 (1992), 70–93; A. JOUSIFFE, "Camel Economy," *Geographical Magazine*, May 1991, 22–25; B. LEWIS, "Rethinking the Middle East," *Foreign Affairs*, Fall 1992, 99–119; P. MANSFIELD, *A History of the Middle East* (Viking, 1991); T. MOSTYN, ed., *The Cambridge Encyclopedia of the Middle East and North Africa* (Cambridge, 1988); "Out of Joint: A Survey of the Middle East," *The Economist*, September 28, 1991, 22 pp.; R. PATAI, "The Middle East as a Culture Area," *Middle East Journal*, 6 (1952), 1–21; B. RUBIN, "Reshaping the Middle East," *Foreign Affairs*, 69 (1990), 131–146; "The Middle East, 1992," *Current History*, January 1992, entire issue (articles on the Gulf War, Kuwait, Iraq, Israel, Palestinians, Lebanon, Egypt, Arab World); see also comparable issues in earlier years; "The Middle East and the Age of Discovery," *Aramco World*, May-June 1992, entire issue.

On the Arab World and Islam, see A. ABU KHALIL, "A New Arab Ideology? The Rejuvenation of Arab Nationalism," *Middle East Journal*, 46 (1992), 22–36; I. R. AL FARUQI and L. L. AL FARUQI, *The Cultural Atlas of Islam* (Macmillan, 1986); S. R. ALI, *Oil, Turmoil, and Islam in the Middle East* (Praeger, 1986); G. W. CHOUDBURY, *Islam and the Contemporary World* (London: Indus Thames Publishers, 1990); F. J. COSTA and A. G. NOBLE, "Planning Arabic Towns," *Geographical Review*, 76 (1986), 160–172; M. DEMPSEY, comp., *The Daily Telegraph Atlas of the Arab World* (London: The Daily Telegraph, 1983);

Horizon, 13, no. 3 (Summer 1971), articles as follows: J. MORRIS, "What Is an Arab?" 4–17; S. DE GRAMONT, "Mohammed: The Prophet Armed," 18–23; and D. DAICHES, "What is a Jew?" 24–35; A. HOURANI, *A History of the Arab Peoples* (Harvard, 1991); E. MORTIMER, "New 'Ism' in the East" (Islamic fundamentalism), *World Monitor*, September 1992, 50–52; F. J. SIMOONS, *Eat Not This Flesh: Food Avoidances in the Old World* (Wisconsin, 1961), especially chap. 3, "Pigs and Pork."

Middle East: African Sector

"North Africa" (Libya, Sudan, Chad, Algeria, Morocco, Western Sahara, Tunisia), *Current History*, April 1990, entire issue.

On Egypt and Sudan, see J. A. ALLAN, "High Aswan Dam Is a Success Story," *Geographical Magazine*, 53 (1981), 393–396; H. ANSARI, *Egypt: The Stalled Society* (SUNY, 1986); K. M. BARBOUR, "The Sudan Since Independence," *Journal of Modern African Studies*, 18 (1980), 73–97; P. K. BECHTOLD, "More Turbulence in Sudan: A New Politics This Time?" *Middle East Journal*, 44 (1990), 579–595; R. E. BENEDICK, "The High Dam and the Transformation of the Nile," *Middle East Journal*, 33 (1979), 119–144; R. BONNER, "Letter from Sudan," *The New Yorker*, July 13, 1992, 70–83; N. J. BROWN, *Peasant Politics in Modern Egypt: The Struggle Against the State* (Yale, 1990); R. O. COLLINS, *The Waters of the Nile: Hydropolitics and the Jonglei, 1980–1988* (Oxford, 1990); S. EL-SHAKHS, "National Factors in the Development of Cairo," *Town Planning Review*, 42 (1971), 233–249; H. M. FAHIM, *Dams, People and Development: The Aswan High Dam Case* (Oxford, 1981); A. T. GROVE, "Egypt Has Too Much Water," *Geographical Magazine*, 54 (1982), 437–441; A. HERACLIDES, "Janus or Sisyphus? The Southern Problem of the Sudan," *Journal of Modern African Studies*, 25 (1987), 213–231; J. J. HOBBS, *Bedouin Life in the Egyptian Wilderness* (Texas, 1989); H. E. Hurst, *The Nile: A General Account of the River and the Utilization of Its Waters*, rev. ed. (Constable, 1957); M. JENNER, "Cairo in Peril" (historic preservation), *Geographical Magazine*, 57 (1985), 474–480; S. KONTOS, "Farmers and the Failure of Agribusiness in Sudan," *Middle East Journal*, 44 (1990), 649–667; A-A I. KASHEF, "Technical and Ecological Impacts of the Aswan High Dam," *Journal of Hydrology*, 53 (1981), 73–84; J. D. PENNINGTON, "The Copts in Modern Egypt," *Middle Eastern Studies*, 18 (1982), 158–179; N. POLLARD, "The Gezira Scheme: A Study in Failure," *The Ecologist*, 11 (1981), 21–31; S. RADWAN and E. LEE, *Agrarian Change in Egypt: An Anatomy of Rural Poverty* (Croom Helm, 1986); A. RICHARDS, "The Agricultural Crisis in Egypt," *Journal of Development Studies*, 16 (1980), 303–321; J. R. ROGGE, *Too Many, Too Long: Sudan's Twenty-Year Refugee Dilemma* (Rowman & Allanheld, 1985); B. WALLACH, "The Nile Valley," *Focus*, 36, Spring 1986, 16–19; also, "The Sudan Gezira," *ibid.*, October 1985, 10–13; also, "Irrigation in Sudan Since Independence" (Gezira and other schemes), *Geographical Review*, 78 (1988), 417–437; J. WATERBURY, *Hydropolitics of the Nile Valley* (Syracuse, 1979); C. WEIL and K. M. KVALE, "Current Research on Geographical Aspects of Schistosomiasis," *Geographical Review*, 75 (1985), 186–216; U. WIKAN, "Living Conditions Among Cairo's Poor: A View from Below," *Middle East Journal*, 39 (1985), 7–26; C. WILLIAMS, "Islamic Cairo: Endangered Legacy," *Middle East Journal* 39 (1985), 231–246.

On the Maghreb and Western Sahara, see T. ARKELL, "The

Decline of Pastoral Nomadism in the Western Sahara," *Geography*, 76 (1991), 162–166; ALLAN M. FINDLAY and ANNE M. FINDLAY, "Regional Disparities and Population Change in Morocco," *Scottish Geographical Magazine*, 102 (1986), 29–41; W. LANGEWIESCHE, "The World in Its Extreme" (the Sahara), *Atlantic Monthly*, November 1991, 105–140; R. I. LAWLESS, "The Concept of *tell* and *sahara* in the Maghreb: A Reappraisal," *Transactions IBG*, 57 (1972), 125–137; N. J. MIDDLETON, "Dust Storms in the Middle East," *Journal of Arid Environments*, 10 (1986), 83–96.

On Ethiopia and Somalia, see C. CLAPHAM, *Transformation and Continuity in Revolutionary Ethiopia* (Cambridge, 1988); R. EVANS, "A Question of Time" (Ethiopia's ethnic problems), *Geographical Magazine*, August 1991, 22–25; M. HAEFLIGER, "War and Death in Somalia," *Swiss Review of World Affairs*, September 1992, 12–13; G. HAILE, "The Unity and Territorial Integrity of Ethiopia," *Journal of Modern African Studies*, 24 (1986), 465–487; G. KEBBEDE and M. J. JACOB, "Drought, Famine and the Political Economy of Environmental Degradation in Ethiopia," *Geography*, 73 (1988), 65–70; H. KLOOS, "Peasant Irrigation Development and Food Production in Ethiopia," *Geographical Journal*, 157 (1991), 295–306; V. LULING, "Wiping Out a Way of Life" (Ethiopia), *Geographical Magazine*, July 1989, 34–37; J. MADELEY, "Ethiopia's New Villagers," *Geographical Magazine*, 58 (1986), 246–249; M. OTTOWAY, "Drought and Development in Ethiopia," *Current History*, 85 (1986), 217–220, 224; also, "Mediation in a Transitional Conflict: Eritrea," *Annals of the American Academy of Political and Social Science*, 518 (November 1991), 69–81; J. D. UNRUH, "Nomadic Pastoralism and Irrigated Agriculture in Somalia: Utilization of Existing Land Use Patterns in Designs for Multiple Access of 'High Potential' Areas of Semi-arid Africa," *GeoJournal*, 25 (1991), 91–108.

Israel, Jordan, Lebanon, Syria

F. AJAMI, "Lebanon and Its Inheritors," *Foreign Affairs*, 63 (1985), 778–799; D. H. K. AMIRAN, "Geographical Aspects of National Planning in Israel: The Management of Limited Resources," *Transactions I3G, n.s.*, 3 (1978), 115–128; *Atlas of Israel*, 3rd ed. (Macmillan, 1985); D. BAHAT, with C. T. RUBENSTEIN, *The Illustrated Atlas of Jerusalem* (Simon & Schuster, 1990); Y. BEN-ARIEH, "Urban Development in the Holy Land," in J. PATTEN, ed., *The Expanding City: Essays in Honour of Professor Jean Gottman* (Academic, 1983), 1–37; H. CATTAN, *The Palestine Question* (Croom Helm, 1988); A CATTANI, "Israel: The Struggle for Existence," *Swiss Review of World Affairs*, June 1988, 8–11; H. COBBAN, *The Making of Modern Lebanon* (Hutchinson, 1985); P. COSSALI et al, "Whose Promised Land? Israel and the Palestinians," *Geographical Magazine*, 60 (1988), 2–19; R. COCKBURN, "A State in Devastation" (Lebanon), *Geographical Magazine*, 55 (1983), 561–568; G. B. CRESSEY, *Crossroads: Land and Life in Southwest Asia* (Lippincott, 1960; a standard geography; dated but still useful for detail), also, "Qanats, Karez and Foggaras," *Geographical Review*, 48 (1958), 27–44; also, "Water in the Desert," *Annals AAG*, 47 (1957), 105–124; A. DE SHERBININ, "Israel," *Population Today*, April 1, 1991, 12; A. DRYSDALE, "Political Conflict and Jordanian Access to the Sea," *Geographical Review*, 77 (1987), 86–102; E. EFRAT, "Israel's Map of Inequality in Spatial Development," *GeoJournal*, 13 (1986), 401–411; A. ELON, "Letter from Israel," *The New Yorker*, February 13, 1989, 74–80; R. EVANS, "Whose Promised Land?" *Geographical Magazine*, July 1990, 38–41; M. GILBERT, *The Arab-Israeli Conflict: Its History in Maps*, 3rd ed. (Weidenfeld & Nicolson, 1979); C. L. HALLOWELL, "The Glory That Was Jerusalem," *Natural History*, 82 (1973), 39–49; L. HALPRIN, "Israel, the Man-Made Landscape," *Landscape*, Winter 1959–1960, 19–23; J. E. HAZLETON, "Land Reform in Jordan: The East Ghor Canal Project," *Middle East Studies*, 15 (1979), 239–257; A. KELLERMAN, "Tel-Aviv," *Cities*, 2 (1985), 98–105; S. KHALAF, *Lebanon's Predicament* (Columbia, 1987); R. G. KHOURI, *The Jordan Valley: Life and Society Below Sea Level* (Longman, 1981); N. KLIOT, "Lebanon—A Geography of Hostages," *Political Geography Quarterly*, 5 (1986), 199–220; A. D. MILLER, "The Arab-Israeli Conflict, 1967–1987: A Retrospective," *Middle East Journal*, 41 (1987), 349–360; J. MUIR, "Lebanon: Arena of Conflict, Crucible of Peace," *Middle East Journal*, 38 (1984), 204–219; D. NEWMAN, *Population, Settlement and Conflict: Israel and the West Bank* (Cambridge: 1991); A. G. NOBLE and E. EFRAT, "Geography of the Intifada," *Geographical Review*, 80 (1990), 288–307; A. A. ODETH, "Two Capitals in an Undivided Jerusalem," *Foreign Affairs*, Spring 1992, 183–188; E. ORNI and E. EFRAT, *Geography of Israel*, 4th ed. (Jerusalem: Israeli Universities Press, 1980); "Peace and Palestine: The Road That is Not Straight," *The Economist*, January 25, 1992, 40–42; D. PERETZ, *Intifada: The Palestinian Uprising* (Westview, 1990); also, *The West Bank: History, Politics, Society, and Economy* (Westview, 1986); also, "Intifadeh: The Palestinian Uprising," *Foreign Affairs*, 66 (1988), 964–980; F. E. PETERS, *Jerusalem and Mecca: The Typology of the Holy City in the Near East* (NYU, 1987); D. PIPES, *Greater Syria: The History of an Ambition* (Oxford, 1990); I. PLOSS and J. RUBENSTEIN, "Water for Peace" (futuristic plan to desalinize Mediterranean water, using hydropower from diversion of Mediterranean water into the Dead Sea), *The New Republic*, September 7 & 14, 1992, 20–22; B. REICH, "Themes in the History of the State of Israel," *American Historical Review*, 96 (1991); 1466–1478; G. ROWLEY, "Developing Perspectives Upon the Areal Extent of Israel: An Outline Evaluation," *GeoJournal*, 19 (1989), 99–111; also, "Divisions in a Holy City," *Geographical Magazine*, 56 (1984), 196–202; E. A. SALEM, "Lebanon's Political Maze: The Search for Peace in a Turbulent Land," *Middle East Journal*, 33 (1979), 444–463; Z. SCHIFF, "Israel After the War," *Foreign Affairs*, Spring 1991, 19–33; S. SHERMAN, "Gaza: A History of Conflict," *Geographical Magazine*, December 1988, 18–23; D. K. SHIPLER, *Arab and Jew: Wounded Spirits in a Promised Land* (Times Books, 1986; by a journalist); C. G. SMITH, "The Disputed Waters of the Jordan," *Transactions IBG*, 40 (December 1966), 111–128; E. STERN, Y. HAYUTH, and Y. GRADUS, "The Negev Continental Bridge: A Chain in an Intermodal Transport System," *Geoforum*, 14 (1983); 461–469; E. STERN and Y. GRADUS, "The Med-Dead Sea Project: A Vision or Reality?" *Geoforum*, 12 (1981), 265–272; G. SZPIRO, "Israel and the Arabs in the Occupied Territories," *Swiss Review of World Affairs*, April 1987, 12–14; T. SZULC, "The Palestinians," *National Geographic*, June 1992, 84–113; M. VIORST, "The Christian Enclave" (Lebanon), *The New Yorker*, October 3, 1988, 40–71; H. VON WISSMAN *et al.*, "On the Role of Nature and Man in Changing the Face of the Dry Belt of Asia," in W. L. THOMAS, JR., ed., *Man's Role in Changing the Face of the Earth* (Chicago, 1956), 278–303; S. WATERMAN, "Ideology and Events in Israeli Human Landscapes," *Geography*, 64 (1979), 171–181; D. WEINTRAUB *et al.*, *Moshava, Kibbutz, and Moshav: Patterns of Jewish Rural Settlement and Development in Palestine* (Cornell, 1969); J. WELLARD, *Samarkand and Beyond: A History of Desert Caravans* (Constable, 1977); M. ZAMIR, *The Foundation of Modern Lebanon* (Croom Helm, 1985).

Arabian Peninsula and Persian/Arabian Gulf

P. ADAMS, "Yemen," *Geographical Magazine*, July 1988, 26–33; F. AL-FARSY, *Saudi Arabia* (Routledge, 1986); M. ALIREZA, "Women of Saudi Arabia," *National Geographic*, 172 (1987), 422–453; H. K. BARTH and F. QUIEL, "Riyadh and Its Development," *GeoJournal*, 15 (1987), 39–46; P. BEAUMONT, "Water and Development in Saudi Arabia," *Geographical Journal*, 143 (1977), 42–60; R. BIDWELL, *The Two Yemens* (Westview, 1983); M. BLACKSELL, "A Crucial State" (Kuwait), *Geographical Magazine*, 58 (1986), 282–287; H. BOWEN-JONES, "Ancient Oman Is Transformed," *Geographical Magazine*, 52 (1980), 286–293; D. CHAMPION *et al.*, "Oil and Gas Reserves of the Middle East and North Africa," *Natural Resources Forum*, 15 (1991), 202–214; A. DE SHERBININ, "Saudi Arabia," *Population Today*, March 1991, 12; P. DRESCH, *Tribes, Government, and History in Yemen* (Oxford, 1989); C. DRAKE, "Oman: Traditional and Modern Adaptations to the Environment," *Focus,* Summer 1988, 15–20; S. A. EL-ARIFI, "The Nature of Urbanization in the Gulf Countries," *GeoJournal*, 13 (1986), 223–235; W. B. Fisher, "The Good Life in Modern Saudi Arabia," *Geographical Magazine*, 51 (1979), 762–768; F. G. GAUSE III, "Yemeni Unity: Past and Future," *Middle East Journal*, 42 (1988), 33–47; D. L. JOHNSON, *The Nature of Nomadism: A Comparative Study of Pastoral Migrations in Southwestern Asia and Northern Africa* (University of Chicago, Department of Geography Research Paper 118, 1969); P. P. KARAN and W. A. BLADEN, "Arabic Cities," *Focus*, January–February 1983, 1–8; M. KEATING, "We Stand United" (Yemen), *Geographical Magazine*, March 1992, 38–43; R. KING, "The Pilgrimage to Mecca: Some Geographical and Historical Aspects," *Erdkunde*, 26 (1972), 61–73; C. M. KORTEPETER, ed., *Oil and the Economic Geography of the Middle East and North Africa: Studies by Alexander Melamid* (Princeton: Darwin Press, 1991); "Lawrence of Arabia" (feature film and video; magnificent desert photography); L. LUXNER, "Jabal Ali: Dubai's Gateway to the World," *Aramco World*, March–April 1992, 32–39; T. R. McHALE, "A Prospect of Saudi Arabia," *International Affairs*, 56 (1980), 622–647; A. MELAMID, "Dubai City," *Geographical Review*, 79 (1989), 345–347; also, "Dhofar," *ibid.*, 74 (1984), 76–79; also, "Qatar," *ibid.*, 77 (1987), 103–105; N. I. NAWWAB, "The Journey of a Lifetime" (pilgrimage to Mecca), *Aramco World*, July/August 1992, 24–35; G. O'REILLY, "Kuwait: Boundaries, Oil, and Shadow Empires," *Geographical Viewpoint* (Dublin, Ireland), 19 (1990–91), 18–30; J. E. PETERSON, "The Arabian Peninsula in Modern Times: A Historiographical Survey," *American Historical Review*, 96 (1991), 1435–1449; B. R. PRIDHAM, ed., *The Arab Gulf and the Arab World* (Croom Helm, 1988); R. K. RAMAZANI, *The Persian Gulf and the Strait of Hormuz* (Sijthoff & Noordhoff, 1979); S. SEARIGHT, "Farmers of the Desert" (Saudi Arabia), *Geographical Magazine*, 58 (1986), 127–131; W. THESIGER, *Arabian Sands* (Dutton, 1959); M. WENNER, *The Yemen Arab Republic: Development and Change in an Ancient Land* (Westview, 1991).

Iran, Iraq, Iran-Iraq War, Gulf War

On Iran and the Iran-Iraq War, see F. AJAMI, "Iran: The Impossible Revolution," *Foreign Affairs*, Winter 1988/1989, 135–155; S. A. ARJOMAND, "Iran's Islamic Revolution in Comparative Perspective," *World Politics*, 38 (1986), 383–414; W. B. FISHER, ed., *The Land of Iran* (*The Cambridge History of Iran*, vol. 1; Cambridge, 1968); P. G. LEWIS, "Iranian Cities," *Focus*, January-February 1983, 12–16; G. SICK, "Trial by Error: Reflections on the Iran-Iraq War," *Middle East Journal*, 43 (1989), 230–245; J. H. SIGLER, "The Iran-Iraq Conflict: The Tragedy of Limited Conventional War," *International Journal*, 41 (1986), 424–456; M. STERNER, "The Iran-Iraq War," *Foreign Affairs*, 63 (1984), 128–134; W. D. SWEARINGEN, "Geopolitical Origins of the Iran-Iraq War," *Geographical Review*, 78 (1988), 405–416; R. WRIGHT, "Tehran Summer," *The New Yorker*, September 5, 1988, 32–72; H. E. WULFF, "The Qanats of Iran," *Scientific American*, 218 (1968), 94–105.

On Iraq, the Gulf War, and the Kurds, see J. BLACKWELL, *Thunder in the Desert. The Strategy and Tactics of the Persian Gulf War* (Bantam Books, 1991); T. Y. CANBY, "After the Storm" (Gulf War), *National Geographic*, August 1991, 2–32; U. S. CENTRAL INTELLIGENCE AGENCY, *Iraq: A Map Folio* (August 1992); S. B. COHEN, "The Geopolitical Aftermath of the Gulf War," *Focus*, Summer 1991, 23–26; S. L. CUTTER, "Ecocide in Babylonia," *Focus*, Summer 1991, 26–31; S. A. EARLE, "Persian Gulf Pollution: Assessing the Damage One Year Later," *National Geographic*, February 1992, 122–134; R. EVANS, "Legacy of Woe" (the Kurds), *Geographical Magazine*, June 1991, 34–38; C. HITCHENS, "Struggle of the Kurds," *National Geographic*, August 1992, 32–61; J. HORGAN, "Up in Flames" (oil well fires in Gulf War), *Scientific American*, May 1991, 17–24; C. KARRER, "The Kurds—At Home and in Exile," *Swiss Review of World Affairs*, August 1991, 17–24; M. SEVERY, "Iraq: Crucible of Civilization," *National Geographic*, May 1991, 102–115.

Turkey, Cyprus, and Afghanistan

On Turkey and Cyprus, see B. W. BEELEY, "The Greek-Turkish Boundary Conflict at the Interface," *Transactions IBG, n.s.*, 3 (1978), 351–366; G. BLAKE, "Turkish Guard on Russian Waters" (Turkish Straits), *Geographical Magazine*, 53 (1981), 950–955; G. D. CAMP, "Greek-Turkish Conflict Over Cyprus," *Political Science Quarterly*, 95 (1980); 43–70; R. COCKBURN, "Two Sides of Cyprus," *Geographical Magazine*, 57 (1985), 144–149; D. A. GILLMORE, "Recent Tourism Development in Cyprus," *Geography*, 74 (1989), 262–265; J. R. GREGORY, "Liquid Asset" (Turkey's Water), *World Monitor*, November 1991, 28–33; M. M. GUNTER, *The Kurds in Turkey: A Political Dilemma* (Westview, 1990); B. R. KUNIHOLM, "Turkey and the West," *Foreign Affairs*, Spring 1991, 34–48; D. LOCKHART and S. ASHTON, "Tourism to Northern Cyprus," *Geography*, 75 (1990), 163–167; N. ROBERTS, "Geopolitics and the Euphrates' Water Resources," *Geography*, 76 (1991), 157–159; S. TALBOTT, "And Now For Some Good News" (Cyprus), *Time*, August 31, 1992, 51; "Turkey: Half Inside, Half Out," *The Economist*, June 18, 1988, 1–30.

On Afghanistan, see N. J. R. ALLAN, "Afghanistan: The End of a Buffer State," *Focus*, Fall 1986, 2–9; E. W. ANDERSON and N. H. DUPREE, *The Cultural Basis of Afghan Nationalism* (London: Pinter Publishers, 1990); M. KEATING, "Afghanistan: Resistance to Assistance," *Geographical Magazine*, November 1991,

36–38; R. B. RAIS, "Afghanistan After the Soviet Withdrawal," *Current History*, March 1992, 123–127; W. B. WOOD, "Long Time Coming: The Repatriation of Afghan Refugees," *Annals AAG*, 79 (1989), 345–369.

PART 4: THE ORIENT

Orient: General

I. ADAMS, "Rice Cultivation in Asia," *American Anthropologist*, 50 (1948), 256–278; C. E. BLACK *et al.*, *The Modernization of Inner Asia* (Sharpe, 1991); H. B. BROUGH, "A New Lay of the Land" (skewed land ownership and environmental breakdown: Orient and world), *World Watch*, January–February 1991, 12–19; G. B. CRESSEY, *Asia's Lands and Peoples* (3rd ed.; McGraw-Hill, 1963; quite dated but still a standard reference); D. DRAKAKIS-SMITH, "Urban Food Distribution in Asia and Africa," *Geographical Journal*, 157 (1991), 51–61; J. S. FEIN and P. L. STEPHENS, eds., *Monsoons* (Wiley, 1987); N. S. GINSBURG, ed., *The Pattern of Asia* (Prentice-Hall, 1958; a major geography text, still very valuable despite its date); J. GOODWIN, *A Time for Tea: Travels Through China and India in Search of Tea* (Knopf, 1991); R. E. HUKE, "The Green Revolution," *Journal of Geography*, 84 (1985), 248–254; O. JIN BEE, "The Tropical Rain Forest: Patterns of Exploitation and Trade," *Singapore Journal of Tropical Geography*, 11 (1990), 117–142; D. F. LACH, *Asia in the Making of Europe*, 2 vols. (Chicago: vol. 1, 1965; vol. 2, 1970, 1978); W. W. LOCK-WOOD, "Asian Triangle: China, India, Japan," *Foreign Affairs*, 52 (1974), 818–838; W. MALENBAUM, "A Gloomy Portrayal of Development Achievements and Prospects: China and India," *Economic Development and Cultural Change*, 38 (1990), 391–406; R. MURPHEY, *The Outsiders: The Western Experience in India and China* (Michigan, 1977); "Rain Forest Priorities" (map), *World Monitor*, August 1991, 10; E. O. REISCHAUER and J. K. FAIRBANK, *A History of East Asian Civilization*, 2 vols. (Houghton Mifflin, vol. 1, *East Asia: The Great Tradition*, 1958, 1960; vol. 2, *East Asia: The Modern Transformation*, 1965); R. REPETTO, "Deforestation in the Tropics," *Scientific American*, April 1990, 36–42; R. A. SCALAPINO, "The United States and Asia: Future Prospects," *Foreign Affairs*, Winter 1991/92, 19–40; J. E. SPENCER, *Oriental Asia: Themes Toward a Geography* (Prentice-Hall, 1973), also, with W. L. THOMAS, *Asia, East by South: A Cultural Geography*, 2nd ed. (Wiley, 1971); G. T. TREWARTHA, "Monsoons: With a Focus on South Asia and Tropical East Africa," *Journal of Geography*, 81 (1982), 4–11; C. TYLER, "Laying Waste" (tropical rain forests), *Geographical Magazine*, January 1990, 26–30; also, "The Sense of Sustainability" (tropical rain forests), *ibid.*, February 1990, 8–13; P. J. VESILIND, "Monsoons," *National Geographic*, 166 (1984), 712–747.

Indian Subcontinent

R. AKHTAR and A. LEARMONTH, "Malaria Returns to India," *Geographical Magazine*, 54 (1982), 135–139; *A Passage to India* (feature film and video; based on the novel by E. M. Forster); H. ARDEN, "Along the Grand Trunk Road: Searching for India," *National Geographic*, May 1990, 116–138; S. BAKER, *Caste: At Home in Hindu India* (Jonathan Cape, 1990); A. L. BASHAM, *The Wonder That Was India*, 3rd ed. (Sidgwick and Jackson, 1967);

C. BAXTER, *Bangladesh: A New Nation in an Old Setting* (Westview, 1984); S. M. BHARDWAJ, *Hindu Places of Pilgrimage in India: A Study in Cultural Geography* (California, 1973); *Cambridge Economic History of India*, 2 vols. (Cambridge, 1982); B. W. BUNTING, "Bhutan: Kingdom in the Clouds," *National Geographic*, May 1991, 78–101; J. CHANG, "The Indian Summer Monsoon," *Geographical Review*, 37 (1967), 373–396; B. CHEL-LANEY, "Drama in Delhi" (Indian debt, ethnic terrorism, military excess), *World Monitor*, December 1991, 40–44; R. CRITCH-FIELD, "Sowing Success, Reaping Guns" (Punjab), *World Monitor*, July 1992, 24–30; B. DOGRA, "Traditional Agriculture in India: High Yields and No Waste," *The Ecologist*, 13 (1983), 84–87; J. P. DORIAN, "The Development of India's Mining Industry," *Geo-Journal*, 19 (1989), 145–160; A. K. DUTT and A. SEN, "Provisional Census of India 1991," *Geographical Review*, 82 (1992), 207–211; A. K. DUTT and M. M. GELB, *Atlas of South Asia, Fully Annotated* (Westview, 1987); A. K. DUTT *et al.*, "Spatial Pattern of Languages in India: A Culture-Historical Analysis," *GeoJournal*, 10 (1985), 51–74; also, DUTT, "Bengal: A Search for Regional Identity," *Focus*, May–June 1984, 1–12; G. ETIENNE, *Food and Poverty: India's Half Won Battle* (Sage Publications, 1988); R. EVANS, "India After the Gandhis," *Geographical Magazine*, September 1991, 28–32; B. H. FARMER, *An Introduction to South Asia* (Methuen, 1983); also, "Perspectives on the 'Green Revolution' in South Asia," *Modern Asian Studies*, 20 (1986), 175–199; A. FRATER, *Chasing the Monsoon* (Knopf, 1991); J. FRYER, "Benefits from the Indian Sacred Cow," *Geographical Magazine*, 53 (1981), 617–624; "Gandhi" (feature film; video); J. H. HUTTON, *Caste in India: Its Nature, Function and Origins*, 4th ed. (Oxford, 1963); "India and South Asia," *Current History*, March 1992, entire issue; P. P. KARAN and S. IIJIMA, "Environmental Stress in the Himalaya," *Geographical Review*, 75 (1985), 71–92; H. KREUTZMANN, "The Karakoram Highway: The Impact of Road Construction on Mountain Societies," *Modern Asian Studies*, 25 (1991), 711–736; R. KRISHNA, "The Economic Development of India," *Scientific American*, 243, no. 3 (September 1980), 166–178; D. LEWIS, "Drowning by Numbers" (dams in India), *Geographical Magazine*, September 1991, 34–38; JOHN MASTERS, historical novels; R. MURPHEY, "The City in the Swamp: Aspects of the Site and Early Growth of Calcutta," *Geographical Journal*, 130 (1964), 241–256; V. S. NAIPAUL, *India: A Wounded Civilization* (Knopf, 1977); A. G. NOBLE and A. K. DUTT, *India: Cultural Patterns and Processes* (Westview, 1982); F. ROBINSON, ed., *The Cambridge Encyclopedia of India, Pakistan, Bangladesh, Sri Lanka, Nepal, Bhutan and the Maldives* (Cambridge, 1989); J. E. SCHWARTZBERG, ed., *An Historical Atlas of South Asia* (Chicago, 1978); F. J. SIMOONS, *Eat Not This Flesh: Food Avoidances in the Old World* (Wisconsin, 1961), especially chap. 4, "Beef," 45–63; J. R. SMITH, "The Man Behind the Mountain" (Sir George Everest), *Geographical Magazine*, November 1990, 38–41; D. E. SOPHER, "Place and Landscape in Indian Tradition," *Landscape*, 29 (1986), no. 2, 1–9; O. H. K. SPATE and A. T. A. LEARMONTH, *India and Pakistan: A General and Regional Geography*, 3rd ed. (Methuen, 1967), with a chapter on Ceylon (now Sri Lanka) by B. H. FARMER—a huge, authoritative textbook that is still indispensable, although aging; B. WALLACH, "India: Learning Curves," *Focus*, Part I, Fall 1991, 28–34; Part II, Winter 1991, 25–28.

Southeast Asia: General

R. W and B. B. DECKER, *Mountains of Fire: The Nature of Volcanoes* (Cambridge, 1991); C. DIXON, *South East Asia in the World-Economy* (Cambridge 1991); A. K. DUTT, ed., *Southeast Asia: Realm of Contrast*, 3rd rev. ed. (Westview, 1985); D. DWYER, ed., *South East Asian Development: Geographical Perspectives* (Longman, 1990); C. A. FISHER, *South-East Asia: A Social, Economic, and Political Geography*, 2nd ed. (Dutton, 1964), a detailed and authoritative survey that is still very valuable despite its age); P. HURST, *Rainforest Politics: Ecological Destruction in South-East Asia* (Zed Books, 1990); L. KONG, "The Malay World in Colonial Fiction," *Singapore Journal of Tropical Geography*, 7 (June 1986), 40–52; R. C. OBERST, "A War Without Winners in Sri Lanka," *Current History*, March 1992, 128–131; J. RIGG, *Southeast Asia: A Region in Transition* (HarperCollins Academic, 1991); J. E. SPENCER, "Southeast Asia," *Progress in Human Geography*, 8 (1984), 284–288; R. ULACK and G. PAUER, *Atlas of Southeast Asia* (Macmillan, 1989); L. YOUNG-LENG, *Southeast Asia: Essays in Political Geography* (Singapore, 1982).

Southeast Asia: Sri Lanka, Burma, Thailand, Cambodia, Laos, Vietnam

R. BIRSEL, "Military Régime Minority Insurgence" (Myanmar), *Geographical Magazine*, February 1990, 28–32; T. L. BROWN, *War and Aftermath in Vietnam* (Routledge, 1991); D. P. CHANDLER, "The Tragedy of Cambodian History," *Pacific Affairs*, 52 (1979), 410–419; T. DE RUBEMPRÉ, "Buddha's Own Country" (Burma), *Geographical Magazine*, 59 (1987), 183–188; B. B. FALL, "Two Thousand Years of War in Viet-Nam," *Horizon*, 9, no. 2 (Spring 1967), 4–22 (a concise summation of Vietnam's internal development through history, as related to foreign invaders and influences); C. A. FISHER, "The Vietnamese Problem in Its Geographical Context," *Geographical Journal*, 131 (1965), 502–5155; G. C. HERRING, "America and Vietnam: The Unending War," *Foreign Affairs*, Winter 1991/92, 104–119; B. IMHASLY, "The Flight of the Rohingya" (Burmese Muslims, fleeing to Bangladesh from Myanmar), *Swiss Review of World Affairs*, June 1992, 20–22; N. KAPLAN, *et al.*, "Indochinese Refugee Families and Academic Achievement," *Scientific American*, February 1992, 36–42; R. C. OBERST, "A War Without Winners in Sri Lanka," *Current History*, March 1992, 128–131; J. SILVERSTEIN, "Burma Through the Prism of Western Novels," *Journal of Southeast Asian Studies*, 16 (1985), 129–140; P. T. WHITE, articles on Hanoi and Saigon, *National Geographic*, November 1989, 558–621.

Southeast Asia: Malaysia, Singapore, Brunei

P. P. COURTENAY, "The Plantation in Malaysian Economic Development," *Journal of Southeast Asian Studies*, 12 (1981), 329–348; B. FIELD and J. SMITH, "Singapore," *Cities*, 3 (1986), 186–199; J. GOODRIDGE, "A New Life for Tin," *Geographical Magazine*, 57 (1985), 424–429; K. GRICE and D. DRAKAKIS-SMITH, "The Role of the State in Shaping Development: Two Decades of Growth in Singapore," *Transactions IBG, n.s.*, 10 (1985), 347–359; B. HODGSON, "Singapore: Mini-Size Superstate," *National Geographic*, 159 (1981), 540–561; W. NEVILLE, "Economy and Employment in Brunei," *Geographical Review*, 75

(1985), 451–461; J. C. RYAN, "Plywood vs. People in Sarawak," *World Watch*, January–February 1991, 8–9; C. TYLER, "Triangular Bonds" (Singapore, Malaysia, and Indonesia), *Geographical Magazine*, March 1992, 34–37.

Southeast Asia: Indonesia, Philippines

M. C. CLEARY and F. J. LIAN, "On the Geography of Borneo," *Progress in Human Geography*, 15 (1991), 163–177; C. DRAKE, "The Spatial Pattern of National Integration in Indonesia," *Transactions IBG, n.s.*, 6 (1981), 471–490; C. A. FISHER, "Indonesia—A Giant Astir," *Geographical Journal*, 138 (1972), 154–165; D. W. FRYER and J. C. JACKSON, *Indonesia* (Ernest Benn, 1977); A. GORDON, "Indonesia, Plantations and the 'Post-colonial' Mode of Production," *Journal of Contemporary Asia*, 12 (1982), 168–187; P. A. KRINKS, "Rural Changes in Java: An End to Involution?" *Geography*, 63 (1978), 31–36; J. T. LINDBLAD, "Economic Aspects of the Dutch Expansion in Indonesia, 1870–1914," *Modern Asian Studies*, 23 (1989), 1–23; S. SCHLOSSTEIN, *Asia's New Little Dragons: The Dynamic Emergence of Indonesia, Thailand and Malaysia* (Contemporary Books, 1991); F. L. WERNSTEDT and J. E. SPENCER, *The Philippine Island World: A Physical, Cultural, and Regional Geography* (California, 1967; comprehensive, authoritative; aging, but still valuable); W. A. WITHINGTON, "Indonesia's Significance in the Study of Regional Geography," *Journal of Geography*, 68 (1969), 227–237; also, "The Major Geographic Regions of Sumatra, Indonesia," *Annals AAG*, 57 (1967), 534–549; W. B. WOOD, "Intermediate Cities on a Resource Frontier" (Indonesia), *Geographical Review*, 76 (1986), 149–159.

China: General

"A New Nuclear Age? The Shadow of the Bomb Still Looms" (includes discussion of China's nuclear role), *World Press Review*, December 1991, 9–15; C. BLUNDEN and M. ELVIN, *Cultural Atlas of China* (Facts on File, 1983); K. BUCHANAN *et al.*, *China: The Land and People* (Crown, 1981); F. BUNGE and R. S. SHINN, eds., *China, A Country Study*, 3rd ed. (Washington, DC: Government Printing Office, 1981; U.S. Department of the Army, Area Handbook Series); T. CANNON and A. JENKINS, eds., *The Geography of Contemporary China: The Impact of Deng Xiaoping's Decade* (Routledge, 1990); "China," *Current History*, September 1991, 1992, entire issues; P. J. M. GEELAN and D. C. TWITCHETT, eds., *The Times Atlas of China* (Van Nostrand, 1984); B. HOOK and D. C. TWICHETT, eds., *The Cambridge Encyclopedia of China* (Cambridge, 1991; a massive, indispensable work); R. G. KNAPP, *China's Traditional Rural Architecture: A Cultural Geography of the Common House* (Hawaii, 1986); also, ed., *China's Island Frontier* (Hawaii, 1980); "Modern China," *Geographical Magazine* 58–59 (1986–1987; articles in separate issues by various authors); D. C. MONEY, *China: The Land and the People*, rev. ed. (London: Evans Bros., 1990); C. L. SALTER, "The New Localism in China's Cultural Landscape," *Landscape*, 25, no. 3 (1981), 10–14; also, "Windows on a Changing China," *Focus*, January 1985, 12–21; C. J. SMITH, *China: People and Places in the Land of One Billion* (Westview, 1991); T. R. TREGEAR, *China: A Geographical Survey* (Hodder and Stoughton/Wiley, 1980); B. WALLACH, "China: Temples of Heaven," Parts I and II, *Focus*, 1990; B. U. WIESER, "Notes from the Middle Kingdom," *Swiss Review of World Affairs*, August

1991, 25–28; D. S. ZAGORIA, "China's Quiet Revolution," *Foreign Affairs*, 62 (1984), 879–904.

China: Historical

B. CATCHPOLE, *A Map History of Modern China* (Heinemann, 1976); J. K. FAIRBANK, *The Great Chinese Revolution 1800;nn1985* (Harper & Row, 1986); also with E. O. REISCHAUER, *China: Tradition and Transformation*, rev. ed. (Houghton Mifflin, 1989); J. GERNET, *A History of Chinese Civilization*, trans. J. R. Foster (Cambridge, 1982); A. HERMANN, *An Historical Atlas of China*; gen. ed., N. GINSBURG; prefatory essay, P. WHEATLEY (Aldine, 1966); "The Last Emperor" (feature film; video) N. B. TUCKER, "China and America: 1941–1991," *Foreign Affairs*, Winter 1991/92, 75–92; A. N. WALDRON, "The Problem of the Great Wall of China," *Harvard Journal of Asiatic Studies*, 43 (1983), 643–663.

China: Environment

A. K. BISWAS, "Long Distance Water Transfer: The Chinese Plans," *GeoJournal*, 6 (1982), 481–487; M. FULLEN and D. MITCHELL, "Taming the Shamo Dragon," (desertification in China), *Geographical Magazine*, November 1991, 26–29; P. HO, "The Loess and the Origin of Chinese Agriculture," *American Historical Review*, 75 (1969), 1–36; L. J. C. MA and A. G. NOBLE, eds., *The Environment: Chinese and American Views* (Methuen, 1981); V. SMIL, "China's Environment," *Current History*, September 1980, 14–18; also, "Controlling the Yellow River," *Geographical Review*, 69 (1979), 253–272; also, *The Bad Earth: Environmental Degradation in China* (Sharpe, 1984); Z. SONGQIAO, *Physical Geography of China* (Wiley, 1986); A. S. WALKER, "Deserts of China," *American Scientist*, 70 (1982), 366–376; D. E. WALLING, "Yellow River Which Never Runs Clear," *Geographical Magazine*, 53 (1981), 568–575; "Water in China" (special issue), *GeoJournal*, 10, no. 2 (1985); C. XUEMIN, "Water Resources Planning on the Yellow River," *Natural Resources Forum*, 4 (1980), 315–324; H. K. YOON, "Loess Cave-Dwellings in Shaanxi Province, China," *GeoJournal*, 21 (1990), 95–102; Z. ZHANG, "Loess in China," *GeoJournal*, 4 (1980), 525–540.

China: Population, Economy, and Development

D. CHEN, "The Economic Development of China," *Scientific American*, 243, no. 3 (September 1980), 152–165; C. COMTOIS, "Transport and Territorial Development in China, 1949–1985," *Modern Asian Studies*, 24 (1990), 777–818; S. JINGZHI, *The Economic Geography of China* (Oxford, 1988); A. J. JOWETT, "China: The Great Leap to Disaster, or China: The Great Famine, or China: The Harvest of Death or . . . ," *Focus*, Fall 1990, 19–23; also, "China: The One, Two, Three, Four and More Child Policy," *ibid.*, Summer 1991, 32–36; L. KRAAR, "The China Bubble Bursts," *Fortune*, July 6, 1987, 86–89; R. W. McCOLL and Y. KOU, "Feeding China's Millions," *Geographical Review*, 80 (1990), 434–443; R. MURPHEY, *The Fading of the Maoist Vision: City and Country in China's Development* (Methuen, 1980); V. NEE and F. W. YOUNG, "Peasant Entrepreneurs in China's 'Second Economy': An Institutional Analysis," *Economic Development and Cultural Change*, 39 (1991), 293–310; C. W. PANNELL, "Recent Chinese Agriculture," *Geographical Review*, 75 (1985), 170–185; also, "Less Land for Chinese Farmers," *Geographical Magazine*, 54 (1982), 324–329; and, with J. C. L. MA, *China: The*

Geography of Development and Modernization (Wiley, 1983); D. R. PHILLIPS, "Oil in Chinese Waters," *Geographical Magazine*, 56 (September 1984), 444–445; J. PRYBYLA, "China's Economic Dynamos," *Current History*, September 1992, 262–267; O. SCHELL, "The Wind of Wanting to Go It Alone" (new economic pragmatism in China), *The New Yorker*, January 23, 1984, 43–85; C. D. STOLTENBERG, "China's Special Economic Zones: Their Development and Prospects," *Asian Survey*, 24 (1984), 637–654; R. TERRILL, "China's Youth Wait for Tomorrow," *National Geographic*, July 1991, 110–136; H. TIEN *et al.*, "China's Demographic Dilemmas," *Population Bulletin*, June 1992, entire issue; S. TONG, "China the Arms Merchant," *World Monitor*, June 1992, 46–50; J. L. and A. S. TYSON, "China's Villages," *Christian Science Monitor*, July 22, 29; August 5, 12, 1992; G. VEECK, ed., *The Uneven Landscape: Geographical Studies in Post-Reform China* (Baton Rouge, LA: Geoscience Publications, 1991); A. WILLIAMS, "Pearl River Powerhouse" (South China's "Special Economic Zones"), *Geographical Magazine*, March 1992, 28–32.

China: Urbanization

D. BRONGER, "Metropolitanization in China?" *GeoJournal*, 8 (1984), 137–146; S. CHANG, "Modernization and China's Urban Development," *Annals AAG*, 71 (1981), 202–219; also, "Peking: The Growing Metropolis of Communist China," *Geographical Review*, 55 (1965), 313–327; D. J. DWYER, "Chengdu, Sichuan: The Modernization of a Chinese City," *Geography*, 71 (1986), 215–227; D. W. EDGINGTON, "Tianjin," *Cities*, 3 (1986), 117–124; R. KIRKBY, "China Goes to Town" (urban growth), *Geographical Magazine*, 58 (1986), 508–511; L. J. C. MA and E. W. HANTEN, *Urban Development in Modern China* (Westview, 1981); C. W. PANNELL, "China's Urban Geography," *Progress in Human Geography*, 14 (1990), 214–236; K. C. TAN, "Small Towns in Chinese Urbanization," *Geographical Review*, 76 (1986), 265–275; M. K. WHYTE, "Urbanism as a Chinese Way of Life," *International Journal of Comparative Sociology*, 24 (1983), 61–85.

Arid China, Hong Kong, Macao, Taiwan, Mongolia

P. ALLEN, "Inside Chinese Tibet," *Geographical Magazine*, 52 (1980), 830–837; T. B. ALLEN, "Time Catches Up with Mongolia," *National Geographic*, 167 (1984), 242–269; W. BELLO and S. ROSENFELD, "High-Speed Industrialization and Environmental Devastation in Taiwan," *The Ecologist*, 20 (1990), 125–132; F. F. CHIEN, "A View from Taipei," *Foreign Affairs*, Winter 1991/92, 92–103; C. DUNCAN, "Macau," *Cities*, 3 (1986), 2–11; D. R. HALL, "Economic and Urban Development in Mongolia," *Geography*, 72 (1987), 73–76; Y. D. HWANG, *The Rise of a New World Economic Power: Postwar Taiwan* (Greenwood, 1991); M. LeVASSEUR, "Destination: Tibet," *Focus*, Spring 1987, 34–35; R. W. McCOLL, "China's Silk Roads: A Modern Journey to China's Western Regions," *Focus*, Summer 1991, 1–6; also, "By Their Dwellings Shall Ye Know Them: Home and Selling Among China's Inner Asian Ethnic Groups," *ibid.*, Winter 1989, 1–6; W. B. OVERHOLT, "China and British Hong Kong," *Current History*, 90 (1991), 270–274; M. ROSSABI, "Mongolia: A New Opening?" *Current History*, September 1992, 278–283; "Slow Thaw in Mongolia," *World Press Review*, May 1992, 42; R. TERRILL, "The Fate of Hong Kong," *World Monitor*, January 1992, 42–50; also, "Hong Kong," *National Geographic*, February 1991, 100–131.

Japan

E. A. ACKERMAN, *Japan's Natural Resources and Their Relation to Japan's Economic Future* (Chicago, 1953; a massive survey, dated in many ways but still useful); ASSOCIATION OF JAPANESE GEOGRAPHERS, ed., *Geography of Japan* (Tokyo: Terikoku-Shoin, 1980); V. J. BUNCE, "The Geography of Japanese Overseas Investment," *Teaching Geography*, 15 (1990), 72–77; W. BURGESS, "Hokkaido: Japan's New Frontier," *Geography*, 67 (1982), 64–68; W. CHAPMAN, *Inventing Japan: The Making of a Postwar Civilization* (Prentice-Hall, 1991); H. CORTAZZI, *The Japanese Achievement* (Sidgwick & Jackson, 1990); J. D. EYRE, "Water Controls in a Japanese Irrigation System," *Geographical Review*, 45 (1955), 197–216; C. A. FISHER, "The Expansion of Japan: A Study in Oriental Geopolitics: Part I, Continental and Maritime Components in Japanese Expansion; Part II, The Greater East Asia Co-Prosperity Sphere," *Geographical Journal*, 115 (1950), 1–19, 179–193; Y. FUNABASHI, "Japan and the New World Order," *Foreign Affairs*, Winter 1991/92, 58–74; C. D. HARRIS, "The Urban and Industrial Transformation of Japan," *Geographical Review*, 72 (1982), 50–89; R. HOLBROOKE, "Japan and the United States: Ending the Unequal Partnership," *Foreign Affairs*, Winter 1991/92, 41–57; J. IMPOCO, "Japan Grows Old on the Farm," *U.S. News and World Report*, August 3, 1992, 39–40; *Japan: A Regional Geography of an Island Nation* (Tokyo: Teikoku Shoin, 1985); "Japan," *Current History*, April 1991, entire issue; "Japan's Troubled Future: Special Report," *Fortune*, March 30, 1987, 21–53; C. KOKUDO, *The National Atlas of Japan* (Tokyo: Japan Map Center, 1977); K. J. S. KOKUSAI, *Atlas of Japan: Physical, Economic, and Social*, 2nd ed. (Tokyo: International Society for Educational Information, 1974); D. KORNHAUSER, *Japan: Geographical Background to Urban-Industrial Development*, 2nd ed. (Longman, 1982; "The World's Landscapes"); D. MacDONALD, *A Geography of Modern Japan* (Paul Norbury, 1985); G. M. MACKLIN, "Time and History in Japan," *American Historical Review*, 85 (1980), 557–571; P. McGILL, "Japan: Anxiety on the Road to World Leadership," *World Press Review*, August 1992, 16–18; I. J. McMULLEN, "How Confucian Is Modern Japan?" *Asian Affairs*, 11 (1980), 276–283; K. MURATA and I. OTA, eds., *An Industrial Geography of Japan* (Bell & Hyman/St. Martin's, 1980); S. MATSUI *et al.*, "Farming at the Pace of Industry," *Geographical Magazine*, 52 (1980), 740–746; M. NISHI, "Regional Variations in Japanese Farmhouses," *Annals AAG*, 57 (1967), 239–266; K. K. OSHIRO, "Mechanization of Rice Production in Japan," *Economic Geography*, 61 (1985), 323–331; J. PEZEU-MASSABUAU, *The Japanese Islands: A Physical and Social Geography* (translated from the French by P. C. Blum; Tuttle, 1978); H. C. REED, "The Ascent of Tokyo As an International Financial Center," *Journal of International Business Studies*, 11 (1980), 19–35; E. O. REISCHAUER, *The Japanese* (Harvard, 1977); J. SARGENT, "Industrial Location in Japan with Special Reference to the Semiconductor Industry," *Geographical Journal*, 153 (1987), 72–85; H. SHITARA, "It Is Cold, It Is Hot, But It Is Seldom Dry" (Japanese climate), *Geographical Magazine*, 52 (1980), 547–554; T. C. SMITH, *The Agrarian Origins of Modern Japan* (Stanford, 1959); H. TANAKA, "Landscape Expression of Buddhism in Japan," *Canadian Geographer*, 28 (1984), 240–257; S. TATSUNO, *The Technopolis Strategy: Japan, High Technology, and the Control of the Twenty-First Century* (Prentice-Hall, 1986); G. T. TREWARTHA, *Japan: A Geography* (Wisconsin, 1965; a standard, authoritative work, now somewhat outdated; MACHIKO YANAGISHITA, "Japan," *Population Bulletin*, April 1992, 12.

Korea

P. M. BARTZ, *South Korea: A Descriptive Geography* (Oxford, 1972); C. A. FISHER, "The Role of Korea in the Far East," *Geographical Journal*, 120 (1954), 282–298; N. JACOBS, *The Korean Road to Modernization and Development* (Illinois, 1985); S. McCUNE, *Korea's Heritage: A Regional and Social Geography* (Charles E. Tuttle, 1964).

Part 5: The Pacific World

Pacific World: General; Pacific Islands

G. BARCLAY, *A History of the Pacific from the Stone Age to the Present Day* (Sidgwick & Jackson, 1978); J. C. BEAGLEHOLE, *The Exploration of the Pacific*, 3rd ed.; (Stanford, 1966); P. BELLWOOD, *Man's Conquest of the Pacific* (Oxford, 1979); also, "The Peopling of the Pacific," *Scientific American*, 243, no. 5 (November 1980), 174–185; also, "The Austronesian Dispersal and the Origin of Languages" (in Oceania), *Scientific American*, July 1991, 88–93; S. G. BRITTON, "The Evolution of a Colonial Space-Economy: The Case of Fiji," *Journal of Historical Geography*, 6 (1980), 251–274; H. C. BROOKFIELD and D. HART, *Melanesia: A Geographical Interpretation of an Island World* (Barnes & Noble, 1971); K. BROWER, *Micronesia: The Land, the People, and the Sea* (Louisiana State, 1981); I. C. CAMPBELL, *A History of the Pacific Islands* (California, 1990); J-H. CHANG, "Sugar Cane in Hawaii and Taiwan: Contrasts in Ecology, Technology, and Economics," *Economic Geography*, 46 (1970), 39–52; H. CLEVELAND, "The Future of the Pacific Basin," *Pacific Viewpoint*, 25 (1984), 1–13; K. B. CUMBERLAND, *Southwest Pacific: A Geography of Australia, New Zealand, and Their Pacific Island Neighbors*, rev. ed. (Praeger, 1968); C. DARWIN, *The Structure and Distribution of Coral Reefs* (California, 1962; first published in 1889); B. R. FINNEY, "Anomalous Westerlies, El Niño, and the Colonization of Polynesia," *American Anthropologist*, 87 (1985), 9–26; O. W. FREEMAN, ed., *Geography of the Pacific* (Wiley, 1951; a standard reference work, out of date in many respects but still informative); H. R. FRIIS, ed., *The Pacific Basin: A History of Its Geographical Exploration* (American Geographical Society, 1967); S. HENNINGHAM, *France and the South Pacific: A Contemporary History* (Hawaii: 1991); M. C. HOWARD, *Mining, Politics, and Development in the South Pacific* (Westview, 1991); E. LEACH, "Ocean of Opportunity" (the Pacific), *Pacific Viewpoint*, 24 (1983), 99–111; D. LEWIS, *We, The Navigators: The Ancient Art of Landfinding in the Pacific* (Hawaii, 1972); G. R. LEWTHWAITE, "Man and Land in Early Tahiti: Polynesian Agriculture Through European Eyes, *Pacific Viewpoint*, 5 (1964), 11–34; also, "Man and the Sea in Early Tahiti," *ibid.*, 7 (1966), 28–53; R. McKIE, "The Island-Hopping Gene" (aboriginal colonization of the islands), *Geographical Magazine*, November 1991, 30–34; W. MENARD, "Coconut," *Pacific Discovery*, 21, no. 2 (March-April 1968), 19–24; A. MOOREHEAD, *The Fatal Impact: An Account of the Invasion of the South Pacific, 1767–1840* (Harper & Row, 1966); R. E. MURPHY, "American Micronesia: A Supplementary Chapter in the Regional Geography of the United States," *Journal of Geography*, 79 (1980), 181–186; also, " 'High' and 'Low' Islands in the Eastern Carolines," *Geographical Review*, 39 (1949), 425–439; D. L. OLIVER, *The Pacific Islands* (Harvard, 1951; reprinted, Doubleday Anchor Books, 1962; a much-cited book by an anthropologist); J. OVERTON, ed., "Fiji Since the Coups" (eight articles, *Pacific Viewpoint*, 30 (1989), 109–216; *Pacific Islands Year Book* (Sydney, Australia: Pacific Publications, annual); *Pacific Viewpoint* (semiannual; a geographical journal of high quality; A. G. PRICE, *The Western Invasion of the Pacific and Its Continents: A Study of Moving Frontiers and Changing Landscapes*, 1513–1958 (Oxford, 1963); R. J. SAGER, "The Pacific Islands: A New Geography," *Focus*, Summer 1988, 10–14; M. D. SAHLINS, *Islands of History* (Oceania) (Chicago, 1985); B. SMITH, *European Vision and the South Pacific*, 2nd ed. (Yale, 1985); O. H. K. SPATE, *Paradise Found and Lost*: Vol. III of Spate, *The Pacific Since Magellan* (Minnesota, 1988); D. R. STODDART, "Catastrophic Human Interference with Coral Island Ecosystems," *Geography*, 53 (1968), 25–40; P. THEROUX, *The Happy Isles of Oceania: Paddling the Pacific* (Putnam, 1992); S. D. THOMAS, "The Puzzle of Micronesian Navigation," *Pacific Discovery*, November-December 1982, 1–12; R. TRUMBULL, *Tin Roofs and Palm Trees: A Report on the New South Seas* (Washington, 1977); H. J. WIENS, *Atoll Environment and Ecology* (Yale, 1962).

Australia

B. ANSELM, "Australians Exploit Their Iron Mountains," *Geographical Magazine*, 53 (1981), 712–718; *Atlas of Australian Resources* (3rd series; issued serially from 1980 by the Australian Division of National Mapping, Canberra); *Australian Geographer* (semiannual); *Australian Geographical Studies* (semiannual); J. S. BEARD, "Some Vegetation Types of Tropical Australia in Relation to Those of Africa and America," *Journal of Ecology*, 55 (1967), 271–290; J. BIRD, *Seaport Gateways of Australia* (Oxford, 1968); B. BLAINEY, *The Tyranny of Distance: How Distance Shaped Australia's History* (St. Martin's, 1968); R. BRITTON, "Australia's Resources Boom," *Focus*, March-April 1982, 1–8; D. J. CARR and S. G. M. CARR, eds., *Plants and Man in Australia* (Sydney: Academic Press, 1981); B. CHALKLEY and H. WINCHESTER, "Australia in Transition," *Geography*, 76 (1991) 97–108; P. P. COURTENAY, *Northern Australia: Patterns and Problems of Tropical Development in an Advanced Country* (Melbourne: Longman Cheshire, 1982); D. CORKE, "Riddles in the Sand" (Australian exploration), *Geographical Magazine*, December 1991, 36–41; P. CRABB, "Managing Australia's Water Resources," *Australian Geographical Studies*, 20 (1982), 96–107; B. D. DAVIDSON, *European Farming in Australia: An Economic History of Australian Farming* (Elsevier, 1981); R. FREESTONE, "Urban Australia: Postwar to Postindustrial," *Focus*, March-April 1982, 9–14; R. HALL, "An Australian Country Townscape," *Landscape* 24, no. 2 (1980), 41–48; R. L. HEATHCOATE and B. G. THOM, eds., *Natural Hazards in Australia* (Canberra: Australian Academy of Science, 1979); J. H. HOLMES, ed., "Queensland: A Geographical Interpretation," *Queensland Geographical Journal*, 4th Series, 1 (1986), 343 pp.; R. HUGHES, *The Fatal Shore: The Epic of Australia's Founding* (Knopf, 1986); D. N. JEANS, ed., *Australia: A Geography, Volume One: The Natural Environment* (Oxford, 1989); T. LANGFORD-SMITH, "New Perspectives on the Australian Deserts," *Australian Geographer*, 15 (1983), 269–284; E. LINACRE and J. HOBBS, *The Australian Climatic Environment* (Wiley, 1977); M. J. LOEFFLER, "Australian-American Interbasin Water Transfer," *Annals AAG*, 60 (1970), 493–516; R. E. LONSDALE and J. H. HOLMES, eds., *Settlement Systems in Sparsely Populated Regions: The United States and Australia* (Oxford, 1981); G. MANNERS, "Unresolved Conflicts in Australian Mineral and Energy Resource Policies," *Geographical Journal*, 158 (1992), 129–144; J. W. McCARTY and C. B. SCHEDVIN, eds., *Australian Capital Cities: Historical Essays* (Sydney, 1978); T. L. McKNIGHT, *Australia's Corner of the World: A Geographical Summation* (Prentice-Hall, 1970); D. J. MULVANEY, "The Prehistory of the Australian Aborigine," *Scientific American*, 214, no. 3 (March 1966), 84–93; D. PARKES, ed., *Northern Australia: The Arenas of Life and Ecosystems on Half a Continent* (Academic, 1984); J. M. POWELL and M. WILLIAMS, eds., *Australian Space, Australian Time: Geographical Perspectives* (Oxford, 1975); *Reader's Digest Atlas of Australia* (Sydney, Australia: Reader's Digest Services, 1978); W. H. RICHMOND and P. C. SHARMA, eds., *Mining and Australia* (Queensland, 1983); G. SIVIOUR, "A New Railway Age in Australia?" *Geography*, 68 (1983), 331–335; T. SORENSON and H. WEINAND, "Regional Well-Being in Australia Revisited," *Australian Geographical Studies*, 29 (1991), 42–70; O. H. K. SPATE, *Australia* (Praeger, 1968); E. STOKES, "Into the Ghastly Blank" (early exploration of Australia), *Geographical Magazine*, November 1991, 6–11; P. STRATHAM, ed., *The Origins of Australia's Capital Cities* (Cambridge, 1989); J. TAYLOR, "Anchorage and Darwin—A Tale of Two (Sister) Cities," *Geography*, 76 (1991), 151–154; M. J. TAYLOR and N. THRIFT, "Large Corporations and Concentrations of Capital in Australia: A Geographical Analysis," *Economic Geography*, 56 (1980), 261–280; J. VESSELS, "The Simpson Outback" (ranching on the desert's edge), *National Geographic*, April 1992, 64–92.

New Zealand

K. B. CUMBERLAND and J. S. WHITELAW, *New Zealand* (Aldine, 1970; "The World's Landscapes"); E. M. K. DOUGLAS, "The Maori," *Pacific Viewpoint*, 20 (1979), 103–109; G. R. HAWKE, *The Making of New Zealand: An Economic History* (Cambridge, 1985); J. I. KELLY, "A Population Cartogram of New Zealand," *New Zealand Journal of Geography*, (October 1985), 7–11; "New Zealand in the 1980s: Market Forces in the Welfare State," *Pacific Viewpoint*, 32, no. 2 (October 1991), entire issue; *New Zealand Geographer* (semiannual); R. P. WILLIS, "Farming in New Zealand and the E. E. C.—The Case of the Dairy Industry," *New Zealand Geographer*, 40 (1984), 3–11.

Part 6: Africa South of the Sahara

Africa: General, Historical, Cultural, Political

Among general, political, and cultural references are *Africa News: The Newspaper of African Affairs* (biweekly); "Africa, 1992", *Current History*, May 1992, entire issue; articles on U. S.–African relations, Zambia, South Africa, AIDS, Kenya, Zimbabwe, Sahel, Somalia; see also the comparable annual issues in earlier years; M. CHEGE, "Remembering Africa," *Foreign Affairs*, 71 (1992), 146–163; S. DENYER, *African Traditional Architecture: An Historical and Geographical Perspective* (Heinemann/Holmes & Meier, 1978); T. FORREST, "Brazil and Africa: Geopolitics, Trade, and Technology in the South Atlantic," *African Affairs*, 81 (1982), 3–20; I. L. L. GRIFFITHS, *An Atlas of African Affairs* (Methuen,

1984); A. T. GROVE, *The Changing Geography of Africa* (Oxford, 1989); T. A. HALE, "Africa and the West: Close Encounters of the Literary Kind," *Comparative Literature Studies*, 20 (1983), 261–275; R. H. JACKSON and C. G. ROSBERG, "Why Africa's Weak States Persist: The Empirical and the Juridical in Statehood," *World Politics*, 35 (1982), 1–24; D. LAMB, *The Africans* (Random House, 1983; A wide-ranging report by a journalist); L. MORROW, "Africa: The Scramble for Existence," *Time*, September 7, 1992, 40–46; J. MURRAY, ed., *Cultural Atlas of Africa* (Facts on File, 1982); A. O'CONNOR, *The African City* (Hutchinson, 1983); R. OLIVER and M. CROWDER, eds., *The Cambridge Encyclopedia of Africa* (Cambridge, 1981; a major one-volume reference); R. M. PRESS, "Africa's Turn" (movement toward democracy), *World Monitor*, February 1992, 36–43; E. RANSDELL, "Africa's Trek to Freedom," *U. S. News & World Report*, August 10, 1992, 28–31; P. RICHARDS, "Spatial Organization As a Theme in African Studies," *Progress in Human Geography*, 8 (1984), 551–561; K. SOMERVILLE, *Foreign Military Intervention in Africa* (London: Pinter, 1990); R. VAN CHI-BONNARDEL, Director, *The Atlas of Africa* (Paris: Jeune Afrique/New York: Hippocrene, 1973); C. WINTERS, "Urban Morphogenesis in Francophone Black Africa," *Geographical Review*, 72 (1982), 139–154.

Historical references include: J. F. AJAYI and M. CROWDER, *Historical Atlas of Africa* (Cambridge, 1985); A. J. CHRISTOPHER, *Colonial Africa* (Rowman & Littlefield/Barnes & Noble, 1984); also, "Continuity and Change of African Capitals, *Geographical Review*, 75 (1985), 44–57; M. CROWDER, ed., *The Cambridge History of Africa*, vol. 8: *c. 1940 to c. 1975* (Cambridge, 1984); B. DAVIDSON, *Africa in History: Themes and Outlines*, 4th ed., rev. (Collier, 1991); J. J. EWALD, "Slavery in Africa and the Slave Trades from Africa" (a review article), *American Historical Review*, 97 (1992), 465–485; G. S. P. FREEMAN-GRENVILLE, *The New Atlas of African History* (Simon & Schuster, 1991); C. HIBBERT, *Africa Explored: Europeans in the Dark Continent* 1769;nn1889 (Allen Lane, 1982); A. M. JOSEPHY, ed., *The Horizon History of Africa*, 2 vols. (American Heritage Publishing Co, 1971); A. MOOREHEAD, *The White Nile* (1961), *The Blue Nile* (1962) (Harper & Row); R. OLIVER, *The African Experience* (HarperCollins, 1991); also, with J. D. FAGE, *A Short History of Africa*, rev. ed. (NYU, 1963); C. PALMER, "African Slave Trade: The Cruelest Commerce," *National Geographic*, September 1992, 62–91; T. PAKENHAM, *The Scramble for Africa: The White Man's Conquest of the Dark Continent from 1876 to 1912* (Random House, 1991); T. RANGER, "White Presence and Power in Africa," *Journal of African History*, 20 (1979), 463–469; T. SEVERIN, *The African Adventure: Four Hundred Years of Exploration in the 'Dangerous Continent'* (Dutton, 1973); "The Scramble for Africa," in *The Colonial Overlords: TimeFrame AD 1850–1990* (Time-Life Books, 1990), ch. 3, 74–102.

Africa: Environment, Population, Economy, Development

A selection of general, environmental and climatic references includes: W. M. ADAMS and F. M. R. HUGHES, "The Environmental Effects of Dam Construction in Tropical Africa: Impacts and Planning Procedures" (features Kenya and Nigeria), *Geoforum*, 17 (1986), 403–410; C. AGNEW and T. O'CONNOR, "The Meteorological Scapegoat" (drought), *Geographical Magazine*, January 1991: "Geographical Analysis," 1–4; D. and J. BARTLETT, "Africa's Skeleton Coast" (Namib Desert), *National Geographic*, January 1992, 54–85; D. BOURN, "Cattle, Rainfall and Tsetse in Africa" (features Ethiopia), *Journal of Arid Environments*, 1 (1978), 49–61; A. BRADSTOCK, "Elephant Saviours in Ivory Towers" (world ivory trade and conservation of elephants), *Geographical Magazine*, December 1990, 14–17; M. CHEATER, "Death of the Ivory Market" (Zimbabwe), *World Watch*, May-June 1991, 34–35; K. CLEAVER and G. SCHREIBER, "Population, Agriculture, and the Environment in Africa," *Finance and Development*, June 1992, 34–35; A. T. GROVE, "Desertification," *Progress in Physical Geography*, 1 (1977), 296–310; L. A. LEWIS and L. BERRY, *African Environments and Resources* (Boston, MA: Unwin Hyman, 1988); R. D. MANN, "Time Running Out: The Urgent Need for Tree Planting in Africa," *The Ecologist*, March/April 1990, 48–53; S. E. NICHOLSON, "Climatic Variations in the Sahel and Other African Regions During the Past Five Centuries," *Journal of Arid Environments*, 1 (1987), 3–24; P. RICHARDS, "The Environmental Factor in African Studies," *Progress in Human Geography*, 4 (1980), 589–600; C. STAGER, "Africa's Great Rift," *National Geographic*, May 1990, 2–14.

On Africa's population, economy, and development, see "Africa: The Economic Challenge," *Finance and Development*, December 1991 (several articles by different authors); D. R. ALTSCHUL, "Transportation in African Development," *Journal of Geography*, 79 (1980), 44–56; A. AKINBODE, "Population Explosion in Africa and Its Implications for Economic Development," *Journal of Geography*, 76 (1977), 28–36; J. BAKER, "Oil and African Development," *Journal of Modern African Studies*, 15 (1977), 175–212; C. K. EICHER, "Facing Up to Africa's Food Crisis," *Foreign Affairs*, 61 (1982), 151–174; J. G. GALATY and P. BONTE, eds., *Herders, Warriors, and Traders: Pastoralism in Africa* (Westview, 1991); M. GRIFFIN, "Dinka and Their Cattle Defy Time," *Geographical Magazine*, 53 (1981), 760–765; A. A. MAZRUI, "The Economic Woman in Africa," *Finance and Development*, June 1992, 42–43; A. MOUNTJOY and D. HILLING, *Africa: Geography and Development* (Rowman & Littlefield/Barnes & Noble, 1987); N. R. L. MWASE, "Role of Transport in Rural Development in Africa," *Impact of Science on Society*, no. 162 (1991), 137–148; A. O'CONNOR, *Poverty in Africa: A Geographical Approach* (Columbia, 1991); R. SANDBROOK and J. BARKER, *Politics of Africa's Economic Stagnation* (Cambridge, 1985); also, "The State and Economic Stagnation in Tropical Africa," *World Development*, 14 (1986), 319–332; D. WESTERN and V. FINCH, "Cattle and Pastoralism: Survival and Production in Arid Lands," *Human Ecology*, 14 (1986), 77–94.

On African diseases, see "AIDS: How Other Nations Suffer and Cope," *World Press Review*, January 1992, 9–14; A. D. CLIFF and M. R. SMALLMAN-RAYNOR, "The AIDS Pandemic: Global Geographical Patterns and Local Spatial Processes," *Geographical Journal*, 158 (1992), 182–198; J. GIBLIN, "Trypanosomiasis Control in African History: An Evaded Issue," *Journal of African History*, 31 (1990), 59–80; H. KLOOS and K. THOMPSON, "Schistosomiasis in Africa: An Ecological Perspective," *Journal of Tropical Geography*, 48 (1979), 31–46; A. LARSON, "The Social Epidemiology of Africa's AIDS Epidemic," *African Affairs*, 89, no. 354 (January 1990), 5–25; R. TURLEY, "Malaria: Worldwide Search for Solutions," *Geographical Magazine*, February 1990, 22–27.

West Africa

C. AGNEW, "Spatial Aspects of Drought in the Sahel," *Journal of Arid Environments*, 18 (1990), 279–293; A. ARECCHI, "Dakar," *Cities*, 2 (1985), 186–197; A. ARMSTRONG, "Ivory Coast: Another New Capital for Africa," *Geography*, 70 (1985), 72–74; M. BARBOUR *et al.*, eds., *Nigeria in Maps* (Holmes & Meier, 1982); T. J. BASSETT, "Fulani Herd Movements," *Geographical Review*, 76 (1986), 233–248; L. C. BECKER, "The Collapse of the Family Farm in West Africa? Evidence from Mali," *Geographical Journal*, 156 (1990), 313–322; B. A. CHOKOR, "Ibadan," *Cities*, 3 (1986), 106–116; R. J. H. CHURCH, *West Africa: A Study of the Environment and Man's Use of It*, 8th ed. (Longman, 1980; a major geographical text); A. DE SHERBININ, "Nigeria," *Population Today*, July/August 1992, 12; BUCHI EMECHETA, *The Bride Price* (1976) and *The Slave Girl* (1977) (Braziller; short novels of Nigerian life by a Nigerian novelist); E. O. ETEJERE and R. B. BHAT, "Traditional Preparation and Uses of Cassava in Nigeria," *Economic Botany*, 39 (1985), 157–164; J. A. GRITZNER, *The West African Sahel: Human Agency and Environmental Change* (University of Chicago Geography Research Paper 226, 1988); A. T. GROVE, *The Niger and Its Neighbours: Environmental History and Hydrobiology, Human Use and Health Hazards of the Major West African Rivers* (Balkema, 1985); C. HAUB, "Nigerian Census Surprises Experts," *Population Today*, June 1992, 3; D. HILLING, "The Evolution of a Port System: The Case of Ghana," *Geography*, 62 (1977), 97–105; A. M. HOWARD, "The Relevance of Spatial Analysis for African Economic History: The Sierra Leone–Guinea System," *Journal of African History*, 17 (1976), 365–388; A. JONES and M. JOHNSON, "Slaves from the Windward Coast" (West Africa), *Journal of African History*, 21 (1980), 17–34; J. G. LOCKWOOD, "The Causes of Drought with Particular Reference to the Sahel," *Progress in Physical Geography*, 10 (1986), 111–119; M. MORTIMORE, *Adapting to Drought: Farmers, Famines and Desertification in West Africa* (Cambridge, 1989); J. C. NWAFOR, "The Relocation of Nigeria's Federal Capital: A Device for Greater Territorial Integration and National Unity," *GeoJournal*, 4 (1980), 359–366; J. K. ONOH, *The Nigerian Oil Economy: From Prosperity to Glut* (Croom Helm, 1983); J. O. C. ONYEMELUKWE and M. O. FILANI, *Economic Geography of West Africa* (Longman, 1983); J. J. PARSONS, "The Canary Islands Search for Stability," *Focus*, April 1985, 22–29; P. RICHARDS, "Farming Systems and Agrarian Change in West Africa," *Progress in Human Geography*, 7 (1983), 1–39; S. ROYLE, "St. Helena: A Geographical Summary," *Geography*, 76 (1991), 266–268; R. D. STERN *et al.*, "The Start of the Rains in West Africa," *Journal of Climatology*, 1 (1981), 59–68; D. E. VERMEER, "Collision of Climate, Cattle, and Culture in Mauritania During the 1970s," *Geographical Review*, 71 (1981), 281–297; M. J. WATTS and T. J. BASSETT, "Politics, the State, and Agrarian Development: A Comparative Study of Nigeria and the Ivory Coast," *Political Geography Quarterly*, 5 (1986), 103–125; G. WENDLER and F. EATON, "On the Desertification of the Sahel Zone," *Climatic Change*, 5 (1983), 365–380.

Equatorial Africa

G. BRUNOLD, "São Tomé e Príncipe: The Way Things Are," *Swiss Review of World Affairs*, June 1992, 23–24; A. COWELL, "Mobutu's Zaire: Magic and Decay," *New York Times Magazine*, April 5, 1992, 30–38; B. EMERSON, *Leopold II of the Belgians: King of Colonialism* (Weidenfeld & Nicolson, 1979); P. FORBATH, *The River Congo* (Secher & Warburg, 1978); N. GORDIMER, "The Congo River," in *The Essential Gesture* (Knopf, 1988); R. W. HARMS, *River of Wealth, River of Sorrow: The Central Zaire Basin in the Era of the Slave and Ivory Trade, 1500–1891* (Yale, 1981); G. C. KABWIT, "Zaire: The Roots of Continuing Crisis," *Journal of Modern African Studies*, 17 (1979), 381–407; J. MADELEY, "Cameroon Grows Its Own," *Geographical Magazine*, 59 (1987), 296–300; V. S. NAIPAUL, *A Bend in the River* (Knopf, 1979; a short, classic novel); E. RANSDELL, "In Zaire, A Big Man Still Rules the Roost," *U. S. News & World Report*, August 10, 1992, 31–34; A. SHOUMATOFF, *In Southern Light: Trekking Through Zaire and the Amazon* (Simon & Schuster, 1986); T. O'TOOLE, *The Central African Republic: The Continent's Hidden Heart* (Westview, 1986); H. WINTERNITZ, *East Along the Equator* (Zaire) (Atlantic Monthly Press, 1987); J. WITTE, "Deforestation in Zaire: Logging and Landlessness," *The Ecologist*, 22 (1992), 58–64; C. YOUNG and T. TURNER, *The Rise and Decline of the Zairian State* (Wisconsin, 1985); "Zaire," *Population Today*, March 1992, 12.

East Africa

R. H. BATES, "The Agrarian Origins of Mau Mau," *Agricultural History*, 61, no. 1 (Winter 1987), 1–28; N. HARMAN, "East Africa: Turning the Corner," *The Economist*, June 20, 1987, 3–18; B. S. HOYLE, *Seaports and Development: The Experience of Kenya and Tanzania* (Gordon & Breach, 1983); E. HUXLEY, *The Flame Trees of Thika: Memories of an African Childhood* (1959), illustrated edition (Weidenfeld & Nicolson, 1987; made into a memorable TV series); J. J. JORGENSEN, *Uganda: A Modern History* (St. Martin's, 1981); R. B. MABELE *et al.*, "The Economic Development of Tanzania," *Scientific American*, 243, no. 3 (September 1980), 182–190; P. J. MARTIN, "The Zanzibar Clove Industry," *Economic Botany*, 45 (1991), 450–459; S. H. OMINDE, ed., *Population and Development in Kenya* (Heinemann, 1984); "Out of Africa" (feature film and video, with spectacular views of Kenya; based on the book by Isak Dinesen).

South Central Africa

P. S. FALK, "Cuba in Africa," *Foreign Affairs*, 65 (1987), 1077–1096; R. HARDY, *Rivers of Darkness* (Putnam, 1979; a gripping novel of tropical medicine and guerrilla warfare in Mozambique); E. P. SCOTT, "Development Through Self-Reliance in Zambia," *Journal of Geography*, 84 (1985), 282–290; J. SHEPHERD, "Zimbabwe: Poised on the Brink," *Atlantic Monthly*, July 1987, 26–31; D. WEINER *et al.*, "Land Use and Agricultural Productivity in Zimbabwe," *Journal of Modern African Studies*, 23 (1985), 251–285; A. WOOD, "When the Bottom Goes Out of Copper" (Zambia), *Geographical Magazine*, 56 (1984), 16–20.

Southern Africa

H. ADAM and K. MOODLEY, "Negotiations About What in South Africa?" *Journal of Modern African Studies*, 27 (1989), 367–381; K. S. O. BEAVON, "Trekking On: Recent Trends in the Human Geography of Southern Africa," *Progress in Human Geography*, 5 (1981), 159–189; A. BRINK, *Writing in a State of Siege* (South Africa) (Simon & Schuster, 1983); A. CHRISTEN, "South Africa:

From Apartheid to Class Conflict?" *Swiss Review of World Affairs*, August 1991, 29–30; A. J. CHRISTOPHER, *South Africa* (Longman, 1982: "The World's Landscapes"); also, "Southern Africa and the United States: A Comparison of Pastoral Frontiers," *Journal of the West*, 20 (1981), 52–59; also *South Africa: The Impact of Past Geographies* (Cape Town: Juta & Co., 1984); B. A. COX and C. M. ROGERSON, "The Corporate Power Elite in South Africa: Interlocking Directorships Among Large Enterprises," *Political Geography Quarterly*, 4 (1985), 219–234; D. CRUSH and J. CRUSH, "A State of Dependence" (Lesotho), *Geographical Magazine*, 55 (1983), 24–29; J. CRUSH and P. WELLINGS, "The Southern African Pleasure Periphery, 1966–83," *Journal of Modern African Studies*, 21 (1983), 673–698; J. DRUMMOND, "Reincorporating the Bantustans into South Africa: The Question of Bophuthatswana," *Geography*, 76 (1991), 369–373; A. DU TOIT, "No Chosen People: The Myth of the Calvinist Origins of Afrikaner Nationalism and Racial Ideology," *American Historical Review*, 88 (1983), 920–952; P. FUHRMAN, "Harry Oppenheimer, African Empire Builder, Is Smiling Again," *Forbes*, September 16, 1991, 130–138; D. HART, "A Literary Geography of Soweto," *GeoJournal*, 12 (1986), 191–195; also, with G. H. PIRIE, "The Sight and Soul of Sophiatown," *Geographical Review*, 74 (1984), 38–47; R. F. HASWELL, "South African Towns on European Plans," *Geographical Magazine*, 51 (1979), 686–694; H. LAMAR and L. THOMPSON, eds., *The Frontier in History: North America and Southern Africa Compared* (Yale, 1981); J. LELYVELD, *Move Your Shadow: South Africa Black and White* (Times Books, 1985); A. LEMON, "Toward the New South Africa," *Journal of Geography*, 90 (1991), 254–263; P. LOPEZ, "Landscapes Open and Closed: A Journey Through Southern Africa," *Harper's Magazine*, July 1987, 51–58; JAMES A. MICHENER, *The Covenant* (Random House, 1980; a vast panoramic novel of South African history; a tremendous store of other fine fiction about South Africa has been created by such able writers as Peter Abrahams, André Brink [see *A Dry White Season*, also a feature film and video], J. M. Coetzee, Nadine Gordimer, Alex La Guma, Doris Lessing, Alan Paton, and many others); S. MUFSON, "South Africa 1990," *Foreign Affairs*, 70, no. 1 (1991), 120–141; G. H. PIRIE, "The Decivilizing Rails: Railways and Underdevelopment in South Africa," *Tijdschrift voor Economische en Sociale Geografie*, 73 (1982), 221–228; P. POOVALINGAN, "The Indians of South Africa: A Century on the Defensive," *Optima*, 28 (1979), 66–91; *Reader's Digest Atlas of Southern Africa* (Cape Town: Reader's Digest, 1984); R. I. ROTBERG *et al.*, *South Africa and Its Neighbors: Regional Security and Self-Interest* (Lexington, 1985); S. P. RULE, "South African Emigration to Australia: Who and Why?" *The South African Geographer*, 17 (1989/1990), 66–75; A. SAMPSON, *Black and Gold* (Pantheon, 1987; a British journalist's view of South Africa); D. M. SMITH, *Apartheid in South Africa* (Cambridge, 1985); also, "Conflict in South African Cities," *Geography*, 72 (1987), 153–158; also, ed., *The Apartheid City and Beyond: Urbanization and Social Change in South Africa* (Routledge, 1992); W. SMITH, "Cape Coloureds Who Survive in Limbo," *Geographical Magazine*, 53 (1980), 188–195; *South African Geographical Journal* (annual); "Southern Africa," *Geographical Magazine*, 58–59 (1986–1987; 13 articles by various authors); A. SPARKS, *The Mind of South Africa* (Heinemann, 1990); W. R. STANLEY, "A Third Port for Southwest Africa/Namibia?" *GeoJournal*, 22 (1990), 363–378; E. STERN, "Competition and Location in the Gaming Industry: The 'Casino States' of Southern Africa," *Geography*, 72 (1987), 140–150; STUDY COMMISSION ON U. S. POLICY TOWARD SOUTHERN AFRICA, *South Africa: Time Running Out* (California, 1981); M. O. SUTCLIFFE, "The Crisis in South Africa: Material Conditions and the Reformist Response," *Geoforum*, 17 (1986), 141–159; L. THOMPSON, *A History of South Africa* (Yale, 1990); B. WALLACH, "South Africa: Seeing for One's Self," *Focus*, Part I, Spring 1990, 21–26; Part II, Summer 1990, 30–35; Part III, Fall 1990, 27–32; G. G. WEIGEND, "Economic Activity Patterns in White Namibia," *Geographical Review*, 75 (1985), 462–81; also, "German Settlement Patterns in Namibia," *ibid.*, 75 (1985), 156–169; P. WELLINGS and A. BLACK, "Industrial Decentralization Under Apartheid: The Relocation of Industry to the South African Periphery," *World Development*, 14 (1986), 1–38; J. WESTERN, *Outcast Cape Town* (Minnesota, 1981; Allen & Unwin, 1982); also, "Undoing the Colonial City" (Cape Town and Tianjin), *Geographical Review*, 75 (1985), 335–357; also, "South African Cities: A Social Geography," *Journal of Geography*, 85 (1986), 249–255; A. ZICH, "Botswana: The Adopted Land," *National Geographic*, December 1990, 70–96.

Indian Ocean Islands

D. W. GADE, "Madagascar and Nondevelopment Culture," *Focus*, October 1985, 14–21; M. GRIFFIN, "The Perfumed Isles" (Comoros), *Geographical Magazine*, 58 (1986), 524–527; A. JOLLY, *A World Like Our Own: Man and Nature in Madagascar* (Yale, 1980); P. LENOIR, "An Extreme Example of Pluralism: Mauritius," *Cultures*, 6 (1979) 63–82; R. TURLEY, "Madagascar," *Geographical Magazine*, 56 (1984), 28–33.

Part 7: Latin America

Latin America: General

J. P. AUGELLI, "Food, Population and Dislocation in Latin America," *Journal of Geography*, 84 (1985), 274–281; M. BARKE, "Different Lifestyle, Same Old Struggle" (women in Latin America), *Geographical Magazine*, October 1990, 4–7; W. L. BERNECKER and H. W. TOBLER, "North and South America: A Comparison," *Swiss Review of World Affairs*, September 1992, 26–27; H. BLAKEMORE and C. T. SMITH, eds., *Latin America: Geographical Perspectives*, 2nd ed. (Methuen, 1983); R. G. BOEHM and S. VISSER, eds., *Latin America: Case Studies* (Kendall/Hunt, 1984); M. BRAWER, *Atlas of South America* (Simon & Schuster, 1991); R. D. F. BROMLEY and R. BROMLEY, *South American Development: A Geographical Introduction* (Cambridge, 1982); S. CORBRIDGE, "Cheap Bananas and Global Capitalism," *Journal of Historical Geography*, 17 (1991), 204–217; J. A. CROW, *The Epic of Latin America*, 3rd ed. (California, 1980); P. D. CURTIN, *The Rise and Fall of the Plantation Complex: Essays in Atlantic History* (Cambridge, 1990); M. DEAS *et al.*, *Latin America in Perspective* (Houghton Mifflin, 1991); S. GREENBLATT, *Marvelous Possessions: The Wonder of the New World* (Chicago, 1991); E. GRIFFIN and L. FORD, "A Model of Latin American City Structure," *Geographical Review*, 70 (1980), 397–422; A. HENNESSY, *The Frontier in Latin American History* (Edward Arnold, 1978); L. A. HOBERMAN and S. M. SOCOLOW, eds., *Cities and Society in Colonial Latin America*

(New Mexico, 1986); J. HUNTER, "The Status of Cacao in the Western Hemisphere," *Economic Botany*, 44 (1990), 425–439; P. E. JAMES and C. W. MINKEL, *Latin America*, 5th ed. (Wiley, 1986; a major textbook); H. S. KLEIN, *African Slavery in Latin America and the Caribbean* (Oxford, 1986); "Latin America," *Current History*, February 1992, entire issue (articles on U.S.–Latin American relations, Mexico, Cuba, Haiti, Central American revolutions, Andrean cocaine, Argentina, Brazil); A. F. LOWENTHAL, "Rediscovering Latin America," *Foreign Affairs*, Fall 1990, 27–41; J. O. MAOS, *The Spatial Organization of New Land Settlement in Latin America* (Westview, 1984); A. S. MORRIS, *South America*, 3rd ed. (Rowman & Littlefield/Barnes & Noble, 1987; a geography); J. H. PARRY, *The Discovery of South America* (Taplinger, 1979); R. S. PLATT, *Latin America: Countrysides and United Regions* (McGraw-Hill, 1942; a classic book based on extended field research); D. A. PRESTON, "Views of Rural Latin America," *Progress in Human Geography*, 4 (1980), 601–610; C. E. REBORATTI, "Human Geography in Latin America," *Progress in Human Geography*, 6 (1982), 397–407; "South America," *Current History*, February 1991, entire issue; "Spain Rediscovers the New World," *The Economist*, July 30, 1988, 17–20.

Caribbean America: General

T. D. ANDERSON, *Geopolitics of the Caribbean: Ministates in a Wider World* (Praeger, 1984); C. N. CAVIEDES, "Five Hundred Years of Hurricanes in the Caribbean: Their Relationship with Global Climatic Variabilities," *GeoJournal*, 23 (1991), 301–310; M. E. CRAHAN, ed., *Africa and the Caribbean: The Legacies of a Link* (Johns Hopkins, 1979); M. W. HELMS, *Middle America: A Culture History of Heartland and Frontier* (Prentice-Hall, 1975); K. R. HOPE, *Economic Development in the Caribbean* (Praeger, 1986); F. W. KNIGHT, *The Caribbean: The Genesis of a Fragmented Nationalism*, 2nd ed. (Oxford, 1990); R. A. PIELKE, *The Hurricane* (Routledge, 1990); B. C. RICHARDSON, *The Caribbean in the Wider World, 1492;nn1992: A Regional Geography* (Cambridge, 1992); C. O. SAUER, *The Early Spanish Main* (California, 1969); R. WAUCHOPE, gen. ed., *Handbook of Middle American Indians* (Texas, 1964–1976), especially vol. 1: R. C. WEST, ed., *Natural Environment and Early Cultures* (1964); R. C. WEST et al., *Middle America: Its Lands and Peoples*, 3rd ed. (Prentice-Hall, 1989; a standard text.)

Mexico

D. D. ARREOLA, "Nineteenth-Century Townscapes of Eastern Mexico," *Geographical Review*, 72 (1982), 1–19; M. BARKE, "Mérida, Yucatán: A Core Within the Periphery," *Scottish Geographical Magazine*, 100 (1984), 160–170; D. M. BRAND, *Mexico, Land of Sunshine and Shadow* (Van Nostrand, 1966); L. B. CASAGRANDE, "The Five Nations of Mexico," *Focus*, Spring 1987, 2–9; P. G. CASANOVA, "The Economic Development of Mexico," *Scientific American*, 243, no. 3 (September 1980), 192–204; J. H. COATSWORTH, "Indispensable Railroads in a Backward Economy: The Case of Mexico," *Journal of Economic History*, 39 (1979), 939–960; W. E. DOOLITTLE, "Aboriginal Agricultural Development in the Valley of Sonora, Mexico," *Geographical Review*, 70 (1980), 328–342; also, "Agricultural Expansion in a Marginal Area of Mexico, *ibid.*, 73 (1983), 301–313; M. and K. FELDSTEIN, "Mexico's Maestro" (recent economic progress), *World Monitor*, July 1992, 42–49; R. A. FERNANDEZ, *The Mexican-American Border Region: Issues and Trends* (Notre Dame, 1989); J. S. HENDERSON, *The World of the Ancient Maya* (Cornell, 1981); L. A. HERZOG, *Where North Meets South: Cities, Space, and Politics on the U.S.–Mexico Border* (Texas, 1990); C. C. LOCKWOOD, *The Yucatán Peninsula* (Louisiana State, 1989); B. McDOWELL, "Mexico City: An Alarming Giant," *National Geographic*, 166 (1984), 138–172; N. J. PERRY, "What's Powering Mexico's Success," *Fortune*, February 10, 1992, 109–115; J. B. PICK et al., *Atlas of Mexico* (Westview, 1989); S. K. PURCELL, "Mexico's New Economic Vitality," *Current History*, February 1992, 54–58; A. RIDING, *Distant Neighbors: A Portrait of the Mexicans* (Knopf, 1985); R. A. SANCHEZ, "Oil Boom, A Blessing for Mexico?" *GeoJournal*, 7 (1983), 229–245; S. SANDERSON, *Land Reform in Mexico: 1910–1980* (Academic, 1984); E. R. STODDARD, R. L. NOSTRAND, and J. P. WEST, eds., *Borderlands Sourcebook: A Guide to the Literature on Northern Mexico and the American Southwest* (Oklahoma, 1983); L. SUAREZ-VILLA, "The Manufacturing Process Cycle and the Industrialization of the United States–Mexico Borderlands," *Annals of Regional Science*, 18 (1984), 1–23.

Central America

J. P. AUGELLI, "Costa Rica's Frontier Legacy," *Geographical Review*, 77 (1987), 1–16; also, "The Panama Canal Area," *Focus*, Spring 1986, 20–29; E. CATER, "Profits from Paradise" ("ecotourism" in Belize), *Geographical Magazine*, March 1992, 16–21; M. COLCHESTER, "Guatemala: The Clamour for Land and the Fate of the Forests," *The Ecologist*, 21 (1991), 177–185; S. L. DRIEVER, "Insurgency in Guatemala: Centuries-Old Conflicts Over Land and Social Inequality Spawn Guerrilla Movements and Hope for Democratic Change," *Focus*, 1985, 2–9; S. FRENKEL, "Geography, Empire, and Environmental Determinism" (illustrated by the Panama Canal Zone), *Geographical Review*, 82 (1992), 143–153; P. GAUPP, "Ecology and Development in the Tropics" (Costa Rica), *Swiss Review of World Affairs*, September 1992, 14–19; C. HALL, *Costa Rica: A Geographical Interpretation in Historical Perspective* (Westview, 1985); W. LaFEBER, *The Panama Canal: The Crisis in Historical Perspective*, updated ed. (Oxford, 1989); T. SAGAWE, "Deforestation and the Behaviour of Households in the Dominican Republic," *Geography*, 76 (1991), 304–314; O. TICKELL, "Uncle Sam Goes Bananas" (Guatemala), *Geographical Magazine*, September 1991, 22–26; R. L. WOODWARD, JR., *Central America: A Nation Divided*, 2nd ed. (Oxford, 1985).

West Indies and Guianas

J. P. AUGELLI, "Nationalization of Dominican Borderlands," *Geographical Review*, 70 (1980), 19–35; *Caribbean Geography* (semiannual); R. CHARDON, "Sugar Plantations in the Dominican Republic," *Geographical Review*, 74 (1984), 441–454; C. E. COBB, JR., "Jamaica: Hard Times, High Hopes," *National Geographic*, 167, no. 1 (January 1985), 114–140; L. A. EYRE, "The Ghettoization of an Island Paradise" (Jamaica), *Journal of Geography*, 82 (1983), 236–239; also "Political Violence and Urban Geography in Kingston, Jamaica," *Geographical Review*, 74 (1984), 24–37; M. FERGUS, "The Turks and Caicos Islands: A Geographical Note," *Geography*, 76 (1990), 66–67; B. FLOYD, "Agricultural Reform in Castro's Cuba," *Geographical Magazine*

50 (1978), 808–815; also, *Jamaica: An Island Microcosm* (St. Martin's, 1979); D. LEWIS and C. WOOD, "French Guiana: Cayman a l'Orange" (environmental opportunities and problems), *Geographical Magazine*, June 1991, 17–20; D. LOWENTHAL, *West Indian Societies* (Oxford, 1972); J. NEWHOUSE, "Socialism or Death" (Cuba), *The New Yorker*, April 27, 1992, 52–83; R. PICÓ, *The Geography of Puerto Rico* (Aldine, 1974); R. B. POTTER, "Tourism and Development: The Case of Barbados, West Indies," *Geography*, 68 (1983), 46–50; S. K. PURCELL, "Collapsing Cuba," *Foreign Affairs*, 71 (1992), 130–145; F. REDMILL, "The Independent Way for St. Kitts," *Geographical Magazine*, 55 (1983), 636–640; B. RICHARDS, "The Uncertain State of Puerto Rico," *National Geographic*, 163 (1983), 516–543; R. J. TATA, *Haiti: Land of Poverty* (University Press of America, 1982); B. WARF, "Anguilla," *Focus*, Fall 1991, 22; B. WALLACH, "Puerto Rico: Growth, Change, Progress, Development," *Focus*, Summer 1989, 27–33; D. WATTS, *The West Indies: Patterns of Development, Culture and Environmental Change Since 1492* (Cambridge, 1987); P. T. WHITE, "Cuba at a Crossroads," *National Geographic, August 1991, 90–121.*

Andean Countries

C. J. ALLEN, "To Be Quechua: The Symbolism of Coca Chewing in Highland Peru," *American Ethnologist*, 8 (1981), 157–171; R. BROMLEY, "The Colonization of Humid Tropical Areas in Ecuador," *Singapore Journal of Tropical Geography*, 2 (1981), 15–26; E. BURCHARD, "Coca Chewing and Diet," *Current Anthropology*, 33, no. 1 (February 1992), 1–24; C. N. CAVIEDES, "El Niño, 1982;nn83," *Geographical Review*, 74 (1984), 267–290; S. L. CUTTER and C. L. TORO, "Colombia: 'Miami Vice' or Terra Incognitae?" *Focus*, Spring 1990, 28–32; D. W. GADE, "Inca and Colonial Settlement, Coca Cultivation and Endemic Disease in the Tropical Forest," *Journal of Historical Geography*, 5 (1979), 263–279; G. A. GEYER, "Democracy Betrayed" (Venezuela), *World Monitor*, September 1992; M. HIRAOKA and S. YAMAMOTO, "Agricultural Development in the Upper Amazon of Ecuador," *Geographical Review*, 70 (1980), 423–445; G. KNAPP, *Andean Ecology: Adaptive Dynamics in Ecuador* (Westview, 1991); P. N. KNOX, "El Niño—A Current Catastrophe," *Earth*, vol. 1 (September 1992), 30–37; E. MORALES, *Cocaine: White Gold Rush in Peru* (Arizona, 1989); J. S. OTTO and N. E. ANDERSON, "Cattle Ranching in the Venezuelan Llanos and the Florida Flatwoods: A Problem in Comparative History," *Comparative Studies in Society and History*, 28 (1986), 672–683; J. J. PARSONS, "Geography as Exploration and Discovery" (Columbia), *Annals AAG*, 67 (1977), 1–16; R. B. SOUTH, "Coca in Bolivia," *Geographical Review*, 67 (1977), 22–33; S. STRONG, "Where the Shining Path Leads" (Maoist guerillas in Peru), *New York Times Magazine*, May 24, 1992, 12–18, 35; C. WEIL, "Migration Among Landholdings by Bolivian Campesinos," *Geographical Review*, 73 (1983), 182–197.

Brazil

J. L. ALEXANDER, "South of São Paulo," *Geographical Magazine*, 57 (1985), 598–600; A. BOTELHO, "Brazil's Independent Computer Strategy," *Technology Review*, May-June 1987, 36–45; J. P. DICKENSON, *Brazil* (Longman, 1982: "The World's Landscapes"); M. EAKIN, "Creating a Growth Pole: The Industrialization of Belo Horizonte, *Brazil*, 1897–1987," *The Americas*, 47

(1991), 383–410; P. FORESTER, "Capital of Dreams" (Brasília), *Geographical Magazine*, 58 (1986), 462–467; J. GREENWOOD, "Riches and Debt in Venezuela," *Geographical Magazine*, 56 (1984), 463–469; P. GREINIER, "The Alcohol Plan and the Development of Northeast Brazil," *GeoJournal*, 11 (1985), 61–68; R. HARVEY, "Clumsy Giant: A Survey of Brazil," *The Economist*, April 25, 1987, 26 pp.; F. I. JOHNSON, "Sugar in Brazil: Policy and Production," *Journal of Developing Areas*, 17 (1983), 243–256; A. J. R. RUSSELL-WOOD, *The Black Man in Slavery and Freedom in Colonial Brazil* (St. Martin's, 1982); M. Y. UNE, "An Analysis of the Effects of Frost on the Principal Coffee Areas of Brazil," *GeoJournal*, 6 (1982), 129–140; L. N. WILLMORE, "The Comparative Performance of Foreign and Domestic Films in Brazil," *World Development*, 14 (1986), 489–502; C. H. WOOD and J. M. DE CARVALHO, *The Democracy of Inequality in Brazil* (Cambridge, 1988).

Southern Mid-Latitude Countries

S. DIBBLE, "Paraguay: Plotting a New Course," *National Geographic*, August 1992, 88–113; J. KING, "Civilization and Barbarism: The Impact of Europe on Argentina," *History Today*, 34 (1984), 16–21; P. McGRATH, "Paraguayan Powerhouse," *Geographical Magazine*, 55 (1983), 192–197; "The Mission" (historical feature film and video, with overpowering evocation of the Iguassu Falls); W. WEISCHET, "Climatic Constraints for the Development of the Far South of Latin America," *GeoJournal*, 11 (1985), 79–87.

Amazonia

"Amazonia: A World Resource at Risk" (map), *National Geographic* separate, August 1992; A. B. ANDERSON, "Smokestacks in the Rainforest: Industrial Development and Deforestation in the Amazon Basin," *World Development*, 18 (1990), 1191–1205; D. L. CLAWSON, "Obstacles to Successful Highlander Colonization of the Amazon and Orinoco Basins," *American Journal of Economics and Sociology*, 41 (1982), 351–362; P. A. COLINVAUX, "The Past and Future Amazon," *Scientific American*, May 1989, 102–108; A. COWELL, *The Decade of Destruction: The Crusade to Save the Amazon Rain Forest* (New York: Henry Holt, 1990); P. CUNNINGHAM, "The Other Side of the Coin" (gold mining), *Geographical Magazine*, September 1990, 30–34; R. E. DICKINSON, *The Geophysiology of Amazonia: Vegetation and Climate Interactions* (Wiley, 1987); P. M. FEARNSIDE, *Human Carrying Capacity of the Brazilian Rainforest* (Columbia, 1986); R. A. FORESTA, "Amazonia and the Politics of Geopolitics," *Geographical Review*, 82 (1992), 128–142; J. GODFREY, "Boom Towns of the Amazon," *Geographical Review*, 80 (1990), 103–117; R. GRIBEL, "The Balbina Disaster: The Need to Ask Why?" (gigantic hydro project in Amazonia), *The Ecologist*, 20 (1990), 133–135; R. B. HAMES and W. T. VICKERS, eds., *Adaptive Responses of Native Amazonians* (Academic, 1983); J. HEMMING, ed., *Change in the Amazon Basin*, vol. 1, *Man's Impact on Forests and Rivers*; vol. 2, *The Frontier After a Decade of Colonization* (Manchester, 1985); also, "Invaded by Gold-Diggers," *Geographical Magazine*, May 1990, 26–30; M. HIRAOKA, "The Development of Amazonia," *Geographical Review*, 72 (1982), 94–98; C. F. JORDAN, "Amazon Rain Forests" *American Scientist*, 70 (1982), 394–401; J. KIRBY, "Agricultural Land-Use and the Settlement of Amazonia," *Pacific Viewpoint*, 17 (1976),

105–132; G. MONBIOT, "Tips for the Trees" (logging in Amazonia), *Geographical Magazine*, November 1991, 22–25; E. F. MORAN, ed., *The Dilemma of Amazonian Development* (Westview, 1983); M. SCHMINK, "Land Conflicts in Amazonia," *American Ethnologist*, 9 (1982), 341–357; also, with C. H. WOOD, eds., *Frontier Expansion in Amazonia* (Florida, 1984); N. J. H. SMITH, *Rainforest Corridor: The Transamazon Colonization Scheme* California, 1982); "The Maracá Rainforest Project" (Brazil), *Geographical Journal*, 156 (1990), 249–296; B. WALLACH, "Manaus," *Focus*, January 1985, 8–11; E. C. WOLF, "Survival of the Rarest" (with world map of rain forest "hotspots"), *World Watch*, March-April 1991, 12–20.

Part 8: North America

Anglo-America: General

W. W. ATWOOD, *The Physiographic Provinces of North America* (Ginn, 1940); B. BAILYN, *The Peopling of British North America: An Introduction* (Knopf, 1986); S. S. BIRDSALL and J. W. FLORIN, *Regional Landscapes of the United States and Canada*, 4th ed. (Wiley, 1991; a geography text); R. A. BURCHELL, ed., *The End of Anglo-America: Historical Essays in the Study of Cultural Divergence* (Manchester, 1991); J. GARREAU, *The Nine Nations of North America* (Houghton Mifflin, 1981; a book for general readers); W. E. GARRET, ed., *Atlas of North America: Space Age Portrait of a Continent* (National Geographic Society, 1985); J. R. GIBSON, *Imperial Russia in Frontier America: The Changing Geography of Supply of Russian America, 1784–1867* (Oxford, 1976); C. B. HUNT, *Natural Regions of the United States and Canada*, rev. ed., (Freeman, 1974); R. D. MITCHELL and P. A. GROVES, eds., *North America: The Historical Geography of Changing Continent* (Rowman & Littlefield, 1987); A. G. NOBLE, *Wood, Brick and Stone: The North American Settlement Landscape* (Massachusetts, 1984); also, ed., *To Build in a New Land: Ethnic Landscapes in North America* (Johns Hopkins, 1992); J. H. PATERSON, *North America: A Geography of the United States and Canada*, 8th ed. (Oxford, 1989); K. B. RAITZ, "Ethnic Maps of North America," *Geographical Review*, 68 (1978), 335–350; J. R. ROONEY, JR., W. ZELINSKY, and D. R. LOUDER, eds., *This Remarkable Continent: An Atlas of United States and Canadian Society and Culture* (Texas A&M, 1982); V. E. SHELFORD, *The Ecology of North America* (Illinois, 1963); M. J. TROUGHTON, "Industrialization of U. S. and Canadian Agriculture," *Journal of Geography*, 84 (1985), 255–263; J. L. VANKAT, *The Natural Vegetation of North America: An Introduction* (Wiley, 1979); C. L. WHITE, E. J. FOSCUE, and T. L. McKNIGHT, *Regional Geography of Anglo-America*, 6th ed. (Prentice-Hall, 1985); W. ZELINSKY, "North America's Vernacular Regions," *Annals AAG*, 70 (1980), 1–16.

Canada: General

"A Survey of Canada," *The Economist*, October 8, 1988, 1–18; "Canada," *Current History*, December 1991, seven articles by different authors, 401–437; "Canada: A Special Issue: (articles and bibliography by numerous authors), *Journal of Geography*, 83, no. 5 (September-October 1984); *Canadian Geographer* (quarterly; a scholarly journal of high quality); *Canadian Geographic* (bimonthly; a well-written magazine for general readers); S. R.

GRAUBARD, ed., *In Search of Canada* (New Brunswick: Transaction Publishers, 1989); C. HARRIS, "Presidential Address: The Pattern of Early Canada," *Canadian Geographer*, 31 (1987), 290–298; also, ed., *Historical Atlas of Canada: From the Beginning to 1800* (Toronto, 1988); and, with J. WARKENTIN, *Canada Before Confederation: A Study in Historical Geography* (Carleton, 1991); D. KEMP, "The Greenhouse Effect and Global Warming: A Canadian Perspective," *Geography*, 76 (1991), 121–130; L. D. McCANN, ed., *Heartland and Hinterland: A Geography of Canada* (Prentice-Hall of Canada, 1982); K. McNAUGHT, *The Penguin History of Canada* (Penguin Books, 1988); J. MORRIS, *O Canada: Travels in an Unknown Country* (HarperCollins, 1992); P. C. NEWMAN, "Three Centuries of the Hudson's Bay Company: Canada's Fur-Trading Empire," *National Geographic*, 172 (1987), 192–229; J. L. ROBINSON, *Concepts and Themes in the Regional Geography of Canada* (Vancouver: Talonbooks, 1983); M. C. STORRIE and C. I. JACKSON, "Canadian Environments," *Geographical Review*, 62 (1972), 309–332; J. WARKENTIN, ed., *Canada: A Geographical Interpretation* (Methuen, 1968; essays, mainly historical, by 23 geographers).

Canada; Ethnicity; Political Geography; French Canada

D. CARTWRIGHT, "Changes in the Patterns of Contact Between Anglophones and Francophones in Quebec," *GeoJournal*, 8 (1984), 109–122; K. J. CROWE, "Why the New Names for Eskimos and Indians?" *Canadian Geographic*, 99, no. 1 (August-September 1979), 68–71; J. L. ELLIOTT, ed., *Two Nations, Many Cultures: Ethnic Groups in Canada* (Prentice-Hall of Canada, 1979); "For Want of Glue: A Survey of Canada," *The Economist*, June 29, 1991, 18 pp.; R. C. HARRIS, *The Seigneurial System in Early Canada: A Geographical Study* (Wisconsin, 1966); also, "Brief Interlude with New France," *Geographical Magazine*, 52 (1980), 274–280; V. KONRAD, "Recurrent Symbols of Nationalism in Canada," *Canadian Geographer*, 30 (1986), 176–180; J. A. LAPONCE, "The French Language in Quebec: Tensions Between Geography and Politics," *Political Geography Quarterly*, 3 (1984), 91–104; P. S. LI, ed., *Race and Ethnic Relations in Canada* (Oxford, 1990); A. H. MALCOLM, *The Canadians* (Times Books, 1985); K. McROBERTS, "Canada and Quebec at a Crossroads," *1992 World Book Year Book*, 113–124; M. RICHTER, *Oh Canada! Oh Quebec! Requiem for a Divided Country* (Knopf, 1992); A. L. SANGUIN, "The Quebec Question and the Political Geography of Canada," *GeoJournal*, 8 (1984), 99–107; D. E. SMITH, "Empire, Crown and Canadian Federalism," *Canadian Journal of Political Science*, 24 (1991), 451–473; G. TOMBS, "Canada Wry," *World Monitor*, January 1991, 58–62; G. WYNN, "Ethnic Migrations and Atlantic Canada: Geographical Perspectives," *Canadian Ethnic Studies*, 18 (1986), 1–15; W. ZELINSKY, "A Sidelong Glance at Canadian Nationalism and Its Symbols," *North American Culture*, 4 (1988), 2–27 (extensive list of references).

Canada; Economy; Cities

M. BARLOW and B. SLACK, "International Cities: Some Geographical Considerations and a Case Study of Montreal," *Geoforum*, 16 (1985), 333–345; A. BLACKBOURN and R. G. PUTNAM, *The Industrial Geography of Canada* (Croom Helm, 1984); R. BOULDING, "Forestry in Canada," *American Forests*, July 1979, 38–43; M. F. FOX, "Regional Changes in Canadian

Agriculture," *Geography*, 71 (1986), 67–70; M. A. GOLDBERG and J. MERCER, *The Myth of the North American City: Continentalism Challenged* (British Columbia, 1986); D. G. HAGLUND, ed., *The New Geopolitics of Minerals: Canada and International Resource Trade* (British Columbia, 1989); G. A. NADER, *Cities of Canada*, 2 vols. (Macmillan of Canada: vol. I, 1975; vol. II, 1976); H. PEARCE, "The Flooding of a Nation" (large Canadian dams and hydro projects), *Geographical Magazine*, November 1991, 18–21; L. A. SANDBERG, "Geographers' Perception of Canada in the World Economic Order," *Progress in Human Geography*, 13, 2 (June 1989), 157–175; G. A. STELTER and A. F. J. ARTIBISE, eds., *Power and Place: Canadian Urban Development in the North American Context* (British Columbia, 1986).

Canada: Provinces and Regions

F. BRUEMMER, "Churchill: Polar Bear Capital of the World," *Canadian Geographic*, 103, no. 6 (December 1983–January 1984), 20–27; W. BURGESS, "Recent Mining Developments in the Canadian Arctic," *Geography*, 68 (1983), 50–53; M. CLAYTON, "Canada's Natives Exercise New Clout on National Scene," *Christian Science Monitor*, July 3, 1992, 10–11; D. EISLER, "Harvests of Ruin: Debt and Low Grain Prices Are Destroying Saskatchewan's Rural Way of Life," *Canadian Geographic*, January/February 1992, 32–43; R. GEORGE, "More Energy for B.C. in Peace River Coal," *Canadian Geographic*, 100, no. 1 (February-March 1980), 26–33; W. HAMLEY, "Tourism in the Northwest Territories," *Geographical Review*, 81 (1991), 389–399; S. J. HORNSBY, "Staple Trades, Subsistence Agriculture, and Nineteenth-Century Cape Breton Island, *Annals AAG*, 79 (1989), 410–434; G. W. LEAHY, "Quebec: Our City of World Heritage Renown," *Canadian Geographic*, 106, no. 6 (December 1986–January 1987), 8–21; J. LEMON, "Toronto," *Cities*, 8 (1991), 258–266; P. MARCHAK, *Green Gold: The Forest Industry in British Columbia* (British Columbia, 1983); H. MILLWARD, "The Development, Decline, and Revival of Mining on the Sydney Coalfield," *Canadian Geographer*, 28 (1984), 180–185; P. C. NEWMAN, "The Beaver and the Bay: From Dreams of Cathay Came an Empire of Furs," *Canadian Geographic*, August-September 1989, 56–64; D. PELLY, "In the Face of Adversity" (Canada's Inuit), *Geographical Magazine*, January 1990, 16–18; J. R. ROGGE, ed., *Developing the Subarctic, Manitoba Geographical Studies*, 1 (1973); J. SCHREINER, "The Port of Vancouver," *Canadian Geographic*, August-September 1987, 10–21; P. M. SLOWE, "The Geography of Borderlands: The Case of the Quebec-US Borderlands," *Geographical Journal*, 157 (1991), 191–198; E. STRUZIK, "Sister Cities North of Sixty— Yellowknife and Whitehorse: Twin Capitals, but Cast in Quite Separate Moulds," *Canadian Geographic*, 106 (1986), 66–71; "The Canada Series" (McGraw-Hill Ryerson: short books on Canada's provinces, 1979); L. TROTIER, gen. ed., *Studies in Canadian Geography* (Toronto, 1972; vol. 1: A. MACPHERSON, ed., *The Atlantic Provinces*; vol 2: F. GRENIER, ed., *Quebec*; vol. 3: R. L. GENTILCORE, ed., *Ontario*; vol. 4: P. J. SMITH, ed., *The Prairie Provinces*; vol. 5: J. L. ROBINSON, ed., *British Columbia*; vol. 6: W. C. Wonders, ed., *The North*); K. TWITCHELL, "The Not-So-Pristine Arctic," *Canadian Geographic*, February-March 1991, 53–60; G. WYNN, "A Province Too Much Dependent on New England" (Nova Scotia), *Canadian Geographer*, 31 (1987), 98–113.

United States: General

J. AGNEW, *The United States in the World Economy: A Regional Geography* (Cambridge, 1987); "America and the World 1990/91," *Foreign Affairs*, 70, no. 1 (1991), numerous articles on world politics and major regions; M. J. BOWDEN, "The Invention of American Tradition," *Journal of Historical Geography*, 18 (1992), 3–26; C. D. HARRIS, *Bibliography of Geography, Part II: Regional*, vol. 1, *The United States of America* (University of Chicago, Department of Geography Research Paper 206, 1984); J. F. HART, ed., *Regions of the United States* (Harper & Row, 1972, published originally as a special issue, *Annals AAG*, 62, no. 2, June 1972; regional essays by geographers, of which the following are especially pertinent to this text: R. W. Durrenberger, "The Colorado Plateau"; J. F. Hart, "The Middle West"; E. C. Mather, "The American Great Plains"; D. W. Meinig, "American West: Preface to a Geographical Introduction"; J. E. Vance, Jr., "California and the Search for the Ideal"); P. L. KNOX *et al.*, *The United States: A Contemporary Human Geography* (Wiley, 1988); "Land of the Eagle" (TV series; videos); K. O. MORGAN, *State Rankings 1992: A Statistical View of the 50 United States* (Lawrence, KS, Morgan Quitno Corporation, 1992); "State Population Estimates, 1991 and Change Since the 1990 Census," *Population Today*, April 1992, 10; U. S. BUREAU OF THE CENSUS, *County and City Data Book* (Washington, DC: Government Printing Office, 1988 and every 5th year); also, *State and Metropolitan Area Data Book 1991* (Government Printing Office); U. S. GEOLOGICAL SURVEY, *The National Atlas of the United States of America* (Washington, DC: 1970); N. WEMMERUS, "People Patterns: Population Gain and Loss in U.S. Counties 1980–1990," *Population Today*, May 1991, 10; J. H. WHEELER, "U.S.A.," chap. 8 (pp. 131–178) in S. GODDARD, ed., *A Guide to Information Sources in the Geographical Sciences* (CromHelm/Barnes & Noble, 1983; chap. 8 is an extensive bibliography of books and articles on the United States through 1981); W. ZELINSKY, *Nation Into State: The Shifting Symbolic Foundations of American Nationalism* (North Carolina, 1988).

United States: Historical Geography

T. H. BREEN, "An Empire of Goods: The Anglicization of Colonial America, 1690–1776," *Journal of British Studies*, 25 (1986), 467–499; R. H. BROWN, *Historical Geography of the United States* (Harcourt Brace, 1948; still a major work); M. CONZEN, ed., *The Making of the American Landscape* (London: Unwin Hyman, 1990); C. V. EARLE, "The First English Towns of North America," *Geographical Review*, 67 (1977), 34–50; "Fashioning the American Landscape," *Geographical Magazine*, 52–53 (1979–1980; 12 articles by various authors); "1491: America Before Columbus," *National Geographic*, October 1991, 2–99; T. G. JORDAN, "Preadaptation and European Colonization in Rural North America," *Annals AAG*, 79 (1989), 489–500; J. KAY, "Landscapes of Men and Women: Rethinking the Regional Historical Geography of the United States and Canada," *Journal of Historical Geography*, 17 (1991), 435–452; J. T. LEMON, "Early Americans and Their Social Environment," *Journal of Historical Geography*, 6 (1980), 115–131; L. LORD and S. BURKE, "America Before Columbus," *U. S. News & World Report*, July 8, 1991, 22–37; D. W. MEINIG, *The Shaping of America: A Geographical Perspective on 500 Years of American History*, vol. 1, *Atlantic America, 1492–1800* (Yale, 1986; a major work of historical geography); also, "The Historical Geography Imperative," *Annals AAG*, 79 (1989), 79–87; NATIONAL GEOGRAPHIC SOCIETY, *Historical Atlas of the United States* (1988); D. H. UBELAKER, "North American Census, 1492" (weighing the evidence for population numbers), *Pacific Discovery*, Winter 1992, 32–35; D. WARD, ed., *Geographic Perspectives on America's Past: Readings on the Historical Geography of the United States* (Oxford, 1979).

United States: Natural Environment
R. G. BAILEY, *Ecoregions of the United States* (map, 1:7,500,000, U. S. Forest Service, 1976); also, comp., *Description of the Ecoregions of the United States* (U. S. Forest Service, 1978); N. M. FENNEMAN, *Physiography of Western United States* (McGraw-Hill, 1931); also, *Physiography of Eastern United States* (McGraw-Hill, 1938); G. J. GRAY and A. ENG, "How Much Old-Growth is Left?" *American Forests*, September/October, 1991, 46–48; R. J. MASON and M. T. MATTSON, *Atlas of United States Environmental Issues* (Macmillan, 1990); M. WILLIAMS, *Americans and Their Forests: A Historical Geography* (Cambridge, 1989).

United States: Cultural Geography (General)
B. BIGELOW, "Roots and Regions: A Summary Definition of the Cultural Geography of America," *Journal of Geography*, 79 (1980), 218–229; J. CONRON, ed., *The American Landscape: A Critical Anthology of Prose and Poetry* (Oxford, 1974); M. CONZEN, "What Makes the American Landscape," *Geographical Magazine*, 53 (1980), 36–41; R. D. GASTIL, *Cultural Regions of the United States* (Washington, 1975); J. F. HART, "The Bypass Strip as an Ideal Landscape," *Geographical Review*, 72 (1982), 218–223; P. LEWIS, "Learning from Looking: Geographic and Other Writing About the American Cultural Landscape," *American Quarterly*, 35 (1983), 242–261; D. B. LUTEN, *Progress Against Growth: On the American Landscape* (Guilford Press, 1986); J. R. SHORTRIDGE, "The Concept of the Place-Defining Novel in American Popular Culture," *Professional Geographer*, 43 (1991), 280–291; extensive reference list; also, "Changing Usage of Four American Regional Labels," *Annals AAG*, 77 (1987), 325–336; G. R. STEWART, *Names on the Land: A Historical Account of Placenaming in the United States*, 4th ed. (Lexikos, 1982); also, *A Concise Dictionary of American Place Names* (Oxford, 1986); W. ZELINSKY, "The Changing Face of Nationalism in the American Landscape," *Canadian Geographer*, 30 (1986), 171–175; also, *The Cultural Geography of the United States* (Prentice-Hall, 1973).

United States: Ethnicity
D. D. ARREOLA, "Urban Mexican Americans" *Focus*, January-February 1984, 7–11; T. D. BOSWELL and T. C. JONES, "A Regionalization of Mexican Americans in the United States," *Geographical Review*, 70 (1980), 88–98; T. D. BOSWELL and M. RIVERS, "Cubans in America: A Minority Group Comes of Age," *Focus*, April 1985, 2–9; W. K. CROWLEY, "Old Order Amish Settlement: Diffusion and Growth," *Annals AAG*, 68 (1978), 294–364; R. DANIELS, *Coming to America: A History of Immigration and Ethnicity in American Life* (HarperCollins, 1990); J. A. DUNLEVY, "On the Settlement Patterns of Recent Caribbean and Latin Immigrants to the United States," *Growth and Change*, 22, no. 1 (Winter 1991), 54–67; D. K. FELLOWS, *A Mosaic of America's Ethnic Minorities* (Wiley, 1972); D. H. FISHER, *Albion's Seed: Four British Folkways in America* (dispersion pattern and effects of British culture-groups) (Oxford, 1989); L. H. GANN and P. J. DUIGNAN, *The Hispanic in the United States: A History* (Westview, 1986); K. E. McHUGH, "Black Migration Reversal in the United States," *Geographical Review*, 77 (1987), 171–182; K. B. RAITZ, "Themes in the Cultural Geography of European Ethnic Groups in the United States," *Geographical Review*, 69 (1979), 45–54; J. R. SHORTRIDGE, "Patterns of Religion in the United States," *Geographical Review*, 66 (1976), 420–434; D. WARD, "The Ethnic Ghetto in the United States: Past and Present," *Transactions IBG*, n.s., 7 (1982), 257–275.

United States: General Economic Geography; Industrial Geography
J. R. BORCHERT, "Major Control Points in American Economic Geography," *Annals AAG*, 68 (1978), 214–232; M. J. BROADWAY and T. WARD, "Recent Changes in the Structure and Location of the U.S. Meatpacking Industry," *Geography*, 76 (1990), 76–79; D. CLARK, *Post-Industrial America: A Geographical Perspective* (Methuen, 1985); S. S. COHEN and J. ZYSMAN, "The Myth of a Post-Industrial Economy." *Technology Review*, 55–62; R. A. ERICKSON and D. J. HAYWARD, "The International Flows of Exports from U. S. Regions," *Annals AAG*, 81 (1991), 371–390; C. FLAVIN, "Conquering U. S. Oil Dependence," *World Watch*, January-February 1991, 28–35; J. F. HART, "Small Towns and Manufacturing," *Geographical Review*, 78 (1988), 272–287; S. P. HUNTINGTON, "The U.S.—Decline or Renewal?" *Foreign Affairs*, 67, no. 2 (Winter 1988/89), 76–96; L. KRAAR, "Japan's Gung-Ho U.S. Car Plants," *Fortune*, January 30, 1989, 98–108; M. MAGNET, "The Resurrection of the Rust Belt," *Fortune*, August 15, 1988, 40–46; D. R. MEYER, "Emergence of the American Manufacturing Belt: An Interpretation," *Journal of Historical Geography*, 9 (1983), 145–174; P. MILLER, "Our Electric Future" (nuclear power), *National Geographic*, August 1991, 60–89; S. NASAR, "America's Competitive Revival," *Fortune*, January 4, 1988, 44–52; J. M. RUBENSTEIN, "The Changing Distribution of U.S. Automobile Plants" *Focus*, Fall 1988, 12–17; J. SZEKELEY, "Can Advanced Technology Save the U.S. Steel Industry?" *Scientific American*, 257 (1987), 34–41; "The Rhine and the Ohio: A Tale of Two Rivers," *The Economist*, May 21, 1988, 21–24; B. L. WEINSTEIN and R. E. FIRESTINE, *Regional Growth and Decline in the United States: The Rise of the Sunbelt and the Decline of the Northeast* (Praeger, 1978).

United States: Agriculture and Land Use
L. R. BROWN, "The Growing Grain Gap (decline in world grain reserves), *World Watch*, September-October 1988, 10–18; W. EBELING, *The Fruited Plain: The Story of American Agriculture* (California, 1979); W. W. COCHRANE, *The Development of American Agriculture: A Historical Analysis* (Minnesota, 1979); C. EARLE and R. HOFFMAN, "The Foundation of the Modern Economy: Agriculture and the Costs of Labor in the United States and England, 1800–60," *American Historical Review*, 85 (1980), 1055–1094; B. GIBBONS, "Do We Treat Our Soil Like Dirt?" *National Geographic*, 166 (1984), 350–389; J. F. HART, *The Land That Feeds Us* (Norton, 1991); also, "Nonfarm Farms," *Geographical Review*, 82 (1992), 166–179; also "The Persistence of Family Farming Areas," *Journal of Geography*, 86 (1987), 198–203; R. H. JACKSON, *Land Use in America* (V. H. Winston, 1981); "Revolution on the Farm: The Plow Is Being Replaced by New Techniques

That Protect the Land and Promise Even More Abundant Crops," *Time*, June 29, 1992, 54–56; E. G. SMITH, JR., "America's Richest Farms and Ranches," *Annals AAG*, 70 (1980), 528–541; I. VOGELER, *The Myth of the Family Farm: Agribusiness Dominance of U. S. Agriculture* (Westview, 1981).

United States: Urbanization

C. ABBOTT, *The New Urban America: Growth and Politics in Sunbelt Cities* (North Carolina, 1987); R. F. ABLER, J. S. ADAMS, and K. S. LEE, eds., *A Comparative Atlas of America's Great Cities: Twenty Metropolitan Regions* (Minnesota, 1976); J. S. ADAMS, ed., *Contemporary Metropolitan America*, 4 vols.: vol. 1, *Cities of the Nation's Historic Metropolitan Core*; vol. 2, *Nineteenth Century Ports*; vol. 3, *Nineteenth Century Inland Cities and Ports*; vol. 4, *Twentieth Century Cities* (Ballinger, 1976); J. BORCHERT, "American Metropolitan Evolution," *Geographical Review*, 57 (1967), 301–332, a classic article of enduring value; also, "Instability in American Metropolitan Growth," *ibid.*, 73 (1983), 127–149; C. E. BROWNING, "The Rise of the Beltways: A Powerful Force for Urban Change," *Focus*, Summer 1990, 18–22; L. R. FORD, "Reading the Skylines of American Cities," *Geographical Review*, 82 (1992), 180–200; J. GARREAU, *Edge City: Life on the New Frontier* (Doubleday, 1991); J. F. HART, ed., *Our Changing Cities* (Johns Hopkins, 1991); R. PALM, *The Geography of American Cities* (Oxford, 1981).

U. S. Northeast

M. M. BELL, "Did New England Go Downhill?" *Geographical Review*, 79 (1989), 450–466; J. E. BENHART and M. E. DUNLOP, "The Iron and Steel Industry of Pennsylvania: Spatial Change and Economic Evolution," *Journal of Geography*, 88 (1989), 173–183; M. J. BOWDEN, "Invented Tradition and Academic Convention in Geographic Thought about New England," *GeoJournal*, 26 (1992), 187–194; K. R. BOWLING, *The Creation of Washington, D.C.: The Idea and Location of the American Capital* (George Mason University Press, 1991); W. CRONON, *Changes in the Land: Indians, Colonists, and the Ecology of New England* (Hill & Wang, 1983); R. A. CYBRIWSKY and T. A. REINER, "Philadelphia in Transition," *Focus*, September-October 1982, 1–16; J. E. DILISIO, *Maryland: A Geography* (Westview, 1983); C. V. EARLE, *The Evolution of a Tidewater Settlement System: All Hallow's Parish, Maryland, 1650;nn1783* (University of Chicago, Department of Geography Research Paper 170, 1975); R. EVANS, "The Big Apple Turns Rotten" (New York), *Geographical Magazine*, February 1992, 10–14; F. FEGLEY, "Plain Pennsylvanians Who Keep Their Faith" (Pennsylvania Germans), *Geographical Magazine*, 53 (1981), 968–975; J. V. FIFER, "Washington, D.C.: The Political Geography of a Federal Capital," *Journal of American Studies*, 15 (1981), 5–26; J. GOTTMAN, *Megalopolis: The Urbanized Northeastern Seaboard of the United States* (Twentieth Century Fund, 1961); G. HALVERSON, "The Real New York, Please Stand Up," *Christian Science Monitor*, July 15, 1992, 10–11; C. L. HEYRMAN, *Commerce and Culture: The Maritime Communities of Colonial Massachusetts, 1690–1750* (Norton, 1984); E. C. HIGBEE, "The Three Earths of New England," *Geographical Review*, 42 (1952), 425–438; J. T. LEMON, *The Best Poor Man's Country: A Geographical Study of Early Southeastern Pennsylvania* (Johns Hopkins, 1972); T. R. LEWIS and J. E. HARMON, *Connecticut: A Geography* (Westview, 1986); D. R. McMA-

NIS, *Colonial New England: A Historical Geography* (Oxford, 1975); B. MARSH, "Continuity and Decline in the Anthracite Towns of Pennsylvania," *Annals AAG*, 77 (1987), 337–352; S. E. MORISON, *The Maritime History of Massachusetts, 1783–860* (Houghton Mifflin, 1921); R. PILLSBURY, "The Pennsylvania Culture Area: A Reappraisal," *North American Culture*, 3, no. 2 (1987), 37–54; extensive reference list on United States cultural geography; H. S. RUSSELL, *A Long, Deep Furrow: Three Centuries of Farming in New England* (University Press of New England, 1976), G. SHANKLAND, "Boston—The Unlikely City," *Geographical Magazine*, 53 (1981), 323–327; T. STEINBERG, *Nature Incorporated: Industrialization and the Waters of New England* (Cambridge, 1991); J. L. SWERDLOW, "Erie Canal: Living Link to Our Past," *National Geographic*, November 1990, 38–65; J. H. THOMPSON, ed., *Geography of New York State*, 2nd ed. (Syracuse, 1977); B. WARF, "Japanese Investments in the New York Metropolitan Region," *Geographical Review*, 78 (1988), 257–271; J. O. WHEELER, "Corporate Role of New York City in the Metropolitan Hierarchy," *Geographical Review*, 80 (1990), 370–381; J. S. WOOD and M. A. STEINITZ, "A World We Have Gained: House, Common, and Village in New England," *Journal of Historical Geography*, 18 (1992), 105–120; W. ZELINSKY, "The Pennsylvania Town: An Overdue Geographical Account," *Geographical Review*, 67 (1977), 127–147.

U. S. South: Agriculture and Rural Life

C. S. Aiken, "New Settlement Pattern of Rural Blacks in the American South," *Geographical Review*, 75 (1985), 383–404; also, "A New Type of Black Ghetto in the Plantation South, *Annals AAG*, 80 (1990), 223–246; P. A. COCLANIS, "Bitter Harvest: The South Carolina Low Country in Historical Perspective," *Journal of Economic History*, 45 (1985), 251–259; A. E. COWDREY, *This Land, This South: An Environmental History* (Kentucky, 1983); P. DANIEL, "The Crossroads of Change: Cotton, Tobacco and Rice Cultures in the Twentieth-Century South," *Journal of Southern History*, 50 (1984), 429–456; also, "The Transformation of the Rural South: 1930 to the Present," *Agricultural History*, 55 (1981), 231–248; C. V. EARLE, "A Staple Interpretation of Slavery and Free Labor," *Geographical Review*, 68 (1978), 51–65; J. F. HART, "Cropland Concentrations in the South," *Annals AAG*, 68 (1978), 505–517; also, "Land Use Change in a Piedmont County," *Annals AAG*, 70 (1980), 492–527; also, "The Demise of King Cotton," *Annals AAG*, 67 (1977), 307–322; also, with E. L. CHESTANG, "Bright Tobacco: A Photo-Essay," *Focus*, Winter 1991, 1–7; T. G. JORDAN, *Trails to Texas: Southern Roots of Western Cattle Ranching* (Nebraska, 1981); I. R. MANNERS "The Persistent Problem of the Boll Weevil: Pest Control in Principle and in Practice," *Geographical Review*, 69 (1979), 25–42; W. A. SCHROEDER, E. S. MUNGER, and D. R. POWARS, "Sickle Cell Anaemia, Genetic Variations, and the Slave Trade to the United States," *Journal of African History*, 31 (1990), 163–180.

U. S. South: Cities and Industries

B. A. BROWNWELL and D. R. GOLDFIELD, eds., *The City in Southern History: The Growth of Urban Civilization in the South* (Kennikat, 1977); C. E. COBB, JR., "Miami," *National Geographic*, January 1992, 86–113; J. R. FEAGIN, "The Global Context of Metropolitan Growth: Houston and the Oil Industry," *American Journal of Sociology*, 90 (1985), 1204–1230; D. R.

GOLDFIELD, "The Urban South: A Regional Framework," *American Historical Review*, 86 (1981), 1009–1034; M. L. JOHNSON, "Postwar Industrial Development in the Southeast and the Pioneer Role of Labor-Intensive Industry," *Economic Geography*, 61 (1985), 46–65; L. H. LARSEN, *The Rise of the Urban South* (Kentucky, 1985).

U. S. South: People and Culture

C. S. AIKEN, "Faulkner's Yoknapatawpha County: A Place in the American South," *Geographical Review*, 69 (1979), 331–348; also "Faulkner's Yoknapatawpha County: Geographical Fact into Fiction," *ibid.*, 67 (1977), 1–21; D. P. ARREOLA, "The Mexican American Cultural Capital" (San Antonio, Texas), *Geographical Review*, 77 (1987), 17–34; R. ARSENAULT, "The End of the Long Hot Summer: The Air Conditioner and Southern Culture," *Journal of Southern History*, 50 (1984), 597–628; W. J. CASH, *The Mind of the South* (Knopf, 1941); P. D. ESCOTT and D. R. GOLDFIELD, eds., *The South For New Southerners* (North Carolina, 1991); D. W. MEINIG, *Imperial Texas: An Interpretive Essay in Cultural Geography* (Texas, 1969); J. P. RADFORD, "Identity and Tradition in the Post-Civil War South," *Journal of Historical Geography*, 18 (1992), 91–103; D. RIEFF, "The Second Havana" (Miami), *The New Yorker*, May 18, 1987, 65–83.

U. S. South: States and Regions

M. J. BRADSHAW, "Public Policy in Appalachia: The Application of a Neglected Geographical Factor?" *Transactions IBG*, n.s., 10 (1985), 385–400; also, "TVA at Fifty," *Geography*, 69 (1984), 209–220; J. W. CLAY et al., *Land of the South* (text, maps, photos, references) (Birmingham, AL: Oxmoor House, 1989); A. B. CRUICKSHANK, "Development of the Deep South: A Reappraisal," *Scottish Geographical Magazine*, 96 (1980), 91–104; J. F. HART, "Land Rotation in Appalachia," *Geographical Review*, 67 (1977), 148–166; T. G. JORDAN et al., *Texas: A Geography* (Westview, 1984); also, JORDAN, "The Imprint of the Upper and Lower South on Mid-Nineteenth-Century Texas," *Annals AAG*, 57 (1967); P. P. KARAN, ed., *Kentucky: A Regional Geography* (Kendall/Hunt, 1973); C. F. KOVACIK and J. J. WINBERRY, *South Carolina: A Geography* (Westview, 1987); also, South Carolina: The Making of a Landscape (South Carolina, 1989); J. MASLOW, "Trade: Blues in the Gulf" (U. S. Gulf Coast seaports), *Atlantic Monthly*, May 1988, 25–31; T. G. MOORE, "Eastern Kentucky as a Model of Appalachia: The Role of Literary Images," *Southeastern Geographer*, 31 (1991), 75–89; V. S. NAIPAUL, *A Turn in the South* (Knopf, 1989; travel account by a distinguished novelist); K. B. RAITZ et al., *Appalachia: A Regional Geography: Land, People, and Development* (Westview, 1984); C. TREPANIER, "The Cajunization of French Louisiana: Forging a Regional Identity," *Geographical Journal*, 157 (1991), 161–171; M. D. WINSBERG, *Florida Weather* (University of Central Florida Press, 1990).

U. S. Midwest

T. J. BAERWALD, "The Twin Cities," *Focus*, Spring 1986, 10–15; B. H. BALTENSBERGER, *Nebraska: A Geography* (Westview, 1985); B. W. BLOUET and F. C. LUEBKE, eds., *The Great Plains: Environment and Culture* (Nebraska, 1979); J. BORCHERT, *America's Northern Heartland* (Minnesota, 1987); C. E. COBB, JR., "The Great Lakes' Troubled Waters," *National Geographic*, July 1987, 2–31; W. CRONON, *Nature's Metropolis: Chicago and the Great West* (Norton, 1991); I. CUTLER, *Chicago: Metropolis of the Mid-Continent*, 3rd ed. (Kendall/Hunt, 1982); R. DALLAS, "The Agricultural Collapse of the Arid Midwest," *Geographical Magazine*, October 1990, 16–22; I. FRAZIER, *Great Plains* (Farrar, Straus & Giroux, 1989); R. L. GERLACH, *Immigrants in the Ozarks: A Study in Ethnic Geography* (Missouri, 1976); also, *Settlement Patterns in Missouri: A Study of Population Origins, with a Wall Map* (Missouri, 1986); F. HAPGOOD, "The Prodigious Soybean," *National Geographic*, July 1987, 66–91; J. F. HART, "Change in the Corn Belt," *Geographical Review*, 76 (1986), 51–72; WILLIAM LEAST HEAT-MOON, *Prairy Earth (a deep map)* (Houghton Mifflin, 1991); A. D. HORSLEY, *Illinois: A Geography* (Westview, 1986); H. B. JOHNSON, *Order Upon the Land: The U. S. Rectangular Land Survey and the Upper Mississippi Country* (Oxford, 1976); T. R. MAHONEY, *River Towns in the Great West: The Structure of Provincial Urbanization in the American Midwest, 1820–1870* (Cambridge, 1990); T. L. McKNIGHT, "Great Circles on the Great Plains: The Changing Geometry of American Agriculture," *Erdkunde*, 33 (1979), 70–79; P. F. MATTINGLY and G. APSBURY, "Devolution of Cotton From the Piedmont," *Tijdschrift voor Economische en Sociale Geografie*, 77 (1986), 197–204; H. M. MAYER and R. C. WADE, *Chicago: Growth of a Metropolis* (Chicago, 1973); "The Midwest," *Journal of Geography*, 85, no. 5, special issue (September-October 1986; seven articles, with extensive reference lists); M. D. RAFFERTY, *Missouri: A Geography* (Westview, 1983); also, *The Ozarks: Land and Life* (Oklahoma, 1980); N. J. ROSENBERG, "Climate of the Great Plains Region of the United States," *Great Plains Quarterly*, 7 (1987), 22–32; W. A. SCHROEDER, *The Eastern Ozarks: A Geographic Interpretation of the Rolla 1:250,000 Topographic Map* (National Council for Geographic Education, 1967); J. R. SHORTRIDGE, "The Emergence of 'Middle West' as an American Regional Label," *Annals AAG*, 74 (1984), 209–220; also, *The Middle West: Its Meaning in American Culture* (Kansas, 1989); also, "The Vernacular Middle West," *Annals AAG*, 75 (1985), 48–57; L. M. SOMMERS, *Michigan: A Geography* (Westview, 1984); I. VOGELER, *Wisconsin: A Geography* (Westview, 1986); B. WALLACH, "The Return of the Prairie," *Landscape*, 28, no. 3 (1985), 1–5; W. P. WEBB, *The Great Plains* (Ginn, 1931); D. WORSTER, *Dust Bowl: The Southern Plains in the 1930's* (Oxford, 1979).

U. S. West: General

J. L. ALLEN, "Horizons of the Sublime: The Invention of the Romantic West," *Journal of Historical Geography*, 18 (1992), 27–40; W. A. BECK and Y. D. HAASE, *Historical Atlas of the American West* (Oklahoma, 1989); J. CARRIER, "The Colorado: A River Drained Dry," *National Geographic*, June 1992, 4–35; B. DeVOTO, *The Course of Empire* (Houghton Mifflin, 1952); M. T. EL-ASHRY and D. C. GIBBONS, eds., *Water and Arid Lands of the Western United States* (Cambridge, 1988); W. L. GRAF, "Science, Public Policy, and Western American Rivers," *Transactions IBG*, n.s., 17 (1992), 5–19; P. HORGAN, *The Heroic Triad* (Heinemann, 1974); W. L. LANG, ed., *Centennial West: Essays on the Northern Tier States* (Washington, 1991); T. L. McKNIGHT, "Irrigation Technology: A Photo-Essay," *Focus*, Summer 1990, 1–6; G. D. NASH, *The American West Transformed: The Impact of the Second World War* (Indiana, 1985); "Privatising

America's West," *The Economist*, October 22, 1988, 21–24; M. REISNER, *Cadillac Desert: The American West and Its Disappearing Water* (Viking, 1986); also, with S. BATES, *Overtapped Oasis: Reform or Revolution for Western Water* (Washington, DC: Island Press, 1990); Z. A. SMITH, *Ground Water in the West* (Academic, 1989); "New West, True West: Interpreting the Region's History," *Western Historical Quarterly*, 18 (1987), 141–156.

U. S. West: California

E. S. BAKKER, *An Island Called California: An Ecological Introduction to Its Natural Communities* (California, 1984); S. L. BOTTLES, *Los Angeles and the Automobile: The Making of the Modern City* (California, 1987); L. M. DILSAVER, "After the Gold Rush," *Geographical Review*, 75 (1985), 1–18; M. W. DONLEY *et al., Atlas of California* (Portland, OR: Professional Book Center, 1979); R. W. DURRENBURGER and R. B. JOHNSON, *California: Patterns on the Land*, 5th ed. (Palo Alto, CA: Mayfield, 1976); N. ETTLINGER, "The Roots of Comparative Advantage in California and Japan," *Annals AAG*, 81 (1991), 391–407; R. L. GENTILCORE, "Missions and Mission Lands of Alta California," *Annals AAG*, 51 (1961), 46–72; R. GOODENOUGH, "The Nature and Implications of Recent Population Growth in California," *Geography*, 77 (1992), 123–133; W. L. KAHRL, ed., *The California Water Atlas* (Sacramento, CA: The Governor's Office of Planning and Research, distributed by William Kaufmann, 1979; a magnificent oversized atlas in color); D. W. LANTIS *et al., California: Land of Contrast*, 3rd ed. (Kendall/Hunt, 1981); E. LIEBMAN, *California Farmland: A History of Large Agricultural Landholdings* (Rowman & Allanheld, 1983); C. LOCKWOOD and C. B. LEINBERGER, "Los Angeles Comes of Age," *Atlantic Monthly*, January 1988, 31–57; T. L. McKNIGHT, "Center Pivot Irrigation in California," *Geographical Review*, 73 (1983), 1–14; J. McPHEE, "Los Angeles Against the Mountains," in *The Control of Nature* (Farrar, Straus, & Giroux, 1989), 181–272; also, "Annals of the Former World: Assembling California" (geologic history), I, II, *The New Yorker*, September 7, 14, 1992, pp. 36–68, 44–84; B. MARCHAND and A. SCOTT, "Los Angeles en 1990: Une Nouvelle Capitale Mondiale," *Annales de Géographie* 100, no. 560 (July-August 1991); in French, with English summary; H. J. NELSON, *The Los Angeles Metropolis* (Kendall/Hunt, 1982); J. J. PARSONS, "A Geographer Looks at the San Joaquin Valley," *Geographical Review*, 76 (1986), 371–389; T. L. PETERS, "Trends in California Viticulture," *Geographical Review*, 74 (1984), 455–467; J. N. RUTGER and D. M. BRANDON, "California Rice Culture," *Scientific American*, February 1981, 42–51; R. A. SAUDER, "Patenting an Arid Frontier: Use and Abuse of the Public Land Laws in Owens Valley, California," *Annals AAG*, 79 (1989), 544–569; A. J. SCOTT, "High Technology Industry and Territorial Development: The Rise of the Orange County Complex, 1955;nn1984," *Urban Geography*, 7 (1986), 3–45; M. SCOTT, *The San Francisco Bay Area: A Metropolis in Perspective* (2nd ed.; California, 1985); R. STEINER, "Large Private Landholdings in California," *Geographical Review*, 72 (1982), 315–326; W. L. THOMAS, JR., ed., "Man, Time and Space in Southern California," *Annals AAG*, 49, Supplement, September 1959, entire issue; B. WALLACH, "The West Side Oil Fields of California," *Geographical Review*, 70 (1980), 50–59; C. WILVERT, "San Diego/San Francisco—Sansan," *Geographical Magazine* 53 (1981), 268–272.

U. S. Interior West

R. H. BROWN, *Wyoming: A Geography* (Westview, 1980); D. B. COLE and J. L. DIETZ, "The Changing Rocky Mountain Region," *Focus*, March-April 1984, 1–11; I. G. CLARK, *Water in New Mexico: A History of Its Management and Use* (New Mexico, 1987); M. L. COMEAUX, *Arizona: A Geography* (Westview, 1981); J. DAVIS, "King Coal in Cattle Country," *Geographical Magazine*, 56 (1984), 368–373; P. L. FRADKIN, *A River No More: The Colorado River and the West* (Knopf, 1981); P. GOBER, "The Retirement Community as a Geographical Phenomenon: The Case of Sun City, Arizona," *Journal of Geography*, 84 (1985), 189–198; W. L. GRAF, "An American Stream" (the Colorado), *Geographical Magazine*, 59 (1987), 504–509; N. HUNDLEY, *Water and the West: The Colorado River Compact and the Politics of Water in the American West* (California, 1975); R. H. JACKSON, "Mormon Perception and Settlement," *Annals AAG*, 68 (1978), 317–334; also, "The Mormon Experience: The Plains as Sinai, the Great Salt Lake as the Dead Sea, and the Great Basin as Desert-cum-Promised Land," *Journal of Historical Geography*, 18 (1992), 41–58; B. LUCKINGHAM, "The American Southwest: An Urban View," *Western Historical Quarterly*, 15 (1984), 261–280; also, *Phoenix: The History of a Southwestern Metropolis* (Arizona, 1989); E. G. McPHERSON and R. A. HAIP, "Emerging Desert Landscape in Tucson," *Geographical Review*, 79 (1989), 435–449; D. W. MEINIG, *Southwest: Three Peoples in Geographical Change, 1600–1970* (Oxford, 1971); also, "The Mormon Culture Region: Strategies and Patterns in the Geography of the American West, 1847–1964," *Annals AAG*, 55 (1965), 191–220; J. A. McPHEE, *Basin and Range* (Farrar, Straus, & Giroux, 1981); JAMES A. MICHENER, *Centennial* (a novel) (Random House, 1974); R. L. NOSTRAND, "The Hispano Homeland in 1900," *Annals AAG*, 70 (1980), 382–396; R. REDFERN, *Corridors of Time: 1,700,000,000 Years of Earth at Grand Canyon* (Times Books, 1980; a spectacular oversized volume in color); B. WALLACH, "Sheep Ranching in the Dry Corner of Wyoming," *Geographical Review*, 71 (1981), 51–63.

U. S. Pacific Northwest

W. A. BOWEN, *The Willamette Valley: Migration and Settlement on the Oregon Frontier* (Washington, 1978); J. R. GIBSON, *Farming the Frontier: The Agricultural Opening of the Oregon Country, 1786–1846* (Washington, 1985); I. HAMILTON, "From Roses to Microchips" (Portland, Oregon), *Geographical Magazine*, 58 (1986), 356–361; A. J. KIMERLING and P. L. JACKSON, eds., *Atlas of the Pacific Northwest* (7th ed.; Oregon State, 1985); G. MACINKO, "The Ebb and Flow of Wheat Farming in the Big Bend, Washington," *Agricultural History*, 59 (1985), 215–228; D. W. MEINIG, *The Great Columbia Plain: A Historical Geography, 1805–1910* (Washington, 1968); D. C. ROSE, "Seattle," *Cities*, 7 (1990), 283–288; J. W. SCOTT and R. L. DELORME, *Historical Atlas of Washington* (Oklahoma, 1988); G. SWAN, "The Beautiful and Dammed" (Columbia River), *Geographical Magazine*, August 1988, 35–40; B. WARF, "Regional Transformation, Everyday Life, and Pacific Northwest Lumber Production," *Annals AAG*, 78 (1988), 326–346.

Alaska

G. KNAPP and T. A. MOREHOUSE, "Alaska's North Slope Borough Revisited," *Polar Record*, 27 (1991), 303–312; D. F. LYNCH *et al.,* "Alaska: Land and Resource Issues," *Focus,* January-February, 1981, 1–16; J. McPHEE, *Coming into the Country* (Alaska) (Farrar-Strauss-Giroux, 1977); J. B. RAY, "Selection of the Marine Terminal for the Trans-Alaska Pipeline," *Journal of Geography,* 78 (1979), 147–151; J. REARDEN, ed., "Alaska's Salmon Fisheries," *Alaska Geographic,* 10, no. 3 (1983), 1–123; B. RICHARDS, "Alaska's Southeast: A Place Apart," *National Geographic,* January 1984, 50–87; J. R. SHORTRIDGE, "The Collapse of Frontier Farming in Alaska," 66 (1976), 583–604.

Hawaii

K. E. KIM and K. LOWRY, "Honolulu," *Cities,* 7 (1990), 274–282; J. R. MORGAN *et al., Hawaii: A Geography* (Westview, 1983); E. C. NORDYKE, *The Peopling of Hawaii* (Hawaii, 1977); T. J. OSBORNE, "Trade or War? America's Annexation of Hawaii Reconsidered," *Pacific Historical Review,* 50 (1981), 285–307; UNIVERSITY OF HAWAII AT MANOA, DEPARTMENT OF GEOGRAPHY, *Atlas of Hawaii,* 2nd ed. (Hawaii, 1983; a handsome atlas in color).

Greenland

A. J. TAAGHOLD, "Greenland and the Future," *Environmental Conservation,* 7 (1980), 295–300; also, "Greenland's Future Development: A Historical and Political Perspective," *Polar Record,* 21, no. 130 (January 1982), 23–32.

ADDITIONAL REFERENCES AND READINGS

This list of books and articles updates the References and Readings section.

The Geographer's Field of Vision

J. BONGAARTS, "Can the Growing Human Population Feed Itself?" *Scientific American,* March 1994, 36–42; F. BRAY, "Agriculture for Developing Nations," *Scientific American,* July 1994, 30–42; "Fish: The Tragedy of the Oceans," *The Economist,* March 12, 1994, 21–24; R. KAPLAN, "The Coming Anarchy," *Atlantic Monthly,* February 1994, 44–76; M. LOWE, *Back on Track: The Global Rail Revival* (Worldwatch Paper 118; Worldwatch Institute, April 1994); M. MONMONIER, *Mapping It Out: Expository Cartography for the Humanities and Social Sciences* (Chicago, 1993); C. RENFREW, "World Linguistic Diversity," *Scientific American,* January 1994, 116–123; "Power to the People: A Survey of Energy," *The Economist,* June 18, 1994, 18 pp.

Part 1: Europe

I. ARMOUR, "Bosnia and Herzegovina in Historical Perspective," 1994 *Britannica Book of the Year,* 425–427; B. BRYSON, "Britain's Hedgerows," *National Geographic,* September 1993, 94–117; F. CARTER and D. TURNOCK, eds., *Environmental Problems in Eastern Europe* (Routledge, 1994); M. CORSON and J. MINGHI, "Reunification of Partitioned Nation-States: Theory Versus Reality in Vietnam and Germany," *Journal of Geography,* May/June 1994, 125–131; W. GÜNTHARDT, "Steaks for America, Potatoes for Europe" [Post-Columbian exchange of foods], *Swiss Review of World Affairs,* January 1993, 26–28; R. HALL, "Europe's Changing Population," *Geography,* 78 (1993), 3–15; C. HARRIS, "New European Countries and Their Minorities," *Geographical Review,* 83 (1993), 301–320; "Minorities: That Other Europe," *The Economist,* December 25, 1993/January 7, 1994, 17–20; P. MÜNSTER, "New Doubts About Dutch Polders," *Swiss Review of World Affairs,* December 1993, 22–23; C. STIEGER, "Greece Strangling Macedonia," *Swiss Review of World Affairs,* June 1994, 2–3; "Switzerland as a Financial Center," Special Section, *Swiss Review of World Affairs,* February 1993, 17–32; C. NEWMAN, "The Light at the End of the Chunnel" [English Channel Tunnel], *National Geographic,* May 1994, 36–47; C. WILVERT, "Spain: Europe's California," *Journal of Geography,* March/April 1994, 74–79.

Part 2: The Former Soviet Region

"Chaos in the Caucasus" [three articles], *Swiss Review of World Affairs,* May 1994, 7–10; "Economist Survey: Ukraine—The Birth and Possible Death of a Country," *The Economist,* May 7, 1994, 18 ff.; C. HARRIS, "Ethnic Tensions in Areas of the Russian Diaspora," *Post-Soviet Geography,* 34 (1993), 233–239; P. HOFHEINZ, "Rising in Russia," *Fortune,* January 24, 1994, 92–97; R. KAISER, *The Geography of Nationalism in Russia and the USSR* (Princeton, 1994); D. PETERSON, *Troubled Lands: The Legacy of Soviet Environmental Destruction* (Westview, 1993); "Return of the Soviet Empire" [six articles], *World Press Review,* April 1994, 8–13; A. RÜESCH, "Turkmenistan's Dream of Rapid Riches" [oil], *Swiss Review of World Affairs,* April 1994, 10–11.

Part 3: The Middle East

T. ALLEN, "Turkey Struggles for Balance," *National Geographic,* May 1994, 2–35; "At Ease in Zion: A Survey of Israel," *The Economist,* January 22, 1994, 22 pp.; M. GILBERT, *Atlas of the Arab-Israeli Conflict* (Oxford, 1994); D. HOWELL *et al.,* "Oil: When Will We Run Out?" *Earth,* March 1993, 26–33; D. JOHNSON, "Nomadism and Desertification in Africa and the Middle East," *GeoJournal,* 31 (1993), 51–66; S. LIBISZEWSKI, "Source of Life, Source of Strife" [Tigris-Euphrates water], *Swiss Review of World Affairs,* June 1994, 8–10; S. REED, "The Battle for Egypt" [Islamic militants], *Foreign Affairs,* September/October 1993, 94–107; J. STANISLAW and D. YERGIN, "Oil: Reopening the Door," *Foreign Affairs,* September/October 1993, 81–93, "The Muslims Are Coming," [five articles], *World Press Review,* May 1994, 8–13; S. ZAIMECHE, "Change, the State and Deforestation: The Algerian Example," *Geographical Journal,* March 1994, 50–56.

Part 4: Monsoon Asia

"China" [seven articles], *Current History,* September 1993, entire issue; "China: Boom or Bust?" [seven articles] *World Press Review,* July 1994, 8–17; "China's Communists: The Road from Tiananmen," *The Economist,* June 4, 1994, 19–21; J. CURRAN, "China's Investment Boom," *Fortune,* March 7, 1994, 116–124;

R. HORNIK and G. SEGAL, "China's Growing Pains," *Foreign Affairs*, May/June 1994, 28–58; L. KRAAR, "Storm Over Hong Kong," *Fortune*, March 8, 1993, 98–106; F. LEEMING, *The Changing Geography of China* (Blackwell, 1993); K. LUDWIG, "Impressions from Old Tibet," *Swiss Review of World Affairs*, May 1994, 18–23; M. MORRISH, "Checking Up on China," *Teaching Geography*, April 1994, 51–57; T. O'NEILL, "The Mekong: A Haunted River's Season of Peace," *National Geographic*, February 1993, 2–35; B. R. SCHLENDER, "Japan: Is It Changing for Good?" *Fortune*, June 13, 1994, 125–134; Z. SONGQIAO, *Geography of China: Environment, Resources, Population, and Development* (Wiley, 1994); S. TEFFT, "Despite Controversy, China Pushes Ahead on Colossal Dam" [Three Gorges Dam], *Christian Science Monitor*, May 11, 1994, 11–13; C. THOMSON, "Political Identity Among Chinese in Thailand," *Geographical Review*, 83 (1993), 397–409; P. WHITE, "Rice: The Essential Harvest," *National Geographic*, May 1994, 48–79. (*See also* M. CORSON under Part II: Europe, p. R-33.)

Part 5: The Pacific World

P. HAGGETT, "The Invasion of Human Epidemic Diseases into Australia, New Zealand, and the Southwest Pacific," *New Zealand Geographer*, 49 (1993), 40–47; D. HOUGHTON, "Long-Distance Commuting: A New Approach to Mining in Australia," *Geographical Journal*, 159 (1993), 281–290; J. OVERTON, "Pacific Futures? Geography and Change in the Pacific Islands," *New Zealand Geographer*, 49 (1993), 48–55.

Part 6: Africa, South of the Sahara

"Africa" [nine articles], *Current History*, May 1994, entire issue; "Africa: A Flicker of Light," *The Economist*, March 5, 1994, 21–24; T. BINNS, "Nigeria: Africa's Restless Giant," *Teaching Geography*, April 1993, 50–56; R. BONNER, "Crying Wolf Over Elephants," *New York Times Magazine*, February 7, 1993, 16–19 ff.; N. CHEGE, "Africa's Non-Timber Forest Economy," *World Watch*, July/August 1994, 19–23; I. GRIFFITHS, *The Atlas of African Affairs* (Routledge, 1994); O. ITEN, "Burundi Between Mistrust and Democracy," *Swiss Review of World Affairs*, January 1993, 14–19; W. MORGAN and J. SOLARZ, "Agricultural Crisis in Sub-Saharan Africa: Development Constraints and Policy Problems," *Geographical Journal*, March 1994, 57–73; K. NEWLAND,

"Refugees: The Rising Flood," [with map], *World Watch*, May-June 1994, 10–20.

Part 7: Latin America

B. BLOUET and O. BLOUET, *Latin America and the Caribbean: A Systematic and Regional Survey* (Wiley, 1993); C. COLLINS and S. SCOTT, "Air Pollution in the Valley of Mexico," *Geographical Review*, 83 (1993), 119–133; J. FIFER, "Chile's Pioneering Location: Pacific Rim and Southern Cone," *Geography*, April 1994, 129–146; B. GODFREY, "Migration to the Gold-Mining Frontier in Brazilian Amazonia," *Geographical Review*, 82 (1992), 458–469; "Into the Spotlight: A Survey of Mexico," *The Economist*, February 13, 1993, 22 pp.

Part 8: North America

D. CHADWICK, "Roots of the Sky: The American Prairie," *National Geographic*, October 1993, 90–119; R. CONNIFF, "California: Desert in Disguise," *National Geographic* [Special Edition: Water], November 1993, 38–53; M. CONZEN et al., *A Scholar's Guide to Geographical Writing on the American and Canadian Past* (University of Chicago, Department of Geography Research Paper 235, 1993); K. DAVIDSON, "Learning from Los Angeles" [faults and earthquakes], *Earth*, September 1994, 40–47; C. EARLE, *Geographical Inquiry and American Historical Problems* (Stanford, 1992); W. ELLIS, "The Mississippi: River Under Siege," *National Geographic*, November 1993, 90–105; J. FARAGHER, "Gunslingers and Bureaucrats: Some Unexpected Truths About the American West," *The New Republic*, December 14, 1992, 29–36; D. FARR, "Creating Nunavut," *World Book Year Book*, 1993, 354–357; H. MEADWELL, "The Politics of Nationalism in Quebec," *World Politics*, 45 (1993), 203–241; D. MEINIG, *The Shaping of America: A Geographical Perspective on 500 Years of Histgory, Vol. 2: Continental America, 1800–1867* (Yale, 1993); J. MITCHELL, James Bay: Where Two Worlds Collide," *National Geographic*, November 1993, 66–75; R. SAUDER, *The Lost Frontier: Water Diversion in the Growth and Destruction of Owens Valley Agriculture* (Arizona, 1994); "Timeline" [recent changes in US railway ownership], *Trains*, November 1990, 21–47; E. WALTER, JR., "The Fight for America's Public Lands," *World Book Year Book*, 1993, Special Report, 164–175; "Water in the West," Special Issue, *Pacific Discovery*, Winter 1993 [includes a "Water Resources" bibliography, p. 52, and a map, "California's Water Geography," p. 26.

CURRENT SOURCES

Chapter 1

Birdsall, Stephen S. "Regard, Respect, and Responsibility: Sketches for a Moral Geography of the Everyday." *Annals of the Association of American Geographers.* 86(4): 619–629, 1996.

Demko, George J. et al. *Why in the World: Adventures in Geography.* New York: Anchorbooks, published by Bantam Doubleday, 1992.

Gaile, Gary L. and Cort J. Wilmott, eds. *Geography in America.* Columbus, Ohio: Merrill Publishing, 1989.

Geography Education Standards Project. *Geography for Life: National Geography Standards 1994.* Washington, D.C.: National Geographic Research and Exploration. 1994.

Lineback, Neal. *Geography in the News.* Southern Pines, NC: Karo Hollow Press, 1995.

Marshall, Bruce, ed. *The Real World.* Boston: Houghton Mifflin Company, 1991.

Rediscovering Geography: New Relevance for Science and Society. Washington, D.C.: National Academy Press, 1997.

Chapter 2

Bennett, Charles F. *Man and Earth's Ecosystems: An Introduction to the Geography of Human Modification of the Earth.* New York: John Wiley & Sons, 1975.

Botkin, Daniel B., Margriet F. Caswell, John E. Estes, and Angelo A. Ario, eds. *Changing the Global Environment.* Boston: Academic Press, 1989.

Brown Lester, ed. *State of the World 1994.* New York: W.W. Norton & Co., 1994.

———. *State of the World 1990.* New York: W.W. Norton & Co., 1990.

Clark, William C. "Managing Planet Earth." *Scientific American* 261(3): 19–26 (1989).

Dasmann, Raymond F. *Environmental Conservation.* New York: John Wiley & Sons, 1984.

———. *Ecodevelopment—An Ecological Perspective: Tropical Ecology and Development, 1331–1335.* Kuala Lumpur: International Society of Tropical Ecology, 1980.

De Souza, Anthony R., and Frederick P. Stutz. *The World Economy: Resources, Location, Trade and Development.* New York: Macmillan, 1994.

Detwyler, Thomas R., ed. *Man's Impact on the Environment.* New York: McGraw-Hill, 1988.

Dubos, Réne. *A God Within.* New York: Charles Scribner's Sons, 1972.

English, Paul, and James A. Miller. *World Regional Geography: A Question of Place.* New York: John Wiley & Sons, 1989.

Glacken, C.J. "Changing Ideas of the Habitable World." In *Man's Role in Changing the Face of the Earth,* edited by W. L. Thomas, 70–92. Chicago: University of Chicago Press, 1956.

Glaeser, Bernhard, ed. *Ecodevelopment: Concepts, Projects, Strategies.* Oxford: Pergamon Press, 1984.

Goudie, Andrew. *The Human Impact on the Natural Environment.* Oxford: Basil Blackwell, 1986.

Greenhouse, Steven. "The Greening of American Diplomacy." *New York Times,* October 9, 1995, A4.

Grossman, Larry. "Man-Environment Relations in Anthropology and Geography." *Annals of the Association of American Geographers* 67: 126–144 (1977).

Gupta, Ajavit. *Ecology and Development in the Third World.* New York: Routledge, Chapman & Hall, 1988.

Hanson, Herbert C. *Dictionary of Ecology.* New York: Philosophical Library, 1962.

Hardesty, Donald L. *Ecological Anthropology.* New York: John Wiley & Sons, 1977.

Hardin, Garrett. "Living on a Lifeboat." In *Readings in Ecology, Energy and Human Society: Contemporary Perspectives,* edited by William R. Burch, Jr. New York: HarperCollins, 1977.

International Union for the Conservation of Nature, United Nations Environmental Program, World Wildlife Fund. *World Conservation Strategy: Living Resource Conservation for Sustainable Development:* IUCN, Gland, 1980.

Kates, Robert W., and Ian Burton, eds. *Geography, Resources, and Environment, Vol. II.* Chicago: University of Chicago Press, 1986.

Kaufman, Les, and Kenneth Mallory. *The Last Extinction.* Cambridge, Mass.: MIT Press, 1986.

Lovelock, J.E. *Gaia.* Oxford: Oxford University Press, 1979.

MacNeill, Jim. "Strategies for Sustainable Economic Development." *Scientific American* 261: 104–113 (1989).

Martin, Geoffrey J., and Preston E. James. *All Possible Worlds: A History of Geographical Ideas.* New York: John Wiley & Sons, 1993.

McLaren, Digby J., and Brian J. Skinner, eds. *Resources and World Development.* New York: John Wiley & Sons, 1987.

Miller, G. Tyler. *Living in the Environment.* Belmont, California: Wadsworth, 1996.

Moran, Emilio F. *Human Adaptability.* Boulder, Colo.: Westview Press, 1982.

Myers, Norman. "Population, Environment and Conflict." *Environmental Conservation* 14: 15–22 (?).

———. *A Wealth of Wild Species: Storehouse for Human Welfare.* Boulder, Colo.: Westview Press, 1983.

Nasar, Sylvia. "Why International Statistical Comparisons Don't Work." *New York Times,* March 8, 1992, E5.

O'Riordan, T. *Environmentalism.* London: Pion, 1976.

Pearce, David W., and R. Kerry Turner, eds. *Economics of Natural Resources and the Environment.* Baltimore: Johns Hopkins University Press, 1990.

Population Reference Bureau. *World Population Data Sheet.* Washington, D.C.: Population Reference Bureau, 1996.

Postel, Sandra. "Carrying Capacity: Earth's Bottom Line." In *State of the World,* edited by Lester Brown, 3–21. New York: W.W. Norton & Co., 1994.

Potter, Thomas Michael. "Social Organization and Environmental Destruction: Capitalism, Socialism and the Environment." In *International Dimensions of the Environmental Crisis,* edited by Richard N. Barrett, 21–29. Boulder, Colo.: Westview Press, 1982.

Repetto, R., ed. *The Global Possible.* New Haven: Yale University Press, 1985.

Roberts, Neil. *The Holocene: An Environmental History.* Oxford: Basil Blackwell, 1989.

Rubenstein, James M. *The Cultural Landscape: An Introduction to Human Geography.* New York: Macmillan, 1994.

Ruckelshaus, William D. "Toward a Sustainable World." *Scientific American* 261: 114–120 (1989).

Sahlins, Marshall. *Stone Age Economics.* Walter DeGruyter Press, 1972.

Simmons, I.G. *Changing the Face of the Earth.* Oxford: Basil Blackwell, 1989.

Soule, Michael E. *Conservation Biology: The Science of Scarcity and Diversity.* Sunderland, Mass.: Sinaeuer Press, 1986.

Stevens, William K. "'95 the Hottest Year on Record as the Global Trend Keeps Up." *New York Times,* January 4, 1996, A1.

———. "Experts Confirm Human Role in Global Warming." *New York Times,* September 10, 1995, A1.

Tivy, Joy, and Greg O'Hare. *Human Impact on the Ecosystem.* Edinburgh: Oliver and Boyd, 1981.

Wilson, Edward O. "Threats to Biodiversity." *Scientific American* 261 (3): 60–66 (1989).

———. *Biodiversity.* Washington, D.C.: Smithsonian Institution Press, 1987.

World Commission on Environment and Development. *Our Common Future.* Oxford: Oxford University Press, 1987.

World Resources Institute. *World Resources, 1994–95.* New York: Oxford University Press, 1994.

———. *The Global Possible: Resources, Development and the New Century.* Washington, D.C.: World Resources Institute, 1984.

Chapter 3

Blouet, Brian W. "The Political Geography of Europe: 1900–2000 A.D." *Journal of Geography.* 95(1): 5–14, 1996.

"European Business." *The Economist.* 340(7974): 22–23, 1996.

Jordan, Terry G. *The European Culture Area: A Systematic Geography.* New York: HarperCollins, 1996.

Renfrew, Colin. "The Origins of Indo-European Languages." *Scientific American.* 261(4): 106–114, 1989.

Chapter 4

"Crossed Fingers in France." *The Economist.* 343(8014): 45–47, 1997.

Katzenstein, Peter J. "United Germany in an Integrated Europe." *Current History.* 96(608): 116–123, 1997

Lueschen, Leila S. "French Agriculture: Trends and Policies." *Agribusiness.* 11(5): 447–462, 1995.

Wild, Trevor and Philip N. Jones. "Spatial Impacts of German Unification." *The Geographical Journal.* 160 (1): 1–16, 1994.

Chapter 5

Clark, Arthur L. *Bosnia: What Every American Should Know.* New York: Berkley Books, 1996.

Murphy, Alexander B. and Anne Hunderi-Ely. "The Geography of the 1994 Nordic Vote on European Union Membership." *Professional Geographer.* 48(3): 2184–2197, 1996.

Silber, Laura and Allan Little. *Yugoslavia: Death of a Nation.* Harmondsworth, U.K.: Penguin, 1995.

Woodward, Susan L. "Bosnia after Dayton: Year Two." *Current History.* v96 n608: 97–103. 1997.

Chapter 6

Burke, Justin. "Volgograd: The Living World War II Memorial." *Christian Science Monitor,* September 10, 1993, 9.

DeBlij, Harm J., and Peter O. Muller. *Geography: Realms, Regions and Concepts.* New York: John Wiley & Sons, 1994.

Ingwerson, Marshall. "Russia's Bear Economy Goes Bullish as Reforms Kick In." *Christian Science Monitor,* October 10, 1995, 6.

Moffett, George. "Assessing a Russia Without Yeltsin at the Helm." *Christian Science Monitor,* November 14, 1995, 1.

Poletz, Lida. "Ukraine and Belarus Elect New Pro-Russian Leaders." *Christian Science Monitor,* July 12, 1994, 3.

"The Selling of Russia." *Economist,* November 18, 1995, 88.

Sneider, Daniel. "Russia's Future Shadowed by Centuries of Conquest." *Christian Science Monitor,* August 10, 1994, 7.

Specter, Michael. "After Lull, War Revives in Chechnya's Ruins." *New York Times,* November 26, 1995, A1.

———. "Chechen Insurgents Take Their Struggle to a Moscow Park." *New York Times,* November 24, 1995, A1.

———. "Pro-Russian Chechen Leader Survives Bombing in Capital." *New York Times,* November 21, 1995, A4.

———. "Russia's Fall Grain Harvest Seen as Worst in 30 Years." *New York Times,* October 10, 1995, A6.

———. "Climb in Russia's Death Rate Sets Off Population Implosion." *New York Times,* A1.

Stanley, Alessandra. "From Repression to Respect, Russian Church in Comeback." *New York Times,* October 3, 1994, A1.

Chapter 7

Barraclough, Colin. "Azerbaijan Keen to Seal Oil Deal with West." *Christian Science Monitor,* August 9, 1994, 9.

———. "Turkmenistan Ripe for Capitalism." *Christian Science Monitor,* June 24, 1992, 7.

Bohlen, Celestine. "Yeltsin and Ukraine's Chief in Pact on Black Sea Fleet." *New York Times,* April 16, 1994, 5.

———. "Poor Region in Russia Lays Claim to Its Diamonds." *New York Times,* November 1, 1992, 3.

———. "Russia Permits Just a Peek at Nature in the Raw." *New York Times,* October 28, 1992, A4.

Broad, William J. "Russians Describe Extensive Dumping of Nuclear Waste." *New York Times,* April 27, 1993, A1.

Burke, Justin. "In Caviar Capital, Cultures Clash." *Christian Science Monitor,* September 14, 1993, 10.

———. "Volga Germans Seek Lost Homeland." *Christian Science Monitor,* September 10, 1993, 8.

———. "Samara's Defense Industry Moves Reluctantly Toward the Free Market." *Christian Science Monitor,* September 9, 1993, 8.

———. "A Part of Russia That Wants Out." *Christian Science Monitor,* September 3, 1993, 6.

———. "The Volga Basin's Merchant Capital Is New Russia's Economic Showcase." *Christian Science Monitor,* September 2, 1993, 8.

————. "A Russian Republic Looks Away." *Christian Science Monitor,* March 10, 1993, 6.

————. "Uzbek Leaders Clamp Down to Keep Peace." *Christian Science Monitor,* January 5, 1993, 8.

————. "Uzbekistan Moves Slowly Toward Economic Reform." *Christian Science Monitor,* January 5, 1993, 8.

————. "Uzbek Leaders Pick Stability Over Reform." *Christian Science Monitor,* December 11, 1992, 10.

————. "Tajiks Struggle for National Identity." *Christian Science Monitor,* September 30, 1992, 10.

————. "Siberia's Buryats Hope for Recovery." *Christian Science Monitor,* September 3, 1992, 7.

Colarusso, John. "Chechnya: The War Without Winners." *Current History* 94(594): 329–336 (1995).

"Communists Lead as Belarus Votes." *Christian Science Monitor,* November 30, 1995, 9.

Dudaev, Dzhohar. "Chechnya is World's Concern." *Christian Science Monitor,* November 1, 1995, 19.

Erlanger, Steven. "U.S. Aid for Huge Russian Lake Is in Jeopardy." *New York Times,* September 3, 1995, A3.

————. "In Russia, Turning Oil into Money Is Actually Hard." *New York Times,* March 20, 1994, E5.

————. "A Cry of Pain Rises From the Cradle of Soviet Science." *New York Times,* November 21, 1993, E18.

————. "Heirs of the Golden Horde Reclaim a Tatar Culture." *New York Times,* August 13, 1993, A3.

————. "To Kazakhstan, With Gold in Mind." *New York Times,* March 10, 1993, A4.

————. "Russia and Ukraine: Condemned to Get Along." *New York Times,* June 21, 1992, E3.

————. "Uzbeks, Free of Soviets, Dethrone Czar Cotton." *New York Times,* June 20, 1992, A4.

Ford, Peter. "Rebels Keep Guns While Russia Tries to Call Shots in Chechnya." *Christian Science Monitor,* September 21, 1995, 8.

————. "Why Georgia (the Country) Tries to Act Big for Its Size." *Christian Science Monitor,* September 6, 1995, 6.

————. "A Remote Republic Rattles Moscow." *Christian Science Monitor,* September 16, 1994, 7.

Gordon, Michael R. "Chechnya Toll Is Far Higher; 80,000 Dead, Lebed Asserts." *New York Times,* September 4, 1996, A3.

————. "Russia Agrees to Closer Links With Three Ex-Soviet Lands." *New York Times,* March 30, 1996, A4.

Greenhouse, Steven. "Ukraine Votes to Become a Nuclear-Free Country." *New York Times,* November 17, 1994, A6.

Greenwald, Igor. "Four Years Old, Ukraine Improvises Its Identity." *Christian Science Monitor,* August 24, 1995, 6.

Ingwerson, Marshall. "Peace with No Honor: Chechnya Pact Leaves Russian Troops Bitter." *Christian Science Monitor,* September 3, 1996, 1.

————. "Marketing Nuclear Plants for an Energy-Hungry World." *Christian Science Monitor,* November 8, 1995, 10.

————. "Foreign Investors in Kazakstan Face Obstacles." *Christian Science Monitor,* October 12, 1995, 6.

————. "Kyrgyz Brokers Dream of a Trading Frenzy." *Christian Science Monitor,* September 20, 1995, 7.

————. "Empire Lost, Russian People Stream Out of Central Asia." *Christian Science Monitor,* September 19, 1995, 6.

————. "New Nation Digs Deep for Its Turkic Roots." *Christian Science Monitor,* September 12, 1995, 1.

————. "'The Turks Are Coming,' Cry Russians in Their Ex-Empire." *Christian Science Monitor,* September 26, 1995, 1.

Ingwerson, Marshall, and Sami Kohen. "At Your Local Gas Pump Soon: Caspian Sea Oil." *Christian Science Monitor,* October 11, 1995, 6.

Kamm, Henry. "With 'No Enemies,' Russia's Baltic Fleet Rusts." *New York Times,* November 27, 1995, A6.

————. "Turks Fear Role in Asia of Russians." *New York Times,* June 18, 1994, A6.

"Kazakhstan: High Stepping." *Economist,* March 16, 1996, 41.

Lee, Rensselaer W. "Post-Soviet Nuclear Trafficking: Myths, Half-Truths, and the Reality." *Current History* 94 (594): 343–348 (1995).

LeVine, Steve. "Caspian's Sun on Rise, Cashing Russian Shadow." *New York Times,* October 27, 1995, A6.

————. "Oil Consortium to Skirt Russia in Its Shipments." *New York Times,* October 7, 1995, A22.

————. "U.S. and Russia at Odds Over Caspian Oil." *New York Times,* October 4, 1995, C2.

————. "After Karl Marx, a 1,000-Year-Old Superman." *New York Times,* August 31, 1995, A4.

Lewison, Arlene. "A Century in Mourning." *Columbia Missourian,* September 17, 1995, G1.

Linden, Eugene. "The Rape of Siberia." *Time,* September 4, 1995, 42–53.

Meyer, Stephen M. "The Devolution of Russian Military Power." *Current History* 94 (594): 322–328 (1995).

Misiunas, Romauld J. "This Tiny Russian Enclave in Europe Could Explode." *Christian Science Monitor,* April 6, 1994, 23.

"The Northern Sea Route: Plain Sailing or Environmental Disaster?" *WWF Arctic Bulletin* 3.94: 10–12 (1994).

Perlez, Jane. "Despite U.S. Help and Hope, Ukraine's Star Fades." *New York Times,* June 27, 1996, A3.

————. "Ukraine Sells Its Companies, But Buyers Are Few." *New York Times,* November 2, 1995, A1.

Poletz, Lida. "Ukraine's Religious Standoff Makes Unlikely Political Allies." *Christian Science Monitor,* July 21, 1995, 6.

————. "Ukraine Between East and West." *Christian Science Monitor,* June 16, 1994, 6.

Rasputin, Valentine. *Siberia on Fire: Stories and Essays by Valentin Rasputin.* De Kalb: Northern Illinois University Press, 1989.

Rosen, Yereth. "USSR Leaves Radioactive Legacy." *Christian Science Monitor,* August 26, 1992, 8.

"Russian Oil: Not a Gusher." *Economist,* October 14, 1995, 78.

Salpukas, Agis. "Kazakhstan and Chevron Try to End Pipeline Impasse." *New York Times,* October 28, 1995, A18.

————. "Siberian Oil Venture by Four Companies." *New York Times,* April 12, 1994, C4.

Schmemann, Serge. "War Bleeds Ex-Soviet Land at Central Asia's Heart." *New York Times,* February 21, 1993, A1.

Shaw, Denis J.B. *The Post Soviet Republics: A Systematic Geography.* Harlow, England: Longman Press, 1995.

Sloane, Wendy. "Mills Quiet in Russia's Manchester." *Christian Science Monitor,* August 5, 1994, 8.

————. "Russia Backs Overthrow of Rebel Leader in Caucasus." *Christian Science Monitor,* August 4, 1994, 6.

————. "Russia-Estonia Summit Seeks Troops Solution." *Christian Science Monitor,* July 27, 1994, 4.

_____. "Baltics Accuse Russia of Reneging on Troop Pullout." *Christian Science Monitor,* April 8, 1994, 5.

_____. "Russia's Bashkir Republic Pushes for Its Autonomy." *Christian Science Monitor,* October 26, 1993, 6.

Sneider, Daniel. "Russian Leaders Watch Closely As Ukraine and Belarus Vote." *Christian Science Monitor,* June 22, 1994, 3.

_____. "Tajikistan, A Tangle of Diverse Identities." *Christian Science Monitor,* May 13, 1994, 6.

_____. "Russian Bear Roams in Battered Tajikistan." *Christian Science Monitor,* May 12, 1994, 7.

_____. "Ukraine's Economy Stumbles." *Christian Science Monitor,* March 30, 1994, 7.

_____. "Turkmenistan, Out from Under Soviet Masters, Reaches to Iran." *Christian Science Monitor,* April 13, 1993, 7.

_____. "Turkmenistan: Slow Reforms, No Dissent." *Christian Science Monitor,* March 25, 1993, 7.

_____. "Estonia Leads Baltic States into New Era." *Christian Science Monitor,* January 25, 1993, 8.

_____. "Oil Fuels Azeris' Hopes for Future." *Christian Science Monitor,* January 13, 1993, 10.

_____. "Central Asians to Build Own Common Market." *Christian Science Monitor,* January 7, 1993, 2.

Specter, Michael. "Decisive Battle, Vague Promises." *New York Times,* September 1, 1996, A7.

_____. "Belarus and Russia Form Union, Reuniting Two Former Soviet Lands." *New York Times,* March 24, 1996, A1.

_____. "Pro-Russian Chechen Leader Survives Bombing in Capital." *New York Times,* November 21, 1995, A4.

_____. "Yeltsin Threatens Action on Warring Secessionist Area." *New York Times,* November 30, 1994, A3.

Stanley, Alessandra. "Yeltsin, Echoing Cold War, Rails Against West." *New York Times,* October 20, 1995, A3.

Sullivan, Walter. "Soviet Nuclear Dumps Disclosed." *New York Times,* November 24, 1992, B9.

Tyler, Patrick F. "Soviets' Secret Nuclear Dumping Raises Fears for Arctic Waters." *New York Times,* May 4, 1992, A1.

Walker, Sam. "No Time Like the Present to Invest in Eastern Siberia's Sakha Republic." *Christian Science Monitor,* March 9, 1994, 9.

Chapter 8

Bates, Daniel G., and Amal Rassam. *Peoples and Cultures of the Middle East.* Englewood Cliffs, N. J.: Prentice-Hall, 1983.

Coon, Carleton S. "Point Four and the Middle East." *Annals of the American Academy of Political and Social Science* 270: 88–92 (1950).

English, Paul Ward. "Geographical Perspectives on the Middle East: The Passing of the Ecological Trilogy." In *Geographers Abroad: Essays on the Problems and Prospects of Research in Foreign Areas,* University of Chicago Development of Geography Research Paper No. 152, edited by Marvin W. Mikesell, 134–164 (1973).

_____. "Urbanities, Peasants, and Nomads: The Middle Eastern Ecological Trilogy." *Journal of Geography* 61: 54–59 (1967).

Held, Colbert. *Middle East Patterns: Places, People and Politics.* Boulder, Colo.: Westview Press, 1994.

Chapter 9

Beaumont, Peter, Gerald H. Blake, and J. Malcolm Wagstaff. *The Middle East: A Geographical Study.* New York: Halstead Press, 1988.

Britannica Book of the Year. 1994.

Bulloch, John, and Adel Darwish. *Water Wars: Coming Conflicts in the Middle East.* London: Victor Gollancz, 1993.

Burns, John F. "The West in Afghanistan, Before and After." *New York Times,* February 18, 1996, E5.

_____. "From Cold War, Afghans Inherit Brutal New Age." *New York Times,* February 14, 1996, A1.

_____. "Afghan Capital Grim As War Follows War." *New York Times,* February 5, 1996, A1.

Crossette, Barbara. "Forgotten Moroccan P.O.W.s Freed From Sandy Squalor." *New York Times,* December 8, 1995, A1.

Drysdale, Alasdair, and Gerald H. Blake. *The Middle East and North Africa: A Political Geography.* Oxford: Oxford University Press, 1985.

Frank, Harry Thomas. *Discovering the Biblical World.* Maplewood, N. J.: Hammond Inc., 1988.

George, Allen. "Dry Them Out." *The Middle East,* May 1993, 17–18.

Girardet, Edward. "Land Mines: Soviets Leave Dangerous Legacy Behind in Afghanistan." *Christian Science Monitor,* June 22, 1988, 7.

Goode's World Atlas. (For Metropolitan Area Population Figures.) New York: Rand McNally, 1995.

Hawkes, Jacquetta, ed. *Atlas of Ancient Archaeology.* New York: McGraw-Hill, 1974.

Hedges, Chris. "In a Remote Southern Marsh, Iraq Is Strangling the Shiites." *New York Times,* November 16, 1993, A1.

Held, Colbert. *Middle East Patterns.* Boulder, Colo.: Westview Press, 1994.

Ibrahim, Youssef M. "Algeria Is Edging Toward Breakup." *New York Times,* April 4, 1994, A1.

"Iraqis Are Said to Wage War on Marsh Arabs." *New York Times,* October 19, 1993, A6.

Landay, Jonathan S. "Bill to Restrict Land Mines May Be Defused on Hill." *Christian Science Monitor,* September 21, 1995, 3.

Lev, Martin. *Traveler's Key to Jerusalem.* New York: Alfred A. Knopf, 1989.

North, Andrew. "Flight of the Marsh Arabs." *The Middle East,* February 1994, 37–39.

_____. "New Evidence Shows Marshlands Draining Away." *The Middle East,* October 1993, 22–23.

Schmemann, Serge. "Israel and the Ethiopians: Welcoming Newcomers Isn't Always So Easy." *New York Times,* February 4, 1996, E3.

Stauffer, T.R. "Water Flows in Qaddafi's Pharaonic Project." *Christian Science Monitor,* January 8, 1996, 1.

Tregenza, Leon Arthur. *Egyptian Years.* London: Oxford University Press, 1958.

Velin, Jo-Anne. "Efforts to Ban Mines Persist After Setback." *Christian Science Monitor,* October 18, 1995, 6.

"Water in the Middle East: As Thick As Blood." *Economist,* December 23, 1995, 53–55.

World Book Encyclopedia Yearbook. New York: World Book Encyclopedia, 1994.

Wren, Christopher S. "Everywhere, Weapons That Keep on Killing." *New York Times,* October 8, 1995, E3.

Chapter 10

Chapman, Graham P., and Kathleen M. Baker. *The Changing Geography of Asia.* London: Routledge, 1992.

Drake, Christine. "National Integration in China and Indonesia." *Geographical Review,* 82: 209–220, 1992.

Lee, Manwoo. "North Korea: The Cold War Continues." *Current History* 95(605): 438–442, 1996.

Shinn, James. "Japan as an 'Ordinary Country.'" *Current History.* 96(605): 401–407, 1996.

Takashi, Inoguchi. "The Coming Pacific Century." *Current History.* 93(597): 25–30, 1994.

Chapter 11

Barr, Cameron. "India's Poor Assess Damage from Rioting." *Christian Science Monitor,* December 21, 1992, 2.

Basham, A.L. *The Wonder That Was India.* New York: Grove Press, 1959.

Bhardwaj, Surinder Mohan. *Hindu Places of Pilgrimage in India.* Berkeley: University of California Press, 1983.

"Blast Kills 53 in Sri Lanka; 1,400 Injured." *New York Times,* February 1, 1996, A1.

Bokhari, Farhan. "Cotton Farmers Face Big Setback." *Christian Science Monitor,* December 29, 1993, 8.

Burns, John F. "India Counts on Vote to Blunt Kashmir Insurgency." *New York Times,* September 7, 1996, A3.

Cressey, George B. *Asia's Lands and Peoples.* New York: McGraw-Hill, 1963.

Gargan, Edward A. "Shackled by Past, Racked by Unrest, India Lurches Toward Uncertain Future." *New York Times,* February 18, 1994, A4.

———. "For Many Brides in India, a Dowry Buys Death." *New York Times,* December 30, 1993, A5.

———. "Though Sikh Rebellion Is Quelled, India's Punjab State Still Seethes." *New York Times,* October 26, 1993, A1, A4.

Ginsburg, Norton, ed. *The Pattern of Asia.* Englewood Cliffs, N.J.: Prentice-Hall, 1958.

Girardet, Edward. "Helping Farmers Shake Poppy Habit." *Christian Science Monitor,* January 12, 1989, 6.

Hazarika, Sanjoy. "Plan to Clean Holy River Fails to Stem Tide of Filth." *New York Times,* October 18, 1994, B7, B10.

———. "India Dam Project Brings a Quandary." *New York Times,* June 2, 1992, A5.

Kour, Samsar Chand. *Beautiful Valleys of Kashmir and Ladakh.* Mysore, India: Wesley Press, 1942.

Lodrick, Deryck O. *Sacred Cows, Sacred Places: Origins and Survivals of Animal Homes in India.* Berkeley: University of California Press, 1981.

Malik, Rajeev. "The Threat to India's Economic Reforms." *Christian Science Monitor,* December 21, 1992, 19.

Miller, G. Tyler. *Living in the Environment,* 9th ed. Belmont, California: Wadsworth Press, 1996.

Moffett, George. "Pakistan's Population Growth Saps Economic Prosperity." *Christian Science Monitor,* September 6, 1994, 7.

Schwartzberg, Joseph E. "South Asia." In *World Geography,* 3rd ed., edited by John W. Morris, 513–552. New York: McGraw-Hill, 1972.

Spencer, J.E., and William L. Thomas. *Asia, East by South: A Cultural Geography.* New York: John Wiley & Sons, 1971.

"Sri Lanka: The Victory Still to Come." *Economist,* December 9, 1995, 39.

Tefft, Sheila. "Caste Dispute Deepens India's Political Crisis." *Christian Science Monitor,* December 28, 1990, 8.

———. "Political and Ethnic Hot Spots in South Asia." *Christian Science Monitor,* May 8, 1990, 8.

"Violence Comes to Shangri-La." *Economist,* October 6, 1990, 40.

Chapter 12

"After the Smoke." *Economist,* September 30, 1995, 40.

Aung-Thwin, Maureen. "Suu Kyi's Release: Act I in Burma's Drama." *Christian Science Monitor,* August 2, 1995, 19.

Barr, Cameron. "A Democracy Fighter's Plight." *Christian Science Monitor,* September 1, 1995, 1.

Browne, Malcolm W. "Crowding and Managerial Gaps Imperil Vietnam." *New York Times,* May 8, 1994, A1.

———. "More Species May Be Discovered in Vietnam." *New York Times,* May 3, 1994, B9.

"Burma and Bangladesh at Odds Over Muslim Refugees." *Christian Science Monitor,* January 28, 1992, 4.

Chenault, Kathy. "Cambodia: Victory at the Ballot Box, But Threat of War Lingers." *Christian Science Monitor,* October 6, 1993, 11.

Cox, C. Berry, and Peter D. Moore. *Biogeography: An Ecological and Evolutionary Approach.* Oxford: Basil Blackwell, 1993.

Erlanger, Steven. "America Opens the Door to a Vietnam It Never Knew." *New York Times,* February 6, 1994, 4A–4.

Erlich, Reese. "Low Wages in Thailand Despite Rapid Growth." *Christian Science Monitor,* August 30, 1995, 9.

———. "Philippines Population Efforts Raise Hackles." *Christian Science Monitor,* January 11, 1994, 10.

Esper, George. "The Rebuilding of Vietnam." *Columbia Missourian,* January 9, 1994, 4.

Gargan, Edward A. "A Boom in Malaysia Reaches for the Sky." *New York Times,* February 2, 1996, C1.

Gillotte, Tony. "Vietnamese Ecologist Assists Village Farmers." *Christian Science Monitor,* August 2, 1995, 14.

Hottelet, Richard C. "Dangerous Isles: Even Barren Rocks Attract Conflict." *Christian Science Monitor,* March 7, 1996, 18.

Kemf, Elizabeth. *The Month of Pure Light: The Regreening of Vietnam.* London: The Women's Press, 1990.

Jones, Clayton. "Economic Cooperation Zones Create New Asian Geometry." *Christian Science Monitor,* December 1, 1993, 12.

———. "Paradise Islands or an Asian Powder Keg?" *Christian Science Monitor,* December 1, 1993, 14.

———. "Search for Security in the Pacific." *Christian Science Monitor,* November 17, 1993, 11.

———. "From Carpet Bombing to Capitalism in Laos." *Christian Science Monitor,* November 10, 1993, 10.

Kamm, Henry. "Decades-Old U.S. Bombs Still Killing and Maiming Laotians." *New York Times,* August 10, 1995, A5.

———. "Laos Capital Defies Time and Change." *New York Times,* August 6, 1995, A5.

———. "Communism in Laos: Poverty and a Thriving Elite." *New York Times,* July 30, 1995, A6.

Mydans, Seth. "New Boat People Exodus: Back to Vietnam." *New York Times,* April 17, 1996, A1.

Peck, Grant. "Asia's Urban Morass: Exhausted Engines of the World Economy." *Columbia Missourian,* March 6, 1994, 4.

Porter, Gareth. "The Environmental Hazards of Asia-Pacific Development: The Southeast Asian Rain Forests." *Current History* 93(587): 430–434 (1994).

Scott, David Clark. "Surging Sales for the Golden Triangle." *Christian Science Monitor,* January 6, 1989, 4.

Shari, Michael. "Indonesia's Brass Polishes Itself." *Christian Science Monitor,* September 18, 1995, 1.

Shenon, Philip. "AIDS Epidemic, Late to Arrive, Now Explodes in Populous Asia." *New York Times,* January 21, 1996, A1.

———. "Filipino Victims May Share in Marcos's Loot." *New York Times,* October 28, 1995, A1.

———. "Burmese Cry Intrusion." *New York Times,* March 29, 1994, A5.

———. "AIDS Onslaught Breaches the Burmese Citadel." *New York Times,* March 11, 1994, A7.

———. "Pol Pot, the Mass Murderer Who Is Still Alive and Well." *New York Times,* February 6, 1994, A1.

Simpson, Beryl Brintnall, and Molly Conner-Ogorzaly. *Economic Botany: Plants in Our World.* New York: McGraw-Hill, 1986.

"Singapore Celebrates Economic Success." *Dodge City* (Kan.) *Daily Globe,* August 12, 1995, 5.

Stanley, Bruce. "Ho's Heirs in Hanoi Divide Power." *Christian Science Monitor,* August 4, 1995, 1.

———. "Vietnam Revels As the World Beats a Path to Its Open Door." *Christian Science Monitor,* July 18, 1995, 7.

Tan, Abby. "Burma's Star Dissident Emerges." *Christian Science Monitor,* July 13, 1995, 6.

Tefft, Sheila. "Boat People Languish: Asia's Mired Huddled Masses." *Christian Science Monitor,* December 7, 1995, 1.

———. "Indonesian Regime Retains Grip but Faces, and Allows, More Dissent." *Christian Science Monitor,* February 25, 1994, 8.

———. "A New Kind of Battle Rages at Angkor Wat." *Christian Science Monitor,* September 18, 1992, 10.

———. "Cambodia's Long, Tough Road Home." *Christian Science Monitor,* May 20, 1992, 9.

———. "Thai Spirit Houses." *Christian Science Monitor,* August 30, 1991, 13.

"Thailand's Tourist Industry: Beached." *Economist,* July 6, 1991, 72.

Thong, Huynh Sanh. *The Heritage of Vietnamese Poetry,* 7, 69, 127, 209. New Haven: Yale University Press, 1979.

Tyler, Patrick E. "China Battles a Spreading Scourge of Illicit Drugs." *New York Times,* November 15, 1995, A1.

"With the Rebels." *Economist,* October 14, 1995, 39.

Chapter 13

"Can a Bear Love a Dragon?" *The Economist.* 343(8014): 19–21, 1997.

Cannon, Terry and Alan Jenkins. *The Geography of Contemporary China: The Impact of Deng Xiaoping's Decade.* London: Routledge, 1992.

Fan, C. Cindy. "Economic Opportunities and Internal Migration: A Case Study of Guangdong Province." *Professional Geographer.* 48(1): 28–45, 1996.

Kaye, Lincoln. "The Grip Slips." *Far Eastern Economic Review* 158(19): 18–20, 1995.

Minqi, Le. "China: Six Years After Tiananmen." *Monthly Review* (20): 1–13, 1996.

"No Truck with Trains: China's Joint Ventures Take to the Highway." *Far Eastern Economic Review.* 159(3): 46, 1996.

Pannell, Clifton W., and Laurence J.C. Ma. *China: The Geography of Development and Modernization.* New York: John Wiley & Sons, 1983.

"Stay Back, China". *Economist.* 338(7957): 39–41, 1996.

Topping, Audrey R. "Ecological Roulette: Damming the Yangtze." *Foreign Affairs.* 74(5): 132–147, 1995.

Chapter 14

Barr, Cameron W. "Tiny Islands Emerge As Big Dispute." *Christian Science Monitor,* February 16, 1996, 6.

Basic Facts on the Nanjing Massacre and the Tokyo War Crimes Trial. Bound Brook, N.J.: New Jersey–Hong Kong Network, 1996 (pamphlet and WWW document).

Blustein, Paul. "Women Need Not Apply." *Washington Post,* National Weekly Edition, August 28–September 3, 1995, 18.

Brauchli, Marcus W., and David P. Hamilton. "Watch Out, Investors: Asia's Hot Spots Are Flaring." *Wall Street Journal,* August 18, 1995, A6.

Broad, William J. "Japan Plans to Conquer Sea's Depths." *New York Times,* October 18, 1994, B7.

———. "New Study Questions Hiroshima Radiation." *New York Times,* October 13, 1992, A6.

Barr, Cameron W. "Kobe's Political Aftershock: Defiant Japanese." *Christian Science Monitor,* July 20, 1995, 1.

Clark, Gregory. "The Confidence Gap Widens in a Shaken Japan." *International Herald Tribune,* February 1, 1995, 8.

Cutler, B.J. "Four Small Islands Prevent Treaty." *Columbia Missourian,* October 21, 1990, 5F.

De Souza, Anthony R., and Frederick P. Stutz. *The World Economy: Resources, Location, Trade and Development.* New York: Macmillan, 1994.

Erlanger, Steven. "As Clinton Visits Changing Asia, Military Concerns Gain Urgency." *New York Times,* April 15, 1996, A1.

Gurdon, Meghan Cox. "Calm in Korea Belies Tension in Nuclear Dispute." *Christian Science Monitor,* February 1, 1994, 1.

Holstein, William J., and Kaxmi Nakarmi. "Korea." *Business Week,* July 31, 1995, 32–38.

"Japan's Unspoken Fears." *Economist,* October 7, 1995, 35–36.

Kambara, Keiko. "Japan's Birthrate Hits Low." *Christian Science Monitor,* March 20, 1990, 4.

Kristof, Nicholas D. "North Korea Wouldn't Invade the South, Would It?" *New York Times,* April 14, 1996, D1.

———. "South Korea's President, in a Rift With North, Delays Talks." *New York Times,* October 15, 1995, A4.

———. "In Japan, Chicken Little Lays the Golden Egg." *New York Times,* July 10, 1995, E5.

MacDonald, Donald. *A Geography of Modern Japan.* Woodchurch, Ashford, Kent (England): Paul Norbury Publications, 1985.

Matthews, Rupert O. *The Atlas of Natural Wonders.* New York: Facts on File, 1988.

McGill, Douglas C. "Scour Technology's Stain With Technology." *New York Times Magazine,* October 4, 1992, 32–60.

Moffett, George. "North Korea Curtails Its Nuclear Program." *New York Times,* September 21, 1995, A1.

Ozawa, Ichiro. *Blueprint for a New Japan: The Rethinking of a Nation.* New York: Kodansha International, 1995.

Palmer, R.R., and Joel Colton. *A History of the Modern World.* New York: Alfred A. Knopf, 1971.

Pollack, Andrew. "Behind North Korea's Barbed Wire: Capitalism." *New York Times,* September 15, 1996, A3.

————. "Okinawans Send Message to Tokyo and U.S. to Cut Bases." *New York Times,* September 9, 1996, A3.

————. "Armed North Korea Troops Again Violate the DMZ." *New York Times,* April 8, 1996, A7.

————. "The Creed on the Farm: Rice Land is Sacred." *New York Times,* February 18, 1993, A4.

————. "Japan's Role in Ecology: Leadership That Has Had a Slow Start at Home." *New York Times,* July 31, 1992, A7.

Powell, Bill. "End of the Age of Hubris." *Newsweek,* January 30, 1995, 33.

Reid, T.R. "Kobe Wakes to a Nightmare." *National Geographic* 188(1): 112–136 (July 1995).

Reid, T.R., and Paul Blustein. "What a Difference a Half Century Makes." *Washington Post,* National Weekly Edition, August 21–27, 1995, 21.

Sanger, David E. "Japan, Bowing to Pressure, Defers Plutonium Projects." *New York Times,* February 23, 1994, A2.

————. "Japan Denies Any Plans to Build Nuclear Bombs." *New York Times,* February 2, 1994, A5.

Shinohara, Makiko. "Scientists Clash Over Whaling." *Christian Science Monitor,* February 27, 1992, 10.

Spaeth, Anthony. "Engineer of Doom." *Time,* June 12, 1995, 57.

Spencer, E.W. "Japan: Stimulus or Scapegoat?" *Foreign Affairs* 62(1): 123–137 (1983).

Sterngold, James. "Life in a Box: Japanese Question Fruits of Success." *New York Times,* January 2, 1994, A1.

"A Striptease in Japan." *Economist,* October 14, 1995, 20.

Wood, Daniel B. "California Rice Growers Anticipate Upturn in Production, Demand." *Christian Science Monitor,* December 13, 1993, 7.

Chapter 15

Chaddock, Gail Russell. "How French Islanders Live Above Ground Zero." *Christian Science Monitor,* August 31, 1995, 6.

————. "Avast, Ye Nuclear Protestors! Prepare to Be Boarded!" *Christian Science Monitor,* August 25, 1995, 6.

————. "French Wallets Take Hit in Protests Over Nuclear Tests." *Christian Science Monitor,* August 25, 1995, 1

————. "France Drops an Economic Bomb on Tahiti." *Christian Science Monitor,* August 14, 1995, 8.

Chaddock, Gail Russell, and David Rohde. "France, Colonies Stay Mum on Nukes." *Christian Science Monitor,* July 12, 1995, 7.

Cox, C. Barry, and Peter D. Moore. *Biogeography: An Ecological and Evolutionary Approach.* Oxford: Basil Blackwell, 1993.

DeBlij, H.J. *Human Geography.* New York: John Wiley & Sons, 1993.

Detwyler, Thomas R., ed. *Man's Impact on the Environment.* New York: McGraw-Hill, 1971.

Diamond, Jared. "Easter's End." *Discover* 16(8): 62–69 (1995).

Foster, Catherine. "War in the Pacific: Legacy of a Copper Mine." *Christian Science Monitor,* July 20, 1994, 11.

Freeman, Otis W. "The Pacific Island World." *Journal of Geography* 44: 25 (1945).

Gabler, Robert E., et al. *Essentials of Physical Geography.* Philadelphia: Saunders College Publishing, 1987.

Jordan, Terry G., and Lester Rowntree. *The Human Mosaic.* New York: Harper and Row, 1986.

Miller, G. Tyler. *Living in the Environment.* Belmont, Calif.: Wadsworth, 1994.

Mitchell, Andrew. *The Fragile South Pacific: An Ecological Odyssey.* Austin: University of Texas Press, 1989.

Oliver, Douglas L. *The Pacific Islands.* New York: Doubleday Anchor Books, 1962.

Scott, David Clark. "Landowners Claim Bigger Slice of Mineral Pie in Papua New Guinea." *Christian Science Monitor,* December 27, 1988, 9.

Shenon, Philip. "A Pacific Island Nation Is Stripped of Everything." *New York Times,* December 10, 1995, A3.

————. "Tahiti's Antinuclear Protests Turn Violent." *New York Times,* September 8, 1995, A4.

Whitney, Craig R. "Under Pressure, France Is Ending Its Nuclear Tests." *New York Times,* January 30, 1996, A1.

Chapter 16

"Australia." *Nature* (film series). Narrated by George Page. Washington, D.C.: Public Broadcasting System, 1988.

Blainey, Geoffrey. *Triumph of the Nomads: A History of Ancient Australia.* Melbourne: Macmillan, 1982.

Chaddock, Gail Russell. "Australian Farmers Consider Commercializing Kangaroo." *Christian Science Monitor,* November 29, 1995, 15.

————. "Australia's Hidden Strength in Asia: Chinese Immigrants." *Christian Science Monitor,* September 21, 1995, 7.

Evans, Howard Ensign, and Mary Alice Evans. *Australia: A Natural History.* Washington, D.C.: Smithsonian Institution Press, 1983.

Foster, Catherine. "Darwin, Aussie Pioneer Town, Seeks Survival in Asian Markets." *Christian Science Monitor,* June 22, 1994, 10.

————. "U.S., New Zealand Mend Ties Despite Differences." *Christian Science Monitor,* February 22, 1994, 6.

————. "Australia's Diamonds Lead Market Uptick." *Christian Science Monitor,* February 2, 1994, 7.

————. "New Gold Mining Activity Surges Ahead in Australia." *Christian Science Monitor,* January 12, 1994, 8.

————. "Australia Grants Aborigines Right to Claim Native Title." *Christian Science Monitor,* December 23, 1993, 3.

————. "Family Setting Fights Addiction." *Christian Science Monitor,* November 17, 1993, 18.

————. "Aborigines Unite to Fight for Australian Land Claims." *Christian Science Monitor,* August 9, 1993, 4.

————. "Aboriginal Families Get Their Land." *Christian Science Monitor,* January 6, 1993, 7.

"Gold Diggers." *Economist,* September 23, 1995, 5.

Grant, Bruce. "Human Rights in Asia: Australia Confronts an Identity Crisis." *New York Times,* March 20, 1994, E5.

MacLeod, Alexander. "New Zealand Shepherds in a Contentious Revolution." *Christian Science Monitor,* November 22, 1995, 6.

Moorehead, Allen. *Cooper's Creek.* New York: Macmillan, 1963.

Pool, Gail. "Traveling in the Dreaming Tracks of Aboriginal Australia." *Christian Science Monitor,* September 2, 1987, 20.

Rohde, David. "Outback and New Market Needs Challenge Australian Cattleman." *Christian Science Monitor,* July 26, 1995, 9.

———. "Aussie Voters, Wary of Rapid Change, Put Heat on Leaders." *Christian Science Monitor,* July 21, 1995, 7.

Scherer, Ron. "Living Off the Sheep's Back." *Christian Science Monitor,* April 16, 1992, 10.

———. "Australia to Improve Native-Rights System." *Christian Science Monitor,* April 1, 1992, 6.

———. "Australian Aborigines Turn to Antidrinking Programs." *Christian Science Monitor,* March 30, 1992, 15.

———. "Australians Confront Poor Treatment of Aborigines." *Christian Science Monitor,* May 22, 1991, 6.

———. "What's Hot, Dry and Big? Australia's Outback." *Christian Science Monitor,* December 19, 1990, 10.

Scott, David Clark. "Knocking Down Trade Walls." *Christian Science Monitor,* August 19, 1988, 7.

Shenon, Philip. "Australian Labor Party Retains Power in Election." *New York Times,* March 14, 1993, A3.

Sullivan, Jack. *Banggaiyerri: The Story of Jack Sullivan As Told to Bruce Shaw.* Canberra: Australian Institute of Aboriginal Studies, 1983.

Terrill, Ross. "The Aborigines' Search For Justice." *World Monitor,* May 1989, 48–55.

———. "Australia's Next Frontier." *World Monitor,* October 1988, 36–46.

"Trade in the Pacific: No Action, No Agenda." *Economist,* November 25, 1995, 75.

Twain, Mark. *Following the Equator.* Reprint. New York: Dover, 1989.

Chapter 17

"Africa's Colonial Favourites." *Economist,* October 28, 1995, 20.

Anderson, Terry L. "Zimbabwe Makes Living With Wildlife Pay." *Wall Street Journal,* October 25, 1991.

Battersby, John. "Severe Drought Threatens Reform in Southern Africa." *Christian Science Monitor,* March 23, 1992, 1.

Butler, Victoria. "Is This the Way to Save Africa's Wildlife?" *International Wildlife,* March/April 1995, 38–43.

Camerapix Publishers International. *Spectrum Guide to Zimbabwe.* Edison, N.J.: Hunter Publishing, Inc., 1991.

French, Howard W. "African Democracies Worry Aid Will Dry Up." *New York Times,* March 19, 1995, A1.

———. "An Ignorance of Africa As Vast As the Continent." *New York Times,* November 20, 1994, E3.

Hanley, Charles J. "Starting From Zero: Self-Sufficiency is a Fantasy for Nations of Sub-Saharan Africa." *Columbia Missourian,* February 28, 1993, 6.

International Union for the Conservation of Nature and Natural Resources. *The Nature of Zimbabwe: A Guide to Conservation and Development.* Gland, Switzerland: IUCN, 1988.

Keller, Bill. "Southern Africa's Old Front Line Ponders Its Future in Mainstream." *New York Times,* November 20, 1994, A1.

———. "Even Short of Horns, Rhinos of Zimbabwe Face Poacher Calamity." *New York Times,* October 11, 1994, B7.

Lederer, Edith. "The Next Famine." *Columbia Missourian,* August 21, 1994, 4.

McPherson, James M. "Involuntary Immigrants." *New York Times Book Review,* 24.

Miller, G. Tyler. *Living in the Environment.* Belmont, Calif.: Wadsworth Press, 1994.

Moffett, George D. "African Elephants to Remain Protected by International Treaty." *Christian Science Monitor,* March 13, 1992, 7.

Nasar, Sylvia. "Political Causes of Famine: It's Never Fair Just to Blame the Weather." *New York Times,* January 17, 1993, A1.

Nelson, Joan M. "Democracy in Africa." *Christian Science Monitor,* March 30, 1993, 18.

Noble, Kenneth B. "Political Chaos in Zaire Disrupts Efforts to Control AIDS Epidemic." *New York Times,* A1.

Perlez, Jane. "Rhino Near Last Stand, Animal Experts Warn." *New York Times,* July 7, 1992, A5.

———. "Zimbabwe Kills Elephants to Help Save Lives." *New York Times,* July 5, 1992, A1.

Press, Robert M. "Mali Elections Break New Ground." *Christian Science Monitor,* February 13, 1992, 4.

Shiner, Cindy. "Wests Finds a Darling in Africa." *Christian Science Monitor,* October 5, 1995, 6.

Stock, Robert. *Africa South of the Sahara: A Geographical Interpretation.* New York: Guilford, 1995.

Varisco, Daniel Martin. "Rhinoceros Horn Is Also the Animal's Achilles' Heel." *Christian Science Monitor,* June 28, 1988, 19.

Weisman, Steven R. "Bluefin Tuna and African Elephants Win Some Help at a Global Meeting." *New York Times,* March 11, 1992, A7.

Zweifel, Thomas D. "New, Genuine Leaders in Africa." *Christian Science Monitor,* September 6, 1995, 19.

Chapter 18

Adams, R. and M., and A. Willens. *Dry Lands: Man and Plants.* London: Architectural Press, 1978.

"After the Hangings." *Economist,* November 18, 1995, 41.

Battersby, John. "Going Home in Southern Africa." *Christian Science Monitor,* March 11, 1994, 6.

———. "Namibian Sovereignty Complete As South Africa Withdraws from Port." *Christian Science Monitor,* March 1, 1994, 7.

———. "Zimbabwe: Unity Pact Papers Over Persistent Tensions." *Christian Science Monitor,* December 28, 1992, 11.

Beran, Pau. "The Only Way to Dislodge Nigeria's Dictator." *Christian Science Monitor,* November 24, 1995, 19.

Biswas, M.R., and A.K. Biswas. *Desertification: Environmental Science and Application,* vol. 12. New York: Pergamon Press, 1980.

Botkin, Daniel B., Margriet F. Caswell, John E. Estes, and Angelo A. Ario, eds. *Changing the Global Environment.* Boston: Academic Press, 1989.

Browne, Malcolm W. "Fish That Dates Back to Age of Dinosaurs Is Verging on Extinction." *New York Times,* April 18, 1995.

"Burundi—the Next Bloodbath?" *Economist,* October 14, 1995, 49.

Crown, Sarah. "In Adopting Black Babies, Whites 'Become African.'" *Christian Science Monitor,* December 26, 1995, 1.
———. "Edgy Whites Still Fleeing South Africa." *Christian Science Monitor,* September 20, 1995, 1.

Daley, Suzanne. "South African Democracy Stumbles in Old Rivalry." *New York Times,* January 7, 1996, A1.
———. "Tradition-Bound Swazis Chafing Under Old Ties." *New York Times,* December 16, 1995, A4.
———. "As Crime Soars, South African Whites Leave." *New York Times,* December 12, 1995, A1.
———. "Foes in Angola Still at Odds Over Diamonds." *New York Times,* September 15, 1995, A1.

Darnton, John. "Intervening with Élan and No Regrets." *New York Times,* June 26, 1994, E3.

De Waal, Alex, and Rakiya Omaar. "The Genocide in Rwanda and the International Response." *Current History* 94(591): 156–161 (1995).

French, Howard W. "Where Proud Moors Rule, Blacks Are Outcasts." *New York Times,* January 11, 1996, A5.
———. "Nigeria Comes on Too Strong." *New York Times,* November 19, 1995, E3.
———. "Repression in Nigeria." *New York Times,* November 12, 1995, 9.
———. "Nigeria Executes Critic of Regime; Nations Protest." *New York Times,* November 11, 1995, A1.
———. "Africa's Ballot Box: Look Out for Sleight of Hand." *New York Times,* October 24, 1995, A3.
———. "Africa's Nations Start to Be Their Brothers' Keepers." *New York Times,* October 15, 1995, E6.
———. "Freetown Journal: Darkness, Not the Diamond's Dazzle." *New York Times,* October 9, 1995, A4.
———. "Nigeria Chief Offers Foes Concessions." *New York Times,* October 2, 1995, A5.
———. "Sleepy Congo, a Poor Land Once Very Rich." *New York Times,* June 18, 1995, 3.

Goldman Environmental Foundation. "Save One of the World's Brave and Honorable Environmental Heroes." *New York Times,* November 3, 1995, A7.

Gritzner, Jeffrey Allman. *West African Sahel: Human Agency and Environmental Change.* University of Chicago Geography Research Paper No. 226, 1988.

Hackel, Joyce. "A Genocide Later, Rwanda Again on Edge." *Christian Science Monitor,* November 28, 1995, 6.
———. "Rwandan Troubles Made Worse by a Shake-Up." *Christian Science Monitor,* August 30, 1995, 1.

Jolly, Alison, and Richard Jolly. "Malagasy Economics and Conservation: A Tragedy Without Villains." In *Key Environments: Madagascar,* 211–217. Oxford: Pergamon Press, 1984.

Kassas, Mohamed A.F. "Ecology and Management of Desertification." In *National Geographic,* 198–211.
———. *Earth '88: Changing Geographic Perspectives.* Washington, D.C.: National Geographic Society, (On reserve in Ellis Library.)

Kazaure, Z.M. "Nigerian Military Government Prepares for Return to Democratic Rule." *New York Times,* October 23, 1995, A9.

Kobani, Badey and Orage Memorial Foundation. "Until Now, Everything You've Read About the Ogoni Situation in Nigeria Has Been One-Sided." *New York Times,* December 6, 1995, A17.

Le Houerou, Henri Noel, and Hubert Gillet. "Conservation versus Desertization in African Arid Lands." In *Conservation Biology: The Science of Scarcity and Diversity,* edited by Michael E. Soule, 444–462. Sinauer Associates, 1986.

Lewis, L.A., and L. Berry. *African Environments and Resources.* Boston: Unwin Hyman Press, 1988.

Lorch, Donatella. "Even With Peace and Rain, Ethiopia Fears Famine." *New York Times,* January 3, 1996, A3.
———. "Ethiopia Deals With Legacy of Kings and Colonels." *New York Times,* December 31, 1995, 3.
———. "Trying to Avert a New Rwanda Refugee Crisis." *New York Times,* November 26, 1995, A6.
———. "Zanzibar Journal: Where Life Has No Spice, a $1 Billion Pick-Me-Up." *New York Times,* November 6, 1995, A4.
———. "A Joyful but Anxious Vote in Tanzania." *New York Times,* October 29, 1995, A8.
———. "Kenya Refuses to Hand Over Suspects in Rwanda Slayings." *New York Times,* October 6, 1995, A3.
———. "Kampala Journal: Cast Out Once, Asians Come Home." *New York Times,* March 22, 1993, A6.

MacArthur, R.H., and E.O. Wilson. *The Theory of Island Biogeography.* Princeton: Princeton University Press, 1967.

McKinley, James C. "Growing Ethnic Strife in Burundi Leads to Fears of New Civil War." *New York Times,* January 14, 1996, A1.

"Madagascar: Money Missing? Who Cares?" *Economist,* October 14, 1995, 50.

Matloff, Judith. "W. Africans Fret Sanctions Will Stir Chaos in Nigeria." *Christian Science Monitor,* December 16, 1995, 6.

McKinley, James C. "Despite Pledge, Burundi Strife Grows." *New York Times,* September 19, 1996, A8.

Moose, George E. "Angola: After Decades of War, Bright Prospects." *Christian Science Monitor,* January 3, 1996, 18.

Nixon, Rob. "The Oil Weapon." *New York Times,* November 17, 1995, A23.

Noble, Kenneth B. "Zaire's Once-Wealthy Mines Now the Prey of Scavengers." *New York Times,* February 21, 1994, A1.
———. "Zaire Is in Turmoil After the Currency Collapses." *New York Times,* December 12, 1993.
———. "Glittering, New Capital Belies Turmoil in Nigeria." *New York Times,* October 31, 1993, 3.
———. "In Zaire, Starvation Is Growing in a Wealthy Land." *New York Times,* May 16, 1993.

Ntuyahaga, Monsignor. "Here Come the Cows." In *A Selection of African Prose,* edited by W.W. Whiteley 142–145. Oxford: Clarendon, 1964.

Perlez, Jane. "In Ethiopia, Islam's Tide Laps at the Rock of Ages." *New York Times,* January 13, 1992, A2.

Sammakia, Nejla. "Revisiting Somalia One Year Later." *Columbia Missourian,* December 12, 1993, 4.

Schulz, William F. "The U.S. Ignores Africa at Its Own Peril." *Christian Science Monitor,* November 6, 1995, 19.

Shiner, Cindy. "Roots of Ex-U.S. Slaves Still Run Deep in Liberia." *Christian Science Monitor,* October 26, 1995, 7.
———. "Liberia Makes a Fresh Start After Six Years of Civil War." *Christian Science Monitor,* September 5, 1995, 6.

Simons, Marlise. "France Invades Comoros to Halt Coup." *New York Times,* October 5, 1995, A6.

Smith, Donald. "Horn of Africa." *St. Louis Post-Dispatch,* December 20, 1992, E1.

Stauffer, Thomas R. "West Africans Skirmish—Over Oil?" *Christian Science Monitor,* March 22, 1994, 9.

Stock, Robert. *Africa South of the Sahara: A Geographical Interpretation.* New York: Guilford, 1995.

Weber, Peter. "Neighbors Under the Gun." *Worldwatch,* July/August 1991, 35–36.

Wilson, Edward O. "Threats to Biodiversity." *Scientific American* 261: 60–66.

Chapter 19

Kopinak, Kathryn. *Desert Capitalism: Maquiladoras in North America's Western Industrial Corridor.* Tucson: University of Arizona Press. 1966.

Price, Marie. "Ecopolitics and Environmental Nongovernmental Organizations in Latin America." *Geographical Review.* 84(1): 42–58, 1994.

Tedeschi, Tony. "Natural Attractions of the Caribbean and Central and South America." *Audubon.* 95(3): 113–119, 1993.

Chapter 20

Goulding, Michael. "Flooded Forests of the Amazon." *Scientific American.* 268(3): 114–120, 1993.

Goulding, Michael et al. *Floods of Fortune: Ecology and Economy Along the Amazon.* New York: Columbia University Press. 1996.

"Fact Sheet: Coca Production and the Environment." *U.S. Department of State Dispatch.* 3(9): 165, 1992.

Franco, Mario de and Ricardo Godoy. "The Economic Consequences of Cocaine Production in Bolivia: Historical, Local, and Macroeconomic Perspectives." *Journal of Latin American Studies.* 24(2): 375–406, 1992.

Manfredo, Jr., Fernando. "The Future of the Panama Canal." *Journal of Interamerican Studies & World Affairs.* 35(3): 103–128, 1993.

Marcus, Joyce. "The Amazon: Divergent Evolution and Divergent View." *National Geographic Research & Exploration.* 10(4): 384–394, 1994.

"Postrevolutionary Central America." *Current History.* v96 n607: 49–86. 1997.

Chapter 21

Barkema, Alan and Mark Drabenston. "A New Agricultural Policy for a New World Market." *Economic Review.* 79(2): 59–72, 1994.

Fox, Annette B. "Environment and Trade: The NAFTA Case." *Political Science Quarterly.* 110(1): 49–68, 1995.

Chapter 22

Parizeau, Jacques. "The Case for a Sovereign Quebec." *Foreign Policy.* 99: 69–76, 1995.

Pelly, David. "Birth of an Inuit Nation." *Geographical Magazine.* 66(4): 23–25, 1994.

Salee, Daniel. "Identities in conflict: The Aboriginal question and the politics of recognition in Quebec." *Ethnic and Racial Studies.* 28(2): 277–314, 1995.

Hamley, Will. "Problems and Challenges in Canada's Northwest Territories." *Geography.* 78(340): 267–280, 1993.

Kaplan, David H. "Population and Politics in a Plural Society: The Changing Geography of Canada's Linguistic Groups." *Annals of the Association of American Geographers.* 84: 46–67, 1994.

Chapter 23

Elazar, Daniel J. *The American Mosaic: The Impact of Space, Time, and Culture on American Politics.* Boulder, CO: Westview Press, 1994.

Fost, Dan. "California Comeback." *American Demographics.* 17(7): 52–53.

Frey, William H. and Alden Speare, Jr. "America at Mid-Decade." *American Demographics.* 17(2): 23–31, 1992.

Garreau, Joel. *Edge City: Life on the New Frontier.* New York: Doubleday, 1991.

Hudson, John C. *Making the Corn Belt: A Geographical History of Middle-Western Agriculture.* Bloomington: Indiana University Press, 1994.

Jakle, John A. and David Wilson. *Derelict Landscapes: The Wasting of America's Built Environment.* Savage, MD: Rowman & Littlefield. 1992.

Miyares, Ines M. "Changing Perceptions of Space and Place as Measures of Hmong Acculturation." *Professional Geographer.* 49(2): 214–224, 1997.

Moody, David W. "Water: Fresh Water Resources of the United States." *National Geographic Research & Exploration.* 9(1): 81–85, 1993.

Page numbers in parentheses refer to places in the text where terms are initially or fully discussed.

Absolute (mathematical) location Determined by the intersection of lines such as latitude and longitude, providing an exact point expressed in degrees, minutes, and seconds. (p. 11)

Acid rain Precipitation that mixes with airborne industrial pollutants, causing the moisture to become highly acidic, and therefore harmful to flora and bodies of water on which it falls. Sulfuric acid is the most common component of this acid precipitation. (p. 78)

Adaptive reuse Finding new uses for older buildings and stores, often accompanied by a shift from decline to steady renewal in an urban neighborhood. (p. 573)

Age of Exploration (also called the Age of Discovery) Era of European seafaring lasting from the 15th century through the early 19th century. (pp. 38, 74)

Age-structure profile (population pyramid) Graphic representation of a country's population by gender and five-year age increments (p. 47)

Agricultural or Neolithic (New Stone Age) Revolution The domestication of plants and animals that began about 10,000 years ago. (p. 36)

Altitudinal zonation In a highland area, the presence of distinctive climatic and associated biotic and economic zones at successively higher elevations. Ethiopia and Bolivia are examples of these kinds of zones. See *tierra caliente, tierra templada, tierra fria.* (p. 464)

anticyclone Atmospheric high-pressure cell. In the cell, the air is descending and becomes warmer. As it warms, its capacity to hold water vapor increases, and the result is minimal precipitation. (p. 26)

Anti-Semitism Anti-Jewish sentiments and activities. (p. 222)

Apartheid The Republic of South Africa's former official policy of "separate development of the races," designed to ensure the racial integrity and political supremacy of the white minority. (p. 441)

Arable Suitable for cultivation.

Archipelago Chain or group of islands. (p. 263)

Arid China In a climatic division of China approximately along the 20° isohyet, Arid China lies to the west of the line. It characterizes more than half of the territory of the country, but accommodates less than 10 percent of China's population. (p. 321)

Atmospheric pollution The modification of the blanket of gases that surround the earth largely through the airborne products of industrial production and the human consumption of fossil fuels. (p. 78)

Atoll Low islands made of coral and usually having an irregular ring shape around a lagoon.

Balkanization Fragmentation of a political area into many smaller independent units, as in former Yugoslavia. (p. 149)

Basin irrigation Ancient Egyptian system of cultivation using fields saturated by seasonal impoundment of Nile floodwaters. (p. 234)

Bazaar (suq) Central market of the traditional Middle Eastern city, characterized by twisting, close-set lanes, and merchant stalls. (p. 218)

Belief systems The set of customs that an individual or culture group has relating to religion, social contracts, and other aspects of cultural organization. (p. 70)

Benelux The name used to collectively refer to the countries of Belgium, Netherlands, and Luxembourg. (p. 113)

Biodiversity hot spots A ranked list of places scientists believe deserve immediate attention for flora and fauna study and conservation. (p. 33)

Biological diversity (biodiversity) The number of plant and animal species and the variety of genetic materials these organisms contain. (p. 32)

Biomass The collective dried weight of organisms in an ecosystem. (p. 35)

Biome Terrestrial ecosystem type categorized by a dominant type of natural vegetation. (p. 28)

Birth rate The annual number of live births per thousand people in a population. (p. 45)

Boat people In this text, the term given to the refugee populations who attempted to flee Cuba and Haiti in the early 1990s by launching themselves from the islands toward Florida in the hope that they would get picked up by the U.S. Coast Guard and given haven. The term had an earlier use by peoples in Southeast Asia making similar efforts to flee Vietnam.

Bourgeoisie In Marxist doctrine, the capitalist class.

Brain drain The exodus of educated or skilled persons from a poor to a rich country, or from a poor to a rich region within a country. (p. 310)

Break-of-bulk point A classic geographic term describing a point in transit when bulk goods must be removed from one mode of transport and installed on another. Trainloads of grain carried to a port for transhipment on cargo boats or barges is a common example. (p. 327)

Buffer state The term used for a generally smaller political unit adjacent to a large, or between several large, political units. Such a role often times enables the smaller state to maintain its independence because of its mutual use to the larger, proximate nations. Uruguay, between Brazil and Argentina, is an example. (p. 119)

Capital goods Goods used to produce other goods. (p. 355)

Carrying capacity The size of a population of any organism that an ecosystem can support. (p. 37)

Cartogram A special map in which an area's shape and size are defined by explicit characteristics of population, economy, or distribution of any stated product. (p. 11)

Cartography   The craft of designing and making maps, the basic language of geography. In recent years, this traditional manual art has been changed profoundly through the use of computers and Geographic Information Systems (GIS), and through major improvements in machine capacity to produce detailed, colored, map products. (p. 9)

Cash (commercial) crops   Crops produced generally for export. (p. 43)

Caste   Hierarchy in the Hindu religion that determines a person's social rank. It is determined by birth and cannot be changed. (p. 290)

Chernobyl   The site in the Ukraine where, in April 1986, the worst nuclear power plant accident in history occurred. It is thought that approximately 5000 people died and a zone with a twenty mile radius is still virtually uninhabitable. 116,000 people were moved from the area and cleanup continues to this day. (p. 78)

Chernozem   A Russian term meaning "black earth." It is a grass-land soil that is exceptionally thick, productive, and durable. (p. 159)

Chestnut soils   Productive soils typical of the Russian steppe and North American Great Plains. (p. 160)

China proper   The relatively well-watered eastern portion of China, which is the home of the great majority of the Chinese population. It is sometimes described as Humid China, as opposed to Arid China that lies to the west, and that has relatively little population. (p. 325)

Chokepoint   A strategic narrow passageway on land or sea that may be closed off by force or threat of force. (p. 226)

Choropleth maps   Maps that are drawn to show the differing distribution of goods or geographic characteristics (including population) across a broad area. Such maps are good for generalizations, but often mask significant local variations in the presence of the item being mapped. (p. 11)

Chunnel (Eurotunnel)   The 23-mile tunnel that links Britain with the European continent. It was completed in 1994 at a cost of more than 15 billion dollars, making it the single most costly project in landscape transformation ever undertaken. (p. 89)

Civilization   The complex culture of urban life. (p. 37)

Climate   The average weather conditions, including temperature, precipitation, and winds, of an area over an extended period of time. (p. 23)

Climatology   The scientific study of patterns and dynamics of climate. (p. 20)

Coal   Residue from organic material compressed for a long period under overlying layers of the earth. Coals vary in hardness and heating value. Anthracite coal burns with a hot flame and almost no smoke. Bituminous coal is used in the largest quantities; some can be used to make coke by baking out the volatile elements. Coke is largely carbon and burns with intense heat when used in blast furnaces to smelt iron ore. Peat, which is coal in the earliest stages of formation, can be burned if dried. Concern over atmospheric pollution by the sulfur in coal smoke has recently increased the demand for low-sulfur coals. (p. 77)

Collective farm   A large-scale farm in the Former Soviet Union that usually incorporated several villages. Workers received shares of the income after the obligations of the collective had been met. (p. 168)

Collectivization   The process of forming collective farms in Communist countries.

Colonization   The European pattern of establishing dependencies abroad to enhance economic development in the home country. (p. 38)

Colored   A South African term referring to persons of mixed racial ancestry. (p. 445)

Common Market   This is an earlier name given to the (current) 15 countries that make up the European Union. In 1957, an initial six countries combined to form the European Economic Community (EEC), and this supranational community has grown to have considerable economic and political importance in Europe. *See* European Union. (p. 81)

Computer cartography   Map making using sophisticated software and computer hardware. It is a new, actively growing career field in geography. (p. 9)

Concentric zone model   A generalized model of a city. A city becomes articulated into contrasting zones arranged as concentric rings around its central business district.

Coniferous vegetation   Needleleaf evergreen trees; most bear seed cones. (p. 29)

Consumer goods   Goods that individuals acquire for short-term use. (p. 355)

Consumer organisms   Animals that cannot produce their own food within a food chain. (p. 35)

Consumption overpopulation   The concept that a few persons, each using a large quantity of natural resources from ecosystems across the world, add up to too many people for the environment to support. (p. 45)

Containerization   Prepackaging of items into larger standardized containers for more efficient transport. (p. 535)

Convectional precipitation   The heavy precipitation that occurs when air is heated by intense surface radiation, then rises and cools rapidly. (p. 24)

Council for Mutual Economic Assistance (COMECON)   A former economic organization consisting of the Soviet Union, Poland, East Germany, Czechoslovakia, Hungary, Romania, Bulgaria, Cuba, Mongolia, and Vietnam, now disbanded. (p. 82)

Crop calendar   The dates by which farmers prepare, plant, and harvest their fields. Farmers who are involved in new agricultural patterns necessitated by greater use of chemical fertilizers and new plant strains are sometimes unable to adjust to the demands of a much more exact crop calendar than has been traditional. (p. 272)

Crop irrigation   Bringing water to the land by artificial methods. (p. 37)

Cultural geography   The study of the ways in which humankind has adopted, adapted to, and modified the face of the earth, with particular attention given to cultural patterns and their associated landscapes. It also includes a culture's influence on environmental perception and assessment. (p. 21)

Cultural landscape   The landscape modified by human transformation, thereby reflecting the cultural patterns of the resident culture at that time. (p. 16)

Cultural mores   The belief systems and customs of a culture group. (p. 70)

Cultural diffusion   One of the most important dynamics in geography, cultural diffusion is the engine of change as crops, languages, culture patterns, and ideas are diffused from one place to other places, often in the course of human migration. (p. 16)

Culture The values, beliefs, aspirations, modes of behavior, social institutions, knowledge, and skills that are transmitted and learned within a group of people. (p. 16)

Culture hearth An area where innovations develop, with subsequent diffusion to other areas. (p. 38)

Culture system A system in which Dutch colonizers required farmers in Java to contribute land and labor for the production of export crops under Dutch supervision. (p. 312)

Cyclone A low-pressure cell that composes an extensive segment of the atmosphere into which different air masses are drawn.

Cyclonic (frontal) precipitation The precipitation generated in traveling low-pressure cells which bring different air masses into contact. (p. 24)

Death rate The annual number of deaths per thousand people in a population. (p. 45)

Debt-for-nature swap An arrangement in which a certain portion of international debt is forgiven in return for the borrower's pledge to invest that amount in nature conservation. (p. 54)

Deciduous trees Broadleaf trees that lose their leaves and cease to grow during the dry or the cold season and resume their foliage and grow vigorously during the hot, wet season. (p. 29)

Delta Landform resulting from the deposition of great quantities of sediment when a stream empties into a larger body of water. (p. 69)

Demographic shift The term for major population redistributions as people move from the countryside to the city, or from one region to another. (p. 139)

Demographic transition A model describing population change within a country. The country initially has a high birth rate, a high death rate, and a low rate of natural increase, moves through a middle stage of high birth rate, low death rate, and high rate of natural increase, and ultimately reaches a third stage of low birth rate, low or medium death rate, and low or negative rate of population increase. (p. 49)

Dependency theory A theory arguing that the world's more developed countries continue to prosper by dominating their former colonies, the now-independent less developed countries. (p. 42)

Desert An area too dry to support a continuous cover of trees or grass. A desert generally receives less than 10 inches (25 cm) of precipitation per year. (p. 29)

Desertification Expansion of a desert brought about by changing environmental conditions or unwise human use. (p. 424)

Detritus Material shed from rock surfaces in the process of disintegration and erosion. (p. 264)

Development A process of improvement in the material conditions of people often linked to the diffusion of knowledge and technology. (p. 39)

Diaspora The scattering of the Jews outside Palestine beginning in the Roman Era. (p. 222)

Dissection Carving of a landscape into erosional forms by running water.

Distributary Stream that results when a river subdivides into branches in a delta. (pp. 69, 327)

Donor fatigue Public or official weariness of extending aid to needy people. (p. 50)

Drought avoidance Adaptations of desert plants and animals to evade dry conditions by migrating (animals) or being active only when wet conditions occur (plants and animals). (p. 207)

Drought endurance Adaptations of desert plants and animals to tolerate dry conditions through water storage and heat loss mechanisms. (p. 207)

Dry farming Planting and harvesting according to the seasonal rainfall cycle. (p. 37)

Ecologically dominant species A species that competes more successfully than others for nutrition and other essentials of life. (p. 37)

Ecology Study of the interrelationships of organisms to one another and to the environment. (p. 34)

Economic development A process that generally includes a major demographic shift toward a more urban population, replacement of major imports with domestically produced goods, expansion of export industries based on manufactured goods, and a general upgrading of domestic patterns of health care, energy use, and literacy. (p. 20)

Ecosphere (biosphere) The vast ecosystem composed of all of Earth's ecosystems. (p. 34)

Ecosystem System composed of interactions between living organisms and nonliving components of the environment. (p. 34)

Ejido An agricultural unit in Mexico characterized by communally farmed land, or common grazing land. It is of particular importance to indigenous villages. (p. 472)

Endemic species A species of plant or animal found exclusively in one area. (pp. 302, 375)

Energy crisis The petroleum shortages and price surges sparked by the 1973 oil embargo. (p. 227)

Environmental assessment The process of determining the condition and value of a particular environmental setting. Used both in aspects of landscape change and in evaluating environmental perception. (p. 4)

Environmental determinism The belief that the physical environment has played a major role in the cultural development of a people or locale. Also called environmentalism. (p. 16)

Environmental perception The systematic evaluation of environmental characteristics in terms of human patterns of settlement, resources, land use, and attitudes toward the earth and its settings. (p. 4)

Environmental possibilism A philosophy seen in contrast to *environmental determinism* that declares that although environmental conditions do have an influence on human and cultural development, people have varied possibilities in how they decide to live within a given environment. (p. 16)

Estuary A deepened ("drowned") river mouth into which the sea has flooded. (p. 69)

Ethnocentrism Regarding one's own group as superior and as setting proper standards for other groups.

European Economic Community Economic organization designed to secure the benefits of large-scale production by pooling resources and markets. The name has been changed to Economic Union. See Common Market and Economic Union. (p. 80)

Evaporation The loss of moisture from the Earth's land surfaces and its water bodies to the air through the ongoing influence of solar radiation and transpiration by plants. (p. 133)

Exotic species A non-native species introduced into a new area. (p. 373)

Exploration The human fascination with new landscapes that leads to the search for new places, new resources, new cultural patterns, and new routes of travel and trade. (p. 74)

Extensive land use   A livelihood, such as hunting and gathering, that requires the use of large land areas. (p. 36)

European Free Trade Association (EFTA)   Organization that maintains free trade among its members but allows each member to set its own tariffs in trading with the outside world. Members are Iceland, Norway, Sweden, Switzerland, Austria, and Finland.

European Union   The current organization begun in the 1950s as the Common Market. It now is made up of fifteen nations, including France, Germany, Italy, Belgium, Luxembourg, the Netherlands, the United Kingdom, Denmark, Ireland, Greece, Portugal, Spain, Austria, Finland, and Sweden. See *Common Market.* (p. 81)

Fall line   Zone of transition in eastern United States where rivers flow from the harder rocks of the Piedmont to the softer rocks of the Atlantic and Gulf Coastal Plain. Falls and/or rapids are characteristic features.

Fertile Crescent   The arc-shaped area stretching from southern Iraq through northern Iraq, southern Turkey, Syria, Lebanon, Israel and western Jordan, where plants and animals were domesticated beginning about 10,000 years ago. (p. 212)

Fjord   Long, narrow extension of the sea into the land usually edged by steep valley walls that have been deepened by glaciation. (p. 126)

flow resource   A resource that can be renewed, hence that extracts a lower cost from the environment in its utilization. (p. 130)

Food chain   The sequence through which energy, in the form of food, passes through an ecosystem. (p. 35)

formal region   *See* Region.

Four Modernizations   The effort of the Chinese after the death of Mao Zedong in 1976 to focus Chinese efforts at economic development in agriculture, industry, science and technology and defense. (p. 332)

Front   Contact zone between unlike air masses. A front is named according to the air mass that is advancing (cold front or warm front). (p. 25)

Frontal precipitation   *See* Cyclonic precipitation.

Fuelwood crisis   Deforestation in the less developed countries caused by subsistence needs. (p. 44)

Functional region   *See* Region.

Gaia Hypothesis   A hypothesis stating that the ecosphere is capable of restoring its equilibrium following any disturbance that is not too drastic. (p. 34)

gentrification   The social and physical process of change in an urban neighborhood by the return of young, often professional, populations to the urban core. These peoples are often attracted by the substantial nature of the original building stock of the place and the proximity to the city center, which generally continues to have major professional opportunities. While this process brings an urban landscape back into a primary role as a tax base, this change does dispossess considerable number of minority peoples who tended to remain in these neighborhoods as the white population moved to the suburbs during the past five or six decades. (p. 574)

Geographic analysis   By giving attention to the spatial aspects of a distribution, geographic analysis helps to explain distribution, density, and flow of a given phenomenon. (p. 4)

Geographic Information Systems (GIS)   The increasingly popular field of computer-assisted geographic analysis and graphic representation of spatial data. It is based on superimposing various data layers that may include everything from soils to hydrology to transportation networks to elevation. Computer software and hardware are steadily improving, enabling GIS to produce ever more detailed and exact output. (p. 21)

Geography   Study of the spatial order and associations of things. Also defined as study of places, study of relationships between people and environment, study of spatial organization. (p. 4)

Geomorphology   The scientific analysis of the landforms of the earth, sometimes called *physiography.* (p. 20)

glacial deposition   In the process of continental and valley glaciation, the deposition of moraines that become lateral or terminal—depending on where they are deposited—in the act of glacial retreat. This same process also leads to glacial scouring as moving ice picks up loose rock and reshapes the landscape as the glacier moves forward or retreats. (p. 67)

glacial scouring   See glacial deposition. (p. 67)

gravity flow   Water flow in an irrigation or a hydroelectric system that allows for the free flow of water from a source to another area without the addition of motor or animal power. In bringing water to a city such a characteristic is of particular importance because of its relative cheapness compared to the need to pump water upslope. (p. 133)

Green Revolution   The transfer of high-yielding seeds, mechanization, irrigation, and massive application of chemical fertilizers to areas where traditional agriculture has been practiced. (pp. 33, 272)

Greenhouse Effect   The observation that increased concentrations of carbon dioxide and other gasses in Earth's atmosphere causes a warmer atmosphere. (p. 51)

Gross domestic product (GDP)   Value of goods and services produced in a country in a given year. Does not include net income earned outside the country. The value is normally given in current prices for the stated year.

Growth national product (GNP)   Value of goods and services produced internally in a given country during a stated year, plus the value resulting from transactions abroad. The value is normally stated in current prices of the stated year. Such data must be used with caution in regard to developing countries because of the broad variance in patterns of data collection and the fact that many people consume a large share of what they produce. (p. 39)

growth pole   The term for a new city, a resource, or some development in a heretofore under settled area that begins to attract population growth and economic development.

Guano   Seabird excrement that is also found in phosphate deposits. (p. 374)

Gulf Stream   Strong ocean current originating in the tropical Atlantic Ocean that skirts the eastern shore of the United States, curves eastward, and reaches Europe as a part of the broader current called the North Atlantic Drift. (p. 63)

Hacienda   A Spanish term for large rural estates owned by the aristocracy in Latin America.

Haj   Pilgrimage to Mecca, the principal holy city in the Islamic religion. Every Muslim is required to make this pilgrimage at least once in a lifetime, if possible.

Han Chinese   The term for the original Chinese peoples who settled in North China, on the margins of the Yellow River (Huang He) and who were central to the development of Chinese culture. Han Chinese make up approximately 94 percent of China's current population. (p. 321)

hierarchichal rule A system by which leadership for a culture group is defined by traditional and sometimes frustrating patterns of age-sex distinction. Evident in most cultures and slowly modified by an increasing role of democratic elections and the establishment of a broader opportunity base for individual advancement. (p. 102)

historical geography Concern with the historical patterns of human settlement, migration, town building, and the human use of the earth. Often the subdiscipline that best blends geography and history as a perspective on human activity. (p. 21)

Holocaust Nazi Germany's attempted extermination of Jews, Gypsies, homosexuals and other minorities during World War II. (p. 223)

Homelands Ten former territorial units in South Africa reserved for native Africans (blacks). They had elected African governments, and some were designated as "independent" republics, although they were not recognized outside of South Africa. Formerly called Bantustans. They were abolished in 1994. (p. 442)

Horizontal migration The movements of pastoral nomads, over relatively flat areas to reach areas of pasture and water. (p. 218)

Hot spot Small area of the Earth's mantle where molten magma is relatively close to the crust. Hot spots are associated with island chains and thermal features. (p. 370)

Household Responsibility System The Chinese system devised in 1978 that enabled their rural population to begin to free itself from the communal structures that had characterized the years of development under Chairman Mao Zedong. This innovation returned much agricultural decisionmaking to the farm household, although there was a required portion of the major crops that had to be sold to the government. (p. 332)

Humboldt (or Peru) Current The cold ocean current that flows northward along the west coast of South and North America. It plays a significant role in regional fish resources.

Humid China The eastern portion of China that is relatively well-watered and where the great majority of the Chinese population has settled. An arc from Kunming in southwest China to Beijing in North China describes the approximate western margin of this zone. (p. 321)

Humid continental Climate with cold winters, warm to hot summers, and sufficient rainfall for agriculture, with the greater part of the precipitation in the summer half-year. (p. 29)

Humid pampa A level to gently rolling area of grassland centered in Argentina with a humid subtropical climate.

Humid subtropical Climate that characteristically occupies the southeastern margins of continents with hot summers, mild to cool winters, and ample precipitation for agriculture. (p. 29)

Humus Decomposed organic soil material. Grasslands characteristically provide more humus than forests do. (p. 159)

Hunting and gathering A mode of livelihood, based on collection of wild plants and hunting of wild animals, characterizing pre-agricultural peoples, generally. (p. 36)

Hydraulic control The ability to control water in irrigation systems, rivers, urban settlements, and for the generation of electricity has been central to the development of major civilizations, and to culture groups throughout human history. (p. 133)

Ideology A system of political and/or economic beliefs such as communism, capitalism, autocracy, or democracy.

Industrial Revolution A period beginning in mid-18th century Britain that saw rapid advances in technology and the use of inanimate power. (p. 36)

Intensive land use A livelihood requiring use of small land areas, such as farming. (p. 37)

Intertillage The growing of two or more crops simultaneously in alternate rows. Also called interplanting. (p. 348)

Irredentism The demand for an international transfer of territory in order to place a minority population in its alleged homeland. (p. 141)

Irrigation Being able to overcome rainfall deficiencies by the control and transport of water to farmland has historically been central to centers of major agricultural production. This act requires engineering skills, political organization all along a water source, and considerable labor investment in the initial creation and the continuing maintenance of such a system. (p. 133)

Island chain Series of islands formed by ocean crust sliding over a stationary "hot spot" in the Earth's mantle. (p. 370)

karst A landscape feature of an area built on limestone or dolomite, which are susceptible to differential erosion, with resultant sinks and mountains that seem to shoot straight up from limestone plains. Although the term comes from the Adriatic Coast in Europe, some of the world's most dramatic karst landscapes occur in southern China. (p. 328)

Keystone species A species which affects many other organisms in an ecosystem. (p. 414)

Lacustrine plain Floor of a former lake where glacial meltwater accumulated and sediments washed in and settled. An extremely flat surface is characteristic.

Law of Return An Israeli law permitting all Jews living in Israel to have Israeli citizenship. (p. 224)

Law of the Sea A United Nations treaty or convention permitting coastal nations to have greater access to marine resources. (p. 359)

Landscape Originally, a term for an artist's view of a setting for human activity but now much more broadly used to define the scene that can be observed from any given point and prospect. (p. 4)

Landscape transformation The human process of making over the earth's surface into a setting seen as more productive, more convenient, and more aesthetically pleasing. From initial human shelters to contemporary massive urban centers, humans have been driven to change their environmental settings, causing negative as well as positive environmental outcomes. (p. 17)

Large-scale map A map constructed to show considerable detail in a small area. (p. 10)

Laterite A material found in tropical regions with highly leached soils; composed mostly of iron and aluminum oxides which harden when exposed and make cultivation difficult.

Latifundia (*sing.* latifundio) Large agricultural Latin American estates with strong commercial orientations. (p. 471)

Latitude Measurement that denotes position with respect to the equator and the poles. Latitude is measured in degrees, minutes, and seconds, which are described as parallels. The low latitudes lie between the Tropics of Cancer and Capricorn; the middle latitudes lie between the Tropics and the Arctic and Antarctic Circles; the high latitudes lie between the Polar Circles and the Poles.

Less developed countries (LDCs) The world's poorer countries. (p. 39)

Lifeboat Ethics   Ecologist's Garrett Hardin's argument that, for ecological reasons, rich countries should not assist poor countries. (p. 50)

Lithospheric plates   The top layer of sediments on the earth and the plates on which human settlement and transport takes place. The -lithic of the Neolithic, for example, comes from the same foot meaning *stone*. Above the lithosphere is the biosphere (layer of life) and above that is the atmosphere (blanket of air). (p. 606)

Location   Central to all geographic analysis is the concept of location. Where something "is" relates to all manner of influences, from climate to migration routes. The classic use of this term comes in the response to the question, "What is most important in real estate?" The answer: "Location, location, location." It is a crucial component in trying to understand patterns of historic and economic development. (p. 9)

Loess   Fine-grained material that has been picked up, transported, and deposited in its present location by wind; it forms an unusually productive soil. (p. 67)

Long March   The 1935–36 flight of approximately 100,000 communist troops under Mao Zedong from south-central China to the mountains of North China. Mao was fleeing attack by General Chiang Kai-shek and the success of the Long March led to the establishment of Mao's troops as a force with the capacity to overcome the much larger and better armed troops of Chiang. (p. 318)

Longitude   Measurement that denotes a position east or west of the prime meridian (Greenwich, England). Longitude is measured in degrees, minutes, and seconds, and meridians of longitude extend from Pole to Pole and intersect parallels of latitude. (p. 10)

Malthusian scenario   The model forecasting that human population growth will outpace growth in food and other resources, with a resulting population die-off. (p. 45)

Map projection   A way to minimize distortion in one or more properties of a map (direction, distance, shape, or area). (p. 10)

Map scale   The actual distance on the earth that is represented by a given linear unit on a map. (p. 10)

Maquiladoras   Operations dedicated to the assembly of manufactured goods generally, in Mexico and Central America, from components initially produced in the United States or other places. With the enactment of NAFTA in 1994, there has been a massive expansion of *maquiladora* operations because of the economic energy generated by this regional economic alliance. (p. 484)

Marginalization   A process by which poor subsistence farmers are pushed onto fragile, inferior or marginal lands which cannot support crops for long and which are degraded by cultivation. (p. 43)

Marine west coast   Climate occupying the western sides of continents in the higher middle latitudes; greatly moderated by the effects of ocean currents that are warm in winter and cool in summer relative to the land. (p. 29)

Marshall Plan   The plan designed largely by the United States after the conclusion of World War II by which U.S. aid was focused on the rebuilding of the very Germany that had been its enemy in the war just concluded. Secretary of State George Marshall (who had been Chief of Staff of U.S. Army from 1939–45), was central to the plan's design and implementation. (p. 81)

Material culture   The items that can be seen and associated with cultural development, such as house architecture, musical instruments, and tools. In study of the cultural landscape, items of material culture add considerable personality to cultural identity. (pp. 16, 74)

Medical geography   The study of patterns of disease diffusion, environmental impact on public health, and the interplay of geographic factors, migration, and population. With the increasing ease of international movement, medical geography is becoming more important as the potential for disease diffusion increases. (p. 20)

Medina   An urban pattern typical of the Middle Eastern city before the 20th century. (p. 218)

Mediterranean (dry-summer subtropical)   Climate that occupies an intermediate location between a marine west coast climate on the poleward side and a steppe or desert climate on the equatorward side. During the high-sun period it is rainless; in the low-sun period it receives precipitation of cyclonic or orographic origin. (p. 29)

Meiji Restoration   The Japanese revolution of 1868 that restored the legitimate sovereignty of the emperor. Meiji means "Enlightened Rule." This event led to a transformation of Japan's society and economy. (p. 352)

Melanesia (Black Islands)   A group of relatively large islands in the Pacific Ocean bordering Australia. (p. 365)

Mental maps   In every individual's mind is a series of locations, access routes, physical and cultural characteristics of places, and often a general sense of the good or bad of locales. The term mental map is used to define such geographies. Education is often aimed at replacing subjective impressions of places and peoples with more objective information in such personal geographies. (p. 7)

Mercantile colonialism   The historical pattern by which Europeans extracted primary products from colonies abroad, particularly in the tropics. (p. 42)

Meridian   *See* Longitude.

Mestizo   In Latin America, a person of mixed European and native Indian ancestry. (p. 469)

Metropolitan area   A city together with suburbs, satellites, and adjacent territory with which the city is functionally interlocked.

Micronesia (Tiny islands)   Thousands of small and scattered islands in the central and western Pacific Ocean, mainly north of the Equator. (p. 368)

Microstate   A political entity that is tiny in area and population and is independent or semi-independent. (p. 58)

Middle Eastern ecological trilogy   The model of mostly symbiotic relations among villagers, pastoral nomads and urbanites in the Middle East. (p. 216)

Migration   A temporary, periodic, or permanent move to a new location. (p. 17, 45)

Milpero   The term for the Latin American farmer who engages in swidden or shifting cultivation in forest lands. The plot is often called the *milpa*. (p. 506)

Minifundia (*sing.* minifundio)   Small Latin American agricultural landholdings, usually with a strong subsistence component. (p. 471)

Mixed forest   Transitional area where both needleleaf and broadleaf trees are present and compete with each other. (p. 29)

Mobility   The pattern of human movement between work and residence, changing because of continual improvement in means of transportation. Increasing mobility has increased the impact that urban space has on adjacent farmlands in most developed countries. (p. 17)

Monsoon   A current of air blowing fairly steadily from a given direction for several weeks or months at a time. Characteristics of a monsoonal climate are a seasonal reversal of wind direction, a strong summer maximum of rainfall, and a long dry season lasting for most or all of the winter months. (p. 282)

Montreal Protocol   A 1989 international treaty to ban chlorofluorocarbons. (p. 53)

Moor   Rainy, deforested upland, covered with grass or heather and often underlain by water-soaked peat. (p. 89)

Moraine   Unsorted material deposited by a glacier during its retreat. Terminal moraines are ridges formed by long-continued deposition at the front of a stationary ice sheet. (p. 67)

More Developed Countries (MDCs)   The world's wealthier countries. (p. 39)

Mulatto   A person of mixed European and black ancestry. (p. 499)

Multiple use   A water system that enables human use of it a number of times, as in stream flow from a river into a turbine and then back into a stream or another, lower, turbine system. Because there is no waste in such a system it is highly desired and leads to sometimes complex engineering in order to achieve multiple use of this resource. (p. 130)

Multiple-cropping   Farming patterns in which several crops are raised on the same plot of land in the course of a year or even a season. Common examples of this are winter wheat and summer corn or soybeans in North America, or several crops of rice in the same plot in Monsoon Asia. (p. 318)

Nation   Commonly, a nation is a term describing the citizens of a state—or that state—but it has additional usage as a collective term for peoples of high cultural homogeneity, existing as a separate political entity. *See* Nation-state. (p. 18)

Nation-state   A political situation in which high cultural and ethnic homogeneity characterizes the political unit in which such people live. The term *ethnic cleansing* has been used to define some of the bloody efforts to achieve nation-state status in the collapse of the former Yugoslavia in the 1990s. (p. 18)

Nationalism   The drive to expand the identity and strength of a political unit which serves as home to a population interested in greater cohesion, and often expanded political power. (p. 78)

Natural levee   Strip of land immediately alongside a stream that has been built up by deposition of sediments during flooding.

Natural resource   Product of the natural environment that can be used to benefit people. Resources are human appraisals.

Near abroad   Russia's name for the now-independent former republics, other than Russia, of the USSR. (p. 173)

Neo-Malthusians   Supporters of forecasts that resources will not be able to keep pace with the needs of growing human populations. (p. 50)

Newly industrializing countries (NICs)   The more prosperous of the world's less developed countries. (p. 39)

Non-black soil zone   Areas of poor soil in cool, humid portions of the Slavic coreland, suitable for cultivation of rye. (p. 180)

North Atlantic Drift   A warm current, originating in tropical parts of the Atlantic Ocean, that drifts north and east, moderating temperatures of western Europe. (p. 63)

North Atlantic Treaty Organization (NATO)   Military alliance formed in 1949 which included the United States, Canada, many European nations, and Turkey. (p. 82)

Oil embargo   The 1973 embargo on oil exports imposed by Arab members of the Organization of Petroleum Exporting Countries (OPEC) against the United States and the Netherlands. (p. 227)

Organization for Economic Cooperation and Development (OECD). *See* Organization for European Economic Cooperation.

Organization for European Economic Cooperation (OEEC)   The Organization for European Economic Cooperation was created in 1948 to organize and facilitate the European response to the 1947 Marshall Plan. In the early 1960s, the OEEC became the Organization for Economic Cooperation and Development (OECD) and this served as a multinational base for continued planning in economic and social development. The Common Market came to overshadow this organization and finally the European Union, in 1993, became the most powerful and influential multinational organization in Europe. (p. 81)

Orographic precipitation   The precipitation that results when moving air strikes a topographic barrier, such as a mountain, and is forced upward. (p. 24)

Outwash   Glacial material carried by sheets of meltwater underneath a glacier and subsequently deposited. (p. 67)

Ozone hole   Areas of depletion of ozone in the Earth's stratosphere, caused by chlorofluorocarbons. (p. 52)

Parallel   Latitude line running parallel to the equator. (p. 10)

Patrilineal descent system   Kinship naming system based on descent through the male line. (p. 218)

People overpopulation   The concept that many persons, each using a small quantity of natural resources to sustain life, add up to too many people for the environment to support. (p. 45)

Per capita   For every people or per people.

Perennial irrigation   Year-round irrigated cultivation of crops, as in the Nile Valley following construction of barrages and dams. (p. 235)

Permafrost   Permanently frozen subsoil. (p. 158)

Photosynthesis   The process by which green plants use the energy of the sun to combine carbon dioxide with water to give off oxygen and produce their own food supply. (p. 34)

Physical geography   The subdiscipline in the field of geography most concerned with the climate, landforms, soils, and physiography of the earth's surface. (p. 20)

Piedmont   A belt of country at an intermediate elevation along the base of a mountain range.

Pillars of Islam   The five fundamental tenets of the faith of Islam. (p. 215)

Place identity   In geographic analysis of a given locale, the nature of place identity becomes a means of understanding people's response to that particular place. Determination of the environmental and cultural characteristics that are most frequently associated with a certain place helps to establish that "sense of place" for a given location. (p. 13)

Planned economy   Government determination of where industrial plants should be located, what products should be produced, and how the labor force should be organized are all aspects of the planned economies found in the Soviet Union, the People's Republic of China, and Eastern Europe from the end of WWII to the early 1990s. From the early 1990s to the present time, this pattern of organization has been replaced with a market economy, or a system that lets markets—both domestic and foreign—determine the organization of resources and industrial activity. (p. 331)

Plantation   Large commercial farming enterprise, generally emphasizing one or two crops and utilizing hired labor. Plantations are commonly found in tropical regions. (p. 464)

Plate tectonics   The dominant force in the creation of the continents, mountain systems, and ocean deeps. The steady, but slow, movement of these massive plates of the earth's mantle and crust has created the positions of the continents that we have, and major patterns of volcanic and seismic activity. Areas where plates are being pulled under other plates are called subduction zones. (p. 606)

Pleistocene Overkill   A hypothesis stating that hunters and gatherers of the Pleistocene Era hunted many species to extinction. (p. 37)

Podzol   Soil with a grayish, bleached appearance when plowed, lacking in well-decomposed organic matter, poorly structured, and very low in natural fertility. Podzols are the dominant soils of the taiga. (p. 159)

Polder   An area reclaimed from the sea and enclosed within dikes in the Netherlands and other countries. Polder soils tend to be very fertile. (p. 117)

Polynesia (Many islands)   Pacific island region that roughly resembles a triangle with its corners at New Zealand, the Hawaiian Islands, and Easter Island. (p. 368)

Population change rate   The birth rate minus the death rate in a population. (p. 45)

Population density   The average number of people living in a square mile or square kilometer. It is a very handy statistic for generalized comparisons but often fails to provide a detailed sense of the real distribution of people. (p. 11)

Population explosion   The surge in Earth's human population which has occurred since the beginning of the Industrial Revolution. (p. 45)

Postindustrial   Just as the Industrial Revolution of the mid-18th century was a monumental time of change in geographic patterns, the current shifts in employment and trade that have taken place in the past two decades are also seen as significant. New importance in postindustrial society is given to information management, financial services, and the service sector. (p. 17)

Prairie   Area of tall grass in the middle latitudes composed of rich soils that have been cleared for agriculture. Original lack of trees may have been due to repeated burnings or periodic drought conditions.

Primary consumers (herbivores)   Consumers of green plants. (p. 35)

Primate city   A city strongly dominant within its country in population, government, and economic activity. Primate cities are generally found in developing countries, although some already developed countries, such as France, have them. (p. 102)

Prime meridian   *See* Latitude.

Producers (Self-feeders)   In a food chain, organisms that produce their own food (mostly, green plants). (p. 35)

Push-pull forces of migration   In non-forced human migration, there is generally a series of influences on the migrant that tend to push him or her away from a place. Economics and social conditions are common push factors, but images of other places (often unsubstantiated) also serve to attract the migrants and influence the future use of migration destinations. (p. 469)

Pyramid of biomass   A diagrammatic representation of decreasing biomass in a food chain. (p. 36)

Pyramid of energy flow   A diagrammatic representation of the loss of high quality, concentrated energy as it passes through the food chain. (p. 35)

Qanat   Tunnel used to carry irrigation and drinking water from an underground source by gravity flow; also called foggara or karez. (p. 240)

Rain shadow   Condition creating dryness in an area located on the lee side of a topographic barrier such as a mountain range. (p. 24)

Region   A "human construct" that is often of considerable size and that has substantial internal unity or homogeneity and that differs in significant respects from adjoining areas. Regions can be classed as formal (homogeneous), functional, or vernacular. The *formal* region, also known as a uniform region, has a unitary quality which derives from a homogeneous characteristic. The United States of America is an example of a formal region. The *functional* region, also called the nodal region, is a coherent structure of areal units organized into a functioning system by lines of movement or influence that converge on a central node or trunk. A major example would be the trading territory served by a large city and bound together by the flow of people, goods, and information over an organized network of transportation and communication lines. *Vernacular* regions are areas that possess regional identity, such as "The Sun Belt," but share less objective criteria in the use of this regional name. *General* regions, such as the major world regions in this text, are recognized on the basis of overall distinctiveness. (p. 559)

Remote sensing   Through the use of aerial and satellite imagery, geographers and other scientists have been able to get vast amounts of data describing places all over the face of the earth. Remote sensing is the science of acquiring and analyzing data without being in contact with the subject. It is used in the study of patterns of land use, seasonal change, agricultural activity, and even human movement along transport lines. This process relates closely to GIS, described above. (p. 21)

Renewable resource   A resource, such as timber, that is grown or renewed so that a continual supply of it is always available. A finite resource is one that, once consumed, cannot be easily used again. Petroleum products are a good example of such a resource—and because it takes too much time to go through the process of creation, they are not seen as renewable. (p. 130)

Resources   Resources become valuable through human appraisal, and their utilization reflects levels of technology, location, and economic ambition. There are few resources that have been universally esteemed (water, land, defensible locales) throughout history, so patterns of resource utilization serve as indices of other levels of cultural development. *See* Natural resource. (p. 17)

Ring of Fire   The long horseshoe-like chain of volcanoes that go from the southern Andes Mountains in South America up the west coast of North America and arch over into the northeast Asian island chains of Japan, the Philippines, and Southeast Asia. This zone of seismic instability and erratic vulcanism is caused by the tension built up in plate tectonics. *See* Plate tectonics. (p. 606)

Riparian   A state containing or bordering a river. (p. 226)

Salinization   The deposition of salts on, and subsequent fertility loss in, soils experiencing a combination of overwatering and high evaporation. (p. 236)

Savanna Low-latitude grassland in an area with marked wet and dry seasons. (p. 29)

Scorched earth The wartime practice of destroying one's own assets to prevent them from falling into enemy hands. (p. 164)

Seamount An underwater volcanic mountain. (p. 370)

Second law of thermodynamics A natural law stating that high quality, concentrated energy is increasingly degraded as it passes through the food chain. (p. 35)

Secondary consumers (carnivores) Consumers of primary consumers. (p. 35)

Sedentarization Voluntary or coerced settling down, particularly pastoral nomads in the Middle East. (p. 218)

Segmentary kinship system Organization of kinship groups in concentric units of membership, as of lineage, clan and tribe among Middle Eastern pastoral nomads. (p. 218)

Service sector The labor sector made up of employees in retail trade and personal services; the sector most likely to increase in employment significance in *postindustrial* society. (p. 17)

Settler colonization The historical pattern by which Europeans sought to create new or "neo-Europes" abroad. (p. 42)

Shatter belt A region having a fragmented and unstable pattern of nationalities and political units. (p. 213)

Shi'a Islam The branch of Islam regarding male descendants of Ali, the cousin and son-in-law of the Prophet Muhammad, as the only rightful successors to the Prophet Muhammad. (p. 215)

Site and situation Site is defined by the specific geographic location of a given place, while situation is defined as the accessability of that site, and the nature of the economic and population characteristics of that locale. The combination term deals broadly with the blend of influences of setting and historical development of a place. (p. 4)

Slavic coreland (Fertile Triangle or Agricultural Triangle) The large area of the western Former Soviet Region containing most of the region's cities, industries and cultivated lands. (p. 175)

Small-scale map A map constructed to give a highly generalized view of a large area. (p. 10)

Soil Earth mantle made of decomposed rock and decayed organic material.

Sovereign state A political unit that has achieved political independence and maintains itself as a separate unit. (p. 18)

Soybean A bean crop that made its way from Asia to the New World during the Age of Discovery and has now become a major crop in the agricultural Midwest of the U.S. It is of particular significance because the plant affixes nitrogen as it grows—making it less demanding on the land—and produces a vegetable crop that is high in protein as well as an oil that can be used as a fuel and as a chemical element in a variety of paint, ink, and other products. It is bound to be a crop of growing significance in this next century as both food and petroleum products become ever more costly.

Spatial analysis *See* Spatial.

Spatial Geography is concerned at all levels with spatial organization—that is, the way in which space has been compartmentalized, modified, utilized. The patterns that are associated with the distribution of both physical and human features of the landscape are central to all geographic analysis—and such analysis is spatial analysis. (p. 6)

Special Economic Zones (SEZs) In China, specific urban areas were given special status in the 1980s to enable Chinese planners and entrepreneurs to begin to anticipate economic changes produced by the 1997 return of Hong Kong from Britain. The most important SEZ is Shenzhen, just adjacent to Hong Kong itself, but all of the SEZs have undergone rapid economic and demographic growth since China's current rapid pace of economic growth began in the early 1980s. These locales have served to give Chinese businesspeople a little more latitude in experimentation with free market and joint-venture activities. (p. 338)

St. Lawrence Seaway The St. Lawrence Seaway is a 450 mi (725 km) water link between Lake Erie and the Atlantic Ocean. It allows major oceanic vessels to reach the Great Lakes, thus enabling ports as far inland as the west side of Lake Superior to serve as international ports. The seaway was completed in 1959 although Canada and the United States began to anticipate such a waterway project in the last years of the 19th century.

State A political unit over which an established government maintains sovereign control. (p. 18)

State farm (sovkhoz) A type of collectivized state-owned agricultural unit in the former Soviet Union; workers receive cash wages in the same manner as industrial workers. (p. 168)

Steppe Semiarid transition zone between humid and arid areas. Grass is the dominant vegetation. (p. 29)

Subarctic High-latitude climate characterized by short, mild summers and long, severe winters. (p. 29)

Subduction zone *See* Plate tectonics.

Subsidence The settling of land by the compression of lower layers of rock and soil, usually accelerated by the removal of water from below the surface. Of particular importance in urban centers, with Mexico City having one of the most persistent problems.

Sun Belt An area of indefinite extent, encompassing most of the South, plus the West at least as far north as Denver, Salt Lake City, and San Francisco. This region has shown a faster growth in population and jobs than the nation as a whole during the past several decades.

Sunni Islam The branch of Islam regarding successorship to the Prophet Muhammad as a matter of consensus among religious elders. (p. 215)

Sustainable development (ecodevelopment) Concepts and efforts to improve the quality of human life while living within the carrying capacity of supporting ecosystems. (p. 53)

Sustainable yield (natural replacement rate) The highest rate at which a renewable resource can be used without decreasing its potential for renewal. (p. 43)

Sweat equity The term used for the actual physical labor expended in remaking some aspect of the cultural landscape. Although it is difficult to give an economic value to this process, the term often reflects the patterns of neighborhood change associated with *gentrification*. (p. 575)

Swidden One of the terms used for slash-and-burn agriculture, or shifting cultivation. In this process, landless peasant farmers move into remote forest lands and cut down forest growth. After a period of drying out, the forest floor is burned and then varied crops are planted in the ash of the burned forest material. Crops may be raised as little as one year, or perhaps several years, and then the farming group moves to adjacent lands in order to take advantage of the fertilizer afforded by the burned tree growth. While swidden farmers do not make up a large percentage of the world's farm peoples, they have an enormous impact on remaining primary forest growth, especially in tropical rainforests. (p. 506)

Teaching water   The Chinese phrase used to define the labors that must be accomplished in order to establish a productive irrigation system. (p. 133)

Taiga   Northern coniferous (needleleaf) forest. (p. 29)

Technocentrists (Cornucopians)   Supporters of forecasts that resources will keep pace with or exceed the needs of growing human populations. (p. 49)

Technological unemployment   Unemployment caused by the replacement of human labor with highly sophisticated technology, even robots. *See* Postindustrialism. (p. 76)

Tertiary consumers   Consumers of secondary consumers. (p. 35)

Theory of Island Biogeography   A theoretical calculation of the relationship between habitat loss and natural species loss, in which a 90 percent loss of natural forest cover results in a loss of half of the resident species. (p. 453)

Tierra caliente   Latin American climatic zone reaching from sea level upward to approximately 3000 feet (914 m). In this hot, wet environment are grown such crops as rice, sugarcane, and cacao. (p. 464)

Tierra fria   Latin American climatic zone found higher than 6000 feet (1829 m) above sea level. Frost occurs and the upper limit of agriculture and tree growth is reached. (p. 464)

Tierra templada   Climatic zone extending from approximately 3000 to 6000 feet (914–1829 m) above sea level. It is a prominent zone of European-induced settlement and commercial agriculture such as coffee-growing. (p. 464)

Togugawa Shogunate   The period 1600 to 1868 in Japan, characterized by a feudal hierarchy and Japan's isolation from the outside world. (p. 351)

Trade winds   Streams of air that originate in semipermanent high-pressure cells on the margins of the tropics and are attracted equatorward by a semipermanent low-pressure cell. (p. 26)

Transportation infrastructure   In efforts of economic development it is always seen that the network of transportation facilities—railroads, highways, ports, airports—has a major impact on a government's or business's efforts to achieve higher levels of productivity. (p. 336)

Triangular trade   Sixteenth to nineteenth century trading links between West Africa, Europe and the Americas, involving guns, alcohol and manufactured goods from Europe to West Africa exchanged for slaves. Slaves to the Americans were exchanged for the gold, silver, tobacco, sugar and rum carried back to Europe. (p. 407)

Trophic (feeding) levels   Stages in the food chain. (p. 35)

Tropical rain forest   Low-latitude broadleaf evergreen forest found where heat and moisture are continuously, or almost continuously, available. (p. 29)

Tundra   A region with a long, cold winter when moisture is unobtainable because it is frozen, and a very short, cool summer. Vegetation includes mosses, lichens, shrubs, dwarf trees, and some grass. (p. 29)

Underground economy   The "black market" typical of the Former Soviet Region and many LDCs. (p. 171)

Undifferentiated highland   Climate that varies with latitude, altitude, and exposure to the sun and moisture-bearing winds. (p. 29)

Vernacular region   *See* Region.

Vertical migration   Movements of pastoral nomads between low winter and high summer pastures. (p. 218)

Virgin and idle lands (New lands)   Steppe areas of Kazakstan and Siberia brought into grain production in the 1950s. (p. 169)

Warsaw Pact   Former military alliance consisting of the Soviet Union and the European countries of Poland, East Germany, Czechoslovakia, Hungary, Romania, and Bulgaria (now dissolved). (p. 83)

Water head   In attempting to control water for the generation of electricity or a gravity flow irrigation project, it is essential to pond up water in relative quantity so that there is an adequate force—water head—to achieve the desired goal. Dam structures are the most common engineering responses to this requirement. (p. 130)

Weather   The atmospheric conditions prevailing at one time and place. (p. 23)

Weathering   The natural process that disintegrates rocks by mechanical or physical means, making soil formation possible.

Westerly winds   Airstream located in the middle latitudes that blows from west to east. Also known as westerlies. (p. 25)

Westernization   The process whereby non-Western societies acquire Western traits, which are adopted with varying degrees of thoroughness. (p. 62)

Zero population growth (ZPG)   The condition of equal birth rates and death rates in a population. (p. 49)

Zionist movement   The political effort, beginning in the late 19th century, to establish a Jewish homeland in Palestine. (p. 223)

Zone of transition   Area where the characteristics of one region change gradually to those of another.

PLACE-NAME PRONUNCIATION GUIDE

The authors acknowledge with thanks the assistance of David Allen, Eastern Michigan University, in preparing an earlier pronunciation guide that the authors expanded and modified for this book. In general, names are those appearing in the text matter; pronunciations of others on maps can be found in Goode's World Atlas (Rand McNally) or standard dictionaries and encyclopedias. Vowel sounds are coded as: ă (uh), ā (ay), ĕ (eh), ē (ee), ī (eye), ĭ (ih), o͞o (oo).

Afghanistan (af-*gann′*-ih-stann′)
  Helmand R. (*hell′*-mund)
  Hindu Kush Mts. (*hinn′*-doo *koosh′*)
  Kabul (*kah′*-b'l)
  Khyber Pass (*keye′*-ber)

Albania (al-*bay′*-nee-a)
  Tirana (tih-*rahn′*-a)

Algeria (al-*jeer′*-ee-a)
  Ahaggar Mts. (uh-*hahg′*-er)
  Sahara Desert (suh-*hahr′*-a)
  Tanezrouft (*tahn′*-ez-rooft′)

American Samoa
  Pago Pago (*pahng′*-oh *pahng′*-oh)
  Tutuila I. (too-too-*ee′*-lah)

Andorra (an-*dohr′*-a)

Angola (ang-*goh′*-luh)
  Benguela R. R. (ben-*gehl′*-a)
  Cabinda (exclave) (kah-*binn′*-duh)
  Luanda (loo-*an′*-duh)
  Lobito (luh-*bee′*-toh)

Anguilla (an-*gwill′*-a)

Antigua and Barbuda (an-*teeg′*-wah, bahr-*bood′*-a)

Argentina (ahr′-jen-*teen′*-a)
  Buenos Aires (*bwane′*-uhs *eye′*-ress)
  Córdoba (*kawrd′*-oh-bah)
  Patagonia (region) (patt′-hu-*goh′*-nee-a)
  Rio de la Plata (*ree′*-oh duh lah *plah′*-tah)
  Rosario (roh-*zahr′*-ee-oh)
  Tierra del Fuego (region) (tee-ehr′-uh dell *fway′*-goh)
  Tucumán (too′-koo-*mahn′*)

Armenia (ahr′-*meeny′*-uh)
  Yerevan (yair′-uh-*vahn′*)

Aruba (ah-*roob′*-a)

Australia (aw-*strayl′*-yuh)

Adelaide (*add′*-el-ade)
Brisbane (*briz′*-bun)
Canberra (*can′*-bur-a)
Melbourne (*mell′*-bern)
Sydney (*sid′*-nih)
Tasmania (state) (tazz-*may′*-nih-uh)

Austria (*aw′*-stree-a)
  Graz (*grahts′*)
  Linz (*lints′*)
  Salzburg (*sawlz′*-burg)
  Vienna (vee-*enn′*-a)

Azerbaijan (ah′-zer-*beye′*-jahn)
  Baku (bah-*koo′*)

Bahamas (buh-*hahm′*-uz)
  Nassau (*nass′*-aw)

Bahrain (bah-*rayn′*)

Bangladesh (bahng′-lah-*desh′*)
  Dacca (Dhaka) (*dahk′*-a)

Barbados (bar-*bay′*-dohss)

Belarus (*bell′*-uh-roos)
  Minsk (*mensk′*)

Belgium
  Antwerp (*ant′*-werp)
  Ardennes Upland (ahr-*den′*)
  Brussels (*brus′*-elz)
  Charleroi (*shahr*-luh-*rwah′*)
  Ghent (*gent′*)
  Liège (lee-*ezh′*)
  Meuse R. (*merz′*)
  Sambre R. (*sohm*-bruh)
  Scheldt R. (*skelt′*)

Belize (buh-*leez′*)

Benin (beh-*neen′*)
  Cotonou (koh-toh-*noo′*)

Bhutan (boo-*tahn′*)
  Thimphu (*thim′*-poo′)

Bolivia (boh-*liv′*-ee-uh)
  Altiplano (region) (al-tih-*plahn′*-oh)
  La Paz (lah *pahz′*)
  Santa Cruz (sahnta *krooz′*)
  Sucre (*soo′*-kray)
  Lake Titicaca (tee′-tee-*kah′*-kah)

Bosnia-Hercegovina (*boz'*-nee-uh,-herts-uh-goh-*veen'*-uh)
  Sarajevo (sahr-uh-*yay'*-voh)

Botswana (baht-*swahn'*-a)
  Gaborone (gahb'-uh-*roh'*-nee)

Brazil (bra-*zill*)
  Amazonia (region) (am-uh-*zohn'*-ih-a)
  Belo Horizonte (*bay'*-loh haw-ruh-*zonn'*-tee)
  Brasília (bruh-*zeel'*-yuh)
  Caatinga (region) (kay-*teeng'*uh)
  Manaus (mah-*noose'*)
  Minas Gerais (*meen'*-us zhuh-*rice*)
  Natal (nah-*tal'*)
  Paraíba Valley (pah-rah-*ee'*-bah)
  Paraná R. (pah-rah-*nah'*)
  Pôrto Alegre (por-too' al-*egg'*-ruh)
  Recife (ruh-*see'*-fee)
  Rio de Janeiro (*ree'*-oh dih juh-*nair'*-oh)
  Salvador (sal'-vah-*dor'*)
  São Paulo (sau *pau'*-loh)

Brunei (*broo'*-neye)

Bulgaria
  Sofia (*soh'*-fee-a)

Burkina Faso (bur-*keena' fah'*-soh)
  Ouagadougou (wah'-guh-*doo'*-goo)

Burundi (buh-*roon'*-dee) ("oo" as in "wood")
  Bujumbura (boo-jem-*boor'*-a) (second "oo" as in "wood")

Cambodia
  Phnom Penh (*nom' pen'*)

Cameroon (kam'-uh-*roon'*)
  Douala (doo-*ahl'*-a)
  Yaoundé (yah-onn-*day'*)

Canada
  Nova Scotia (*noh'*-vuh *scoh'*-shuh)
  Montreal (mon'-tree-*awl'*)
  Ottawa (*aht'*-ah-wah)
  Quebec (kwih-*beck'*; Fr., kay-*beck'*)
  Saguenay R. (*sag'*-uh-nay)

Cape Verde (*verd'*)

Central African Republic
  Bangui (*bahng'*-ee)

Chile (*chill'*-ee)
  Atacama Desert (ah-tah-*kay'*-mah)
  Chuquicamata (choo-kee-*kah'*-mah-tah)
  Concepción (con-sep'-*syohn'*)
  Santiago (sahn'-tih-*ah'*-goh)
  Valparaíso (vahl'-pah-rah-*ee'*-soh)

China
  Altai Mts. (al-*teye*)
  Amur R. (ah-*moor'*) ("oo" as in "wood")
  Anshan (ahn-*shahn'*)
  Baotou (*bau'*-toh')
  Beijing (bay-*zhing'*)
  Canton (kan'-*tahn'*)
  Chang Jiang (chung jee-*ahng'*)
  Chengdu (chung-doo')

Chongqing (chong-*ching'*)
Dzungarian Basin (zun-*gair'*-ree-uhn)
Guangdong (gwahng-*dung'*)
Guangzhou (gwahng-*joh'*)
Hainan I. (*heye'*-nahn')
Harbin (*hahr'*-ben)
Huang He (*hwahng' huh'*)
Lanzhou (lahn-*joh'*)
Lhasa (*lah'*-suh)
Li R. (*lee'*)
Liao (lih-ow')
Lüda (*loo'*-dah)
Nanjing (nahn-*zhing'*)
Peking (pea-*king'*)
Shandong (shahn-*dung'*)
Shanghai (shang-*heye'*)
Shenyang (shun-*yahng'*)
Sichuan (zehch-*wahn'*)
Taklamakan Desert (*tah'*-kla-mah-*kahn'*)
Tarim Basin (*tah'*-reem')
Tiananmen (*tyahn'*-an-men)
Tianjin (tyahn-jeen')
Tien Shan (*tih'*-en *shahn'*)
Tientsin (*tinn'*-*tsinn'*)
Tsinling Shan (*chinn'*-leeng *shahn'*)
Ürümqi (oo-*room'*-chee)
Wuhan (*woo'*-hahn')
Xian (shee-*ahn'*)
Xinjiang (shin-jee-*ahng'*)
Xizang (shee-*dzahng'*)
Yangtze R. (*Yang'*-see)
Yunnan (yoo-*nahn'*) ("oo" as in "wood")

Colombia (koh-*lomm'*-bih-uh)
  Barranquilla (bah-rahn-*keel'*-yah)
  Bogotá (boh-ga-*tah'*)
  Cali (*kah'*-lee)
  Cartagena (kahrt'-a-*hay'*-na)
  Cauca R. (kow'-kah)
  Magdalena R. (mahg'-thah-*lay'*-nah)
  Medellín (med-deh-*yeen'*)

Comoros (*kahm'*-o-rohs')

Congo
  Brazzaville (*brazz'*-uh-vill'; Fr.: brah-zah-*veel'*)
  Pointe Noire (pwahnt nwahr')

Costa Rica (*kohs'*-tuh *ree'*-kuh)
  San José (sahn hoh-*zay'*)

Croatia (kroh-ay'-shuh)
  Zagreb (*zah'*-grebb)

Cyprus (*seye'*-pruhs)
  Nicosia (nik'-o-*see'*-uh)

Czechoslovakia (check'-oh-sloh-*vah'*-kee-a)
  Bratislava (bratt'-ih-*slahv'*-a, braht-)
  Brno (*ber'*-noh)
  Moravia (muh-*rayve'*-ih-uh)
  Ostrava (*aw'*-strah-vah)
  Prague (*prahg'*)

Denmark
  Copenhagen (koh′-pen-hahg′-gen)

Djibouti (jih-*boot′*-ee)

Dominica (dahm′-i-*nee′*-ka)

Dominican Republic (duh-*min′*-i-kan)
  Santo Domingo (sant′-oh d-*min′*-goh)

Ecuador (ec′-wa-dawr)
  Guayaquil (gweye′-uh-*keel′*)
  Quito (*kee′*-toh)

Egypt
  Aswan (*ahs′*-wahn)
  Cairo (*keye′*-roh)
  Port Said (sah-*eed′*)

El Salvador (el-*sal′*-va-dor)

Equatorial Guinea (*gin′*-ee)

Estonia (ess-*toh′*-nee-a)
  Tallin (*tahl′*-in)

Ethiopia (ee′-thee-*oh′*-pea-a)
  Addis Ababa (*ad′*-iss *ab′*-a-ba)
  Asmara (az-mahr′-a)

Fiji (*fee′*-jee)
  Suva (*soo′*-va)

Finland
  Helsinki (*hel′*-sing-kee)
  Karelian Isthmus (kuh-*ree′*-lih-uhn)

France
  Alsace (region) (al-*sass′*)
  Ardennes (region) (ahr-*den′*)
  Bordeaux (bawr-*doh′*)
  Boulogne (boo-*lohn′*-yuh)
  Calais (kah-*lay′*)
  Carcassonne (kahr-kuh-*suhn′*)
  Champagne (sham-*pahn*-yuh)
  Garonne R. (guh-*rahn′*)
  Jura Mts. (*joo′*-ruh)
  Le Havre (luh ah′-vruh)
  Lille (*leel′*)
  Loire R. (lwahr′)
  Lorraine (loh-*rane′*)
  Lyons (Fr., Lyon), both (*lee*-aw)
  Marseilles (mahr-*say′*)
  Massif Central (upland) (mah-seef′ saw-*trahl*)
  Meuse R. (*merz′*)
  Moselle R. (moh-*zell′*)
  Nancy (naw-*see′*)
  Nantes (*nawnt′*)
  Nice (*neece′*)
  Oise r. (*wahz′*)
  Pyrenees (peer′-uh-*neez′*)
  Riviera (riv′-ih-*ehr′*-a)
  Rouen (roo-*aw′*)
  Saône R. (*sohn′*)
  Seine R. (*senn′*)
  Strasbourg (strahz-*boor′*)

  Toulon (too-*law′*)
  Toulouse (too-*looz′*)
  Vosges Mts. (*vohzh′*)

French Guiana (gee′-*ahn*-a)
  Cayenne (*keye′*-enn)

Gabon (gah-*baw′*)

Gambia (*gamm′*-bih-uh)
  Banjul (*bahn*-jool)

Georgia
  Abkhazia (ahb-*kahz′*-zee-uh)
  Adjaria (ah-*jahr′*-ree-uh)
  Caucasus Mts. (kaw′-kuh-suhs)
  South Ossetia (oh-see′-shee-uh)
  Tbilisi (tuh-*blee′*-see)

Germany
  Aachen (*ah′*-ken)
  Bremen (*bray′*-men)
  Chemnitz (*kemm′*-nits)
  Cologne (kuh-*lohn′*)
  Dresden (*drez′*-den)
  Düsseldorf (*doo′*-sel-dorf)
  Elbe R. (*ell′*-buh)
  Erzgebirge (ehrts-guh-beer-guh)
  Frankfurt (*frahngk′*-foort)
  Karlsruhe (*kahrls′*-roo-uh)
  Leipzig (*leyep′*-sig)
  Main R. (*mane′*; Ger., *mine′*)
  Mannheim-Ludwïgshafen (*man′*-hime, *loot′*-vikhs-hah-fen)
  Munich (*myoo′*-nik)
  Neisse R. (*nice′*-uh)
  Oder R. (*oh′*-der)
  Ruhr (region) (*roor′*)
  Stuttgart (*stuht′*-gahrt)
  Weser R. (*vay′*-zer)
  Wiesbaden-Mainz (*vees′*-bahd-en *meyents*)
  Wuppertal (*voop′*-er-tahl) ("oo" as in "wood")

Ghana (*gahn′*-a)
  Accra (a-*krah′*)
  Kumasi (koo-*mahs′*-ee)
  Tema (*tay′*-muh)
  Volta R. (*vohl′*-tuh)

Greece
  Aegean Sea (uh-jee′-un)
  Piraeus (peye-*ree′*-us)
  Thessaloniki (thess′-uh-loh-*nee′*-kih)

Grenada (gruh-*nayd′*-a)

Guadeloupe (gwah′-duh-*loop′*)

Guatemala (gwah′-tuh-*mah′*-luh)

Guinea (*gin′*-ee)
  Conakry (*kahn′*-a-kree)

Guinea-Bissau (biw-*ow′*)

Guyana (geye-*an′*-a; geye-*ahn*-a′)

Haiti (*hayt′*-ee)
  Port-au-Prince (pohrt′-oh-*prants′*)

Honduras (hahn-*dur'*-as)
  Tegucigalpa (tuh-goo'-sih-*gahl'*-pah)

Hungary (*hun'*-guh-rih)
  Budapest (boo-duh-*pesht'*)

Iceland
  Akureyri (ah'-koor-*ray'*-ree)
  Reykjavik (*ray'*-kyuh-vik')

India
  Agra (*ah'*-gruh)
  Ahmedabad (ah'-mud-uh-*bahd'*)
  Assam (state) (uh-*sahm'*)
  Bengal (region) (benn-*gahl'*)
  Bihar (state) (bih-*hahr'*)
  Brahmaputra R. (brah'-muh-*poo'*-truh)
  Deccan (region) (*deck'*-un)
  Delhi (*deh'*-lih)
  Ganges R. (*gan'*-jeez)
  Eastern, Western Ghats (mountains) (*gahts'*)
  Himalaya Mts. (himm-*ah'*-luh-yuh, -ah-*lay*-yuh)
  Jaipur (*jeye'*-poor)
  Jamshedpur (juhm-*shayd'*-poor)
  Kashmir (region) (*cash'*-meer)
  Kerala (state) (*kay'*-ruh-luh)
  Madras (muh-*drass'*, -*drahs'*)
  Punjab (region) (*pun'*-jahb, -jab; pun-*jahb'*, -*jab'*)
  Rajasthan (state) (*rah'*-juh-stahn)
  Uttar Pradesh (state) (*oo'*-tahr pruh-*desh'*)

Indonesia
  Bali (*bah'*-lih)
  Bandung (*bahn'*-doong)
  Celebes (Sulawesi) I. (*sell'*-uh-beez; sool-uh-*way'*-see)
  Irian Jaya (ihr'-ee-ahn-*jeye'*-uh)
  Jakarta (yah-*kahr'*-tah)
  Java (*jah'*-vuh)
  Kalimantan (kal-uh-*mann'*-tann)
  Medan (muh-*dahn'*)
  Palembang (pah-lemm-*bahng'*)
  Sumatra (suh-*mah'*-truh)
  Surabaya (soo'-ruh-*bah'*-yah)
  Ujung Pandang (oo'-jung pahn-*dahng'*)

Iran (ih-*rahn'*)
  Abadan (ah-buh-*dahn'*)
  Elburz Mts. (ell-*boorz'*)
  Isfahan (izz-fah-*hahn'*)
  Meshed (muh-*shed'*)
  Tabriz (tah-*breez'*)
  Tehran (teh-huh-*rahn'*)
  Zagros Mts. (*zah'*-gruhs)

Iraq (i-*rahk'*)
  Baghdad (*bag'*-dad)
  Basra (*bahs'*-ra)
  Euphrates R. (yoo-frate'-eez)
  Kirkuk (Kihr-*kook'*) ("oo" as in "wood")
  Mosul (moh-*sool'*)
  Shatt al Arab (R.) (*shaht'* ahl ah-*rahb'*)
  Tigris R. (*teye'*-griss)

Italy
  Adriatic Sea (ay-drih-*at'*-ik)
  Apennines (Mts.) (*app'*-uh-nines)
  Bologna (boh-*lohn'*-ya)
  Genoa (*jen'*-o-a)
  Milan (mih-*lahn'*)
  Naples (*nay'*-pels)
  Palermo (pah-*ler'*-moh)
  Turin (*toor'*-in)

Ivory Coast (Côte d'Ivoire) (koht-d'-vwahr)
  Abidjan (abb'-ih-jahn')

Israel
  Gaza Strip (*gahz'*-uh)
  Eilat (*ay'*-laht)
  Haifa (*heye'*-fa)
  Tel Aviv (tel'-a-*veev'*)

Jamaica (ja-*may'*-ka)

Japan
  Fukuoka (foo'-koo-*oh'*-kah)
  Hiroshima (heer'-a-*shee'*-mah)
  Hokkaido (island) (hah-*keye'*-doh)
  Honshu (island) (*honn'*-shoo)
  Kitakyushu (kee-*tah'*-*kyoo'*-shoo)
  Kobe (*koh'*-bay)
  Kyoto (kee-*oht'*-oh)
  Kyushu (island) (kee-*yoo'*-shoo)
  Kyoto (kee-*oht'*-oh')
  Nagoya (nay-*goy'*-ya)
  Osaka (oh-*sahk'*-a)
  Sapporo (sahp-*pohr'*-oh')
  Shikoku (island) (shih-*koh'*-koo)
  Tokyo (*toh'*-kee-oh)
  Yokohama (yoh'-ko-*hahm'*-a)

Jordan (*jor'*-dan)
  Amman (a-*mahn'*)

Kazakhstan (kuh-*zahk'*-stahn)
  Alma Ata (*al'*-mah ah-tah')
  Aral Sea (uh-*rahl'*)
  Karaganda (kahr'-rah-*gahn'*-dah)
  Lake Balkhash (bul-*kahsh'*)
  Syr Darya (river) (*seer'* dahr-*yah'*)

Kenya (*ken'*-ya)
  Mombasa (mahm-*bahs*-a)
  Nairobi (neye-*roh'*-bee)

Kiribati (keer'-ih-*bah'*-tih)

Korea, North
  Pyongyang (pea-ong-*yahng'*)
  Yalu River (*yah'*-loo)

Korea, South
  Inchon (*inn'*-chonn)
  Pusan (*poo'*-sahn)
  Seoul (*sohl'*)
  Taegu (teye-goo')

Kuwait (koo-*wayt'*)

Kyrgyzstan (keer'-geez-*stahn'*)

Laos (*lah'*-ohs)
   Vientiane (vyen-*tyawn'*)

Latvia
   Riga (*reeg'*-a)

Lebanon (*leb'*-a-non)
   Beirut (bay-*root'*)

Lesotho (luh-*soh'*-toh)

Libya (*lib'*-yah)
   Benghazi (benn-*gah'*-zee)
   Tripoli (*tripp'*-uh-lee)

Liechtenstein (*lick'*-tun-stine')

Lithuania (lith'-oo-*ane'*-ee-uh)
   Vilnyus (*vill'*-nee-us)

Luxembourg (*luk'*-sem-burg)

Madagascar
   Antananarivo (an-ta-nan'-a-*ree'*-voh)

Malawi (muh-*lah'*-wee)
   Lilongwe (lih-*lawng'*-way)

Malaysia
   Kuala Lumpur (*kwah'*-luh *loom'*-poor) ("oo" as in "wood")
   Sabah (*sah'*-bah)
   Sarawak (suh-*rah'*-wahk)
   Strait of Malacca (muh-*lahk'*-uh)

Maldives (*mawl'*-deevz)

Mali (*mah'*-lee)
   Bamako (*bam'*-ah-koh')

Malta (*mawl'*-tuh)

Martinique (mahr'-tih-*neek'*)

Mauritania (mahr-ih-*tay'*-nee-a)

Mauritius (muh-*rish'*-us)

Mexico
   Ciudad Juárez (see-yoo-*dahd' wahr'*-ez)
   Guadalajara (gwad'-a-la-*har'*-a)
   Monterrey (mahnt-uh-*ray'*)
   Puebla (poo-*eb'*-la)
   Tijuana (tee-*hwah'*-nah)

Moldova (mohl-*doh'*-vah)
   Kishinev (kish'-in-*yeff'*)

Monaco (*mon'*-a-koh')

Mongolia (mon-*gohl'*-yuh)
   Ulan Bator (oo-*lahn' bah'*tor)

Montserrat (mont'-suh-*rat'*)

Morocco (moh-*rah'*-koh)
   Casablanca (kas-a-*blang'*-ka)
   Rabat (rah-*baht'*)

Mozambique (moh'-zamm-*beek'*)
   Beira (*bay'*-rah)
   Cabora Bassa Dam (kah-*bore'*-ah *bah'*-sah)
   Maputo (mah-*poot'*-oh')

Namibia (nah-*mib'*-ee-a)

Nauru (nah-*oo*-roo)

Nepal (nuh-*pawl'*)
   Kathmandu (kath'-man-*doo'*)

Netherlands
   Ijsselmeer (*eye'*-sul-mahr')
   The Hague (*hayg'*)
   Utrecht (*yoo'*-trekt')
   Zuider Zee (*zeye'*-der-zay')

Netherlands Antilles
   Curaçao (*koo'*-rahs-ow')

New Zealand
   Auckland (*aw'*-klund)

Nicaragua (nik'-a-*rahg'*-wah)

Niger (*neye'*-jer; Fr., nee-*zhere'*)
   Niamey (nee-*ah'*-may)

Nigeria (neye-*jeer*-ee-a)
   Abuja (a-*boo'*-ja)
   Ibadan (ee-*bahd'*-n)
   Kano (*kahn'*-oh)
   Lagos (*lahg'*-us)

Norway
   Bergen (*berg'*-n)
   Oslo (*ahz'*-loh)
   Stavanger (stah-*vahng'*-er)
   Trondheim (*trahn'*-haym')

Oman (oh-*mahn'*)
   Muscat (*mus'*-kat')

Pakistan (pahk-i-*stahn'*)
   Hyderabad (*heyed'*-er-a-bahd')
   Islamabad (is-*lahm*-a-bahd')
   Karachi (kah-*rahch'*-ee')
   Karakoram (mountains) (kahr'-a-*kohr*-um)
   Lahore (lah-*hohr'*)
   Peshawar (puh-*shah'*-wahr)
   Punjab (see India)

Papua New Guinea (*pap'*-yoo-uh new *gin'*-ee)
   Port Moresby (*morz'*-bee)

Paraguay (*pair'*-uh-gweye')
   Asuncíon (a-*soon'*-see-*ohn'*)

Peru (puh-*roo'*)
   Callao (kah-*yow'*)
   Chimbote (chimm-*boh'*-te)
   Cerro de Pasco (*sehr'*-roh day *pahs'*-koh)
   Cuzco (*koos'*-koh)
   Iquitos (ee-*kee'*-tohs)
   Lima (*lee'*-ma)

Philippines (*fil*-i-peenz')
   Luzon (island) (loo-*zahn'*)
   Mindanao (island) (minn'-dah-*nah'*-oh)
   Mount Pinatubo (pea'-nah-*too'*-buh)
   Visayan Islands (vih-*sah'*-y'n)

Poland
   Gdansk (ga-*dahntsk'*)
   Katowice (kaht-uh-*veet'*-sih)
   Krakow (*krahk'*-ow)

Lodz (*looj'*)
Posnan (pohz'-nan-ya)
Silesia (region) (sih-*lee'*-shuh)
Szczecin (*shchet'*-seen')
Wroclaw (*vrawt'*-slahf)

Portugal (*pohr'*-chi-gal)
Lisbon (*liz'*-bon)
Oporto (oh-*pohrt'*-oh)

Qatar (*kaht'*-ar)

Romania (roh-*main'*-yuh)
Bucharest (boo'-kuh-*rest'*)
Carpathian Mts. (kahr-*pay'*-thih-un)

Russia
Altai Mts. (*al'*-teye)
Amur R. (ah-*moor'*) ("oo" as in "wood")
Angara R. (ahn'-guh-*rah'*)
Arkhangelsk (ahr-*kann'*-jelsk)
Astrakhan (as-trah-*khahn'*)
Baikal, Lake (beye-*kahl'*)
Bashkiria (bahsh-*keer'*-ee-uh)
Bratsk (*brahtsk'*)
Buryatia (boor-*yaht'*-ee-uh)
Caucasus (kaw'-kah-suss)
Chechen-Ingushia (*chehch'*-yenn, enn-*gooshia'*)
Chelyabinsk (*chell'*-yah-binnsk)
Cherepovets (*chehr'*-yuh-puh-vyetz')
Chuvashia (choo-*vahsh'*-i-a)
Dagestan Rep. (dag'-uh-*stahn'*)
Irkutsk (ihr-*kootsk'*)
Ivanovo (ih-*vahn'*-uh-voh)
Izhevsk (ih-*zhehvsk'*)
Kama R. (kah'-mah)
Karelia (kah-*reel'*-yuh)
Kazan (kuh-*zahn'*)
Khabarovsk (kah-*bah'*-rawfsk)
Kola Pen. (*koh'*-lah)
Kolyma R. (kuh-*lee'*-muh)
Krasnoyarsk (kras'-nuh-*yahrsk'*)
Kuril Is. (*koo'*-rill)
Kuznetsk Basin (kooz'-*netsk'*)
Ladoga, Lake (*lad'*-uh-guh)
Lena R. (lee-nah, lay-nah)
Lipetsk (lyee'-*petsk'*)
Magnitogorsk (mag-*nee'*-toh-gorsk')
Moscow (*mahs'*-koh or *mahs'*-kow)
Murmansk (moor'-*mahntsk'*)
Nakhodka (nuh-kawt-kah)
Nizhniy Novgorod (*nizh'*-nee nahv-guh-*rahd'*)
Nizhniy Tagil (tah-*geel*)
Novokuznetsk (naw'-voh-*kooznetsk'*)
Novaya Zemlya (*naw'*-vuh-yuh zem-*lyah'*)
Novosibirsk (naw-vuh-suh-*beersk'*)
Oka R. (oh-kah')
Okhotsk, Sea of (oh-*kahtsk'*)
Omsk (*ahmsk'*)
Pechora R. (peh-*chore'*-a)
Rostov (rahs-*tawf'*)

Sakhalin (sahk'-uh-*leen'*)
Samara (suh-*mahr'*-a)
Saratov (suh-*raht'*-uff)
Sayan Mts. (sah-yahn')
Tatarstan (Tatar Rep.) (Tah-*tahr'*-stahn')
Togliatti (tawl-*yah'*-tee)
Tuva Rep. (*too'*-va)
Ufa (oo-*fah'*)
Ussuri R. (oo-soor'-i)
Vladivostok (vlah'-dih-vahs-*tawk'*)
Volga R. (vahl'-guh, *vohl'*-guh)
Volgograd (vahl-guh-*grahd'*, vohl'-)
Voronezh (vah-*raw'*-nyesh)
Yakutsk (yah-*kootsk'*)
Yekaterinburg (yeh-*kahta'*-rinn-berg)

Rwanda (roo-*ahn*-da)

St. Kitts and Nevis (*nee'*-vis)

St. Lucia (*loo'*-sha)

St. Vincent-Grenadines (gren'-a-*deenz'*)

São Tomé and Príncipe (sau too-meh, pren-see'-puh)

Saudi Arabia (*saud'*-ee)
Jiddah (*jid'*-a)
Riyadh (ree-*adh'*)

Senegal (sen'-i-*gahl'*)
Dakar (da'-*kahr'*)

Seychelles (say'-*shelz'*)

Sierra Leone (see-*ehr'*-a lee-*ohn'*)

Slovenia (sloh-*veen'*-i-a)
Ljubljana (*lyoo'*-b'l-yah-nah)

Somalia (so-mahl'-ee-a)
Mogadishu (mahg'-uh-*dish'*-oo)

South Africa
Bloemfontein (*bloom'*-fonn-*tayn'*)
Bophuthatswana (homeland) (boh-poot'-aht-*swahna'*)
Drakensberg (escarpment) (*drahk*-n's-burg')
Durban (durr-bun)
Johannesburg (joh-*hann'*-iss-burg)
Kwazulu (homeland) (kwah-zoo'-loo)
Natal (province) (nuh-*tahl'*)
Pretoria (prih-*tohr'*-ee-uh)
Transkei (homeland) (trans-*keye'*)
Transvaal (province) (trans-*vahl'*)

Spain
Barcelona (bahr'-suh-*lohn*-uh)
Bilbao (bill-*bah'*-oh)
Catalonia (region) (katt'-uh-*lohn'*-ee-a)
Madrid (muh-*dridd'*)
Meseta (plateau) (muh-*say'*-tuh)
Seville (suh-*vill'*)
Valencia (vull-*enn'*-shee-uh)

Sri Lanka (sree *lahnkah'*)

Sudan (*soo'*-dann)
Gezira (guh-*zeer'*-uh)
Juba (*joo'*-bah)

Khartoum (kahr′-*toom*′)
Omdurman (ahm-*durr*′-man)
Wadi Halfa (*wahdi*′ *hawl*′-fah)

Suriname (*soor*′-i-nahm′)

Swaziland (*swah*′-zih-land′)

Sweden
Göteborg (yort′-e-*bor*′)
Skåne (*skohn*′)
Småland (*smoh*′-lund′)

Switzerland
Basel (*bah*′-z′l)
Zürich (*zoor*′-ick)

Syria (seer′-i-uh)
Aleppo (a-*lep*-oh′)
Damascus (da-*mas*′-kus)

Tajikistan (tah-*jick*′-ih-stahn′)

Tanzania (tan′-za-*nee*′-a)
Dar es Salaam (dahr′ es sa-*lahm*′)

Thailand (*teye*′-land′)
Bangkok (*bang*′-kok′)

Togo (*toh*′-goh)
Lomé (loh-*may*′)

Tonga (*tawng*′-a)

Trinidad and Tobago (to-*bay*′-goh)

Tunisia (too-*nee*′-zhee-a)
Tunis (*too*′-nis)

Turkey
Anatolian Peninsula (ann′-ah-*tohl*′-yun)
Ankara (*ang*′-kah-rah)
Bosporus (strait) (*bahs*′-puhr-us)
Dardanelles (strait) (dahr′-duh-*nelz*′)
Istanbul (iss′-tann-*bool*′)
Izmir (izz′-*meer*′)
Sea of Marmara (*mahr*′-muh-ruh)

Turkmenistan (turk′-menn-ih-*stahn*′)
Ashkhabad (*ash*′-kuh-bahd′)

Ukraine (yoo′-*krane*′)
Crimea (cry′-*mee*′-uh)
Dnepropetrovsk (d′nyepp-pruh-pay-*trawfsk*′)
Dnieper R. (duh-*nyepp*′-er)
Donetsk (duhn-*yehtsk*′)
Kharkov (*kahr*′-kawf)
Kiev (*kee*′-yeff)
Krivoy Rog (*kree*′-voi *rohg*′)
Sevastopol (sih-*vass*′-tuh-*pohl*′)
Yalta (*yahl*′-tuh)
Zaporozhye (zah-puh-*rawzh*′-yuh)

United Arab Emirates
Abu Dhabi (*ah*′-boo dah′-bee)
Dubai (doo-*beye*′)

United Kingdom
Edinburgh (*ed*′-in-buh-ruh)
Glasgow (glass′-goh)

Thames R. (*timms*′)

United States
Albuquerque (al′-buh-*ker*′-kih)
Aleutian Is. (uh-*loo*′-sh′n)
Allegheny R. (*al*′-ih-gay′-nih)
Appalachian Highlands (ap′-uh-*lay*′-chih-un)
Coeur d'Alene (*kurr*′ duh-*layn*′)
Des Moines (duh *moin*′)
Grand Coulee Dam (*koo*′-lee)
Juneau (*joo*′-noh)
Mesabi Range (muh-*sahb*′-ih)
Monongahela R. (muh-*nahng*′-guh-*hee*′-luh)
Omaha (oh′-mah-*hah*′)
Ouachita Mts. (*wosh*′-ih-taw′)
Palouse (region) (pah-*loose*′)
Phoenix (*fee*′-nix)
Provo (*proh*′-voh)
Rio Grande (*ree*′-oh *grahn*′-day)
San Joaquin R. (sann wah-*keen*′)
Spokane (spoh′-*kann*′)
Tucson (*too*′-sonn)
Valdez (val-*deez*′)

Uruguay (*yoor*′-uh-gweye′)
Montevideo (mahnt′-e-vi-*day*′-oh)

Uzbekistan (ooz-*beck*′-ih-stahn′) ("oo" as in "wood")
Amu Darya (river) (ah-moo′ *dahr*′-yuh, dahr-*yah*′)
Bukhara (buh-*kah*′-ruh)
Fergana Valley (fair-*gahn*′-uh)
Samarkand (*samm*′-ur-*kand*′; Rus., suh-mur-*kahnt*′)
Tashkent (tash-*kent*′, tahsh-)

Vanuatu (van′-oo-*ay*′-too)
Port Vila (*vee*′-la)

Venezuela (ven′-ez-*way*′-la)
Caracas (ka-*rak*′-as)
Maracaibo (mar′-a-*keye*′-boh)
Orinoco R. (oh-rih-*noh*′-koh)
Valencia (va-*len*′-see-a)

Vietnam (vee-et′-*nahm*′)
Annam (region) (uh-*namm*′)
Cochin China (*koh*′-chin)
Hanoi (ha-*noi*′)
Ho Chi Minh City (Saigon) (hoe chee min′; *seye*′-gahn)
Mekong R. (may-*kahng*′)

Western Samoa (sa-*moh*′-a)
Apia (ah-pea-uh)

Yemen
Aden (*ah*-d′n)
S'ana (sah-*na*′)

Yugoslavia (yoo′-goh-*slahv*′-ee-a)
Belgrade (*bell*′-grade, -grahd)
Kosovo (*kaw*′-suh-voh′)
Montenegro (republic) (mahn′-tee-*nay*′-groh)

Zaire (*zeye*′-eer′)
Kananga (kay-*nahng*′-guh)
Kasai R. (kah-*seye*′)

PG-8   Place-Name Pronunciation Guide

Kinshasa (*keen'*-shah-suh)
Kisangani (kee'-sahn-*gahn'*-ih)
Lake Kivu (*kee'*-voo)
Lualaba R. (loo'-ah-*lahb'*-a)
Lubumbashi (loo-boom-*bah'*-shee)
Matadi (muh-tah'-dee)
Shaba (region) (*shah'*-bah)

Zambia (*zam'*-bee-a)
Kariba Dam (kah-*ree'*-buh)

Lusaka (loo-*sahk'*-a)
Zambezi R. (zamm-*bee'*-zee)

Zimbabwe (zim-*bahb'*-way)
Bulawayo (bool'-uh-*way'*-oh)
Harare (hah-*rahr'*-ee)
Que Que (*kway' kway'*)
Wankie (*wahn'*-kee)

INDEX